CHEMICAL THERMODYNAMICS OF SELENIUM

CHEMICAL THERMODYNAMICS

Vol. 1. Chemical Thermodynamics of Uranium (H. Wanner and I. Forest, eds.)
Vol. 2. Chemical Thermodynamics of Americium (R.J. Silva et al.)
Vol. 3. Chemical Thermodynamics of Technetium (M.C.A. Sandino and E. Östhols, eds.)
Vol. 4. Chemical Thermodynamics of Neptunium and Plutonium (OECD Nuclear Energy Agency, ed.)
Vol. 5. Update on the Chemical Thermodynamics of Uranium, Neptunium, Plutonium, Americium and Technetium (OECD Nuclear Energy Agency, ed.)
Vol. 6. Chemical Thermodynamics of Nickel (OECD Nuclear Energy Agency, ed.)
Vol. 7. Chemical Thermodynamics of Selenium (OECD Nuclear Energy Agency, ed.)

CHEMICAL THERMODYNAMICS 7

Chemical Thermodynamics of Selenium

Åke OLIN (Chairman)
Department of Analytical Chemistry
Uppsala University
Uppsala (Sweden)

Bengt NOLÄNG
The Ångström Laboratory
Department of Materials Chemistry
Uppsala University
Uppsala (Sweden)

Evgeniy G. OSADCHII
Institute of Experimental Mineralogy RAS
Chernogolovka (Russian Federation)

Lars-Olof ÖHMAN
Department of Chemistry
Umeå University
Umeå (Sweden)

Erik ROSÉN
Department of Chemistry
Umeå University
Umeå (Sweden)

Edited by
Federico J. MOMPEAN (Series Editor and Project Co-ordinator)
Jane PERRONE (Volume Editor) and Myriam ILLEMASSÈNE
OECD Nuclear Energy Agency, Data Bank
Issy-les-Moulineaux (France)

2005

ELSEVIER

AMSTERDAM • BOSTON • HEIDELBERG • LONDON • NEW YORK • OXFORD

PARIS • SAN DIEGO • SAN FRANCISCO • SINGAPORE • SYDNEY • TOKYO

Published by
ELSEVIER B.V.
Radarweg 29
P.O. Box 211, 1000 AE Amsterdam, The Netherlands

ISBN: 0-444-51403-1
ISBN: 0-444-51974-2 (CD-ROM)

This volume has been reproduced by the publisher from the formatted pages prepared by the editors.

∞ The paper used in this publication meets the requirements of ANSI/NISO Z39.48-1992 (Permanence of Paper).

Printed in The Netherlands

Preface

This is the seventh volume of the series "Chemical Thermodynamics" edited by the OECD Nuclear Energy Agency (NEA). It is the first volume that deals primarily with a non-metal, selenium. It contains an evaluation of experimental thermodynamic data of binary inorganic selenium compounds in the solid and gaseous state and in aqueous solution. Thus organic selenium compounds and metal complexes with selenium-containing organic ligands are not included. Ternary systems are also excluded except when the proton is one of the components. The book is based on publications found in a broad electronic search of the literature in the spring of 1999. Since then information has been retrieved occasionally of missing data on specific species and a restricted general search was made this autumn for the period 1999 – 2004. It is thus hoped that the relevant information available up to the end of year 2003 has been included in the review.

At the NEA Data Bank the responsibility for the overall co-ordination of the Project was placed with Eric Östhols (from its initiation in 1998 to February 2000), with Stina Lundberg (from March 2000 to September 2000) and with Federico Mompean (since September 2000). Federico Mompean was in charge of the preparation of the successive drafts, updating the NEA thermodynamic database and editing the book to its present final form, with assistance from Myriam Illemassène, Jane Perrone, Katy Ben Said and Cristina Domènech-Ortí.

The first meetings of the review team took place in October 1998 and March 1999 at which the scope of the review and its delimitations were set up. The areas for review were also allotted to the members. Additional meetings were held in May 2001 and March 2002. The discussions were mainly centred around problems with the strict application of the TDB Guidelines to the review and the bringing together and presentation of the material in the book. The active participation and constructive ideas of Kastriot Spahiu, Chairman of the NEA-TDB Executive Group, and the Project co-ordinators in solving our problems have been most valuable.

After having accepted to partake in the evaluation of solution data, I suddenly found myself assigned as chairman of the project. In that position I recruited the other initial members of the team Bengt Noläng and Lars-Olof Öhman. After the search of the literature had been done it became evident that a substantial part of the information originated from work carried out in the former Soviet Union and it was not available in translations. This problem was overcome by the addition of Eugeniy Osadchii to the team. Erik Rosén joined the group at a later date when it became obvious to the initial members of the team that the volume of information was too large to be handled only by them.

The team as a whole is responsible for the final result but various parts of the review have been the main responsibility of the individual members. Thus Bengt Noläng has focused on the thermodynamic properties of the element and solid selenides. He also made the statistical-mechanical calculations on spectroscopic data for gaseous molecules. I would like here to acknowledge his painstaking efforts to penetrate the literature on elemental selenium in the gaseous phase. Lars-Olof Öhman and Erik Rosén have evaluated the material on hydrogen selenide and the aqueous chemistry of the selenides. They also took on the main responsibility for compounds formed with the non-metals in Groups 14 to 17 of the Periodic Table except oxygen. Evgeniy Osadchii collected the "Russian papers" and made a preliminary evaluation of their contents. The essential parts of the papers that contained thermodynamic data were translated into English and sent to Sweden. His knowledge also of Belorussian and Ukrainian was indispensable in this context. The short communications published in Zhur. Fiz. Khim., which often contained the final results only, meant that he had to spend time in the archives to extract the primary data that were necessary for the assessment of the papers. The compilation of selenium minerals is also his work. I am mainly responsible for the assessment of the selenium oxides and their aqueous chemistry, the metal selenites and selenates both in aqueous solution and as solids.

Various arrangements of the presentation of the material in Chapter V have been considered. A sub-division according to type of compound such as selenide, selenite etc. would lead to the best overview of a particular type of species. This arrangement was tried but abandoned and the system used in previous reviews was adhered to, *i.e.* the compounds and complexes formed by selenium with a certain element are presented together.

Selenium forms compounds with most elements of the Periodic Table. NEA-TDB selected data for a number of these elements are not available for use as auxiliary data in the evaluation of formation data and entropies of their selenium compounds from reaction data. It would not have been a realistic task for the selenium project to assess all non-selected auxiliary data needed according to the NEA-TDB Guidelines. Instead the information required was obtained from compilations of thermochemical data such as "The NBS tables of chemical thermodynamic properties" and "JANAF thermochemical tables" supplemented by "Critical reviews of ..." published in chemical journals. The review of the literature on selenium and its compounds has thus resulted in two thermochemical data sets. One set of data is in accord with the NEA-TDB Guidelines and compatible with the requirements for addition to the NEA-TDB Data Bank. The other set, obtained with non-TDB auxiliary data, does not fulfil the requirements of the Data Bank. These facts created a problem in the presentation of the results of the selenium project, which was solved by the inclusion of an Appendix E at a suggestion by Federico Mompean. This appendix presents the same type of information as the ordinary Chapter III (selected data) and Chapter IV (selected auxiliary data). The experimental sources of the entries in Table E-1 in Appendix E have been reviewed according to the NEA-TDB Guidelines but the formation data and entropies could not be fully evaluated using the set of auxiliary data currently accepted by the NEA-TDB Project.

The thermodynamic information available for a large number of the compounds and systems originates from one experimental source only. The assessment of the accuracy and extrapolation of the data to standard conditions are then virtually impossible to make in an indisputable way. The compilers are fully aware that the uncertainties of the results obtained in these cases are highly subjective although this may not be mentioned in the context. A number of the quantities derived in Chapter V would probably have been given a low weight or even rejected if a richer crop of experimental data had been available. Such entries in Chapter V, often subjected to recalculations of the original data, should be regarded as an effort by the team to present a reasonably complete account, within the restrictions of the project, of the present knowledge of the chemical thermodynamics of selenium and its compounds. Thereby areas where further work is needed will be highlighted. It is hoped that this approach will not distract those readers who would prefer a more stringent selection and validation of the material at hand.

Uppsala, Sweden, November 2004 Åke Olin, Chairman

Acknowledgements

For the preparation of this book, the authors have received financial support from the NEA TDB Phase II Project. The following organisations take part in the Project:

ANSTO, Australia
ONDRAF/NIRAS, Belgium
RAWRA, Czech Republic
POSIVA, Finland
ANDRA, France
IPSN (now IRSN), France
FZK, Germany
JNC, Japan
ENRESA, Spain
SKB, Sweden
SKI, Sweden
HSK, Switzerland
NAGRA, Switzerland
PSI, Switzerland
BNFL, UK
Nirex, UK
DoE, USA

The authors would like to express their sincere gratitude to the Peer Reviewers for their numerous comments. These comments have greatly improved the scientific and linguistic quality of the book. The suggestion by Jean Fuger and Malcolm H. Rand to include also thermodynamic data obtainable from spectroscopic work should especially be mentioned here. The latter reviewer also provided computer code for the calculations that served as a check of our programs.

Åke Olin wants to thank Peter M. May, Murdoch University, for discussions, now and earlier, on the art of 'stability constant determinations in aqueous solution' and for putting the selenium data in the JESS Thermodynamic Database at his disposal. His thanks also go to Christomir Christov, University of California at San Diego, for providing data prior to their publication and for generous information on the application of the Pitzer approach to these data. Help with the assessment of voltammetric work was obtained from Leif Nyholm, Uppsala University and it is gratefully acknowledged. Finally his thanks extend to Jean Pettersson, Uppsala University, for his readiness to sort out an old timer's problems with computer and computer codes.

The coordination of the project was initially in the hands of Erik Östhols. After he left the NEA this responsibility was taken over by Stina Lundberg and shortly after by Federico Mompean. The team is grateful for his continuous help and support in finalising this project. The seemingly never-ending revisions of the original manuscript have been taken care of by Katy Ben Said, Myriam Illemassène, and Jane Perrone. The authors wish to express their sincere gratitude to them.

At the NEA Data Bank, Pierre Nagel and Eric Lacroix have provided excellent software and advice, which have eased the editorial and database work. Cynthia Picot, Solange Quarmeau and Amanda Costa from NEA Publications have provided considerable help in editing the present series. Their contributions and the support of many NEA staff members are highly appreciated.

The authors and editors are grateful to Dr. Ludmila Shelinova, RAS, and to the following publishers for their permission to produce some figures found in this volume:

Academic Press,
ASM International,
Elsevier,
Kluwer Academic/Plenum Publishers.

The entire manuscript of this book has undergone a peer review by an independent international group of reviewers, according to the procedures in the TDB-6 Guideline, available from the NEA. The peer reviewers have seen and approved the modifications made by the authors in response to their comments. The peer review comment records may be obtained on request from the OECD Nuclear Energy Agency. The peer reviewers were:

Prof Jean Fuger, University of Liège, Liège, Belgium,
Prof Yizhak Marcus, The Hebrew University of Jerusalem, Israel
Prof Peter May, Murdoch University, Murdoch, Australia
Dr Malcolm Rand, Winters Hill Consultancy, Oxfordshire, UK

Their contributions are gratefully acknowledged.

Note from the Chairman of the NEA-TDB Project Phase II

The need to make available a comprehensive, internationally recognised and quality-assured chemical thermodynamic database that meets the modeling requirements for the safety assessment of radioactive waste disposal systems prompted the Radioactive Waste Management Committee (RWMC) of the OECD Nuclear Energy Agency (NEA) to launch in 1984 the Thermochemical Database Project (NEA-TDB) and to foster its continuation as a semi-autonomous project known as NEA-TDB Phase II in 1998.

The RWMC assigned a high priority to the critical review of relevant chemical thermodynamic data of inorganic species and compounds of the actinides uranium, neptunium, plutonium and americium, as well as the fission product technetium. The first four books in this series on the chemical thermodynamics of uranium, americium, neptunium and plutonium, and technetium originated from this initiative.

The organisation of Phase II of the TDB Project reflects the interest in many OECD/NEA member countries for a timely compilation of the thermochemical data that would meet the specific requirements of their developing national waste disposal programmes.

The NEA-TDB Phase II Review Teams, comprising internationally recognised experts in the field of chemical thermodymanics, exercise their scientific judgement in an independent way during the preparation of the review reports. The work of these Review Teams has also been subjected to further independent peer review.

Phase II of the TDB Project consisted of: (i) updating the existing, CODATA-compatible database for inorganic species and compounds of uranium, neptunium, plutonium, americium and technetium, (ii) extending it to include selected data on inorganic species and compounds of nickel, selenium and zirconium, (iii) and further adding data on organic complexes of citrate, oxalate, EDTA and iso-saccharinic acid (ISA) with uranium, neptunium, plutonium, americium, technetium, nickel, selenium, zirconium and some other competing cations.

The NEA-TDB Phase II objectives were formulated by the 17 participating organisations coming from the fields of radioactive waste management and nuclear regulation. The TDB Management Board is assisted for technical matters by an Executive Group of experts in chemical thermodynamics. In this second phase of the Project, the NEA acts as coordinator, ensuring the application of the Project Guidelines and liaising with the Review Teams.

The present volume is the third one published within the scope of NEA-TDB Phase II and contains a database for inorganic species and compounds of selenium. We trust that the efforts of the reviewers, the peer reviewers and the NEA Data Bank staff merit the same high recognition from the broader scientific community as received for previous volumes of this series.

Mehdi Askarieh
United Kingdom Nirex limited
Chairman of TDB Project Phase II Management Board
On behalf of the NEA TDB Project Phase II Participating Organisations:
ANSTO, Australia
ONDRAF/NIRAS, Belgium
RAWRA, Czech Republic
POSIVA, Finland
ANDRA, France
IPSN (now IRSN), France
FZK, Germany
JNC, Japan
ENRESA, Spain
SKB, Sweden
SKI, Sweden
HSK, Switzerland
PSI, Switzerland
BNFL, UK
Nirex, UK
DoE, USA

Editor's note

This is the seventh volume of a series of expert reviews of the chemical thermodynamics of key chemical elements in nuclear technology and waste management. This volume is devoted to the inorganic species and compounds of selenium. The tables contained in Chapters III and IV list the currently selected thermodynamic values within the NEA-TDB Project. The database system developed at the NEA Data Bank, see Section II.6, assures consistency among all the selected and auxiliary data sets.

The recommended thermodynamic data are the result of a critical assessment of published information. The values in the auxiliary data set, see Tables IV-1 and IV-2 have been adopted from CODATA key values or have been critically reviewed in this or earlier volumes of the series.

How to contact the NEA-TDB Project

Information on the NEA and the TDB Project, on-line access to selected data and computer programs, as well as many documents in electronic format are available at

www.nea.fr.

To contact the TDB project coordinator and the authors of the review reports, send comments on the TDB reviews, or to request further information, please send e-mail to tdb@nea.fr. If this is not possible, write to:

TDB project coordinator
OECD Nuclear Energy Agency, Data Bank
Le Seine-St. Germain
12, boulevard des Îles
F-92130 Issy-les-Moulineaux
FRANCE

The NEA Data Bank provides a number of services that may be useful to the reader of this book.

- The recommended data can be obtained via internet directly from the NEA Data Bank.
- The NEA Data Bank maintains a library of computer programs in various areas. This includes geochemical codes such as PHREEQE, EQ3/6, MINEQL, MINTEQ and PHRQPITZ, in which chemical thermodynamic data like those presented in this book are required as the basic input data. These computer codes can be obtained on request from the NEA Data Bank.

Contents

List of Figures

List of Tables

Chapter I

Introduction

I.1 Background

The modelling of the behaviour of hazardous materials under environmental conditions is among the most important applications of natural and technical sciences for the protection of the environment. In order to assess, for example, the safety of a waste deposit, it is essential to be able to predict the eventual dispersion of its hazardous components in the environment (geosphere, biosphere). For hazardous materials stored in the ground or in geological formations, the most probable transport medium is the aqueous phase. An important factor is therefore the quantitative prediction of the reactions that are likely to occur between hazardous waste dissolved or suspended in ground water, and the surrounding rock material, in order to estimate the quantities of waste that can be transported in the aqueous phase. It is thus essential to know the relative stabilities of the compounds and complexes that may form under the relevant conditions. This information is often provided by speciation calculations using chemical thermodynamic data. The local conditions, such as ground water and rock composition or temperature, may not be constant along the migration paths of hazardous materials, and fundamental thermodynamic data are the indispensable basis for dynamic modelling of the chemical behaviour of hazardous waste components.

In the field of radioactive waste management, the hazardous material consists to a large extent of actinides and fission and activation products from nuclear reactors (such is the case of the fission product ^{79}Se). The scientific literature on thermodynamic data, mainly on equilibrium constants and redox potentials in aqueous solution, has been contradictory in a number of cases. A critical and comprehensive review of the available literature is necessary in order to establish a reliable thermochemical database that fulfils the requirements of a proper modelling of the behaviour of the actinide and fission and activation products in the environment.

The Radioactive Waste Management Committee (RWMC) of the OECD Nuclear Energy Agency recognised the need for an internationally acknowledged, high-quality thermochemical data base for the application in the safety assessment of radioactive waste disposal, and undertook the development of the NEA Thermochemical Data

Base (TDB) project [85MUL], [88WAN], [91WAN]. The RWMC assigned a high priority to the critical review of relevant chemical thermodynamic data of compounds and complexes for this area containing the actinides uranium, neptunium, plutonium and americium, as well as the fission product technetium. In 1998, Phase II of the TDB Project (TDB-II) was started to provide for the further needs of the radioactive waste management programs by updating the existing database and applying the TDB review methodology to other elements (occurring in waste as fission or activation products) and to simple organic complexes. In TDB-II the overall objectives are set by a Management Board, integrated by representatives of 17 organisations from the field of radioactive waste management. These participating organisations, together with the NEA, provide financial support for TDB-II. The TDB-II Management Board is assisted in technical matters by a group of experts in chemical thermodynamics (the Executive Group). The NEA acts in this phase as Project Co-ordinator ensuring the implementation of the Project Guidelines and liaising with the Review Teams. After the books on chemical thermodynamics of uranium [92GRE/FUG], americium [95SIL/BID], technetium [99RAR/RAN], neptunium and plutonium, [2001LEM/FUG], the Update Review on U, Np, Pu, Am and Tc [2003GUI/FAN], and nickel [2005GAM/BUG], the present report on the fission product selenium is the seventh volume in the series. Two additional volumes are in preparation as this book goes to press dealing with the inorganic species and compounds of zirconium and with compounds and complexes formed by the organic ligands oxalate, citrate, EDTA and iso-saccharinic acid with U, Np, Pu, Am, Tc, Zr, Ni and Se.

I.2 Focus of the review

The first and most important step in the modelling of chemical reactions is to decide whether they are controlled by chemical thermodynamics or kinetics, or possibly by a combination of the two. This also applies to the modelling of more complex chemical systems and processes, such as waste repositories of various kinds, the processes describing transport of toxic materials in ground and surface water systems, the global geochemical cycles, *etc.*

As outlined in the previous section, the focus of the critical review presented in this report is on the thermodynamic data of selenium relevant to the safety assessment of radioactive waste repositories in the geosphere. This includes the release of waste components from the repository into the geosphere (*i.e.*, its interaction with the waste container and the other near-field materials) and their migration through the geological formations and the various compartments of the biosphere. As ground waters and pore waters are the transport media for the waste components, the knowledge of the thermodynamics of the corresponding elements in waters of various compositions is of fundamental importance.

The present review therefore puts much weight on the assessment of the low-temperature thermodynamics of selenium in aqueous solution and makes independent analyses of the available literature in this area. The standard method used for the analysis of ionic interactions between components dissolved in water (see Appendix B) allows the general and consistent use of the selected data for modelling purposes, regardless of the type and composition of the ground water, within the ionic strength limits given by the experimental data used for the data analyses in the present review.

The interactions between solid compounds, such as the rock materials, and the aqueous solution and its components are as important as the interactions within the aqueous solution, because the solid materials in the geosphere control the chemistry of the ground water, and they also contribute to the overall solubilities of key elements. The present review therefore also considers the chemical behaviour of solid compounds containing selenium. A list of selenium containing minerals is presented in Appendix D, but no thermodynamic data are available for the minerals. It is, however, difficult to assess the relative importance of the solid phases for performance assessment purposes, particularly since their interactions with the aqueous phase are in many cases known to be subject to quantitatively unknown kinetic constraints. Furthermore, in some circumstances sorption of aqueous ions at mineral water interfaces may be a more important factor in determining migration of selenium than dissolution and precipitation phenomena.

This book contains a summary and a critical review of the thermodynamic data on compounds and complexes containing selenium in its various oxidation states as reported in the available chemical literature up to the end of 2003. A comparatively large number of primary references are discussed separately in Appendix A. No attempt has been made to estimate the relative importance of a certain complex or solid with respect to its behaviour under environmental or geological conditions. Thus the review tries to present an as complete account as possible of the thermodynamic quantities of all binary inorganic selenium species for which such data are available. Data for ternary species with O, H (selenites, selenates, *etc.*) and the halogens (oxyhalides and mixed halides) are included as well as some sulphur analogues such as selenocyanates. Other ternary or higher species and systems are in general not considered in the review.

The present review does not include any compounds or complexes containing organic ligands. This class of compounds is the subject of a separate review in the NEA-TDB series. Neither are the thermodynamic properties of vitreous forms of the compounds considered in the review.

Although the focus of this review is on selenium, it is necessary to use data on a number of other species during the evaluation process that lead to the recommended data. These so-called auxiliary data are taken both from the publication of CODATA Key Values [89COX/WAG] and from the evaluation of additional auxiliary data in the volume on uranium [92GRE/FUG] and the succeeding volumes noted in Section I.1, and their use is recommended by this review. Selenium forms compounds with most of the elements of the Periodic Table. The sources of auxiliary data mentioned do not contain all data that were needed, particularly for the recalculation of calorimetric work carried out in the former Soviet Union. Additional auxiliary data were often obtained from "The NBS tables of chemical thermodynamic properties" [82WAG/EVA]. These data appear, together with their estimated uncertainties, in Appendix E; the recalculations are presented in Appendix A and Chapter V. They permit a re-evaluation with new ancillary data when needed. Care has been taken that all the selected thermodynamic data at standard conditions are internally consistent. For this purpose, special software has been developed at the NEA Data Bank that is operational in conjunction with the NEA-TDB data base system, *cf.* Section II.6. In order to maintain consistency in the application of the values selected by this review, it is essential to use these auxiliary data.

I.3 Review procedure and results

The objective of the present review is to present an assessment of the sources of published thermodynamic data in order to decide on the most reliable values that can be recommended. Experimental measurements, as well as quantities for simple molecules calculated by statistical mechanics methods, published in the scientific literature are the main source for the selection of recommended data. When necessary, experimental source data are re-evaluated by using chemical models which are found more realistic than those used by the original author. Re-evaluation of literature values might be also necessary to correct for known systematic errors (for example, if the junction potentials are neglected in the original publication) or to make extrapolations to standard state conditions ($I = 0$) by using the specific ion interaction (SIT) equations (*cf.* Appendix B). For convenience, these SIT equations are referred to in some places in the text as "the SIT". A great number of recalculations have also been made to ensure that the final result is consistent with selected data for basic reactions, such as vapour pressure - temperature relationships and the protonation constants of the various selenium anions. The great majority of studies on equilibria in aqueous solution have been carried out with the final result presented on the molar scale (M). In the presentation of the primary result this scale has been kept and no recalculation to the molal scale (m) performed. Extrapolation to $I = 0$ by SIT thus involves this recalculation, although it is not explicitly mentioned. All pressures are expressed with the bar as unit and the necessary transformation has been made by the review. In some instances, this rule was not followed for various reasons. The pressure unit is then explicitly mentioned.

In order to ensure that consistent procedures are used for the evaluation of primary data, a number of guidelines have been developed. They have been updated and improved since 1987, and their most recent versions are available at the NEA [2000OST/WAN], [2000GRE/WAN], [99WAN/OST], [2000WAN/OST], [99WAN].

Some of these procedures are also outlined in this volume, *cf.* Chapter II, Appendix B, and Appendix C.

Once the critical review process in the NEA-TDB project is completed, the resulting manuscript is reviewed independently by qualified experts nominated by the NEA. The independent peer review is performed according to the procedures outlined in the TDB-6 guideline [99WAN]. The purpose of the additional peer review is to receive an independent view of the judgements and assessments made by the primary reviewers, to verify assumptions, results and conclusions, and to check whether the relevant literature has been exhaustively considered. The independent peer review is performed by personnel having technical expertise in the subject matter to be reviewed, to a degree at least equivalent to that needed for the original review.

The thermodynamic data compiled in the present review (see Chapters III and IV and Appendix E) refer to the reference temperature of 298.15 K and to standard conditions, *cf.* Section II.3. For the modelling of real systems it is, in general, necessary to recalculate the standard thermodynamic data to non-standard state conditions. For aqueous species a procedure for the calculation of the activity factors is thus required. This review uses the approximate specific ion interaction method (SIT) for the extrapolation of experimental data to the standard state in the data evaluation process, and in some cases this requires the re-evaluation of original experimental values (solubilities, emf data, *etc.*). For maximum consistency, this method, as described in Appendix B, should always be used in conjunction with the selected data presented in this review. However, some solubility data for highly soluble selenates were evaluated in the original papers by the Pitzer approach. No attempt was made to re-evaluate these data by the SIT method.

As discussed in Appendix C, there is no unique way to assign uncertainties. The assignments made in this review of the uncertainties representing the 95% confidence level are to a large extent based on the subjective choice by the reviewers, supported by their scientific and technical experience in the corresponding area. The thermodynamic data for a considerable number of the selenium compounds or species are based on a single source only. In addition, the experimental material might be sparse or details of the experiments and methods described in little detail. The reviewers have therefore found it difficult to apply, in a consistent way, the standards set by the NEA guidelines in the assessment of the thermodynamic quantities of a considerable number of selenium compounds or systems. In order to provide an as complete picture as possible of what is known about the thermochemistry of selenium, the review has approached the task as follows. A result is accepted (after recalculation) if there is no obvious reason to reject it and the accepted quantity is presented in Chapter V. Accepted results that can be confidently extrapolated to standard conditions are then used to derive the thermodynamic parameters in the Tables of Chapter III and Appendix E. The confidence limits assigned are thought to be valid if data from several independent sources or methods are available. Confidence intervals are also assigned to values for which only one accepted datum is available. These intervals are often based on little statistical information and therefore highly subjective. The background material in Chapter V and/or Appendix A should therefore be consulted. Papers that have been found faulty have been rejected and the reasons are given in Appendix A.

The quality of thermodynamic models cannot be better than the quality of the data used in the model and the "completeness" of the chemical model used, *e.g.*, the inclusion of all the relevant dissolved chemical species and solid phases. For the user it is important to consider that the data sets presented in this review (Chapters III and IV and Appendix E) may not be "complete" with respect to all the conceivable systems and conditions; there are gaps in the information. The gaps and the uncertainties about the reliability of the data are pointed out in Chapter V, and this information may be used as a basis for the assignment of research priorities.

Chapter II

Standards, Conventions, and Contents of the Tables

This chapter outlines and lists the symbols, terminology and nomenclature, the units and conversion factors, the order of formulae, the standard conditions, and the fundamental physical constants used in this volume. They are derived from international standards and have been specially adjusted for the TDB publications.

II.1 Symbols, terminology and nomenclature

II.1.1 Abbreviations

Abbreviations are mainly used in tables where space is limited. Abbreviations for methods of measurement are listed in Table II-1.

Table II-1: Abbreviations for experimental methods

AIX	Anion exchange
AES	Atomic Emission Spectroscopy
CAL	Calorimetry
CHR	Chromatography
CIX	Cation exchange
COL	Colorimetry
CON	Conductivity
COR	Corrected
COU	Coulometry
CRY	Cryoscopy
DIS	Distribution between two phases
DSC	Differential Scanning Calorimetry
DTA	Differential Thermal Analysis
EDS	Energy Dispersive Spectroscopy

(Continued on next page)

Table II-1: (continued)

EM	Electromigration
EMF	Electromotive force, not specified
EPMA	Electron Probe Micro Analysis
EXAFS	Extended X-ray Absorption Fine Structure
FTIR	Fourier Transform Infra Red
IDMS	Isotope Dilution Mass-Spectroscopy
IR	Infrared
GL	Glass electrode
ISE-X	Ion selective electrode with ion X stated
IX	Ion exchange
KIN	Rate of reaction
LIBD	Laser Induced Break Down
MVD	Mole Volume Determination
NMR	Nuclear Magnetic Resonance
PAS	Photo Acoustic Spectroscopy
POL	Polarography
POT	Potentiometry
PRX	Proton relaxation
QH	Quinhydrone electrode
RED	Emf with redox electrode
REV	Review
SEM	Scanning Electron Microscopy
SP	Spectrophotometry
SOL	Solubility
TC	Transient Conductivity
TGA	Thermo Gravimetric Analysis
TLS	Thermal Lensing Spectrophotometry
TRLFS	Time Resolved Laser Fluorescence Spectroscopy
UV	Ultraviolet
VLT	Voltammetry
XANES	X-ray Absorption Near Edge Structure
XRD	X-ray Diffraction
?	Method unknown to the reviewers

Other abbreviations may also be used in tables, such as SHE for the standard hydrogen electrode or SCE for the saturated calomel electrode. The abbreviation NHE has been widely used for the "normal hydrogen electrode", which is by definition identical to the SHE. It should nevertheless be noted that NHE customarily refers to a standard state pressure of 1 atm, whereas SHE always refers to a standard state pressure of 0.1 MPa (1 bar) in this review.

II.1.2 Symbols and terminology

The symbols for physical and chemical quantities used in the TDB review follow the recommendations of the International Union of Pure and Applied Chemistry, IUPAC [79WHI], [88MIL/CVI]. They are summarised in Table II-2.

Table II-2: Symbols and terminology.

Symbols and terminology	
length	l
height	h
radius	r
diameter	d
volume	V
mass	m
density (mass divided by volume)	ρ
time	t
frequency	ν
wavelength	λ
internal transmittance (transmittance of the medium itself, disregarding boundary or container influence)	T_i
internal transmission density, (decadic absorbance): $\log_{10}(1/T)$	A
molar (decadic) absorption coefficient: $A / c_B l$	ε
relaxation time	τ
Avogadro constant	N_A
relative molecular mass of a substance[(a)]	M_r
thermodynamic temperature, absolute temperature	T
Celsius temperature	t
(molar) gas constant	R
Boltzmann constant	k
Faraday constant	F
(molar) entropy	S_m
(molar) heat capacity at constant pressure	$C_{p,m}$
(molar) enthalpy	H_m
(molar) Gibbs energy	G_m
chemical potential of substance B	μ_B
pressure	p
partial pressure of substance B: $x_B p$	p_B
fugacity of substance B	f_B
fugacity coefficient: f_B/p_B	$\gamma_{f,B}$

(Continued on next page)

Table II-2: (continued)

Symbols and terminology	
amount of substance[(b)]	n
mole fraction of substance B:	x_B
molarity or concentration of a solute substance B (amount of B divided by the volume of the solution) [(c)]	c_B, [B]
molality of a solute substance B (amount of B divided by the mass of the solvent) [(d)]	m_B
factor for the conversion of molarity to molality of a solution: m_B/c_B	ϱ
mean ionic molality [(e)], $m_{\pm}^{(\nu_+ + \nu_-)} = m_+^{\nu_+} m_-^{\nu_-}$	$m_{\pm}$
activity of substance B	a_B
activity coefficient, molality basis: a_B / m_B	γ_B
activity coefficient, concentration basis: a_B / c_B	y_B
mean ionic activity [(e)], $a_{\pm}^{(\nu_+ + \nu_-)} = a_B = a_+^{\nu_+} a_-^{\nu_-}$	$a_{\pm}$
mean ionic activity coefficient [(e)], $\gamma_{\pm}^{(\nu_+ + \nu_-)} = \gamma_+^{\nu_+} \gamma_-^{\nu_-}$	$\gamma_{\pm}$
osmotic coefficient, molality basis	ϕ
ionic strength: $I_m = \frac{1}{2}\sum_i m_i z_i^2$ or $I_c = \frac{1}{2}\sum_i c_i z_i^2$	I
SIT ion interaction coefficient between substance B_1 and substance B_2 stoichiometric	$\varepsilon(B_1, B_2)$
coefficient of substance B (negative for reactants, positive for products)	
stoichiometric coefficient	ν_B
general equation for a chemical reaction	$0 = \sum_B \nu_B B$
equilibrium constant [(f)]	K
charge number of an ion B (positive for cations, negative for anions)	z_B
charge number of a cell reaction	n
electromotive force	E
$\text{pH} = -\log_{10}[a_H/(\text{mol}\cdot\text{kg}^{-1})]$	pH
electrolytic conductivity	κ
superscript for standard state [(g)]	$^\circ$

a ratio of the average mass per formula unit of a substance to $\frac{1}{12}$ of the mass of an atom of nuclide ^{12}C.

b *cf.* Sections 1.2 and 3.6 of the IUPAC manual [79WHI].

c This quantity is called "amount-of-substance concentration" in the IUPAC manual [79WHI]. A solution with a concentration equal to 0.1 mol·dm^{-3} is called a 0.1 molar solution or a 0.1 M solution.

d A solution having a molality equal to 0.1 mol·kg^{-1} is called a 0.1 molal solution or a 0.1 m solution.

e For an electrolyte $N_{\nu_+}X_{\nu_-}$ which dissociates into $\nu_{\pm}(= \nu_+ + \nu_-)$ ions, in an aqueous solution with concentration m, the individual cationic molality and activity coefficient are $m_+ (= \nu_+ m)$ and $\gamma_+ (= a_+ / m_+)$. A similar definition is used for the anionic symbols. Electrical neutrality requires that $\nu_+ z_+ = \nu_- z_-$

f Special notations for equilibrium constants are outlined in Section II.1.6. In some cases, K_c is used to indicate a concentration constant in molar units, and K_m a constant in molal units

g See Section II.3.1.

II.1.3 Chemical formulae and nomenclature

This review follows the recommendations made by IUPAC [71JEN], [77FER], [90LEI] on the nomenclature of inorganic compounds and complexes, except for the following items:

- The formulae of coordination compounds and complexes are not enclosed in square brackets [71JEN] (Rule 7.21). Exceptions are made in cases where square brackets are required to distinguish between coordinated and uncoordinated ligands.
- The prefixes "oxy-" and "hydroxy-" are retained if used in a general way, *e.g.*, "gaseous uranium oxyfluorides". For specific formula names, however, the IUPAC recommended citation [71JEN] (Rule 6.42) is used, *e.g.*, "uranium(IV) difluoride oxide" for $UF_2O(cr)$.

An IUPAC rule that is often not followed by many authors [71JEN] (Rules 2.163 and 7.21) is recalled here: the order of arranging ligands in coordination compounds and complexes is the following: central atom first, followed by ionic ligands and then by the neutral ligands. If there is more than one ionic or neutral ligand, the alphabetical order of the symbols of the ligating atoms determines the sequence of the ligands. For example, $(UO_2)_2CO_3(OH)_3^-$ is standard, $(UO_2)_2(OH)_3CO_3^-$ is non-standard and is not used.

Abbreviations of names for organic ligands appear sometimes in formulae. Following the recommendations by IUPAC, lower case letters are used, and if necessary, the ligand abbreviation is enclosed within parentheses. Hydrogen atoms that can be replaced by the metal atom are shown in the abbreviation with an upper case "H", for example: H_3edta^-, Am(Hedta)(s) (where edta stands for ethylenediaminetetraacetate).

II.1.4 Phase designators

Chemical formulae may refer to different chemical species and are often required to be specified more clearly in order to avoid ambiguities. For example, UF_4 occurs as a gas, a solid, and an aqueous complex. The distinction between the different phases is made by phase designators that immediately follow the chemical formula and appear in parentheses. The only formulae that are not provided with a phase designator are aqueous ions. They are the only charged species in this review since charged gases are not considered. The use of the phase designators is described below.

- The designator (l) is used for pure liquid substances, *e.g.*, $H_2O(l)$.
- The designator (aq) is used for undissociated, uncharged aqueous species, *e.g.*, $U(OH)_4(aq)$, $CO_2(aq)$. Since ionic gases are not considered in this review, all ions may be assumed to be aqueous and are not designed with (aq). If a chemical reaction refers to a medium other than H_2O (*e.g.*, D_2O, 90% etha-

nol/10% H_2O), then (aq) is replaced by a more explicit designator, *e.g.*, "(in D_2O)" or "(sln)". In the case of (sln), the composition of the solution is described in the text.

- The designator (sln) is used for substances in solution without specifying the actual equilibrium composition of the substance in the solution. Note the difference in the designation of H_2O in Eqs.(II.2) and (II.3). $H_2O(l)$ in Reaction (II.2) indicates that H_2O is present as a pure liquid, *i.e.*, no solutes are present, whereas Reaction (II.3) involves a HCl solution, in which the thermodynamic properties of $H_2O(sln)$ may not be the same as those of the pure liquid $H_2O(l)$. In dilute solutions, however, this difference in the thermodynamic properties of H_2O can be neglected, and $H_2O(sln)$ may be regarded as pure $H_2O(l)$.

  Example:

  $$UO_2Cl_2(cr) + 2HBr(sln) \rightleftharpoons UO_2Br_2(cr) + 2HCl(sln) \quad (II.1)$$

  $$UO_2Cl_2{\cdot}3H_2O(cr) \rightleftharpoons UO_2Cl_2{\cdot}H_2O(cr) + 2H_2O(l) \quad (II.2)$$

  $$UO_3(\gamma) + 2HCl(sln) \rightleftharpoons UO_2Cl_2(cr) + H_2O(sln) \quad (II.3)$$

- The designators (cr), (am), (vit), and (s) are used for solid substances. (cr) is used when it is known that the compound is crystalline, (am) when it is known that it is amorphous, and (vit) for glassy substances. Otherwise, (s) is used.

- In some cases, more than one crystalline form of the same chemical composition may exist. In such a case, the different forms are distinguished by separate designators that describe the forms more precisely. If the crystal has a mineral name, the designator (cr) is replaced by the first four characters of the mineral name in parentheses, *e.g.*, SiO_2(quar) for quartz and SiO_2(chal) for chalcedony. If there is no mineral name, the designator (cr) is replaced by a Greek letter preceding the formula and indicating the structural phase, *e.g.*, α-UF_5, β-UF_5.

Phase designators are also used in conjunction with thermodynamic symbols to define the state of aggregation of a compound to which a thermodynamic quantity refers. The notation is in this case the same as outlined above. In an extended notation (*cf.* [82LAF]) the reference temperature is usually given in addition to the state of aggregation of the composition of a mixture.

Example:

$\Delta_f G_m^\circ$ (Na^+, 298.15 K)	standard molar Gibbs energy of formation of aqueous Na^+ at 298.15 K
S_m° ($UO_2(SO_4)\cdot 2.5H_2O$, cr. 298.15 K)	standard molar entropy of $UO_2(SO_4)\cdot 2.5H_2O$(cr) at 298.15 K
$C_{p,m}^\circ$ (UO_3, α, 298.15 K)	standard molar heat capacity of α-UO_3 at 298.15 K
$\Delta_f H_m$ (HF, sln, HF·7.8H_2O)	enthalpy of formation of HF diluted 1:7.8 with water.

II.1.5 Processes

Chemical processes are denoted by the operator Δ, written before the symbol for a property, as recommended by IUPAC [82LAF]. An exception to this rule is the equilibrium constant, *cf.* Section II.1.6. The nature of the process is denoted by annotation of the Δ, *e.g.*, the Gibbs energy of formation, $\Delta_f G_m$, the enthalpy of sublimation, $\Delta_{sub} H_m$, *etc.* The abbreviations of chemical processes are summarised in Table II-3.

Table II-3: Abbreviations used as subscripts of Δ to denote the type of chemical process.

Subscript of Δ	Chemical process
at	separation of a substance into its constituent gaseous atoms (atomisation)
dehyd	elimination of water of hydration (dehydration)
dil	dilution of a solution
f	formation of a compound from its constituent elements
fus	melting (fusion) of a solid
hyd	addition of water of hydration to an unhydrated compound
mix	mixing of fluids
r	chemical reaction (general)
sol	process of dissolution
sub	sublimation (evaporation) of a solid
tr	transfer from one solution or liquid phase to another
trs	transition of one solid phase to another
vap	vaporisation (evaporation) of a liquid

The most frequently used symbols for processes are $\Delta_f G$ and $\Delta_f H$, the Gibbs energy and the enthalpy of formation of a compound or complex from the elements in their reference states (*cf.* Table II-6).

II.1.6 Equilibrium constants

The IUPAC has not explicitly defined the symbols and terminology for equilibrium constants of reactions in aqueous solution. The NEA has therefore adopted the conventions that have been used in the work *Stability Constants of Metal ion Complexes* by Sillén and Martell [64SIL/MAR], [71SIL/MAR]. An outline is given in the paragraphs below. Note that, for some simple reactions, there may be different correct ways to index an equilibrium constant. It may sometimes be preferable to indicate the number of the reaction to which the data refer, especially in cases where several ligands are discussed that might be confused. For example, for the equilibrium:

$$m\mathrm{M} + q\mathrm{L} \rightleftharpoons \mathrm{M}_m\mathrm{L}_q \qquad \text{(II.4)}$$

both $\beta_{q,m}$ and β (II.4) would be appropriate, and $\beta_{q,m}$ (II.4) is accepted, too. Note that, in general, K is used for the consecutive or stepwise formation constant, and β is used for the cumulative or overall formation constant. In the following outline, charges are only given for actual chemical species, but are omitted for species containing general symbols (M, L).

II.1.6.1 Protonation of a ligand

$$\mathrm{H}^+ + \mathrm{H}_{r-1}\mathrm{L} \rightleftharpoons \mathrm{H}_r\mathrm{L} \qquad K_{1,r} = \frac{[\mathrm{H}_r\mathrm{L}]}{[\mathrm{H}^+][\mathrm{H}_{r-1}\mathrm{L}]} \qquad \text{(II.5)}$$

$$r\mathrm{H}^+ + \mathrm{L} \rightleftharpoons \mathrm{H}_r\mathrm{L} \qquad \beta_{1,r} = \frac{[\mathrm{H}_r\mathrm{L}]}{[\mathrm{H}^+]^r[\mathrm{L}]} \qquad \text{(II.6)}$$

This notation has been proposed and used by Sillén and Martell [64SIL/MAR], but it has been simplified later by the same authors [71SIL/MAR] from $K_{1,r}$ to K_r. This review retains, for the sake of consistency, *cf.* Eqs.(II.7) and (II.8), the older formulation of $K_{1,r}$.

For the addition of a ligand, the notation shown in Eq.(II.7) is used.

$$\mathrm{HL}_{q-1} + \mathrm{L} \rightleftharpoons \mathrm{HL}_q \qquad K_q = \frac{[\mathrm{HL}_q]}{[\mathrm{HL}_{q-1}][\mathrm{L}]} \qquad \text{(II.7)}$$

Eq.(II.8) refers to the overall formation constant of the species $\mathrm{H}_r\mathrm{L}_q$.

$$r\mathrm{H}^+ + q\mathrm{L} \rightleftharpoons \mathrm{H}_r\mathrm{L}_q \qquad \beta_{q,r} = \frac{[\mathrm{H}_r\mathrm{L}_q]}{[\mathrm{H}^+]^r[\mathrm{L}]^q}. \qquad \text{(II.8)}$$

In Eqs.(II.5), (II.6) and (II.8), the second subscript r can be omitted if $r = 1$, as shown in Eq.(II.7).

Example:

$$H^+ + PO_4^{3-} \rightleftharpoons HPO_4^{2-} \qquad \beta_{1,1} = \beta_1 = \frac{[HPO_4^{2-}]}{[H^+][PO_4^{3-}]}$$

$$2\,H^+ + PO_4^{3-} \rightleftharpoons H_2PO_4^- \qquad \beta_{1,2} = \frac{[H_2PO_4^-]}{[H^+]^2[PO_4^{3-}]}.$$

II.1.6.2 Formation of metal ion complexes

$$ML_{q-1} + L \rightleftharpoons ML_q \qquad K_q = \frac{[ML_q]}{[ML_{q-1}][L]} \qquad (II.9)$$

$$M + {}_qL \rightleftharpoons ML_q \qquad \beta_q = \frac{[ML_q]}{[M][L]^q} \qquad (II.10)$$

For the addition of a metal ion, *i.e.*, the formation of polynuclear complexes, the following notation is used, analogous to Eq.(II.5):

$$M + M_{m-1}L \rightleftharpoons M_mL \qquad K_{1,m} = \frac{[M_mL]}{[M][M_{m-1}L]}. \qquad (II.11)$$

Eq.(II.12) refers to the overall formation constant of a complex M_mL_q

$$mM + qL \rightleftharpoons M_mL_q \qquad \beta_{q,m} = \frac{[M_mL_q]}{[M]^m[L]^q}. \qquad (II.12)$$

The second index can be omitted if it is equal to 1, *i.e.*, $\beta_{q,m}$ becomes β_q if $m = 1$. The formation constants of mixed ligand complexes are not indexed. In this case, it is necessary to list the chemical reactions considered and to refer the constants to the corresponding reaction numbers.

It has sometimes been customary to use negative values for the indices of the protons to indicate complexation with hydroxide ions, OH^-. This practice is not adopted in this review. If OH^- occurs as a reactant in the notation of the equilibrium, it is treated like a normal ligand L, but in general formulae the index variable n is used instead of q. If H_2O occurs as a reactant to form hydroxide complexes, H_2O is considered as a protonated ligand, HL, so that the reaction is treated as described below in Eqs.(II.13) to (II.15) using n as the index variable. For convenience, no general form is used for the stepwise constants for the formation of the complex $M_mL_qH_r$. In many experiments, the formation constants of metal ion complexes are determined by adding a ligand in its protonated form to a metal ion solution. The complex formation reactions thus involve a deprotonation reaction of the ligand. If this is the case, the equilibrium constant is supplied with an asterisk, as shown in Eqs.(II.13) and (II.14) for mononuclear and in Eq.(II.15) for polynuclear complexes.

$$\mathrm{ML}_{q-1} + \mathrm{HL} \rightleftharpoons \mathrm{ML}_q + \mathrm{H}^+ \qquad {}^*K_q = \frac{[\mathrm{ML}_q][\mathrm{H}^+]}{[\mathrm{ML}_{q-1}][\mathrm{HL}]} \tag{II.13}$$

$$\mathrm{M} + q\mathrm{HL} \rightleftharpoons \mathrm{ML}_q + q\mathrm{H}^+ \qquad {}^*\beta_q = \frac{[\mathrm{ML}_q][\mathrm{H}^+]^q}{[\mathrm{M}][\mathrm{HL}]^q} \tag{II.14}$$

$$m\mathrm{M} + q\mathrm{HL} \rightleftharpoons \mathrm{M}m\mathrm{L}q + q\mathrm{H}^+ \qquad {}^*\beta_{q,m} = \frac{[\mathrm{M}_m\mathrm{L}_q][\mathrm{H}^+]^q}{[\mathrm{M}]^m[\mathrm{HL}]^q} \tag{II.15}$$

Example:

$$\mathrm{UO_2^{2+}} + \mathrm{HF(aq)} \rightleftharpoons \mathrm{UO_2F^+} + \mathrm{H^+} \qquad {}^*K_1 = {}^*\beta_1 = \frac{[\mathrm{UO_2F^+}][\mathrm{H^+}]}{[\mathrm{UO_2^{2+}}][\mathrm{HF\ (aq)}]}$$

$$3\mathrm{UO_2^{2+}} + 5\mathrm{H_2O(l)} \rightleftharpoons \mathrm{(UO_2)_3(OH)_5^+} + 5\mathrm{H^+} \qquad {}^*\beta_{5,3} = \frac{[\mathrm{(UO_2)_3(OH)_5^+}][\mathrm{H^+}]^5}{[\mathrm{UO_2^{2+}}]^3}.$$

Note that an asterisk is only assigned to the formation constant if the protonated ligand that is added is deprotonated during the reaction. If a protonated ligand is added and coordinated as such to the metal ion, the asterisk is to be omitted, as shown in Eq.(II.16)

$$\mathrm{M} + q\mathrm{H}_r\mathrm{L} \rightleftharpoons \mathrm{M(H}_r\mathrm{L)}q \qquad \beta_q = \frac{[\mathrm{M(H}_r\mathrm{L)}_q]}{[\mathrm{M}][\mathrm{H}_r\mathrm{L}]^q}. \tag{II.16}$$

Example:

$$\mathrm{UO_2^{2+}} + 3\mathrm{PO_4^-} \rightleftharpoons \mathrm{UO_2(H_2PO_4)_3^-} \qquad \beta_3 = \frac{[\mathrm{UO_2(H_2PO_4)_3^-}]}{[\mathrm{UO_2^{2+}}][\mathrm{H_2PO_4^-}]^3}.$$

II.1.6.3 Solubility constants

Conventionally, equilibrium constants involving a solid compound are denoted as "solubility constants" rather than as formation constants of the solid. An index "s" to the equilibrium constant indicates that the constant refers to a solubility process, as shown in Eqs.(II.17) to (II.19)

$$\mathrm{M}_a\mathrm{L}_b(\mathrm{s}) \rightleftharpoons a\mathrm{M} + b\mathrm{L} \qquad K_{s,0} = [\mathrm{M}]^a\,[\mathrm{L}]^b. \qquad \text{(II.17)}$$

$K_{s,0}$ is the conventional solubility product, and the subscript "0" indicates that the equilibrium reaction involves only uncomplexed aqueous species. If the solubility constant includes the formation of aqueous complexes, a notation analogous to that of Eq.(II.12) is used:

$$\frac{m}{a}\mathrm{M}_a\mathrm{L}_b(\mathrm{s}) \rightleftharpoons \mathrm{M}_m\mathrm{L}_q + \left(\frac{mb}{a} - q\right)\mathrm{L} \qquad K_{s,q,m} = \left[\mathrm{M}_m\mathrm{L}_q\right]\left[\mathrm{L}\right]^{(\frac{mb}{a}-q)}. \qquad \text{(II.18)}$$

Example:

$$\mathrm{UO_2F_2(cr)} \rightleftharpoons \mathrm{UO_2F^+} + \mathrm{F^-} \qquad K_{s,1,1} = K_{s,1} = \left[\mathrm{UO_2F^+}\right]\left[\mathrm{F^-}\right].$$

Similarly, an asterisk is added to the solubility constant if it simultaneously involves a protonation equilibrium:

$$\frac{m}{a}\mathrm{M}_a\mathrm{L}_b(\mathrm{s}) + \left(\frac{mb}{a} - q\right)\mathrm{H}^+ \rightleftharpoons \mathrm{M}_m\mathrm{L}_q + \left(\frac{mb}{a} - q\right)\mathrm{HL}$$

$$^*K_{s,q,m} = \frac{\left[\mathrm{M}_m\mathrm{L}_q\right]\left[\mathrm{HL}\right]^{(\frac{mb}{a}-q)}}{\left[\mathrm{H}^+\right]^{(\frac{mb}{a}-q)}}. \qquad \text{(II.19)}$$

Example:

$$\mathrm{U(HPO_4)_2 \cdot 4H_2O(cr)} + \mathrm{H^+} \rightleftharpoons \mathrm{UHPO_4^{2+}} + \mathrm{H_2PO_4^-} + 4\,\mathrm{H_2O(l)}$$

$$^*K_{s,1,1} = {^*K_{s,1}} = \frac{\left[\mathrm{UHPO_4^{2+}}\right]\left[\mathrm{H_2PO_4^-}\right]}{\left[\mathrm{H^+}\right]}.$$

II.1.6.4 Equilibria involving the addition of a gaseous ligand

A special notation is used for constants describing equilibria that involve the addition of a gaseous ligand, as outlined in Eq.(II.20):

$$\mathrm{ML}_{q-1} + \mathrm{L(g)} \rightleftharpoons \mathrm{ML}_q \qquad K_{\mathrm{p},q} = \frac{\left[\mathrm{ML}_q\right]}{\left[\mathrm{ML}_{q-1}\right]p_\mathrm{L}}. \qquad \text{(II.20)}$$

The subscript "p" can be combined with any other notations given above.

Example:

$$\mathrm{CO_2(g)} \rightleftharpoons \mathrm{CO_2(aq)} \qquad K_\mathrm{p} = \frac{\left[\mathrm{CO_2(aq)}\right]}{p_{\mathrm{CO_2}}},$$

$$3\,\mathrm{UO_2^{2+}} + 6\,\mathrm{CO_2(g)} + 6\,\mathrm{H_2O(l)} \rightleftharpoons \mathrm{(UO_2)_3(CO_3)_6^{6-}} + 12\,\mathrm{H^+}$$

$$^{*}\beta_{p,6,3} = \frac{[(UO_2)_3(CO_3)_6^{6-}][H^+]^{12}}{[UO_2^{2+}]^3\, p_{CO_2}^6},$$

$$UO_2CO_3(cr) + CO_2(g) + H_2O(l) \rightleftharpoons UO_2(CO_3)_2^{2-} + 2\,H^+$$

$$^{*}K_{p,s,2} = \frac{[UO_2(CO_3)_2^{2-}][H^+]^2}{p_{CO_2}}.$$

In cases where the subscripts become complicated, it is recommended that K or β be used with or without subscripts, but always followed by the equation number of the equilibrium to which it refers.

II.1.6.5 Redox equilibria

Redox reactions are usually quantified in terms of their electrode (half cell) potential, E, which is identical to the electromotive force (emf) of a galvanic cell in which the electrode on the left is the standard hydrogen electrode, SHE[1], in accordance with the "1953 Stockholm Convention" [88MIL/CVI]. Therefore, electrode potentials are given as reduction potentials relative to the standard hydrogen electrode, which acts as an electron donor. In the standard hydrogen electrode, $H_2(g)$ is at unit fugacity (an ideal gas at unit pressure, 0.1 MPa), and H^+ is at unit activity. The sign of the electrode potential, E, is that of the observed sign of its polarity when coupled with the standard hydrogen electrode. The standard electrode potential, E°, *i.e.*, the potential of a standard galvanic cell relative to the standard hydrogen electrode (all components in their standard state, *cf.* Section II.3.1, and with no liquid junction potential) is related to the standard Gibbs energy change $\Delta_r G_m^\circ$ and the standard (or thermodynamic) equilibrium constant K° as outlined in Eq.(II.21).

$$E^\circ = -\frac{1}{nF}\Delta_r G_m^\circ = \frac{RT}{nF}\ln K^\circ \qquad \text{(II.21)}$$

and the potential, E, is related to E° by:

$$E = E^\circ - (RT/nF)\Sigma \nu_i \ln a_i\,. \qquad \text{(II.22)}$$

For example, for the hypothetical galvanic cell:

$$\text{Pt} \,\Big|\, \text{H}_2(\text{g}, p = 1\text{ bar}) \,\Big|\, \text{HCl(aq}, a_\pm = 1, f_{H_2} = 1) \,\vdots\, \begin{matrix}\text{Fe(ClO}_4)_2\text{ (aq}, a_{Fe^{2+}} = 1)\\ \text{Fe(ClO}_4)_3\text{ (aq}, a_{Fe^{3+}} = 1)\end{matrix} \,\Big|\, \text{Pt} \qquad \text{(II.23)}$$

where $\vdots$ denotes a liquid junction and $|$ a phase boundary, the reaction is:

[1] The definitions of SHE and NHE are given in Section II.1.1.

$$\mathrm{Fe^{3+} + \tfrac{1}{2}H_2(g) \rightleftharpoons Fe^{2+} + H^+}. \quad \text{(II.24)}$$

Formally Reaction (II.24) can be represented by two half cell reactions, each involving an equal number of electrons, (designated "e^-"), as shown in the following equations:

$$\mathrm{Fe^{3+} + e^- \rightleftharpoons Fe^{2+}} \quad \text{(II.25)}$$

$$\mathrm{\tfrac{1}{2}H_2(g) \rightleftharpoons H^+ + e^-}. \quad \text{(II.26)}$$

The terminology is useful, although it must be emphasised that "e^-" here does not represent the hydrated electron.

The equilibrium constants of the two half cell reactions may be written:

$$K^\circ\text{(II.25)} = \frac{a_{\mathrm{Fe^{2+}}}}{a_{\mathrm{Fe^{3+}}} \cdot a_{\mathrm{e^-}}} \quad \text{(II.27)}$$

$$K^\circ\text{(II.26)} = \frac{a_{\mathrm{H^+}} \cdot a_{\mathrm{e^-}}}{\sqrt{f_{\mathrm{H_2}}}} = 1. \quad \text{(II.28)}$$

The "Stockholm Convention" implies that the constant K°(II.26) = 1, which also defines the numerical value of a_{e^-} to 1 for the standard hydrogen electrode. In addition, $\Delta_r G_m^\circ$(II.26) = 0, $\Delta_r H_m^\circ$(II.26) = 0, $\Delta_r S_m^\circ$(II.26) = 0 by definition, at all temperatures, and therefore $\Delta_r G_m^\circ$(II.25) = $\Delta_r G_m^\circ$(II.24).

The equilibrium constant of the more general cell reaction:

$$\mathrm{Ox} + \tfrac{1}{2}n\,\mathrm{H_2(g)} \rightleftharpoons \mathrm{Red} + n\,\mathrm{H^+} \quad \text{(II.29)}$$

and of the half cell reaction

$$\mathrm{Ox} + n\,\mathrm{e^-} \rightleftharpoons \mathrm{Red} \quad \text{(II.30)}$$

together with K°(II.26) yield

$$K^\circ\text{(II.29)} = K^\circ\text{(II.30)} \times K^\circ\text{(II.26)}^n = \frac{a_{\mathrm{Red}}}{a_{\mathrm{Ox}}} \frac{1}{(a_{\mathrm{e^-}}/(a_{\mathrm{e^-}}(\mathrm{SHE}){=}1))^n} = \frac{a_{\mathrm{Red}}}{a_{\mathrm{Ox}}} \frac{1}{a_{\mathrm{e^-}}^n}. \quad \text{(II.31)}$$

The activity scale is thus fixed by the above convention. Further

$$E\text{(II.29)} = E^\circ\text{(II.29)} - \frac{RT\ln(10)}{nF} \log_{10}\left(\frac{a_{\mathrm{Red}}}{a_{\mathrm{Ox}}}\right) \quad \text{(II.32)}$$

Combination of Eqs.((II.21), (II.31) and (II.32)) yields the relationship between a_{e^-} and the electrode potential *vs*. SHE:

$$-\log_{10} a_{\mathrm{e^-}} = \frac{F}{RT\ln(10)} E\text{ (II.29)} \quad \text{(II.33)}$$

The splitting of redox reactions into two half cell reactions by introducing the symbol "e^-" is highly useful. It should be noted that the e^- notation does not in any way refer to solvated electrons. When calculating the equilibrium composition of a chemical system, both "e^-", and H^+ can be chosen as components and they can be treated numerically in a similar way: equilibrium constants, mass balances, *etc.* may be defined for both. However, while H^+ represents the hydrated proton in aqueous solution, the above equations use only the activity of "e^-", and never the concentration of "e^-". Concentration to activity conversions (or activity coefficients) are never needed for the electron (*cf.* Appendix B, Example B.3).

In the literature on geochemical modelling of natural waters, it is customary to represent the "electron activity" of an aqueous solution with the symbol "pe" or "pε"($= -\log_{10} a_{e^-}$) by analogy with pH ($= -\log_{10} a_{H^+}$), and the redox potential of an aqueous solution relative to the standard hydrogen electrode is usually denoted by either "Eh" or "E_H" (see for example [81STU/MOR], [82DRE], [84HOS], [86NOR/MUN]). The two representations are interrelated via Eq.(II.33) yielding the expression:

$$\text{pe} = \frac{F}{RT\ln(10)} E_{\text{H}}.$$

The symbol $E^{'o}$ is used to denote the so-called "formal potential" [74PAR]. The formal (or "conditional") potential can be regarded as a standard potential for a particular medium in which the activity coefficients are independent (or approximately so) of the reactant concentrations [85BAR/PAR] (the definition of $E^{'o}$ parallels that of "concentration quotients" for equilibria). Therefore, from

$$E = E^{'o} - \frac{RT}{nF} \sum \nu_i \ln c_i \qquad \text{(II.34)}$$

$E^{'o}$ is the potential E for a cell when the ratio of the *concentrations* (not the activities) on the right–hand side and the left–hand side of the cell reaction is equal to unity, and

$$E^{'o} = E^o - \frac{RT}{nF} \sum \nu_i \ln \varrho \gamma_i = -\frac{\Delta_r G_m}{nF} \qquad \text{(II.35)}$$

where the γ_i are the molality activity coefficients and ϱ is (m_i/c_i), the ratio of molality to molarity (*cf.* Section 0). The medium must be specified.

II.1.7 pH

Because of the importance that potentiometric methods have in the determination of equilibrium constants in aqueous solutions, a short discussion on the definition of "pH" and a simplified description of the experimental techniques used to measure pH will be given here.

The acidity of aqueous solutions is often expressed in a logarithmic scale of the hydrogen ion activity. The definition of pH as:

$$\mathrm{pH} = -\log_{10} a_{\mathrm{H^+}} = -\log_{10}(m_{\mathrm{H^+}}\gamma_{\mathrm{H^+}})$$

can only be strictly used in the limiting range of the Debye-Hückel equation (that is, in extremely dilute solutions). In practice the use of pH values requires extra assumptions on the values for single ion activities. In this review values of pH are used to describe qualitatively ranges of acidity of experimental studies, and the assumptions described in Appendix B are used to calculate single ion activity coefficients.

The determination of pH is often performed by emf measurements of galvanic cells involving liquid junctions [69ROS], [73BAT]. A common setup is a cell made up of a reference half cell (*e.g.* Ag(s)/AgCl(s) in a solution of constant chloride concentration), a salt bridge, the test solution, and a glass electrode (which encloses a solution of constant acidity and an internal reference half cell):

$$\mathrm{Pt(s)}\;\Big|\;\mathrm{Ag(s)}\;\Big|\;\mathrm{AgCl(s)}\;\Big|\;\mathrm{KCl(aq)}\;\underset{a}{\vdots}\;\begin{matrix}\text{salt}\\ \text{bridge}\end{matrix}\;\underset{b}{\vdots}\;\begin{matrix}\text{test}\\ \text{solution}\end{matrix}\;\wr\;\mathrm{KCl(aq)}\;\Big|\;\mathrm{AgCl(s)}\;\Big|\;\mathrm{Ag(s)}\;\Big|\;\mathrm{Pt(s)} \qquad \text{(II.36)}$$

where $\wr$ stands for a glass membrane.

The emf of such a cell (assuming Nernstian behaviour of the glass electrode) is given by:

$$E = E^* + \frac{RT}{F}\ln a_{\mathrm{H^+}} + E_{\mathrm{j}}$$

where E^* is a constant, and E_{j} is the liquid junction potential. The purpose of the salt bridge is to minimise the junction potential in junction "*b*", while keeping constant the junction potential for junction "*a*". Two methods are most often used to reduce and control the value of E_{j}. An electrolyte solution of *high* concentration (the "salt bridge") is a requirement of both methods. In the first method, the salt bridge is a saturated (or nearly saturated) solution of potassium chloride. A problem with a bridge of high potassium concentration, is that potassium perchlorate might precipitate[2] inside the liquid junction when the test solution contains a high concentration of perchlorate ions.

In the other method the salt bridge contains the same *high* concentration of the same inert electrolyte as the test solution (for example, 3 M $NaClO_4$). However, if the concentration of the background electrolyte in the salt bridge and test solutions is reduced, the values of E_{j} are dramatically increased. For example, if both the bridge and the test solution have $[ClO_4^-] = 0.1$ M as background electrolyte, the dependence of the

---

[2] $KClO_4$(cr) has a solubility of ≈ 0.15 M in pure water at 25°C

liquid junction at "b" on acidity is $E_j \approx -440 \times [H^+]$ mV·dm^3·mol^{-1} at 25°C [69ROS] (p.110), which corresponds to a correction at pH = 2 of ≥ 0.07 pH units.

Because of the problems in eliminating the liquid junction potentials and in defining individual ionic activity coefficients, an "operational" definition of pH is given by IUPAC [88MIL/CVI]. This definition involves the measurement of pH differences between the test solution and standard solutions of known pH and similar ionic strength (in this way similar values of γ_{H^+} and E_j cancel each other when emf values are subtracted).

In order to deduce the stoichiometry and equilibrium constants of complex formation reactions and other equilibria, it is necessary to vary the concentrations of reactants and products over fairly large concentration ranges under conditions where the activity coefficients of the species are either known, or constant. Only in this manner is it possible to use the mass balance equations for the various components together with the measurement of one or more free concentrations to obtain the information desired [61ROS/ROS], [90BEC/NAG], [97ALL/BAN], p. 326 – 327. For equilibria involving hydrogen ions, it is necessary to use concentration units, rather than hydrogen ion activity. For experiments in an ionic medium, where the concentration of an "inert" electrolyte is much larger than the concentration of reactants and products we can ensure that, as a first approximation, their trace activity coefficients remain constant even for moderate variations of the corresponding total concentrations. Under these conditions of fixed ionic strength the free proton concentration may be measured directly, thereby defining it in terms of $-\log_{10}[H^+]$ rather than on the activity scale as pH, and the value of $-\log_{10}[H^+]$ and pH will differ by a constant term, *i.e.*, $\log_{10}\gamma_{H^+}$. Equilibrium constants deduced from measurements in such ionic media are therefore *conditional* constants, because they refer to the given medium, not to the standard state. In order to compare the magnitude of equilibrium constants obtained in different ionic media it is necessary to have a method for estimating activity coefficients of ionic species in mixed electrolyte systems to a *common* standard state. Such procedures are discussed in Appendix B.

Note that the precision of the measurement of $-\log_{10}[H^+]$ and pH is virtually the same, in very good experiments, ± 0.001. However, the accuracy is generally considerably poorer, depending in the case of glass electrodes largely on the response of the electrode (linearity, age, pH range, *etc.*), and to a lesser extent on the calibration method employed.

II.1.8 Spectroscopic constants and statistical mechanics calculations for gaseous species

The thermal functions for a number of gaseous species have been calculated by statistical-mechanical relations (see for example Chapter 27 of [61LEW/RAN]). The parameters defining the vibrational and rotational energy levels of the molecule in terms of the rotational (J) and vibrational (v) quantum numbers, and thus many of its thermodynamic properties, are:

- for diatomic molecules (non-rigid rotator, anharmonic oscillator approximation): ω (vibrational frequency in wavenumber units), x (anharmonicity constant), B (rotational constant for equilibrium position), D (centrifugal distortion constant), α (rotational constant correction for excited vibrational states), and σ (symmetry number), where the energy levels with quantum numbers v and J are given by:

$$E_{(v,J)}/hc = \omega\,(v+½) - \omega\,x(v+½)^2 + B\,J(J+1) - D\,J^2(J+1)^2 - \alpha(v+½)\,J\,(J+1) \quad \text{(II.37)}$$

- for linear polyatomic molecules, the parameters are the same as those for diatomic molecules, except that the contributions for anharmonicity are usually neglected.

- for non-linear polyatomic molecules (rigid rotator, harmonic oscillator approximation): $I_x\,I_y\,I_z$, the product of the principal moments of inertia (readily calculated from the geometrical structure of the molecule), v(i), the vibration frequencies and σ, the symmetry number. While the vibrational energy levels for polyatomic molecules are given approximately by the first term of Eq.(II.37) for each of the normal vibrations, the rotational energy levels cannot be expressed as a simple general formula. However, the required rotational partition function can be expressed with sufficient accuracy simply in terms of the product of the principal moments of inertia. As for linear polyatomic molecules, anharmonic contributions are usually neglected. In each case, the symmetry number σ, the number of indistinguishable positions into which the molecule can be turned by simple rotations, is required to calculate the correct entropy.

The relations for calculating the thermal functions from the partition function defined by the energy levels are well-known – again, see Chapter 27 of [61LEW/RAN], for a simple description. In each case, the relevant translational and electronic contributions (calculated from the molar mass and the electronic energy levels and degeneracies) must be added. Except where accurate spectroscopic data exist, the geometry and parameters of the excited states are assumed to be the same as those for the ground state

II.1.9 Order of formulae

To be consistent with CODATA, the data tables are given in "*Standard Order of Arrangement*" [82WAG/EVA]. This scheme is presented in Figure II-1 below, and shows the sequence of the ranks of the elements in this convention. The order follows the ranks of the elements.

Figure II-1: Standard order of arrangement of the elements and compounds based on the periodic classification of the elements (from Ref.[82WAG/EVA])

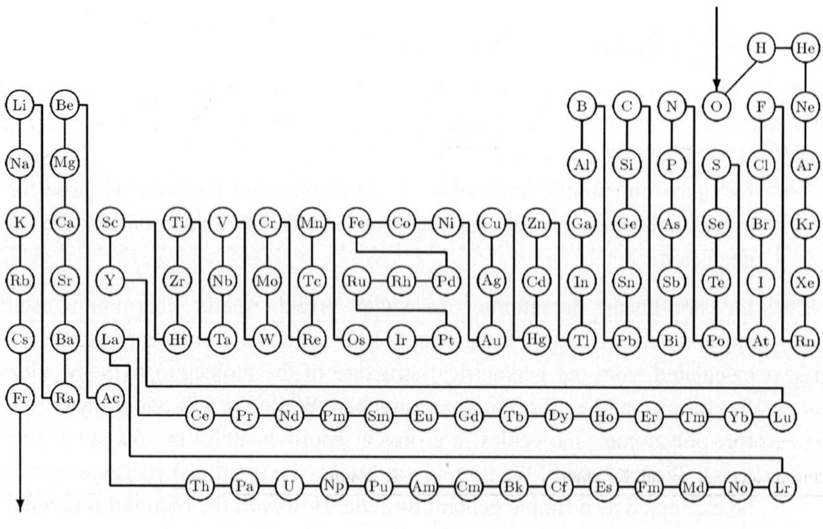

For example, for uranium, this means that, after elemental uranium and its monoatomic ions (*e.g.*, U^{4+}), the uranium compounds and complexes with oxygen would be listed, then those with hydrogen, then those with oxygen and hydrogen, and so on, with decreasing rank of the element and combinations of the elements. Within a class, increasing coefficients of the lower rank elements go before increasing coefficients of the higher rank elements. For example, in the U–O–F class of compounds and complexes, a typical sequence would be $UOF_2(cr)$, $UOF_4(cr)$, $UOF_4(g)$, $UO_2F(aq)$, UO_2F^+, $UO_2F_2(aq)$, $UO_2F_2(cr)$, $UO_2F_2(g)$, $UO_2F_3^-$, $UO_2F_4^{2-}$, $U_2O_3F_6(cr)$, *etc.* [92GRE/FUG]. Formulae with identical stoichiometry are in alphabetical order of their designators. An exception is made for ammonium salts, which are included in group 1 before the alkali metals.

II.1.10 Reference codes

The references cited in the review are ordered chronologically and alphabetically by the first two authors within each year, as described by CODATA [87GAR/PAR]. A reference code is made up of the final two digits of the year of appearance (if the publication is not from the 20th century, the year will be put in full). The year is followed by the first three letters of the surnames of the first two authors, separated by a slash.

If there are multiple reference codes, a "2" will be added to the second one, a "3" to the third one, and so forth. Reference codes are always enclosed in square brackets.

II.2 Units and conversion factors

Thermodynamic data are given according to the *Système International d'unités* (SI units). The unit of energy is the joule. Some basic conversion factors, also for non-thermodynamic units, are given in Table II-4.

Table II-4: Unit conversion factors

To convert from (non-SI unit symbol)	to (SI unit symbol)	multiply by
ångström (Å)	metre (m)	1×10^{-10} (exactly)
standard atmosphere (atm)	pascal (Pa)	1.01325×10^{5} (exactly)
bar (bar)	pascal (Pa)	1×10^{5} (exactly)
thermochemical calorie (cal)	joule (J)	4.184 (exactly)
entropy unit e.u. $\hat{=}$ $\mathrm{cal \cdot K^{-1} \cdot mol^{-1}}$	$\mathrm{J \cdot K^{-1} \cdot mol^{-1}}$	4.184 (exactly)

Since a large part of the NEA-TDB project deals with the thermodynamics of aqueous solutions, the units describing the amount of dissolved substance are used very frequently. For convenience, this review uses "M" as an abbreviation of "$\mathrm{mol \cdot dm^{-3}}$" for molarity, *c*, and, in Appendices B and C, "m" as an abbreviation of "$\mathrm{mol \cdot kg^{-1}}$" for molality, *m*. It is often necessary to convert concentration data from molarity to molality and vice versa. This conversion is used for the correction and extrapolation of equilibrium data to zero ionic strength by the specific ion interaction theory, which works in molality units (*cf.* Appendix B). This conversion is made in the following way. Molality is defined as m_B moles of substance B dissolved in 1 kilogram of pure water. Molarity is defined as c_B moles of substance B dissolved in $(\rho - c_\mathrm{B}M)$ kilogram of pure water, where ρ is the density of the solution in $\mathrm{kg \cdot dm^{-3}}$ and M_B the molar weight of the solute in $\mathrm{kg \cdot mol^{-1}}$.

From this it follows that:

$$m_\mathrm{B} = \frac{c_\mathrm{B}}{\rho - c_\mathrm{B}M}.$$

When the ionic strength is kept high and constant by an inert electrolyte, I, the ratio $m_\mathrm{B}/c_\mathrm{B}$ can be approximated by:

$$\frac{m_B}{c_B} \approx \frac{1}{\rho - c_I M_I}$$

where c_I is the concentration of the inert electrolyte in mol·dm^{-3} and M_I its molar mass in kg·mol^{-1}.

Baes and Mesmer [76BAE/MES], (*p*.439) give a table with conversion factors (from molarity to molality) for nine electrolytes and various ionic strengths. Conversion factors at 298.15 K for twenty four electrolytes, calculated using the density equations reported by Söhnel and Novotný [85SOH/NOV], are reported in Table II-5.

Example:

1.00 M $NaClO_4$ ≙ 1.05 m $NaClO_4$

2.00 M NaCl ≙ 1.02 m NaCl

4.00 M $NaClO_4$ ≙ 4.95 m $NaClO_4$

6.00 M $NaNO_3$ ≙ 7.55 m $NaNO_3$

It should be noted that equilibrium constants need also to be converted if the concentration scale is changed from molarity to molality or vice versa. For a general equilibrium reaction, $0 = \sum_B \nu_B B$, the equilibrium constants can be expressed either in molarity or molality units, K_c or K_m, respectively:

$$\log_{10} K_c = \sum_B \nu_B \log_{10} c_B$$

$$\log_{10} K_m = \sum_B \nu_B \log_{10} m_B$$

With $(m_B / c_B) = \varrho$, or $(\log_{10} m_B - \log_{10} c_B) = \log_{10} \varrho$, the relationship between K_c and K_m becomes very simple, as shown in Eq.(II.38)

$$\log_{10} K_m = \log_{10} K_c + \sum_B \nu_B \log_{10} \varrho . \qquad \text{(II.38)}$$

$\sum_B \nu_B$ is the sum of the stoichiometric coefficients of the reaction, *cf.* Eq.(II.54) and the values of ϱ are the factors for the conversion of molarity to molality as tabulated in Table II-5 for several electrolyte media at 298.15 K. In the case of very dilute solutions, these factors are approximately equal to the reciprocal of the density of the pure solvent. Then, if the solvent is water, molarity and molality may be used interchangeably, and $K_c \approx K_m$. The differences between the values in Table II-5 and the values listed in the uranium NEA-TDB review [92GRE/FUG] (*p*.23) are found at the highest concentrations, and are no larger than ± 0.003 dm^3·kg^{-1}, reflecting the accuracy expected in this type of conversion. The uncertainty introduced by the use of Eq. (II.38) in the values of $\log_{10} K_m$ will be no larger than ± 0.001 $\sum_B \nu_B$.

Table II-5: Factors ϱ for the conversion of molarity, c_B, to molality, m_B, of a substance B, in various media at 298.15 K (calculated from densities in [85SOH/NOV])

	$\varrho = m_B / c_B$ (dm^3 of solution per kg of H_2O)							
c (M)	$HClO_4$	$NaClO_4$	$LiClO_4$	NH_4ClO_4	$Ba(ClO_4)_2$	HCl	NaCl	LiCl
0.10	1.0077	1.0075	1.0074	1.0091	1.0108	1.0048	1.0046	1.0049
0.25	1.0147	1.0145	1.0141	1.0186	1.0231	1.0076	1.0072	1.0078
0.50	1.0266	1.0265	1.0256	1.0351	1.0450	1.0123	1.0118	1.0127
0.75	1.0386	1.0388	1.0374	1.0523	1.0685	1.0172	1.0165	1.0177
1.00	1.0508	1.0515	1.0496	1.0703	1.0936	1.0222	1.0215	1.0228
1.50	1.0759	1.0780	1.0750	1.1086	1.1491	1.0324	1.0319	1.0333
2.00	1.1019	1.1062	1.1019		1.2125	1.0430	1.0429	1.0441
3.00	1.1571	1.1678	1.1605		1.3689	1.0654	1.0668	1.0666
4.00	1.2171	1.2374	1.2264			1.0893	1.0930	1.0904
5.00	1.2826	1.3167				1.1147	1.1218	1.1156
6.00	1.3547	1.4077				1.1418		1.1423
c (M)	KCl	NH_4Cl	$MgCl_2$	$CaCl_2$	NaBr	HNO_3	$NaNO_3$	$LiNO_3$
0.10	1.0057	1.0066	1.0049	1.0044	1.0054	1.0056	1.0058	1.0059
0.25	1.0099	1.0123	1.0080	1.0069	1.0090	1.0097	1.0102	1.0103
0.50	1.0172	1.0219	1.0135	1.0119	1.0154	1.0169	1.0177	1.0178
0.75	1.0248	1.0318	1.0195	1.0176	1.0220	1.0242	1.0256	1.0256
1.00	1.0326	1.0420	1.0258	1.0239	1.0287	1.0319	1.0338	1.0335
1.50	1.0489	1.0632	1.0393	1.0382	1.0428	1.0478	1.0510	1.0497
2.00	1.0662	1.0855	1.0540	1.0546	1.0576	1.0647	1.0692	1.0667
3.00	1.1037	1.1339	1.0867	1.0934	1.0893	1.1012	1.1090	1.1028
4.00	1.1453	1.1877	1.1241	1.1406	1.1240	1.1417	1.1534	1.1420
5.00		1.2477		1.1974	1.1619	1.1865	1.2030	1.1846
6.00					1.2033	1.2361	1.2585	1.2309
c (M)	NH_4NO_3	H_2SO_4	Na_2SO_4	$(NH_4)_2SO_4$	H_3PO_4	Na_2CO_3	K_2CO_3	NaSCN
0.10	1.0077	1.0064	1.0044	1.0082	1.0074	1.0027	1.0042	1.0069
0.25	1.0151	1.0116	1.0071	1.0166	1.0143	1.0030	1.0068	1.0130
0.50	1.0276	1.0209	1.0127	1.0319	1.0261	1.0043	1.0121	1.0234
0.75	1.0405	1.0305	1.0194	1.0486	1.0383	1.0065	1.0185	1.0342
1.00	1.0539	1.0406	1.0268	1.0665	1.0509	1.0094	1.0259	1.0453
1.50	1.0818	1.0619	1.0441	1.1062	1.0773	1.0170	1.0430	1.0686
2.00	1.1116	1.0848		1.1514	1.1055	1.0268	1.0632	1.0934
3.00	1.1769	1.1355		1.2610	1.1675		1.1130	1.1474
4.00	1.2512	1.1935		1.4037	1.2383		1.1764	1.2083
5.00	1.3365	1.2600			1.3194		1.2560	1.2773
6.00	1.4351	1.3365			1.4131			1.3557

II.3 Standard and reference conditions

II.3.1 Standard state

A precise definition of the term "standard state" has been given by IUPAC [82LAF]. The fact that only changes in thermodynamic parameters, but not their absolute values, can be determined experimentally, makes it important to have a well-defined standard state that forms a base line to which the effect of variations can be referred. The IUPAC [82LAF] definition of the standard state has been adopted in the NEA-TDB project. The standard state pressure, $p^\circ = 0.1$ MPa (1 bar), has therefore also been adopted, *cf.* Section II.3.2. The application of the standard state principle to pure substances and mixtures is summarised below. It should be noted that the standard state is always linked to a reference temperature, *cf.* Section II.3.3.

- The standard state for a gaseous substance, whether pure or in a gaseous mixture, is the pure substance at the standard state pressure and in a (hypothetical) state in which it exhibits ideal gas behaviour.
- The standard state for a pure liquid substance is (ordinarily) the pure liquid at the standard state pressure.
- The standard state for a pure solid substance is (ordinarily) the pure solid at the standard state pressure.
- The standard state for a solute B in a solution is a hypothetical liquid solution, at the standard state pressure, in which $m_B = m^\circ = 1 \text{ mol} \cdot \text{kg}^{-1}$, and in which the activity coefficient γ_B is unity.

It should be emphasised that the use of superscript, $^\circ$, *e.g.*, in $\Delta_\text{f} H^\circ_\text{m}$, implies that the compound in question is in the standard state and that the elements are in their reference states. The reference states of the elements at the reference temperature (*cf.* Section II.3.3) are listed in Table II-6.

Table II-6: Reference states for some elements at the reference temperature of 298.15 K and standard pressure of 0.1 MPa [82WAG/EVA], [89COX/WAG], [91DIN].

O_2	gaseous	Al	crystalline, cubic
H_2	gaseous	Zn	crystalline, hexagonal
He	gaseous	Cd	crystalline, hexagonal
Ne	gaseous	Hg	liquid
Ar	gaseous	Cu	crystalline, cubic
Kr	gaseous	Ag	crystalline, cubic
Xe	gaseous	Fe	crystalline, cubic, bcc
F_2	gaseous	Tc	crystalline, hexagonal
Cl_2	gaseous	V	crystalline, cubic
Br_2	liquid	Ti	crystalline, hexagonal
I_2	crystalline, orthorhombic	Am	crystalline, dhcp
S	crystalline, orthorhombic	Pu	crystalline, monoclinic
Se	crystalline, trigonal (*)	Np	crystalline, orthorhombic
Te	crystalline, hexagonal	U	crystalline, orthorhombic
N_2	gaseous	Th	crystalline, cubic
P	crystalline, cubic ("white")	Be	crystalline, hexagonal
As	crystalline, rhombohedral ("grey")	Mg	crystalline, hexagonal
Sb	crystalline, rhombohedral	Ca	crystalline, cubic, fcc
Bi	crystalline, rhombohedral	Sr	crystalline, cubic, fcc
C	crystalline, hexagonal (graphite)	Ba	crystalline, cubic
Si	crystalline, cubic	Li	crystalline, cubic
Ge	crystalline, cubic	Na	crystalline, cubic
Sn	crystalline, tetragonal ("white")	K	crystalline, cubic
Pb	crystalline, cubic	Rb	crystalline, cubic
B	β, crystalline, rhombohedral	Cs	crystalline, cubic

* This review

II.3.2 Standard state pressure

The standard state pressure chosen for all selected data is 0.1 MPa (1 bar) as recommended by the International Union of Pure and Applied Chemistry IUPAC [82LAF].

However, the majority of the thermodynamic data published in the scientific literature and used for the evaluations in this review, refer to the old standard state pressure of 1 "standard atmosphere" (= 0.101325 MPa). The difference between the thermodynamic data for the two standard state pressures is not large and lies in most cases within the uncertainty limits. It is nevertheless essential to make the corrections for the change in the standard state pressure in order to avoid inconsistencies and propagation of errors. In practice the parameters affected by the change between these two standard state pressures are the Gibbs energy and entropy changes of all processes that involve gaseous species. Consequently, changes occur also in the Gibbs energies of formation of species that consist of elements whose reference state is gaseous (H, O, F, Cl, N, and the noble gases). No other thermodynamic quantities are affected significantly. A large part of the following discussion has been taken from the NBS tables of chemical thermodynamic properties [82WAG/EVA], see also Freeman [84FRE].

The following expressions define the effect of pressure on the properties of all substances:

$$\left(\frac{\partial H}{\partial p}\right)_T = V - T\left(\frac{\partial V}{\partial T}\right)_p = V(1-\alpha T) \tag{II.39}$$

$$\left(\frac{\partial C_p}{\partial p}\right)_T = -T\left(\frac{\partial^2 V}{\partial T^2}\right)_p \tag{II.40}$$

$$\left(\frac{\partial S}{\partial p}\right)_T = -V\alpha = -\left(\frac{\partial V}{\partial T}\right)_p \tag{II.41}$$

$$\left(\frac{\partial G}{\partial p}\right)_T = V \tag{II.42}$$

$$\text{where } \alpha \equiv \frac{1}{V}\left(\frac{\partial V}{\partial T}\right)_p \tag{II.43}$$

For ideal gases, $V = \frac{RT}{p}$ and $\alpha = \frac{R}{pV} = \frac{1}{T}$. The conversion equations listed below (Eqs. (II.44) to (II.51)) apply to the small pressure change from 1 atm to 1 bar (0.1 MPa). The quantities that refer to the old standard state pressure of 1 atm are assigned the superscript $^{(\text{atm})}$ while those that refer to the new standard state pressure of 1 bar carry the superscript $^{(\text{bar})}$.

For all substances the changes in the enthalpy of formation and heat capacity are much smaller than the experimental accuracy and can be disregarded. This is exactly true for ideal gases.

$$\Delta_f H^{(bar)}(T) - \Delta_f H^{(atm)}(T) = 0 \qquad \text{(II.44)}$$

$$C_p^{(bar)}(T) - C_p^{(atm)}(T) = 0 \qquad \text{(II.45)}$$

For gaseous substances, the entropy difference is:

$$S^{(bar)}(T) - S^{(atm)}(T) = R \ln\left(\frac{p^{(atm)}}{p^{(bar)}}\right) = R \ln 1.01325$$

$$= 0.1094 \text{ J}\cdot\text{K}^{-1}\cdot\text{mol}^{-1} \qquad \text{(II.46)}$$

This is exactly true for ideal gases, as follows from Eq.(II.41) with $\alpha = \frac{R}{pV}$. The entropy change of a reaction or process is thus dependent on the number of moles of gases involved:

$$\Delta_r S^{(bar)} - \Delta_r S^{(atm)} = \delta \cdot R \ln\left(\frac{p^{(atm)}}{p^{(bar)}}\right)$$

$$= \delta \times 0.1094 \text{ J}\cdot\text{K}^{-1}\cdot\text{mol}^{-1} \qquad \text{(II.47)}$$

where δ is the net increase in moles of gas in the process.

Similarly, the change in the Gibbs energy of a process between the two standard state pressures is:

$$\Delta_r G^{(bar)} - \Delta_r G^{(atm)} = -\delta \cdot RT \ln\left(\frac{p^{(atm)}}{p^{(bar)}}\right)$$

$$= -\delta \cdot 0.03263 \text{ kJ}\cdot\text{mol}^{-1} \text{ at } 298.15 \text{ K.} \qquad \text{(II.48)}$$

Eq.(II.48) applies also to $\Delta_f G^{(bar)} - \Delta_f G^{(atm)}$, since the Gibbs energy of formation describes the formation process of a compound or complex from the reference states of the elements involved:

$$\Delta_f G^{(bar)} - \Delta_f G^{(atm)} = -\delta \times 0.03263 \text{ kJ}\cdot\text{mol}^{-1} \text{ at } 298.15 \text{ K.} \qquad \text{(II.49)}$$

The changes in the equilibrium constants and cell potentials with the change in the standard state pressure follow from the expression for Gibbs energy changes, Eq.(II.48)

$$\log_{10} K^{(\text{bar})} - \log_{10} K^{(\text{atm})} = -\frac{\Delta_r G^{(\text{bar})} - \Delta_r G^{(\text{atm})}}{RT \ln 10}$$

$$= \delta \cdot \frac{\ln\left(\dfrac{p^{(\text{atm})}}{p^{(\text{bar})}}\right)}{\ln 10} = \delta \cdot \log_{10}\left(\frac{p^{(\text{atm})}}{p^{(\text{bar})}}\right)$$

$$= \delta \cdot 0.005717 \qquad \text{(II.50)}$$

$$E^{(\text{bar})} - E^{(\text{atm})} = -\frac{\Delta_r G^{(\text{bar})} - \Delta_r G^{(\text{atm})}}{nF}$$

$$= \delta \cdot \frac{RT \ln\left(\dfrac{p^{(\text{atm})}}{p^{(\text{bar})}}\right)}{nF}$$

$$= \delta \cdot \frac{0.0003382}{n} \text{V} \quad \text{at } 298.15\text{K}. \qquad \text{(II.51)}$$

It should be noted that the standard potential of the hydrogen electrode is equal to 0.00 V exactly, by definition.

$$H^+ + e^- \rightleftharpoons ½H_2(g) \qquad E^\circ \stackrel{\text{def}}{=} 0.00\text{V} \qquad \text{(II.52)}$$

This definition will not be changed, although a gaseous substance, $H_2(g)$, is involved in the process. The change in the potential with pressure for an electrode potential conventionally written as:

$$Ag^+ + e^- \rightleftharpoons Ag\ (cr)$$

should thus be calculated from the balanced reaction that includes the hydrogen electrode,

$$Ag^+ + ½H_2\ (g) \rightleftharpoons Ag\ (cr) + H^+.$$

Here $\delta = -0.5$. Hence, the contribution to δ from an electron in a half cell reaction is the same as the contribution of a gas molecule with the stoichiometric coefficient of 0.5. This leads to the same value of δ as the combination with the hydrogen half cell.

Example:

$Fe(cr) + 2H^+ \rightleftharpoons Fe^{2+} + H_2(g)$	$\delta = 1$	$E^{(\text{bar})} - E^{(\text{atm})} = 0.00017$ V
$CO_2(g) \rightleftharpoons CO_2(aq)$	$\delta = -1$	$\log_{10} K^{(\text{bar})} - \log_{10} K^{(\text{atm})} = -0.0057$
$NH_3(g) + \frac{5}{4} O_2(g) \rightleftharpoons NO(g) + \frac{3}{2} H_2O(g)$	$\delta = 0.25$	$\Delta_r G^{(\text{bar})} - \Delta_r G^{(\text{atm})} = -0.008$ kJ·mol^{-1}
$\frac{1}{2} Cl_2(g) + 2O_2(g) + e^- \rightleftharpoons ClO_4^-$	$\delta = -3$	$\Delta_f G^{(\text{bar})} - \Delta_f G^{(\text{atm})} = 0.098$ kJ·mol^{-1}

II.3.3 Reference temperature

The definitions of standard states given in Section II.3 make no reference to fixed temperature. Hence, it is theoretically possible to have an infinite number of standard states of a substance as the temperature varies. It is, however, convenient to complete the definition of the standard state in a particular context by choosing a reference temperature. As recommended by IUPAC [82LAF], the reference temperature chosen in the NEA-TDB project is $T = 298.15$ K or $t = 25.00$°C. Where necessary for the discussion, values of experimentally measured temperatures are reported after conversion to the IPTS-68 [69COM]. The relation between the absolute temperature T (K, kelvin) and the Celsius temperature t (°C) is defined by $t = (T - T_0)$ where $T_0 = 273.15$ K.

II.4 Fundamental physical constants

To ensure the consitency with other NEA-TDB Reviews, the fundamental physical constants are taken from a publication by CODATA [86COD]. Those relevant to this review are listed in Table II-7. Note that updated values of the fundamental constants can be obtained from CODATA, notably through its Internet site. In most cases, recalculation of the NEA-TDB database entries with the updated values of the fundamental constants will not introduce significant (with respect to their quoted uncertainties) excursions from the current NEA-TDB selections.

Table II-7: Fundamental physical constants. These values have been taken from CODATA [86COD]. The digits in parentheses are the one-standard-deviation uncertainty in the last digits of the given value.

Quantity	Symbol	Value	Units
speed of light in vacuum	c	299 792 458	$m \cdot s^{-1}$
permeability of vacuum	μ_0	$4\pi \times 10^{-7} = 12.566\ 370\ 614...$	$10^{-7}\ N \cdot A^{-2}$
permittivity of vacuum	ϵ_0	$1/\mu_0 c^2 = 8.854\ 187\ 817...$	$10^{-12}\ C^2 \cdot J^{-1} \cdot m^{-1}$
Planck constant	h	6.626 0755(40)	10^{-34} J·s
elementary charge	e	1.602 177 33(49)	10^{-19} C
Avogadro constant	N_A	6.022 1367(36)	$10^{-23}\ mol^{-1}$
Faraday constant	F	96 485.309(29)	$C \cdot mol^{-1}$
molar gas constant	R	8.314 510(70)	$J \cdot K^{-1} \cdot mol^{-1}$
Boltzmann constant, R/N_A	k	1.380 658(12)	$10^{-23}\ J \cdot K^{-1}$
Non-SI units used with SI:			
electron volt, (e/C) J	eV	1.602 177 33(49)	10^{-19} J
atomic mass unit, $1u = m_u = \frac{1}{12} m(^{12}C)$	u	1.660 5402(10)	10^{-27} kg

II.5 Uncertainty estimates

One of the principal objectives of the NEA-TDB development effort is to provide an idea of the uncertainties associated with the data selected in the reviews. In general the uncertainties should define the range within which the corresponding data can be reproduced with a probability of 95%. In many cases, a full statistical treatment is limited or impossible due to the availability of only one or few data points. Appendix C describes in detail the procedures used for the assignment and treatment of uncertainties, as well as the propagation of errors and the standard rules for rounding.

II.6 The NEA-TDB system

A data base system has been developed at the NEA Data Bank that allows the storage of thermodynamic parameters for individual species as well as for reactions. The structure of the data base system allows consistent derivation of thermodynamic data for individual species from reaction data at standard conditions, as well as internal recalculations of data at standard conditions. If a selected value is changed, all the dependent values will be recalculated consistently. The maintenance of consistency of all the selected data, including their uncertainties (*cf.* Appendix C), is ensured by the software developed for this purpose at the NEA Data Bank. The literature sources of the data are also stored in the data base.

The following thermodynamic parameters, valid at the reference temperature of 298.15 K and at the standard pressure of 1 bar, are stored in the data base:

$\Delta_f G_m^\circ$ the standard molar Gibbs energy of formation from the elements in their reference states (kJ·mol^{-1})

$\Delta_f H_m^\circ$ the standard molar enthalpy of formation from the elements in their reference states (kJ·mol^{-1})

S_m° the standard molar entropy (J·K^{-1}·mol^{-1})

$C_{p,m}^\circ$ the standard molar heat capacity (J·K^{-1}·mol^{-1}).

For aqueous neutral species and ions, the values of $\Delta_f G_m^\circ$, $\Delta_f H_m^\circ$, S_m° and $C_{p,m}^\circ$ correspond to the standard partial molar quantities, and for individual aqueous ions they are relative quantities, defined with respect to the aqueous hydrogen ion, according to the convention [89COX/WAG] that $\Delta_f H_m^\circ(H^+, T) = 0$ and that $S_m^\circ(H^+, T) = 0$. Furthermore, for an *ionised solute* B containing any number of different cations and anions:

$$\Delta_f H_m^\circ(B_\pm, aq) = \sum_+ \nu_+ \, \Delta_f H_m^\circ(\text{cation, aq}) + \sum_- \nu_- \, \Delta_f H_m^\circ(\text{anion, aq})$$

$$S_m^\circ(B_\pm, aq) = \sum_+ \nu_+ \, S_m^\circ(\text{cation, aq}) + \sum_- \nu_- \, S_m^\circ(\text{anion, aq})$$

As the thermodynamic parameters vary as a function of temperature, provision is made for including the compilation of the coefficients of empirical temperature functions for these data, as well as the temperature ranges over which they are valid. In many cases the thermodynamic data measured or calculated at several temperatures were published for a particular species, rather than the deduced temperature functions. In these cases, a non-linear regression method is used in this review to obtain the most significant coefficients of the following empirical function for a thermodynamic parameter, X:

$$\begin{aligned} X(T) &= a_X + b_X \cdot T + c_X \cdot T^2 + d_X \cdot T^{-1} + e_X \cdot T^{-2} + f_X \cdot \ln T + g_X \cdot T \ln T \\ &\quad + h_X \cdot \sqrt{T} + \frac{i_X}{\sqrt{T}} + j_X \cdot T^3 + k_X \cdot T^{-3}. \end{aligned} \qquad \text{(II.53)}$$

Most temperature variations can be described with three or four parameters, a, b and e being the ones most frequently used. In the present review, only $C^{\circ}_{p,\mathrm{m}}(T)$, *i.e.*, the thermal functions of the heat capacities of individual species, are considered and stored in the data base. They refer to the relation:

$$C^{\circ}_{p,\mathrm{m}}(T) = a + b\times T + c\times T^2 + d\times T^{-1} + e\times T^{-2} + i\times T^{-1/2}$$

(where the subindices for the coefficients have been dropped) and are listed in Table III-3.

The pressure dependence of thermodynamic data has not been the subject of critical analysis in the present compilation. The reader interested in higher temperatures and pressures, or the pressure dependency of thermodynamic functions for geochemical applications, is referred to the specialised literature in this area, *e.g.*, [82HAM], [84MAR/MES], [88SHO/HEL], [88TAN/HEL], [89SHO/HEL], [89SHO/HEL2], [90MON], [91AND/CAS].

Selected standard thermodynamic data referring to chemical reactions are also compiled in the data base. A chemical reaction "r", involving reactants and products 'B", can be abbreviated as:

$$0 = \sum_{\mathrm{B}} \nu^{r}_{\mathrm{B}}\,\mathrm{B} \qquad \text{(II.54)}$$

where the stoichiometric coefficients ν^{r}_{B} are positive for products, and negative for reactants. The reaction parameters considered in the NEA-TDB system include:

$\log_{10} K^{\circ}_{r}$	the equilibrium constant of the reaction, logarithmic
$\Delta_{\mathrm{r}} G^{\circ}_{\mathrm{m}}$	the molar Gibbs energy of reaction ($\mathrm{kJ\cdot mol^{-1}}$)
$\Delta_{\mathrm{r}} H^{\circ}_{\mathrm{m}}$	the molar enthalpy of reaction ($\mathrm{kJ\cdot mol^{-1}}$)
$\Delta_{\mathrm{r}} S^{\circ}_{\mathrm{m}}$	the molar entropy of reaction ($\mathrm{J\cdot K^{-1}\cdot mol^{-1}}$)
$\Delta_{\mathrm{r}} C^{\circ}_{p,\mathrm{m}}$	the molar heat capacity of reaction ($\mathrm{J\cdot K^{-1}\cdot mol^{-1}}$)

The temperature functions of these data, if available, are stored according to Eq.(II.53).

The equilibrium constant, K_r°, is related to $\Delta_r G_m^\circ$ according to the following relation:

$$\log_{10} K_r^\circ = -\frac{\Delta_r G_m^\circ}{RT \ln(10)}$$

and can be calculated from the individual values of $\Delta_f G_m^\circ(B)$ (for example, those given in Table III-1 and Table IV-1), according to:

$$\log_{10} K_r^\circ = -\frac{1}{RT \ln(10)} \sum_B \nu_B^r \, \Delta_f G_m^\circ(B) \,. \qquad (II.55)$$

II.7 Presentation of the data

The selenium data are presented in Chapter III and Appendix E. Unless otherwise indicated, they refer to standard conditions (*cf.* Section II.3) and 298.15K (25°C) and are provided with an uncertainty which should correspond to the 95% confidence level (see Appendix C). Thermodynamic parameters (formation data and entropies) that could be evaluated from reaction data with selected TDB auxiliary data in Chapter IV are denoted as selected and presented in Chapter III. When use of non-TDB auxiliary data had to be resorted to in the evaluation, the result is denoted as adopted and presented in Appendix E. The difference in the status between a selected and an adopted value thus depends entirely on the background of the auxiliary data used in the assessment.

Chapter III contains a table of selected thermodynamic data for individual compounds and complexes of selenium (Table III-1), a table of selected reaction data (Table III-2) for reactions concerning selenium species and a table containing selected thermal functions of the heat capacities of individual species of selenium (Table III-3). The selection of these data is discussed in Chapter V.

Chapter IV contains, for auxiliary compounds and complexes that do not contain selenium, a table of the thermodynamic data for individual species (Table IV-1) and a table of reaction data (Table IV-2). Most of these values are the CODATA Key Values [89COX/WAG]. The selection of the remaining auxiliary data is discussed in [92GRE/FUG], and other preceding volumes of this series. As just mentioned, Appendix E contains a table, Table E-1, of the adopted thermodynamic data for individual selenium compounds calculated with auxiliary data not contained in Chapter IV. The non-TDB auxiliary data, usually from [82WAG/EVA], are presented in Table E-2 and the evaluations are discussed in Chapter V and Appendix A.

All the selected data presented in Table III-1, Table III-2, Table IV-1 and Table IV-2 are internally consistent. This consistency is maintained by the internal consistency verification and recalculation software developed at the NEA Data Bank in conjunction with the NEA-TDB data base system, *cf.* Section II.6. Therefore, when using the selected data for selenium species, the auxiliary data of Chapter IV must be used together with the data in Chapter III to ensure internal consistency of the data set.

It is important to note that Table III-2 and Table IV-2 include only those species for which the primary selected data are reaction data. The formation data derived from them and listed in Table III-1 are obtained using auxiliary data, and their uncertainties are propagated accordingly. In order to maintain the uncertainties originally assigned to the selected data in this review, the user is advised to make direct use of the reaction data presented in Table III-2 and Table IV-2, rather than taking the derived values in Table III-1 and Table IV-1 to calculate the reaction data with Eq.(II.55). The later approach would imply a twofold propagation of the uncertainties and result in reaction data whose uncertainties would be considerably larger than those originally assigned.

The thermodynamic data in the selected set refer to a temperature of 298.15 K (25.00°C), but they can be recalculated to other temperatures if the corresponding data (enthalpies, entropies, heat capacities) are available [97PUI/RAR]. For example, the temperature dependence of the standard reaction Gibbs energy as a function of the standard reaction entropy at the reference temperature (T_0= 298.15 K), and of the heat capacity function is:

$$\Delta_r G_m^\circ(T) = \Delta_r H_m^\circ(T_0) + \int_{T_0}^{T} \Delta_r C_{p,m}^\circ(T)\,dT - T\left(\Delta_r S_m^\circ(T_0) + \int_{T_0}^{T} \frac{\Delta_r C_{p,m}^\circ(T)}{T}dT\right),$$

and the temperature dependence of the standard equilibrium constant as a function of the standard reaction enthalpy and heat capacity is:

$$\log_{10} K^\circ(T) = \log_{10} K^\circ(T_0) - \frac{\Delta_r H_m^\circ(T_0)}{R\ln(10)}\left(\frac{1}{T} - \frac{1}{T_0}\right) - \frac{1}{RT\ln(10)}\int_{T_0}^{T} \Delta_r C_{p,m}^\circ(T)\,dT + \frac{1}{R\ln(10)}\int_{T_0}^{T} \frac{\Delta_r C_{p,m}^\circ(T)}{T}dT,$$

where R is the gas constant (*cf.* Table II-7).

In the case of aqueous species, for which enthalpies of reaction are selected or can be calculated from the selected enthalpies of formation, but for which there are no selected heat capacities, it is in most cases possible to recalculate equilibrium constants to temperatures up to 100 to 150°C, with an additional uncertainty of perhaps about 1 to 2 logarithmic units, due to neglecting the heat capacity contributions to the temperature correction. However, it is important to observe that "new" aqueous species, *i.e.*, species not present in significant amounts at 25°C and therefore not detected, may be significant at higher temperatures, see for example the work by Ciavatta *et al.* [87CIA/IUL]. Additional high-temperature experiments may therefore be needed in order to ascertain that proper chemical models are used in the modelling of hydrothermal systems. For many species, experimental thermodynamic data are not available to allow a selection of parameters describing the temperature dependence of equilibrium constants and Gibbs energies of formation. The user may find information on various procedures to estimate the temperature dependence of these thermodynamic parameters in [97PUI/RAR]. The thermodynamic data in the selected set refer to infinite dilution for soluble species. Extrapolation of an equilibrium constant K, usually measured at high ionic strength, to K° at $I = 0$ using activity coefficients γ, is explained in Appendix B. The corresponding Gibbs energy of dilution is:

$$\Delta_{\text{dil}}G_{\text{m}} = \Delta_{\text{r}}G^{\circ}_{\text{m}} - \Delta_{\text{r}}G_{\text{m}} \quad \text{(II.56)}$$

$$= -\ RT\ \Delta_{\text{r}} \ln \gamma_{\pm} \quad \text{(II.57)}$$

Similarly $\Delta_{\text{dil}}S_{\text{m}}$ can be calculated from $\ln \gamma_{\pm}$ and its variations with T, while:

$$\Delta_{\text{dil}}H_{\text{m}} = RT^2 \frac{\partial}{\partial T}(\Delta_{\text{r}} \ln \gamma_{\pm})_p \quad \text{(II.58)}$$

depends only on the variation of γ with T, which is neglected in this review, when no data on the temperature dependence of γ's are available. In this case the Gibbs energy of dilution $\Delta_{\text{dil}}G_{\text{m}}$ is entirely assigned to the entropy difference. This entropy of reaction is calculated using $\Delta_{\text{r}}G^{\circ}_{\text{m}} = \Delta_{\text{r}}H^{\circ}_{\text{m}} - T\Delta_{\text{r}}S^{\circ}_{\text{m}}$, the above assumption $\Delta_{\text{dil}}H_{\text{m}} = 0$, and $\Delta_{\text{dil}}G_{\text{m}}$.

Chapter III

Selected selenium data

This chapter presents the chemical thermodynamic data set for selenium species which has been selected in this review. Table III-1 contains the recommended thermodynamic data of the selenium species, Table III-2 the recommended thermodynamic data of chemical equilibrium reactions by which the selenium compounds and complexes are formed, and Table III-3 the temperature coefficients of the heat capacity data of Table III-1where available (see Appendix E for additional selenium data, *cf.* Section II.7).

The species and reactions in the tables appear in standard order of arrangement. Table III-2 contains information only on those reactions for which primary data selections are made in Chapter V of this review. These selected reaction data are used, together with data for key selenium species and auxiliary data selected in this review, to derive the corresponding formation data in Table III-1. The uncertainties associated with values for key selenium species and the auxiliary data are in some cases substantial, leading to comparatively large uncertainties in the formation quantities derived in this manner.

The values of $\Delta_r G_m^\circ$ for many reactions are known more accurately than would be calculated directly from the uncertainties of the $\Delta_f G_m^\circ$ values in Table III-1 and auxiliary data. The inclusion of a table for reaction data (Table III-2) in this report allows the use of equilibrium constants with total uncertainties that are based directly on the experimental accuracies. This is the main reason for including both Table III-1 and Table III-2.

The selected thermal functions of the heat capacities, listed in Table III-3, refer to the relation

$$C_{p,m}^\circ(T) = a + b \times T + c \times T^2 + d \times T^{-1} + e \times T^{-2}.$$

A detailed discussion of the selection procedure is presented in Chapter V. It may be noted that this chapter contains data on more species or compounds than are present in the tables of Chapter III. The main reasons for this situation are the lack of information for a proper extrapolation of the primary data to standard conditions in some systems and lack of solid primary data in others.

A warning: The addition of any aqueous species and their data to this internally consistent data base can result in a modified data set, which is no longer rigorous and can lead to erroneous results. The situation is similar when gases or solids are added.

A further set of formation data for selenium species and is available in Table E-1, Appendix E. Although the experimental sources leading to these selections have been critically reviewed according to the NEA-TDB Guidelines, the corresponding values for some of their formation thermodynamic properties cannot be fully evaluated using the set of auxiliary data currently accepted by the NEA-TDB Project. Table E-2 in Appendix E details the auxiliary data (other than NEA-TDB Auxiliary Data) that have been used in the evaluation of the formation values listed in Table E-1.

Table III-1: Selected thermodynamic data for selenium compounds and complexes. All ionic species listed in this table are aqueous species. Unless noted otherwise, all data refer to the reference temperature of 298.15 K and to the standard state, *i.e.*, a pressure of 0.1 MPa and, for aqueous species, infinite dilution ($I = 0$). The uncertainties listed below each value represent total uncertainties and correspond in principle to the statistically defined 95% confidence interval. Values obtained from internal calculation, *cf.* footnotes (a) and (b), are rounded at the third digit after the decimal point and may therefore not be exactly identical to those given in Part V. Systematically, all the values are presented with three digits after the decimal point, regardless of the significance of these digits. The data presented in this table are available on computer media from the OECD Nuclear Energy Agency.

Compound	$\Delta_f G_m^o$ (kJ·mol^{-1})	$\Delta_f H_m^o$ (kJ·mol^{-1})	S_m^o (J·K^{-1}·mol^{-1})	$C_{p,m}^o$ (J·K^{-1}·mol^{-1})
Se(cr) (= trigonal)	0.000	0.000	42.090 ± 0.330	25.090 (c) ± 0.300
Se(mono)	1.281 (a) ± 0.184	2.140 ± 0.100	44.970 ± 0.400	25.090 (c) ± 0.800
Se(l)				(c)
Se(g)	195.927 (a) ± 1.524	236.070 (b) ± 1.521	176.729 ± 0.006	20.819 (c) ± 0.001
Se^{2-}	128.600 (b) ± 3.000			
$Se_2(g)$	92.442 (a) ± 3.009	141.100 ± 3.000	247.380 ± 0.400	41.710 (c) ± 0.200
Se_2^{2-}	112.670 (b) ± 6.294			

(Continued on next page)

Table III-1: (continued)

Compound	$\Delta_f G_m^o$ (kJ·mol^{-1})	$\Delta_f H_m^o$ (kJ·mol^{-1})	S_m^o (J·K^{-1}·mol^{-1})	$C_{p,m}^o$ (J·K^{-1}·mol^{-1})
$Se_3(g)$	123.549 [(a)] ± 10.439	178.000 ± 10.000	308.900 ± 10.000	
Se_3^{2-}	100.590 [(b)] ± 9.198			
$Se_4(g)$	111.647 [(a)] ± 13.406	163.000 ± 12.000	340.600 ± 20.000	
Se_4^{2-}	97.580 [(b)] ± 12.149			
$Se_5(g)$	88.393 [(a)] ± 5.256	144.400 ± 4.300	398.300 ± 10.000	
$Se_6(g)$	83.638 [(a)] ± 4.866	136.100 ± 3.800	428.500 ± 10.000	
$Se_7(g)$	92.845 [(a)] ± 4.891	150.200 ± 4.500	487.000 ± 6.000	
$Se_8(g)$	100.307 [(a)] ± 3.580	156.800 ± 3.000	526.200 ± 6.000	
$SeO(g)$	30.855 [(a)] ± 6.218	57.400 ± 6.210	233.700 ± 1.000	32.520 [(c)] ± 0.250
$SeO_2(cr)$	− 171.797 [(a)] ± 0.620	− 225.390 ± 0.600	67.490 ± 0.400	58.230 [(c)] ± 0.180
$SeO_2(g)$	− 115.166 [(a)] ± 2.600	− 110.590 [(b)] ± 2.571	262.590 ± 1.250	43.360 [(c)] ± 0.150
$SeO_3(cr)$	− 86.154 [(a)] ± 2.222	− 163.100 ± 2.200	91.740 ± 1.000	77.240 [(c)] ± 0.790
SeO_3^{2-}	− 362.392 [(b)] ± 1.756	− 507.160 ± 1.130	− 5.055 [(a)] ± 7.011	
SeO_4^{2-}	− 439.485 [(b)] ± 1.431	− 603.500 [(b)] ± 3.500	32.965 [(b)] ± 12.687	
$Se_2O_5(cr)$		− 414.590 [(b)] ± 2.577		
HSe^-	43.471 [(b)] ± 2.024			
$H_2Se(g)$	15.217 [(a)] ± 2.003	29.000 ± 2.000	219.000 ± 0.100	34.700 [(c)] ± 0.100
$H_2Se(aq)$	21.495 [(b)] ± 2.003	14.300 [(b)] ± 2.022	148.637 [(b)] ± 1.029	
$HSeO_3^-$	− 410.112 [(b)] ± 1.166	− 512.330 ± 1.010	137.656 [(a)] ± 5.184	

(Continued on next page)

Table III-1: (continued)

Compound	$\Delta_f G_m^o$ (kJ·mol⁻¹)	$\Delta_f H_m^o$ (kJ·mol⁻¹)	S_m^o (J·K⁻¹·mol⁻¹)	$C_{p,m}^o$ (J·K⁻¹·mol⁻¹)
$HSeO_4^-$	– 449.474 (b) ± 1.312	– 582.700 (d) ± 4.700	136.232 (a) ± 16.370	
H_2SeO_3(cr)		– 524.720 (b) ± 0.650		
H_2SeO_3(aq)	– 425.181 (b) ± 0.849	– 505.320 ± 0.650	211.710 (a) ± 3.601	
H_2SeO_4(cr)		– 530.500 ± 1.880		
SeF_2(g)			271.280 ± 0.050	47.660 (c) ± 0.040
SeF_4(l)	– 775.070 (a) ± 24.067	– 851.000 ± 24.000	193.000 (d) ± 6.000	
SeF_4(g)	– 765.549 (b) ± 24.047	– 806.000 (b) ± 24.047	312.000 ± 0.100	82.030 (c) ± 0.090
SeF_6(g)	– 1017.445 (a) ± 0.438	– 1118.000 ± 0.400	313.200 ± 0.500	110.440 (c) ± 0.050
$SeOF_2$(l)	– 526.563 (a) ± 16.003	– 575.000 ± 16.000	185.000 (d) ± 1.000	
$SeOF_2$(g)	– 515.418 (b) ± 16.032	– 530.700 (b) ± 16.031	296.200 ± 0.100	63.330 (c) ± 0.050
Se_2F_2(g)			331.120 ± 0.100	72.230 (c) ± 0.040
SeF_5Cl(g)			341.000 ± 0.200	116.300 (c) ± 0.100
$SeCl_2$(g)	– 25.947 (b) ± 4.473	– 17.000 (b) ± 4.472	295.180 ± 0.060	53.570 (c) ± 0.030
$SeCl_4$(cr)	– 85.902 (a) ± 3.114	– 179.000 ± 2.000	176.000 (d) ± 8.000	
Se_2Cl_2(l)	– 45.846 (a) ± 4.273	– 82.000 ± 4.000	186.000 (b) ± 5.000	
Se_2Cl_2(g)	– 34.935 (a) ± 10.002	– 21.000 ± 10.000	354.000 ± 0.200	77.600 (c) ± 0.100
$SeOCl_2$(l)	– 132.107 (a) ± 2.089	– 180.600 ± 2.000	205.100 (d) ± 2.000	
$SeOCl_2$(g)	– 114.223 (b) ± 2.502	– 128.000 (b) ± 2.500	321.540 ± 0.100	70.200 (c) ± 0.050
$SeBr_2$(g)	– 4.583 (a) ± 20.226	32.000 (b) ± 20.224	317.000 ± 1.000	55.420 (c) ± 0.500

(Continued on next page)

Table III-1: (continued)

Compound	$\Delta_f G_m^o$ (kJ·mol⁻¹)	$\Delta_f H_m^o$ (kJ·mol⁻¹)	S_m^o (J·K⁻¹·mol⁻¹)	$C_{p,m}^o$ (J·K⁻¹·mol⁻¹)
$SeBr_4$(cr)		− 70.000 ±5.000		
Se_2Br_2(l)		− 31.000 ±3.000		
Se_2Br_2(g)			377.400 ± 0.200	79.500 (c) ± 0.100
$SeOBr_2$(cr)		− 140.000 ±5.000		
SeS(g)	92.356 (a) ± 6.871	143.327 (b) ± 6.869	245.100 ± 0.500	35.800 (c) ± 0.100
TeSe(g)			257.900 ± 0.500	37.800 (c) ± 0.100
Se_4N_4(cr)		679.000 ± 35.000		
α-P_4Se_3			239.600 ± 1.200	166.200 ± 1.000
As_2Se_3(cr)			206.000 ± 3.000	121.400 (c) ± 3.000
As_4Se_3(cr)				172.500 (c) ± 4.000
As_4Se_4(cr)				196.600 (c) ± 4.000
SbSe(g)	134.874 (a) ± 20.020	177.000 ± 20.000	228.900 ± 3.000	
Sb_2Se_3(cr)	− 127.415 (a) ± 3.723	− 128.700 ± 3.600	213.000 ± 3.000	124.900 (c) ± 1.500
BiSe(cr)	− 46.807 ± 5.720	− 46.100 ± 5.700	101.200 ± 1.500	55.500 ± 0.900
BiSe(g)			263.900 ± 6.000	36.460 (c) ± 0.500
Bi_2Se_3(cr)	− 146.618 (a) ± 10.065	− 145.500 ± 10.000	243.500 ± 3.600	132.600 (c) ± 2.100
CSe(g)	309.848 (a) ± 9.202	362.000 ± 9.200	222.750 ± 0.500	30.580 (c) ± 0.200
CSe_2(cr)				(c)
CSe_2(l)	196.020 (a) ± 8.820	219.300 ± 7.600	168.000 ± 15.000	88.530 (c) ± 0.900

(Continued on next page)

Table III-1: (continued)

Compound	$\Delta_f G^o_m$ ($kJ\cdot mol^{-1}$)	$\Delta_f H^o_m$ ($kJ\cdot mol^{-1}$)	S^o_m ($J\cdot K^{-1}\cdot mol^{-1}$)	$C^o_{p,m}$ ($J\cdot K^{-1}\cdot mol^{-1}$)
$CSe_2(g)$	203.937 [(a)] ± 7.611	255.600 [(b)] ± 7.603	263.200 ± 1.000	49.900 [(c)] ± 0.100
$SeCN^-$	136.051 [(b)] ± 3.800			
SiSe(g)	98.119 [(a)] ± 10.501	150.200 ± 10.500	235.580 ± 0.200	33.620 [(c)] ± 0.100
$SiSe_2(cr)$		− 177.600 ± 3.200		
GeSe(cr)	− 52.226 [(a)] ± 1.473	− 50.300 [(b)] ± 1.449	79.640 ± 0.800	49.880 [(c)] ± 0.500
GeSe(g)	70.450 [(a)] ± 1.770	122.400 [(b)] ± 1.761	247.420 ± 0.500	35.250 [(c)] ± 0.200
$GeSe_2(cr)$	− 93.239 [(a)] ± 12.007	− 93.200 ± 12.000	115.400 ± 1.200	71.700 ± 0.700
$GeSe_2(g)$		124.800 [(b)] ± 21.633		
α-SnSe		− 109.500 ± 15.000		51.910 [(c)] ± 2.000
SnSe(g)	61.861 [(a)] ± 15.693	110.200 [(b)] ± 15.689	255.400 ± 1.000	35.930 [(c)] ± 0.500
$SnSe_2(cr)$		− 114.900 ± 10.000		76.700 [(c)] ± 0.500
$SnSe_2(g)$		78.000 ± 16.100		
PbSe(cr)	− 97.936 [(a)] ± 7.694	− 98.500 ± 7.600	105.000 ± 4.000	50.610 [(c)] ± 1.000
PbSe(g)	82.866 [(a)] ± 8.689	129.500 [(b)] ± 8.683	263.300 ± 1.000	36.360 [(c)] ± 0.500
$PbSeO_3(cr)$	− 457.981 [(b)] ± 5.985	− 532.080 ± 1.800	166.087 [(a)] ± 20.968	
$PbSeO_4(cr)$	− 503.108 [(b)] ± 2.060	− 607.300 [(b)] ± 4.326	167.734 [(b)] ± 16.027	
$BSe_2(g)$	131.400 [(a)] ± 20.871	187.100 ± 20.000	276.900 ± 20.000	
AlSe(g)	206.569 [(a)] ± 35.000	258.000 ± 35.000	242.890 ± 0.500	34.750 [(c)] ± 0.200
$Al_2Se_3(cr)$		− 532.000 ± 40.000		

(Continued on next page)

Table III-1: (continued)

Compound	$\Delta_f G_m^o$ (kJ·mol^{-1})	$\Delta_f H_m^o$ (kJ·mol^{-1})	S_m^o (J·K^{-1}·mol^{-1})	$C_{p,m}^o$ (J·K^{-1}·mol^{-1})
GaSe(cr)		− 145.000 ± 10.000	70.400 ± 0.700	48.600 ± 0.900
Ga_2Se_3(cr)		− 384.000 ± 40.000		
InSe(cr)			83.890 ± 0.900	49.750 (c) ± 1.060
α-In_2Se_3		− 318.000 ± 20.000	203.100 ± 2.000	124.100 (c) ± 4.100
β-In_2Se_3				(c)
α-TlSe			103.000 ± 1.300	50.100 ± 0.800
Tl_2Se(cr)			174.800 ± 3.000	82.200 ± 2.500
Tl_2SeO_4(cr)	− 526.546 (b) ± 1.772			
TlSeCN(aq)	93.662 (b) ± 4.156			
α-ZnSe	− 172.488 (a) ± 4.013	− 176.000 ± 4.000	71.940 ± 1.000	47.740 (c) ± 1.500
ZnSe(g)		239.300 ± 20.000		
α-$ZnSeO_3$		− 636.800 ± 11.300		
β-$ZnSeO_3$		− 652.700 ± 10.900		
$ZnSeO_4$(cr)		− 664.730 ± 5.600		
$ZnSeO_4$(aq)	− 599.017 (b) ± 1.493	− 752.290 (b) ± 5.319	− 20.053 (b) ± 18.507	
$ZnSeO_4 \cdot H_2O$(cr)		− 983.070 ± 5.600		
$ZnSeO_4 \cdot 6H_2O$(cr)	− 2018.309 (b) ± 1.524	− 2458.540 ± 5.610	417.016 (a) ± 19.502	
$ZnSe_2O_5$(cr)		− 888.300 ± 25.000		
$Zn(HSeO_3)_2 \cdot 2H_2O$(cr)		− 1775.300 ± 25.000		

(Continued on next page)

Table III-1: (continued)

Compound	$\Delta_f G_m^o$ (kJ·mol^{-1})	$\Delta_f H_m^o$ (kJ·mol^{-1})	S_m^o (J·K^{-1}·mol^{-1})	$C_{p,m}^o$ (J·K^{-1}·mol^{-1})
$ZnSeCN^+$	– 18.058 [(b)] ± 3.824			
$Zn(SeCN)_2(aq)$	115.311 [(b)] ± 7.630			
α-CdSe	– 140.897 [(a)] ± 1.918	– 143.100 ± 1.900	86.500 ± 0.800	49.430 [(c)] ± 0.480
$CdSeO_3(cr)$	– 493.432 [(a)] ± 6.467	– 576.800 ± 2.500	122.000 ± 20.000	
$CdSeO_4(cr)$		– 634.100 ± 5.630		
$CdSeO_4(aq)$	– 530.175 [(b)] ± 1.652	– 671.120 [(b)] ± 5.349	31.463 [(b)] ± 18.561	
$CdSeO_4{\cdot}H_2O(cr)$		– 943.050 ± 5.620		
$CdSe_2O_5(cr)$		– 828.000 ± 17.000		
$Cd_3H_2(SeO_3)_4(cr)$		– 2306.000 ± 45.000		
$CdSeCN^+$	45.533 [(b)] ± 3.889			
$Cd(SeCN)_2(cr)$	161.834 [(b)] ± 8.153			
$Cd(SeCN)_2(aq)$	175.305 [(b)] ± 7.668			
$Cd(SeCN)_3^-$	308.674 [(b)] ± 11.488			
$Cd(SeCN)_4^{2-}$	440.216 [(b)] ± 15.232			
α-HgSe	– 51.200 ± 4.000	– 57.011 [(a)] ± 4.100	98.500 ± 3.000	54.400 [(c)] ± 0.700
$HgSe_2^{2-}$	65.334 [(b)] ± 5.641			
$HgSeO_3(cr)$	– 290.195 [(b)] ± 5.980	– 373.600 ± 5.000	151.732 [(a)] ± 26.108	
$Hg(SeO_3)_2^{2-}$	– 644.882 [(b)] ± 6.291			
$Hg_2SeO_3(cr)$	– 295.587 [(b)] ± 5.998			

(Continued on next page)

Table III-1: (continued)

Compound	$\Delta_f G_m^o$ (kJ·mol^{-1})	$\Delta_f H_m^o$ (kJ·mol^{-1})	S_m^o (J·K^{-1}·mol^{-1})	$C_{p,m}^o$ (J·K^{-1}·mol^{-1})
$Hg_3H_2(SeO_3)_4(cr)$		− 1652.800 ± 32.500		
$Hg(SeCN)_2(aq)$	309.481 [(b)] ± 9.510			
$Hg(SeCN)_3^-$	419.846 [(b)] ± 12.754			
$Hg(SeCN)_4^{2-}$	541.627 [(b)] ± 15.470			
α-CuSe	− 36.835 [(a)] ± 0.563	− 39.500 ± 0.500	66.300 ± 0.800	50.300 ± 0.600
β-CuSe	− 34.910 [(a)] ± 0.563	− 36.800 ± 0.500	68.900 ± 0.800	
CuSe(g)		320.800 ± 15.000		
α-Cu_2Se		− 59.300 ± 0.500		
$CuSe_2(cr)$		− 41.200 ± 5.900		
$Cu_3Se_2(cr)$		− 96.800 ± 7.100		
$CuSeO_3(cr)$		− 386.900 ± 1.520		
$CuSeO_3·2H_2O(cr)$		− 998.600 ± 1.510		
$CuSeO_4(cr)$		− 480.330 ± 2.810		
$CuSeO_4·5H_2O(cr)$	− 1574.100 [(b)] ± 2.420	− 1973.330 ± 2.830	312.800 [(a)] ± 12.494	
AgSe(g)		309.000 ± 6.000		
α-Ag_2Se	− 46.900 ± 1.300	− 40.129 [(a)] ± 1.318	149.900 ± 0.500	81.150 ± 0.900
$Ag_2SeO_3(cr)$	− 298.400 [(b)] ± 2.440	− 363.440 ± 1.020	216.773 [(a)] ± 8.885	
$Ag_2SeO_4(cr)$	− 330.150 [(b)] ± 3.200	− 422.510 ± 2.120	227.717 [(a)] ± 12.885	
AgSeCN(cr)	133.247 [(b)] ± 4.783			

(Continued on next page)

Table III-1: (continued)

Compound	$\Delta_f G^o_m$ (kJ·mol^{-1})	$\Delta_f H^o_m$ (kJ·mol^{-1})	S^o_m (J·K^{-1}·mol^{-1})	$C^o_{p,m}$ (J·K^{-1}·mol^{-1})
$Ag(SeCN)_3^{2-}$	406.193 [(b)] ± 11.529			
$Ni_{0.88}Se(cr)$	– 69.628 [(a)] ± 1.621	– 69.800 ± 1.600	67.800 ± 0.800	48.260 ± 0.500
$NiSe_2(cr)$	– 112.355 [(a)] ± 7.017	– 115.500 ± 7.000	103.500 ± 1.500	74.500 [(c)] ± 1.000
$Ni_3Se_2(cr)$		– 179.500 ± 6.000		
$NiSeO_3\cdot 2H_2O(cr)$	– 915.553 [(b)] ± 6.022	– 1109.330 ± 2.830	196.268 [(a)] ± 22.320	
$NiSeO_4(aq)$	– 500.499 [(b)] ± 1.650			
$NiSeO_4\cdot 6H_2O(cr)$	– 1915.983 [(b)] ± 1.664			
$NiSeCN^+$	80.175 [(b)] ± 3.893			
$Ni(SeCN)_2(aq)$	213.544 [(b)] ± 7.681			
$Co_{0.84}Se(cr)$	– 56.300 [(a)] ± 6.500	– 55.400 ± 6.000		
$CoSe_2(cr)$		– 105.500 ± 15.000	97.200 ± 2.000	75.600 ± 1.000
$Co_9Se_8(cr)$		– 617.600 ± 40.000		
δ-$Fe_{0.875}Se$				[(c)]
$FeSe_2(cr)$		– 108.700 ± 15.000	86.800 ± 1.000	72.900 [(c)] ± 1.000
β-$Fe_{1.04}Se$		– 69.600 ± 4.000	72.100 ± 0.800	57.100 [(c)] ± 0.700
γ-Fe_3Se_4		– 235.000 ± 30.000	279.800 ± 3.000	220.100 ± 2.000
α-Fe_7Se_8		– 463.500 ± 20.000	613.800 ± 5.000	442.100 ± 4.000
$Pd_4Se(cr)$			200.100 ± 3.000	137.000 ± 2.000
$RuSe_2(cr)$			80.800 ± 0.800	69.870 [(c)] ± 1.000

(Continued on next page)

Table III-1: (continued)

Compound	$\Delta_f G_m^o$ (kJ·mol^{-1})	$\Delta_f H_m^o$ (kJ·mol^{-1})	S_m^o (J·K^{-1}·mol^{-1})	$C_{p,m}^o$ (J·K^{-1}·mol^{-1})
α-MnSe			90.800 ± 3.000	51.040 ± 0.300
α-Cr_2Se_3			187.700 ± 1.500	120.800 ± 0.900
γ-Cr_2Se_3			178.000 ± 6.000	122.900 ± 0.900
$MoSe_2$(cr)			89.100 ± 0.500	70.584 [c] ± 0.500
WSe_2(cr)				72.600 [c] ± 1.000
$NbSe_2$(cr)			103.500 ± 0.600	73.520 [c] ± 0.300
EuSe(cr)				51.000 [c] ± 3.000
Sm_2Se_3(cr)			221.200 ± 3.000	128.200 [c] ± 1.500
Nd_2Se_3(cr)			224.100 ± 2.500	130.200 ± 2.000
Nd_3Se_4(cr)			359.600 ± 5.800	187.060 ± 1.470
Pr_3Se_4(cr)			371.000 ± 6.000	183.760 ± 1.450
LaSe(cr)				49.200 ± 1.000
La_2Se_3(cr)			202.200 ± 1.800	125.600 [c] ± 1.000
PuSe(cr)			92.100 ± 1.800	59.700 ± 1.200
USe(cr)	– 273.255 [a] ± 18.006	– 272.000 ± 18.000	96.500 ± 1.500	54.800 ± 0.300
β-USe_2			133.800 ± 1.500	79.160 ± 0.300
U_3Se_4(cr)		– 1023.000 ± 92.000		
UO_2SeO_3(cr)		– 1522.200 ± 1.100		
UO_2SeO_4(cr)		– 1539.300 ± 3.300		

(Continued on next page)

Table III-1: (continued)

Compound	$\Delta_f G_m^o$ (kJ·mol^{-1})	$\Delta_f H_m^o$ (kJ·mol^{-1})	S_m^o (J·K^{-1}·mol^{-1})	$C_{p,m}^o$ (J·K^{-1}·mol^{-1})
UO_2SeO_4(aq)	– 1407.676 [(b)] ± 2.672			
$MgSeO_3$(cr)		– 900.300 ± 2.500		
$MgSeO_3·6H_2O$(cr)	– 2273.830 [(b)] ± 2.600	– 2707.210 ± 1.300	328.460 [(a)] ± 9.756	
$MgSeO_4$(aq)	– 907.418 [(b)] ± 2.266			
$MgSeO_4$(cr)		– 971.320 ± 2.620		
$MgSeO_4·H_2O$(cr)		– 1298.240 ± 2.580		
$MgSeO_4·4H_2O$(cr)		– 2192.670 ± 2.650		
$MgSeO_4·6H_2O$(cr)	– 2324.200 [(b)] ± 2.000	– 2781.400 ± 2.400	351.144 [(a)] ± 10.484	
CaSe(cr)		– 414.400 ± 35.000		
$CaSeO_3·H_2O$(cr)	– 1188.870 [(b)] ± 2.495			
$CaSeO_4$(cr)		– 1112.140 ± 2.580		
$CaSeO_4$(aq)	– 1003.707 [(b)] ± 1.865			
$CaSeO_4·2H_2O$(cr)	– 1481.900 [(b)] ± 2.300	– 1709.000 ± 2.600	198.799 [(a)] ± 11.654	
SrSe(cr)		– 406.000 ± 35.000		
$SrSeO_3$(cr)	– 962.200 [(b)] ± 3.400	– 1051.900 ± 2.700	104.663 [(a)] ± 14.567	
$BaSeO_3$(cr)	– 957.200 [(b)] ± 3.400	– 1036.700 ± 2.900	145.594 [(a)] ± 15.016	
$BaSeO_4$(cr)	– 1040.294 [(b)] ± 3.007	– 1144.000 ± 5.000	166.983 [(a)] ± 19.590	
NH_4HSe(cr)	– 24.162 [(b)] ± 5.344	– 132.040 [(b)] ± 4.046	102.770 [(b)] ± 11.702	
$(NH_4)_2SeO_4$(cr)	– 593.100 [(b)] ± 1.600	– 878.450 ± 3.500	209.654 [(a)] ± 12.912	

(Continued on next page)

Table III-1: (continued)

Compound	$\Delta_f G_m^o$ (kJ·mol^{-1})	$\Delta_f H_m^o$ (kJ·mol^{-1})	S_m^o (J·K^{-1}·mol^{-1})	$C_{p,m}^o$ (J·K^{-1}·mol^{-1})
$Li_2Se(cr)$		– 434.100 ± 15.000		
$Li_2Se_2O_5(cr)$		– 1310.870 ± 4.700		
$Li_2SeO_3(cr)$		– 1031.810 ± 2.460		
$Li_2SeO_3{\cdot}H_2O(cr)$		– 1328.600 ± 2.430		
$Li_2SeO_4(cr)$		– 1127.690 ± 5.600		
$Li_2SeO_4{\cdot}H_2O(cr)$	– 1252.400 (b) ± 1.500	– 1420.000 ± 5.600	181.757 (a) ± 19.452	
$LiHSeO_3(cr)$		– 817.850 ± 1.840		
$LiH_3(SeO_3)_2(cr)$		– 1371.700 ± 3.700		
$Na_2Se(cr)$		– 337.700 ± 10.000		
$Na_2SeO_3(cr)$	– 868.679 (a) ± 0.967	– 959.620 ± 0.950	147.400 ± 0.300	126.100 ± 0.300
$Na_2SeO_3{\cdot}5H_2O(cr)$		– 2404.670 ± 1.000		
$Na_2SeO_4(cr)$		– 1076.500 ± 3.500		
$Na_2SeO_4{\cdot}10H_2O(cr)$	– 3338.700 (b) ± 1.600	– 4013.600 ± 3.600	623.928 (a) ± 13.224	
$Na_2Se_2O_5(cr)$		– 1219.640 ± 4.710		
$Na_2Se_2O_7(cr)$		– 1322.000 ± 5.000		
$NaHSeO_3(cr)$		– 760.100 ± 2.500		
$NaHSeO_3{\cdot}3H_2O(cr)$		– 1642.550 ± 2.500		
$NaHSeO_4(cr)$		– 822.230 ± 5.590		
$NaH_3(SeO_3)_2(cr)$		– 1316.640 ± 4.710		

(Continued on next page)

Table III-1: (continued)

Compound	$\Delta_f G_m^o$ (kJ·mol^{-1})	$\Delta_f H_m^o$ (kJ·mol^{-1})	S_m^o (J·K^{-1}·mol^{-1})	$C_{p,m}^o$ (J·K^{-1}·mol^{-1})
$K_2Se(cr)$		– 398.700 ± 10.000		
$K_2SeO_3(cr)$		– 981.090 ± 2.360		
$K_2SeO_4(cr)$	– 999.300 [(b)] ± 1.500	– 1112.497 [(a)] ± 1.537	202.090 ± 1.000	142.640 ± 1.000
$K_2Se_2O_5(cr)$		– 1250.050 ± 4.720		
$KHSeO_3(cr)$		– 773.490 ± 1.840		
$KH_3(SeO_3)_2(cr)$		– 1329.080 ± 4.710		
$Rb_2Se(cr)$				74.700 ± 2.000
$Rb_2SeO_3(cr)$		– 973.170 ± 2.400		
$Rb_2SeO_4(cr)$	– 1005.100 [(b)] ± 1.500	– 1116.500 ± 2.200	232.317 [(a)] ± 8.957	
$Cs_2Se(cr)$				76.000 ± 2.000
$Cs_2SeO_3 \cdot H_2O(cr)$		– 1311.830 ± 2.550		
$Cs_2SeO_4(cr)$	– 1018.800 [(b)] ± 1.800	– 1138.700 ± 3.500	220.707 [(a)] ± 13.229	

a Value calculated internally using the equation $\Delta_f G_m^o = \Delta_f H_m^o - T \sum_i S_{m,i}^o$

b Value calculated internally from reaction data (see Table III-2)

c Temperature coefficients of this function are listed in Table III-3

d Value calculated internally from reaction data for a different key species

Table III-2: Selected thermodynamic data for reactions involving selenium compounds and complexes. All ionic species listed in this table are aqueous species. Unless noted otherwise, all data refer to the reference temperature of 298.15 K and to the standard state, *i.e.*, a pressure of 0.1 MPa and, for aqueous species, infinite dilution (I = 0). The uncertainties listed below each value represent total uncertainties and correspond in principle to the statistically defined 95% confidence interval. Values obtained from internal calculation, *cf.* footnote (a), are rounded at the third digit after the decimal point and may therefore not be exactly identical to those given in Part V. Systematically, all the values are presented with three digits after the decimal point, regardless of the significance of these digits. The data presented in this table are available on computer media from the OECD Nuclear Energy Agency.

Species	Reaction			
	$\log_{10} K^{\circ}$	$\Delta_r G_m^{\circ}$ (kJ·mol^{-1})	$\Delta_r H_m^{\circ}$ (kJ·mol^{-1})	$\Delta_r S_m^{\circ}$ (J·K^{-1}·mol^{-1})
Se(g)	$0.5Se_2(g) \rightleftharpoons Se(g)$			
			165.520 ± 0.250	
Se^{2-}	$Se(cr) + 2 e^- \rightleftharpoons Se^{2-}$			
	– 22.530 ± 0.526	128.600 ± 3.000		
Se_2^{2-}	$2Se^{2-} \rightleftharpoons Se_2^{2-} + 2 e^-$			
	25.320 ± 0.333	– 144.530 ± 1.900		
Se_3^{2-}	$3Se^{2-} \rightleftharpoons Se_3^{2-} + 4 e^-$			
	49.966 ± 0.333	– 285.210 ± 1.900		
Se_4^{2-}	$4Se^{2-} \rightleftharpoons Se_4^{2-} + 6 e^-$			
	73.023 ± 0.333	– 416.820 ± 1.900		
$SeO_2(g)$	$SeO_2(cr) \rightleftharpoons SeO_2(g)$			
			114.800 ± 2.500	
SeO_3^{2-}	$HSeO_3^- \rightleftharpoons H^+ + SeO_3^{2-}$			
	– 8.360 ± 0.230	47.719 ± 1.313		
SeO_4^{2-}	$HSeO_4^- \rightleftharpoons H^+ + SeO_4^{2-}$			
	– 1.750 ± 0.100	9.989 ± 0.571	– 20.800 ± 3.200	– 103.267 [(a)] ± 10.902

(Continued on next page)

Table III-2: (continued)

Species	Reaction $\log_{10} K^{\circ}$	$\Delta_r G_m^{\circ}$ $(kJ\cdot mol^{-1})$	$\Delta_r H_m^{\circ}$ $(kJ\cdot mol^{-1})$	$\Delta_r S_m^{\circ}$ $(J\cdot K^{-1}\cdot mol^{-1})$
$Se_2O_5(cr)$	$SeO_2(cr) + SeO_3(cr) \rightleftharpoons Se_2O_5(cr)$			
			– 26.100 ± 1.200	
HSe^-	$H_2Se(aq) \rightleftharpoons H^+ + HSe^-$			
	– 3.850 ± 0.050	21.976 ± 0.285		
$H_2Se(aq)$	$H_2Se(g) \rightleftharpoons H_2Se(aq)$			
	– 1.100 ± 0.010	6.279 ± 0.057	– 14.700 ± 0.300	– 70.363 [(a)] ± 1.024
$HSeO_3^-$	$H_2SeO_3(aq) \rightleftharpoons H^+ + HSeO_3^-$			
	– 2.640 ± 0.140	15.069 ± 0.799		
$HSeO_4^-$	$H_2O(l) + H_2SeO_3(aq) \rightleftharpoons 3H^+ + HSeO_4^- + 2\,e^-$			
	– 37.289 [(c)] ± 0.175	212.847 ± 1.000		
$H_2SeO_3(cr)$	$H_2O(l) + SeO_2(cr) \rightleftharpoons H_2SeO_3(cr)$			
			– 13.500 ± 0.180	
$H_2SeO_3(aq)$	$3H_2O(l) + Se(cr) \rightleftharpoons 2H_2(g) + H_2SeO_3(aq)$			
	– 50.147 ± 0.147	286.240 ± 0.840		
$SeF_4(g)$	$SeF_4(l) \rightleftharpoons SeF_4(g)$			
	– 1.668 ± 0.409	9.520 ± 2.335	45.000 ± 1.500	119.000 ± 6.000
$SeOF_2(g)$	$SeOF_2(l) \rightleftharpoons SeOF_2(g)$			
	– 1.953 ± 0.183	11.146 ± 1.044	44.300 ± 1.000	111.200 ± 1.000
$SeCl_2(g)$	$SeCl_4(cr) \rightleftharpoons Cl_2(g) + SeCl_2(g)$			
	– 10.517 ± 0.816	60.033 ± 4.657	162.000 ± 4.000	342.000 ± 8.000
$Se_2Cl_2(l)$	$Se(cr) + SeCl_2(g) \rightleftharpoons Se_2Cl_2(l)$			
	3.150 [(c)] ± 0.437	– 17.979 [(a)] ± 2.494	– 63.000 ± 2.000	– 151.000 ± 5.000

(Continued on next page)

Table III-2: (continued)

Species	Reaction $\log_{10} K^{\circ}$	$\Delta_r G_m^{\circ}$ (kJ·mol⁻¹)	$\Delta_r H_m^{\circ}$ (kJ·mol⁻¹)	$\Delta_r S_m^{\circ}$ (J·K⁻¹·mol⁻¹)
$SeOCl_2(g)$	$SeOCl_2(l) \rightleftharpoons SeOCl_2(g)$			
	− 3.135 ± 0.283	17.895 ± 1.614	52.600 ± 1.500	116.400 ± 2.000
$SeBr_2(g)$	$Se_2Br_2(l) \rightleftharpoons Se(cr) + SeBr_2(g)$			
			63.000 ± 20.000	
$SeS(g)$	$0.5S_2(g) + 0.5Se_2(g) \rightleftharpoons SeS(g)$			
	− 1.104 [(c)] ± 1.174	6.300 ± 6.700	8.476 [(a)] ± 6.702	7.300 ± 0.500
$TeSe(g)$	$0.5Se_2(g) + 0.5Te_2(g) \rightleftharpoons TeSe(g)$			
	− 0.140 [(c)] ± 0.543	0.800 ± 3.100	2.200 ± 3.000	4.700 ± 1.000
$CSe_2(g)$	$CSe_2(l) \rightleftharpoons CSe_2(g)$			
			36.300 ± 0.200	
$SeCN^-$	$HCN(aq) + Se(mono) \rightleftharpoons H^+ + SeCN^-$			
	− 3.574 ± 0.497	20.400 ± 2.838		
$GeSe(cr)$	$GeF_4(g) + SeF_6(g) \rightleftharpoons 5F_2(g) + GeSe(cr)$			
			2257.900 ± 1.300	
$GeSe(g)$	$GeSe(cr) \rightleftharpoons GeSe(g)$			
			172.700 ± 1.000	
$GeSe_2(g)$	$GeSe_2(cr) \rightleftharpoons GeSe_2(g)$			
			218.000 ± 18.000	
$SnSe(g)$	α-SnSe $\rightleftharpoons SnSe(g)$			
			219.700 ± 4.600	
$PbSe(g)$	$PbSe(cr) \rightleftharpoons PbSe(g)$			
			228.000 ± 4.200	

(Continued on next page)

Table III-2: (continued)

Species	Reaction $\log_{10} K^\circ$	$\Delta_r G_m^\circ$ (kJ·mol^{-1})	$\Delta_r H_m^\circ$ (kJ·mol^{-1})	$\Delta_r S_m^\circ$ (J·K^{-1}·mol^{-1})
$PbSeO_3(cr)$	$Pb^{2+} + SeO_3^{2-} \rightleftharpoons PbSeO_3(cr)$			
	12.500 ± 1.000	− 71.351 ± 5.708		
$PbSeO_4(cr)$	$Pb^{2+} + SeO_4^{2-} \rightleftharpoons PbSeO_4(cr)$			
	6.900 ± 0.250	− 39.385 ± 1.427	− 4.720 ± 2.530	116.269 [a] ± 9.742
$Ga_2(SeO_3)_3{\cdot}6H_2O(cr)$	$2Ga^{3+} + 6H_2O(l) + 3SeO_3^{2-} \rightleftharpoons Ga_2(SeO_3)_3{\cdot}6H_2O(cr)$			
	37.000 ± 2.000	− 211.198 ± 11.416		
$In_2(SeO_3)_3{\cdot}6H_2O(cr)$	$6H_2O(l) + 2In^{3+} + 3SeO_3^{2-} \rightleftharpoons In_2(SeO_3)_3{\cdot}6H_2O(cr)$			
	39.000 ± 2.000	− 222.614 ± 11.416		
$Tl_2SeO_4(cr)$	$SeO_4^{2-} + 2Tl^+ \rightleftharpoons Tl_2SeO_4(cr)$			
	3.900 ± 0.150	− 22.261 ± 0.856	− 43.000 ± 5.000	− 69.558 [a] ± 17.014
$TlSeCN(aq)$	$SeCN^- + Tl^+ \rightleftharpoons TlSeCN(aq)$			
	1.750 ± 0.290	− 9.989 ± 1.655		
$ZnSeO_4(aq)$	$SeO_4^{2-} + Zn^{2+} \rightleftharpoons ZnSeO_4(aq)$			
	2.160 ± 0.060	− 12.329 ± 0.342	4.600 ± 4.000	56.781 [a] ± 13.465
$ZnSeO_4{\cdot}6H_2O(cr)$	$6H_2O(l) + SeO_4^{2-} + Zn^{2+} \rightleftharpoons ZnSeO_4{\cdot}6H_2O(cr)$			
	1.538 ± 0.068	− 8.779 ± 0.388		
$ZnSeCN^+$	$SeCN^- + Zn^{2+} \rightleftharpoons ZnSeCN^+$			
	1.210 ± 0.060	− 6.907 ± 0.342	− 5.950 ± 0.340	3.209 [a] ± 1.619
$Zn(SeCN)_2(aq)$	$2SeCN^- + Zn^{2+} \rightleftharpoons Zn(SeCN)_2(aq)$			
	1.680 ± 0.110	− 9.590 ± 0.628	− 6.350 ± 1.200	10.865 [a] ± 4.542
$CdSeO_4(aq)$	$Cd^{2+} + SeO_4^{2-} \rightleftharpoons CdSeO_4(aq)$			
	2.270 ± 0.060	− 2.957 ± 0.342	8.300 ± 4.000	71.297 [a] ± 13.465

(Continued on next page)

Table III-2: (continued)

Species	Reaction $\log_{10} K^{\circ}$	$\Delta_r G_m^{\circ}$ (kJ·mol^{-1})	$\Delta_r H_m^{\circ}$ (kJ·mol^{-1})	$\Delta_r S_m^{\circ}$ (J·K^{-1}·mol^{-1})
$CdSeCN^{+}$	$Cd^{2+} + SeCN^{-} \rightleftharpoons CdSeCN^{+}$			
	2.240 ± 0.060	− 12.786 ± 0.342	− 10.050 ± 0.200	9.177 [(a)] ± 1.330
$Cd(SeCN)_3^{-}$	$Cd^{2+} + 3SeCN^{-} \rightleftharpoons Cd(SeCN)_3^{-}$			
	3.810 ± 0.210	− 21.748 ± 1.199	− 3.350 ± 14.000	61.706 [(a)] ± 47.128
$Cd(SeCN)_4^{2-}$	$Cd^{2+} + 4SeCN^{-} \rightleftharpoons Cd(SeCN)_4^{2-}$			
	4.600 ± 0.110	− 26.257 ± 0.628	− 40.000 ± 14.000	− 46.094 [(a)] ± 47.003
$Cd(SeCN)_2(cr)$	$Cd^{2+} + 2SeCN^{-} \rightleftharpoons Cd(SeCN)_2(cr)$			
	5.700 ± 0.500	− 32.536 ± 2.854		
$Cd(SeCN)_2(aq)$	$Cd^{2+} + 2SeCN^{-} \rightleftharpoons Cd(SeCN)_2(aq)$			
	3.340 ± 0.120	− 19.065 ± 0.685	− 26.350 ± 2.000	− 24.434 [(a)] ± 7.091
$HgSe_2^{2-}$	$HSe^{-} + \alpha\text{-}HgSe \rightleftharpoons H^{+} + HgSe_2^{2-}$			
	− 12.800 ± 0.600	73.063 ± 3.425		
$HgSeO_3(cr)$	$Hg^{2+} + SeO_3^{2-} \rightleftharpoons HgSeO_3(cr)$			
	16.200 ± 1.000	− 92.470 ± 5.708		
$Hg(SeO_3)_2^{2-}$	$HgSeO_3(cr) + SeO_3^{2-} \rightleftharpoons Hg(SeO_3)_2^{2-}$			
	− 1.350 ± 0.150	7.706 ± 0.856		
$Hg_2SeO_3(cr)$	$Hg_2^{2+} + SeO_3^{2-} \rightleftharpoons Hg_2SeO_3(cr)$			
	15.200 ± 1.000	− 86.762 ± 5.708		
$Hg(SeCN)_2(aq)$	$Hg^{2+} + 2SeCN^{-} \rightleftharpoons Hg(SeCN)_2(aq)$			
	22.300 ± 1.000	− 127.289 ± 5.708		
$Hg(SeCN)_3^{-}$	$Hg^{2+} + 3SeCN^{-} \rightleftharpoons Hg(SeCN)_3^{-}$			
	26.800 ± 1.000	− 152.976 ± 5.708		

(Continued on next page)

Table III-2: (continued)

Species	Reaction $\log_{10} K^{\circ}$	$\Delta_r G_m^{\circ}$ (kJ·mol^{-1})	$\Delta_r H_m^{\circ}$ (kJ·mol^{-1})	$\Delta_r S_m^{\circ}$ (J·K^{-1}·mol^{-1})
$Hg(SeCN)_4^{2-}$	$Hg^{2+} + 4SeCN^- \rightleftharpoons Hg(SeCN)_4^{2-}$			
	29.300 ± 0.500	− 167.246 ± 2.854	− 193.000 ± 3.000	− 86.381 (a) ± 13.888
$CuSeO_4{\cdot}5H_2O(cr)$	$Cu^{2+} + 5H_2O(l) + SeO_4^{2-} \rightleftharpoons CuSeO_4{\cdot}5H_2O(cr)$			
	2.440 ± 0.200	− 13.928 ± 1.142		
$Ag_2SeO_3(cr)$	$2Ag^+ + SeO_3^{2-} \rightleftharpoons Ag_2SeO_3(cr)$			
	15.800 ± 0.300	− 90.187 ± 1.712		
$Ag_2SeO_4(cr)$	$2Ag^+ + SeO_4^{2-} \rightleftharpoons Ag_2SeO_4(cr)$			
	7.860 ± 0.500	− 44.865 ± 2.854		
$AgSeCN(cr)$	$Ag^+ + SeCN^- \rightleftharpoons AgSeCN(cr)$			
	13.998 (c) ± 0.508	− 79.900 ± 2.900	− 108.100 ± 1.700	− 94.583 (a) ± 11.275
$Ag(SeCN)_3^{2-}$	$Ag^+ + 3SeCN^- \rightleftharpoons Ag(SeCN)_3^{2-}$			
	13.850 ± 0.300	− 79.056 ± 1.712		
$NiSeO_3{\cdot}2H_2O(cr)$	$2H_2O(l) + Ni^{2+} + SeO_3^{2-} \rightleftharpoons NiSeO_3{\cdot}2H_2O(cr)$			
	5.800 ± 1.000	− 33.107 ± 5.708		
$NiSeO_4(aq)$	$Ni^{2+} + SeO_4^{2-} \rightleftharpoons NiSeO_4(aq)$			
	2.670 ± 0.050	− 15.240 ± 0.285		
$NiSeO_4{\cdot}6H_2O(cr)$	$6H_2O(l) + Ni^{2+} + SeO_4^{2-} \rightleftharpoons NiSeO_4{\cdot}6H_2O(cr)$			
	1.381 ± 0.045	− 7.883 ± 0.257		
$NiSeCN^+$	$Ni^{2+} + SeCN^- \rightleftharpoons NiSeCN^+$			
	1.770 ± 0.060	− 10.103 ± 0.342	− 12.800 ± 0.400	− 9.045 (a) ± 1.766
$Ni(SeCN)_2(aq)$	$Ni^{2+} + 2SeCN^- \rightleftharpoons Ni(SeCN)_2(aq)$			
	2.240 ± 0.140	− 12.786 ± 0.799	− 25.000 ± 4.000	− 40.966 (a) ± 13.681

(Continued on next page)

Table III-2: (continued)

Species	Reaction $\log_{10} K^{\circ}$	$\Delta_r G_m^{\circ}$ (kJ·mol^{-1})	$\Delta_r H_m^{\circ}$ (kJ·mol^{-1})	$\Delta_r S_m^{\circ}$ (J·K^{-1}·mol^{-1})
$CoSeO_3 \cdot 2H_2O(cr)$	$Co^{2+} + 2H_2O(l) + SeO_3^{2-} \rightleftharpoons CoSeO_3 \cdot 2H_2O(cr)$			
	7.900	– 45.094		
	± 0.400	± 2.283		
$CoSeO_4(aq)$	$Co^{2+} + SeO_4^{2-} \rightleftharpoons CoSeO_4(aq)$			
	2.700	– 15.412		
	± 0.050	± 0.285		
$CoSeO_4 \cdot 6H_2O(cr)$	$Co^{2+} + 6H_2O(l) + SeO_4^{2-} \rightleftharpoons CoSeO_4 \cdot 6H_2O(cr)$			
	1.759	– 10.040		
	± 0.043	± 0.245		
$CoSeCN^+$	$Co^{2+} + SeCN^- \rightleftharpoons CoSeCN^+$			
	1.500	– 8.562	– 6.000	8.593 [a]
	± 0.300	± 1.712	± 4.000	± 14.594
$FeSeO_3^+$	$Fe^{3+} + SeO_3^{2-} \rightleftharpoons FeSeO_3^+$			
	11.150	– 63.645		
	± 0.110	± 0.628		
$Fe_2(SeO_3)_3 \cdot 6H_2O(cr)$	$2Fe^{3+} + 6H_2O(l) + 3SeO_3^{2-} \rightleftharpoons Fe_2(SeO_3)_3 \cdot 6H_2O(cr)$			
	41.580	– 237.340		
	± 0.110	± 0.628		
$MnSeO_3 \cdot 2H_2O(cr)$	$2H_2O(l) + Mn^{2+} + SeO_3^{2-} \rightleftharpoons MnSeO_3 \cdot 2H_2O(cr)$			
	7.600	– 43.381		
	± 1.000	± 5.708		
$MnSeO_4(aq)$	$Mn^{2+} + SeO_4^{2-} \rightleftharpoons MnSeO_4(aq)$			
	2.430	– 13.871		
	± 0.050	± 0.285		
$UO_2SeO_4(aq)$	$SeO_4^{2-} + UO_2^{2+} \rightleftharpoons UO_2SeO_4(aq)$			
	2.740	– 15.640		
	± 0.250	± 1.427		
$Th(SeO_3)_2(cr)$	$Th(SeO_3)_2 \cdot H_2O(cr) \rightleftharpoons H_2O(g) + Th(SeO_3)_2(cr)$			
	– 6.499	37.094	94.100	191.200
	± 0.560	± 3.195	± 2.900	± 4.500
$MgSeO_3 \cdot 6H_2O(cr)$	$6H_2O(l) + Mg^{2+} + SeO_3^{2-} \rightleftharpoons MgSeO_3 \cdot 6H_2O(cr)$			
	5.820	– 33.221		
	± 0.250	± 1.427		

(Continued on next page)

Table III-2: (continued)

Species	Reaction $\log_{10} K^\circ$	$\Delta_r G_m^\circ$ (kJ·mol^{-1})	$\Delta_r H_m^\circ$ (kJ·mol^{-1})	$\Delta_r S_m^\circ$ (J·K^{-1}·mol^{-1})
$MgSeO_4(aq)$	$Mg^{2+} + SeO_4^{2-} \rightleftharpoons MgSeO_4(aq)$			
	2.200 ± 0.200	− 12.558 ± 1.142		
$MgSeO_4{\cdot}6H_2O(cr)$	$6H_2O(l) + Mg^{2+} + SeO_4^{2-} \rightleftharpoons MgSeO_4{\cdot}6H_2O(cr)$			
	1.133 ± 0.044	− 6.467 ± 0.251		
$CaSeO_3{\cdot}H_2O(cr)$	$Ca^{2+} + H_2O(l) + SeO_3^{2-} \rightleftharpoons CaSeO_3{\cdot}H_2O(cr)$			
	6.400 ± 0.250	− 36.531 ± 1.427		
$CaSeO_4(aq)$	$Ca^{2+} + SeO_4^{2-} \rightleftharpoons CaSeO_4(aq)$			
	2.000 ± 0.100	− 11.416 ± 0.571		
$CaSeO_4{\cdot}2H_2O(cr)$	$Ca^{2+} + 2H_2O(l) + SeO_4^{2-} \rightleftharpoons CaSeO_4{\cdot}2H_2O(cr)$			
	2.680 ± 0.250	− 15.298 ± 1.427		
$SrSeO_3(cr)$	$SeO_3^{2-} + Sr^{2+} \rightleftharpoons SrSeO_3(cr)$			
	6.300 ± 0.500	− 35.961 ± 2.854		
$BaSeO_3(cr)$	$Ba^{2+} + SeO_3^{2-} \rightleftharpoons BaSeO_3(cr)$			
	6.500 ± 0.250	− 37.102 ± 1.427		
$BaSeO_4(cr)$	$Ba^{2+} + SeO_4^{2-} \rightleftharpoons BaSeO_4(cr)$			
	7.560 ± 0.100	− 43.153 ± 0.571		
$NH_4HSe(cr)$	$H_2Se(g) + NH_3(g) \rightleftharpoons NH_4HSe(cr)$			
	4.024 [(c)] ± 0.866	− 22.972 [(a)] ± 4.942	− 115.100 ± 3.500	− 309.000 ± 11.700
$(NH_4)_2SeO_4(cr)$	$2NH_4^+ + SeO_4^{2-} \rightleftharpoons (NH_4)_2SeO_4(cr)$			
	− 0.911 ± 0.065	5.200 ± 0.371		
$Li_2SeO_4{\cdot}H_2O(cr)$	$H_2O(l) + 2Li^+ + SeO_4^{2-} \rightleftharpoons Li_2SeO_4{\cdot}H_2O(cr)$			
	− 1.762 ± 0.065	10.058 ± 0.371		

(Continued on next page)

Table III-2: (continued)

Species	Reaction $\log_{10} K^{\circ}$	$\Delta_r G_m^{\circ}$ $(kJ \cdot mol^{-1})$	$\Delta_r H_m^{\circ}$ $(kJ \cdot mol^{-1})$	$\Delta_r S_m^{\circ}$ $(J \cdot K^{-1} \cdot mol^{-1})$
$LiH_3(SeO_3)_2(cr)$	$H_2O(l) + 3LiHSeO_3(cr) \rightleftharpoons Li_2SeO_3 \cdot H_2O(cr) + LiH_3(SeO_3)_2(cr)$			
	− 0.163 [(c)] ± 0.018	0.930 ± 0.100		
$LiHSeO_3(cr)$	$LiH_3(SeO_3)_2(cr) \rightleftharpoons H_2SeO_3(cr) + LiHSeO_3(cr)$			
	− 2.689 [(c)] ± 0.158	15.350 ± 0.900		
$Na_2SeO_4 \cdot 10H_2O(cr)$	$10H_2O(l) + 2Na^+ + SeO_4^{2-} \rightleftharpoons Na_2SeO_4 \cdot 10H_2O(cr)$			
	0.681 ± 0.087	− 3.887 ± 0.497		
$NaHSeO_3(cr)$	$NaH_3(SeO_3)_2(cr) \rightleftharpoons H_2SeO_3(cr) + NaHSeO_3(cr)$			
	− 3.162 [(c)] ± 0.193	18.050 ± 1.100		
$NaH_3(SeO_3)_2(cr)$	$3NaHSeO_3(cr) \rightleftharpoons Na_2SeO_3(cr) + NaH_3(SeO_3)_2(cr)$			
	− 1.109 [(c)] ± 0.070	6.330 ± 0.400		
$NaH_7(SeO_3)_4(cr)$	$2H_2SeO_3(cr) + NaH_3(SeO_3)_2(cr) \rightleftharpoons NaH_7(SeO_3)_4(cr)$			
	1.086 [(c)] ± 0.070	− 6.200 ± 0.400		
$K_2SeO_3(cr)$	$3KHSeO_3(cr) \rightleftharpoons K_2SeO_3(cr) + KH_3(SeO_3)_2(cr)$			
	− 1.093 [(c)] ± 0.070	6.240 ± 0.400		
$K_2SeO_4(cr)$	$2K^+ + SeO_4^{2-} \rightleftharpoons K_2SeO_4(cr)$			
	− 0.904 ± 0.065	5.160 ± 0.371		
$Rb_2SeO_3(cr)$	$3RbHSeO_3(cr) \rightleftharpoons Rb_2SeO_3(cr) + RbH_3(SeO_3)_2(cr)$			
	− 1.498 ± 0.088	8.550 ± 0.500		
$Cs_2SeO_3 \cdot H_2O(cr)$	$3CsHSeO_3(cr) + H_2O(l) \rightleftharpoons Cs_2SeO_3 \cdot H_2O(cr) + CsH_3(SeO_3)_2(cr)$			
	− 2.533 [(c)] ± 0.158	14.460 ± 0.900		
$Cs_2SeO_4(cr)$	$2Cs^+ + SeO_4^{2-} \rightleftharpoons Cs_2SeO_4(cr)$			
	− 0.636 ± 0.065	3.630 ± 0.371		

(Continued on next page)

Table III-2: (continued)

Species	Reaction $\log_{10} K^{\circ}$	$\Delta_r G_m^{\circ}$ (kJ·mol⁻¹)	$\Delta_r H_m^{\circ}$ (kJ·mol⁻¹)	$\Delta_r S_m^{\circ}$ (J·K⁻¹·mol⁻¹)
$CsHSeO_3(cr)$	$CsH_3(SeO_3)_2(cr) \rightleftharpoons CsHSeO_3(cr) + H_2SeO_3(cr)$			
	− 3.372 [(c)] ± 0.210	19.250 ± 1.200		
$CsH_3(SeO_3)_2(cr)$	$CsH_5(SeO_3)_3(cr) \rightleftharpoons CsH_3(SeO_3)_2(cr) + H_2SeO_3(cr)$			
	− 1.349 ± 0.088	7.700 ± 0.500		

(a) Value calculated internally using the equation $\Delta_r G_m^{\circ} = \Delta_r H_m^{\circ} - T\,\Delta_r S_m^{\circ}$

(b) Value calculated from a selected standard potential.

(c) Value of $\log_{10} K^{\circ}$ calculated internally from $\Delta_r G_m^{\circ}$.

Table III-3: Selected temperature coefficients for heat capacities marked with [(c)] in Table III-1 according to the form $C^{\circ}_{p,m}(T) = a + bT + cT^2 + dT^{-1} + eT^{-2}$. The functions are valid between the temperatures T_{min} and T_{max} (in K). The notation E±*nm* indicates the power of 10. Units for $C^{\circ}_{p,m}$ are J·K^{-1}·mol^{-1}.

Compound	*a*	*b*	*c*	*d*	*e*	T_{max}	T_{min}
Se(cr) (= trigonal)	2.48014E+01	1.28590E–03	9.92730E–06		– 8.71300E+04	298	500
Se(mono)	2.56980E+01	– 2.16540E–03	1.36408E–05		– 1.04720E+05	298	413
Se(l)	5.24080E+01	– 4.97660E–02	3.26080E–05			303	1000
Se(g)	1.15228E+01	1.68482E–02	– 6.47162E–06	1.66200E+03	– 6.42570E+04	298	1100
Se(g)	1.11256E+01	3.11880E–03	– 2.58660E–07	1.92957E+04	– 9.80328E+06	1100	5000
Se_2(g)	1.93485E+01	1.24903E–02	– 2.07010E–06	1.09846E+04	– 1.60249E+06	298	1300
Se_2(g)	5.81884E+01	– 4.10150E–03	4.78900E–07	– 2.82796E+04	1.297335E+07	1300	3000
SeO(g)	4.20600E+01	– 2.75700E–03	1.20100E–06	– 4.11300E+03	4.41300E+05	300	2000
SeO_2(cr)	6.95000E+01	3.89000E–03			– 1.10500E+06	290	305
SeO_2(g)	6.66380E+01	– 2.70100E–03		– 8.88400E+03	6.44900E+05	300	1500
SeO_3(cr)	1.76400E+01	2.00000E–01				300	390
H_2Se(g)	2.32030E+01	3.23156E–02	– 8.89220E–06	3.05500E+02	1.46472E+05	298	1500
SeF_2(g)	6.50600E+01	– 3.94790E–03	8.19000E–07	– 5.20470E+03	1.02790E+05	298	2000
SeF_4(g)	1.22798E+02	– 8.48620E–03	1.76310E–06	– 1.11220E+04	– 9.65900E+04	298	2000
SeF_6(g)	1.96298E+02	– 2.58210E–02	6.44800E–06	– 2.57220E+04	6.70400E+05	298	1500
$SeOF_2$(g)	9.67972E+01	– 7.47720E–03	1.49940E–06	– 1.11580E+04	5.38060E+05	298	2000
Se_2F_2(g)	8.82590E+01	– 2.98520E–03	6.25000E–07	– 3.79710E+03	– 2.18640E+05	298	2000
SeF_5Cl(g)	1.41600E+02	2.23830E–02	– 8.25000E–06		– 2.78740E+06	298	1400
$SeCl_2$(g)	5.94678E+01	– 7.54700E–04	1.60400E–07	– 9.12300E+02	– 2.33210E+05	298	2000
Se_2Cl_2(g)	8.19200E+01	2.29300E–03	– 1.19500E–06		– 4.34600E+05	298	1000
$SeOCl_2$(g)	9.10154E+01	– 4.26780E–03	8.49300E–07	– 6.51540E+03	1.98530E+05	298	2000
$SeBr_2$(g)	5.81200E+01	4.43500E–05			– 2.41000E+05	298	2000
Se_2Br_2(g)	8.28500E+01	3.59900E–04	– 9.80000E–08		– 3.11200E+05	298	1000
SeS(g)	3.20010E+01	4.90020E–03	– 5.35000E–07	1.86320E+03	– 3.45960E+05	298	2000
TeSe(g)	1.19809E+02	– 7.49930E–02	2.42400E–05	– 2.95990E+04	3.33040E+06	298	950
TeSe(g)	– 1.4324E+01	2.48440E–02	– 3.82760E–06	5.14240E+04	– 1.53300E+07	950	2000
As_2Se_3(cr)	1.21230E+02	– 3.81700E–02	1.30021E–04			273	634
As_4Se_3(cr)	5.15536E+01	2.30751E–01	– 2.88980E–05	2.43452E+04	– 2.39391E+06	130	440
As_4Se_4(cr)	6.98425E+03	– 1.14877E+01	7.27981E–03	– 1.75572E+06	1.67046E+08	300	540

(Continued next page)

Table III-3: (continued)

Compound	a	b	c	d	e	T_{max}	T_{min}
$Sb_2Se_3(cr)$	1.21700E+02	1.07300E–02				298	885
BiSe(g)	3.73251E+01	4.16100E–04	– 4.68000E–08		– 8.77670E+04	298	1500
$Bi_2Se_3(cr)$	1.26950E+02	1.89400E–02				298	985
CSe(g)	4.42955E+01	– 3.42560E–03	6.91000E–07	– 6.01840E+03	6.60504E+05	298	2000
$CSe_2(cr)$	5.17100E+01	5.62300E–02			– 3.10000E–04	78	229
$CSe_2(l)$	6.87000E+01	6.65000E–02				230	328
$CSe_2(g)$	7.00729E+01	– 3.39950E–03	5.68700E–07	– 8.13830E+03	7.22560E+05	298	2000
SiSe(g)	3.59540E+01	2.00500E–03	– 7.42700E–06		– 2.56238E+05	298	1500
GeSe(cr)	4.67770E+01	1.50990E–02	– 3.16000E–08		– 1.23100E+05	298	940
GeSe(g)	3.78410E+01	1.34000E–04	– 3.08000E–08	– 3.77200E+02	– 1.21800E+05	298	3000
α-SnSe	5.26200E+01	7.28000E–03	– 1.32900E–06		– 2.45690E+05	298	796
SnSe(g)	3.76190E+01	2.10400E–04	– 1.51000E–08	– 1.78700E+02	– 1.02881E+05	298	3000
$SnSe_2(cr)$	7.53900E+01	1.15000E–02			– 1.92000E+05	230	570
PbSe(cr)	4.69600E+01	1.22600E–02				298	1349
PbSe(g)	3.73380E+01	4.34300E–04	– 2.69000E–08		– 9.80680E+04	298	2000
AlSe(g)	3.80600E+01	3.18300E–04	– 6.33300E+02		– 1.14300E+05	298	2000
InSe(cr)	4.64960E+01	7.28100E–03			9.63140E+04	298	913
α-In_2Se_3	1.09190E+02	5.00000E–02				298	471
β-In_2Se_3	1.13420E+02	3.59900E–02				471	700
α-ZnSe	5.31550E+01	2.65170E–03	1.07890E–06		– 5.59680E+05	298	1500
α-CdSe	4.58700E+01	1.19400E–02				298	1500
α-HgSe	5.16600E+01	9.52000E–03				293	513
$NiSe_2(cr)$	1.03642E+02	– 5.51057E–02	4.53165E–05		– 1.48810E+06	298	1000
δ-$Fe_{0.875}Se$	4.70600E+01	– 1.48700E–02	2.39200E–05		4.51900E+06	660	1020
$FeSe_2(cr)$	3.70140E+01	8.02730E–02			9.57991E+05	298	600
$FeSe_2(cr)$	–8.35808E+02	1.050076E+00			1.056967E+08	600	853
β-$Fe_{1.04}Se$	5.62200E+01	2.31760E–02	– 9.70000E–07		– 5.34290E+05	298	730
$RuSe_2(cr)$	7.22980E+01	9.07620E–03	– 8.00200E–07		– 4.65942E+05	298	900
$RuSe_2(cr)$	– 4.3518E+01	1.405064E–01	– 3.88342E–05		2.248624E+07	900	1500
$MoSe_2(cr)$	6.91120E+01	1.42210E–02			– 2.46210E+05	298	1800
$WSe_2(cr)$	7.10860E+01	1.46270E–02			– 2.53246E+05	298	1800
$NbSe_2(cr)$	6.87630E+01	2.35000E–02			– 2.00000E+05	1180	1656

(Continued next page)

Table III-3: (continued)

Compound	*a*	*b*	*c*	*d*	*e*	T_{max}	T_{min}
EuSe(cr)	4.87600E+01	7.37600E–03				298	1600
Sm_2Se_3(cr)	1.21850E+02	3.66380E–02			– 4.03880E+04	1180	1900
La_2Se_3(cr)	1.16450E+02	2.76400E–02			8.06310E+04	298	2000

Chapter IV

Selected auxiliary data

This chapter presents the chemical thermodynamic data for auxiliary compounds and complexes which are used within the NEA's TDB project. Most of these auxiliary species are used in the evaluation of the recommended selenium data in Table III-1, Table III-2, and Table III-3. It is therefore essential to always use these auxiliary data in conjunction with the selected data for selenium. The use of other auxiliary data can lead to inconsistencies and erroneous results. Additional auxiliary data used in this review are found in Appendix E, Table E-2 (*cf.* Section II.7).

The values in the tables of this chapter are either CODATA Key Values, taken from [89COX/WAG], or were evaluated within the NEA's TDB project, as described in the corresponding chapters of the uranium review [92GRE/FUG], the technetium review [99RAR/RAN], the Update review [2003GUI/FAN], the nickel review [2005GAM/BUG] and the present review.

Table IV-1 contains the selected thermodynamic data of the auxiliary species and Table IV-2 the selected thermodynamic data of chemical reactions involving auxiliary species. The reason for listing both reaction data and entropies, enthalpies and Gibbs energies of formation is, as described in Chapter III, that uncertainties in reaction data are often smaller than the derived S°_m, $\Delta_f H^\circ_m$ and $\Delta_f G^\circ_m$, due to uncertainty accumulation during the calculations.

All data in Table IV-1 and Table IV-2 refer to a temperature of 298.15 K, the standard state pressure of 0.1 MPa and, for aqueous species and reactions, to the infinite dilution standard state ($I = 0$).

The uncertainties listed below each reaction value in Table IV-2 are total uncertainties, and correspond mainly to the statistically defined 95% confidence interval. The uncertainties listed below each value in Table IV-1 have the following significance:

- for CODATA values from [89COX/WAG], the ± terms have the meaning: "it is probable, but not at all certain, that the true values of the thermodynamic quantities differ from the recommended values given in this report by no more than twice the ± terms attached to the recommended values".

- for values from [92GRE/FUG], [99RAR/RAN], [2003GUI/FAN], [2005GAM/BUG] and the present Review, the ± terms are derived from total uncertainties in the corresponding equilibrium constant of reaction (*cf.* Table IV-2), and from the ± terms listed for the necessary CODATA key values.

CODATA [89COX/WAG] values are available for $CO_2(g)$, HCO_3^-, CO_3^{2-}, $H_2PO_4^-$ and HPO_4^{2-}. From the values given for $\Delta_f H_m^\circ$ and S_m° the values of $\Delta_f G_m^\circ$ and, consequently, all the relevant equilibrium constants and enthalpy changes can be calculated. The propagation of errors during this procedure, however, leads to uncertainties in the resulting equilibrium constants that are significantly higher than those obtained from experimental determination of the constants. Therefore, reaction data for $CO_2(g)$, HCO_3^-, CO_3^{2-}, which were absent from the corresponding Table IV-2 in [92GRE/FUG], are included in this volume.

Note that the values in Table IV-1 and Table IV-2 may contain more digits than those listed in either [89COX/WAG] or in the discussions on auxiliary data selection of NEA-TDB reviews, because the data in the present chapter are retrieved directly from the computerised data base and rounded to three digits after the decimal point throughout.

Table IV-1: Selected thermodynamic data for auxiliary compounds and complexes adopted in the NEA-TDB project, including the CODATA Key Values [89COX/WAG]. All ionic species listed in this table are aqueous species. Unless noted otherwise, all data refer to 298.15 K and a pressure of 0.1 MPa and, for aqueous species, a standard state of infinite dilution (I = 0). The uncertainties listed below each value represent total uncertainties and correspond in principle to the statistically defined 95% confidence interval. Values in bold typeface are CODATA Key Values and are taken directly from Ref. [89COX/WAG] without further evaluation. Values obtained from internal calculation, *cf.* footnotes (a) and (b), are rounded at the third digit after the decimal point. Systematically, all the values are presented with three digits after the decimal point, regardless of the significance of these digits. The reference listed for each entry in this table indicates the NEA-TDB Review where the corresponding data have been adopted as NEA-TDB auxiliary data. The data presented in this table are available on computer media from the OECD Nuclear Energy Agency.

Compound and NEA-TDB Review	$\Delta_f G_m^\circ$ (kJ·mol^{-1})	$\Delta_f H_m^\circ$ (kJ·mol^{-1})	S_m° (J·K^{-1}·mol^{-1})	$C_{p,m}^\circ$ (J·K^{-1}·mol^{-1})
O(g) [92GRE/FUG]	**231.743** [(a)] **± 0.100**	**249.180** **± 0.100**	**161.059** **± 0.003**	**21.912** **± 0.001**
O_2(g) [92GRE/FUG]	**0.000**	**0.000**	**205.152** **± 0.005**	**29.378** **± 0.003**
H(g) [92GRE/FUG]	**203.276** [(a)] **± 0.006**	**217.998** **± 0.006**	**114.717** **± 0.002**	**20.786** **± 0.001**
H^+ [92GRE/FUG]	**0.000**	**0.000**	**0.000**	**0.000**
H_2(g) [92GRE/FUG]	**0.000**	**0.000**	**130.680** **± 0.003**	**28.836** **± 0.002**
OH^- [92GRE/FUG]	**– 157.220** [(a)] **± 0.072**	**– 230.015** **± 0.040**	**– 10.900** **± 0.200**	
H_2O(g) [92GRE/FUG]	**– 228.582** [(a)] **± 0.040**	**– 241.826** **±0.040**	**188.835** **± 0.010**	**33.609** **± 0.030**
H_2O(l) [92GRE/FUG]	**– 237.140** [(a)] **± 0.041**	**– 285.830** **± 0.040**	**69.950** **± 0.030**	**75.351** **± 0.080**
H_2O_2(aq) [92GRE/FUG]		– 191.170 ± 0.100		
He(g) [92GRE/FUG]	**0.000**	**0.000**	**126.153** **± 0.002**	**20.786** **± 0.001**
Ne(g) [92GRE/FUG]	**0.000**	**0.000**	**146.328** **± 0.003**	**20.786** **± 0.001**
Ar(g) [92GRE/FUG]	**0.000**	**0.000**	**154.846** **± 0.003**	**20.786** **± 0.001**

(Continued on next page)

Table IV-1: (continued)

Compound and NEA-TDB Review	$\Delta_f G_m^\circ$ (kJ·mol^{-1})	$\Delta_f H_m^\circ$ (kJ·mol^{-1})	S_m° (J·K^{-1}·mol^{-1})	$C_{p,m}^\circ$ (J·K^{-1}·mol^{-1})
Kr(g) [92GRE/FUG]	**0.000**	**0.000**	**164.085 ± 0.003**	**20.786 ± 0.001**
Xe(g) [92GRE/FUG]	**0.000**	**0.000**	**169.685 ± 0.003**	**20.786 ± 0.001**
F(g) [92GRE/FUG]	**62.280** (a) **± 0.300**	**79.380 ± 0.300**	**158.751 ± 0.004**	**22.746 ± 0.002**
F^- [92GRE/FUG]	**– 281.523** (a) **± 0.692**	**– 335.350 ± 0.650**	**– 13.800 ± 0.800**	
F_2(g) [92GRE/FUG]	**0.000**	**0.000**	**202.791 ± 0.005**	**31.304 ± 0.002**
HF(aq) [92GRE/FUG]	– 299.675 ± 0.702	– 323.150 ± 0.716	88.000 (a) ± 3.362	
HF(g) [92GRE/FUG]	**– 275.400** (a) **± 0.700**	**– 273.300 ± 0.700**	**173.779 ± 0.003**	**29.137 ± 0.002**
HF_2^- [92GRE/FUG]	– 583.709 ± 1.200	– 655.500 ± 2.221	92.683 (a) ± 8.469	
Cl(g) [92GRE/FUG]	**105.305** (a) **± 0.008**	**121.301 ± 0.008**	**165.190 ± 0.004**	**21.838 ± 0.001**
Cl^- [92GRE/FUG]	**– 131.217** (a) **± 0.117**	**– 167.080 ± 0.100**	**56.600 ± 0.200**	
Cl_2(g) [92GRE/FUG]	**0.000**	**0.000**	**223.081 ± 0.010**	**33.949 ± 0.002**
ClO^- This review	– 37.669 (b) ± 0.962	– 108.300 (b) ± 1.500	42.560 (b) ± 9.238	
ClO_2^- [92GRE/FUG]	10.250 ± 4.044			
ClO_3^- [92GRE/FUG]	– 7.903 (a) ± 1.342	– 104.000 ± 1.000	162.300 ± 3.000	
ClO_4^- [92GRE/FUG]	**– 7.890** (a) **± 0.600**	**– 128.100 ± 0.400**	**184.000 ± 1.500**	
HCl(g) [92GRE/FUG]	**– 95.298** (a) **± 0.100**	**– 92.310 ± 0.100**	**186.902 ± 0.005**	**29.136 ± 0.002**
HClO(aq) This review	– 80.023 (a) ± 0.613	– 125.700 ± 1.500	126.255 ± 5.434	
$HClO_2$(aq) [92GRE/FUG]	– 0.938 ± 4.043			

(Continued on next page)

Table IV-1: (continued)

Compound and NEA-TDB Review	$\Delta_f G_m^o$ (kJ·mol^{-1})	$\Delta_f H_m^o$ (kJ·mol^{-1})	S_m^o (J·K^{-1}·mol^{-1})	$C_{p,m}^o$ (J·K^{-1}·mol^{-1})
Br(g) [92GRE/FUG]	**82.379** [(a)] **± 0.128**	**111.870 ± 0.120**	**175.018 ± 0.004**	**20.786 ± 0.001**
Br^- [92GRE/FUG]	**– 103.850** [(a)] **± 0.167**	**– 121.410 ± 0.150**	**82.550 ± 0.200**	
Br_2(aq) [92GRE/FUG]	4.900 ± 1.000			
Br_2(g) [92GRE/FUG]	**3.105** [(a)] **± 0.142**	**30.910 ± 0.110**	**245.468 ± 0.005**	**36.057 ± 0.002**
Br_2(l) [92GRE/FUG]	**0.000**	**0.000**	**152.210 ± 0.300**	
BrO^- [92GRE/FUG]	– 32.095 ± 1.537			
BrO_3^- [92GRE/FUG]	19.070 [(a)] ± 0.634	– 66.700 ± 0.500	161.500 ± 1.300	
HBr(g) [92GRE/FUG]	**– 53.361** [(a)] **± 0.166**	**– 36.290 ± 0.160**	**198.700 ± 0.004**	**29.141 ± 0.003**
HBrO(aq) [92GRE/FUG]	– 81.356 [(b)] ± 1.527			
I(g) [92GRE/FUG]	**70.172** [(a)] **± 0.060**	**106.760 ± 0.040**	**180.787 ± 0.004**	**20.786 ± 0.001**
I^- [92GRE/FUG]	**– 51.724** [(a)] **± 0.112**	**– 56.780 ± 0.050**	**106.450 ± 0.300**	
I_2(cr) [92GRE/FUG]	**0.000**	**0.000**	**116.140 ± 0.300**	
I_2(g) [92GRE/FUG]	**19.323** [(a)] **± 0.120**	**62.420 ± 0.080**	**260.687 ± 0.005**	**36.888 ± 0.002**
IO_3^- [92GRE/FUG]	– 126.338 [(a)] ± 0.779	– 219.700 ± 0.500	118.000 ± 2.000	
HI(g) [92GRE/FUG]	**1.700** [(a)] **± 0.110**	**26.500 ± 0.100**	**206.590 ± 0.004**	**29.157 ± 0.003**
HIO_3(aq) [92GRE/FUG]	– 130.836 ± 0.797			
S(cr)(orthorhombic) [92GRE/FUG]	**0.000**	**0.000**	**32.054 ± 0.050**	**22.750 ± 0.050**
S(g) [92GRE/FUG]	**236.689** [(a)] **± 0.151**	**277.170 ± 0.150**	**167.829 ± 0.006**	**23.674 ± 0.001**
S^{2-} [92GRE/FUG]	120.695 ± 11.610			

(Continued on next page)

Table IV-1: (continued)

Compound and NEA-TDB Review	$\Delta_f G_m^\circ$ (kJ·mol^{-1})	$\Delta_f H_m^\circ$ (kJ·mol^{-1})	S_m° (J·K^{-1}·mol^{-1})	$C_{p,m}^\circ$ (J·K^{-1}·mol^{-1})
S_2(g) [92GRE/FUG]	**79.686** [a] **± 0.301**	**128.600 ± 0.300**	**228.167 ± 0.010**	**32.505 ± 0.010**
SO_2(g) [92GRE/FUG]	**– 300.095** [a] **± 0.201**	**– 296.810 ± 0.200**	**248.223 ± 0.050**	**39.842 ± 0.020**
SO_3^{2-} [92GRE/FUG]	– 487.472 ± 4.020			
$S_2O_3^{2-}$ [92GRE/FUG]	– 519.291 ± 11.345			
SO_4^{2-} [92GRE/FUG]	**– 744.004** [a] **± 0.418**	**– 909.340 ± 0.400**	**18.500 ± 0.400**	
HS^- [92GRE/FUG]	**12.243** [a] **± 2.115**	**– 16.300 ± 1.500**	**67.000 ± 5.000**	
H_2S(aq) [92GRE/FUG]	**– 27.648** [a] **± 2.115**	**– 38.600 ± 1.500**	**126.000 ± 5.000**	
H_2S(g) [92GRE/FUG]	**– 33.443** [a] **± 0.500**	**– 20.600 ± 0.500**	**205.810 ± 0.050**	**34.248 ± 0.010**
HSO_3^- [92GRE/FUG]	– 528.684 ± 4.046			
$HS_2O_3^-$ [92GRE/FUG]	– 528.366 ± 11.377			
H_2SO_3(aq) [92GRE/FUG]	– 539.187 ± 4.072			
HSO_4^- [92GRE/FUG]	**– 755.315** [a] **± 1.342**	**– 886.900 ± 1.000**	**131.700 ± 3.000**	
HSO_4^- [2001LEM/FUG]	– 755.315 ± 1.342	– 886.900 ± 1.000	131.700 ± 3.000	
Te(cr) [92GRE/FUG]	0.000	0.000	49.221 ± 0.050	25.550 ± 0.100
TeO_2(cr) [2003GUI/FAN]	– 265.996 [a] ± 2.500	– 321.000 ± 2.500	69.890 ± 0.150	60.670 ± 0.150
N(g) [92GRE/FUG]	**455.537** [a] **± 0.400**	**472.680 ± 0.400**	**153.301 ± 0.003**	**20.786 ± 0.001**
N_2(g) [92GRE/FUG]	**0.000**	**0.000**	**191.609 ± 0.004**	**29.124 ± 0.001**
N_3^- [92GRE/FUG]	348.200 ± 2.000	275.140 ± 1.000	107.710 [a] ± 7.500	
NO_3^- [92GRE/FUG]	**– 110.794** [a] **± 0.417**	**– 206.850 ± 0.400**	**146.700 ± 0.400**	

(Continued on next page)

Table IV-1: (continued)

Compound and NEA-TDB Review	$\Delta_f G_m^o$ (kJ·mol^{-1})	$\Delta_f H_m^o$ (kJ·mol^{-1})	S_m^o (J·K^{-1}·mol^{-1})	$C_{p,m}^o$ (J·K^{-1}·mol^{-1})
$HN_3(aq)$ [92GRE/FUG]	321.372 ± 2.051	260.140 ± 10.050	147.381 ± 34.403	
$NH_3(aq)$ [92GRE/FUG]	– 26.673 ± 0.305	– 81.170 ± 0.326	109.040 ± 0.913	
$NH_3(g)$ [92GRE/FUG]	**– 16.407** (a) **± 0.350**	**– 45.940** **± 0.350**	**192.770** **± 0.050**	**35.630** **± 0.005**
NH_4^+ [92GRE/FUG]	**– 79.398** (a) **± 0.278**	**– 133.260** **± 0.250**	**111.170** **± 0.400**	
P(am)(red) [92GRE/FUG]		– 7.500 ± 2.000		
P(cr)(white, cubic) [92GRE/FUG]	**0.000**	**0.000**	**41.090** **± 0.250**	**23.824** **± 0.200**
P(g) [92GRE/FUG]	**280.093** (a) **± 1.003**	**316.500** **± 1.000**	**163.199** **± 0.003**	**20.786** **± 0.001**
$P_2(g)$ [92GRE/FUG]	**103.469** (a) **± 2.006**	**144.000** **± 2.000**	**218.123** **± 0.004**	**32.032** **± 0.002**
$P_4(g)$ [92GRE/FUG]	**24.419** (a) **± 0.448**	**58.900** **± 0.300**	**280.010** **± 0.500**	**67.081** **± 1.500**
PO_4^{3-} [92GRE/FUG]	– 1025.491 ± 1.576	– 1284.400 ± 4.085	– 220.970 ± 12.846	
$P_2O_7^{4-}$ [92GRE/FUG]	– 1935.503 ± 4.563			
HPO_4^{2-} [92GRE/FUG]	**– 1095.985** (a) **± 1.567**	**– 1299.000** **± 1.500**	**– 33.500** **± 1.500**	
$H_2PO_4^-$ [92GRE/FUG]	**– 1137.152** (a) **± 1.567**	**– 1302.600** **± 1.500**	**92.500** **± 1.500**	
$H_3PO_4(aq)$ [92GRE/FUG]	– 1149.367 ± 1.576	– 1294.120 ± 1.616	161.912 ± 2.575	
$HP_2O_7^{3-}$ [92GRE/FUG]	– 1989.158 ± 4.482			
$H_2P_2O_7^{2-}$ [92GRE/FUG]	– 2027.117 ± 4.445			
$H_3P_2O_7^-$ [92GRE/FUG]	– 2039.960 ± 4.362			
$H_4P_2O_7(aq)$ [92GRE/FUG]	– 2045.668 ± 3.299	– 2280.210 ± 3.383	274.919 ± 6.954	
As(cr) [92GRE/FUG]	0.000	0.000	35.100	24.640 ± 0.500

(Continued on next page)

Table IV-1: (continued)

Compound and NEA-TDB Review	$\Delta_f G_m^\circ$ (kJ·mol^{-1})	$\Delta_f H_m^\circ$ (kJ·mol^{-1})	S_m° (J·K^{-1}·mol^{-1})	$C_{p,m}^\circ$ (J·K^{-1}·mol^{-1})
AsO_2^- [92GRE/FUG]	– 350.022 [(a)] ± 4.008	– 429.030 ± 4.000	40.600 ± 0.600	
AsO_4^{3-} [92GRE/FUG]	– 648.360 [(a)] ± 4.008	– 888.140 ± 4.000	– 162.800 ± 0.600	
As_2O_5(cr) [92GRE/FUG]	– 782.449 [(a)] ± 8.016	– 924.870 ± 8.000	105.400 ± 1.200	116.520 ± 0.800
As_4O_6(cubic) [92GRE/FUG]	– 1152.445 [(a)] ± 16.032	– 1313.940 ± 16.000	214.200 ± 2.400	191.290 ± 0.800
As_4O_6(monoclinic) [92GRE/FUG]	– 1154.009 [(a)] ± 16.041	– 1309.600 ± 16.000	234.000 ± 3.000	
As_4O_6(g) [2005GAM/BUG]	– 1092.716 [(a)] ± 16.116	– 1196.250 ± 16.000	408.600 ± 6.000	
$HAsO_2$(aq) [92GRE/FUG]	– 402.925 [(a)] ± 4.008	– 456.500 ± 4.000	125.900 ± 0.600	
$H_2AsO_3^-$ [92GRE/FUG]	– 587.078 [(a)] ± 4.008	– 714.790 ± 4.000	110.500 ± 0.600	
H_3AsO_3(aq) [92GRE/FUG]	– 639.681 [(a)] ± 4.015	– 742.200 ± 4.000	195.000 ± 1.000	
$HAsO_4^{2-}$ [92GRE/FUG]	– 714.592 [(a)] ± 4.008	– 906.340 ± 4.000	– 1.700 ± 0.600	
$H_2AsO_4^-$ [92GRE/FUG]	– 753.203 [(a)] ± 4.015	– 909.560 ± 4.000	117.000 ± 1.000	
H_3AsO_4(aq) [92GRE/FUG]	– 766.119 [(a)] ± 4.015	– 902.500 ± 4.000	184.000 ± 1.000	
$(As_2O_5)_3\cdot 5\,H_2O$(cr) [92GRE/FUG]		– 4248.400 ± 24.000		
Sb(cr) [92GRE/FUG]	0.000	0.000	45.520 ± 0.210	25.260 ± 0.200
Bi(cr) [2001LEM/FUG]	0.000	0.000	56.740 ± 0.420	25.410 ± 0.200
C(cr) [92GRE/FUG]	**0.000**	**0.000**	**5.740 ± 0.100**	**8.517 ± 0.080**
C(g) [92GRE/FUG]	**671.254** [(a)] **± 0.451**	**716.680 ± 0.450**	**158.100 ± 0.003**	**20.839 ± 0.001**
CO(g) [92GRE/FUG]	**– 137.168** [(a)] **± 0.173**	**– 110.530 ± 0.170**	**197.660 ± 0.004**	**29.141 ± 0.002**
CO_2(aq) [92GRE/FUG]	**– 385.970** [(a)] **± 0.270**	**– 413.260 ±0.200**	**119.360 ± 0.600**	

(Continued on next page)

Table IV-1: (continued)

Compound and NEA-TDB Review	$\Delta_f G^\circ_m$ (kJ·mol^{-1})	$\Delta_f H^\circ_m$ (kJ·mol^{-1})	S°_m (J·K^{-1}·mol^{-1})	$C^\circ_{p,m}$ (J·K^{-1}·mol^{-1})
CO_2(g) [92GRE/FUG]	**– 394.373** (a) **± 0.133**	**– 393.510 ± 0.130**	**213.785 ± 0.010**	**37.135 ± 0.002**
CO_3^{2-} [92GRE/FUG]	**– 527.900** (a) **± 0.390**	**– 675.230 ± 0.250**	**– 50.000 ± 1.000**	
HCO_3^- [92GRE/FUG]	**– 586.845** (a) **± 0.251**	**– 689.930 ± 0.200**	**98.400 ± 0.500**	
CN^- This review	166.939 (b) ± 2.519	147.350 (b) ± 3.541	101.182 (b) ± 8.475	
HCN(aq) This review	114.368 (b) ± 2.517	103.750 (b) ± 3.536	131.271 (b) ± 8.440	
HCN(g) This review	119.517 (a) ± 2.500	129.900 ± 2.500	201.710 ± 0.100	
SCN^- [92GRE/FUG]	92.700 ± 4.000	76.400 ± 4.000	144.268 (a) ± 18.974	
Si(cr) [92GRE/FUG]	**0.000**	**0.000**	**18.810 ± 0.080**	**19.789 ± 0.030**
Si(g) [92GRE/FUG]	**405.525** (a) **± 8.000**	**450.000 ± 8.000**	**167.981 ± 0.004**	**22.251 ± 0.001**
SiO_2(α– quartz) [92GRE/FUG]	**– 856.287** (a) **± 1.002**	**– 910.700 ± 1.000**	**41.460 ± 0.200**	**44.602 ± 0.300**
$SiO_2(OH)_2^{2-}$ [92GRE/FUG]	– 1175.651 ± 1.265	– 1381.960 ± 15.330	– 1.488 ± 51.592	
$SiO(OH)_3^-$ [92GRE/FUG]	– 1251.740 ± 1.162	– 1431.360 ± 3.743	88.024 ± 13.144	
$Si(OH)_4$(aq) [92GRE/FUG]	– 1307.735 (b) ± 1.156	– 1456.960 (b) ±3.163	189.973 (b) ± 11.296	
$Si_2O_3(OH)_4^{2-}$ [92GRE/FUG]	– 2269.878 ± 2.878			
$Si_2O_2(OH)_5^-$ [92GRE/FUG]	– 2332.096 ± 2.878			
$Si_3O_6(OH)_3^{3-}$ [92GRE/FUG]	– 3048.536 ± 3.870			
$Si_3O_5(OH)_5^{3-}$ [92GRE/FUG]	– 3291.955 ± 3.869			
$Si_4O_8(OH)_4^{4-}$ [92GRE/FUG]	– 4075.179 ± 5.437			
$Si_4O_7(OH)_5^{3-}$ [92GRE/FUG]	– 4136.826 ± 4.934			

(Continued on next page)

Table IV-1: (continued)

Compound and NEA-TDB Review	$\Delta_f G^\circ_m$ (kJ·mol^{-1})	$\Delta_f H^\circ_m$ (kJ·mol^{-1})	S°_m (J·K^{-1}·mol^{-1})	$C^\circ_{p,m}$ (J·K^{-1}·mol^{-1})
SiF_4(g) [92GRE/FUG]	**– 1572.773** [(a)] **± 0.814**	**– 1615.000 ± 0.800**	**282.760 ± 0.500**	**73.622 ± 0.500**
Ge(cr) [92GRE/FUG]	**0.000**	**0.000**	**31.090 ± 0.150**	**23.222 ± 0.100**
Ge(g) [92GRE/FUG]	**331.209** [(a)] **± 3.000**	**372.000 ± 3.000**	**167.904 ± 0.005**	**30.733 ± 0.001**
GeO_2(tetragonal) [92GRE/FUG]	**– 521.404** [(a)] **± 1.002**	**– 580.000 ± 1.000**	**39.710 ± 0.150**	**50.166 ± 0.300**
GeF_4(g) [92GRE/FUG]	**– 1150.018** [(a)] **± 0.584**	**– 1190.200 ± 0.500**	**301.900 ± 1.000**	**81.602 ± 1.000**
Sn(cr) [92GRE/FUG]	**0.000**	**0.000**	**51.180 ± 0.080**	**27.112 ± 0.030**
Sn(g) [92GRE/FUG]	**266.223** [(a)] **± 1.500**	**301.200 ± 1.500**	**168.492 ± 0.004**	**21.259 ± 0.001**
Sn^{2+} [92GRE/FUG]	**– 27.624** [(a)] **± 1.557**	**– 8.900 ± 1.000**	**– 16.700 ± 4.000**	
SnO(tetragonal) [92GRE/FUG]	**– 251.913** [(a)] **± 0.220**	**– 280.710 ± 0.200**	**57.170 ± 0.300**	**47.783 ± 0.300**
SnO_2(cassiterite, tetragonal) [92GRE/FUG]	**– 515.826** [(a)] **± 0.204**	**– 577.630 ± 0.200**	**49.040 ± 0.100**	**53.219 ± 0.200**
Pb(cr) [92GRE/FUG]	**0.000**	**0.000**	**64.800 ± 0.300**	**26.650 ± 0.100**
Pb(g) [92GRE/FUG]	**162.232** [(a)] **± 0.805**	**195.200 ± 0.800**	**175.375 ± 0.005**	**20.786 ± 0.001**
Pb^{2+} [92GRE/FUG]	**– 24.238** [(a)] **± 0.399**	**0.920 ± 0.250**	**18.500 ± 1.000**	
$PbSO_4$(cr) [92GRE/FUG]	**– 813.036** [(a)] **± 0.447**	**– 919.970 ± 0.400**	**148.500 ± 0.600**	
B(cr) [92GRE/FUG]	**0.000**	**0.000**	**5.900 ± 0.080**	**11.087 ± 0.100**
B(g) [92GRE/FUG]	**521.012** [(a)] **± 5.000**	**565.000 ± 5.000**	**153.436 ± 0.015**	**20.796 ± 0.005**
B_2O_3(cr) [92GRE/FUG]	**– 1194.324** [(a)] **± 1.404**	**– 1273.500 ± 1.400**	**53.970 ± 0.300**	**62.761 ± 0.300**
$B(OH)_3$(aq) [92GRE/FUG]	**– 969.268** [(a)] **± 0.820**	**– 1072.800 ± 0.800**	**162.400 ± 0.600**	
$B(OH)_3$(cr) [92GRE/FUG]	**– 969.667** [(a)] **± 0.820**	**– 1094.800 ± 0.800**	**89.950 ± 0.600**	**86.060 ± 0.400**

(Continued on next page)

Table IV-1: (continued)

Compound and NEA-TDB Review	$\Delta_f G_m^o$ (kJ·mol⁻¹)	$\Delta_f H_m^o$ (kJ·mol⁻¹)	S_m^o (J·K⁻¹·mol⁻¹)	$C_{p,m}^o$ (J·K⁻¹·mol⁻¹)
BF_3(g) [92GRE/FUG]	**– 1119.403** [(a)] **± 0.803**	**– 1136.000 ± 0.800**	**254.420 ± 0.200**	**50.463 ± 0.100**
Al(cr) [92GRE/FUG]	**0.000**	**0.000**	**28.300 ± 0.100**	**24.200 ± 0.070**
Al(g) [92GRE/FUG]	**289.376** [(a)] **± 4.000**	**330.000 ± 4.000**	**164.554 ± 0.004**	**21.391 ± 0.001**
Al^{3+} [92GRE/FUG]	**– 491.507** [(a)] **± 3.338**	**– 538.400 ± 1.500**	**– 325.000 ± 10.000**	
Al_2O_3(corundum) [92GRE/FUG]	**– 1582.257** [(a)] **± 1.302**	**– 1675.700 ± 1.300**	**50.920 ± 0.100**	**79.033 ± 0.200**
AlF_3(cr) [92GRE/FUG]	**– 1431.096** [(a)] **± 1.309**	**– 1510.400 ± 1.300**	**66.500 ± 0.500**	**75.122 ± 0.400**
Tl^+ [99RAR/RAN]	– 32.400 ± 0.300			
Zn(cr) [92GRE/FUG]	**0.000**	**0.000**	**41.630 ± 0.150**	**25.390 ± 0.040**
Zn(g) [92GRE/FUG]	**94.813** [(a)] **± 0.402**	**130.400 ± 0.400**	**160.990 ± 0.004**	**20.786 ± 0.001**
Zn^{2+} [92GRE/FUG]	**– 147.203** [(a)] **± 0.254**	**– 153.390 ± 0.200**	**– 109.800 ± 0.500**	
ZnO(cr) [92GRE/FUG]	**– 320.479** [(a)] **± 0.299**	**– 350.460 ± 0.270**	**43.650 ± 0.400**	
Cd(cr) [92GRE/FUG]	**0.000**	**0.000**	**51.800 ± 0.150**	**26.020 ± 0.040**
Cd(g) [92GRE/FUG]	**77.230** [(a)] **± 0.205**	**111.800 ± 0.200**	**167.749 ± 0.004**	**20.786 ± 0.001**
Cd^{2+} [92GRE/FUG]	**– 77.733** [(a)] **± 0.750**	**– 75.920 ± 0.600**	**– 72.800 ± 1.500**	
CdO(cr) [92GRE/FUG]	**– 228.661** [(a)] **± 0.602**	**– 258.350 ± 0.400**	**54.800 ± 1.500**	
$CdSO_4 \cdot 2.667\ H_2O$(cr) [92GRE/FUG]	**– 1464.959** [(a)] **± 0.810**	**– 1729.300 ± 0.800**	**229.650 ± 0.400**	
Hg(g) [92GRE/FUG]	**31.842** [(a)] **± 0.054**	**61.380 ± 0.040**	**174.971 ± 0.005**	**20.786 ± 0.001**
Hg(l) [92GRE/FUG]	**0.000**	**0.000**	**75.900 ± 0.120**	
Hg^{2+} [92GRE/FUG]	**164.667** [(a)] **± 0.313**	**170.210 ± 0.200**	**– 36.190 ± 0.800**	

(Continued on next page)

Table IV-1: (continued)

Compound and NEA-TDB Review	$\Delta_f G^\circ_m$ (kJ·mol^{-1})	$\Delta_f H^\circ_m$ (kJ·mol^{-1})	S°_m (J·K^{-1}·mol^{-1})	$C^\circ_{p,m}$ (J·K^{-1}·mol^{-1})
Hg_2^{2+} [92GRE/FUG]	**153.567** [a] **± 0.559**	**166.870 ± 0.500**	**65.740 ± 0.800**	
HgO(montroydite, red) [92GRE/FUG]	**– 58.523** [a] **± 0.154**	**– 90.790 ± 0.120**	**70.250 ± 0.300**	
Hg_2Cl_2(cr) [92GRE/FUG]	**– 210.725** [a] **± 0.471**	**– 265.370 ± 0.400**	**191.600 ± 0.800**	
Hg_2SO_4(cr) [92GRE/FUG]	**– 625.780** [a] **± 0.411**	**– 743.090 ± 0.400**	**200.700 ± 0.200**	
Cu(cr) [92GRE/FUG]	**0.000**	**0.000**	**33.150 ± 0.080**	**24.440 ± 0.050**
Cu(g) [92GRE/FUG]	**297.672** [a] **± 1.200**	**337.400 ± 1.200**	**166.398 ± 0.004**	**20.786 ± 0.001**
Cu^{2+} [92GRE/FUG]	**65.040** [a] **± 1.557**	**64.900 ± 1.000**	**– 98.000 ± 4.000**	
CuCl(g) [2003GUI/FAN]		76.800 ± 10.000		
$CuSO_4$(cr) [92GRE/FUG]	**– 662.185** [a] **± 1.206**	**– 771.400 ± 1.200**	**109.200 ± 0.400**	
Ag(cr) [92GRE/FUG]	**0.000**	**0.000**	**42.550 ± 0.200**	**25.350 ± 0.100**
Ag(g) [92GRE/FUG]	**246.007** [a] **± 0.802**	**284.900 ± 0.800**	**172.997 ± 0.004**	**20.786 ± 0.001**
Ag^+ [92GRE/FUG]	**77.096** [a] **± 0.156**	**105.790 ± 0.080**	**73.450 ± 0.400**	
AgCl(cr) [92GRE/FUG]	**– 109.765** [a] **± 0.098**	**– 127.010 ± 0.050**	**96.250 ± 0.200**	
Ni(cr) [2005GAM/BUG]	0.000	0.000	29.870 ± 0.200	26.070 ± 0.100
Ni^{2+} [2005GAM/BUG]	– 45.773 [b] ± 0.771	– 55.012 [a] ± 0.878	– 131.800 [d] ± 1.400	– 46.100 ± 7.500
Ti(cr) [92GRE/FUG]	**0.000**	**0.000**	**30.720 ± 0.100**	**25.060 ± 0.080**
Ti(g) [92GRE/FUG]	**428.403** [a] **± 3.000**	**473.000 ± 3.000**	**180.298 ± 0.010**	**24.430 ± 0.030**
TiO_2(rutile) [92GRE/FUG]	**– 888.767** [a] **± 0.806**	**– 944.000 ± 0.800**	**50.620 ± 0.300**	**55.080 ± 0.300**
$TiCl_4$(g) [92GRE/FUG]	**– 726.324** [a] **± 3.229**	**– 763.200 ± 3.000**	**353.200 ± 4.000**	**95.408 ± 1.000**

(Continued on next page)

Table IV-1: (continued)

Compound and NEA-TDB Review	$\Delta_f G^o_m$ (kJ·mol^{-1})	$\Delta_f H^o_m$ (kJ·mol^{-1})	S^o_m (J·K^{-1}·mol^{-1})	$C^o_{p,m}$ (J·K^{-1}·mol^{-1})
Th(cr) [92GRE/FUG]	**0.000**	**0.000**	**51.800 ± 0.500**	**26.230 ± 0.050**
Th(g) [92GRE/FUG]	**560.745 ± 6.002** [(a)]	**602.000 ± 6.000**	**190.170 ± 0.050**	**20.789 ± 0.100**
ThO_2(cr) [92GRE/FUG]	**– 1169.238 ± 3.504** [(a)]	**– 1226.400 ± 3.500**	**65.230 ± 0.200**	
Be(cr) [92GRE/FUG]	**0.000**	**0.000**	**9.500 ± 0.080**	**16.443 ± 0.060**
Be(g) [92GRE/FUG]	**286.202 ± 5.000** [(a)]	**324.000 ± 5.000**	**136.275 ± 0.003**	**20.786 ± 0.001**
BeO(bromellite) [92GRE/FUG]	**– 580.090 ± 2.500** [(a)]	**– 609.400 ± 2.500**	**13.770 ± 0.040**	**25.565 ± 0.100**
Mg(cr) [92GRE/FUG]	**0.000**	**0.000**	**32.670 ± 0.100**	**24.869 ± 0.020**
Mg(g) [92GRE/FUG]	**112.521 ± 0.801** [(a)]	**147.100 ± 0.800**	**148.648 ± 0.003**	**20.786 ± 0.001**
Mg^{2+} [92GRE/FUG]	**– 455.375 ± 1.335** [(a)]	**– 467.000 ± 0.600**	**– 137.000 ± 4.000**	
MgO(cr) [92GRE/FUG]	**– 569.312 ± 0.305** [(a)]	**– 601.600 ± 0.300**	**26.950 ± 0.150**	**37.237 ± 0.200**
MgF_2(cr) [92GRE/FUG]	**– 1071.051 ± 1.210** [(a)]	**– 1124.200 ± 1.200**	**57.200 ± 0.500**	**61.512 ± 0.300**
Ca(cr) [92GRE/FUG]	**0.000**	**0.000**	**41.590 ± 0.400**	**25.929 ± 0.300**
Ca(g) [92GRE/FUG]	**144.021 ± 0.809** [(a)]	**177.800 ± 0.800**	**154.887 ± 0.004**	**20.786 ± 0.001**
Ca^{2+} [92GRE/FUG]	**– 552.806 ± 1.050** [(a)]	**– 543.000 ± 1.000**	**– 56.200 ± 1.000**	
CaO(cr) [92GRE/FUG]	**– 603.296 ± 0.916** [(a)]	**– 634.920 ± 0.900**	**38.100 ± 0.400**	**42.049 ± 0.400**
CaF(g) [2003GUI/FAN]	– 302.118 ± 5.104	– 276.404 ± 5.100	229.244 ± 0.500	33.671 ± 0.500
CaCl(g) [2003GUI/FAN]	– 129.787 ± 5.001	– 103.400 ± 5.000	241.634 ± 0.300	35.687 ± 0.010
Sr(cr) [92GRE/FUG]	0.000	0.000	55.700 ± 0.210	
Sr^{2+} [92GRE/FUG]	– 563.864 ± 0.781 [(a)]	– 550.900 ± 0.500	– 31.500 ± 2.000	

(Continued on next page)

Table IV-1: (continued)

Compound and NEA-TDB Review	$\Delta_f G_m^\circ$ (kJ·mol^{-1})	$\Delta_f H_m^\circ$ (kJ·mol^{-1})	S_m° (J·K^{-1}·mol^{-1})	$C_{p,m}^\circ$ (J·K^{-1}·mol^{-1})
SrO(cr) [92GRE/FUG]	– 559.939 [(a)] ± 0.914	– 590.600 ± 0.900	55.440 ± 0.500	
$SrCl_2$(cr) [92GRE/FUG]	– 784.974 [(a)] ± 0.714	– 833.850 ± 0.700	114.850 ± 0.420	
$Sr(NO_3)_2$(cr) [92GRE/FUG]	– 783.146 [(a)] ± 1.018	– 982.360 ± 0.800	194.600 ± 2.100	
Ba(cr) [92GRE/FUG]	0.000	0.000	62.420 ± 0.840	
Ba(g) [2003GUI/FAN]	152.852 ± 5.006	185.000 ± 5.000	170.245 ± 0.010	20.786 ± 0.001
Ba^{2+} [92GRE/FUG]	– 557.656 [(a)] ± 2.582	– 534.800 ± 2.500	8.400 ± 2.000	
BaO(cr) [92GRE/FUG]	– 520.394 [(a)] ± 2.515	– 548.100 ± 2.500	72.070 ± 0.380	
BaF(g) [2003GUI/FAN]	– 349.569 ± 6.705	– 324.992 ± 6.700	246.219 ± 0.210	34.747 ± 0.300
$BaCl_2$(cr) [92GRE/FUG]	– 806.953 [(a)] ± 2.514	– 855.200 ± 2.500	123.680 ± 0.250	
Li(cr) [92GRE/FUG]	**0.000**	**0.000**	**29.120 ± 0.200**	**24.860 ± 0.200**
Li(g) [92GRE/FUG]	**126.604** [(a)] **± 1.002**	**159.300 ± 1.000**	**138.782 ± 0.010**	**20.786 ± 0.001**
Li^+ [92GRE/FUG]	**– 292.918** [(a)] **± 0.109**	**– 278.470 ± 0.080**	**12.240 ± 0.150**	
Na(cr) [92GRE/FUG]	**0.000**	**0.000**	**51.300 ± 0.200**	**28.230 ± 0.200**
Na(g) [92GRE/FUG]	**76.964** [(a)] **± 0.703**	**107.500 ± 0.700**	**153.718 ± 0.003**	**20.786 ± 0.001**
Na^+ [92GRE/FUG]	**– 261.953** [(a)] **± 0.096**	**– 240.340 ± 0.060**	**58.450 ± 0.150**	
NaF(cr) [2001LEM/FUG]	– 546.327 [(a)] ± 0.704	– 576.600 ± 0.700	51.160 ± 0.150	
NaCl(cr) [2001LEM/FUG]	– 384.221 ± 0.147	– 411.260 ± 0.120	72.150 ± 0.200	50.500
$NaNO_3$(cr) [2003GUI/FAN]		– 467.580 ± 0.410		
K(cr) [92GRE/FUG]	**0.000**	**0.000**	**64.680 ± 0.200**	**29.600 ± 0.100**

(Continued on next page)

Table IV-1: (continued)

Compound and NEA-TDB Review	$\Delta_f G_m^o$ (kJ·mol⁻¹)	$\Delta_f H_m^o$ (kJ·mol⁻¹)	S_m^o (J·K⁻¹·mol⁻¹)	$C_{p,m}^o$ (J·K⁻¹·mol⁻¹)
K(g) [92GRE/FUG]	**60.479** (a) **± 0.802**	**89.000 ± 0.800**	**160.341 ± 0.003**	**20.786 ± 0.001**
K^+ [92GRE/FUG]	**– 282.510** (a) **± 0.116**	**– 252.140 ± 0.080**	**101.200 ± 0.200**	
KCl(cr) [2005GAM/BUG]		– 436.461 ± 0.129		
KBr(cr) [2005GAM/BUG]		– 393.330 ± 0.188		
KI(cr) [2005GAM/BUG]		– 329.150 ± 0.137		
Rb(cr) [92GRE/FUG]	**0.000**	**0.000**	**76.780 ± 0.300**	**31.060 ± 0.100**
Rb(g) [92GRE/FUG]	**53.078** (a) **± 0.805**	**80.900 ± 0.800**	**170.094 ± 0.003**	**20.786 ± 0.001**
Rb^+ [92GRE/FUG]	**– 284.009** (a) **± 0.153**	**– 251.120 ± 0.100**	**121.750 ± 0.250**	
Cs(cr) [92GRE/FUG]	**0.000**	**0.000**	**85.230 ± 0.400**	**32.210 ± 0.200**
Cs(g) [92GRE/FUG]	**49.556** (a) **± 1.007**	**76.500 ± 1.000**	**175.601 ± 0.003**	**20.786 ± 0.001**
Cs^+ [92GRE/FUG]	**– 291.456** (a) **± 0.535**	**– 258.000 ± 0.500**	**132.100 ± 0.500**	
CsCl(cr) [2001LEM/FUG]	– 413.807 (a) ± 0.208	– 442.310 ± 0.160	101.170 ± 0.200	52.470
CsBr(cr) [2001LEM/FUG]	– 391.171 ± 0.305	– 405.600 ± 0.250	112.940 ± 0.400	52.930

a Value calculated internally using $\Delta_f G_m^o = \Delta_f H_m^o - T\sum_i S_{m,i}^o$.

b Value calculated internally from reaction data (see Table V-2).

Table IV-2: Selected thermodynamic data for reactions involving auxiliary compounds and complexes used in the evaluation of thermodynamic data for the NEA-TDB Project data. All ionic species listed in this table are aqueous species. Unless noted otherwise, all data refer to 298.15 K and a pressure of 0.1 MPa and, for aqueous species, a standard state of infinite dilution ($I = 0$). The uncertainties listed below each value represent total uncertainties and correspond in principle to the statistically defined 95% confidence interval. Systematically, all the values are presented with three digits after the decimal point, regardless of the significance of these digits. The reference listed for each entry in this table indicates the NEA-TDB Review where the corresponding data have been adopted as NEA-TDB Auxiliary data. The data presented in this table are available on computer media from the OECD Nuclear Energy Agency.

Species and NEA-TDB Review	Reaction $\log_{10} K^\circ$	$\Delta_r G_m^\circ$ (kJ·mol⁻¹)	$\Delta_r H_m^\circ$ (kJ·mol⁻¹)	$\Delta_r S_m^\circ$ (J·K⁻¹·mol⁻¹)
$HF(aq)$	$F^- + H^+ \rightleftharpoons HF(aq)$			
[92GRE/FUG]	3.180	− 18.152	12.200	101.800 [a]
	± 0.020	± 0.114	± 0.300	± 1.077
HF_2^-	$F^- + HF(aq) \rightleftharpoons HF_2^-$			
[92GRE/FUG]	0.440	− 2.511	3.000	18.486 [a]
	± 0.120	± 0.685	± 2.000	± 7.090
ClO^-	$HClO(aq) \rightleftharpoons ClO^- + H^+$			
This review	− 7.420	42.354	17.400	− 83.695 [a]
	± 0.130	± 0.742	± 2.100	± 7.470
ClO_2^-	$HClO_2(aq) \rightleftharpoons ClO_2^- + H^+$			
[92GRE/FUG]	− 1.960	11.188		
	± 0.020	± 0.114		
$HClO(aq)$	$Cl_2(g) + H_2O(l) \rightleftharpoons Cl^- + H^+ + HClO(aq)$			
[92GRE/FUG]	− 4.537	25.900		
	± 0.105	± 0.600		
$HClO_2(aq)$	$H_2O(l) + HClO(aq) \rightleftharpoons 2H^+ + HClO_2(aq) + 2e^-$			
[92GRE/FUG]	− 55.400 [b]	316.230		
	± 0.700	± 3.996		
BrO^-	$HBrO(aq) \rightleftharpoons BrO^- + H^+$			
[92GRE/FUG]	− 8.630	49.260	30.000	− 64.600 [a]
	± 0.030	± 0.171	± 3.000	± 10.078
$HBrO(aq)$	$Br_2(aq) + H_2O(l) \rightleftharpoons Br^- + H^+ + HBrO(aq)$			
[92GRE/FUG]	− 8.240	47.034		
	± 0.200	± 1.142		

(Continued on next page)

Table IV-2: (continued)

Species and NEA-TDB Review	Reaction $\log_{10} K^{\circ}$	$\Delta_r G^{\circ}_m$ (kJ·mol^{-1})	$\Delta_r H^{\circ}_m$ (kJ·mol^{-1})	$\Delta_r S^{\circ}_m$ (J·K^{-1}·mol^{-1})
$HIO_3(aq)$	$H^+ + IO_3^- \rightleftharpoons HIO_3(aq)$			
[92GRE/FUG]	0.788 ± 0.029	− 4.498 ± 0.166		
S^{2-}	$HS^- \rightleftharpoons H^+ + S^{2-}$			
[92GRE/FUG]	− 19.000 ± 2.000	108.450 ± 11.416		
SO_3^{2-}	$H_2O(l) + SO_4^{2-} + 2\,e^- \rightleftharpoons 2OH^- + SO_3^{2-}$			
[92GRE/FUG]	− 31.400 (b) ± 0.700	179.230 ± 3.996		
$S_2O_3^{2-}$	$3H_2O(l) + 2SO_3^{2-} + 4\,e^- \rightleftharpoons 6OH^- + S_2O_3^{2-}$			
[92GRE/FUG]	− 39.200 (b) ± 1.400	223.760 ± 7.991		
$H_2S(aq)$	$H_2S(aq) \rightleftharpoons H^+ + HS^-$			
[92GRE/FUG]	− 6.990 ± 0.170	39.899 ± 0.970		
HSO_3^-	$H^+ + SO_3^{2-} \rightleftharpoons HSO_3^-$			
[92GRE/FUG]	7.220 ± 0.080	− 41.212 ± 0.457	66.000 ± 30.000	359.590 (a) ± 100.630
$HS_2O_3^-$	$H^+ + S_2O_3^{2-} \rightleftharpoons HS_2O_3^-$			
[92GRE/FUG]	1.590 ± 0.150	− 9.076 ± 0.856		
$H_2SO_3(aq)$	$H^+ + HSO_3^- \rightleftharpoons H_2SO_3(aq)$			
[92GRE/FUG]	1.840 ± 0.080	− 10.503 ± 0.457	16.000 ± 5.000	88.891 (a) ± 16.840
HSO_4^-	$H^+ + SO_4^{2-} \rightleftharpoons HSO_4^-$			
[92GRE/FUG]	1.980 ± 0.050	− 11.302 ± 0.285		
$HN_3(aq)$	$H^+ + N_3^- \rightleftharpoons HN_3(aq)$			
[92GRE/FUG]	4.700 ± 0.080	− 26.828 ± 0.457	− 15.000 ± 10.000	39.671 (a) ± 33.575
$NH_3(aq)$	$NH_4^+ \rightleftharpoons H^+ + NH_3(aq)$			
[92GRE/FUG]	− 9.237 ± 0.022	52.725 ± 0.126	52.090 ± 0.210	− 2.130 (a) ± 0.821

(Continued on next page)

Table IV-2 (continued)

Species and NEA-TDB Review	Reaction $\log_{10} K^{\circ}$	$\Delta_r G_m^{\circ}$ ($kJ\cdot mol^{-1}$)	$\Delta_r H_m^{\circ}$ ($kJ\cdot mol^{-1}$)	$\Delta_r S_m^{\circ}$ ($J\cdot K^{-1}\cdot mol^{-1}$)
$HNO_2(aq)$	$H^+ + NO_2^- \rightleftharpoons HNO_2(aq)$			
[92GRE/FUG]	3.210	– 18.323	– 11.400	23.219 [(a)]
	± 0.160	± 0.913	± 3.000	± 10.518
PO_4^{3-}	$HPO_4^{2-} \rightleftharpoons H^+ + PO_4^{3-}$			
[92GRE/FUG]	– 12.350	70.494	14.600	– 187.470 [(a)]
	± 0.030	± 0.171	± 3.800	± 12.758
$P_2O_7^{4-}$	$HP_2O_7^{3-} \rightleftharpoons H^+ + P_2O_7^{4-}$			
[92GRE/FUG]	– 9.400	53.656		
	± 0.150	± 0.856		
$H_2PO_4^-$	$H^+ + HPO_4^{2-} \rightleftharpoons H_2PO_4^-$			
[92GRE/FUG]	7.212	– 41.166	– 3.600	126.000 [(a)]
	± 0.013	± 0.074	± 1.000	± 3.363
$H_3PO_4(aq)$	$H^+ + H_2PO_4^- \rightleftharpoons H_3PO_4(aq)$			
[92GRE/FUG]	2.140	– 12.215	8.480	69.412 [(a)]
	± 0.030	± 0.171	± 0.600	± 2.093
$HP_2O_7^{3-}$	$H_2P_2O_7^{2-} \rightleftharpoons H^+ + HP_2O_7^{3-}$			
[92GRE/FUG]	– 6.650	37.958		
	± 0.100	± 0.571		
$H_2P_2O_7^{2-}$	$H_3P_2O_7^- \rightleftharpoons H^+ + H_2P_2O_7^{2-}$			
[92GRE/FUG]	– 2.250	12.843		
	± 0.150	± 0.856		
$H_3P_2O_7^-$	$H_4P_2O_7(aq) \rightleftharpoons H^+ + H_3P_2O_7^-$			
[92GRE/FUG]	– 1.000	5.708		
	± 0.500	± 2.854		
$H_4P_2O_7(aq)$	$2H_3PO_4(aq) \rightleftharpoons H_2O(l) + H_4P_2O_7(aq)$			
[92GRE/FUG]	– 2.790	15.925	22.200	21.045 [(a)]
	± 0.170	± 0.970	± 1.000	± 4.673
$CO_2(aq)$	$H^+ + HCO_3^- \rightleftharpoons CO_2(aq) + H_2O(l)$			
[92GRE/FUG]	6.354	– 36.269		
	± 0.020	± 0.114		
$CO_2(g)$	$CO_2(aq) \rightleftharpoons CO_2(g)$			
[92GRE/FUG]	1.472	– 8.402		
	± 0.020	± 0.114		

(Continued on next page)

Table IV-2 (continued)

Species and NEA-TDB Review	Reaction $\log_{10} K^\circ$	$\Delta_r G_m^\circ$ (kJ·mol⁻¹)	$\Delta_r H_m^\circ$ (kJ·mol⁻¹)	$\Delta_r S_m^\circ$ (J·K⁻¹·mol⁻¹)
HCO_3^-	$CO_3^{2-} + H^+ \rightleftharpoons HCO_3^-$			
[92GRE/FUG]	10.329 ± 0.020	– 58.958 ± 0.114		
CN^-	$HCN(aq) \rightleftharpoons CN^- + H^+$			
This review	– 9.210 ± 0.020	52.571 ± 0.114	43.600 ± 0.200	– 30.089 [(a)] ± 0.772
$HCN(aq)$	$HCN(g) \rightleftharpoons HCN(aq)$			
This review	0.902 ± 0.050	– 5.149 ± 0.285	– 26.150 ± 2.500	– 70.439 [(a)] ± 8.440
$SiO_2(OH)_2^{2-}$	$Si(OH)_4(aq) \rightleftharpoons 2H^+ + SiO_2(OH)_2^{2-}$			
[92GRE/FUG]	– 23.140 ± 0.090	132.080 ± 0.514	75.000 ± 15.000	– 191.460 [(a)] ± 50.340
$SiO(OH)_3^-$	$Si(OH)_4(aq) \rightleftharpoons H^+ + SiO(OH)_3^-$			
[92GRE/FUG]	– 9.810 ± 0.020	55.996 ± 0.114	25.600 ± 2.000	– 101.950 [(a)] ± 6.719
$Si(OH)_4(aq)$	$2H_2O(l) + SiO_2(quar) \rightleftharpoons Si(OH)_4(aq)$			
[92GRE/FUG]$	– 4.000 ± 0.100	22.832 ± 0.571	25.400 ± 3.000	8.613 [(a)] ± 10.243
$Si_2O_3(OH)_4^{2-}$	$2Si(OH)_4(aq) \rightleftharpoons 2H^+ + H_2O(l) + Si_2O_3(OH)_4^{2-}$			
[92GRE/FUG]	– 19.000 ± 0.300	108.450 ± 1.712		
$Si_2O_2(OH)_5^-$	$2Si(OH)_4(aq) \rightleftharpoons H^+ + H_2O(l) + Si_2O_2(OH)_5^-$			
[92GRE/FUG]	– 8.100 ± 0.300	46.235 ± 1.712		
$Si_3O_6(OH)_3^{3-}$	$3Si(OH)_4(aq) \rightleftharpoons 3H^+ + 3H_2O(l) + Si_3O_6(OH)_3^{3-}$			
[92GRE/FUG]	– 28.600 ± 0.300	163.250 ± 1.712		
$Si_3O_5(OH)_5^{3-}$	$3Si(OH)_4(aq) \rightleftharpoons 3H^+ + 2H_2O(l) + Si_3O_5(OH)_5^{3-}$			
[92GRE/FUG]	– 27.500 ± 0.300	156.970 ± 1.712		
$Si_4O_8(OH)_4^{4-}$	$4Si(OH)_4(aq) \rightleftharpoons 4H^+ + 4H_2O(l) + Si_4O_8(OH)_4^{4-}$			
[92GRE/FUG]	– 36.300 ± 0.500	207.200 ± 2.854		

(Continued on next page)

Table IV-2 (continued)

Species and NEA-TDB Review	Reaction $\log_{10} K^\circ$	$\Delta_r G_m^\circ$ (kJ·mol^{-1})	$\Delta_r H_m^\circ$ (kJ·mol^{-1})	$\Delta_r S_m^\circ$ (J·K^{-1}·mol^{-1})
$Si_4O_7(OH)_5^{3-}$	$4Si(OH)_4(aq) \rightleftharpoons 3H^+ + 4H_2O(l) + Si_4O_7(OH)_5^{3-}$			
[92GRE/FUG]	– 25.500 ± 0.300	145.560 ± 1.712		

a Value calculated internally using $\Delta_r G_m^\circ = \Delta_r H_m^\circ - T\,\Delta_r S_m^\circ$.

b Value calculated from a selected standard potential.

c Value of $\log_{10} K^\circ$ calculated internally from $\Delta_r G_m^\circ$.

Chapter V

Discussion of data selection for selenium

V.1 Elemental selenium

The chemistry of elemental selenium is complex and it has been the subject of a large number of studies. Two well-defined polymorphs, trigonal and monoclinic selenium, exist in the solid state. The trigonal form of selenium is often improperly denoted as "hexagonal" selenium. Liquid selenium is easily supercooled and undergoes a glass transition at 303.4 K. The trigonal form of selenium is thermodynamically stable in the temperature range from 0 K up to the melting point at 494.2 K. The monoclinic form is metastable and melts at 413 K. In the gas phase, all species $Se_n(g)$, n = 1 to 12, have been detected. These properties taken together render a thermodynamic evaluation of the system difficult and the results of various investigations differ. In particular, the assessment of data for the gaseous species varies depending on the method of evaluation and the method for recalculation to 298.15 K from high temperatures. The proper evaluation of the data for the gaseous species requires the knowledge of the thermodynamic properties of the trigonal and liquid phases because most of the experimental results have been derived from vapour pressure measurements involving equilibria with solid and liquid selenium. In addition, heat capacities and entropies of the gaseous species are needed for second and third law evaluations. These quantities were, except for Se(g) and $Se_2(g)$, unknown at the time of the investigations and various estimates were used in the evaluation procedures. These circumstances have resulted in different data sets that may serve to describe the thermodynamic properties of the system as long as data from one source only are consistently selected. However, the thermodynamic quantities for individual gaseous species differ from one evaluation to another, making it hazardous to select or compare quantities from different evaluations. In the following, the available experimental information from the various sources has been re-evaluated using a single set of heat capacity and entropy data, focussing on the experimental values and minimising the use of estimates.

The procedure for the selection of data for selenium has been the following:

- The selection of entropy and heat capacity data for trigonal selenium in the temperature range 298.15 to 494.2 K and for the liquid in the temperature range 494.2 to 1500 K. The thermodynamic properties of the metastable monoclinic phase and the supercooled liquid are assessed for use as auxiliary data. The major source of information is the review by Gaur, Shu, Mehta, and Wunderlich [81GAU/SHU] which has been combined with other information for phase transformations.
- The selection of heat capacities and entropies at 298.15 K for the gaseous species as well as expressions of the temperature dependence of the heat capacity for the same species. Experimental information for the proper calculation of these quantities from statistical mechanics are only available for the species Se(g) and Se_2(g). For all the other selenium gas molecules, the published values are based on estimates made by comparison with gaseous sulphur and other selenium molecules. The selected values rely mostly on the work of Pupishev and Rusin [84PUP/RUS] who have made an elaborate analysis of the Se_2(g)-Se_8(g) and S_2(g)-S_8(g) molecules using the LCAO method.
- The selection of enthalpies of formation for the gaseous species by re-evaluating previous investigations using the data selected in the previous steps. This evaluation order allows the third law to be used with a consistent set of heat capacities and entropies in deriving enthalpies of formation. The third law has rarely been applied in selenium gas phase studies and it reduces the scatter between derived experimental enthalpy values from various investigations by 50 to 80%.

A complete set of thermodynamic data for the polymorphs of solid selenium, the liquid, and the gas molecules Se_n(g), n = 1 to 8, was derived in the review by Mills [74MIL]. This work has been a valuable source of information, but several important later studies have appeared. A compilation of data for the equilibrium between liquid selenium and saturated selenium vapour was made by Rao [82RAO]. The work was mainly based on the contribution by Rau [74RAU] and the selection of data sources was limited.

There are two major sources of experimental information on selenium gas equilibria; total vapour pressure measurements and mass spectroscopic investigations. The composition of the gas phase is complex, see Figure V-1, and limited thermodynamic information can be obtained from the total vapour pressure measurements except at high temperatures where Se_2(g) dominates. The mass spectrometric investigations provide information about individual gaseous species. However, there are serious problems due to fragmentation in the mass spectrometer. It has been found that a large part of the observed intensities of $Se_3^+(g)$, $Se_4^+(g)$, and to some extent $Se_2^+(g)$ ions are caused by fragmentation of the species Se_5(g) and Se_6(g). Thermodynamic evaluations have therefore been omitted for the former species in most studies, and where data have been derived, the vapour pressures are too large and the enthalpies of formation too small. Similarly, the vapour pressures of the $Se_5(g)$ and $Se_6(g)$ species become too low and the corresponding enthalpies of formation too large. The saturated vapour pressure measurements are useful for the normalisation of the saturated total pressure obtained from mass spectrometric investigations.

Figure V-1: Mole fractions of species in saturated selenium vapour. Abrupt changes in the partial pressures occur at the melting point 494 K of selenium for some of the species.

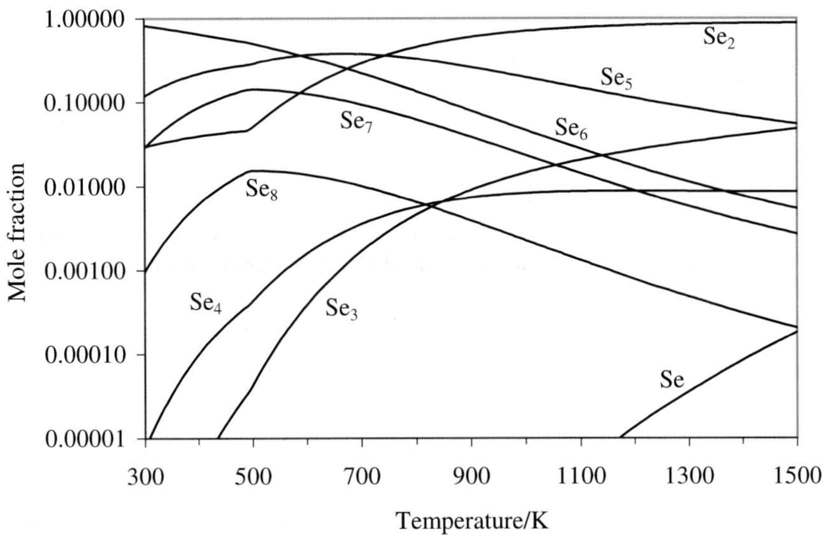

The selected thermodynamic data have been used to calculate the total selenium vapour pressure above solid and liquid selenium at equilibrium. The quotient between the observed and calculated pressures as a function of the inverse temperature is shown in Figure V-2.

Figure V-2: Comparison between the saturated pressure of selenium calculated from the selected thermodynamic quantities and the total pressure measured in various investigations. A logarithmic unit of 0.1 corresponds approximately to a relative error of 25%.

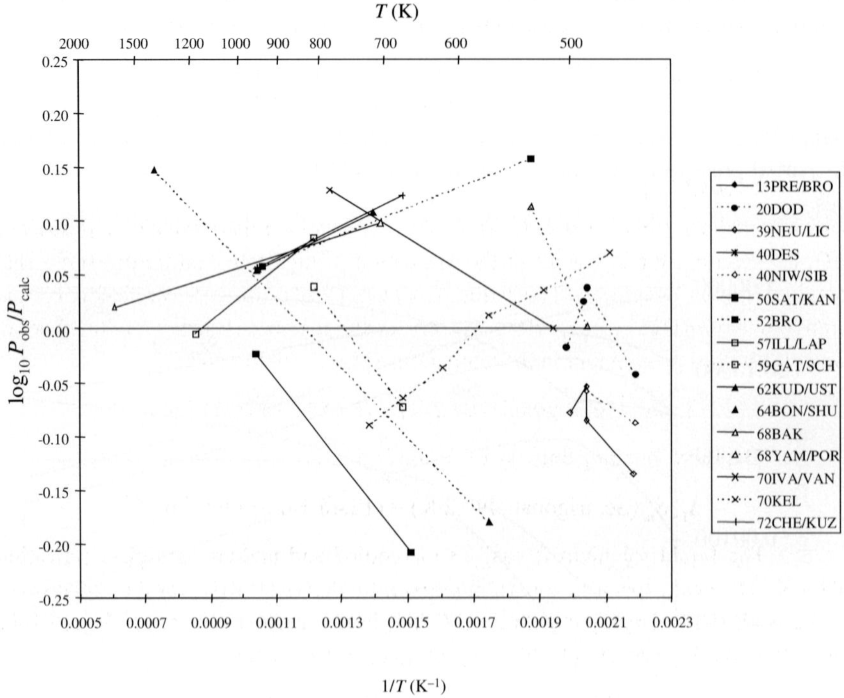

V.1.1 Se(trigonal)

The thermodynamic properties of the condensed forms of selenium were reviewed and re-assessed in a very thorough study by Gaur, Shu, Mehta, and Wunderlich [81GAU/SHU]. This was complemented by a discussion of the zero-point entropy of the metastable forms by Judovits and Wunderlich [85JUD/WUN]. The thermodynamic properties were found to be in excellent agreement with inelastic neutron scattering studies of the selenium polymorphs made by Phillips, Buchenau, Nucker, Dianoux and Petry [89PHI/BUC]. No changes to the data have appeared in the literature since then and the data in [81GAU/SHU] have been selected for this work. The selected entropy at 298.15 K is:

$$S^{\circ}_{m}\,(\text{Se, trigonal, 298.15 K}) = (42.09 \pm 0.33)\ \text{J}\cdot\text{K}^{-1}\cdot\text{mol}^{-1}$$

and

$$C^{\circ}_{p,\mathrm{m}}\,(\mathrm{Se, trigonal, 298.15\ K}) = (25.09 \pm 0.30)\ \mathrm{J\cdot K^{-1}\cdot mol^{-1}}.$$

[81GAU/SHU] described the temperature dependence of the heat capacity using an unusual analytical expression and for the present work the heat capacity expression was re-evaluated using the standard form. The selected expression is:

$$C^{\circ}_{p,\mathrm{m}}\,(\mathrm{Se, trigonal, (298.15 - 500)\ K}) = (24.8014 + 1.2859 \times 10^{-3}\,T + 9.9273 \times 10^{-6}\,T^{2} - 0.8713 \times 10^{5}\,T^{-2})\ \mathrm{J\cdot K^{-1}\cdot mol^{-1}}.$$

V.1.2 Se(liquid)

Selenium melts at (494.2 ± 0.2) K [80WUN/SHU] and the thermodynamic properties of Se(l) are required for this work in the evaluation of high temperature measurements of selenium vapour pressures. The value accepted for the enthalpy of fusion is that of Grønvold [73GRO2], who performed very accurate measurements of the thermodynamic properties of selenium in the temperature range 298 to 1000 K:

$$\Delta_{\mathrm{fus}}H^{\circ}_{\mathrm{m}}\,(\mathrm{Se, trigonal, 494.2\ K}) = (6.159 \pm 0.012)\ \mathrm{kJ\cdot mol^{-1}}.$$

The corresponding entropy of fusion is:

$$\Delta_{\mathrm{fus}}S^{\circ}_{\mathrm{m}}\,(\mathrm{Se, trigonal, 494.2\ K}) = (12.46 \pm 0.03)\ \mathrm{J\cdot K^{-1}\cdot mol^{-1}}.$$

The liquid selenium is easily supercooled and undergoes a glass transition at 303.4 K. The heat capacity expression derived by [81GAU/SHU] for the liquid is based on the work in [73GRO2] and [80SHU/GAU] in the temperature ranges 500 to 1000 K and 330 to 520 K, respectively. The expression is selected here,

$$C^{\circ}_{p,\mathrm{m}}\,(\mathrm{Se, l, (303.4 - 1000)\ K}) = (52.408 - 49.766 \times 10^{-3}\,T + 3.2608 \times 10^{-5}\,T^{2})\ \mathrm{J\cdot K^{-1}\cdot mol^{-1}}.$$

No experimental values are available for the heat capacity of liquid selenium above 1000 K. For this work a constant value equal to the value at 1000 K was used in evaluations of data above 1000 K:

$$C^{\circ}_{p,\mathrm{m}}\,(\mathrm{Se, l, (1000 - 1500)\ K}) = 35.250\ \mathrm{J\cdot K^{-1}\cdot mol^{-1}}.$$

V.1.3 Se(monoclinic)

Monoclinic selenium is metastable and its enthalpy of formation and entropy are needed as auxiliary data for some evaluations. They were derived from a thermodynamic cycle involving the enthalpy and entropy of fusion at the melting point 413 K, the selected data for trigonal and liquid selenium, and the heat capacity of monoclinic selenium. The selected enthalpy of fusion is that in [81GAU/SHU]:

$$\Delta_{\mathrm{fus}}H^{\circ}_{\mathrm{m}}\,(\mathrm{Se, monoclinic, 413\ K}) = 3.24\ \mathrm{kJ\cdot mol^{-1}}.$$

The corresponding entropy of fusion is:

$$\Delta_{\text{fus}}S^{\circ}_{\text{m}}\text{(Se, monoclinic, 413 K)} = 7.86\ \text{J·K}^{-1}\text{·mol}^{-1}.$$

Similarly, the heat capacity at 298.15 K assessed in [81GAU/SHU] is selected:

$$C^{\circ}_{p,\text{m}}\text{(Se, monoclinic, 298.15 K)} = (25.09 \pm 0.80)\ \text{J·K}^{-1}\text{·mol}^{-1}.$$

The expression

$$C^{\circ}_{p,\text{m}}\text{(Se, monoclinic, (298 – 413) K)} = (25.698 - 2.1654 \times 10^{-3}\,T + 13.6408 \times 10^{-6}\,T^{2} - 1.0472 \times 10^{5}\,T^{-2})\ \text{J·K}^{-1}\text{·mol}^{-1}$$

was derived by the review from the heat capacity values assessed in [81GAU/SHU].

The enthalpy of formation and entropy of monoclinic selenium from this cycle are $\Delta_{\text{f}}H^{\circ}_{\text{m}}$ (Se, monoclinic, 298.15 K) = (2.14 ± 0.10) kJ·mol^{-1} and S°_{m} (Se, monoclinic, 298.15 K) = (44.97 ± 0.40) J·K^{-1}·mol^{-1}. The corresponding values from the data in [81GAU/SHU] are $\Delta_{\text{f}}H^{\circ}_{\text{m}}$ (Se, monoclinic, 298.15 K) = 2.18 kJ·mol^{-1} and S°_{m} (Se, monoclinic, 298.15 K) = 45.07 J·K^{-1}·mol^{-1}. The differences between the values originate from the use of the enthalpy of fusion of trigonal selenium determined in [80WUN/SHU], $\Delta_{\text{fus}}H^{\circ}_{\text{m}}$ (Se, trigonal, 494.2 K) = 6.20 kJ·mol^{-1}, in the latter case, rather than the value $\Delta_{\text{fus}}H^{\circ}_{\text{m}}$ (Se, trigonal, 494.2 K) = (6.159 ± 0.012) kJ·mol^{-1} selected by the review. The selected enthalpy of formation and entropy of monoclinic selenium are:

$$\Delta_{\text{f}}H^{\circ}_{\text{m}}\text{(Se, monoclinic, 298.15 K)} = (2.14 \pm 0.10)\ \text{kJ·mol}^{-1}$$

and

$$S^{\circ}_{\text{m}}\text{(Se, monoclinic, 298.15 K)} = (44.97 \pm 0.40)\ \text{J·K}^{-1}\text{·mol}^{-1}.$$

The Gibbs energy of formation is calculated from the selected enthalpy of formation and entropy to be:

$$\Delta_{\text{f}}G^{\circ}_{\text{m}}\text{(Se, monoclinic, 298.15 K)} = (1.28 \pm 0.18)\ \text{kJ·mol}^{-1}.$$

V.1.4 Se(g)

Kelley and King [61KEL/KIN] calculated the standard entropy of Se(g) to be S°_{m} (Se, g, 298.15 K) = (176.61 ± 0.05) J·K^{-1}·mol^{-1} from atomic constant data. This value (for 1 atm standard pressure) was selected by Mills [74MIL]. The entropy was recalculated by the review using statistical mechanics employing the energy levels given by Kelley [87KEL] and a molar mass of 78.96 g·mol^{-1} yielding the selected value:

$$S^{\circ}_{\text{m}}\text{(Se, g, 298.15 K)} = (176.729 \pm 0.006)\ \text{J·K}^{-1}\text{·mol}^{-1}.$$

The major contribution to the error limits is the uncertainty in the molar mass of Se(g) due to the natural variation in the isotopic composition of selenium. The selected value of the heat capacity was calculated from the same data as the entropy yielding:

$$C^{\circ}_{p,\mathrm{m}}\,(\mathrm{Se, g, 298.15\ K}) = (20.819 \pm 0.001)\ \mathrm{J{\cdot}K^{-1}{\cdot}mol^{-1}}.$$

The heat capacity expressions

$$C^{\circ}_{p,\mathrm{m}}\,(\mathrm{Se, g, (298.15 - 1100)\ K}) = (11.5228 + 16.8482 \times 10^{-3}\,T - 6.47162 \times 10^{-6}\,T^{2} - 1.6620 \times 10^{3}\,T^{-1} - 0.64257 \times 10^{5}\,T^{-2})\ \mathrm{J{\cdot}K^{-1}{\cdot}mol^{-1}},$$

and

$$C^{\circ}_{p,\mathrm{m}}\,(\mathrm{Se, g, (1100 - 5000)\ K}) = (11.1256 + 3.1188 \times 10^{-3}\,T - 0.25866 \times 10^{-6}\,T^{2} + 19.2957 \times 10^{3}\,T^{-1} - 98.0328 \times 10^{5}\,T^{-2})\ \mathrm{J{\cdot}K^{-1}{\cdot}mol^{-1}}$$

were derived by the review and describe the heat capacity calculated from statistical mechanics within $\pm$ 0.03 $\mathrm{J{\cdot}K^{-1}{\cdot}mol^{-1}}$.

The enthalpy of formation of Se(g) is calculated from determinations of the dissociation energy of Se_2(g). The available studies of this quantity were reviewed and discussed thoroughly by Drowart and Smoes [77DRO/SMO]. The determinations of the dissociation energy were also re-assessed by Grønvold, Drowart, and Westrum [84GRO/DRO]. Both reviews selected the highest of three spectroscopically possible values originating from the work by Barrow, Chandler, and Meyer [66BAR/CHA]. However, it can be noted that these authors themselves preferred the lowest of the three possible values. No later investigations changing the conclusions in [77DRO/SMO] and [84GRO/DRO] have been found and the value for the atomisation energy of Se_2(g) as recalculated to 298.15 K from 0 K in [84GRO/DRO] is selected:

$$\Delta_{\mathrm{at}} H^{\circ}_{\mathrm{m}}\,(\mathrm{Se_2, g, 298.15\ K}) = (331.04 \pm 0.50)\ \mathrm{kJ{\cdot}mol^{-1}}.$$

This value is combined with the selected enthalpy of formation of Se_2(g) to give the selected enthalpy of formation of Se(g):

$$\Delta_{\mathrm{f}} H^{\circ}_{\mathrm{m}}\,(\mathrm{Se, g, 298.15\ K}) = (236.1 \pm 1.5)\ \mathrm{kJ{\cdot}mol^{-1}}.$$

The Gibbs energy of formation is calculated from the selected enthalpy of formation and entropy to be:

$$\Delta_{\mathrm{f}} G^{\circ}_{\mathrm{m}}\,(\mathrm{Se, g, 298.15\ K}) = (196.0 \pm 1.5)\ \mathrm{kJ{\cdot}mol^{-1}}.$$

V.1.5 Se_2(g)

The calculation of the entropy of Se_2(g) from spectroscopic data has been done with and without inclusion of the splitting of the ground state $X^3\Sigma^+$ of the Se_2(g) molecule. Stull and Sinke [56STU/SIN] calculated a value of S°_{m} (Se_2, g, 298.15 K) = (251.99 $\pm$ 0.90) $\mathrm{J{\cdot}K^{-1}{\cdot}mol^{-1}}$ by treating the state as a triplet, *i.e.* using a statistical weight of 3. In the handbook by Hultgren *et al.* [73HUL/DES] and in the review by Mills [74MIL], the splitting energy was estimated resulting in the values S°_{m} (Se_2, g, 298.15 K) = (249.46 $\pm$ 0.90) $\mathrm{J{\cdot}K^{-1}{\cdot}mol^{-1}}$ and S°_{m} (Se_2, g, 298.15 K) = (243.6 $\pm$ 3.8) $\mathrm{J{\cdot}K^{-1}{\cdot}mol^{-1}}$, respectively. A reliable experimental value for the splitting energy became available through the work of Greenwood and Barrow [76GRE/BAR] and it was used in Vendrikh, Gorbov, and Pashinkin [82VEN/GOR] to calculate an accurate value of S°_{m} (Se_2, g, 298.15 K) = (247.42 ± 0.05) $J \cdot K^{-1} \cdot mol^{-1}$. Grønvold, Drowart and Westrum [84GRO/DRO] reviewed the spectroscopic data of Se_2(g) and calculated the entropy to be S°_{m} (Se_2, g, 298.15 K) = (247.38 ± 0.05) $J \cdot K^{-1} \cdot mol^{-1}$ from the experimental work performed primarily by Barrow, Chandler and Meyer [66BAR/CHA], and by Prosser, Barrow, Effantin, d'Incan, and Vergès [82PRO/BAR]. The calculations in [84GRO/DRO] differ mainly from those in [82VEN/GOR] by using a slightly different value for the splitting of the $X^3\Sigma^+$ state, 512.81 cm^{-1} instead of 509.95 cm^{-1}, and a different estimated value of the unobserved $a^1\Delta_g$ state, 4500 cm^{-1} instead of 3000 cm^{-1}. The energy of the $a^1\Delta_g$ state was calculated to be 4618 cm^{-1} by Heinemann, Koch, Lindner, Reinen and Widmark [96HEI/KOC] using *ab initio* molecular-orbital methods. Their work gives slightly higher energies for the low-lying electronic states than those determined experimentally and the estimate made in [84GRO/DRO] for the $a^1\Delta_g$ state (4500 ± 500) cm^{-1}, appears to be preferable to that in [82VEN/GOR].

For the purpose of this review, experimental entropies were calculated from a number of investigations involving Se_2(g) equilibria using the second law. Such evaluations were not made in the original work and the details are discussed in Appendix A for each paper. The results are summarised in Table V-1.

The scatter is considerable among the second law values due to the often very limited temperature ranges of the investigations. The value in [68YAM/POR] is especially low, a fact which is probably due to contributions from fragmentation of higher molecules to the registered Se_2^+(g) ion intensity. The entropy calculated in [84GRO/DRO] is selected:

$$S^{\circ}_{m}(\mathrm{Se_2, g, 298.15\ K}) = (247.38 \pm 0.40)\ \mathrm{J \cdot K^{-1} \cdot mol^{-1}}.$$

Similarly, the heat capacity at 298.15 K calculated by the same authors is selected:

$$C^{\circ}_{p,\mathrm{m}}(\mathrm{Se_2, g, 298.15\ K}) = (41.71 \pm 0.20)\ \mathrm{J \cdot K^{-1} \cdot mol^{-1}}.$$

The heat capacity expressions

$$C^{\circ}_{p,\mathrm{m}}(\mathrm{Se_2, g, (298.15 - 1300)\ K}) = (19.3485 + 12.4903 \times 10^{-3}\, T - 2.0701 \times 10^{-6}\, T^{2} + 1.09846 \times 10^{4}\, T^{-1} - 1.602488 \times 10^{6}\, T^{-2})\ \mathrm{J \cdot K^{-1} \cdot mol^{-1}},$$

and

$$C^{\circ}_{p,\mathrm{m}}(\mathrm{Se_2, g, (1300 - 3000)\ K}) = (58.1884 - 4.1015 \times 10^{-3}\, T + 0.4789 \times 10^{-6}\, T^{2} - 2.82796 \times 10^{4}\, T^{-1} + 1.297335 \times 10^{7}\, T^{-2})\ \mathrm{J \cdot K^{-1} \cdot mol^{-1}}$$

were derived by the review and describe the heat capacity calculated in [84GRO/DRO] from statistical mechanics to within ± 0.01 $J \cdot K^{-1} \cdot mol^{-1}$.

Table V-1: Entropy of $Se_2(g)$ at 298.15 K as (1) calculated from molecular constants and (2) evaluated using the second law.

Reference	S°_{m} (298.15 K)$^{(1)}$ (J·K^{-1}·mol^{-1})	S°_{m} (298.15 K)$^{(2)}$ (J·K^{-1}·mol^{-1})	Temperature (K)
[56STU/SIN]	(252.0 ± 0.9)$^{(a)}$		
[73HUL/DES]	(249.4 ± 0.9)$^{(a)}$		
[74MIL]	(243.6 ± 3.8)$^{(a)}$		
[82VEN/GOR]	(247.42 ± 0.05)$^{(a)}$		
[84GRO/DRO]	(247.38 ± 0.05)		
[62RAT/RIC]		(249.4 ± 6.3)	523 – 673
[66BER/CHU]		(243.9 ± 1.4)	437 – 469, 420 – 546$^{(b)}$
[66BER/CHU]		(232.6 ± 3.2)	505 – 546, 420 – 546$^{(c)}$
[68YAM/POR]		(216.5 ± 10.5)	578 – 667
[70KEL]		(250.0 ± 4.1)	573 – 723
[82GRI/GRI]		(245.2 ± 8.5), (231.1 ± 8.5)	463 – 483
[84HUA/GIL]		(263.6 ± 14.2)	485 – 503
Average		(241.5 ± 28.3)	

a Error limits as given in the reference.

b The first temperature range pertains to the equilibrium 6Se(trigonal) $\rightleftharpoons$ $Se_6(g)$ and the second temperature interval to the equilibrium $Se_6(g) \rightleftharpoons 3Se_2(g)$.

c The first temperature range pertains to the equilibrium 6Se(l) $\rightleftharpoons$ $Se_6(g)$ and the second temperature interval to the equilibrium $Se_6(g) \rightleftharpoons 3Se_2(g)$.

The enthalpy of formation of $Se_2(g)$ has been determined from mass spectrometric investigations of selenium vapour and from measurements of the total selenium pressure and the vapour density. The enthalpy of formation of $Se_2(g)$ cannot be determined from the latter two methods without additional information because of the complexity of the gas phase. $Se_2(g)$ is a minority species at temperatures below approximately 800 K and other species such as $Se_5(g)$, $Se_6(g)$, and $Se_7(g)$ dominate. Nevertheless, the enthalpy of formation of $Se_2(g)$ has been derived from vapour density measurements at these low temperatures in a number of investigations by assuming the presence of particular selenium species in addition to $Se_2(g)$. Such evaluations are prone to be unreliable, in particular in cases where the evaluation was made with the wrong major species, because the properties determined for $Se_2(g)$ are essentially dependent on the small difference between the total vapour pressure and the vapour pressures of the majority species. At temperatures above 800 K the mole fraction of $Se_2(g)$ exceeds 50% and by calculating the mole fractions of the other selenium species from data determined in mass spectrometric investigations at low temperatures, the mole fraction of

$Se_2(g)$ can be found to within ± 10% at high temperatures. This allows the evaluation of fairly reliable values of the enthalpy of formation of $Se_2(g)$ from high temperature total vapour pressure data. The mass spectrometric investigations are treated first and are followed by a discussion of the high temperature total pressure measurements.

The determinations of the enthalpy of formation of $Se_2(g)$ from mass spectrometric investigations of selenium vapour using Knudsen cells can be separated into the approximate temperature ranges 420 to 494 K for equilibria with solid selenium and 494 to 700 K for equilibria with liquid selenium. The various investigations have used different methods and auxiliary data for deriving enthalpies of formation at 298.15 K from the measurements and for the purpose of this review the measurements were re-evaluated using the selected values for heat capacities and entropies of the selenium gaseous species and condensed phases. The details of the evaluations are given in Appendix A for each reference and the enthalpies of formation of $Se_2(g)$ are summarised in Table V-2. The values of the enthalpy of formation of $Se_2(g)$ derived in the review by Grønvold, Drowart, and Westrum [84GRO/DRO] are included for comparison.

The large differences between the second and third law values for the measurements in [64MAS/WEI] and [68YAM/POR] indicate serious systematic errors in the measurements. The values in [66BER/CHU] were derived in two steps. First the enthalpy of formation was evaluated for $Se_6(g)$ from the reactions $6Se(s) \rightleftharpoons Se_6(g)$ in the temperature range 446 to 469 K and $6Se(l) \rightleftharpoons Se_6(g)$ in the temperature range 505 to 546 K. The enthalpy of formation of $Se_2(g)$ was then obtained from the reaction $Se_6(g) \rightleftharpoons 3Se_2(g)$ studied in the temperature range 420 to 543 K. The experimental results for the first two reactions lead to different values for the enthalpy of formation of $Se_6(g)$ and an enthalpy of fusion $\Delta_{fus}H^{\circ}_{m}$ (Se, trigonal) = 9.623 $J{\cdot}K^{-1}{\cdot}mol^{-1}$. This value is about 50% larger than the selected value of 6.159 $J{\cdot}K^{-1}{\cdot}mol^{-1}$ and explains the difference between the two second law values of Berkowitz and Chupka [66BER/CHU]. Yamdagni and Porter [68YAM/POR] concluded that the $Se_2^+(g)$ ions were formed primarily from $Se_2(g)$ and not from fragmentation of larger molecules because no evidence for a relationship between the intensities of the ion beams of the $Se_2^+(g)$ ions and the larger molecules could be found. However, they noted the substantial difference between their second and third law evaluations. In the light of later investigations, the $Se_2^+(g)$ ion intensity in their investigations seems to contain a considerable contribution caused by fragmentation of larger molecules after all. As such, the values derived from their data will constitute lower limits for the enthalpy of formation because the partial pressures calculated from these ion intensities would be higher than the real equilibrium values. The second and third law values of Keller [70KEL] agree well with each other but the total pressure of selenium is too low in the upper part of the temperature range where the data for $Se_2(g)$ were evaluated. The enthalpy values are therefore too large, and the correct values should be lower. Grimley, Grindstaff, DeMercurio, and Forsman [82GRI/GRI] noted that the ion intensity of $Se_2^+(g)$ was dependent on the properties of the ion source and the uncertainties are therefore large. This is shown by the discrepancy between the two third law enthalpies calculated by the review from the ion intensity data and the supposedly corrected mole fractions in Tables I and V of the paper. The results of Huang and Gilles were obtained for a temperature interval of 17 K only and it is not surprising that the second law value differs from the third law value.

Table V-2: Re-evaluated second and third law enthalpy of formation of $Se_2(g)$ at 298.15 K as obtained from mass spectrometric investigations and determinations of the magnetic susceptibility of selenium gas. The last column contains the values evaluated by Grønvold, Drowart, and Westrum [84GRO/DRO] from the cited references. The value attributed to [84GRO/DRO] in the first column was evaluated from the measurements of Tobisawa [60TOB] who did not originally evaluate partial pressures or thermodynamic properties.

Reference	$\Delta_f H_m^\circ$ (298.15 K), 2nd law (kJ·mol^{-1})	$\Delta_f H_m^\circ$ (298.15 K), 3rd law (kJ·mol^{-1})	Temperature (K)	$\Delta_f H_m^\circ$ (298.15 K), 3rd law [84GRO/DRO] (kJ·mol^{-1})
[62RAT/RIC]	(146.1 ± 3.7)	(144.7 ± 0.7)	523 – 673	(144.6 ± 0.7)
[64MAS/WEI]	(121.1 ± 4.6)[(a)]	(143.5 ± 3.4)	900 – 1100	(144.4 ± 2.6)
[66BER/CHU]	(139.4 ± 0.7)	(141.2 ± 1.4)	437 – 469, 420 – 546[(b)]	141.0
[66BER/CHU]	(133.8 ± 1.7)	(141.2 ± 1.4)	505 – 546, 420 – 546[(c)]	
[68YAM/POR]	(113.0 ± 6.3)[(a)]	(131.6 ± 6.3)[(a)]	578 – 667	142.7
[70KEL]	(145.8 ± 2.6)	(144.0 ± 1.0)	573 – 723	(144.1 ± 1.3)
[75HOA/REY]	(138.9 ± 8.4)		446 – 549	
[82GRI/GRI]	(135.5 ± 4.1)	(136.5 ± 4.1)[(d)], (143.3 ± 4.1)	463 – 483	143.1
[84GRO/DRO]		(143.1 ± 0.9)	1073 – 1373	(143.1 ± 0.9)
[84HUA/GIL]	(148.6 ± 7.1)	(140.6 ± 10.0)	485 – 503	
Average	(141.2 ± 11.5)	(142.7 ± 2.9)		

a Not included in the average.

b The first temperature range pertains to the equilibrium 6Se(trigonal) $\rightleftharpoons$ $Se_6(g)$ and the second temperature interval to the equilibrium $Se_6(g) \rightleftharpoons 3Se_2(g)$.

c The first temperature range pertains to the equilibrium 6Se(l) $\rightleftharpoons$ $Se_6(g)$ and the second temperature interval to the equilibrium $Se_6(g) \rightleftharpoons 3Se_2(g)$.

d Based on raw intensity data and not included in the average.

The thermal functions of condensed selenium and $Se_2(g)$ employed by the review and in [84GRO/DRO] are almost identical and the differences between the third law values derived in the two reviews deserve some comments. The difference between the third law values derived by the review and [84GRO/DRO] from [64MAS/WEI] is due to the different temperature intervals employed. The complete experimental interval of 800 to 1100 K was used by the review and the interval 900 to 1100 K in [84GRO/DRO]. The difference between the two evaluations is well within the assigned uncertainty limits. A large discrepancy exists for the values derived from the measurements in [68YAM/POR]. The third law value of the enthalpy of formation of $Se_2(g)$ determined in [68YAM/POR] was given as the range 104.6 to 115.5 $kJ \cdot mol^{-1}$ at 625 K in the original paper. It was obtained by estimating the mole fraction of $Se_2(g)$ to be in the range 10% to 100% at 625 K from the mass spectrometric measurements and corresponds to the range 123.7 to 134.6 $kJ \cdot mol^{-1}$ at 298.15 K. The value recommended in [68YAM/POR] for the enthalpy of formation of $Se_2(g)$ is (106.7 ± 10.5) $kJ \cdot mol^{-1}$ at 625 K, corresponding to (125.8 ± 10.5) $kJ \cdot mol^{-1}$ at 298.15 K, by also taking into account their second law value. The value $\Delta_f H_m^\circ$ (Se_2, g, 298.15 K) = 142.7 $kJ \cdot mol^{-1}$ attributed to [68YAM/POR] in [84GRO/DRO] was derived by considering only the lower limit of the mole fraction, 10%, and using the total pressures in [63NES] rather than those determined in the original paper. The value $\Delta_f H_m^\circ$ (Se_2, g, 298.15 K) = 131.61 $kJ \cdot mol^{-1}$ obtained by the review was calculated using the mole fraction x_{Se_2} = 0.43 corresponding to the relative ion intensity at 600 K and the total pressure as measured in the original paper.

The enthalpy of formation of $Se_2(g)$ can also be obtained from the total pressure of selenium vapour in equilibrium with the liquid at higher temperatures. Above about 800 K the mole fraction of $Se_2(g)$ exceeds 50% and third law evaluations will result in reliable values since the total pressure can be measured with good accuracy. For the purpose of this review, a number of such evaluations were carried out (see Appendix A) and have been summarised in Table V-3. The mole fractions of $Se_2(g)$ applied in the evaluations were calculated from the mass spectrometric measurements discussed above rather than setting the mole fraction to unity. The fractions calculated in this way are probably correct to within ± 0.1 and are consistent with the adopted thermochemical data. A change in the mole fraction by 0.1 shifts the derived enthalpy value by about 1.2 $kJ \cdot mol^{-1}$. None of the original papers included this type of evaluation.

The values derived from the measurements in [13PRE/BRO] and [57ILL/LAP] are 3 to 4 $kJ \cdot mol^{-1}$ larger in [84GRO/DRO] than in the current review. The difference originates from the assumption made in the evaluations in [84GRO/DRO] that $Se_6(g)$ is the major species besides $Se_2(g)$ in the gas phase while in fact $Se_5(g)$ is more abundant than $Se_6(g)$ at these temperatures. This fact leads to too low mole fractions of $Se_2(g)$ and results in larger values of the enthalpy of formation. The value of the enthalpy of formation of $Se_2(g)$ selected in [84GRO/DRO], $\Delta_f H_m^\circ$ (Se_2, g, 298.15 K) = (144.14 ± 0.63) kJ·mol^{-1} is therefore too large and the error limits are too small. The present review selects the weighted mean of the results in Tables V-2 and V-3:

$$\Delta_f H_m^\circ (Se_2, g, 298.15 K) = (141.1 \pm 3.0) kJ\cdot mol^{-1}$$

and

$$\Delta_f G_m^\circ (Se_2, g, 298.15 K) = (92.4 \pm 3.0) kJ\cdot mol^{-1},$$

where the uncertainties have been increased to account for unknown additional systematic errors that may be present.

Table V-3: Enthalpy of formation of $Se_2(g)$ at 298.15 K evaluated from total pressure measurements by the third law. The values derived in [84GRO/DRO] are included for comparison. The mole fraction and temperature columns do not apply to these values.

Reference	$\Delta_f H_m^\circ$ (298.15 K) (kJ·mol^{-1})	Temperature (K)	Mole fraction $Se_2(g)$	$\Delta_f H_m^\circ$ (298.15 K), [84GRO/DRO] (kJ·mol^{-1})
[13PRE/BRO]	(136.7 ± 2.5)	950	0.64	(143.3 ± 1.1)
[52BRO]	(140.1 ± 1.6)	950	0.64	
[57ILL/LAP]	(141.1 ± 2.1)	1150	0.80	(144.01 ± 0.25)
[62KUD/UST]	(140.1 ± 1.6)	950	0.64	
[64BON/SHU]	(139.9 ± 3.4)	1150	0.80	
[68BAK]	(140.5 ± 2.2)	1500	0.90	
[72CHE/KUZ]	(140.8 ± 1.5)	940	0.63	
[74RAU]	(138.9 ± 1.9)	1373	0.88	
Average	(139.8 ± 2.7)			

V.1.6 $Se_3(g)$

No experimental information is available on the entropy and heat capacity of $Se_3(g)$. The published values are based on estimates made by comparison with gaseous sulphur and other selenium molecules.

The entropy of the $Se_3(g)$ molecule was calculated from estimated molecular data by Keller [70KEL]. Mills [74MIL] proposes a value cited from a private communication with Drowart and Smoes. Pupishev and Rusin [84PUP/RUS] calculated the entropy and heat capacity for the temperature range 298.15 to 1500 K from estimated molecular data, which were obtained from a scaling procedure motivated by a comparison of the $Se_2(g)$-$Se_8(g)$ and $S_2(g)$-$S_8(g)$ molecules using the LCAO method. For this review, second law entropies were calculated from investigations involving $Se_3(g)$ equilibria as discussed in Appendix A ([66BER/CHU], [70KEL], [82GRI/GRI], and [84HUA/GIL]). Such evaluations were not made in the original papers. The entropy values are summarised in Table V-4.

Table V-4: Entropy of $Se_3(g)$ at 298.15 K as (1) calculated from estimated molecular constants and (2) evaluated using the second law.

Reference	S°_{m} (Se_3, g, 298.15 K)$^{(1)}$ (J·K^{-1}·mol^{-1})	S°_{m} (Se_3, g, 298.15 K)$^{(2)}$ (J·K^{-1}·mol^{-1})
[70KEL]	315.2$^{(a)}$	
[74MIL]	315.0	
[84PUP/RUS]	308.9	
[66BER/CHU]		(292.2 ± 4.3)
[66BER/CHU]		(275.4 ± 9.6)
[70KEL]		(313.3 ± 3.0)
[82GRI/GRI]		(302.6 ± 8.5)
[84HUA/GIL]		(288.2 ± 8.4)
Average		(294.3 ± 28.2)

a 300 K

The values in [70KEL] and [74MIL] seem to be based on the same set of molecular parameters although those in [74MIL] are undocumented. The second law entropies are in favour of the lower of the values calculated from molecular data and the selected entropy value is that in [84PUP/RUS],

$$S^{\circ}_{m}(\text{Se}_3, \text{g}, 298.15\ \text{K}) = (308.9 \pm 10.0)\ \text{J·K}^{-1}\text{·mol}^{-1}.$$

All available heat capacity values of $Se_3(g)$ at 298.15 K are based on estimated quantities and are summarised in Table V-5. The heat capacity value attributed to [84PUP/RUS] was calculated by the review from the estimated molecular parameters reported in the paper.

The heat capacity expression $C^{\circ}_{p,m}$(Se_3, g, (298.15 – 1500) K) = (57.6443 + $0.7762 \times 10^{-3}\ T - 0.2902 \times 10^{-6}\ T^{2} - 3.34573 \times 10^{5}\ T^{-2}$) J·K^{-1}·mol^{-1} was evaluated by the review by fitting the expression to heat capacity values calculated from the molecular parameters reported in [84PUP/RUS]. The heat capacity calculated from this expression is 2 to 5 J·K^{-1}·mol^{-1} smaller than that in [74MIL] in the whole temperature range 298.15 to 1500 K while it is smaller than that in [74RAU] by less than 1 J·K^{-1}·mol^{-1} in the temperature range 400 to 1100 K. The heat capacity expression derived by the review was employed in recalculations and evaluations wherever the thermodynamic properties of $Se_3(g)$ were required at other temperatures than the standard temperature. The heat capacity at 298.15 K calculated from the expression is $C^{\circ}_{p,m}$ (Se_3, g, 298.15 K) = (54.1 ± 2.0) $J \cdot K^{-1} \cdot mol^{-1}$. No value of the heat capacity is selected becase all tentative values are based on estimated quantities only.

Table V-5: Heat capacity values for $Se_3(g)$ at 298.15 K. The value attributed to [84PUP/RUS] was calculated by the review from the molecular parameters given in the reference.

Reference	$C^{\circ}_{p,m}$ (Se_3, g, 298.15 K) ($J \cdot K^{-1} \cdot mol^{-1}$)
[74RAU]	56.4
[74MIL]	56.6
[84PUP/RUS]	54.1

$Se_3(g)$ is a minor species in selenium vapour. Nevertheless, the ion $Se_3^+(g)$ is found in relatively large quantities in most of the mass spectrometric investigations due to fragmentation of the molecules $Se_5(g)$ and $Se_6(g)$. The evaluation of thermodynamic data for $Se_3(g)$ has therefore been avoided except in a few studies, summarised in Table V-6, where the investigators thought they were able to obtain ion concentrations free from fragmentation effects. A necessary requirement is that the investigation is carried out at such low selenium activities that the concentrations of $Se_5(g)$ and $Se_6(g)$ are low. The investigation of Keller [70KEL] is the only one in which this condition has been monitored properly and, after a re-assessment using the currently adopted thermal functions of the $Se_n(g)$ species, the review selects from this study:

$$\Delta_f H^{\circ}_m (Se_3, g, 298.15 K) = (178.0 \pm 10.0) kJ \cdot mol^{-1}$$

and

$$\Delta_f G^{\circ}_m (Se_3, g, 298.15 K) = (123.5 \pm 10.4) kJ \cdot mol^{-1},$$

where the uncertainties have been increased to indicate the considerable uncertainties of these values.

Table V-6: Re-evaluated second and third law values of the enthalpy of formation of $Se_3(g)$ at 298.15 K obtained from mass spectrometric investigations. Determinations that yield too large $Se_3^+(g)$ concentrations result in too low enthalpies of formation.

Reference	$\Delta_f H_m^\circ$ (298.15 K), 2nd law (kJ·mol^{-1})	$\Delta_f H_m^\circ$ (298.15 K), 3rd law (kJ·mol^{-1})	Temperature (K)
[66BER/CHU]	(157.3 ± 2.1)	(170.0 ± 7.0)	437 – 469, 694 – 1087 (a)
[66BER/CHU]	(149.0 ± 5.1)	(170.0 ± 7.0)	505 – 546, 694 – 1087 (b)
[70KEL]	(181.0 ± 2.1)	(178.0 ± 0.8)	673 – 723
[82GRI/GRI]	(143.6 ± 4.1)	(146.7 ± 4.1)	463 – 483
[84HUA/GIL]	(138.8 ± 4.2)	(149.3 ± 7.7)	485 – 503
Average	(153.9 ± 32.6)	(162.8 ± 27.3)	

a The first temperature range pertains to the equilibrium $6Se(trigonal) \rightleftharpoons Se_6(g)$ and the second temperature interval to the equilibrium $Se_6(g) \rightleftharpoons 2Se_3(g)$.

b The first temperature range pertains to the equilibrium $6Se(l) \rightleftharpoons Se_6(g)$ and the second temperature interval to the equilibrium $Se_6(g) \rightleftharpoons 2Se_3(g)$.

V.1.7 $Se_4(g)$

No experimental information is available about the entropy and heat capacity of $Se_4(g)$. All published values are based on estimates made by comparison with gaseous sulphur and other selenium molecules.

The entropy of the $Se_4(g)$ molecule was calculated from estimated molecular data by Keller [70KEL]. Mills [74MIL] proposes a value cited from a private communication with Drowart and Smoes. Pupishev and Rusin [84PUP/RUS] calculated the entropy and heat capacity for the temperature range 298.15 to 1500 K from estimated molecular data, which were obtained from a scaling procedure motivated by a comparison of the $Se_2(g)$-$Se_8(g)$ and $S_2(g)$-$S_8(g)$ molecules using the LCAO method. Second law entropies were calculated as discussed in Appendix A from those investigations in which the temperature dependence could be measured for equilibria involving $Se_4(g)$. Such evaluations were not made in the original papers. The entropy values are summarised in Table V-7.

The values in [70KEL] and [74MIL] seem to be based on the same set of molecular parameters although those in [74MIL] are undocumented. Although not decisive, the second law entropies are in favour of the lower of the values calculated from molecular data. The entropy value in [84PUP/RUS] is selected,

$$S_m^\circ(Se_4, g, 298.15\ K) = (340.6 \pm 20.0)\ J \cdot K^{-1} \cdot mol^{-1},$$

being also the value that is based on the most elaborate comparisons. The error limits have been set to reflect the uncertainty in the molecular structure.

Table V-7: Entropy of $Se_4(g)$ at 298.15 K as (1) calculated from estimated molecular constants and (2) evaluated using the second law.

Reference	S°_{m} (Se_4, g, 298.15 K)$^{(1)}$ ($J \cdot K^{-1} \cdot mol^{-1}$)	S°_{m} (Se_4, g, 298.15 K)$^{(2)}$ ($J \cdot K^{-1} \cdot mol^{-1}$)
[70KEL]	379.8$^{(a)}$	
[74MIL]	379.2	
[84PUP/RUS]	340.6	
[82GRI/GRI]		(323.7 ± 8.5), (309.8 ± 8.5)
[84HUA/GIL]		(326.9 ± 8.0)

a 300 K

All available heat capacity values of $Se_4(g)$ at 298.15 K are based on estimated quantities and are summarised in Table V-8. The heat capacity value attributed to [84PUP/RUS] was calculated by the review from the estimated molecular parameters reported in the paper.

Table V-8: Heat capacity values of $Se_4(g)$ at 298.15 K. The value attributed to [84PUP/RUS] was calculated by the review from the molecular parameters given in the reference.

Reference	$C^{\circ}_{p,m}$ (Se_4, g, 298.15 K) ($J \cdot K^{-1} \cdot mol^{-1}$)
[74RAU]	81.1
[74MIL]	80.3
[84PUP/RUS]	76.1

Expressions for the variation of the heat capacity with temperature have been given by [74RAU] and [74MIL], and derived by the review from the molecular parameters estimated in [84PUP/RUS] yielding $C^{\circ}_{p,m}$ (Se_4, g, (298.15 – 1500) K) = (82.1759 + $1.3493 \times 10^{-3}\,T - 0.5043 \times 10^{-6}\,T^{2} - 5.68525 \times 10^{5}\,T^{-2}$) $J \cdot K^{-1} \cdot mol^{-1}$. The heat capacity calculated from this expression is 0.8 to 5.0 $J \cdot K^{-1} \cdot mol^{-1}$ smaller than that in [74MIL] in the whole temperature range 298.15 to 1500 K while it is smaller than that in [74RAU] by less than 1 $J \cdot K^{-1} \cdot mol^{-1}$ above 650 K. The heat capacity expression derived by the review was employed in recalculations and evaluations wherever the thermodynamic properties of $Se_4(g)$ were required at other temperatures than the standard temperature. The heat capacity at 298.15 K calculated from the expression is $C^{\circ}_{p,m}$ (Se_4, g, 298.15 K) = (76.1 ± 2.0) $J \cdot K^{-1} \cdot mol^{-1}$. No value of the heat capacity is selected because all the tentative values are based on estimated quantities only.

$Se_4(g)$ is a minor species in selenium vapour and the situation is even worse than for $Se_3(g)$ with respect to fragmentation problems. The evaluation of thermodynamic data for $Se_4(g)$ has therefore been avoided except in three studies, summarised in Table V-9, where the investigators thought the fragmentation was under control. The study of Keller [70KEL] is the only one in which the fragmentation has been monitored properly and the review selects from this study:

$$\Delta_f H_m^\circ (Se_4, g, 298.15 K) = (163.0 \pm 12.0) kJ\cdot mol^{-1}$$

and

$$\Delta_f G_m^\circ (Se_4, g, 298.15 K) = (111.6 \pm 13.4) kJ\cdot mol^{-1},$$

where the uncertainties have been increased to indicate the considerable uncertainties of these values.

Table V-9 Re-evaluated second and third law values of the enthalpy of formation of $Se_4(g)$ at 298.15 K obtained from mass spectrometric investigations. Determinations that yield too large $Se_4^+(g)$ concentrations result in too low enthalpies of formation.

Reference	$\Delta_f H_m^\circ$ (298.15 K), 2nd law (kJ·mol^{-1})	$\Delta_f H_m^\circ$ (298.15 K), 3rd law (kJ·mol^{-1})	Temperature (K)
[70KEL]		(163.0 ± 3.4)	673
[82GRI/GRI]	(133.7 ± 4.1)	(141.9 ± 4.1), (148.6 ± 4.1)	463 – 483
[84HUA/GIL]	(134.4 ± 3.8)	(141.4 ± 11.4)	485 – 503

V.1.8 $Se_5(g)$

No experimental information is available on the entropy and heat capacity of $Se_5(g)$. All published values are based on estimates made by comparison with gaseous sulphur and other selenium molecules.

The entropy of the $Se_5(g)$ molecule was calculated from estimated molecular data by Keller [70KEL]. Mills [74MIL] proposes a value cited from a private communication with Drowart and Smoes. Steudel [81STE] calculated the heat capacity and entropy by analogy with sulphur molecules as discussed in Appendix A. Pupishev and Rusin [84PUP/RUS] calculated the entropy and heat content for the temperature range 298.15 to 1500 K from estimated molecular data, which were obtained from a scaling procedure motivated by a comparison of the $Se_2(g)$-$Se_8(g)$ and $S_2(g)$-$S_8(g)$ molecules using the LCAO method. Second law entropies were calculated by the review from those investigations in which the temperature dependence could be measured for equilibria involving $Se_5(g)$. These calculations are discussed for each individual investigation in Appendix A. Such evaluations were not made in the original papers. The entropy values are summarised in Table V-10.

Table V-10: Entropy of $Se_5(g)$ at 298.15 K as (1) calculated from estimated molecular constants and (2) evaluated from vapour pressure measurements using the second law.

Reference	S°_{m} (Se_5, g, 298.15 K)$^{(1)}$ (J·K^{-1}·mol^{-1})	S°_{m} (Se_5, g, 298.15 K)$^{(2)}$ (J·K^{-1}·mol^{-1})	Temperature (K)
[70KEL]	386.0[a]		
[74MIL]	385.4		
[81STE]	383.5		
[84PUP/RUS]	398.3		
[66BER/CHU]		(430.6 ± 3.1)	437 – 469, 420 – 546[b]
[66BER/CHU]		(402.5 ± 7.8)	505 – 546, 420 – 546[c]
[66FUJ/WES]		(421.0 ± 9.2)	375 – 460
[68YAM/POR]		(394.2 ± 13.9)	578 – 667
[70KEL]		(392.6 ± 7.2)	573 – 673
[74RAU]		(411.2 ± 18.8)	485 – 575
[75HOA/REY]		(373.2 ± 15.9)	429 – 464
[75HOA/REY]		(402.5 ± 8.5), (409.8 ± 8.5)	463 – 483
[82GRI/GRI]		(419.7 ± 8.8)	430 – 490
[84HUA/GIL]		(422.8 ± 8.8)	485 – 503
Average		(407.3 ± 32.5)	

a 300 K

b The first temperature range pertains to the equilibrium 6Se(trigonal) $\rightleftharpoons$ $Se_6(g)$ and the second temperature interval to the equilibrium $5Se_6(g) \rightleftharpoons 6Se_5(g)$.

c The first temperature range pertains to the equilibrium 6Se(l) $\rightleftharpoons$ $Se_6(g)$ and the second temperature interval to the equilibrium $5Se_6(g) \rightleftharpoons 6Se_5(g)$.

The values in [70KEL] and [74MIL] seem to be based on the same set of molecular parameters although those in [74MIL] are undocumented. The second law entropies are in favour of the higher of the values calculated from molecular data and the selected entropy value is that in [84PUP/RUS],

$$S^{\circ}_{m}(\text{Se}_5, \text{g}, 298.15\ \text{K}) = (398.3 \pm 10.0)\ \text{J·K}^{-1}\text{·mol}^{-1}.$$

The reported values of the heat capacity of $Se_5(g)$ at 298.15 K are summarised in Table V-11. The heat capacity of $Se_5(g)$ assigned to [84PUP/RUS] was calculated as the mean of the heat capacity of Se_4(g) and Se_6(g). Similarly, the heat capacity expression $C^{\circ}_{p,m}$ (Se_5, g, (298.15 – 1500) K) = $(107.3452 + 1.039 \times 10^{-3}\ T - 0.3887 \times 10^{-6}\ T^2 - 6.02604 \times 10^5\ T^{-2})$ $J \cdot K^{-1} \cdot mol^{-1}$ was derived by the review from the mean of the heat capacity expressions for Se_4(g) and Se_6(g) (cf. Sections V.1.7 and V.1.9). The heat capacity calculated from this expression and that in [74MIL] differ by less than 0.1 $J \cdot K^{-1} \cdot mol^{-1}$ in the temperature range 750 to 1500 K and amounts to a maximum of 0.45 $J \cdot K^{-1} \cdot mol^{-1}$ at 298.15 K. The expression was employed in recalculations and evaluations wherever the thermodynamic properties of Se_5(g) were required at other temperatures than the standard temperature. The heat capacity at 298.15 K calculated from the expression is $C^{\circ}_{p,m}$ (Se_5, g, 298.15 K) = (100.8 ± 3.0) $J \cdot K^{-1} \cdot mol^{-1}$. No value of the heat capacity is selected because all tentative values are based on estimated quantities only.

Table V-11: Heat capacity values of Se_5(g) at 298.15 K. The value attributed to [84PUP/RUS] was not explicitly stated in the paper and was therefore calculated as the mean of their values for Se_4(g) and Se_6(g) in agreement with their heat content calculations.

Reference	$C^{\circ}_{p,m}$ (Se_5, g, 298.15 K) ($J \cdot K^{-1} \cdot mol^{-1}$)
[74RAU]	103.5
[74MIL]	101.3
[81STE]	107.5
[84PUP/RUS]	100.8

The enthalpy of formation of Se_5(g) has been determined from mass spectrometric investigations of selenium vapour using Knudsen cells. The measurements can be separated into the approximate temperature ranges 420 to 494 K for equilibria with solid selenium and 494 to 700 K for equilibria with liquid selenium. The various investigations have used different methods and auxiliary data for deriving the enthalpy of formation at 298.15 K from the measurements. This review has re-evaluated the measurements using the adopted values for heat capacities and entropies of the gaseous species and condensed phases. The details of the evaluations are given in Appendix A and the values of the enthalpies of formation of Se_5(g) are summarised in Table V-12.

The third law evaluations made in this review reduce the scatter of the enthalpy values by more than 50%. The scatter of the values obtained from the second law is an effect of the small temperature ranges used in most of the studies. The second law value in [75HOA/REY] for the equilibrium with liquid selenium is not included in the average in Table V-12 because it stems from measurements which result in an enthalpy of fusion of selenium that is twice as large as the correct value.

Table V-12: Re-evaluated second and third law values of the enthalpy of formation of $Se_5(g)$ at 298.15 K obtained from mass spectrometric investigations.

Reference	$\Delta_f H^\circ_m$ (298.15 K), 2nd law (kJ·mol^{-1})	$\Delta_f H^\circ_m$ (298.15 K), 3rd law (kJ·mol^{-1})	Temperature (K)
[66BER/CHU]	(160.4 ± 1.4)	(145.3 ± 3.1)	437 – 469, 420 – 546[a]
[66BER/CHU]	(146.7 ± 4.1)	(145.3 ± 3.1)	505 – 546, 420 – 546[b]
[66FUJ/WES]	(155.5 ± 4.1)	(145.3 ± 4.1)	405 – 438
[68YAM/POR]	(140.2 ± 8.4)	(142.7 ± 8.4)	578 – 667
[70KEL]	(145.2 ± 4.5)	(148.6 ± 1.6)	498 – 573
[75HOA/REY]	(129.3 ± 8.4)[c]	(142.5 ± 8.4)	485 – 575
[75HOA/REY]	(151.2 ± 8.4)	(145.5 ± 5.9)	429 – 464
[82GRI/GRI]	(149.1 ± 4.1)	(147.0 ± 4.1), (143.5 ± 4.1)	463 – 483
[84DRO/SMO]	(155.4 ± 4.2)	(145.3 ± 4.2)	430 – 490
[84HUA/GIL]	(152.2 ± 4.2)	(140.1 ± 8.0)	485 – 503
Average	(150.7 ± 12.0)	(144.6 ± 4.6)	

a The first temperature range pertains to the equilibrium $6Se(trigonal) \rightleftharpoons Se_6(g)$ and the second temperature interval to the equilibrium $5Se_6(g) \rightleftharpoons 6Se_5(g)$.

b The first temperature range pertains to the equilibrium $6Se(l) \rightleftharpoons Se_6(g)$ and the second temperature interval to the equilibrium $5Se_6(g) \rightleftharpoons 6Se_5(g)$.

c Not included in the average.

It was observed in the mass spectrometric studies that $Se_3(g)$ and $Se_4(g)$ were formed due to the fragmentation of $Se_5(g)$ and $Se_6(g)$ in the spectrometer and in most of the investigations data for $Se_3(g)$ and $Se_4(g)$ were not evaluated because of this problem. However, in all studies the partial pressures of $Se_5(g)$ and $Se_6(g)$ were evaluated without compensating for fragmentation. The partial pressures of these species are therefore too low and the enthalpies of formation should be somewhat lower than the values calculated from the experimental data. This conclusion is supported by the fact that the total pressure becomes too small, when it is calculated from the above data for $Se_5(g)$ and $Se_6(g)$ and the selected data of the other selenium species in the temperature range where $Se_5(g)$ and $Se_6(g)$ are present with large mole fractions (see Figure V-1). In order to fit the calculated total pressures better to the measured total pressures, the selected quantities have been shifted by about – 1.0 kJ·mol^{-1} from the weighted mean of the second and third law values. The review thus selects:

$$\Delta_f H^\circ_m (Se_5, g, 298.15 K) = (144.4 \pm 4.3) kJ·mol^{-1}$$

and

$$\Delta_f G^\circ_m (Se_5, g, 298.15 K) = (88.4 \pm 5.3) kJ·mol^{-1}.$$

V.1.9 $Se_6(g)$

No experimental information is available about the entropy and heat capacity of $Se_6(g)$. The published values are based on estimates made by comparison with gaseous sulphur and other selenium molecules.

The entropy of the $Se_6(g)$ molecule was calculated from estimated molecular data by Keller [70KEL]. Mills [74MIL] proposes a value cited from a private communication with Drowart and Smoes. Steudel [81STE] calculated the heat capacity and entropy by analogy with sulphur molecules as discussed in Appendix A. Pupishev and Rusin [84PUP/RUS] calculated the entropy and heat content for the temperature range 298.15 to 1500 K from estimated molecular data, which were obtained from a scaling procedure motivated by a comparison of the $Se_2(g)$-$Se_8(g)$ and $S_2(g)$-$S_8(g)$ molecules using the LCAO method. Second law entropies were calculated by the review as discussed in Appendix A from those investigations in which the temperature dependence could be measured for equilibria involving $Se_6(g)$. Such evaluations were not made in the original papers. The entropy values are summarised in Table V-13.

Table V-13: Entropy of $Se_6(g)$ at 298.15 K as (1) calculated from estimated molecular constants and (2) evaluated using the second law.

Reference	S°_{m} (Se_6, g, 298.15 K)$^{(1)}$ ($J\cdot K^{-1}\cdot mol^{-1}$)	S°_{m} (Se_6, g, 298.15 K)$^{(2)}$ ($J\cdot K^{-1}\cdot mol^{-1}$)	Temperature (K)
[70KEL]	433.9 [(a)]		
[74MIL]	433.6		
[81STE]	433.1		
[84PUP/RUS]	428.5		
[66BER/CHU]		(444.3 ± 1.7)	437 – 469
[66BER/CHU]		(410.6 ± 4.6)	505 – 546
[66FUJ/WES]		(446.9 ± 9.2)	405 – 438
[68YAM/POR]		(429.1 ± 13.9)	578 – 667
[70KEL]		(441.7 ± 14.2)	473 – 673
[75HOA/REY]		(434.6 ± 18.8)	485 – 575
[75HOA/REY]		(395.6 ± 15.9) [(b)]	429 – 464
[82GRI/GRI]		(420.6 ± 8.5), (422.8 ± 8.5)	463 – 483
[84DRO/SMO]		(443.3 ± 8.0)	430 – 490
[84HUA/GIL]		(449.3 ± 7.1)	485 – 503
Average		(434.3 ± 25.5)	

a 300 K

b Not included in the average

The difference between values calculated from molecular data is small. Despite the fact that the average value calculated from the second law is slightly in favour of the larger entropy values calculated from molecular data, the entropy value in [84PUP/RUS] is selected since it is the result of the best documented molecular calculations,

$$S^{\circ}_{m}(Se_6, g, 298.15 K) = (428.5 \pm 10.0) J \cdot K^{-1} \cdot mol^{-1}.$$

All values of the heat capacity of Se_6(g) at 298.15 K are based on estimated quantities and are summarised in Table V-14. The heat capacity value attributed to [84PUP/RUS] was calculated by the review from the estimated molecular parameters reported in the paper. Similarly, the heat capacity expression $C^{\circ}_{p,m}$ (Se_6, g, (298.15 – 1500) K) = (132.511 + 0.7288 × 10^{-3} T – 0.2731 × 10^{-6} T^2 – 6.36682 × 10^5 T^{-2}) $J \cdot K^{-1} \cdot mol^{-1}$ was obtained by fitting the expression to heat capacity values calculated from the molecular parameters reported in [84PUP/RUS]. The difference between the heat capacity calculated from this expression and that calculated from the expression in [74MIL] amounts to 0.05 $J \cdot K^{-1} \cdot mol^{-1}$ in the temperature range 1000 to 1500 K and reaches a maximum of 0.7 $J \cdot K^{-1} \cdot mol^{-1}$ at 298.15 K. The derived heat capacity expression was employed by the review in all recalculations and evaluations wherever the thermodynamic properties of Se_6(g) were required at other temperatures than the standard temperature. The heat capacity at 298.15 K calculated from the expression is $C^{\circ}_{p,m}$ (Se_6, g, 298.15 K) = (125.5 ± 2.0) $J \cdot K^{-1} \cdot mol^{-1}$. No value of the heat capacity is selected because all tentative values are based on estimated quantities only.

The heat capacities calculated from the expressions in [74MIL] and [84PUP/RUS] differ by less than 0.6 $J \cdot K^{-1} \cdot mol^{-1}$ in the whole temperature range 298.15 to 1500 K.

Table V-14: Heat capacity values of Se_6(g) at 298.15 K. The value attributed to [84PUP/RUS] was calculated by the review from the molecular parameters given in the reference.

Reference	$C^{\circ}_{p,m}$ (Se_6, g, 298.15 K) ($J \cdot K^{-1} \cdot mol^{-1}$)
[74RAU]	128.6
[74MIL]	126.3
[81STE]	129.7
[84PUP/RUS]	125.5

The enthalpy of formation of Se_6(g) has been determined from mass spectrometric investigations of selenium vapour using Knudsen cells. The measurements can be separated into the approximate temperature ranges 420 to 494 K for equilibria with solid selenium and 494 to 700 K for equilibria with liquid selenium. The various investigations have used different methods and auxiliary data for deriving the enthalpy of formation at 298.15 K from the measurements. This review has re-evaluated the measurements using the adopted values for the heat capacities and entropies of the gaseous species and condensed phases. The details of the evaluations are given in Appendix A. The values of the enthalpy of formation of $Se_6(g)$ are summarised in Table V-15.

Table V-15: Re-evaluated second and third law values of the enthalpy of formation of $Se_6(g)$ at 298.15 K from mass spectrometric investigations.

Reference	$\Delta_f H^\circ_m$ (298.15 K), 2nd law (kJ·mol^{-1})	$\Delta_f H^\circ_m$ (298.15 K), 3rd law (kJ·mol^{-1})	Temperature (K)
[66BER/CHU]	(145.5 ± 0.8)	(138.3 ± 0.4)	437 – 469
[66BER/CHU]	(128.9 ± 2.4)	(138.3 ± 0.5)	505 – 546
[66FUJ/WES]	(145.6 ± 4.1)	(137.4 ± 4.1)	405 – 438
[68YAM/POR]	(134.8 ± 8.4)	(134.4 ± 8.4)	578 – 667
[70KEL]	(142.5 ± 8.3)	(134.3 ± 2.4)	473 – 673
[75HOA/REY]	(118.9 ± 8.4) [(a)]	(136.2 ± 8.4)	485 – 575
[75HOA/REY]	(140.5 ± 8.4)	(137.8 ± 5.9)	429 – 464
[82GRI/GRI]	(135.5 ± 4.1)	(139.3 ± 4.1), (138.3 ± 4.1)	463 – 483
[84DRO/SMO]	(144.0 ± 3.8)	(137.0 ± 3.8)	430 – 490
[84HUA/GIL]	(143.2 ± 3.3)	(133.0 ± 7.1)	485 – 503
Average	(140.1 ± 11.3)	(136.8 ± 4.0)	

a Not included in the average.

The third law evaluations made in this review reduce the scatter of the enthalpy values by more than 65%. The scatter of the values obtained from the second law is an effect of the small temperature ranges used in most of the studies. The second law value in [75HOA/REY] for the equilibrium with liquid selenium is not included in the average in Table V-15 because it stems from measurements which result in an enthalpy of fusion of selenium that is twice as large as the correct value.

It was observed in the mass spectrometric studies that $Se_3(g)$ and $Se_4(g)$ were formed due to the fragmentation of $Se_5(g)$ and $Se_6(g)$ in the spectrometer and in most of the investigations $Se_3(g)$ and $Se_4(g)$ were omitted from evaluation because of this problem. However, in all studies the partial pressures of $Se_5(g)$ and $Se_6(g)$ were evaluated without compensating for fragmentation. The partial pressures of these species are therefore too low and the enthalpies of formation should be somewhat lower than the values calculated from the experimental data. As discussed in the previous sub-section the selected quantities have been shifted by about – 1.0 kJ·mol^{-1} from the weighted mean of the second and third law values. The review thus selects:

$$\Delta_f H_m^\circ (Se_6, g, 298.15 K) = (136.1 \pm 3.8) kJ·mol^{-1}$$

and

$$\Delta_f G_m^\circ (Se_6, g, 298.15 K) = (83.6 \pm 4.9) kJ·mol^{-1}.$$

V.1.10 Se_7(g)

No experimental information is available about the entropy and heat capacity of Se_7(g). The published values are based on estimates made by comparison with gaseous sulphur and other selenium molecules.

The entropy of the Se_7(g) molecule was calculated from estimated molecular data by Keller [70KEL]. Mills [74MIL] proposes a value cited from a private communication with Drowart and Smoes. Steudel [81STE] calculated the heat capacity and entropy by analogy with sulphur molecules as discussed in Appendix A. Pupishev and Rusin [84PUP/RUS] calculated the entropy and heat content for the temperature range 298.15 to 1500 K from estimated molecular data, which were obtained from a scaling procedure motivated by a comparison of the Se_2(g)-Se_8(g) and S_2(g)-S_8(g) molecules using the LCAO method.

Second law entropies were calculated by the review as discussed in Appendix A from those investigations in which the temperature dependence could be measured for equilibria involving Se_7(g). Such evaluations were not made in the original papers. The values in [66BER/CHU] were derived in two steps. First the entropy of Se_6(g) was determined from the equilibria Se_6(g)/liquid and Se_6(g)/solid and then the entropy of Se_7(g) was evaluated from the reaction $7Se_6(g) \rightleftharpoons 6Se_7(g)$. The first step resulted in two different values for Se_6(g) and hence also in two separate entropies of Se_7(g). The entropy values are summarised in Table V-16.

The difference between the average of the second law values and those calculated from molecular data is not large, but the second law values are widely scattered due to the small temperature ranges of the measurements. The entropy value in [84PUP/RUS] is selected since it is the result of the best documented molecular calculations,

$$S_m^\circ (Se_7, g, 298.15 K) = (487.0 \pm 6.0) J·K^{-1}·mol^{-1}.$$

Table V-16: Entropy of $Se_7(g)$ at 298.15 K as (1) calculated from estimated molecular constants and (2) evaluated using the second law.

Reference	S^{o}_{m} (Se_7, g, 298.15 K) [(1)] (J·K^{-1}·mol^{-1})	S^{o}_{m} (Se_7, g, 298.15 K) [(2)] (J·K^{-1}·mol^{-1})	Temperature (K)
[70KEL]	486.7 [(a)]		
[74MIL]	486.5		
[81STE]	482.7		
[84PUP/RUS]	487.0		
[66BER/CHU]		(505.1 ± 4.2)	437 – 469, 442 – 592 [(b)]
[66BER/CHU]		(465.8 ± 10.9)	505 – 546, 442 – 592 [(c)]
[66FUJ/WES]		(519.3 ± 9.3)	405 – 438
[68YAM/POR]		(477.4 ± 7.0)	578 – 667
[70KEL]		(505.2 ± 14.8)	473 – 673
[75HOA/REY]		(494.1 ± 9.4)	485 – 575
[75HOA/REY]		(438.7 ± 8.0) [(d)]	429 – 464
[82GRI/GRI]		(479.8 ± 8.5), (477.8 ± 8.5)	463 – 483
[84DRO/SMO]		(495.0 ± 11.5)	430 – 490
[84HUA/GIL]		(513.5 ± 8.0)	485 – 503
Average		(493.3 ± 34.6)	

a 300 K

b The first temperature range pertains to the equilibrium 6Se(trigonal) $\rightleftharpoons$ $Se_6(g)$ and the second temperature interval to the equilibrium $7Se_6(g) \rightleftharpoons 6Se_7(g)$.

c The first temperature range pertains to the equilibrium 6 Se(l) $\rightleftharpoons$ $Se_6(g)$ and the second temperature interval to the equilibrium $7Se_6(g) \rightleftharpoons 6Se_7(g)$.

d Not included in the average.

The reported values of the heat capacity of $Se_7(g)$ at 298.15 K are summarised in Table V-17. The heat capacity value attributed to [84PUP/RUS] was calculated by the review from the estimated molecular parameters reported in the paper. Similarly, the heat capacity expression $C^{o}_{p,m}$ (Se_7, g, (298.15 – 1500) K) = $(157.344 + 0.8809 \times 10^{-3}\ T - 0.3297 \times 10^{-6}\ T^{2} - 7.51088 \times 10^{5}\ T^{-2})$ J·K^{-1}·mol^{-1} was obtained by fitting the expression to heat capacity values calculated from the molecular parameters reported in [84PUP/RUS]. The difference between the heat capacity calculated from this expression and that calculated from the expression in [74MIL] amounts to 0.1 J·K^{-1}·mol^{-1} in the temperature range 800 to 1500 K and reaches a maximum of 0.6 J·K^{-1}·mol^{-1} at 298.15 K. The derived heat capacity expression was employed by the review in all recalculations and evaluations wherever the thermodynamic properties of $Se_7(g)$ were required at other temperatures than the standard temperature. The heat capacity at 298.15 K calculated from the expression is $C^{\circ}_{p,m}$ (Se_7, g, 298.15 K) = (149.1 ± 2.0) $J \cdot K^{-1} \cdot mol^{-1}$. No value of the heat capacity is selected because all tentative values are based on estimated quantities only.

Table V-17: Heat capacity values of Se_7(g) at 298.15 K. The value attributed to [84PUP/RUS] was calculated by the review from the molecular parameters given in the reference.

Reference	$C^{\circ}_{p,m}$ (Se_7, g, 298.15 K) ($J \cdot K^{-1} \cdot mol^{-1}$)
[74RAU]	148.0
[74MIL]	148.5
[81STE]	151.9
[84PUP/RUS]	149.1

The enthalpy of formation of Se_7(g) has been determined from mass spectrometric investigations of selenium vapour using Knudsen cells. The measurements can be separated into the approximate temperature ranges 420 to 494 K for equilibria with solid selenium and 494 to 700 K for equilibria with liquid selenium. The various investigations have used different methods and auxiliary data for deriving enthalpies of formation at 298.15 K from the measurements and for the purpose of this review the measurements were re-evaluated using the adopted values for heat capacities and entropies of the gaseous species and condensed phases. The details of the evaluations are given in Appendix A. The values of the enthalpies of formation of Se_7(g) are summarised in Table V-18.

The third law evaluation of the experimental data made in this review reduces the scatter of the enthalpy values by 50%. The scatter of the values obtained from the second law is an effect of the small temperature ranges used in most of the studies. The second law value in [75HOA/REY] for the equilibrium with the liquid selenium is not included in the average in Table V-18 because it stems from measurements which result in an enthalpy of fusion of selenium that is twice as large as the correct value. The review selects the weighted average of the second and third law values:

$$\Delta_f H^{\circ}_m (Se_7, g, 298.15 K) = (150.2 \pm 4.5)\ kJ \cdot mol^{-1}$$

and

$$\Delta_f G^{\circ}_m (Se_7, g, 298.15 K) = (92.8 \pm 4.9)\ kJ \cdot mol^{-1}.$$

Table V-18: Re-evaluated second and third law enthalpy of formation of $Se_7(g)$ at 298.15 K from mass spectrometric investigations.

Reference	$\Delta_f H_m^\circ$ (298.15 K), 2nd law (kJ·mol^{-1})	$\Delta_f H_m^\circ$ (298.15 K), 3rd law (kJ·mol^{-1})	Temperature (K)
[66BER/CHU]	(159.4 ± 1.9)	(151.2 ± 0.9)	437 – 469, 420 – 546$^{(a)}$
[66BER/CHU]	(140.1 ± 5.7)	(151.2 ± 1.3)	505 – 546, 420 – 546$^{(b)}$
[66FUJ/WES]	(165.2 ± 4.1)	(150.7 ± 4.1)	375 – 460
[68YAM/POR]	(143.3 ± 8.4)	(149.1 ± 8.4)	578 – 667
[70KEL]	(158.8 ± 8.5)	(148.0 ± 2.4)	573 – 673
[75HOA/REY]	(121.7 ± 8.4)$^{(c)}$	(147.0 ± 8.4)	485 – 575
[75HOA/REY]	(151.7 ± 8.4)	(148.6 ± 5.9)	429 – 464
[82GRI/GRI]	(149.3 ± 4.1)	(152.8 ± 4.1), (153.7 ± 4.1)	463 – 483
[84DRO/SMO]	(154.8 ± 5.4)	(151.0 ± 5.4)	430 – 490
[84HUA/GIL]	(159.5 ± 3.8)	(146.2 ± 6.2)	485 – 503
Average	(153.6 ± 16.1)	(150.0 ± 4.6)	

a The first temperature range pertains to the equilibrium 6Se(trigonal) $\rightleftharpoons$ $Se_6(g)$ and the second temperature interval to the equilibrium $6Se_7(g) \rightleftharpoons 7Se_6(g)$.

b The first temperature range pertains to the equilibrium 6Se(l) $\rightleftharpoons$ $Se_6(g)$ and the second temperature interval to the equilibrium $6Se_7(g) \rightleftharpoons 7Se_6(g)$.

c Not included in the average.

V.1.11 $Se_8(g)$

No experimental information is available on the entropy and heat capacity of $Se_8(g)$. All published values are based on estimates made by comparison with gaseous sulphur and other selenium molecules, and in particular with molecular parameters for Se_8 rings known from the solid state.

The entropy of the $Se_8(g)$ molecule was calculated from estimated molecular data by Keller [70KEL]. Mills [74MIL] proposes a value cited from a private communication with Drowart and Smoes. Steudel [81STE] calculated the heat capacity and entropy by analogy with sulphur molecules as discussed in Appendix A. Pupishev and Rusin [84PUP/RUS] calculated the entropy and heat content for the temperaure range 298.15 to 1500 K from estimated molecular data which were obtained from a scaling procedure motivated by a comparison of the $Se_2(g)$-$Se_8(g)$ and $S_2(g)$-$S_8(g)$ molecules using the LCAO method.

Second law entropies were calculated by the review as discussed in Appendix A from those investigations in which the temperature dependence could be measured for equilibria involving $Se_8(g)$. Such evaluations were not made in the original papers. The values in [66BER/CHU] were derived in two steps. First the entropy of $Se_6(g)$ was determined from the equilibria $Se_6(g)$/liquid and $Se_6(g)$/solid and then the entropy of $Se_8(g)$ was evaluated from the reaction $8Se_6(g) \rightleftharpoons 6Se_8(g)$. The first step resulted in two different values for $Se_6(g)$ and hence also in two separate entropies of $Se_8(g)$. The entropy values are summarised in Table V-19.

Table V-19: Entropy of $Se_8(g)$ at 298.15 K as (1) calculated from estimated molecular constants and (2) evaluated using the second law.

Reference	S^{o}_{m} (Se_8, g, 298.15 K) (1) (J·K^{-1}·mol^{-1})	S^{o}_{m} (Se_8, g, 298.15 K) (2) (J·K^{-1}·mol^{-1})	Temperature (K)
[70KEL]	531.9 (a)		
[74MIL]	531.3		
[81STE]	532.3		
[84PUP/RUS]	526.2		
[66BER/CHU]		(553.7 ± 4.8)	437 – 469, 472 – 606 (b)
[66BER/CHU]		(508.7 ± 12.4)	505 – 546, 472 – 606 (c)
[66FUJ/WES]		(561.7 ± 12.8)	405 – 438
[68YAM/POR]		(528.2 ± 27.9)	578 – 667
[70KEL]		(537.7 ± 14.4)	498 – 573
[82GRI/GRI]		(511.0 ± 8.5), (508.6 ± 8.5)	463 – 483
[84HUA/GIL]		(541.2 ± 7.1)	485 – 503
Average		(531.4 ± 40.7)	

a 300 K

b The first temperature range pertains to the equilibrium $6Se(\text{trigonal}) \rightleftharpoons Se_6(g)$ and the second temperature interval to the equilibrium $8Se_6(g) \rightleftharpoons 6Se_8(g)$.

c The first temperature range pertains to the equilibrium $6Se(l) \rightleftharpoons Se_6(g)$ and the second temperature interval to the equilibrium $8Se_6(g) \rightleftharpoons 6Se_8(g)$.

The difference between the entropy values calculated from molecular data is small while the second law values are widely scattered mainly due to the small temperature ranges of the measurements. The entropy value in [84PUP/RUS] is well within their range and it is selected since it is the result of the best documented molecular calculations,

$$S^{o}_{m}(Se_8, g, 298.15\ K) = (526.2 \pm 6.0)\ J{\cdot}K^{-1}{\cdot}mol^{-1}.$$

The reported values of the heat capacity of $Se_8(g)$ at 298.15 K are summarised in Table V-20. The heat capacity value attributed to [84PUP/RUS] was calculated by the review from the estimated molecular parameters reported in the paper. Similarly, the heat capacity expression $C^{\circ}_{p,m}$ (Se_8, g, (298.15 – 1500) K) = (182.1676 + 1.0499 × 10^{-3} T – 0.3934 × 10^{-6} T^2 – 8.53742 × 10^5 T^{-2}) J·K^{-1}·mol^{-1} was obtained by fitting the expression to heat capacity values calculated from the molecular parameters reported in [84PUP/RUS]. The largest difference between the heat capacity calculated from this expression and that calculated from the expression in [74MIL] amounts to 0.8 J·K^{-1}·mol^{-1} in the temperature range 298.15 to 1500 K. The derived heat capacity expression was employed by the review in all recalculations and evaluations wherever the thermodynamic properties of Se_8(g) were required at other temperatures than the standard temperature. The heat capacity at 298.15 K calculated from the expression is $C^{\circ}_{p,m}$ (Se_8, g, 298.15 K) = (172.8 ± 2.0) J·K^{-1}·mol^{-1}. No value of the heat capacity is selected because all tentative values are based on estimated quantities only.

Table V-20: Heat capacity values of Se_8(g) at 298.15 K. The value attributed to [84PUP/RUS] was calculated by the review from the molecular parameters given in the reference.

Reference	$C^{\circ}_{p,m}$ (Se_8, g, 298.15 K) (J·K^{-1}·mol^{-1})
[74RAU]	177.3
[74MIL]	173.9
[81STE]	174.0
[84PUP/RUS]	172.8

The enthalpy of formation of Se_8(g) has been determined from mass spectrometric investigations of selenium vapour using Knudsen cells. The measurements can be separated into the approximate temperature ranges 420 to 494 K for equilibria with solid selenium and 494 to 700 K for equilibria with liquid selenium. The various investigations have used different methods and auxiliary data for deriving enthalpies of formation at 298.15 K from the measurements. For the purpose of this review the measurements were recalculated (see Appendix A) using the adopted values for heat capacities and entropies of the gaseous species and condensed phases. The values of the enthalpies of formation of Se_8(g) are summarised in Table V-21.

The third law evaluations made in this review reduce the scatter of experimental values by more than 80%, which indicates that the temperature ranges of the investigations are in general too small for second law evaluations. The discrepancy between the two second law values in [66BER/CHU] stem from an unusually large enthalpy of fusion determined in their investigation. The value in [75HOA/REY] is omitted from the averaging because the enthalpy of fusion determined in the investigation is two times larger than the correct value. In addition to the values above, Rau [74RAU] derived $\Delta_f H_m^\circ$ (Se_8, g, 298.15 K) = 169.410 $kJ \cdot mol^{-1}$ by fitting thermochemical data to total pressure measurements. However, $Se_8(g)$ is a minor species in the whole temperature range and the results obtained in this way are therefore very uncertain for this species. The review selects the weighted average of the second and the third law values:

$$\Delta_f H_m^\circ (Se_8, g, 298.15\ K) = (156.8 \pm 3.0)\ kJ \cdot mol^{-1}$$

and

$$\Delta_f G_m^\circ (Se_8, g, 298.15\ K) = (100.3 \pm 3.6)\ kJ \cdot mol^{-1}.$$

Table V-21: Re-evaluated second and third law values of the enthalpy of formation of $Se_8(g)$ at 298.15 K from mass spectrometric investigations.

Reference	$\Delta_f H_m^\circ$ (298.15 K), 2nd law ($kJ \cdot mol^{-1}$)	$\Delta_f H_m^\circ$ (298.15 K), 3rd law ($kJ \cdot mol^{-1}$)	Temperature (K)
[66BER/CHU]	(169.8 ± 2.2)	(157.0 ± 1.2)	437 – 469, 472 – 606[(a)]
[66BER/CHU]	(147.8 ± 6.6)	(157.0 ± 1.6)	505 – 546, 472 – 606[(b)]
[66FUJ/WES]	(171.8 ± 5.7)	(155.8 ± 5.7)	405 – 438
[68YAM/POR]	(156.8 ± 8.4)	(155.6 ± 8.4)	578 – 667
[70KEL]	(161.7 ± 7.7)	(155.4 ± 2.3)	498 – 573
[75HOA/REY]	(119.8 ± 8.4)[(c)]		465 – 571
[82GRI/GRI]	(150.9 ± 4.1)	(158.3 ± 4.1), (159.4 ± 4.1)	463 – 483
[84HUA/GIL]	(162.4 ± 3.3)	(154.9 ± 5.8)	485 – 503
Average	(160.2 ± 17.6)	(156.7 ± 3.1)	

a The first temperature range pertains to the equilibrium 6Se(trigonal) $\rightleftharpoons$ $Se_6(g)$ and the second temperature interval to the equilibrium $8Se_6(g) \rightleftharpoons 6Se_8(g)$.

b The first temperature range pertains to the equilibrium 6Se(l) $\rightleftharpoons$ $Se_6(g)$ and the second temperature interval to the equilibrium $8Se_6(g) \rightleftharpoons 6Se_8(g)$.

c Not included in the average.

V.1.12 $Se_9(g)$

The ion $Se_9^+(g)$ has been observed in the mass spectrometric investigations by Goldfinger and Jeunehomme [63GOL/JEU], Fujisaki and Westmore [66FUJ/WES], Yamdagni and Porter [68YAM/POR] and Hoareau, Reymond, Cabaud and Uzan [75HOA/REY]. The measured signals were large enough for a quantitative determination only in the last investigation and $\Delta_r H_m^\circ$(530 K) = (49.0 ± 8.4) $kJ \cdot mol^{-1}$ was obtained for the reaction $2Se_9(g) \rightleftharpoons 3Se_6(g)$. Considering the low concentrations measured and the unreliable values obtained for other selenium species in the investigation, the enthalpy of reaction is not selected for a calculation of the enthalpy of formation of $Se_9(g)$ at 298.15 K.

The entropy and heat capacity of $Se_9(g)$ at 298.15 K were calculated by Steudel [81STE] by analogy with sulphur molecules as discussed in Appendix A. Considering the uncertainties in the molecular structure of the $Se_9(g)$ molecule the proposed values are not selected by this review.

V.1.13 $Se_{10}(g)$

The ion Se_{10}^{+} (g) has been observed in the mass spectrometric investigations by Goldfinger and Jeunehomme [63GOL/JEU], Fujisaki and Westmore [66FUJ/WES], Yamdagni and Porter [68YAM/POR] and Hoareau, Reymond, Cabaud and Uzan [75HOA/REY]. The measured signals were large enough for a quantitative determination only in the last investigation and $\Delta_r H_m^\circ$(530 K) = (33.5 ± 16.8) kJ·mol^{-1} was obtained for the reaction $4Se_{10}(g) \rightleftharpoons 5Se_8(g)$. Considering the low concentrations measured and the unreliable values obtained for the other selenium species in the investigation, the enthalpy of reaction is not selected for a calculation of the enthalpy of formation of $Se_{10}(g)$ at 298.15 K.

The entropy and heat capacity of $Se_{10}(g)$ at 298.15 K were calculated by Steudel [81STE] by analogy with sulphur molecules as discussed in Appendix A. Considering the uncertainties in the molecular structure of the $Se_{10}(g)$ molecule the proposed values are not selected by this review.

V.1.14 $Se_{11}(g)$ and $Se_{12}(g)$

The entropies and heat capacities of $Se_{11}(g)$ and $Se_{12}(g)$ at 298.15 K were calculated by Steudel [81STE] by analogy with sulphur molecules as discussed in Appendix A. The $Se_{11}(g)$ and $Se_{12}(g)$ molecules have not been reported in experimental investigations and the values are not selected by this review.

V.2 Selenium oxides and hydrides

V.2.1 Selenium monoxide

The thermodynamic quantities selected for selenium monoxide, SeO(g), stem from the work by Smoes and Drowart [84SMO/DRO]. Earlier work has been reviewed by Mills [74MIL]. The equilibrium reaction:

$$SeO(g) + SeS(g) \rightleftharpoons Se_2(g) + SO(g) \qquad (V.1)$$

was set up in a Knudsen cell. The partial pressures of the species were assumed to be proportional to the ionisation currents measured by the mass spectrometer. The equilibrium constant was determined in the temperature range 1300 to 1800 K and found to follow the relationship:

$$\log_{10} K^\circ((\text{V.1}), (1300 - 1800)\ \text{K}) = -(0.376 \pm 0.049) + (2753 \pm 77)\ T^{-1}.$$

The main purpose of the work was the determination of the energy of dissociation of SeO(g) by a joint evaluation of the above thermochemical data and spectroscopic information as sketched in Appendix A. The enthalpy and Gibbs functions of SeO(g), SeS(g), Se_2(g), and SO(g) required for the process were calculated by statistical thermodynamic methods in the temperature interval 298.15 to 2000 K. The value adopted from the evaluation by the review is $\Delta_{\text{at}} H^\circ_{\text{m}}$ (SeO, g, 0 K) = (424.7 ± 6.0) kJ·mol^{-1}.

The standard enthalpy of formation of SeO(g) was calculated from the dissociation energy and the auxiliary data in Appendix A. The selected result is:

$$\Delta_{\text{f}} H^\circ_{\text{m}} (\text{SeO, g, 298.15 K}) = (57.4 \pm 6.2)\ \text{kJ·mol}^{-1}.$$

The selected standard entropy of the monoxide was calculated from the molecular parameters in [84SMO/DRO] to be:

$$S^\circ_{\text{m}} (\text{SeO, g, 298.15 K}) = (233.7 \pm 1.0)\ \text{J·K}^{-1}\text{·mol}^{-1}.$$

This value agrees well with the value obtained by Papousek [62PAP], 234.1 J·K^{-1}·mol^{-1}, from a statistical-mechanical calculation. The uncertainty is added here.

The standard entropy and Gibbs energy of formation can now be calculated with selected auxiliary data to be:

$$\Delta_{\text{f}} S^\circ_{\text{m}} (\text{SeO, g, 298.15 K}) = (89.0 \pm 1.1)\ \text{J·K}^{-1}\text{·mol}^{-1},$$

$$\Delta_{\text{f}} G^\circ_{\text{m}} (\text{SeO, g, 298.15 K}) = (30.9 \pm 6.2)\ \text{kJ·mol}^{-1}.$$

The following relationship for the heat capacity was obtained from the molecular parameters in [84SMO/DRO]

$$C^\circ_{p,\text{m}} (\text{SeO, g, (300 - 2000) K}) = (42.06 - 2.757 \times 10^{-3}\ T + 1.201 \times 10^{-6}\ T^2 - 4.113 \times 10^3\ T^{-1} + 4.413 \times 10^5\ T^{-2})\ \text{J·K}^{-1}\text{·mol}^{-1}$$

which yields the selected value:

$$C^\circ_{p,\text{m}} (\text{SeO, g, 298.15 K}) = (32.52 \pm 0.25)\ \text{J·K}^{-1}\text{·mol}^{-1}.$$

The fit to the data is very good but the $C^\circ_{p,\text{m}}$ values are on the average 0.8 J·K^{-1}·mol^{-1} greater than those calculated in [62PAP].

V.2.2 Selenium dioxide

Selenium reacts with oxygen, but not readily, when heated in air forming selenium dioxide. Under standard conditions the compound is a solid, which sublimes.

V.2.2.1 SeO_2(cr)

The low temperature heat capacity of selenium dioxide has been measured by Mal'tsev , Pashinkin, Bakeeva and Zhdanov [68MAL/PAS] between 53.5 and 305 K and by Amitin *et al.* [87AMI/MIN] between 7 and 304 K. The thermodynamic properties of SeO_2(cr) obtained for 298.15 K agree reasonably well, although there are differences between the $C^{\circ}_{p,m}$ values at $T < 100$ K. The results obtained in [87AMI/MIN] will be selected, since the measurements were carried out to lower temperatures and great care exercised in the removal of impurities, in particular moisture, from the selenium dioxide specimen:

$$C^{\circ}_{p,m}(SeO_2, cr, 298.15\ K) = (58.23 \pm 0.18)\ J{\cdot}K^{-1}{\cdot}mol^{-1},$$

$$S^{\circ}_{m}(SeO_2, cr, 298.15\ K) = (67.49 \pm 0.40)\ J{\cdot}K^{-1}{\cdot}mol^{-1}.$$

The paper also tabulates the values of the thermodynamic functions of SeO_2(cr) between 0 and 305 K.

The values $C^{\circ}_{p,m}(SeO_2$, cr, 298.15 K) = 58.33 $J{\cdot}K^{-1}{\cdot}mol^{-1}$ and $S^{\circ}_{m}(SeO_2$, cr, 298.15 K) = (66.69 ± 1.67) $J{\cdot}K^{-1}{\cdot}mol^{-1}$ accepted by Mills [74MIL] are those obtained in [68MAL/PAS].

The heat capacity in the range 293 to 413 K was measured by Pashinkin *et al.* [71PAS/ARO]. Their results are presented as $C^{\circ}_{p,m}(SeO_2, cr, T) = (35.94 + 7.53 \times 10^{-2}\ T)\ J{\cdot}K^{-1}{\cdot}mol^{-1}$. The value of $C^{\circ}_{p,m}$ at 298.15 K, 58.39 $J{\cdot}K^{-1}{\cdot}mol^{-1}$, calculated by the equation agrees with the selected value. In [68MAL/PAS] the relationship (slightly changed to yield the selected value at 298.15 K):

$$C^{\circ}_{p,m}(SeO_2, cr, T) = (69.50 + 3.89 \times 10^{-3}\ T - 1.105 \times 10^{6}\ T^{-2})\ J{\cdot}K^{-1}{\cdot}mol^{-1}$$

was derived from the data in the range 290 to 305 K. The equation, which was suggested for extrapolations up to 600 K, was tested and accepted by Behrens, Lemons and Rosenblatt [74BEH/LEM]. It is also accepted and selected by the review.

The enthalpy of formation of SeO_2(cr) is obtained from the three accepted measurements of the heat of combustion of Se(trigonal) listed in Table V-22 (references to earlier experimental results or selected values in compilations can be found in [62GAT]). Selenium reacts only partly with oxygen in the bomb calorimeter unless mixed with a combustion aid. Benzoic acid was used for this purpose and the heat evolved by the combustion aid was about 50% in the static bomb and about 90% in the rotating bomb experiments.

Reference [62GAT] contains the experimental results of Gattow, which were later recalculated by Gattow and Dräger in [66GAT/DRA] and corrected mainly for the formation of H_2SeO_3(cr). Gadzhiev [66GAD] was the first to point out the need for a correction for the enthalpy of the reaction between SeO_2(cr) and H_2O(l) (from the combustion of benzoic acid) with formation of H_2SeO_3(cr). The two references [66GAD] and [66HAJ] refer to the same author and contain essentially the same information. The calorimeter used by Barnes and Mortimer [73BAR/MOR] contained 50 ml of water and thus the heat of the reaction Se(cr) + O_2(g) + 400H_2O → SeO_2(aq, 1:400) was measured, – (221.67 ± 1.88) $kJ \cdot mol^{-1}$. The total enthalpy of dissolution of selenium dioxide in water to the concentration (1:400) was also determined, (3.85 ± 0.08) $kJ \cdot mol^{-1}$. The two data combine to the value in Table V-22.

Table V-22: Accepted experimental determinations of the enthalpy of formation of SeO_2(cr), $\Delta_f H_m^\circ$ (SeO_2, cr, 298.15 K).

Reference	Method	$\Delta_f H_m^\circ$ (SeO_2, cr, 298.15 K) $kJ \cdot mol^{-1}$
[62GAT], [66GAT/DRA]	Static bomb	– (224.01 ± 0.84)
[66GAD], [66HAJ]	Static bomb	– (225.73 ± 0.42)
[73BAR/MOR]	Rotating bomb	– (225.52 ± 1.88)

The review selects the weighted mean, with suitable uncertainty limits:

$$\Delta_f H_m^\circ (SeO_2, \text{cr}, 298.15 \text{ K}) = -(225.39 \pm 0.60) \text{ kJ} \cdot \text{mol}^{-1}.$$

In [92GRE/FUG], $\Delta_f H_m^\circ$ (SeO_2, cr, 298.15 K) = – (225.1 ± 2.1) $kJ \cdot mol^{-1}$ was adopted from Mills [74MIL]. With the selected data the Gibbs energy of formation is found to be:

$$\Delta_f G_m^\circ (SeO_2, \text{cr}, 298.15 \text{ K}) = -(171.80 \pm 0.62) \text{ kJ} \cdot \text{mol}^{-1}.$$

The solubility of selenium dioxide in water will be treated in Section V.3.2 as H_2SeO_3(cr) is the stable solid phase in contact with water below 343 K.

V.2.2.2 SeO_2(l)

No heat capacity measurements of liquid selenium dioxide are available.

Makatun and Pechkovskii [70MAK/PEC] measured the vapour pressure of SeO_2(g) over supercooled SeO_2(l) (SeO_2(cr) sublimes) and SeO_2(cr) in the temperature interval 530 to 660 K. No primary data are available and the results were represented by the relationships:

$$\log_{10} p\,(SeO_2, \text{s}, (530 - 660) \text{ K}) = 9.635 - 5740\, T^{-1}$$

and

$$\log_{10} p\,(SeO_2, \text{l}, (530 - 660) \text{ K}) = 7.635 - 4420\, T^{-1}.$$

The calculated triple point temperature is 660 K, which is close to the visually observed melting point of (662.5 ± 0.5) K in the same reference. The triple point pressure is evaluated to be (9.0 ± 0.3) bar. Pashinkin *et al.* [71PAS/ARO] observed a melting point of (657 ± 10) K. The melting point at 613 K suggested in [74MIL] hence appears to be in error.

The enthalpy changes obtained from the variation of the vapour pressure with temperature are $\Delta_{sub}H^{o}_{m}$ (SeO_2, cr, 600 K) = (109.90 ± 1.50) $kJ\cdot mol^{-1}$ and $\Delta_{vap}H^{o}_{m}$ (SeO_2, l, 600 K) = (84.60 ± 1.50) $kJ\cdot mol^{-1}$ with the uncertainties added by the review. The enthalpy of fusion obtained from these data is (25.3 ± 2.1) $kJ\cdot mol^{-1}$. From a DTA curve it was determined to be (17.5 ± 6.3) $kJ\cdot mol^{-1}$ in [71PAS/ARO] and 23.0 $kJ\cdot mol^{-1}$ in [82NIS/SOK]. The review accepts the average of these three values of the enthalpy of fusion with a large estimated uncertainty, thus:

$$\Delta_{fus}H^{o}_{m}(SeO_2, cr, 660 K) = (22.0 \pm 5.0) kJ\cdot mol^{-1},$$

$$\Delta_{vap}H^{o}_{m}(SeO_2, l, 600 K) = (84.60 \pm 1.50) kJ\cdot mol^{-1}.$$

There are no data available for the calculation of the thermodynamic properties of SeO_2(l) under standard conditions.

V.2.2.3 SeO_2(g)

The thermodynamic properties of SeO_2(g) have been calculated from molecular properties by Nagarajan [63NAG2], Pashinkin *et al.* [71PAS/ARO], and Behrens *et al.* [74BEH/LEM]. The agreement between the data is fairly good. The data in [74BEH/LEM] will be selected as somewhat inaccurate molecular data were used by Nagarajan [63NAG2] and Pashinkin *et al.* [71PAS/ARO], who report $C^{o}_{p,m}$ (SeO_2, g, 298.15 K) = 42.64 and 43.10 $J\cdot K^{-1}\cdot mol^{-1}$ and S^{o}_{m} (SeO_2, g, 298.15 K) = 265.00 and 265.15 $J\cdot K^{-1}\cdot mol^{-1}$, respectively. The selected values are:

$$C^{o}_{p,m}(SeO_2, g, 298.15 K) = (43.36 \pm 0.15) J\cdot K^{-1}\cdot mol^{-1},$$

$$S^{o}_{m}(SeO_2, g, 298.15 K) = (262.6 \pm 1.3) J\cdot K^{-1}\cdot mol^{-1}.$$

The uncertainties have been estimated by the review.

The following temperature dependence of the heat capacity was derived from the data in [74BEH/LEM]:

$$C^{o}_{p,m}(SeO_2, g, T) = (66.638 - 2.701 \times 10^{-3}\, T - 8.884 \times 10^{3}\, T^{-1} + 6.449 \times 10^{5}\, T^{-2})\ J\cdot K^{-1}\cdot mol^{-1}.$$

The relationship reproduces the data to better than 0.07 $J\cdot K^{-1}\cdot mol^{-1}$ in the temperature range 300 to 1500 K with a mean absolute deviation of 0.04 $J\cdot K^{-1}\cdot mol^{-1}$. This is somewhat better than obtained with the relationship proposed by Brebrick [2000BRE] for the same data.

Solid selenium dioxide sublimes:

$$SeO_2(cr) \rightleftharpoons SeO_2(g) \quad (V.2)$$

and the vapour consists of SeO_2 molecules at least up to 1000 K according to the vapour density measurements by D'yachkova, Vigdorovich, Ustyugov and Kudryavtsev [69DYA/VIG]. Sonin and Polyachenok [71SON/POL] made molecular weight determinations in the presence of Se(l) ($T < 600$ K) or O_2 ($T < 1000$ K) and found no evidence for the formation of SeO or SeO_3, respectively. The thermodynamic data from vapour pressure measurements have all been evaluated on the assumption that SeO_2 is the sole species in the gaseous phase.

The experimental determinations of the saturated vapour pressure over SeO_2(cr) are entered in Table V-23. The results by Gadzhiev, Dobrotin, and Suvorov [65GAD/DOB] have been excluded since they are apparently based on the data by the same authors in [68DOB/SUV]. The values of $\Delta_{sub}H^{\circ}_{m}$ (SeO_2, cr, T) and $\Delta_{sub}S^{\circ}_{m}$ (SeO_2, cr, T) were calculated from the experimental results presented as $\log_{10} p_{SeO_2} = AT^{-1} + B$.

Table V-23: Experimental values of the enthalpy and entropy of sublimation of SeO_2(cr) from vapour pressure measurements over the temperature intervals in column 2.

Reference	Interval (K)	Method	$\Delta_{sub}H^{\circ}_{m}$ (SeO_2, cr, T) (kJ·mol^{-1})	$\Delta_{sub}S^{\circ}_{m}$ (SeO_2, cr, T) (J·K^{-1}·mol^{-1})
[44AME/BEL]	446 – 500	Transp.	106.1	175.2
[62MAR/GET]	403 – 505	Transp.	94.50	150.3
[64GLE/MUL]	420 – 492	Transp.	95.63	155.2
[68DOB/SUV]	473 – 583	Static	92.37	150.5
[69DYA/VIG]	500 – 603	Static	107.1	177.6
[69FIC/THO]	361 – 461	Knudsen	94.86	150.5
[69PIA/MAL]	360 – 450	Knudsen	99.16	–
[69SON/NOV]	473 – 593	Static	109.2	182.6
[70MAK/PEC]	523 – 663	Static	109.9	184.5
[74BEH/LEM]	374 – 427	Knudsen	110.8	182.7
[96KAZ/BOL]	349 – 394	Knudsen	110.9	190.1
[2000BRE]	416 – 505	Static	109.2	182.8

The transpiration experiments used oxygen, air, or water vapour as transporting medium. The effusion experiments used mass spectrometric detection except in [74BEH/LEM] that registered the weight loss with a recording microbalance. A membrane null manometer measured the pressure in the static method except in [2000BRE]. In this work, the pressure was evaluated from a measurement of the optical density of the gaseous phase.

The standard enthalpy of sublimation was estimated by the second law approach from $\Delta_{\text{sub}}H^{\circ}_{\text{m}}$ (SeO_2, cr, T) and $\Delta_{\text{r}}C^{\circ}_{p,\text{m}}$ ((V.2), (300 K < T < 600) K) = – (16.3 ± 1.0) J·K^{-1}·mol^{-1} calculated from the data in [74BEH/LEM]. The latter figure is an average value of a slightly varying $\Delta_{\text{r}}C^{\circ}_{p,\text{m}}$ in the temperature interval. The results are found in Table V-24. No uncertainties have been estimated, as only $\Delta_{\text{sub}}H^{\circ}_{\text{m}}$ (SeO_2, cr, 298.15 K) calculated by the third law approach will be used by the review.

Table V-24: $\Delta_{\text{sub}}H^{\circ}_{\text{m}}$ (SeO_2, cr, 298.15 K) calculated by the second and third, and $\Delta_{\text{sub}}S^{\circ}_{\text{m}}$ (SeO_2, cr, 298.15 K) by the second law approach.

Reference	$\Delta_{\text{sub}}H^{\circ}_{\text{m}}$ (SeO_2) (kJ·mol^{-1})		$\Delta_{\text{sub}}S^{\circ}_{\text{m}}$ (SeO_2) (J·K^{-1}·mol^{-1})
	2nd law	3rd law	2nd law
[44AME/BEL]	109.0	(115.3 ± 0.4)	182.7
[62MAR/GET]	97.08	(114.6 ± 1.4)	157.2
[64GLE/MUL]	98.29	(113.8 ± 0.8)	162.1
[68DOB/SUV]	96.23	115.2	159.8
[69DYA/VIG]	111.3	(115.9 ± 0.3)	187.7
[69FIC/THO]	96.74	(113.1 ± 1.4)[(a)]	155.7
[69PIA/MAL]	101.0	–	–
[69SON/NOV]	113.1	115.1	192.1
[70MAK/PEC]	114.8	114.8	195.7
[74BEH/LEM]	112.5	(115.8 ± 0.4)	187.5
[96KAZ/BOL]	112.1	(112.6 ± 0.7)	193.7
[2000BRE]	111.8	112.7[(b)]	189.8

a Systematic trend

b Calculated in the paper

The data obtained by the third law approach in Table V-24 were calculated from primary vapour pressures, when available, or from the estimated pressure at the mean temperature of the interval, when only the A and B coefficients were published. The Gibbs energy function data were extracted from [74BEH/LEM].

The review selects:

$$\Delta_{\text{sub}}H^{\circ}_{\text{m}}(SeO_2, \text{cr}, 298.15\ \text{K}) = (114.8 \pm 2.5)\ \text{kJ·mol}^{-1}$$

based on all third law data except those in [69FIC/THO].

The value of $\Delta_{\text{sub}}S^{\circ}_{\text{m}}$ (SeO_2, cr, 298.15 K) calculated from the selected values of the entropies of solid and gaseous SeO_2 is:

$$\Delta_{\text{sub}} S^{\circ}_{\text{m}} (\text{SeO}_2, \text{cr}, 298.15 \text{ K}) = (195.1 \pm 0.4) \text{ J}\cdot\text{K}^{-1}\cdot\text{mol}^{-1}.$$

As can be seen from Table V-24 the experimental values are consistently lower than the expected value. Several workers have observed the discrepancy. The great hygroscopicity of selenium dioxide has been suggested as causing errors, particularly in low temperature measurements. This view is also adopted by the review, which selects the value of $\Delta_{\text{sub}} S^{\circ}_{\text{m}}$ (SeO_2, cr, 298.15 K) calculated above. With the selected data one obtains the enthalpy and Gibbs energy of formation of SeO_2(g) as:

$$\Delta_{\text{f}} H^{\circ}_{\text{m}} (\text{SeO}_2, \text{g}, 298.15 \text{ K}) = -(110.6 \pm 2.6) \text{ kJ}\cdot\text{mol}^{-1},$$

$$\Delta_{\text{f}} G^{\circ}_{\text{m}} (\text{SeO}_2, \text{g}, 298.15 \text{ K}) = -(115.2 \pm 2.6) \text{ kJ}\cdot\text{mol}^{-1}.$$

The vapour pressure expression derived from the selected data is:

$$\log_{10} p\,(\text{SeO}_2, \text{g}, T) = 15.893 - 6250.2\ T^{-1} - 1.960 \times \log_{10} T\ .$$

Figure V-3 shows the vapour pressure curve and some experimental points from sources containing primary data.

Figure V-3: Vapour pressure of SeO_2(cr).

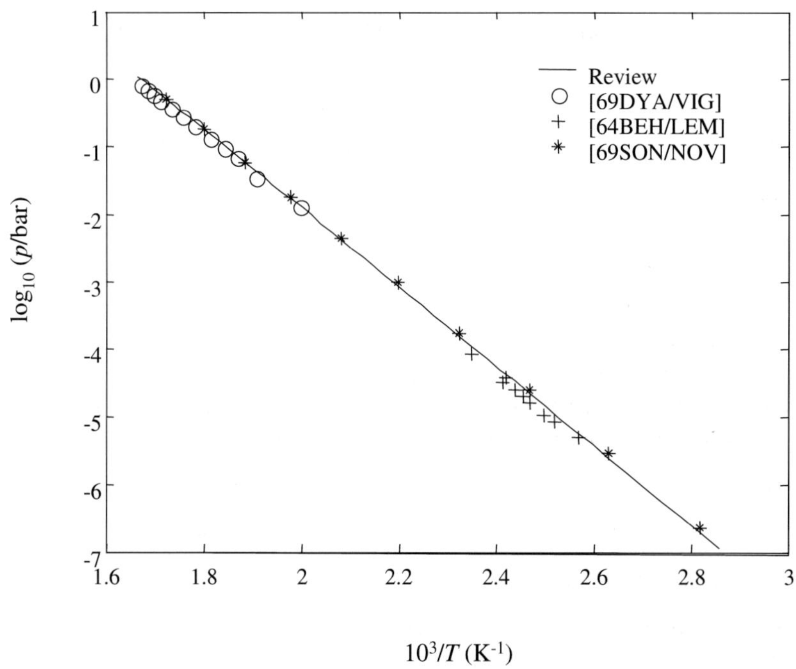

V.2.3 Selenium trioxide

Selenium trioxide cannot be prepared directly in a pure state by reaction between the elements. Dehydration of selenic acid by P_4O_{10} under stringent conditions [51TOU/DOS] and reaction of sulphur trioxide with potassium selenate [52LEH/KRU], followed by vacuum distillation, are two preferred methods of preparation. It is an extremely reactive solid at room temperature.

V.2.3.1 SeO_3(cr)

The entropy of SeO_3(cr) was estimated by Hart [52HAR] to be S°_{m} (SeO_3, cr, 298.15 K) = 84.1 $J\cdot K^{-1}\cdot mol^{-1}$ from an extrapolation within the periodic table. This datum will be given zero weight.

Kondrat'ev, Strizhov, and Doronin [84KON/STR] measured the heat capacity of SeO_3(cr) in the temperature range 55 to 307 K and at 12 and 46 K. Extrapolation to 0 K was made by a graphical method, but no details of the extrapolation procedure are presented. The results were:

$$C^{\circ}_{p,m}(\mathrm{SeO_3, cr, 298.15\ K}) = (77.24 \pm 0.79)\ \mathrm{J\cdot K^{-1}\cdot mol^{-1}},$$

$$S^{\circ}_{m}(\mathrm{SeO_3, cr, 298.15\ K}) = (91.7 \pm 1.0)\ \mathrm{J\cdot K^{-1}\cdot mol^{-1}}.$$

The uncertainties have been added by the review as twice the estimated relative standard deviations indicated by the authors. No other low temperature studies of SeO_3(cr) have been found and the above results are selected.

Trzil [58TRZ] calculated from experimental mean heat capacity measurements by Tuma [57TUM] (original data not available) in the temperature range 293 to 373 K:

$$C^{\circ}_{p,m}(\mathrm{SeO_3, cr, (293 - 373)\ K}) = (14.60 + 0.205\ T)\ \mathrm{J\cdot K^{-1}\cdot mol^{-1}}.$$

The estimated uncertainty is $\pm 5\%$. The value at 298.15 K is then (75.72 ± 3.80) $J\cdot K^{-1}\cdot mol^{-1}$, which agrees within the error limits with [84KON/STR]. The temperature derivative of the heat capacity calculated from [84KON/STR] is 0.19 $J\cdot K^{-2}\cdot mol^{-1}$ at 295 K. The relationship:

$$C^{\circ}_{p,m}(\mathrm{SeO_3, cr}, T) = (77.24 + 0.20 \times (T - 298))\ \mathrm{J\cdot K^{-1}\cdot mol^{-1}}$$

will be selected as an estimate of $C^{\circ}_{p,m}$ ($\pm$ 5%) in the range 300 to 390 K (melting point).

The enthalpy of formation of SeO_3(cr) can not be determined directly. It is found from the enthalpy of formation of selenic acid and the enthalpy of dissolution of SeO_3(cr). The latter quantity was determined for a final dilution of (aq, 1:2000) by Tuma, cited in [58TRZ], to be $-(143.89 \pm 0.80)$ $kJ\cdot mol^{-1}$ and by Jerschkewitz and Menning [62JER/MEN] to be $-(134.52 \pm 0.80)$ $kJ\cdot mol^{-1}$. The two estimates differ significantly, which may be caused by experimental difficulties from the extreme hygroscopicity of SeO_3(cr). Moisture content of 1% diminishes the enthalpy of dissolution by about 5 $kJ\cdot mol^{-1}$ according to [58TRZ]. The review adopts $-(143.89 \pm 0.80)$ $kJ\cdot mol^{-1}$

for this enthalpy of dissolution, since the specimen used by Tuma is judged to be of better quality.

The enthalpy of formation of H_2SeO_4(aq, 1:1200) is evaluated in Section V.3.3.1 to be – (590.80 ± 1.70) $kJ\cdot mol^{-1}$. The corresponding value at (aq, 1:2000) is – (592.80 ± 2.00) $kJ\cdot mol^{-1}$ obtained with heats of dilution of selenic acid presented in [58TRZ]. The data combined with the selected enthalpy of formation of H_2O(l) lead to the selected value:

$$\Delta_f H_m^\circ (\mathrm{SeO_3, cr, 298.15\ K}) = -(163.1 \pm 2.2)\ \mathrm{kJ\cdot mol^{-1}}.$$

This result is reasonably coherent with the value – 166.9 $kJ\cdot mol^{-1}$ in [82WAG/EVA] but not the value selected by [74MIL], – 170.3 $kJ\cdot mol^{-1}$, largely based on the original data of Thomsen [1882THO]. Mijlhoff [63MIJ] calculated the standard enthalpy of formation of SeO_3(cr) essentially using data from the same reference and arrived at – 184.5 $kJ\cdot mol^{-1}$. This result will be ignored, *cf.* Appendix A.

The selected data yield:

$$\Delta_f G_m^\circ (\mathrm{SeO_3, cr, 298.15\ K}) = -(86.2 \pm 2.2)\ \mathrm{kJ\cdot mol^{-1}}.$$

V.2.3.2 SeO_3(l)

Selenium trioxide melts at (392.5 ± 1.5) K, [62JER/MEN], [63MIJ/GER], to a colourless liquid. The decomposition of SeO_3(l) on heating was studied by Jerschkewitz and Menning [62JER/MEN] and found to be dependent on the purity of the specimen. The presence of a selenate salt, selenium dioxide, or water greatly enhances the rate of the decomposition reaction:

$$2\mathrm{SeO_3(l)} \rightarrow \mathrm{Se_2O_5(s)} + 0.5\mathrm{O_2(g)} \qquad \text{(V.3)}$$

The stability range of SeO_3(l) is somewhat difficult to assess but an upper limit of 425 to 435 K is indicated by the decomposition curves in [62JER/MEN] although vapour pressure measurements of SeO_3(g) over SeO_3(l) are reported up to 510 K, Section V.2.3.3.

Mijlhoff and Gerding [63MIJ/GER] estimated the enthalpy of fusion of SeO_3(cr) from a comparison of the time taken to melt a specimen with the time taken for two other substances with well-known heats of fusion and obtained:

$$\Delta_{fus} H_m^\circ (\mathrm{SeO_3, cr, 394\ K}) = (7.10 \pm 1.00)\ \mathrm{kJ\cdot mol^{-1}}.$$

No other thermodynamic datum has been found for SeO_3(l).

V.2.3.3 SeO_3(g)

Selenium trioxide forms tetramers in the solid state according to the crystal structure determination by Mijlhoff [65MIJ]. The same author [65MIJ2] inferred from electron diffraction studies that the tetramers persisted in the gaseous state perhaps with about

30% of the monomer present. The temperature of the investigation was 393 K. This proposal was supported by infrared absorption measurements on gaseous SeO_3 at 413 K by Paetzold and Amoulong [65PAE/AMO] as far as these measurements gave no evidence for the presence of a monomer. Ficalora, Thompson, and Margrave [69FIC/THO] observed no polymers of SeO_3 in their mass spectroscopic measurements. Since the position of the equilibrium between a tetramer and a monomer is highly pressure dependent, a polymer might dissociate under the experimental conditions in the mass spectrometer. Brassington, Edwards, and Fawcett [87BRA/EDW] measured the vibrational Raman spectra of selenium trioxide in the solid and gaseous state. The data confirmed the $(SeO_3)_4$ tetramer in the solid state. The much simpler spectrum of the gas at 410 K indicated that selenium trioxide in the vapour phase is essentially monomeric. Brisdon and Ogden [87BRI/OGD] carried out matrix isolation IR spectroscopic and mass spectroscopic studies of the vapour above heated selenium trioxide in the temperature interval 380 to 410 K. The experimental results were interpreted to show that $(SeO_3)_4$ was predominantly present and the degree of dissociation was estimated to be 2 to 3% at about 390 K. Higher temperatures did not significantly increase the proportion of the monomer, but instead resulted in extensive decomposition.

Since the review can not conclude on the nature of the selenium(VI) oxide species in the gaseous phase, the notation SeO_3(vap) will be used.

The sources of experimental data for Reactions ((V.4) and (V.5)):

$$SeO_3(cr) \rightleftharpoons SeO_3(vap) \qquad (V.4)$$

$$SeO_3(l) \rightleftharpoons SeO_3(vap) \qquad (V.5)$$

are listed in Table V-25. The values of $\Delta_r H_m^\circ$ and $\Delta_r S_m^\circ$ were derived from the A and B coefficients of the relationship $\log_{10} p_{SeO_3} = AT^{-1} + B$. No error limits are presented in the table since the systematic errors are evidently much larger than the random errors. The primary experimental results for Reaction (V.4) are shown in Figure V-4.

The result obtained by Mijlhoff and Block [64MIJ/BLO] is discussed in Appendix A and the data in the table originate from this discussion. The experimental difficulties in the determination of these equilibria are illustrated in the paper by Bleijerveld and Mijlhoff [75BLE/MIJ]. They repeated the previous experiments in the same apparatus. Different pressures were obtained when the sample was heated to or cooled to the set temperature. An irreversible decomposition of SeO_3(vap) with formation of oxygen was suggested. The correction for the partial pressure of oxygen was found from the residual pressure at temperatures below 273 K. The tacit assumption was made that the highest temperature of the experiment determined the amount of oxygen formed and only pressures measured after decrease of temperature were used in the evaluation of the experiment. The correction for the presence of oxygen is substantial and amounts from 80 to 40% of the total pressure measured. The mass spectrometric data by Ficalora *et al.* [69FIC/THO] yield a smooth $\log_{10} p_{SeO_3}$ *versus* $1/T$ curve, but the pressures are smaller by a factor of almost ten compared with the static data.

Table V-25: Thermodynamic data derived from vapour pressure measurements of SeO_3(cr) and SeO_3(l).

Reference	State	Method	Range (K)	$T_{average}$ (K)	$\Delta_r H^\circ_m$ (kJ·mol^{-1})	$\Delta_r S^\circ_m$ (J·K^{-1}·mol^{-1})
[64MIJ/BLO]	s	static	350 – 390	370	53.3	92.7
[69FIC/THO]	s	dynamic	302 – 365	334	40.7	42
[75BLE/MIJ]	s	static	370 – 393	382	74.0	142
[64MIJ/BLO]	l	static	390 – 405	398	55.3	99.3
[66GAD/DOB]	l	static	400 – 453	426	49.1	83
[72SID/KON]	l	boil. pt.	450 – 510	480	76.6	147
[75BLE/MIJ]	l	static	393 – 413	403	64.3	118

The available data for Reaction (V.5) are also entered in Table V-25. Results obtained by Gadzhiev *et al.* and Dobrotin *et al.* from measurements with a membrane null manometer are entered only once as [65GAD/DOB], [66GAD/DOB], [68DOB/SUV], and [69GAD/SAT] are apparently based on the same experimental material. Sidorova *et al.* [72SID/KON] determined the boiling point of SeO_3(l) from thermograms obtained at pressures between 0.067 to 0.53 bar. Although the method is anticipated to yield somewhat imprecise results, a well-behaved plot of $\log_{10} p_{SeO_3}$ *versus* $1/T$ was obtained.

The scatter of the data leads the review to the conclusion that no thermodynamic quantities can be selected with confidence for Reactions (V.4) and (V.5). Although consistency can be found between certain sets of enthalpy data, for instance, between the data below 380 K in [64MIJ/BLO] and the mass spectroscopic results, this consistency is lost when vapour pressures are compared. Hence no data selection will be made for gaseous selenium trioxide by the review.

Figure V-4: Vapour pressure of $SeO_3(g)$ over $SeO_3(cr)$.

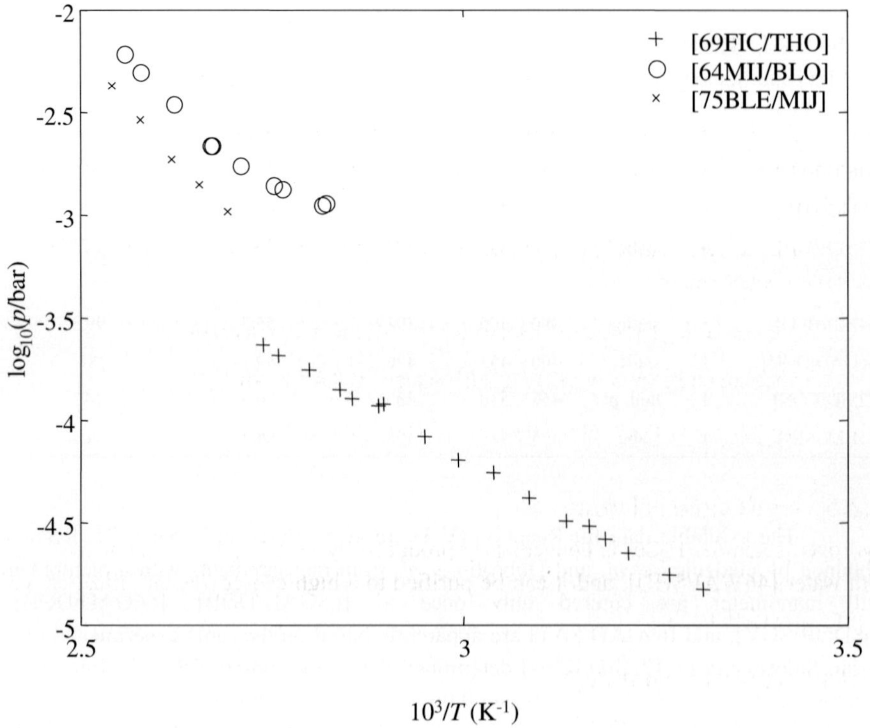

V.2.4 Diselenium pentoxide

Diselenium pentoxide can be prepared by thermal decomposition of selenium trioxide at about 450 K, Reaction (V.3). It is a white, hygroscopic solid that starts to decompose slowly above 455 K according to:

$$Se_2O_5(cr) \rightarrow 2SeO_2(cr) + 0.5O_2(g). \quad (V.6)$$

The compound melts at (497 ± 1) K under violent disintegration, [62JER/MEN], [72SID/KON]. The crystal structure of Se_2O_5 was determined by Zak [80ZAK]. The structure is composed of alternating SeO and SeO_2 units joined by O bridges into infinite chains, $[–Se^{IV}(O)–O–Se^{VI}(O_2)–O–]_n$.

Jerschkewitz and Menning [62JER/MEN] measured the enthalpy of dissolution of $Se_2O_5(cr)$ in water to a concentration of 0.018 M to be – (114.2 ± 0.4) $kJ \cdot mol^{-1}$. This datum has been combined with the enthalpies of dissolution to the same concentration of $SeO_2(cr)$, (5.06 ± 0.24) $kJ \cdot mol^{-1}$, found in [61SEL/PAK] and of $SeO_3(cr)$,

– (145.4 ± 1.1) kJ·mol^{-1}, calculated from the data in [58TRZ] and [62JER/MEN] (see Section V.2.3.1) to give the enthalpy change of the reaction:

$$SeO_2(cr) + SeO_3(cr) \rightleftharpoons Se_2O_5(cr) \quad (V.7)$$

The result is:

$$\Delta_r H^\circ_m((V.7), 298.15\text{ K}) = -(26.1 \pm 1.2)\text{ kJ·mol}^{-1}.$$

Changes in $\Delta_r H^\circ_m$((V.7)) between 293 and 298 K have been neglected and a correction, calculated to be (0.60 ± 0.20) kJ·mol^{-1}, for different degrees of protolysis has been applied.

The selected enthalpies of formation of SeO_2(cr) and SeO_3(cr), and $\Delta_r H^\circ_m$(V.7) give the selected value:

$$\Delta_f H^\circ_m(Se_2O_5, \text{cr}, 298.15\text{ K}) = -(414.6 \pm 2.6)\text{ kJ·mol}^{-1}.$$

The value reported in [82WAG/EVA] is – 408.4 kJ·mol^{-1}.

No other thermochemical data have been found for Se_2O_5.

V.2.5 Hydrogen selenide

Hydrogen selenide, H_2Se, is conveniently produced by reacting an excess of Al_2Se_3(s) with water [46WAI/SHU], and it can be purified to a high degree via solidification and sublimation [37KRU/CLU].

V.2.5.1 H_2Se(cr) and H_2Se(l)

The low-temperature properties of H_2Se were studied by Kruis and Clusius [37KRU/CLU] who found that the solid phase was characterised by two phase transitions at 80 to 87 K and at 172.5 K, respectively. The temperature span of the lower transition is indicative of a higher order transition, and the higher transition takes place between two isotropic phases. The heats of transformation for the two transitions are (1.574 ± 0.002) and (1.116 ± 0.002) kJ·mol^{-1}, respectively. H_2Se(cr) melts at (207.430 ± 0.018) K with an enthalpy of fusion of:

$$\Delta_{fus} H^\circ_m(H_2Se, \text{cr}, 207.43\text{ K}) = (2.515 \pm 0.002)\text{ kJ·mol}^{-1}.$$

The triple point pressure is 0.270 bar. The boiling point of H_2Se at 1 bar is 231 to 232 K and the critical point is at 410 to 411 K and 91 bar.

V.2.5.2 H_2Se(g)

The heat capacity expression

$$C^\circ_{p,m}(H_2Se, \text{g}, (298.15 - 1500)\text{ K}) = (23.203 + 3.23156 \times 10^{-2}\,T - 8.8922 \times 10^{-6}\,T^2 + 3.055 \times 10^{2}\,T^{-1} + 1.46472 \times 10^{5}\,T^{-2})\text{ J·K}^{-1}\text{·mol}^{-1}$$

was derived by the review from molecular parameters in [84LAN/EDW]. This expression leads to the selected value:

$$C^{\circ}_{p,\mathrm{m}}(\mathrm{H_2Se, g, 298.15\ K}) = (34.7 \pm 0.1)\ \mathrm{J\cdot K^{-1}\cdot mol^{-1}}$$

which is in harmony with the values presented by Gordon [69GOR] and Mills [74MIL], 34.75 $\mathrm{J\cdot K^{-1}\cdot mol^{-1}}$ and 34.65 $\mathrm{J\cdot K^{-1}\cdot mol^{-1}}$, respectively.

Equilibria characterising the formation of $H_2Se(g)$ at higher temperatures have been studied utilising liquid selenium [1898PEL], [1899BOD], [66RAW/TOG], [69UST/VIG3], [69YUS/DEV] as well as various selenide compounds [67RAU], [69FLO], [72FLO] in combination with hydrogen gas.

For the reaction:

$$\mathrm{H_2(g) + Se(l) \rightleftharpoons H_2Se(g)} \qquad (V.8)$$

Rawling and Toguri [66RAW/TOG] (see Appendix A) presented an expression for the temperature dependent equilibrium constant:

$\log_{10} K^{\circ}((V.8), T) = (13.07 + 2.97 \times 10^{-4}\ T - 3.69 \log_{10} T - 2209\ T^{-1} - 4.81 \times 10^{3}\ T^{-2})$.

This expression also closely describes the data in [1899BOD] and [69YUS/DEV].

The experimental data presented by Rawling and Toguri in the temperature range of 800 to 900 K have been analysed by the third-law method using the selected heat capacity functions for $H_2Se(g)$, $H_2(g)$ and Se(cr, l). The derived enthalpy of formation for $H_2Se(g)$ from this analysis is $\Delta_f H^{\circ}_{m}(\mathrm{H_2Se, g, 298.15\ K}) = (29.5 \pm 1.5)$ $\mathrm{kJ\cdot mol^{-1}}$.

For the reaction:

$$\mathrm{H_2(g) + \tfrac{1}{2}Se_2(g) \rightleftharpoons H_2Se(g)} \qquad (V.9)$$

Flögel [72FLO] (see Appendix A) summarised the various investigations made and reached a "best" expression for the temperature dependent equilibrium constant: $\log_{10} K^{\circ}((V.9), T) = (4.574)^{-1} \times ((10.2 \pm 0.2) \times 10^{3}\ T^{-1} - 8.9 - (2.9 - 1.5 \log_{10}(T/300)) \times (7 \times 10^{-4}\ T - 0.2))$. This equation was covered by experimental data in the temperature range 500 to 1450 K. A corresponding third-law analysis using this expression yielded the value $\Delta_f H^{\circ}_{m}(\mathrm{H_2Se, g, 298.15\ K}) = (28.0 \pm 1.5)\ \mathrm{kJ\cdot mol^{-1}}$.

The standard enthalpy of formation of $H_2Se(g)$ was also directly measured by Gunn [64GUN] by exploding mixtures of $H_2Se(g)$ and $H_2Te(g)$. Depending on the H_2Te:H_2Se ratio, values in the range of 15 $\mathrm{kJ\cdot mol^{-1}}$ to 27 $\mathrm{kJ\cdot mol^{-1}}$ were gained and, on the extrapolation of these data to a ratio of zero, an approximate value of 33 $\mathrm{kJ\cdot mol^{-1}}$ was obtained.

The review selects:

$$\Delta_f H_m^\circ (H_2Se, g, 298.15 K) = (29.0 \pm 2.0) kJ\cdot mol^{-1}.$$

Mills [74MIL] selected the value (29.3 ± 0.8) $kJ\cdot mol^{-1}$. Mayer *et al.* [97MAY/GAL] reached a $\Delta_f H_m^\circ (H_2Se, g, 298.15 K)$ value of 34.1 $kJ\cdot mol^{-1}$ from *ab initio* molecular orbital calculations.

The standard entropy of $H_2Se(g)$ was calculated from molecular parameters in [84LAN/EDW] to yield the selected value:

$$S_m^\circ (H_2Se, g, 298.15 K) = (219.0 \pm 0.1) J\cdot K^{-1}\cdot mol^{-1}.$$

The selected entropy of formation is derived from the calculated entropy value of $H_2Se(g)$ and accepted standard entropy data of $H_2(g)$ and Se(cr):

$$\Delta_f S_m^\circ (H_2Se, g, 298.15 K) = (46.2 \pm 0.2) J\cdot K^{-1}\cdot mol^{-1}.$$

The selected Gibbs energy of formation is calculated from the enthalpy of formation and the entropy:

$$\Delta_f G_m^\circ (H_2Se, g, 298.15 K) = (15.2 \pm 2.0) kJ\cdot mol^{-1}.$$

In [82WAG/EVA], the reported values for $\Delta_f H_m^\circ$ and $\Delta_f G_m^\circ$ are 29.7 and 15.9 $kJ\cdot mol^{-1}$, respectively.

V.3 Selenium oxides and hydrides in aqueous solution. Solid selenious acid and selenic acid.

V.3.1 Hydrogen selenide-selenide system

The temperature dependence of the aqueous solubility of $H_2Se(g)$ was studied by McAmis and Felsing [25AMI/FEL] between 288 to 308 K and by Dubeau, Sisi and Ozanne [71DUB/SIS] between 298 to 343 K. In both studies precipitation of $Ag_2Se(s)$ was used for the analysis of the solution and implies that the sum of the concentrations of $H_2Se(aq)$ and HSe^- was measured. At 298.15 K, and a partial pressure of $H_2Se(g)$ of 1 atm, the solubilities reported are $(84.15 \pm 0.18) \times 10^{-3}$ M [25AMI/FEL] and $(83.36 \pm 0.30) \times 10^{-3}$ M [71DUB/SIS], respectively. From these data, the average value of $(83.8 \pm 0.4) \times 10^{-3}$ M is selected which, with a first dissociation constant for $H_2Se(aq)$ of $10^{-(3.85 \pm 0.05)}$ (see below), corresponds to a $H_2Se(aq)$ concentration of $(80.5 \pm 0.5) \times 10^{-3}$ M.

For the reaction:

$$H_2Se(g) + aq \rightleftharpoons H_2Se(aq) \quad (V.10)$$

and at a partial pressure of 1 bar, this corresponds to an equilibrium constant:

$$\log_{10} K^\circ ((V.10), 298.15 K) = -(1.10 \pm 0.01)$$

and to

$$\Delta_r G^\circ_m((V.10), 298.15\ K) = (6.28 \pm 0.06)\ kJ\cdot mol^{-1}.$$

From the recorded temperature dependence of the aqueous solubility the enthalpy of Reaction (V.10) was calculated to be $\Delta_r H^\circ_m((V.10), 298.15\ K) = -10.17$ $kJ\cdot mol^{-1}$ in [25AMI/FEL] and – 14.74 $kJ\cdot mol^{-1}$ in [71DUB/SIS], respectively. The value reported in [71DUB/SIS] is more credible considering that a much wider temperature range was employed in this study. However, the dissociation of $H_2Se(aq)$ was not taken into account, which adds to the uncertainty in the reaction enthalpy. The review selects the value reported in [71DUB/SIS]:

$$\Delta_r H^\circ_m((V.10), 298.15\ K) = -(14.7 \pm 0.3)\ kJ\cdot mol^{-1}.$$

The uncertainty has been estimated by this review from a comparison with the temperature dependence of the dissociation constant of $H_2S(aq)$.

The first dissociation constant for $H_2Se(aq)$:

$$H_2Se(aq) \rightleftharpoons H^+ + HSe^- \qquad (V.11)$$

has been investigated in two early conductometric studies [13BRU], [23HLA], one potentiometric study [41HAG] in dilute solution, and one indirect solubility study in 1 M $NaClO_4$ medium [70MEH/GUB]. The study by Mehra and Gubeli [70MEH/GUB] was based on data comparing the solubility of $Ag_2Se(s)$ in different regions of pH (*cf.* [71MEH/GUB] in Appendix A) and has, due to its inherent uncertainty, not been further considered. In the two conductometric studies, pure water was saturated with $H_2Se(g)$ at a pressure of 0.1 to 1 atm. The degree of dissociation was obtained from the conductance of the resulting solution and, together with an assumed $H_2Se(aq)$ solubility of 0.1 M at $p_{H_2Se} = 1$ atm, used to calculate K(V.11). With the revised data for the total solubility of $H_2Se(g)$ (see above) these data yield $\log_{10} K^\circ((V.11), 298.15\ K) = -3.85$ and $-(3.82 \pm 0.02)$, respectively, see Appendix A. These data compare relatively well with the value – 3.89 determined by potentiometry in [41HAG]. This review selects an average of the three determinations. It differs little from the value selected in [92GRE/FUG] but is associated with a considerably lower uncertainty:

$$\log_{10} K^\circ((V.11), 298.15\ K) = -(3.85 \pm 0.05).$$

The second dissociation constant of $H_2Se(aq)$

$$HSe^- \rightleftharpoons H^+ + Se^{2-} \qquad (V.12)$$

has been the subject of a series of studies but, due to the sensitivity of Se(– II) towards oxidation and the very weak acidity of HSe^-, many of these results are connected with large uncertainties.

Hagisawa [41HAG] estimated $\log_{10} K\ (V.12) = -11$ from the pH of a dilute solution of NaHSe. This procedure may introduce an uncertainty of several orders of magnitude depending on the purity of the reagents employed. Lingane and Niedrach [48LIN/NIE] studied the half-wave potential of the reaction:

$$HgSe(s) + 2H^+ + 2e^- \rightleftharpoons Hg(l) + H_2Se(aq) \qquad (V.13)$$

and noticed a continuous shift with pH in the whole pH range investigated, which included 1 M NaOH as an upper limit. From this information, they chose $\log_{10} K$ (V.12) = – 14 "as the most probable value". Wood [58WOO] added limited amounts of H_2Se to a solution of KOH and measured the change in pH with a two-point KOH(aq)-calibrated glass electrode. He also determined the solubility of $Na_2Se(s)$ in a series of solutions with known initial concentrations of sodium hydroxide and sodium acetate. Wood evaluated the equilibrium constant of the reaction:

$$Se^{2-} + H_2O(l) \rightleftharpoons HSe^- + OH^- \qquad (V.14)$$

from the first set of data, at I = 0.2 M and 273.6 K, to be K(V.14) = (1.1 ± 0.4). The second set of data, at I = 4.2 M and 295 K, were used to simultaneously determine the solubility product of $Na_2Se(s)$ and K(V.14) = (0.25 ± 0.09). The transformation of these constants to $\log_{10} K^\circ$ ((V.12), 295 K) employed estimated values of both activity coefficients and entropies and yielded the values: – (14.5 ± 0.4) and – (15.9 ± 0.8), respectively. Details of this study are provided in Appendix A.

Mehra and Gubeli [70MEH/GUB] reported $\log_{10} K$ ((V.12), I = 1 M, 298 K) = – 11.6 from a study [68MEH] in which the solubility of $Ag_2Se(s)$ in the pH-range of 4.4 to 10.0 was compared to the corresponding quantity at pH > 12.6 (*cf.* [71MEH/GUB] in Appendix A). As discussed in Section V.10.2.1, however, this approach cannot be accepted and this datum is therefore refuted.

Lyons and Young [86LYO/YOU] measured the UV-spectra of HSe^- and Se^{2-} at high pH and 304 K, and noted that "substantial concentrations of both ions were present in 1 M KOH". This technique was later employed by Levy and Myers [90LEV/MYE], who systematically varied the sodium hydroxide concentration in the test solutions over the range of 0.1 to 11.6 molal. From the absorbance data they evaluated an expression for $\log_{10} K$ ((V.14), 298.15 K) = 0.997 – 0.793 × $m^{1/2}$, where m denotes the molality of sodium hydroxide in the test solution. By combining this equation with activity coefficient functions and $\log_{10} K^\circ_w$ = – 13.997, $\log_{10} K^\circ$ ((V.12), 298.15 K) = – (15.05 ± 0.04) was deduced, see Appendix A. This value is by far the most reliable value of $\log_{10} K^\circ$ (V.12) reported in the literature and it is accepted by this review. However, the uncertainty reported by Levy and Myers is only based on internal data extrapolations and does not consider systematic errors, and is therefore considerably increased by this review:

$$\log_{10} K^\circ ((V.12), 298.15\ K) = -(15.05 \pm 0.20).$$

The enthalpy change on dissolution of $H_2Se(g)$ in 0.25 M lithium, sodium, or potassium hydroxide was measured by Fabre [1887FAB] with concordant results. From these measurements carried out at about 287 K the review estimates in Appendix A the enthalpy of formation of HSe^- to be (14.3 ± 3.2) $kJ \cdot mol^{-1}$. This result has not been selected since a proper correction from 287 K to standard conditions could not be made.

The value in [82WAG/EVA] is 15.9 $kJ \cdot mol^{-1}$, probably estimated from the same experimental data. The temperature dependences of the equilibrium constants presented in [68DYA/KHO] are based on various empirical relations and assumptions and not on experimental data, and they are therefore not accepted by this review.

Based on the selected data for $H_2Se(g)$, Section V.2.5.2, the following thermodynamic quantities for the $H_2Se(aq)$-HSe^--Se^{2-} system are obtained:

$$\Delta_f H_m^\circ (H_2Se, aq, 298.15\ K) = (14.3 \pm 2.0)\ kJ \cdot mol^{-1},$$

$$\Delta_f S_m^\circ (H_2Se, aq, 298.15\ K) = -(24.1 \pm 9.5)\ J \cdot K^{-1} \cdot mol^{-1},$$

$$\Delta_f G_m^\circ (H_2Se, aq, 298.15\ K) = (21.5 \pm 2.0)\ kJ \cdot mol^{-1},$$

$$\Delta_f G_m^\circ (HSe^-, 298.15\ K) = (43.5 \pm 2.0)\ kJ \cdot mol^{-1},$$

$$\Delta_f G_m^\circ (Se^{2-}, 298.15\ K) = (129.4 \pm 2.3)\ kJ \cdot mol^{-1}.$$

The results for $H_2Se(aq)$ and HSe^- are selected. The value of $\Delta_f G_m^\circ (Se^{2-}$, 298.15 K) is compatible with the corresponding value independently evaluated from redox potentials of mono- and polyselenide ions in Section V.3.4.3, (127.9 ± 1.9) $kJ \cdot mol^{-1}$. The finally selected value is chosen as the mean of these values

$$\Delta_f G_m^\circ (Se^{2-}, 298.15\ K) = (128.6 \pm 3.0)\ kJ \cdot mol^{-1}$$

and corresponds to a value of

$$\log_{10} K^\circ ((V.12), 298.15\ K) = -(14.91 \pm 0.20).$$

For $H_2Se(aq)$, the values listed in [82WAG/EVA] of $\Delta_f H_m^\circ$ and $\Delta_f G_m^\circ$ are 19.2 and 22.2 $kJ \cdot mol^{-1}$, respectively, while for HSe^- and Se^{2-}, $\Delta_f G_m^\circ$ values of 44.0 $kJ \cdot mol^{-1}$ and 129.3 $kJ \cdot mol^{-1}$ are reported.

V.3.2 Selenious acid-selenite system

V.3.2.1 $H_2SeO_3(cr)$

The enthalpy of formation of $H_2SeO_3(cr)$ is obtained indirectly. Selivanova and Pakhorukov [61SEL/PAK] measured the enthalpy changes of the reactions:

$$SeO_2(cr) + H_2O(l) + aq \rightleftharpoons H_2SeO_3(aq, 1{:}200) \qquad (V.15)$$

$$\Delta_r H_m^\circ ((V.15), 298.15\ K) = (4.18 \pm 0.17)\ kJ \cdot mol^{-1}$$

and

$$H_2SeO_3(cr) + aq \rightleftharpoons H_2SeO_3(aq, 1{:}200) \qquad (V.16)$$

$$\Delta_r H_m^\circ ((V.16), 298.15\ K) = (17.68 \pm 0.05)\ kJ \cdot mol^{-1}$$

which yield for the enthalpy of hydration of $SeO_2(cr)$ by one mole of $H_2O(l)$ to $H_2SeO_3(cr)$:

$$\Delta_{\text{hyd}}H_{\text{m}}^{\circ}(\text{SeO}_2, \text{cr}, 298.15 \text{ K}) = -(13.50 \pm 0.18) \text{ kJ·mol}^{-1}.$$

Jannek and Meyer [13JAN/MEY] made similar measurements at "room temperature" in the dilution range 1:400 to 1:500 and obtained $\Delta_{\text{hyd}}H_{\text{m}}^{\circ}(\text{SeO}_2, \text{cr}) = -(13.35 \pm 0.25)$ kJ·mol^{-1}. The more recent value by Selivanova and Pakhorukov [61SEL/PAK] is preferred by the review.

With the selected value of the enthalpy of formation of SeO_2(cr) and the enthalpy of formation of H_2O(l) from [89COX/WAG] $\Delta_{\text{f}}H_{\text{m}}^{\circ}(\text{H}_2\text{SeO}_3, \text{cr}, 298.15 \text{ K}) = -(524.72 \pm 0.65)$ kJ·mol^{-1} is obtained. Reference [82WAG/EVA] reports – 524.46 kJ·mol^{-1}. The review selects:

$$\Delta_{\text{f}}H_{\text{m}}^{\circ}(\text{H}_2\text{SeO}_3, \text{cr}, 298.15 \text{ K}) = -(524.72 \pm 0.65) \text{ kJ·mol}^{-1}.$$

No low temperature heat capacity measurements of H_2SeO_3(cr) have been found. An estimated value of the standard entropy $S_{\text{m}}^{\circ}(\text{H}_2\text{SeO}_3, \text{cr}, 298.15 \text{ K}) = (116.0 \pm 4.0)$ J·K^{-1}·mol^{-1} is derived below from measurements of Reaction (V.17).

The equilibrium constant of the dehydration reaction:

$$\text{H}_2\text{SeO}_3(\text{cr}) \rightleftharpoons \text{SeO}_2(\text{cr}) + \text{H}_2\text{O}(\text{g}) \qquad (\text{V.17})$$

has been measured by Jannek and Meyer [13JAN/MEY], Manchot and Ortner [22MAN/ORT], Ishikawa and Abe [38ISH/ABE], and Dobrotin *et al.* [68DOB/SUV]. The last reference does not contain primary data but a graph indicates good agreement between the measured pressures and those in [38ISH/ABE]. The results in [22MAN/ORT] differ widely from the others and will therefore not be considered. The measurements, shown in Figure V-5, indicate a phase transition at 343 K:

$$\text{H}_2\text{SeO}_3(\text{cr}) \rightarrow \text{SeO}_2(\text{cr}) + \text{H}_2\text{O}(\text{l, sat SeO}_2(\text{cr})). \qquad (\text{V.18})$$

In [68DOB/SUV] it was shown that the vapour pressure at 338 K kept constant when the solid phase changed from the composition H_2SeO_3 to SeO_2. Thus formation of polymeric selenious acids is not indicated.

Ishikawa and Abe [38ISH/ABE] used two different equipments, one with a null-point membrane manometer and the other with a directly attached manometer, and report:

$$\log_{10} p_{\text{H}_2\text{O}}((\text{V.17}), T) = -(3510 \pm 40)\, T^{-1} + (8.965 \pm 0.120)$$

respectively,

$$\log_{10} p_{\text{H}_2\text{O}}((\text{V.17}), T) = -(3120 \pm 30)\, T^{-1} + (7.810 \pm 0.085).$$

The uncertainties entered here are based on the statistical errors from the least squares fit to the data. The corresponding enthalpies of reaction, $\Delta_{\text{r}}H_{\text{m}}^{\circ}((\text{V.17}), (300 - 340) \text{ K})$, are (67.02 ± 0.77) and (59.74 ± 0.57) kJ·mol^{-1}, respectively. Reference [68DOB/SUV] reports (66.9 ± 8.4) kJ·mol^{-1}. The expected enthalpy change of Reaction (V.17), calculated from $\Delta_{\text{hyd}}H_{\text{m}}^{\circ}(\text{SeO}_2, \text{cr}, 298.15 \text{ K})$ and the heat of evaporation of water, is (57.51 ± 0.20) kJ·mol^{-1}. The observed values are thus too high and might indicate difficulties with the establishment of equilibrium in the temperature range 300 to 320 K. Ishikawa and Abe [38ISH/ABE] observed reversibility above 313 K, while the other authors found hysteresis effects below the transition point. Jannek and Meyer [13JAN/MEY] suggested that the H_2SeO_3(s) formed on lowering the temperature created a skin on the sample, which prevented true equilibrium to be established.

Figure V-5: Vapour pressure of H_2O above H_2SeO_3(cr) and SeO_2(aq, saturated).

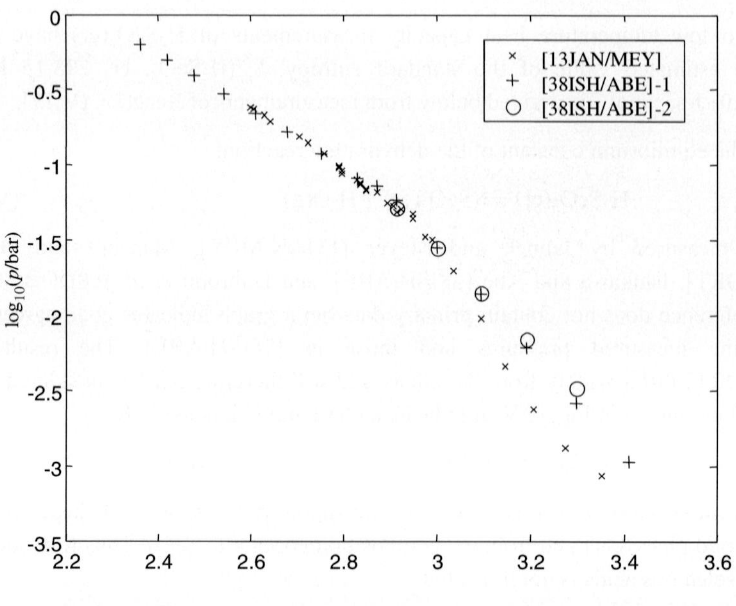

Since the vapour pressure measurements in [38ISH/ABE] and [68DOB/SUV] appear to be consistent an extrapolation to 298.15 K has been made. The vapour pressure relationships yield the Gibbs energy of Reaction (V.17) as (16.03 ± 1.03) and (15.15 ± 0.75) kJ·mol^{-1}, respectively, at this temperature. With the accepted value of the enthalpy of reaction $\Delta_r S^\circ_m$ ((V.17), 298.15 K) becomes (140.6 ± 3.5) J·K^{-1}·mol^{-1}. From this figure S°_m (H_2SeO_3, cr, 298.15 K) = (116.0 ± 4.0) J·K^{-1}·mol^{-1} is obtained. This result is not selected since the assumptions made in its derivation are of uncertain validity.

Extended solubility measurements of H_2SeO_3(cr) in water have been made by Manchot and Ortner [22MAN/ORT], Ishikawa and Abe [38ISH/ABE], and Rustamov and Mardakhaev [63RUS/MAR]. The results are shown in Figure V-6. The data in [22MAN/ORT], [63RUS/MAR] agree reasonably well, as do the data in [38ISH/ABE], [38NEA/CRO]. It has not been possible to find a reason for the large difference between the sets. Sabbah and Perinet [65SAB/PER] and Micka and Ebert [86MIC/EBE2] report a solubility of H_2SeO_3 at 298.15 K of (31.70 ± 0.10) m corresponding to 69.1 mass-% of SeO_2. This result is consistent with the set showing the lower solubility.

Figure V-6: Solubility in the system SeO_2-H_2O.

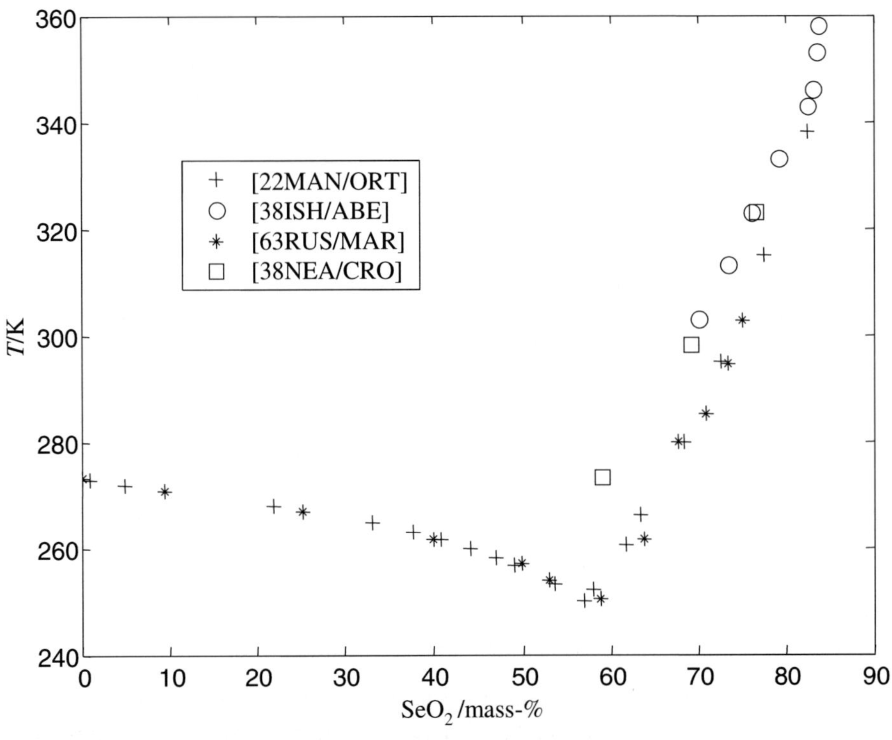

V.3.2.2 H_2SeO_3(aq)

Most work on the determination of the protonation constants of the selenite ion:

$$SeO_3^{2-} + H^+ \rightleftharpoons HSeO_3^- \quad (V.19)$$

$$HSeO_3^- + H^+ \rightleftharpoons H_2SeO_3(aq) \quad (V.20)$$

has been carried out with potentiometric methods. Extensive concentration ranges, up to 1 to 3 M, were studied by Sabbah and Carpeni [66SAB/CAR], and Barcza and Sillén [71BAR/SIL] in various salt backgrounds using the glass electrode. The binuclear species $H_4(SeO_3)_2$, $H_3(SeO_3)_2^-$, $H_2(SeO_3)_2^{2-}$, and $H(SeO_3)_2^{3-}$ were suggested to form in addition to $HSeO_3^-$ and H_2SeO_3. Similar measurements were made by Ganelina, Kuz'micheva, and Krasnopol'skaya [73GAN/KUZ] but without addition of an inert salt background. On the other hand, the measurements by Ozeki, Yagasaki, Ichida and Sasaki [88OZE/YAG] did not indicate the formation of higher complexes. They made a potentiometric study in the ranges $6.9 > -\log_{10}[H^+] > 2$ and 10.6×10^{-3} M < $[Se(IV)]_{tot}$ < 34.5×10^{-3} M. The investigation was carried out in 1 M Na(Cl) and the titration curves did not show any dependence on selenite concentration.

Most determinations of the protonation constants have been carried out at moderate total selenite concentration, say 0.025 M, with or without an added salt background. The constants were evaluated on the assumption that only mononuclear species were formed. A speciation calculation at a total selenite concentration of 0.025 M and the constants presented in [71BAR/SIL] for 1 M $NaClO_4$ shows that the concentrations of binuclear complexes are substantial. Hence the results of the simple evaluation model may be doubted.

The proposed reaction scheme with binuclear proton complexes is supported by the calorimetric titrations by Arnek and Barcza [72ARN/BAR]. Their measurements could not be satisfactorily interpreted unless all six complexes mentioned above were included. Additional support for the presence of binuclear species comes from kinetic investigations of ligand exchange reactions on Co(III) and Rh(III) in [77FOW/STR], [77FOW/STR2], oxidation of selenite to selenate by Np(VII) in [76COO/WOO], and oxygen exchange between selenite ion and water in [73OKU/OKA]. Oganesyan *et al.* [84OGA/KAP] also suggested formation of polymeric species at concentrations above 0.01 M from dialysis experiments. At low concentrations the predominating species at pH 3 would be $Se(OH)_4$, which was proposed to react with H^+ to form $Se(OH)_{4-n}^{n+}$ and finally yield Se^{4+} in 3 M $HClO_4$. This proposal and the derived equilibrium constants are not accepted by the review, since they are in conflict with other evidence.

Walrafen [62WAL2] measured the intensities of the Raman bands at 692 and 898 cm^{-1} of 0.8 to 8.0 M solutions of selenious acid. The intensities were directly proportional to the total concentrations and Walrafen concluded that the acid was present as H_2SeO_3 molecules. In very concentrated solution (> 8 M) shifts in the positions of the bands were noted and formation of $(H_2SeO_3)_2$ was suggested. In a later work, [62WAL], which also comprised studies of molten selenious acid, he summarises the measurements as follows. "The results of Raman studies of aqueous solutions of selenious acid and of molten selenious acid indicate a continuous sequence of species ranging from H_2SeO_3, possibly hydrated in aqueous solution, to $(H_2SeO_3)_2$ and $H_2Se_2O_5$ in concentrated aqueous solution and in the melt, with $(H_2SeO_3)_2$ predominating near 50 mol-% SeO_2 in H_2O and $H_2Se_2O_5$ predominating near 66.7 mol-% in H_2O".

Milne [87MIL] measured quantitative Raman spectra of selenious acid in the concentration range 0.048 to 1.425 M. The band at 890 cm^{-1} clearly shows a peak at 890 cm^{-1} from H_2SeO_3 and a shoulder at 850 cm^{-1} caused by a peak at this frequency from $HSeO_3^-$. The band was decomposed into the contributions from H_2SeO_3 and $HSeO_3^-$. The protonation constant evaluated in this way varied, somewhat irregularly, from $\log_{10} K_{2,1} = 2.12$ to 1.99 with increasing concentration of Se(IV). An approximate analysis has been made here with addition of the dominant binuclear species, $(H_2SeO_3)_2$. A dimerisation constant of 1 to 2 M^{-1} from [71BAR/SIL] and [66SAB/CAR] was used. The calculations resulted in a very pronounced trend in $K_{2,1}$.

Milne [87MIL] also measured 0.38 to 0.95 M solutions of $CsHSeO_3$. The peak intensities of the $HSeO_3^-$ bands at 850 and 610 cm^{-1} were directly proportional to the total concentration. No indication of the formation of $H_2(SeO_3)_2^{2-}$ was found in the concentration range. Simon and Paetzold [60SIM/PAE2] had earlier reached similar conclusions. These authors showed that the band at 540 cm^{-1} due to the Se–O–Se bridge appeared first above a selenite concentration of 3 M.

Baes and Mesmer [76BAE/MES] rejected the dimeric species and suggested that experimental artefacts caused their inclusion in the equilibrium model. The large medium changes made are likely to create uncompensated variations in liquid junction potentials and activity coefficients. This position was also taken in [92GRE/FUG]. The conclusion drawn here from the discussion of the evidence for binuclear species is that the equilibrium constants proposed for these species are most likely too large and need confirmation.

The evaluation of the protonation constants will be made from sources, which based the calculations on Reactions (V.19) and (V.20). The following sources are available [25WIL/PRI], [31BRI/ROB], [32BRI/ROB], [33RUM], [39HAG], [67KAW/MAN], [68SEK/IWA], [69SAL/HAK], [73NAZ/SHI], [73OKU/OKA], [75VES/KOS], [77FOW/STR], [82FRA/ZUM], [88OZE/YAG], and [90DAS/NAR]. Some of the references are only of historical interest and others do not contain enough information on experimental details to be considered. The accepted references are found in Table V-26. The results from a recent extrapolation of data at high ionic strengths to $I = 0$ by Seby *et al.* [2001SEB/POT] have been omitted, see Appendix A, although they support the selection made below.

The extrapolated values are given equal weights and

$$\log_{10} K_1^\circ((\text{V.19}), 298.15\ \text{K}) = (8.36 \pm 0.23),$$

$$\Delta_r G_m^\circ((\text{V.19}), 298.15\ \text{K}) = -(47.72 \pm 1.31)\ \text{kJ·mol}^{-1},$$

$$\log_{10} K_{2,1}^\circ((\text{V.20}), 298.15\ \text{K}) = (2.64 \pm 0.14),$$

$$\Delta_r G_m^\circ((\text{V.20}), 298.15\ \text{K}) = -(15.07 \pm 0.80)\ \text{kJ·mol}^{-1}$$

will be selected. They differ little from those selected in [92GRE/FUG].

The Gibbs energy of formation for $H_2SeO_3(aq)$ obtained in Section V.3.4.1 further yields:

$$\Delta_f G^\circ_m(H_2SeO_3, aq, 298.15 K) = -(425.18 \pm 0.85) kJ\cdot mol^{-1},$$

$$\Delta_f G^\circ_m(HSeO_3^-, 298.15 K) = -(410.1 \pm 1.7) kJ\cdot mol^{-1},$$

$$\Delta_f G^\circ_m(SeO_3^{2-}, 298.15 K) = -(362.4 \pm 1.8) kJ\cdot mol^{-1}.$$

The last value disagrees with NBS [82WAG/EVA].

Table V-26: Accepted protonation constants of SeO_3^{2-} at 298.15 K.

Medium	$\log_{10} K_1$	$\log_{10} K_{2,1}$	Reference
$\to 0$	(8.50 ± 0.04)	(2.75 ± 0.04)	[67KAW/MAN]
0.1M $NaClO_4$	–	2.33	[68SEK/IWA]
$\to 0^{(a)}$	–	2.55	– „ –
$\to 0$	8.25	2.62	[69SAL/HAK]
$\to 0$ (H_2O)	–	(2.648 ± 0.001)	[75VES/KOS]
$\to 0$ (D_2O)	–	(3.015 ± 0.013)	– „ –
1 M $NaClO_4$	(8.06 ± 0.05)	(2.36 ± 0.05)	[77FOW/STR]
1 M NaCl	(8.28 ± 0.04)	(2.36 ± 0.04)	[88OZE/YAG]
0	(8.33 ± 0.12)	(2.64 ± 0.08)	[90DAS/NAR]

a Extrapolation to $I = 0$ made by the review

Thomsen [1882THO] measured the heat of neutralisation of selenious acid. The enthalpy changes $\Delta_r H^\circ_m$((V.19), 298.15 K) = –4.99 and $\Delta_r H^\circ_m$((V.20), 298.15 K) = 6.94 kJ·mol^{-1} have been estimated here from the data. The values of the protonation enthalpies calculated from the enthalpies of formation in [82WAG/EVA] are $\Delta_r H^\circ_m$((V.19), 298.15 K) = – 5.35 and $\Delta_r H^\circ_m$((V.20), 298.15 K) = 7.07 kJ·mol^{-1}, respectively. It is concluded that the NBS data most likely originate from [1882THO].

Arnek and Barcza [72ARN/BAR] made calorimetric measurements on the H^+-SeO_3^{2-} system with 3 M $NaClO_4$, 1 M $NaClO_4$, and 3 M $LiClO_4$ as salt backgrounds. The data from 3 M $LiClO_4$ will not be considered as they were obtained at Se(IV) concentrations, which are too large for a proper determination of the current protonation enthalpies. The values $\Delta_r H^\circ_m$((V.19), 298.15 K) = – 5.27 and – 5.02, and $\Delta_r H^\circ_m$((V.20), 298.15 K) = 5.27 and 6.28 kJ·mol^{-1} were found in 3 M and 1 M medium, respectively.

Vesala and Koskinen [75VES/KOS] measured the second protonation constant at 9 temperatures between 288 and 304 K. The value of the constant shows a minimum at about 294 K and varies by + and – 0.025 logarithmic unit in the temperature interval.

The enthalpy change was estimated from the data to be (3.6 ± 0.6) kJ·mol^{-1} at 298.15 K. The empirical relationship of Harned and Embree [34HAR/EMB] yields (0.85 ± 0.35) kJ·mol^{-1} from the same data. As data from calorimetric measurements are preferred, since they are more accurate generally, the review finds no evidence for a major revision of the protonation enthalpies obtained from the NBS [82WAG/EVA] data.

Selivanova and Pakhorukov [61SEL/PAK] measured the integral heat of dissolution of H_2SeO_3(cr), see Appendix A. The enthalpy change of:

$$H_2SeO_3(cr) + aq \rightarrow H_2SeO_3(aq, 1{:}\infty) \qquad (V.21)$$

was obtained by extrapolation of their data with allowance for protolysis as:

$$\Delta_r H^\circ_m((V.21), 298.15\text{ K}) = (19.40 \pm 0.60)\text{ kJ·mol}^{-1}.$$

The enthalpy of formation of H_2SeO_3(aq) is then calculated as $\Delta_f H^\circ_m(H_2SeO_3$, aq, 298.15 K) $= -(505.32 \pm 0.88)$ kJ·mol^{-1}. The value in [82WAG/EVA] is -507.4 kJ·mol^{-1}.

The enthalpies of formation selected in the H^+-SeO_3^{2-} system from the present evaluation are:

$$\Delta_f H^\circ_m(H_2SeO_3, aq, 298.15\text{ K}) = -(505.32 \pm 0.65)\text{ kJ·mol}^{-1},$$

$$\Delta_f H^\circ_m(HSeO_3^-, 298.15\text{ K}) = -(512.3 \pm 1.0)\text{ kJ·mol}^{-1},$$

$$\Delta_f H^\circ_m(SeO_3^{2-}, 298.15\text{ K}) = -(507.2 \pm 1.1)\text{ kJ·mol}^{-1}.$$

These estimates are then compatible with the following enthalpies of protonation:

$$\Delta_r H^\circ_m((V.19), 298.15\text{ K}) = -(5.2 \pm 0.5)\text{ kJ·mol}^{-1},$$

$$\Delta_r H^\circ_m((V.20), 298.15\text{ K}) = (7.0 \pm 0.5)\text{ kJ·mol}^{-1}.$$

The enthalpy and Gibbs energy of formation of H_2SeO_3(aq) yield:

$$\Delta_f S^\circ_m(H_2SeO_3, aq, 298.15\text{ K}) = -(268.79 \pm 3.66)\text{ J·K}^{-1}\text{·mol}^{-1}$$

and further with auxiliary data from [89COX/WAG] and [92GRE/FUG]:

$$S^\circ_m(H_2SeO_3, aq, 298.15\text{ K}) = (211.7 \pm 3.6)\text{ J·K}^{-1}\text{·mol}^{-1},$$

$$S^\circ_m(HSeO_3^-, 298.15\text{ K}) = (137.7 \pm 5.2)\text{ J·K}^{-1}\text{·mol}^{-1},$$

$$S^\circ_m(SeO_3^{2-}, 298.15\text{ K}) = (5.1 \pm 7.0)\text{ J·K}^{-1}\text{·mol}^{-1}.$$

V.3.3 Selenic acid-selenate system

V.3.3.1 H_2SeO_4(cr)

Thomsen [1882THO] determined the enthalpy change of Reaction (V.22) to be – (125.03 ± 0.40) kJ·mol^{-1},

$$H_2SeO_3(\text{aq, 1:500}) + HClO(\text{aq, 1:700}) \rightarrow H_2SeO_4(\text{aq, 1:1200}) + HCl(\text{aq, 1:1200}). \quad (V.22)$$

The result of this measurement seems to form the basis of later evaluations of the standard enthalpy of formation of selenic acid.

Trzil [58TRZ] calculated $\Delta_f H_m^\circ$ (H_2SeO_4, aq 1:1200) by the thermochemical cycle shown in [58TRZ] in Appendix A to be – 601.65 kJ·mol^{-1}. This cycle has also been employed by the review to obtain this datum. The auxiliary data introduced in the calculation are shown in the same reference. The result is – (590.80 ± 1.67) kJ·mol^{-1}.

The value reported in [82WAG/EVA] is – 587.0 kJ·mol^{-1}. The more negative value calculated by Trzil is mainly caused by an earlier accepted value of the enthalpy of formation of SeO_2(cr), which was about 10 kJ·mol^{-1} too exothermic. The review adopts:

$$\Delta_f H_m^\circ (H_2SeO_4, \text{aq 1:1200, 298.15 K}) = -(590.80 \pm 1.70)\ \text{kJ·mol}^{-1}.$$

The standard enthalpy of formation of H_2SeO_4(cr) is obtained by combining this value with heats of dissolution of the acid.

Tuma, cited in [58TRZ], measured integral heats of dissolution of H_2SeO_4(cr) and obtained (uncertainty added here):

$$\Delta_{dis} H_m^\circ (H_2SeO_4(\text{cr}), \text{aq 1:1200, 293 K}) = -(60.30 \pm 0.80)\ \text{kJ·mol}^{-1}.$$

If the temperature effects between 293 and 298 K are neglected this value results in $\Delta_f H_m^\circ$ (H_2SeO_4, cr, 298.15 K) = – (530.5 ± 1.9) kJ·mol^{-1}. The value in [82WAG/EVA] is – 530.1 kJ·mol^{-1}. The review selects:

$$\Delta_f H_m^\circ (H_2SeO_4, \text{cr, 298.15 K}) = -(530.5 \pm 1.9)\ \text{kJ·mol}^{-1}.$$

No other thermochemical data have been found for H_2SeO_4(cr).

V.3.3.2 H_2SeO_4(aq)

Kremann and Hofmeier [08KRE/HOF], Meyer and Aulich [28MEY/AUL], Kapustinskii and Zdhanova [51KAP/ZHD3], and Vuillard [56VUI], [57VUI] have contributed to the establishment of the phase diagram of the system H_2SeO_4-H_2O. Much of the data were obtained from heating curves due to high viscosity and severe supercooling in certain concentration regions. The phase diagram in Figure V-7 is essentially constructed from data in [56VUI]. Because of the slow formation of, in particular $H_2SeO_4·2H_2O$ and

$H_2SeO_4{\cdot}6H_2O$, metastable equilibria develop. These are not indicated in the phase diagram but contained in Table V-27, which lists the invariant points in the diagram.

Figure V-7: The system H_2SeO_4-H_2O.

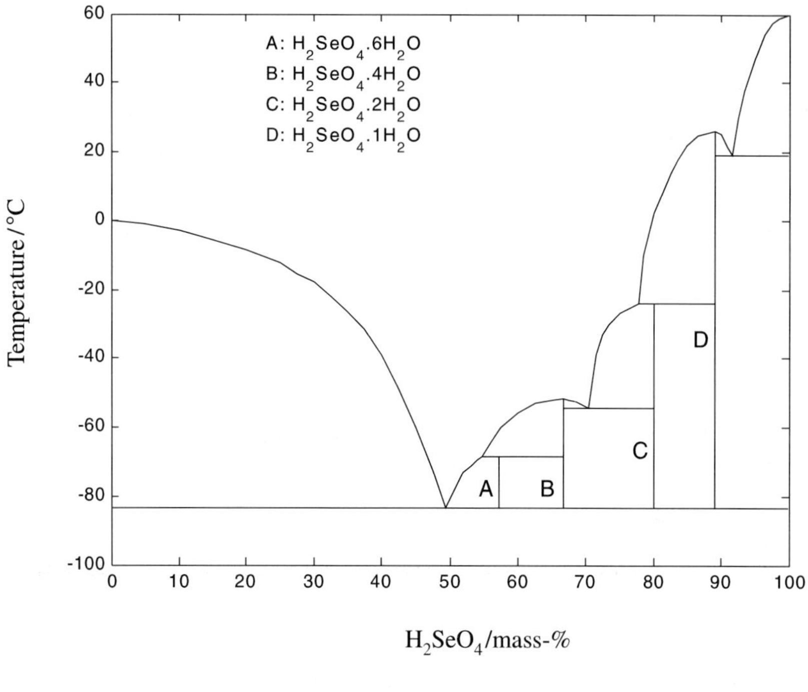

A cryoscopic measurement by Wasif, Nour, and Hussein [74WAS/NOU] resulted in:

$$\Delta_{\text{fus}} H^{\circ}_{\text{m}} (\text{H}_2\text{SeO}_4, \text{cr}, 335.10 \text{ K}) = (14.32 \pm 0.30) \text{ kJ·mol}^{-1}.$$

Dostal [55DOS] has established the phase diagram in the composition range H_2SeO_4 to SeO_3. This region is, however, considered to be outside the scope of the review.

Table V-27: Invariant points in the H_2SeO_4-H_2O phase diagram.

Phases (s)	t (°C)	H_2SeO_4mass-%	Remark
H_2O	0	0	m.p.
$H_2O + H_2SeO_4{\cdot}6H_2O$	– 83.4	49.4	eutectic
$H_2SeO_4{\cdot}6H_2O + H_2SeO_4{\cdot}4H_2O$	– 68.5	54.8	peritectic
$H_2SeO_4{\cdot}4H_2O$	– 51.7	66.8	m.p.
$H_2SeO_4{\cdot}4H_2O + H_2SeO_4{\cdot}2H_2O$	– 54.5	70.3	eutectic
$H_2SeO_4{\cdot}2H_2O + H_2SeO_4{\cdot}H_2O$	– 24	77.7	peritectic
$H_2SeO_4{\cdot}H_2O$	26	89.0	m.p.
$H_2SeO_4{\cdot}H_2O + H_2SeO_4$	19	91.5	eutectic
H_2SeO_4	62.0	100	m.p. [74WAS/NOU]
$H_2O + H_2SeO_4{\cdot}4H_2O$	– 88	50.3	eutectic (metastable)
$H_2SeO_4{\cdot}4H_2O + H_2SeO_4{\cdot}H_2O$	– 58.5	74.1	eutectic (metastable)
$H_2SeO_4{\cdot}6H_2O$	– 68.5	57.3	m.p. (incongruent)
$H_2SeO_4{\cdot}2H_2O$	– 24	80.1	m.p. (incongruent)

Walrafen [63WAL] made a Raman spectroscopic study of the system H_2SeO_4-H_2O at 298.15 K (the solutions at the largest concentrations were supercooled). The species H_2SeO_4, $HSeO_4^-$, and SeO_4^{2-} were identified and their concentrations as a function of the total concentration of selenic acid are shown in Figure V-8. The highest total concentration, about 18 M, corresponds to almost neat H_2SeO_4. Selenic acid is a strong acid and is completely protolysed at total concentrations below 11 to 12 M.

Kapustinskii and Zhdanova [51KAP/ZHD] measured the density of selenic acid solutions between 3 and 98 w-% selenic acid at 298.15 K and Macalpine and Sayce [32ALP/SAY] the aqueous vapour pressure at 298.15 K of solutions containing 0 to 93.5 mass-% H_2SeO_4.

The only protonation reaction of the selenate ion that needs to be considered is

$$SeO_4^{2-} + H^+ \rightleftharpoons HSeO_4^- . \quad (V.23)$$

Pamfilov and Agafonova [50PAM/AGA], Nair [64NAI], and Gosh and Nair [70GHO/NAI] studied Reaction (V.23) by potentiometric measurements of cells without a liquid junction. The test solutions contained low concentrations of the reactants, typically below 0.01 m, and the equilibrium was studied in the temperature interval 273 to 318 K. The Russian workers used a quinhydrone electrode and the other measurements used the hydrogen electrode. Earlier potentiometric work by Sherrill and Lyons [32SHE/LYO], and by Gelbach and King [42GEL/KIN] will be ignored, as experimental artefacts affect the results, see Appendix A.

Figure V-8: Speciation of $H_2SeO_4(aq)$ as a function of total concentration.

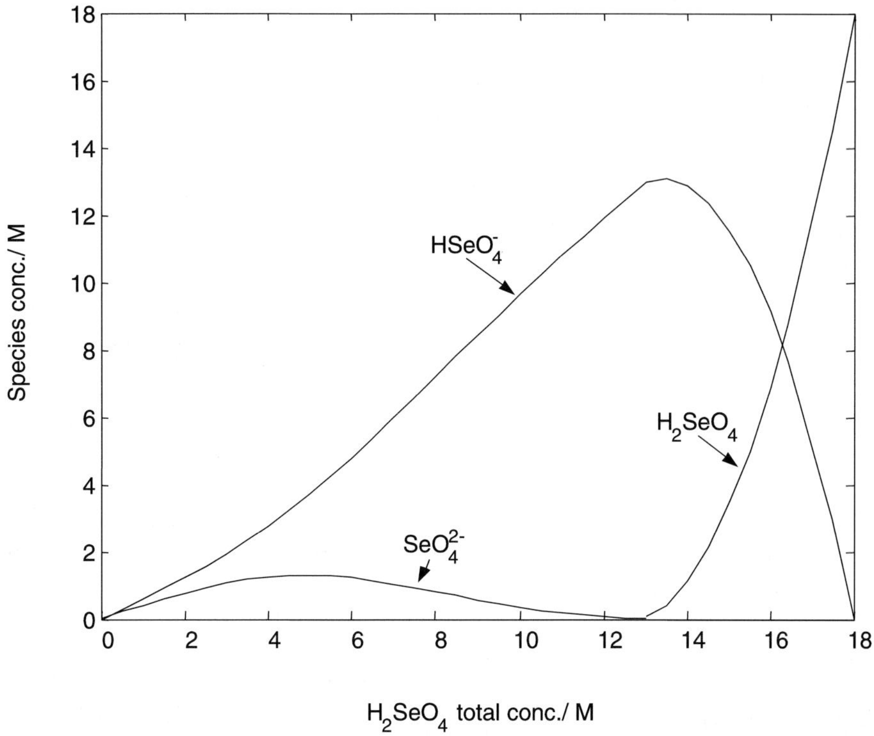

The equilibrium constant K_1° (V.23) is small. The problem is the evaluation of the concentration quotient from a measurement of the hydrogen ion activity rather than the extrapolation of this quotient to $I = 0$. The value obtained for K_1° (V.23) is thus sensitive to the expression chosen for the calculation of the activity coefficients. The data in [64NAI], measured by the cell:

$$H_2, Pt \mid HCl(m_1), Na_2SeO_4(m_2) \mid AgCl, Ag$$

at fifteen compositions with $0.31 < m_1 / m_2 < 1.39$, have been recalculated a number of times. The outcome of the different approaches is summarised in Table V-28 for the measurements at 298.15 K. The SIT calculation made by the review introduced the approximations $\varepsilon(H^+, HSeO_4^-) = \varepsilon(H^+, Cl^-) = 0.12$ kg·mol^{-1}, $\varepsilon(Na^+, HSeO_4^-) = \varepsilon(Na^+, HSO_4^-) = -0.01$ kg·mol^{-1}, and $\varepsilon(Na^+, SeO_4^{2-}) = \varepsilon(Na^+, SO_4^{2-}) = -0.12$ kg·mol^{-1}. The numerical values were extracted from [92GRE/FUG].

Table V-28: Results from recalculations of the data in [64NAI] at 298.15 K with different estimates of the activity coefficients.

Reference	Remark	$\log_{10} K_1^{\circ}$
[64NAI]	Davies equation with – 0.2 I	1.66
[65COV/DOB]	Debye-Hückel with the ion size parameter 1.0	1.70
[65COV/DOB]	Debye-Hückel with the ion size parameter 1.7	1.78
[70GHO/NAI]	Davies equation with – 0.3 I	1.68
[70GHO/NAI]	As above and ion pairing between Na^+ and SeO_4^{2-} (K = 8.1 m^{-1})	1.70
[76BAE/MES]	Brönsted-Guggenheim	1.66
Review	SIT	1.72

Ghosh and Nair [70GHO/NAI] repeated the earlier experiments on 6 test solutions at 298.15 K. The data were evaluated with the approach used in the fifth entry in Table V-28. The result was $\log_{10} K_1^{\circ}$(V.23) = (1.72 ± 0.03) and the same result was obtained here by the SIT model. The main measurements in [70GHO/NAI] were made on 8 test solutions of composition $HCl(m_1) + H_2SeO_4(m_3)$ with $0.050 < m_1 / m_3 < 0.174$. The data were evaluated with the method used in the fourth entry in Table V-28. The result at 298.15 K was $\log_{10} K_1^{\circ}$(V.23) = (1.745 ± 0.019). The SIT approach yielded $\log_{10} K_1^{\circ}$(V.23) = (1.781 ± 0.011).

Pamfilov and Agafonova [50PAM/AGA] report $\log_{10} K_1^{\circ}$(V.23) = 1.88 at 298.15 K. Since the Debye-Hückel limiting law expression was used, the data were recalculated to yield $\log_{10} K_1^{\circ}$(V.23) = (1.92 ± 0.02) by Covington and Dobson [65COV/DOB]. Although the difference between the values of $\log_{10} K_1^{\circ}$(V.23) at 298.15 K in [64NAI] and [50PAM/AGA] is significant, the big difference between the two sources is the temperature dependence of the protonation constant. The enthalpy of Reaction (V.23) calculated by Pamfilov and Agafonova was 7.4 kJ·mol^{-1}. A simplified calculation from the slightly curved $\log_{10} K_1^{\circ}(V.23) = f(1/T)$ graph yields (5.0 ± 0.3) kJ·mol^{-1}. These values are much lower than expected from an analogy with the sulphate ion with $\Delta_r H_m^{\circ}$ = (22.4 ± 1.1) kJ·mol^{-1} at 298.15 K in [92GRE/FUG]. The selected value of $\log_{10} K_1^{\circ}$(V.23) will therefore be based solely on the data by Nair [64NAI], and Gosh and Nair [70GHO/NAI]:

$$\log_{10} K_1^{\circ}((V.23), 298.15\ K) = (1.75 \pm 0.10),$$

$$\Delta_r G_m^{\circ}((V.23), 298.15\ K) = -(9.99 \pm 0.57)\ kJ \cdot mol^{-1}.$$

This value differs slightly from the value (1.80 ± 0.14) selected in [92GRE/FUG].

Treinin and Wilf [70TRE/WIL] made a spectrophotometric determination in UV of $\log_{10} K_1$ (V.23). The value obtained at 293 K was (1.84 ± 0.04). This direct measurement of the concentrations of the selenate species is interesting in view of the previous discussion. However, since the paper contains insufficient information and the experiment was carried out at high ionic strengths, it could not be considered in the selection.

The $\log_{10} K_1^\circ$ (V.23) determinations at nine temperatures between 273 and 318 K made in [64NAI] form a slightly curved $\log_{10} K_1^\circ$ (V.23) = f(1/T) plot. Nair fitted a polynomial of the second order to the data from which $\Delta_r H_m^\circ$((V.23), 298.15 K) = 24.0 kJ·mol^{-1} was obtained. The relationship $\log_{10} K_1^\circ$ (V.23) = ($a + b\,T^{-1} + c \ln T$) was fitted to the data by the review and yielded a slightly better fit resulting in $\Delta_r H_m^\circ$((V.23), 298.15 K) = 24.2 kJ·mol^{-1}. The seven temperatures between 273 and 318 K studied in [70GHO/NAI] yield a linear $\log_{10} K_1^\circ$ (V.23) = f(1/T) plot from which $\Delta_r H_m^\circ$((V.23), 298.15 K) = (24.0 ± 1.5) kJ·mol^{-1} is found. An analogous investigation of the protonation of the sulphate ion by Nair and Nancollas [58NAI/NAN] yielded an enthalpy value of 23.4 kJ·mol^{-1}, which is only slightly higher than the selected value of 22.4 kJ·mol^{-1} from [89COX/WAG]. The determination of $\Delta_r H_m^\circ$((V.23), 298.15 K) from the temperature variation of the protonation constant thus appears to result in a correct value despite the sensitivity of the method to the evaluation model used and experimental errors. On the other hand a value of 17.5 kJ·mol^{-1} is indicated by the data in [82WAG/EVA]. The smaller value is supported by the calorimetric measurements by Thomsen [1882THO]. The review will regard 24.0 kJ·mol^{-1} as an upper limit and 17.5 kJ·mol^{-1} as a lower limit of the enthalpy change and selects the mean of the two values. Thus:

$$\Delta_r H_m^\circ((\text{V.23}), 298.15\text{ K}) = (20.80 \pm 3.20)\text{ kJ·mol}^{-1}.$$

The review has calculated the standard enthalpy of formation of the selenate ion to be – (600.4 ± 3.3) kJ·mol^{-1} from $\Delta_f H_m^\circ$ (H_2SeO_4, aq 1:5000, 298.15 K) = – (591.2 ± 3.3) kJ·mol^{-1} cited in [77COR/OUW2] and an estimate of the heat of dilution to the aqueous standard state of – 9.2 kJ·mol^{-1}. The estimate is based on the selected data for Reaction (V.23). Another and similar estimate was made from the heats of dilution published by Trzil [58TRZ] and the selected value of $\Delta_f H_m^\circ$ (H_2SeO_4, aq 1:1200, 298.15 K). The result is – (605.7 ± 2.8) kJ·mol^{-1}. The weighted average of the two estimates is selected with error bars assigned by the review. The non-referenced standard enthalpy of formation of the selenate ion in [82WAG/EVA] is – 599.1 kJ·mol^{-1}.

The selected thermodynamic data for hydrogen selenate and selenate ions, with $\Delta_f G_m^\circ$ ($HSeO_4^-$, 298.15 K) derived in Section V.3.4.2, would then be:

$$\Delta_f G_m^\circ(\text{HSeO}_4^-, 298.15\text{ K}) = -(449.5 \pm 1.3)\text{ kJ·mol}^{-1},$$

$$\Delta_f H_m^\circ(\text{HSeO}_4^-, 298.15\text{ K}) = -(582.7 \pm 4.7)\text{ kJ·mol}^{-1},$$

$$S_m^\circ(\text{HSeO}_4^-, 298.15\text{ K}) = (136.2 \pm 16.4)\text{ J·K}^{-1}\text{·mol}^{-1},$$

$$\Delta_f G_m^\circ(\text{SeO}_4^{2-}, 298.15\text{ K}) = -(439.5 \pm 1.4)\text{ kJ·mol}^{-1},$$

$$\Delta_f H^\circ_m(\text{SeO}_4^{2-}, 298.15\ \text{K}) = -(603.5 \pm 3.5)\ \text{kJ·mol}^{-1},$$

$$S^\circ_m(\text{SeO}_4^{2-}, 298.15\ \text{K}) = (33.0 \pm 12.7)\ \text{J·K}^{-1}\text{·mol}^{-1}.$$

V.3.4 Redox potentials

Redox equilibria between the various oxidation states of selenium have been little studied. This is apparently a consequence of slow reaction rates. For instance, it has been repeatedly demonstrated that the redox potential measured by a platinum electrode is not affected by the ratio of Se(VI) / Se(IV) present in solution, [87RUN/LIN]. The values of the standard electrode potentials of the important redox couples Se(VI) / Se(IV) and Se(IV) / Se(0) are essentially based on only one experimental investigation each. A more detailed discussion than would normally be required will therefore be made of the two investigations.

V.3.4.1 The redox couple Se(IV)-Se(0)

Schott, Swift, and Yost [28SCH/SWI] measured the equilibrium reaction:

$$\text{Se(s)} + 2\text{I}_2\text{(cr)} + 3\text{H}_2\text{O(l)} \rightleftharpoons \text{H}_2\text{SeO}_3\text{(aq)} + 4\text{H}^+ + 4\text{I}^- \qquad (\text{V.24})$$

at 298.15 K. The conditional equilibrium constant was calculated from a determination of the composition of the test solution by chemical analysis. Corrections for side reactions were made with auxiliary data pertaining to the solubility of iodine, tri-iodide formation, and dissociation of H_2SeO_3. For the calculation of the thermodynamic constant, the activity coefficient was assumed to be equal to the activity coefficient of hydriodic acid at a concentration equal to the ionic strength of the equilibrium mixture. The activity coefficients were obtained from [23PEA/FOR]. The value of $\log_{10} K^\circ(\text{V.24}) = -(13.836 \pm 0.018)$ was evaluated from 5 measurements. The small statistical error is remarkable considering the large exponents in the expression for the equilibrium constant.

The data have been reconsidered by the review and the following observations were made, see also Appendix A. The solid selenium used in the experiments was obtained by reduction of a selenite solution by thiosulphate. The selenium might therefore not be in its standard state. The activity of the specimen is most likely close enough to the standard state activity, however, since the precipitate was kept at boiling temperature for several hours. A recalculation of the side-reactions with more recent values of the auxiliary equilibrium constants made little difference to the result. The analytical data are not always consistent with the stoichiometry of Reaction (V.24) and the known initial composition of the test solution. The authors also observed this and held oxidation of iodide by initially present oxygen responsible for the discrepancies. However, in some instances the deviations from the expected concentrations are remarkably large. The deviations do not invalidate the results if equilibrium prevails, which was tested.

A major uncertainty in $\log_{10} K^\circ$ (V.24) arises from the activity coefficient correction. This correction has been applied under the initial assumption, but with more recent data from Harned and Robinson [41HAR/ROB]. An estimate of the activity coefficient was also made with the SIT approach using $\varepsilon(H^+, ClO_4^-) = 0.12$ kg·mol^{-1} from [92GRE/FUG], and $\varepsilon(H^+, I^-) = 0.18$ and $\varepsilon(K^+, I^-) = 0.014$ kg·mol^{-1} calculated from data in [59ROB/STO]. The results of the calculations were:

$$\log_{10} K^\circ ((\text{V.24}), 298.15\ \text{K}) = -(13.831 \pm 0.021);$$

new auxiliary equilibrium constants, γ from [23PEA/FOR],

$$\log_{10} K^\circ ((\text{V.24}), 298.15\ \text{K}) = -(13.928 \pm 0.016);$$

new auxiliary equilibrium constants, γ from [41HAR/ROB],

$$\log_{10} K^\circ ((\text{V.24}), 298.15\ \text{K}) = -(13.988 \pm 0.045);$$

new auxiliary equilibrium constants, γ from SIT.

The review selects:

$$\log_{10} K^\circ ((\text{V.24}), 298.15\ \text{K}) = -(13.90 \pm 0.10).$$

With the Gibbs energy of formation of I^- from [92GRE/FUG] the Gibbs energy change of the reaction:

$$H_2SeO_3(aq) + 2H_2(g) \rightleftharpoons Se(cr) + 3H_2O(l) \qquad (\text{V.25})$$

$$\Delta_r G_m^\circ ((\text{V.25}), 298.15\ \text{K}) = -(286.24 \pm 0.84)\ \text{kJ·mol}^{-1}$$

is obtained. The standard electrode potential of the redox couple:

$$H_2SeO_3(aq) + 4H^+ + 4e^- \rightleftharpoons Se(cr) + 3H_2O(l) \qquad (\text{V.26})$$

is thus

$$E^\circ((\text{V.26}), 298.15\ \text{K}) = (0.742 \pm 0.002)\ \text{V}$$

which agrees with Latimer's value of 0.740 V, [56LAT].

Osman-Zade and Vagramyan [66OSM/VAG] measured the cell:

$$\text{Hg, } Hg_2Cl_2 \mid 0.1\ \text{M KCl} \mid \text{KCl (sat.)} \mid C\ \text{M } H_2SeO_3 \mid \text{Se, Pt}$$

at 298.15 K for $C = 0.005$, 0.02, 0.07, 0.2, and 0.6 M. The selenium electrode was obtained by plating selenium onto platinum by electrolysis. Their data have been recalculated with the selected value of $\log_{10} K_{1,2}^\circ$ for the selenite ion and an approximate activity coefficient correction. The result is $E^\circ((\text{V.26}), 298.15\ \text{K}) = (0.745 \pm 0.004)$ V, which is almost the same value as reported by the authors but without this correction. The potentiometric measurements thus confirm the value of the standard electrode potential obtained above. A similar study in the alkaline range by Globa, Lomakina, and Sklyarov [86GLO/LOM] seemingly provides additional support, but it is not accepted by the review as valid confirmation, see Appendix A.

The Gibbs energy of formation of $H_2SeO_3(aq)$ was calculated by combining $\Delta_r G_m^o$ ((V.25), 298.15 K) with data for $H_2O(l)$ in [92GRE/FUG] to be:

$$\Delta_f G_m^o (H_2SeO_3, aq, 298.15 K) = -(425.18 \pm 0.85) \text{ kJ·mol}^{-1}.$$

The selected result agrees with [82WAG/EVA].

V.3.4.2 The redox couple Se(VI)-Se(IV)

Sherrill and Izard [28SHE/IZA] studied the reactions:

$$Cl_2(g) + H_2SeO_3(aq) + H_2O(l) \rightleftharpoons 3H^+ + HSeO_4^- + 2Cl^- \qquad (V.27)$$

$$K^\circ(V.27) = \frac{m_{H^+}^3 m_{HSeO_4^-} m_{Cl^-}^2}{p_{Cl_2} m_{H_2SeO_3} a_{H_2O}} \frac{\gamma_{H^+}^3 \gamma_{Cl^-}^2 \gamma_{HSeO_4^-}}{\gamma_{H_2SeO_3}}$$

and

$$Br_2(aq) + H_2SeO_3(aq) + H_2O(l) \rightleftharpoons 3H^+ + HSeO_4^- + 2Br^- \qquad (V.28)$$

$$K^\circ(V.28) = \frac{m_{H^+}^3 m_{HSeO_4^-} m_{Br^-}^2}{m_{Br_2} m_{H_2SeO_3} a_{H_2O}} \frac{\gamma_{H^+}^3 \gamma_{Br^-}^2 \gamma_{HSeO_4^-}}{\gamma_{Br_2} \gamma_{H_2SeO_3}}$$

at 298.15 K. Equilibrium was approached from both directions and the approach to equilibrium took 4 to 7 days for Reaction (V.27) and 50 to 70 days for Reaction (V.28). The former reaction was carried out in a medium of about 7 to 8 m hydrochloric acid and the other reaction in a medium dominated by hydrobromic acid in the concentration range 0.2 to 0.4 m.

The concentrations at equilibrium were found by chemical analysis with the exception of m_{H^+}, which was calculated from the total halide concentration. Corrections were applied for the formation of the Cl_3^- and Br_3^- ions. Formation of chloroselenate(IV) or bromoselenate(IV) species was not considered.

The strongly acid medium used in Reaction (V.27) ensures that selenium (IV) and (VI) are present as H_2SeO_3 and $HSeO_4^-$, respectively. It was assumed that the acidity constants of H_2SeO_3 and $HSeO_4^-$ were approximately equal in the evaluation of K(V.28). Consequently no correction factors were applied for the dissociation of the acids as the factors would cancel. However, this assumption is not valid as the acidity constant of $HSeO_4^-$ is about 8 times larger than for H_2SeO_3. A calculation showed that a correction for the dissociation of $HSeO_4^-$ changes the K(V.28) values by less than 2% because the decrease in the concentration of $HSeO_4^-$ is largely compensated by the increase in the hydrogen ion concentration.

The activity coefficients of uncharged species were assumed to be unity in the evaluation of the equilibrium constants. The activity coefficients of all monovalent ions were set equal and taken as the mean activity coefficient of hydrochloric or hydrobromic acid, respectively, at the concentration of the hydrogen ions in the equilibrium mixture. The water activity was found from vapour pressure data. The values used agree well with later data in [37AKE/TEA] and [36HAR/KES].

The values of the equilibrium constants were:

$$\log_{10} K^\circ((\text{V.27}), 298.15\ \text{K}) = (9.148 \pm 0.057),$$

$$\log_{10} K^\circ((\text{V.28}), 298.15\ \text{K}) = -(0.0560 \pm 0.0256).$$

The change in Gibbs energy of the redox reaction:

$$HSeO_4^- + 3H^+ + 2e^- \rightleftharpoons H_2SeO_3(aq) + H_2O(l) \qquad (\text{V.29})$$

is obtained by combining K°(V.27) and K°(V.28) with $\Delta_r G_m^\circ$ (½Cl_2(g)/Cl^-, 298.15 K) = – (131.22 ± 0.12) kJ·mol^{-1}, respectively, $\Delta_r G_m^\circ$ (½Br_2(aq)/Br^-, 298.15 K) = – (106.30 ± 0.53) kJ·mol^{-1} calculated from [92GRE/FUG]. The result is E°(V.29) = (1.089 ± 0.002) V for hydrochloric acid medium and E°(V.29) = (1.103 ± 0.006) V for hydrobromic acid medium.

The agreement between the two sets of data was perfect in the original paper. However, the calculation for the bromide medium was based on E°(½ Br_2(aq)/Br^-, 298.15 K) = 1.087 V. With the new selected thermodynamic data used here, this datum equals 1.101 V.

The results are difficult to reconcile with the more recent Raman measurements by Milne and co-workers in [79MIL/HAI], [85MIL/HAI], [87MIL]. In solutions of selenious acid in hydrochloric acid of a concentration greater than 4 M, a new band appears which was assigned to the formation of a Se–Cl bond. In their first interpretation of the data, [79MIL/HAI], the concentration of free H_2SeO_3(aq) would decrease by less than 5% in the concentration range 6 to 7 M used in [28SHE/IZA]. The re-interpretation of the data in [87MIL] suggested a much stronger complexation by chloride. The concentrations of H_2SeO_3(aq) would then be much lower and amount to 11.6 and 3.7% of the total Se(IV) concentration in 6 and 7 M HCl, respectively, according to Table 3 in the cited reference. A variation in the value of K°(V.27) reflecting the variation in $m_{H_2SeO_3}$ is not seen, however (*cf.* Section V.4.2.5).

Similarly selenious acid was found to be complexed by bromide in [85MIL/HAI]. The first complex suggested to form is $HSeO_2Br$,

$$H_2SeO_3(aq) + Br^- + H^+ \rightleftharpoons HSeO_2Br(aq) + H_2O(l) \qquad (\text{V.30})$$

with $\log_{10} K$ (V.30) = (0.70 ± 0.05). Under the experimental conditions used in [28SHE/IZA] the fraction of Se(IV) present as H_2SeO_3(aq) would then vary between 75 and 90% (*cf.* Section V.4.3.6).

The review concludes that the good agreement between the data from Reactions (V.27) and (V.28), despite the high powers in the expression for the equilibrium constant, and the absence of trends indicate that unaccounted side-reactions are absent.

On the other hand, the Raman spectroscopic work suggests that such reactions are present. Evidently further work is needed to settle this issue. The review will use, without corrections, the equilibrium constant from the hydrobromic acid medium, in which the corrections for side-reactions would be fairly small. It should also be noted that the data from the hydrochloric acid medium have been subjected to an activity correction varying between 8.9×10^3 to 1.1×10^5. The review thus selects:

$$E^{\circ}((\text{V.29}), 298.15\text{ K}) = (1.103 \pm 0.006)\text{ V},$$

$$\Delta_r G_m^{\circ}((\text{V.29}), 298.15\text{ K}) = -(212.85 \pm 1.00)\text{ kJ}\cdot\text{mol}^{-1}.$$

With the Gibbs energy of formation of $H_2SeO_3(aq)$, $-(425.18 \pm 0.85)$, and $H_2O(l)$, -237.14 kJ·mol^{-1}, and the protonation constant of the selenate ion, ($\log_{10} K_1^{\circ} = (1.75 \pm 0.10)$), there results:

$$\Delta_f G_m^{\circ}(\text{HSeO}_4^-, 298.15\text{ K}) = -(449.5 \pm 1.3)\text{ kJ}\cdot\text{mol}^{-1},$$

$$\Delta_f G_m^{\circ}(\text{SeO}_4^{2-}, 298.15\text{ K}) = -(439.5 \pm 1.4)\text{ kJ}\cdot\text{mol}^{-1}.$$

The corresponding values in [82WAG/EVA] are – 452.2 and – 441.3 kJ·mol^{-1}.

V.3.4.3 The redox couples Se(0) - Se_n^{2-} ; $1 \leq n \leq 4$.

The presence of polyselenide ions, Se_n^{2-}, is well documented in solid phases with the alkali metal ions Na^+ and K^+ [39KLE/SOD]. They can be prepared by mixing stoichiometric proportions of the elements in liquid ammonia [97RAY/DIC]. In this solvent, molecular complexes of composition Na_2Se_3, Na_2Se_4 and Na_2Se_6 have been indicated from UV-visible spectroscopic measurements [77SHA/KOE], while ^{77}Se-NMR spectra of isolated Se_3^{2-}, Se_4^{2-}, and Se_6^{2-} ions were recorded after complexation of the sodium ions with Cryptand 2,2,2 [91BJO/SCH]. With respect to their formal oxidation states, these ions represent intermediate values between 0 and – II.

In aqueous solution, polyselenide ions can be prepared by reacting sodium or potassium selenide and elemental selenium under strict oxygen-free conditions [49PAP/HAI], [67ZAI/GRE], [82GLA/DUB], [86LYO/YOU], [95LIC/FOR]. The early papers indicate that the saturated solution contains somewhat less than four times as much selenium as present in the original selenide solution. This has been taken to indicate that the highest polyselenide ion formed in aqueous solution is Se_4^{2-}.

Only one study has been published on the protonation of polyselenide ions. Schöneshöfer, Karmann, and Henglein [69SCH/KAR] exposed $H_2Se(aq)$ to radiolysis with 1.5 MeV electrons and identified H_2Se_2 as the end product of a series of radical reactions. The acidity constants of this compound were determined from spectrophotometric measurements, but as no information concerning the pH determination is provided, the protonation constants reported, $\log_{10} K_{1,2} = 1.7$ and $\log_{10} K_1 = 9.3$, are of limited value. It seems clear, however, that the ordinary trend that polynuclear compounds are more acidic than their mononuclear counterparts is fulfilled also for the polyselenides. It is probably also safe to conclude that no protonation occurs under the conditions employed in the studies of polyselenide equilibria in [86LYO/YOU] and [95LIC/FOR].

Lyons and Young [86LYO/YOU] have made the most extensive study of the system Se(cr)-Se(– II). They reported molar absorptivity curves for the ions Se^{2-}, Se_2^{2-}, Se_3^{2-}, and Se_4^{2-} together with formal redox potentials (*vs.* SCE) of the couples ½ Se_2^{2-} / Se^{2-}, ⅓ Se_3^{2-} / Se^{2-}, and ¼ Se_4^{2-} / Se^{2-} from UV-visible spectroscopic and redox potential measurements at 300 to 304 K in a 1 M KOH medium. Details are presented in Appendix A. The formal potentials determined were recalculated in this review to the SHE scale with E^{o} = 0.2369 V at 304 K for the saturated calomel electrode. The potentials were further corrected for the protonation of the selenide ion with $\log_{10} K_1$ = 13.97 and $\log_{10} K_w$ = – 13.79 valid at I = 1 M from [90LEV/MYE]. The recalculated results are estimated to carry uncertainties of ± 0.006 V and they are summarised below:

$$1/2Se_2^{2-} + e^- \rightleftharpoons Se^{2-} \quad (V.31)$$

E^{o}((V.31), 1 M KOH, 304 K) = – (0.722 ± 0.006) V

$\log_{10} K$ ((V.31), 1 M KOH, 304 K) = – (11.97 ± 0.10);

$$1/3Se_3^{2-} + 4/3e^- \rightleftharpoons Se^{2-} \quad (V.32)$$

E^{o}((V.32), 1 M KOH, 304 K) = – (0.712 ± 0.006) V

$\log_{10} K$ ((V.32), 1 M KOH, 304 K) = – (15.74 ± 0.14);

$$1/4Se_4^{2-} + 3/2e^- \rightleftharpoons Se^{2-} \quad (V.33)$$

E^{o}((V.33), 1 M KOH, 304 K) = – (0.693 ± 0.006) V

$\log_{10} K$ ((V.33), 1 M KOH, 304 K) = – (17.24 ± 0.15).

Licht and Forouzan [95LIC/FOR] later performed a similar study in 0.5, 1, and 2 M KOH but without defining the temperature used. From spectrophotometric data they deduced the equilibrium constants for the reactions:

$$Se^{2-} + Se_3^{2-} \rightleftharpoons 2Se_2^{2-} \quad (V.34)$$

$$Se^{2-} + 2Se_4^{2-} \rightleftharpoons 3Se_3^{2-} \quad (V.35)$$

to be $\log_{10} K$ (V.34) = 0.65 and $\log_{10} K$ (V.35) = 4.0 or 4.2 (two different values reported). These values were found to be independent of the ionic medium employed and they compare well with those calculated from the data obtained from [86LYO/YOU], *viz.* $\log_{10} K$ (V.34) = (0.66 ± 0.6) and $\log_{10} K$ (V.35)= (3.74 ± 1.75), respectively.

Licht and Forouzan [95LIC/FOR] also measured the redox potentials of solutions in which 0.01 M and 1 M K_2Se in 1 M KOH were gradually supplied with Se(0), 0 < $[Se(0)]_{tot}$ / $[Se(-II)]_{tot}$ < 2.8. From the data collected at 0.01 M Se(– II), a best fit to the measured potentials was obtained with E°((V.31), 1 M KOH) = – 0.716 V *vs.* SHE after due correction for the protonation of the selenide ion. The recorded potential was shifted by – 0.05 V when $[Se(-II)]_{tot}$ was increased from 0.01 to 1 M at a constant $[Se(0)]_{tot} / [Se(-II)]_{tot}$ ratio. This is in acceptable agreement with the proposed model of Lyons and Young [86LYO/YOU], which indicates a practically constant shift of – 0.06 V for this change in total selenide concentration.

The review therefore accepts the equilibrium constants presented by Lyons and Young [86LYO/YOU] at I = 1 M KOH and 304 K.

An experimental observation not exploited in either [86LYO/YOU] or [95LIC/FOR] is the fact that these polyselenide solutions reach equilibrium with respect to elemental Se(cr) at a $[Se(0)]_{tot} / [Se(-II)]_{tot}$ ratio of 2.8. This piece of information can, in combination with the polyselenide equilibrium constants just discussed, be used to evaluate an equilibrium constant for the reaction

$$Se(cr) + 2e^- \rightleftharpoons Se^{2-}. \qquad (V.36)$$

This can be achieved by solving the mass balance equations for $[Se(-II)]_{tot}$ and $[Se(0)]_{tot}$ under the condition that $a_{Se(cr)}$ reaches unity at $[Se(0)]_{tot} / [Se(-II)]_{tot}$ = 2.8. The resulting equation of the third degree yielded $\log_{10} K$ ((V.36), 1 M KOH, 304 K) = – 21.1.

An alternative approach is to vary the equilibrium constant of Reaction (V.36) and identify the value for which the above conditions are fulfilled. The computer programme WinSGW [http://www.chem.umu.se/dep/inorgchem/] was used together with the data matrix in Table V-29.

Table V-29: Formation constants for polyselenide $H_pe_qSe_r$ complexes.

Species	$\log_{10} K$	p	q	r
H^+	0	1	0	0
e^-	0	0	1	0
Se^{2-}	0	0	0	1
OH^-	– 13.79	– 1	0	0
HSe^-	13.97	1	0	1
Se_2^{2-}	23.95	0	– 2	2
Se_3^{2-}	47.23	0	– 4	3
Se_4^{2-}	68.95	0	– 6	4
Se(cr)	Varied	0	– 2	1

In these calculations, a strong dependence between $\log_{10} K$ (V.36) and the point of Se(cr) saturation was registrered. The condition $a_{\text{Se(cr)}} = 1$ at $[\text{Se(0)}]_{\text{tot}} / [\text{Se}(-\text{II})]_{\text{tot}} = (2.8 \pm 0.05)$ was satisfied for $\log_{10} K$ ((V.36), 1 M KOH, 304 K) = – (21.08 ± 0.15).

From these calculations the review accepts:

$$E^{\circ}((\text{V.36}), 1\ \text{M KOH}, 304\ \text{K}) = -(0.636 \pm 0.004)\ \text{V}.$$

The distribution of the selenium species during the titration experiment performed by Lyons and Young [86LYO/YOU] is illustrated in Figure V-9.

Figure V-9: Distribution of selenium species for $[\text{Se}(-\text{II})]_{\text{tot}} = 0.01$ M as a function of $[\text{Se(0)}]_{\text{tot}}/[\text{Se}(-\text{II})]_{\text{tot}}$ in 1 M KOH at 304 K.

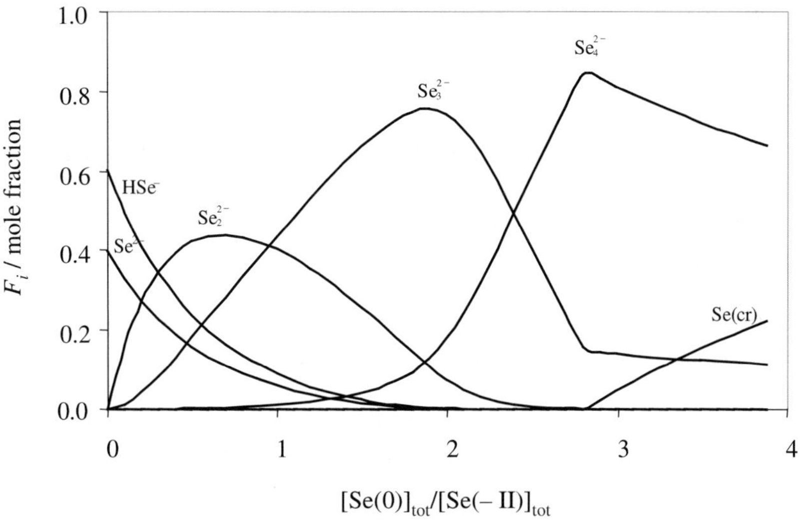

The Figure depicts an experiment in which an increasing amount of Se(0) is added to a 10^{-2} M solution of HSe^- / Se^{2-}. Up to a total concentration of Se(0) of 2.8×10^{-2} mol·dm^{-3}, all of the elemental selenium added is incorporated into polyselenide ions. At larger Se(0) additions, an increasing fraction of the addition remains as Se(cr) and $[\text{Se(0)}]_{\text{tot}}$ then represents a hypothetical concentration. The diagram shows that Se_3^{2-} and Se_4^{2-} predominate in solutions saturated with Se(cr). Hence, the calculated value of K(V.36) is practically independent of the protonation constant of the Se^{2-} ion.

The accepted redox potentials were recalculated to $I = 0$ by SIT and the assumption $\varepsilon(\mathrm{Se}_n^{2-}, \mathrm{K}^+) \approx -0.06$ kg·mol^{-1}. No ion interaction coefficients for Se_n^{2-}, K^+ are available in the literature and the approximation chosen represents an average of the $\varepsilon(\mathrm{X}^{2-}, \mathrm{K}^+)$ data presented in Table B-5. No information about the reaction enthalpies of these reactions are available in the literature. Therefore, the resulting values valid at 304 K were also the values accepted at standard conditions, but with an increase in the estimated uncertainties. The result for E°((V.36), 298.15 K) = – (0.663 ± 0.010) V corresponds to $\Delta_f G_m^\circ(\mathrm{Se}^{2-}$, 298.15 K) = (127.9 ± 1.9) kJ·mol^{-1}. The selected standard potential is based on $\Delta_f G_m^\circ(\mathrm{Se}^{2-}$, 298.15 K) = (128.6 ± 3.0) kJ·mol^{-1} selected in Section V.3.1. The selected potentials are:

$$E^\circ((\mathrm{V.36}), 298.15\ \mathrm{K}) = -(0.666 \pm 0.010)\ \mathrm{V},$$

$$E^\circ((\mathrm{V.31}), 298.15\ \mathrm{K}) = -(0.749 \pm 0.010)\ \mathrm{V},$$

$$E^\circ((\mathrm{V.32}), 298.15\ \mathrm{K}) = -(0.739 \pm 0.010)\ \mathrm{V},$$

$$E^\circ((\mathrm{V.33}), 298.15\ \mathrm{K}) = -(0.720 \pm 0.010)\ \mathrm{V}.$$

From these standard potentials, the corresponding Gibbs energies of formation were calculated to be:

$$\Delta_f G_m^\circ(\mathrm{Se}_2^{2-}, 298.15\ \mathrm{K}) = (112.7 \pm 6.3)\ \mathrm{kJ\cdot mol^{-1}},$$

$$\Delta_f G_m^\circ(\mathrm{Se}_3^{2-}, 298.15\ \mathrm{K}) = (100.6 \pm 9.2)\ \mathrm{kJ\cdot mol^{-1}},$$

$$\Delta_f G_m^\circ(\mathrm{Se}_4^{2-}, 298.15\ \mathrm{K}) = (97.6 \pm 12.1)\ \mathrm{kJ\cdot mol^{-1}}.$$

V.3.5 Summary of the H^+-e^--Se system

Based on the selected data presented in Sections V.3.1 to V.3.4, a series of predominance area diagrams at different $[\mathrm{Se}]_{\mathrm{tot}}$ were constructed. These diagrams are presented in Figure V-10. The major difference compared to previously presented predominance area diagrams for the system, *e.g.* in [2001SEB/POT], is the identification of predominance areas for the polyselenide ions. Since these ions are formed from Se(cr) without proton exchange, they appear as horizontal bands beneath the stability field of Se(cr), and they diminish in importance with the diminishing field of Se(cr) at low total concentration. No quantitative information concerning the protonation of these ions has been found in the literature. It is therefore a remaining and open question if these fields should be complemented in the acidic direction with areas for HSe_n^- and $\mathrm{H_2Se}_n$(aq).

Figure V-10: Predominance areas in the pH-pe-diagram for selenium at standard conditions and 298.15 K for three total concentrations.

a) $[Se]_{tot} = 10^{-3}$ M

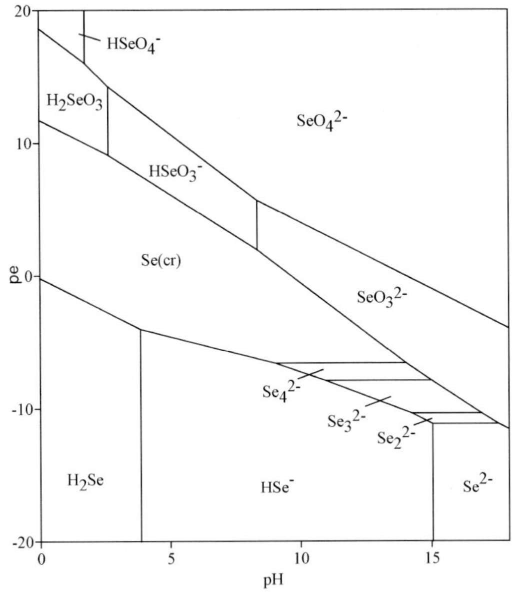

b) $[Se]_{tot} = 10^{-6}$ M

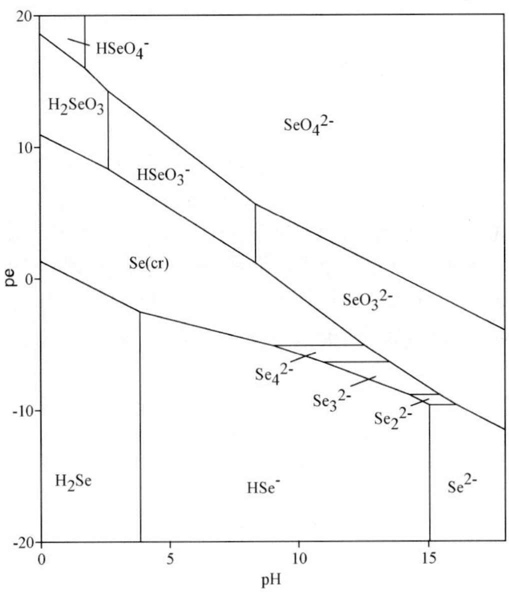

c) $[\mathrm{Se}]_{\mathrm{tot}} = 10^{-9}$ M

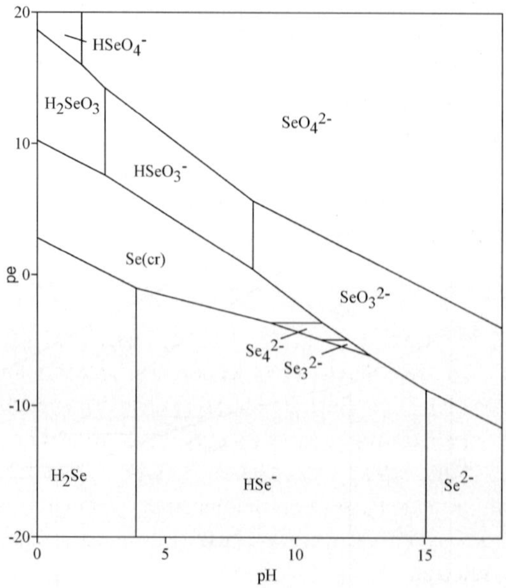

V.4 Group 17 compounds and complexes

V.4.1 Fluorine compounds

V.4.1.1 Se_2F_2(g)

There are two structural isomers of Se_2F_2(g), *viz.* FSeSeF and SeSeFF, of which the former is thermodynamically more stable according to Haas and Willner [79HAA/WIL]. High temperature thermodynamic data for the stable isomeric form FSeSeF, were calculated by the review based on spectroscopic data by Haas and Willner [79HAA/WIL]. The following selected standard entropy was then obtained:

$$S^{\circ}_{\mathrm{m}}(\mathrm{FSeSeF, g, 298.15\ K}) = (331.12 \pm 0.10)\ \mathrm{J \cdot K^{-1} \cdot mol^{-1}}$$

and the heat capacity expression in the temperature range 298.15 to 2000 K:

$$C^{\circ}_{p,\mathrm{m}}(\mathrm{FSeSeF, g}, T) = (88.259 - 2.9852 \times 10^{-3}\ T + 6.2500 \times 10^{-7}\ T^{2} - 3.7971 \times 10^{3}\ T^{-1} - 2.1864 \times 10^{5}\ T^{-2})\ \mathrm{J \cdot K^{-1} \cdot mol^{-1}}.$$

The selected heat capacity value at 298.15 K is:

$$C^{\circ}_{p,\mathrm{m}}(\mathrm{FSeSeF, g, 298.15\ K}) = (72.23 \pm 0.04)\ \mathrm{J \cdot K^{-1} \cdot mol^{-1}}.$$

The only original value reported of the heat of formation for $Se_2F_2(g)$, $\Delta_f H^\circ_m(Se_2F_2, g, 298.15\ K) = -(62 \pm 25)\ kJ \cdot mol^{-1}$, was estimated by O'Hare [68OHA] by a procedure equivalent to the assumption that the reaction enthalpy of the dissociation of $Se_2F_2(g)$ to Se(g) and $SeF_2(g)$ was equal to zero. By applying the currently presented values for the enthalpies of formation of Se(g), see Table III-1, and of $SeF_2(g)$, *cf.* Section V.4.1.2 below, the adjusted value $\Delta_f H^\circ_m(Se_2F_2, g, 298.15\ K) = -(128 \pm 30)\ kJ \cdot mol^{-1}$ is calculated with the uncertainty assigned by the review. Considering the rough estimates made, this value of the standard enthalpy of formation for $Se_2F_2(g)$ cannot be selected particularly as it is not clear which of the isomeric forms (see above) of $Se_2F_2(g)$ was considered by O'Hare.

V.4.1.2 $SeF_2(g)$

The heat of formation for $SeF_2(g)$, $\Delta_f H^\circ_m(SeF_2, g, 298.15\ K) = -(293 \pm 22)\ kJ \cdot mol^{-1}$, was estimated by O'Hare [68OHA] by a procedure equivalent to the assumption that the reaction $SeOF_2(g) + SeO(g) \rightleftharpoons SeO_2(g) + SeF_2(g)$ has a zero enthalpy change. By applying the currently selected values for the enthalpies of formation of $SeOF_2(g)$, SeO(g) and $SeO_2(g)$, see Table III-1, the adjusted value $\Delta_f H^\circ_m(SeF_2, g, 298.15\ K) = -(364 \pm 30)\ kJ \cdot mol^{-1}$ is calculated with the uncertainty added by the review. Since no experimental data are available, this value of the standard enthalpy of formation of $SeF_2(g)$ will not be selected.

Thermal thermodynamic quantities of $SeF_2(g)$ were calculated by the review based on spectroscopic data by Willner [81WIL]. In the temperature range 298.15 to 2000 K, the heat capacity expression:

$$C^\circ_{p,m}(SeF_2, g, T) = (65.060 - 3.9479 \times 10^{-3}\,T + 8.1900 \times 10^{-7}\,T^2 - 5.2047 \times 10^{3}\,T^{-1} + 1.0279 \times 10^{5}\,T^{-2})\ J \cdot K^{-1} \cdot mol^{-1}$$

was obtained. The selected heat capacity and the selected standard entropy at 298.15 K were calculated to be:

$$C^\circ_{p,m}(SeF_2, g, 298.15\ K) = (47.66 \pm 0.04)\ J \cdot K^{-1} \cdot mol^{-1}$$

$$S^\circ_m(SeF_2, g, 298.15\ K) = (271.28 \pm 0.05)\ J \cdot K^{-1} \cdot mol^{-1}.$$

V.4.1.3 SeF_4

V.4.1.3.1 $SeF_4(g)$

The equilibrium reaction:

$$SeF_4(l) \rightleftharpoons SeF_4(g) \qquad (V.37)$$

was studied by vapour pressure measurements by Peacock [53PEA] and by Dagron [62DAG], with experimental data collected between 260 to 340 K and 315 to 380 K, respectively. Their concordant results have been used to derive the following relationship for the temperature dependence of the equilibrium constant:

$$\log_{10} K^\circ((\text{V.37}), T) = -(2350 \pm 70)\, T^{-1} + (6.21 \pm 0.30).$$

This equation yields the boiling point 377.9 K of SeF_4 at 1.0 bar, in good agreement with the observed value 379 K [53PEA].

From this equation the enthalpy and entropy of reaction at the mean temperature, 320 K, are calculated to be (45.0 ± 1.5) kJ·mol^{-1} and (119 ± 6) J·K^{-1}·mol^{-1}, respectively. These results are accepted to be valid at standard conditions, *i.e.* $\Delta_r C^\circ_{p,m}$ (V.37) = 0 J·K^{-1}·mol^{-1}, thus:

$$\Delta_r H^\circ_m((\text{V.37}), 298.15\ \text{K}) = (45.0 \pm 1.5)\ \text{kJ·mol}^{-1},$$

$$\Delta_r S^\circ_m((\text{V.37}), 298.15\ \text{K}) = (119 \pm 6)\ \text{J·K}^{-1}\text{·mol}^{-1},$$

$$\Delta_r G^\circ_m((\text{V.37}), 298.15\ \text{K}) = (9.5 \pm 2.3)\ \text{kJ·mol}^{-1}.$$

From the relationship $\Delta_r H^\circ_m$ ((V.37), 298.15 K) = $\Delta_f H^\circ_m$ (SeF_4, g, 298.15 K) – $\Delta_f H^\circ_m$ (SeF_4, l, 298.15 K) and the selected standard enthalpy of formation of SeF_4(l) below:

$$\Delta_f H^\circ_m(\text{SeF}_4, \text{g}, 298.15\ \text{K}) = -(806.0 \pm 24.0)\ \text{kJ·mol}^{-1}$$

is obtained.

The entropy value 296.1 J·K^{-1}·mol^{-1} for SeF_4(g) was calculated by O'Hare [68OHA] from molecular constants. By use of spectroscopic data by Bowater *et al.* [68BOW/BRO] and by Adams and Downs [72ADA/DOW], the review obtained the following selected standard entropy value:

$$S^\circ_m(\text{SeF}_4, \text{g}, 298.15\ \text{K}) = (312.00 \pm 0.10)\ \text{J·K}^{-1}\text{·mol}^{-1}.$$

The review also calculated the heat capacity expression in the temperature range 298.15 to 2000 K:

$$C^\circ_{p,m}(\text{SeF}_4, \text{g}, T) = (122.798 - 8.4862 \times 10^{-3}\, T + 1.7631 \times 10^{-6}\, T^2 - 1.1122 \times 10^4\, T^{-1} - 9.6590 \times 10^4\, T^{-2})\ \text{J·K}^{-1}\text{·mol}^{-1}.$$

The selected heat capacity value at 298.15 K is:

$$C^\circ_{p,m}(\text{SeF}_4, \text{g}, 298.15\ \text{K}) = (82.03 \pm 0.09)\ \text{J·K}^{-1}\text{·mol}^{-1}.$$

The selected data further yield:

$$\Delta_f G^\circ_m(\text{SeF}_4, \text{g}, 298.15\ \text{K}) = -(765.5 \pm 24.0)\ \text{kJ·mol}^{-1}.$$

V.4.1.3.2 SeF_4(l)

The enthalpies of dissolution of SeO_2(cr) and SeF_4(l) in aqueous 1 M NaOH were measured by Carre *et al.* [79CAR/GER]. The calorimetric results thus obtained were combined with pertinent literature data on standard enthalpies of formation and heats of dilution to yield the standard enthalpy of formation for SeF_4(l), $\Delta_f H^\circ_m$ (SeF_4, l,

298.15 K) = – (849.4 ± 24.4) kJ·mol^{-1}. This value was slightly corrected by Woolf [80WOO] to be – 851.4 kJ·mol^{-1} using CODATA auxiliary values. The following standard enthalpy of formation for SeF_4(l) is selected:

$$\Delta_f H^\circ_m (SeF_4, l, 298.15 K) = -(851.0 \pm 24.0) kJ\cdot mol^{-1}$$

with the uncertainty assigned by this review.

From the relationship $\Delta_r S^\circ_m$ ((V.37), 298.15 K) = S°_m (SeF_4, g, 298.15 K) – S°_m (SeF_4, l, 298.15 K) and the entropy values given above:

$$S^\circ_m (SeF_4, l, 298.15 K) = (193 \pm 6) J\cdot K^{-1}\cdot mol^{-1}$$

is obtained with the uncertainty assigned by the present review.

The selected data further yield:

$$\Delta_f G^\circ_m (SeF_4, l, 298.15 K) = -(775.1 \pm 24.1) kJ\cdot mol^{-1}.$$

V.4.1.3.3 SeF_4(s)

The melting point of SeF_4(s) is 263.7 K as determined by Aynsley *et al.* [52AYN/PEA]. No other thermodynamic data are available.

V.4.1.4 SeF_5(g)

The existence of the compound SeF_5(g) seems to be doubtful.

O'Hare [68OHA] calculated the enthalpy of formation of SeF_5(g) to be – (941 ± 15) kJ·mol^{-1} from estimated values of bond energies. Considering the rough estimations made, no value of the standard enthalpy of formation for SeF_5(g) is selected.

Mills [74MIL] calculated thermal thermodynamic quantities of the compound from estimated molecular parameters by O'Hare [68OHA]. In the temperature range 298.15 to 2000 K, the heat capacity is presented as:

$$C^\circ_{p,m} (SeF_5, g, T) = (127.67 + 2.812 \times 10^{-3}\, T - 2.940 \times 10^{6}\, T^{-2}) J\cdot K^{-1}\cdot mol^{-1}.$$

The heat capacity and standard entropy at 298.15 K are calculated to be:

$$C^\circ_{p,m} (SeF_5, g, 298.15 K) = (95.4 \pm 1.0) J\cdot K^{-1}\cdot mol^{-1}$$

$$S^\circ_m (SeF_5, g, 298.15 K) = (322 \pm 2) J\cdot K^{-1}\cdot mol^{-1}.$$

The data further yield the standard Gibbs energy of formation at 298.15 K, $\Delta_f G^\circ_m$ (SeF_5, g, 298.15 K) = – (872 ± 30) kJ·mol^{-1}.

Since no experimental data are available to support these quantities, the thermodynamic data for SeF_5(g) have not been selected.

V.4.1.5 $SeF_6(g)$

The reaction between solid selenium and gaseous fluorine to form $SeF_6(g)$ was studied by O'Hare *et al.* [66OHA/SET] and by O'Hare [93OHA] using bomb calorimetry. The two studies are in good agreement and the following re-determined value by O'Hare [93OHA] of the standard enthalpy of formation for $SeF_6(g)$ is selected:

$$\Delta_f H_m^\circ(SeF_6, g, 298.15 K) = -(1118.0 \pm 0.4) kJ \cdot mol^{-1}.$$

Ohta *et al.* [96OHT/YAM] carried out low-temperature calorimetric measurements on SeF_6 in the temperature range 6 to 240 K. Experimental data were evaluated and the value (312.9 ± 0.5) $J \cdot K^{-1} \cdot mol^{-1}$ was determined for the standard entropy of $SeF_6(g)$ at 298.15 K. By statistical thermodynamic methods, Nagarajan and Brinkley [71NAG/BRI] obtained the value 313.7 $J \cdot K^{-1} \cdot mol^{-1}$ and later O'Hare [93OHA] the value 312.5 $J \cdot K^{-1} \cdot mol^{-1}$. With the molecular parameters in the latter reference the review obtained 313.55 $J \cdot K^{-1} \cdot mol^{-1}$. Considering all contributions, the following standard entropy for $SeF_6(g)$ is selected:

$$S_m^\circ(SeF_6, g, 298.15 K) = (313.2 \pm 0.5) J \cdot K^{-1} \cdot mol^{-1}$$

with the uncertainty assigned by the present review.

The review also calculated heat capacities for $SeF_6(g)$. The expression:

$$C_{p,m}^\circ(SeF_6, g, T) = (196.298 - 2.5821 \times 10^{-2} T + 6.4480 \times 10^{-6} T^2 - 2.5722 \times 10^4 T^{-1} + 6.7040 \times 10^5 T^{-2}) J \cdot K^{-1} \cdot mol^{-1}$$

was obtained from a least squares calculation on the data in the temperature range 298.15 to 1500 K. The selected value at 298.15 K is:

$$C_{p,m}^\circ(SeF_6, g, 298.15 K) = (110.44 \pm 0.05) J \cdot K^{-1} \cdot mol^{-1}.$$

By using the selected entropy of $SeF_6(g)$, and selected auxiliary entropy data of Se(cr) and $F_2(g)$, the following entropy of formation is obtained $\Delta_f S_m^\circ(SeF_6, g, 298.15 K) = -(337.3 \pm 0.6)$ $J \cdot K^{-1} \cdot mol^{-1}$. From the selected standard enthalpy of formation and the derived entropy of formation of $SeF_6(g)$, the following calculated value of the Gibbs energy of formation at 298.15 K is selected:

$$\Delta_f G_m^\circ(SeF_6, g, 298.15 K) = -(1017.4 \pm 0.4) kJ \cdot mol^{-1}.$$

The values reported in [82WAG/EVA] are $\Delta_f H_m^\circ(SeF_6, g, 298.15 K) = -1117$ $kJ \cdot mol^{-1}$, $S_m^\circ(SeF_6, g, 298.15 K) = 313.87$ $J \cdot K^{-1} \cdot mol^{-1}$ and $\Delta_f G_m^\circ(SeF_6, g, 298.15 K) = -1017$ $kJ \cdot mol^{-1}$, respectively.

V.4.1.6 $SeF_5Cl(g)$

IR-frequencies were determined by Christe *et al.* [72CHR/SCH] and, together with molecular constant data, they computed thermodynamic properties of $SeF_5Cl(g)$ using the rigid-rotor, harmonic-oscillator approximation. The following temperature dependence of the heat capacity was derived in the temperature range 298 to 1400 K from the data in [72CHR/SCH]:

$$C^{\circ}_{p,\mathrm{m}}(\mathrm{SeF_5Cl, g}, T) = (141.60 + 22.383 \times 10^{-3}\,T - 8.25 \times 10^{-6}\,T^2 - 2.7874 \times 10^{6}\,T^{-2})\ \mathrm{J{\cdot}K^{-1}{\cdot}mol^{-1}}.$$

The selected value at 298.15 K is:

$$C^{\circ}_{p,\mathrm{m}}(\mathrm{SeF_5Cl, g, 298.15\ K}) = (116.3 \pm 0.1)\ \mathrm{J{\cdot}K^{-1}{\cdot}mol^{-1}}.$$

The standard entropy value for $SeF_5Cl(g)$ calculated by Christe *et al.* [72CHR/SCH] is selected:

$$S^{\circ}_{\mathrm{m}}(\mathrm{SeF_5Cl, g, 298.15\ K}) = (341.0 \pm 0.2)\ \mathrm{J{\cdot}K^{-1}{\cdot}mol^{-1}}$$

with the uncertainty added by the review.

V.4.1.7 $SeOF_2$

V.4.1.7.1 $SeOF_2(g)$

The equilibrium reaction:

$$\mathrm{SeOF_2(l)} \rightleftharpoons \mathrm{SeOF_2(g)} \qquad \text{(V.38)}$$

was studied by Dagron [62DAG] in the temperature range 320 to 400 K by vapour pressure measurements. Evaluation of his results yields $\log_{10} K^{\circ}((\mathrm{V.38}), T) = (5.81 \pm 0.05) - (2316 \pm 50)\,T^{-1}$ for the temperature dependence of the equilibrium constant. From this equation, the following thermodynamic quantities are derived at 360 K: $\Delta_{\mathrm{r}}H^{\circ}_{\mathrm{m}}((\mathrm{V.38}), 360\ \mathrm{K}) = (44.3 \pm 1.0)\ \mathrm{kJ{\cdot}mol^{-1}}$ and $\Delta_{\mathrm{r}}S^{\circ}_{\mathrm{m}}((\mathrm{V.38}), 360\ \mathrm{K}) = (111.2 \pm 1.0)\ \mathrm{J{\cdot}K^{-1}{\cdot}mol^{-1}}$ respectively. These values are accepted to be valid also at 298.15 K, *i.e.* $\Delta_{\mathrm{r}}C^{\circ}_{p,\mathrm{m}}(\mathrm{V.38}) = 0\ \mathrm{J{\cdot}K^{-1}{\cdot}mol^{-1}}$:

$$\Delta_{\mathrm{r}}H^{\circ}_{\mathrm{m}}((\mathrm{V.38}), 298.15\ \mathrm{K}) = (44.3 \pm 1.0)\ \mathrm{kJ{\cdot}mol^{-1}},$$

$$\Delta_{\mathrm{r}}S^{\circ}_{\mathrm{m}}((\mathrm{V.38}), 298.15\ \mathrm{K}) = (111.2 \pm 1.0)\ \mathrm{J{\cdot}K^{-1}{\cdot}mol^{-1}}.$$

The data further yield:

$$\Delta_{\mathrm{r}}G^{\circ}_{\mathrm{m}}((\mathrm{V.38}), 298.15\ \mathrm{K}) = (11.1 \pm 1.0)\ \mathrm{kJ{\cdot}mol^{-1}}.$$

By applying the enthalpy relationship $\Delta_{\mathrm{r}}H^{\circ}_{\mathrm{m}}((\mathrm{V.38}), 298.15\ \mathrm{K}) = \Delta_{\mathrm{f}}H^{\circ}_{\mathrm{m}}(\mathrm{SeOF_2, g, 298.15\ K}) - \Delta_{\mathrm{f}}H^{\circ}_{\mathrm{m}}(\mathrm{SeOF_2, l, 298.15\ K})$ and inserting the selected standard enthalpy of formation for $SeOF_2(l)$ presented below, the following selected value:

$$\Delta_{\mathrm{f}}H^{\circ}_{\mathrm{m}}(\mathrm{SeOF_2, g, 298.15\ K}) = -(531 \pm 16)\ \mathrm{kJ{\cdot}mol^{-1}}$$

is obtained.

By statistical thermodynamic methods, Nagarajan and Müller [66NAG/MUL] calculated the standard entropy value of $SeOF_2(g)$ to be 296.6 $\mathrm{J{\cdot}K^{-1}{\cdot}mol^{-1}}$. Spectroscopic data by Bowater *et al.* [68BOW/BRO2] and by Alexander and Beattie [72ALE/BEA] were used by the review to calculate the selected value:

$$S^{\circ}_{m}(\text{SeOF}_2, \text{g}, 298.15\text{ K}) = (296.2 \pm 0.1)\text{ J}\cdot\text{K}^{-1}\cdot\text{mol}^{-1}.$$

The review also calculated the heat capacity expression in the temperature range 298.15 to 2000 K:

$$C^{\circ}_{p,m}(\text{SeOF}_2, \text{g}, T) = (96.7972 - 7.4772 \times 10^{-3}\, T + 1.4994 \times 10^{-6}\, T^{2} - 1.1158 \times 10^{4}\, T^{-1} + 5.3806 \times 10^{5}\, T^{-2})\text{ J}\cdot\text{K}^{-1}\cdot\text{mol}^{-1}.$$

The selected heat capacity value at 298.15 K is:

$$C^{\circ}_{p,m}(\text{SeOF}_2, \text{g}, 298.15\text{ K}) = (63.33 \pm 0.05)\text{ J}\cdot\text{mol}^{-1}\cdot\text{K}^{-1}.$$

The selected data further yield:

$$\Delta_f G^{\circ}_{m}(\text{SeOF}_2, \text{g}, 298.15\text{ K}) = -(515 \pm 16)\text{ kJ}\cdot\text{mol}^{-1}.$$

V.4.1.7.2 $SeOF_2$(l)

The standard enthalpy of formation for $SeOF_2$(l) was calorimetrically determined by Carre *et al.* [79CAR/GER] to be – (574.0 ± 15.9) kJ·mol^{-1}. Woolf [80WOO] corrected this result to be – 575.3 kJ·mol^{-1} using CODATA auxiliary values and the following standard enthalpy of formation for $SeOF_2$(l) is selected:

$$\Delta_f H^{\circ}_{m}(\text{SeOF}_2, \text{l}, 298.15\text{ K}) = -(575 \pm 16)\text{ kJ}\cdot\text{mol}^{-1}.$$

Use of the general relationship $\Delta_r S^{\circ}_{m}$ ((V.38), 298.15 K) = S°_{m} ($SeOF_2$, g, 298.15 K) – S°_{m} ($SeOF_2$, l, 298.15 K) gives the selected standard entropy for $SeOF_2$(l):

$$S^{\circ}_{m}(\text{SeOF}_2, \text{l}, 298.15\text{ K}) = (185.0 \pm 1.0)\text{ J}\cdot\text{K}^{-1}\cdot\text{mol}^{-1}$$

The selected Gibbs energy of formation for $SeOF_2$(l) is thus:

$$\Delta_f G^{\circ}_{m}(\text{SeOF}_2, \text{l}, 298.15\text{ K}) = -(527 \pm 16)\text{ kJ}\cdot\text{mol}^{-1}$$

V.4.1.8 Aqueous complexes

The formation of aqueous selenium(IV)-fluoride complexes in solutions of high HF(aq) concentration was detected by Milne [87MIL]. The study was based on Raman spectroscopic measurements and showed, at 28 M HF, the presence of new bands in addition to those of H_2SeO_3(aq). The bands were interpreted as originating from a species $HSeO_2F$(aq). At even higher HF concentrations, > 50 M, additional Raman bands appeared and these were claimed to be due to a species $SeOF_2$(aq). No attempts were made to elucidate thermodynamic information from the measurements.

V.4.2 Chlorine compounds

V.4.2.1 Se_2Cl_2

V.4.2.1.1 $Se_2Cl_2(l)$

The equilibrium reaction

$$Se_2Cl_2(l) \rightleftharpoons Se(cr) + SeCl_2(g) \qquad (V.39)$$

was studied by Lundkvist and Lellep [68LUN/LEL] in the temperature range 293 to 348 K. The vapour pressure of $SeCl_2(g)$ was determined by a flow method with N_2 as carrier gas and is described by the equation $\log_{10} p\,(SeCl_2, g, T) = 6.73 - 2947\,T^{-1}$. As $SeCl_2(l)$ was found to dissolve some selenium, especially at higher temperatures, the equilibrium constant of Reaction (V.39) should be written:

$$K^\circ(V.39) = \frac{p_{SeCl_2(g)}}{a_{Se_2Cl_2(sln)}}$$

The notation $Se_2Cl_2(sln)$ represents $Se_2Cl_2(l)$ saturated with Se(cr). The activity of $Se_2Cl_2(sln)$ was estimated from determinations of the solubility of Se(cr). The temperature dependence of the equilibrium constant was recalculated by this review to be:

$$\log_{10} K^\circ((V.39), T) = 7.89 - 3284\,T^{-1}.$$

From this equation, the following thermodynamic quantities are calculated ($\Delta_r C^\circ_{p,m} = 0$):

$$\Delta_r H^\circ_m((V.39), 298.15\text{ K}) = (63 \pm 2)\text{ kJ·mol}^{-1},$$

$$\Delta_r S^\circ_m((V.39), 298.15\text{ K}) = (151 \pm 5)\text{ J·K}^{-1}\text{·mol}^{-1},$$

$$\Delta_r G^\circ_m((V.39), 298.15\text{ K}) = (18.0 \pm 2.5)\text{ kJ·mol}^{-1}.$$

The vapour pressures of $SeCl_2(g)$ obtained in a similar study by Oppermann *et al.* [86OPP/HAN] agree satisfactorily at lower temperatures with those in [68LUN/LEL]. Their vapour pressure expression, $\log_{10} p\,(SeCl_2, g, T) = 4.54 - 2260\,T^{-1}$, yields $\Delta_r G^\circ_m((V.39), 298.15\text{ K}) = (17.4 \pm 1.0)\text{ kJ·mol}^{-1}$.

By applying the enthalpy relationship $\Delta_r H^\circ_m((V.39), 298.15\text{ K}) = \Delta_f H^\circ_m(SeCl_2, g, 298.15\text{ K}) - \Delta_f H^\circ_m(Se_2Cl_2, l, 298.15\text{ K})$, and inserting the selected standard enthalpy of formation for $SeCl_2(g)$, see Section V.4.2.2, the following value $\Delta_f H^\circ_m(Se_2Cl_2, l, 298.15\text{ K}) = -(80 \pm 4)\text{ kJ·mol}^{-1}$ is obtained. From a re-assessment of solution calorimetric data determined by Thomsen [1882THO] and Petersen [1891PET], Mills presents the value $\Delta_f H^\circ_m(Se_2Cl_2, l, 298.15\text{ K}) = -(84 \pm 4)\text{ kJ·mol}^{-1}$. The present review selects:

$$\Delta_f H^\circ_m(Se_2Cl_2, l, 298.15\text{ K}) = -(82 \pm 4)\text{ kJ·mol}^{-1}.$$

By use of the relationship $\Delta_r S_m^o$ ((V.39), 298.15) = S_m^o (Se, cr, 298.15 K) + S_m^o ($SeCl_2$, g, 298.15 K) – S_m^o (Se_2Cl_2, l, 298.15 K) and the value derived above for $\Delta_r S_m^o$, and S_m^o ($SeCl_2$, g, 298.15 K) presented in Section V.4.2.2, as well as the selected standard entropy of Se(cr):

$$S_m^o(\mathrm{Se_2Cl_2, l, 298.15\ K}) = (186 \pm 5)\ \mathrm{J \cdot K^{-1} \cdot mol^{-1}}$$

is obtained, with the uncertainty assigned by the present review.

The selected data further yield the standard Gibbs energy of formation:

$$\Delta_f G_m^o(\mathrm{Se_2Cl_2, l, 298.15\ K}) = -(46 \pm 4)\ \mathrm{kJ \cdot mol^{-1}}.$$

V.4.2.1.2 Se_2Cl_2(g)

Thermodynamic properties for Se_2Cl_2(g) were calculated by Ramaswamy and Jayaraman [71RAM/JAY] by statistical thermodynamic methods. The heat capacity expression:

$$C_{p,m}^o(\mathrm{Se_2Cl_2, g}, T) = (81.92 + 2.293 \times 10^{-3}\ T - 1.195 \times 10^{-6}\ T^2 - 0.4346 \times 10^6\ T^{-2})\ \mathrm{J \cdot K^{-1} \cdot mol^{-1}}$$

was obtained by the review from a least squares calculation on the data given in the temperature range 298.15 to 1000 K. The selected value at 298.15 K is:

$$C_{p,m}^o(\mathrm{Se_2Cl_2, g, 298.15\ K}) = (77.6 \pm 0.1)\ \mathrm{J \cdot K^{-1} \cdot mol^{-1}}.$$

The calculated standard entropy value of Se_2Cl_2(g) is selected:

$$S_m^o(\mathrm{Se_2Cl_2, g, 298.15\ K}) = (354.0 \pm 0.2)\ \mathrm{J \cdot K^{-1} \cdot mol^{-1}}.$$

From voltammetric measurements with a Se(l) electrode in a melt of LiCl-KCl at around 700 K, Bodewig and Plambeck [70BOD/PLA] estimated the value of the Gibbs energy of reaction of:

$$2\mathrm{Se(l)} + \mathrm{Cl_2(g)} \rightleftharpoons \mathrm{Se_2Cl_2(g)} \quad (V.40)$$

to be $\Delta_r G_m^o$ ((V.40), 700 K) = – 46 $kJ \cdot mol^{-1}$. This quantity equals the Gibbs energy of formation of Se_2Cl_2(g) at 700 K. Using the third law method, the adopted standard enthalpy of formation for Se_2Cl_2(g) is derived to be:

$$\Delta_f H_m^o(\mathrm{Se_2Cl_2, g, 298.15\ K}) = -(21 \pm 10)\ \mathrm{kJ \cdot mol^{-1}}$$

with the uncertainty assigned by the present review. In this analysis, Gibbs energy functions were calculated with standard auxiliary data for Se(l) and Cl_2(g) and with data from [71RAM/JAY] for Se_2Cl_2(g).

The data further yield the standard Gibbs energy of formation:

$$\Delta_f G_m^o(\mathrm{Se_2Cl_2, g, 298.15\ K}) = -(35 \pm 10)\ \mathrm{kJ \cdot mol^{-1}}.$$

V.4.2.2 $SeCl_2(g)$

The equilibrium reaction:

$$SeCl_4(cr) \rightleftharpoons SeCl_2(g) + Cl_2(g) \qquad (V.41)$$

was studied by vapour pressure measurements by Yost and Kircher [30YOS/KIR], and by Oppermann *et al.* [86OPP/HAN]. Experimental data were collected between 380 to 500 K and 350 to 480 K, respectively. Their results are mainly consistent and the following relationship for the temperature dependence is used from [86OPP/HAN]:

$$\log_{10} K^\circ((V.41), T) = 17.101 - 8200\ T^{-1}.$$

From this equation the enthalpy and entropy of reaction at the mean temperature, 415 K, are calculated to be $(157 \pm 3)\ kJ\cdot mol^{-1}$ and $(327 \pm 6)\ J\cdot K^{-1}\cdot mol^{-1}$, respectively. By use of the second-law method with the estimated $\Delta_r C^\circ_{p,m} = -46\ J\cdot K^{-1}\cdot mol^{-1}$, the following values were obtained for standard conditions:

$$\Delta_r H^\circ_m((V.41), 298.15\ K) = (162 \pm 4)\ kJ\cdot mol^{-1},$$

$$\Delta_r S^\circ_m((V.41), 298.15\ K) = (342 \pm 8)\ J\cdot K^{-1}\cdot mol^{-1},$$

$$\Delta_r G^\circ_m((V.41), 298.15\ K) = (60 \pm 5)\ kJ\cdot mol^{-1},$$

with all uncertainties assigned by the present review.

Applying the relationship $\Delta_r H^\circ_m((V.41), 298.15\ K) = \Delta_f H^\circ_m(SeCl_2, g, 298.15\ K) - \Delta_f H^\circ_m(SeCl_4, cr, 298.15\ K)$ and inserting the value of the selected standard enthalpy of formation for $SeCl_4(cr)$, $-(179 \pm 2)\ kJ\cdot mol^{-1}$ (see Section V.4.2.3), the following selected value:

$$\Delta_f H^\circ_m(SeCl_2, g, 298.15\ K) = -(17 \pm 4)\ kJ\cdot mol^{-1}$$

is obtained.

The enthalpy value of $SeCl_2(g)$ reported by Mills [74MIL], and later used in the Barin Tables [95BAR], is $\Delta_f H^\circ_m(SeCl_2, g, 298.15\ K) = -(33.5 \pm 12.0)\ kJ\cdot mol^{-1}$. This value was evaluated by Mills [74MIL] mainly based upon the vapour pressure measurements by Yost and Kircher [30YOS/KIR] and the enthalpy of formation for $SeCl_4(cr)$, $\Delta_f H^\circ_m(SeCl_4, cr, 298.15\ K) = -188.7\ kJ\cdot mol^{-1}$ (see Section V.4.2.3).

From spectroscopic data by Ozin and Van der Voet [70OZI/VOE], Mills [74MIL] calculated the standard entropy value of $SeCl_2(g)$ to be $295.7\ J\cdot K^{-1}\cdot mol^{-1}$. Considering the molecular parameters by Fernholt *et al.* [83FER/HAA], a recalculation by the review yielded the selected entropy value:

$$S^\circ_m(SeCl_2, g, 298.15\ K) = (295.18 \pm 0.06)\ J\cdot K^{-1}\cdot mol^{-1}$$

and the heat capacity expression in the temperature range 298.15 to 2000 K:

$$C^\circ_{p,m}(SeCl_2, g, T) = (59.4678 - 7.5470 \times 10^{-4}\ T + 1.6040 \times 10^{-7}\ T^2$$

$$- 9.1230 \times 10^{2}\, T^{-1} - 2.3321 \times 10^{5}\, T^{-2})\ \mathrm{J{\cdot}K^{-1}{\cdot}mol^{-1}}.$$

The selected heat capacity at 298.15 K is:

$$C^{\circ}_{p,\mathrm{m}}\,(\mathrm{SeCl_2, g, 298.15\ K}) = (53.57 \pm 0.03)\ \mathrm{J{\cdot}K^{-1}{\cdot}mol^{-1}}.$$

The selected data further yield the standard Gibbs energy of formation at 298.15 K:

$$\Delta_{\mathrm{f}}G^{\circ}_{\mathrm{m}}\,(\mathrm{SeCl_2, g, 298.15\ K}) = -\,(26 \pm 4)\ \mathrm{kJ{\cdot}mol^{-1}}.$$

V.4.2.3 $SeCl_4$(cr)

The enthalpies of dissolution of $SeCl_4$(cr) and SeO_2(cr) in 4.02 M HCl were measured by Westland and Makhija [78WES/MAK] whereas Oppermann *et al.* [85OPP/HAN] measured the dissolution enthalpies of $SeCl_4$(cr) and SeO_2(cr) in 1 M NaOH. The calorimetric results obtained by Westland and Makhija [78WES/MAK] were combined with present data on the standard enthalpies of formation for SeO_2(cr), HCl(g) and H_2O(l), as well as heats of solution of HCl(g). The standard enthalpy of formation they derived for $SeCl_4$(cr) was $\Delta_{\mathrm{f}}H^{\circ}_{\mathrm{m}}$ ($SeCl_4$, cr, 298.15 K) = – (179 ± 2) $\mathrm{kJ{\cdot}mol^{-1}}$. On the other hand, the corresponding value reported by Oppermann *et al.* [85OPP/HAN], $\Delta_{\mathrm{f}}H^{\circ}_{\mathrm{m}}$ ($SeCl_4$, cr, 298.15 K) = – (184 ± 4) $\mathrm{kJ{\cdot}mol^{-1}}$, has been recalculated by the review using the selected standard enthalpies of formation for SeO_2(cr) and H_2O(l), as well as the derived partial enthalpies of formation of 1 M NaOH, $\Delta_{\mathrm{f}}H^{\circ}_{\mathrm{m}}$ (NaOH, sln, 298.15 K) = – (470.32 ± 0.07) and 1 M NaCl, $\Delta_{\mathrm{f}}H^{\circ}_{\mathrm{m}}$ (NaCl, sln, 298.15 K) = – (408.50 ± 0.13) from [89COX/WAG] and [82WAG/EVA]. Hence, the corrected enthalpy value from the measurements by Oppermann *et al.* [85OPP/HAN] is $\Delta_{\mathrm{f}}H^{\circ}_{\mathrm{m}}$ ($SeCl_4$, cr, 298.15 K) = – (179.3 ± 1.2) $\mathrm{kJ{\cdot}mol^{-1}}$. Considering this latter value and that given by Westland and Makhija [78WES/MAK], the following standard enthalpy of formation for $SeCl_4$(cr) is selected:

$$\Delta_{\mathrm{f}}H^{\circ}_{\mathrm{m}}\,(\mathrm{SeCl_4, cr, 298.15\ K}) = -\,(179.0 \pm 2.0)\ \mathrm{kJ{\cdot}mol^{-1}}.$$

Mills [74MIL] calculated a value of the enthalpy of formation for $SeCl_4$(cr), amounting to – 188.7 $\mathrm{kJ{\cdot}mol^{-1}}$, mainly based on calorimetric measurements by Thomsen [1882THO]. Mills [74MIL] also estimated the heat capacity at 298.15 K to be $C^{\circ}_{p,\mathrm{m}}$ ($SeCl_4$, cr, 298.15 K) = 133.9 $\mathrm{J{\cdot}K^{-1}{\cdot}mol^{-1}}$.

Based on the vapour pressures determined in the study of Reaction (V.41) described in Section V.4.2.2, the value of the entropy of Reaction (V.41) at 298.15 K was derived to be (342 ± 8) $\mathrm{J{\cdot}K^{-1}{\cdot}mol^{-1}}$. By use of the relationship $\Delta_{\mathrm{r}}S^{\circ}_{\mathrm{m}}$ ((V.41), 298.15 K) = S°_{m} ($SeCl_2$, g, 298.15 K) + S°_{m} (Cl_2, g, 298.15 K) – S°_{m} ($SeCl_4$, cr, 298.15 K) and inserting the selected standard entropy values of Cl_2(g) and $SeCl_2$(g), the selected standard entropy of $SeCl_4$(cr):

$$S^{\circ}_{\mathrm{m}}\,(\mathrm{SeCl_4, cr, 298.15\ K}) = (176 \pm 8)\ \mathrm{J{\cdot}K^{-1}{\cdot}mol^{-1}}$$

is obtained, with the uncertainty assigned by the present review.

The selected data further yield:

$$\Delta_f G_m^\circ(\text{SeCl}_4, \text{cr}, 298.15\ \text{K}) = -(85.9 \pm 3.1)\ \text{kJ·mol}^{-1}.$$

V.4.2.4 $SeOCl_2$

V.4.2.4.1 $SeOCl_2(g)$

The equilibrium reaction:

$$\text{SeOCl}_2(\text{l}) \rightleftharpoons \text{SeOCl}_2(\text{g}) \qquad \text{(V.42)}$$

was studied in the temperature range 310 to 430 K by vapour pressure measurements by Oppermann *et al.* [86OPP/HAN]. They presented the following expression for the temperature dependence of the equilibrium constant, $\log_{10} K^\circ$((V.42), (310 – 430) K) = 6.08 – 2750 T^{-1}. From this equation, the thermodynamic quantities $\Delta_r H_m^\circ$((V.42), 370 K) = (52.6 ± 1.5) kJ·mol^{-1}, $\Delta_r S_m^\circ$((V.42), 370 K) = (116.4 ± 2.0) J·K^{-1}·mol^{-1} are derived at the mean temperature, 370 K. These values are accepted to be valid also at 298.15 K, *i.e.* $\Delta_r C_{p,m}^\circ$((V.42)) = 0 J·K^{-1}·mol^{-1}:

$$\Delta_r H_m^\circ((\text{V.42}), 298.15\ \text{K}) = (52.6 \pm 1.5)\ \text{kJ·mol}^{-1},$$

$$\Delta_r S_m^\circ((\text{V.42}), 298.15\ \text{K}) = (116.4 \pm 2.0)\ \text{J·K}^{-1}\text{·mol}^{-1}.$$

These data at 298.15 K further yield:

$$\Delta_r G_m^\circ((\text{V.42}), 298.15\ \text{K}) = (17.9 \pm 1.6)\ \text{kJ·mol}^{-1}.$$

By applying the enthalpy relationship $\Delta_r H_m^\circ$((V.42), 298.15 K) = $\Delta_f H_m^\circ$($SeOCl_2$, g, 298.15 K) – $\Delta_f H_m^\circ$($SeOCl_2$, l, 298.15 K) and inserting the selected standard enthalpy of formation for $SeOCl_2$(l) presented below, the following selected value:

$$\Delta_f H_m^\circ(\text{SeOCl}_2, \text{g}, 298.15\ \text{K}) = -(128 \pm 3)\ \text{kJ·mol}^{-1}$$

is derived.

By statistical thermodynamic methods, Nagarajan and Müller [66NAG/MUL] calculated the standard entropy value of $SeOCl_2$(g) to be 323.7 J·K^{-1}·mol^{-1}. Spectroscopic data by Westphal and Rosenberger [80WES/ROS] were used by the review to calculate the selected value:

$$S_m^\circ(\text{SeOCl}_2, \text{g}, 298.15\ \text{K}) = (321.54 \pm 0.10)\ \text{J·K}^{-1}\text{·mol}^{-1}.$$

The review also calculated the heat capacity expression in the temperature range 298.15 to 2000 K:

$$C_{p,m}^\circ(\text{SeOCl}_2, \text{g}, T) = (91.0154 - 4.2678 \times 10^{-3}\,T + 8.4930 \times 10^{-7}\,T^2 - 6.5154 \times 10^3\,T^{-1} + 1.9853 \times 10^5\,T^{-2})\ \text{J·K}^{-1}\text{·mol}^{-1}.$$

The selected heat capacity value at 298.15 K is:

$$C^{\circ}_{p,\text{m}}\,(\text{SeOCl}_2,\ \text{g},\ 298.15\ \text{K}) = (70.20 \pm 0.05)\ \text{J}\cdot\text{K}^{-1}\cdot\text{mol}^{-1}.$$

The selected data further yield:

$$\Delta_\text{f}G^{\circ}_\text{m}\,(\text{SeOCl}_2,\ \text{g},\ 298.15\ \text{K}) = -\,(114 \pm 3)\ \text{kJ}\cdot\text{mol}^{-1}.$$

V.4.2.4.2 $SeOCl_2$(l)

Oppermann *et al.* [85OPP/HAN] measured the dissolution enthalpies of $SeOCl_2$(l) and SeO_2(cr) in 1 M NaOH. In the evaluation of the calorimetric results, they obtained the following value of the standard enthalpy of formation for $SeOCl_2$(l), $\Delta_\text{f}H^{\circ}_\text{m}$ ($SeOCl_2$, l, 298.15 K) = – (182.9 ± 2.9) kJ·mol^{-1}. However, a recalculation by the review using the selected standard enthalpies of formation for SeO_2(cr) and H_2O(l), as well as the derived partial enthalpies of formation of 1 M NaOH, $\Delta_\text{f}H^{\circ}_\text{m}$ (NaOH, 1 M, 298.15 K) = – (470.32 ± 0.07) and 1 M NaCl, $\Delta_\text{f}H^{\circ}_\text{m}$ (NaCl, 1 M, 298.15 K) = – (408.50 ± 0.13) from [89COX/WAG] and [82WAG/EVA], gave the selected value:

$$\Delta_\text{f}H^{\circ}_\text{m}\,(\text{SeOCl}_2,\ \text{l},\ 298.15\ \text{K}) = -\,(180.6 \pm 2.0)\ \text{kJ}\cdot\text{mol}^{-1}$$

with the uncertainty assigned by the present review.

By use of the relationship $\Delta_\text{r}S^{\circ}_\text{m}$ ((V.42), 298.15) = S°_m ($SeOCl_2$, g, 298.15 K) – S°_m ($SeOCl_2$, l, 298.15 K) and inserting the above entropy values, the selected standard entropy for $SeOCl_2$(l):

$$S^{\circ}_\text{m}\,(\text{SeOCl}_2,\ \text{l},\ 298.15\ \text{K}) = (205.1 \pm 2.0)\ \text{J}\cdot\text{K}^{-1}\cdot\text{mol}^{-1}$$

is obtained.

The selected standard Gibbs energy of formation for $SeOCl_2$(l) thus amounts to:

$$\Delta_\text{f}G^{\circ}_\text{m}\,(\text{SeOCl}_2,\ \text{l},\ 298.15\ \text{K}) = -\,(132.1 \pm 2.1)\ \text{kJ}\cdot\text{mol}^{-1}.$$

V.4.2.5 Aqueous complexes

The formation of aqueous selenium(IV)-chloride complexes in solutions of high hydrochloric acid concentration has been studied by Milne and LaHaie [79MIL/HAI] and by Milne [87MIL]. The studies are based on Raman spectroscopic measurements and, in [79MIL/HAI], the occurrence of bands at 152, 265, and 336 cm^{-1} were interpreted as originating from a species $SeOCl_2$(aq). The intensities of these bands were significant at HCl concentrations of 7 M and above, and were used to deduce an equilibrium constant for the reaction

$$\text{H}_2\text{SeO}_3(\text{aq}) + 2\text{H}^+ + 2\text{Cl}^- \rightleftharpoons \text{SeOCl}_2(\text{aq}) + 2\text{H}_2\text{O(l)}$$

with

$$K = ([\text{SeOCl}_2] \times a_\text{w}^2)\ /\ ([\text{H}_2\text{SeO}_3] \times a_\text{HCl}^2).$$

In the evaluation, the activity coefficients of $SeOCl_2$ and H_2SeO_3 were assumed to be equal. The molar intensities of the Se–Cl bands were adjusted to values that resulted in the smallest variation in K over the HCl concentration range 7.1 to 12.2 M. The value of K so obtained was $(3.2 \pm 1.1) \times 10^{-6}$ M^{-2}.

In [87MIL], a significant re-interpretation of these measurements was made. The formation of a species $HSeO_2Cl(aq)$ was postulated from a slight increase in the intensity of a Se–OH stretching band at 690 cm^{-1} over the range of 4 to 6 M HCl. Additional support was gathered from a minor increase in intensity over the same concentration interval of a broad band at 330 cm^{-1}. The equilibrium constant for the reaction

$$H_2SeO_3(aq) + H^+ + Cl^- \rightleftharpoons HSeO_2Cl(aq) + H_2O(l),$$

$$K = ([HSeO_2Cl] \times a_w) / ([H_2SeO_3] \times a_{\pm HCl}^2),$$

was evaluated to 0.015 M^{-2} by varying the molar intensity of the assumed Se–OH band of $HSeO_2Cl(aq)$ at 690 cm^{-1}. However, since the resulting molar intensity practically coincides with the molar intensity of the 690 cm^{-1} band of H_2SeO_3, this value must be judged as highly uncertain.

At HCl concentrations exceeding 7 M data were, as in [79MIL/HAI], interpreted as due to the formation of $SeOCl_2(aq)$. Here, however, the equilibrium constant was correctly formulated, $K = ([SeOCl_2] \times a_w^2)/([H_2SeO_3] \times a_{\pm HCl}^4)$, and new values for a_w and $a_{\pm HCl}$ were employed. Again, the molar intensities of the Se–Cl bands at 100 to 450 cm^{-1} were changed until the minimum of the variation in K was reached, $K = 1.4 \times 10^{-7}$ M^{-4}.

To conclude, the formation of a soluble Se(IV)-Cl^- adduct at hydrochloric acid concentration higher than 7 M seems well documented from the appearance of specific Raman bands, while the formation of a species at lower HCl concentrations appears much more speculative. However, due to the lack of information concerning the temperature employed and uncertainties in the evaluation technique, the equilibrium constants are not selected by the review.

Elemental selenium can be dissolved in solutions containing Se(IV) and 10 M HCl or more. The resulting strongly yellow solutions were studied at 11.8 M HCl by Mahadevan and Milne [83MAH/MIL] using gravimetric and Raman spectroscopic techniques. Both data sets indicated the simultaneous formation of $Se_2Cl_2(aq)$ and $SeCl_2(aq)$. The two techniques yielded concordant results for the values at 298 K of the equilibrium constants describing the formation of these species from Se(cr) and $SeOCl_2(aq)$. The results were:

$$3Se(cr) + SeOCl_2(aq) + 2HCl \rightleftharpoons 2Se_2Cl_2(aq) + H_2O;$$

$$K = ([Se_2Cl_2]^2 \times a_w) / ([SeOCl_2] \times a_{HCl}^2) = (1.47 \pm 0.15) \times 10^{-9}\ M^{-1}$$

and

$$\text{Se(cr)} + \text{SeOCl}_2\text{(aq)} + 2\text{HCl} \rightleftharpoons 2\text{SeCl}_2\text{(aq)} + \text{H}_2\text{O};$$

$$K = ([\text{Se}_2\text{Cl}_2]^2 \times a_\text{w}) / ([\text{SeOCl}_2] \times a_{\text{HCl}}{}^2) = (0.75 \pm 0.10) \times 10^{-9}\ \text{M}^{-1}.$$

The conditions for the existence of these complexes are, however, judged to fall outside the limits of this review and the equilibrium constants are therefore not selected.

V.4.3 Bromine compounds

V.4.3.1 Se_2Br_2

V.4.3.1.1 Se_2Br_2(l)

Tsvetkov [78TSV] mixed liquid bromine and crystalline selenium at various Br/Se ratios in a sealed glass ampoule and measured the heat of reaction in an adiabatic calorimeter. The standard enthalpy of formation for Se_2Br_2(l) thus obtained:

$$\Delta_\text{f} H^\circ_\text{m} (\text{Se}_2\text{Br}_2, \text{l}, 298.15\ \text{K}) = -(31 \pm 3)\ \text{kJ·mol}^{-1}$$

is selected with the uncertainty assigned by the present review.

V.4.3.1.2 Se_2Br_2(g)

Thermodynamic properties for Se_2Br_2(g) were calculated by Ramaswamy and Jayaraman [71RAM/JAY] by statistical thermodynamic methods. The heat capacity expression of Se_2Br_2(g)

$$C^\circ_{p,\text{m}} (\text{Se}_2\text{Br}_2, \text{g}, T) = (82.85 + 0.3599 \times 10^{-3}\ T - 0.098 \times 10^{-6}\ T^2 - 0.3112 \times 10^{6}\ T^{-2})\ \text{J·K}^{-1}\text{·mol}^{-1}$$

was obtained by the review from a least squares calculation on the data given in the temperature range 298.15 to 1000 K. The selected value at 298.15 K is:

$$C^\circ_{p,\text{m}} (\text{Se}_2\text{Br}_2, \text{g}, 298.15\ \text{K}) = (79.5 \pm 0.1)\ \text{J·K}^{-1}\text{·mol}^{-1}.$$

The calculated standard entropy value of Se_2Br_2(g) is selected:

$$S^\circ_\text{m} (\text{Se}_2\text{Br}_2, \text{g}, 298.15\ \text{K}) = (377.4 \pm 0.2)\ \text{J·K}^{-1}\text{·mol}^{-1}$$

with the uncertainty assigned by the review.

The value of the standard enthalpy of formation for Se_2Br_2(g) reported in [82WAG/EVA], $\Delta_\text{f} H^\circ_\text{m}$ (Se_2Br_2, g, 298.15 K) = 29 kJ·mol^{-1} is adopted, but not selected, as its provenance is not known.

V.4.3.1.3 Se_2Br_2(cr)

From a phase diagram study of the system Se-Br, Born *et al.* [79BOR/KNI] identified an intermediate phase of composition Se_2Br_2 with a melting point at 278 K. No thermodynamic data have been found for this compound.

V.4.3.2 $SeBr_2$(g)

The equilibrium reaction:

$$Se_2Br_2(l) \rightleftharpoons Se(cr) + SeBr_2(g) \quad (V.43)$$

was studied by Högberg and Lundqvist [70HOG/LUN] in the temperature range 293 to 348 K. The vapour pressure of $SeBr_2$(g) in equilibrium with Se(cr) and Se_2Br_2(l), saturated with Se, was determined by a flow method with N_2 as carrier gas. The partial pressure of $SeBr_2$(g) is described by the equation $\log_{10} p\,(SeBr_2, g, (293 - 348)\ K) = 6.43 - 3278\ T^{-1}$. No solubilities of Se(cr) in Se_2Br_2(l) are presented (*cf.* Section V.4.2.1.1 concerning determined solubilities in Se_2Cl_2(l) by Lundkvist and Lellep [68LUN/LEL]). The solubility is assumed to be small so that the activities of the condensed phases are unity in the temperature range studied. The temperature dependence of the equilibrium constant can be written:

$$\log_{10} K\ ((V.43), (293 - 348)\ K) = (6.43 \pm \text{constant}) - 3278\ T^{-1}.$$

From this equation, the enthalpy of reaction is derived to be:

$$\Delta_r H_m^\circ\ ((V.43), 298.15\ K) = (63 \pm 20)\ kJ\cdot mol^{-1}$$

with the uncertainty estimated by the present review. Reaction (V.43) has also been studied by Oppermann *et al.* [2001OPP/SCH] by direct pressure measurements. The reaction enthalpy reported is $(49 \pm 16)\ kJ\cdot mol^{-1}$. The reason for the selection of the value in [70HOG/LUN] is presented in Appendix A.

By applying the enthalpy relationship $\Delta_r H_m^\circ$ ((V.43), 298.15 K) = $\Delta_f H_m^\circ$ ($SeBr_2$, g, 298.15 K) − $\Delta_f H_m^\circ$ (Se_2Br_2, l, 298.15 K) and inserting the selected standard enthalpy of formation for Se_2Br_2(l), see Section V.4.3.1.1, the following selected value of the standard enthalpy of formation for $SeBr_2$(g) is obtained:

$$\Delta_f H_m^\circ\ (SeBr_2, g, 298.15\ K) = (32 \pm 20)\ kJ\cdot mol^{-1}.$$

Oppermann *et al.* [2001OPP/SCH] report a widely different value for the enthalpy of formation of $SeBr_2$(g). The origin of the discrepancy is discussed in Appendix A.

Mills [74MIL] calculated thermal thermodynamic quantities from estimated molecular parameters. In the temperature range 298.15 to 2000 K, the heat capacity for $SeBr_2$(g) is presented as:

$$C_{p,m}^\circ\ (SeBr_2, g, T) = (58.12 + 0.04435 \times 10^{-3}\ T - 0.2410 \times 10^{6}\ T^{-2})\ J\cdot K^{-1}\cdot mol^{-1}.$$

The heat capacity at 298.15 K is:

$$C^{\circ}_{p,\mathrm{m}}(\mathrm{SeBr_2, g, 298.15\ K}) = (55.42 \pm 0.50)\ \mathrm{J{\cdot}K^{-1}{\cdot}mol^{-1}}$$

and the standard entropy for $SeBr_2$(g), is

$$S^{\circ}_{\mathrm{m}}(\mathrm{SeBr_2, g, 298.15\ K}) = (317 \pm 1)\ \mathrm{J{\cdot}K^{-1}{\cdot}mol^{-1}}$$

with the uncertainties assigned by the review.

The data further yield the standard Gibbs energy of formation at 298.15 K:

$$\Delta_{\mathrm{f}}G^{\circ}_{\mathrm{m}}(\mathrm{SeBr_2, g, 298.15\ K}) = -(5 \pm 20)\ \mathrm{kJ{\cdot}mol^{-1}}.$$

These data are selected.

V.4.3.3 SeBr₄

V.4.3.3.1 SeBr₄(cr)

The enthalpies of dissolution of $SeBr_4$(cr) and SeO_2(cr) in 4.02 M HCl were measured by Westland and Makhija [78WES/MAK] and in 1 M HBr by Oppermann *et al.* [89OPP/RIT]. The calorimetric results obtained in these studies were combined with literature data on standard enthalpies of formation for SeO_2(cr), HBr(g), H_2O(l) and heat of solution of HBr(g). Although the standard enthalpies of formation they used deviate slightly from the present values given in Table IV-1, their reported values $\Delta_{\mathrm{f}}H^{\circ}_{\mathrm{m}}(\mathrm{SeBr_4, cr, 298.15\ K}) = -(74.9 \pm 2.1)$ and $= -(69.7 \pm 6.6)\ \mathrm{kJ{\cdot}mol^{-1}}$, respectively, are quite representative considering the given uncertainties. From measurements in an adiabatic calorimeter, Tsvetkov [78TSV] obtained the value $\Delta_{\mathrm{f}}H^{\circ}_{\mathrm{m}}(\mathrm{SeBr_4, cr, 298.15\ K}) = -(65.3 \pm 1.3)\ \mathrm{kJ{\cdot}mol^{-1}}$. Considering all these studies, the review selects:

$$\Delta_{\mathrm{f}}H^{\circ}_{\mathrm{m}}(\mathrm{SeBr_4, cr, 298.15\ K}) = -(70.0 \pm 5.0)\ \mathrm{kJ{\cdot}mol^{-1}}.$$

V.4.3.3.2 SeBr₄(g)

At room temperatures there is still some evaporation of $SeBr_4$(cr) to $SeBr_4$(g). However, at temperatures higher than 340 K, $SeBr_4$(g) dissociates to Se_2Br_2(l, g) and Br_2(g). No thermodynamic data for $SeBr_4$(g) are available.

V.4.3.4 SeOBr₂(cr)

The enthalpies of dissolution of $SeOBr_2$(cr) and SeO_2(cr) in aqueous hydrobromic acid were measured by Oppermann *et al.* [89OPP/RIT]. The calorimetric results obtained were combined with literature data on standard enthalpies of formation for SeO_2(cr), HBr(g), H_2O(l) and heat of solution of HBr(g). The standard enthalpy of formation thus derived for $SeOBr_2$(cr) is selected:

$$\Delta_{\mathrm{f}}H^{\circ}_{\mathrm{m}}(\mathrm{SeOBr_2, cr, 298.15\ K}) = -(140 \pm 5)\ \mathrm{kJ{\cdot}mol^{-1}}$$

with the uncertainty assigned by the present review.

V.4.3.5 $SeOBr_2(g)$

Thermodynamic data for $SeOBr_2(g)$ are reported by Oppermann *et al.* [2001OPP/SCH] from pressure measurements of equilibrium reactions involving $SeOBr_2(g)$. As discussed in Appendix A the derived formation data could not be selected by the review.

V.4.3.6 Aqueous complexes

Milne and LaHaie [85MIL/HAI] have studied the interactions occurring between Se(IV) and Br^- in 1 to 8.9 M aqueous solutions of HBr. The experimental techniques used were Raman and UV/Vis spectroscopy, and the appearance of changing band intensities and/or new bands was interpreted as being due to the formation of a series of Se(IV)-Br^- adducts.

At 1 and 3 M HBr, the species $HSeO_2Br(aq)$ was postulated to form according to

$$H_2SeO_3(aq) + H^+ + Br^- \rightleftharpoons HSeO_2Br(aq) + H_2O(l);$$

$$K = ([HSeO_2Br] \times a_w) / ([H_2SeO_3] \times [H^+] \times [Br^-] \times \gamma_{\pm HBr}^2).$$

The molar intensities of some assumed $HSeO_2Br$ Raman bands at 100 to 300 cm^{-1}, 700 cm^{-1}, and 900 cm^{-1} were varied and the corresponding value of K was evaluated from the total intensities measured. The value of the molar intensity that gave reasonable agreement in K for both HBr concentrations was selected and the corresponding value of K accepted as the best value of the equilibrium constant. A critical perusal of the procedure used reveals, however, that the molar intensities of the $HSeO_2Br$ bands obtained in this manner are quite similar to the corresponding bands for H_2SeO_3 in exactly the same wave-number regions. This implies that the evaluated value of K, 4.9 M^{-2}, must be judged as being highly uncertain and it is therefore not accepted by this review.

At 3.5 to 5 M HBr, a significant increase in the 100 to 300 cm^{-1} band area was registered, and this was postulated to be due to the formation of a species $SeOBr_3^-$. Again, the molar intensity of the band was varied to yield an invariant equilibrium constant and

$$K = ([SeOBr_3^-] \times a_w^2) / ([H_2SeO_3] \times [H^+]^2 \times [Br^-]^3 \times \gamma_{\pm HBr}^4) = 4.7 \times 10^{-3}\ M^{-5}$$

was evaluated. However, the evidence presented to support the stoichiometry of this species is very weak and, therefore, this equilibrium constant is not accepted by the review. Finally, at above 5.5 M HBr, a shoulder in the UV/Vis spectrum at 385 nm was recorded and taken, without further proof, as being due to the simultaneous formation of $SeBr_5^-$ and $SeBr_6^{2-}$. Again, equilibrium constants were calculated by varying the molar absorbtivities and, again, the values are refuted due to the lack of evidence for the stoichiometries of the postulated complexes.

In solutions containing high HBr concentrations and Se(IV), elemental Se(cr) can be dissolved. Milne [89MIL] studied these reactions at 4.8 and 8.72 M HBr (293 K) by gravimetric and Raman spectroscopic measurements.

At 8.72 M HBr, a linear relationship between the concentration of Se(IV) and the solubility of Se(0) was recorded for [Se(IV)] < 0.0165 M, and a constant Se(0) solubility above this value. The linear part of the curve was used to conclude that the average oxidation state of selenium in these solutions was (1.49 ± 0.03). On this basis and the positions of three Raman bands, an equilibrium constant for the reaction

$$Se_2Br_2(aq) + 2Br^- \rightleftharpoons SeBr_4^{2-} + Se(cr),$$

$$K = ([SeBr_4^{2-}]) / ([Se_2Br_2] \times [Br^-]^2),$$

was calculated to be 0.025 M^{-2}. The value of the constant solubility at [Se(0)] = 0.0287 M was combined with this equilibrium constant to conclude that the solubility of $Se_2Br_2(cr)$ is 0.0116 M in 8.72 M HBr. These findings are accepted by the review but, since the data cannot be extrapolated to standard conditions, they are not selected by the review. Also in 4.8 M HBr, a gradual increase in the solubility of Se(cr) with increasing concentrations of Se(IV) was recorded. The effect was considerably weaker than at 8.72 M HBr and significantly higher concentrations of Se(IV) were needed. The effect was modelled by assuming the presence of Se_2Br_2 and $SeBr_4^{2-}$ in combination with $HSeO_2Br$ and $SeOBr_3^-$.

V.4.4 Iodine compounds

A few Se-I compounds, for example Se_8I, have been identified by Gopal and Milne [92GOP/MIL] in solutions of CS_2 or CCl_4. In air, no stable compounds seem to be formed. No thermodynamic data are available for these compounds.

V.5 Group 16 compounds and complexes

V.5.1 Sulphur compounds

Due to the chemical and structural similarities between selenium and sulphur, this is a research field, which has received considerable interest over the years.

V.5.1.1 The selenium-sulphur system

The binary phase diagram was studied by Ringer [02RIN] and by Boctor and Kullerud [87BOC/KUL] with rather different results. Although the liquidus temperature curves reported are quite similar, the interpretation of the solidus and subsolidus relationships differ considerably. Thus, while Ringer reported on the existence of an eutectoid at 348 K and 16.5 mol-% Se, Boctor and Kullerud could not confirm this finding. Also, the peritectic isotherm in the Se-rich part of the system was considerably expanded in [87BOC/KUL] compared to that in [02RIN]. Since the phase relation diagram presented by Boctor and Kullerud is based on modern experimental techniques, it is judged as far more trustworthy and it is presented in Figure V-11. Observe that this diagram is schematic only since it lacks the (possibly narrow) diphasic regions and invariant reactions characteristic of a true phase diagram. It should be noted that the review presented in [96SHA/CHA2] only covered the literature up until 1985. Its conclusions are therefore based on the study by Ringer only.

Figure V-11: Phase relations in the condensed Se-S system. Reprinted from [87BOC/KUL], Copyright (1987), with permission from Elsevier.

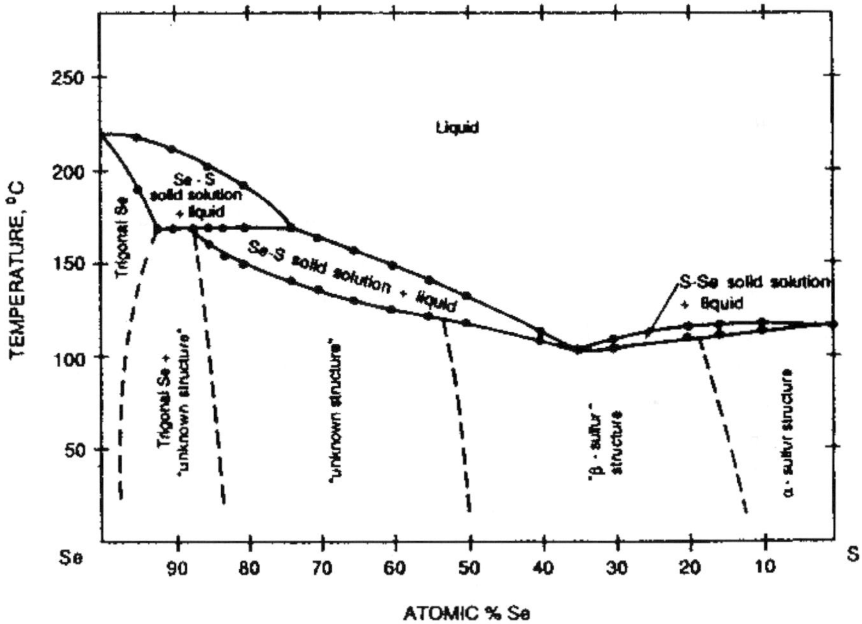

X-ray diffraction has been used to conclude that S_8- and Se_8-ring molecules exist in the S- and Se-rich regions of the diagram, respectively, while the structures in the intermediate region are unknown and the solids characterised by poor crystallinity. These solids have been the subject of a number of studies in which they were extracted with various solvents and recrystallised [67COO/CUL], [78LAI/NII], [79LAI/NII], [79EYS/SUN], [80LAI/STE], [83STE/STR]. In all of these studies, the presence of a series of different Se_nS_{8-n} molecules was indicated. As pointed out by Bocter and Kullerud, however, these data do not necessarily prove the existence of such molecules in the "unknown structure region", since they may actually have been formed during the extraction process.

No thermodynamic data characterising the mixed Se_nS_{8-n} phases have been identified in the literature, except for some enthalpies of fusion presented in [78LAI/NII].

Bocter and Kullerud [87BOC/KUL] also studied Se-S melts, and found no evidence of liquid immiscibility. They found, however, that the temperature for polymerisation in the melt significantly decreased, and could principally be described by an equilibrium copolymerisation model [84KEN/WHE]. The only explicit thermodynamic datum available is the enthalpy of mixing, which was studied by Maekawa, Yokokawa and Niwa [73MAE/YOK] at 620 and 730 K. Data were collected in the range of $0.15 \le x_{Se} \le 0.85$ and could, at both temperatures, be described by the equation:

$$\Delta_{mix} H^{\circ}_{m} = ((5.03 \pm 0.33) \times x_{Se} \times (1 - x_{Se}))\ kJ \cdot mol^{-1}$$

where x_{Se} is the mole fraction of selenium in the melt.

Devyatykh and Umilin *et al.* [63DEV/UMI], [65DEV/UMI] determined the relative amounts of selenium and sulphur at 700 to 950 K in the gas phase above these melts. They found that the mole fraction of selenium was lower in the gaseous phase than in the melt. In [63DEV/UMI], the deviations from Raoult's law were utilised to derive thermodynamic information of an assumed $S_nSe(l)$ molecule in the melt. However, since it was later shown by the same group in [65DEV/UMI] that the apparent fractionation on evaporation strongly depended on the experimental technique employed for the measurements, this review has not accepted the calculations presented in [63DEV/UMI].

V.5.1.2 SeS(g)

From molecular parameters presented in [87FIN/KRU], the heat capacity expression

$$C^{\circ}_{p,m}(SeS, g, (298 - 2000)\ K) = 32.001 + 4.9002 \times 10^{-3}\ T - 5.35 \times 10^{-7}\ T^{2} + 1.8632 \times 10^{3}\ T^{-1} - 3.4596 \times 10^{5}\ T^{-2})\ J \cdot K^{-1} \cdot mol^{-1}$$

was derived by the review. This expression leads to the selected value:

$$C^{\circ}_{p,m}(SeS, g, 298.15\ K) = (35.8 \pm 0.1)\ J \cdot K^{-1} \cdot mol^{-1}.$$

The molecular parameters were also used to calculate the selected standard entropy of SeS(g):

$$S^{\circ}_{m}(SeS, g, 298.15\ K) = (245.1 \pm 0.5)\ J \cdot K^{-1} \cdot mol^{-1}.$$

The reaction:

$$\tfrac{1}{2}Se_2(g) + \tfrac{1}{2}S_2(g) \rightleftharpoons SeS(g) \qquad (V.44)$$

has been studied by Drowart and Smoes [77DRO/SMO], and Huang, Meschi and Searcy [79HUA/MES] employing a Knudsen effusion cell with EuS/VSe_2 and In_2S_3/In_2Se_3, respectively. From mass spectrometric measurements in the temperature range of 1760 to 2110 K, Drowart and Smoes [77DRO/SMO] obtained $\log_{10} K^\circ((V.44), T) = (0.20 \pm 0.04) - (260 \pm 80)\, T^{-1}$. Huang *et al.* [79HUA/MES] performed measurements at 1120 to 1200 K and found the value of K° (V.44) obtained to vary irregularly with the temperature and to depend on the diameter of the orifice in the Knudsen cell. Two values of the constant are reported, namely $\log_{10} K^\circ(V.44) = -(0.08 \pm 0.08)$ and (0.24 ± 0.09), depending on the orifice diameter employed. Extrapolation of the equation presented in [77DRO/SMO] to the temperature range of 1120 to 1200 K yields $\log_{10} K^\circ(V.44) = -0.02$ to -0.03, a result which falls between the determinations in [79HUA/MES].

The present review selects the data in [77DRO/SMO]. A third law evaluation using these data results in:

$$\Delta_r H_m^\circ((V.44), 298.15\text{ K}) = (8.5 \pm 6.7)\text{ kJ·mol}^{-1},$$

$$\Delta_r G_m^\circ((V.44), 298.15\text{ K}) = (6.3 \pm 6.7)\text{ kJ·mol}^{-1},$$

$$\Delta_r S_m^\circ((V.44), 298.15\text{ K}) = (7.3 \pm 0.5)\text{ J·K}^{-1}\text{·mol}^{-1}.$$

The selected reaction data yield with accepted auxiliary data:

$$\Delta_f H_m^\circ(\text{SeS, g, 298.15 K}) = (143.3 \pm 6.9)\text{ kJ·mol}^{-1},$$

$$\Delta_f G_m^\circ(\text{SeS, g, 298.15 K}) = (92.4 \pm 6.7)\text{ kJ·mol}^{-1}.$$

V.5.1.3 Aqueous selenothionates

Greiver and Zaitseva [67GRE/ZAI] observed that elemental selenium dissolves in a concentrated, highly alkaline aqueous solution of mono- and polysulphides under oxygen-free conditions. The colour of the solution changes from colourless/yellow over to intense red-brown. Based on some relatively crude stoichiometric data, the authors claimed that the dominating species in such solutions is SSe_2^{2-}. The standard potential of the couple Se, S/SSe_2^{2-} was reported to be -0.383 V at 298.15 K in the same reference. However, the electrode reaction of the cell employed for this determination was never established and neither was its reversibility proved. This datum (2Se(s) + S(s) + $2e^- \rightleftharpoons SSe_2^{2-}$; $\log_{10} K = -12.95$) is therefore not accepted by the present review.

In a later paper by the same authors, [71GRE/ZAI], highly concentrated sodium sulphide solutions were heated in the presence of various amounts of elemental selenium. The crystals formed upon cooling were analysed for their sodium, selenium, and sulphur contents. The existence of non-stoichiometric compounds in the system Na_2S-Na_2SSe_2 was claimed from the analytical data and a series of X-ray powder diffractograms. The aqueous solubilities of these compounds at room temperature were reported to fluctuate between 3.5 to 4 M. It was also noted that the colour of these solutions gradually changed from brownish yellow to rose-yellow at higher sulphide contents. The change in colour was hypothetically assumed to indicate the simultaneous existence of the species SSe^{2-} and SSe_2^{2-}. On this basis and from spectrophotometric data at 390 nm, the equilibrium constant of the reaction $SSe_2^{2-} + S^{2-} \rightleftharpoons 2SSe^{2-}$ was determined to be $\log_{10} K = (0.9 \pm 0.1)$ at 291 K and 1 M NaOH medium. No verification of the hypothesis was, however, presented in the paper and the equilibrium constant is therefore not accepted by the review.

In conclusion, the review finds no reliable thermodynamic data for selenothionates in aqueous solution.

V.5.1.4 Aqueous selenosulphates

The selenosulphate ion, $SeSO_3^{2-}$, is formed when an oxygen-free alkali sulphite solution is reacted with a finely divided powder of elemental selenium [60KLE/OST], [95BAL/MIL]. It is characterised by a single NMR resonance at 646.3 ppm *vs.* Me_2Se and by a strong Raman band at 310 cm^{-1} [95BAL/MIL]. Klebanov and Ostapkevich [60KLE/OST] studied the reaction:

$$Se(s) + SO_3^{2-} \rightleftharpoons SeSO_3^{2-} \qquad (V.45)$$

in solutions of sodium and potassium sulphite (concentrations not reported). The reported values of the equilibrium constant at 293 K were 0.49 and 0.75, respectively. In the article, it is hypothesised that the difference is due to a difference in solution pH between the sodium and the potassium sulphite experiments, but the cause of such a difference cannot be deduced from the experimental protocol presented.

Ball and Milne [95BAL/MIL] employed molar solutions of sodium sulphite and obtained an equilibrium constant of 0.43 (triplicate) at 293 K, *i.e.* in relatively good agreement with the value found by Klebanov and Ostapkevich. In both articles, Reaction (V.45) was found to be favoured by an increase in temperature. The reported equilibrium constants in sodium sulphite solution in the temperature range 273 to 371 K have, by this review, been utilised to calculate a reaction enthalpy of:

$$\Delta_r H_m^\circ((V.45), I = 3\ M, 298.15\ K) = (19.7 \pm 1.2)\ kJ\cdot mol^{-1}.$$

This datum was used to estimate from the data in [95BAL/MIL]:

$$\log_{10} K((V.45), I = 3\ M, 298.15\ K) = -(0.28 \pm 0.20),$$

corresponding to:

$$\Delta_r G_m^\circ((V.45), I = 3\ M, 298.15\ K) = (1.6 \pm 1.2)\ kJ\cdot mol^{-1}.$$

Extrapolation of the value of the equilibrium constant to standard conditions cannot be made, but the value might as a first approximation be considered independent of *I*.

The entropy change of the reaction becomes:

$$\Delta_r S^\circ_m ((V.45), I = 3 \text{ M}, 298.15 \text{ K}) = (61 \pm 6) \text{ J·K}^{-1}\text{·mol}^{-1}.$$

Combined with $\Delta_f G^\circ_m (SO_3^{2-}, 298.15 \text{ K}) = -(487.5 \pm 4.0) \text{ kJ·mol}^{-1}$ from [92GRE/FUG], the Gibbs energy of formation for $SeSO_3^{2-}$ is estimated to be:

$$\Delta_f G^\circ_m (SeSO_3^{2-}, 298.15 \text{ K}) = -(485.9 \pm 4.2) \text{ kJ·mol}^{-1}.$$

Fofanov and Kitaev [69FOF/KIT] employed a strongly bound Pb(II)-Trilon B complex (the identity of "Trilon B" is not explicitly given) to study the disproportionation of the selenosulphate ion:

$$SeSO_3^{2-} + H_2O(l) \rightleftharpoons H_2Se(aq) + SO_4^{2-} \qquad (V.46)$$

The Pb(II) complex reacted with $SeSO_3^{2-}$ in a 0.2 to 1.2 M OH^- medium at an undefined temperature forming PbSe(s). The composition of the reaction mixture at equilibrium was found from a determination of the concentration of Pb(II) remaining in solution. The concentrations of the other members of the reaction:

$$SeSO_3^{2-} + PbX^{2-x} + 2OH^- \rightleftharpoons PbSe(s) + SO_4^{2-} + X^{x-} + H_2O(l) \qquad (V.47)$$

were calculated from the stoichiometry of the reaction. The equilibrium constant of the reaction was evaluated from these data to be $\log_{10} K$ (V.47) = (1.65 ± 0.43). Combined with the reported stability constant of the Trilon B complex, $\log_{10} K$ = 18.04, and accepted data for the acidity constants of H_2Se and the solubility product of PbSe(cr), $\log_{10} K$ ((V.46), I = 0.6 – 1.6 M) = – (23 ± 2) is obtained. This value is not selected due to the undefined temperature and other conditions of the experiments.

V.5.2 Tellurium compounds

V.5.2.1 The selenium-tellurium system

The phase diagram of the Se-Te system is characterised by the occurrence of a continuous series of solid solutions and complete liquid miscibility. The Se-rich melts are prone to strong supercooling. The phase diagram has been repeatedly studied (thermal analysis, electroconductivity, thermopower) with conforming results as reviewed by Chizhevskaya and Shelimova [97CHI/SHE]. Ghosh *et al.* [88GHO/LUK] have successfully modelled the solidus and liquidus curves by using the Redlich-Kistner polynomial to describe the excess Gibbs energy functions.

The Se-Te melt shows a complex behaviour with respect to structural, thermodynamic [73MAE/YOK] and electrical properties. The heat capacity, and the enthalpy and Gibbs energy of mixing of these melts have been recently modelled [95CLA/COM], [96AMZ/GIL], [97COM/MOR] by assuming the presence of polymeric species with twofold coordinated selenium atoms and two- and threefold coordinated tellurium atoms.

In two papers, [69KOZ/VAN] and [69UST/VIG4], the possibility of refining tellurium from 0.01 to 0.6 mass-% of selenium impurities by distillation has been studied. Both papers indicate that selenium is enriched in the vapour by a factor of 2 to 3 as compared with its fraction in the liquid phase. Furthermore, in [69KOZ/VAN] it is shown that the separation is more efficient at atmospheric pressure than at reduced pressure. No attempts are made in these papers to rationalise the findings in thermodynamic terms.

V.5.2.2 TeSe(g)

From molecular parameters presented in [89FIN/SET], the heat capacity expression

$$C^{\circ}_{p,\mathrm{m}}(\mathrm{TeSe, g, (298-950)\ K}) = 119.809 - 7.4993 \times 10^{-2}\, T + 2.424 \times 10^{-5}\, T^2 - 2.9599 \times 10^{4}\, T^{-1} + 3.3304 \times 10^{6}\, T^{-2})\ \mathrm{J{\cdot}K^{-1}{\cdot}mol^{-1}}$$

and

$$C^{\circ}_{p,\mathrm{m}}(\mathrm{TeSe, g, (950-2000)\ K}) = (-14.324 + 2.4844 \times 10^{-2}\, T - 3.8276 \times 10^{-6}\, T^2 + 5.1424 \times 10^{4}\, T^{-1} - 1.533 \times 10^{7}\, T^{-2})\ \mathrm{J{\cdot}K^{-1}{\cdot}mol^{-1}}$$

were derived by the review. The former expression leads to the selected value:

$$C^{\circ}_{p,\mathrm{m}}(\mathrm{TeSe, g, 298.15\ K}) = (37.8 \pm 0.1)\ \mathrm{J{\cdot}K^{-1}{\cdot}mol^{-1}}.$$

The molecular parameters were also used to calculate the standard entropy of TeSe(g):

$$S^{\circ}_{\mathrm{m}}(\mathrm{TeSe, g, 298.15\ K}) = (257.9 \pm 0.5)\ \mathrm{J{\cdot}K^{-1}{\cdot}mol^{-1}}.$$

The formation of gaseous TeSe(g) was studied by Ustyugov, Vigdorovich and Kudryavtsev [68UST/VIG] at 1090 to 1270 K and by Drowart and Smoes [77DRO/SMO] at 1400 to 1640 K. In both articles, the data were interpreted in terms of the reaction

$$\tfrac{1}{2}\mathrm{Se_2(g)} + \tfrac{1}{2}\mathrm{Te_2(g)} \rightleftharpoons \mathrm{TeSe(g)} \qquad \text{(V.48)}$$

Ustyugov *et al.* measured the total unsaturated vapour pressure of selenium-tellurium mixtures with a quartz membrane manometer. The partial pressures of $Se_2(g)$, $Te_2(g)$, and TeSe(g) were calculated from a knowledge of the total pressure, the amounts of substance taken, and the equilibrium constants for the side-reactions with formation of the Se_4, Se_6, Se_8, Te, and Te_4 gaseous species. From the partial pressures presented it is found that the minor species contribute, in total, to less than 10% of the total pressure. The results can be expressed by $\log_{10} K^{\circ}((\mathrm{V.48}), T) = 0.6270 - 368.4\ T^{-1}$. A third law evaluation of these data yields $\Delta_{\mathrm{r}} H^{\circ}_{\mathrm{m}}((\mathrm{V.48}), 298.15\ \mathrm{K}) = (0.1 \pm 0.2)$ kJ·mol^{-1}.

Drowart and Smoes set up the equilibrium in a Knudsen cell containing VSe_2 and TiTe. The relative partial pressures were calculated from the ion intensities measured by a mass spectrometer. The equilibrium constant followed the relationship $\log_{10} K^{\circ}((\mathrm{V.48}), T) = (0.25 \pm 0.07) - (40 \pm 100)\ T^{-1}$. The third law evaluation of the enthalpy of reaction yields $\Delta_r H_m^\circ$((V.48), 298.15 K) = (4.4 ± 1.0) kJ·mol^{-1}. The review adopts the average of these results and obtains:

$$\Delta_r H_m^\circ((\text{V.48}), 298.15\text{ K}) = (2.2 \pm 3.0)\text{ kJ·mol}^{-1},$$

$$\Delta_r G_m^\circ((\text{V.48}), 298.15\text{ K}) = (0.8 \pm 3.1)\text{ kJ·mol}^{-1},$$

$$\Delta_r S_m^\circ((\text{V.48}), 298.15\text{ K}) = (4.7 \pm 1.0)\text{ J·K}^{-1}\text{·mol}^{-1}$$

and

$$\Delta_f H_m^\circ(\text{TeSe, g, 298.15 K}) = (154.4 \pm 3.4)\text{ kJ·mol}^{-1},$$

$$\Delta_f G_m^\circ(\text{TeSe, g, 298.15 K}) = (104.7 \pm 3.4)\text{ kJ·mol}^{-1}.$$

The formation data were calculated using $\Delta_f H_m^\circ$(Te_2, g, 298.15 K) = (163.2 ± 1.0) kJ·mol^{-1} from [95DAV/RAN] and are compiled in Appendix E, since they were calculated with non-TDB auxiliary data.

The values reported in [82WAG/EVA] are: $\Delta_f H_m^\circ$ = 159.0 kJ·mol^{-1}, S_m° = 265.8 J·K^{-1}·mol^{-1} and $\Delta_f G_m^\circ$ = 108.8 kJ·mol^{-1}, respectively.

V.5.2.3 Aqueous complexes

Zaitseva *et al.* [71ZAI/GRE] recorded the dissolution of tellurium when elemental tellurium was reacted at 353 K with concentrated aqueous solutions of strictly oxygen-free sodium mono- and polyselenides. The maximum Te:Se ratio recorded in the experiments ranged between 0.3 and 0.4 and these figures were interpreted as being due to the formation of $TeSe^{2-}$ and $TeSe_3^{2-}$ ions. The former species was also reported to crystallise as $Na_2TeSe·10H_2O$(s) from the aqueous phase. No experimentally verified thermodynamic data characterising these species have been found.

V.5.2.4 Selenium dioxide-tellurium dioxide

By reacting equimolar amounts of SeO_2(cr) and TeO_2(cr) under sealed and inert conditions at 620 K, $TeSeO_4$(cr) is formed [82CAS/JER], [83DIM/LAK], [84BAR/PET], [84CAS/PIC]. According to Bart and Petrini [84BAR/PET], two modifications of the compound exist, an α-$TeSeO_4$ phase with a triclinic unit cell with a volume of 181.7 Å^3 and a β-$TeSeO_4$ phase with an orthorhombic unit cell with a volume of 335.8 Å^3 described in [83DIM/LAK] and [82CAS/JER].

The compound is highly hygroscopic and a range of phases can be identified after exposure to moist air [84BAR/PET]. When heated in an inert but open atmosphere, Castro, Pico, Jerez and Veiga [84CAS/PIC] recorded, as did [84BAR/PET], that $TeSeO_4$(cr) started to decompose into SeO_2(g) and TeO_2(cr) at 590 K. Under closed conditions, Dimitriev, Lakov and Ivanova [83DIM/LAK] observed a strong endothermic effect in DTA data at 748 K and interpreted this effect as due to the melting of $TeSeO_4$(cr). Based on this effect, and a second endothermic effect at 593 K for mixtures of higher SeO_2-content, Dimitriev *et al.* constructed a partial phase diagram for the

SeO_2-TeO_2 system containing an eutectic point at about 75 mol-% SeO_2 and (593 ± 5) K.

Castro *et al.* [82CAS/JER], [84CAS/PIC] have also indicated the presence of compounds with other stoichiometries in the system, and characterised them by differential scanning calorimetry. However, the existence of these compounds do not find support in the studies by Dimitriev *et al.* [83DIM/LAK] and Bart and Petrini [84BAR/PET], and they have therefore not been accepted by the review.

V.6 Group 15 compounds and complexes

V.6.1 Nitrogen compounds

V.6.1.1 Selenium nitride

Selenium nitride, Se_4N_4(cr), can be prepared by reacting SeO_2, $SeCl_4$ or $SeBr_4$ with NH_3 at 345 to 355 K and 40 to 50 bar [60JAN/DOE]. It is characterised by an intense orange-red colour, it is practically insoluble in water and it can spontaneously explode when in pure form [58JAN/DOE]. An early report on its heat of decomposition, based on two data only and without experimental detail, is that of Berthelot and Vieille [1884BER/VIE], – 708 kJ·mol^{-1}. More recently, Barker *et al.* [65BAR/COR] studied the system calorimetrically with a high standard deviation in the results, ± 70 kJ·mol^{-1}, n = 8. From the argument that low values could be indicative of partial decomposition of the sample before the calorimetric measurement, the "best" value was taken as the average of the four most exothermic runs, – (679 ± 17) kJ·mol^{-1}, of the reaction:

$$Se_4N_4(cr) \rightarrow 4Se(cr) + 2N_2(g). \qquad (V.49)$$

The review accepts this argument and selects:

$$\Delta_f H_m^\circ (Se_4N_4, cr, 298.15 K) = (679 \pm 35) kJ \cdot mol^{-1}.$$

The value reported in [82WAG/EVA] is 708 kJ·mol^{-1}, and is obviously based on the early [1884BER/VIE] study.

V.6.2 Phosphorus compounds

V.6.2.1 Phosphorus selenides

The compounds P_4Se_3(cr), P_4Se_4(cr), P_4Se_5(cr), and P_4Se_{10}(cr) were reported by Blachnik and Hoppe [79BLA/HOP]. Thermodynamic information is only available for P_4Se_3(cr). Phase diagrams can be found in [79BLA/HOP] and [93BLA].

V.6.2.1.1 P_4Se_3(cr)

P_4Se_3(cr) exist in the polymorphs α, β, γ, and α′. The α′ phase is metastable at all temperatures while the other phases have a given range of stability. The enthalpies and temperatures of transition reported in [79BLA/HOP] are given in Table V-30.

Table V-30: Enthalpies and temperatures of transition of the P_4Se_3(cr) polymorphs reported in [79BLA/HOP].

Transition	Temperature (K)	$\Delta_{tr}H^{\circ}_{m}(T)$ (kJ·mol^{-1})
$\alpha \rightarrow \beta$	356	10.98
$\beta \rightarrow \gamma$	479	1.3
$\gamma \rightarrow$ liquid	522	2.79

The heat capacity of α-P_4Se_3 was measured in the temperature range 5 to 350 K by Clever and Westrum [65CLE/WES]. The reported values of the heat capacity and entropy are selected by the review:

$$C^{\circ}_{p,m}(P_4Se_3, \alpha, 298.15\ K) = (166.2 \pm 1.0)\ J{\cdot}K^{-1}{\cdot}mol^{-1}$$

and

$$S^{\circ}_{m}(P_4Se_3, \alpha, 298.15\ K) = (239.6 \pm 1.2)\ J{\cdot}K^{-1}{\cdot}mol^{-1}.$$

Heat capacity measurements in the temperature range 125 to 650 K were also reported in [79BLA/HOP]. However, the result was only presented in the form of a graphical format not easily amenable to re-evaluation.

V.6.2.2 Selenophosphates

Selenophosphates, $H_nSePO_3^{n-3}$ with $0 < n < 3$, are labile selenium donors required for the biochemical synthesis of selenium-dependent enzymes and seleno-RNAs. In the biological fluids, it is formed from ATP and selenide by selenophosphate synthetase [97KAM/GLA]. Aqueous solutions containing $H_nSePO_3^{n-3}$ can be prepared by reacting tris(trimethylsilyloxide)monoselenophosphate [93GLA/SIN] or diisopropylethylammonium O,O-bis(trimethylsilyl)monoselenophosphate [97KAM/GLA] with oxygen-free water. Data presented by Kaminski *et al.* in [97KAM/GLA] show that $HSePO_3^{2-}$ is thermodynamically unstable in aqueous solution and decomposes under first-order reaction kinetics to $H_2PO_4^-$ and HSe^-. The activation parameters for this process are presented in the paper.

Glass *et al.* [93GLA/SIN] recorded the pH-dependent ^{31}P-NMR chemical shift of H_3SePO_3 and concluded that the protonation constants, $\log_{10} K_1$ and $\log_{10} K_2$, of the selenophosphate ion could be estimated to be 8.8 and 4.6, respectively. However, these measurements were performed in a highly complex ionic medium consisting of 0.1 M Tricine + 0.02 M Dithiothreitol + 0.06 M Mg^{2+} with no information on the performance of the pH measurements or on the temperature employed. Therefore, these equilibrium constants are not accepted by the review.

V.6.3 Arsenic compounds

V.6.3.1 Arsenic selenides

Equilibrium conditions in the selenium-arsenic system are very difficult to reach experimentally. The reactions are slow and annealing times of months may not be sufficient because supercooled glasses, that do not crystallise, are easily formed. In addition, the gas phase consists of all the species $Se_n(g)$, n = 1 to 8, $As_m(g)$, m = 1 to 4, together with AsSe(g), $As_2Se_2(g)$, $As_4Se_3(g)$, and $As_4Se_4(g)$. These conditions explain the different and often contradictory results reported in investigations of the system. The phase diagram of the selenium-arsenic system is shown in Figure V-12.

Figure V-12: The phase diagram selenium-arsenic as optimised in [97DEG/PEL]. Reprinted from [97DEG/PEL] with permission from ASM International.

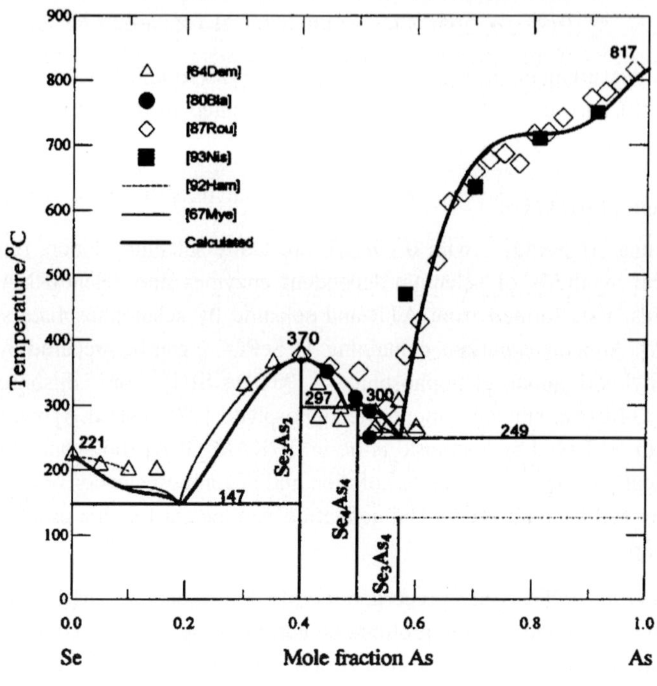

The thermodynamic properties of the condensed phases and the gaseous species were calculated by Degterov, Pelton, and L'Ecuyer [97DEG/PEL] by a least squares optimisation. All available thermodynamic data and phase relationships were treated simultaneously in order to obtain the best agreement between calculated and observed quantities. However, many estimates had to be introduced as auxiliary parameters because of insufficient experimental information. For the solids in the system, experimental information is essentially only available for As_2Se_3(cr) and the properties of As_4Se_3(cr), and As_4Se_4(cr) (referred to as AsSe(cr) by some authors) were estimated from the phase diagram and the properties of As_2Se_3(cr). Similarly, the thermodynamic properties of the liquid phase is dependent on the enthalpy of formation of As_2Se_3(cr) determined by O'Hare, Lewis, Susman and Volin [90OHA/LEW] using fluorine combustion calorimetry. Unfortunately, it was not observed in [97DEG/PEL] that the enthalpy of formation of As_2Se_3(cr) was revised by O'Hare [93OHA] due to a change in the standard enthalpy of formation of AsF_5(g) to $-(1234.2 \pm 0.7)$ $kJ \cdot mol^{-1}$. The thermodynamic optimisation in [97DEG/PEL] therefore needs a revision and no enthalpies of formation or entropies can be selected from the work.

The vapour phase has been studied in a number of investigations, see Table V-31, but it was concluded in [97DEG/PEL] that the measured vapour pressures are accurate only to within one or two orders of magnitude. In the optimisation, all heat capacity and entropy data of the gaseous species were based on estimates. The enthalpies of formation were obtained by matching the calculated and observed total pressures of the system. Considering the strong interplay between the entropy and enthalpy values, the low accuracy of the measured vapour pressures, and the many estimations involved, none of the thermodynamic properties derived for the gaseous species are selected. No more recent experimental thermodynamic information exists.

Table V-31: Experimental investigations of the vapour phase in the system selenium-arsenic.

Reference	Method	Temperature (K)
[68GOR/KRE]	Knudsen effusion and mass spectrometry	540 – 626
[68KNO/BAN]	Laser/mass spectrometry	~ 1300
[73POL/ZAK]	Laser/mass spectrometry	Not reported
[74GOS/MOL]	Knudsen effusion and mass spectrometry	469 – 543
[84DRO/SMO]	Knudsen effusion and mass spectrometry	470
[86STE/ALI]	Knudsen effusion and mass spectrometry	570 – 690

V.6.3.1.1 AsSe(g)

The values of the thermodynamic properties of AsSe(g) reported from the thermodynamic optimisation in [97DEG/PEL] are $C^{\circ}_{p,\mathrm{m}}$ (AsSe, g, 298.15 K) = 35.7 J·K^{-1}·mol^{-1}, $C^{\circ}_{p,\mathrm{m}}$ (AsSe, g, (298.15 – 1800) K) = (35.271 + 1.556 × 10^{-3} T) J·K^{-1}·mol^{-1}, S°_{m} (AsSe, g, 298.15 K) = 206.8 J·K^{-1}·mol^{-1}, and $\Delta_{\mathrm{f}}H^{\circ}_{\mathrm{m}}$ (AsSe, g, 298.15 K) = 86.6 kJ·mol^{-1}. These values need a revision as discussed in Section V.6.3.1. In addition, they are all essentially based on estimates and are consequently not selected in the review.

V.6.3.1.2 As_2Se_2(g)

The values of the thermodynamic properties of As_2Se_2(g) reported from the thermodynamic optimisation in [97DEG/PEL] are $C^{\circ}_{p,\mathrm{m}}$ (As_2Se_2, g, 298.15 K) = 80.8 J·K^{-1}·mol^{-1}, $C^{\circ}_{p,\mathrm{m}}$ (As_2Se_2, g, (298.15 – 1000) K) = (88.75 – 1.910 × 10^{-3} T – 127.51 $T^{-0.5}$) J·K^{-1}·mol^{-1}, S°_{m} (As_2Se_2, g, 298.15 K) = 394.1 J·K^{-1}·mol^{-1}, and $\Delta_{\mathrm{f}}H^{\circ}_{\mathrm{m}}$ (As_2Se_2, g, 298.15 K) = 143.733 kJ·mol^{-1}. These values need a revision as discussed in Section V.6.3.1. In addition, they are all essentially based on estimates and are consequently not selected in the review.

V.6.3.1.3 As_2Se_3(cr)

The optimisation in [97DEG/PEL] accepted the numerical values of the thermodynamic properties of As_2Se_3(cr), including the standard enthalpy, entropy, and low- and high-temperature heat capacities, from the review by O'Hare, Lewis, Susman, and Volin [90OHA/LEW]. Unfortunately, it was not observed in [97DEG/PEL] that the enthalpy of formation of As_2Se_3(cr) was revised by O'Hare [93OHA] due to a change in the enthalpy of formation of AsF_5(g). The thermodynamic optimisation in [97DEG/PEL] therefore needs a revision as discussed in Section V.6.3.1 and only the value of the entropy and heat capacity of As_2Se_3(cr) can be selected from these reviews.

The selected heat capacity expression is:

$$C^{\circ}_{p,\mathrm{m}}(\mathrm{As_2Se_3, cr, (273.15 - 634)\ K}) = (121.23 - 38.170 \times 10^{-3}\,T + 130.21 \times 10^{-6}\,T^2)\ \mathrm{J{\cdot}K^{-1}{\cdot}mol^{-1}}$$

and was derived in [90OHA/LEW] from the heat capacity measurements made by Zigel and Orlova [73ZIG/ORL], Rasulov and Medzhidov [77RAS/MED], and Zhdanov and Maltsev [68ZHD/MAL]. The selected value of the heat capacity at 298.15 K is obtained from the heat capacity expression to be:

$$C^{\circ}_{p,\mathrm{m}}(\mathrm{As_2Se_3, cr, 298.15\ K}) = (121.4 \pm 3.0)\ \mathrm{J{\cdot}K^{-1}{\cdot}mol^{-1}},$$

where the error limits were set by the present review.

The selected value of the entropy is:

$$S^{\circ}_{\mathrm{m}}(\mathrm{As_2Se_3, cr, 298.15\ K}) = (206.0 \pm 3.0)\ \mathrm{J{\cdot}K^{-1}{\cdot}mol^{-1}},$$

and was derived in [90OHA/LEW] from the heat capacity measurements made in the temperature range 53 to 310 K by Zhdanov and Maltsev [68ZHD/MAL]. The heat capacity in the temperature range 0 to 53 K was obtained by an extrapolation procedure based on a quantum-theory function suggested by Nemilov [82NEM].

The enthalpy of formation of As_2Se_3(cr) has been determined in a number of investigations. Blachnik and Schneider [70BLA/SCH] determined the value $\Delta_f H^\circ_m(As_2Se_3, cr, 298.15\ K) = -(103.0 \pm 21.0)\ kJ\cdot mol^{-1}$ using direct synthesis calorimetry. Steblevski, Alikhanyan, Gorgoraki, and Pashinkin [86STE/ALI] determined a number of values in the range $\Delta_f H^\circ_m(As_2Se_3, cr, 298.15\ K) = -95$ to $-49\ kJ\cdot mol^{-1}$ from various gas phase reactions in a mass spectrometric Knudsen-effusion study. Neverov, Rybakova, and Panus [89NEV/RYB] reported the value $\Delta_f H^\circ_m(As_2Se_3, cr, 298.15\ K) = -(122.0 \pm 10.0)\ kJ\cdot mol^{-1}$ from dissolution experiments in "bromine-saturated alkali". The value was recalculated in [90OHA/LEW] using more recent auxiliary data in [82WAG/EVA] yielding $\Delta_f H^\circ_m(As_2Se_3, cr, 298.15\ K) = -(94.0 \pm 10.0)\ kJ\cdot mol^{-1}$. The investigation in [90OHA/LEW] using fluorine combustion calorimetry is the most accurate study and yielded the value $\Delta_f H^\circ_m(As_2Se_3, cr, 298.15\ K) = -(86.1 \pm 4.1)\ kJ\cdot mol^{-1}$. It was later revised by O'Hare [93OHA] due to a change in the enthalpy of formation of AsF_5(g) and the new value is accepted and included in Appendix E since it was calculated using a non-TDB auxiliary datum:

$$\Delta_f H^\circ_m(As_2Se_3, cr, 298.15\ K) = -(81.1 \pm 4.1)\ kJ\cdot mol^{-1}.$$

The Gibbs energy of formation of As_2Se_3(cr) is calculated from the selected values of the enthalpy of formation and entropy, the selected entropy of selenium, and the entropy of arsenic in [92GRE/FUG] yielding:

$$\Delta_f G^\circ_m(As_2Se_3, cr, 298.15\ K) = -(83.9 \pm 4.2)\ kJ\cdot mol^{-1}.$$

V.6.3.1.4 As_4Se_3(cr)

The enthalpy of formation and entropy of As_4Se_3(cr) have not been measured. Degterov *et al.* [97DEG/PEL] estimated the entropy to be $S^\circ_m(As_4Se_3, cr, 298.15\ K) = (250 \pm 10)\ J\cdot K^{-1}\cdot mol^{-1}$ by comparison with the value of As_2Se_3(cr) and the already estimated value of As_4Se_4(cr). The value of the enthalpy of formation was then obtained from the optimisation procedure in [97DEG/PEL] yielding $\Delta_f H^\circ_m(As_4Se_3, cr, 298.15\ K) = -111.496\ kJ\cdot mol^{-1}$. No thermodynamic properties of As_4Se_3(cr) are selected from this optimisation because the assessment in [97DEG/PEL] needs a revision as discussed in Section V.6.3.1.

Blachnik, Hoppe, and Wickel [80BLA/HOP] measured the heat capacity in the temperature range 130 to 650 K. The expression

$$C^\circ_{p,m}(As_4Se_3, cr, (130-440)\ K) = (51.5536 + 230.751 \times 10^{-3}\ T - 28.898 \times 10^{-6}\ T^2 + 2.43452 \times 10^4\ T^{-1} - 2.393908 \times 10^6\ T^{-2})\ J\cdot K^{-1}\cdot mol^{-1}$$

was derived by the review from the values listed in Table 2 of the paper and describes the measured values within ± 1.4 $J\cdot K^{-1}\cdot mol^{-1}$. The relatively large error limit is caused by the scatter of the experimental values. The corresponding heat capacity at 298.15 K is selected:

$$C^{\circ}_{p,m}(\mathrm{As_4Se_3, cr, 298.15\ K}) = (172.5 \pm 4.0)\ \mathrm{J\cdot K^{-1}\cdot mol^{-1}}.$$

V.6.3.1.5 $As_4Se_3(g)$

The values of the thermodynamic properties of $As_4Se_3(g)$ reported from the thermodynamic optimisation in [97DEG/PEL] are $C^{\circ}_{p,m}(\mathrm{As_4Se_3, g, 298.15\ K}) = 148.5\ \mathrm{J\cdot K^{-1}\cdot mol^{-1}}$, $C^{\circ}_{p,m}(\mathrm{As_4Se_3, g, (298.15 - 1000)\ K}) = (157.8107 - 8.26039 \times 10^{5}\ T^{-2})\ \mathrm{J\cdot K^{-1}\cdot mol^{-1}}$, $S^{\circ}_{m}(\mathrm{As_4Se_3, g, 298.15\ K}) = 532.116\ \mathrm{J\cdot K^{-1}\cdot mol^{-1}}$, and $\Delta_f H^{\circ}_{m}(\mathrm{As_4Se_3, g, 298.15\ K}) = 83.11\ \mathrm{kJ\cdot mol^{-1}}$. These values need a revison as discussed in Section V.6.3.1. In addition, they are all essentially based on estimates and are consequently not selected by the review.

V.6.3.1.6 $As_4Se_4(cr)$

The thermodynamic optimisation in [97DEG/PEL] adopted the entropy value $S^{\circ}_{m}(\mathrm{As_4Se_4, cr, 298.15\ K}) = (304.6 \pm 18.0)\ \mathrm{J\cdot K^{-1}\cdot mol^{-1}}$, which was estimated by Mills [74MIL] in a comparison with the entropy of $As_2Se_3(cr)$. The enthalpy of formation of $As_4Se_4(cr)$ was estimated by Mills to be $\Delta_f H^{\circ}_{m}(\mathrm{As_4Se_4, cr, 298.15\ K}) = -(154.0 \pm 50.0)\ \mathrm{kJ\cdot mol^{-1}}$ and the value was adjusted by 3.2 $kJ\cdot mol^{-1}$ in the optimisation procedure in order to reproduce the phase diagram. These values need a revison as discussed in Section V.6.3.1. In addition, they are all essentially based on estimates and are consequently not selected by the review.

Blachnik, Hoppe, and Wickel [80BLA/HOP] measured the heat capacity in the temperature range 300 to 540 K. The expression

$$C^{\circ}_{p,m}(\mathrm{As_4Se_4, cr, (300 - 540)\ K}) = (6.984249 \times 10^{3} - 11.48774\ T + 7.279814 \times 10^{-3}\ T^{2} - 1.755719 \times 10^{6}\ T^{-1} + 1.67046 \times 10^{8}\ T^{-2})\ \mathrm{J\cdot K^{-1}\cdot mol^{-1}}$$

was derived by the review from the values listed in Table 2 of the paper and describes the measured values within ± 1.3 $J\cdot K^{-1}\cdot mol^{-1}$. The relatively large error limit is caused by the scatter of the experimental values. The corresponding heat capacity at 298.15 K is selected:

$$C^{\circ}_{p,m}(\mathrm{As_4Se_4, cr, 298.15\ K}) = (196.6 \pm 4.0)\ \mathrm{J\cdot K^{-1}\cdot mol^{-1}}$$

V.6.3.1.7 $As_4Se_4(g)$

The values of the thermodynamic properties of $As_4Se_4(g)$ reported from the thermodynamic optimisation in [97DEG/PEL] are $C^{\circ}_{p,m}(\mathrm{As_4Se_4, g, 298.15\ K}) = 178.2\ \mathrm{J\cdot K^{-1}\cdot mol^{-1}}$, $C^{\circ}_{p,m}(\mathrm{As_4Se_4, g, (298.15 - 1000)\ K}) = (199.8215 - 9.0928856 \times 10^{-3}\ T - 2.1245953 \times 10^{-6}\ T^{2} - 330.16855\ T^{-0.5})\ \mathrm{J\cdot K^{-1}\cdot mol^{-1}}$, $S^{\circ}_{m}(\mathrm{As_4Se_4, g, 298.15\ K}) = 653\ \mathrm{J\cdot K^{-1}\cdot mol^{-1}}$, and

$\Delta_f H_m^\circ$ (As_4Se_4, g, 298.15 K) = 81.234 kJ·mol^{-1}. These values need a revison as discussed in Section V.6.3.1. In addition, they are all essentially based on estimates and are consequently not selected by the review.

V.6.4 Antimony compounds

V.6.4.1 Antimony selenides

The phase diagram and the thermodynamic properties of the liquid phase in the antimony-selenium system were assessed by Ghosh, Lukas, and Delaey [89GHO/LUK] and by Ghosh [93GHO]. Sb_2Se_3(cr) is the only intermediary compound in the system.

The gas phase in the antimony-selenium system is complex and the species SbSe(g), Sb_2Se_2(g), Sb_2Se_3(g), Sb_2Se_4(g), Sb_3Se(g), Sb_3Se_2(g), Sb_3Se_3(g), Sb_4Se_3(g), and Sb_4Se_4(g) were all found to be present in non-negligible amounts in the mass-spectroscopic study by Sullivan, Prusaczyk, Miller, and Carlson [79SUL/PRU]. This finding invalidates the previous determinations of thermodynamic properties of gaseous antimony selenides by Gospodinov, Pashinkin, Boncheva-Mladenova, and Novoselova [70GOS/PAS] and Hoareau, Cabaud, and Uzan [72HOA/CAB]. The selected thermodynamic data of the gaseous species were derived from the work in [79SUL/PRU] as discussed in Appendix A.

V.6.4.1.1 SbSe(g)

The value of the entropy of SbSe(g) calculated in [79SUL/PRU] from a combination of experimental and estimated spectroscopic data is selected:

$$S_m^\circ(\text{SbSe, g, 298.15 K}) = (228.9 \pm 3.0)\ \text{J·K}^{-1}\text{·mol}^{-1}.$$

The enthalpy of formation of SbSe(g) calculated by the review as discussed in Appendix A from the work in [79SUL/PRU] is selected:

$$\Delta_f H_m^\circ(\text{SbSe, g, 298.15 K}) = (177.0 \pm 20.0)\ \text{kJ·mol}^{-1}.$$

The Gibbs energy of formation of SbSe(g) is calculated from the selected values of the enthalpy of formation and entropy, the selected entropy of selenium, and the entropy of antimony in [92GRE/FUG] yielding:

$$\Delta_f G_m^\circ(\text{SbSe, g, 298.15 K}) = (134.9 \pm 20.0)\ \text{kJ·mol}^{-1}.$$

V.6.4.1.2 Sb_2Se_2(g)

The standard enthalpy of formation of Sb_2Se_2(g) is calculated by the review as discussed in Appendix A from the work in [79SUL/PRU] to be $\Delta_f H_m^\circ$ (Sb_2Se_2, g, 298.15 K) = (127.2 ± 30.0) kJ·mol^{-1}. The result is not selected since it is partly based on estimated thermal properties of Sb_2Se_2(g) of unknown accuracy.

V.6.4.1.3 $Sb_2Se_3(cr)$

The heat capacity of $Sb_2Se_3(cr)$ was measured in the temperature range 53 to 300 K by Zhdanov [71ZHD]. The heat capacity was extrapolated to 0 K and the entropy at 298.15 K evaluated. The values of the heat capacity and entropy obtained in [71ZHD] are selected:

$$C^{\circ}_{p,m}(Sb_2Se_3, cr, 298.15 K) = (124.9 \pm 1.5) J \cdot K^{-1} \cdot mol^{-1}$$

and

$$S^{\circ}_{m}(Sb_2Se_3, cr, 298.15 K) = (213.0 \pm 3.0) J \cdot K^{-1} \cdot mol^{-1}.$$

Blachnik and Schneider [71BLA/SCH] measured the heat capacity in the temperature range 417 to 1041 K across the melting point at 885 K using drop calorimetry. The heat capacity of the liquid was determined to be a constant value of $C^{\circ}_{p,m}(Sb_2Se_3, l, (885 - 1041) K) = 230.1 J \cdot K^{-1} \cdot mol^{-1}$. No heat capacity expression was evaluated for $Sb_2Se_3(cr)$ and the heat capacity expression

$$C^{\circ}_{p,m}(Sb_2Se_3, cr, (298.15 - 885) K) = (121.70 + 10.73 \times 10^{-3}\, T) J \cdot K^{-1} \cdot mol^{-1}$$

was evaluated by the review from a combination of the heat capacity values listed in Table 2 of the paper and the selected heat capacity at 298.15 K.

The enthalpy of fusion was determined to be $\Delta_{fus}H^{\circ}_{m}(Sb_2Se_3, cr, 885 K) = (54.3 \pm 2.4) kJ \cdot mol^{-1}$ by Glata and Cordo [66GLA/COR], $(52.3 \pm 5.0) kJ \cdot mol^{-1}$ by Blachnik and Schneider [70BLA/SCH], $(55.2 \pm 4.0) kJ \cdot mol^{-1}$ by Blachnik and Schneider [71BLA/SCH], and $(77.4 \pm 2.5) kJ \cdot mol^{-1}$ by Glazov, Kuliev, and Krestovnikov [71GLA/KUL]. The value in the last paper is rejected and the value $\Delta_{fus}H^{\circ}_{m}(Sb_2Se_3, cr, 885 K) = (54.2 \pm 1.5) kJ \cdot mol^{-1}$ is calculated as a weighted average from the other three investigations.

The enthalpy of mixing of antimony and selenium in the liquid state at 935 K was determined by Maekawa, Yokokawa, and Niwa [72MAE/YOK]. On combining their value of the enthalpy of mixing for the composition $Sb_{0.40}Se_{0.60}$, $\Delta_{mix}H^{\circ}_{m}(Sb_{0.40}Se_{0.60}, l, 935 K) = -(29.3 \pm 0.5) kJ \cdot mol^{-1}$, with the heat capacities of solid and liquid Sb_2Se_3, the enthalpy of fusion of $Sb_2Se_3(cr)$, the data for antimony in [73HUL/DES], and the selected data of selenium, the enthalpy of formation of $Sb_2Se_3(cr)$ is calculated to be $\Delta_{f}H^{\circ}_{m}(Sb_2Se_3, cr, 298.15 K) = -(133.0 \pm 8.0) kJ \cdot mol^{-1}$. The enthalpy of formation was also determined by Howlett, Misra, and Bever [64HOW/MIS] using liquid metal calorimetry to be $\Delta_{f}H^{\circ}_{m}(Sb_2Se_3, cr, 298.15 K) = -(127.6 \pm 4.0) kJ \cdot mol^{-1}$. The selected value of the enthalpy of formation is calculated as a weighted average of the value derived from the work in [72MAE/YOK] and the investigation in [64HOW/MIS] yielding:

$$\Delta_{f}H^{\circ}_{m}(Sb_2Se_3, cr, 298.15 K) = -(128.7 \pm 3.6) kJ \cdot mol^{-1}.$$

The Gibbs energy of formation is calculated from the selected enthalpy of formation and entropy, the selected entropy of selenium, and the entropy of antimony in [92GRE/FUG] to be:

$$\Delta_f G_m^\circ(Sb_2Se_3, cr, 298.15 K) = -(127.4 \pm 3.7) kJ\cdot mol^{-1}.$$

In the aqueous phase, Umland and Wallmeier [80UML/WAL] attempted to determine the solubility product of Sb_2Se_3(cr) by the polarographic method outlined in Appendix A. Their result, $\log_{10} K_{s,0}^\circ$ (Sb_2Se_3, cr, 298.15 K) = – (113 ± 2), is not accepted since the measurements were conducted in presence of 0.5 M tartaric acid and, therefore, the conditions for the evaluation of the constant from the measurements are not fulfilled.

V.6.4.1.4 Sb_2Se_3(g), Sb_2Se_4(g), Sb_3Se(g), Sb_3Se_2(g), Sb_3Se_3(g), Sb_4Se_3(g), and Sb_4Se_4(g)

The enthalpies of formation of the species enumerated above were calculated by the review as discussed in Appendix A from the work in [79SUL/PRU] to be $\Delta_f H_m^\circ$(Sb_2Se_3, g, 298.15 K) = (83.5 ± 30.0) kJ·mol^{-1}, $\Delta_f H_m^\circ$(Sb_2Se_4, g, 298.15 K) = (58.3 ± 30.0) kJ·mol^{-1}, $\Delta_f H_m^\circ$(Sb_3Se, g, 298.15 K) = (171.3 ± 40.0) kJ·mol^{-1}, $\Delta_f H_m^\circ$(Sb_3Se_2, g, 298.15 K) = (141.4 ± 30.0) kJ·mol^{-1}, $\Delta_f H_m^\circ$(Sb_3Se_3, g, 298.15 K) = (88.6 ± 30.0) kJ·mol^{-1}, $\Delta_f H_m^\circ$(Sb_4Se_3, g, 298.15 K) = (81.2 ± 40.0) kJ·mol^{-1}, $\Delta_f H_m^\circ$(Sb_4Se_4, g, 298.15 K) = (35.6 ± 40.0) kJ·mol^{-1}. These results are not selected since they are partly based on estimated thermal properties and entropies of the species

V.6.4.2 Antimony selenites

Gospodinov [93GOS] investigated the system Sb_2O_3-SeO_2-H_2O at 298.15 K by solubility measurements over the concentration range 1.7 to 64.8 mass-% SeO_2 in the liquid phase. The primary solubility data and the phase diagram are presented. The crystalline phases of composition $Sb_2O_3\cdot SeO_2$, $Sb_2O_3\cdot 2SeO_2$, $Sb_2(SeO_3)_3$, and $Sb_2(SeO_3)_3\cdot SeO_2$ were identified from the tie lines and by chemical and X-ray analysis. A large crystallisation field is present only for $Sb_2(SeO_3)_3\cdot SeO_2$, while the fields of the other phases are very narrow and characterised by only one experimental point each. The solubility of Sb(III) is low over the whole concentration interval in selenious acid.

No thermodynamic data have been found for antimony selenites.

V.6.4.3 Antimony selenates

No thermodynamic data have been found for antimony selenates.

V.6.5 Bismuth compounds

V.6.5.1 Bismuth selenides

The phase diagram and the thermodynamic data of the system bismuth-selenium were reviewed by Chizhevskaya, Shelimova, and Zaitseva [94CHI/SHE] and Antipov, Rudnyi, and Dobrokhotova [2001ANT/RUD] thermodynamically modelled and assessed the complete system. There are severe experimental problems in the system and the alloys in the system require several months of homogenising annealing to attain equilibrium and this fact explains much of the diversity among the experimental results. The solid phases in the system are Bi_3Se_2(cr), Bi_2Se_3(cr), and BiSe(cr). The first two compounds have narrow homogeneity ranges, while the last one is a solid solution. The phase diagram bismuth-selenium is shown in Figure V-13.

Figure V-13: The phase diagram bismuth-selenium. After [94CHI/SHE]. Reprinted from [94CHI/SHE] with permission from the authors and Kluger/Plenum Publishers.

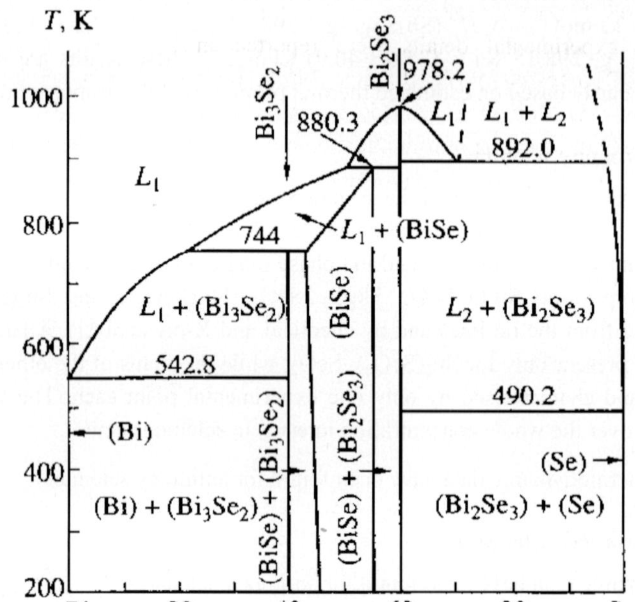

V.6.5.1.1 BiSe(cr)

BiSe(cr) is a solid solution with a homogeneity range extending from approximately $Bi_{0.57}Se_{0.43}$ to $Bi_{0.45}Se_{0.55}$.

Stolyarova, Gavrilov, and Nekrasov [90STO/GAV] measured the heat capacity of the composition $Bi_{0.5}Se_{0.5}$ in the temperature range 6 to 300 K and reported the values $C^{\circ}_{p,m}$(BiSe, cr, 298.15 K) = (55.5 ± 0.3) $J\cdot K^{-1}\cdot mol^{-1}$ and S°_{m}(BiSe, cr, 298.15 K) = (101.2 ± 0.5) $J\cdot K^{-1}\cdot mol^{-1}$. The heat capacity value at 298.15 K was accepted in the assessment in [2001ANT/RUD], but the entropy value was considered too large to be consistent with the thermodynamic modelling. The entropy was instead assigned the value 90.8 $J\cdot K^{-1}\cdot mol^{-1}$ rather than the value in [90STO/GAV] in order to better fit the model to the phase diagram. The review selects the experimental values determined in [90STO/GAV], but increases the uncertainty limits to incorporate additional systematic errors:

$$C^{\circ}_{p,m}(\text{BiSe, cr, 298.15 K}) = (55.5 \pm 0.9)\ \text{J}\cdot\text{K}^{-1}\cdot\text{mol}^{-1}$$

and

$$S^{\circ}_{m}(\text{BiSe, cr, 298.15 K}) = (101.2 \pm 1.5)\ \text{J}\cdot\text{K}^{-1}\cdot\text{mol}^{-1}.$$

The enthalpy of formation of BiSe(cr) at 298.15 K was determined by Bros, Castanet, and Kehiaian [76BRO/CAS] to be – (44 ± 4) $kJ\cdot mol^{-1}$ using solution calorimetry where small amounts of the solid were dissolved in liquid bismuth at 560 K. Direct synthesis techniques, using evacuated silica tubes in the calorimeter, were employed in [90STO/GAV] yielding $\Delta_f H^{\circ}_{m}$(BiSe, cr, 298.15 K) = – (48.1 ± 1.20) $kJ\cdot mol^{-1}$. No experimental details were reported in this work. Antipov *et. al.* [2001ANT/RUD] assessed $\Delta_f H^{\circ}_{m}$(BiSe, cr, 298.15 K) = – 51.8 $kJ\cdot mol^{-1}$ based on their thermodynamic modelling. The value is higher than the experimental values and it is consistent with their selection of a smaller value for the entropy of BiSe(cr). The review selects the average of the experimental values,

$$\Delta_f H^{\circ}_{m}(\text{BiSe, cr, 298.15 K}) = -(46.1 \pm 5.7)\ \text{kJ}\cdot\text{mol}^{-1}.$$

The Gibbs energy of formation is calculated from the selected enthalpy of formation and entropy, the selected entropy of selenium, and the entropy of bismuth in Table IV-1 to be:

$$\Delta_f G^{\circ}_{m}(\text{BiSe, cr, 298.15 K}) = -(46.8 \pm 5.7)\ \text{kJ}\cdot\text{mol}^{-1}.$$

V.6.5.1.2 BiSe(g)

The entropy and enthalpy functions of BiSe(g) were calculated by Gorbov and Krestovnikov [66GOR/KRE] from molecular parameters using the vibrational parameters ω_e = 264.7 cm^{-1} and $\omega_e x_e$ = 0.4 cm^{-1} determined by Sharma [54SHA] from gas phase studies and the estimated molecular parameters B_e = 0.505 cm^{-1}, $\alpha_e = 0.105 \times 10^{-3}$ cm^{-1}, $D_0 = 0.735 \times 10^{-8}$ cm^{-1}, and r_e = 2.41Å obtained from a comparison of molecular properties of the sulphides, selenides, and tellurides of arsenic, antimony, and bismuth. These parameters were used by Uy and Drowart [69UY/DRO] in their calculation of the dissociation energy of BiSe(g) from mass-spectrometric measurements at high temperatures.

The ground state frequencies were determined to be ω_e = 264.0 cm^{-1} and $\omega_e x_e$ = 0.61 cm^{-1} by Ahmed [85FAK] from fluorescence spectra of BiSe molecules isolated in an argon matrix. The first excited electronic state lies above 13000 cm^{-1} according to the absorption studies of Barrow, Stobart, and Vaughan [67BAR/STO] and does not contribute significantly to the heat capacity and the entropy below 1500 K. The following selected values were calculated by the review using the usual non-rigid rotator, anharmonic oscillator approximation and the parameters above given in [66GOR/KRE]:

$$C^{\circ}_{p,\mathrm{m}}(\mathrm{BiSe, g, 298.15\ K}) = (36.46 \pm 0.50)\ \mathrm{J{\cdot}K^{-1}{\cdot}mol^{-1}}$$

and

$$S^{\circ}_{\mathrm{m}}(\mathrm{BiSe, g, 298.15\ K}) = (263.9 \pm 6.0)\ \mathrm{J{\cdot}K^{-1}{\cdot}mol^{-1}}.$$

The frequency parameters obtained from the matrix isolation studies in [85FAK] lead to values for the heat capacity and entropy which are larger by approximately 0.1 and 6 J·K^{-1}·mol^{-1}, respectively.

The expression

$$C^{\circ}_{p,\mathrm{m}}(\mathrm{BiSe, g, (298-1500)\ K}) = (37.3251 + 0.4161 \times 10^{-3}\,T - 4.68 \times 10^{-8}\,T^{2} - 8.7767 \times 10^{4}\,T^{-2})\ \mathrm{J{\cdot}K^{-1}{\cdot}mol^{-1}}$$

describes the heat capacity obtained from statistical mechanics calculations to within ± 0.005 J·K^{-1}·mol^{-1}.

Uy and Drowart [69UY/DRO] studied the reaction Pb(g) + BiSe(g) $\rightleftharpoons$ Bi(g) + PbSe(g) in the temperature range 864 to 1050 K using Knudsen effusion and mass spectrometry. The results were re-evaluated by the review as discussed in Appendix A yielding $\Delta_f H^{\circ}_{m}$(BiSe, g, 298.15 K) = (172.5 ± 7.2) kJ·mol^{-1} and $\Delta_f H^{\circ}_{m}$(BiSe, g, 298.15 K) = (174.8 ± 7.2) kJ·mol^{-1} from the second and the third law, respectively. Similarly, Porter and Spencer [60POR/SPE] studied the reaction Bi_2(g) + Se_2(g) $\rightleftharpoons$ 2BiSe(g) at 1050 K using Knudsen effusion and mass spectrometry and determined the equilibrium constant. The result was re-evaluated by the review as discussed in Appendix A yielding $\Delta_f H^{\circ}_{m}$(BiSe, g, 298.15 K) = (167.5 ± 9.0) kJ·mol^{-1} from the third law. The adopted value of the enthalpy of formation is calculated as the weighted average of the values from [69UY/DRO] and [60POR/SPE] yielding:

$$\Delta_f H^{\circ}_{\mathrm{m}}(\mathrm{BiSe, g, 298.15\ K}) = (172.2 \pm 4.4)\ \mathrm{kJ{\cdot}mol^{-1}}$$

and is included in Appendix E since it was calculated with non-TDB auxiliary data.

V.6.5.1.3 Bi_2Se_3(cr)

The low temperature heat capacity of Bi_2Se_3(cr) was measured in the temperature range 6 to 300 K in [90STO/GAV]. The heat capacity at 298.15 K was determined to be $C^{\circ}_{p,m}$(Bi_2Se_3, cr, 298.15 K) = (132.6 ± 2.1) J·K^{-1}·mol^{-1}. Melekh and Semenkovich [67MEL/SEM] and Blachnik and Schneider [71BLA/SCH] determined the heat capacity in the temperature range 398 to 773 K and 452 to 985 K, respectively, both using drop calorimetry. Rasulov [88RAS] reported the heat capacity in the temperature range 300 to 1100 K. The heat capacity measurements are summarised in Figure V-14. Heat capacity values derived from enthalpy measurements using drop calorimetry at low temperatures are less accurate and the value at 298.15 K determined in [90STO/GAV] is selected although it is somewhat larger than the values extrapolated from the drop calorimetric determinations. Similarly, the assessment in [2001ANT/RUD] selected a value, 132.0 $J \cdot K^{-1} \cdot mol^{-1}$, close to that of the low temperature measurements.The selected value is:

$$C^{\circ}_{p,m}(Bi_2Se_3, cr, 298.15 K) = (132.6 \pm 2.1)\ J \cdot K^{-1} \cdot mol^{-1}.$$

The temperature dependence of the heat capacity varies considerably between the investigations in [67MEL/SEM], [71BLA/SCH], and [88RAS] as shown in Figure V-14. The heat capacity expression

$$C^{\circ}_{p,m}(Bi_2Se_3, cr, (298.15 - 985) K) = (126.95 + 18.94 \times 10^{-3}\ T)\ J \cdot K^{-1} \cdot mol^{-1}.$$

was derived by the review by assuming a linear relationship and selecting a weighted average of $C^{\circ}_{p,m}(Bi_2Se_3, cr, 900 K) = 144.0\ J \cdot K^{-1} \cdot mol^{-1}$ in addition to the selected value of the heat capacity at 298.15 K.

The entropy of formation of Bi_2Se_3(cr) was determined to be $\Delta_f S^{\circ}_m(Bi_2Se_3, cr, 298.15 K) = -(7.5 \pm 29.0)\ J \cdot K^{-1} \cdot mol^{-1}$ by Malekh and Semenkovich [68MAL/SEM] and to $-(5.2 \pm 15.6)\ J \cdot K^{-1} \cdot mol^{-1}$ by Vasiliev, Somov, Nikolskaya, and Gerassimov [68VAS/SOM] from electrochemical cell measurements in the temperature range 553 to 583 K and 310 to 411 K, respectively. The third law entropy derived from the heat capacity measurements in [90STO/GAV] is $S^{\circ}_m(Bi_2Se_3, cr, 298.15 K) = (243.5 \pm 3.6)\ J \cdot K^{-1} \cdot mol^{-1}$ and corresponds to an entropy of formation of $\Delta_f S^{\circ}_m(Bi_2Se_3, cr, 298.15 K) = (3.76 \pm 3.60)\ J \cdot K^{-1} \cdot mol^{-1}$. The thermodynamic modelling in [2001ANT/RUD] resulted in an entropy value of 167.5 $J \cdot K^{-1} \cdot mol^{-1}$ corresponding to $\Delta_f S^{\circ}_m(Bi_2Se_3, cr, 298.15 K) = -72.3\ J \cdot K^{-1} \cdot mol^{-1}$. The review finds this value much too low and selects the third law value in [90STO/GAV]

$$S^{\circ}_m(Bi_2Se_3, cr, 298.15 K) = (243.5 \pm 3.6)\ J \cdot K^{-1} \cdot mol^{-1}.$$

Figure V-14: The heat capacity of Bi_2Se_3(cr) as measured in various investigations.

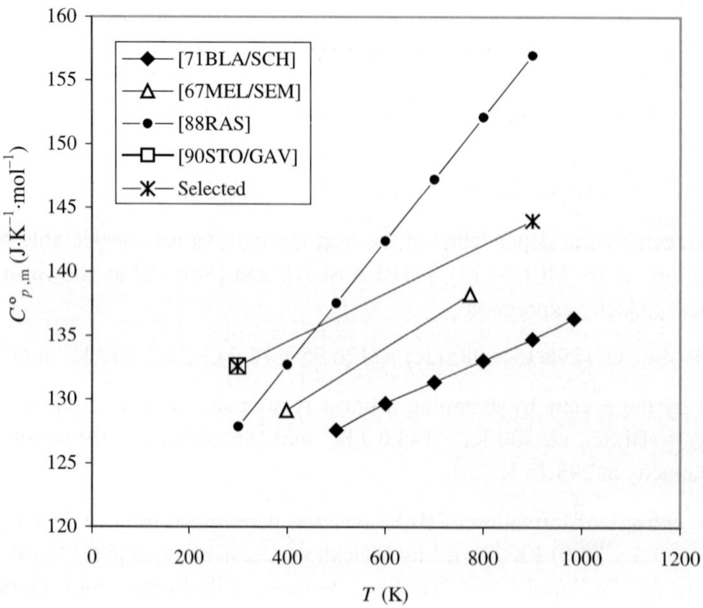

Measurements of the total pressure of the vapour in equilibrium with Bi_2Se_3 are summarised in Table V-32. No mass spectrometric investigations are available for equilibria involving solid bismuth selenides, but Porter and Spencer [60POR/SPE] identified the species Se_2(g), BiSe(g), and Bi_2(g) in the vapour in equilibrium with liquid bismuth-selenium alloys of various compositions. It was found that selenium-rich melts primarily vaporise by losing Se_2(g) and that bismuth-rich melts primarily vaporise by losing Bi_2(g). The information available about the species in the gas phase is too meagre for a proper evaluation of thermodynamic data from the vapour pressure measurements. The vaporisation studies in [68BON/PAS] are further discussed in Appendix A.

The melting point of Bi_2Se_3(cr) is 978 K and the measurements of the enthalpy of fusion are summarised in Table V-33. Premelting phenomena are pronounced and explain the scatter in the measured values. The selected value of the enthalpy of fusion is:

$$\Delta_{\text{fus}}H^{\circ}_{\text{m}}(\text{Bi}_2\text{Se}_3, \text{cr}, 978\ \text{K}) = (81.6 \pm 7.0)\ \text{kJ}\cdot\text{mol}^{-1}.$$

Table V-32: Summary of the measurements of the total vapour pressure in equilibrium with Bi_2Se_3 using various experimental techniques. Note that some of the measurements were made above the melting point of Bi_2Se_3(cr) at 978 K.

Reference	Temperature (K)	Method
[66OHA/KOZ]	773 – 1073	Dew point
[67KRE/GOR]	747 – 896	Knudsen effusion
[68BON/PAS]	735 – 879	Knudsen effusion
	794 – 900	Langmuir
[69UST/VIG]	956 – 1288	Static
[70GOS/PAS]	677 – 822	Knudsen effusion

Table V-33: Summary of determinations of the enthalpy of fusion of Bi_2Se_3(cr).

Reference	$\Delta_{fus}H^{\circ}_{m}$ (Bi_2Se_3, cr, 978 K) ($kJ \cdot mol^{-1}$)
[66GLA/COR]	(74.0 ± 4.5)
[70BLA/SCH]	(85.8 ± 4.2)
[76BRO/CAS]	(85 ± 5)
Average	(81.6 ± 7.0)

The experimental determinations of the enthalpy of formation of Bi_2Se_3(cr) are summarised in Table V-34. The value of the enthalpy of formation determined in [55GAT/SCH] was considered tentative by the authors themselves and is far out of line with the other determinations. The value is therefore rejected. Similarly, the value determined in [68BON/PAS] is rejected because it stems from an evaluation using an inappropriate chemical reaction as discussed in Appendix A. The value of the enthalpy of formation attributed to [72MAE/YOK] was calculated by the review from the enthalpies of mixing as discussed in Appendix A. The thermodynamic modelling in [2001ANT/RUD] resulted in an enthalpy of formation of Bi_2Se_3(cr) of – 158.5 $kJ \cdot mol^{-1}$, a value which is 13.0 $kJ \cdot mol^{-1}$ more negative than the average of the experimental values in Table V-34. The value in this table is selected:

$$\Delta_f H^{\circ}_{m} (Bi_2Se_3, \text{cr}, 298.15 \text{ K}) = -(145.5 \pm 10.0) \text{ kJ} \cdot \text{mol}^{-1}.$$

The Gibbs energy of formation of Bi_2Se_3(cr) is calculated from the selected enthalpy of formation, the selected entropy of selenium, and the entropy of bismuth in [2001LEM/FUG] yielding:

$$\Delta_f G^{\circ}_{m} (Bi_2Se_3, \text{cr}, 298.15 \text{ K}) = -(146.6 \pm 10.1) \text{ kJ} \cdot \text{mol}^{-1}.$$

Table V-34: Summary of determinations of the enthalpy of formation of Bi_2Se_3(cr) using various experimental techniques. The values denoted by (*) were not included in the calculation of the average.

Reference	$\Delta_f H^\circ_m$ (Bi_2Se_3, cr 298.15 K) (kJ·mol^{-1})	Temperature (K)	Method
[55GAT/SCH]	– (58.2 ± 6.3) (*)		Oxygen combustion
[64HOW/MIS]	– (140.0 ± 1.9)	623 – 873	Liquid metal solution calorimetry
[65AND/KUD]	– (152.3 ± 6.0)		Oxygen combustion
[68BON/PAS]	– (141.8 ± 6.0) (*)	735 – 879	Vapour pressure
[68MAL/SEM]	– (154.0 ± 6.0)	553 – 583	Electrochemical cell
[68VAS/SOM]	– (136.4 ± 1.9)	310 – 411	Electrochemical cell
[72MAE/YOK]	– 188.8 (*)	998	Liquid metal solution calorimetry
[76BRO/CAS]	– (155.0 ± 10.0)	888	Liquid metal solution calorimetry
[90STO/GAV]	– (135.3 ± 4.2)		Direct synthesis
Average	– (145.5 ± 10.0)		

V.6.5.1.4 Bi_3Se_2(cr)

No experimental thermodynamic properties have been reported for Bi_3Se_2(cr). However, the thermodynamic modelling in [2001ANT/RUD] yielded the values $C^\circ_{p,m}$(Bi_3Se_2, cr, 298.15 K) = 127 J·K^{-1}·mol^{-1}, S°_m(Bi_3Se_2, cr, 298.15 K) = 277.5 J·K^{-1}·mol^{-1}, and $\Delta_f H^\circ_m$(Bi_3Se_2, cr, 298.15 K) = – 92 kJ·mol^{-1}. The values are not based on experimental data for this phase and are not selected by the review. In addition, since the values selected by the review for BiSe(cr) and Bi_2Se_3(cr) are quite different from those resulting from the phase diagram modelling, the values in [2001ANT/RUD] for Bi_3Se_2(cr) should not be combined with those selected by the review for the other two phases.

V.6.5.2 Bismuth selenites

Ojkova and Gospodinov [80OJK/GOS] studied the phase equilibria in the system Bi_2O_3-SeO_2-H_2O at (373.2 ± 1.0) K by solubility measurements in the range 0 to 83.0 mass-% SeO_2. The incongruently soluble phases $Bi_2(SeO_3)_3$ and $Bi_2(SeO_3)_3{\cdot}H_2SeO_3$ were formed. A phase diagram and the solubility data are presented. Basic salts are formed below 1.6 mass-% of SeO_2 in the aqueous phase and the solubility is less than about 3×10^{-3} mol(Bi(III))·kg^{-1} below 70 mass-% SeO_2.

Andreeva and Karapet'yants [65AND/KAR], [66AND/KAR] performed calorimetric measurements from which the standard enthalpies of formation of $Bi_2(SeO_3)_3$(cr) and $(BiO)_2SeO_3$(cr) were found. Their data have been re-evaluated with new auxiliary data in Appendix A. The result is $\Delta_f H^\circ_m$($Bi_2(SeO_3)_3$, cr, 298.15 K) =

– (1401.81 ± 6.77) kJ·mol^{-1} and $\Delta_f H_m^\circ$ ((BiO)$_2$SeO$_3$, cr, 298.15 K) = – (959.49 ± 5.99) kJ·mol^{-1}.

Bismuth selenite dissociates thermally in two steps

$$Bi_2(SeO_3)_3(cr) \rightleftharpoons (BiO)_2SeO_3(cr) + 2SeO_2(g) \quad (V.50)$$

$$(BiO)_2SeO_3(cr) \rightleftharpoons Bi_2O_3(cr) + SeO_2(g) \quad (V.51)$$

according to Marcovskii, Sapozhnikov, and Boev [64MAR/SAP]. Bakeeva, Pashinkin, and Bakeev [72BAK/PAS] measured the vapour pressure of SeO_2(g) by the dew point method above a mixture of $Bi_2(SeO_3)_3$(cr) and $(BiO)_2SeO_3$(cr) over the temperature range 786 to 885 K and between 786 to 869 K above a mixture of $(BiO)_2(SeO_3)_3$(cr) and Bi_2O_3(cr). The pressures were evaluated from the dew point temperatures with the relationship in [69SON/NOV]. The data have been recalculated with the selected relationship for the saturated vapour pressure to yield:

$$\log_{10} p_{SeO_2} ((V.50), (786 – 885)\ K) = (9.464 \pm 0.081) – (8236 \pm 63)\ T^{-1}$$

$$\log_{10} p_{SeO_2} ((V.51), (786 – 869)\ K) = (7.978 \pm 0.235) – (7868 \pm 71)\ T^{-1}.$$

However, according to Oppermann *et al.* [99OPP/GOB] the decomposition of $(BiO)_2SeO_3$(cr) to Bi_2O_3(cr) and SeO_2(g) comprises additional steps. Therefore no thermochemical data have been calculated by the review from these relationships.

Oppermann *et al.* [99OPP/GOB] performed a comprehensive study of the system Bi_2O_3-SeO_2. The standard enthalpies of formation of $Bi_2(SeO_3)_3$(cr) and $(BiO)_2SeO_3$(cr) prepared by high temperature synthesis were found to be – (1404.6 ± 7.1) kJ·mol^{-1} and – (902.5 ± 5.4) kJ·mol^{-1}, respectively, from solution calorimetry. The corresponding values from vapour pressure measurements of the decomposition reactions were reported to be – (1389.9 ± 33.5) kJ·mol^{-1} and – (884.5 ± 12.6) kJ·mol^{-1}. No thermodynamic quantities can be recommended for $(BiO)_2SeO_3$ due to the scatter of the present data. For $Bi_2(SeO_3)_3$ the agreement is good and the mean of the values obtained by calorimetry will be adopted and included in Appendix E since it was evaluated with non-TDB auxiliary data:

$$\Delta_f H_m^\circ (Bi_2(SeO_3)_3, cr, 298.15\ K) = – (1403.2 \pm 5.0)\ kJ \cdot mol^{-1}.$$

The entropy change of the decomposition reaction yields S_m° ($Bi_2(SeO_3)_3$, cr, 298.15 K) = (392 ± 38) J·K^{-1}·mol^{-1} in [99OPP/GOB]. This value is not selected.

V.6.5.3 Bismuth selenates

No thermodynamic data have been found for bismuth selenates.

V.7 Group 14 compounds and complexes

V.7.1 Carbon compounds

V.7.1.1 Carbon selenides

V.7.1.1.1 CSe(g)

The heat capacity and entropy of CSe(g) were calculated by the review from the ground state properties given by Lebreton, Bosser and Marsigny [73LEB/BOS]. The lowest excited state lies at 24396 cm^{-1} according to Bosser and Lebreton [81BOS/LEB] and does not contribute significantly to the thermal properties below 2000 K. The selected values are:

$$C^{\circ}_{p,\mathrm{m}}(\mathrm{CSe, g, 298.15\ K}) = (30.58 \pm 0.20)\ \mathrm{J{\cdot}K^{-1}{\cdot}mol^{-1}}$$

and

$$S^{\circ}_{\mathrm{m}}(\mathrm{CSe, g, 298.15\ K}) = (222.75 \pm 0.50)\ \mathrm{J{\cdot}K^{-1}{\cdot}mol^{-1}}.$$

The expression

$$C^{\circ}_{p,\mathrm{m}}(\mathrm{CSe, g, (298.15 - 2000)\ K}) = (44.2955 - 3.4256 \times 10^{-3}\ T + 6.91 \times 10^{-7}\ T^{2} - 6.0184 \times 10^{3}\ T^{-1} + 6.60504 \times 10^{5}\ T^{-2})\ \mathrm{J{\cdot}K^{-1}{\cdot}mol^{-1}}$$

describes the heat capacity obtained from the statistical mechanics calculations to within $\pm 0.02\ \mathrm{J{\cdot}K^{-1}{\cdot}mol^{-1}}$.

The atomisation energy of CSe(g) was determined to be $\Delta_{\mathrm{at}}H^{\circ}_{\mathrm{m}}(\mathrm{CSe, g, 0\ K}) = (578.9 \pm 23.0)\ \mathrm{kJ{\cdot}mol^{-1}}$ by Drowart [64DRO] and to be $\Delta_{\mathrm{at}}H^{\circ}_{\mathrm{m}}(\mathrm{CSe, g, 298.15\ K}) = (590.8 \pm 9.0)\ \mathrm{kJ{\cdot}mol^{-1}}$ by Smoes and Drowart [77SMO/DRO] from mass spectrometric measurements. The selected value of the enthalpy of formation is calculated by combining the latter value of the enthalpy of atomisation with the selected value of the enthalpy of formation of Se(g) and the corresponding value for C(g) in [89COX/WAG] yielding:

$$\Delta_{\mathrm{f}}H^{\circ}_{\mathrm{m}}(\mathrm{CSe, g, 298.15\ K}) = (362.0 \pm 9.2)\ \mathrm{kJ{\cdot}mol^{-1}}.$$

The Gibbs energy of formation is calculated from the selected enthalpy of formation, the selected entropy of selenium, and the entropy of carbon in [89COX/WAG] to be:

$$\Delta_{\mathrm{f}}G^{\circ}_{\mathrm{m}}(\mathrm{CSe, g, 298.15\ K}) = (309.9 \pm 9.2)\ \mathrm{kJ{\cdot}mol^{-1}}.$$

V.7.1.1.2 CSe_2(cr, l, g)

Carbon diselenide, CSe_2, can be prepared by reacting dichloromethane, CH_2Cl_2, with liquid selenium at 870 K. The gaseous reaction mixture is solidified by cooling to 195 K and CSe_2 is purified from unreacted CH_2Cl_2 by fractional distillation [66GAT/DRA2].

The melting and boiling points of CSe_2 are (229.47 ± 0.05) K and (398.2 ± 0.5) K, respectively, at a pressure of 1 bar. At room temperature, the golden yellow liquid soon darkens and produces what is believed to be polymeric $(CSe_2)_n$(s) [68BRO/WHA] while, at 195 K, the solid could be kept for several months without deterioration [66GAT/DRA2].

Gattow and Dräger [66GAT/DRA2], [66GAT/DRA3] have made the most comprehensive and careful studies of the properties and thermochemistry of CSe_2. They utilised vapour pressure measurements and calorimetric measurements to conclude that the enthalpy and the entropy of fusion are:

$$\Delta_{\text{fus}}H^\circ_{\text{m}}(\text{CSe}_2, \text{cr}, 229.47\ \text{K}) = (6.4 \pm 0.2)\ \text{kJ}\cdot\text{mol}^{-1}$$

$$\Delta_{\text{fus}}S^\circ_{\text{m}}(\text{CSe}_2, \text{cr}, 229.47\ \text{K}) = (27.8 \pm 1.0)\ \text{J}\cdot\text{K}^{-1}\cdot\text{mol}^{-1}.$$

The heat capacity of CSe_2(cr, l) was studied within the temperature range of 78 to 328 K with the drop calorimetry method, and the data were reproduced by the equations:

$$C^\circ_{p,\text{m}}(\text{CSe}_2, \text{cr}, (78-229)\ \text{K}) = (51.71 + 5.623 \times 10^{-2}\ T - 3.1 \times 10^{-4}\ T^{-2})\ \text{J}\cdot\text{K}^{-1}\cdot\text{mol}^{-1}$$

$$C^\circ_{p,\text{m}}(\text{CSe}_2, \text{l}, (230-328)\ \text{K}) = (68.70 + 6.65 \times 10^{-2}\ T)\ \text{J}\cdot\text{K}^{-1}\cdot\text{mol}^{-1}.$$

From the second equation, the selected value:

$$C^\circ_{p,\text{m}}(\text{CSe}_2, \text{l}, 298.15\ \text{K}) = (88.53 \pm 0.90)\ \text{J}\cdot\text{K}^{-1}\cdot\text{mol}^{-1}$$

is obtained.

Gattow and Dräger [66GAT/DRA] determined the vapour pressures over crystalline and liquid CSe_2, respectively, by an isoteniscope. The measurements were found to obey the equations:

$$\log_{10} p(\text{CSe}_2, \text{cr}, (219-229)\ \text{K}) = 6.915 - 2418\ T^{-1}$$

and

$$\log_{10} p(\text{CSe}_2, \text{l}, (230-337)\ \text{K}) = 22.87 - 2702\ T^{-1} - 6.2345 \log_{10} T.$$

The equation presented by Ives, Pittman and Wardlaw [47IVE/PIT] for liquid CSe_2 is $\log_{10} p(\text{CSe}_2, \text{l}, (273-333)\ \text{K}) = 5.04 - 1987\ T^{-1}$. Since the CSe_2 specimen employed by Gattow and Dräger is judged by this review to be of significantly higher purity, the former equation is given preference. It is used to calculate the selected enthalpy of vaporisation for CSe_2(l)

$$\Delta_{\text{vap}}H^\circ_{\text{m}}(\text{CSe}_2, \text{l}, 298.15\ \text{K}) = (36.3 \pm 0.2)\ \text{kJ}\cdot\text{mol}^{-1}.$$

The heat of combustion of CSe_2(l) at 298.15 K was measured by bomb calorimetry in [66GAT/DRA3]. Corrections were made for the formation of minor quantities of elemental carbon and selenium. For the reaction CSe_2(l) + $3O_2$(g) → CO_2(g) + $2SeO_2$(cr) a reaction enthalpy of $\Delta_{\text{r}}H^\circ_{\text{m}} = -(1064.0 \pm 7.5)\ \text{kJ}\cdot\text{mol}^{-1}$ was obtained.

Merten and Schluter [36MER/SCH] found the value $-1008\ \text{kJ}\cdot\text{mol}^{-1}$ with the same method. Since Gattow and Dräger considered the formation of by-products in the experiments, their value is selected. It combines with the selected enthalpies of formation of SeO_2(cr) and CO_2(g) to:

$$\Delta_f H_m^\circ(\text{CSe}_2, \text{l}, 298.15\ \text{K}) = (219.3 \pm 7.6)\ \text{kJ}\cdot\text{mol}^{-1}.$$

This value combines with the selected enthalpy of vaporisation for CSe_2(l) to:

$$\Delta_f H_m^\circ(\text{CSe}_2, \text{g}, 298.15\ \text{K}) = (255.6 \pm 7.6)\ \text{kJ}\cdot\text{mol}^{-1}.$$

In [82WAG/EVA], the enthalpy of formation for CSe_2(l) at 298.15 K is reported to be $164.8\ \text{kJ}\cdot\text{mol}^{-1}$. This value is apparently based on the data in [36MER/SCH].

The standard entropy of CSe_2(g) was calculated from molecular parameters in [88BUR/WIL] to yield the selected value:

$$S_m^\circ(\text{CSe}_2, \text{g}, 298.15\ \text{K}) = (263.2 \pm 1.0)\ \text{J}\cdot\text{K}^{-1}\cdot\text{mol}^{-1}.$$

The molecular constant data were also employed to calculate the heat capacity of CSe_2(g)

$$C_{p,m}^\circ(\text{CSe}_2, \text{g}, (298-2000)\ \text{K}) = (70.0729 - 3.3995\times10^{-3}\,T + 5.687\times10^{-7}\,T^2 - 8.1383\times10^{3}\,T^{-1} + 7.2256\times10^{5}\,T^{-2})\ \text{J}\cdot\text{K}^{-1}\cdot\text{mol}^{-1}.$$

This relationship yields:

$$C_{p,m}^\circ(\text{CSe}_2, \text{g}, 298.15\ \text{K}) = (49.9 \pm 0.1)\ \text{J}\cdot\text{K}^{-1}\cdot\text{mol}^{-1}.$$

The value of $S_m^\circ(\text{CSe}_2, \text{l}, 298.15\ \text{K})$ was calculated by Gattow and Dräger [66GAT/DRA3] from the heat content measurements. It was also calculated from $S_m^\circ(\text{CSe}_2, \text{g}, 298.15\ \text{K})$ combined with $\Delta_{vap}S_m^\circ(\text{CSe}_2, \text{l}, 298.15\ \text{K})$ obtained from the vapour pressure measurements. The two values were in acceptable agreement, (165.3 ± 9.2) and $(172.4 \pm 8.4)\ \text{J}\cdot\text{K}^{-1}\cdot\text{mol}^{-1}$, respectively. They also agree with the value calculated from vapour pressure measurements by Ives *et al.* [47IVE/PIT], $(166.9 \pm 12.6)\ \text{J}\cdot\text{K}^{-1}\cdot\text{mol}^{-1}$. On this basis, the review selects

$$S_m^\circ(\text{CSe}_2, \text{l}, 298.15\ \text{K}) = (168 \pm 15)\ \text{J}\cdot\text{K}^{-1}\cdot\text{mol}^{-1}.$$

The selected data yield:

$$\Delta_f G_m^\circ(\text{CSe}_2, \text{l}, 298.15\ \text{K}) = (196.0 \pm 8.8)\ \text{kJ}\cdot\text{mol}^{-1},$$

$$\Delta_f G_m^\circ(\text{CSe}_2, \text{g}, 298.15\ \text{K}) = (204.0 \pm 7.6)\ \text{kJ}\cdot\text{mol}^{-1}.$$

V.7.1.2 Selenocyanate

The selenocyanate ion, $SeCN^-$, is an ambidentate ion which may coordinate to metal ions via either the nitrogen atom (to “hard” metal ions) or the selenium atom (to “soft” metal ions) [74AHR/AVS], [75NOR], [79KAR/CRE]. Its potassium salt can be prepared by reacting a suspension of elemental selenium and potassium cyanide in liquid ammonia or in boiling absolute ethanol [65GOL/SKO]. The compound is sensitive to light and moisture and can be purified via recrystallisation from acetone/diethylether solution [66BOU/KEL].

Boughton and Keller [66BOU/KEL] measured the change in pH when a pure specimen of KSeCN was added to water and found it to be practically negligible. From this, the authors concluded that the $\mathrm{p}K_\mathrm{a}$ of HSeCN is, at least, below 1. Attempts to titrate a solution of KSeCN with hydrochloric acid failed due to the rapid decomposition into elemental selenium and hydrogen cyanide. The thermodynamics of the reaction

$$\mathrm{SeCN^-} + \mathrm{H^+} \rightleftharpoons \mathrm{Se(monoclinic)} + \mathrm{HCN(aq)} \qquad \text{(V.52)}$$

was studied in dilute solution between 273.5 and 285.7 K by Hamada [61HAM]. As discussed in Appendix A, the results of this investigation could only in part be selected with

$$\log_{10} K^\circ((\text{V.52}), 298.15\ \text{K}) = (3.6 \pm 0.5)$$

and

$$\Delta_\mathrm{f} G^\circ_\mathrm{m}(\mathrm{SeCN^-}, 298.15\ \mathrm{K}) = (136.1 \pm 3.8)\ \mathrm{kJ{\cdot}mol^{-1}}.$$

V.7.2 Silicon compounds

V.7.2.1 Silicon selenides

Tomaszkiewicz, Susman, Volin, and O'Hare in [94TOM/SUS] measured the enthalpy of formation of $\mathrm{SiSe_2(cr)}$ and thoroughly reviewed the thermodynamic properties of silicon selenides. No new experimental information is available and in general the values reported in [94TOM/SUS] are accepted by the review. Thermodynamic information is available for $\mathrm{SiSe_2(cr)}$ and SiSe(g).

V.7.2.1.1 SiSe(g)

The heat capacity and entropy were calculated in [94TOM/SUS] using statistical mechanics and experimental values of the molecular constants. The values obtained are selected,

$$C^\circ_{p,\mathrm{m}}(\mathrm{SiSe, g}, 298.15\ \mathrm{K}) = (33.6 \pm 0.1)\ \mathrm{J{\cdot}K^{-1}{\cdot}mol^{-1}}$$

and

$$S^\circ_\mathrm{m}(\mathrm{SiSe, g}, 298.15\ \mathrm{K}) = (235.58 \pm 0.20)\ \mathrm{J{\cdot}K^{-1}{\cdot}mol^{-1}}.$$

The heat capacity expression

$$C^\circ_{p,\mathrm{m}}(\mathrm{SiSe, g}, (298.15 - 1500)\ \mathrm{K}) = (35.954 + 2.005 \times 10^{-3}\, T - 0.7427 \times 10^{-6}\, T^2 - 2.56238 \times 10^{5}\, T^{-2})\ \mathrm{J{\cdot}K^{-1}{\cdot}mol^{-1}}$$

was derived by the review from the heat capacity values given in Table 5 in [94TOM/SUS].

The enthalpy of the reaction

$$\tfrac{1}{2}SiSe_2(cr) + \tfrac{1}{2}Si(cr) \rightleftharpoons SiSe(g) \quad (V.53)$$

was assessed in [94TOM/SUS] to be $\Delta_r H_m^\circ((V.53), 298.15\ K) = (239 \pm 10)\ kJ\cdot mol^{-1}$ by critically examining the vapour pressure measurements by Emons and Theisen [72EMO/THE] and Ustyugov, Bezobrazov, Mareev, and Yu [70UST/BEZ], the spectroscopic studies by Huber and Herzberg [79HUB/HER], and the mass-spectrometric observations of Drowart [64DRO]. The enthalpy of formation of SiSe(g) calculated from the reaction enthalpy and the selected enthalpy of formation of $SiSe_2(cr)$ is selected,

$$\Delta_f H_m^\circ(SiSe, g, 298.15\ K) = (150.2 \pm 10.5)\ kJ\cdot mol^{-1}.$$

The Gibbs energy of formation is calculated from the selected enthalpy of formation, the selected entropy of selenium, and the entropy of silicon in [89COX/WAG],

$$\Delta_f G_m^\circ(SiSe, g, 298.15\ K) = (98.1 \pm 10.5)\ kJ\cdot mol^{-1}.$$

V.7.2.1.2 $SiSe_2(cr)$

The enthalpy of formation of $SiSe_2(cr)$ was determined to be $\Delta_f H_m^\circ(SiSe_2, cr, 298.15\ K) = (5 \pm 2)\ kJ\cdot mol^{-1}$ by Gabriel and Alvarez-Tostado [52GAB/ALV] by measuring the enthalpy of dissolution of $SiSe_2(cr)$ in concentrated NaOH(aq). The reaction $SiSe_2(cr) + 6NaOH(aq) \rightleftharpoons 2Na_2Se(aq) + Na_2SiO_3(aq) + 3H_2O(l)$ was assumed to occur in the calorimeter. The result differs completely from the very accurate fluorine combustion calorimetry work in [94TOM/SUS]. The study determined the reaction enthalpy for

$$SiSe_2(cr) + 8F_2(g) \rightleftharpoons SiF_4(g) + 2SeF_6(g) \quad (V.54)$$

to be $\Delta_r H_m^\circ((V.54), 298.15\ K) = -(3673.4 \pm 3.0)\ kJ\cdot mol^{-1}$, yielding $\Delta_f H_m^\circ(SiSe_2, cr, 298.15\ K) = -(177.6 \pm 3.2)\ kJ\cdot mol^{-1}$ using the CODATA [89COX/WAG] enthalpy value of $SiF_4(g)$ and the selected enthalpy value of $SeF_6(g)$. The experimental procedure in [52GAB/ALV] was thoroughly examined and discussed in [94TOM/SUS] but all attempts to explain the large difference between the two values failed and it was concluded that the reaction actually taking place in the study in [52GAB/ALV] was different from that anticipated. The value determined in [94TOM/SUS] is selected:

$$\Delta_f H_m^\circ(SiSe_2, cr, 298.15\ K) = -(177.6 \pm 3.2)\ kJ\cdot mol^{-1}.$$

V.7.3 Germanium compounds

V.7.3.1 Germanium selenides

The thermodynamics of the compounds in the system germanium-selenium were reviewed and critically assessed by O'Hare, Zywocinski, and Curtiss [96OHA/ZYW]. No new experimental information is available and in general the values reported in [96OHA/ZYW] are accepted by this review. Thermodynamic information is available for GeSe(cr), $GeSe_2$(cr), GeSe(g), $GeSe_2$(g) and Ge_2Se_2(g).

V.7.3.1.1 GeSe(cr)

The heat capacity and entropy of GeSe(s) were assessed in [96OHA/ZYW] from the heat capacities reported by Zhdanov [69ZHD], Wiedemeier, Siemers, Gaur, and Wunderlich [78WIE/SIE], Rasulov and Medzhidov [76RAS/MED], and Rasulov [88RAS]. The assessed values are selected:

$$C^{\circ}_{p,\mathrm{m}}(\mathrm{GeSe, cr, 298.15\ K}) = (49.9 \pm 0.5)\ \mathrm{J{\cdot}K^{-1}{\cdot}mol^{-1}}$$

and

$$S^{\circ}_{\mathrm{m}}(\mathrm{GeSe, cr, 298.15\ K}) = (79.6 \pm 0.8)\ \mathrm{J{\cdot}K^{-1}{\cdot}mol^{-1}}.$$

The heat capacity expression

$$C^{\circ}_{p,\mathrm{m}}(\mathrm{GeSe, cr, (298.15-940)\ K}) = (46.777 + 15.099 \times 10^{-3}\,T - 0.0316 \times 10^{-6}\,T^2 - 1.231 \times 10^{5}\,T^{-2})\ \mathrm{J{\cdot}K^{-1}{\cdot}mol^{-1}}$$

was evaluated by the review from the values tabulated in Table 6 in [96OHA/ZYW].

The enthalpy of reaction for

$$\mathrm{GeSe(cr)} + 5\mathrm{F_2(g)} \rightleftharpoons \mathrm{GeF_4(g)} + \mathrm{SeF_6(g)} \qquad \text{(V.55)}$$

was determined to be $\Delta_r H^{\circ}_{\mathrm{m}}((\mathrm{V.55}), 298.15\ \mathrm{K}) = -(2257.9 \pm 1.3)\ \mathrm{kJ{\cdot}mol^{-1}}$ by O'Hare, Susman, and Volin [89OHA/SUS] using fluorine combustion calorimetry. The selected value of the enthalpy of formation of GeSe(cr) is obtained by combining this enthalpy of reaction with the CODATA [89COX/WAG] enthalpy value of GeF_4(g) and the selected enthalpy value of SeF_6(g) yielding:

$$\Delta_f H^{\circ}_{\mathrm{m}}(\mathrm{GeSe, cr, 298.15\ K}) = -(50.3 \pm 1.4)\ \mathrm{kJ{\cdot}mol^{-1}}.$$

The Gibbs energy of formation is calculated from the selected enthalpy of formation and entropy of GeSe(cr), the selected entropy of selenium, and the entropy of germanium in [89COX/WAG] to be:

$$\Delta_f G^{\circ}_{\mathrm{m}}(\mathrm{GeSe, cr, 298.15\ K}) = -(52.2 \pm 1.4)\ \mathrm{kJ{\cdot}mol^{-1}}.$$

V.7.3.1.2 GeSe(g)

The heat capacity and entropy of GeSe(g) were calculated by the review from the molecular parameters of $^{74}Ge^{80}Se(g)$ given by Herzberg [79HUB/HER] using the usual non-rigid rotator, anharmonic oscillator approximation. The parameters were scaled to the mean molecular mass by the procedures outlined in [50HER]. No excited states were considered since they all lie above 30,000 cm^{-1} and do not affect the properties below 3000 K. The calculated values are selected:

$$C^{\circ}_{p,\mathrm{m}}(\mathrm{GeSe, g, 298.15\ K}) = (35.25 \pm 0.20)\ \mathrm{J{\cdot}K^{-1}{\cdot}mol^{-1}}$$

and

$$S^{\circ}_{\mathrm{m}}(\mathrm{GeSe, g, 298.15\ K}) = (247.42 \pm 0.50)\ \mathrm{J{\cdot}K^{-1}{\cdot}mol^{-1}}.$$

The heat capacity expression

$$C^{\circ}_{p,\mathrm{m}}(\mathrm{GeSe, g, (298.15 - 3000)\ K}) = (37.841 + 0.1340 \times 10^{-3}\ T - 0.0308 \times 10^{-6}\ T^{2} - 3.772 \times 10^{2}\ T^{-1} - 1.218 \times 10^{5}\ T^{-2})\ \mathrm{J{\cdot}K^{-1}{\cdot}mol^{-1}}$$

describes the statistical mechanics results to within 0.01 J·K^{-1}·mol^{-1}.

GeSe(cr) sublimes congruently according to the reaction GeSe(cr) $\rightleftharpoons$ GeSe(g) and the enthalpy of sublimation of GeSe(g) was assessed to be $\Delta_{\mathrm{sub}}H^{\circ}_{\mathrm{m}}$(GeSe, cr, 298.15 K) = (172.7 ± 1.0) kJ·mol^{-1} in [96OHA/ZYW] from the vapour pressure measurements by Chun-Hua, Pashinkin, and Novoselova [62CHU/PAS], Karbanov, Zlomanov, and Novoselova [68KAR/ZLO], Pashinkin, Ukhlinov, and Novoselova [69PAS/UKH], Wiedmeier and Irene [74WIE/IRE], and Ferro, Piacente, and Bardi [86FER/PIA]. The sublimation enthalpy was combined with the selected value of the enthalpy of formation of GeSe(cr) to give the selected value of the enthalpy of formation of GeSe(g):

$$\Delta_{\mathrm{f}}H^{\circ}_{\mathrm{m}}(\mathrm{GeSe, g, 298.15\ K}) = (122.4 \pm 1.8)\ \mathrm{kJ{\cdot}mol^{-1}}.$$

The Gibbs energy of formation is calculated from the selected enthalpy of formation and entropy of GeSe(g), the selected entropy of selenium, and the entropy of germanium in [89COX/WAG] to be:

$$\Delta_{\mathrm{f}}G^{\circ}_{\mathrm{m}}(\mathrm{GeSe, g, 298.15\ K}) = (70.5 \pm 1.8)\ \mathrm{kJ{\cdot}mol^{-1}}.$$

V.7.3.1.3 $GeSe_2$(cr)

The heat capacity and entropy of $GeSe_2$(cr) were assessed to be (71.3 ± 0.7) J·K^{-1}·mol^{-1} and (114.3 ± 1.2) J·K^{-1}·mol^{-1}, respectively, in [96OHA/ZYW] mainly from the heat capacity measurements made in the temperature range 52 to 310 K by Tarasov, Zhdadov, and Maltsev [68TAR/ZHD]. Since then, the heat capacity of $GeSe_2$(cr) has been measured in the temperature range 13 to 900 K using adiabatic calorimetry by Stølen *et al.* [99STO/JOH]. Their values of the heat capacity and entropy at 298.15 K are within the uncertainty ranges of the assessed values in [96OHA/ZYW] and are preferred. The selected values are:

$$C^{\circ}_{p,\mathrm{m}}(\mathrm{GeSe_2, cr, 298.15\ K}) = (71.7 \pm 0.7)\ \mathrm{J{\cdot}K^{-1}{\cdot}mol^{-1}}$$

and

$$S^{\circ}_{\mathrm{m}}(\mathrm{GeSe_2, cr, 298.15\ K}) = (115.4 \pm 1.2)\ \mathrm{J{\cdot}K^{-1}{\cdot}mol^{-1}}.$$

The heat capacity of $GeSe_2$(cr) shows a sharp λ-type effect at 423 K and no analytical expression for the temperature dependence of the heat capacity was therefore evaluated.

The enthalpy of reaction for

$$\mathrm{GeSe_2(cr) + 8F_2(g) \rightleftharpoons GeF_4(g) + 2SeF_6(g)} \qquad \text{(V.56)}$$

was determined to be $\Delta_{\mathrm{r}}H^{\circ}_{\mathrm{m}}$((V.56), 298.15 K) = – (3325.0 ± 2.3) kJ·mol^{-1} by O'Hare [86OHA] and to $\Delta_{\mathrm{r}}H^{\circ}_{\mathrm{m}}$((V.56), 298.15 K) = – (3323.5 ± 2.9) kJ·mol^{-1} by O'Hare, Susman, and Volin [87OHA/SUS] using fluorine combustion calorimetry. The values were combined with the CODATA [89COX/WAG] enthalpy value of GeF_4(g) and the selected enthalpy value of SeF_6(g) to yield $\Delta_{\mathrm{f}}H^{\circ}_{\mathrm{m}}$($GeSe_2$, cr, 298.15 K) = – (101.2 ± 2.3) kJ·mol^{-1} and $\Delta_{\mathrm{f}}H^{\circ}_{\mathrm{m}}$($GeSe_2$, cr, 298.15 K) = – (102.7 ± 2.9) kJ·mol^{-1}, respectively. Boone and Kleppa [92BOO/KLE] determined $\Delta_{\mathrm{f}}H^{\circ}_{\mathrm{m}}$($GeSe_2$, cr, 298.15 K) = – (84.4 ± 1.8) kJ·mol^{-1} using direct synthesis calorimetry. The discrepancy between the result obtained in the two methods is much larger than the claimed accuracy of the measurements. All three determinations seem to have been carried out with utmost care and under very precise conditions. Nevertheless, there must be unknown systematic errors hidden in at least one of the methods. This problem was analysed in detail in the review in [96OHA/ZYW] without finding any conclusive reason why the results differ. Nevertheless it was argued in favour of the fluorine combustion technique because the vapour pressures of GeSe(g) and $GeSe_2$(g) in equilibrium with a mixture of GeSe(cr) and $GeSe_2$(cr) as calculated from the enthalpy value in [92BOO/KLE] are inconsistent with vapour pressure measurements. However, the enthalpy of formation of GeSe(cr) employed in the calculations was determined by fluorine combustion calorimetry (*cf* V.7.3.1.1) and may eventually be subject to similar systematic errors. At present, there is no conclusive reason to dismiss the result from either method and the review selects the mean obtained from the methods

$$\Delta_{\mathrm{f}}H^{\circ}_{\mathrm{m}}(\mathrm{GeSe_2, cr, 298.15\ K}) = -(93.2 \pm 12.0)\ \mathrm{kJ{\cdot}mol^{-1}}.$$

The Gibbs energy of formation is calculated from the selected enthalpy of formation and entropy of $GeSe_2$(cr), the selected entropy of selenium, and the entropy of germanium in [89COX/WAG] to be:

$$\Delta_{\mathrm{f}}G^{\circ}_{\mathrm{m}}(\mathrm{GeSe_2, cr, 298.15\ K}) = -(93.2 \pm 12.0)\ \mathrm{kJ{\cdot}mol^{-1}}.$$

V.7.3.1.4 $GeSe_2(g)$

The heat capacity and entropy were calculated from statistical mechanics employing molecular parameters obtained by quantum mechanical calculations in [96OHA/ZYW] to be $C^{\circ}_{p,m}$ ($GeSe_2$, g, 298.15 K) = (57.2 ± 1.8) $J \cdot K^{-1} \cdot mol^{-1}$ and S°_{m} ($GeSe_2$, g, 298.15 K) = (285.3 ± 9.0) $J \cdot K^{-1} \cdot mol^{-1}$.

The heat capacity expression $C^{\circ}_{p,m}$ ($GeSe_2$, g, (298.15 – 3000) K) = $(61.888 + 0.4049 \times 10^{-3}\, T - 0.0881 \times 10^{-6}\, T^2 - 4.302 \times 10^5\, T^{-2})$ $J \cdot K^{-1} \cdot mol^{-1}$ was evaluated by the review from the values tabulated in Table 3 in [96OHA/ZYW]. The entropy and heat capacity values are not selected because they are based on quantum mechanical calculations only and not on experimental data.

The enthalpy of sublimation of $GeSe_2$(cr) was assessed to be $\Delta_{sub}H^{\circ}_{m}$ ($GeSe_2$, cr, 298.15 K) = (218 ± 18) $kJ \cdot mol^{-1}$ in [96OHA/ZYW] from the vapour pressure measurements by Karbanov, Zlomanov, and Novoselova [68KAR/ZLO], Wiedemeier and Irene [74WIE/IRE], and Chirulli and Piacente [84CHI/PIA]. This value is adopted since the additional uncertainty originating from the recalculation of the high temperature data to standard conditions is judged to be moderate compared to the estimated experimental uncertainty. The sublimation enthalpy was combined with the selected value of the enthalpy of formation of $GeSe_2$(cr) to give the selected value of the enthalpy of formation of $GeSe_2$(g)

$$\Delta_f H^{\circ}_{m}(GeSe_2, g, 298.15\ K) = (124.8 \pm 21.6)\ kJ \cdot mol^{-1}.$$

V.7.3.1.5 $Ge_2Se_2(g)$

The heat capacity and entropy were calculated from statistical mechanics employing molecular parameters obtained by quantum mechanical calculations in [96OHA/ZYW] to be $C^{\circ}_{p,m}$ (Ge_2Se_2, g, 298.15 K) = (79.0 ± 2.4) $J \cdot K^{-1} \cdot mol^{-1}$ and S°_{m} (Ge_2Se_2, g, 298.15 K) = (357.8 ± 11.0) $J \cdot K^{-1} \cdot mol^{-1}$. The heat capacity expression $C^{\circ}_{p,m}$ (Ge_2Se_2, g, (298.15 – 3000) K) = $(83.025 + 0.1039 \times 10^{-3}\, T - 0.0224 \times 10^{-6}\, T^2 - 3.594 \times 10^5\, T^{-2})$ $J \cdot K^{-1} \cdot mol^{-1}$ was evaluated by the review from the values tabulated in Table 4 in [96OHA/ZYW]. These results are not selected.

The enthalpy of the dissociation

$$Ge_2Se_2(g) \rightleftharpoons 2GeSe(g) \qquad (V.57)$$

was assessed in [96OHA/ZYW] by a third law evaluation of the mass spectrometric studies by Zlomanov, Novozhilov, and Makarov [80ZLO/NOV] and Balducci, Sacconi, and Gigli [91BAL/SAC]. to be $\Delta_r H^{\circ}_{m}$((V.57), 298.15 K) = (109 ± 5) $kJ \cdot mol^{-1}$. The value of the enthalpy of formation of Ge_2Se_2(g), (135.8 ± 6.4) $kJ \cdot mol^{-1}$, is calculated from this value and the selected enthalpy of formation of GeSe(g). The results have not been adopted since the accuracy of the thermal data and entropy of Ge_2Se_2(g) needed in the calculations could not be ascertained by the review.

V.7.3.2 Germanium selenites

Slavtscheva, Popova, and Gospodinov [84SLA/POP] prepared $Ge(SeO_3)_2$(cr) and measured its solubility in dilute mineral acids. The solubility product calculated from these measurements is not accepted, see Appendix A.

Gospodinov and Bogdanov [89GOS/BOG] studied the thermal decomposition of germanium selenite. The heat capacity of $Ge(SeO_3)_2$ was determined by differential scanning calorimetry to be $C^{\circ}_{p,m}$ ($Ge(SeO_3)_2$, cr, (400 – 550) K) = (– 22.47 + 0.8109 T – 1.653 × 10^{-4} T^2) $J \cdot K^{-1} \cdot mol^{-1}$. The relationship is not recommended since concurrent measurements on $PbSeO_3$(cr) led to dubious results.

No other thermodynamic data have been found for germanium selenites.

V.7.3.3 Germanium selenates

No thermodynamic data have been found for germanium selenates.

V.7.4 Tin compounds

V.7.4.1 Tin selenides

The phase diagram and the thermodynamic properties of the condensed phases in the system tin-selenium were assessed by Feutelais, Majid, Legendre, and Fries [96FEU/MAJ]. The solid phases reported in the system were α-SnSe, β-SnSe, and $SnSe_2$. No new experimental thermodynamic information is available and the selected thermodynamic properties of the solid phases were in general calculated from the Gibbs energy expressions evaluated in [96FEU/MAJ].

The gaseous species SnSe(g), $SnSe_2$(g), and Sn_2Se_2(g) have been reported in mass spectrometric investigations of the vaporisation of the solid compounds.

V.7.4.1.1 α-SnSe

α-SnSe is the stable form of crystalline SnSe below (796 ± 3) K at which temperature it transforms into β-SnSe.

The heat capacity of α-SnSe was evaluated in [96FEU/MAJ] from the heat capacity and heat content measurements summarised in Table V-35.

The expression

$$C^{\circ}_{p,m}(\text{SnSe}, \alpha, (298.15 - 796)\text{ K}) = (52.62 + 7.28 \times 10^{-3}\,T - 1.329 \times 10^{-6}\,T^2 - 2.4569 \times 10^{5}\,T^{-2})\text{ J}\cdot\text{K}^{-1}\cdot\text{mol}^{-1}$$

was calculated by the review from the Gibbs energy expression selected in [96FEU/MAJ]. The heat capacity at 298.15 K calculated from this expression is selected:

$$C^{\circ}_{p,\mathrm{m}}(\mathrm{SnSe}, \alpha, 298.15\ \mathrm{K}) = (51.9 \pm 2.0)\ \mathrm{J{\cdot}K^{-1}{\cdot}mol^{-1}}.$$

Table V-35: Heat capacity and heat content measurements for α-SnSe and β-SnSe. The transition α → β occurs at (796 ± 3) K.

Reference	Temperature range (K)	Method
[74BLA/IGE2]	298 – 787	Drop calorimetry
[81BAL/LEG]	375 – 1135	Drop calorimetry
[81WIE/PUL]	230 – 580	Differential scanning calorimetry
[91RAS]	320 – 1146	Drop calorimetry
[94YAM/KAM]	700 – 1400	Drop calorimetry

The standard entropy of α-SnSe was evaluated to be 86.93 $\mathrm{J{\cdot}K^{-1}{\cdot}mol^{-1}}$, corresponding to $\Delta_{\mathrm{f}}S^{\circ}_{\mathrm{m}}(\mathrm{SnSe}, \alpha, 298.15\ \mathrm{K}) = -\,6.3\ \mathrm{J{\cdot}K^{-1}{\cdot}mol^{-1}}$, in the thermodynamic optimisation and assessment of the Sn-Se system in [96FEU/MAJ]. The value originates mainly from the modelling and assumptions made about the liquid phase in the system and the recalculation to 298.15 K by the use of enthalpies of phase transformations and heat capacities. The only experimental determination of the entropy at low temperatures was made by Melekh, Stepanova, Fomina, and Semenkovich [71MEL/STE] who performed emf measurements on the galvanic cells

$$(-)\ \mathrm{Sn(cr)}\ |\ \mathrm{SnCl_2}\ (\text{in } \mathrm{AlCl_3} + \mathrm{NaCl}\ \text{melt})\ |\ \mathrm{SnSe_2(cr)}, \mathrm{Se(trig)}\ (+)$$

$$(-)\ \mathrm{Sn(cr)}\ |\ \mathrm{SnCl_2}\ (\text{in } \mathrm{AlCl_3} + \mathrm{NaCl}\ \text{melt})\ |\ \mathrm{SnSe(cr)}, \mathrm{SnSe_2(cr)}\ (+)$$

in the temperature range 435 to 485 K resulting in $\Delta_{\mathrm{f}}S^{\circ}_{\mathrm{m}}(\mathrm{SnSe}, \alpha, 298.15\ \mathrm{K}) = (5.0 \pm 4.6)\ \mathrm{J{\cdot}K^{-1}{\cdot}mol^{-1}}$. The results in [96FEU/MAJ] and [71MEL/STE] disagree and since both values are based on a number of estimates no value is selected for the entropy at 298.15 K.

The enthalpy of formation of α-SnSe was evaluated in [96FEU/MAJ] by considering the results of the direct synthesis calorimetry measurements of Gadzhiev and Sharifov [60GAD/SHA] and Boone and Kleppa [92BOO/KLE], the Knudsen effusion measurements of Colin and Drowart [64COL/DRO], and the emf measurements in [71MEL/STE]. The review accepts the assessment in [96FEU/MAJ] and obtains from the data presented there the selected value:

$$\Delta_{\mathrm{f}}H^{\circ}_{\mathrm{m}}(\mathrm{SnSe}, \alpha, 298.15\ \mathrm{K}) = -\,(109.5 \pm 15.0)\ \mathrm{kJ{\cdot}mol^{-1}}$$

where the error limits were set by the review.

V.7.4.1.2 β-SnSe

β-SnSe is the stable form of crystalline SnSe from 796 K up to 1146 K where it melts.

No heat capacity or heat content measurements have been made for β-SnSe below 796 K and consequently no value can be selected at 298.15K. [96FEU/MAJ] employed the heat capacity expression of α-SnSe to describe the heat capacity of β-SnSe in the thermochemical modelling of the tin-selenium system.

The enthalpy of the transformation α-SnSe → β-SnSe was estimated to be 1.28 $kJ\cdot mol^{-1}$ in [96FEU/MAJ] from the heat content measurements summarised in Table V-35.

V.7.4.1.3 SnSe(g)

The heat capacity and entropy of SnSe(g) were calculated by the review from the molecular parameters of $^{120}Sn^{80}Se(g)$ given by Herzberg [79HUB/HER] using the usual non-rigid rotator, anharmonic oscillator approximation. The parameters were scaled to the mean molecular mass by the procedures outlined in [50HER]. No excited states were considered since they all lie above 27,000 cm^{-1} and do not affect the properties below 3000 K. The calculated values are selected:

$$C^{\circ}_{p,m}(\text{SnSe, g, 298.15 K}) = (35.93 \pm 0.50)\ \text{J}\cdot\text{K}^{-1}\cdot\text{mol}^{-1}$$

and

$$S^{\circ}_{m}(\text{SnSe, g, 298.15 K}) = (255.4 \pm 1.0)\ \text{J}\cdot\text{K}^{-1}\cdot\text{mol}^{-1}.$$

The heat capacity expression

$$C^{\circ}_{p,m}(\text{SnSe, g, (298.15 – 3000) K}) = (37.619 + 0.2104 \times 10^{-3}\,T - 0.0151 \times 10^{-6}\,T^{2} - 1.787 \times 10^{2}\,T^{-1} - 1.02881 \times 10^{5}\,T^{-2})\ \text{J}\cdot\text{K}^{-1}\cdot\text{mol}^{-1}$$

describes the results obtained by statistical mechanics to within 0.005 $J\cdot K^{-1}\cdot mol^{-1}$.

The mass spectrometric investigations of the vaporisation of SnSe(cr) by Colin and Drowart [64COL/DRO] and by Kulyukhina, Zlomanov, and Novoselova [79KUL/ZLO] have shown that SnSe(g) is the major product formed in the vaporisation process. The species Sn(g), Se_2(g), $SnSe_2$(g) and Sn_2Se_2(g) were only found in minor quantities. The thermodynamic properties of SnSe(g) can thus be evaluated from the measurements of the saturated total pressure in equilibrium with SnSe(cr) assuming the reaction α-SnSe ⇌ SnSe(g). The results of such studies evaluated as discussed in Appendix A are summarised in Table V-36. The weighted average $\Delta_{sub}H^{\circ}_{m}$(SnSe, α, 298.15 K) = (219.7 ± 4.6) $kJ\cdot mol^{-1}$ was combined with the selected enthalpy of formation of α-SnSe to yield the selected enthalpy of formation of SnSe(g):

$$\Delta_{f}H^{\circ}_{m}(\text{SnSe, g, 298.15 K}) = (110.2 \pm 15.7)\ \text{kJ}\cdot\text{mol}^{-1}.$$

Table V-36: Determinations of the enthalpy and entropy of sublimation of α-SnSe.

Reference	$\Delta_{sub}H^{\circ}_{m}$ (SnSe, α 298.15 K) (kJ·mol^{-1})	$\Delta_{sub}S^{\circ}_{m}$ (SnSe, α, 298.15 K) (J·K^{-1}·mol^{-1})	T (K)	Measurement method
[61NES/PAS]	(188.2 ± 20.0)	(133.0 ± 25.0)	862 – 920	Knudsen effusion
[63HIR/ICH]	(210.8 ± 12.0)	(158.9 ± 13.0)	773 – 898	Knudsen effusion
[64COL/DRO]	(150.8 ± 45.0)	(92.7 ± 40.0)	792 – 979	Knudsen effusion, mass spectroscopy
[70BLA/MUN]	(217.0 ± 7.0)	(164.7 ± 8.0)	789 – 975	Torsion effusion
[79KUL/ZLO]	(209.1 ± 25.0)	(198.7 ± 30.0)	550 – 693	Knudsen effusion, mass spectroscopy
[83WIE/PUL]	(219.7 ± 10.0)	(166.7 ± 7.0)	736 – 967	Knudsen effusion
[95ZOC/PIA]	(222.5 ± 3.4)	(168.9 ± 4.2)	793 – 1040	Torsion effusion

The Gibbs energy of formation is calculated from the selected enthalpy of formation and entropy of SnSe(g), the selected entropy of selenium, and the entropy of tin in [89COX/WAG] to be:

$$\Delta_f G^{\circ}_{m}(\text{SnSe, g, 298.15 K}) = (61.9 \pm 15.7)\ \text{kJ·mol}^{-1}.$$

V.7.4.1.4 $SnSe_2$(cr)

The heat capacity of $SnSe_2$(cr) was measured in the temperature range 230 to 570 K in a careful study by Wiedemeier, Pultz, Gaur, and Wunderlich [81WIE/PUL]. Their results are selected:

$$C^{\circ}_{p,m}(\text{SnSe}_2\text{, cr, (230 – 570) K}) = (75.39 + 11.5 \times 10^{-3}\ T - 1.92 \times 10^{5}\ T^{-2})\ \text{J·K}^{-1}\text{·mol}^{-1}$$

and

$$C^{\circ}_{p,m}(\text{SnSe}_2\text{, cr, 298.15K}) = (76.7 \pm 0.5)\ \text{J·K}^{-1}\text{·mol}^{-1}.$$

The standard entropy of $SnSe_2$(cr) was evaluated to 139.8 J·K^{-1}·mol^{-1}, corresponding to $\Delta_f S^{\circ}_{m}$ ($SnSe_2$, cr, 298.15 K) = 4.4 J·K^{-1}·mol^{-1}, in the thermodynamic optimisation and assessment of the Sn-Se system in [96FEU/MAJ]. The value originates mainly from the modelling and assumptions made about the liquid phase in the system and the recalculation to 298.15 K by the use of enthalpies of phase transformations and heat capacities. The only experimental information about the entropy at low temperatures originates from the emf measurements on the galvanic cell

$$(-)\ \text{Sn(cr)} \mid \text{SnCl}_2\ (\text{in AlCl}_3 + \text{NaCl melt}) \mid \text{SnSe}_2\text{(cr), Se(trig)}\ (+)$$

made in the temperature range 435 to 485 K by Melekh, Stepanova, Fomina, and Semenkovich [71MEL/STE]. The resulting $\Delta_f S^{\circ}_{m}$ ($SnSe_2$, cr, 298.15 K) = – (17.2 ± 2.5)

$J \cdot K^{-1} \cdot mol^{-1}$ disagrees with the value in [96FEU/MAJ] and because of the large discrepancy no value is selected for the entropy at 298.15 K.

The enthalpy of formation of $SnSe_2(cr)$ was evaluated in [96FEU/MAJ] by considering the results of the direct synthesis calorimetry measurements of Boone and Kleppa [92BOO/KLE], the vapour pressure measurements of Karakhanova, Pashinkin, and Novoselova [67KAR/PAS] and Kulyukhina, Zlomanov, and Novoselova [77KUL/ZLO], and the emf measurements in [71MEL/STE]. The review accepts the assessment in [96FEU/MAJ] and obtains from the data presented there the selected value:

$$\Delta_f H_m^\circ (SnSe_2, cr, 298.15 K) = -(114.9 \pm 10.0) kJ \cdot mol^{-1}$$

where the error limits were set by the review.

V.7.4.1.5 $SnSe_2(g)$

The molecule $SnSe_2(g)$ was found to be present in very low concentrations in the mass spectrometric investigation of the vaporisation of $SnSe_2(cr)$ in the temperature range 550 to 690 K in [79KUL/ZLO] but not in a similar study of the vaporisation of solid SnSe in [64COL/DRO]. The enthalpy of the reaction

$$SnSe_2(cr) \rightleftharpoons SnSe_2(g) \quad (V.58)$$

was reported in [79KUL/ZLO] to be $\Delta_r H_m^\circ ((V.58), 298.15 K) = (190.5 \pm 20.9) kJ \cdot mol^{-1}$ from the second law and $\Delta_r H_m^\circ ((V.58), 298.15 K) = (193.8 \pm 12.6) kJ \cdot mol^{-1}$ from the third law. No experimental heat capacities or entropies are available for a proper recalculation of the high temperature results to the standard temperature and no presentation of the data used in the paper is made. Since the additional uncertainties originating from the fairly short extrapolation to the standard temperature are small in relation to the uncertainties of the reaction enthalpy, the review accepts the procedure employed in [79KUL/ZLO]. The weighted average of the reaction enthalpy values is combined with the selected enthalpy of formation of $SnSe_2(cr)$ to yield the selected enthalpy of formation of $SnSe_2(g)$:

$$\Delta_f H_m^\circ (SnSe_2, g, 298.15 K) = (78.0 \pm 16.1) kJ \cdot mol^{-1}.$$

V.7.4.1.6 $Sn_2Se_2(g)$

Small amounts of the molecule $Sn_2Se_2(g)$ were found in the mass spectrometric investigations in the temperature range 792 to 979 K in [64COL/DRO] and in the range 550 to 693 K in [79KUL/ZLO]. The values reported for the dimerisation reaction

$$2SnSe(g) \rightleftharpoons Sn_2Se_2(g) \quad (V.59)$$

were $\Delta_r H_m^\circ ((V.59), 298.15 K) = -(194.6 \pm 22.0) kJ \cdot mol^{-1}$ and $\Delta_r H_m^\circ ((V.59), 298.15 K) = -(196.4 \pm 12.6) kJ \cdot mol^{-1}$ from the third law in [64COL/DRO] and [79KUL/ZLO], respectively. The weighted average of the reaction enthalpy values,

– (195.9 ± 11.1) kJ·mol^{-1}, is combined with the selected enthalpy of formation of SnSe(g) to yield the enthalpy of formation of Sn_2Se_2(g), $\Delta_f H_m^\circ$ (Sn_2Se_2, g, 298.15 K) = (16.3 ± 12.2) kJ·mol^{-1}.

The extrapolation of the high temperature results to 298.15 K in [79KUL/ZLO] and [64COL/DRO] employs estimated values of the Gibbs energy function. The result for the enthalpy of formation of Sn_2Se_2(g) is not selected since the accuracy of the extrapolations cannot be ascertained.

V.7.4.2 Tin selenites

Slavtscheva *et al.* [84SLA/POP] prepared $Sn(SeO_3)_2$(cr) and measured its solubility in dilute mineral acids. The solubility product calculated from these measurements is not accepted, see Appendix A.

Gospodinov and Bogdanov [89GOS/BOG] studied the thermal decomposition of $Sn(SeO_3)_2$ and determined the heat capacity of the compound by differential scanning calorimetry: $C_{p,m}^\circ$ ($Sn(SeO_3)_2$, cr, (400 – 550) K) = (204.1 + 2.544 × 10^{-2} T + 2.477 × 10^{-6} T^2) J·K^{-1}·mol^{-1}. The relationship is not recommended since concurrent measurements on $PbSeO_3$(cr) led to dubious results.

No other thermodynamic data have been found for tin selenites.

V.7.4.3 Tin selenates

No thermodynamic data have been found for tin selenates.

V.7.5 Lead compounds

V.7.5.1 Lead selenides

Lead monoselenide PbSe(cr) has a very narrow homogeneity range, melts congruently at 1354 K, and is the only solid compound in the lead-selenium system. Temperature-composition and temperature-pressure phase diagrams are available in the review by Novoselova, Zlomanov, Karbanov, Matveyev, and Gas'kov [72NOV/ZLO].

V.7.5.1.1 PbSe(cr)

References to the heat capacity measurements of PbSe(cr) are found in Table V-37.

The heat capacity reported in [83ELS/ABO] is rejected because it is about 60 times too small. A linear heat capacity expression for the remaining data and the temperature range 240 to 1349 K was generated by forcing the regression line to pass through the value of the heat capacity at 240 K in [54PAR/QUA] and adapting it to the measured values at high temperatures as shown in Figure V-15. The resulting expression is

$$C^{\circ}_{p,\mathrm{m}}(\mathrm{PbSe, cr, (298.15 - 1349)\ K}) = (46.96 + 12.26 \times 10^{-3}\ T)\ \mathrm{J{\cdot}K^{-1}{\cdot}mol^{-1}}$$

and the selected heat capacity at 298.15 K is calculated to be:

$$C^{\circ}_{p,\mathrm{m}}(\mathrm{PbSe, cr, 298.15\ K}) = (50.6 \pm 1.0)\ \mathrm{J{\cdot}K^{-1}{\cdot}mol^{-1}}.$$

Table V-37: Heat capacity measurements of PbSe(cr).

Reference	Temperature range (K)
[54PAR/QUA]	25 – 240
[74BLA/IGE2]	300 – 1349
[78RAS/MED]	300 – 1349
[82LYK/CHE]	1 – 20
[83ELS/ABO]	300 – 700
[94PAS/ZLO]	350 – 580
[94YAM/KAM]	800 – 1349

Figure V-15: The heat capacity of PbSe(cr) as a function of temperature.

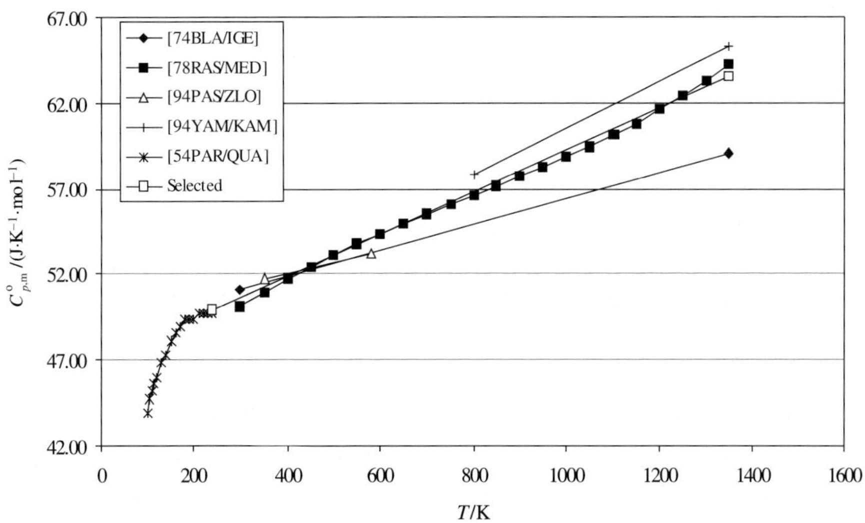

A third law value of the entropy of PbSe(cr), S°_{m}(PbSe, cr, 298.15 K) = (102.4 ± 4.0) J·K^{-1}·mol^{-1}, was calculated by the review by integrating the heat capacity resulting from a combination of the selected heat capacity expression and the measurements in [54PAR/QUA] and [82LYK/CHE]. Second law entropies were calculated by the review as discussed in Appendix A from the electrochemical cell measurements by Finch and Wagner [60FIN/WAG], Sadikov and Semenkovich [66SAD/SEM], and Shamsuddin and Misra [73SHA/MIS]. The results are summarised in Table V-38. All second law entropies are larger than the value obtained from the third law. Considering the uncertainties in the heat capacities at low temperatures and the missing heat capacity data in the temperature range 240 to 300 K, the selected value is:

$$S^{\circ}_{m}(\text{PbSe, cr, 298.15 K}) = (105.0 \pm 4.0)\ \text{J}\cdot\text{K}^{-1}\cdot\text{mol}^{-1}.$$

Second law values of the enthalpy of formation calculated as discussed in Appendix A from the electrochemical measurements in [60FIN/WAG], [66SAD/SEM], and [73SHA/MIS] are tabulated in Table V-39 together with the results of direct synthesis calorimetry investigations made by Hajiev [70HAJ] and Boone and Kleppa [92BOO/KLE].

Table V-38: Second law entropies at 298.15 K calculated by the review from electrochemical cell measurements.

Reference	Temperature (K)	Reaction	S°_{m} (PbSe, cr, 298.15 K) ($\text{J}\cdot\text{K}^{-1}\cdot\text{mol}^{-1}$)
[60FIN/WAG]	473 – 573	Pb(cr) + Se(l) $\rightleftharpoons$ PbSe(cr)	(108.7 ± 6.0)
[66SAD/SEM]	673 – 773	Pb(l) + Se(l) $\rightleftharpoons$ PbSe(cr)	(106.7 ± 4.0)
[73SHA/MIS]	490 – 600	Pb(cr) + Se(l) $\rightleftharpoons$ PbSe(cr)	(109.6 ± 4.0)
[73SHA/MIS]	650 – 858	Pb(l) + Se(l) $\rightleftharpoons$ PbSe(cr)	(107.9 ± 4.0)

Table V-39: Values of the enthalpy of formation of PbSe(cr) at 298.15 K determined using electrochemical cells ([60FIN/WAG], [66SAD/SEM], [73SHA/MIS]) and direct synthesis calorimetry ([70HAJ], [92BOO/KLE]).

Reference	$\Delta_f H^{\circ}_{m}$ (PbSe, cr, 298.15 K) ($\text{kJ}\cdot\text{mol}^{-1}$)
[60FIN/WAG]	– (99.5 ± 3.0)
[66SAD/SEM]	– (95.1 ± 3.0)
[73SHA/MIS]	– (100.3 ± 3.0), – (94.9 ± 4.0)
[70HAJ]	– (105.0 ± 4.0)
[92BOO/KLE]	– (96.3 ± 3.0)

The mean of the enthalpy of formation values listed in Table V-39 is selected:

$$\Delta_f H^{\circ}_{m}(\text{PbSe, cr, 298.15 K}) = -(98.5 \pm 7.6)\ \text{kJ}\cdot\text{mol}^{-1}.$$

The Gibbs energy of formation is calculated from the selected enthalpy of formation and entropy of PbSe(cr), the selected entropy of selenium, and the entropy of lead in [89COX/WAG] to be:

$$\Delta_f G_m^\circ(\text{PbSe, cr, 298.15 K}) = -(97.9 \pm 7.7)\ \text{kJ·mol}^{-1}.$$

In the aqueous phase, Umland and Wallmeier [80UML/WAL] studied the reduction of selenite at the mercury electrode in the presence of Pb^{2+} by polarography, see Appendix A. The solubility product of PbSe(s) was obtained from the position of the half-wave potential of the second reduction step, $HgSe(s) + Pb^{2+} + 2e^- \rightleftharpoons PbSe(s) + Hg(l)$, to be $\log_{10} K_{s,0}^\circ$ (PbSe, s, 298.15 K) = – (33.6 ± 2.0). Combined with CODATA for Pb^{2+} and the selected value for Se^{2-}, it corresponds to $\Delta_f G_m^\circ$ (PbSe, cr, 298.15 K) = – (87.4 ± 11.8) kJ·mol^{-1}.

The value is not in full harmony with the value obtained from an extrapolation of high temperature data. Since the validity of the polarographic method has not been documented, the value of $\Delta_f G_m^\circ$ (PbSe, cr, 298.15 K) obtained from the extrapolation of the high temperature data has been selected.

V.7.5.1.2 PbSe(g)

Uy and Drowart [69UY/DRO] found PbSe(g) to be the major species in gaseous mixtures of lead and selenium using mass spectrometry. The vaporisation of PbSe(cr) occurs according to the reaction PbSe(cr) $\rightleftharpoons$ PbSe(g).

The heat capacity and entropy of PbSe(g) were calculated using the molecular data of Herzberg [50HER] by Kelley [60KEL] and Kelley and King [61KEL/KIN], respectively. The calculations were repeated by the review using the molecular parameters of $^{208}Pb^{80}Se(g)$ given by Ndikumana, Carleer, and Colin [90NDI/CAR] employing the usual non-rigid rotator, anharmonic oscillator approximation. The parameters were scaled to the mean molecular mass by the procedures outlined in [50HER]. No excited states were considered. All of them lie above 18,000 cm^{-1} and do not affect the properties below 2000 K. The calculated values are selected:

$$C_{p,m}^\circ(\text{PbSe, g, 298.15 K}) = (36.36 \pm 0.50)\ \text{J·K}^{-1}\text{·mol}^{-1}$$

and

$$S_m^\circ(\text{PbSe, g, 298.15 K}) = (263.3 \pm 1.0)\ \text{J·K}^{-1}\text{·mol}^{-1}.$$

The heat capacity expression

$$C_{p,m}^\circ(\text{PbSe, g, (298.15 – 2000) K}) = (37.338 + 0.4343 \times 10^{-3}\,T - 0.0269 \times 10^{-6}\,T^2 - 0.98068 \times 10^{5}\,T^{-2})\ \text{J·K}^{-1}\text{·mol}^{-1}$$

describes the results obtained by statistical mechanics to within 0.008 J·K^{-1}·mol^{-1}.

Second law entropies were calculated by the review as discussed in Appendix A from the evaporation studies summarised in Table V-40. The second laws values agree well with the value calculated from molecular data.

Table V-40: Second law entropy values for PbSe(g) at 298.15 K derived from vapour pressure measurements as discussed in Appendix A.

Reference	S°_{m} (PbSe, g, 298.15 K) ($J \cdot K^{-1} \cdot mol^{-1}$)	Temperature (K)	Measurement method
[59ZLO/POP]	(259.8 ± 9.0)	774 – 991	Langmuir, Knudsen weight loss
[69HAN/MUN]	(274.7 ± 7.0)	936 – 1106	Torsion effusion
[69SOK/PAS]	(270.5 ± 7.0)	936 – 1016	Knudsen weight loss
[93BRU/PIA]	(260.6 ± 7.0)	573 – 823	Torsion effusion
Average	(266.4 ± 14.4)		

Second and third law enthalpies of sublimation were evaluated by the review as discussed in Appendix A from the evaporation studies summarised in Table V-41. The selected value of the enthalpy of sublimation of PbSe(cr) is calculated as the weighted average of the second and third law mean values, $\Delta_{sub}H^{\circ}_{m}$ (PbSe, cr, 298.15 K) = (228.0 ± 4.2) $kJ \cdot mol^{-1}$. The selected value of the enthalpy of formation of PbSe(g) is calculated from the selected value of the sublimation enthalpy and the enthalpy of formation of PbSe(cr) yielding:

$$\Delta_{f}H^{\circ}_{m}(\text{PbSe, g, 298.15 K}) = (129.5 \pm 8.7)\ \text{kJ}\cdot\text{mol}^{-1}.$$

Table V-41: Second and third law enthalpies of sublimation of PbSe(cr) at 298.15 K evaluated by the review as discussed in Appendix A.

Reference	$\Delta_{sub}H^{\circ}_{m}$ (PbSe, cr, 298.15 K), 2nd law ($kJ \cdot mol^{-1}$)	$\Delta_{sub}H^{\circ}_{m}$ (PbSe, crs, 298.15 K), 3rd law ($kJ \cdot mol^{-1}$)
[59ZLO/POP]	(222.6 ± 9.0)	(226.0 ± 3.0)
[69HAN/MUN]	(242.6 ± 7.0)	(231.0 ± 4.0)
[69SOK/PAS]	(234.9 ± 7.0)	(227.1 ± 3.0)
[93BRU/PIA]	(224.4 ± 7.0)	(227.1 ± 3.0)
Average	(231.1 ± 18.4)	(227.8 ± 4.3)

The Gibbs energy of formation is calculated from the selected enthalpy of formation and entropy of PbSe(g), the selected entropy of selenium, and the entropy of lead in [89COX/WAG] to be:

$$\Delta_f G_m^\circ(\text{PbSe, g, 298.15 K}) = (82.9 \pm 8.7)\ \text{kJ·mol}^{-1}.$$

V.7.5.2 Lead selenites

Oikova and Gospodinov [81OIK/GOS] studied the system PbO-SeO_2-H_2O at (373.2 ± 1.0) K over the concentration range 0 to 71 mass-% SeO_2 in the liquid phase. The incongruently soluble phases $PbSeO_3$ and $PbSeO_3 \cdot H_2SeO_3$ were formed. The solubility data and a phase diagram are available in the paper. The solubility of Pb(II) is low over the whole concentration range in H_2SeO_3.

Chukhlantsev and Tomashevsky [57CHU/TOM] prepared lead selenite by mixing an acidified 0.2 M solution of lead acetate and a 0.1 M sodium selenite solution in stoichiometric amounts. The precipitate was crystalline. Chemical analysis confirmed the 1:1 ratio between Pb(II) and Se(IV). No X-ray diffraction measurements were performed. The solubility of the specimen in dilute solution of nitric or hydrochloric acid was measured at 293 K. The experiments were performed and the data recalculated as described in Appendix A, [56CHU]. The result for:

$$PbSeO_3(cr) \rightleftharpoons Pb^{2+} + SeO_3^{2-} \qquad (V.60)$$

is $\log_{10} K_{s,0}^\circ$ ((V.60), 293 K) = – (12.10 ± 0.09). The reported conditional constant was – (11.49 ± 0.17). Slavtscheva *et al.* [84SLA/POP] made similar measurements in dilute nitric acid at 293 K. As no primary data are available, the constant, $\log_{10} K_{s,0}^\circ$ = – (12.50 ± 0.06), cannot be recalculated. The review accepts and selects:

$$\log_{10} K_{s,0}^\circ((\text{V.60}), 298.15\ \text{K}) = -(12.5 \pm 1.0).$$

Wide uncertainty limits are applied since the experimental information is somewhat meagre.

Selivanova and Leshchinskaya [62SEL/LES3] measured the enthalpy change of the reaction between Na_2SeO_3(cr) and a lead nitrate solution with formation of crystalline lead selenite. From their data the standard enthalpy change of the reaction $Na_2SeO_3(cr) + Pb^{2+} \rightarrow PbSeO_3(cr) + 2Na^+$ was calculated to be – (54.06 ± 1.50) kJ·mol^{-1} in Appendix A. Combined with selected standard enthalpies of formation this datum yields:

$$\Delta_f H_m^\circ(\text{PbSeO}_3\text{, cr, 298.15 K}) = -(532.1 \pm 1.8)\ \text{kJ·mol}^{-1}.$$

The selected result differs from the value in [82WAG/EVA], – 537.6 kJ mol^{-1}. The causes of the difference are discussed in Appendix A.

From the selected data:

$$\Delta_f G_m^\circ(\text{PbSeO}_3\text{, cr, 298.15 K}) = -(458.0 \pm 6.0)\ \text{kJ·mol}^{-1},$$

$$S^{\circ}_{m}(\mathrm{PbSeO_3, cr, 298.15\ K}) = (166.1 \pm 21.0)\ \mathrm{J \cdot K^{-1} \cdot mol^{-1}}.$$

were calculated.

Bakeeva, Buketov, and Pashinkin [71BAK/BUK] measured the vapour pressure of SeO_2(g) above lead selenite. All measurements were carried out above the eutectic temperature of the $PbSeO_3$-$2PbO \cdot PbSeO_3$ sector of the $PbSeO_3$-PbO phase diagram [69GME]. For $PbSeO_3$(cr) and the temperature interval 893 to 968 K, $\log_{10} p(\mathrm{SeO_2, g}, T) = (10.73 \pm 0.22) - (12223 \pm 628)\ T^{-1}$, and for $PbSeO_3$(l) and the interval 981 to 1023 K, $\log_{10} p(\mathrm{SeO_2, g}, T) = (7.71 \pm 0.96) - (9274 \pm 302)\ T^{-1}$ were obtained.

Gospodinov and Bogdanov [89GOS/BOG] determined the melting point of $PbSeO_3$(cr) to be (948 ± 5) K and its enthalpy of fusion by differential scanning calorimetry to be:

$$\Delta_{\mathrm{fus}} H^{\circ}_{m}(\mathrm{PbSeO_3, cr, 948\ K}) = (36.28 \pm 0.13)\ \mathrm{kJ \cdot mol^{-1}}.$$

By the same technique the following heat capacities were found for the temperature range 400 to 550 K: $C^{\circ}_{p,m}(\mathrm{PbSeO_3, cr, (400-550)\ K}) = (514.4 - 1.359\ T + 2.511 \times 10^{-3}\ T^2)\ \mathrm{J \cdot K^{-1} \cdot mol^{-1}}$, $C^{\circ}_{p,m}(\mathrm{PbSeO_3 \cdot 2PbO, cr, (400-550)\ K}) = (-16.77 + 0.8958\ T - 7.075 \times 10^{-4}\ T^2)\ \mathrm{J \cdot K^{-1} \cdot mol^{-1}}$, $C^{\circ}_{p,m}(\mathrm{PbSeO_3 \cdot 4PbO, cr, (400-550)\ K}) = (17.64 + 1.446\ T - 1.432 \times 10^{-3}\ T^2)\ \mathrm{J \cdot K^{-1} \cdot mol^{-1}}$. The unit of the heat capacity is not mentioned in the paper, but it is likely to be $\mathrm{cal \cdot K^{-1} \cdot mol^{-1}}$ since the calorie has been used for other data. The coefficients have been recalculated accordingly. The relationship for $PbSeO_3$(cr) results in $C^{\circ}_{p,m}$ values that are in total disagreement with estimates in [73BAR/KNA], [73KLU/LES]. The above equations for the heat capacities are therefore not recommended.

V.7.5.3 Lead selenates

A phase diagram and related information of the system PbO-SeO_3-H_2O at 373.15 K are available in [81OIK/GOS]. Only $PbSeO_4$(cr) is formed.

Selivanova and Boguslavskii [55SEL/BOG] measured the solubility of $PbSeO_4$(cr) at 298.15 K in alkali metal selenate solutions of concentration (0, 0.1, 0.5, 1, 10, 100) × 10^{-3} M (rounded figures). No pH measurements were performed. A calculation with hydrolysis constants from [76BAE/MES] indicated that the test solutions should be slightly acidic, but the concentrations of $PbOH^+$ and $HSeO_4^-$ should be negligible compared to the total concentrations. The solubility product of $PbSeO_4$(cr) obtained from the data and extrapolated to $I = 0$ by SIT with $\Delta\varepsilon = 0\ \mathrm{kg \cdot mol^{-1}}$ is selected:

$$\mathrm{PbSeO_4(cr)} \rightleftharpoons \mathrm{Pb^{2+}} + \mathrm{SeO_4^{2-}} \qquad \text{(V.61)}$$

$$\log_{10} K^{\circ}_{s,0}((\text{V.61}), 298.15\ \mathrm{K}) = -(6.90 \pm 0.25).$$

The uncertainty is twice the standard deviation. There is no indication in the data of the formation of $PbSeO_4$(aq) at higher selenate concentrations.

Feroci, Fini, Badiello, and Breccia [97FER/FIN] studied the complex formation between Pb^{2+} and SeO_4^{2-} by classical polarography in 0.15 M $NaNO_3$. They found no evidence for an interaction, see Appendix A. The review estimates from an assumed accuracy of the polarographic measurement of ±2 mV, the equilibrium constant of the reaction:

$$Pb^{2+} + SeO_4^{2-} \rightleftharpoons PbSeO_4(aq) \qquad (V.62)$$

to be:

$$\log_{10} \beta_1^\circ\ ((V.62),\ 298.15\ K) < 2.2.$$

Selivanova, Kapustinskii, and Zubova [59SEL/KAP] determined the enthalpy of Reaction (V.61) from the temperature dependence of the solubility product and by a calorimetric measurement of the enthalpy of the reaction:

$$Pb(NO_3)_2(aq,\ 1{:}850) + H_2SeO_4(7.07\ M) \rightarrow PbSeO_4(cr) + 2HNO_3(aq,\ 1{:}\ 425)$$

The review has reconsidered the data and selects from a calculation presented in Appendix A [59SEL/KAP]:

$$\Delta_r H_m^\circ((V.61),\ 298.15\ K) = (4.72 \pm 2.53)\ kJ\cdot mol^{-1}.$$

With the selected enthalpies of formation of the ions, the standard enthalpy of formation of $PbSeO_4$(cr) becomes:

$$\Delta_f H_m^\circ(PbSeO_4,\ cr,\ 298.15\ K) = -(607.3 \pm 4.3)\ kJ\cdot mol^{-1}.$$

The value reported in [82WAG/EVA] is – 609.2 $kJ\cdot mol^{-1}$. The difference between the two data can be traced to the new values of the ionic enthalpy of Pb^{2+} and $\Delta_f H_m^\circ(Pb(NO_3)_2$, aq 1:850, 298.15 K) as discussed in [59SEL/KAP] in Appendix A. The value reported in the paper is – 622.2 $kJ\cdot mol^{-1}$. No details of the calculations are available, but neglect of enthalpies of dilution is probably the major cause of the discrepancy.

The standard Gibbs energy of formation of lead selenate is calculated from the Gibbs energy of Reaction (V.61) and the selected values of $\Delta_f G_m^\circ$ for the ions to be:

$$\Delta_f G_m^\circ(PbSeO_4,\ cr,\ 298.15\ K) = -(503.1 \pm 2.1)\ kJ\cdot mol^{-1}.$$

The selected value agrees with [82WAG/EVA].

The selected data further yield:

$$S_m^\circ(PbSeO_4,\ cr,\ 298.15\ K) = (167.7 \pm 16.1)\ J\cdot K^{-1}\cdot mol^{-1}.$$

V.7.5.4 Lead selenocyanates

Golub and Kosmatyi [59GOL/KOS3], and Golub [59GOL] studied the complex formation between Pb^{2+} and $SeCN^-$ by potentiometric and solubility measurements. Both sets of experiments, carried out with the ligand in the concentration range of 0.7 to 3.6 M, were interpreted to show that the species $Pb(SeCN)_6^{4-}$ was formed. This conclusion and the equilibrium constants calculated from the measurements are not accepted by the review for reasons developed in [59GOL/KOS3] in Appendix A.

V.8 Group 13 compounds and complexes

V.8.1 Boron compounds

V.8.1.1 Boron selenides

B_2Se_3(cr) is the only solid compound in the system boron-selenium. A phase diagram for the system B_2Se_3-Se is available in [72HIL/CUE]. Melucci and Wahlbeck [70MEL/WAH] identified the boron selenides BSe(g), BSe_2(g), BSe_3(g), B_2Se_2(g), and B_2Se_3(g) in studies of the vaporisation behaviour of solid B_2Se_3. It was concluded that the vaporisation is similar to that of B_2S_3(cr). Some intensity values were reported but no thermodynamic quantities can be derived from the measurements. Thermodynamic information is only available for the gaseous species BSe(g) and BSe_2(g).

V.8.1.1.1 BSe(g)

Uy and Drowart [70UY/DRO] determined the enthalpy change of the reaction

$$B(g) + YSe(g) \rightleftharpoons Y(g) + BSe(g) \qquad (V.63)$$

to $\Delta_r H_m^\circ$((V.63), 0 K) = – (25.0 ± 12.1) kJ·mol^{-1} in a mass spectrometric study at temperatures in the range 2231 to 2368 K. Their evaluation was based on the third law using Gibbs energy functions calculated from estimated parameters for the molecules YSe(g) and BSe(g). The dissociation enthalpy of BSe(g) was calculated to be $\Delta_{at} H_m^\circ$(BSe, g, 0 K) = (457.6 ± 14.8) kJ·mol^{-1} using an undocumented value of the dissociation enthalpy of YSe(g), $\Delta_{at} H_m^\circ$(YSe, g, 0 K) = (432.6 ± 10.5) kJ·mol^{-1}. A recalculation to 298.15 K using the molecular parameters of YSe(g) and BSe(g) estimated in [70UY/DRO] for the reaction above yields $\Delta_r H_m^\circ$((V.63), 298.15K) = – (25.2 ± 12.1) kJ·mol^{-1}. The value was combined with the selected enthalpy of formation of Se(g), the enthalpy of formation of B(g) in [89COX/WAG], and the enthalpy of formation of YSe(g) in Section V.11.23.1.1 to give $\Delta_f H_m^\circ$(BSe, g, 298.15 K) = (342.5 ± 20.0) kJ·mol^{-1}.

The auxiliary data calculated from estimated molecular parameters for a diatomic molecule are expected to be quite accurate. The uncertainty in the recalculation to the standard temperature of the result for this iso-molecular reaction is anticipated to be moderate compared to the experimental uncertainty. The value for the enthalpy of formation of BSe(g) is therefore adopted

$$\Delta_f H_m^\circ(\text{BSe, g, 298.15 K}) = (342.5 \pm 20.0)\ \text{kJ·mol}^{-1}.$$

and included in Appendix E since it was calculated with non-TDB auxiliary data.

V.8.1.1.2 $BSe_2(g)$

Binnewies [90BIN] performed a mass spectrometric investigation of the reaction between solid boron and $Se_2(g)$ in the temperature range 1020 to 1220 K. A second law evaluation of the reaction:

$$B(cr) + Se_2(g) \rightleftharpoons BSe_2(g) \qquad (V.64)$$

resulted in an enthalpy of reaction of $\Delta_r H_m^\circ$((V.64), 1100 K) = 47.6 kJ·mol^{-1} and an entropy of reaction of $\Delta_r S_m^\circ$((V.64), 1100 K) = 26.5 J·K^{-1}·mol^{-1}. The values were recalculated to 298.15 K using the heat capacity change for the analogous reaction C(cr) + $S_2(g) \rightleftharpoons CS_2(g)$ as obtained from data tabulated in [95BAR]. The values derived were $\Delta_r H_m^\circ$((V.64), 298.15 K) = 46.0 kJ·mol^{-1} and $\Delta_r S_m^\circ$((V.64), 298.15 K) = 23.6 J·K^{-1}·mol^{-1}. No error limits were established.

The combination of the reaction entropy at 298.15 K with the entropy of B(cr) in [89COX/WAG] and the selected entropy value of $Se_2(g)$ yields the selected value of the standard entropy:

$$S_m^\circ(BSe_2, g, 298.15 K) = (276.9 \pm 20.0) J·K^{-1}·mol^{-1}.$$

The selected value for the enthalpy of formation is obtained by combining the enthalpy of reaction at 298.15 K with the selected enthalpy of formation of $Se_2(g)$ yielding:

$$\Delta_f H_m^\circ(BSe_2, g, 298.15 K) = (187.1 \pm 20.0) kJ·mol^{-1}.$$

The error limits of the entropy and the enthalpy of formation were estimated by the review. The selected Gibbs energy of formation is calculated from the selected enthalpy of formation and entropy, the selected entropy of selenium, and the entropy of boron in [89COX/WAG] to be:

$$\Delta_f G_m^\circ(BSe_2, g, 298.15K) = (131.4 \pm 20.9) kJ·mol^{-1}.$$

V.8.2 Aluminium compounds

V.8.2.1 Aluminium selenides

$Al_2Se_3(cr)$ is the only intermediate solid compound in the aluminium-selenium system. The gaseous molecules AlSe(g), $Al_2Se(g)$, and $Al_2Se_2(g)$ have been observed in mass spectrometric investigations of gaseous aluminium-selenium mixtures.

V.8.2.1.1 AlSe(g)

The heat capacity and entropy of AlSe(g) were calculated by the review using the usual non-rigid rotator, anharmonic oscillator approximation and the molecular parameters of $^{27}Al^{80}Se(g)$ determined by Lavendy and Pinchemel [79LAV/PIN] in a rotational analysis of high-resolution spectra of the $A^2\Sigma^+$ - $X^2\Sigma^+$ system. The parameters were scaled to the mean molecular mass by the procedures outlined in [50HER]. No excited states were considered since they all lie above 23,000 cm^{-1} and do not affect the properties below 2000 K. The calculated values are selected:

$$C^{\circ}_{p,m}(\text{AlSe, g, 298.15 K}) = (34.75 \pm 0.20)\ \text{J}\cdot\text{K}^{-1}\cdot\text{mol}^{-1}$$

and

$$S^{\circ}_{m}(\text{AlSe, g, 298.15 K}) = (242.89 \pm 0.50)\ \text{J}\cdot\text{K}^{-1}\cdot\text{mol}^{-1}.$$

The heat capacity expression

$$C^{\circ}_{p,m}(\text{AlSe, g, (298.15 – 2000) K}) = (38.060 + 0.3183 \times 10^{-3}\ T - 6.333 \times 10^{2}\ T^{-1} - 1.143 \times 10^{5}\ T^{-2})\ \text{J}\cdot\text{K}^{-1}\cdot\text{mol}^{-1}$$

describes the statistical mechanics results to within 0.03 $J\cdot K^{-1}\cdot mol^{-1}$.

The abundances of the species AlSe(g), Al_2Se(g), and Al_2Se_2(g) in gaseous aluminium-selenium mixtures have been studied by Ficalora, Hastie, and Margrave [68FIC/HAS] and Uy and Drowart [71UY/DRO] using mass spectroscopy and Knudsen effusion cells. The investigations are discussed in Appendix A and the results concerning AlSe(g) as recalculated by the review are summarised in Table V-42. The enthalpies of the reactions were combined with the selected properties of the selenium species and the CODATA [89COX/WAG] values of aluminium to obtain the enthalpy of formation of AlSe(g). The second and third law values in [68FIC/HAS] differ substantially, because the second law entropy of the reaction is quite different from that calculated from the entropies of the species. This indicates experimental errors and the results in [68FIC/HAS] were assigned zero weights. The second and third law values in [71UY/DRO] were combined to calculate a mean value, which is selected:

$$\Delta_f H^{\circ}_{m}(\text{AlSe, g, 298.15 K}) = (258 \pm 35)\ \text{kJ}\cdot\text{mol}^{-1}.$$

Table V-42: Second and third law enthalpies of formation of AlSe(g) and the derived values of the enthalpy of formation of AlSe(g). The uncertainties are those of the references.

Reference	Reaction	$\Delta_r H^{\circ}_{m}$ (298.15 K) 2nd law, ($kJ\cdot mol^{-1}$)	$\Delta_r H^{\circ}_{m}$ (298.15 K) 3rd law, ($kJ\cdot mol^{-1}$)	$\Delta_f H^{\circ}_{m}$ (AlSe, g, 298.15 K) 2nd law, ($kJ\cdot mol^{-1}$)	$\Delta_f H^{\circ}_{m}$ (AlSe, g, 298.15 K) 3rd law, ($kJ\cdot mol^{-1}$)
[68FIC/HAS]	Al(g) + ½Se_2(g) ⇌ AlSe(g)	– (220 ± 13)	– (138 ± 9)	(180 ± 13)	(262 ± 9)
[71UY/DRO]	AlSe(g) + Se(g) ⇌ Al(g) + Se_2(g)	– (6 ± 25)	– (39 ± 20)	(241 ± 25)	(274 ± 20)

The Gibbs energy of formation was calculated from the selected enthalpy of formation and entropy of AlSe(g), the selected standard entropy of trigonal selenium, and the standard entropy of Al(cr) in [89COX/WAG] to be:

$$\Delta_f G_m^\circ(\text{AlSe, g, 298.15K}) = (207 \pm 35)\ \text{kJ·mol}^{-1}.$$

V.8.2.1.2 Al_2Se(g)

The heat capacity and entropy of Al_2Se(g) were calculated by the review using statistical mechanics and the molecular parameters estimated in [71UY/DRO]. The calculated values are $C_{p,m}^\circ$(Al_2Se, g, 298.15 K) = 52.2 J·K^{-1}·mol^{-1} and S_m°(Al_2Se, g, 298.15 K) = 289.3 J·K^{-1}·mol^{-1}. The heat capacity expression $C_{p,m}^\circ$(Al_2Se, g, (298.15 – 2000) K) = $(59.479 + 0.4008 \times 10^{-3}\, T - 1.2535 \times 10^{3}\, T^{-1} - 2.633 \times 10^{5}\, T^{-2})$ J·K^{-1}·mol^{-1} describes the statistical mechanics results to within 0.05 J·K^{-1}·mol^{-1}. These properties were used in the evaluations described below but are not selected since they are calculated with estimated molecular parameters.

The abundances of the species AlSe(g), Al_2Se(g), and Al_2Se_2(g) in gaseous aluminium-selenium mixtures have been studied by Ficalora, Hastie, and Margrave [68FIC/HAS] and Uy and Drowart [71UY/DRO] using mass spectroscopy and Knudsen effusion cells. The investigations are discussed in Appendix A and the results concerning Al_2Se(g) as recalculated by the review are summarised in Table V-43. The enthalpies of the reactions were combined with the selected properties of the selenium species, the CODATA [89COX/WAG] values of aluminium, and the selected values of AlSe(g) to obtain the enthalpy of formation of Al_2Se(g). These values have large assigned uncertainties, differ greatly, and because of the estimations involved, no value for the enthalpy of formation Al_2Se(g) is selected.

Table V-43: Second and third law enthalpies of formation of Al_2Se(g) and the derived values of the enthalpy of formation of Al_2Se(g). The uncertainties are those of the references.

Reference	Reaction	$\Delta_r H_m^\circ$ (298.15 K) 2nd law, (kJ·mol^{-1})	$\Delta_r H_m^\circ$ (298.15 K) 3rd law, (kJ·mol^{-1})	$\Delta_f H_m^\circ$ (Al_2Se, g, 298.15 K) 2nd law, (kJ·mol^{-1})	$\Delta_f H_m^\circ$ (Al_2Se, g, 298.15 K) 3rd law, (kJ·mol^{-1})
[68FIC/HAS]	2Al(g) + ½Se_2(g) $\rightleftharpoons$ Al_2Se(g)	– (507 ± 30)	– (433 ± 25)	(224 ± 30)	(297 ± 25)
[71UY/DRO]	Al_2Se(g) + Se(g) $\rightleftharpoons$ 2AlSe(g)	(3 ± 13)	– (4 ± 18)	(277 ± 71)	(284 ± 72)

V.8.2.1.3 $Al_2Se_2(g)$

The heat capacity and entropy of $Al_2Se_2(g)$ were calculated by the review using statistical mechanics and the molecular parameters estimated in [71UY/DRO]. The calculated values are $C^{\circ}_{p,m}(Al_2Se_2, g, 298.15 K) = 78.2 J·K^{-1}·mol^{-1}$ and $S^{\circ}_{m}(Al_2Se_2, g, 298.15 K) = 311.0 J·K^{-1}·mol^{-1}$. The heat capacity expression $C^{\circ}_{p,m}(Al_2Se_2, g, (298.15 – 2000) K) = (89.088 – 0.5604 \times 10^{-3} T – 1.7536 \times 10^{3} T^{-1} – 4.27693 \times 10^{5} T^{-2}) J·K^{-1}·mol^{-1}$ describes the statistical mechanics results to within $0.07 J·K^{-1}·mol^{-1}$. These properties were used in the evaluations described below but are not selected since they are calculated with estimated molecular parameters.

The abundances of the species AlSe(g), $Al_2Se(g)$, and $Al_2Se_2(g)$ in gaseous aluminium-selenium mixtures have been studied by Ficalora, Hastie, and Margrave [68FIC/HAS] and Uy and Drowart [71UY/DRO] using mass spectroscopy and Knudsen effusion cells. The investigations are discussed in Appendix A and the results concerning $Al_2Se_2(g)$ as recalculated by the review are summarised in Table V-44. The enthalpies of the reactions were combined with the selected properties of the selenium species, the CODATA [89COX/WAG] values of aluminium, and the selected values of AlSe(g) to obtain the enthalpy of formation of $Al_2Se_2(g)$. The uncertainties are large and because the accuracy of the estimations involved cannot be ascertained, no value for the enthalpy of formation $Al_2Se_2(g)$ is selected.

Table V-44: Second and third law enthalpies of formation of $Al_2Se_2(g)$ and the derived values of the enthalpy of formation of $Al_2Se_2(g)$. The uncertainties are those of the references.

Reference	Reaction	$\Delta_r H^{\circ}_m$ (298.15 K) 2nd law, ($kJ·mol^{-1}$)	$\Delta_r H^{\circ}_m$ (298.15 K) 3rd law, ($kJ·mol^{-1}$)	$\Delta_f H^{\circ}_m$ (Al_2Se_2, g, 298.15 K) 2nd law, ($kJ·mol^{-1}$)	$\Delta_f H^{\circ}_m$ (Al_2Se_2, g, 298.15 K) 3rd law, ($kJ·mol^{-1}$)
[68FIC/HAS]	$2Al(g) + Se_2(g) \rightleftharpoons Al_2Se_2(g)$	– (768 ± 45)	– (978 ± 60)	(33 ± 45)	– (177 ± 60)
[71UY/DRO]	$Al_2Se_2(g) + 2Se(g) \rightleftharpoons 2AlSe(g) + Se_2(g)$	(58 ± 38)	(43 ± 50)	(127 ± 80)	(142 ± 86)

V.8.2.1.4 $Al_2Se_3(cr)$

No experimental information is available on the entropy and heat capacity of $Al_2Se_3(cr)$. The entropy was estimated to be $S^{\circ}_{m}(Al_2Se_3, cr, 298.15 K) = (154.8 \pm 30.0) J·K^{-1}·mol^{-1}$ and the heat capacity to be $C^{\circ}_{p,m}(Al_2Se_3, cr, (298 – 2000) K) = (107.74 + 34.31 \times 10^{-3} T) J·K^{-1}·mol^{-1}$ by Mills [74MIL]. No values of the heat capacity and entropy have been selected.

The enthalpy of formation has been measured by synthesis calorimetry to be $\Delta_f H_m^\circ$ (Al_2Se_3, cr, 298.15 K) = – (566.9 ± 13.0) kJ·mol^{-1} by Kapustinskii and Golutvin [51KAP/GOL]. Schneider and Gattow [54SCH/GAT2] used combustion calorimetry to determine the enthalpy of the reaction $Al_2Se_3(cr) + 4\frac{1}{2}O_2(g) \rightleftharpoons Al_2O_3(cr) + 3SeO_2(cr)$ to be – (1855.2 ± 7.4) kJ·mol^{-1}. The reaction products were identified using X-ray diffraction and only a small residue of Al_2Se_3(cr) was found in addition to the two oxides. From this reaction, the review calculates $\Delta_f H_m^\circ$ (Al_2Se_3, cr, 298.15 K) = – (496.7 ± 7.7) kJ·mol^{-1} using the selected enthalpy of SeO_2(cr) and the CODATA [89COX/WAG] enthalpy of Al_2O_3(cr). The confidence intervals of the two values of the enthalpy of formation of Al_2Se_3(cr) do not overlap and assigning equal weights to the values, the review selects:

$$\Delta_f H_m^\circ (Al_2Se_3, \text{cr}, 298.15\ \text{K}) = -(532 \pm 40)\ \text{kJ·mol}^{-1}$$

where the error limits have been adjusted to incorporate the two experimental values.

V.8.2.2 Aluminium selenites

Gospodinov [91GOS] established the phase diagram at 373.2 K of the system Al_2O_3-SeO_2-H_2O from solubility measurements. The tie lines and chemical analysis of the solids showed the formation of crystalline phases of composition $Al_2(SeO_3)_3 \cdot 6H_2O$, $Al_2(SeO_3)_3 \cdot H_2SeO_3$, and $Al_2(SeO_3)_3 \cdot 3H_2SeO_3$. The solubility data, which cover the range 1.6 to 89.3 mass-% SeO_2 in the liquid phase, are presented together with the phase diagram.

Popova, Slavtscheva, and Gospodinov [86POP/SLA] determined the solubility of aluminium selenite in dilute nitric acid. The preparation and composition of the selenite is not discussed. It is assumed to have the formula $Al_2(SeO_3)_3 \cdot 6H_2O$(cr) by the review. The solubility product calculated in the paper, see Appendix A, for Reaction (V.65),

$$Al_2(SeO_3)_3 \cdot 6H_2O(cr) \rightleftharpoons 2Al^{3+} + 3SeO_3^{2-} + 6H_2O(l) \qquad (V.65)$$

is $\log_{10} K_{s,0}^\circ$ ((V.65), 298.15 K) = – 31.1. No value of the solubility of the salt is presented in the paper, but the reported solubility product indicates that it is fairly large. The corresponding ionic strength of the solution is estimated to be greater than 0.2 M. The extrapolation to standard conditions of the conditional equilibrium constant, which involves activity coefficients of doubly and triply charged ions, is sensitive to the unknown composition of the equilibrium solution. The reported value can hence not be accepted by the review.

No other thermodynamic data have been found for aluminium selenites.

V.8.2.3 Aluminium selenates

Most investigators [50BAS], [63SEL/SAM], [69ROS/PER], [87OJK/STA] conclude that the aluminium selenate crystallising at room temperature contains 16 molecules of water. Two references [56ERD/PAU], [73SEL/ZAL] report 18 molecules of water of crystallisation. In no instance has the water content been analysed directly. The number of water molecules of crystallisation cannot be regarded as definitely established and the varying results may indicate the formation of phases with 16 and 18 molecules of water as has been reported for aluminium sulphate [27KRA/FRI]. The solubility of a specimen with the assigned formula $Al_2(SeO_4)_3{\cdot}16H_2O$ was measured to be 0.603 mol($Al_2(SeO_4)_3$)·kg^{-1} in [87OJK/STA]. The review accepts:

$$m(Al_2(SeO_4)_3{\cdot}16H_2O(cr), \text{aq sat}, 298.15\text{ K}) = 0.603\text{ mol}(Al_2(SeO_4)_3){\cdot}kg^{-1}.$$

No uncertainty can be assigned to this value.

A phase diagram and related information are available for the system: $Al_2(SeO_4)_3$-$MnSeO_4$-H_2O at 298.15 K in [87OJK/STA]; $Al_2(SeO_4)_3$-$MgSeO_4$-H_2O at 298.15 K in [87OJK/STA].

Selivanova, Zalogina, and Gorokhov [73SEL/ZAL] determined the enthalpy of formation of aluminium selenate. The specimen used in the experiment was assigned the composition $Al_2(SeO_4)_3{\cdot}18H_2O$, which "satisfactorily" agreed with the chemical analysis. This composition is thus at variance with most other reports. The reaction studied in [73SEL/ZAL] was $Al_2(SeO_4)_3{\cdot}18H_2O(cr)$ + $3BaCl_2$(aq, 1:1600) → $3BaSeO_4$(cr) + $2AlCl_3$(aq, 1:2400) + $18H_2O$(l). The review calculated the enthalpy change of the reaction:

$$Al_2(SeO_4)_3{\cdot}18H_2O(cr) + 3Ba^{2+} \rightarrow 3BaSeO_4(cr) + 2Al^{3+} + 18H_2O(l) \qquad (V.66)$$

from the data in Appendix A and obtained:

$$\Delta_r H^\circ_m((V.66), 298.15\text{ K}) = -(79.32 \pm 0.87)\text{ kJ}{\cdot}mol^{-1}.$$

This datum is combined with selected enthalpies of formation to:

$$\Delta_f H^\circ_m(Al_2(SeO_4)_3{\cdot}18H_2O, \text{cr}, 298.15\text{ K}) = -(7970.0 \pm 17.1)\text{ kJ}{\cdot}mol^{-1}.$$

The enthalpies of dissolution of $Al_2(SeO_4)_3{\cdot}18H_2O$(cr) and $Al_2(SeO_4)_3$(cr) in 0.5 M KOH were also measured. These measurements yield:

$$\Delta_f H^\circ_m(Al_2(SeO_4)_3, \text{cr}, 298.15\text{ K}) = -(2561.7 \pm 17.1)\text{ kJ}{\cdot}mol^{-1}.$$

The evaluation in Appendix A resulted in enthalpies of formation considerably different from those obtained in the paper, which reports – 8000.6 and – 2592.4 kJ·mol^{-1}, respectively. It is not possible to analyse the causes of the difference as no auxiliary data are disclosed in the publication. Since no other data for aluminium selenates are available, the values calculated by the review will be accepted but not selected as they partly rest on nonselected auxiliary values. They will thus be included in Appendix E. It may be noted that the standard enthalpy of formation of the anhydrous salt is almost independent of the number of water molecules of crystallisation ascribed to the hydrate.

V.8.3 Gallium compounds

V.8.3.1 Gallium selenides

The phase diagram of the Ga-Se system was studied in detail by Dieleman, Sanders, and Dommelen [82DIE/SAN]. Stable solid compounds form at the nominal compositions GaSe and Ga_2Se_3. GaSe(cr) has a narrow homogeneity range and melts congruently at 1210 K. Ga_2Se_3(cr) exists as the polymorphs α-Ga_2Se_3 with compositions in the range $Ga_{0.413}Se_{0.587}$ to $Ga_{0.39}Se_{0.61}$ and β-Ga_2Se_3 with compositions in the range $Ga_{0.398}Se_{0.602}$ to $Ga_{0.393}Se_{0.607}$. In addition, a metastable α'-Ga_2Se_3 phase was found. Reactions and transitions between the polymorphs are slow and equilibrium is difficult to obtain. β-Ga_2Se_3 is stable at room temperature and transforms into the α-Ga_2Se_3 modification at temperatures in the range 898 to 968 K. α-Ga_2Se_3 melts congruently at 1280 K.

The mass spectrometric investigations of the vaporisation of GaSe(cr) by Piacente, Bardi, Paolo, and Ferro [76PIA/BAR] and of Ga_2Se_3(cr) by Uy, Muenow, Ficalora, and Margrave [68UY/MUE] and Berger, Strelchenko, Bondar, Molodyk, Balanevskaya, and Lebedev [69BER/STR] have shown that Ga_2Se(g) is the only important gaseous gallium selenide.

V.8.3.1.1 GaSe(cr)

The heat capacity of GaSe(cr) has been measured by Mamedov and Kerimov [67MAM/KER2], Mamedov, Kerimov, Kostrykov, and Mekhtiev [70MAM/KER], and Mamedov, Aldchanov, Kerimov, and Mekhtiev [78MAM/ALD] in the temperature ranges 60 to 300 K, 12 to 300 K, and 2 to 60 K, respectively. The heat capacity at 298.15 K was reported to be $C^{\circ}_{p,m}$(GaSe, cr, 298.15 K) = (48.92 ± 1.50) $J \cdot K^{-1} \cdot mol^{-1}$ in [67MAM/KER2] and (48.24 ± 0.50) $J \cdot K^{-1} \cdot mol^{-1}$ in [70MAM/KER]. The selected value of the heat capacity is:

$$C^{\circ}_{p,m}(\text{GaSe, cr, 298.15 K}) = (48.6 \pm 0.9)\ \text{J} \cdot \text{K}^{-1} \cdot \text{mol}^{-1}.$$

The heat capacity expression $C^{\circ}_{p,m}$(GaSe, cr, (298.15 – 1200) K) = (44.64 + 12.97 × 10^{-3} T) $J \cdot K^{-1} \cdot mol^{-1}$ was estimated by Mills [74MIL] and has been used by the review in evaluations wherever the heat capacity of GaSe(cr) above room temperature has been required.

The entropy of GaSe(cr) at 298.15 K reported from the heat capacity measurements in [70MAM/KER] is S°_{m}(GaSe, cr, 298.15 K) = (70.25 ± 0.80) $J \cdot K^{-1} \cdot mol^{-1}$. A combination of the entropies derived from the heat capacity measurements in [67MAM/KER2] and [78MAM/ALD] yields S°_{m}(GaSe, cr, 298.15 K) = (71.05 ± 1.50) $J \cdot K^{-1} \cdot mol^{-1}$. The selected value of the standard entropy is calculated as the weighted average yielding:

$$S^{\circ}_{m}(\text{GaSe, cr, 298.15 K}) = (70.4 \pm 0.7)\ \text{J} \cdot \text{K}^{-1} \cdot \text{mol}^{-1}.$$

The enthalpy of formation of GaSe(cr) has been measured by combustion calorimetry by Hahn and Burow [56HAH/BUR] and by direct synthesis calorimetry by Sharifov, Azizov, Aliev, and Abbasov [73SHA/AZI]. The values reported are $\Delta_f H_m^\circ$(GaSe, cr, 298.15 K) = – (146.4 ± 13.0) kJ·mol^{-1} and – (143.5 ± 5.0) kJ·mol^{-1}, respectively. The selected value of the enthalpy of formation is:

$$\Delta_f H_m^\circ(\text{GaSe, cr, 298.15 K}) = -(145.0 \pm 10.0)\ \text{kJ·mol}^{-1}.$$

The enthalpy of fusion of GaSe(cr) was reported to be $\Delta_{fus} H_m^\circ$(GaSe, cr, 1210 K) = (56.0 ± 3.5) kJ·mol^{-1} in [82DIE/SAN].

The Gibbs energy of formation is calculated from the selected enthalpy of formation and entropy, the selected entropy of selenium, and the entropy of gallium in [82WAG/EVA] to be:

$$\Delta_f G_m^\circ(\text{GaSe, cr, 298.15 K}) = -(141.2 \pm 15.0)\ \text{kJ·mol}^{-1}.$$

The result is entered in the Appendix E as it was calculated with non-TDB auxiliary data.

V.8.3.1.2 Ga_2Se(g)

The heat capacity and entropy of Ga_2Se(g) was calculated in [74MIL] from the molecular constants estimated by Uy, Muenow and Ficalora [68UY/MUE]. The calculations were repeated by the review yielding $C_{p,m}^\circ$(Ga_2Se, g, 298.15 K) = 54.9 J·K^{-1}·mol^{-1}, S_m°(Ga_2Se, g, 298.15 K) = 315.8 J·K^{-1}·mol^{-1}, and the expression $C_{p,m}^\circ$(Ga_2Se, g, (298.15 – 2000) K) = (58.00 + 0.1236 × 10^{-3} T – 2.7723 × 10^5 T^{-2}) J·K^{-1}·mol^{-1}. The results were used in the evaluations made by the review wherever these thermodynamic quantities were needed. No values are selected since they are all based on estimated quantities.

The formation of Ga_2Se(g) has been studied in mass spectrometric vaporisation investigations of GaSe(cr) in [76PIA/BAR] and of Ga_2Se_3(cr) in [68UY/MUE] and [69BER/STR]. In addition, Bendrikh and Pashinkin [79BEN/PAS] have measured the saturated total pressure of the vapour in equilibrium with Ga_2Se_3(cr) using the weight loss Knudsen method. The vaporisation was concluded to be incongruent for GaSe(cr) and congruent for Ga_2Se_3(cr) and the vaporisation reactions are

$$4\text{GaSe(cr)} \rightleftharpoons \text{Ga}_2\text{Se(g)} + \text{Ga}_2\text{Se}_3\text{(cr)} \qquad \text{(V.67)}$$

and

$$\text{Ga}_2\text{Se}_3\text{(cr)} \rightleftharpoons \text{Ga}_2\text{Se(g)} + \text{Se}_2\text{(g)}. \qquad \text{(V.68)}$$

The experimental conditions are summarised in Table V-45.

Table V-45: Reactions and methods used for the vaporisation studies involving $Ga_2Se(g)$.

Reference	Reaction	Temperature (K)	Measurement method
[68UY/MUE]	(V.68)	931 – 1094	Mass spectrometry and Knudsen effusion cells
[69BER/STR]	(V.68)	970 – 1180	Mass spectrometry and Knudsen effusion cells
[76PIA/BAR]	(V.67)	1053 – 1143	Mass spectrometry, thermogravimetry, and torsion-effusion Knudsen cells
[79BEN/PAS]	(V.68)	1121 – 1269	Weight loss Knudsen cells

The measurements in [76PIA/BAR] were re-evaluated by the review as discussed in Appendix A yielding $\Delta_r H^\circ_m$((V.67), 298.15 K) = 400.8 kJ·mol^{-1} and $\Delta_r S^\circ_m$((V.67), 298.15 K) = 241.4 J·K^{-1}·mol^{-1} from the second law and $\Delta_r H^\circ_m$((V.67), 298.15 K) = 334.3 kJ·mol^{-1} from the third law for Reaction (V.67). Similarly, the investigations involving Reaction (V.68) were re-evaluated by the review as discussed in Appendix A yielding the results presented in Table V-46. The reaction quantities were combined with the selected standard enthalpy of $Se_2(g)$ and the values of the enthalpies of formation of GaSe(cr) and $Ga_2Se_3(cr)$ in Sections V.8.3.1.1 and V.8.3.1.3 to obtain values of the enthalpy of formation of $Ga_2Se(g)$. These values can be found in Table V-46. The average of the second and third law values of the enthalpy of formation of $Ga_2Se(g)$ is (130.0 ± 28.9) kJ·mol^{-1} excluding the second law value attributed to [76PIA/BAR]. No value is selected because the reaction schemes are uncertain and a number of estimates are involved in the evaluations.

Table V-46: The enthalpy of the evaporation reactions at 298.15 K and the standard enthalpy of formation of $Ga_2Se(g)$.

Reference	Reaction	$\Delta_r H^\circ_m$ (298.15 K) (kJ·mol^{-1})		$\Delta_f H^\circ_m$ (Ga_2Se, g, 298.15 K) (kJ·mol^{-1})	
		2nd law	3rd law	2nd law	3rd law
[68UY/MUE]	(V.68)	(649.6 ± 20.0)		(124.2 ± 50.0)	
[69BER/STR]	(V.68)	(678.6 ± 42.0)	(636.9 ± 30.0)	(153.2 ± 63.0)	(111.5 ± 54.0)
[76PIA/BAR]	(V.67)	(400.8 ± 43.0)	(334.3 ± 20.0)	(205.1 ± 50.0)	(138.6 ± 46.0)
[79BEN/PAS]	(V.68)	(657.6 ± 46.0)	(645.6 ± 30.0)	(132.2 ± 65.0)	(120.2 ± 54.0)

V.8.3.1.3 $Ga_2Se_3(cr)$

No experimental information on the heat capacity of $Ga_2Se_3(cr)$ is available. Mills [74MIL] estimated the expression $C^\circ_{p,m}$(Ga_2Se_3, cr, (298.15 – 1300) K) = (105.73 + 35.31 × 10^{-3} T) J·K^{-1}·mol^{-1}.

The heat capacity at 298.15 K calculated from this expression, $C^{\circ}_{p,\mathrm{m}}$ (Ga_2Se_3, cr, 298.15 K) = (116.3 ± 1.0) $J\cdot K^{-1}\cdot mol^{-1}$, results in $\Delta_f C^{\circ}_{p,\mathrm{m}}$ (Ga_2Se_3, cr, 298.15 K) = – 10.7 $J\cdot K^{-1}\cdot mol^{-1}$ for the formation of Ga_2Se_3(cr) from the elements. The value is reasonable in comparison with the corresponding value for GaSe(cr), $\Delta_f C^{\circ}_{p,\mathrm{m}}$ (GaSe, cr, 298.15 K) = – 2.71 $J\cdot K^{-1}\cdot mol^{-1}$, considering the change in composition. The expression of the heat capacity above was used by the review in evaluations wherever the heat capacity of Ga_2Se_3(cr) was required.

The entropy of Ga_2Se_3(cr) at 298.15 K was estimated in [74MIL] to be (179.9 ± 17.0) by comparison with related compounds and a second law evaluation of galvanic cell data recorded by Mamedov [69MAM]. The value is not selected since it is in effect based on estimates.

The enthalpy of formation of Ga_2Se_3(cr) has been measured using bomb calorimetry and galvanic cells. The results of the investigations are summarised in Table V-47.

Table V-47: The enthalpy of formation of Ga_2Se_3(cr) determined in various investigations. The values denoted by (a) and (b) were evaluated by Mills [74MIL] according to the second and third law, respectively.

Reference	$\Delta_f H^{\circ}_{\mathrm{m}}$ (Ga_2Se_3, cr, 298.15 K) ($kJ\cdot mol^{-1}$)	Technique
[56HAH/BUR]	– (439.3 ± 13.0)	Combustion calorimetry
[61GAD/SHA]	– (460.2 ± 21.0)	Combustion calorimetry
[65EFE/KAR]	– (376.6 ± 21.0)	Direct synthesis calorimetry
[67SHA/AZI]	– (368.6 ± 13.0)	Direct synthesis calorimetry
[69MAM]	– (405.0 ± 21.0)[(a)], – (408.8 ± 13.0)[(b)]	Galvanic cell measurements

The enthalpy of formation of Ga_2Se_3(cr) must not be more negative than – 435 $kJ\cdot mol^{-1}$ to be consistent with the phase diagram and the selected value of the enthalpy of formation of GaSe(cr). Mills [74MIL] selected the value $\Delta_f H^{\circ}_{\mathrm{m}}$ (Ga_2Se_3, cr, 298.15 K) = – (408.8 ± 13.0) $kJ\cdot mol^{-1}$. Burylev [77BUR] compared trends in the enthalpy of formation of the gallium sulphides, selenides, and tellurides and selected the value $\Delta_f H^{\circ}_{\mathrm{m}}$ (Ga_2Se_3, cr, 298.15 K) = – (355.6 ± 13.0) $kJ\cdot mol^{-1}$. The mean of the values in [65EFE/KAR], [67SHA/AZI], [69MAM] is selected:

$$\Delta_f H^{\circ}_{\mathrm{m}} (Ga_2Se_3, \text{cr}, 298.15\ \text{K}) = -(384 \pm 40)\ \text{kJ}\cdot\text{mol}^{-1}.$$

The enthalpy of fusion of Ga_2Se_3(cr) was reported to be $\Delta_{fus} H^{\circ}_{\mathrm{m}}$ (Ga_2Se_3, cr, 1280 K) = (84 ± 7) $kJ\cdot mol^{-1}$ in [82DIE/SAN] and to be $\Delta_{fus} H^{\circ}_{\mathrm{m}}$ (Ga_2Se_3, cr, 1293 K) = 48.1 $kJ\cdot mol^{-1}$ in [76BAB/KUL].

V.8.3.2 Gallium selenites

Savchenko, Tananaev, and Volodina [68SAV/TAN] studied the system Ga_2O_3-SeO_2-H_2O at 298.15 K by solubility measurements in the concentration range 3.3 to 40.7 mass-% SeO_2. Slow reaction rates resulted in few equilibrium data. The presented phase diagram contains only the crystallisation field of $Ga_2(SeO_3)_3{\cdot}6H_2O(cr)$. Similar measurements were performed by Gospodinov [84GOS4] at (373.2 ± 1.0) K over the concentration range 1.0 to 74.4 mass-% SeO_2 in the liquid phase. The compositions of the solids obtained from the tie-lines were $Ga_2(SeO_3)_3{\cdot}6H_2O$ and $Ga_2(SeO_3)_3{\cdot}3H_2SeO_3$. The latter crystallises at SeO_2 concentration greater than about 35 mass-%, which may account for its absence in [68SAV/TAN]. Both compounds are incongruently soluble and the solubility data show that the concentration of Ga(III) is small over the whole concentration range of SeO_2.

Popova *et al.* [86POP/SLA] determined the solubility of gallium selenite in dilute nitric or perchloric acid with concordant results. The preparation and composition of the selenite is not discussed. It is assumed to have the formula $Ga_2(SeO_3)_3{\cdot}6H_2O$ by the review. The solubility product calculated from the total concentration of Ga(III) in the saturated solution is accepted and selected, *cf.* Appendix A.

$$Ga_2(SeO_3)_3{\cdot}6H_2O(cr) \rightleftharpoons 2Ga^{3+} + 3SeO_3^{2-} + 6H_2O(l) \qquad (V.69)$$

$$\log_{10} K^{\circ}_{s,0}((V.69), 298.15\text{ K}) = -(37.0 \pm 2.0).$$

The uncertainty limits have been widened considerably by the review in view of the uncertainty in the values of the activity coefficients and the high powers of the activities in the expression for the solubility product.

The Gibbs energy of formation

$$\Delta_f G^{\circ}_m(Ga_2(SeO_3)_3{\cdot}6H_2O, \text{cr}, 298.15\text{ K}) = -(3039.2 \pm 14.7)\text{ kJ·mol}^{-1}$$

was calculated with $\Delta_f G^{\circ}_m(Ga^{3+}, 298.15\text{ K}) = -(159.0 \pm 4.0)\text{ kJ·mol}^{-1}$ [82WAG/EVA], with the uncertainty estimated by the review. Since the calculation of the formation datum from the experiments involved non-TDB auxiliary data it is included in Appendix E.

Tananaev, Volodina, Selivanova and Bol'shakova, [76TAN/VOL], [76TAN/VOL2] prepared a number of gallium selenites and determined their standard enthalpies of formation from calorimetric measurements of enthalpies of dissolution in HCl(aq, 1:8). The experimental data have been re-evaluated, see Appendix A, to yield:

$$\Delta_f H^{\circ}_m(Ga_2(SeO_3)_3{\cdot}6H_2O, \text{cr}, 298.15\text{ K}) = -(3682.3 \pm 6.6)\text{ kJ·mol}^{-1},$$

$$\Delta_f H^{\circ}_m(Ga_2(SeO_3)_3{\cdot}H_2O, \text{cr}, 298.15\text{ K}) = -(2138.4 \pm 6.6)\text{ kJ·mol}^{-1},$$

$$\Delta_f H^{\circ}_m(Ga_2(SeO_3)_3, \text{cr}, 298.15\text{ K}) = -(1823.7 \pm 6.5)\text{ kJ·mol}^{-1},$$

$$\Delta_f H^{\circ}_m(GaH(SeO_3)_2{\cdot}2H_2O, \text{cr}, 298.15\text{ K}) = -(1788.4 \pm 3.5)\text{ kJ·mol}^{-1},$$

$$\Delta_f H_m^\circ (\text{GaH(SeO}_3)_2, \text{cr}, 298.15 \text{ K}) = -(1176.7 \pm 3.5) \text{ kJ·mol}^{-1}.$$

The recalculated values differ substantially from the published values for reasons discussed in Appendix A. Since the calculation of the formation enthalpies from the experiments involved non-TDB auxiliary data, they are included in Appendix E.

V.8.3.3 Gallium selenates

Tananaev, Gorokhov, Selivanova and Bol'shakova, [68TAN/GOR] measured the enthalpy change of the reaction $Ga_2(SeO_4)_3 \cdot 18H_2O(cr) + 3BaCl_2(aq, 1{:}1600) \rightarrow 3BaSeO_4(cr) + 2GaCl_3(aq, 1{:}2400) + 18H_2O(l)$ by calorimetry. The review calculated the enthalpy change of the reaction:

$$Ga_2(SeO_4)_3 \cdot 18H_2O(cr) + 3Ba^{2+} \rightarrow 3BaSeO_4(cr) + 2Ga^{3+} + 18H_2O(l) \qquad (V.70)$$

from the data to be $\Delta_r H_m^\circ ((V.70), 298.15 \text{ K}) = -(64.90 \pm 3.19) \text{ kJ·mol}^{-1}$. This datum is combined with accepted enthalpies of formation to:

$$\Delta_f H_m^\circ (Ga_2(SeO_4)_3 \cdot 18H_2O, \text{cr}, 298.15 \text{ K}) = -(7331.1 \pm 17.8) \text{ kJ·mol}^{-1}.$$

The enthalpies of dissolution of $Ga_2(SeO_4)_3 \cdot 18H_2O(cr)$ and $Ga_2(SeO_4)_3(cr)$ in 0.2 M KOH were also measured. These measurements yield:

$$\Delta_f H_m^\circ (Ga_2(SeO_4)_3, \text{cr}, 298.15 \text{ K}) = -(1908.4 \pm 17.9) \text{ kJ·mol}^{-1}.$$

The enthalpies of formation reported in [68TAN/GOR] are – 7401.1 and – 1978.2 kJ·mol^{-1} respectively. An analysis of the evaluation there shows that the difference from the present results is mainly caused by the different selections of the standard enthalpy of formation of $BaSeO_4(cr)$, see Appendix A. The values calculated by the review are adopted, although the experimental difficulties with highly hydrated salts are well known. This matter is further discussed in Appendix A, [68TAN/GOR]. Since the calculation of the formation enthalpies from the experiments involved non-TDB auxiliary data, they are included in Appendix E.

V.8.4 Indium compounds

V.8.4.1 Indium selenides

The thermodynamic properties of the system indium-selenium have been reviewed by Chatillon [93CHA] and Vassilev, Daouchi, Record, and Tedenac [98VAS/DAO]. The system is difficult to study because of slow reactions and the identification of the solid phases has been controversial. Similarly, the gas phase has been difficult to study since, in addition to gaseous indium selenide molecules, it also comprises all the selenium species Se(g)-Se_n(g), n = 1 to 8. Proper heat capacities and entropies are missing for most condensed phases and gaseous species and the investigators have employed a number of assumptions and estimations in their evaluations making it difficult to sort out the quantities actually measured. The measurements of the thermodynamic properties of the compounds in the indium-selenium system are in general of poor quality.

The reviews in [98VAS/DAO] and [93CHA] report the stable compounds In_4Se_3(cr), InSe(cr), In_6Se_7(cr), In_5Se_7(cr), and In_2Se_3(cr). Experimental thermodynamic information is essentially only available for InSe(cr) and In_2Se_3(cr). Thermodynamic properties have also been reported for the phases In_2Se(s) and In_5Se_6(s), but these phases do not actually exist. Mass spectrometric investigations have shown that the molecules InSe(g), In_2Se(g), and In_2Se_2(g) form in the gas phase.

After the assessment of the thermodynamic properties of indium selenides was made by the review, the thermodynamic modelling of the In-Se system by Li, Record, and Tedenac [2003LI/REC] has appeared. The paper brings no new experimental information and does not affect the selection of the thermodynamic properties made by the review. However, the modelling resulted in thermodynamic properties of all condensed phases in the system and the values are given in the pertinent sections. Most of these data are not based on experimental thermodynamic information for individual phases, but are a result of the modelling of the thermodynamics and the phase diagram of the system as a whole. The data are included in the review for information only.

V.8.4.1.1 InSe(cr)

The low temperature heat capacity of InSe(cr) has been measured in the temperature range 55 to 300 K by Mamedov, Kerimov, Kostryukov, and Guseinov [67MAM/KER] and in the range 12 to 304 K by Mamedov, Kerimov, Kostryukov, and Mekthiev [70MAM/KER]. The heat capacity reported in [70MAM/KER] for the temperature range 55 to 304 K is identical to that reported in [67MAM/KER] and the values obviously stem from the same measurements. Mills [76MIL] measured the heat capacity of InSe(cr) in the temperature range 129 to 920 K. The heat capacity values reported at 298.15 K were $C^{\circ}_{p,\mathrm{m}}$(InSe, cr, 298.15 K) = (50.3 ± 1.5) $\mathrm{J{\cdot}K^{-1}{\cdot}mol^{-1}}$ and (49.2 ± 1.5) $\mathrm{J{\cdot}K^{-1}{\cdot}mol^{-1}}$ in [67MAM/KER] and [76MIL], respectively. The average of the two heat capacity values is selected:

$$C^{\circ}_{p,\mathrm{m}}(\mathrm{InSe, cr, 298.15\ K}) = (49.8 \pm 1.1)\ \mathrm{J{\cdot}K^{-1}{\cdot}mol^{-1}}.$$

The heat capacity expression in [76MIL] was modified by the review to fit the selected value of the heat capacity at 298.15 K, yielding:

$$C^{\circ}_{p,\mathrm{m}}(\mathrm{InSe, cr, (298.15 - 913)\ K}) = (46.496 + 7.281 \times 10^{-3}\ T + 0.96314 \times 10^{5}\ T^{-2})\ \mathrm{J{\cdot}K^{-1}{\cdot}mol^{-1}}.$$

An attempt to determine the enthalpy and entropy of formation of InSe(cr) was made by Mustafaev, Azizov, Aliev, and Abbasov [74MUS/AZI] from electrochemical cell measurements in the temperature range 510 to 710 K. However, the reported values were obtained using thermochemical data of the non-existent phase In_5Se_6(s) and are therefore rejected. The third law entropy of InSe(cr) at 298.15 K derived from the values of the low-temperature heat capacity reported in [70MAM/KER] is selected:

$$S^{\circ}_{m}(\text{InSe, cr, 298.15 K}) = (83.9 \pm 0.9)\ \text{J}\cdot\text{K}^{-1}\cdot\text{mol}^{-1}.$$

Hahn and Burow [56HAH/BUR] determined the enthalpy of formation of InSe(cr) to be $-(118.0 \pm 13.0)$ kJ·mol^{-1} using combustion calorimetry in oxygen. The experimental conditions are summarised shortly and auxiliary data used are not mentioned in the paper. The quality of the measurements can therefore not be assessed and neither be recalculated using the auxiliary data set employed by the review. The value can therefore not be selected.

The thermodynamic modelling in [2003LI/REC], *cf.* Section V.8.4.1, resulted in the values $\Delta_f H^{\circ}_{m}$(InSe, cr, 298.15 K) = $-$ 120.6 kJ·mol^{-1}, S°_{m}(InSe, cr, 298.15 K) = 76.1 J·K^{-1}·mol^{-1}, $C^{\circ}_{p,m}$(InSe, cr, 298.15 K) = 49.16 J·K^{-1}·mol^{-1}, and the heat capacity expression $C^{\circ}_{p,m}$(InSe, cr, (298.15 – 850) K) = $(49.633 + 3.447 \times 10^{-3}\ T - 1.33339 \times 10^{5}\ T^{-2})$ J·K^{-1}·mol^{-1}, which are given for information only.

V.8.4.1.2 InSe(g)

Chatillon [93CHA] calculated the heat capacity and entropy of InSe(g) to be $C^{\circ}_{p,m}$(InSe, g, 298.15 K) = (36.1 ± 1.5) J·K^{-1}·mol^{-1} and S°_{m}(InSe, g, 298.15 K) = (254.7 ± 3.0) J·K^{-1}·mol^{-1}, respectively, from estimated molecular parameters given by Colin and Drowart [68COL/DRO]. The values can be compared with $C^{\circ}_{p,m}$(InSe, g, 298.15 K) = 35.2 J·K^{-1}·mol^{-1} and S°_{m}(InSe, g, 298.15 K) = 257.8 J·K^{-1}·mol^{-1} calculated by the review from the molecular parameters obtained by Erkoc, Katircioglu, and Yilmaz [2001ERK/KAT] in quantum mechanical studies of In-Se clusters. None of the values are based on experimental information on the InSe(g) molecule and no values are selected.

The atomisation enthalpy of InSe(g) was determined by Colin and Drowart [68COL/DRO] from mass spectrometric measurements in the temperature range 1495 to 1815 K. The evaluation employed Gibbs energy functions calculated from estimated molecular parameters and the experimental results were therefore re-evaluated as discussed in Appendix A using the results in [2001ERK/KAT]. The calculated values are $\Delta_f H^{\circ}_{m}$(InSe, g, 298.15 K) = 216.7 kJ·mol^{-1} from the second law and $\Delta_f H^{\circ}_{m}$(InSe, g, 298.15 K) = 226.6 kJ·mol^{-1} from the third law. The molecular properties calculated from quantum mechanics and statistical mechanics for a diatomic molecule can be expected to be quite accurate and the average of the enthalpy values is accepted with an uncertainty estimated by the review

$$\Delta_f H^{\circ}_{m}(\text{InSe, g, 298.15 K}) = (221.6 \pm 20.0)\ \text{kJ}\cdot\text{mol}^{-1}.$$

The value is included in Appendix E since it was calculated with non-TDB auxiliary data.

V.8.4.1.3 $In_2Se(g)$

Greenberg, Borjakova, and Shevelkov [73GRE/BOR] calculated the heat capacity and entropy of $In_2Se(g)$ to be $C^\circ_{p,m}$ (In_2Se, g, 298.15 K) = 56.9 $J\cdot K^{-1}\cdot mol^{-1}$ and S°_m (In_2Se, g, 298.15 K) = 333.4 $J\cdot K^{-1}\cdot mol^{-1}$, respectively, from estimated molecular parameters. The values can be compared with $C^\circ_{p,m}$ (In_2Se, g, 298.15 K) = 55.1 $J\cdot K^{-1}\cdot mol^{-1}$ and S°_m (In_2Se, g, 298.15 K) = 322.4 $J\cdot K^{-1}\cdot mol^{-1}$ calculated by the review from the molecular parameters obtained by Erkoc, Katircioglu, and Yilmaz [2001ERK/KAT] in quantum mechanical studies of In-Se clusters. None of the values are based on experimental information on the $In_2Se(g)$ molecule and no values are selected.

The atomisation enthalpy of $In_2Se(g)$ was determined by Colin and Drowart [68COL/DRO] from mass spectrometric measurements in the temperature range 1495 to 1815 K. The evaluation employed Gibbs energy functions calculated from estimated molecular parameters and the experimental results were therefore re-evaluated as discussed in Appendix A. The results were $\Delta_f H^\circ_m$ (In_2Se, g, 298.15 K) = 167.8 $kJ\cdot mol^{-1}$ from the second law and $\Delta_f H^\circ_m$ (In_2Se, g, 298.15 K) = 147.3 $kJ\cdot mol^{-1}$ from the third law. No value is selected because the calculations depend on free energy functions derived from non-experimental information only.

V.8.4.1.4 $In_2Se_2(g)$

Chatillon [93CHA] calculated the heat capacity and entropy of $In_2Se_2(g)$ to be $C^\circ_{p,m}$ (In_2Se_2, g, 298.15 K) = 82.3 $J\cdot K^{-1}\cdot mol^{-1}$ and S°_m (In_2Se_2, g, 298.15 K) = 395.7 $J\cdot K^{-1}\cdot mol^{-1}$, respectively, from estimated molecular parameters obtained in a comparison of similar molecules with known or partly known molecular parameters. The values can be compared with $C^\circ_{p,m}$ (In_2Se_2, g, 298.15 K) = 82.5 $J\cdot K^{-1}\cdot mol^{-1}$ and S°_m (In_2Se_2, g, 298.15 K) = 389.5 $J\cdot K^{-1}\cdot mol^{-1}$ calculated by the review from the molecular parameters obtained by Erkoc, Katircioglu, and Yilmaz [2001ERK/KAT] in quantum mechanical studies of In-Se clusters. None of the values are based on experimental information for the $In_2Se_2(g)$ molecule and no values are selected.

The atomisation enthalpy of $In_2Se_2(g)$ was determined by Colin and Drowart [68COL/DRO] from mass spectrometric measurements in the temperature range 1030 to 1134 K. The evaluation employed Gibbs energy functions calculated from estimated molecular parameters and the experimental results were therefore re-evaluated as discussed in Appendix A. The results were $\Delta_f H^\circ_m$ (In_2Se_2, g, 298.15 K) = 139.8 $kJ\cdot mol^{-1}$ from the second law and $\Delta_f H^\circ_m$ (In_2Se_2, g, 298.15 K) = 159.4 $kJ\cdot mol^{-1}$ from the third law. No value is selected because the calculations depend on free energy functions derived from non-experimental information only.

V.8.4.1.5 $In_2Se_3(cr)$

Several modifications of $In_2Se_3(cr)$ are known but the stability ranges of the various polymorphs are not well known. The stability sequence α', α, β, γ, and δ with the transition temperatures 148 K, 473 K, 793 K, and 1003 K was suggested by Popovic, Tonejc, Grzeta-Plenkovic, Celustka, and Trojko [79POP/TON]. The α phase can be obtained in two modifications, hexagonal α-In_2Se_3 and rhombohedral α-In_2Se_3, depending on the synthesis method. Vassilev, Daouchi, Record, and Tedenac [98VAS/DAO] concluded that it is actually the γ-In_2Se_3 phase and not α-In_2Se_3 that is stable at 298.15 K. Equilibrium is difficult to attain and annealing times vary from days to months. A further complication in the evaluation of thermodynamic data for solid In_2Se_3 is the poor X-ray characterisation of the phases in many of the investigations. The distinct forms of In_2Se_3(cr) were in general not characterised in the thermodynamic investigations and unless otherwise stated all forms are treated as being identical in this review.

The low temperature heat capacity of α-In_2Se_3 has been measured by Demidenko, Koshchenko, Griberg, Boryakova, and Gastev, [75DEM/KOS], Mills [76MIL], Koshchenko, Grinberg, Demidenko, and Zhelagina [81KOS/GRI], and Boehnke, Kühn, Berezovskii and Spassov [87BOE/KUH]. The reported heat capacity and entropy values at 298.15 K are summarised in Table V-48. The heat capacity measurements all agree up 260 K, but above this temperature the heat capacity measured in [87BOE/KUH] remains almost constant while it continues to increase in the other investigations. The heat capacity value at 298.15 K is therefore somewhat lower in [87BOE/KUH] than in the other investigations. The mean of the heat capacity values is selected:

$$C^{\circ}_{p,\mathrm{m}}(\mathrm{In_2Se_3}, \alpha, 298.15\ \mathrm{K}) = (124.1 \pm 4.1)\ \mathrm{J\cdot K^{-1}\cdot mol^{-1}}.$$

Table V-48: Experimental heat capacity and third law entropy values of α-In_2Se_3 at 298.15 K. The values attributed to [87BOE/KUH] were evaluated by the review as discussed in Appendix A.

Reference	$C^{\circ}_{p,\mathrm{m}}$ (In_2Se_3, α, 298.15 K) ($\mathrm{J\cdot K^{-1}\cdot mol^{-1}}$)	S°_{m} (In_2Se_3, α, 298.15 K) ($\mathrm{J\cdot K^{-1}\cdot mol^{-1}}$)	Temperature (K)
[75DEM/KOS]	(125.5 ± 1.0)	(197.6 ± 1.8)	55 – 330
[76MIL]	(124.1 ± 1.0)		129 – 700
[81KOS/GRI]	(125.5 ± 0.3)	(203.2 ± 0.6)	5 – 303
[87BOE/KUH]	(121.1 ± 0.4)	(203.0 ± 0.6)	7 – 285

Mills [76MIL] measured the heat capacity of In_2Se_3(cr) up to 700 K and the heat capacity expressions

$$C^{\circ}_{p,\mathrm{m}}(\mathrm{In_2Se_3}, \alpha, (298.15 - 471)\ \mathrm{K}) = (109.19 + 50.0 \times 10^{-3}\ T)\ \mathrm{J\cdot K^{-1}\cdot mol^{-1}}.$$

and

$$C^{\circ}_{p,\mathrm{m}}(\mathrm{In_2Se_3}, \beta, (471-700)\ \mathrm{K}) = (113.42 + 35.99 \times 10^{-3}\ T)\ \mathrm{J{\cdot}K^{-1}{\cdot}mol^{-1}}.$$

were derived by the review from the values listed in Table 1 of the paper. The transition temperature was changed from the original value of 486 K in [76MIL] to 471 K determined in [98VAS/DAO].

The entropy of α-In_2Se_3 at 298.15 K evaluated in [75DEM/KOS] is based on an estimated heat capacity in the range 0 to 55 K and the value is lower than in the other investigations, see Table V-48. A second law entropy of formation of In_2Se_3(cr), presumably α-In_2Se_3, was determined to be $\Delta_f S^{\circ}_{m}(\mathrm{In_2Se_3}, \alpha, 298.15\ \mathrm{K}) = -(50.6 \pm 8.8)\ \mathrm{J{\cdot}K^{-1}{\cdot}mol^{-1}}$ by Mustafaev, Azizov, Aliev, and Abbasov [74MUS/AZI] in electrochemical cell measurements in the temperature range 510 to 710 K. The entropy value $S^{\circ}_{m}(\mathrm{In_2Se_3}, \alpha, 298.15\ \mathrm{K}) = (191.3 \pm 9.0)\ \mathrm{J{\cdot}K^{-1}{\cdot}mol^{-1}}$ is obtained from this investigation on combining the entropy of formation of α-In_2Se_3 with the selected standard entropy of selenium and the entropy of In(cr) in [82WAG/EVA]. The mean of the values of the standard entropy in [81KOS/GRI] and, from [87BOE/KUH] as evaluated by this review, is selected:

$$S^{\circ}_{m}(\mathrm{In_2Se_3}, \alpha, 298.15\ \mathrm{K}) = (203.1 \pm 2.0)\ \mathrm{J{\cdot}K^{-1}{\cdot}mol^{-1}}.$$

The transition temperatures and enthalpies of the various polymorphic transformations of In_2Se_3(cr) are discussed in detail in the reviews in [93CHA] and [98VAS/DAO]. The values recommended in the two reviews are similar and those in [98VAS/DAO] are:

$$\Delta_{\mathrm{trs}}H^{\circ}_{m}(\mathrm{In_2Se_3}, \alpha \rightarrow \beta, 471\ \mathrm{K}) = (3.95 \pm 0.30)\ \mathrm{kJ{\cdot}mol^{-1}},$$

$$\Delta_{\mathrm{trs}}H^{\circ}_{m}(\mathrm{In_2Se_3}, \beta \rightarrow \gamma, 913\ \mathrm{K}) = (3.25 \pm 0.30)\ \mathrm{kJ{\cdot}mol^{-1}},$$

$$\Delta_{\mathrm{trs}}H^{\circ}_{m}(\mathrm{In_2Se_3}, \gamma \rightarrow \delta, 1021\ \mathrm{K}) = (0.95 \pm 0.30)\ \mathrm{kJ{\cdot}mol^{-1}}.$$

The enthalpy of formation of α-In_2Se_3 was determined to be $\Delta_f H^{\circ}_{m}(\mathrm{In_2Se_3}, \alpha, 298.15\ \mathrm{K}) = -(344 \pm 13)\ \mathrm{kJ{\cdot}mol^{-1}}$ by Hahn and Burow [56HAH/BUR] using combustion calorimetry in oxygen. The experimental conditions are only briefly described and the auxiliary data used in calculations are not given. The quality of these measurements can therefore not be assessed and neither be recalculated using the auxiliary data set employed by the review. Sharifov and Azizov [67SHA/AZI] determined the enthalpy of formation to be $\Delta_f H^{\circ}_{m}(\mathrm{In_2Se_3}, \alpha, 298.15\ \mathrm{K}) = -(318.0 \pm 10.0)\ \mathrm{kJ{\cdot}mol^{-1}}$ using direct synthesis in a bomb calorimeter. Unreacted selenium was distilled off and the final product was X-rayed. This is the better documented investigation of the two.

The vaporisation of In_2Se_3(cr) has been studied in a number of investigations, usually by measuring the total pressure employing Knudsen effusion techniques. The results are summarised in Table V-49 and the individual investigations are discussed in Appendix A. Mass spectrometry was used in [69BER/STR] and [79VEN/PAS] to also record the partial pressures of In_2Se(g) and Se_2(g). In all investigations, except that by Srinavasa and Edwards [87SRI/EDW], the assumed reaction was $\mathrm{In_2Se_3(cr)} \rightleftharpoons \mathrm{In_2Se(g)}$

+ $Se_2(g)$ without actually verifying that this was the case. In [87SRI/EDW] the vaporisation of a number of samples of different compositions in the indium-selenium system were investigated and by analysing the residues with X-ray techniques, it was concluded that the samples vaporised incongruently under the formation of liquids. Nevertheless, it was possible to combine results for selected reactions involving liquids to form the hypothetical reaction $In_2Se_3(cr) \rightleftharpoons In_2Se(g) + Se_2(g)$ and thereby to reduce the effects of the uncertain compositions of the liquids. The procedure is discussed in more detail in Appendix A. Unfortunately, the numerical values presented in the paper are only given as evaluated at 298.15 K without stating the auxiliary data used in the evaluations. The results can therefore not be recalculated using the auxiliary data of the review. The probably incongruent vaporisation of $In_2Se_3(cr)$ makes the results from the vaporisation studies doubtful and since no reliable data exist for $In_2Se(g)$, the thermodynamic properties of $In_2Se_3(cr)$ cannot be determined with confidence from these investigations.

Table V-49: Determinations of thermochemical properties of the anticipated reaction α-$In_2Se_3 \rightleftharpoons In_2Se(g) + Se_2(g)$. Note that the actual vaporisation of α-In_2Se_3 is probably incongruent, a condition, which was only taken into account in the study in [87SRI/EDW]. Most of the results are therefore doubtful.

Reference	$\Delta_r S^\circ_m$ (298.15 K), ($J \cdot K^{-1} \cdot mol^{-1}$)	$\Delta_r H^\circ_m$ (298.15 K), 2nd law, ($kJ \cdot mol^{-1}$)	$\Delta_r H^\circ_m$ (298.15 K), 3rd law, ($kJ \cdot mol^{-1}$)	Temperature (K)	Method
[69BER/STR]	168.4	400.2	598.4	900 – 1120	Knudsen effusion and mass spectrometry.
[72GRI/BOR]	373.6	451.8	444.8	923 – 1013	Bourdon gauge.
[79VEN/PAS]	319.4	536.2	588.1	1055 – 1167	Knudsen effusion and mass spectrometry.
[87SRI/EDW]			604.2, 600.4	1004 – 1156	Knudsen and torsion effusion.
[88BAR/PIA]	332	549	596	984 – 1159	Knudsen effusion
[88PAS/ZHA]	360	(546.4 ± 16.0)	(563.2 ± 14.0)	980 – 1100	Knudsen effusion

The results in [56HAH/BUR] are rejected by the review because of the rudimentary documentation and the review selects the value of the enthalpy of formation determined in [67SHA/AZI],

$$\Delta_f H^\circ_m (In_2Se_3, \alpha, 298.15 \text{ K}) = -(318 \pm 20) \text{ kJ} \cdot \text{mol}^{-1},$$

but the uncertainty was increased since systematic errors cannot be excluded.

The Gibbs energy of formation of α-In_2Se_3 is calculated from the selected values of the enthalpy of formation and entropy, the selected entropy of selenium, and the entropy of indium in [82WAG/EVA] yielding:

$$\Delta_f G_m^\circ(\mathrm{In_2Se_3}, \alpha, 298.15\,\mathrm{K}) = -(306 \pm 20)\ \mathrm{kJ\cdot mol^{-1}}.$$

The datum is included in Appendix E since it was calculated with non-TDB auxiliary data.

The thermodynamic modelling in [2003LI/REC], *cf.* Section V.8.4.1, resulted in the values $\Delta_f H_m^\circ(\mathrm{In_2Se_3}, \alpha, 298.15\ \mathrm{K}) = -270.8\ \mathrm{kJ\cdot mol^{-1}}$, $S_m^\circ(\mathrm{In_2Se_3}, \alpha, 298.15\ \mathrm{K}) = 223.1\ \mathrm{J\cdot K^{-1}\cdot mol^{-1}}$, $C_{p,m}^\circ(\mathrm{In_2Se_3}, \alpha, 298.15\ \mathrm{K}) = 124.5\ \mathrm{J\cdot K^{-1}\cdot mol^{-1}}$, $\Delta_f H_m^\circ(\mathrm{In_2Se_3}, \beta, 298.15\ \mathrm{K}) = -268.5\ \mathrm{kJ\cdot mol^{-1}}$, $S_m^\circ(\mathrm{In_2Se_3}, \beta, 298.15\ \mathrm{K}) = 231.9\ \mathrm{J\cdot K^{-1}\cdot mol^{-1}}$, $C_{p,m}^\circ(\mathrm{In_2Se_3}, \beta, 298.15\ \mathrm{K}) = 121.1\ \mathrm{J\cdot K^{-1}\cdot mol^{-1}}$, $\Delta_f H_m^\circ(\mathrm{In_2Se_3}, \gamma, 298.15\ \mathrm{K}) = -267.7\ \mathrm{kJ\cdot mol^{-1}}$, $S_m^\circ(\mathrm{In_2Se_3}, \gamma, 298.15\ \mathrm{K}) = 229.8\ \mathrm{J\cdot K^{-1}\cdot mol^{-1}}$, $C_{p,m}^\circ(\mathrm{In_2Se_3}, \gamma, 298.15\ \mathrm{K}) = 122.9\ \mathrm{J\cdot K^{-1}\cdot mol^{-1}}$, $\Delta_f H_m^\circ(\mathrm{In_2Se_3}, \delta, 298.15\ \mathrm{K}) = -262.5\ \mathrm{kJ\cdot mol^{-1}}$, $S_m^\circ(\mathrm{In_2Se_3}, \delta, 298.15\ \mathrm{K}) = 234.9\ \mathrm{J\cdot K^{-1}\cdot mol^{-1}}$, $C_{p,m}^\circ(\mathrm{In_2Se_3}, \delta, 298.15\ \mathrm{K}) = 122.9\ \mathrm{J\cdot K^{-1}\cdot mol^{-1}}$, and the heat capacity expressions $C_{p,m}^\circ(\mathrm{In_2Se_3}, \alpha, (298.15 - 474)\ \mathrm{K}) = (123.06 + 17.857 \times 10^{-3}\ T - 3.4857 \times 10^{5}\ T^{-2})\ \mathrm{J\cdot K^{-1}\cdot mol^{-1}}$, $C_{p,m}^\circ(\mathrm{In_2Se_3}, \beta, (471 - 1170)\ \mathrm{K}) = (109.5 + 38.73 \times 10^{-3}\ T)\ \mathrm{J\cdot K^{-1}\cdot mol^{-1}}$, $C_{p,m}^\circ(\mathrm{In_2Se_3}, \gamma, (474 - 1018)\ \mathrm{K}) = (111.38 + 38.73 \times 10^{-3}\ T)\ \mathrm{J\cdot K^{-1}\cdot mol^{-1}}$, and $C_{p,m}^\circ(\mathrm{In_2Se_3}, \delta, (1018 - 1073)\ \mathrm{K}) = (111.38 + 38.73 \times 10^{-3}\ T)\ \mathrm{J\cdot K^{-1}\cdot mol^{-1}}$, which are given for information only.

V.8.4.1.6 In_4Se_3(cr)

No experimental thermodynamic information exists for this phase, but the thermodynamic modelling in [2003LI/REC], *cf.* Section V.8.4.1, resulted in the values $\Delta_f H_m^\circ(\mathrm{In_4Se_3}, \mathrm{cr}, 298.15\ \mathrm{K}) = -367.1\ \mathrm{kJ\cdot mol^{-1}}$, $S_m^\circ(\mathrm{In_4Se_3}, \mathrm{cr}, 298.15\ \mathrm{K}) = 285.1\ \mathrm{J\cdot K^{-1}\cdot mol^{-1}}$, $C_{p,m}^\circ(\mathrm{In_4Se_3}, \mathrm{cr}, 298.15\ \mathrm{K}) = 179.3\ \mathrm{J\cdot K^{-1}\cdot mol^{-1}}$, and the heat capacity expression $C_{p,m}^\circ(\mathrm{In_4Se_3}, \mathrm{cr}, (298.15 - 850)\ \mathrm{K}) = (191.58 + 5.78 \times 10^{-3}\ T - 12.4493 \times 10^{5}\ T^{-2})\ \mathrm{J\cdot K^{-1}\cdot mol^{-1}}$, which are given for information only.

V.8.4.1.7 In_5Se_7(cr)

No experimental thermodynamic information exists for this phase, but the thermodynamic modelling in [2003LI/REC], *cf.* Section V.8.4.1, resulted in the values $\Delta_f H_m^\circ(\mathrm{In_5Se_7}, \mathrm{cr}, 298.15\ \mathrm{K}) = -670.2\ \mathrm{kJ\cdot mol^{-1}}$, $S_m^\circ(\mathrm{In_5Se_7}, \mathrm{cr}, 298.15\ \mathrm{K}) = 528.6\ \mathrm{J\cdot K^{-1}\cdot mol^{-1}}$, $C_{p,m}^\circ(\mathrm{In_5Se_7}, \mathrm{cr}, 298.15\ \mathrm{K}) = 298.7\ \mathrm{J\cdot K^{-1}\cdot mol^{-1}}$, and the heat capacity expression $C_{p,m}^\circ(\mathrm{In_5Se_7}, \mathrm{cr}, (298.15 - 1150)\ \mathrm{K}) = (295.34 + 42.856 \times 10^{-3}\ T - 8.36568 \times 10^{5}\ T^{-2})\ \mathrm{J\cdot K^{-1}\cdot mol^{-1}}$, which are given for information only.

V.8.4.1.8 In_6Se_7(cr)

No experimental thermodynamic information exists for this phase, but the thermodynamic modelling in [2003LI/REC], *cf.* Section V.8.4.1, resulted in the values

$\Delta_f H_m^\circ$ (In_6Se_7, cr, 298.15 K) = – 773.8kJ·mol^{-1}, S_m° (In_6Se_7, cr, 298.15 K) = 523.9 J·K^{-1}·mol^{-1}, $C_{p,m}^\circ$ (In_6Se_7, cr, 298.15 K) = 321.9 J·K^{-1}·mol^{-1}, and the heat capacity expression $C_{p,m}^\circ$ (In_6Se_7, cr, (298.15 – 943) K) = (313.18 + 53.27 × 10^{-3} T – 6.34328 × 10^5 T^{-2}) J·K^{-1}·mol^{-1}, which are given for information only.

V.8.4.1.9 In_9Se_{11}(cr)

The phase In_9Se_{11}(cr) is not mentioned in the reviews of the system In-Se in [93CHA] and [98VAS/DAO] and no experimental thermodynamic information exists for the phase. Nevertheless, the thermodynamic modelling in [2003LI/REC], *cf.* Section V.8.4.1, reports the values $\Delta_f H_m^\circ$ (In_9Se_{11}, cr, 298.15 K) = – 1173 kJ·mol^{-1}, S_m° (In_9Se_{11}, cr, 298.15 K) = 826.9 J·K^{-1}·mol^{-1}, $C_{p,m}^\circ$ (In_9Se_{11}, cr, 298.15 K) = 495.3 J·K^{-1}·mol^{-1}, and the heat capacity expression $C_{p,m}^\circ$ (In_9Se_{11}, cr, (298.15 – 1000) K) = (481.8 + 81.96 × 10^{-3} T – 9.7589 × 10^5 T^{-2}) J·K^{-1}·mol^{-1}, which are included for information only.

V.8.4.2 Indium selenites

Savchenko, Tananaev, and Volodina [68SAV/TAN] studied the system In_2O_3-SeO_2-H_2O at 298.15 K by solubility measurements in the concentration range 0.7 to 35.3 mass-% SeO_2. Two crystalline phases were observed namely $In_2(SeO_3)_3·6H_2O$ and $In(HSeO_3)_3·3H_2O$. Gospodinov [84GOS6] made a similar investigation at (373.2 ± 1.0) K in the concentration interval 1.4 to 78.1 mass-% SeO_2 in the liquid phase. The tie lines were used to identify the incongruently soluble solid phases $In_2(SeO_3)_3·6H_2O$, $In_2(SeO_3)_3·H_2SeO_3$, and $In_2(SeO_3)_3·2H_2SeO_3·H_2O$. The reported solubility of In(III) is small over the whole concentration range in SeO_2.

Mityureva [59MIT] measured the solubility of $In_2(SeO_3)_3·6H_2O$(cr) in 0.1 M H_2SO_4 and 0.05 M HCl at 293.2 K. The reported solubility product, $\log_{10} K_{s,0}$ = – 32.61, is not accepted as side reactions of In^{3+} were not considered. No attempt has been made to re-evaluate the few data presented in the paper as discussed in Appendix A. Popova, Slavtscheva, and Gospodinov [86POP/SLA] determined the solubility of indium selenite in dilute nitric acid. The preparation and composition of the selenite is not discussed. It is assumed to have the formulae $In_2(SeO_3)_3·6H_2O$(cr) by the review. The solubility product calculated in the paper is accepted and selected, *cf.* Appendix A,

$$In_2(SeO_3)_3·6H_2O(cr) \rightleftharpoons 2In^{3+} + 3SeO_3^{2-} + 6H_2O(l) \quad (V.71)$$

$$\log_{10} K_{s,0}^\circ ((V.71), 298.15\ K) = -(39.0 \pm 2.0)$$

but the uncertainty has been increased considerably by the review due to the uncertainty in the values of the activity coefficients and the high powers of the activities in the expression for the solubility product.

No other thermodynamic data have been found for indium selenites.

V.8.4.3 Indium selenates

Tananaev *et al.* have determined the solubility of indium selenate in water at 293 K to be 1.87 mol($In_2(SeO_4)_3$)·kg^{-1} in a number of papers (see below). The composition of the solid phase is reported to be $In_2(SeO_4)_3\cdot9H_2O$(cr). The review accepts

$$m(In_2(SeO_4)_3\cdot9H_2O(cr), \text{aq sat}, 293.15\text{ K}) = 1.87\text{ mol}(In_2(SeO_4)_3)\cdot\text{kg}^{-1}.$$

No uncertainty can be assigned to this figure.

A phase diagram or related information is available for the systems: $In_2(SeO_4)_3$-H_2SeO_4-H_2O at 293 K in [72DEI/TAN]; $In_2(SeO_4)_3$-$(NH_4)_2SeO_4$-H_2O at 293 K in [80KAD/TAN]; $In_2(SeO_4)_3$-Li_2SeO_4-H_2O at 293 K in [82TAN/KAD]; $In_2(SeO_4)_3$-Na_2SeO_4-H_2O at 293 K in [73DEI/TAN]; $In_2(SeO_4)_3$-K_2SeO_4-H_2O at 293 K in [75KAD/DEI]; $In_2(SeO_4)_3$-Rb_2SeO_4-H_2O at 293 K in [77KAD/DEI]; $In_2(SeO_4)_3$-Cs_2SeO_4-H_2O at 293 K in [77KAD/DEI3].

No other thermodynamic data have been found for indium selenates.

V.8.5 Thallium compounds

V.8.5.1 Thallium selenides

Römermann, Feutelais, Fries and Blachnik [2000ROM/FEU] investigated the Tl-Se system by differential scanning calorimetry and X-ray diffraction techniques and assessed and modelled the thermodynamic properties of the condensed phases in the system. The resulting phase diagram is presented in Figure V-16 and it comprises the solid phases Tl_2Se(cr), α-TlSe, β-TlSe, all of fixed or nearly fixed stoichiometry, and a solution phase γ with an upper limit of the selenium content corresponding to the compound Tl_5Se_3(cr). In addition, the phase Tl_2Se_3(cr) has been reported by Terpilowski, Zaleska, and Gawel [68TER/ZAL]. The phase has not been found in later investigations and it probably does not exist. It was not included in the phase diagram in [2000ROM/FEU]. The saturated vapour pressure in equilibrium with solid Tl_2Se(cr) and TlSe(cr) has been measured by Shakhtakhtinskii, Kuliev, and Abdullaev [63SHA/KUL] and by Genov, Nesmeyanov, and Priselkov [62GEN/NES]. However, the species in the gas phase have not been identified and the measured pressures cannot reliably be attributed to specific molecules. No thermodynamic data are therefore selected for gaseous thallium selenides.

Figure V-16: Phase diagram of the selenium-thallium system.

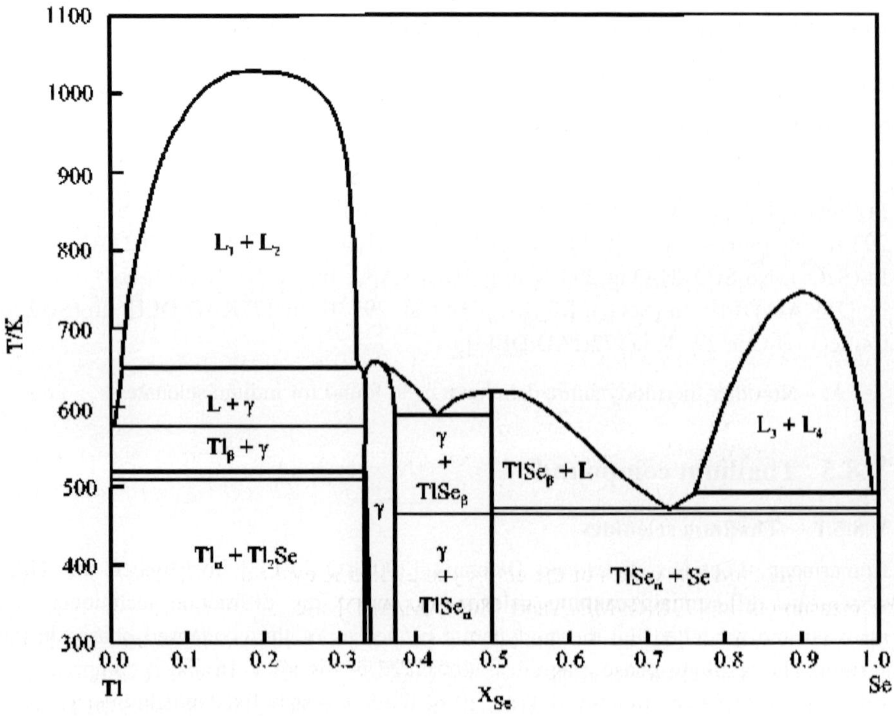

V.8.5.1.1 α-TlSe and β-TlSe

There are two solid polymorphs of the composition TlSe; α-TlSe, which is stable below 461 K, and β-TlSe, which is stable from 472K up to the congruent melting point at 620 K. The transformation $\alpha \rightarrow \beta$ occurs in the temperature range 461 to 472 K with a tiny change in composition. The change was neglected in the modelling of the phases in [2000ROM/FEU] which used the average temperature 466 K for the transition.

The heat capacity of α-TlSe was measured in the temperature range 3 to 300 K using a heat pulse technique and in the temperature range 190 to 461 K using a heat content method by Brekow, Meissner, Scheiba, Tausend, and Wobig [73BRE/MEI], and in the temperature range 2 to 300 K by Mamedov, Aldzhanov, Mekhtiev, and Kerimov [80MAM/ALD]. The values reported at 298.15 K were $C^{\circ}_{p,m}$(TlSe, α, 298.15 K) = (50.6 ± 1.5) $J\cdot K^{-1}\cdot mol^{-1}$ and (49.9 ± 1.0) $J\cdot K^{-1}\cdot mol^{-1}$, respectively. The weighted average is selected:

$$C^{\circ}_{p,m}(\text{TlSe}, \alpha, 298.15\ \text{K}) = (50.1 \pm 0.8)\ \text{J}\cdot\text{K}^{-1}\cdot\text{mol}^{-1}.$$

Brekow *et al.* [73BRE/MEI] also determined the heat capacity of β-TlSe in the temperature range (466 to 600) K and reported the heat capacity expression $C^{\circ}_{p,\mathrm{m}}$(TlSe, cr, (190 – 600) K) = (42.8 + 26.3 × 10^{-3} T) J·K^{-1}·mol^{-1} to describe the heat capacity of the TlSe polymorphs to within 3%. The expression describes the heat capacity across the α → β transformation and the transition enthalpy is too small to be determinable by their technique. Morgant, Legendre, and Souleau [82MOR/LEG] used drop calorimetry in the temperature range 298 to 775 K to determine the heat capacities of α-TlSe, β-TlSe, and TlSe(l) yielding the constant values $C^{\circ}_{p,\mathrm{m}}$(TlSe, α, (298 – 461) K) = 50.3 J·K^{-1}·mol^{-1}, $C^{\circ}_{p,\mathrm{m}}$(TlSe, β, (472 – 619) K) = 51.5 J·K^{-1}·mol^{-1}, and $C^{\circ}_{p,\mathrm{m}}$(TlSe, l, (619 – 775) K) = 78.8 J·K^{-1}·mol^{-1}. No value was determined for the enthalpy of the transition α → β, but the enthalpy and entropy of fusion were determined to be $\Delta_{\mathrm{fus}}H^{\circ}_{\mathrm{m}}$(β-TlSe, 619 K) = (22.19 ± 0.20) kJ·mol^{-1} and $\Delta_{\mathrm{fus}}S^{\circ}_{\mathrm{m}}$(β-TlSe, 619 K) = 35.85 J·K^{-1}·mol^{-1}, respectively. The assessment in [2000ROM/FEU] combined the results in [73BRE/MEI], [80MAM/ALD], and [82MOR/LEG] and the following quantities: $C^{\circ}_{p,\mathrm{m}}$(TlSe, α, (298.15 – 466) K) = (42.742 + 26.28 × 10^{-3} T) J·K^{-1}·mol^{-1}, $C^{\circ}_{p,\mathrm{m}}$(TlSe, α, (466 – 620) K) = (42.836 + 26.11 × 10^{-3} T) J·K^{-1}·mol^{-1}, and $\Delta_{\mathrm{trs}}H^{\circ}_{\mathrm{m}}$(α-TlSe→β-TlSe, 466 K) = 0.27 kJ·mol^{-1} are reported. The heat capacity of α-TlSe at 298.15 K calculated from the expression is $C^{\circ}_{p,\mathrm{m}}$(TlSe, α, 298.15 K) = 50.62 J·K^{-1}·mol^{-1} which is within the uncertainty limits of the selected value.

The third law values of the entropy at 298.15 K evaluated from the heat capacity measurents in [73BRE/MEI] and [80MAM/ALD] are S°_{m}(TlSe, α, 298.15 K) = (104.5 ± 3.0) J·K^{-1}·mol^{-1} and (102.6 ± 1.5) J·K^{-1}·mol^{-1}, respectively. The assessment in [2000ROM/FEU] selected the value S°_{m}(TlSe, α, 298.15 K) = 104.3 J·K^{-1}·mol^{-1}. The review selects the weighted average of the measurements in [73BRE/MEI] and [80MAM/ALD]:

$$S^{\circ}_{\mathrm{m}}(\mathrm{TlSe}, \alpha, 298.15\ \mathrm{K}) = (103.0 \pm 1.3)\ \mathrm{J\cdot K^{-1}\cdot mol^{-1}}.$$

The enthalpy of formation of α-TlSe was determined to be $\Delta_{\mathrm{f}}H^{\circ}_{\mathrm{m}}$(TlSe, α, 298.15 K) = – (61.0 ± 2.0) kJ·mol^{-1} by Vasiliev, Nikolskaya, Bachinskaya, and Gerasimov [67VAS/NIK] from electrochemical measurements in the temperature range 313 to 413 K using the cell (–) Tl(cr) | TlCl, KCl (in glycerol) | α-TlSe, Se(trig) (+).

Morgant, Feutelais, Schnitter, and Legendre [91MOR/FEU] measured the cell

(–) Tl(l) | 1 weight-% TlCl (in LiCl + KCl melt) | $Tl_{1-x}Se_x$(l) (+)

in the temperature range 663 to 763 K. Partial and integral enthalpies of formation of the solutions were derived. Their value for the enthalpy of mixing of the composition $Tl_{0.5}Se_{0.5}$ at 738 K, $\Delta_{\mathrm{f}}H^{\circ}_{\mathrm{m}}$($Tl_{0.5}Se_{0.5}$, l, 738 K) = – 25.13 kJ·mol^{-1}, was combined by the review with the heat content measurements in [82MOR/LEG], the selected properties of selenium, and the SGTE [91DIN] thermodynamic properties of thallium yielding $\Delta_{\mathrm{f}}H^{\circ}_{\mathrm{m}}$(TlSe, α, 298.15 K) = – 57.0 kJ·mol^{-1}. Maekawa, Yokokawa, and Niwa [71MAE/YOK] measured the enthalpy of mixing using calorimetry at 753 K. The results were recalculated to 298.15 K by the review using the same procedure as for the results in [91MOR/FEU] yielding $\Delta_f H_m^\circ$(TlSe, α, 298.15 K) = – 56.4 kJ·mol^{-1}. The Gibbs energy of formation of α-TlSe at 433 K was determined to be $\Delta_f G_m^\circ$(TlSe, α, 433 K) = – (56.5 ± 1.2) kJ·mol^{-1} by Terpilowski, Zaleska, and Gawel [68TER/ZAL] using cells similar to those used in [67VAS/NIK]. However, the results were obtained via measurements on Tl_2Se_3(cr) which probably does not exist. Their results are therefore dubious and were rejected in [2000ROM/FEU] and by the review. The activity of selenium in liquid thallium-selenium mixtures at 860 K was determined by Predel, Piehl, and Pool [75PRE/PIE] by measuring the equilibrium selenium pressure. Partial and integral free energies were calculated. The calculations were later revised by Predel, Gerdes, and Gerling [79PRE/GER] using new thermodynamic data for the gaseous selenium species. No quantities were calculated at standard conditions. Since the primary data from the measurements are not presented, a recalculation to standard conditions is not possible even if necessary thermal functions were available. Predel *et al.* [75PRE/PIE] also measured the enthalpy of mixing calorimetrically at 860 K and the results were recalculated to 298.15 K by the review using the same procedure as for the results in [91MOR/FEU] yielding $\Delta_f H_m^\circ$(TlSe, α, 298.15 K) = – 52.2 kJ·mol^{-1}. Römermann *et al.* [2000ROM/FEU] assessed the value to be $\Delta_f H_m^\circ$(TlSe, α, 298.15/K) = – 63.2 kJ·mol^{-1}, but the review calculates and accepts the average obtained on assigning equal weights to the values derived from [67VAS/NIK], [91MOR/FEU], [71MAE/YOK], and [75PRE/PIE]

$$\Delta_f H_m^\circ(\text{TlSe}, \alpha, 298.15\text{ K}) = -(56.7 \pm 7.1)\text{ kJ·mol}^{-1}.$$

The datum is included in Appendix E since it was calculated with non-TDB auxiliary data.

The Gibbs energy of formation is calculated from the accepted enthalpy of formation and entropy of α-TlSe, the selected entropy of selenium, and the entropy of thallium in [82WAG/EVA] to be:

$$\Delta_f G_m^\circ(\text{TlSe}, \alpha, 298.15\text{K}) = -(55.7 \pm 7.1)\text{ kJ·mol}^{-1}.$$

The datum is included in Appendix E since it was calculated with non-TDB auxiliary data.

V.8.5.1.2 Tl_2Se(cr)

The heat capacity of Tl_2Se(cr) was measured in the temperature range 3 to 640 K using a heat pulse method by Brekow, Meissner, Scheiba, Tausend, and Wobig [75BRE/MEI]. Mamedov, Yangirov, Guseinov, and Abdullaev [88MAM/YAN] presented measured heat capacity values in a graph together with the data in [75BRE/MEI] and concluded that the two series of data agreed. No numerical values were reported. The heat capacity and the entropy at 298.15 K derived from the heat capacity data in [75BRE/MEI] are selected:

$$C^{\circ}_{p,\mathrm{m}}(\mathrm{Tl_2Se, cr, 298.15\ K}) = (82.2 \pm 2.5)\ \mathrm{J{\cdot}K^{-1}{\cdot}mol^{-1}},$$

and

$$S^{\circ}_{\mathrm{m}}(\mathrm{Tl_2Se, cr, 298.15\ K}) = (174.8 \pm 3.0)\ \mathrm{J{\cdot}K^{-1}{\cdot}mol^{-1}}.$$

The heat capacity expression $C^{\circ}_{p,\mathrm{m}}(\mathrm{Tl_2Se, cr, (190-640)\ K}) = (75.2 + 27.6 \times 10^{-3}\ T)\ \mathrm{J{\cdot}K^{-1}{\cdot}mol^{-1}}$ was also reported in [75BRE/MEI].

Vasiliev, Nikolskaya, and Gerasimov [71VAS/NIK] extended the work in [67VAS/NIK] on α-TlSe, *cf.* Section V.8.5.1.1, to lower selenium contents and performed electrochemical measurements in the temperature range 313 to 413 K using cells of the type

$$(-)\ \mathrm{Tl(cr)}\ |\ \mathrm{TlCl, KCl\ (in\ glycerol)}\ |\ \mathrm{Tl_{1-\mathit{x}}Se_{\mathit{x}}}\ (+)$$

with x in the range 0.375 to 0.333. The enthalpy of formation of $\mathrm{Tl_2Se(cr)}$ was determined to be $\Delta_{\mathrm{f}}H^{\circ}_{\mathrm{m}}(\mathrm{Tl_2Se, cr, 298.15\ K}) = -(92.6 \pm 2.0)\ \mathrm{kJ{\cdot}mol^{-1}}$ by combining their earlier results for α-TlSe with the new measurements.

In Section V.8.5.1.1 the enthalpy of formation of α-TlSe at standard conditions was evaluated by the review from the measurements of the enthalpy of mixing in the liquid state at high temperatures made in [91MOR/FEU], [71MAE/YOK], and [75PRE/PIE] by combining their results with the heat content measurements in [82MOR/LEG]. The procedure was repeated for the composition $\mathrm{Tl_2Se}$ yielding the enthalpy values $-$ 82.7 $\mathrm{kJ{\cdot}mol^{-1}}$, $-$ 86.4 $\mathrm{kJ{\cdot}mol^{-1}}$, and $-$ 60.2 $\mathrm{kJ{\cdot}mol^{-1}}$ from the measurements in [91MOR/FEU], [71MAE/YOK], and [75PRE/PIE], respectively. The result in [75PRE/PIE] deviates from the other two values because the shape of the curve of the enthalpy of mixing as a function of selenium content is different in this investigation. The assessment by Römermann *et al.* [2000ROM/FEU] rejected the value because of this condition and selected the value $\Delta_{\mathrm{f}}H^{\circ}_{\mathrm{m}}(\mathrm{Tl_2Se, cr, 298.15\ K}) = -95.6\ \mathrm{kJ{\cdot}mol^{-1}}$ in their thermodynamic modelling of the phase diagram. The review accepts the average enthalpy value calculated by assigning equal weights to the values derived from [71VAS/NIK], [91MOR/FEU], and [71MAE/YOK],

$$\Delta_{\mathrm{f}}H^{\circ}_{\mathrm{m}}(\mathrm{Tl_2Se, cr, 298.15\ K}) = -(87.2 \pm 9.8)\ \mathrm{kJ{\cdot}mol^{-1}}.$$

The datum is included in Appendix E since it was calculated with non-TDB auxiliary data.

The Gibbs energy of formation is calculated from the accepted enthalpy of formation and entropy of $\mathrm{Tl_2Se(cr)}$, the selected entropy of selenium, and the entropy of thallium in [82WAG/EVA] to be:

$$\Delta_{\mathrm{f}}G^{\circ}_{\mathrm{m}}(\mathrm{Tl_2Se, s, 298.15K}) = -(88.5 \pm 9.8)\ \mathrm{kJ{\cdot}mol^{-1}}.$$

The datum is included in Appendix E since it was calculated with non-TDB auxiliary data.

In the aqueous phase, Umland and Wallmeier [80UML/WAL] studied the reduction of selenite at the mercury electrode in the presence of Tl^+ by polarography, see Appendix A. The solubility product of $Tl_2Se(s)$ was obtained from the position of the half-wave potential of the second reduction step, $HgSe(s) + 2Tl^+ + 2e^- \rightleftharpoons Tl_2Se(s) + Hg(l)$, to be $\log_{10} K^\circ_{s,0}$ (Tl_2Se, s, 298.15 K) = – (33.3 ± 2.0). Combined with the $\Delta_f G^\circ_m$ value for Tl^+ (– 32.40 kJ·mol^{-1}) in [82WAG/EVA] and the selected value for Se^{2-}, it corresponds to $\Delta_f G^\circ_m$ (Tl_2Se, s, 298.15 K) = – (126.3 ± 13.2) kJ·mol^{-1}. The value is in disagreement with the value extrapolated from high temperature measurements. Since the validity of the polarographic method has not been documented, the value of $\Delta_f G^\circ_m$ (Tl_2Se, cr, 298.15 K) obtained from the extrapolation of the high temperature data has been adopted.

V.8.5.1.3 Tl_2Se_3(cr)

According to the assessment of the phase diagram information by Römermann *et al.* [2000ROM/FEU], Tl_2Se_3(cr) probably does not exist and the thermodynamic properties estimated by Mills [74MIL] and reported by Terpilowski, Zaleska, and Gawel [68TER/ZAL] are therefore irrelevant.

V.8.5.2 Thallium selenites

V.8.5.2.1 Tl(I)

Micka, Ebert, France, and Ptacek [86MIC/EBE] established the phase diagram of the system Tl_2SeO_3-H_2SeO_3-H_2O at 298.15 K from solubility measurements. The crystalline phases Tl_2SeO_3, $Tl_2Se_2O_5$, and $TlH_3(SeO_3)_2$ were identified. Few primary solubility data are presented and the paper is largely devoted to the thermal analysis and spectrochemical characterisation of the compounds.

V.8.5.2.2 Tl(III)

Gospodinov [84GOS5] studied the phase diagram of the system Tl_2O_3-SeO_2-H_2O at (373.2 ± 1.0) K over the concentration range 0.2 to 78.8 mass-% SeO_2 in the liquid phase. The compositions of the solids obtained from the tie lines were $Tl_2(SeO_3)_3$ and $Tl_2(SeO_3)_3 \cdot H_2SeO_3$, both of which dissolve incongruently. The published solubility data show that the solubility of Tl(III) is small in the whole interval of SeO_2 concentrations investigated.

Mityureva [59MIT] measured the solubility of thallium(III) selenite, $Tl_2(SeO_3)_3$(cr), in 0.1 M H_2SO_4 and 0.1 M HNO_3 at 293.2 K. The reported solubility product, $\log_{10} K_{s,0}$ = – 38.72, is not accepted as side reactions of Tl^{3+} were not considered. No attempt has been made to re-evaluate the few data presented in the paper as discussed in Appendix A. Popova, Slavtscheva, and Gospodinov [86POP/SLA] determined the solubility of thallium(III) selenite in dilute sulphuric acid. The solubility product, $\log_{10} K_{s,0}$ = – 46.69, calculated from the total concentration of Tl(III) in the saturated solution is not accepted since the strong hydrolysis of Tl^{3+} was not taken into account. A recalculation is not possible, see Appendix A.

No other thermodynamic data have been found for thallium selenites.

V.8.5.3 Thallium selenates

V.8.5.3.1 Tl(I)

Vojtisek and Ebert [89VOJ/EBE] determined the solubility of Tl_2SeO_4(cr) in water at 298.15 K to be 0.0588 mol·kg^{-1} as part of a study of the system Tl_2SeO_4-$MgSeO_4$-H_2O.

Selivanova [58SEL] calculated thermodynamic quantities for Tl_2SeO_4(cr) at 298.15 K from solubilities interpolated in an unknown way from data in the literature, see Appendix A. No experimental value of the solubility at 298.15 K was available then. As the solubility data surrounding the standard temperature were obtained at 293 and 353 K, her results were checked by calculating the solubility product from data available at 282.5, 285.2, and 293.2 K. Activity coefficients were estimated by SIT with $\varepsilon(Tl^+, SeO_4^{2-}) = \varepsilon(K^+, SeO_4^{2-}) = -0.06$ kg·mol^{-1}. The linear $\log_{10} K^\circ_{s,0} = f(1/T)$ plot was extrapolated to 298.15 K and the solubility product of the reaction

$$Tl_2SeO_4(cr) \rightleftharpoons 2Tl^+ + SeO_4^{2-} \quad (V.72)$$

was obtained to be $\log_{10} K^\circ_{s,0} = -3.90$. The value calculated from [89VOJ/EBE] is – 3.88 and:

$$\log_{10} K^\circ_{s,0}((V.72), 298.15\text{ K}) = -(3.90 \pm 0.15)$$

will be selected. The value proposed by Selivanova was – 4.00.

The standard Gibbs energy of formation of thallium(I) selenate is calculated from the solubility product and the Gibbs energies of the ions ($\Delta_f G^\circ_m(Tl^+, 298.15\text{ K}) = -(32.40 \pm 0.30)$ kJ·mol^{-1} [99RAR/RAN]) to be:

$$\Delta_f G^\circ_m(Tl_2SeO_4, \text{cr}, 298.15\text{ K}) = -(526.5 \pm 1.8)\text{ kJ·mol}^{-1}.$$

Selivanova calculated the enthalpy change of Reaction (V.72) from the interpolated solubilities in the range 283.15 to 313.15 K to be (43.10 ± 2.90) kJ·mol^{-1}. The review obtained the same result from the original data at the temperatures mentioned above, and (41.67 ± 2.95) kJ·mol^{-1} for the whole temperature range between 282.5 and 373.2 K. The value of $\Delta_r H^\circ_m((V.72), 298.15\text{ K})$ is uncertain as there is no experimental value of the solubility between 293.2 and 353.2 K. In addition, the data used originate from two sources. The selected value:

$$\Delta_r H^\circ_m((V.72), 298.15\text{ K}) = (43.0 \pm 5.0)\text{ kJ·mol}^{-1}$$

reflects the uncertainty and, with $\Delta_f H^\circ_m(Tl^+, 298.15\text{ K}) = (5.36 \pm 1.00)$ kJ·mol^{-1} in [82WAG/EVA], leads to

$$\Delta_f H_m^\circ(\text{Tl}_2\text{SeO}_4, \text{cr}, 298.15 \text{ K}) = -(635.8 \pm 6.4) \text{ kJ·mol}^{-1}.$$

The value reported in [82WAG/EVA] is – 632 kJ·mol^{-1}.

The data yield:

$$S_m^\circ(\text{Tl}_2\text{SeO}_4, \text{cr}, 298.15 \text{ K}) = (214.4 \pm 22.3) \text{ J·K}^{-1}\text{·mol}^{-1}.$$

Since the calculation of the formation data and entropy from the experimental data involved non-TDB auxiliary data they are included in Appendix E.

V.8.5.3.2 Tl(III)

Gospodinov and Barkov [92GOS/BAR] studied the system Tl_2O_3-H_2SeO_4-H_2O at 373.15 K. $(TlO)_2SeO_4$ is formed when the SeO_3 concentration in the liquid phase exceeds about 20 mass-%.

The standard Gibbs energy of formation of thallium(III) selenate is reported as $\Delta_f G_m^\circ(\text{Tl}_2(\text{SeO}_4)_3, \text{cr}, 298.15 \text{ K}) = -901.1 \text{ kJ·mol}^{-1}$ in [82WAG/EVA]. The source of this datum could not be traced and it has therefore not been selected. No other data have been found for this compound.

V.8.5.4 Thallium selenocyanates

Ramakrishna and Thuraisingham [73RAM/THU] studied the complex formation between Tl^+ and $SeCN^-$ at 300 K in 2 M $NaNO_3$ medium by polarographic measurements. The evaluation of the $E_{1/2}$ *vs.* [$SeCN^-$] data relied on graphical techniques and indicated the formation of three mononuclear complexes:

$$\text{Tl}^+ + \text{SeCN}^- \rightleftharpoons \text{TlSeCN(aq)} \quad \text{(V.73)}$$

$$\text{Tl}^+ + 2\text{SeCN}^- \rightleftharpoons \text{Tl(SeCN)}_2^- \quad \text{(V.74)}$$

$$\text{Tl}^+ + 3\text{SeCN}^- \rightleftharpoons \text{Tl(SeCN)}_3^{2-} \quad \text{(V.75)}$$

with $\log_{10} K$ (V.73) = 0.97; $\log_{10} K$ (V.74) = 0.88 and $\log_{10} K$ (V.75) = 1.24. The values of the constants form a somewhat unexpected sequence, which would partly jeopardise the evaluation method used. Of the three complexes, only the mono-selenocyanate species is formed in appreciable amounts within the allowed range for the constant ionic medium method, *i.e.*, a 10% replacement of the medium ions. Only the stability constant of this complex is therefore accepted by this review. Its value at $I = 0$ and 298.15 K has been estimated by SIT with $\varepsilon(\text{Tl}^+, \text{NO}_3^-) \approx \varepsilon(\text{Tl}^+, \text{ClO}_4^-) = -(0.21 \pm 0.10) \text{ kg·mol}^{-1}$ and $\varepsilon(\text{Na}^+, \text{SeCN}^-) \approx \varepsilon(\text{Na}^+, \text{SCN}^-) = (0.05 \pm 0.01) \text{ kg·mol}^{-1}$. A correction of the order of 0.01 logarithmic unit to 298.15 K from 300 K was made with a $\Delta_r H_m^\circ$ of – 10 kJ·mol^{-1} estimated from other Me-$SeCN^-$ systems. The result is

$$\log_{10} K^\circ((\text{V.73}), 298.15 \text{ K}) = (1.75 \pm 0.29).$$

The uncertainty limits include an assumed uncertainty in the original value of ± 0.20.

The selection corresponds to:

$$\Delta_r G_m^\circ((\text{V.73}), 298.15\ \text{K}) = -(10.0 \pm 1.7)\ \text{kJ}\cdot\text{mol}^{-1}.$$

Combined with the value for $\Delta_f G_m^\circ(\text{SeCN}^-, 298.15\ \text{K})$ selected in this review, and the [99RAR/RAN] value of $\Delta_f G_m^\circ(\text{Tl}^+, 298.15\ \text{K})$ ($-32.40\ \text{kJ}\cdot\text{mol}^{-1}$), this corresponds to:

$$\Delta_f G_m^\circ(\text{TlSeCN, aq}, 298.15\ \text{K}) = (93.7 \pm 4.2)\ \text{kJ}\cdot\text{mol}^{-1}.$$

V.9 Group 12 compounds and complexes

V.9.1 Zinc compounds

V.9.1.1 Zinc selenides

The zinc-selenium phase diagram was assessed and modelled by Sharma and Young [96SHA/CHA] and Brebrick and Liu [96BRE/LIU]. ZnSe is the only solid compound in the system and three different polymorphs have been characterised. According to [96SHA/CHA] the α form, which crystallises in the zinc blende (sphalerite) structure, is the stable form up to the melting point at 1799 K. The β form crystallises in the sodium chloride structure and is stable at high pressures only. The third form of ZnSe crystallises in the wurtzite structure and is metastable at all temperatures and pressures [96SHA/CHA]. In contrast to these statements, the DTA studies by Okada, Kawanaka, and Ohmoto [96OKA/KAW] indicate that α-ZnSe transforms to the wurtzite structure type at 1684 K.

Experimental thermodynamic information is only available for α-ZnSe and ZnSe(g).

V.9.1.1.1 α-ZnSe

The low temperature heat capacity of α-ZnSe has been measured by Irwin and LaCombe [74IRW/LAC], Birch [75BIR], and by Sirota, Petrova, and Sokolovski [80SIR/PET] in the temperature ranges 15 to 150 K, 2 to 25 K, and 4 to 300 K, respectively. The heat capacity measured in [74IRW/LAC] and [80SIR/PET] differ by $-(0.24 \pm 1.32)\ \text{J}\cdot\text{K}^{-1}\cdot\text{mol}^{-1}$ in the temperature range 15 to 150 K and in [80SIR/PET] the discrepancy was attributed to different degrees of pulverisation of the samples. The heat capacity at 298.15 K in [80SIR/PET] is selected by the review:

$$C_{p,m}^\circ(\text{ZnSe}, \alpha, 298.15\ \text{K}) = (47.74 \pm 1.50)\ \text{J}\cdot\text{K}^{-1}\cdot\text{mol}^{-1}.$$

The high temperature heat capacity was determined by Yamaguchi, Kameda, Takeda, and Itagaki [94YAM/KAM] from heat content measurements in the temperature range 800 to 1500 K using drop calorimetry. The analytical expression $C^{\circ}_{p,m}$ (ZnSe, α, (800 – 1500) K) = (50.4 + 5.94 × 10^{-3} T – 0.27 × 10^{5} T^{-2}) J·K^{-1}·mol^{-1} was derived for the heat capacity. This expression and the selected heat capacity at 298.15 K were combined by the review to give a heat capacity expression for the complete temperature range:

$$C^{\circ}_{p,m}(\text{ZnSe}, \alpha, (298.15-1500)\ \text{K}) = (53.155 + 2.6517 \times 10^{-3}\ T + 1.0789 \times 10^{-6}\ T^{2} - 5.5968 \times 10^{5}\ T^{-2})\ \text{J}\cdot\text{K}^{-1}\cdot\text{mol}^{-1}.$$

The expression was employed in the re-evaluations made by the review in all cases where the heat capacity of α-ZnSe was required at temperatures above 298.15 K.

The vaporisation of α-ZnSe takes place according to the reaction α-ZnSe $\rightleftharpoons$ Zn(g) + ½Se_2(g) and has been studied by different techniques in a number of investigations. The methods of evaluation and the auxiliary data used vary and in order to obtain a consistent set of thermochemical quantities, all investigations were re-evaluated by the review using the second and third laws and an identical set of auxiliary data. This is discussed in Appendix A. The results are summarised in Table V-50 and Table V-51.

The Gibbs energy of formation of α-ZnSe was studied in the temperature range 360 to 420 K using the cell

(–) Zn(cr) | $ZnCl_2$(in glycerol) | α-ZnSe, Se(trig) (+)

by Terpilowski and Ratajczak [67TER/RAT] and in the temperature range 635 to 825 K using the cell

(–) Zn(cr, l) | 5 weight-% $ZnCl_2$ (in LiCl + KCl melt) | α-ZnSe, Se(l) (+)

by Nasar and Shamsuddin [90NAS/SHA]. Similarly to the vaporisation studies, these investigations were re-evaluated by the review using the second and third laws and an identical set of auxiliary data as discussed in Appendix A. The results are included in Table V-50 and Table V-51.

A third law value of the entropy at 298.15 K, S°_{m} (ZnSe, α, 298.15 K) = (71.94 ± 1.00) J·K^{-1}·mol^{-1}, was derived by the review from the low temperature heat capacity measurements in [80SIR/PET] as discussed in Appendix A. The second law entropies are widely scattered, but when obviously poor measurements are neglected, the weighted average given in Table V-50 agrees well with the third law value. The latter value is selected:

$$S^{\circ}_{m}(\text{ZnSe}, \alpha, 298.15\ \text{K}) = (71.94 \pm 1.00)\ \text{J}\cdot\text{K}^{-1}\cdot\text{mol}^{-1}.$$

Table V-50: Second law entropies of α-ZnSe at 298.15 K evaluated by the review as discussed in Appendix A. Values denoted by (*) were not included in the average.

Reference	S°_{m} (ZnSe, α, 298.15 K) ($J \cdot K^{-1} \cdot mol^{-1}$)	Method	Temperature (K)
[60KOR/SOK]	(31.6 ± 20.0) (*)	Knudsen effusion weight loss	913 – 1093
[62WOS/GEE]	(58.9 ± 14.0)	Gas flow method	1060 – 1393
[66SED/AGU]	(62.2 ± 20.0)	Static method	900 – 1200
[67TER/RAT]	(69.9 ± 6.0)	Electrochemical cell	360 – 420
[69BOE/BEN]	– (58 ± 43) (*)	Gas flow method	1150 – 1240
[69FLO]	(54.9 ± 12.0)	Gas flow method	1173 – 1413
[73HAS/MUN]	(88.7 ± 12.0)	Knudsen torsion effusion	952 – 1209
[76KIR/RAV]	(86.2 ± 9.0)	Mass spectrometry and Knudsen effusion weight loss	923 – 1123
[78ROL/HOC]	(25.4 ± 16.0) (*)	Mass spectrometry and Knudsen effusion weight loss	875 – 1100
[90BAR/TRI]	(66.1 ± 9.0)	Knudsen effusion weight loss	1025 – 1288
[90NAS/SHA]	(59.4 ± 12.0)	Electrochemical cell	635 – 693
[90NAS/SHA]	(63.9 ± 9.0)	Electrochemical cell	693 – 825
[96BRE/LIU2]	(74.0 ± 6.0)	Optical density	1260 – 1410
[96SCH/FRE]	(68.1 ± 14.0)	Knudsen effusion weight loss	1190 – 1310
Weighted average	(70.4 ± 2.8)		

In addition to the previously mentioned measurements of the enthalpy of formation, determinations have also been made by direct synthesis DTA calorimetry by Charlot, Tikhomiroff , and Lafitte [70CHA/TIK] and in [90NAS/SHA]. The results are included in Table V-51.

The weighted average of the second and third law values of the enthalpy of formation is selected:

$$\Delta_f H^{\circ}_{m} (\text{ZnSe}, \alpha, 298.15 \text{ K}) = -(176.0 \pm 4.0) \text{ kJ} \cdot \text{mol}^{-1}.$$

The error limits were increased in order to reflect that the error limits of the second and third law values do not overlap.

The Gibbs energy of formation is calculated from the selected enthalpy of formation and the standard entropies to be:

$$\Delta_f G^{\circ}_{m} (\text{ZnSe}, \alpha, 298.15 \text{ K}) = -(172.5 \pm 4.0) \text{ kJ} \cdot \text{mol}^{-1}.$$

Table V-51: Second and third law enthalpies of formation of α-ZnSe at 298.15 K evaluated by the review as discussed in Appendix A. Values denoted by (*) were not included in the average.

Reference	$\Delta_f H_m^\circ$ (ZnSe, α, 298.15 K) (kJ·mol^{-1})		Temperature (K)
	2nd law	3rd law	
[60KOR/SOK]	– (215.9 ± 35.0) (*)	– (175.2 ± 12.0) (*)	913 – 1093
[62WOS/GEE]	– (198.6 ± 19.0)	– (182.8 ± 6.0)	1060 – 1393
[66SED/AGU]	– (185.8 ± 16.0)		900 – 1200
[67TER/RAT]	– (166.1 ± 6.0)		360 – 420
[69BOE/BEN]	– (383 ± 51) (*)	– (193.0 ± 18.0) (*)	1150 – 1240
[69FLO]	– (192.5 ± 23.0)	– (170.5 ± 4.9)	1173 – 1413
[70CHA/TIK]	– (156.0 ± 12.0)		300 – 800
[73HAS/MUN]	– (157.8 ± 13.0)	– (175.9 ± 6.6)	952 – 1209
[76KIR/RAV]	– (164.5 ± 6.0)	– (179.1 ± 5.2)	923 – 1123
[78ROL/HOC]	– (215.5 ± 19.0) (*)	– (169.7 ± 15.0) (*)	875 – 1100
[90BAR/TRI]	– (183.4 ± 14.0)	– (181.8 ± 7.0)	1025 – 1288
[90NAS/SHA]	– (177.5 ± 6.0)	– (177.3 ± 4.2)	635 – 693
[90NAS/SHA]	– (175.0 ± 6.0)	– (177.5 ± 4.2)	693 – 825
[90NAS/SHA]	– (177.1 ± 8.0)		707
[96BRE/LIU2]	– (175.4 ± 8.0)	– (178.2 ± 3.6)	1260 – 1410
[96SCH/FRE]	– (182.4 ± 14.0)	– (177.6 ± 3.8)	1190 – 1310
Weighted average	– (172.5 ± 2.4)	– (177.5 ± 1.6)	

In the aqueous phase, Umland and Wallmeier [80UML/WAL] studied the reduction of selenite at the mercury electrode in the presence of Zn^{2+} by polarography, see Appendix A. The solubility product of ZnSe(s) was obtained from the position of the half-wave potential of the second reduction step, HgSe(s) + Zn^{2+} + 2e$^-$ $\rightleftharpoons$ ZnSe(s) + Hg(l), to be $\log_{10} K_{s,0}^\circ$ (ZnSe, s, 298.15 K) = – (23.2 ± 2.0). Combined with CODATA for Zn^{2+} and the selected value for Se^{2-}, it corresponds to $\Delta_f G_m^\circ$ (ZnSe, s, 298.15 K) = – (151.0 ± 11.8) kJ·mol^{-1}. The value is in disagreement with the value obtained by extrapolation of high temperature data. Since the validity of the polarographic method has not been documented, the value of $\Delta_f G_m^\circ$ (ZnSe, α, 298.15 K) obtained from the extrapolation of the high temperature data has been selected.

A comparison between $\log_{10} K^\circ$ for the reaction α-ZnSe $\rightleftharpoons$ Zn(g) + ½Se_2(g) calculated from the selected thermodynamic quantities and the experimental results is made in Figure V-17.

Figure V-17: The difference between calculated and experimental values of $\log_{10} K$ for the reaction α-ZnSe $\rightleftharpoons$ Zn(g) + ½Se_2(g). The calculated values were derived from the selected thermodynamic quantities.

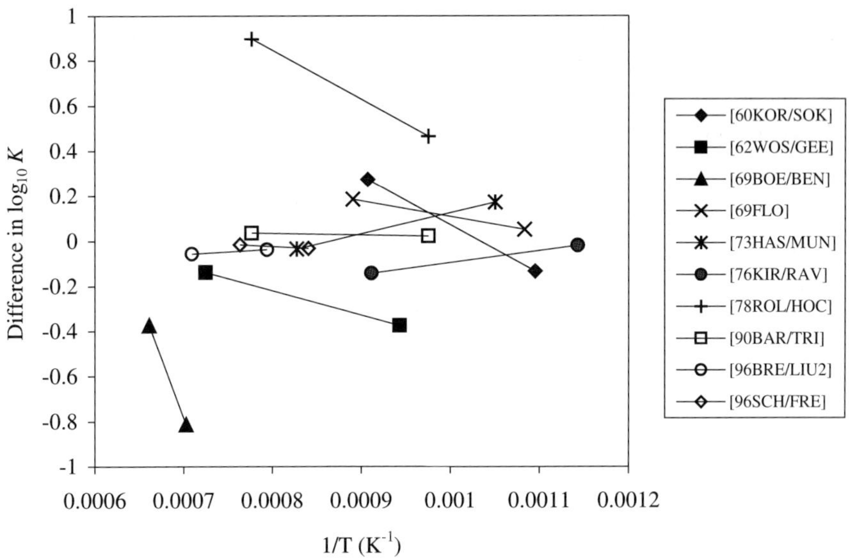

The enthalpy of fusion of α-ZnSe was determined to be $\Delta_{\text{fus}}H^{\circ}_{\text{m}}$ (ZnSe, α, 1788 K) = (66.9 ± 5.0) kJ·mol^{-1} by Lakeenkov and Pelevin [77LAK/PEL].

After the assessment of the thermodynamic properties of α-ZnSe was made by the review, the heat capacity measurements made by Pashinkin and Malkova [2003PAS/MAL] have appeared. The measurements were performed in the temperature range 400 to 600 K and fill a gap in the experimental values. The new values of the heat capacity are between 0.9 to 1.7 J·K^{-1}·mol^{-1} smaller than the corresponding values calculated from the heat capacity expression used by the review. On combining the new results with the previous measurements in [94YAM/KAM] for the temperature range 800 to 1500 K, a heat capacity expression results which for typical temperatures (800 to 1300 K) of the measurements in Table V-51 yields enthalpy and entropy increments smaller by about 0.5 kJ·mol^{-1} and 1.1 J·K^{-1}·mol^{-1}, respectively. If the new results had been taken into account, the second and third law values of the enthalpy of formation in Table V-51 would have been about 0.5 kJ·mol^{-1} less negative and about 0.5 kJ·mol^{-1} more negative, respectively. The difference between the second and the third law results would thus have increased. Similarly, the second law entropies in Table V-50 would have been larger by about 1.1 J·K^{-1}·mol^{-1}.

V.9.1.1.2 ZnSe(g)

The mass spectrometric ion intensity of $ZnSe^+$(g) was measured in the temperature range 1240 to 1376 K in the α-ZnSe evaporation studies made by De Maria, Goldfinger, Malaspina, and Piacente [65MAR/GOL]. The ion intensity was of the order of 10^4 to 10^5 times smaller than the ion intensities of Zn^+(g) and Se_2^+(g). The two simultaneous equilibria α-ZnSe $\rightleftharpoons$ Zn(g) + ½Se_2(g) and Zn(g) + ½Se_2(g) $\rightleftharpoons$ ZnSe(g) were combined with the pressure independent equilibrium ZnSe(g) $\rightleftharpoons$ ⅓α-ZnSe + ⅔Zn(g) + ⅓Se_2(g). The data were evaluated by the third law using Gibbs energy functions calculated with estimated molecular parameters for ZnSe(g). The result was presented as the atomisation enthalpy of ZnSe(g), $\Delta_{at}H_m^\circ$ (ZnSe, g, 0 K) = (132.6 ± 13.6) kJ·mol^{-1}.

The experimental work was evaluated by the review using the selected properties of α-ZnSe and Se_2(g), the properties of Zn(g) in [89COX/WAG], and thermal functions of ZnSe(g) calculated from the estimated molecular parameters in [65MAR/GOL] yielding $\Delta_f H_m^\circ$ (ZnSe, g, 298.15 K) = (225.7 ± 50.0) kJ·mol^{-1} from the second law and $\Delta_f H_m^\circ$ (ZnSe, g, 298.15 K) = (241.5 ± 20.0) kJ·mol^{-1} from the third law. The weighted average is (239.3 ± 18.6) kJ·mol^{-1}. The corresponding dissociation enthalpy is $\Delta_{at}H_m^\circ$ (ZnSe, g, 298.15 K) = (127.2 ± 19.0) kJ·mol^{-1}.

The derived value of the enthalpy of formation of ZnSe(g) is selected:

$$\Delta_f H_m^\circ (\text{ZnSe, g, 298.15 K}) = (239.3 \pm 20.0)\ \text{kJ·mol}^{-1}$$

since the additional uncertainty originating from the recalculation to standard conditions is judged to be moderate compared to the estimated experimental uncertainty, which has been increased to include a possible error from the calculation of the equilibrium constant from the ion intensities.

V.9.1.2 Zinc selenites

The system $ZnSeO_3$-SeO_2-H_2O has been investigated by solubility measurements at 273.15 and 298.15 K by Ebert, Micka, and Uchytilova [84EBE/MIC]. They identified the solid phases $ZnSeO_3$ and $Zn(HSeO_3)_2 \cdot 2H_2O$ at 273.15 K, and $ZnSeO_3$ and $ZnSe_2O_5$ at 298.15 K in the phase diagrams. The formation of a hydrogen selenite and a diselenite, respectively, was confirmed by infrared spectroscopy. No primary solubility data are published. A similar study by Gospodinov [84GOS3] performed at 373.15 K in the concentration range 1.2 to 80.1 mass-% SeO_2 in the liquid phase yielded $ZnSeO_3$ and $ZnSeO_3 \cdot H_2SeO_3$. The empirical formula corresponds to either $Zn(HSeO_3)_2$ or $ZnSe_2O_5 \cdot H_2O$. A selection between these alternatives can not be made from the presented data, which include the results of the solubility measurements.

V.9.1.2.1 $ZnSeO_3$(cr)

Zinc selenite exits in two crystalline forms, α- and β-$ZnSeO_3$. The α form appears to be metastable at ordinary temperatures. Sapozhnikov and Markovskii [66SAP/MAR] determined the enthalpies of dissolution of α- and β-$ZnSeO_3$ in mineral acid at 298.15 K. These measurements yielded $\Delta_{trs}H_m^\circ$($ZnSeO_3$, α, 298.15 K) = – (15.82 ± 1.41) $kJ\cdot mol^{-1}$. Gospodinov and Bogdanov [83GOS/BOG] studied the behaviour of α-$ZnSeO_3$ on heating. A transition of the α form to the β form occurs at (600 ± 5) K according to measurements with differential scanning calorimetry. The enthalpy of the transition observed was $\Delta_{trs}H_m^\circ$($ZnSeO_3$, α, 600 K) = (13.22 ± 0.13) $kJ\cdot mol^{-1}$. Fusion occurs at (913 ± 5) K with $\Delta_{fus}H_m^\circ$($ZnSeO_3$, β, 913 K) = (31.21 ± 0.13) $kJ\cdot mol^{-1}$. The uncertainties in the temperatures have been added by the review. The two measurements of the transition enthalpy of α-$ZnSeO_3$ are difficult to reconcile and the review selects:

$$\Delta_{trs}H_m^\circ(\mathrm{ZnSeO_3}, \alpha, 298.15\ \mathrm{K}) = -(15.82 \pm 1.41)\ \mathrm{kJ\cdot mol^{-1}},$$

$$\Delta_{fus}H_m^\circ(\mathrm{ZnSeO_3}, \beta, 913\ \mathrm{K}) = (31.21 \pm 0.13)\ \mathrm{kJ\cdot mol^{-1}}.$$

Bakeeva, Buketov, and Bakeev [64BAK/BUK], and Polukarov and Makhmetov [72POL/MAK] measured the dissociation pressure of β-$ZnSeO_3$ by the dew point method. The dew point temperatures measured in the latter reference are available in the deposited VINITI document (No. 3973-72). There are inconsistencies in the primary data in [64BAK/BUK]. The dew point temperatures derived from the pressures obtained with the vapour pressure relationship in [44AME/BEL] in the paper were deemed most reliable. The results obtained after recalculation with the accepted relationship for the saturated vapour pressure were:

$$\beta\text{-}\mathrm{ZnSeO_3} \rightleftharpoons \mathrm{ZnO(cr)} + \mathrm{SeO_2(g)} \qquad (V.76)$$

[64BAK/BUK]: $\log_{10} p_{SeO_2}$ ((V.76), (738 – 881) K) = (9.80 ± 0.63) – (9060 ± 510) T^{-1};

[72POL/MAK]: $\log_{10} p_{SeO_2}$ ((V.76), (742 – 892) K) = (8.52 ± 0.15) – (8239 ± 115) T^{-1};

$$\mathrm{ZnSeO_3(l)} \rightleftharpoons \mathrm{ZnO(cr)} + \mathrm{SeO_2(g)} \qquad (V.77)$$

[64BAK/BUK]: $\log_{10} p_{SeO_2}$ ((V.77), (901 – 997) K) = (8.47 ± 0.86) – (7875 ± 805) T^{-1};

[72POL/MAK]: $\log_{10} p_{SeO_2}$ ((V.77), (920 – 992) K) = (7.23 ± 0.43) – (7130 ± 430) T^{-1}.

The agreement between the two investigations is poor, particularly at high temperatures. The results for Reaction (V.77) will not be accepted.

An extrapolation of the enthalpy and entropy changes derived from the vapour pressure expressions to standard conditions was made with $\Delta_r C_{p,m}^\circ$ = (24.6 – 47.0 × 10^{-3} T – 19.0 × 10^5 T^{-2}) $J\cdot K^{-1}\cdot mol^{-1}$calculated from data in [73BAR/KNA] and resulted in, respectively, $\Delta_f H_m^\circ$($ZnSeO_3$, β, 298.15 K) = – (639.5 ± 10.1) and – (623.9 ± 3.4) $kJ\cdot mol^{-1}$, and S_m°($ZnSeO_3$, β, 298.15 K) = (110.0 ± 12.1) and (134.3 ± 2.9) $J\cdot K^{-1}\cdot mol^{-1}$. The value of $\Delta_f H_m^\circ$ of β-$ZnSeO_3$ calculated in Appendix A from [66SAP/MAR] is – (652.7 ± 10.9) $kJ\cdot mol^{-1}$. The review selects the calorimetric result from [66SAP/MAR]:

$$\Delta_f H_m^\circ(\mathrm{ZnSeO_3}, \alpha, 298.15\ \mathrm{K}) = -(636.8 \pm 11.3)\ \mathrm{kJ\cdot mol^{-1}},$$

$$\Delta_f H^\circ_m (\text{ZnSeO}_3, \beta, 298.15 \text{ K}) = -(652.7 \pm 10.9) \text{ kJ·mol}^{-1}.$$

An estimate of the standard entropy of β-$ZnSeO_3$ is calculated as the average from [64BAK/BUK] and [72POL/MAK] to be S°_m ($ZnSeO_3$, β, 298.15 K) = (122.2 ± 15.0) $J·K^{-1}·mol^{-1}$. Being an estimate this value does not qualify for selection.

The heat capacity of $ZnSeO_3$ was determined by Gospodinov and Bogdanov [83GOS/BOG] in the temperature range 350 to 500 K to be $C^\circ_{p,m} = (-180.2 + 1.166\,T - 8.37 \times 10^{-4}\,T^2)$ $J·K^{-1}·mol^{-1}$. This result is not accepted since the value of $C^\circ_{p,m}$ and its temperature variation appear to be too large.

V.9.1.2.2 $ZnSeO_3·H_2O$(cr)

The phase diagram studies show that anhydrous zinc selenite is the phase in equilibrium with the saturated solution at 298.15 K. However, $ZnSeO_3·H_2O$ is precipitated from dilute solution below about 330 K [64LES/SEL]. Above this temperature β-$ZnSeO_3$ is formed. For his solubility studies Chukhlantsev [56CHU] prepared zinc selenite by mixing a 0.1 M sodium selenite solution with the equivalent amount of a zinc sulphate solution at 323 to 333 K. The precipitate crystallised on prolonged keeping in the mother liquor. Chemical analysis confirmed the 1:1 ratio between Zn and Se(IV). No X-ray diffraction measurements were performed. It is thus not known whether the specimen investigated was $ZnSeO_3·H_2O$ or β-$ZnSeO_3$. Its solubility in dilute solution of nitric or sulphuric acid was measured at 293 K. The data have been recalculated as described in Appendix A, [56CHU]. The result for

$$\text{ZnSeO}_3[\cdot\text{H}_2\text{O}](\text{cr}) \rightleftharpoons \text{Zn}^{2+} + \text{SeO}_3^{2-} + [\text{H}_2\text{O(l)}] \qquad (\text{V.78})$$

was $\log_{10} K^\circ_{s,0}$ ((V.78), 293.15 K) = – (7.25 ± 0.29). The reported conditional constant is – (6.59 ± 0.22). The [H_2O] notation is used to point out the uncertainty about the composition of the solid phase.

As described in Appendix A, Ripan and Vericeanu [68RIP/VER] studied the solubility of zinc selenite by conductivity measurements. A calculation based on the equilibrium constants in [76BAE/MES] and the total concentrations provided in the paper, mean 1.40×10^{-4} M, leads to the conclusion that about 30% of dissolved Zn(II) would be present in hydrolysed forms, if the solvent were conductivity water. No reliable solubility product can therefore be derived from the conductivity data. The value $\log_{10} K^\circ_{s,0}$ = – (7.71 ± 0.05) was estimated by Masson, Lutz and Engelen [86MAS/LUT] from the data, but the composition of the solid phase is apparently not known.

Sharmasakar, Reddy, and Vance [96SHA/RED] obtained $\log_{10} K^\circ_{s,0}$ ((V.78), 298.2 K) = – 7.70 from measurements of the solubility of crystalline $ZnSeO_3·H_2O$ in aqueous media. For reasons presented in Appendix A, this result has been given zero weight.

From the thermodynamic data for β-$ZnSeO_3$ in the previous sub-Section $\log_{10} K^{\circ}_{s,0}$ ($ZnSeO_3$, β, 298.15 K) = – (11.0 ± 2.0) is obtained.

In view of the scarce and conflicting information available the review cannot select a value of the solubility product of zinc selenite, but $\log_{10} K^{\circ}_{s,0}$ ((V.78), 298.15 K) ≤ – 7.25 is indicated for an aged precipitate of zinc selenite.

Leshchinskaya and Selivanova [64LES/SEL] measured the enthalpy change of the reaction $ZnSO_4{\cdot}7H_2O$(cr) + Na_2SeO_3(aq, 1:1000) → $ZnSeO_3{\cdot}H_2O$(cr) + Na_2SO_4(aq, 1:1000) + $6H_2O$(l) in an electrically calibrated calorimeter. The formation of $ZnSeO_3{\cdot}H_2O$(cr) appears well established although the equilibrium studies cited above indicate that the anhydrous form would be the stable phase under the conditions of the calorimetric experiment. The standard enthalpy of formation of $ZnSeO_3{\cdot}H_2O$(cr) is evaluated from the data in Appendix A to be – (931.01 ± 2.73) kJ·mol^{-1}. Sapozhnikov and Markovskii [66SAP/MAR] measured the enthalpy change of the dissolution of $ZnSeO_3{\cdot}H_2O$ in H_2SO_4(aq, 1:50). The review calculates $\Delta_f H^{\circ}_m$ = – (943.9 ± 10.5) kJ·mol^{-1} from their data, see Appendix A. The review adopts the value obtained in [64LES/SEL]:

$$\Delta_f H^{\circ}_m (ZnSeO_3{\cdot}H_2O, cr, 298.15 K) = -(931.0 \pm 2.7) kJ{\cdot}mol^{-1},$$

in close agreement with the evaluation in [82WAG/EVA]. It is included in Appendix E as it was partly calculated using non selected auxiliary data.The other thermodynamic data in this reference for $ZnSeO_3{\cdot}H_2O$ are apparently based on [56CHU].

V.9.1.2.3 $Zn(HSeO_3)_2{\cdot}2H_2O$(cr) and $ZnSe_2O_5$(cr)

The standard enthalpies of formation of $Zn(HSeO_3)_2{\cdot}2H_2O$(cr) and $ZnSe_2O_5$(cr) are selected from [66SAP/MAR] in Appendix A to be:

$$\Delta_f H^{\circ}_m (Zn(HSeO_3)_2{\cdot}2H_2O, cr, 298.15 K) = -(1775.3 \pm 25.0) kJ{\cdot}mol^{-1}$$

and

$$\Delta_f H^{\circ}_m (ZnSe_2O_5, cr, 298.15 K) = -(888.3 \pm 25.0) kJ{\cdot}mol^{-1}.$$

V.9.1.3 Zinc selenates

A table and a graph of the solubility of zinc selenate as a function of temperature were established by Klein [40KLE]. He found that the hexahydrate was the stable phase below 308 K, the pentahydrate between 308 and 317 K, the monohydrate between 317 and 335.7 K, and the anhydrous salt above 335.7 K. The solubility of $ZnSeO_4{\cdot}6H_2O$(cr) at 298.15 K has been recorded in [84GOS], [87VOJ/EBE3], [93OJK/STO]. The reported composition of the saturated solution in mass-% recalculated to the molality of dissolved $ZnSeO_4$ is 3.189, 3.256, and 3.053 mol·kg^{-1}, respectively. The mean of the three values is selected:

$$m(ZnSeO_4{\cdot}6H_2O(cr), aq\ sat, 298.15 K) = (3.17 \pm 0.20) mol(ZnSeO_4){\cdot}kg^{-1}.$$

A phase diagram and related information can be found for the systems: $ZnSeO_4$-H_2SeO_4-H_2O at 298.15 and 373.15 K in [84GOS]; $ZnSeO_4$-$NiSeO_4$-H_2O at 298.15 K in [93OJK/STO]; $ZnSeO_4$-$CoSeO_4$-H_2O at 298.15 K in [94OJK/STO]; $ZnSeO_4$-$MgSeO_4$-H_2O at 298.15 K in [90OJK/BAL]; $ZnSeO_4$-$(NH_4)_2SeO_4$-H_2O at 298.15 K in [87VOJ/EBE3]; $ZnSeO_4$-Na_2SeO_4-H_2O at 298.15 K in [91OYK/MIH]; $ZnSeO_4$-K_2SeO_4-H_2O at 298.15 K in [87VOJ/EBE].

The value of solubility product of $ZnSeO_4{\cdot}6H_2O(cr)$ defined by the reaction:

$$ZnSeO_4{\cdot}6H_2O(cr) \rightleftharpoons Zn^{2+} + SeO_4^{2-} + 6H_2O(l) \qquad (V.79)$$

at 298.15 K has been reported by Christov [97CHR] to be $\log_{10} K^{\circ}_{s,0} = -1.572$ and by Kumov and Batyreva [90KUM/BAT] to be -1.49, see Appendix A. The results are based on the same isopiestic data by Ojkova and Staneva [89OJK/STA]. The former value will be selected as it has been evaluated with the Pitzer ion interaction approach. With the selected solubility of zinc selenate this value is recalculated to be:

$$\log_{10} K^{\circ}_{s,0}\ ((V.79),\ 298.15\ K) = -(1.538 \pm 0.068).$$

The uncertainty added by the review includes an uncertainty in the mean activity coefficient of 5%.

Banks [34BAN] made careful measurements of the conductivity of zinc selenate solutions at 298.15 K. The concentration range used was approximately 2×10^{-4} to 1×10^{-3} M. On the assumption that only $ZnSeO_4(aq)$ was formed the data were evaluated by an iterative procedure in which the inter-ionic attraction was corrected for using the Debye-Hückel (activity coefficient) and Onsager (ionic mobility) equations. The result for:

$$Zn^{2+} + SeO_4^{2-} \rightleftharpoons ZnSeO_4(aq) \qquad (V.80)$$

was $\log_{10} \beta_1^{\circ}$ ((V.80), 298.15 K) = (2.19 ± 0.06). The uncertainty added is twice the standard deviation derived from 8 measurements.

Moriya and Sekine [74MOR/SEK] studied the complex formation by a chelate extraction method and found evidence for the formation of $Zn(SeO_4)_2^{2-}$ also. The measurements were made at 298.15 K in 1 M $NaClO_4$ with the following results, $\log_{10} \beta_1$ ((V.80), 1 M $NaClO_4$, 298.15 K) = (0.71 ± 0.05) and for:

$$Zn^{2+} + 2SeO_4^{2-} \rightleftharpoons Zn(SeO_4)_2^{2-} \qquad (V.81)$$

$\log_{10} \beta_2$ ((V.81), 1 M $NaClO_4$, 298.15 K) = (1.34 ± 0.05). The uncertainties have been added here. The value of $\log_{10} \beta_1$ at $I = 0$ has been estimated with $\varepsilon(Zn^{2+}, ClO_4^-) = 0.33$ and $\varepsilon(SeO_4^{2-}, Na^+) \approx \varepsilon(SO_4^{2-}, Na^+) = -0.12$ from [92GRE/FUG] to be (2.13 ± 0.09) in good agreement with the value from the conductivity measurement. The mean of the two results will be selected:

$$\log_{10} \beta_1^{\circ}\ ((V.80),\ 298.15\ K) = (2.16 \pm 0.06).$$

The correspondingly extrapolated value of β_2 with $\varepsilon(\text{Zn(SeO}_4)_2^{2-}, \text{Na}^+) \approx \varepsilon(\text{SO}_4^{2-}, \text{Na}^+)$ and $\Delta\varepsilon = -(0.21 \pm 0.11)$ results in $\log_{10} \beta_2^\circ$ (V.81) = (2.76 ± 0.12). This value will not be selected due to the lack of data supporting the extrapolation.

Aruga [78ARU] made calorimetric titrations of 0.17 M zinc nitrate solution with 0.17 M tetraethylammonium selenate, see Appendix A. The concentration of $ZnSeO_4(aq)$ formed was calculated with the equilibrium constant determined in [34BAN] corrected to I = 0.5 M by Davies' equation. After experimentally determined corrections for heats of dilution had been applied, the enthalpy change of Reaction (V.80) was calculated to be $\Delta_r H_m$((V.80), I = 0.5 M, 298.15 K) = (0.20 ± 0.05) kJ·mol^{-1}. The recalculation of the equilibrium constant to I = 0.5 M is uncertain. The uncertainty assigned by the author to $\Delta_r H_m^\circ$ has therefore been increased to include the uncertainty in the value of the equilibrium constant.

The calculated difference between the enthalpy change at I = 0 and 0.5 M was 4.05 kJ·mol^{-1}, while the estimate by the review is (4.40 ± 0.2) kJ·mol^{-1}. With the latter value,

$$\Delta_r H_m^\circ((\text{V.80}), 298.15\ \text{K}) = (4.6 \pm 4.0)\ \text{kJ·mol}^{-1}$$

is obtained and selected. The error limits are estimated here as discussed in Appendix A.

This selection for Reaction (V.80) results in:

$$\Delta_f G_m^\circ(\text{ZnSeO}_4(\text{aq}), 298.15\ \text{K}) = -(599.0 \pm 1.5)\ \text{kJ·mol}^{-1},$$

$$\Delta_f H_m^\circ\ \text{ZnSeO}_4(\text{aq}), 298.15\ \text{K}) = -(752.3 \pm 5.3)\ \text{kJ·mol}^{-1},$$

$$S_m^\circ(\text{ZnSeO}_4(\text{aq}), 298.15\ \text{K}) = -(20.1 \pm 18.5)\ \text{J·K}^{-1}\text{·mol}^{-1}.$$

Selivanova, Maier, and Luk'yanova [63SEL/MAI3] have investigated the enthalpies of formation of zinc selenates. They prepared crystalline $ZnSeO_4{\cdot}6H_2O$, $ZnSeO_4{\cdot}H_2O$, and $ZnSeO_4$ and measured their enthalpies of dissolution to (aq, 1:1200). The investigation also included a determination of the enthalpy change of the reaction $ZnSeO_4{\cdot}H_2O(cr) + BaCl_2(aq, 1{:}1200) \rightarrow BaSeO_4(cr) + ZnCl_2(aq, 1{:}1200) + H_2O(l)$. The review has recalculated the data in Appendix A and arrived at the following results:

$$\text{ZnSeO}_4{\cdot}\text{H}_2\text{O(cr)} + \text{Ba}^{2+} \rightarrow \text{BaSeO}_4\text{(cr)} + \text{Zn}^{2+} + \text{H}_2\text{O(l)} \qquad \text{(V.82)}$$

$$\Delta_r H_m^\circ((\text{V.82}), 298.15\ \text{K}) = -(65.35 \pm 0.40)\ \text{kJ·mol}^{-1},$$

$$\Delta_f H_m^\circ(\text{ZnSeO}_4, \text{cr}, 298.15\ \text{K}) = -(664.7 \pm 5.6)\ \text{kJ·mol}^{-1},$$

$$\Delta_f H_m^\circ(\text{ZnSeO}_4{\cdot}\text{H}_2\text{O}, \text{cr}, 298.15\ \text{K}) = -(983.1 \pm 5.6)\ \text{kJ·mol}^{-1},$$

$$\Delta_f H_m^\circ(\text{ZnSeO}_4{\cdot}6\text{H}_2\text{O}, \text{cr}, 298.15\ \text{K}) = -(2458.5 \pm 5.6)\ \text{kJ·mol}^{-1}.$$

The values of the standard enthalpies of formation overlap those reported in [82WAG/EVA]. It is observed that the selected enthalpies of formation lead to an unexpectedly large enthalpy of dilution of $ZnSeO_4$(aq, 1:1200) to the aqueous standard state as compared with other selenates (and sulphates) of divalent metal ions.

The standard Gibbs energy of formation of $ZnSeO_4{\cdot}6H_2O$(cr) is calculated from the Gibbs energy of Reaction (V.79) and the selected values of $\Delta_f G_m^\circ$ for the ions and water to be:

$$\Delta_f G_m^\circ(\text{ZnSeO}_4{\cdot}6\text{H}_2\text{O, cr, 298.15 K}) = -(2018.3 \pm 1.5)\ \text{kJ}{\cdot}\text{mol}^{-1}.$$

V.9.1.4 Zinc selenocyanates

Aqueous complexation between Zn^{2+} and $SeCN^-$ has been studied in two works, a careful and comprehensive potentiometric and calorimetric study by Ahrland, Avsar, and Kullberg [74AHR/AVS], and a rather restricted polarographic study by Humffray, Bond, and Forrest [67HUM/BON]. The latter study is based on 6 experimental data only and solutions containing high concentrations of potassium selenocyanate, 0.2 to 1.4 M, in addition to a 2 M potassium nitrate background. The activity coefficients of the species formed cannot, therefore, be considered constant in these solutions, and the evaluated equilibrium constants ($\log_{10} K$ (V.83) = 0.76, $\log_{10} K$ (V.84) = 1.0 at 303 K) are accordingly not accepted by the review.

In the former study, at 298.15 K and 1 M $NaClO_4$, potentiometric (zinc amalgam electrode) titration data were used to conclude that two mononuclear complexes were formed:

$$\text{Zn}^{2+} + \text{SeCN}^- \rightleftharpoons \text{ZnSeCN}^+ \qquad \text{(V.83)}$$

$$\text{Zn}^{2+} + 2\text{SeCN}^- \rightleftharpoons \text{Zn(SeCN)}_2\text{(aq)}. \qquad \text{(V.84)}$$

The corresponding equilibrium constants were evaluated from a numerical minimisation of the function $\Sigma(E_{\text{calc}} - E_{\text{obs}})^2$ with the result:

$$\log_{10} K\,((\text{V.83}), I = 1\ \text{M NaClO}_4, 298.15\ \text{K}) = (0.44 \pm 0.02)$$

$$\log_{10} K\,((\text{V.84}), I = 1\ \text{M NaClO}_4, 298.15\ \text{K}) = (0.64 \pm 0.04).$$

These constants have been extrapolated to standard conditions by SIT using the $\Delta\varepsilon$ values suggested in [97BAH/PAR] in Appendix A to the selected values:

$$\log_{10} K^\circ((\text{V.83}), 298.15\ \text{K}) = (1.21 \pm 0.06),$$

$$\log_{10} K^\circ((\text{V.84}), 298.15\ \text{K}) = (1.68 \pm 0.11).$$

In the evaluation of calorimetric data, the stability constants were treated as "known" and reaction enthalpies were found by a least squares optimisation of the function $\Sigma(Q_{\text{calc}} - Q_{\text{obs}})^2$. Q_{calc} and Q_{obs} denote the calculated and observed enthalpy changes in the calorimetric titration, respectively. The review estimates that the ionic strength dependence of the reported values is small and the values selected at $I = 0$ are therefore assumed to be the same as those reported in [74AHR/AVS] for 1 M $NaClO_4$. The uncertainties reported there are doubled by this review to account for the uncertainty in this assumption. The selected result is:

$$\Delta_r H_m^\circ((V.83), 298.15\ K) = -(5.95 \pm 0.34)\ kJ\cdot mol^{-1}$$

$$\Delta_r H_m^\circ((V.84), 298.15\ K) = -(6.35 \pm 1.20)\ kJ\cdot mol^{-1}$$

$$\Delta_r G_m^\circ((V.83), 298.15\ K) = -(6.91 \pm 0.34)\ kJ\cdot mol^{-1}$$

$$\Delta_r G_m^\circ((V.84), 298.15\ K) = -(9.59 \pm 0.63)\ kJ\cdot mol^{-1}$$

$$\Delta_r S_m^\circ((V.83), 298.15\ K) = (3.2 \pm 1.6)\ J\cdot K^{-1}\cdot mol^{-1}$$

$$\Delta_r S_m^\circ((V.84), 298.15\ K) = (10.9 \pm 4.5)\ J\cdot K^{-1}\cdot mol^{-1}.$$

The values selected for $\Delta_r G_m^\circ$ combines with the selected values of $\Delta_f G_m^\circ(\text{SeCN}^-, 298.15\ \text{K})$ and $\Delta_f G_m^\circ(\text{Zn}^{2+}, 298.15\ \text{K})$ to yield:

$$\Delta_f G_m^\circ(\text{ZnSeCN}^+, 298.15\ \text{K}) = -(18.1 \pm 3.8)\ \text{kJ}\cdot\text{mol}^{-1}$$

$$\Delta_f G_m^\circ(\text{Zn(SeCN)}_2, \text{aq}, 298.15\ \text{K}) = (115.3 \pm 7.6)\ \text{kJ}\cdot\text{mol}^{-1}.$$

V.9.2 Cadmium compounds

V.9.2.1 Cadmium selenides

The cadmium-selenium phase diagram was assessed and modelled by Sharma and Chang in [89SHA/CHA] and [96SHA/CHA3]. CdSe is the only solid compound in the system and two different polymorphs have been characterised. The α form, which crystallises in the wurtzite structure, is the stable form up to the melting point at 1537 K. The β form crystallises in the sodium chloride structure and is stable at high pressures only. The homogeneity range of α-CdSe is small, *i.e.* $x \le 0.002$ in CdSe_{1-x} [72SIG/WIE]. Experimental thermodynamic information is only available for α-CdSe.

V.9.2.1.1 α-CdSe

The low temperature heat capacity of α-CdSe has been measured by Demidenko [69DEM], Shaulov and Kostina [69SHA/KOS], Petkova and Kofman [76PET/KOF], and Sirota, Gavaleshko, Novikova, and Novikov [92SIR/GAV]. The heat capacity values and the corresponding third law entropies at 298.15 K are presented in Table V-52. The third law entropy derived from the measurements in [69DEM] was reported in the work of Demidenko and Maltsev [69DEM/MAL]. The entropy value in Table V-52 and attributed to [76PET/KOF] was derived by the review from the heat capacity reported in the paper.

Table V-52. Heat capacity and third law entropy of α-CdSe at 298.15 K. Original values including an estimated entropy term for the temperature range 0 to 50 K are denoted by (a), values corrected using the experimental mean entropy at 50 K from [76PET/KOF] and [92SIR/GAV] are denoted by (b), and values derived from estimated Debye-Einstein functions are denoted by (c).

Reference	$C^{\circ}_{p,m}$ (CdSe, α, 298.15 K) ($J \cdot K^{-1} \cdot mol^{-1}$)	S°_{m} (CdSe, α, 298.15 K) ($J \cdot K^{-1} \cdot mol^{-1}$)	S°_{m} (CdSe, α, 50 K) ($J \cdot K^{-1} \cdot mol^{-1}$)	T (K)
[69DEM]	(49.42 ± 0.4)			55 – 300
[69DEM/MAL]		(82.34 ± 2.1)$^{(a)}$, (85.28 ± 2.1)$^{(b)}$	(13.23 ± 1.7)$^{(c)}$	55 – 300
[69SHA/KOS]	(49.76 ± 0.3)	(84.19 ± 1.7)$^{(a)}$, (87.13 ± 1.7)$^{(b)}$	(13.89 ± 0.84)$^{(c)}$	55 – 310
[76PET/KOF]	(49.39 ± 0.3)	(86.76 ± 1.7)	(16.72 ± 0.4)	5 – 300
[92SIR/GAV]	(49.16 ± 0.6)	(85.60 ± 2.5)	(16.27 ± 0.5)	5 – 300

The heat capacity values at 298.15 K agree well and the selected value is:

$$C^{\circ}_{p,m}(\text{CdSe, α, 298.15 K}) = (49.43 \pm 0.48)\ \text{J·K}^{-1}\text{·mol}^{-1}.$$

The heat capacity above room temperature has been measured using differential scanning calorimetry in the temperature range 320 to 760 K by Sigai and Wiedemeier [72SIG/WIE] and in the temperature range 360 to 760 K by Glazov, Pashinkin, Malkova [89GLA/PAS].

The heat capacity was also derived from heat content measurements using drop calorimetry in the temperature range 800 to 1512 K by Yamaguchi, Kameda, Takeda, and Itagaki [94YAM/KAM]. The high temperature heat capacity results are summarised in Figure V-18. The selected heat capacity expression is:

$$C^{\circ}_{p,m}(\text{CdSe, α, (298.15 – 1500) K}) = (45.87 + 11.94 \times 10^{-3}\ T)\ \text{J·K}^{-1}\text{·mol}^{-1}$$

which matches the selected heat capacity at 298.15 K and follows that in [94YAM/KAM] at high temperatures as shown in Figure V-18.

The third law entropy values at 298.15 K in [69DEM/MAL] and [69SHA/KOS] involved the evaluation of the entropy for the temperature range 0 to 50 K using estimated Debye-Einstein functions. The procedure results in entropy values that are approximately 3 $\text{J·K}^{-1}\text{·mol}^{-1}$ smaller than the corresponding experimental values in [76PET/KOF] and [92SIR/GAV]. A correction for this discrepancy brings the entropy values at 298.15 K in [69DEM/MAL] and [69SHA/KOS] on a par with the values in [76PET/KOF] and [92SIR/GAV]. Second law entropies are given in Table V-53 and the weighted average of the second and third law values is selected:

$$S^{\circ}_{m}(\text{CdSe}, \alpha, 298.15\text{ K}) = (86.5 \pm 0.8)\ \text{J·K}^{-1}\text{·mol}^{-1}.$$

Figure V-18: Experimental and selected heat capacity of α-CdSe above 298.15 K.

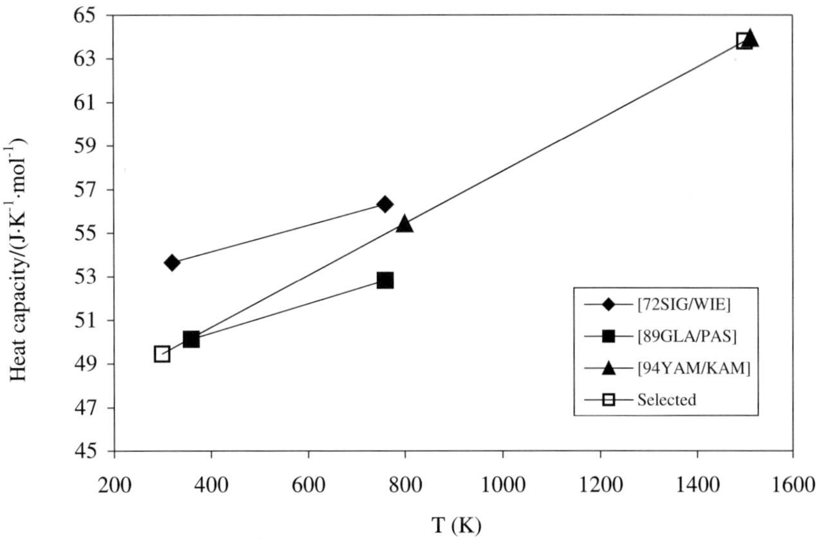

Table V-53: Second law entropies of α-CdSe at 298.15 K evaluated by the review as discussed in Appendix A. The value denoted by (*) was not included in the average.

Reference	S°_{m} (CdSe, α, 298.15 K) (J·K^{-1}·mol^{-1})	Method	Temperature (K)
[60KOR/SOK]	(90.9 ± 10.0)	Knudsen effusion	913 – 1093
[61WOS]	(75.4 ± 15.0)	Gas flow method	975 – 1210
[61SOM]	(84.5 ± 8.0)$^{(*)}$	Bourdon gauge and dew point	1016 – 1170
[64TER/RAT]	(86.3 ± 2.0)	Electrochemical cell	600 – 700
[69FLO]	(85.0 ± 12.0)	Gas flow method	973 – 1373
[71SEA/MUN]	(84.1 ± 8.0)	Knudsen torsion effusion	942 – 1041
[72SIG/WIE]	(85.6 ± 5.0)	Knudsen effusion weight loss	866 – 1035
[88BAR/IER]	(90.5 ± 4.0)	Knudsen effusion weight loss	830 – 1115
Weighted average	(86.8 ± 1.6)		

The vaporisation of α-CdSe takes place according to the reaction: α-CdSe $\rightleftharpoons$ Cd(g) + ½Se_2(g) and it has been studied by different techniques in a number of investigations. The methods of evaluation and the auxiliary data used vary from one investigation to another and in order to obtain a consistent set of thermochemical quantities, all investigations were re-evaluated by the review using the second and third laws and an identical set of auxiliary data. The vapour pressure measurements in [69BOE/BEN2] were rejected for reasons given in Appendix A. The individual investigations are discussed in Appendix A. The results are summarised in Table V-53 and Table V-54.

Table V-54: Second and third law enthalpies of formation of α-CdSe at 298.15 K evaluated by the review as discussed in Appendix A. Values denoted by (*) were not included in the average.

Reference	$\Delta_f H_m^\circ$ (CdSe, α, 298.15 K) (kJ·mol^{-1})		Temperature (K)
	2nd law	3rd law	
[60KOR/SOK]	– (139.0 ± 15.0)	– (141.9 ± 8.0)	913 – 1093
[61SOM]	– (121.2 ± 8.0)$^{(*)}$	– (119.4 ± 6.0)$^{(*)}$	975 – 1210
[61WOS]	– (158.0 ± 18.0)	– (146.2 ± 3.0)	1016 – 1170
[64TER/RAT]	– (136.6 ± 4.0)	– (136.4 ± 2.0)	600 – 700
[69FLO]	– (143.8 ± 15.0)	– (142.3 ± 3.0)	973 – 1373
[71SEA/MUN]	– (144.1 ± 12.0)	– (142.0 ± 2.0)	942 – 1041
[72SIG/WIE]	– (147.7 ± 5.0)	– (147.1 ± 2.0)	866 – 1035
[88BAR/IER]	– (142.6 ± 3.0)	– (146.8 ± 2.0)	830 – 1115
Weighted average	– (142.1 ± 2.1)	– (143.3 ± 0.9)	

The Gibbs energy of formation of α-CdSe was studied in the temperature range 603 to 703 K using the electrochemical cell

(–) Cd(l) | 5 weight-% $CdBr_2$ (in LiBr + KBr melt) | α-CdSe, Se(l) (+)

by Terpilowski and Ratajczak [64TER/RAT]. Similarly to the vaporisation studies, this investigation was re-evaluated by the review using the second and third laws and an identical set of auxiliary data as discussed in Appendix A. The results are included in Table V-53 and Table V-54. The values of the enthalpy of formation derived from the measurements in [61SOM] are substantially less negative than those of the other investigations. This is an effect of the significantly larger total pressures measured. It has been suggested in [72SIG/WIE] that this could be due to the use of closed systems in Somorjai's method. A slight excess of an elemental component can cause serious errors in closed systems, especially at low pressures. The results of [61SOM] are not considered in the calculation of averages in Table V-53 and Table V-54. The experimental sublimation pressures obtained in the various investigations are given in Figure V-19.

The selected value of the enthalpy of formation is:

$$\Delta_f H_m^\circ(\text{CdSe, } \alpha\text{, 298.15 K}) = -(143.1 \pm 1.9)\ \text{kJ·mol}^{-1}.$$

The Gibbs energy of formation is calculated from the selected enthalpy of formation and the standard entropies to be:

$$\Delta_f G_m^\circ(\text{CdSe, } \alpha\text{, 298.15 K}) = -(140.9 \pm 1.9)\ \text{kJ·mol}^{-1}.$$

Figure V-19: Sublimation pressures of α-CdSe obtained in experimental studies.

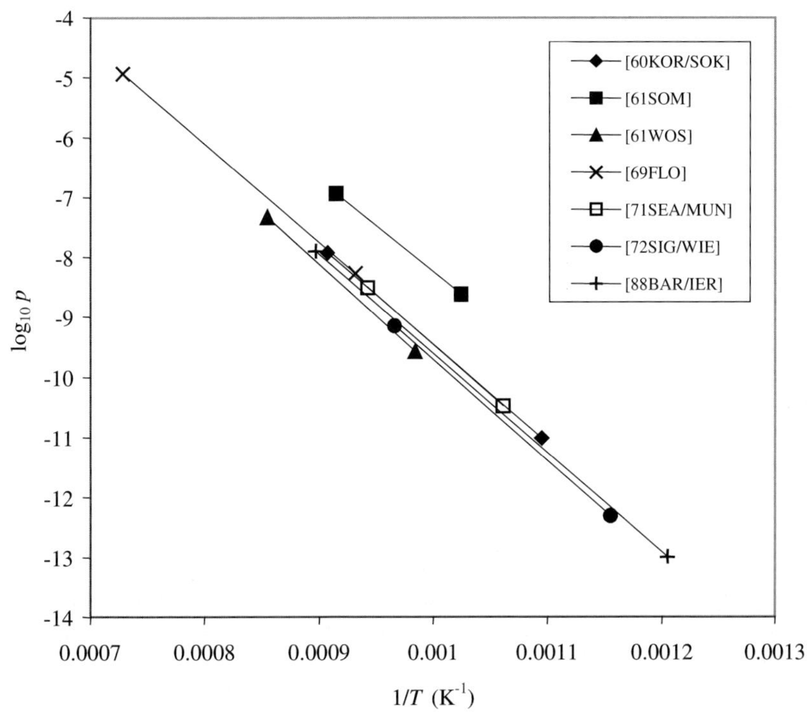

In the aqueous phase, Umland and Wallmeier [80UML/WAL] studied the reduction of selenite at the mercury electrode in the presence of Cd^{2+} by polarography, see Appendix A. The solubility product of CdSe(s) was obtained from the position of the half-wave potential of the second reduction step, $HgSe(s) + Cd^{2+} + 2e^- \rightleftharpoons CdSe(s) +$

Hg(l), to be $\log_{10} K^{\circ}_{s,0}$ (CdSe, s, 298.15 K) = – (31.2 ± 2.0). Combined with CODATA for Cd^{2+} and the selected value for Se^{2-}, it corresponds to $\Delta_f G^{\circ}_m$ (CdSe, s, 298.15 K) = – (127.2 ± 11.8) kJ·mol^{-1}. The value is not in harmony with the value obtained by extrapolation of high temperature data. Since the validity of the polarographic method has not been documented, the value of $\Delta_f G^{\circ}_m$ (CdSe, α, 298.15 K) obtained from the extrapolation of the high temperature data has been selected.

After the assessment of the thermodynamic properties of α-CdSe was made by the review, the heat capacity measurements made by Pashinkin, Malkova, and Mikhailova [2002PAS/MAL] have appeared. The measurements were performed in the temperature range 500 to 760 K and the result does not differ significantly from the values calculated from the selected heat capacity expression. The results therefore do not affect any of the selected thermodynamic properties of α-CdSe.

V.9.2.2 Cadmium selenites

Micka, Uchytilova, and Ebert [84MIC/UCH] present the phase diagram of the system $CdSeO_3$-SeO_2-H_2O at 298 K from solubility measurements. The solid phases $CdSeO_3$, $Cd_3H_2(SeO_3)_4$, and $CdSe_2O_5$ were identified by chemical analysis and infrared spectroscopy. No primary solubility data are published. Cadmium selenite crystallises in three modifications and the enthalpies of transition between the different forms are within 3.0 kJ·mol^{-1} at 298.15 K according to Sapozhnikov and Markovskii [66SAP/MAR].

Chukhlantsev [56CHU] prepared cadmium selenite by mixing a 0.1 M sodium selenite solution with the equivalent amount of a cadmium chloride solution. The product was amorphous. Chemical analysis confirmed the 1:1 ratio between Cd and Se(IV). The solubility of the specimen in dilute solution of nitric or sulphuric acid was measured at 293 K. The data have been recalculated as described in Appendix A, [56CHU]. The result for

$$CdSeO_3(am) \rightleftharpoons Cd^{2+} + SeO_3^{2-} \quad (V.85)$$

is $\log_{10} K_{s,0}$ ((V.85), 293.15 K) = – (9.62 ± 0.29). The reported conditional constant was – (8.90 ± 0.19). Ripan and Vericeanu [68RIP/VER] studied the solubility of $CdSeO_3$(s) by conductivity measurements as described in Appendix A. A calculation based on the equilibrium constants in [76BAE/MES] and the total concentrations provided in the paper leads to the conclusion that about 5% of dissolved Cd(II) would be present in hydrolysed forms, if the solvent were conductivity water. The solubility product estimated with allowance for the hydrolysis of Cd^{2+} and SeO_3^{2-} is $\log_{10} K_{s,0}$ ((V.85), 291 K) = – (8.39 ± 0.10). The uncertainty reflects the variance from the solubility measurements. The value of $\log_{10} K_{s,0}$ obtained can only be regarded as indicating the order of magnitude of the constant. The selection of a value of the solubility product will be made later.

Toropova [57TOR] measured the shift of the polarographic half-wave potential of Cd^{2+} in 0.15 to 0.50 M sodium selenite solutions. The solutions also contained

$NaNO_3$, but it is not possible to find out from the text whether the sodium ion concentration or the ionic strength was kept constant. As no experimental data are provided to support the reported equilibrium constant for the proposed complexation according to the reaction:

$$Cd^{2+} + 2SeO_3^{2-} \rightleftharpoons Cd(SeO_3)_2^{2-} \qquad (V.86)$$

the value of $\log_{10} K$ ((V.86), 1 M $NaNO_3$ (?), 298.15 K) = 5.15 cannot be selected.

Leshchinskaya and Selivanova have reported on the enthalpy of formation of cadmium selenite in three publications [63LES/SEL5], [63LES/SEL6], [63SEL/LES5]. The enthalpy change of the reaction Na_2SeO_3(cr) + $CdCl_2$(aq, 1:1000) → $CdSeO_3$(am) + 2NaCl(aq, 1:500) was measured in an electrically calibrated calorimeter [63SEL/LES5]. The result from this experiment is combined in Appendix A with determinations of the enthalpies of dissolution of $CdSeO_3$(cr) and $CdSeO_3$(am) in nitric acid reported in [63LES/SEL6] to yield $\Delta_f H_m^\circ$ ($CdSeO_3$, cr, 298.15 K) = – (576.00 ± 1.18) $kJ\cdot mol^{-1}$ and $\Delta_f H_m^\circ$ ($CdSeO_3$, am, 298.15 K) = – (574.60 ± 1.17) $kJ\cdot mol^{-1}$. The measurements of the enthalpy of dissolution of $CdSeO_3$(cr) in dilute mineral acid in [66SAP/MAR], which are re-evaluated in Appendix A, result in $\Delta_f H_m^\circ$ ($CdSeO_3$, cr, 298.15 K) = – (585.1 ± 4.0) $kJ\cdot mol^{-1}$ as a mean value for the two common polymorphs denoted α-$CdSeO_3$ and β-$CdSeO_3$. The review selects:

$$\Delta_f H_m^\circ (CdSeO_3, cr, 298.15\ K) = -(576.80 \pm 2.50)\ kJ\cdot mol^{-1}.$$

Gospodinov and Bogdanov [83GOS/BOG] studied the thermal properties of the two common polymorphs of $CdSeO_3$(cr) by differential scanning calorimetry with the sample in a closed vessel. A transition from the α to the β form occurred at (667 ± 5) K with $\Delta_{trs} H_m^\circ$ = (6.11 ± 0.13) $kJ\cdot mol^{-1}$. The β-$CdSeO_3$ melts at (970 ± 5) K with $\Delta_{fus} H_m^\circ$ = (29.0 ± 0.2) $kJ\cdot mol^{-1}$. The transition enthalpy at 298.15 K obtained in [66SAP/MAR] is – (3.13 ± 1.05) $kJ\cdot mol^{-1}$. The review selects:

$$\Delta_{trs} H_m^\circ (CdSeO_3, \alpha, 298.15\ K) = -(3.13 \pm 1.05)\ kJ\cdot mol^{-1},$$

$$\Delta_{fus} H_m^\circ (CdSeO_3, \beta, 970\ K) = (29.0 \pm 0.2)\ kJ\cdot mol^{-1}.$$

Bakeeva, Buketov, and Pashinkin [68BAK/BUK3] measured the equilibrium pressure of SeO_2(g) above $CdSeO_3$(β or l) by the dew point method. Their primary data have been recalculated as described in [71PAS/ARO] in Appendix A with the following results:

$$\beta\text{-}CdSeO_3 \rightleftharpoons CdO(cr) + SeO_2(g) \qquad (V.87)$$

$$\log_{10} p_{SeO_2} ((V.87), (881 - 947)\ K) = (10.91 \pm 0.41) - (12030 \pm 370)\ T^{-1};$$

$$CdSeO_3(l) \rightleftharpoons CdO(cr) + SeO_2(g) \qquad (V.88)$$

$$\log_{10} p_{SeO_2} ((V.88), (963 - 1083)\ K) = (7.65 \pm 0.28) - (8880 \pm 280)\ T^{-1}.$$

These data yield T_{fus} = 960 K and $\Delta_{fus}H^\circ_m$ = (60.2 ± 8.2) kJ·mol^{-1}. The latter datum disagrees substantially with the directly measured value in [83GOS/BOG] and has been given zero weight.

The extrapolation of the data to standard conditions was made with the estimated heat capacities in [73BAR/KNA]. The derived standard enthalpy and entropy values are $\Delta_f H^\circ_m$ ($CdSeO_3$, β, 298.15 K) = – (608.3 ± 8.4) kJ·mol^{-1} and S°_m ($CdSeO_3$, β, 298.15 K) = (108.2 ± 8.0) J·K^{-1}·mol^{-1}. The recalculation performed by Pashinkin *et al.* [71PAS/ARO] with another set of auxiliary data agrees within the uncertainties with the result obtained by the review. The extrapolated enthalpy of formation of $CdSeO_3$ from the high temperature measurements has been given zero weight.

The solubility product of $CdSeO_3$(am) determined in [56CHU] yields S°_m ($CdSeO_3$, am, 298.15 K) = (127.4 ± 13.0) J·K^{-1}·mol^{-1}, while the extrapolated S°_m value (108.2 J·K^{-1}·mol^{-1}) corresponds to $\log_{10} K^\circ_{s,0}$ ($CdSeO_3$, cr, 298.15 K) = – (8.6 ± 0.6). In order to make $\log_{10} K^\circ_{s,0} \leq -9.6$, the entropy of $CdSeO_3$(cr) should exceed about 130 J·K^{-1}·mol^{-1}. With the experimental data at hand, the inconsistency between the entropy of cadmium selenite obtained from the high temperature data and the solubility measurement cannot be resolved. The review makes the selection:

$$S^\circ_m (CdSeO_3, \text{cr}, 298.15 \text{ K}) = (122.0 \pm 20.0) \text{ J·K}^{-1}\text{·mol}^{-1}.$$

The selection corresponds to a solubility product for $CdSeO_3$(cr) of

$$\log_{10} K^\circ_{s,0} (CdSeO_3, \text{cr}, 298.15 \text{ K}) = -(9.3 \pm 1.2).$$

The corresponding data in [82WAG/EVA] seem to originate from the conditional solubility product published in [56CHU].

The standard enthalpies of formation of $CdSe_2O_5$(cr) and $Cd_3H_2(SeO_3)_4$(cr) are selected from [66SAP/MAR] in Appendix A to be:

$$\Delta_f H^\circ_m (CdSe_2O_5, \text{cr}, 298.15 \text{ K}) = -(828 \pm 17) \text{ kJ·mol}^{-1}$$

and

$$\Delta_f H^\circ_m (Cd_3H_2(SeO_3)_4, \text{cr}, 298.15 \text{ K}) = -(2306 \pm 45) \text{ kJ·mol}^{-1}.$$

V.9.2.3 Cadmium selenates

Klein [40KLE] made a thorough investigation of the solubility of cadmium selenates in water as a function of temperature. The results are presented in a table and as a graph. The dihydrate, $CdSeO_4·2H_2O$, seems to invariably form by crystallisation at room temperature although it is metastabile with respect to $CdSeO_4·H_2O$. Metastable equilibria with $CdSeO_4·2H_2O$ could be established up to 343 K, thereafter a transformation to the monohydrate occurred. The solubility at 298.15 K of the dihydrate interpolated from the tabled data is 2.62 mol($CdSeO_4$)·kg^{-1}. This quantity was determined by Vojtisek and

Ebert [87VOJ/EBE3] and by Oy(j)kova and Mihov [92OYK/MIH], both investigations reporting 2.56 $mol(CdSeO_4)\cdot kg^{-1}$. The review selects:

$$m(CdSeO_4\cdot 2H_2O(cr), \text{aq sat}, 298.15 \text{ K}) = (2.56 \pm 0.02) \text{ mol}(CdSeO_4)\cdot kg^{-1}.$$

The monohydrate was the stable phase within the temperature range 263.7 to 371.7 K investigated in [40KLE]. The review selects the interpolated solubility:

$$m(CdSeO_4\cdot H_2O(cr), \text{aq sat}, 298.15 \text{ K}) = (2.42 \pm 0.05) \text{ mol}(CdSeO_4)\cdot kg^{-1}.$$

A phase diagram or related information is available for the systems: $CdSeO_4$-H_2SeO_4-H_2O at 373.15 K in [85GOS]; $CdSeO_4$-$(NH_4)_2SeO_4$-H_2O at 298.15 K in [87VOJ/EBE3]; $CdSeO_4$-Na_2SeO_4-H_2O at 298.15 K in [92OYK/MIH]; $CdSeO_4$-K_2SeO_4-H_2O at 298.15 K in [87VOJ/EBE3].

The value of solubility product of $CdSeO_4\cdot 2H_2O(cr)$:

$$CdSeO_4\cdot 2H_2O(cr) \rightleftharpoons Cd^{2+} + SeO_4^{2-} + 2H_2O(l) \qquad (V.89)$$

at 298.15 K has been reported by Kumov and Batyreva [90KUM/BAT] to be $\log_{10} K^{\circ}_{s,0}$ ((V.89), 298.15 K) = – (1.85 ± 0.09), see Appendix A. As the reviewers failed to assess the accuracy of this datum, it will not be included amongst the selected data.

The stability constant of $CdSeO_4(aq)$ was determined by Banks [34BAN] concurrently with the same datum for $ZnSeO_4(aq)$ as described in Section V.9.1.3. The result for

$$Cd^{2+} + SeO_4^{2-} \rightleftharpoons CdSeO_4(aq) \qquad (V.90)$$

was $\log_{10} \beta_1^{\circ} = 2.27$.

Badiello, Feroci, and Fini [96BAD/FER] made polarographic measurements of 1×10^{-5} M Cd^{2+} in 0.15 M $NaNO_3$ containing selenate concentrations from 0 to 1×10^{-2} M. The half-wave potential, $E_{1/2}$, did not change with selenate concentration and it was concluded that no complex formation occurred. The total expected change in $E_{1/2}$ would be less than 3 mV if the value of the stability constant determined by Banks is correct. It is doubtful if the experimental accuracy of the polarographic work is sufficient to detect it. The value of the stability constant from the conductivity measurements will be selected:

$$\log_{10} \beta_1^{\circ} ((V.90), 298.15 \text{ K}) = (2.27 \pm 0.06).$$

Aruga [78ARU] made calorimetric titrations of 0.17 M cadmium nitrate solution with 0.17 M tetraethylammonium selenate, see Appendix A. The evaluation of the experiment was made as described for zinc in Section V.9.1.3. The selected result is:

$$\Delta_r H_m^{\circ} ((V.90), 298.15 \text{ K}) = (8.3 \pm 4.0) \text{ kJ}\cdot\text{mol}^{-1}.$$

The selections for Reaction (V.90) result in:

$$\Delta_f G_m^{\circ} (CdSeO_4, \text{aq}, 2985.15) = -(530.2 \pm 1.7) \text{ kJ}\cdot\text{mol}^{-1},$$

$$\Delta_f H^\circ_m (\text{CdSeO}_4, \text{aq}, 2985.15) = -(671.1 \pm 5.3) \text{ kJ·mol}^{-1},$$

$$S^\circ_m (\text{CdSeO}_4, \text{aq}, 2985.15) = (31.5 \pm 18.6) \text{ J·K}^{-1}\text{·mol}^{-1}.$$

Selivanova, Maier, and Luk'yanova [63SEL/MAI] have investigated the enthalpies of formation of solid cadmium selenates. They prepared crystalline $CdSeO_4·H_2O$ and $CdSeO_4$ and measured their enthalpy of dissolution to (aq, 1:1200). The investigation also included a determination of the enthalpy change of the reaction $CdSeO_4·H_2O(cr) + BaCl_2(aq, 1:800) \rightarrow BaSeO_4(cr) + CdCl_2(aq, 1:800) + H_2O(l)$. The review has recalculated the data in Appendix A and arrived at the following results:

$$\text{CdSeO}_4\text{·H}_2\text{O(cr)} + \text{Ba}^{2+} \rightarrow \text{BaSeO}_4\text{(cr)} + \text{Cd}^{2+} + \text{H}_2\text{O(l)} \quad \text{(V.91)}$$

$$\Delta_r H^\circ_m ((\text{V.91}), 298.15 \text{ K}) = -(27.90 \pm 0.16) \text{ kJ·mol}^{-1},$$

$$\Delta_f H^\circ_m (\text{CdSeO}_4, \text{cr}, 298.15 \text{ K}) = -(634.12 \pm 5.63) \text{ kJ·mol}^{-1},$$

$$\Delta_f H^\circ_m (\text{CdSeO}_4\text{·H}_2\text{O}, \text{cr}, 298.15 \text{ K}) = -(943.05 \pm 5.62) \text{ kJ·mol}^{-1}.$$

The selected values of the standard enthalpies of formation overlap those reported in [82WAG/EVA]. This reference also contains the Gibbs energy of formation of the anhydrous salt, $\Delta_f G^\circ_m(\text{CdSeO}_4, \text{cr}, 298.15 \text{ K}) = -531.7 \text{ kJ·mol}^{-1}$. The source of this datum could not be traced.

V.9.2.4 Cadmium selenocyanates

The complex formation between Cd^{2+} and $SeCN^-$ in aqueous solution has been studied in five papers [56TOR], [62GOL/AND], [67HUM/BON], [74AHR/AVS], [75SAT/SAH]. Of these papers, the paper by Ahrland, Avsar, and Kullberg [74AHR/AVS] is of superior quality and should be given a premium weight in the evaluation. Toropova [56TOR] based her findings on data obtained from solutions containing 0.25 to 1 M KSeCN added to a 0.8 M ionic medium of $NaNO_3$. Golub and Andreichenko [62GOL/AND] evaluated the equilibrium constants from data obtained in solutions in which the ionic medium had been exchanged by up to 100%. The results of Humffray *et al.* [67HUM/BON] and Satyanarayana *et al.* [75SAT/SAH] are both based on a limited number of data, $n = 8$ and 6, respectively. In addition, the latter study was made in the absence of supporting electrolyte and the activity coefficients of $SeCN^-$ and $CdSeCN^+$ were assumed to cancel in the calculation of the formation constant. In all five papers, the complex formation is described in terms of a series of mononuclear complexes

$$\text{Cd}^{2+} + n\text{SeCN}^- \rightleftharpoons \text{Cd(SeCN)}_n^{2-n} \quad (n = 1 \text{ to } 4); \text{(V.92)}$$

and Table V-55 lists the equilibrium constants reported.

For reasons given above, the present review rejects all studies in favour of [74AHR/AVS]. The constants in the paper have been recalculated and extrapolated to standard conditions by SIT using Δε values in [97BAH/PAR] in Appendix A. The selected results are:

$$\log_{10} K^\circ ((V.92), n = 1, 298.15 \text{ K}) = (2.24 \pm 0.06),$$

$$\log_{10} K^\circ ((V.92), n = 2, 298.15 \text{ K}) = (3.34 \pm 0.12),$$

$$\log_{10} K^\circ ((V.92), n = 3, 298.15 \text{ K}) = (3.81 \pm 0.21),$$

$$\log_{10} K^\circ ((V.92), n = 4, 298.15 \text{ K}) = (4.60 \pm 0.11).$$

Table V-55: Stability constants for complexes of Cd^{2+} with $SeCN^-$.

Method	Ionic medium	t (°C)	$\log_{10}\beta_1$	$\log_{10}\beta_2$	$\log_{10}\beta_3$	$\log_{10}\beta_4$	$\log_{10}\beta_5$	Reference
pol	0.8 M $NaNO_3$	25				3.6		[56TOR]
pot	1.5 M KNO_3	20	1.34	2.26	3.00	3.24	3.89	[62GOL/AND]
pol	2 M KNO_3	30	(1.30 ± 0.04)	(2.00 ± 0.02)	(2.64 ± 0.01)	(3.00 ± 0.01)		[67HUM/BON]
pot	1 M $NaClO_4$	25	(1.47 ± 0.01)	(2.30 ± 0.07)	(2.85 ± 0.15)	(4.04 ± 0.04)		[74AHR/AVS]
pot	0	25	(1.98 ± 0.04)					[75SAT/SAH]

Comprehensive calorimetric data for the formation of cadmium(II)-selenocyanate complexes in 1 M $NaClO_4$ were also collected in [74AHR/AVS]. Their evaluation relied on the stability constants determined by potentiometry and reaction enthalpies were obtained by a least squares optimisation of the function $\Sigma(Q_{calc} - Q_{obs})^2$. Q_{calc} and Q_{obs} denote the calculated and the observed enthalpy changes during the calorimetric titration, respectively. The review accepts these values as valid also at $I = 0$, but doubles the uncertainties reported to account for the uncertainty of this assumption. The selected values are:

$$\Delta_r H_m^\circ ((V.92), n = 1, 298.15 \text{ K}) = -(10.05 \pm 0.20) \text{ kJ·mol}^{-1},$$

$$\Delta_r H_m^\circ ((V.92), n = 2, 298.15 \text{ K}= -(26.35 \pm 2.00) \text{ kJ·mol}^{-1},$$

$$\Delta_r H_m^\circ ((V.92), n = 3, 298.15 \text{ K}) = -(3.35 \pm 14.00) \text{ kJ·mol}^{-1},$$

$$\Delta_r H_m^\circ ((V.92), n = 4, 298.15 \text{ K}) = -(40.00 \pm 14.00) \text{ kJ·mol}^{-1},$$

$$\Delta_r G_m^\circ ((V.92), n = 1, 298.15 \text{ K}) = -(12.79 \pm 0.34) \text{ kJ·mol}^{-1},$$

$$\Delta_r G_m^\circ ((V.92), n = 2, 298.15 \text{ K}) = -(19.06 \pm 0.68) \text{ kJ·mol}^{-1},$$

$$\Delta_r G^\circ_m \text{((V.92)}, n = 3, 298.15\text{ K)} = -(21.75 \pm 1.20)\text{ kJ·mol}^{-1},$$

$$\Delta_r G^\circ_m \text{((V.92)}, n = 4, 298.15\text{ K)} = -(26.26 \pm 0.63)\text{ kJ·mol}^{-1},$$

$$\Delta_r S^\circ_m \text{((V.92)}, n = 1, 298.15\text{ K)} = (9.2 \pm 1.3)\text{ J·K}^{-1}\text{·mol}^{-1},$$

$$\Delta_r S^\circ_m \text{((V.92)}, n = 2, 298.15\text{ K)} = -(24.4 \pm 7.1)\text{ J·K}^{-1}\text{·mol}^{-1},$$

$$\Delta_r S^\circ_m \text{((V.92)}, n = 3, 298.15\text{ K)} = (61.7 \pm 47.1)\text{ J·K}^{-1}\text{·mol}^{-1},$$

$$\Delta_r S^\circ_m \text{((V.92)}, n = 4, 298.15\text{ K)} = -(46.1 \pm 47.0)\text{ J·K}^{-1}\text{·mol}^{-1}.$$

It can be noted that the value of $\Delta_r H^\circ_m$((V.92), n = 1, 298.15 K) is in fair agreement with the value determined in [75SAT/SAH], $-(11.1 \pm 0.4)$ kJ·mol^{-1}, also obtained by calorimetry.

Golub and Andreichenko [62GOL/AND] also measured at 293 K the solubility of $Cd(SeCN)_2$(cr) in pure water. The solubility reported is 0.248 M, but from the calculations provided in the paper it should obviously read 0.0248 M. This value is accepted and combined with the selected complex formation model to yield a solubility product for the reaction:

$$Cd(SeCN)_2(cr) \rightleftharpoons Cd^{2+} + 2SeCN^- \qquad (V.93)$$

of $\log_{10} K^\circ_{s,0} = -5.7$. This result is accepted but the uncertainty limit is set at $\pm$ 0.5 in the selected value to account for uncertainties in the experimental set-up, in the activity coefficients and in the temperature extrapolation from 293 K:

$$\log_{10} K^\circ_{s,0} \text{((V.93)}, 298.15\text{ K)} = -(5.7 \pm 0.5).$$

The selection corresponds to:

$$\Delta_r G^\circ_m \text{((V.93)}, 298.15\text{ K)} = (32.5 \pm 2.9)\text{ kJ·mol}^{-1}.$$

The selected values for $\Delta_r G^\circ_m$ combines with the selected value of $\Delta_f G^\circ_m$($SeCN^-$, 298.15 K) and the auxiliary value of $\Delta_f G^\circ_m$(Cd^{2+}, 298.15 K) to yield:

$$\Delta_f G^\circ_m (CdSeCN^+, 298.15\text{ K}) = (45.5 \pm 3.9)\text{ kJ·mol}^{-1},$$

$$\Delta_f G^\circ_m (Cd(SeCN)_2, \text{aq}, 298.15\text{ K}) = (175.3 \pm 7.7)\text{ kJ·mol}^{-1},$$

$$\Delta_f G^\circ_m (Cd(SeCN)_3^-, 298.15\text{ K}) = (308.7 \pm 11.5)\text{ kJ·mol}^{-1},$$

$$\Delta_f G^\circ_m (Cd(SeCN)_4^{2-}, 298.15\text{ K}) = (440.2 \pm 15.2)\text{ kJ·mol}^{-1},$$

$$\Delta_f G^\circ_m (Cd(SeCN)_2, \text{cr}, 298.15\text{ K}) = (161.8 \pm 8.2)\text{ kJ·mol}^{-1}.$$

V.9.3 Mercury compounds

V.9.3.1 Mercury selenides

The mercury-selenium phase diagram was assessed and modelled by Sharma and Chang in [92SHA/CHA] and [93SHA/CHA]. HgSe(s) is the only solid compound in the system but four different polymorphs have been characterised. The α form, which crystallises in the zincblende structure, is the stable form up to the melting point at 1072 K. The β, γ, and δ forms are all high-pressure polymorphs. The homogeneity range of α-HgSe is not well established and deviations from stoichiometry in the range $HgSe_{0.953}$ to $HgSe_{1.030}$ have been reported [92SHA/CHA]. The reaction between mercury and selenium appears to be slow in the final stages and the equilibration time is important. Experimental thermodynamic information is only available for α-HgSe.

V.9.3.1.1 α-HgSe

The low temperature heat capacity of α-HgSe has been measured in the temperature range 80 to 300 K by Gultyaev and Petrov [59GUL/PET] and in the temperature range 15 to 290 K by Bogdanov and Bezborodova [70BOG/BEZ] yielding the values $C^{\circ}_{p,m}$(HgSe, α, 298.15 K) = (49.8 ± 2.5) J·K^{-1}·mol^{-1} and (54.3 ± 1.0) J·K^{-1}·mol^{-1} (see Appendix A), respectively. The heat capacity above room temperature was measured by Kelemen, Cruceanu and Niculescu [65KEL/CRU] (see Appendix A) in the temperature range 293 to 513 K yielding the expression $C^{\circ}_{p,m}$(HgSe, α, (293 – 513) K) = (52.31 + 9.52 × 10^{-3} T) J·K^{-1}·mol^{-1} and $C^{\circ}_{p,m}$(HgSe, α, 298.15 K) = (55.2 ± 1.0) J·K^{-1}·mol^{-1}. The weighted average of the three values of the heat capacity is selected:

$$C^{\circ}_{p,m}(\text{HgSe}, \alpha, 298.15\ \text{K}) = (54.4 \pm 0.7)\ \text{J·K}^{-1}\text{·mol}^{-1}.$$

The heat capacity expression in [65KEL/CRU] was adjusted to match the selected heat capacity at 298.15 K yielding:

$$C^{\circ}_{p,m}(\text{HgSe}, \alpha, (293 - 513)\ \text{K}) = (51.66 + 9.52 \times 10^{-3}\ T)\ \text{J·K}^{-1}\text{·mol}^{-1}.$$

This expression has been used by the review wherever the heat capacity of α-HgSe has been needed in second and third law evaluations. It was also extrapolated to higher temperatures where necessary.

The mass spectrometric studies of the vaporisation of α-HgSe by Goldfinger and Jeunehomme [63GOL/JEU], Berkowitz and Chupka [66BER/CHU], and Grade and Hirschwald [80GRA/HIR] have shown that the vaporisation is considerably more complex than for the compounds α-ZnSe and α-CdSe due to the formation of significant amounts of polymeric selenium species other than the Se_2(g) molecule. The vaporisation cannot be simplified to the reaction α-HgSe $\rightleftharpoons$ Hg(g) + ½Se_2(g) and in principle all the selenium species in the series Se(g)-Se_8(g) have to be considered. This problem has been treated in various ways in vaporisation studies of α-HgSe and various sets of auxiliary thermodynamic data for the selenium species have been used. In order to obtain a consistent set of thermodynamic quantities, all the vaporisation studies were re-evaluated by the review as discussed in Appendix A using the second and third laws and the selected set of thermodynamic data for condensed selenium and the selenium gaseous species. No experimental evidence for the existence of the molecule HgSe(g) is available.

The second law entropy values from the various investigations are summarised in Table V-56. A third law value, S°_{m} (HgSe, α, 298.15 K) = (100.7 ± 4.0) $J\cdot K^{-1}\cdot mol^{-1}$, was derived by the review as discussed in Appendix A from the heat capacity measurements in [70BOG/BEZ]. The entropies derived from the measurements in [68REZ/ISA] and [72MUN/MES] deviate strongly from the other values and were rejected. The selected value of the entropy was calculated as the weighted average of the second and third law values where, in the absence of proper error limits for all values, the second law values were assigned unit weight and the third law value was given a weight factor of four. The selected value is:

$$S^{\circ}_{m}\text{ (HgSe, }\alpha\text{, 298.15 K)} = (98.5 \pm 3.0)\ J\cdot K^{-1}\cdot mol^{-1}.$$

Table V-56: Second law entropy values for α-HgSe at 298.15 K derived from vapour pressure and emf measurements. Values denoted by (*) were not considered in calculating the average.

Reference	S°_{m} (HgSe, α, 298.15 K) ($J\cdot K^{-1}\cdot mol^{-1}$)	Temperature (K)	Measurement method
[65BRE]	103.3	723 – 873	Optical density
[65SIL/KAR]	94.4	613 – 753	Gas flow
[68RAT/TER]	(93.7 ± 9.0)	320 – 420	Electrochemical cell
[68REZ/ISA]	155.5(*)	573 – 823	Gas flow
[69PAS/UST]	(96.0 ± 10.0)	756 – 951	Quartz membrane manometer
[72MUN/MES]	(62.6 ± 22.0)(*)	530 – 570	Torsion effusion
[84IMP/PIA]	(96.6 ± 12.0)	541 – 602	Torsion effusion
Average	96.8		

The second and third law enthalpies of formation are summarised in Table V-57. The second and third law values of the enthalpy of formation derived from the measurements in [68REZ/ISA] differ substantially indicating serious systematic errors. The third law values are considerably less scattered than the second law values. The average $\Delta_f H^{\circ}_{m}$ (HgSe, α, 298.15 K) = – (58.0 ± 4.0) $kJ\cdot mol^{-1}$ is considered to summarise the high temperature measurements in Table V-57. The error limits were set by considering the scatter of the third law values. The corresponding Gibbs energy of formation is calculated to be $\Delta_f G^{\circ}_{m}$ (HgSe, α, 298.15 K) = – (52.2 ± 4.0) $kJ\cdot mol^{-1}$.

In four studies, the solubility product of mercury(II) selenide in the aqueous phase has been estimated from polarographic measurements of the reduction potential of the couple HgSe(s) + 2e$^-$ $\rightleftharpoons$ Hg(l) + Se^{2-}. This datum combines with the standard electrode potential of Hg^{2+}/Hg(l) to the solubility product. The results are collected in Table V-58.

Table V-57: Second and third law enthalpies of formation of α-HgSe at 298.15 K evaluated by the review as discussed in Appendix A. Values denoted by (*) were not considered in calculating the average.

Reference	$\Delta_f H^\circ_m$ (HgSe, α, 298.15 K) (kJ·mol^{-1})		Temperature (K)
	2nd law	3rd law	
[65BRE]	– 51.2	– 55.0	723 – 873
[65SIL/KAR]	– 62.9	– 60.0	613 – 753
[68RAT/TER]	– (45.2 ± 9.0)		320 – 420
[68REZ/ISA]	– 17.9$^{(*)}$	– 57.6$^{(*)}$	573 – 823
[69PAS/UST]	– (63.3 ± 10.0)	– (61.1 ± 10.0)	756 – 951
[72MUN/MES]	– (73.5 ± 30.0)	– (57.0 ± 30.0)	530 – 570
[84IMP/PIA]	– (58.5 ± 8.0)	– (57.0 ± 8.0)	541 – 602
Average	– 59.1	– 58.0	

Table V-58: Values of the solubility product of HgSe(s) obtained by polarographic measurement at 298.15 K.

Reference	$\log_{10} K_{s,0}$	Ionic medium
[48LIN/NIE]	– 59	Variable + 3 × 10^{-4} % gelatin
[72GLA/KIR]	– 58	1 M Na_2SO_4
[80UML/WAL]	– 61	0.1 M $(NH_4)_2SO_4$
[89GLA/KOV]	– 57.3	1 M NaOH

All results are based on relatively few measurements and none of the studies can be given preference. It might have been preferable to recalculate the solubility products to standard conditions with accepted protonation constants of the selenide ion. This has not been done since the necessary means for doing so are lacking and, in addition, it can be anticipated that the calculation would not reduce the scatter amongst the data.

Mehra and Gubeli [71MEH/GUB] studied the solubility of HgSe(s) in 1 M $NaClO_4$ medium at 298.15 K, see Appendix A. The potentiometric measurements provided the concentration of Hg^{2+} in solutions saturated with HgSe(s) in the pH range of 0.5 to 10 and with a total concentration of Se(– II) of 0.015 to 0.08 M. The reported solubility product, $\log_{10} K_{s,0}$ (HgSe(s), 1 M $NaClO_4$, 298.15 K) = – (56.6 ± 0.2), is not accepted since it was obtained by using erroneous values for the protonation constant of Se^{2-}. Most of the experimental data were collected below pH 3. The solubility equilibrium is then best described by the reaction:

$$HgSe(s) + 2H^+ \rightleftharpoons Hg^{2+} + H_2Se(aq). \tag{V.94}$$

The review obtained from experimental data presented in [68MEH] $\log_{10} K$ ((V.94), 1 M $NaClO_4$, 298.15 K) = – (41.65 ± 0.45) from the data below pH 3 and $\log_{10} \gamma_{H^+}$ estimated by SIT to be – 0.06. After extrapolation to standard condition with SIT and $\Delta\varepsilon$ = 0.06 kg·mol^{-1}, and insertion of the accepted protonation constants of Se^{2-}, $\log_{10} K^{\circ}_{s,0}$ (HgSe, s, 298.15 K) = – (60.7 ± 0.5) was obtained. The different estimates of the solubility product thus agree with a mean value of $\log_{10} K^{\circ}_{s,0}$ = – (60.0 ± 2.0). Combined with the CODATA values for Hg^{2+} and the selected data for Se^{2-} this corresponds to $\Delta_f G^{\circ}_m$ (HgSe, s, 298.15 K) = – (49.2 ± 11.8) kJ·mol^{-1}.

The $\Delta_f G^{\circ}_m$ (HgSe, s, 298.15 K) from the high temperature measurements and from solution measurements agree within the uncertainty limits. These data were combined (α-HgSe is assumed to be present in the solubility measurement) with the weights 2:1 to the selected value:

$$\Delta_f G^{\circ}_m \text{(HgSe, α, 298.15 K)} = – (51.2 ± 4.0) \text{ kJ·mol}^{-1}.$$

The selected standard enthalpy of formation is calculated with the selected entropy to be:

$$\Delta_f H^{\circ}_m \text{(HgSe, α, 298.15 K)} = – (57.0 ± 4.1) \text{ kJ·mol}^{-1}.$$

The solubility product calculated from the Gibbs energy of formation of α-HgSe is $\log_{10} K^{\circ}_{s,0}$ = – (60.35 ± 0.70) and overlaps with the values from the experimental measurements.

V.9.3.1.2 Aqueous mercury(II) selenide complexes

From the solubility measurements conducted with radioactively labelled mercury, Mehra and Gubeli [71MEH/GUB] recorded significantly higher concentrations of Hg(II) than those corresponding to the measured Hg^{2+} activity. They suggested that this corresponded to the formation of soluble complexes, with Hg(HSe)(OH)(aq) predominating in the pH region 0 to 3, $Hg(HSe)_2(OH)^-$ predominating between pH 4 to 5.5, and $Hg(HSe)_2(OH)_2^{2-}$ predominating at pH > 7.0. The compositions of these species were derived from the dependence of the solubility on total selenide concentration and pH and their formation constants were calculated.

The review prefers to denote the species by the equivalent formulas HgSe(aq), $HgSe(HSe)^-$, and $HgSe_2^{2-}$ and has recalculated the data to yield:

$$HgSe(s) \rightleftharpoons HgSe(aq) \qquad (V.95)$$

$$\log_{10} K\ ((V.95),\ 1\ M\ NaClO_4,\ 298.15\ K) = -(7.84 \pm 0.44);$$

$$HgSe(s) + HSe^- \rightleftharpoons HgSe(HSe)^- \qquad (V.96)$$

$$\log_{10} K\ ((V.96),\ 1\ M\ NaClO_4,\ 298.15\ K) = -(6.30 \pm 0.60);$$

$$HgSe(s) + HSe^- \rightleftharpoons HgSe_2^{2-} + H^+ \qquad (V.97)$$

$$\log_{10} K\ ((V.97),\ 1\ M\ NaClO_4,\ 298.15\ K) = -(12.00 \pm 0.60).$$

The three equilibrium constants are obtained directly from the experimental data and may be combined with other constants to furnish the stability constants of the mercury(II)-selenide complexes.

The increase in the solubility of HgSe(s) with increasing hydrogen selenide and selenide activities is well documented at high pH and the third equilibrium constant is accepted by the review. The first constant, which equals an intrinsic solubility of HgSe(s) is most likely too large due to the rather simple method used for the separation of solid material from the aqueous phase. It is tentatively included due to its potential importance in connection with the mobility of mercury from repositories. Further investigations with a more efficient phase separation technique would be welcome here.

The value of the third constant at standard conditions is estimated with SIT and $\Delta\varepsilon = 0.04\ kg{\cdot}mol^{-1}$ to be:

$$\log_{10} K^{o}\ ((V.97),\ 298.15\ K) = -(12.8 \pm 0.60).$$

Combined with the selected data for α-HgSe and HSe^- this corresponds to:

$$\Delta_f G_m^{o}(HgSe_2^{2-},\ 298.15\ K) = (65.3 \pm 5.6)\ kJ{\cdot}mol^{-1}.$$

V.9.3.2 Mercury selenites

V.9.3.2.1 Hg(I)

Chukhlantsev and Tomashevsky [57CHU/TOM] prepared mercury(I) selenite by mixing 0.1 M solutions of mercury(I) nitrate and sodium selenite in stoichiometric amounts. Chemical analysis confirmed the 2:1 ratio between Hg(I) and Se(IV). No X-ray diffraction measurements were performed. The solubility of the specimen in dilute solution of nitric or sulphuric acid was measured at 293 K. The experiments were performed and the data recalculated as described in Appendix A, [56CHU]. The result for:

$$Hg_2SeO_3(s) \rightleftharpoons Hg_2^{2+} + SeO_3^{2-} \qquad (V.98)$$

was $\log_{10} K^{\circ}_{s,0}$ ((V.98), 293.15 K) = – (15.23 ± 0.25). The reported conditional constant is – (14.64 ± 0.11) after a correction has been applied by the review for a mistake in the original publication. The result is selected and assumed to be valid at standard conditions:

$$\log_{10} K^{\circ}_{s,0} \text{ ((V.98), 298.15 K)} = -(15.2 \pm 1.0),$$

but the uncertainty limits have been increased as the quality of the examined specimen is unknown.

V.9.3.2.2 Hg(II)

Micka, Uchytilova, and Ebert [84MIC/UCH] present the phase diagram of the system $HgSeO_3$-SeO_2-H_2O at 298 K from solubility measurements. The solid phases $HgSeO_3$ and $Hg_3H_2(SeO_3)_4$ were identified by chemical analysis and infrared spectroscopy. No primary solubility data are published. Gospodinov and Barkov [91GOS/BAR2] studied the system at 298.15 and 373.15 K. They identified $HgSeO_3$ and $Hg_3H_2(SeO_3)_4$ as the only solids formed in agreement with the previous investigation. The primary solubility data are available. Two forms of mercury(II) selenite were observed, α-$HgSeO_3$ formed in the experiments at 298 K and β-$HgSeO_3$ formed at 373 K. The α-form transformed to the β form on heating of the solid to 675 to 725 K.

Toropova [57TOR] measured the solubility of $HgSeO_3$(cr) in acidified 1 M $NaNO_3$ medium. The data were used to calculate the mixed conditional equilibrium constant of the reaction:

$$HgSeO_3(cr) + 2H^+ \rightleftharpoons Hg^{2+} + H_2SeO_3(aq) \qquad (V.99)$$

to be $\log_{10} K$ ((V.99), 1 M $NaNO_3$, 298.15 K) = – (4.07 ± 0.02) in Appendix A. Extrapolation to $I = 0$ by the SIT expression and combination with the protonation constants of the selenite ion yielded for the solubility product of $HgSeO_3$(cr), $\log_{10} K^{\circ}_{s,0}$ ((V.100), 298.15 K) = – (15.98 ± 0.30). The solubility product is defined by the reaction:

$$HgSeO_3(cr) \rightleftharpoons Hg^{2+} + SeO_3^{2-} \qquad (V.100)$$

Rosenheim and Pritze [09ROS/PRI] measured the solubility of $HgSeO_3$(cr) in sodium selenite solution in the concentration range 0.0625 to 2.00 M. The solubility increases with selenite concentration and an interpretation of the data in terms of the equilibrium:

$$HgSeO_3(cr) + SeO_3^{2-} \rightleftharpoons Hg(SeO_3)_2^{2-} \qquad (V.101)$$

was attempted. The influence of the activity coefficients can be expected to largely cancel in the expression for the equilibrium constant. The approximate constancy of $\log_{10} K^{\circ}$ ((V.101), 298.15 K) at – (1.35 ± 0.15) for selenite concentrations ≥ 0.50 M provide evidence for the proposed reaction. The increase in the value of the constant at lower selenite concentration cannot be explained from the data provided in the paper, but might indicate a change in the composition of the solid phase. An almost identical value of the constant can be calculated from the data in [57TOR]. The review accepts and selects:

$$\log_{10} K^\circ((\text{V.101}), 298.15\ \text{K}) = -(1.35 \pm 0.15).$$

Sapozhnikov and Markovskii [65SAP/MAR] prepared three modifications of mercury selenite by mixing solutions containing selenite and mercury(II). They were, however, unable to delineate the conditions for the formation of each modification, which were denoted as α, β and γ-$HgSeO_3$. The close resemblance between the three forms is evidenced by the small differences in the enthalpies of dissolution in hydrochloric acid and the transition enthalpy from the α to the β modification was determined to be $-(2.60 \pm 1.05)$ kJ·mol^{-1} in [66SAP/MAR]. Gospodinov and Bogdanov [83GOS/BOG] determined the temperature of the transition of α-$HgSeO_3$ to β-$HgSeO_3$ to be (650 ± 5) K and the corresponding enthalpy change to be $\Delta_{trs}H^\circ_m$($HgSeO_3$, α, 650 K) = (1.72 ± 0.13) kJ·mol^{-1}. Fusion occurs at (739 ± 5) K with $\Delta_{fus}H^\circ_m$($HgSeO_3$, β, 739 K) = (14.81 ± 0.13) kJ·mol^{-1}. The data were obtained by differential scanning calorimetry and the uncertainties in the transition temperatures have been added by the review. The review selects:

$$\Delta_{trs}H^\circ_m(\text{HgSeO}_3, \alpha, 298.15\ \text{K}) = -(2.60 \pm 1.05)\ \text{kJ·mol}^{-1},$$

$$\Delta_{fus}H^\circ_m(\text{HgSeO}_3, \beta, 739\ \text{K}) = (14.81 \pm 0.13)\ \text{kJ·mol}^{-1}.$$

Pashinkin *et al.* [72PAS/UST] studied the thermal decomposition of $HgSeO_3$ by measurements of the total vapour pressure by a membrane manometer. Experiments over the temperature interval 856 to 974 K with unsaturated vapours showed that the decomposition proceeds by the reaction:

$$\text{HgSeO}_3(\text{l}) \rightleftharpoons \text{Hg(g)} + \text{SeO}_2(\text{g}) + \tfrac{1}{2}\text{O}_2(\text{g}). \quad (\text{V.102})$$

Measurements over the temperature interval 778 to 847 K of the saturated vapour pressure resulted in: $\Delta_rH^\circ_m$((V.102), 812 K) = (325.5 ± 11.4) kJ·mol^{-1}, and $\Delta_rS^\circ_m$((V.102), 812 K) = (346.9 ± 14.0) J·K^{-1}·mol^{-1}, calculated from the temperature dependence of the equilibrium constant of Reaction (V.102). The thermodynamic quantities at standard conditions were evaluated by the review to be $\Delta_fH^\circ_m$($HgSeO_3$, β, 298.15 K) = – (406.1 ± 12.1) kJ·mol^{-1} and S°_m($HgSeO_3$, α, 298.15 K) = (149.9 ± 15.0) J·K^{-1}·mol^{-1} with $C^\circ_{p,m}$ values from [73BAR/KNA]. Pashinkin and Gospodinov [94PAS/GOS] estimated the value of S°_m($HgSeO_3$, α, 298.15 K) to be (142.6 ± 25.1) J·K^{-1}·mol^{-1} from the same experimental data. A comparison between the two estimates cannot be made since the $C^\circ_{p,m}$ values used in [94PAS/GOS] are not presented explicitly. The extrapolated standard enthalpy of formation of $HgSeO_3$ does not agree with values calculated from [66SAP/MAR] in Appendix A. The mean value for the three polymorphs determined by calorimetry will be selected and yields:

$$\Delta_f H_m^\circ(\text{HgSeO}_3, \text{cr}, 298.15 \text{ K}) = -(373.6 \pm 5.0) \text{ kJ·mol}^{-1}.$$

The estimate of the entropy of $HgSeO_3$(cr) together with the selected value of $\Delta_f H_m^\circ(\text{HgSeO}_3, \text{cr}, 298.15 \text{ K})$ yield $\log_{10} K_{s,0}^\circ((\text{V.100}), 298.15 \text{ K}) = -(16.4 \pm 1.2)$. In view of the experimental result in [57TOR] the review selects:

$$\log_{10} K_{s,0}^\circ((\text{V.100}), 298.15 \text{ K}) = -(16.2 \pm 1.0)$$

for mercury(II) selenite.

The corresponding data in [82WAG/EVA] seem to derive from the conditional solubility product in [57TOR].

Sapozhnikov and Markovskii [66SAP/MAR] determined, see Appendix A,

$$\Delta_f H_m^\circ(\text{Hg}_3\text{H}_2(\text{SeO}_3)_4, \text{cr}, 298.15 \text{ K}) = -(1652.8 \pm 32.5) \text{ kJ·mol}^{-1},$$

which is selected.

V.9.3.3 Mercury selenates

Gospodinov and Barkov [91GOS/BAR] determined the phase diagram of the system $HgSeO_4$-H_2SeO_4-H_2O at 373.15 K. Only $HgSeO_4·H_2O$ and $HgSeO_4$ were observed as solid phases.

No other data pertaining to the review have been found.

V.9.3.4 Mercury selenocyanates

The formation of $Hg(SeCN)_4^{2-}$ is well established by the potentiometric work of Toropova [56TOR], while her experimental data pertaining to the formation and the formation constant of $Hg(SeCN)_3^-$ only comprise a few points. In their polarographic work Murayama and Takayanagi [72MUR/TAK] studied the anodic mercury wave in the presence of 0.001 to 0.003 M $SeCN^-$. The electrode process was assumed to comprise the charge transfer Hg(l) $\rightleftharpoons$ Hg^{2+} + $2e^-$ combined with the formation of $Hg(SeCN)_2$(aq) and $Hg(SeCN)_3^-$. No primary data are provided and the evaluation procedure is rather involved, which makes the assessment difficult. The results are mixed equilibrium constants, since an activity coefficient correction was applied to the Hg^{2+} ion. The following complexes are thus proposed to prevail in the Hg^{2+}-$SeCN^-$ system:

$$\text{Hg}^{2+} + 2\text{SeCN}^- \rightleftharpoons \text{Hg(SeCN)}_2(\text{aq}), \quad (\text{V.103})$$

$$\text{Hg}^{2+} + 3\text{SeCN}^- \rightleftharpoons \text{Hg(SeCN)}_3^-, \quad (\text{V.104})$$

$$\text{Hg}^{2+} + 4\text{SeCN}^- \rightleftharpoons \text{Hg(SeCN)}_4^{2-}. \quad (\text{V.105})$$

The equilibrium constants reported are given in Table V-59.

Recalculation and extrapolation of the data in the table to standard conditions by SIT with $\Delta\varepsilon$ from [97BAH/PAR] in Appendix A resulted in the accepted values:

$$\log_{10} K^\circ((\text{V.103}), 298.15\text{ K}) = (22.3 \pm 1.0),$$

$$\log_{10} K^\circ((\text{V.104}), 298.15\text{ K}) = (26.8 \pm 1.0),$$

$$\log_{10} K^\circ((\text{V.105}), 298.15\text{ K}) = (29.3 \pm 0.5).$$

The uncertainties assigned by this review to the first two constants are based on the discrepancy between the two reported values of $\log_{10} K^\circ((\text{V.104}), 298.15\text{ K})$. The uncertainty limits of the final complex is based on the difference (0.2 logarithmic units) between the extrapolated values from 0.3 and 0.8 M $NaNO_3$, respectively.

Table V-59: Formation constants for complexes of Hg^{2+} with $SeCN^-$.

Method	Ionic medium	t (°C)	$\log_{10} \beta_2$	$\log_{10} \beta_3$	$\log_{10} \beta_4$	Reference
pot	0.3 M $NaNO_3$	25		26.4	28.9	[56TOR]
pot	0.8 M $NaNO_3$	25			28.7	[56TOR]
pot	0.3 M $NaNO_3$	15			30.1	[56TOR]
pot	0.3 M $NaNO_3$	20			29.5	[56TOR]
pot	0.3 M $NaNO_3$	30			28.3	[56TOR]
pol	0.3 M KNO_3	25	21.4	25.6		[72MUR/TAK]
pol	0.3 M KNO_3	35	20.4	24.4		[72MUR/TAK]

Kullberg [74KUL2] determined the enthalpy change of Reaction (V.105) at $I = 0.3$ and 1 M from measurements presented in Appendix A to be $-(191.9 \pm 1.5)$ kJ·mol^{-1}. This value compares favourably with the value, $-(194 \pm 3)\ \text{kJ·mol}^{-1}$, derived from the temperature dependence of the equilibrium constant in [56TOR] at $I = 0.3$ M.

The present review accepts the average of these values as valid also at $I = 0$, and assigns an uncertainty of $\pm 3\ \text{kJ·mol}^{-1}$ to account for the uncertainty of this assumption. The selection yields:

$$\Delta_r G^\circ_m((\text{V.103}), 298.15\text{ K}) = -(127.3 \pm 5.7)\ \text{kJ·mol}^{-1},$$

$$\Delta_r G^\circ_m((\text{V.104}), 298.15\text{ K}) = -(153.0 \pm 5.7)\ \text{kJ·mol}^{-1},$$

$$\Delta_r H^\circ_m((\text{V.105}), 298.15\text{ K}) = -(193.0 \pm 3.0)\ \text{kJ·mol}^{-1},$$

$$\Delta_r G^\circ_m((\text{V.105}), 298.15\text{ K}) = -(167.2 \pm 3.0)\ \text{kJ·mol}^{-1},$$

$$\Delta_r S^\circ_m((\text{V.105}), 298.15\text{ K}) = -(86.4 \pm 13.9)\ \text{J·K}^{-1}\text{·mol}^{-1}.$$

Combined with the selected value for $\Delta_f G^\circ_m(SeCN^-, 298.15\text{ K})$ and auxiliary data for $\Delta_f G^\circ_m(Hg^{2+}, 298.15\text{ K})$, this yields:

$$\Delta_f G^\circ_m(\text{Hg(SeCN)}_2, \text{aq}, 298.15\text{ K}) = (309.5 \pm 9.5)\ \text{kJ·mol}^{-1},$$

$$\Delta_f G^\circ_m(\text{Hg(SeCN)}_3^-, 298.15 \text{ K}) = (419.8 \pm 12.8) \text{ kJ·mol}^{-1},$$

$$\Delta_f G^\circ_m(\text{Hg(SeCN)}_4^{2-}, 298.15 \text{ K}) = (541.6 \pm 15.5) \text{ kJ·mol}^{-1}.$$

V.10 Group 11 compounds and complexes

V.10.1 Copper compounds

V.10.1.1 Copper selenides

The solid compounds formed in the system copper-selenium are $Cu_{2-x}Se(cr)$, $Cu_3Se_2(cr)$, CuSe(cr), and $CuSe_2(cr)$. The homogeneity range of $Cu_{2-x}Se(cr)$ and the polymorphic transitions of the phases are shown in the phase diagram by Massalski [86MAS] in Figure V-20. The phase diagram critically evaluated from literature data by Glazov, Pashinkin, and Fedorov [2000GLA/PAS] differs only in minor details from that in [86MAS].

Figure V-20: The phase diagram copper-selenium; after Massalski [86MAS]. Reprinted from [86MAS], Copyright (1986), with permission from ASM International.

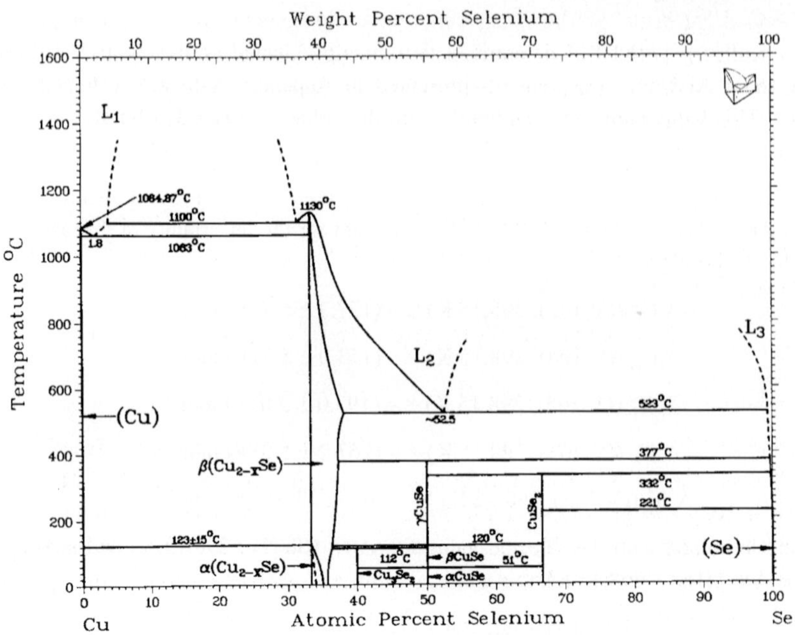

V.10.1.1.1 α-CuSe

Stølen, Fjellvåg, Grønvold, Sipowska, and Westrum [96STO/FJE] measured the heat capacity of α-CuSe in the temperature range 5 to 653 K using adiabatic shield calorimetry. The heat capacity determined at 298.15 K is selected:

$$C^{\circ}_{p,\mathrm{m}}(\mathrm{CuSe}, \alpha, 298.15\ \mathrm{K}) = (50.3 \pm 0.6)\ \mathrm{J{\cdot}K^{-1}{\cdot}mol^{-1}}.$$

The entropy at 298.15 K was determined to be S°_{m}(CuSe, α, 298.15 K) = (66.3 ± 0.8) $\mathrm{J{\cdot}K^{-1}{\cdot}mol^{-1}}$ and corresponds to an entropy of formation of $\Delta_{\mathrm{f}}S^{\circ}_{\mathrm{m}}$(CuSe, α, 298.15 K) = – (8.9 ± 0.8) $\mathrm{J{\cdot}K^{-1}{\cdot}mol^{-1}}$. The electrochemical measurements in the temperature range 320 to 420 K by Askerova, Alieva, Azizov, Abbasov, and Mustafayev [76ASK/ALI] using the cell

$$(-)\ \mathrm{Cu(cr)}\ |\ \mathrm{CuBr, KBr(in\ glycerol)}\ |\ \mathrm{CuSe(cr), CuSe_2(cr)}\ (+)$$

yielded the value $\Delta_{\mathrm{f}}S^{\circ}_{\mathrm{m}}$(CuSe, α, 298.15 K) = – (4.5 ± 4.2) $\mathrm{J{\cdot}K^{-1}{\cdot}mol^{-1}}$ after a correction for the entropy of the α → β transition (*cf.* Section V.10.1.1.2). The confidence intervals of the values of the entropy of formation of α-CuSe in [96STO/FJE] and [76ASK/ALI] thus overlap and the standard entropy in [96STO/FJE] is selected:

$$S^{\circ}_{\mathrm{m}}(\mathrm{CuSe}, \alpha, 298.15\ \mathrm{K}) = (66.3 \pm 0.8)\ \mathrm{J{\cdot}K^{-1}{\cdot}mol^{-1}}.$$

Gattow and Schneider [56GAT/SCH] determined the enthalpy of formation of α-CuSe to be $\Delta_{\mathrm{f}}H^{\circ}_{\mathrm{m}}$(CuSe, α, 298.15 K) = – (39.5 ± 0.5) $\mathrm{kJ{\cdot}mol^{-1}}$ (error estimated by the review) using direct synthesis calorimetry. The measurements in [76ASK/ALI] result in a value of $\Delta_{\mathrm{f}}H^{\circ}_{\mathrm{m}}$(CuSe, α, 298.15 K) = – (33.5 ± 10.0) $\mathrm{kJ{\cdot}mol^{-1}}$ after a correction for the enthalpy of the α → β transition (*cf.* Section V.10.1.1.2). The weighted mean of the enthalpy values is selected:

$$\Delta_{\mathrm{f}}H^{\circ}_{\mathrm{m}}(\mathrm{CuSe}, \alpha, 298.15\ \mathrm{K}) = -(39.5 \pm 0.5)\ \mathrm{kJ{\cdot}mol^{-1}}.$$

The Gibbs energy of formation of α-CuSe is calculated from the selected enthalpy of formation and entropy, the selected entropy of selenium and the entropy of copper in [89COX/WAG] yielding:

$$\Delta_{\mathrm{f}}G^{\circ}_{\mathrm{m}}(\mathrm{CuSe}, \alpha, 298.15\ \mathrm{K}) = -(36.8 \pm 0.6)\ \mathrm{kJ{\cdot}mol^{-1}}.$$

V.10.1.1.2 β-CuSe

The enthalpy of transformation of the α → β transition was determined to be $\Delta_{\mathrm{trs}}H^{\circ}_{\mathrm{m}}$(CuSe, α → β, 327 K) = (0.855 ± 0.006) $\mathrm{kJ{\cdot}mol^{-1}}$ in [96STO/FJE]. The corresponding entropy change is (2.61 ± 0.02) $\mathrm{J{\cdot}K^{-1}{\cdot}mol^{-1}}$. Due to closeness of the transition and standard temperatures, selected values of the enthalpy of formation and entropy of β-CuSe were calculated by adding the enthalpy of transformation and the corresponding entropy to the values of the enthalpy of formation and the entropy of α-CuSe. The selected values are:

$$S^{\circ}_{\mathrm{m}}(\mathrm{CuSe}, \beta, 298.15\ \mathrm{K}) = (68.9 \pm 0.8)\ \mathrm{J{\cdot}K^{-1}{\cdot}mol^{-1}}$$

and

$$\Delta_f H_m^\circ(\text{CuSe}, \beta, 298.15\text{ K}) = -(36.8 \pm 0.5)\text{ kJ·mol}^{-1}.$$

The Gibbs energy of formation of β-CuSe is calculated from the selected enthalpy of formation and entropy, the selected entropy of selenium and the entropy of copper in [89COX/WAG] yielding:

$$\Delta_f G_m^\circ(\text{CuSe}, \beta, 298.15\text{ K}) = -(34.91 \pm 0.56)\text{ kJ·mol}^{-1}.$$

V.10.1.1.3 CuSe(g)

The dissociation energy of the CuSe(g) molecule was determined by Smoes, Mandy, Auwera-Mahieu and Drowart [72SMO/MAN] from high temperature mass spectrometric measurements of ion intensities of the participating species. The dissociation energy was obtained from the reactions CuSe(g) $\rightleftharpoons$ ½Cu_2(g) + ½Se_2(g) studied in the temperature range 1302 to 1793 K and CuSe(g) + Se(g) $\rightleftharpoons$ Cu(g) + Se_2(g) studied at 1707 K giving the values $\Delta_{at}H_m^\circ$(CuSe, g, 0 K) = (253.1 ± 5.4) kJ·mol^{-1} and (249.8 ± 8.4) kJ·mol^{-1}, respectively. The authors considered the average value $\Delta_{at}H_m^\circ$(CuSe, g, 0 K) = (251.5 ± 14.6) kJ·mol^{-1} to properly describe their experiments. The evaluation involved Gibbs energy functions of CuSe(g) calculated from estimated molecular parameters. The results were re-evaluated by the review using the selected properties of Se(g) and Se_2(g), the thermodynamic properties of Cu(g) in [89COX/WAG], the thermodynamic properties of Cu_2(g) in [98CHA], and thermal properties of CuSe(g) calculated from the parameters estimated in [72SMO/MAN] (properties at standard conditions calculated to be $C_{p,m}^\circ$(CuSe, g, 298.15 K) = 36.3 J·K^{-1}·mol^{-1} and S_m°(CuSe, g, 298.15K) = 255.1 J·K^{-1}·mol^{-1}), yielding $\Delta_f H_m^\circ$(CuSe, g, 298.15 K) = 324.5 kJ·mol^{-1} from the third law and the first reaction and $\Delta_f H_m^\circ$(CuSe, g, 298.15 K) = 317.5 kJ·mol^{-1} from the second law, and $\Delta_f H_m^\circ$(CuSe, g, 298.15 K) = 320.3 kJ·mol^{-1} from the third law and the last reaction. The reasonable agreement between the results obtained in the temperature extrapolations with the second and the third laws indicate that the estimated thermal data and the entropies are sufficiently accurate to permit the average of the enthalpy values to be accepted with error bars from [72SMO/MAN]. The review thus selects:

$$\Delta_f H_m^\circ(\text{CuSe}, \text{g}, 298.15\text{ K}) = (320.8 \pm 15.0)\text{ kJ·mol}^{-1}.$$

V.10.1.1.4 CuSe$_2$(cr)

Gattow and Schneider [56GAT/SCH] determined the enthalpy of formation of $CuSe_2$(cr) to be $\Delta_f H_m^\circ$($CuSe_2$, cr, 298.15 K) = – 43.1 kJ·mol^{-1} using direct synthesis calorimetry. Electrochemical cell measurements in the temperature range 320 to 420 K in [76ASK/ALI] using the cell

(–) Cu(cr) | CuBr, KBr (in glycerol) | $CuSe_2$(cr), Se(trig) (+)

yielded the value $\Delta_f H^\circ_m$ ($CuSe_2$, cr, 298.15 K) = – (39.3 ± 4.0) $kJ \cdot mol^{-1}$. The mean of the enthalpy values is selected:

$$\Delta_f H^\circ_m (CuSe_2, cr, 298.15 K) = -(41.2 \pm 5.9)\ kJ \cdot mol^{-1},$$

where the results in the two investigations have been given equal weights.

V.10.1.1.5 α-Cu_2Se

The composition of α-Cu_2Se varies within a small range at room temperature, see the phase diagram in Figure V-20. Kubaschewski [73KUB] measured the heat capacity of $Cu_{1.968}Se$(cr) in the temperature range 180 to 560 K. The heat capacity above 250 K is strongly dependent on the exact composition of the α-Cu_2Se phase and no reliable value for the stoichiometric composition Cu_2Se can be derived from these measurements. Blachnik and Gunia [78BLA/GUN] measured the enthalpy content in the temperature range 350 to 1500 K using drop calorimetry and tabulated the heat capacity as a function of temperature. The heat capacity at 300 K was reported to be $C^\circ_{p,m}$ (Cu_2Se, α, 300 K) = 75.23 $J \cdot K^{-1} \cdot mol^{-1}$ and was obtained from an extrapolation of the values measured in the temperature range 350 to 395 K. Considering the rapid changes in the heat capacity within the range, the estimated value at 300 K is of doubtful quality and not selected by the review.

Glazov and Korenchuk [71GLA/KOR] measured the saturated vapour pressure of α-Cu_2Se in the temperature range 1133 to 1293 K yielding $\Delta_f H^\circ_m$ (Cu_2Se, α, 298.15 K) = – (26.5 ± 10.0) $kJ \cdot mol^{-1}$ and S°_m (Cu_2Se, α, 298.15 K) = (90.1 ± 20.0) $J \cdot K^{-1} \cdot mol^{-1}$ after a recalculation as discussed in Appendix A. Gattow and Schneider [56GAT/SCH] determined the enthalpy of formation to be $\Delta_f H^\circ_m$ (Cu_2Se, α, 298.15 K) = – 59.3 $kJ \cdot mol^{-1}$ using direct synthesis calorimetry. Electrochemical cell measurements in the temperature range 320 to 420 K by Askerova, Alieva, Azizov, Abbasov, and Mustafayev [76ASK/ALI] using the cell

$$(-)\ Cu(cr)\ |\ CuBr, KBr\ (in\ glycerol)\ |\ Cu_2Se(cr), Cu_3Se_2(cr)\ (+)$$

yielded the values $\Delta_f H^\circ_m$ (Cu_2Se, α, 298.15 K) = – (65.7 ± 6.5) $kJ \cdot mol^{-1}$ and $\Delta_f S^\circ_m$ (Cu_2Se, α, 298.15 K) = – (28.9 ± 6.0) $J \cdot K^{-1} \cdot mol^{-1}$ where the error limits are those proposed by the authors. However, the measurements are less reliable than the error limits indicate because they were made across the α → β transition at 395 K. No value is selected for the entropy of α-Cu_2Se from the measurements in [71GLA/KOR] and [76ASK/ALI] because of the doubtful experimental conditions. The value of the enthalpy of formation in [56GAT/SCH] is selected:

$$\Delta_f H^\circ_m (Cu_2Se, \alpha, 298.15 K) = -(59.3 \pm 0.5)\ kJ \cdot mol^{-1},$$

where the error limits were set by the review.

V.10.1.1.6 β-Cu_2Se

The β-Cu_2Se phase has a variable composition and is unstable at the stoichiometric composition at 298.15 K, see the phase diagram in Figure V-20.

The enthalpy change of the α → β transition was determined to be $\Delta_{trs}H_m^\circ$(Cu_2Se, α → β, 395 K) = (6.4 ± 2.0) kJ·mol^{-1} by Murray and Heyding [75MUR/HEY] and 6.5 kJ·mol^{-1} in [78BLA/GUN]. The review estimates the enthalpy of formation of β-Cu_2Se at standard conditions by adding the transformation enthalpy to the enthalpy of formation of α-Cu_2Se to be $\Delta_fH_m^\circ$(Cu_2Se, β, 298.15 K) = – (52.9 ± 2.1) kJ·mol^{-1}. This value is not selected.

V.10.1.1.7 Cu_2Se(g)

The atomisation enthalpy of the Cu_2Se(g) molecule was determined by Smoes, Mandy, Auwera-Mahieu and Drowart [72SMO/MAN] from mass spectrometric measurements of ion intensities of the species in the reaction Cu_2Se(g) + Se(g) $\rightleftharpoons$ 2CuSe(g) in the temperature range 1508 to 1707 K. The results were evaluated using only the third law involving Gibbs energy functions of CuSe(g) and Cu_2Se(g) calculated from estimated molecular parameters, since the experimental data were too scattered to provide a meaningful second law value. The results were recalculated by the review using the selected properties of Se(g), the properties of CuSe(g), as calculated in Section V.10.1.1.3, and the thermal properties of Cu_2Se(g) calculated from the parameters estimated in [72SMO/MAN] (properties at standard conditions calculated to be $C_{p,m}^\circ$(Cu_2Se, g, 298.15K) = 55.3 J·K^{-1}·mol^{-1} and S_m°(Cu_2Se, g, 298.15K) = 311.9 J·K^{-1}·mol^{-1}), yielding $\Delta_fH_m^\circ$(Cu_2Se, g, 298.15 K) = (388.3 ± 18.3) kJ·mol^{-1} from the third law. The value is not selected by the review because of the difficulties with the second law value and the uncertainties originating from the tentative properties of the Cu_2Se(g) molecule.

V.10.1.1.8 Cu_3Se_2(cr)

Gattow and Schneider [56GAT/SCH] determined the enthalpy of formation of Cu_3Se_2(cr) to be $\Delta_fH_m^\circ$(Cu_3Se_2, cr, 298.15 K) = – 98.9 kJ·mol^{-1} using direct synthesis calorimetry. Electrochemical cell measurements in the temperature range 320 to 420 K in [76ASK/ALI] using the cell

(–) Cu(cr) | CuBr, KBr (in glycerol) | Cu_3Se_2(cr), CuSe(cr) (+)

yielded the value $\Delta_fH_m^\circ$(Cu_3Se_2, cr, 298.15 K) = – (94.6 ± 7.1) kJ·mol^{-1}. The selected value of the enthalpy of formation is

$$\Delta_fH_m^\circ(Cu_3Se_2, \text{cr}, 298.15\text{ K}) = -(96.8 \pm 7.1)\text{ kJ·mol}^{-1}$$

where the results in the two investigations have been given equal weights.

V.10.1.2 Copper selenites

Ojkova and Gospodinov [82OJK/GOS] studied the phase equilibria in the system CuO-SeO_2-H_2O at 373.2 K by solubility measurements. The solubility data are presented together with the phase diagram. The diagram contains crystallisation fields of $CuSeO_3{\cdot}2H_2O$ and $CuSeO_3{\cdot}H_2SeO_3$. The solubility of Cu(II) is small over the whole range from 0 to 70 mass-% SeO_2 investigated. Micka, Cermak, and Niznansky [90MIC/CER] studied this system at 298.15 K and obtained solids with the same empirical formulas. The compound $CuSeO_3{\cdot}H_2SeO_3$ was shown by infrared spectroscopy to be $Cu(HSeO_3)_2$. The phase diagram is the only result presented of the solubility measurements.

Chukhlantsev [56CHU] prepared copper selenite by mixing 0.1 M solutions of copper sulphate and sodium selenite in the cold. The product was aged for 24 hours. Chemical analysis confirmed the 1:1 ratio between Cu(II) and Se(IV). The solubility of the specimen in dilute solution of nitric or sulphuric acid was measured at 293 K. No X-ray diffraction measurements were performed and the solubility equilibrium will be written on the assumption that the composition of the solid phase is $CuSeO_3{\cdot}2H_2O$:

$$CuSeO_3{\cdot}2H_2O(s) \rightleftharpoons Cu^{2+} + SeO_3^{2-} + 2H_2O(l) \qquad (V.106)$$

The data have been recalculated as described in Appendix A, [56CHU]. The result was $\log_{10} K^{\circ}_{s,0}$ ((V.106), 293.15 K) = – (8.26 ± 0.30). The reported conditional constant is – (7.68 ± 0.35).

Ripan and Vericeanu [68RIP/VER] studied the solubility of $CuSeO_3{\cdot}2H_2O(s)$ by conductivity measurements as described in Appendix A. An equilibrium analysis performed as outlined there on the total concentrations provided in the paper, about 1.8×10^{-4} M, leads to the conclusion that almost 80% of the copper(II) would be present as hydroxo complexes at equilibrium. Hence, a conductivity measurement is unsuitable for the determination of the solubility of copper selenite and no reliable value of the solubility product can be calculated from the data in [68RIP/VER]. Masson *et al.* [86MAS/LUT] calculated $\log_{10} K^{\circ}_{s,0} = -(7.49 \pm 0.10)$ from the data neglecting hydrolysis of Cu^{2+}.

Feroci *et al.* [97FER/FIN] studied the system Cu^{2+}-SeO_3^{2-} by polarography. No shift in $E_{1/2}$, but a steady decrease in the limiting current, was observed with increasing selenite concentration. It was assumed from these results that solid copper selenite was formed, although no precipitate was observed in the electrode vessel. The solubility product was reported as $\log_{10} K_{s,0}$ ((V.106), 0.15 M $NaNO_3$, ? K) = – 8.42. As discussed in Appendix A, the result of this investigation can not be considered as reliable.

Slavtscheva, Popova, and Gospodinov [93SLA/POP] determined the solubility of copper selenite in dilute nitric acid and calculated its solubility product from measurements of pH and total Cu(II) concentration in the equilibrium solution. No primary data are provided, which precludes a recalculation of their result, $\log_{10} K^{\circ}_{s,0}$ ((V.106), 298.15 K) = – (10.82 ± 0.10), with selected auxiliary data.

In view of the scatter between the reported values of the solubility product and the limited experimental information the review can only regard the results in [56CHU] and [93SLA/POP] as upper and lower limits, which leads to the not selected estimate $\log_{10} K^{\circ}_{s,0}$ ((V.106), 298.15 K) = – (9.5 ± 1.5).

The standard Gibbs energy of formation of $CuSeO_3$(cr) in [82WAG/EVA] is apparently based on the conditional solubility product in [56CHU] and cannot be recommended.

Leshchinskaya and Selivanova [65LES/SEL] measured the enthalpy change of the reaction Na_2SeO_3(cr) + $CuSO_4$(aq, 1:800) + $2H_2O$(l) → $CuSeO_3\cdot 2H_2O$(cr) + Na_2SO_4(aq, 1:800). They also measured the integral enthalpies of dissolution of $CuSeO_3\cdot 2H_2O$(cr) and $CuSeO_3$(cr) in 8% HNO_3. The calorimetric data have been used to calculate the standard enthalpies of formation of the copper selenites in Appendix A to be:

$$\Delta_f H^{\circ}_m (\mathrm{CuSeO_3\cdot 2H_2O, cr, 298.15\ K}) = -(998.60 \pm 1.51)\ \mathrm{kJ\cdot mol^{-1}}$$

and

$$\Delta_f H^{\circ}_m (\mathrm{CuSeO_3, cr, 298.15\ K}) = -(386.90 \pm 1.52)\ \mathrm{kJ\cdot mol^{-1}}.$$

These data are selected by the review.

Thermal analysis [82OJK/GOS], [84GOS/BOG] indicates that under equilibrium conditions the dissociation of copper selenite would follow the scheme:

$$2\mathrm{CuSeO_3(cr)} \rightleftharpoons \mathrm{CuO\cdot CuSeO_3(cr)} + \mathrm{SeO_2(g)} \qquad \text{(V.107)}$$

followed by:

$$\mathrm{CuO\cdot CuSeO_3(cr)} \rightleftharpoons 2\mathrm{CuO(cr)} + \mathrm{SeO_2(g)}. \qquad \text{(V.108)}$$

The vapour pressures over $CuSeO_3$(cr) and $CuO\cdot CuSeO_3$(cr) were measured with a membrane null manometer and with a Knudsen effusion cell with mass spectrometric detection by Makatun, Rat'kovskii, and Pechkovskii [71MAK/RAT]. The dew point method was used by Bakeeva *et al.* [72BAK/BUK] for the same purpose.

The results obtained for Reaction (V.108) differ substantially between the two investigations. For instance, the observed pressure of SeO_2(g) at 830 K was 4 times larger in [71MAK/RAT] than in [72BAK/BUK]. The variation of the equilibrium constant with temperature is also different and corresponds to $\Delta_r H^{\circ}_m$ ((V.108), 820 K) = 176 kJ·mol^{-1} in [71MAK/RAT] and 216 kJ·mol^{-1} at 860 K in [72BAK/BUK].

Both research groups observed a break in the plot of $\log_{10} p_{SeO_2}$ against $1/T$ for the experiments with $CuSeO_3$(cr). There is, however, a difference of about 65 K between the break point temperatures. Bakeeva *et al.* assigned the change in slope to a phase transition of $CuSeO_3$, which would then have the notably high transition enthalpy of about 65.5 $kJ \cdot mol^{-1}$. Makatun *et al.* suggested that the break was due to a change in the dissociation reaction. It was found by chemical and X-ray analysis that CuO was the reaction product below 700 K, while at higher temperatures $CuO \cdot CuSeO_3$(cr) was present. This proposal is not supported by the presented enthalpy changes.

Gospodinov and Bogdanov [84GOS/BOG] report the total enthalpy change of the reaction $CuSeO_3(cr) \rightarrow CuO(cr) + SeO_2(g)$ from a measurement by differential scanning calorimetry (DSC) to be 36 $kJ \cdot mol^{-1}$, which is an improbably low value.

The conflicting results of the investigations indicate unsolved experimental difficulties in establishing representative conditions. No thermodynamic quantities of the thermal dissociation of $CuSeO_3$ will be extracted from the references.

The heat capacities of $CuSeO_3$(cr) and $CuO \cdot CuSeO_3$(cr) were determined in the temperature range 593 to 673 K by DSC in [84GOS/BOG] to be $C^{\circ}_{p,m}$($CuSeO_3$, cr, (593 – 673) K) = $(1.21 + 0.306\,T + 1.79 \times 10^{-5}\,T^2)$ $J \cdot K^{-1} \cdot mol^{-1}$, and $C^{\circ}_{p,m}$($CuO \cdot CuSeO_3$, cr, (593 – 673) K) = $(148.4 - 2.66 \times 10^{-2}\,T + 1.78 \times 10^{-4}\,T^2)$ $J \cdot K^{-1} \cdot mol^{-1}$.

The result for $CuSeO_3$(cr) differs substantially from the estimate in [73BAR/KNA] and the heat capacities will not be accepted.

V.10.1.3 Copper selenates

Klein [40KLE] and Gospodinov [84GOS2] studied the temperature dependence of the solubility of copper selenates and found $CuSeO_4 \cdot 5H_2O$ to be the composition of the solid phase below 363 K. According to Klein this compound is only stable to about 320 K in the presence of water. Above this temperature a transformation to basic copper selenates starts. This phenomenon was not noted by Gospodinov who presents a table of solubilities from 273 to 373 K. The agreement between the two sets of solubility data is good in the overlapping temperature range 273 to 353 K.

The solubility of $CuSeO_4 \cdot 5H_2O$(cr) at 298.15 K is available in [32ALP/SAY], [45KIN/BEC], [84GOS2], [95STO/OJK], which report the (recalculated) values 0.915, 0.929, 0.903, and 0.941 mol($CuSeO_4$)$\cdot kg^{-1}$, respectively. The review selects the mean value:

$$m(CuSeO_4 \cdot 5H_2O(cr), \text{aq sat}, 298.15\ K) = (0.922 \pm 0.033)\ mol(CuSeO_4) \cdot kg^{-1}.$$

A phase diagram and related information are available for the systems: $CuSeO_4$-H_2SeO_4-H_2O at 298.15 K in [32ALP/SAY], [84GOS2]; $CuSeO_4$-$MgSeO_4$-H_2O at 298.15 K in [95STO/OJK]; $CuSeO_4$-Na_2SeO_4-H_2O at 298.15 K in [91OYK/MIH]; $CuSeO_4$-K_2SeO_4-H_2O at 298.15 K in [87VOJ/EBE].

The value of the solubility product of $CuSeO_4 \cdot 5H_2O$(cr)

$$CuSeO_4 \cdot 5H_2O(cr) \rightleftharpoons Cu^{2+} + SeO_4^{2-} + 5H_2O(l) \qquad (V.109)$$

at 298.15 K has been reported by Kumov and Batyreva [90KUM/BAT] to be $\log_{10} K^{\circ}_{s,0}$ (V.109) = – (2.44 ± 0.09). The value is probably derived from the isopiestic data by Ojkova and Staneva [89OJK/STA]. The review calculated the value $\log_{10} K^{\circ}_{s,0}$ = – 2.44 from the data in [89OJK/STA] allotted to the saturated solution. It is to be noted that the mean activity factors and water activities of copper selenate solutions in [89OJK/STA] are based on very few experimental data. An attempt by the review to calculate the activities in saturated solution by the Pitzer approach from the tabulated data for non-saturated solutions in [89OJK/STA] failed. The value:

$$\log_{10} K^{\circ}_{s,0} \text{ ((V.109), 298.15 K)} = -(2.44 \pm 0.20)$$

will be accepted and selected with an increased uncertainty added here.

The standard Gibbs energy of formation of $CuSeO_4{\cdot}5H_2O$(cr) is calculated from the Gibbs energy of Reaction (V.109) and the selected values of $\Delta_f G^{\circ}_m$ for the ions and water to be:

$$\Delta_f G^{\circ}_m (CuSeO_4{\cdot}5H_2O, \text{cr, 298.15 K}) = -(1574.10 \pm 2.42) \text{ kJ·mol}^{-1}.$$

Feroci, Fini, Badiello, and Breccia [97FER/FIN] studied the complex formation between Cu^{2+} and SeO_4^{2-} by classical polarography and found no evidence for an interaction, see Appendix A. The review estimates from an assumed accuracy of the polarographic measurement of ± 2 mV the equilibrium constant of

$$Cu^{2+} + SeO_4^{2-} \rightleftharpoons CuSeO_4(aq) \qquad (V.110)$$

to be $\log_{10} \beta_1^{\circ}$ ((V.110), 298.15 K) < 2.2.

The standard enthalpies of formation of $CuSeO_4$(cr) and $CuSeO_4{\cdot}5H_2O$(cr) have been obtained from the calorimetric measurements by Selivanova, Samplavskaya, and Maier [62SEL/SAM]. The recalculated values:

$$\Delta_f H^{\circ}_m (CuSeO_4, \text{cr, 298.15 K}) = -(480.33 \pm 2.81) \text{ kJ·mol}^{-1},$$

$$\Delta_f H^{\circ}_m (CuSeO_4{\cdot}5H_2O, \text{cr, 298.15 K}) = -(1973.33 \pm 2.83) \text{ kJ·mol}^{-1}$$

are selected. The first value agrees with [82WAG/EVA] but the second disagrees by about 20 kJ·mol^{-1}. It appears likely that the NBS evaluation entered the enthalpy of dissolution of $CuSeO_4{\cdot}5H_2O$(cr) with the wrong sign.

V.10.1.4 Copper selenocyanates

Golub and Skopenko [60GOL/SKO] measured the emf of the cell

$$\text{Cu, CuSeCN} \mid H_2O, \text{satd.} \mid 0.1 \text{ M KCl} \mid 0.1 \text{ M KCl} \mid Hg_2Cl_2, \text{Hg}$$

at 293 K. It is not stated whether the left half-cell was separately saturated with CuSeCN(s) or if the solution in immediate contact with the electrode was only assumed to be saturated with the solid phase. The cell is stated to yield reproducible emf readings and from the value registered, E = – 0.0972 V, the concentration of Cu^+ was calculated to be $\log_{10} [Cu^+] = -4.869$ and, subsequently, $\log_{10} K_{s,0}$ (CuSeCN, s, 293.15 K) = -9.738. However, these calculations are invalidated by the disproportionation properties of Cu^+ which, under the conditions, can be calculated to correspond to a Cu^{2+} concentration of 6.3×10^{-4} M.

Kilmartin and Wright [97KIL/WRI] recently studied the development of thin CuSeCN(s) layers on metallic copper in 0.1 M $SeCN^-$ solutions by cyclic voltammetry. The peaks observed were assumed to relate to the formation and removal of CuSeCN(s) according to the reaction Cu(cr) + $SeCN^- \rightleftharpoons$ CuSeCN(s) + e^-. The shapes of the voltammograms do, however, indicate that the electrode reaction exhibits non-Nernstian behaviour. The authors also noted that during their experiments, a possible formation of elemental selenium occurred on the electrode surface. Therefore, the recorded electrode potential characteristics cannot be regarded as well established and these data cannot be used for calculating the standard electrode potential of the above redox couple and the solubility product of CuSeCN(cr).

Kullberg [74KUL] found it impossible to study the aqueous Cu^{2+}-$SeCN^-$ system since a brown precipitate was immediately formed upon addition of selenocyanate ions to a copper(II) solution. Toropova [56TOR] prepared CuSeCN(s) by adding KSeCN to a solution of copper(I) sulphite or thiosulphate complexes, and tried to study the complex formation by measuring the solubility of CuSeCN in selenocyanate solutions. The experiment failed since the complexes initially formed were found to decompose into elementary selenium and copper(I) cyanide compounds.

In conclusion, the review finds that no sound information on the thermodynamic properties of copper selenocyanates is available.

The compound CuSeCN(cr) is without doubt sparingly soluble in water, and additions of Cu^+ can be used to scavenge $SeCN^-$ ions from crude oil processing waters and wastewaters [97MAN/GAL]. With time, however, it seems to decompose into Se(cr) and CuCN(aq).

V.10.2 Silver compounds

V.10.2.1 Silver selenides

The silver-selenium binary phase diagram was reviewed by Karakaya and Thompson [90KAR/THO]. Ag_2Se(cr) is the only intermediary phase in the system and exists in two polymorphs, α and β. The α form is stable below 406 K. The melting point is 1170 K.

V.10.2.1.1 AgSe(g)

The dissociation energy of the AgSe(g) molecule was determined by Smoes, Mandy, Auwera-Mahieu and Drowart [72SMO/MAN] from high temperature mass spectrometric measurements of the ion intensities of the participating species. The dissociation energy was obtained from the reactions AgSe(g) $\rightleftharpoons$ ½Ag_2(g) + ½Se_2(g) studied in the temperature range 1445 to 1544 K and AgSe(g) + Se(g) $\rightleftharpoons$ Ag(g) + Se_2(g) studied in the temperature range 1295 to 1490 K giving the values $\Delta_{at}H^{\circ}_{m}$ (AgSe, g, 0 K) = (203.8 ± 6.3) kJ·mol^{-1} and $\Delta_{at}H^{\circ}_{m}$ (AgSe, g, 0 K) = (208.8 ± 5.0) kJ·mol^{-1}, respectively. The authors considered the average value $\Delta_{at}H^{\circ}_{m}$ (AgSe, g, 0 K) = (206.3 ± 14.6) kJ·mol^{-1} to properly describe their experiments. The evaluation involved Gibbs energy functions of AgSe(g) calculated from estimated molecular parameters. The results were re-evaluated by the review using the selected properties of Se(g) and Se_2(g), the thermal functions of Ag(g) in [89COX/WAG], the enthalpy and entropy of Ag(g) in [89COX/WAG], the enthalpy and entropy of Ag_2(g) in [82WAG/EVA], thermal properties of Ag_2(g) calculated from the molecular properties in [86MOR3], and thermal properties of AgSe(g) calculated from the parameters estimated in [72SMO/MAN], yielding $\Delta_{f}H^{\circ}_{m}$ (AgSe, g, 298.15 K) = 304.6 kJ·mol^{-1} from the second law and $\Delta_{f}H^{\circ}_{m}$ (AgSe, g, 298.15 K) = 311.8 kJ·mol^{-1} from the third law and the first reaction and $\Delta_{f}H^{\circ}_{m}$ (AgSe, g, 298.15 K) = 310.4 kJ·mol^{-1} from the second law and $\Delta_{f}H^{\circ}_{m}$ (AgSe, g, 298.15 K) = 309.3 kJ·mol^{-1} from the third law and the last reaction. The values correspond to the dissociation enthalpies $\Delta_{at}H^{\circ}_{m}$ (AgSe, g, 298.15 K) = 216.1 kJ·mol^{-1}, 208.8 kJ·mol^{-1}, 210.3 kJ·mol^{-1}, and 211.4 kJ·mol^{-1}, respectively. The derived values of the heat of formation of AgSe(g) agree well. Since the entropy and heat capacity of a diatomic molecule calculated from estimated molecular parameters are expected to be quite accurate, the review selects the mean of the four results and obtains:

$$\Delta_{f}H^{\circ}_{m} \text{(AgSe, g, 298.15 K)} = (309.0 \pm 6.0)\ \text{kJ·mol}^{-1}.$$

The error bars have been increased to account for a possible systematic error.

V.10.2.1.2 Ag_2Se(cr)

The heat capacity measurements made on Ag_2Se and the determinations of the enthalpy of transition from α-Ag_2Se to β-Ag_2Se are summarised in Table V-60.

The weighted mean of the heat capacity values is selected:

$$C^{\circ}_{p,m} (\text{Ag}_2\text{Se}, \alpha, 298.15\ \text{K}) = (81.15 \pm 0.90)\ \text{J·K}^{-1}\text{·mol}^{-1}$$

where the error limits have been increased because the overlap of the uncertainty ranges of the individual values is poor.

The temperature dependence of the heat capacity above 298.15 K was only presented as graphs in all publications except in [78BLA/GUN] and [2003GRO/STO] in which also numerical values were presented. No analytical expression for the temperature dependence of the heat capacity was derived by the review because the numerical values in the two series of measurements differ by several J·K^{-1}·mol^{-1}.

Table V-60: Heat capacity measurements and determinations of the enthalpy of transition from α-Ag_2Se to β-Ag_2Se.

Reference	Temperature (K)	$C^{o}_{p,m}$ (Ag_2Se, α, 298.15 K) ($J\cdot K^{-1}\cdot mol^{-1}$)	$\Delta_{trs}H^{o}_{m}$ (Ag_2Se, α → β, 406K) ($kJ\cdot mol^{-1}$)
[59GUL/PET]	80 – 300	(79.2 ± 5.0)	
[62BAE/BUS]	293 – 463	(83.7 ± 3.0)	(7.0 ± 0.6)
[62WAL/ART]	16 – 300	(81.46 ± 0.30)	
[68JOS/KUB]	200 – 630	(82.4 ± 0.8)	
[78BLA/GUN]	300 – 1500	(80.95 ± 0.50)	(6.80 ± 0.14)
[82SHU/SEN]			(7.68 ± 0.10)
[87HON/IID]	120 – 520	(83.7 ± 3.0)	(6.7 ± 1.0)
[2003GRO/STO]	300 – 900	(80.35 ± 0.40)	(6.82 ± 0.02)

The determinations of the entropy at 298.15 K are summarised in Table V-61. The value determined from low temperature heat capacity measurements in [62WAL/ART] agrees well with the second law entropy value obtained from the electrochemical measurements in the temperature range 433 to 643 K made by Kiukkola and Wagner [57KIU/WAG]. The other studies were made at high temperatures and the second law entropies scatter considerably due to the small temperature ranges of the measurements. The weighted mean of the values in [62WAL/ART] and [57KIU/WAG] is selected:

$$S^{o}_{m}(Ag_2Se, \alpha, 298.15\ K) = (149.9 \pm 0.5)\ J\cdot K^{-1}\cdot mol^{-1}.$$

Table V-61: Determinations of the entropy of α-Ag_2Se at 298.15 K. All values were evaluated using the second law except that in [62WAL/ART] which was obtained from low temperature heat capacity measurements. The error limits represent the uncertainties in the linear regressions where applicable.

Reference	S^{o}_{m} (Ag_2Se, α, 298.15 K) ($J\cdot K^{-1}\cdot mol^{-1}$)	Temperature (K)
[57KIU/WAG]	(148.5 ± 1.0)	433 – 643
[62WAL/ART]	(150.2 ± 0.5)	16 – 300
[77DUT/TIS]	(175.8 ± 6.0)	1170 – 1270
[77DUT/TIS]	(153.1 ± 6.0)	1070 – 1170
[90SCA/PIA]	(131.9 ± 5.0)	1045 – 1166
[90SCA/PIA]	(127.6 ± 4.0)	1187 – 1305

The enthalpy of transition from α-Ag_2Se to β-Ag_2Se is dependent on the exact stoichiometry of the selenide as shown by the studies of Shukla, Sen and Sarma [82SHU/SEN]. Measured values for the enthalpy of transition were 6.28, 7.26, and 7.74 $kJ\cdot mol^{-1}$ for the deviations δ = 0.00069, 0.00038, and – 0.000095 in the formula $Ag_{2+\delta}Se$, respectively. Considering these facts, the other transformation enthalpies in Table V-60 are within the range of the determinations made in [82SHU/SEN]. The review selects from this reference:

$$\Delta_{trs}H^{\circ}_{m}(Ag_2Se, \alpha \rightarrow \beta, 406\ K) = (7.68 \pm 0.50)\ kJ\cdot mol^{-1},$$

where the error limits have been increased to reflect the possibility of systematic errors.

The enthalpy of formation has been determined from vapour pressure and galvanic cell measurements. Each determination has been re-evaluated as discussed in Appendix A using the second and third laws, the selected heat capacity, the entropy of $Ag_2Se(cr)$, the selected properties of selenium, and the CODATA [89COX/WAG] values of silver. The results are summarised in Table V-62.

Table V-62: Re-evaluated enthalpy of formation of α-Ag_2Se at 298.15 K obtained from high temperature measurements using (a) the second and (b) the third law. The error limits of the second law values represent the uncertainties in the linear regressions.

Reference	$\Delta_f H^{\circ}_{m}$ (Ag_2Se, α, 298.15 K) (a) ($kJ\cdot mol^{-1}$)	$\Delta_f H^{\circ}_{m}$ (Ag_2Se, α, 298.15 K) (b) ($kJ\cdot mol^{-1}$)	Temperature (K)	Method
[57KIU/WAG]	– (43.6 ± 1.6)	– 40.0	433 – 643	emf
[77DUT/TIS]	– (37.8 ± 8.0)	– 40.9	1070 – 1170	Mass spectrometry
[77DUT/TIS]	– (14.5 ± 8.0)	– 40.8	1170 – 1270	Mass spectrometry
[90SCA/PIA]	– (58.9 ± 6.0)	– 38.7	1045 – 1166	Torsion effusion
[90SCA/PIA]	– (67.9 ± 6.0)	– 40.9	1187 – 1305	Torsion effusion

The second law values obtained from the high temperature measurements are widely scattered, but the third law evaluations made by the review reduce the differences between different investigations substantially. The study of Kiukkola and Wagner [57KIU/WAG] is the only one in which the second and third law values agree and the result conforms to the third law values of the other investigations. The mean of the third law values $\Delta_f H^{\circ}_{m}$ (Ag_2Se, α, 298.15 K) = – (40.3 ± 1.9) $kJ\cdot mol^{-1}$ is considered to summarise the high temperature measurements in Table V-62. The error limit was set by considering the scatter of the third law values. The corresponding Gibbs energy of formation is calculated to be $\Delta_f G_m^\circ$ (Ag_2Se, α, 298.15 K) = – (47.2 ± 2.0) kJ·mol^{-1}.

In aqueous solution, Mehra and Gubeli [70MEH/GUB] studied the solubility of Ag_2Se(s) at 298.15 K in a 1 M $NaClO_4$ medium as a function of pH and selenide concentration, see [71MEH/GUB] in Appendix A. The potentiometric measurements yielded the concentration of Ag^+ in solutions saturated with Ag_2Se(s) and containing a total selenide concentration between 0.01 and 0.1 M in the pH range of 1.3 to 13.6. The primary data in [68MEH] at pH > 12.6 were used to calculate the concentration product Q = $[Ag^+]^2$[Se(– II)], where [Se(– II)] denotes the sum of the concentrations of H_2Se(aq), HSe^- and Se^{2-}. A constant value of Q was obtained and Q was identified with the solubility product of silver selenide and a value of – (53.8 ± 0.1) was assigned to the logarithm of this equilibrium constant. However, since the protonation constant of Se^{2-} in 1 M $NaClO_4$ is estimated from [90LEV/MYE] to be $\log_{10} K_1$ = (13.97 ± 0.10), the constant value of Q appears to be fortuitous and the proposed value of the solubility product is not accepted.

From the data in the pH range of 4.4 to 10.0, as reported in [68MEH], the review calculates the equilibrium constant of the reaction

$$Ag_2Se(s) + H^+ \rightleftharpoons 2Ag^+ + HSe^- \qquad (V.111)$$

to be $\log_{10} K$ (V.111) = – (42.2 ± 0.4). In this pH range HSe^- predominates and the glass electrode measurements are expected to yield reasonably accurate hydrogen ion activities. The value of $\log_{10} \gamma_{H^+}$ is estimated from SIT to be – 0.06, which leads to the accepted equilibrium constant:

$$\log_{10} K \text{ ((V.111), 1 M NaClO}_4\text{, 298.15 K)} = -(42.2 \pm 0.4).$$

The value extrapolated to standard conditions with SIT and $\Delta\varepsilon$ = – 0.09 kg·mol^{-1} is – (42.7 ± 0.4). Combined with the accepted first protonation constant of Se^{2-} this results in $\log_{10} K_{s,0}^\circ$ (Ag_2Se, s, 298.15 K) = – (57.6 ± 0.4).

From the measurement in the pH range of 1.3 to 2.2, the review calculates for the equilibrium

$$Ag_2Se(s) + 2H^+ \rightleftharpoons 2Ag^+ + H_2Se(aq), \qquad (V.112)$$

$\log_{10} K$ ((V.112), 1 M $NaClO_4$, 298.15 K) = – (38.8 ± 0.4).

The value of the constant at standard conditions is estimated with SIT and $\Delta\varepsilon$ = – 0.14 kg·mol^{-1} to be – (38.95 ± 0.40). This value combines with the protonation constants of Se^{2-} to $\log_{10} K_{s,0}^\circ$ (Ag_2Se, s, 298.15 K) = – (57.7 ± 0.4). The two estimates of the solubility product thus agree with a mean value of $\log_{10} K_{s,0}^\circ$ = – (57.65 ± 0.50). Combined with the CODATA values for Ag^+ and the selected data for Se^{2-} this corresponds to $\Delta_f G_m^\circ$ (Ag_2Se, s, 298.15 K) = – (46.3 ± 4.1) kJ·mol^{-1}.

The $\Delta_f G^\circ_m$ (Ag_2Se, α, 298.15 K) from the high temperature measurements and from solution measurements agree within the uncertainty limits. These data were combined (α-Ag_2Se is assumed to be present in the solubility measurement) with the weights 2:1 to the selected value:

$$\Delta_f G^\circ_m (Ag_2Se, \alpha, 298.15\ K) = -(46.9 \pm 1.3)\ kJ\cdot mol^{-1}.$$

The selected standard enthalpy of formation is calculated with the selected entropy to be:

$$\Delta_f H^\circ_m (Ag_2Se, \alpha, 298.15\ K) = -(40.1 \pm 1.3)\ kJ\cdot mol^{-1}.$$

The estimate of the solubility product from the Gibbs energy of formation of α-Ag_2Se is $\log_{10} K^\circ_{s,0} = -(57.76 \pm 0.57)$ which overlaps the values from the experimental measurements.

V.10.2.1.3 Aqueous silver(I) selenide complexes

In the measurements performed by Mehra and Gubeli [70MEH/GUB], the solubility measurements conducted with radioactively labelled silver showed a "constant dissolved" silver selenide concentration of about 6×10^{-8} to 8×10^{-9} M at pH < 10. The solubility was independent of the total Se(– II) concentration in the test solutions. From this finding it was concluded that a species of the formula $Ag_2(HSe)(OH)(aq)$ was formed and its stability constant was evaluated from the data. The composition of this complex is formally equivalent to the composition $Ag_2Se(aq)$ and therefore not unique. The observed constant solubility would then correspond to an intrinsic solubility of $Ag_2Se(s)$

$$Ag_2Se(s) \rightleftharpoons Ag_2Se(aq) \qquad (V.113)$$

and the data in [68MEH] have been recalculated to yield $\log_{10} K$ ((V.113), 1 M $NaClO_4$, 298.15 K) = – (7.66 ± 0.48).

However, an alternative explanation to the minute solubility registered might be the presence of $Ag_2Se(s)$ particles of colloidal size, since only a sintered glass filter was used to separate the aqueous phase from the suspension. Therefore, this review does not accept the proposed species and the equilibrium constant presented, although perhaps a more random solubility would be expected from a faulty experimental technique. Obviously further investigations are needed in this area with a better separation technique, *cf.* [95RAI/FEL3].

At pH values well above 11, the data show an increase in the Ag(I) concentration with pH and $[Se(-II)]_{tot}$. The formation of the species $Ag(Se)_2(OH)^{4-}$ was suggested. The stoichiometry of this species is unlikely, since Ag(I) is well known to prefer a linear co-ordination. Moreover, the derivation of its stoichiometry assumed that Se^{2-} predominates in the experimental pH range, which is not correct. The composition of the complex and its formation constant are thus refuted, although soluble Ag(I)-species of unknown stoichiometry apparently form at increasing selenide concentration and high pH. No attempt was made to re-evaluate these data.

V.10.2.2 Silver selenites

Ojkova and Gospodinov [82OJK/GOS] performed a solubility study of the system Ag_2O-SeO_2-H_2O at 373.2 K. Only the phase diagram is presented and it contains Ag_2SeO_3(cr) as the sole phase formed in the investigated concentration range from 0 to 70 mass-% SeO_2. The solubility of Ag(I) is small over the whole interval.

The accepted experimental determinations of the solubility product of silver selenite defined by the equilibrium:

$$Ag_2SeO_3(cr) \rightleftharpoons 2Ag^+ + SeO_3^{2-} \quad (V.114)$$

are collected in Table V-63.

The data in [56CHU], [62SEL/LES], [76CHA/CHE] were recalculated by the review to $\log_{10} K^{\circ}_{s,0}$ values, see Appendix A. The measurements by Mehra and Gubeli [68MEH], [69MEH/GUB] are also discussed in this Appendix. The value in the Table from this source is a mixed conditional constant on the molar scale. Lin and Pan [61LIN/PAN] obtained $\log_{10} K_{s,0}$ ((V.114), 298.15 K) = – 14.74 from potentiometric measurements with a silver/silver selenite electrode. This result is not accepted for reasons that are presented in Appendix A. The rejection is supported by the analysis of the same data made by Masson *et al.* [86MAS/LUT].

Table V-63: Accepted experimental values of the solubility product of Ag_2SeO_3(cr)

Method	Medium	t (°C)	$-\log_{10} K^{\circ}_{s,0}$	Reference
sol	dilute acid	20	(15.71 ± 0.35)	[56CHU]
sol	Water	25	(15.87 ± 0.02)	[62SEL/LES]
pot	pH 11, I = 0.1	20	(15.85 ± 0.30)	[76CHA/CHE]
sol	dilute acid	25	(15.76 ±0.10)	[93SLA/POP]
pot, sol	1 M $NaClO_4$	25	(15.40 ± 0.35)	[69MEH/GUB]

The review selects the average of the first four $\log_{10} K^{\circ}_{s,0}$ values in the Table and obtains:

$$\log_{10} K^{\circ}_{s,0} \text{ ((V.114), 298.15 K)} = -(15.80 \pm 0.30).$$

The uncertainty has been chosen rather large as systematic errors cannot be excluded in some of the measurements.

Mehra and Gubeli [68MEH], [69MEH/GUB] made extensive measurements of the solubility of $Ag_2SeO_3(s)$ in aqueous solutions at 298.15 K as a function of pH and total selenite concentration. From these measurements, which are presented and evaluated in Appendix A, the review accepts the results $\log_{10} K_{s,0}$ ((V.114), 1 M $NaClO_4$, 298.15 K) = – (15.40 ± 0.35) and the stability constant of the reaction:

$$Ag^{+} + SeO_3^{2-} \rightleftharpoons AgSeO_3^{-} \qquad (V.115)$$

$\log_{10} \beta_1$ ((V.115), 1 M $NaClO_4$, 298.15 K) = (2.39 ± 0.25). The value of the stability constant of the reaction:

$$Ag^{+} + 2SeO_3^{2-} \rightleftharpoons Ag(SeO_3)_2^{3-} \qquad (V.116)$$

$\log_{10} \beta_2$ ((V.116), I = 1 – 3 M ?, 298.15 K) = (3.76 ± 0.05), is not accepted since it may be affected by variations in the medium. An attempt to extrapolate the data to standard condition has not been made.

Selivanova, Leshchinskaya, and Klushina [62SEL/LES] measured the enthalpy change when crystalline silver selenite was formed from $AgNO_3$(cr) and a solution of sodium selenite in a calorimetric experiment. Their result is used in Appendix A to estimate the standard enthalpy of formation of Ag_2SeO_3(cr) to be $\Delta_f H_m^\circ$ (Ag_2SeO_3, cr, 298.15 K) = – (363.44 ± 1.02) kJ·mol^{-1}. This value agrees well with [82WAG/EVA] but differs appreciably from the result in the paper, – 345.0 kJ·mol^{-1}.

Silver selenite decomposes on heating following the reaction scheme:

$$Ag_2SeO_3(\text{cr or l}) \rightleftharpoons 2Ag(\text{cr}) + SeO_2(\text{g}) + 0.5O_2(\text{g}) \qquad (V.117)$$

according to [64BUK/PAS], [84GOS/BOG]. In the latter reference the melting point was determined to be (810 ± 5) K and $\Delta_{fus} H_m^\circ$ (Ag_2SeO_3, cr, 810 K) = (38.7 ± 4.0) kJ·mol^{-1}. The uncertainties have been added by the review.

Bakeeva, Buketov, and Pashinkin [68BAK/BUK] measured the vapour pressure of SeO_2(g) above Ag_2SeO_3(cr or l) by the dew point method. Their primary data have been recalculated as described in Appendix A, [71PAS/ARO]. The following results were obtained from the temperature dependence of the equilibrium constant: Temperature range 757 to 797 K, $\Delta_r H_m^\circ$ ((V.117), cr, 777 K) = (231.5 ± 19.6) kJ·mol^{-1}, $\Delta_r S_m^\circ$ ((V.117), cr, 777 K) = (224.4 ± 25.2) J·K^{-1}·mol^{-1}; temperature range 809 to 951 K, $\Delta_r H_m^\circ$ ((V.117), l, 880 K) = (184.0 ± 5.0) kJ·mol^{-1}, $\Delta_r S_m^\circ$ ((V.117), l, 880 K) = (164.1 ± 5.7) J·K^{-1}·mol^{-1}. These data yield $\Delta_{fus} H_m^\circ$ (Ag_2SeO_3, cr, 790 K) = (47.5 ± 20.2) kJ·mol^{-1}. The directly measured value is selected:

$$\Delta_{fus} H_m^\circ (Ag_2SeO_3, \text{cr}, 810\ \text{K}) = (38.7 \pm 4.0)\ \text{kJ·mol}^{-1}.$$

The vapour pressure is low before the melting point. The narrow temperature range thus accessible to measurement causes the large uncertainties of the thermodynamic quantities for Ag_2SeO_3(cr). No calculation of $\Delta_f H_m^\circ$ (Ag_2SeO_3, cr, 298.15 K) was therefore attempted by the review. Pashinkin *et al.* [71PAS/ARO] report

$\Delta_f H_m^\circ$ (Ag_2SeO_3, cr, 298.15 K) = – (352.3 ± 16.3) $kJ \cdot mol^{-1}$ from an extrapolation of recalculated data in [68BAK/BUK]. This result agrees with the value obtained by the calorimetric measurement in [62SEL/LES]. The review selects:

$$\Delta_f H_m^\circ (Ag_2SeO_3, \text{cr}, 298.15 \text{ K}) = -(363.44 \pm 1.02) \text{ kJ} \cdot \text{mol}^{-1}.$$

The standard entropy of formation of Ag_2SeO_3(cr) was calculated in [71PAS/ARO] to be (207.9 ± 23.8) and in [94PAS/GOS] to be (205.5 ± 23.8) $J \cdot K^{-1} \cdot mol^{-1}$ from the data in [68BAK/BUK]. The Gibbs energy of formation of Ag_2SeO_3(cr) calculated from the selected value of $\log_{10} K_{s,0}^\circ$ ((V.114), 298.15 K) together with $\Delta_f H_m^\circ$ (Ag_2SeO_3, cr, 298.15 K) from [62SEL/LES] yield S_m° (Ag_2SeO_3, cr, 298.15 K) = (216.7 ± 9.6) $J \cdot K^{-1} \cdot mol^{-1}$. The high and low temperature data are thus in agreement and the review selects:

$$\Delta_f G_m^\circ (Ag_2SeO_3, \text{cr}, 298.15 \text{ K}) = -(298.40 \pm 2.44) \text{ kJ} \cdot \text{mol}^{-1},$$

$$S_m^\circ (Ag_2SeO_3, \text{cr}, 298.15 \text{ K}) = (216.8 \pm 8.9) \text{ J} \cdot \text{K}^{-1} \cdot \text{mol}^{-1}.$$

Gospodinov and Bogdanov [84GOS/BOG] report $C_{p,m}^\circ$ (Ag_2SeO_3, cr, (593 – 673) K) = (73.68 + 0.312 T – 2.88 × 10^{-4} T^2) $J \cdot K^{-1} \cdot mol^{-1}$.

V.10.2.3 Silver selenates

Four observations of the solubility of Ag_2SeO_4(cr) in water are available. The result obtained by Meyer and Hinke [32MEY/HIN] is two orders of magnitude lower than the others and will be ignored. The other results in [25LEN/KAO], [42GEL/KIN], [59SEL/ZUB] vary between 1.24 × 10^{-3} and 3.3 × 10^{-3} M, which may indicate difficulties in the preparation of a crystalline silver selenate specimen. The review estimates that the solubility of Ag_2SeO_4(cr) is situated in the interval (1.2 – 2.4) × 10^{-3} $mol \cdot kg^{-1}$.

Selivanova, Zubova, and Finkel'stein [59SEL/ZUB] prepared the salt from silver nitrate and selenic acid. The product was shown to be crystalline by X-ray diffraction. The solubility product of

$$Ag_2SeO_4(\text{cr}) \rightleftharpoons 2Ag^+ + SeO_4^{2-} \qquad \text{(V.118)}$$

found from the measured solubility, (1.24 ± 0.01) × 10^{-3} m, is $\log_{10} K_{s,0}^\circ$ ((V.118), 298.15 K) = – (8.26 ± 0.15). The uncertainty added reflects the uncertainty in the extrapolation to I = 0. This value agrees with the data in [82WAG/EVA] but not with the published value, which is too small by a factor of 4 due to an arithmetic mistake (*cf.* [59SEL/ZUB] in Appendix A).

Gelbach and King [42GEL/KIN] prepared their specimen of Ag_2SeO_4(s) by slow precipitation from 10% silver nitrate and magnesium selenate solutions. The solubility in six solutions of selenic acid in the concentration range from 0.00 to 0.12 M was determined. Equilibrium was approached from under- as well as supersaturation. The solubility in water was found to be 2.42 × 10^{-3} M. No solubility product was calculated from the data. This has been done here with $\log_{10} K_1^\circ$ = 1.75 for the protonation constant of the selenate ion and activity coefficients from SIT with $\varepsilon = 0$ kg·mol^{-1}. The result was $\log_{10} K^{\circ}_{s,0}$ ((V.118), 298.15 K) = – (7.46 ± 0.03). The uncertainty corresponds to the standard deviation from the 6 measurements.

The authors also measured the cell (Q, QH_2 = quinhydrone)

$$\text{Pt} \mid \text{Q, QH}_2,\ c\ \text{H}_2\text{SeO}_4 \mid c\ \text{H}_2\text{SeO}_4(\text{sat'd Ag}_2\text{SeO}_4) \mid \text{Ag}_2\text{SeO}_4, \text{Ag}$$

in order to determine the standard electrode potential of

$$\text{Ag}_2\text{SeO}_4(\text{cr}) + 2\text{e}^- \rightleftharpoons 2\text{Ag} + \text{SeO}_4^{2-}. \qquad \text{(V.119)}$$

The value of E°((V.119), 298.15 K), – 0.8289 V, presented is in error due to an arithmetic mistake. The correct value is 0.5699 V. Combination with the standard electrode potential of the silver electrode yields $\log_{10} K^{\circ}_{s,0}$ ((V.118), 298.15 K) = – 7.75.

Apart from analytical errors, a possible explanation for the difference between the two sets of solubility data would be that the specimen used by Gelbach and King was not perfectly crystalline and hence had a higher solubility. As there is no obvious ground for rejecting one of the values, the review selects the mean of the two solubility products with a large uncertainty:

$$\log_{10} K^{\circ}_{s,0}\ ((\text{V.118}), 298.15\ \text{K}) = -(7.86 \pm 0.50).$$

The standard Gibbs energy of formation of silver(I) selenate is calculated from the solubility product and the selected Gibbs energies of the ions to be:

$$\Delta_f G^{\circ}_m\ (\text{Ag}_2\text{SeO}_4, \text{cr}, 298.15\ \text{K}) = -(330.15 \pm 3.20)\ \text{kJ·mol}^{-1}.$$

This selection is just outside the value, – 334.2 kJ·mol^{-1}, in [82WAG/EVA] obviously calculated from the data in [59SEL/ZUB].

Selivanova, Zubova, and Finkel'stein [59SEL/ZUB] also made a calorimetric study of the reaction between a silver nitrate and a selenic acid solution with formation of Ag_2SeO_4(cr). The standard enthalpy of formation of the compound has been evaluated from the data in Appendix A. The value obtained there:

$$\Delta_f H^{\circ}_m\ (\text{Ag}_2\text{SeO}_4, \text{cr}, 298.15\ \text{K}) = -(422.51 \pm 2.12)\ \text{kJ·mol}^{-1}$$

is selected and in reasonable agreement with the value in [82WAG/EVA], – 420.5 kJ·mol^{-1}, but differs substantially from the value derived in [59SEL/ZUB], – 439.5 kJ·mol^{-1}.

The selected data yield:

$$S^{\circ}_m\ (\text{Ag}_2\text{SeO}_4, \text{cr}, 298.15\ \text{K}) = (227.7 \pm 12.9)\ \text{J·K}^{-1}\text{·mol}^{-1}.$$

V.10.2.4 Silver selenocyanates

The formation of aqueous silver(I)-selenocyanate complexes has been characterised in two restricted studies by Toropova [56TOR], and by Golub and Pomerants

[59GOL/POM]. Both studies are based on ion-selective (Ag^+/Ag) electrode data and the stoichiometry of the dominating reaction

$$Ag^+ + 3SeCN^- \rightleftharpoons Ag(SeCN)_3^{2-} \qquad (V.120)$$

was deduced from the constancy of the corresponding equilibrium constant with a variation of the $SeCN^-$ concentration between 0.06 to 0.44 M. The two equilibrium constants reported, $\log_{10} K$ (V.120) = (13.91 ± 0.06) (n = 5) in [56TOR] and (13.80 ± 0.11) (n = 7) in [59GOL/POM], show good agreement despite that the temperatures employed were 298 and 293 K, respectively. The ionic strength employed by Toropova was specified, I = 0.3 M ($NaNO_3$?), while Golub and Pomerants only stated that their data were collected at "constant ionic strength of potassium nitrate". The review selects the average of the two values reported and assigns it an uncertainty of ± 0.3 logarithmic units for the extrapolation to standard conditions

$$\log_{10} K^\circ((V.120), 298.15 \text{ K}) = (13.85 \pm 0.30).$$

The selection corresponds to:

$$\Delta_r G_m^\circ((V.120), 298.15 \text{ K}) = -(79.1 \pm 1.7) \text{ kJ·mol}^{-1}.$$

A salt of low solubility is also formed in the system according to the reaction:

$$Ag^+ + SeCN^- \rightleftharpoons AgSeCN(cr). \qquad (V.121)$$

This phase was first described by Birkenbach and Huttner [30BIR/HUT], who also estimated its aqueous solubility product from a one-point determination of [Ag^+] in a saturated AgSeCN suspension to be $\log_{10} K_{s,0}$ (AgSeCN, cr, 291 K) = – 15.4.

Das, Sahu, Satyanarayama and Misra [74DAS/SAH] have more recently conducted a systematic study of the potential of the cell

$$\text{Ag,AgCl} \mid x \text{ m KCl}, (y-x) \text{ m KNO}_3 \parallel x \text{ m KSeCN}, (y-x) \text{ m KNO}_3 \mid \text{Ag,AgSeCN}.$$

The emf of this cell, E, was measured as a function of x and y, and its E° values were deduced in the temperature range 308 to 323 K via linear extrapolations of $E_T = f(x, y)$ to $(x, y) = 0$, *i.e.* $I = 0$.

This review selects the data extrapolated to 298.15 K reported by Das *et al.* for the cell reaction AgCl(cr) + $SeCN^-$ → AgSeCN(cr) + Cl^-; $\Delta_r G_m^\circ$ (298.15 K) = – (24.3 ± 2.9) kJ·mol^{-1}, $\Delta_r H_m^\circ$ (298.15 K) = – (42.4 ± 1.7) kJ·mol^{-1}, and $\Delta_r S_m^\circ$ (298.15 K) = – (60.7 ± 15.0) J·K^{-1}·mol^{-1}. These results combine with the CODATA for the reaction: $Ag^+ + Cl^-$ → AgCl(cr) ($\Delta_r G_m^\circ$ = – (55.64 ± 0.22) kJ·mol^{-1}, $\Delta_r H_m^\circ$ = – (65.72 ± 0.14) kJ·mol^{-1}, and $\Delta_r S_m^\circ$ = – (33.80 ± 0.49) J·K^{-1}·mol^{-1}) to:

$$\Delta_r G_m^\circ((V.121), 298.15 \text{ K}) = -(79.9 \pm 2.9) \text{ kJ·mol}^{-1}$$

$$\Delta_r H_m^\circ((V.121), 298.15 \text{ K}) = -(108.1 \pm 1.7) \text{ kJ·mol}^{-1}$$

$$\Delta_r S_m^\circ((V.121), 298.15 \text{ K}) = -(94.6 \pm 11.3) \text{ J·K}^{-1}\text{·mol}^{-1}.$$

The Gibbs energy change corresponds to a solubility product for AgSeCN of:

$$\log_{10} K^{\circ}_{s,0}\,(\text{AgSeCN, cr, 298.15 K}) = -\,(14.0 \pm 0.5).$$

The $\Delta_r G^{\circ}_m$ values selected combine with the selected value of $\Delta_f G^{\circ}_m(\text{SeCN}^-$, 298.15 K) and $\Delta_f G^{\circ}_m(\text{Ag}^+$, 298.15 K) to yield:

$$\Delta_f G^{\circ}_m(\text{Ag(SeCN)}_3^{2-}, 298.15\text{ K}) = (406.2 \pm 11.5)\text{ kJ·mol}^{-1},$$

$$\Delta_f G^{\circ}_m(\text{AgSeCN, cr, 298.15 K}) = (133.2 \pm 4.8)\text{ kJ·mol}^{-1}.$$

V.10.3 Gold compounds

V.10.3.1 Gold selenides

The phase diagram of the gold-selenium system was investigated by Rabenau, Rau, and Rosenstein [71RAB/RAU]. Two polymorphs, α and β of the composition AuSe were found to form. The α polymorph decomposes eutectoidically at 698 K and the β modification is metastable. No other solid compounds were found.

V.10.3.1.1 α-AuSe

The total vapour pressure of selenium in equilibrium with a mixture of Au(cr) and α-AuSe was measured in the temperature range 505 to 602 K using the Knudsen effusion method in [71RAB/RAU]. The result is presented as the partial pressure of Se_2(g) at equilibrium and the enthalpy of formation and entropy at 298.15 were evaluated to be $\Delta_f H^{\circ}_m$(AuSe, α, 298.15 K) = – 7.9 kJ·mol^{-1} and S°_m(AuSe, α, 298.15 K) = 80.8 J·K^{-1}·mol^{-1}, respectively. However, Se_2(g) is not the major species in the gas phase at the temperatures and total pressures of the study, and it is not clearly stated how the partial pressure was derived from the experiments. The result is therefore questionable and impossible to re-evaluate using the selected thermodynamic properties of the gaseous selenium species. No values were selected by the review.

V.10.3.1.2 AuSe(g)

The dissociation enthalpy of AuSe(g) was determined by Smoes, Mandy, Aumera-Mahieu, and Drowart in [72SMO/MAN] from mass spectrometric measurements of ion intensities of the species participating in the reaction

$$\text{AuSe(g)} + \text{Se(g)} \rightleftharpoons \text{Au(g)} + \text{Se}_2\text{(g)} \qquad \text{(V.122)}$$

at temperatures in the range 1466 to 1904 K. The dissociation enthalpy of AuSe(g) was evaluated to be $\Delta_r H^{\circ}_m$((V.122), 0 K) = (247.3 ± 14.6) kJ·mol^{-1} using Gibbs energy functions calculated from estimated molecular parameters of AuSe(g). The results were re-evaluated by the review using the selected properties of Se(g) and Se_2(g), the thermal functions of Au(g) in [73HUL/DES], the enthalpy of Au(g) in [82WAG/EVA], and thermal properties of AuSe(g) calculated from the parameters estimated in

[72SMO/MAN], yielding $\Delta_f H_m^\circ$ (AuSe, g, 298.15 K) = 338.0 kJ·mol^{-1} from the second law and $\Delta_f H_m^\circ$ (AuSe, g, 298.15 K) = 351.5 kJ·mol^{-1} from the third law. These values correspond to the dissociation enthalpies $\Delta_{at} H_m^\circ$ (AuSe, g, 298.15 K) = 264.2 kJ·mol^{-1} and $\Delta_{at} H_m^\circ$ (AuSe, g, 298.15 K) = 250.7 kJ·mol^{-1}, respectively.

The derived values of the heat of formation of AuSe(g) agree reasonably well. Since the entropy and heat capacity of a diatomic molecule calculated from estimated molecular parameters are expected to be quite accurate, the review adopts the mean of the two results and obtains:

$$\Delta_f H_m^\circ (\text{AuSe, g, 298.15 K}) = (344.8 \pm 14.6)\ \text{kJ·mol}^{-1}.$$

The error bars are those in [72SMO/MAN] and the result is included in Appendix E since it was obtained with non-TDB auxiliary data.

V.10.3.2 Gold selenites

The thermal dissociation of gold(III) selenite proceeds by the reaction

$$Au_2(SeO_3)_3(cr) \rightleftharpoons 2Au(cr) + 3SeO_2(g) + 1.5O_2(g). \qquad (V.123)$$

Bakeeva, Pashinkin, Bakeev, and Buketov [73BAK/PAS] measured the selenium dioxide pressure over gold selenite in the interval 489 to 599 K by the dew point method. The pressure was calculated from the dew point temperature by the relationship for the saturated vapour pressure in [69SON/NOV]. The data in the deposited VINITI document (No. 4959-72) have been recalculated with the relationship selected by the review. The enthalpy and entropy changes obtained from the temperature variation of the equilibrium constant are $\Delta_r H_m^\circ$ ((V.123), 544 K) = (576.8 ± 13.0) kJ·mol^{-1} and $\Delta_r S_m^\circ$ ((V.123), 544 K) = (899.4 ± 24.0) J·K^{-1}·mol^{-1}. The uncertainties are entered here as twice the standard deviations from the least-squares calculation.

No experimental data have been found, which would permit a proper calculation of the enthalpy and entropy changes at 298.15 K. Pashinkin and Gospodinov [94PAS/GOS] report S_m° ($Au_2(SeO_3)_3$, cr, 298.15 K) = (290.0 ± 28.0) J·K^{-1}·mol^{-1} from an evaluation of the data. The auxiliary data used are not furnished. Bakeeva *et al.* estimated $\Delta_r C_{p,m}^\circ$ to be – 29.3 J·K^{-1}·mol^{-1} and assumed that it was constant in the temperature interval in their extrapolation of the data. On this assumption the review obtains $\Delta_f H_m^\circ$ ($Au_2(SeO_3)_3$, cr, 298.15 K) = – (916.4 ± 15.7) kJ·mol^{-1} and S_m° ($Au_2(SeO_3)_3$, cr, 298.15 K) = (273.3 ± 29.0) J·K^{-1}·mol^{-1} with selected auxiliary data. The results calculated by the review are not selected due to the uncertainty about the extrapolation procedure.

The value of the standard enthalpy of formation of gold(III) selenite reported in [73BAK/PAS] is a misprint.

No other data for gold selenites have been found.

V.10.3.3 Gold selenates

No thermodynamic data have been found for gold selenates.

V.11 Group 10 to group 3 compounds and complexes

V.11.1 Nickel compounds

V.11.1.1 Nickel selenides

The solid compounds Ni_3Se_2(cr), Ni_6Se_5(cr), $Ni_{21}Se_{20}$(cr), $Ni_{1-x}Se$(cr), and $NiSe_2$(cr) have been reported to form in the system nickel-selenium. The phase diagram of the system is presented in [86MAS]. Experimental thermodynamic information is available for Ni_3Se_2(cr), $Ni_{1-x}Se$(cr), and $NiSe_2$(cr).

V.11.1.1.1 Ni_3Se_2(cr)

Morozova, Vladimirova, Stolyarova, and Nepomnyashchaya [74MOR/VLA] determined the enthalpy of formation of Ni_3Se_2(cr) using direct reaction calorimetry. The obtained value is selected:

$$\Delta_f H_m^\circ(\text{Ni}_3\text{Se}_2, \text{cr}, 298.15\ \text{K}) = -(179.5 \pm 6.0)\ \text{kJ}\cdot\text{mol}^{-1}.$$

V.11.1.1.2 $Ni_{1-x}Se$(cr)

$Ni_{1-x}Se$(cr) is a solid solution and the composition can vary from approximately $Ni_{0.80}Se$(cr) to $Ni_{0.95}Se$(cr). The thermodynamic properties change with the exact composition and because the modelling of solid solutions is out of the scope of this review, the $Ni_{1-x}Se$(cr) phase is chosen to be represented by the single composition $Ni_{0.88}Se$(cr).

The heat capacity of $Ni_{1-x}Se$(cr) samples with the compositions $Ni_{0.80}Se$(cr), $Ni_{0.875}Se$(cr), and $Ni_{0.95}Se$(cr) were measured in the temperature range 5 to 350 K by Grønvold, Thurmann-Moe, Westrum, and Levitin [60GRO/THU]. The values reported of the heat capacity and the entropy were $C_{p,m}^\circ(\text{Ni}_{0.80}\text{Se, cr, 298.15 K}) = (45.46 \pm 0.50)$ $\text{J}\cdot\text{K}^{-1}\cdot\text{mol}^{-1}$, $C_{p,m}^\circ(\text{Ni}_{0.875}\text{Se, cr, 298.15 K}) = (48.12 \pm 0.50)\ \text{J}\cdot\text{K}^{-1}\cdot\text{mol}^{-1}$, $C_{p,m}^\circ(\text{Ni}_{0.95}\text{Se, cr, 298.15 K}) = (50.21 \pm 0.50)\ \text{J}\cdot\text{K}^{-1}\cdot\text{mol}^{-1}$, $S_m^\circ(\text{Ni}_{0.80}\text{Se, cr, 298.15 K}) = (64.1 \pm 0.8)$ $\text{J}\cdot\text{K}^{-1}\cdot\text{mol}^{-1}$, $S_m^\circ(\text{Ni}_{0.875}\text{Se, cr, 298.15 K}) = (67.5 \pm 0.8)\ \text{J}\cdot\text{K}^{-1}\cdot\text{mol}^{-1}$, and $S_m^\circ(\text{Ni}_{0.95}\text{Se, cr, 298.15 K}) = (71.4 \pm 0.5)\ \text{J}\cdot\text{K}^{-1}\cdot\text{mol}^{-1}$. The selected values for the composition $Ni_{0.88}Se$(cr) were obtained from linear interpolations between the values for $Ni_{0.875}Se$(cr) and $Ni_{0.95}Se$(cr):

$$C_{p,m}^\circ(\text{Ni}_{0.88}\text{Se, cr, 298.15 K}) = (48.26 \pm 0.50)\ \text{J}\cdot\text{K}^{-1}\cdot\text{mol}^{-1},$$

and

$$S_m^\circ(\text{Ni}_{0.88}\text{Se, cr, 298.15 K}) = (67.8 \pm 0.8)\ \text{J}\cdot\text{K}^{-1}\cdot\text{mol}^{-1}.$$

The heat capacity of $Ni_{1-x}Se(cr)$ above room temperature was measured for various compositions in the temperature range 298 to 1050 K by Grønvold [70GRO] and in the temperature range 298 to 750 K by Grønvold, Kveseth, and Sveen [75GRO/KVE]. Several order-disorder transformations, strongly dependent on the exact composition, were found. No heat capacity expression was selected because of the complex variation with temperature and composition.

Grønvold [72GRO] determined the enthalpy of formation of the compositions $Ni_{0.95}Se(cr)$, $Ni_{0.88}Se(cr)$, and $Ni_{0.80}Se(cr)$ by direct synthesis calorimetry at 1050 K. The values reported after a recalculation to 298.15 K were $\Delta_f H_m^\circ$ ($Ni_{0.95}Se$, cr, 298.15 K) = – (71.2 ± 1.6) $kJ \cdot mol^{-1}$, $\Delta_f H_m^\circ$ ($Ni_{0.88}Se$, cr, 298.15 K) = – (69.8 ± 1.6) $kJ \cdot mol^{-1}$, and $\Delta_f H_m^\circ$ ($Ni_{0.80}Se$, cr, 298.15 K) = – (66.2 ± 1.6) $kJ \cdot mol^{-1}$.

The selected value of the enthalpy of formation is:

$$\Delta_f H_m^\circ (Ni_{0.88}Se, cr, 298.15 K) = -(69.8 \pm 1.6)\ kJ \cdot mol^{-1}.$$

The Gibbs energy of formation is calculated from the selected enthalpy of formation and entropy, the selected entropy of selenium, and the entropy of nickel in [2005GAM/BUG] to be:

$$\Delta_f G_m^\circ (Ni_{0.88}Se, cr, 298.15 K) = -(69.8 \pm 1.6)\ kJ \cdot mol^{-1}.$$

V.11.1.1.3 $NiSe_2$(cr)

Grønvold and Westrum [62GRO/WES] measured the heat capacity of $NiSe_2(cr)$ in the temperature range 5 to 350 K. The reported values of the heat capacity and entropy at 298.15 K are selected:

$$C_{p,m}^\circ (NiSe_2, cr, 298.15 K) = (74.5 \pm 1.0)\ J \cdot K^{-1} \cdot mol^{-1}$$

and

$$S_m^\circ (NiSe_2, cr, 298.15 K) = (103.5 \pm 1.5)\ J \cdot K^{-1} \cdot mol^{-1}.$$

Grønvold [75GRO] measured the heat capacity of $NiSe_2(cr)$ in the temperature range 298.15 to 1000 K. The expression

$$C_{p,m}^\circ (NiSe_2, cr, (298.15 - 1000) K) = (103.642 - 55.1057 \times 10^{-3}\ T + 4.53165 \times 10^{-5}\ T^2 - 1.488097 \times 10^{6}\ T^{-2})\ J \cdot K^{-1} \cdot mol^{-1}$$

was evaluated by the review from these measurements.

The value of the enthalpy of formation of $NiSe_2(cr)$ determined by Morozova, Vladimirova, Stolyarova, and Nepomnyashchaya [74MOR/VLA] using direct reaction calorimetry is selected:

$$\Delta_f H_m^\circ (NiSe_2, cr, 298.15 K) = -(115.5 \pm 7.0)\ kJ \cdot mol^{-1}.$$

The Gibbs energy of formation is calculated from the selected enthalpy of formation and entropy, the selected entropy of selenium, and the entropy of nickel in [2005GAM/BUG] to be:

$$\Delta_f G^\circ_m (NiSe_2, cr, 298.15 K) = -(112.4 \pm 7.0) kJ\cdot mol^{-1}.$$

V.11.1.2 Nickel selenites

The phase diagram of the system $NiSeO_3$-SeO_2-H_2O at 298.15 K was established by Ebert, Micka, and Pekova [82EBE/MIC2]. The diagram is presented and contains the three solid phases $NiSeO_3\cdot 2H_2O$, $Ni(HSeO_3)_2\cdot 2H_2O$, and H_2SeO_3. The formula of the incongruently soluble acid salt was verified by an infrared spectroscopic measurement. No solubility data except for the compositions of the invariant points are published.

V.11.1.2.1 $NiSeO_3\cdot 2H_2O$(cr)

Chukhlantsev and Tomashevsky [57CHU/TOM] prepared nickel selenite by mixing 0.1 M solutions of nickel sulphate and sodium selenite in stoichiometric amounts. The precipitate was aged for 24 hours in the mother liquor. Chemical analysis confirmed the 1:1 ratio between Ni(II) and Se(IV). No X-ray diffraction measurements were performed. The solubility of the specimen, which is assumed by the review to have the composition $NiSeO_3\cdot 2H_2O$, in dilute solution of nitric or sulphuric acid was measured at 293 K. The experiments were performed and the data recalculated as described in Appendix A, [56CHU]. The result for

$$NiSeO_3\cdot 2H_2O(cr) \rightleftharpoons Ni^{2+} + SeO_3^{2-} + 2H_2O(l) \qquad (V.124)$$

was $\log_{10} K^\circ_{s,0}$ ((V.124), 293.15 K) = – (5.77 ± 0.08). The reported conditional constant is – (5.00 ± 0.06). Ripan and Vericeanu [68RIP/VER] studied the solubility of $NiSeO_3$(s) by conductivity measurements as described in Appendix A. A calculation based on the assumptions made in the Appendix and the solubility, 2.25×10^{-3} M, provided in the paper leads to the conclusion that about 20% of dissolved Ni(II) is present in hydrolysed forms. The solubility product estimated with allowance for the hydrolysis of Ni^{2+} and SeO_3^{2-} is $\log_{10} K^\circ_{s,0}$ ((V.124), 291 K) = – (5.90 ± 0.20). This value can only be regarded as indicating the order of magnitude of the constant. The uncertainty reflects the variance from the solubility measurements.

The review proposes and selects the value:

$$\log_{10} K^\circ_{s,0} ((V.124), 298.15 K) = -(5.8 \pm 1.0).$$

Selivanova *et al.* [63SEL/LES3] made calorimetric measurements for the determination of the standard enthalpy of formation of nickel selenite in the same way as described for $CoSeO_3\cdot 2H_2O$(cr) in [63LES/SEL4] in Appendix A. The precipitate of $NiSeO_3\cdot 2H_2O$ formed in the calorimeter was amorphous, however. The standard enthalpy of formation of this phase is calculated to be $\Delta_f H^\circ_m$ ($NiSeO_3\cdot 2H_2O$, am, 298.15 K) = – (1107.2 ± 2.8) kJ·mol^{-1}. Selivanova and Leshchinskaya published the same data in [64SEL/LES]. In this paper crystalline $NiSeO_3\cdot 2H_2O$ was prepared and its enthalpy of dissolution in 70 mass-% H_2SO_4 was measured. The data are evaluated in Appendix A.

The selected result is:

$$\Delta_f H_m^\circ (NiSeO_3\cdot 2H_2O, cr, 298.15 K) = -(1109.33 \pm 2.83) kJ\cdot mol^{-1}.$$

The value calculated in the paper and in [82WAG/EVA], – 1134.3 $kJ\cdot mol^{-1}$, is not accepted, see Appendix A. The selected solubility product yields:

$$\Delta_f G_m^\circ (NiSeO_3\cdot 2H_2O, cr, 298.15 K) = -(915.56 \pm 6.02) kJ\cdot mol^{-1},$$

$$S_m^\circ (NiSeO_3\cdot 2H_2O, cr, 298.15 K) = (196.3 \pm 22.3) J\cdot K^{-1}\cdot mol^{-1}.$$

V.11.1.2.2 $NiSeO_3$(cr)

The equilibrium:

$$NiSeO_3(cr) \rightleftharpoons NiO(cr) + SeO_2(g) \qquad (V.125)$$

was studied with the dew point method by Bakeeva, Buketov, and Pashinkin [69BAK/BUK] over the temperature interval 817 to 987 K. The relationship used to calculate p_{SeO_2} from the dew point temperatures is not disclosed. From data in [71PAS/ARO] it could be deduced that the vapour pressure expression in [62MAR/GET] had been used for this purpose. The published pressures have been re-calculated accordingly with the selected expression for the saturated vapour pressure of SeO_2(cr), see [71PAS/ARO] in Appendix A. The result is $\log_{10} p(SeO_2, g, T) = (9.671 \pm 0.094) - (9880 \pm 84)\, T^{-1}$ for the temperature interval 817 to 987 K. The uncertainties are entered here as twice the standard deviations from the least-squares calculation. The high temperature data have been extrapolated to standard conditions with $C_{p,m}^\circ$ values in [73BAR/KNA] leading to the following estimates $\Delta_f H_m^\circ (NiSeO_3, cr, 298.15 K) = -(547.2 \pm 5.0) kJ\cdot mol^{-1}$ and $S_m^\circ (NiSeO_3, cr, 298.15 K) = (103.7 \pm 5.0) J\cdot K^{-1}\cdot mol^{-1}$.

The corresponding values calculated in [71PAS/ARO] with unknown auxiliary data are $-(543.9 \pm 6.3) kJ\cdot mol^{-1}$ and $(100.4 \pm 8.3) J\cdot K^{-1}\cdot mol^{-1}$, respectively. The entropy was estimated to be $(98.2 \pm 8.4) J\cdot K^{-1}\cdot mol^{-1}$ in [94PAS/GOS] also with unknown auxiliary data. The value estimated from $S_m^\circ (NiSeO_3\cdot 2H_2O, cr, 298.15 K)$ by subtraction of the entropy contribution from water of crystallisation, 44.7 $J\cdot K^{-1}\cdot mol^{-1}$ in [97PUI/RAR], is $(110.3 \pm 15.0) J\cdot K^{-1}\cdot mol^{-1}$. The agreement between the estimates of $S_m^\circ (NiSeO_3, cr, 298.15 K)$ from the high and low temperature measurements may be fortuitous since it is not obtained for $CoSeO_3$. The extrapolated values of the entropies of $NiSeO_3$ and $CoSeO_3$ are, however, quite similar.

The two sets of thermodynamic quantities for $NiSeO_3\cdot 2H_2O$ and $NiSeO_3$ are not quite consistent, since they imply that the vapour pressure of the reaction

$NiSeO_3{\cdot}2H_2O(cr) \rightleftharpoons NiSeO_3(cr) + 2H_2O(g)$ would be $\log_{10} p(H_2O, g, 298.15\ K) = (0.58 \pm 0.70)$. This result contradicts the observed stability of the hydrate. The contradiction cannot be removed with the experimental information at hand and the data for $NiSeO_3(cr)$ are therefore not selected.

V.11.1.3 Nickel selenates

Klein [40KLE] measured the solubility of nickel selenate in water as a function of temperature and presents the results in a table and a graph. The hexahydrate, $NiSeO_4{\cdot}6H_2O$, is the stable phase below 355 K and $NiSeO_4{\cdot}4H_2O$ above this temperature.

Recent solubility determinations at 298.15 K fall into two groups. Vojtisek, Ebert, and Leblova [87VOJ/EBE4] and Ojkova, Christov, and Mihov [98OJK/CHR] report a solubility of 1.929 $mol(NiSeO_4){\cdot}kg^{-1}$, while Ojkova, Stoilova, and Staneva [93OJK/STO], [95STO/OJK] found the solubility to be 1.785 and 1.744 $mol(NiSeO_4){\cdot}kg^{-1}$, respectively.

The review selects the greater value from the fact that Ojkova, who is the co-author of papers belonging to both sets, uses this datum in her later publications:

$$m(NiSeO_4{\cdot}6H_2O(cr),\ aq\ sat,\ 298.15\ K) = (1.93 \pm 0.05)\ mol(NiSeO_4){\cdot}kg^{-1}.$$

A phase diagram and similar information are available for the system: $NiSeO_4$-H_2SeO_4-H_2O at 303.15 K in [50ROH/FRO]; $NiSeO_4$-$CoSeO_4$-H_2O at 303.15 K in [85MES/MAR]; $NiSeO_4$-$MgSeO_4$-H_2O at 298.15 K in [95STO/OJK]; $NiSeO_4$-$(NH_4)_2SeO_4$-H_2O at 298.15 K in [87VOJ/EBE4]; $NiSeO_4$-K_2SeO_4-H_2O at 298.15 K in [85VOJ/EBE]; $NiSeO_4$-Cs_2SeO_4-H_2O at 298.15 K in [2000OJK/BAR].

The value of solubility product of $NiSeO_4{\cdot}6H_2O(cr)$ at 298.15 K,

$$NiSeO_4{\cdot}6H_2O(cr) \rightleftharpoons Ni^{2+} + SeO_4^{2-} + 6H_2O(l) \qquad (V.126)$$

has been reported by Christov and co-workers in [95STO/CHR], [97CHR], [98CHR/OJK], [98OJK/CHR], [2001BAR/CHR], see [98OJK/CHR] in Appendix A, to be $\log_{10} K^{\circ}_{s,0} = -1.381$ and by Kumok and Batyreva [90KUM/BAT] to be -1.45. The results are based on the same isopiestic data by Ojkova and Staneva [89OJK/STA]. The former value:

$$\log_{10} K^{\circ}_{s,0}((V.126),\ 298.15\ K) = -(1.381 \pm 0.045)$$

is selected as it has been evaluated with the Pitzer ion interaction approach. The uncertainty added by the review corresponds to an uncertainty in the mean activity coefficient of 5%.

The standard Gibbs energy of formation of $NiSeO_4{\cdot}6H_2O(cr)$ is calculated from the Gibbs energy of Reaction (V.126) and the values of $\Delta_f G^{\circ}_m$ for the ions and water to be:

$$\Delta_f G^{\circ}_m(NiSeO_4{\cdot}6H_2O,\ cr,\ 298.15\ K) = -(1916.0 \pm 1.7)\ kJ{\cdot}mol^{-1}.$$

Ghosh and Nair [70GHO/NAI] determined the equilibrium constant of the reaction:

$$Ni^{2+} + SeO_4^{2-} \rightleftharpoons NiSeO_4(aq) \quad (V.127)$$

in the temperature range 273.15 to 318.15 K by potentiometric measurements, see Appendix A. The enthalpy of the reaction was found from the variation of the stability constant with temperature. The following results are selected:

$$\log_{10} \beta_1^\circ ((V.127), 298.15 \text{ K}) = (2.67 \pm 0.05),$$

$$\Delta_r G_m^\circ ((V.127), 298.15 \text{ K}) = -(15.24 \pm 0.29) \text{ kJ·mol}^{-1},$$

while $\Delta_r H_m^\circ$((V.127), 298.15 K) = (14.70 ± 1.30) kJ·mol^{-1} is not accepted for reasons discussed in Appendix A.

The uncertainty in $\log_{10} \beta_1^\circ$ has been increased considerably from the value in the reference to include the uncertainty from the estimate of the activity coefficients.

V.11.1.4 Nickel selenocyanates

Two studies characterising the formation of soluble Ni^{2+}-$SeCN^-$ complexes have been published. The study by Kullberg [74KUL] was performed at 298.15 K and I = 1 M $NaClO_4$ and the results rely on comprehensive spectrophotometric and calorimetric data sets, while the potentiometric and calorimetric study by Satyanarayana, Sahu and Das [75SAT/SAH] is much more limited with respect to the number of data. The potentiometric experiments using an Ag/AgSeCN electrode were conducted in the absence of a supporting electrolyte and the correction for the influence of activity coefficients is somewhat dubious. The review therefore focuses on the data collected by Kullberg. He utilised spectrophotometric data at several wavelengths and at considerable ligand excess to evaluate the equilibrium constant for $NiSeCN^+$, and calorimetric titration data for $NiSeCN^+$ and $Ni(SeCN)_2(aq)$. The value of β_1 obtained by spectrophotometry will be selected.

The complex formation is thus described in terms of mononuclear complexes

$$Ni^{2+} + n\,SeCN^- \rightleftharpoons Ni(SeCN)_n^{2-n} \quad (n = 1 \text{ to } 3);(V.128)$$

and Table V-64 lists the equilibrium constants reported.

Extrapolation of the data from 1 M $NaClO_4$ to standard conditions by SIT and $\Delta\varepsilon$ from [97BAH/PAR] in Appendix A resulted in the selected values:

$$\log_{10} K^\circ ((V.128), n = 1, 298.15 \text{ K}) = (1.77 \pm 0.06),$$

$$\log_{10} K^\circ ((V.128), n = 2, 298.15 \text{ K}) = (2.24 \pm 0.14).$$

In the data material, the $Ni(SeCN)_3^-$ complex never amounted to more than a few per cent of the total nickel(II) concentration and the value of the corresponding equilibrium constant has to be regarded as very approximate only. It has therefore not been accepted by this review.

Table V-64: Formation constants for complexes of Ni^{2+} with $SeCN^-$.

Method	Ionic medium	t (°C)	$\log_{10}\beta_1$	$\log_{10}\beta_2$	$\log_{10}\beta_3$	Reference
sp	1 M $NaClO_4$	25	(0.99 ± 0.01)			[74KUL]
cal	1 M $NaClO_4$	25		(1.20 ± 0.10)	≈1.9	[74KUL]
pot	→ 0	25	(1.66 ± 0.2)			[75SAT/SAH]

The calorimetric data were optimised with respect to the enthalpies of reaction and equilibrium constants. Various calculation models were tested with similar result. The result of the optimisation with $\log_{10} K$ ((V.128), n = 2), $\Delta_r H^\circ_m$ ((V.128), n = 1), and $\Delta_r H^\circ_m$ ((V.128), n = 2) as unknowns and $\log_{10} K$ ((V.128), n = 1) = 0.99 as fixed is selected. The reaction enthalpies found at I = 1 M $NaClO_4$ are assumed as valid also at I = 0. The uncertainties reported are doubled to account for the uncertainty of this assumption. This yields the selected results:

$$\Delta_r H^\circ_m((\text{V.128}), n = 1, 298.15 \text{ K}) = -(12.8 \pm 0.4) \text{ kJ·mol}^{-1},$$

$$\Delta_r H^\circ_m((\text{V.128}), n = 2, 298.15 \text{ K}) = -(25 \pm 4) \text{ kJ·mol}^{-1},$$

$$\Delta_r G^\circ_m((\text{V.128}), n = 1, 298.15 \text{ K}) = -(10.10 \pm 0.34) \text{ kJ·mol}^{-1},$$

$$\Delta_r G^\circ_m((\text{V.128}), n = 2, 298.15 \text{ K}) = -(12.79 \pm 0.80) \text{ kJ·mol}^{-1},$$

$$\Delta_r S^\circ_m((\text{V.128}), n = 1, 298.15 \text{ K}) = -(9.1 \pm 1.8) \text{ J·K}^{-1}\text{·mol}^{-1},$$

$$\Delta_r S^\circ_m((\text{V.128}), n = 2, 298.15 \text{ K}) = -(41.0 \pm 13.7) \text{ J·K}^{-1}\text{·mol}^{-1}.$$

The value of $\Delta_r H^\circ_m$ ((V.128), n = 1, 298.15 K) disagrees with the value in [75SAT/SAH], – (8.7 ± 1.1) kJ·mol^{-1}, also obtained by calorimetry. The disagreement is somewhat surprising since the equilibrium constants at standard conditions approximately agree.

Combined with the value for $\Delta_f G^\circ_m$ ($SeCN^-$, 298.15 K) selected in this review, and the value of $\Delta_f G^\circ_m$ (Ni^{2+}, 298.15 K) (– 45.77 kJ·mol^{-1}) from [2005GAM/BUG], this corresponds to the selected values:

$$\Delta_f G^\circ_m(\text{NiSeCN}^+, 298.15 \text{ K}) = (80.2 \pm 3.9) \text{ kJ·mol}^{-1},$$

$$\Delta_f G^\circ_m(\text{Ni(SeCN)}_2, \text{aq}, 298.15 \text{ K}) = (213.5 \pm 7.7) \text{ kJ·mol}^{-1}.$$

V.11.2 Cobalt compounds

V.11.2.1 Cobalt selenides

The solid compounds $Co_{1-x}Se(cr)$, $CoSe_2(cr)$, and $Co_9Se_8(cr)$ have been reported to form in the system cobalt-selenium. The phase diagram of the system is presented in [72KOM/WES].

V.11.2.1.1 $Co_9Se_8(cr)$

The $Co_9Se_8(cr)$ phase has a narrow homogeneity range and is often denoted $CoSe_{0.89}(cr)$.

The enthalpy of formation of $Co_9Se_8(cr)$ was determined to be $\Delta_f H_m^\circ$ (Co_9Se_8, cr, 298.15 K) = – (617.6 ± 40.0) $kJ\cdot mol^{-1}$ by Morozova, Vladimirova, Grudetskii, and Malysheva [77MOR/VLA] using direct synthesis calorimetry. The value is selected:

$$\Delta_f H_m^\circ (Co_9Se_8, cr, 298.15\ K) = -(617.6 \pm 40.0)\ kJ\cdot mol^{-1}.$$

V.11.2.1.2 $Co_{1-x}Se(cr)$

The $Co_{1-x}Se(cr)$ phase is a solid solution where the composition can vary in the approximate range $Co_{0.72}Se$ to $Co_{0.96}Se$ at temperatures below 1200 K. These compositions can also be written $Co_{0.42}Se_{0.58}$ and $Co_{0.49}Se_{0.51}$. The thermodynamic properties of $Co_{1-x}Se(cr)$ at different temperatures and compositions have been studied by vapour pressure and electrochemical cell techniques. The results are summarised in Table V-65. The individual papers are discussed in Appendix A, except [74MIL].

Table V-65: Enthalpy and entropy of formation of $Co_{1-x}Se(s)$. All values except that in [74MIL], which is a re-assessment of the experimental work in [69LAF/CER], were recalculated by the review from high temperature measurements as discussed in Appendix A.

Reference	Composition	$\Delta_f H_m^\circ$ ($Co_{1-x}Se$, cr, 298.15 K) ($kJ\cdot mol^{-1}$)	$\Delta_f S_m^\circ$ ($Co_{1-x}Se$, cr, 298.15 K) ($J\cdot K^{-1}\cdot mol^{-1}$)
[74MIL]	$Co_{0.90}Se$	– 58.8	
[72MAT/GEI]	$Co_{0.75}Se$	– (54.5 ± 4.0)	(2.3 ± 6.0)
[72MAT/GEI]	$Co_{0.93}Se$	– (52.7 ± 4.0)	(8.3 ± 6.0)
[72MAT/GEI2]	$Co_{0.72}Se$	– (53.8 ± 4.0)	(0.6 ± 4.0)
[74JEL/KOM]	$Co_{0.75}Se$	– 71.4	– 2.1
[74JEL/KOM]	$Co_{0.93}Se$	– 57.4	20.9

The phase $Co_{1-x}Se$(cr) is chosen to be represented by the single intermediate composition $Co_{0.84}Se$(cr) by the review. The selected enthalpy of formation is chosen to incorporate all experimental values and compositions except the most negative value in [74JEL/KOM] which deviates substantially from the others:

$$\Delta_f H^\circ_m (Co_{0.84}Se, cr, 298.15 K) = -(55.4 \pm 6.0) kJ\cdot mol^{-1}.$$

The adopted value of the entropy is calculated from a selected value of the entropy of formation of $\Delta_f S^\circ_m (Co_{0.84}Se, cr, 298.15 K) = (3.0 \pm 8.0) J\cdot K^{-1}\cdot mol^{-1}$, the selected entropy of selenium, and the entropy of cobalt in [98CHA] yielding:

$$S^\circ_m (Co_{0.84}Se, cr, 298.15 K) = (70.3 \pm 8.0) J\cdot K^{-1}\cdot mol^{-1}.$$

This value is included in Appendix E since it was calculated with a non-TDB auxiliary datum.

The Gibbs energy of formation is calculated from the selected enthalpy and entropy of formation to be:

$$\Delta_f G^\circ_m (Co_{0.84}Se, cr, 298.15 K) = -(56.3 \pm 6.5) kJ\cdot mol^{-1}.$$

V.11.2.1.3 $CoSe_2$(cr)

The heat capacity of $CoSe_2$(cr) at 298.15 K was derived by the review by extrapolating the heat capacity measurements made by Ogawa and Nishihara [77OGA/NIS] as discussed in Appendix A. The value obtained is selected:

$$C^\circ_{p,m} (CoSe_2, cr, 298.15 K) = (75.6 \pm 1.0) J\cdot K^{-1}\cdot mol^{-1}.$$

The thermodynamic properties of $CoSe_2$(cr) were investigated in the temperature range 343 to 431 K by Matlasevich and Geiderikh [72MAT/GEI2] using the electrochemical cell

$$(-) Co(cr) | NaCl, CoCl_2(in glycerol) | CoSe_2(cr), Se(cr) (+).$$

The reported values of the enthalpy and entropy of formation of $CoSe_2$(cr) were $\Delta_f H^\circ_m (CoSe_2, cr, 298.15 K) = -92.5 kJ\cdot mol^{-1}$ and $\Delta_f S^\circ_m (CoSe_2, cr, 298.15 K) = -5.1 J\cdot K^{-1}\cdot mol^{-1}$, respectively. The corresponding entropy of $CoSe_2$(cr) is calculated to be $S^\circ_m (CoSe_2, cr, 298.15 K) = 109.1 J\cdot K^{-1}\cdot mol^{-1}$ employing the selected entropy of selenium and the data for cobalt in [98CHA]. The third law entropy value $S^\circ_m (CoSe_2, cr, 298.15 K) = (97.2 \pm 2.0) J\cdot K^{-1}\cdot mol^{-1}$ was derived by the review from the measurements in [77OGA/NIS]. The latter value is preferred and selected:

$$S^\circ_m (CoSe_2, cr, 298.15 K) = (97.2 \pm 2.0) J\cdot K^{-1}\cdot mol^{-1}.$$

Morozova, Vladimirova, Grudetskii, and Malysheva [77MOR/VLA] determined the enthalpy of formation of $CoSe_2$(cr) to be $\Delta_f H^\circ_m (CoSe_2, cr, 298.15 K) = -(118.4 \pm 5.0) kJ\cdot mol^{-1}$ using direct synthesis calorimetry. The selected value is chosen as the mean of the values in [72MAT/GEI2] and [77MOR/VLA] yielding:

$$\Delta_f H_m^o (CoSe_2, cr, 298.15 K) = -(105.5 \pm 15.0) kJ\cdot mol^{-1}.$$

The Gibbs energy of formation is calculated from the selected enthalpy of formation and entropy, the selected entropy of selenium, and the entropy of cobalt in [98CHA] to be:

$$\Delta_f G_m^o (CoSe_2, cr, 298.15 K) = -(100.4 \pm 15.0) kJ\cdot mol^{-1}.$$

This value is included in Appendix E since it was calculated with a non-TDB auxiliary datum.

V.11.2.2 Cobalt selenites

The phase diagram of the system $CoSeO_3$-SeO_2-H_2O at 298.15 K was established by Ebert, Micka, and Pekova [82EBE/MIC]. The diagram is presented and contains the solid phases $CoSeO_3\cdot 2H_2O$, $Co(HSeO_3)_2\cdot 2H_2O$, and H_2SeO_3. The formula of the incongruently soluble acid salt was verified by an infrared spectroscopic measurement. No solubility data except for the compositions of the invariant points are published.

V.11.2.2.1 $CoSeO_3\cdot 2H_2O$(cr)

Chukhlantsev and Tomashevsky [57CHU/TOM] prepared cobalt selenite by mixing 0.1 M solutions of cobalt nitrate and sodium selenite in stoichiometric amounts. The precipitate was amorphous and aged for a long time in the mother liquor. Chemical analysis confirmed the 1:1 ratio between Co(II) and Se(IV). It is assumed to have the composition $CoSeO_3\cdot 2H_2O$ by the review. The solubility of the specimen in dilute solution of nitric or sulphuric acid was measured at 293 K. The experiments were performed and the data recalculated as outlined in Appendix A, [56CHU]. The result for

$$CoSeO_3\cdot 2H_2O(cr) \rightleftharpoons Co^{2+} + SeO_3^{2-} + 2H_2O(l) \qquad (V.129)$$

was $\log_{10} K_{s,0}^o$ ((V.129), 293.15 K) = – (7.62 ± 0.27). The reported conditional constant is – (6.84 ± 0.22). Ripan and Vericeanu [68RIP/VER] studied the solubility of $CoSeO_3$(s) by conductivity measurements as described in Appendix A. A calculation based on the assumptions made in the Appendix and the solubility, 3.40×10^{-4} M, provided in the paper leads to the conclusion that about 15% of dissolved Co(II) is present in hydrolysed forms. The solubility product estimated with allowance for the hydrolysis of Co^{2+} and SeO_3^{2-} is $\log_{10} K_{s,0}^o$ ((V.129), 291 K) = – (7.30 ± 0.20). This value can only be regarded as indicating the order of magnitude of the constant. The uncertainty only reflects the variance in the solubility measurements. Pyatnitskii and Durdyev [66PYA/DUR] measured the solubility of $CoSeO_3\cdot 2H_2O$(cr) in water and in selenite solution. The pH of the saturated solutions was also determined. The results from the determination of the solubility in water have been recalculated as described in [68RIP/VER] in Appendix A with the protonation constant of the selenite ion selected here. The result is $\log_{10} K_{s,0}$ ((V.129), 291.2K) = – (8.17 ± 0.05).

The review selects:

$$\log_{10} K^{\circ}_{s,0}\ ((\text{V.129}), 298.15\ \text{K}) = -(7.9 \pm 0.4)$$

from the results in [57CHU/TOM] and [66PYA/DUR].

Pyatnitskii and Durdyev [66PYA/DUR] attempted to determine the stability constants of cobalt-selenite complexes from solubility measurements in selenite solution. As discussed in Appendix A, their equilibrium model is likely to be incorrect. The proposed equilibrium constant of the reaction:

$$Co^{2+} + 2SeO_3^{2-} \rightleftharpoons Co(SeO_3)_2^{2-}, \qquad (\text{V.130})$$

$\log_{10} K$ ((V.130), $I = 0.01$, 291.2 K) = 5.2 cannot therefore be selected.

Leshchinskaya *et al.* [63LES/SEL4] measured the enthalpy change of the reaction between $Co(NO_3)_2{\cdot}6H_2O$ and a solution of sodium selenite with formation of $CoSeO_3{\cdot}2H_2O$(cr). The calorimetric result is used in Appendix A to calculate the standard enthalpy of formation of $CoSeO_3{\cdot}2H_2O$(cr) to be:

$$\Delta_f H^{\circ}_m (CoSeO_3{\cdot}2H_2O, \text{cr}, 298.15\ \text{K}) = -(1115.56 \pm 2.81)\ \text{kJ}\cdot\text{mol}^{-1}.$$

The value agrees well with [82WAG/EVA]. The selected solubility product yields:

$$\Delta_f G^{\circ}_m (CoSeO_3{\cdot}2H_2O, \text{cr}, 298.15\ \text{K}) = -(936.28 \pm 4.80)\ \text{kJ}\cdot\text{mol}^{-1},$$

$$S^{\circ}_m (CoSeO_3{\cdot}2H_2O, \text{cr}, 298.15\ \text{K}) = (245.1 \pm 18.6)\ \text{J}\cdot\text{K}^{-1}\cdot\text{mol}^{-1}.$$

Since the calculation of the formation and entropy values from the experimental results involved non-TDB auxiliary data, they are included in Appendix E.

V.11.2.2.2 $CoSeO_3$(cr)

Bakeeva, Buketov, and Pashinkin [69BAK/BUK2] investigated the equilibrium:

$$CoSeO_3(\text{cr, l}) \rightleftharpoons CoO(\text{cr}) + SeO_2(\text{g}) \qquad (\text{V.131})$$

by the dew point technique in the temperature range 839 to 991 K. Cobalt selenite melts at (923 ± 5) K. The vapour pressure relationship used to calculate p_{SeO_2} is not presented in the paper. From data in [71PAS/ARO] it could be deduced that the vapour pressure expression in [62MAR/GET] had been used for this purpose. The published pressures have been recalculated accordingly with the selected expression for the saturated vapour pressure of SeO_2(cr), see [71PAS/ARO] in Appendix A. The result for $CoSeO_3$(cr) and the temperature interval 839 to 923 K is $\log_{10} p(SeO_2, \text{g}, T) = (10.38 \pm 0.26) - (10640 \pm 230)\ T^{-1}$ and for $CoSeO_3$(l) $\log_{10} p(SeO_2, \text{g}, T) = (9.509 \pm 0.376) - (9843 \pm 366)\ T^{-1}$ in the temperature interval 923 to 991 K.

The uncertainties are entered here as twice the standard deviations from the least-squares calculation. The data yield $\Delta_{fus} H^{\circ}_m (CoSeO_3, \text{cr}, 923\ \text{K}) = (15.2 \pm 8.3)$ kJ·mol^{-1}.

The high temperature data have been extrapolated to standard conditions with $C^{\circ}_{p,\mathrm{m}}$ values in [73BAR/KNA] (no phase transitions assumed) leading to the estimates $\Delta_{\mathrm{f}}H^{\circ}_{\mathrm{m}}(\mathrm{CoSeO_3, cr, 298.15\ K}) = -(558.2 \pm 5.5)\ \mathrm{kJ\cdot mol^{-1}}$ and $S^{\circ}_{\mathrm{m}}(\mathrm{CoSeO_3, cr, 298.15\ K}) = (107.8 \pm 5.9)\ \mathrm{J\cdot K^{-1}\cdot mol^{-1}}$.

The corresponding values calculated in [71PAS/ARO] with unknown auxiliary data are $-(560.7 \pm 7.9)\ \mathrm{kJ\cdot mol^{-1}}$ and $(99.2 \pm 10.5)\ \mathrm{J\cdot K^{-1}\cdot mol^{-1}}$, respectively. The value estimated from $S^{\circ}_{\mathrm{m}}(\mathrm{CoSeO_3{\cdot}2H_2O, cr, 298.15\ K})$ by subtraction of the entropy contribution from water of crystallisation, 44.7 $\mathrm{J\cdot K^{-1}\cdot mol^{-1}}$ in [97PUI/RAR], is $(159.3 \pm 15.0)\ \mathrm{J\cdot K^{-1}\cdot mol^{-1}}$. The estimates of $S^{\circ}_{\mathrm{m}}(\mathrm{CoSeO_3, cr, 298.15\ K})$ from the high and low temperature experiments hence differ considerably, *cf.* the result for nickel in Section V.11.1.2.

The two sets of thermodynamic quantities for $CoSeO_3{\cdot}2H_2O$ and $CoSeO_3$, respectively, lead to an equilibrium constant for the reaction $\mathrm{CoSeO_3{\cdot}2H_2O(cr)} \rightleftharpoons \mathrm{CoSeO_3(cr)} + \mathrm{2H_2O(g)}$ of $\log_{10} K = -(0.4 \pm 1.1)$ at 298.15 K. The mutual consistency between the sets thus appears to be reasonably satisfactory, but the data for $CoSeO_3$(cr) have not been selected as the accuracy of the extrapolation cannot be assessed.

V.11.2.3 Cobalt selenates

Klein [40KLE] measured the solubility of cobalt selenate in water as a function of temperature and presented the results in a table and a graph. Several hydrates are formed and the transition temperatures observed were; heptahydrate → hexahydrate 284.6 K, hexahydrate → tetrahydrate 306.7 K, and tetrahydrate → monohydrate 346.7 K. $CoSeO_4{\cdot}6H_2O$ readily forms metastable saturated solutions up to about 340 K.

The solubility of $CoSeO_4{\cdot}6H_2O$ in water at 298.15 K has been measured by Vojtisek, Ebert, and Leblova [87VOJ/EBE4], Ojkova, Stoikova, and Staneva [94OJK/STO], and Ojkova, Barkov, and Popov [2000OJK/BAR] with consistent results. The review selects:

$$m(\mathrm{CoSeO_4{\cdot}6H_2O(cr)},\ \text{aq sat},\ 298.15\ \mathrm{K}) = (2.82 \pm 0.02)\ \mathrm{mol(CoSeO_4)\cdot kg^{-1}}.$$

This figure is also obtained by interpolation from the measurements by Klein.

A phase diagram and similar information are available for the system: $CoSeO_4$-$MgSeO_4$-H_2O at 298.15 K in [92OJK/STO]; $CoSeO_4$-$(NH_4)_2SeO_4$-H_2O at 298.15 K in [87VOJ/EBE4]; $CoSeO_4$-Li_2SeO_4-H_2O at 298.15 K in [93OJK/MIC]; $CoSeO_4$-Na_2SeO_4-H_2O at 298.15 K in [93OJK/MIC]; $CoSeO_4$-K_2SeO_4-H_2O at 298.15 K in [85VOJ/EBE]; $CoSeO_4$-Rb_2SeO_4-H_2O at 298.15 K in [2003BAR/CHR]; $CoSeO_4$-Cs_2SeO_4-H_2O at 298.15 K in [2000OJK/BAR].

The value of solubility product of $CoSeO_4{\cdot}6H_2O$(cr) at 298.15 K,

$$\mathrm{CoSeO_4{\cdot}6H_2O(cr)} \rightleftharpoons \mathrm{Co^{2+}} + \mathrm{SeO_4^{2-}} + \mathrm{6H_2O(l)} \qquad \text{(V.132)}$$

has been reported by Christov [97CHR] to be $\log_{10} K^{\circ}_{s,0} = -1.759$ and by Kumov and Batyreva [90KUM/BAT] to be -1.54, see Appendix A. The results are based on the same isopiestic data by Ojkova and Staneva [89OJK/STA]. The former value:

$$\log_{10} K^{\circ}_{s,0}((\text{V.132}), 298.15\text{ K}) = -(1.759 \pm 0.043)$$

is selected as it has been evaluated with the Pitzer ion interaction approach. The uncertainty added by the review corresponds to an uncertainty in the mean activity coefficient of 5%.

The standard Gibbs energy of formation of $CoSeO_4{\cdot}6H_2O(cr)$ is calculated from the Gibbs energy of Reaction (V.132) and the values of $\Delta_f G^{\circ}_m$ for the ions and water to be:

$$\Delta_f G^{\circ}_m(CoSeO_4{\cdot}6H_2O, \text{cr}, 298.15\text{ K}) = -(1926.9 \pm 4.2)\text{ kJ}{\cdot}\text{mol}^{-1}.$$

Since the calculation of this value from the experimental results involved a non-TDB auxiliary datum, it is included in Appendix E.

Ghosh and Nair [70GHO/NAI] determined the equilibrium constant of the reaction:

$$Co^{2+} + SeO_4^{2-} \rightleftharpoons CoSeO_4(aq) \qquad (\text{V.133})$$

in the temperature range 273.15 to 318.15 K by potentiometric measurements, see Appendix A. The enthalpy of the reaction was found from the variation of the stability constant with temperature. The following results are selected:

$$\log_{10} \beta_1^{\circ}((\text{V.133}), 298.15\text{ K}) = (2.70 \pm 0.05),$$

$$\Delta_r G^{\circ}_m((\text{V.133}), 298.15\text{ K}) = -(15.41 \pm 0.29)\text{ kJ}{\cdot}\text{mol}^{-1},$$

while $\Delta_r H^{\circ}_m((\text{V.133}), 298.15\text{ K}) = (12.20 \pm 0.40)\text{ kJ}{\cdot}\text{mol}^{-1}$ is not accepted for reasons discussed in Appendix A.

The uncertainty in $\log_{10} \beta_1^{\circ}$ has been increased considerably from the value in the reference to include the uncertainty from the estimate of the activity coefficients.

V.11.2.4 Cobalt selenocyanates

Three papers have reported on the formation of soluble Co^{2+}-$SeCN^-$ complexes. Toropova [56TOR] concluded from spectrophotometric measurements that the interactions in water were too weak to be quantified by this technique. Golub and Skopenko [62GOL/SKO] studied the system in water, aqueous acetone, acetone, and methanol by means of spectrophotometric, potentiometric (Ag/AgSeCN-electrode), and conductometric methods. The potentiometric data in water ($n = 5$) were collected at an undefined temperature with $[Co^{2+}]_{tot} = [SeCN^-]_{tot} = 0.02$ to 0.04 M and in the absence of a supporting electrolyte. The calibration of the electrode system is not described. Data were analysed on the assumption that only the first complex was formed according to the reaction

$$Co^{2+} + SeCN^- \rightleftharpoons CoSeCN^+. \quad (V.134)$$

The analysis resulted in a concentration dependent equilibrium constant, from K((V.134), 298.15 K) = 22 M^{-1} to 13 M^{-1} at the total concentrations of 0.02 and 0.04 M, respectively.

Also the work by Satyanarayana, Sahu and Das [75SAT/SAH] is based on potentiometric measurements with the Ag/AgSeCN-electrode and conducted in the absence of a supporting electrolyte. The Davies equation was applied to the calculation of the activity coefficient of Co^{2+}, 0.03 to 0.08 M, and the activity coefficients of $SeCN^-$ and $CoSeCN^+$ were assumed to cancel in the calculation of the formation constant. On this basis, K(V.134) = (30.9 ± 3.8) M^{-1} was calculated from 6 experimental data at 298.15 K.

Considering the inherent uncertainties associated with the evaluation of K(V.134) in both papers, the review can only accept the value by Satyanarayana *et al.* [75SAT/SAH] assigned with an uncertainty of ± 0.3 logarithmic units:

$$\log_{10} K^\circ((V.134), 298.15\ K) = (1.5 \pm 0.3).$$

Satyanarayana *et al.* [75SAT/SAH] also determined the enthalpy change of Reaction (V.134) by direct calorimetry. The equilibrium constant evaluated from the potentiometric measurements was used to calculate concentration changes. This implies that a close connection exists between the two quantities evaluated and hence, that the reaction enthalpy reported is connected with a considerable uncertainty. The review therefore accepts the reaction enthalpy reported assigned with a considerably larger uncertainty than given in the paper:

$$\Delta_r H^\circ_m((V.134), 298.15\ K) = -(6.0^{+2.0}_{-4.0})\ kJ\cdot mol^{-1}.$$

The unsymmetrical uncertainty limits are suggested by the results obtained for the Ni^{2+}-$SeCN^-$ system in the same paper, *cf.* Section V.11.1.4.

The selection yields:

$$\Delta_r G^\circ_m((V.134), 298.15\ K) = -(8.56 \pm 1.71)\ kJ\cdot mol^{-1},$$

$$\Delta_r S^\circ_m((V.134), 298.15\ K) = (8.6^{+8.8}_{-14.6})\ J\cdot K^{-1}\cdot mol^{-1}.$$

Combined with the value for $\Delta_f G^\circ_m(SeCN^-, 298.15\ K)$ selected in this review, and the [82WAG/EVA] value of $\Delta_f G^\circ_m(Co^{2+}, 298.15\ K)$ (– 54.4 $kJ\cdot mol^{-1}$), this corresponds to:

$$\Delta_f G^\circ_m(CoSeCN^+, 298.15\ K) = (73.0 \pm 5.7)\ kJ\cdot mol^{-1}.$$

Since the calculation of this value from the experiments involved a non-TDB auxiliary datum, it is included in Appendix E.

V.11.3 Iron compounds

V.11.3.1 Iron selenides

The phase diagram iron-selenium was assessed by Okamoto [91OKA] and it is shown in Figure V-21. The system contains several solid solution phases with stability ranges delimited by temperature. The naming of the phases is confusing and has varied over the years. The names used in [91OKA] are adopted by the review.

Figure V-21: The phase diagram iron-selenium (after [91OKA]). Reprinted from [91OKA], Copyright (1991), with permission from ASM International.

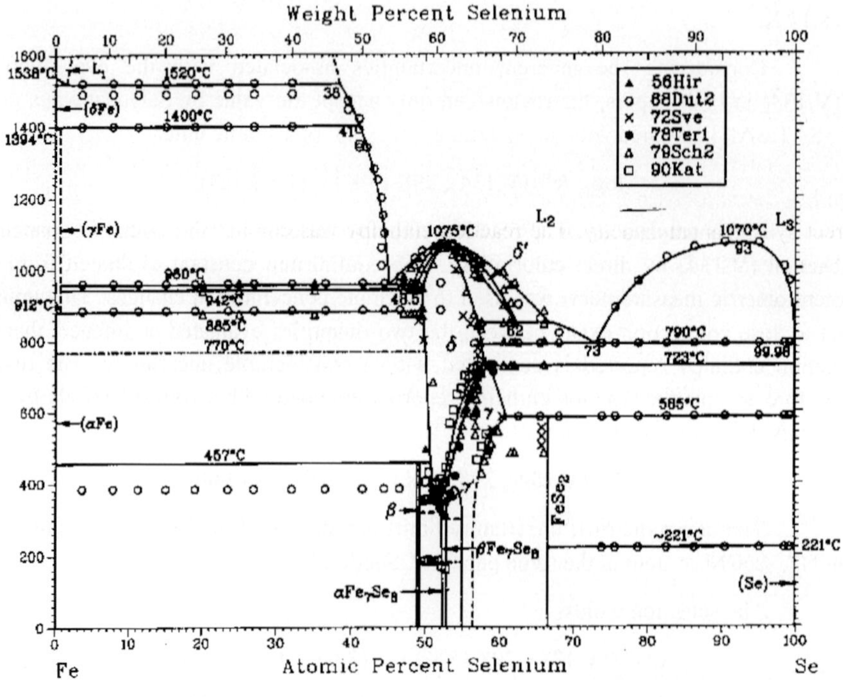

V.11.3.1.1 β-FeSe

The composition of β-FeSe is approximately $Fe_{1.04}Se$ and it has a homogeneity range extending from 49.0 atom-% to 49.4 atom-% Se. The phase is stable below 730 K.

The heat capacity of β-$Fe_{1.04}Se$ was measured in the temperature range 5 to 350 K by Grønvold and Westrum [59GRO/WES]. The heat capacity and entropy values at 298.15 K are selected:

$$C^{\circ}_{p,\mathrm{m}}\,(\mathrm{Fe_{1.04}Se}, \beta, 298.15\ \mathrm{K}) = (57.1 \pm 0.7)\ \mathrm{J{\cdot}K^{-1}{\cdot}mol^{-1}},$$

and

$$S^{\circ}_{\mathrm{m}}\,(\mathrm{Fe_{1.04}Se}, \beta, 298.15\ \mathrm{K}) = (72.1 \pm 0.8)\ \mathrm{J{\cdot}K^{-1}{\cdot}mol^{-1}}.$$

The heat capacity of a sample with the composition $Fe_{1.04}Se$ was measured in the temperature range 298 to 1050 K by Grønvold [68GRO] using adiabatic shield calorimetry.

The heat capacity expression

$$C^{\circ}_{p,\mathrm{m}}\,(\mathrm{Fe_{1.04}Se}, \beta, (298.15 - 730)\ \mathrm{K}) = (56.22 + 23.176 \times 10^{-3}\,T - 0.9700 \times 10^{-6}\,T^{2} - 5.3429 \times 10^{5}\,T^{-2})\ \mathrm{J{\cdot}K^{-1}{\cdot}mol^{-1}}$$

was evaluated by the review from the heat capacity values below 730 K listed in Table 4 of the paper.

Grønvold [72GRO] measured the enthalpy change for the formation of the equilibrium phases from the pure elements at the ratio 1.04Fe:1.00Se and 1050 K using direct synthesis calorimetry. The reaction enthalpy was combined with the heat content measurements of the same composition in [68GRO] yielding the selected enthalpy of formation of β-FeSe at 298.15 K:

$$\Delta_{\mathrm{f}}H^{\circ}_{\mathrm{m}}\,(\mathrm{Fe_{1.04}Se}, \beta, 298.15\ \mathrm{K}) = -\,(69.6 \pm 4.0)\ \mathrm{kJ{\cdot}mol^{-1}}.$$

The Gibbs energy of formation is calculated from the selected enthalpy of formation and the standard entropies to be:

$$\Delta_{\mathrm{f}}G^{\circ}_{\mathrm{m}}\,(\mathrm{Fe_{1.04}Se}, \beta, 298.15\ \mathrm{K}) = -\,(70.1 \pm 4.0)\ \mathrm{kJ{\cdot}mol^{-1}}.$$

This value is included in Appendix E since it was calculated with a non-TDB auxiliary datum.

V.11.3.1.2 δ-FeSe

δ-FeSe is a solid solution of the composition $Fe_{1-x}Se$ ($0 < x < 0.28$) and it is stable in the temperature range 623 to 1348 K. Samples of β-Fe_7Se_8 (composition $Fe_{0.875}Se$) transform on heating into δ-FeSe above approximately 650 K.

The heat capacity of a sample of the composition $Fe_{0.875}Se$ was measured in the temperature range 298 to 1050 K by Grønvold [68GRO] using adiabatic shield calorimetry. The heat capacity expression

$$C^{\circ}_{p,\mathrm{m}}\,(\mathrm{Fe_{0.875}Se}, \delta, (660 - 1020)\ \mathrm{K}) = (47.06 - 14.87 \times 10^{-3}\,T + 23.92 \times 10^{-6}\,T^{2} + 45.19 \times 10^{5}\,T^{-2})\ \mathrm{J{\cdot}K^{-1}{\cdot}mol^{-1}}$$

was evaluated by the review from the heat capacity values above 660 K listed in Table 4 of the paper.

The enthalpy of formation of δ-FeSe of stoichiometric composition (*i.e.*, FeSe) was determined from vapour pressure measurements by Rumyantsev, Zhitenva, and Bolandz [62RUM/ZHI], Svendsen [72SVE], and Piacente, Scardala, and Fontana [92PIA/SCA]. The reaction studied was δ-FeSe $\rightleftharpoons$ Fe(cr) + ½Se_2(g) and the composition FeSe corresponds to the iron rich limit of the homogeneity range of the δ-FeSe phase. Grønvold [72GRO] determined the enthalpy of formation of δ-FeSe of the composition $Fe_{0.875}Se$ to be $\Delta_f H_m^\circ$ ($Fe_{0.875}Se$, δ, 1050 K) = – (58.0 ± 2.0) kJ·mol^{-1} at 1050 K using direct synthesis calorimetry. A strict recalculation of these measurements to 298.15 K is not possible because the δ-FeSe phase is unstable at this temperature and auxiliary data are missing. The reported values of the enthalpies of formation of FeSe at 298.15 K were $\Delta_f H_m^\circ$(FeSe, δ, 298.15 K)= – 77.0 kJ·mol^{-1}, – 74.1 kJ·mol^{-1}, and – (69.0 ± 4.0) kJ·mol^{-1} in [62RUM/ZHI], [72SVE], and [92PIA/SCA], respectively. In order to be compatible with the phase diagram and the enthalpies of formation of β-FeSe and γ-FeSe, the enthalpy of formation of δ-FeSe must not be more negative than – 67.8 kJ·mol^{-1}. This argument neglects the often small change in entropy between neighbouring phases. No enthalpy value is selected since all values are outside the valid range.

V.11.3.1.3 α-Fe_7Se_8

The α-Fe_7Se_8 phase is formed in the low temperature part of the iron rich branch of the γ-FeSe field.

The heat capacity of α-Fe_7Se_8 was measured in the temperature range 5 to 350 K by Grønvold and Westrum [59GRO/WES]. The heat capacity and entropy values at 298.15 K are selected:

$$C_{p,m}^\circ(Fe_7Se_8, \alpha, 298.15\ K) = (442.1 \pm 4.0)\ J\cdot K^{-1}\cdot mol^{-1}$$

and

$$S_m^\circ(Fe_7Se_8, \alpha, 298.15\ K) = (613.8 \pm 5.0)\ J\cdot K^{-1}\cdot mol^{-1}.$$

The heat capacity of a sample of the composition Fe_7Se_8 was measured in the temperature range 298 to 1050 K by Grønvold [68GRO] using adiabatic shield calorimetry. The values measured below 450 K can be attributed to the α-Fe_7Se_8 phase.

Grønvold [72GRO] determined the enthalpy of formation of δ-FeSe of the composition Fe_7Se_8 at 1050 K to be $\Delta_f H_m^\circ$ (Fe_7Se_8, δ, 1050 K) = – (464.0 ± 16.0) kJ·mol^{-1} using direct synthesis calorimetry. The value was combined with the heat content measurements of the same composition in [68GRO] yielding $\Delta_f H_m^\circ$ (Fe_7Se_8, α, 298.15 K) = – (463.5 ± 20.0) kJ·mol^{-1}. Svendsen [72SVE] determined the Gibbs energy change for the reaction 7Fe(cr) + 4Se_2(g) $\rightleftharpoons$ α-Fe_7Se_8 to be $\Delta_r G_m^\circ$ (934 K) =

– (632.6 ± 25.0) kJ·mol^{-1} from vapour pressure measurements using Knudsen effusion cells. The value $\Delta_f H_m^\circ$ (Fe_7Se_8, α, 298.15 K) = – (536.0 ± 42.0) kJ·mol^{-1} was calculated using an estimated free energy function for the composition Fe_7Se_8. An adjustment to the selected enthalpy of formation of Se_2(g) yields $\Delta_f H_m^\circ$ (Fe_7Se_8, cr, 298.15 K) = – (528.8 ± 42.0) kJ·mol^{-1}. The value of the enthalpy of formation of α-Fe_7Se_8 reported in [72GRO] is preferred since it is based only on measured quantities. The selected value is:

$$\Delta_f H_m^\circ (Fe_7Se_8, \alpha, 298.15 \text{ K}) = -(463.5 \pm 20.0) \text{ kJ·mol}^{-1}.$$

The Gibbs energy of formation is calculated from the selected enthalpy of formation and the standard entropies to be:

$$\Delta_f G_m^\circ (Fe_7Se_8, \alpha, 298.15 \text{ K}) = -(489 \pm 20) \text{ kJ·mol}^{-1}.$$

This value is included in Appendix E since it was calculated with a non-TDB auxiliary datum.

V.11.3.1.4 γ-Fe_3Se_4

The γ-Fe_3Se_4 solid solution phase is also known as γ-FeSe. The heat capacity was measured in the temperature range 5 to 350 K by Grønvold and Westrum [59GRO/WES] and by Guseinov, Aldzhanov, Sadykov [89GUS/ALD]. The heat capacity values differ substantially between the investigations as shown in Figure V-22. The values reported at 298.15 K are $C_{p,m}^\circ$ (Fe_3Se_4, γ, 298.15 K) = (220.1 ± 2.0) J·K^{-1}·mol^{-1} and S_m° (Fe_3Se_4, γ, 298.15 K) = (279.8 ± 3.0) J·K^{-1}·mol^{-1} in [59GRO/WES] and $C_{p,m}^\circ$ (Fe_3Se_4, γ, 298.15 K) = (201.0 ± 2.0) J·K^{-1}·mol^{-1} and S_m° (Fe_3Se_4, γ, 298.15 K) = (295.0 ± 3.0) J·K^{-1}·mol^{-1} in [89GUS/ALD]. The samples in [59GRO/WES] were equilibrated for 60 days altogether while those in [89GUS/ALD] were heat treated for 3 days only. The selected heat capacity and entropy values at 298.15 K are:

$$C_{p,m}^\circ (Fe_3Se_4, \gamma, 298.15 \text{ K}) = (220.1 \pm 2.0) \text{ J·K}^{-1}\text{·mol}^{-1}$$

and

$$S_m^\circ (Fe_3Se_4, \gamma, 298.15 \text{ K}) = (279.8 \pm 3.0) \text{ J·K}^{-1}\text{·mol}^{-1}.$$

Grønvold [72GRO] determined the enthalpy of formation of γ-Fe_3Se_4 at 1050 K using direct synthesis calorimetry. The value was combined with the heat content measurements in [68GRO] yielding $\Delta_f H_m^\circ$ (Fe_3Se_4, γ, 298.15 K) = – (212.1 ± 13.0) kJ·mol^{-1}. Svendsen [72SVE] determined the enthalpy of formation of γ-Fe_3Se_4 to be $\Delta_f H_m^\circ$ (Fe_3Se_4, γ, 298.15 K) = – (257.3 ± 13.0) kJ·mol^{-1} from vapour pressure measurements using Knudsen effusion cells. The selected value is the mean of the two investigations:

$$\Delta_f H_m^\circ (Fe_3Se_4, \gamma, 298.15 \text{ K}) = -(235 \pm 30) \text{ kJ·mol}^{-1}.$$

The Gibbs energy of formation is calculated from the selected enthalpy of formation and the standard entropies to be:

$$\Delta_f G_m^\circ (Fe_3Se_4, \gamma, 298.15 \text{ K}) = -(244 \pm 30) \text{ kJ·mol}^{-1}.$$

This value is included in Appendix E since it was calculated with a non-TDB auxiliary datum.

Figure V-22: Low temperature heat capacity of γ-Fe_3Se_4.

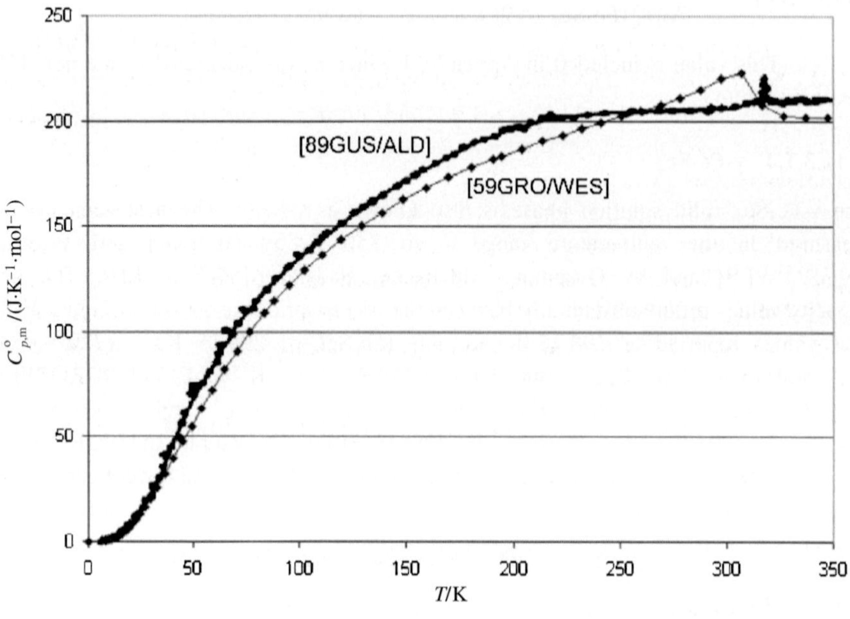

V.11.3.1.5 $FeSe_2$(cr)

Grønvold and Westrum [62GRO/WES] measured the heat capacity of $FeSe_2$(cr) in the temperature range 5 to 350 K. The heat capacity and entropy values at 298.15 K are selected:

$$C_{p,m}^\circ (FeSe_2, \text{cr}, 298.15 \text{ K}) = (72.9 \pm 1.0) \text{ J·K}^{-1}\text{·mol}^{-1}$$

and

$$S_m^\circ (FeSe_2, \text{cr}, 298.15 \text{ K}) = (86.8 \pm 1.0) \text{ J·K}^{-1}\text{·mol}^{-1}.$$

Svendsen [72SVE2] and Grønvold [75GRO] measured the heat capacity of $FeSe_2$(cr) in the temperature range 300 to 853 K. The results are very similar and the heat capacity expressions:

$$C^{\circ}_{p,\mathrm{m}}(\mathrm{FeSe_2, cr, (298.15 - 600)\ K}) = (37.014 + 80.273 \times 10^{-3}\ T + 9.57991 \times 10^{5}\ T^{-2})\ \mathrm{J{\cdot}K^{-1}{\cdot}mol^{-1}}$$

and

$$C^{\circ}_{p,\mathrm{m}}(\mathrm{FeSe_2, cr, (600 - 853)\ K}) = (-835.808 + 1.050076\ T + 1.056967 \times 10^{8}\ T^{-2})\ \mathrm{J{\cdot}K^{-1}{\cdot}mol^{-1}}$$

were derived by the review from the combined heat capacity data reported in the two investigations.

The enthalpy of formation of $FeSe_2$(cr) was determined to be $\Delta_f H^{\circ}_m$($FeSe_2$, cr, 298.15 K) = – (123.0 ± 13.0) kJ·mol^{-1} by Svendsen [72SVE] from static vapour pressure measurements. This enthalpy value is too negative to be consistent with the selected thermodynamic data of the other phases and the phase diagram. The enthalpy of decomposition of $FeSe_2$(cr) into γ-Fe_3Se_4 and liquid selenium was determined to be (30.93 ± 0.05) kJ·mol^{-1} in [75GRO] at 857 K. On combining the value with the selected values of selenium, the heat content measurement of γ-Fe_3Se_4 in [68GRO], and the heat capacity of $FeSe_2$(cr) the enthalpy of formation of $FeSe_2$(cr) is calculated to be $\Delta_f H^{\circ}_m$($FeSe_2$, cr, 298.15 K) = – (108.7 ± 15.0) kJ·mol^{-1}. This value is consistent with the selected thermodynamic data of the other iron selenides and the phase diagram and it is selected by the review:

$$\Delta_f H^{\circ}_m(\mathrm{FeSe_2, s, 298.15\ K}) = -(108.7 \pm 15.0)\ \mathrm{kJ{\cdot}mol^{-1}}.$$

The Gibbs energy of formation is calculated from the selected enthalpy of formation and the standard entropies to be:

$$\Delta_f G^{\circ}_m(\mathrm{FeSe_2, cr\ 298.15\ K}) = -(101.3 \pm 15.0)\ \mathrm{kJ{\cdot}mol^{-1}}.$$

This value is included in Appendix E since it was calculated with a non-TDB auxiliary datum.

V.11.3.2 Iron selenite

Rai, Felmy, and Moore [95RAI/FEL3] made a serious attempt to determine the solubility product of iron(III) selenite as described in Appendix A. Their work shows that $Fe_2(SeO_3)_3{\cdot}6H_2O$ is the stable phase in contact with an aqueous phase with a pH below 4. Above this value the selenite starts to transform into other, basic phases of unknown composition.

The interpretation of careful experimental work resulted in an equilibrium model, which included the two equilibria:

$$\mathrm{Fe_2(SeO_3)_3{\cdot}6H_2O(cr)} \rightleftharpoons 2\mathrm{Fe^{3+}} + 3\mathrm{SeO_3^{2-}} + 6\mathrm{H_2O(l)} \qquad (V.135)$$

$$\mathrm{Fe}^{3+} + \mathrm{SeO}_3^{2-} \rightleftharpoons \mathrm{FeSeO}_3^{+} \qquad \text{(V.136)}$$

and the equilibrium constants:

$$\log_{10} K^{\circ}_{s,0}\ ((\text{V.135}), 296\ \text{K}) = -(41.58 \pm 0.11)$$

$$\log_{10} \beta^{\circ}_{1}\ ((\text{V.136}), 296\ \text{K}) = (11.15 \pm 0.11)$$

The uncertainties are those of the paper. The stability constant of $FeSeO_3^+$ appears exceptionally large although there is known to be a great affinity between Fe(III) and Se(IV). The value of the solubility product is not greatly affected by the value of $\log_{10} \beta^{\circ}_1$ (V.136), however. These results are selected by the review and assumed valid at standard conditions.

Chukhlantsev and Tomashevsky [57CHU/TOM] prepared iron(III) selenite by mixing a 0.2 M solution of iron(III) sulphate and a 0.1 M sodium selenite solution. The pH of the reaction mixture was 5 to 6. The precipitate was aged for 24 hours in the mother liquor. No information on the quality of the specimen is available but it is most likely amorphous. Chemical analysis showed a 2:3 ratio between Fe(III) and Se(IV). The solubility of the specimen in dilute solution of nitric or sulphuric acid was measured at 293 K. The experiments were performed and the data recalculated as described in Appendix A, [56CHU]. Two experimental points were deleted, as they would be much affected by the hydrolysis of Fe^{3+}. The result for

$$\mathrm{Fe_2(SeO_3)_3(am)} \rightleftharpoons 2\mathrm{Fe}^{3+} + 3\mathrm{SeO}_3^{2-} \qquad \text{(V.137)}$$

was $\log_{10} K^{\circ}_{s,0}$ ((V.137), 293.15 K) = $-(33.77 \pm 0.15)$. The reported value was $-(30.91 \pm 0.25)$. The uncertainties reflect only the standard deviation in the measurement. This result, which is not selected, would indicate that the solubility of freshly precipitated iron(III) selenite is 1 to 2 orders of magnitude greater than for the stable form.

No calorimetric determination of the standard enthalpy of formation of iron(III) selenites from aqueous solution has been found.

Pinaev and Volkova [76PIN/VOL] have prepared iron(III) selenites by hydrothermal synthesis from $Fe_2(SeO_3)_3 \cdot 7H_2O$, SeO_2, and H_2O in the temperature range 393 to 513 K. Crystallisation fields were established for a number of selenites written as $Fe_2O_3 \cdot m SeO_2 \cdot n H_2O$. The compounds appear to have been crystalline but some of the *m* and *n* values in the empirical formulas are significantly different from whole numbers. Pinaev, Gornostaeva, Volkova, and Glybin [74PIN/GOR] measured the heat of dissolution in 7.5 M HCl of four of the compounds at 333 K (not 433 K as written in the paper). The enthalpies of dissolution of α-Fe_2O_3, SeO_2(cr), and H_2O(l) in the same solvent were also measured and the data combined to yield the enthalpy change of the reaction

$$\mathrm{Fe_2O_3}\cdot m\mathrm{SeO_2}\cdot n\mathrm{H_2O(cr)} \rightarrow \alpha\text{-}\mathrm{Fe_2O_3(cr)} + m\mathrm{SeO_2(cr)} + n\mathrm{H_2O(l)} \qquad \text{(V.138)}$$

The $\Delta_r H^{\circ}_m$(V.138) values so obtained (available in VINITI No. 2968-2112-74) could be reproduced by the equation $\Delta_r H^{\circ}_m = (m \times (21.3 \pm 0.8) + n \times (12.6 \pm 0.8))$

$kJ\cdot mol^{-1}$. Since only condensed phases are involved in the reaction, it was proposed that $\Delta C^{\circ}_{p,m}$ is close to zero and therefore the relationship would be valid at 298.15 K also. Together with the standard enthalpies of formation of α-Fe_2O_3(cr), SeO_2(cr), and H_2O(l) the equation can then be used to calculate standard enthalpies of formation of iron(III) selenites. The standard enthalpy of formation of $Fe_2(SeO_3)_3$(cr) estimated in this way is $-(1565.0 \pm 3.0)$ $kJ\cdot mol^{-1}$.

According to Bakeeva, Pashinkin, Bakeev, and Buketov [73BAK/PAS2] the thermal dissociation of iron(III) selenite proceeds by the reactions:

$$Fe_2(SeO_3)_3(cr) \rightleftharpoons Fe_2O_3{\cdot}2SeO_2(cr) + SeO_2(g) \qquad (V.139)$$

$$4Fe_2O_3{\cdot}2SeO_2(cr) \rightleftharpoons (Fe_2O_3)_4\, SeO_2(cr) + 7SeO_2(g) \qquad (V.140)$$

$$(Fe_2O_3)_4{\cdot}SeO_2(cr) \rightleftharpoons 4Fe_2O_3(cr) + SeO_2(g). \qquad (V.141)$$

The compositions of the solids were established by chemical analysis and they were shown to be crystalline by X-ray diffraction. The vapour pressure of SeO_2(g) over the selenites was measured by the dew point method and the equilibrium pressures calculated by the relationship in [69SON/NOV]. The original pressure data furnished in the deposited VINITI document (No. 4965-72) have been recalculated with the relationship for the saturated vapour pressure over SeO_2(cr) selected by the review; this leads only to small differences from the original results. The result is for Reaction (V.139) and the temperature interval 679 to 791 K: $\log_{10} p(SeO_2, g, T) = (10.11 \pm 0.20) - (8239 \pm 146)\, T^{-1}$; for Reaction (V.140) and the temperature interval 749 to 909 K: $\log_{10} p(SeO_2, g, T) = (9.840 \pm 0.113) - (8560 \pm 90)\, T^{-1}$; for Reaction (V.141) and the temperature interval 773 to 912 K: $\log_{10} p(SeO_2, g, T) = (9.028 \pm 0.188) - (8100 \pm 160)\, T^{-1}$. The uncertainties are entered here as twice the standard deviations from the least-squares calculation.

No experimental data have been found which would permit a proper calculation of the enthalpy and entropy changes at 298.15 K. The authors estimated standard enthalpies of formation on the assumption that the change in the heat capacity per mole of SeO_2(g) formed in the reaction is $\Delta_r C^{\circ}_{p,m} = C^{\circ}_{p,m}(SeO_2(g)) - C^{\circ}_{p,m}(SeO_2(cr)) \approx -16.0$ $J\cdot K^{-1}\cdot mol^{-1}$. The result obtained for $Fe_2(SeO_3)_3$ was $\Delta_f H^{\circ}_m(Fe_2(SeO_3)_3$, cr, 298.15 K$) = -(1660.0 \pm 10.0)$ $kJ\cdot mol^{-1}$ and $S^{\circ}_m(Fe_2(SeO_3)_3$, cr, 298.15 K$) = (260.0 \pm 12.0)$ $J\cdot K^{-1}\cdot mol^{-1}$, which represent the averages from the original data and those obtained by the review.

No values of the standard entropies or enthalpies of formation of iron(III) selenites can be selected in view of the substantial difference between the value of $\Delta_f H^{\circ}_m(Fe_2(SeO_3)_3$, cr, 298.15 K) obtained in [74PIN/GOR] and [73BAK/PAS2]. The data from these references are included as the only source of information on the subject.

V.11.3.3 Iron selenates

Wells and Salam [68WEL/SAL] attempted to determine the stability constant of the reaction:

$$Fe^{2+} + SeO_4^{2-} \rightleftharpoons FeSeO_4(aq) \qquad (V.142)$$

from a kinetic study of the oxidation of Fe^{2+} by hydrogen peroxide in the presence of various concentrations of the selenate ion. The interpretation of the kinetic data involved the formation of a 1:1 complex between Fe^{2+} and selenate. No primary data are provided and $\beta_1 = 12\ M^{-1}$ at 298.15 K and $I = 1$ M is reported.

The value of the stability constant is not accepted. An analysis of the method used to determine the parameters of the kinetic model made in Appendix A indicates that the method used to calculate β_1 is in error.

No other thermodynamic data for iron selenates have been identified.

V.11.4 Palladium compounds

V.11.4.1 Palladium selenides

A number of compounds with various compositions in the palladium-selenium system have been reported. Experimental thermodynamic information is available for $Pd_4Se(cr)$ only. Additional thermodynamic quantities have been estimated by Mills [74MIL].

V.11.4.1.1 $Pd_4Se(cr)$

The heat capacity of $Pd_4Se(cr)$ was measured in the temperature range 5 to 350 K by Grønvold, Westrum, and Radebaugh [69GRO/WES]. The heat capacity at 298.15 K and the standard entropy were selected:

$$C^{\circ}_{p,m}(Pd_4Se, cr, 298.15\ K) = (137.0 \pm 2.0)\ J\cdot K^{-1}\cdot mol^{-1}$$

and

$$S^{\circ}_{m}(Pd_4Se, cr, 298.15\ K) = (200.1 \pm 3.0)\ J\cdot K^{-1}\cdot mol^{-1}.$$

V.11.5 Rhodium compounds

V.11.5.1 Rhodium selenides

A number of solid rhodium selenides have been reported, but no experimental thermodynamic information is available. Mills [74MIL] has estimated the enthalpy of formation and the entropy of $RhSe_2(cr)$. No values are selected by the review.

V.11.6 Ruthenium compounds

V.11.6.1 Ruthenium selenides

$RuSe_2$(cr) is the only reported compound in the system ruthenium-selenium.

V.11.6.1.1 $RuSe_2$(cr)

Murray and Heyding [67MUR/HEY] studied the decomposition of $RuSe_2$(cr) in the temperature range 1006 to 1127 K by measuring the saturated vapour pressure using the torsion effusion technique. The data were evaluated according to the reaction $RuSe_2(cr) \rightleftharpoons Ru(cr) + Se_2(g)$ using the second law and assuming $\Delta_r C^\circ_{p,m} = 0$ for the reaction. No evaluation was made by the third law due to the lack of entropy information for $RuSe_2$(cr) at the time of the work. Svendsen, Grønvold, and Westrum [87SVE/GRO] measured the heat capacity of $RuSe_2$(cr) in the temperature range 5 to 988 K using adiabatic low- and high-temperature calorimeters and the enthalpy increment of the compound in the temperature range 882 to 1483 K using a drop calorimeter. In addition, the saturated vapour pressure in equilibrium with a mixture of Ru(cr) and $RuSe_2$(cr) was measured in the temperature range 1270 to 1480 K using a silica-glass manometer. Unfortunately, the evaluation of the vapour pressures in [87SVE/GRO] contains errors and the investigation was re-evaluated by the review. Similarly, the results in [67MUR/HEY] were re-evaluated using the heat capacity and entropy determined in [87SVE/GRO].

The experimental heat capacity values determined in [87SVE/GRO] were smoothed using a least-squares procedure involving the fitting of analytical expressions to the values in four overlapping temperature intervals. The expressions derived were not given but smoothed values were tabulated at regular temperature intervals. The heat capacity determined at 298.15 K is selected:

$$C^\circ_{p,m}(\mathrm{RuSe_2, cr, 298.15\ K}) = (69.9 \pm 1.0)\ \mathrm{J\cdot K^{-1}\cdot mol^{-1}}.$$

The following heat capacity expressions were evaluated by the review from the values in Table 8 of the paper :

$$C^\circ_{p,m}(\mathrm{RuSe_2, cr, (298.15-900)\ K}) = (72.298 + 9.0762 \times 10^{-3}\ T - 8.002 \times 10^{-7}\ T^2 - 4.65942 \times 10^5\ T^{-2})\ \mathrm{J\cdot K^{-1}\cdot mol^{-1}}$$

and

$$C^\circ_{p,m}(\mathrm{RuSe_2, cr, (900-1500)\ K}) = (-\,43.518 + 140.5064 \times 10^{-3}\ T - 3.88342 \times 10^{-5}\ T^2 + 2.248624 \times 10^7\ T^{-2})\ \mathrm{J\cdot K^{-1}\cdot mol^{-1}}.$$

The entropy of $RuSe_2$(cr) at 298.15 K was evaluated using the third law from the heat capacity in the temperature range 5 to 298.15 K in [87SVE/GRO]. Second law entropy values were derived by the review from vapour pressure measurements in

[67MUR/HEY] and [87SVE/GRO]. The values are summarised in Table V-66. The third law value determined in [87SVE/GRO] is selected:

$$S^{\circ}_{m}(\text{RuSe}_2, \text{cr}, 298.15\ \text{K}) = (80.8 \pm 0.8)\ \text{J·K}^{-1}\text{·mol}^{-1}.$$

Table V-66: The standard entropy of $RuSe_2$(cr) determined in different investigations. The second law values were derived by the review in evaluations of experimental data given in the references.

Reference	S°_{m} ($RuSe_2$, cr, 298.15 K) 2nd law, ($J·K^{-1}·mol^{-1}$)	S°_{m} ($RuSe_2$, cr, 298.15 K) 3rd law, ($J·K^{-1}·mol^{-1}$)	Temperature range 2nd law, (K)
[67MUR/HEY]	(75.3 ± 4.0)		1006 – 1127
[87SVE/GRO]	(77.2 ± 3.0)	(80.8 ± 0.8)	1271 – 1480

The re-evaluation of the data in [67MUR/HEY] made by the review using the second law and the selected heat capacity yields an enthalpy of formation that is 6.5 kJ·mol^{-1} less negative than obtained in the original work. Svendsen *et al.* combined their vapour pressure measurements with those in [67MUR/HEY] and performed a third law evaluation on the whole dataset using auxiliary data for the species Se(g)-Se_8(g). The procedure yielded an enthalpy of formation of $\Delta_f H^{\circ}_{m}$ ($RuSe_2$, cr, 298.15 K) = – (162.1 ± 2.2) kJ·mol^{-1}. However, the constants given in Table 4 of the paper, supposedly representing the equilibria between the selenium species, were calculated incorrectly and do not reproduce the equilibria in the selenium gas phase. This has been confirmed in a private communication between the review and Svendsen. The results of the experiments with excess of ruthenium reported in Table 10 of the paper were therefore re-evaluated using the selected data for the gaseous selenium species. The Se_2(g) molecule dominates in the gas phase and other selenium gaseous species can be disregarded. The logarithm of the total pressure of selenium (essentially Se_2(g)) is plotted against the inverse absolute temperature in Figure V-23.

The values obtained for the enthalpy of formation of $RuSe_2$(cr) at 298.15 K from the second and third laws are summarised in Table V-67. The weighted average is:

$$\Delta_f H^{\circ}_{m}(\text{RuSe}_2, \text{cr}, 298.15\ \text{K}) = -(175.3 \pm 8.6)\ \text{kJ·mol}^{-1}.$$

The value is included in Appendix E since it was calculated with a non-TDB auxiliary datum.

Figure V-23: Logarithm of the total pressure of selenium (essentially $Se_2(g)$) as a function of temperature. The diagram is based on experimental series 8 to 13 reported in Table 10 in [87SVE/GRO].

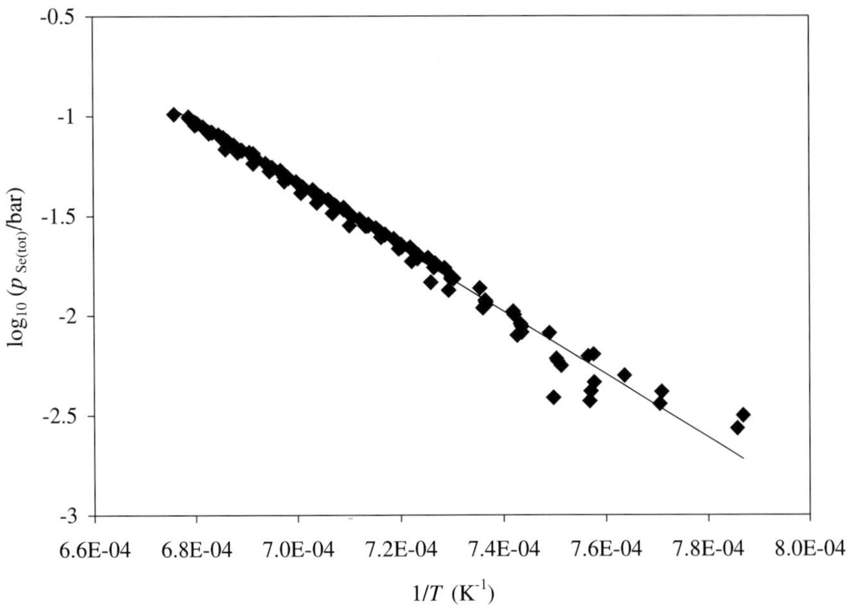

Table V-67: The standard enthalpy of formation of $RuSe_2(cr)$ determined in different investigations. The values were derived by the review in evaluations based on the experimental data of the references.

Reference	$\Delta_f H_m^\circ$ ($RuSe_2$, cr, 298.15 K) ($kJ\cdot mol^{-1}$)		Temperature range (K)
	2nd law	3rd law	
[67MUR/HEY]	– (173.0 ± 15.0)	– (181.1 ± 3.0)	1006 – 1127
[87SVE/GRO]	– (173.2 ± 15.0)	– (169.7 ± 3.0)	1271 – 1480

V.11.7 Platinum compounds

V.11.7.1 Platinum selenides

The platinum-selenium phase diagram was investigated by Richter and Ipser [94RIC/IPS] and the compounds $Pt_5Se_4(cr)$ and $PtSe_2(cr)$ were reported. Experimental thermodynamic information is only available for Pt_5Se_4(cr). Mills [74MIL] has estimated some additional thermodynamic data.

V.11.7.1.1 Pt_5Se_4(cr)

The total pressure of selenium (mainly Se_2(g)) in equilibrium with Pt(cr) and Pt_5Se_4(cr) was measured in the temperature range 943 to 1033 K by Rau [67RAU] using a transport method. The vapour pressure was given as $\log_{10} p(\mathrm{Se_2, g}, T) = 10100 - 14574\, T^{-1}$ and the enthalpy of formation and entropy at 298.15 K were calculated to be $\Delta_f H^\circ_m$ (Pt_5Se_4, cr, 298.15 K) = – (265.3 ± 25.0) kJ·mol^{-1} and S°_m (Pt_5Se_4, cr, 298.15 K) = (327.2 ± 22.0) J·K^{-1}·mol^{-1}, respectively. The procedure for deriving the values from the given pressure expression is not clearly stated. The quantities were therefore recalculated by the review using the selected thermodynamic data for selenium, the data for platinum in [91DIN] and assuming a $\Delta_f C^\circ_{p,m}$ (Pt_5Se_4, cr) = 0 J·K^{-1}·mol^{-1} at all temperatures. The second law results are $\Delta_f H^\circ_m$ (Pt_5Se_4, cr, 298.15 K) = – 308.6 kJ·mol^{-1} and S°_m (Pt_5Se_4, cr, 298.15 K) = 265.3 J·K^{-1}·mol^{-1}. The new values differ substantially from those calculated in the original paper for unknown reasons. No values are selected by the review.

V.11.8 Iridium compounds

V.11.8.1 Iridium selenides

The compounds Ir_2Se_3(cr), $IrSe_2$(cr), and $IrSe_3$(cr) have been reported, but no experimental thermodynamic information is available. Mills [74MIL] has estimated thermodynamic properties of the compounds. No values were selected by the review.

V.11.9 Osmium compounds

V.11.9.1 Osmium selenides

$OsSe_2$(cr) is the only reported compound in the system osmium-selenium.

V.11.9.1.1 $OsSe_2$(cr)

Murray and Heyding [67MUR/HEY] studied the decomposition of $OsSe_2$(cr) in the temperature range 892 to 960 K by measuring the saturated vapour pressure using the torsion effusion technique. The data were evaluated according to the reaction $OsSe_2(cr) \rightleftharpoons Os(cr) + Se_2(g)$ using the second law and assuming $\Delta_r C^\circ_{p,m} = 0$ for the reaction. The derived enthalpy of formation at 298.15 K was $\Delta_f H^\circ_m$ ($OsSe_2$, cr, 298.15 K) = – (136.0 ± 21.0) kJ·mol^{-1}. In the absence of proper heat capacity expressions and an entropy value of $OsSe_2$(cr), the review performed an evaluation using the second and third laws together with the selected heat capacity and entropy of $RuSe_2$(cr). The results were $\Delta_f H^\circ_m$ ($OsSe_2$, cr, 298.15 K) = – (131.5 ± 8.0) kJ·mol^{-1} and

– (125.6 ± 8.0) kJ·mol^{-1} from the second and third laws, respectively. The mean of the last two values is calculated to be:

$$\Delta_f H_m^\circ (\text{OsSe}_2, \text{cr}, 298.15 \text{ K}) = -(128.6 \pm 11.0) \text{ kJ·mol}^{-1}.$$

The value is included in Appendix E since it was calculated with non-TDB auxiliary data.

V.11.10 Manganese compounds

V.11.10.1 Manganese selenides

The manganese-selenium system was reviewed by Schlesinger in [98SCH]. It was concluded that the phase diagram provided in [86MAS] is largely conjectural and based on an analogy with the Mn-Te system. Intermediary stable phases at 1 bar and 298.15 K are α-MnSe and $MnSe_2$(cr). α-MnSe transforms to δ-MnSe below a transition temperature of about 240 K. The phases β-MnSe and γ-MnSe are both metastable at 1 bar. Thermodynamic information is available for the solid phases α-MnSe and $MnSe_2$(cr) and the gaseous species MnSe(g), Mn_2Se(g), $MnSe_2$(g) and Mn_2Se_2(g).

V.11.10.1.1 α-MnSe

The low-temperature heat capacity of solid MnSe was measured in the temperature range 54 to 287 K by Kelley [39KEL]. A pronounced peak centred around 247 K apparently corresponds to the δ → α transition. The heat capacity determined at 298.15 K is selected:

$$C_{p,m}^\circ (\text{MnSe}, \alpha, 298.15 \text{ K}) = (51.04 \pm 0.30) \text{ J·K}^{-1}\text{·mol}^{-1}.$$

The temperature dependence of the heat capacity of α-MnSe was estimated by Wiedemeier and Goyette [68WIE/GOY] to be $C_{p,m}^\circ$(MnSe, α, (298.15 – 1600) K) = $(49.036 + 6.736 \times 10^{-3}\ T)$ J·K^{-1}·mol^{-1} and the expression was employed by the review in the evaluation of high temperature investigations as discussed below and in Appendix A.

The standard entropy of α-MnSe was evaluated to be S_m°(MnSe, α, 298.15 K) = (90.8 ± 3.0) J·K^{-1}·mol^{-1} in [39KEL] from the heat capacity measurements by extrapolating the heat capacity into the range 0 to 54 K using Debye and Einstein functions. A second law evaluation of the vaporisation of α-MnSe in the temperature range 1414 to 1610 K made by Wiedemeier and Goyette [68WIE/GOY] yielded S_m°(MnSe, α, 298.15 K) = (94.3 ± 9.0) J·K^{-1}·mol^{-1} as discussed in Appendix A. The entropy of formation of α-MnSe was determined to be $\Delta_f S_m^\circ$(MnSe, α, 298.15 K) = (26.7 ± 10.0) J·K^{-1}·mol^{-1} by Lukashenko and Polotskaya [78LUK/POL] from emf measurements in the range 630 to 760 K using the electrochemical cell

$$(-)\ \alpha\text{-Mn} \mid 2 \text{ weight-\% MnCl}_2 \text{ (in LiCl + KCl melt)} \mid \alpha\text{-MnSe, MnSe}_2\text{(cr)}\ (+).$$

The standard entropy value S°_{m}(MnSe, α, 298.15 K) = (100.8 ± 10.0) J·K^{-1}·mol^{-1} is obtained by combining the value of the entropy of formation in [78LUK/POL] with the the entropy of α-Mn in [82WAG/EVA] and the selected entropy of Se(trig). The third law value in [39KEL] is selected:

$$S^{\circ}_{m}(\text{MnSe, α, 298.15 K}) = (90.8 \pm 3.0)\ \text{J·K}^{-1}\text{·mol}^{-1}.$$

The total pressure of the saturated vapour in equilibrium with α-MnSe was measured in the temperature range 1414 to 1610 K using Knudsen effusion cells in [68WIE/GOY] and in the temperature range 1280 to 1779 K by Smoes, Pattje, and Drowart [78SMO/PAT] by measuring the partial pressures of the gas species using Knudsen effusion cells and mass spectroscopy. The congruent vaporisation of α-MnSe was experimentally verified in both investigations, see Appendix A. The results were evaluated taking both Se(g) and Se_2(g) into account using the reaction α-MnSe ⇌ Mn(g) + x Se(g) + ½(1 – x)Se_2(g). The consideration of the formation of both Se(g) and Se_2(g) is essential because the two species form in approximately equal amounts at the experimental conditions. The treatment of this problem is discussed in Appendix A for each study. The second and third law values of the enthalpy of formation of α-MnSe derived from [68WIE/GOY] are $\Delta_f H^{\circ}_{m}$(MnSe, α, 298.15 K) = – (169.5 ± 10.0) kJ·mol^{-1} and – (174.7 ± 8.0) kJ·mol^{-1}, respectively. The value of the enthalpy of formation of α-MnSe was calculated to be $\Delta_f H^{\circ}_{m}$(MnSe, α, 298.15 K) = – (180.7 ± 10.0) kJ·mol^{-1} from the measurements in [78SMO/PAT] as discussed in Appendix A. The enthalpy of formation of α-MnSe was determined to be $\Delta_f H^{\circ}_{m}$(MnSe, α, 298.15 K) = – (172.0 ± 12.0) kJ·mol^{-1} in [78LUK/POL] from emf studies. The weighted average $\Delta_f H^{\circ}_{m}$(MnSe, α, 298.15 K) = – (174.2 ± 10.0) kJ·mol^{-1} is considered to summarise these high temperature measurements. The confidence interval was set by considering the scatter of the determinations. The corresponding Gibbs energy of formation is calculated to be $\Delta_f G^{\circ}_{m}$(MnSe, α, 298.15 K) = – (179.2 ± 12.0) kJ·mol^{-1}.

In the aqueous phase, Mehra and Gubeli [70MEH/GUB2] measured the solubility of radioactively labelled MnSe(s) as described in [71MEH/GUB] in Appendix A. From the original data in [68MEH], in the pH region 4.2 to 5.5, the review calculates for the reaction

$$\text{MnSe(s)} + \text{H}^{+} \rightleftharpoons \text{Mn}^{2+} + \text{HSe}^{-}, \qquad (V.143)$$

$$\log_{10} K\ ((V.143),\ 1\ \text{M NaClO}_4,\ 298.15\ \text{K}) = -(0.52 \pm 0.70).$$

Extrapolation to standard conditions with SIT and Δε = 0.25 kg·mol^{-1} and combination with the selected protonation constant of Se^{2-} yields the value $\log_{10} K^{\circ}_{s,0}$(MnSe, s, 298.15 K) = – (16.0 ± 0.8). Combined with the [82WAG/EVA] data for Mn^{2+} (– 228.1 kJ·mol^{-1}) and the selected data for Se^{2-} this corresponds to $\Delta_f G^{\circ}_{m}$(MnSe, s, 298.15 K) = – (190.8 ± 6.8) kJ·mol^{-1}.

The $\Delta_f G^{\circ}_{m}$(MnSe, s, 298.15 K) from the high temperature measurements and from solution measurement do not agree within the uncertainty limits. The data were nevertheless combined (α-MnSe is assumed to be present in the solubility measurement) with the weights 2:1 to yield:

$$\Delta_f G_m^\circ(\text{MnSe}, \alpha, 298.15\text{ K}) = -(182.9 \pm 12.0)\text{ kJ·mol}^{-1}$$

where the error limits were increased to allow for possible systematic errors.

The standard enthalpy of formation is calculated with the selected entropy to be:

$$\Delta_f H_m^\circ(\text{MnSe}, \alpha, 298.15\text{ K}) = -(177.9 \pm 12.0)\text{ kJ·mol}^{-1}.$$

These values are included in Appendix E since they were calculated with non-TDB auxiliary data.

The estimate of the solubility product from the Gibbs energy of formation of α-MnSe is:

$$\log_{10} K_{s,0}^\circ(\text{MnSe}, \alpha, 298.15\text{ K}) = -(14.6 \pm 2.1).$$

V.11.10.1.2 MnSe(g)

The heat capacity and entropy of MnSe(g) were derived by the review as discussed in Appendix A from the work by Smoes, Pattje, and Drowart [78SMO/PAT]. The values are S_m°(MnSe, g, 298.15 K) = (262 ± 5) J·K^{-1}·mol^{-1}, $C_{p,m}^\circ$(MnSe, g, 298.15 K) = (35.8 ± 1.0) J·K^{-1}·mol^{-1}, $C_{p,m}^\circ$(MnSe, g, (298.15 – 2000) K) = $(37.215 + 0.7302 \times 10^{-3}\,T - 6.9 \times 10^{-8}\,T^2 - 1.48257 \times 10^5\,T^{-2})$ J·K^{-1}·mol^{-1}. These properties are not selected since they were calculated with estimated molecular parameters but used in the evaluation of the enthalpy of formation of MnSe(g) as described in Appendix A yielding:

$$\Delta_f H_m^\circ(\text{MnSe, g}, 298.15\text{ K}) = (278.4 \pm 10.0)\text{ kJ·mol}^{-1}.$$

This value is adopted since the heat capacity and entropy calculated from estimated molecular parameters can be expected to be quite accurate for a diatomic molecule. The additional uncertainty originating from the recalculation to the standard temperature is hence moderate in relation to the uncertainty in the reaction enthalpy. The result is included in Appendix E since it was calculated with non-TDB auxiliary data.

V.11.10.1.3 $MnSe_2$(cr)

Lukashenko and Polotskaya [78LUK/POL] measured the emf of the cell

$$(-)\ \alpha\text{-Mn} \mid 2\text{ weight-\% MnCl}_2\text{ (in LiCl + KCl melt)} \mid \text{MnSe}_2\text{(cr), Se(l)}\ (+)$$

in the temperature range 630 to 760 K. Their formation data were recalculated to 298.15 K by the review using the selected data for selenium, data for α-Mn in [82WAG/EVA], and a mean heat capacity of $MnSe_2$(cr) estimated to be 85.0 J·K^{-1}·mol^{-1}. The following results were obtained:

$$S_m^\circ(\text{MnSe}_2, \text{cr}, 298.15\text{ K}) = (140.3 \pm 12.0)\text{ J·K}^{-1}\text{·mol}^{-1},$$

$$\Delta_f H_m^\circ(\text{MnSe}_2, \text{cr}, 298.15 \text{ K}) = -(180.5 \pm 10.0) \text{ kJ·mol}^{-1}.$$

The Gibbs energy of formation is calculated from the above data of $MnSe_2$(cr) to be:

$$\Delta_f G_m^\circ(\text{MnSe}_2, \text{cr}, 298.15 \text{ K}) = -(187.7 \pm 10.6) \text{ kJ·mol}^{-1}.$$

These data are included in Appendix E since they were calculated with a non-TDB datum.

V.11.10.1.4 $MnSe_2$(g)

The heat capacity and entropy of $MnSe_2$(g) were derived by the review as discussed in Appendix A from the work by Smoes, Pattje, and Drowart [78SMO/PAT]. The values are S_m°($MnSe_2$, g, 298.15 K) = (326 ± 10) $J·K^{-1}·mol^{-1}$, $C_{p,m}^\circ$($MnSe_2$, g, 298.15 K) = (54.3 ± 2.0) $J·K^{-1}·mol^{-1}$, $C_{p,m}^\circ$($MnSe_2$, g, (298.15 – 2000) K) = $(57.795 + 0.4579 \times 10^{-3}\ T - 1.41 \times 10^{-7}\ T^2 - 3.25757 \times 10^5\ T^{-2})$ $J·K^{-1}·mol^{-1}$. These properties are not selected since they were calculated with estimated molecular parameters but used in the evaluation of the enthalpy of formation of $MnSe_2$(g) as described in Appendix A yielding $\Delta_f H_m^\circ$($MnSe_2$, g, 298.15 K) = (231.5 ± 20.0) $kJ·mol^{-1}$.

This result is not selected because the calculations depend on free energy functions derived from non-experimental information only.

V.11.10.1.5 Mn_2Se(g)

The heat capacity and entropy of Mn_2Se(g) were derived by the review as discussed in Appendix A from the work by Smoes, Pattje, and Drowart [78SMO/PAT]. The values are S_m°(Mn_2Se, g, 298.15 K) = (307 ± 8) $J·K^{-1}·mol^{-1}$, $C_{p,m}^\circ$(Mn_2Se, g, 298.15 K) = (54.2 ± 2.0) $J·K^{-1}·mol^{-1}$, $C_{p,m}^\circ$(Mn_2Se, g, (298.15 – 2000) K) = $(57.792 + 0.4718 \times 10^{-3}\ T - 1.42 \times 10^{-7}\ T^2 - 3.29577 \times 10^5\ T^{-2})$ $J·K^{-1}·mol^{-1}$. These properties are not selected since they were calculated with estimated molecular parameters but used in the evaluation of the enthalpy of formation of Mn_2Se(g) as described in Appendix A yielding $\Delta_f H_m^\circ$(Mn_2Se, g, 298.15 K) = (264.3 ± 20.0) $kJ·mol^{-1}$.

This result is not selected because the calculations depend on free energy functions derived from non-experimental information only.

V.11.10.1.6 Mn_2Se_2(g)

The heat capacity and entropy of Mn_2Se_2 (g) were derived by the review as discussed in Appendix A from the work by Smoes, Pattje, and Drowart [78SMO/PAT]. The values are S_m°(Mn_2Se_2, g, 298.15 K) = (365 ± 12) $J·K^{-1}·mol^{-1}$, $C_{p,m}^\circ$(Mn_2Se_2, g, 298.15 K) = (78.6 ± 3.0) $J·K^{-1}·mol^{-1}$, $C_{p,m}^\circ$(Mn_2Se_2, g, (298.15 – 2000) K) = $(82.846 + 0.3437 \times 10^{-3}\ T - 1.033 \times 10^{-7}\ T^2 - 3.90602 \times 10^5\ T^{-2})$ $J·K^{-1}·mol^{-1}$. These properties are not selected since they were calculated with estimated molecular parameters but used in the evaluation of the enthalpy of formation of Mn_2Se_2(g) as described in Appendix A yielding $\Delta_f H_m^\circ$ (Mn_2Se_2, g, 298.15 K) = (135.5 ± 25.0) kJ·mol^{-1}.

This result is not selected because the calculations depend on free energy functions derived from non-experimental information only.

V.11.10.1.7 Aqueous manganese(II) selenide complexes

In the solubility measurements employing radioactively labelled manganese, and at 6 < pH < 13.6, Mn(II)(aq) concentrations ranging between 1.5×10^{-6} and 1.5×10^{-7} M were indicated. Mehra and Glubeli proposed that this corresponded to the formation of a soluble species Mn(HSe)(OH)(aq) and calculated the stability constant of the complex. The review prefers to write:

$$\text{MnSe(s)} \rightleftharpoons \text{MnSe(aq)} \qquad (V.144)$$

and the data in [68MEH] yield an intrinsic solubility corresponding to $\log_{10} K$ ((V.144), 1 M $NaClO_4$, 298.15 K) = – (6.42 ± 0.52). However, as discussed in Appendix A, these minute "concentrations" might originate from MnSe(s) particles of colloidal size, since only a sintered glass disk was used for phase separation. Therefore, this equilibrium constant is not accepted by the review.

V.11.10.2 Manganese selenites

No phase diagram of the system $MnSeO_3$-H_2O-SeO_2 has been found.

Chukhlantsev and Tomashevsky [57CHU/TOM] prepared manganese selenite by mixing 0.2 M solutions of manganese sulphate and sodium selenite in stoichiometric amounts. The precipitate was amorphous and aged for a long time, which led to the formation of a crystalline product but no X-ray diffraction measurements were performed. Chemical analysis confirmed the 1:1 ratio between Mn(II) and Se(IV) and the specimen is assigned the composition $MnSeO_3 \cdot 2H_2O$ by the review from the preparative work in [66LES/SEL3]. The solubility of the specimen in dilute solution of hydrochloric or sulphuric acid was measured at 293 K. The experiments were performed and the data recalculated as described in Appendix A, [56CHU]. The result for:

$$\text{MnSeO}_3\cdot 2\text{H}_2\text{O(cr)} \rightleftharpoons \text{Mn}^{2+} + \text{SeO}_3^{2-} + 2\text{H}_2\text{O(l)} \qquad (V.145)$$

was $\log_{10} K_{s,0}^\circ$ ((V.145), 293.15 K) = – (7.64 ± 0.08). The reported conditional constant is – (6.93 ± 0.13). Sharmasakar, Reddy, and Vance [96SHA/RED] obtained $\log_{10} K_{s,0}^\circ$ ((V.145), 298.15 K) = – 7.11 from measurements of the solubility of crystalline $MnSeO_3$ in aqueous media. For reasons presented in Appendix A, this result has been given zero weight. The review accepts and selects:

$$\log_{10} K_{s,0}^\circ \text{((V.145), 298.15 K)} = -(7.6 \pm 1.0).$$

Leshchinskaya and Selivanova [66LES/SEL3] measured the enthalpies of dissolution of crystalline $MnSeO_3{\cdot}2H_2O$, $MnSO_4$ and H_2SeO_3 in H_2SO_4(aq, 1:30). Their data have been recalculated in Appendix A to the value:

$$\Delta_f H_m^\circ (MnSeO_3{\cdot}2H_2O, cr, 298.15 K) = -(1280.7 \pm 2.6) kJ{\cdot}mol^{-1}.$$

The data yield:

$$\Delta_f G_m^\circ (MnSeO_3{\cdot}2H_2O, cr, 298.15 K) = -(1108.2 \pm 6.7) kJ{\cdot}mol^{-1},$$

$$S_m^\circ (MnSeO_3{\cdot}2H_2O, cr, 298.15 K) = (269.3 \pm 24.1) J{\cdot}K^{-1}{\cdot}mol^{-1}.$$

Since the calculation of the formation values and entropy from the experiments involved non-TDB auxiliary data, they are included in Appendix E.

V.11.10.3 Manganese selenates

The stable manganese selenate in contact with water at 298.15 K is the pentahydrate, $MnSeO_4{\cdot}5H_2O$. However, the system is quickly oxidised by the oxygen in air unless precautions are taken according to References [89VOJ/EBE2] and [99KOL/STO]. Koleva and Stoilova [99KOL/STO] determined the solubility of manganese selenate in the temperature range 293.15 to 333.15 K. $MnSeO_4{\cdot}2H_2O$ is reported to be the stable phase above about 313 K.

The solubility of the salt in water at the standard temperature was determined by Vojtisek, Ebert, and Sieglova [89VOJ/EBE2] to be 2.949, by Oy(j)kova and Mihov [92OYK/MIH] to be 2.851, and by Koleva and Stoikova [99KOL/STO] to be 2.786 $mol(MnSeO_4){\cdot}kg^{-1}$. The review selects the mean of the three determinations:

$$m(MnSeO_4{\cdot}5H_2O(cr), aq sat, 298.15 K) = (2.86 \pm 0.08) mol(MnSeO_4){\cdot}kg^{-1}.$$

A phase diagram and related information are available for the system: $MnSeO_4$-H_2SeO_4-H_2O at 293.15 K in [99KOL/STO]; $MnSeO_4$-UO_2SeO_4-H_2O at 298.15 K in [83SER/SER]; $MnSeO_4$-$(NH_4)_2SeO_4$-H_2O at 298.15 K in [89VOJ/EBE2]; $MnSeO_4$-Na_2SeO_4-H_2O at 298.15 K in [92OYK/MIH]; $MnSeO_4$-K_2SeO_4-H_2O at 298.15 K in [89VOJ/EBE2].

The value of solubility product of $MnSeO_4{\cdot}5H_2O$(cr),

$$MnSeO_4{\cdot}5H_2O(cr) \rightleftharpoons Mn^{2+} + SeO_4^{2-} + 5H_2O(l) \qquad (V.146)$$

at 298.15 K has been reported by Kumov and Batyreva [90KUM/BAT] to be $\log_{10} K_{s,0}^\circ$ ((V.146), 298.15 K) = $-(2.05 \pm 0.03)$, see Appendix A. As the reviewers failed to assess the accuracy of this datum, it is not included amongst the selected data.

Ghosh and Nair [70GHO/NAI] determined the equilibrium constant of the reaction:

$$Mn^{2+} + SeO_4^{2-} \rightleftharpoons MnSeO_4(aq) \qquad (V.147)$$

in the temperature range 273.15 to 318.15 K by potentiometric measurements, see Appendix A. The enthalpy of the reaction was found from the variation of the stability constant with temperature. The following results are selected:

$$\log_{10} \beta_1^\circ \text{((V.147), 298.15 K)} = (2.43 \pm 0.05),$$

$$\Delta_r G_m^\circ \text{((V.147), 298.15 K)} = -(13.87 \pm 0.29)\ \text{kJ·mol}^{-1},$$

while $\Delta_r H_m^\circ$((V.147), 298.15 K) = (14.70 ± 1.70) kJ·mol^{-1} is not accepted for reasons discussed in Appendix A.

The uncertainty in $\log_{10} \beta_1^\circ$ has been increased considerably from the value in the reference to include the uncertainty from the estimate of the activity coefficients.

V.11.11 Technetium compounds

V.11.11.1 Technetium selenides

No thermodynamic information is available for technetium selenides (*cf.* [99RAR/RAN]).

V.11.12 Rhenium compounds

V.11.12.1 Rhenium selenides

The compounds $ReSe_2$(cr) and Re_2Se_7(cr) have been reported. No experimental thermodynamic information is available but enthalpies of formation and standard entropies have been estimated by Mills [74MIL]. No values are selected by the review.

V.11.13 Chromium compounds

V.11.13.1 Chromium selenides

The compounds CrSe(cr), Cr_3Se_4(cr) and Cr_2Se_3(cr) have been reported in the system chromium-selenium. Cr_2Se_3(cr) exists in two modifications: the low-temperature γ phase, which is stable up to 1084 K and the high-temperature α phase. Experimental thermodynamic information is only available for the two polymorphs of Cr_2Se_3(cr).

V.11.13.1.1 α-Cr_2Se_3

The heat capacity of α-Cr_2Se_3 was measured in the temperature range 6 to 300 K by Koshchenko, Yachmenev, Zhegalina, Pashinkin, and Danilenko in [90KOS/YAC]. The values determined for the heat capacity and entropy at 298.15 K are selected:

$$C_{p,m}^\circ (Cr_2Se_3, \alpha, 298.15\ \text{K}) = (120.8 \pm 0.9)\ \text{J·K}^{-1}\text{·mol}^{-1}$$

and

$$S^\circ_m(\text{Cr}_2\text{Se}_3, \alpha, 298.15 \text{ K}) = (187.7 \pm 1.5) \text{ J}\cdot\text{K}^{-1}\cdot\text{mol}^{-1}.$$

V.11.13.1.2 γ-Cr_2Se_3

The heat capacity of γ-Cr_2Se_3 was measured in the temperature range 60 to 300 K by Koshchenko, Ivanyas, Zhegalina, and Danilenko in [92KOS/IVA]. The value determined for the heat capacity at 298.15 K is selected:

$$C^\circ_{p,m}(\text{Cr}_2\text{Se}_3, \gamma, 298.15 \text{ K}) = (122.9 \pm 0.9) \text{ J}\cdot\text{K}^{-1}\cdot\text{mol}^{-1}.$$

The entropy of γ-Cr_2Se_3 at 298.15 K was evaluated in [92KOS/IVA] using their measured heat capacity for the temperature range 60 to 298.15 K combined with the heat capacity of α-Cr_2Se_3 in [90KOS/YAC] for the temperature interval 0 to 60 K. The effect of the antiferromagnetic-paramagnetic transition in α-Cr_2Se_3 at 43.3 K was subtracted from the latter. The entropy value of γ-Cr_2Se_3 obtained from this procedure is selected, but it is less accurate than for α-Cr_2Se_3,

$$S^\circ_m(\text{Cr}_2\text{Se}_3, \gamma, 298.15 \text{ K}) = (178.0 \pm 6.0) \text{ J}\cdot\text{K}^{-1}\cdot\text{mol}^{-1}.$$

The Gibbs energy of formation of γ-Cr_2Se_3 was measured in the temperature range 413 to 493 K using the electrochemical cell

$$(-)\ \text{Cr(cr)}\ |\ 0.3 \text{ weight-\% CrCl}_2\text{(in melt)}\ |\ \gamma\text{-Cr}_2\text{Se}_3, \text{Se(cr)}\ (+)$$

by Goncharuk and Lukashenko [86GON/LUK]. The melt was made up from mixtures of ($AlCl_3$, NaCl, KCl) or ($ZnCl_2$, NaCl, KCl). The Gibbs energy, enthalpy, and entropy of formation were evaluated at the average temperature 453 K to be $\Delta_f G^\circ_m(\text{Cr}_2\text{Se}_3, \gamma, 453 \text{ K}) = -(248.6 \pm 3.3) \text{ kJ}\cdot\text{mol}^{-1}$, $\Delta_f H^\circ_m(\text{Cr}_2\text{Se}_3, \gamma, 453 \text{ K}) = -(209.5 \pm 6.5) \text{ kJ}\cdot\text{mol}^{-1}$, and $\Delta_f S^\circ_m(\text{Cr}_2\text{Se}_3, \gamma, 453 \text{ K}) = (86.2 \pm 16.7) \text{ J}\cdot\text{K}^{-1}\cdot\text{mol}^{-1}$, respectively. The entropy of formation is unreasonably large compared to that obtained from the selected value of the entropy and those of the elements ($S^\circ_m(\text{Cr, cr, 298.15 K}) = 23.77 \text{ J}\cdot\text{K}^{-1}\cdot\text{mol}^{-1}$, [82WAG/EVA]), $\Delta_f S^\circ_m(\text{Cr}_2\text{Se}_3, \gamma, 298.15 \text{ K}) = (4.2 \pm 6.0) \text{ J}\cdot\text{K}^{-1}\cdot\text{mol}^{-1}$. The heat capacity change on forming γ-Cr_2Se_3 is small, $(1.4 \pm 0.9) \text{ J}\cdot\text{K}^{-1}\cdot\text{mol}^{-1}$, and the Gibbs energy of formation at 453 K can therefore be converted to 298.15 K using this entropy change without significant loss of accuracy. The following results are thus adopted and included in Appendix E, since they were calculated with non-TDB auxiliary data:

$$\Delta_f H^\circ_m(\text{Cr}_2\text{Se}_3, \gamma, 298.15 \text{ K}) = -(246.7 \pm 20.0) \text{ kJ}\cdot\text{mol}^{-1}$$

and

$$\Delta_f G^\circ_m(\text{Cr}_2\text{Se}_3, \gamma, 298.15\text{K}) = -(247.9 \pm 20.0) \text{ kJ}\cdot\text{mol}^{-1},$$

but because of the dubious experimental entropy change of the formation reaction, the error limits were increased to allow for possible unknown systematic errors.

V.11.14 Molybdenum compounds

V.11.14.1 Molybdenum selenides

Spiesser, Rouxel, Kerriou and Goureaux [69SPI/ROU] report that $MoSe_2$(cr) is the only phase formed below 1463 K in the system molybdenum-selenium, and that the homogeneity range of $MoSe_2$(cr) extends to $MoSe_{1.91}$ at 1523 K. Above 1463 K the phase Mo_3Se_4(cr) forms with a homogeneity range extending from $MoSe_{1.27}$ to $MoSe_{1.45}$. Despite these experimental findings, Brewer and Lamoreaux [80BRE/LAM], with reference to [69SPI/ROU], constructed a phase diagram where the Mo_3Se_4(cr) phase is stable down to 273 K. In addition, thermodynamic properties were derived for all phases in the system. The phase diagram, which thus does not agree with experimental results, has been copied into the compilations by Moffatt [81MOF] and Massalski [86MAS]. Metastable phases, such as Mo_6Se_6 and $Mo_{15}Se_{19}$ [87OHA/TAS], have been synthesised but not by direct combination of the elements.

V.11.14.1.1 $MoSe_2$(cr)

The heat capacity of $MoSe_2$(cr) has been measured in the temperature range 6 to 342 K by Kiwia and Westrum [75KIW/WES3]. No anomalies in the heat capacity were found. The heat capacity at 298.15 K is selected:

$$C^{\circ}_{p,\mathrm{m}}(\mathrm{MoSe_2, cr, 298.15\ K}) = (70.6 \pm 0.5)\ \mathrm{J{\cdot}K^{-1}{\cdot}mol^{-1}}.$$

Bolgar, Trofimova, Yanaki, Kopylova, and Zaletilo [87BOL/TRO] derived analytical expressions for the temperature dependence of the heat capacity and enthalpy from enthalpy measurements in the temperature range 1219 to 1806 K. However, the expressions are in error and the experimental enthalpy was re-evaluated by the review as discussed in Appendix A. The heat capacity expression is: $C^{\circ}_{p,\mathrm{m}}$ ($MoSe_2$, cr, (298.15 – 1806) K) = $(69.112 + 0.014221\ T - 2.4621 \times 10^5\ T^{-2})\ \mathrm{J{\cdot}K^{-1}{\cdot}mol^{-1}}$.

Second law entropies were evaluated by the review as discussed in Appendix A from the investigations in [76GLA/MIK] and [86VIK/GOR]. The result is presented in Table V-68. The entropies disagree substantially and the entropy at 298.15 K obtained from the integration of the low temperature heat capacity in [75KIW/WES3] is selected:

$$S^{\circ}_{\mathrm{m}}(\mathrm{MoSe_2, cr, 298.15\ K}) = (89.1 \pm 0.5)\ \mathrm{J{\cdot}K^{-1}{\cdot}mol^{-1}}.$$

The vapour pressure of the thermal decomposition of $MoSe_2$(cr) has been studied by Glazunov, Mikhailov, Piskarev, and Chupakin [76GLA/MIK], Viksman, Gordienko, Yanaki, and Klochkov [86VIK/GOR], and Semenov-Kobzar, Kulikov, Bogatyr, and Zatelio [92SEM/KUL]. The results in [76GLA/MIK] and [86VIK/GOR] were re-evaluated as discussed in Appendix A and the results are summarised in Table V-68. The measurements in [92SEM/KUL] were made at high temperatures where the decomposition products are unknown and therefore an evaluation of thermodynamic properties could not be made.

Table V-68: Re-evaluated enthalpy of formation and entropy of $MoSe_{1.91}(cr)$ at 298.15 K from studies of the reaction $MoSe_{1.91}(cr) \rightleftharpoons Mo(cr) + 0.955Se_2(g)$ as obtained from (a) the second and (b) the third law. The error limits are those of the original papers.

Reference	$\Delta_f H_m^\circ$ ($MoSe_{1.91}$, cr, 298.15 K) ($kJ\cdot mol^{-1}$)		S_m° ($MoSe_{1.91}$, cr, 298.15 K) ($J\cdot K^{-1}\cdot mol^{-1}$)	Temperature (K)
	(a)	(b)	(a)	
[76GLA/MIK]	– (264.1 ± 13.0)	– (209.4 ± 13.0)	(42.0 ± 10.0)	1233 – 1353
[86VIK/GOR]	– 70.2	– (259.2 ± 35.0)	210.2	1459 – 1570

The second law results from the study in [86VIK/GOR] are obviously in error. Fluorine combustion calorimetry was used in a careful study by O'Hare, Tasker, and Tarascon [87OHA/TAS] to determine the enthalpy of formation of $MoSe_2(cr)$ from the reaction

$$MoSe_2(cr) + 9F_2(g) \rightleftharpoons MoF_6(g) + 2SeF_6(g). \qquad (V.148)$$

The combustion enthalpy was determined to be $\Delta_r H_m^\circ$((V.148), 298.15 K) = – (3559.5 ± 3.0) $kJ\cdot mol^{-1}$ and combined with their value $\Delta_f H_m^\circ$ (MoF_6, g, 298.15 K) = – (1557.9 ± 0.9) $kJ\cdot mol^{-1}$ and the selected value of the enthalpy of formation of $SeF_6(g)$, the enthalpy of formation of $MoSe_2(cr)$ is calculated to be:

$$\Delta_f H_m^\circ (MoSe_2, cr, 298.15 K) = -(234.4 \pm 3.5)\ kJ\cdot mol^{-1}.$$

This result is preferred to those in Table V-68 and it is included in Appendix E since it was calculated with a non-TDB auxiliary datum.

V.11.14.1.2 $Mo_3Se_4(cr)$

The heat capacity of $Mo_3Se_4(cr)$ has been measured in the temperature range 7 to 300 K in [90MIN/AMI]. Unfortunately, the results were only discussed and neither primary data nor values at 298.15 K were given.

V.11.14.1.3 $Mo_6Se_6(cr)$

Fluorine combustion calorimetry was used in a careful study by O'Hare, Tasker, and Tarascon [87OHA/TAS] to determine the enthalpy of formation of $Mo_6Se_6(cr)$ from the reaction

$$Mo_6Se_6(cr) + 36F_2(g) \rightleftharpoons 6MoF_6(g) + 6SeF_6(g). \qquad (V.149)$$

The combustion enthalpy was determined to be $\Delta_r H_m^\circ$((V.149), 298.15 K) = – (15577.0 ± 7.9) kJ·mol^{-1} and combined with their value $\Delta_f H_m^\circ$ (MoF_6, g, 298.15 K) = – (1557.9 ± 0.9) kJ·mol^{-1} and the selected value of the enthalpy of formation of SeF_6(g), the enthalpy of formation of Mo_6Se_6(cr) is calculated to be:

$$\Delta_f H_m^\circ (\mathrm{Mo_6Se_6, cr, 298.15\ K}) = -(478.4 \pm 8.6)\ \mathrm{kJ \cdot mol^{-1}}.$$

The value is included in Appendix E since it was calculated with a non-TDB auxiliary datum.

V.11.14.2 Molybdenum selenites

Sasaki *et al.* [88OZE/YAG] studied the H^+ - MoO_4^{2-} - SeO_3^{2-} system at 298.15 K by potentiometric and ^{77}Se NMR measurements in 1 M NaCl. The equilibrium analysis based on the general reaction scheme:

$$p\mathrm{H^+} + q\,\mathrm{MoO_4^{2-}} + r\,\mathrm{SeO_3^{2-}} \rightleftharpoons (\mathrm{H^+})_p(\mathrm{MoO_4^{2-}})_q(\mathrm{SeO_3^{2-}})_r$$

resulted in the model and formation constants in Table V-69.

The formation of the heteropolyanions $Se_2Mo_5O_{21}^{4-}$ and $HSe_2Mo_5O_{21}^{3-}$ is well established and accepted, whereas the formation of the (14,8,2) and (15,8,2) species is tentative, see Appendix A. No attempt to extrapolate the data to standard conditions has been made.

Table V-69: Formation constants of heteropoly anions in the system H^+ - MoO_4^{2-} - SeO_3^{2-}.

(p,q,r)	$\log_{10}\beta_{pqr}$
(10,5,2)	(71.38 ± 0.02)
(11,5,2)	(72.73 ± 0.04)
(14,8,2)	(97.97 ± 0.02)
(15,8,2)	(101.04 ± 0.03)

V.11.15 Tungsten compounds

V.11.15.1 Tungsten selenides

Thermodynamic information is only available for the WSe_2(cr) phase in the system tungsten-selenium.

V.11.15.1.1 WSe_2(cr)

Bolgar, Trofimova, and Yanaki [90BOL/TRO] measured the heat capacity of WSe_2(cr) in the temperature range 60 to 300 K and the enthalpy in the range 516 to 1765 K. No anomalies in the heat capacity were found. The heat capacity at 298.15 K:

$$C^{\circ}_{p,\mathrm{m}}(\mathrm{WSe_2, cr, 298.15\ K}) = (72.6 \pm 1.0)\ \mathrm{J \cdot K^{-1} \cdot mol^{-1}}$$

and the heat capacity expression derived from the enthalpy measurements:

$$C^{\circ}_{p,\mathrm{m}}(\mathrm{WSe_2, cr, (298.15 - 1800)\ K}) = (71.086 + 14.627 \times 10^{-3}\ T - 2.53246 \times 10^{5}\ T^{-2})\ \mathrm{J \cdot K^{-1} \cdot mol^{-1}}$$

are selected after correction of some errors mentioned in Appendix A.

Second law entropies were evaluated by the review as discussed in Appendix A from a number of investigations listed in Table V-70 of the reaction $WSe_2(cr) \rightleftharpoons W(cr) + Se_2(g)$. The values are very scattered and of little significance. The entropy at 298.15 K was derived to be $S^{\circ}_{\mathrm{m}}(\mathrm{WSe_2, cr, 298.15\ K}) = (89.9 \pm 2.0)\ \mathrm{J \cdot K^{-1} \cdot mol^{-1}}$ in [90BOL/TRO] from an integration of the heat capacity using an extrapolation for the temperature range 0 to 60 K. The value is not selected because of the uncertainties in this range.

Table V-70: Re-evaluated enthalpy of formation and entropy of WSe_2(cr) at 298.15 K obtained from vapour pressure measurements using (a) the second and (b) the third law.

Reference	$\Delta_f H^{\circ}_{\mathrm{m}}(\mathrm{WSe_2, cr, 298.15\ K})$ ($\mathrm{kJ \cdot mol^{-1}}$) (a)	(b)	$S^{\circ}_{\mathrm{m}}(\mathrm{WSe_2, cr, 298.15\ K})$ ($\mathrm{J \cdot K^{-1} \cdot mol^{-1}}$) (a)
[77MIK/GLA]	– (133.9 ± 10.0), (14.0 ± 25.0)	– (209.7 ± 10.0)	(245.9 ± 20.0)
[77PIS/MIK]	– 252.4	– 198.3	40.8
[82SCH/FRA]	– (175.0 ± 12.0)	– (125.6 ± 12.0)	(53.2 ± 5.0)
[83ZEL/KOL]	– (98.0 ± 50.0)	– (190.2 ± 10.0)	(146.3 ± 30.0)
[86VIK/GOR]		– (265.2 ± 35.0)	
[92SEM/KUL]	– (234.7 ± 50.0)	– (256.1 ± 50.0)	(103.2 ± 33.0)

The enthalpy of formation of WSe_2(cr) has been determined in a number of vaporisation investigations using various methods. The studies have been re-evaluated by the review as discussed in Appendix A and the results are summarised in Table V-70. There is a considerable scatter among the values probably due to difficulties in attaining chemical equilibrium and to a considerable homogeneity range of the WSe_2 phase as indicated by the large shifts in the cell parameters of WSe_2 observed in most of the studies. However, this has not been thoroughly investigated. Direct synthesis calorimetry was used by Zelikman, Kolchin, and Golutvin [83ZEL/KOL] and the enthalpy of formation was determined to be $\Delta_f H_m^\circ$ (WSe_2, cr, 298.15 K) = – (154.2 ± 2.0) $kJ\cdot mol^{-1}$. The reaction was incomplete and 10 to 15% of the selenium remained unreacted. Fluorine combustion calorimetry was used in a careful study by O'Hare, Lewis, and Parkinson [88OHA/LEW2] to determine the enthalpy of formation of WSe_2(cr) from the reaction

$$WSe_2(cr) + 9F_2(g) \rightleftharpoons WF_6(g) + 2SeF_6(g). \quad (V.150)$$

The combustion enthalpy was determined to be $\Delta_r H_m^\circ$((V.150), 298.15 K) = – (3772.2 ± 5.1) $kJ\cdot mol^{-1}$ and combined with their value $\Delta_f H_m^\circ$ (WF_6, g, 298.15 K) = – (1721.7 ± 1.7) $kJ\cdot mol^{-1}$ and the selected value of the enthalpy of formation of SeF_6(g), the enthalpy of formation of WSe_2(cr) is calculated to be:

$$\Delta_f H_m^\circ (WSe_2, cr, 298.15 K) = -(185.5 \pm 5.5)\ kJ\cdot mol^{-1}.$$

This is the most precise work and the value is included in Appendix E since it was calculated with a non-TDB auxiliary datum.

V.11.16 Vanadium compounds

V.11.16.1 Vanadium selenides

The vanadium selenides in general have considerable homogeneity ranges and partial molar enthalpies for various compositions in the vanadium-selenium system have been determined by vapour pressure measurements by Hayashi and Nakahira [78HAY/NAK] and Miyauchi, Hayashi, and Nakahira [83MIY/HAY]. However, additional information is required to obtain integral quantities. Mills [74MIL] has estimated the entropies at 298.15 K for VSe(cr), V_2Se_3(cr), and VSe_2(cr). The values are not based on experimental information and are not selected here.

V.11.16.1.1 VSe(g)

The heat capacity of VSe(g) was calculated from estimated molecular parameters in [74MIL] and the heat capacity expression and heat capacity at 298.15 K are: $C_{p,m}^\circ$ (VSe, g, (298.15 – 2000) K) = (37.2502 + 0.0879 × 10^{-3} T – 2.0543 × 10^5 T^{-2}) $J\cdot K^{-1}\cdot mol^{-1}$ and $C_{p,m}^\circ$ (VSe, g, 298.15 K) = (35.0 ± 2.0) $J\cdot K^{-1}\cdot mol^{-1}$, respectively.

The entropy of VSe(g) was calculated in [74MIL] from estimated molecular parameters to be: S_m° (VSe, g, 298.15 K) = (260.8 ± 10.0) $J\cdot K^{-1}\cdot mol^{-1}$.

The atomisation enthalpy of VSe(g) was determined by measuring the partial pressures of the gas species using mass spectroscopy and Knudsen effusion cells by Bergman, Coppens, Drowart and Smoes [70BER/COP]. The mean of the second and third law values of the enthalpy of atomisation at 0 K was the only value reported, $\Delta_{at}H^{\circ}_{m}$ (VSe, g, 0 K) = (341.6 ± 5.5) kJ·mol^{-1}. The evaluations involved the Gibbs energy function of VSe(g) which was calculated from undocumented estimated molecular properties. In absence of a proper value of the heat capacity, the value at 298.15 K is estimated by the review to be approximately 3 kJ·mol^{-1} larger than the value at 0 K yielding $\Delta_{at}H^{\circ}_{m}$ (VSe, g, 298.15 K) = (345 ± 15) kJ·mol^{-1}. The error limits were increased to reflect the estimated overall uncertainty. The enthalpy of formation of VSe(g) is calculated to be $\Delta_{f}H^{\circ}_{m}$ (VSe, g, 298.15 K) = (405 ± 16) kJ·mol^{-1} on combining the atomisation enthalpy with the enthalpy of formation of V(g) in [82WAG/EVA] and the selected enthalpy of formation of Se(g). The reasonable agreement between the values of $\Delta_{at}H^{\circ}_{m}$ (VSe, g, 0 K) obtained by the second and third law extrapolations suggests that this value can be accepted despite the use of undocumented thermal data. Thus the review adopts:

$$\Delta_{f}H^{\circ}_{m}\text{(VSe, g, 298.15 K)} = (405 \pm 16)\ \text{kJ·mol}^{-1}.$$

This value is included in Appendix E since it was calculated with non-TDB auxiliary data.

V.11.17 Niobium compounds

V.11.17.1 Niobium selenides

The compounds Nb_5Se_4(cr), Nb_3Se_4(cr), Nb_2Se_3(cr), $NbSe_2$(cr), and $NbSe_4$(cr) have been reported in the niobium-selenium system [79IGA/NIS]. The thermal decomposition and partial pressure of selenium have been studied for $NbSe_2$(cr) by Mikhailov, Piskarev, and Chupakhin [77MIK/PIS] and for compositions $NbSe_x$ in the range $1.48 \leq x \leq 2.00$ by Igaki and Nishine [79IGA/NIS]. The investigations result in partial quantities for selenium and no integral values were obtained.

V.11.17.1.1 $NbSe_2$(cr)

The heat capacity in the temperature range 4.7 to 314 K was measured by Naumov, Rakhmenkulov, Fedorov, and Mischenko [83NAU/RAK]. Their results at 298.15 K are selected:

$$C^{\circ}_{p,m}(\text{NbSe}_2\text{, cr, 298.15 K}) = (73.52 \pm 0.30)\ \text{J·K}^{-1}\text{·mol}^{-1},$$

and

$$S^{\circ}_{m}(\text{NbSe}_2\text{, cr, 298.15 K}) = (103.5 \pm 0.6)\ \text{J·K}^{-1}\text{·mol}^{-1}.$$

The high temperature enthalpy was measured in the temperature interval 1180 to 1656 K in [87BOL/TRO]. The authors combined their results with the heat capacity at 298.15 K in [83NAU/RAK], and derived expressions for the heat capacity and the enthalpy. However, the numerical expressions given in [87BOL/TRO] are in error and do not reproduce the reported experimental values. The heat capacity expression was re-evaluated by the review as discussed in Appendix A giving the expression: $C^{\circ}_{p,\mathrm{m}}$ ($NbSe_2$, cr, (1180 – 1656) K) = $(68.763 + 23.5 \times 10^{-3}\ T - 2.0 \times 10^{5}\ T^{-2})\ \mathrm{J{\cdot}K^{-1}{\cdot}mol^{-1}}$.

V.11.17.1.2 $NbSe_3$(cr)

The heat capacity of $NbSe_3$(cr) has been studied in the temperature range 50 to 160 K in [81TOM/BIL] and [81TOM/BIL2]. The temperatures are too low for deriving the value at 298.15 K.

V.11.18 Tantalum compounds

V.11.18.1 Tantalum selenides

The compounds $TaSe_2$(cr) and $TaSe_3$(cr) have been reported. Thermodynamic measurements have only been performed on $TaSe_2$(cr).

V.11.18.1.1 $TaSe_2$(cr)

Glazunov, Mikhailov, Piskarev, and Chupakhin [76GLA/MIK] studied the thermal decomposition of $TaSe_2$(cr) in the temperature range 1243 to 1413 K by mass spectroscopy and Knudsen effusion cells. Se_2(g) was the only gaseous species detected and it was concluded that the selenide vaporises according to the reaction $TaSe_2(cr) \rightleftharpoons Ta(cr) + Se_2(g)$. A second law evaluation resulted in an enthalpy of evaporation of $\Delta_r H^{\circ}_{m}$(1350 K) = (246.9 ± 13.0) kJ·mol^{-1}. A re-evaluation made by the review of the partial pressures given in Table 1 of the paper yields $\Delta_r H^{\circ}_{m}$(1350 K) = (164.4 ± 40.0) kJ·mol^{-1} and $\Delta_r S^{\circ}_{m}$(1350 K) = (22.6 ± 10.0) J·K^{-1}·mol^{-1}. A third law evaluation of the same data set using the entropy of Ta(cr) in [98CHA] and the selected data for Se_2(g) and of $NbSe_2$(cr) yields $\Delta_r H^{\circ}_{m}$(1350 K) = (337 ± 20) kJ·mol^{-1}. Data for $NbSe_2$(cr) were used due to the lack of data for $TaSe_2$(cr). The large difference between the derived enthalpies of reaction makes this investigation very dubious and no value is selected by the review.

V.11.19 Titanium compounds

V.11.19.1 Titanium selenides

The titanium-selenium system was reviewed by Murray [86MUR]. It was found that no phase diagram could be constructed from the available experimental information. Hirota, Ueda, and Kosuge [88HIR/UED] performed X-ray and DTA measurements in the composition range $TiSe_x$, $0.95 < x < 2.00$ and presented a phase diagram for this part of the system. $TiSe_{0.96}$(cr) with an orthorhombic unit cell was found to be the compound closest to the composition TiSe. In contrast, Pankratova and Zvinchuk [97PAN/ZVI] determined the homogeneity range of a hexagonal titanium monoselenide to extend from $TiSe_{0.95}$ to $TiSe_{1.04}$ by X-ray powder diffraction. It can be concluded that the system is complex and contains several ordered and disordered phases based on the NiAs and the CdI_2 structures. Reactions are slow and establishment of true equilibrium is problematic. The identification and the exact composition of the phases are doubtful in several studies because of the severe experimental difficulties. Mills [74MIL] estimated the enthalpies of formation and the entropies for the phases TiSe(cr), Ti_2Se_3(cr) and $TiSe_2$(cr) but, considering the uncertainties in the phase identification, these data are not selected in this review.

V.11.19.1.1 TiSe(cr)

As mentioned above, the homogeneity range and crystal structure of TiSe are uncertain. A provisional value for the enthalpy of formation $\Delta_f H_m^\circ$(TiSe, cr, 298.15 K) = – (222 ± 42) kJ·mol^{-1} was obtained in [97PAN/ZVI] from calorimetric measurements described briefly in the paper. Considering the general experimental uncertainties no value is selected in this review.

V.11.19.1.2 $TiSe_2$(cr)

The heat capacity of $TiSe_2$(cr) single crystals was measured in the temperature range 1.3 to 300 K by Craven, Di Salvo and Hsu [78CRA/SAL]. Unfortunately, the measured values are only presented in a small graph and no attempt was made to evaluate the absolute entropy at 298.15 K. A second order transformation at 200 K complicates the evaluation. No value is selected in this review.

V.11.20 Zirconium compounds

V.11.20.1 Zirconium selenides

The existence of the zirconium selenides ZrSe(cr), $ZrSe_{1.5}$(cr), $ZrSe_2$(cr), and $ZrSe_3$(cr) have been reported. No experimental thermodynamic data are available except for $ZrSe_3$(cr) for which the heat capacity has been measured in the temperature range 8 to 200 K [86PRO/AYA]. These temperatures are too low for a derivation of thermodynamic quantities at 298.15 K. Mills [74MIL] has estimated some thermodynamic values by comparison with the corresponding sulphides and tellurides.

V.11.20.2 Zirconium selenites

Pinaev, Nesterenko, Glybin, and Pechkovskii [73PIN/NES] determined the standard enthalpy of formation of $Zr(SeO_3)_2$(cr) from enthalpies of dissolution of ZrO_2(cr), SeO_2(cr), and $Zr(SeO_3)_2$(cr) in 11.28 M HF, see Appendix A. The result is $\Delta_f H_m^\circ$($Zr(SeO_3)_2$, cr, 298.15 K) = – (1609.6 ± 2.7) kJ·mol^{-1}.

Nesterenko, Pinaev, and Pechkovskii [73NES/PIN] measured the vapour pressure of SeO_2(g) over zirconium(IV) selenite, $Zr(SeO_3)_2$, in the temperature range 763 to 813 K by a membrane manometer. The equilibria were sluggish, but appear to be reversible, and 7 to 30 days were needed to reach a steady state. From the data furnished in the deposited VINITI document (No. 5021-72) it is seen that the specimen was contaminated, probably by SeO_2 as the selenite was prepared by mixing solutions of dioxozirconium chloride and selenious acid. No attempt was made to remove the contamination from the crystalline specimen, which makes the interpretation of the data somewhat uncertain. The results for the reaction:

$$Zr(SeO_3)_2(cr) \rightleftharpoons ZrO_2(cr) + 2SeO_2(g) \quad (V.151)$$

in the temperature range 763 to 813 K were presented as $\Delta_r H^\circ_m$((V.151), 788 K) = (272.0 ± 20.0) kJ·mol^{-1} and $\Delta_r S^\circ_m$((V.151), 788 K) = (327.6 ± 20.0) J·K^{-1}·mol^{-1}.

No data are available for a proper extrapolation of these data to standard conditions. On the assumption that $\Delta_r C^\circ_{p,m}$ for the reaction equals $2(C^\circ_{p,m}(SeO_2(g)) - C^\circ_{p,m}(SeO_2(cr))) \approx -32.0$ J·K^{-1}·mol^{-1}, the review obtains $\Delta_f H^\circ_m(Zr(SeO_3)_2$, cr, 298.15 K) = $-(1610.0 \pm 20.0)$ kJ·mol^{-1} and $S^\circ_m(Zr(SeO_3)_2$, cr, 298.15 K) = (217.0 ± 21.0) J·K^{-1}·mol^{-1}. The standard enthalpy of formation agrees with the calorimetric value, which is preferred, adopted and included in Appendix E:

$$\Delta_f H^\circ_m(Zr(SeO_3)_2, cr, 298.15 K) = -(1609.6 \pm 2.7) kJ·mol^{-1}.$$

Pashinkin and Gospodinov [94PAS/GOS] calculated the standard entropy of $Zr(SeO_3)_2$ to be (196.5 ± 33.5) J·K^{-1}·mol^{-1} from the same data. The auxiliary data used are not included in the paper. The mean value of the two entropy estimates is $S^\circ_m(Zr(SeO_3)_2$, cr, 298.15 K) = (210.0 ± 25.0) J·K^{-1}·mol^{-1}. Being an estimate, this value is not selected.

V.11.21 Hafnium compounds

V.11.21.1 Hafnium selenides

The compound Hf_2Se_3(cr) has been reported in the hafnium-selenium system. No thermodynamic information is available.

V.11.22 Scandium compounds

V.11.22.1 Scandium selenides

Experimental thermodynamic information is not available about solid scandium selenides, but Mills [74MIL] has estimated thermodynamic properties for ScSe(cr) and Sc_2Se_3(cr). The former selenide has been reported as non-existent in [65ELL].

V.11.22.1.1 ScSe(g)

The dissociation reaction ScSe(g) $\rightleftharpoons$ Sc(g) + Se(g) was thoroughly studied by Ni and Wahlbeck [72NI/WAH] in the temperature range 1980 to 2240 K by measuring the partial pressures of the gas species using mass spectroscopy and Knudsen effusion cells. The starting material was Sc_2Se_3(cr) which was found to vaporise congruently. The results were evaluated to be $\Delta_{at}H^{\circ}_{m}$ (ScSe, g, 298.15 K) = (382.0 ± 3.5) kJ·mol^{-1} from the second law and to be $\Delta_{at}H^{\circ}_{m}$ (ScSe, g, 298.15 K) = (395.3 ± 36.7) kJ·mol^{-1} from the third law giving a weighted mean of $\Delta_{at}H^{\circ}_{m}$ (ScSe, g, 298.15 K) = (382.1 ± 13.6) kJ·mol^{-1}. The error limits have been increased to account for the estimated overall uncertainty. The evaluation involved the Gibbs energy function of ScSe(g) calculated from estimated molecular parameters. The dissociation enthalpy of ScSe(g) was also determined in a similar study by Bergman, Coppens, Drowart and Smoes [70BER/COP] to be $\Delta_{at}H^{\circ}_{m}$ (ScSe, g, 0 K) = (375.6 ± 5.5) kJ·mol^{-1}. The starting material was given as ScSe(s), a compound that has been reported to be non-existent in [65ELL], and because the experimental conditions were sparsely reported, the result is considered less reliable than the work in [72NI/WAH]. The total vapour pressure of the congruent vaporisation of Sc_2Se_3(cr) was studied in the temperature range 1858 to 2092 K using Knudsen effusion cells by Leary and Wahlbeck [69LEA/WAH]. The authors concluded that it was not possible to derive thermodynamic quantities from these experiments because the vaporisation occurs via two simultaneous reactions, Sc_2Se_3(cr) $\rightleftharpoons$ Se(g) + 2ScSe(g) and Sc_2Se_3(cr) $\rightleftharpoons$ 2Sc(g) + 3Se(g). The dissociation enthalpy in [72NI/WAH] obtained from the third law was recalculated to 298.15 K using their estimated molecular parameters of ScSe(g), the enthalpy of Sc(g) in [82WAG/EVA] and the selected properties of Se(g) yielding $\Delta_{r}H^{\circ}_{m}$ (ScSe, g, 298.15 K) = (385.8 ± 13.6) kJ·mol^{-1}. The value was combined with the enthalpy of formation of Sc(g) in [82WAG/EVA] and the selected enthalpy of formation of Se(g) yielding $\Delta_{f}H^{\circ}_{m}$ (ScSe, g, 298.15 K) = (228.1 ± 13.7) kJ·mol^{-1}. The substantial difference between the results obtained by the temperature extrapolation with the second law and the third law, respectively, suggests experimental errors or insufficient accuracy of the estimated thermal data. However, the result in [70BER/COP] supports the result in [72NI/WAH] and the review adopts:

$$\Delta_{f}H^{\circ}_{m}\text{(ScSe, g, 298.15 K)} = (228.1 \pm 13.7)\ \text{kJ·mol}^{-1}$$

This value is included in Appendix E since it was calculated with non-TDB auxiliary data.

V.11.22.2 Scandium selenites

Maier, Suponitskii, and Karapet'yants [68MAI/SUP] calculated the enthalpy of formation of scandium selenite from measurements of the enthalpy change of the reaction between Na_2SeO_3(cr) and a solution of $ScCl_3$. The precipitated $Sc_2(SeO_3)_3$ was amorphous and chemical analysis showed that it contained 10 molecules of water. The review failed to obtain agreement between the enthalpy changes reported for the studied reactions and the primary data presented as discussed in Appendix A. The re-evaluation therefore results in two values for the enthalpy of formation of $Sc_2(SeO_3)_3\cdot10H_2O$(am) namely, – 5567.1 kJ·mol^{-1} from the reported enthalpy changes and – 5615.6 kJ·mol^{-1} with enthalpy changes calculated from the primary data. The not referenced value in [82WAG/EVA] is $\Delta_{f}H^{\circ}_{m}$ ($Sc_2(SeO_3)_3\cdot10H_2O$, cr, 298.15 K) = – 5550.1 kJ·mol^{-1}. The reaction most affected by the uncertainty in the experimental data was replaced in a second evaluation by additional auxiliary data. The result then obtained was – 5646.5 $kJ\cdot mol^{-1}$.

In view of the disagreement between the primary and the reported experimental data and the scatter amongst the derived values of $\Delta_f H_m^\circ$ ($Sc_2(SeO_3)_3\cdot 10H_2O$, am, 298.15 K), no value of the enthalpy of formation of scandium selenite will be selected.

No other thermodynamic data have been found for scandium selenites.

V.11.22.3 Scandium selenates

Kolosov *et al.* [67KOL/IVA] studied the complex formation between Sc^{3+} and SeO_4^{2-} at pH 4.0 to 4.5 by distribution measurements of Sc^{3+} between the aqueous phase and a cation exchanger. No interference from hydrolysis of Sc^{3+} was observed although the data in [76BAE/MES] predict at least a minor interference. The ligand concentration was changed from 0 to 0.167 M, while the ionic strength was kept constant at 0.5 M by means of $NaClO_4$. Measurements were performed at 298.15, 308.15, and 318.15 K. The data were interpreted in terms of the equilibria:

$$Sc^{3+} + SeO_4^{2-} \rightleftharpoons ScSeO_4^+ \qquad (V.152)$$

$$Sc^{3+} + 2SeO_4^{2-} \rightleftharpoons Sc(SeO_4)_2^- . \qquad (V.153)$$

As discussed in Appendix A, only β_1((V.152), 298.15 K) will be accepted due to the large change in the medium and the accompanying change in the Na^+ concentration, which will affect the evaluation method. The value accepted is:

$$\log_{10} \beta_1 \text{((V.152), 0.5 M NaClO}_4\text{, 298.15 K)} = (1.65 \pm 0.20).$$

A correction to the standard state cannot be made with confidence.

No other thermodynamic data have been identified for scandium selenates. The data for $ScSeO_4^+$ and $Sc(SeO_4)_2^-$ in [82WAG/EVA] most likely derive from [67KOL/IVA] and are therefore subject to the same criticism as the original data.

V.11.23 Yttrium compounds

V.11.23.1 Yttrium selenides

Experimental thermodynamic information is not available on solid yttrium selenides, but Mills [74MIL] has estimated thermodynamic properties for YSe(cr) and Y_2Se_3(cr) by comparison with the yttrium sulphides and lanthanum selenides. No values for solid selenides were selected by the review.

V.11.23.1.1 YSe(g)

Ni and Wahlbeck [72NI/WAH] studied the atomisation of YSe(g) in the temperature range 1980 to 2240 K by measurement of the partial pressures of the gas species using mass spectroscopy and Knudsen effusion cells. The starting material was YSe(cr) which was found to vaporise congruently. The results were evaluated according to the second and third laws giving a mean value of $\Delta_{at}H^{\circ}_{m}$ (YSe, g, 0 K) = (427.6 ± 15.3) kJ·mol^{-1}. The third law treatment involved the Gibbs energy function of YSe(g) calculated from estimated molecular parameters. A recalculation to 298.15 K using the estimated molecular parameters of YSe(g) given in [72NI/WAH] yields $\Delta_{at}H^{\circ}_{m}$ (YSe, g, 298.15 K) = (431.1 ± 15.3) kJ·mol^{-1}. This value was combined with the enthalpy of formation of Y(g), (421.3 ± 2.5 (estimated)) kJ·mol^{-1}, in [82WAG/EVA] and the selected enthalpy of formation of Se(g) to give $\Delta_{f}H^{\circ}_{m}$ (YSe, g, 298.15 K) = (226.3 ± 15.6) kJ·mol^{-1}. The substantial difference between the results obtained by the temperature extrapolation with the second law and the third law, respectively, suggests experimental errors or insufficient accuracy of the estimated thermal data. However, the heat capacity and entropy of a diatomic molecule calculated from estimated molecular properties are expected to be quite accurate and therefore the review has adopted with a large error bar

$$\Delta_{f}H^{\circ}_{m}\text{(YSe, g, 298.15 K)} = (226.3 \pm 15.6)\ \text{kJ·mol}^{-1}.$$

This value is included in Appendix E since it was calculated with non-TDB auxiliary data.

V.11.23.2 Yttrium selenites

Gospodinov and Stancheva [94GOS/STA] investigated the system Y_2O_3-SeO_2-H_2O at 373.15 K by solubility measurements over the interval 0 to 85 mass-% of selenious acid in the liquid phase. Solids of composition $Y_2(SeO_3)_3·4H_2O$ and $Y_2(SeO_3)_3·H_2SeO_3·5H_2O$ were identified in the phase diagram from the tie lines and by chemical and X-ray analysis. The primary solubility data are available.

Maier, Suponitskii, and Karapet'yants [68MAI/SUP] prepared crystalline $Y_2(SeO_3)_3$ and made a calorimetric determination of its enthalpy of dissolution in HCl(aq, 1:18.5). The review has recalculated the data and obtains $\Delta_{f}H^{\circ}_{m}$ ($Y_2(SeO_3)_3$, cr, 298.15 K) = – (2900 ± 19) kJ·mol^{-1}. Zhang-Presse and Oppermann obtained for this quantity the value – (2828.0 ± 8.4) kJ·mol^{-1} from calorimetric measurements in [2002ZHA/OPP] and – (2893 ± 50) kJ·mol^{-1} from measurements of equilibria at high temperatures in [2002ZHA/OPP2]. The two calorimetric values differ by about 70 kJ·mol^{-1} and the two values obtained by Zhang-Presse and Oppermann with different techniques by 65 kJ·mol^{-1}. The review concludes that no value can be selected for $\Delta_{f}H^{\circ}_{m}$ ($Y_2(SeO_3)_3$, cr, 298.15 K) particularly as the latter authors did not discuss the discrepancy in their results.

V.12 Lanthanide compounds and complexes

V.12.1 Lutetium compounds

V.12.1.1 Lutetium selenides

V.12.1.1.1 LuSe(g)

The entropy of LuSe(g) at 298.15 K was calculated to be S°_{m}(LuSe, g, 298.15 K) = 265.8 $J \cdot K^{-1} \cdot mol^{-1}$ by Kovalevskaya, Sidorenko, Lysenko, and Fesenko [77KOV/SID]. The entropy value is not selected because all parameters required for the calculation using statistical mechanics were estimated.

The enthalpy change of the reaction

$$\text{LuSe(g)} + \text{Sc(g)} \rightleftharpoons \text{Lu(g)} + \text{ScSe(g)} \qquad \text{(V.154)}$$

was determined by Bergman, Coppens, Drowart, and Smoes [70BER/COP] in mass spectrometric measurements of ion intensities of the species in the temperature range 1852 to 2320 K. The results were evaluated to be $\Delta_r H^{\circ}_{m}$((V.154), 0 K) = (41.7 ± 4.5) $kJ \cdot mol^{-1}$ from the second law and $\Delta_r H^{\circ}_{m}$((V.154), 0 K) = (38.3 ± 1.2) $kJ \cdot mol^{-1}$ from the third law, respectively, using Gibbs energy functions of LuSe(g) and ScSe(g) calculated from estimated molecular parameters. The weighted average, $\Delta_r H^{\circ}_{m}$((V.154), 0 K) = (38.5 ± 1.2) $kJ \cdot mol^{-1}$, was recalculated to 298.15 K by the review employing molecular parameters for LuSe(g) and ScSe(g) identical to those in [70BER/COP] and the properties of Sc(g) and Lu(g) in [82WAG/EVA] yielding $\Delta_r H^{\circ}_{m}$((V.154), 298.15 K) = (37.7 ± 1.2) $kJ \cdot mol^{-1}$. The enthalpy of formation of LuSe(g) was calculated to be:

$$\Delta_f H^{\circ}_{m}(\text{LuSe, g, 298.15 K}) = (240.2 \pm 14.9)\ \text{kJ} \cdot \text{mol}^{-1}$$

by combining the reaction enthalpy with the enthalpy values of Sc(g) and Lu(g) in [82WAG/EVA] and the Appendix E value for the enthalpy of formation of ScSe(g). The good agreement between the values of $\Delta_r H^{\circ}_{m}$((V.154), 0 K) obtained by the second and third law extrapolations suggests that this value can be accepted despite the use of estimated thermal data. The formation datum is included in Appendix E since it was calculated with non-TDB auxiliary data.

V.12.1.2 Lutetium selenites

Smolyakova and Serebrennikov [78SMO/SER] measured the enthalpy change of the reaction between an aqueous solution of lutetium chloride and Na_2SeO_3(cr). The description of the experiments and calculations is somewhat meagre. Their data have been re-evaluated as described in Appendix A and yielded the result $\Delta_f H^{\circ}_{m}$($Lu_2(SeO_3)_3 \cdot 8H_2O$, am, 298.15 K) = – (5176 ± 12) $kJ \cdot mol^{-1}$.

As an amorphous phase is apparently formed when lutetium selenite is precipitated from aqueous solution, the result of this investigation is included in the review but not selected.

No other thermodynamic data have been found for lutetium selenites.

V.12.2 Ytterbium compounds

V.12.2.1 Ytterbium selenides

V.12.2.1.1 YbSe(cr)

The congruent vaporisation of YbSe(cr) was studied in the temperature range 1744 to 2044 K by Marx and Petzel [90MAR/PET] by measuring the total vapour pressure using the Knudsen weight loss technique. The results were evaluated assuming to the reaction

$$\mathrm{YbSe(cr)} \rightleftharpoons \mathrm{Yb(g)} + \mathrm{Se(g)} \quad \text{(V.155)}$$

yielding the reaction enthalpies $\Delta_r H_m^\circ$((V.155), 298.15 K) = (831.4 ± 32.2) kJ·mol^{-1} and $\Delta_r H_m^\circ$((V.155), 298.15 K) = (833.1 ± 26.0) kJ·mol^{-1} from the second and third laws, respectively. Lacking heat capacity and entropy data for YbSe(cr), values for EuSe(cr) in [74MCM/GSC] were used to obtain the reaction enthalpy values at 298.15 K. The enthalpy of formation of YbSe(cr) was calculated from the mean of the second and third law values, $\Delta_r H_m^\circ$((V.155), 298.15 K) = (832.3 ± 29.1) kJ·mol^{-1}, by combination with the selected value of the enthalpy of formation of Se(g) and the enthalpy of formation of Yb(g), (152.3 ± 4.0(estimated)) kJ·mol^{-1}, in [82WAG/EVA] yielding:

$$\Delta_f H_m^\circ (\mathrm{YbSe, cr, 298.15\ K}) = -(443.9 \pm 31.3)\ \mathrm{kJ \cdot mol^{-1}}.$$

The datum is included in Appendix E since it was calculated with non-TDB auxiliary data.

V.12.2.1.2 YbSe(g)

The entropy of YbSe(g) at 298.15 K was calculated to be S_m°(YbSe, g, 298.15 K) = 271.1 J·K^{-1}·mol^{-1} by Kovalevskaya, Sidorenko, Lysenko, and Fesenko [77KOV/SID]. The entropy value is not selected because all parameters required for the calculation using statistical mechanics were estimated.

V.12.2.1.3 Yb_2Se_3(cr)

The enthalpy of formation of Yb_2Se_3(cr) was reported to be – (965.7 ± 30.0) kJ·mol^{-1} by Azizov, Abbasov, Mustafaev, and Aliev [77AZI/ABB] as a result of direct synthesis calorimetry. The experimental information is minimal and the quality of the work cannot be validated. No value of the enthalpy of formation of Yb_2Se_3(cr) is selected.

V.12.2.2 Ytterbium selenites

Smolyakova and Serebrennikov [78SMO/SER] measured the enthalpy change of the reaction between an aqueous solution of ytterbium chloride and Na_2SeO_3(cr). The description of the experiments and calculations is somewhat meagre. Their data have been re-evaluated as described in Appendix A and yielded the result $\Delta_f H_m^\circ$ ($Yb_2(SeO_3)_3 \cdot 8H_2O$, am, 298.15 K) = – (5115 ± 7) kJ·mol^{-1}.

As an amorphous phase is apparently formed when ytterbium selenite is precipitated from aqueous solution, the result of this investigation is included in the review but not selected.

Oppermann *et al.* [2002OPP/ZHA2] report the enthalpy of formation of $Yb_2(SeO_3)_3$(cr) at 298.15 K to be – (2779 ± 42) kJ·mol^{-1} from equilibrium measurements of decomposition reactions in the temperature range 825 to 1050 K, see also [2002OPP/ZHA2] in Appendix A. This is the only result found and the review adopts the rounded value

$$\Delta_f H_m^\circ (Yb_2(SeO_3)_3, \text{cr}, 298.15\ \text{K}) = -(2780 \pm 50)\ \text{kJ·mol}^{-1}.$$

Since this quantity was evaluated using non-TDB auxiliary data it has been included in Table E-2 of Appendix E.

V.12.2.3 Ytterbium selenates

The measurements of Serebrennikov, Batyreva, and Tsybukova [81SER/BAT], which established the phase diagram of the system $Yb_2(SeO_4)_3$-$Nd_2(SeO_4)_3$-H_2O at 298.15 K, indicate that $Yb_2(SeO_4)_3 \cdot 8H_2O$ is the stable ytterbium selenate in contact with water. The reported solubility:

$$m(Yb_2(SeO_4)_3 \cdot 8H_2O(\text{cr}), \text{aq sat}, 298.15\ \text{K}) = 1.289\ \text{mol}(Yb_2(SeO_4)_3) \cdot \text{kg}^{-1}$$

is accepted but an estimate of the uncertainty cannot be made.

No other thermodynamic data of ytterbium selenates have been found.

V.12.3 Thulium compounds

V.12.3.1 Thulium selenides

The phase diagram of the system thulium-selenium has been studied by Fritzler, Kaldis, and Jilek [82FRI/KAL].

V.12.3.1.1 TmSe(cr)

The enthalpies of formation of solid solutions with compositions in the range $Tm_{0.87}Se$-$Tm_{1.05}Se$ were determined by Spychiger, Kaldis, and Fritzler [85SPY/KAL] at temperatures around 2400 K using fluorine combustion calorimetry. The phase relationships around the stoichiometric composition are unclear and two phases seem to form at or near the 1:1 composition. No enthalpy of formation of TmSe(cr) at 298.15 K was calculated.

V.12.3.1.2 TmSe(g)

The entropy of TmSe(g) at 298.15 K was calculated to be S°_{m}(TmSe, g, 298.15 K) = 272.0 $J\cdot K^{-1}\cdot mol^{-1}$ by Kovalevskaya, Sidorenko, Lysenko, and Fesenko [77KOV/SID]. The entropy value is not selected because all parameters required for the calculation using statistical mechanics were estimated.

The dissociation enthalpy of TmSe(g) was estimated to be $\Delta_{at}H^{\circ}_{m}$(TmSe, g, 0 K) = (270 ± 40) $kJ\cdot mol^{-1}$ by Bergman, Coppens, Drowart, and Smoes [70BER/COP] by comparison with other lanthanide selenides. No value of the enthalpy of formation of TmSe(g) is selected.

V.12.4 Erbium compounds

V.12.4.1 Erbium selenides

V.12.4.1.1 ErSe(g)

The entropy of ErSe(g) at 298.15 K was calculated to be S°_{m}(ErSe, g, 298.15 K) = 278.8 $J\cdot K^{-1}\cdot mol^{-1}$ by Kovalevskaya, Sidorenko, Lysenko, and Fesenko [77KOV/SID]. The entropy value is not selected because all parameters required for the calculation using statistical mechanics were estimated.

The dissociation enthalpy of ErSe(g) was estimated to be $\Delta_{at}H^{\circ}_{m}$(ErSe, g, 0 K) = (322 ± 20) $kJ\cdot mol^{-1}$ by Bergman, Coppens, Drowart, and Smoes [70BER/COP] by comparison with other lanthanide selenides. No value of the enthalpy of formation of ErSe(g) is selected.

V.12.4.2 Erbium selenites

Smolyakova and Serebrennikov [78SMO/SER] measured the enthalpy change of the reaction between an aqueous solution of erbium chloride and Na_2SeO_3(cr). The description of the experiments and calculations is somewhat meagre. Their data have been re-evaluated as described in Appendix A and yielded the result $\Delta_{f}H^{\circ}_{m}$($Er_2(SeO_3)_3\cdot 8H_2O$, am, 298.15 K) = – (5176 ± 11) $kJ\cdot mol^{-1}$.

As an amorphous phase is apparently formed when erbium selenite is precipitated from aqueous solution, the result of this investigation is included in the review but not selected.

No other thermodynamic data have been found for erbium selenites.

V.12.4.3 Erbium selenates

The only study of the complex formation between selenate and trivalent lanthanides found is an X-ray diffraction investigation of some lanthanide selenate solutions by Johansson, Niinistö, and Wakita [85JOH/NII]. The radial distribution function was interpreted to show that inner-sphere complexes were formed. For Er^{3+} an average of 0.4 SeO_4^{2-} per Er^{3+} was estimated for a solution containing 0.78 M Er^{3+} and 1.55 M SeO_4^{2-}. The conditional equilibrium constant for the inner-sphere complex formation:

$$Er^{3+} + SeO_4^{2-} \rightleftharpoons ErSeO_4^{+} \text{ (inner-sphere)} \quad (V.156)$$

would then be β_1((V.156), conditional, 293 K ?) = 0.5 M^{-1}.

This value will not be included in the data selection.

V.12.5 Holmium compounds

V.12.5.1 Holmium selenides

V.12.5.1.1 HoSe(g)

The entropy of HoSe(g) at 298.15 K was calculated to be S_m° (HoSe, g, 298.15 K) = 280.2 $J \cdot K^{-1} \cdot mol^{-1}$ by Kovalevskaya, Sidorenko, Lysenko, and Fesenko [77KOV/SID]. The entropy value is not selected because all parameters required for the calculation using statistical mechanics were estimated.

The enthalpy change of the reaction

$$HoSe(g) + V(g) \rightleftharpoons Ho(g) + VSe(g) \quad (V.157)$$

was determined by Bergman, Coppens, Drowart, and Smoes [70BER/COP] in mass spectrometric measurements of ion intensities of the species in the temperature range 2038 to 2196 K. The results were evaluated to be $\Delta_r H_m^\circ$((V.157), 0 K) = – (12.4 ± 0.5) $kJ \cdot mol^{-1}$ from the third law using Gibbs energy functions of HoSe(g) and VSe(g) calculated from estimated molecular parameters. Unfortunately, the molecular data used for VSe(g) were not mentioned and the reaction enthalpy was recalculated to 298.15 K by the review employing a value of 9.5 $kJ \cdot mol^{-1}$ for the enthalpy increment of VSe(g) from 0 to 298.15 K, the molecular parameters for HoSe(g) identical to those in [70BER/COP], and the enthalpy increments of Ho(g) and V(g) in [82WAG/EVA] yielding $\Delta_r H_m^\circ$((V.157), 298.15 K) = – (14.1 ± 1.0) $kJ \cdot mol^{-1}$. The enthalpy of formation of HoSe(g) was calculated to be $\Delta_f H_m^\circ$ (HoSe, g, 298.15 K) = (205.7 ± 17.0) $kJ \cdot mol^{-1}$ by combining the reaction enthalpy with the enthalpy values of V(g) and Ho(g) in [82WAG/EVA] and the Appendix E value for the enthalpy of formation of VSe(g).

A rounded value with increased error bars is adopted by the review:

$$\Delta_f H_m^\circ \text{(HoSe, g, 298.15 K)} = (206 \pm 20)\ kJ \cdot mol^{-1}$$

and included in Appendix E since it was calculated with non-TDB auxiliary data.

V.12.5.2 Holmium selenites

Smolyakova and Serebrennikov [78SMO/SER] measured the enthalpy change of the reaction between an aqueous solution of holmium chloride and Na_2SeO_3(cr). The description of the experiments and calculations is somewhat meagre. Their data have been re-evaluated as described in Appendix A and yielded the result $\Delta_f H_m^\circ$ ($Ho_2(SeO_3)_3\cdot 8H_2O$, am, 298.15 K) = – (5174 ± 9) kJ·mol^{-1}.

As an amorphous phase is apparently formed when holmium selenite is precipitated from aqueous solution, the result of this investigation is included in the review but not selected.

No other thermodynamic data have been found for holmium selenites.

V.12.6 Dysprosium compounds

V.12.6.1 Dysprosium selenides

V.12.6.1.1 DySe(g)

The entropy of DySe(g) at 298.15 K was calculated to be S_m° (DySe, g, 298.15 K) = 281.6 J·K^{-1}·mol^{-1} by Kovalevskaya, Sidorenko, Lysenko, and Fesenko [77KOV/SID]. The entropy value is not selected because all parameters required for the calculation using statistical mechanics were estimated.

The dissociation enthalpy of DySe(g) was estimated to be $\Delta_{at} H_m^\circ$ (DySe, g, 0 K) = (318 ± 20) kJ·mol^{-1} by Bergman, Coppens, Drowart, and Smoes [70BER/COP] by comparison with other lanthanide selenides. No value of the enthalpy of formation of DySe(g) is selected.

V.12.6.1.2 Dy_2Se_3(cr)

The enthalpy of formation of Dy_2Se_3(cr) was reported to be – (964 ± 30) kJ·mol^{-1} by Azizov, Abbasov, Mustafaev, and Aliev [77AZI/ABB] as a result of direct synthesis calorimetry. The experimental information is minimal and the quality of the work cannot be validated. No value of the enthalpy of formation of Dy_2Se_3(cr) is selected.

V.12.6.2 Dysprosium selenites

Gospodinov and Stancheva [97GOS/STA] investigated the system Dy_2O_3-SeO_2-H_2O at 373.15 K by solubility measurements over the concentration interval 0.06 to 68.4 mass-% SeO_2 in the liquid phase. The formation of $Dy_2(SeO_3)_3\cdot 4H_2O$ and $DyH(SeO_3)_2\cdot 2H_2O$ were found from the tie lines and by chemical analysis. The solubility data and a phase diagram are presented.

Smolyakova and Serebrennikov [78SMO/SER] measured the enthalpy change of the reaction between an aqueous solution of dysprosium chloride and Na_2SeO_3(cr). The description of the experiments and calculations is somewhat meagre. Their data have been re-evaluated as described in Appendix A and yielded the result $\Delta_f H^\circ_m$ ($Dy_2(SeO_3)_3\cdot8H_2O$, am, 298.15 K) = – (5158 ± 9) kJ·mol^{-1}.

Note that the composition assigned to the amorphous phase disagrees with the composition of the equilibrium phase at 373 K. As an amorphous phase is apparently formed when dysprosium selenite is precipitated from aqueous solution at 298 K, the result of this investigation is included in the review but not selected.

No other thermodynamic data have been found for dysprosium selenites.

V.12.6.3 Dysprosium selenates

Smolyakova, Efremov, and Serebrennikov [73SMO/EFR] determined the enthalpy change of the reaction between $Dy_2(SeO_4)_3\cdot8H_2O$(cr) and a solution of barium chloride with formation of $BaSeO_4$(cr) by a calorimetric measurement. The review calculated:

$$\Delta_f H^\circ_m (Dy_2(SeO_4)_3\cdot8H_2O, \text{cr}, 298.15\ \text{K}) = -(5371.4 \pm 17.5)\ \text{kJ·mol}^{-1}$$

from this measurement, see Appendix A. The result has been adopted although the source is of uncertain quality and included in Appendix E since it was calculated with non-TDB auxiliary data.

No other thermodynamic data have been found for dysprosium selenates.

V.12.7 Terbium compounds

V.12.7.1 Terbium selenides

V.12.7.1.1 TbSe(g)

The entropy of TbSe(g) at 298.15 K was calculated to be S°_m(TbSe, g, 298.15 K) = 282.3 J·K^{-1}·mol^{-1} by Kovalevskaya, Sidorenko, Lysenko, and Fesenko [77KOV/SID]. The entropy value is not selected because all parameters required for the calculation using statistical mechanics were estimated.

The dissociation enthalpy of TbSe(g) was estimated to be $\Delta_{at} H^\circ_m$(TbSe, g, 0 K) = (419 ± 20) kJ·mol^{-1} by Bergman, Coppens, Drowart, and Smoes [70BER/COP] by comparison with other lanthanide selenides. No value of the enthalpy of formation of TbSe(g) is selected.

V.12.7.2 Terbium selenites

Smolyakova and Serebrennikov [78SMO/SER] measured the enthalpy change of the reaction between an aqueous solution of terbium chloride and Na_2SeO_3(cr). The description of the experiments and calculations is somewhat meagre. Their data have been re-evaluated as described in Appendix A and yielded the result $\Delta_f H_m^\circ$ ($Tb_2(SeO_3)_3 \cdot 8H_2O$, am, 298.15 K) = – (5156 ± 9) $kJ \cdot mol^{-1}$.

As an amorphous phase is apparently formed when terbium selenite is precipitated from aqueous solution, the result of this investigation is included in the review but not selected.

No other thermodynamic data have been found for terbium selenites.

V.12.7.3 Terbium selenates

Smolyakova, Efremov, and Serebrennikov [73SMO/EFR] determined the enthalpy change of the reaction between $Tb_2(SeO_4)_3 \cdot 8H_2O$(cr) and a solution of barium chloride with formation of $BaSeO_4$(cr) by a calorimetric measurement. The review calculated:

$$\Delta_f H_m^\circ (Tb_2(SeO_4)_3 \cdot 8H_2O, \text{cr}, 298.15\text{ K}) = -(5426.1 \pm 17.5)\text{ kJ}\cdot\text{mol}^{-1}$$

from this measurement, see Appendix A. The result is adopted although the source is of uncertain quality and included in Appendix E since it was calculated with non-TDB auxiliary data.

No other thermodynamic data have been found for terbium selenates.

V.12.8 Gadolinium compounds

V.12.8.1 Gadolinium selenides

The phase diagram of the system gadolinium-selenium has been studied by Pribylskii, Vasilieva, and Gamidov, [82PRI/VAS]. The compounds GdSe(cr), Gd_2Se_3(cr), Gd_3Se_4(cr), and $GdSe_2$(cr) have been reported.

Azizov, Agaev, Guseinov, Abbasov, Aliev, and Mustafaev [80AZI/AGA] measured the emf of the cells

(–) Gd(cr) | $GdCl_3$, KCl (in glycerol) | $GdSe_2$(cr), Se(cr) (+);

(–) Gd(cr) | $GdCl_3$, KCl (in glycerol) | Gd_2Se_3(cr), $GdSe_2$(cr) (+);

(–) Gd(cr) | $GdCl_3$, KCl (in glycerol) | Gd_3Se_4(cr), Gd_2Se_3(cr) (+)

and (–) Gd(cr) | $GdCl_3$, KCl (in glycerol) | GdSe(cr), Gd_3Se_4(cr) (+)

in the temperature range 320 to 420 K. Enthalpies and entropies of formation of the compounds were calculated. The determined values conflict strongly with the enthalpy of formation $\Delta_f H_m^\circ$ (Gd_2Se_3, cr, 298.15 K) = – (1068.2 ± 35.0) $kJ \cdot mol^{-1}$ determined by Agaev and Sharifov [72AGA/SHA] using direct synthesis calorimetry. No other thermodynamic studies of solid gadolinium selenides have been made. Until further experimental information is available it is not possible to judge the quality of these investigations. Consequently, no thermodynamic properties of solid gadolinium selenides were selected.

V.12.8.1.1 GdSe(g)

The entropy of GdSe(g) at 298.15 K was calculated to be S°_{m}(GdSe, g, 298.15 K) = 283.2 $J\cdot K^{-1}\cdot mol^{-1}$ by Kovalevskaya, Sidorenko, Lysenko, and Fesenko [75KOV/SID]. The entropy value is not selected because all parameters required for the calculation using statistical mechanics were estimated.

The enthalpy change of the reaction

$$GdSe(g) + Sc(g) \rightleftharpoons Gd(g) + ScSe(g) \qquad (V.158)$$

was determined by Bergman, Coppens, Drowart, and Smoes [70BER/COP] in mass spectrometric measurements of ion intensities of the species in the temperature range 2065 to 2428 K. The results were evaluated to be $\Delta_r H^{\circ}_{m}$((V.158), 0 K) = (36.2 ± 4.8) $kJ\cdot mol^{-1}$ from the second law and $\Delta_r H^{\circ}_{m}$((V.158), 0 K) = (49.9 ± 1.3) $kJ\cdot mol^{-1}$ from the third law, respectively, using Gibbs energy functions of GdSe(g) and ScSe(g) calculated from estimated molecular parameters. The weighted average, $\Delta_r H^{\circ}_{m}$((V.158), 0 K) = (49.0 ± 1.3) $kJ\cdot mol^{-1}$, was recalculated to 298.15 K by the review employing molecular parameters for GdSe(g) and ScSe(g) identical to those in [70BER/COP] and the properties of Sc(g) and Gd(g) in [82WAG/EVA] yielding $\Delta_r H^{\circ}_{m}$((V.158), 298.15 K) = (49.6 ± 1.6) $kJ\cdot mol^{-1}$. The enthalpy of formation of GdSe(g) was calculated to be:

$$\Delta_f H^{\circ}_{m}(\text{GdSe, g, 298.15 K}) = (198.2 \pm 14.9)\ kJ\cdot mol^{-1}$$

by combining the reaction enthalpy with the enthalpy values of Sc(g) and Gd(g) in [82WAG/EVA] and the Appendix E value for the enthalpy of formation of ScSe(g). The reasonable agreement between the values of $\Delta_r H^{\circ}_{m}$((V.158), 0 K) obtained by the second and third law extrapolations suggests that this value can be accepted despite the use of estimated thermal data. The formation datum is included in Appendix E since it was calculated with non-TDB auxiliary data.

V.12.8.2 Gadolinium selenites

Smolyakova and Serebrennikov [78SMO/SER] measured the enthalpy change of the reaction between an aqueous solution of gadolinium chloride and Na_2SeO_3(cr). The description of the experiments and calculations is somewhat meagre. Their data have been re-evaluated as described in Appendix A and yielded the result $\Delta_f H^{\circ}_{m}$($Gd_2(SeO_3)_3\cdot 8H_2O$, am, 298.15 K) = – (5168 ± 7) $kJ\cdot mol^{-1}$.

As an amorphous phase is apparently formed when gadolinium selenite is precipitated from aqueous solution, the result of this investigation is included in the review but not selected.

No other thermodynamic data have been found for gadolinium selenites.

V.12.8.3 Gadolinium selenates

Smolyakova, Efremov, and Serebrennikov [73SMO/EFR] determined the enthalpy change of the reaction between $Gd_2(SeO_4)_3{\cdot}8H_2O(cr)$ and a solution of barium chloride with formation of $BaSeO_4(cr)$ by a calorimetric measurement. The review calculated:

$$\Delta_f H_m^\circ (Gd_2(SeO_4)_3{\cdot}8H_2O, \text{cr}, 298.15\ \text{K}) = -(5444.5 \pm 17.5)\ \text{kJ}{\cdot}\text{mol}^{-1}$$

from this measurement, see Appendix A. The result has been adopted although the source is of uncertain quality and included in Appendix E, since it was calculated with non-TDB auxiliary data.

No other thermodynamic data have been found for gadolinium selenates.

V.12.9 Europium compounds

V.12.9.1 Europium selenides

The phase diagram of the system europium-selenium has been studied by Sadovskaya, Yarembashh, and Naginaev, [71SAD/YAR]. The compounds EuSe(cr), Eu_2Se_3(cr) and Eu_4Se_7(cr) were reported.

V.12.9.1.1 EuSe(cr)

The heat capacity of EuSe(cr) below 16 K was measured in [95LI/HAG], [72WHI/MCC], and in [80JOH/MCC]. McMasters, Gschneider, Kaldis, and Sampietro [74MCM/GSC] measured the heat content of EuSe(cr) in the temperature range 298.15 to 1680 K using drop calorimetry. The heat capacity at 298.15 K derived in [74MCM/GSC] is selected:

$$C_{p,m}^\circ (\text{EuSe, cr}, 298.15\ \text{K}) = (51.0 \pm 3.0)\ \text{J}{\cdot}\text{K}^{-1}{\cdot}\text{mol}^{-1}.$$

The heat capacity expression

$$C_{p,m}^\circ (\text{EuSe, cr}, (298.15 - 1600)\ \text{K}) = (48.76 + 7.376 \times 10^{-3}\ T)\ \text{J}{\cdot}\text{K}^{-1}{\cdot}\text{mol}^{-1}$$

was derived by the review from the values in Table 4 of the paper.

The sublimation thermodynamics of EuSe was studied in the temperature range 1808 to 2131 K by Hariharan and Eick [74HAR/EIC] using mass spectrometry and Knudsen cells. The evaporation reaction is EuSe(cr) $\rightleftharpoons$ Eu(g) + Se(g). The second law value

$$S_m^\circ (\text{EuSe, cr}, 298.15\ \text{K}) = (99.6 \pm 7.3)\ \text{J}{\cdot}\text{K}^{-1}{\cdot}\text{mol}^{-1}$$

was re-evaluated by the review from the measurements using the heat capacity expression of EuSe(cr) given above, the selected properties of Se(g), and the heat capacity of Eu(g) in [73HUL/DES]. The adopted datum is included in Appendix E since it was calculated with non-TDB auxiliary data.

The enthalpy of atomisation of EuSe(cr) was measured to be $\Delta_{at}H^{\circ}_{m}$ (EuSe, cr, 1788 K) = (826 ± 10) kJ·mol^{-1} by Gordienko, Fesenko, Fenochka, and Lysenko [71GOR/FES], to be $\Delta_{at}H^{\circ}_{m}$ (EuSe, cr, 1970 K) = (864 ± 10) kJ·mol^{-1} in [74HAR/EIC], and to be $\Delta_{at}H^{\circ}_{m}$ (EuSe, cr, 0 K) = (854 ± 20) kJ·mol^{-1} by Nagai, Shinmei, and Yokokawa [74NAG/SHI], all using mass spectrometry and Knudsen effusion cells. The temperature ranges of the measurements were 1665 to 1910 K and 1808 to 2131 K in [71GOR/FES] and [74HAR/EIC], respectively. The temperature range for the measurements in [74NAG/SHI] was not mentioned. The atomisation enthalpy in [71GOR/FES] and [74HAR/EIC] was converted to 298.15 K by the review using the heat capacity expression of EuSe(cr) given above, the selected heat capacity of Se(g), and the heat capacity of Eu(g) in [73HUL/DES] yielding the values $\Delta_{at}H^{\circ}_{m}$ (EuSe, cr, 298.15 K) = (842 ± 13) kJ·mol^{-1} and (884 ± 13) kJ·mol^{-1}. The value in [74NAG/SHI] was converted to 0 K by the authors of the paper without stating the start temperature of the extrapolation and the procedure for the recalculation. The temperature effect is small and the enthalpy change at 298.15 K is estimated by the review to be 2 kJ·mol^{-1} more positive than at 0 K yielding $\Delta_{at}H^{\circ}_{m}$ (EuSe, cr, 298.15 K) = (856 ± 20) kJ·mol^{-1}. The weighted average of the three values of the enthalpy of atomisation of EuSe(cr) is $\Delta_{at}H^{\circ}_{m}$ (EuSe, cr, 298.15 K) = (861.8 ± 8.4) kJ·mol^{-1}. The value was combined with the enthalpy of Eu(g) in [82WAG/EVA] and the selected enthalpy of Se(g) to yield:

$$\Delta_{f}H^{\circ}_{m}(\text{EuSe, cr, 298.15 K}) = -(450.4 \pm 10.0)\ \text{kJ·mol}^{-1}.$$

The datum is included in Appendix E since it was calculated with non-TDB auxiliary data.

V.12.9.1.2 EuSe(g)

The entropy of EuSe(g) at 298.15 K was calculated to be S°_{m} (EuSe, g, 298.15 K) = 284.7 J·K^{-1}·mol^{-1} by Kovalevskaya, Sidorenko, Lysenko, and Fesenko [75KOV/SID]. The entropy value is not selected because all parameters required for the calculation using statistical mechanics were estimated.

The dissociation enthalpy of EuSe(g) was determined to be $\Delta_{at}H^{\circ}_{m}$ (EuSe, g, 0 K) = (299 ± 15) kJ·mol^{-1} by Bergman, Coppens, Drowart, and Smoes [70BER/COP] from a study of the reaction EuSe(g) + Se(g) $\rightleftharpoons$ Eu(g) + Se_2(g) in the temperature range 1809 to 2116 K using mass spectrometry and Knudsen effusion cells. The recalculation to 0 K from high temperatures employed Gibbs energy functions calculated from estimated molecular parameters of EuSe(g). The dissociation enthalpy at 298.15 K is calculated by the review to be 2.8 kJ·mol^{-1} larger than the value at 0 K using the molecular parameters estimated for EuSe(g) in [70BER/COP], the enthalpy of Eu(g) in [82WAG/EVA] and the selected properties of Se(g) yielding $\Delta_{at}H^{\circ}_{m}$ (EuSe, g, 298.15 K) = (301.8 ± 15.0) kJ·mol^{-1}

Nagai, Shinmei, and Yokokawa [74NAG/SHI] reported $\Delta_{at}H^{\circ}_{m}$ (EuSe, g, 298.15 K) = (302 ± 20) kJ·mol^{-1} also from mass spectrometric measurements together with Knudsen effusion cells. The experimental conditions and the assumptions made in the recalculation of the high temperature results to standard conditions are poorly documented and a recalculation by the review is not possible because, among other details, no temperatures are given. The mean of the two values was combined with the enthalpy of formation of Eu(g) in [82WAG/EVA] and the selected enthalpy of Se(g) to yield:

$$\Delta_f H_m^\circ(\text{EuSe, g, 298.15 K}) = (108.9 \pm 15.6)\ \text{kJ}\cdot\text{mol}^{-1}.$$

The datum is adopted and included in Appendix E since it was calculated with non-TDB auxiliary data.

V.12.9.2 Europium selenites

No thermodynamic data have been found for europium selenites.

V.12.9.3 Europium selenates

The measurements of Serebrennikov and Tsybukova [79SER/TSY], which established the phase diagram of the system $Eu_2(SeO_4)_3$-$La_2(SeO_4)_3$-H_2O, indicate that $Eu_2(SeO_4)_3\cdot 8H_2O$ is the stable europium selenate in contact with water at 298.15 K. The reported solubility is accepted:

$$m(\text{Eu}_2(\text{SeO}_4)_3\cdot 8\text{H}_2\text{O, aq sat, 298.15 K}) = 0.608\ \text{mol}(\text{Eu}_2(\text{SeO}_4)_3)\cdot\text{kg}^{-1}.$$

No uncertainty can be assigned to this value.

Smolyakova, Efremov, and Serebrennikov [73SMO/EFR] determined the enthalpy change of the reaction between $Eu_2(SeO_4)_3\cdot 8H_2O$(cr) and a solution of barium chloride with formation of $BaSeO_4$(cr) by a calorimetric measurement. The review calculated:

$$\Delta_f H_m^\circ(\text{Eu}_2(\text{SeO}_4)_3\cdot 8\text{H}_2\text{O, cr, 298.15 K}) = -(5235.2 \pm 17.5)\ \text{kJ}\cdot\text{mol}^{-1}$$

from this measurement, see Appendix A. The result has been adopted although the source is of uncertain quality and included in Appendix E, since it was calculated with non-TDB auxiliary data.

No other thermodynamic data have been found for europium selenates.

V.12.10 Samarium compounds

V.12.10.1 Samarium selenides

V.12.10.1.1 SmSe(cr)

The enthalpy of atomisation of SmSe(cr) was determined to be $\Delta_{at} H_m^\circ$ (SmSe, cr, 1900 K) = (1006 ± 15) kJ·mol^{-1} by Gordienko, Fesenko, Fenochka, and Lysenko [71GOR/FES] and to be $\Delta_{at} H_m^\circ$ (SmSe, cr, 0 K) = (887 ± 19) kJ·mol^{-1} by Nagai, Shinmei, and Yokokawa [74NAG/SHI] by measuring the partial pressures of the species in the gas phase using mass spectrometry and Knudsen effusion cells. The temperature range of the measurements in [71GOR/FES] was 1805 to 2010 K and that in [74NAG/SHI] was not mentioned. The results in [74NAG/SHI] were extrapolated to 0 K using estimated Gibbs energy functions, while Gordienko *et al.* did not recalculate their atomisation enthalpy to standard conditions because of the lack of data for the heat capacity and entropy of SmSe(cr). The discrepancy between the two measurements is much larger than the claimed experimental uncertainties and the effect of the different temperatures. The atomisation enthalpies of SmSe(g) and NdSe(g) have been studied in [71GOR/FES] and [74NAG/SHI], and that of NdSe(g) in [70BER/COP] by Bergman, Coppens, Drowart, and Smoes. The results in [71GOR/FES] deviate strongly from the values obtained in the other two studies, which agree well. The result in [71GOR/FES] for solid SmSe(cr) is therefore rejected. The value of the enthalpy of formation of SmSe(cr) was obtained from the enthalpy of atomisation in [74NAG/SHI] by adding 2 $kJ\cdot mol^{-1}$, estimated by the review, to convert from 0 K to 298.15 K, the selected value of the enthalpy of formation of Se(g), and the enthalpy of formation of Sm(g), (206.7 ± 4.0 (estimated)) $kJ\cdot mol^{-1}$, in [82WAG/EVA] yielding:

$$\Delta_f H_m^\circ(\text{SmSe, cr, 298.15 K}) = -(447 \pm 20)\ \text{kJ}\cdot\text{mol}^{-1}.$$

The datum is included in Appendix E, since it was calculated with non-TDB auxiliary data.

V.12.10.1.2 SmSe(g)

The entropy of SmSe(g) at 298.15 K was calculated to be S_m°(SmSe, g, 298.15 K) = 281.8 $J\cdot K^{-1}\cdot mol^{-1}$ by Kovalevskaya, Sidorenko, Lysenko, and Fesenko [75KOV/SID]. The entropy value is not selected because all parameters required for the calculation using statistical mechanics were estimated.

The dissociation enthalpy of SmSe(g) was determined to be $\Delta_{at}H_m^\circ$(SmSe, g, 1950 K) = (475 ± 15) $kJ\cdot mol^{-1}$ by Gordienko, Fesenko, Fenochka, and Lysenko [71GOR/FES], and to be $\Delta_{at}H_m^\circ$(SmSe, g, 0 K) = (327 ± 15) $kJ\cdot mol^{-1}$ by Nagai, Shinmei, and Yokokawa [74NAG/SHI] by mesuring the partial pressures of the species in the gas phase using mass spectrometry and Knudsen effusion cells. The temperature range of the measurements in [71GOR/FES] was 1890 to 2010 K and that in [74NAG/SHI] was not mentioned. The results in [74NAG/SHI] were extrapolated to 0 K using estimated Gibbs energy functions while Gordienko *et al.* did not recalculate their atomisation enthalpy to standard conditions because of the lack of thermal functions. The discrepancy between the two measurements is much larger than the claimed accuracy and the effect of different temperatures. The atomisation enthalpy was estimated to be $\Delta_{at}H_m^\circ$(SmSe, g, 298.15 K) = (313 ± 45) $kJ\cdot mol^{-1}$ by Bergman, Coppens, Drowart, and Smoes [70BER/COP] by comparison with other lanthanide selenides. This value is in reasonable agreement with the value in [74NAG/SHI] and the value in

[71GOR/FES] is likely to be in error. The enthalpy of formation of SmSe(g) was calculated from the enthalpy of atomisation in [74NAG/SHI] (corrected to 298.15 K by adding 3 kJ·mol^{-1}, estimated by the review), the selected value of the enthalpy of formation of Se(g), and the enthalpy of formation of Sm(g) in [82WAG/EVA] yielding $\Delta_f H_m^\circ$ (SmSe, g, 298.15 K) = (112.2 ± 15.6) kJ·mol^{-1}. Since the accuracy of the estimated thermal data cannot be ascertained the review adopts a rounded value for the enthalpy of formation of SmSe(g) with increased error bars

$$\Delta_f H_m^\circ (\text{SmSe, g, 298.15 K}) = (112 \pm 20)\ \text{kJ·mol}^{-1}$$

The datum is included in Appendix E since it was calculated with non-TDB auxiliary data.

V.12.10.1.3 Sm_2Se_3(cr)

The heat capacity of Sm_2Se_3(cr) was measured in the temperature range 5 to 300 K by Bolgar, Kopan, and Kriklya [2001BOL/KOP]. The heat capacity and entropy resulting from the measurements are selected,

$$C_{p,m}^\circ (\text{Sm}_2\text{Se}_3, \text{cr, 298.15 K}) = (128.2 \pm 1.5)\ \text{J·K}^{-1}\text{·mol}^{-1}$$

and

$$S_m^\circ (\text{Sm}_2\text{Se}_3, \text{cr, 298.15 K}) = (221.2 \pm 3.0)\ \text{J·K}^{-1}\text{·mol}^{-1}.$$

In addition, the heat content was measured in the temperature range 1181 to 2095 K using drop calorimetry. The result was described analytically by the expression

$$C_{p,m}^\circ (\text{Sm}_2\text{Se}_3, \text{cr, (1180 – 1900) K}) = (121.85 + 36.638 \times 10^{-3}\ T - 4.0388 \times 10^4\ T^{-2})\ \text{J·K}^{-1}\text{·mol}^{-1}.$$

V.12.10.2 Samarium selenites

Aleksandrovich and Serebrennikov [62ALE/SER] measured the solubility of $Sm_2(SeO_3)_3$(s) at (298.15 ± 0.05) K in HCl, HNO_3, and H_2SO_4 of unspecified concentration. The experiments and the evaluation of the solubility product of the reaction:

$$\text{Sm}_2(\text{SeO}_3)_3(\text{s}) \rightleftharpoons 2\text{Sm}^{3+} + 3\text{SeO}_3^{2-} \qquad \text{(V.159)}$$

are described in Appendix A. The result is $\log_{10} K_{s,0}^\circ$ ((V.159), 298.15 K) = – (34.67 ± 0.28). The reported conditional value is – 32.6. The measurements probably refer to a hydrated amorphous phase. The review accepts $\log_{10} K_{s,0}^\circ$ ((V.159), 298.15 K) = – (35.0 ± 2.0), for use with a freshly precipitated samarium selenite although the quality of the datum is unproved. This datum is not selected.

Two experimental values of the enthalpy of formation of $Sm_3(SeO_3)_3$(cr) are available. The result $\Delta_f H_m^\circ$ ($Sm_2(SeO_3)_3$, cr, 298.15 K) = – (2921 ± 41) kJ·mol^{-1} was obtained from measurements of equilibria at high temperatures, see [2001OPP/ZHA] in Appendix A for details, and – (2827.1 ± 5.9) kJ·mol^{-1} from calorimetric measurements in [2002ZHA/OPP]. A short account of the calorimetric work can be found in Appendix A. Since the calorimetric result should be more accurate than the result based on a reaction datum extrapolated to standard conditions from 1000 K the review adopts:

$$\Delta_f H_m^\circ (Sm_2(SeO_3)_3, cr, 298.15 K) = -(2827 \pm 6) kJ\cdot mol^{-1}$$

and it is included in Appendix E, since it was obtained using non-TDB auxiliary data.

V.12.10.3 Samarium selenates

Serebrennikov and Tsybukova [79SER/TSY] established the phase diagram of the system $Sm_2(SeO_4)_3$-$La_2(SeO_4)_3$-H_2O and the same authors made a similar study of the system $Sm_2(SeO_4)_3$-$Nd_2(SeO_4)_3$-H_2O together with Velikov [84SER/TSY]. The measurements indicate that $Sm_2(SeO_4)_3\cdot 8H_2O$ is the stable samarium selenate in contact with water at 298.15 K. The reported solubility is accepted:

$$m(Sm_2(SeO_4)_3\cdot 8H_2O(cr), aq sat, 298.15 K) = 0.582 mol(Sm_2(SeO_4)_3)\cdot kg^{-1}.$$

No uncertainty can be assigned to this value.

Suponitskii *et al.* [69SUP/MAI] determined the enthalpy change of the reaction between $Sm_2(SeO_4)_3\cdot 8H_2O(cr)$ and a solution of barium chloride with formation of $BaSeO_4(cr)$ by a calorimetric measurement. The review calculated the enthalpy of formation:

$$\Delta_f H_m^\circ (Sm_2(SeO_4)_3\cdot 8H_2O, cr, 298.15 K) = -(5395.0 \pm 17.8) kJ\cdot mol^{-1}$$

from this measurement, see Appendix A. Since the calculation of the enthalpy of formation from the experimental results involved non-TDB auxiliary data, it is included in Appendix E.

No other thermodynamic data have been found for samarium selenates.

V.12.11 Promethium compounds

V.12.11.1 Promethium selenides

The compounds PmSe(cr) and Pm_2Se_3(cr) have been reported, but no thermodynamic information is available.

V.12.12 Neodymium compounds

V.12.12.1 Neodymium selenides

V.12.12.1.1 NdSe(cr)

The enthalpy of atomisation of NdSe(cr) was determined to be $\Delta_{at} H_m^\circ$ (NdSe, cr, 2090 K) = (1158 ± 17) kJ·mol^{-1} by Gordienko, Fesenko, Fenochka, and Lysenko [71GOR/FES] and to be $\Delta_{at} H_m^\circ$ (NdSe, cr, 0 K) = (971 ± 19) kJ·mol^{-1} by Nagai, Shinmei, and Yokokawa [74NAG/SHI] by measuring the partial pressures of the species in the gas phase using mass spectrometry and Knudsen effusion cells. The temperature range of the measurements in [71GOR/FES] was 1990 to 2190 K and that in [74NAG/SHI] was not mentioned. The results in [74NAG/SHI] were extrapolated to 0 K using estimated Gibbs energy functions while Gordienko *et al.* did not recalculate their atomisation enthalpy to standard conditions because experimental values of the heat capacity and entropy of NdSe(cr) are missing. The discrepancy between the measurements is much larger than the claimed experimental uncertainties and the effect of the different temperatures. The atomisation enthalpy of NdSe(g) has been studied in [71GOR/FES] and [74NAG/SHI], and in [70BER/COP] by Bergman, Coppens, Drowart, and Smoes. The results in [71GOR/FES] deviate strongly from the values obtained in the other two studies, which agree well. The result in [71GOR/FES] for NdSe(cr) is therefore rejected. The value of the enthalpy of formation of NdSe(cr) was calculated from the enthalpy of atomisation in [74NAG/SHI] by adding 2 $kJ\cdot mol^{-1}$, estimated by the review, to convert from 0 K to 298.15 K, the selected enthalpy of formation of Se(g), the enthalpy of formation of Nd(g), (327.6 ± 4.0 (estimated)) $kJ\cdot mol^{-1}$, in [82WAG/EVA], yielding:

$$\Delta_f H_m^\circ(\text{NdSe, cr 298.15 K}) = -(409.9 \pm 19.8)\ kJ\cdot mol^{-1}.$$

The datum is included in Appendix E since it was calculated with non-TDB auxiliary data.

V.12.12.1.2 NdSe(g)

The entropy of NdSe(g) at 298.15 K was calculated to be S_m°(NdSe, g, 298.15 K) = 277.5 $J\cdot K^{-1}\cdot mol^{-1}$ by Kovalevskaya, Sidorenko, Lysenko, and Fesenko [75KOV/SID]. The entropy value is not selected because all parameters required for the calculation using statistical mechanics were estimated.

The enthalpy change of the reaction

$$\text{NdSe(g)} + \text{Sc(g)} \rightleftharpoons \text{Nd (g)} + \text{ScSe(g)} \quad (V.160)$$

was determined by Bergman, Coppens, Drowart, and Smoes [70BER/COP] in mass spectrometric measurements of ion intensities of the gaseous species in the temperature range 1968 to 2447 K. The results were evaluated to be $\Delta_r H_m^\circ$((V.160), 0 K) = (4.7 ± 3.8) $kJ\cdot mol^{-1}$ from the second law and $\Delta_r H_m^\circ$((V.160), 0 K) = (3.7 ± 1.0) $kJ\cdot mol^{-1}$ from the third law, respectively, using Gibbs energy functions of NdSe(g) and ScSe(g) calculated from estimated molecular parameters. The weighted average, $\Delta_r H_m^\circ$((V.160), 0 K) = (3.8 ± 1.0) $kJ\cdot mol^{-1}$, was recalculated to 298.15 K by the review employing molecular parameters for NdSe(g) and ScSe(g) identical to those in [70BER/COP] and the properties of Sc(g) and Nd(g) in [82WAG/EVA] yielding $\Delta_r H_m^\circ$((V.160), 298.15 K) = (3.1 ± 1.2) $kJ\cdot mol^{-1}$. The enthalpy of formation of NdSe(g) is calculated to be $\Delta_f H_m^\circ$(NdSe, g, 298.15 K) = (174.1 ± 14.9) $kJ\cdot mol^{-1}$ by combining the reaction enthalpy with the enthalpy values of Sc(g) and Nd(g) in [82WAG/EVA] and the Appendix E value for the enthalpy of formation of ScSe(g). The corresponding atomisation enthalpy of NdSe(g) at 298.15 K is (389.6 ± 15.5) $kJ \cdot mol^{-1}$.

The dissociation enthalpy of NdSe(g) was determined to be $\Delta_{at}H^{\circ}_{m}$ (NdSe, g, 2050 K) = (544 ± 17) $kJ \cdot mol^{-1}$ by Gordienko, Fesenko, Fenochka, and Lysenko [71GOR/FES] and to $\Delta_{at}H^{\circ}_{m}$ (NdSe, g, 0 K) = (401 ± 18) $kJ \cdot mol^{-1}$ by Nagai, Shinmei, and Yokokawa [74NAG/SHI] by measuring the partial pressures of the species in the gas phase using mass spectrometry and Knudsen effusion cells. The temperature range of the measurements was 1990 to 2110 K in [71GOR/FES] while the temperature range was not mentioned in [74NAG/SHI]. The results in [74NAG/SHI] were converted to 0 K using estimated Gibbs energy functions while Gordienko *et al.* did not recalculate their atomisation enthalpy to standard conditions. The discrepancy between the values of the atomisation enthalpy in [71GOR/FES] on the one hand and those in [70BER/COP] and [74NAG/SHI] on the other is much larger than the claimed experimental uncertainties and the effect of the different temperatures. The result in [71GOR/FES] is therefore rejected. The dissociation energy at 0 K in [74NAG/SHI] was recalculated by the review to 298.15 K and the enthalpy of formation was calculated using the same thermodynamic data as for [70BER/COP] yielding $\Delta_{f}H^{\circ}_{m}$ (NdSe, g, 298.15 K) = (159.7 ± 18.5) $kJ \cdot mol^{-1}$. The weighted average of the values of the enthalpy of formation of NdSe(g) from the measurements in [70BER/COP] and [74NAG/SHI] is

$$\Delta_{f}H^{\circ}_{m}(\text{NdSe, g, 298.15 K}) = (168.4 \pm 15.0)\ \text{kJ}\cdot\text{mol}^{-1}.$$

The datum is adopted since the heat capacity and entropy calculated for diatomic molecules from estimated molecular parameters can be expected to be quite accurate. The error bars have been increased somewhat and the result included in Appendix E since it was calculated with non-TDB auxiliary data.

V.12.12.1.3 Nd_2Se_3(cr)

The heat capacity of Nd_2Se_3(cr) was measured in the temperature range 12 to 299 K by Nogteva, Paukov, and Yarembash [68NOG/PAU]. The heat capacity and entropy values at 298.15 K obtained from the investigation are selected:

$$C^{\circ}_{p,m}(\text{Nd}_2\text{Se}_3, \text{cr, 298.15 K}) = (130.2 \pm 2.0)\ \text{J}\cdot\text{K}^{-1}\cdot\text{mol}^{-1}$$

and

$$S^{\circ}_{m}(\text{Nd}_2\text{Se}_3, \text{cr, 298.15 K}) = (224.1 \pm 2.5)\ \text{J}\cdot\text{K}^{-1}\cdot\text{mol}^{-1}.$$

V.12.12.1.4 Nd_3Se_4(cr)

The heat capacity of Nd_3Se_4(cr) was measured in the temperature range 56 to 298 K by Bolgar, Kopan, and Gorbachuk [2002BOL/KOP] using adiabatic calorimetry. Their value of the heat capacity at 298.15 K is selected

$$C^{\circ}_{p,\mathrm{m}}(\mathrm{Nd_3Se_4, cr, 298.15\ K}) = (187.06 \pm 1.47)\ \mathrm{J{\cdot}K^{-1}{\cdot}mol^{-1}}.$$

The heat content of Nd_3Se_4(cr) was measured in the temperature range 488 to 2296 K by Bolgar and Kopan [2003BOL/KOP]. The authors combined their results with the heat capacity value at 298.15 K in [2002BOL/KOP] and described the heat capacity of the solid by the expression: $C^{\circ}_{p,\mathrm{m}}(\mathrm{Nd_3Se_4, cr, (298.15 - 1971)\ K}) = (156.74 + 70.04 \times 10^{-3}\ T + 8.39088 \times 10^{5}\ T^{-2})\ \mathrm{J{\cdot}K^{-1}{\cdot}mol^{-1}}$. The temperature range is limited by the melting point of Nd_3Se_4(cr) which was determined to be at 1971 K.

The entropy of Nd_3Se_4(cr) at 298.15 K was evaluated in [2002BOL/KOP] by extrapolation of the measured heat capacity to 0 K using Debye functions. The value obtained is selected by the review, but the uncertainty limits have been increased in order to account for unknown systematic errors

$$S^{\circ}_{\mathrm{m}}(\mathrm{Nd_3Se_4, cr, 298.15\ K}) = (359.6 \pm 5.8)\ \mathrm{J{\cdot}K^{-1}{\cdot}mol^{-1}}.$$

V.12.12.2 Neodymium selenites

Aleksandrovich and Serebrennikov [62ALE/SER] measured the solubility of $Nd_2(SeO_3)_3$(s) at (298.15 ± 0.05) K in HCl, HNO_3, and H_2SO_4 of unspecified concentration. The experiments and the evaluation of the solubility product of the reaction:

$$\mathrm{Nd_2(SeO_3)_3(s) \rightleftharpoons 2Nd^{3+} + 3SeO_3^{2-}} \qquad (V.161)$$

are described in Appendix A. The result is $\log_{10} K^{\circ}_{s,0}$((V.161), 298.15 K) = – (30.06 ± 0.83). The reported conditional value is – 27.8. The measurements probably refer to a hydrated amorphous phase. The review accepts $\log_{10} K^{\circ}_{s,0}$((V.161), 298.15 K) = – (30.0 ± 2.0) for use with freshly precipitated neodymium selenite although the quality of the datum is difficult to assess. This datum is not selected.

Maier, Suponitskii, and Karapet'yants [67MAI/SUP] made a calorimetric determination of the enthalpy change of the reaction between an aqueous solution of neodymium(III) chloride and Na_2SeO_3(cr). The neodymium selenite formed was amorphous. The calorimetric measurement is re-evaluated in Appendix A on the assumption that the selenite contains 8 molecules of water of crystallisation. The result is $\Delta_f H^{\circ}_{\mathrm{m}}(\mathrm{Nd_2(SeO_3)_3{\cdot}8H_2O, am, 298.15\ K}) = -(5141.9 \pm 5.3)\ \mathrm{kJ{\cdot}mol^{-1}}$. Smolyakova and Serebrennikov [78SMO/SER] measured the same reaction, but the description of the experiments and calculations is meagre. Their data have been re-evaluated as described in Appendix A and yielded $\Delta_f H^{\circ}_{\mathrm{m}}(\mathrm{Nd_2(SeO_3)_3{\cdot}8H_2O, am, 298.15\ K}) = -(5141 \pm 8)\ \mathrm{kJ{\cdot}mol^{-1}}$. As an amorphous phase is apparently formed when neodymium selenite is precipitated from aqueous solution, the result of these investigations,

$\Delta_f H_m^\circ$ ($Nd_2(SeO_3)_3 \cdot 8H_2O$, am, 298.15 K) = – (5141.5 ± 5.0) kJ·mol^{-1}, is included in the review but not selected.

Maier, Suponitskii, and Karapet'yants [71MAI/SUP] prepared crystalline $Nd_2(SeO_3)_3$ and subjected it to calorimetric measurements, see Appendix A. The re-evaluated data resulted in the value $\Delta_f H_m^\circ$ ($Nd_2(SeO_3)_3$, cr, 298.15 K) = – (2848.2 ± 5.8) kJ·mol^{-1}. Oppermann *et al.* obtained for this quantity the value –(2833.8 ± 5.4) kJ·mol^{-1} from calorimetric measurements in [2002ZHA/OPP] and – (2846 ± 41) kJ·mol^{-1} from measurements of equilibria at high temperatures in [2002OPP/ZHA].

The calorimetric results should be the most accurate but they are significantly different. The review adopts their mean value with suitable error bars:

$$\Delta_f H_m^\circ (Nd_2(SeO_3)_3, \text{cr}, 298.15\ \text{K}) = -(2841.0 \pm 15.0)\ \text{kJ·mol}^{-1}.$$

Since the calculation of the enthalpy of formation from the experimental results involved non-TDB auxiliary data, it is included in Appendix E.

V.12.12.3 Neodymium selenates

The measurements of Serebrennikov, Batyreva, and Tsybukova [78SER/BAT], [81SER/BAT] and by Serebrennikov, Tsybukova, and Velikov [84SER/TSY] indicate that at 298.15 K $Nd_2(SeO_4)_3 \cdot 5H_2O$ is the stable neodymium selenate in contact with water. The reported solubility is 0.917 and 0.91(4) mol($Nd_2(SeO_4)_3$)·kg^{-1}, respectively. It is not clear if the two values refer to independent measurements. They are widely different from the result of an earlier investigation by Friend [31FRI], who reports 0.546 mol($Nd_2(SeO_4)_3$)·kg^{-1}. He also reports on difficulties in the solubility measurements from formation of basic salts. Such problems are not mentioned in the Russian papers on the solubility of lanthanide selenates, [78SER/BAT], [79SER/TSY], [81SER/BAT], [84SER/TSY].

The solubilities of other lanthanide selenates (penta- and octahydrates) are more in line with the result obtained by Friend, while the result by Serebrennikov *et al.* appears excessive. The review cannot find the reason for the discrepancy and hence no value of the solubility of $Nd_2(SeO_4)_3 \cdot 5H_2O$ will be selected.

Suponitskii *et al.* [69SUP/MAI] and Smolyakova, Efremov, and Serebrennikov [73SMO/EFR] determined the enthalpy change of the reaction between $Nd_2(SeO_4)_3 \cdot 5H_2O$(cr) and a solution of barium chloride with formation of $BaSeO_4$(cr) by calorimetric measurements. The review calculated, *cf.* Appendix A, $\Delta_f H_m^\circ$ ($Nd_2(SeO_4)_3 \cdot 5H_2O$, cr, 298.15 K) = – (4511.1 ± 17.5) kJ·mol^{-1} from [69SUP/MAI] and – (4524.2 ± 17.5) kJ·mol^{-1} from [73SMO/EFR] and adopts:

$$\Delta_f H_m^\circ (Nd_2(SeO_4)_3 \cdot 5H_2O, \text{cr}, 298.15\ \text{K}) = -(4517.7 \pm 15.0)\ \text{kJ·mol}^{-1}.$$

Since the calculation of the enthalpy of formation from the experimental results involved non-TDB auxiliary data, it is included in Appendix E.

No other thermodynamic data have been found for neodymium selenates.

V.12.13 Praseodymium compounds

V.12.13.1 Praseodymium selenides

The phase diagram of the praseodymium-selenium system has been studied by Kalitin, Yarembash, and Luzhnaya [66KAL/YAR]. The compounds PrSe(cr), Pr_5Se_6(cr), Pr_3Se_4(cr), Pr_4Se_7(cr), $PrSe_{1.9}$(cr) and Pr_3Se_7(cr) were reported. Experimental thermodynamic information is only available for solid and gaseous PrSe and Pr_3Se_4(cr).

V.12.13.1.1 PrSe(cr)

The enthalpy of atomisation of PrSe(cr) was determined to be $\Delta_{at}H^{\circ}_{m}$ (PrSe, cr, 0 K) = (1020 ± 27) kJ·mol^{-1} by Nagai, Shinmei, and Yokokawa [74NAG/SHI] and $\Delta_{at}H^{\circ}_{m}$ (PrSe, cr, 298.15 K) = (1014.6 ± 39.3) kJ·mol^{-1} by Gordienko, Fenochka, Viksman, Klochkov, and Mikhlina [82GOR/FEN] by measuring the partial pressures of the species in the gas phase using mass spectrometry and Knudsen effusion cells. The temperature range of the measurements in [82GOR/FEN] was 2104 to 2498 K but not mentioned in [74NAG/SHI]. The conversion of the results from high to low temperatures was made using unspecified estimated Gibbs energy functions. In the absence of low temperature heat capacity data of PrSe(cr), the dissociation enthalpy is estimated by the review to be 3 kJ·mol^{-1} larger at 298.15 K than at 0 K yielding $\Delta_{at}H^{\circ}_{m}$ (PrSe, cr, 298.15 K) = (1023 ± 27) kJ·mol^{-1} from the measurements in [74NAG/SHI]. The weighted average of the results from [74NAG/SHI] and [82GOR/FEN], $\Delta_{at}H^{\circ}_{m}$ (298.15 K) = (1020.3 ± 22.3) kJ·mol^{-1}, was combined with the enthalpy of Pr(g) in [82WAG/EVA] and the selected enthalpy of Se(g) to yield:

$$\Delta_f H^{\circ}_{m} \text{(PrSe, cr, 298.15 K)} = -(428.6 \pm 22.7)\ \text{kJ·mol}^{-1}.$$

The datum is included in Appendix E, since it was calculated with non-TDB auxiliary data.

V.12.13.1.2 PrSe(g)

The entropy of PrSe(g) at 298.15 K was calculated to be S°_{m} (PrSe, g, 298.15 K) = 269.7 J·K^{-1}·mol^{-1} by Kovalevskaya, Sidorenko, Lysenko, and Fesenko [75KOV/SID]. The entropy value is not selected because all parameters required for the calculation using statistical mechanics were estimated.

The enthalpy of atomisation of PrSe(g) was determined to be $\Delta_{at}H^{\circ}_{m}$ (PrSe, g, 0 K) = (443 ± 23) kJ·mol^{-1} by Nagai, Shinmei, and Yokokawa [74NAG/SHI] and to be $\Delta_{at}H^{\circ}_{m}$ (PrSe, g, 298.15 K) = (447.4 ± 17.4) kJ·mol^{-1} by Gordienko, Fenochka, Viksman, Klochkov, and Mikhlina [82GOR/FEN] by measurements of the partial pressures of the species in the gas phase using mass spectrometry and Knudsen effusion cells. The temperature range of the measurements in [82GOR/FEN] was 2104 to

2498 K but not mentioned in [74NAG/SHI]. The conversion of the results from high to low temperatures was made using unspecified estimated Gibbs energy functions. In the absence of molecular properties of PrSe(g) the dissociation enthalpy is estimated by the review to be 2 kJ·mol^{-1} larger at 298.15 K than at 0 K yielding $\Delta_{at}H^{\circ}_{m}$ (PrSe, g, 298.15 K) = (445 ± 23) kJ·mol^{-1} from the measurements in [74NAG/SHI]. The weighted average of the results from [74NAG/SHI] and [82GOR/FEN], $\Delta_{at}H^{\circ}_{m}$ (PrSe, g, 298.15 K) = (446.5 ± 13.9) kJ·mol^{-1}, was combined with the enthalpy of Pr(g) in [82WAG/EVA] and the selected enthalpy of Se(g) to yield $\Delta_{f}H^{\circ}_{m}$ (PrSe, g, 298.15 K) = (145.2 ± 14.6) kJ·mol^{-1}. Although the results from the two determinations agree well, they have been obtained by the use of undocumented auxiliary thermal data. The review therefore adopts a rounded value for the enthalpy of formation of PrSe(g) with increased error bars

$$\Delta_{f}H^{\circ}_{m}\,(\text{PrSe, g, 298.15 K}) = (145 \pm 20)\ \text{kJ·mol}^{-1}.$$

The datum is included in Appendix E since it was calculated with non-TDB auxiliary data.

V.12.13.1.3 Pr_3Se_4(cr)

The heat capacity of Pr_3Se_4(cr) was measured in the temperature range 53 to 300 K by Bolgar, Kopan, and Gorbachuk [2002BOL/KOP] using adiabatic calorimetry. Their value of the heat capacity at 298.15 K is selected:

$$C^{\circ}_{p,m}\,(\text{Pr}_3\text{Se}_4\text{, cr, 298.15 K}) = (183.76 \pm 1.45)\ \text{J·K}^{-1}\text{·mol}^{-1}.$$

The heat content of Pr_3Se_4(cr) was measured in the temperature range 476 to 2350 K by Bolgar and Kopan [2003BOL/KOP]. The authors combined their results with the heat capacity value at 298.15 K in [2002BOL/KOP] and described the heat capacity of the solid by the expression : $C^{\circ}_{p,m}$ (Pr_3Se_4, cr, (298.15 – 1890) K) = (153.20 + 69.58 × 10^{-3} T + 8.72562 × 10^{5} T^{-2}) J·K^{-1}·mol^{-1}. The temperature range is limited by the melting point of Pr_3Se_4(cr) which was determined to be at 1890 K.

The entropy of Pr_3Se_4(cr) at 298.15 K was evaluated in [2002BOL/KOP] by extrapolation of the measured heat capacity to 0 K using Debye functions. The value obtained is selected by the review, but the uncertainty limits have been increased in order to account for unknown systematic errors:

$$S^{\circ}_{m}\,(\text{Pr}_3\text{Se}_4\text{, cr, 298.15 K}) = (371.0 \pm 6.0)\ \text{J·K}^{-1}\text{·mol}^{-1}.$$

V.12.13.2 Praseodymium selenites

Aleksandrovich and Serebrennikov [62ALE/SER] measured the solubility of $Pr_2(SeO_3)_3$(s) at (298.15 ± 0.05) K in HCl, HNO_3, and H_2SO_4 of unspecified concentration. The experiments and the evaluation of the solubility product of the reaction:

$$\text{Pr}_2(\text{SeO}_3)_3(\text{s}) \rightleftharpoons 2\text{Pr}^{3+} + 3\text{SeO}_3^{2-} \qquad \text{(V.162)}$$

are described in Appendix A. The result is $\log_{10} K^{\circ}_{s,0}$ ((V.162), 298.15 K) = – (34.19 ± 0.42). The reported conditional value is – 31.6. The measurements most likely refer to a hydrated amorphous phase. The review accepts $\log_{10} K^{\circ}_{s,0}$ ((V.162), 298.15 K) = – (34.0 ± 2.0) for use with freshly precipitated praseodymium selenite although the quality of the datum cannot be assessed. This datum is not selected.

Maier *et al.* [69MAI/SOL], [70KAR/MAI], [71MAI/SUP] prepared a number of praseodymium selenites. The enthalpies of dissolution of the selenites, $PrCl_3{\cdot}7H_2O$(cr), and H_2SeO_3(cr) in HCl(aq, 1:18.5) were measured and have been used in Appendix A to calculate the following enthalpies of formation:

$\Delta_f H^{\circ}_m$ ($Pr_2(SeO_3)_3$, cr, 298.15 K) = – (2856.8 ± 5.0) kJ·mol^{-1}; [71MAI/SUP]

$\Delta_f H^{\circ}_m$ ($Pr(HSeO_3)_3$, cr, 298.15 K) = – (2212.5 ± 2.9) kJ·mol^{-1}; [70KAR/MAI]

$\Delta_f H^{\circ}_m$ ($Pr(HSeO_3)_3{\cdot}2H_2O$, cr, 298.15 K) = – (2823.5 ± 3.0) kJ·mol^{-1}; [70KAR/MAI]

$\Delta_f H^{\circ}_m$ ($Pr_2H_2(SeO_3)_4$, cr, 298.15 K) = – (3366.2 ± 5.2) kJ·mol^{-1}; [69MAI/SOL]

$\Delta_f H^{\circ}_m$ ($Pr_2H_2(SeO_3)_4{\cdot}5H_2O$, cr, 298.15 K) = – (4864.5 ± 5.2) kJ·mol^{-1}; [69MAI/SOL].

Since the calculation of the formation enthalpies from the experimental results involved non-TDB auxiliary data, they are included in Appendix E.

The standard enthalpies of formation of $Pr_2(SeO_3)_3{\cdot}3SeO_2$, (– 3545.2), $Pr_2(SeO_3)_3{\cdot}SeO_2$, (– 3108.3), and $Pr_2O_3{\cdot}SeO_2$, (– 2302.4) kJ·mol^{-1} evaluated from [70KAR/MAI] are not accepted as the solid phases were not sufficiently well characterised in the paper.

V.12.13.3 Praseodymium selenates

The measurements of Serebrennikov, Batyreva, and Tsybukova [78SER/BAT], which established the phase diagram of the system $Pr_2(SeO_4)_3$-$Nd_2(SeO_4)_3$-H_2O, indicate that $Pr_2(SeO_4)_3{\cdot}5H_2O$ is the stable praseodymium selenate in contact with water at 298.15 K. The reported solubility is accepted:

$$m(Pr_2(SeO_4)_3{\cdot}5H_2O(cr), \text{aq sat}, 298.15\ K) = 0.459\ mol(Pr_2(SeO_4)_3){\cdot}kg^{-1}.$$

No uncertainty can be assigned to this value.

Suponitskii *et al.* [69SUP/MAI] determined the enthalpy change of the reaction between $Pr_2(SeO_4)_3{\cdot}5H_2O$(cr) and a solution of barium chloride with formation of $BaSeO_4$(cr) by a calorimetric measurement. The review calculated the enthalpy of formation:

$$\Delta_f H^{\circ}_m (Pr_2(SeO_4)_3{\cdot}5H_2O, cr, 298.15\ K) = -(4545.0 \pm 17.6)\ kJ{\cdot}mol^{-1}$$

from this measurement, see Appendix A. Since the calculation of the formation enthalpy from the experimental results involved non-TDB auxiliary data, it is included in Appendix E.

No other thermodynamic data have been found for praseodymium selenates.

V.12.14 Cerium compounds

V.12.14.1 Cerium selenides

V.12.14.1.1 CeSe(cr)

The enthalpy of atomisation of CeSe(cr) was determined to be $\Delta_{at}H_m^\circ$ (CeSe, cr, 0 K) = (1070 ± 19) kJ·mol^{-1} by Nagai, Shinmei, and Yokokawa [74NAG/SHI] by measuring the partial pressures of the species in the gas phase using mass spectrometry and Knudsen effusion cells. The temperature range of the measurements was not mentioned but the extrapolation of the results from high temperatures to 0 K was made using undocumented estimated Gibbs energy functions. In the absence of low temperature heat capacity data of CeSe(cr), the atomisation enthalpy is estimated by the review to be 3 kJ·mol^{-1} larger at 298.15 K than at 0 K yielding $\Delta_{at}H_m^\circ$ (CeSe, cr, 298.15 K) = (1073 ± 20) kJ·mol^{-1}. The enthalpy of formation of CeSe(cr) was calculated from the enthalpy of atomisation, the selected enthalpy of formation of Se(g), the enthalpy of formation of Ce(g), (423 ± 10(estimated)) kJ·mol^{-1}, in [82WAG/EVA], to be:

$$\Delta_f H_m^\circ(\text{CeSe, cr, 298.15 K}) = -(414 \pm 22)\ \text{kJ·mol}^{-1}.$$

The datum is included in Appendix E since it was calculated with non-TDB auxiliary data.

V.12.14.1.2 CeSe(g)

The entropy of CeSe(g) at 298.15 K was calculated to be S_m° (CeSe, g, 298.15 K) = 266.7 J·K^{-1}·mol^{-1} by Kovalevskaya, Sidorenko, Lysenko, and Fesenko [75KOV/SID]. The entropy value is not selected because all parameters required for the calculation using statistical mechanics were estimated.

The dissociation enthalpy of CeSe(g) was determined to be $\Delta_{at}H_m^\circ$ (CeSe, g, 0 K) = (491 ± 15) kJ·mol^{-1} by Nagai, Shinmei, and Yokokawa [74NAG/SHI] by measuring the partial pressures of the species in the gas phase using mass spectrometry and Knudsen effusion cells. The temperature range of the measurements was not mentioned but the extrapolation of the results from high temperatures to 0 K was made using undocumented estimated Gibbs energy functions. The dissociation enthalpy was estimated to be $\Delta_{at}H_m^\circ$ (CeSe, g, 0 K) = (473 ± 20) kJ·mol^{-1} by Bergman, Coppens, Drowart, and Smoes [70BER/COP] by comparison with other gaseous lanthanide monoselenides. In the absence of low temperature heat capacity data of CeSe(g), the atomisation enthalpy is estimated by the review to be 3 kJ·mol^{-1} larger at 298.15 K than at 0 K yielding $\Delta_{at}H_m^\circ$ (CeSe, g, 298.15 K) = (494 ± 21) kJ·mol^{-1}. The enthalpy of formation of CeSe(g) was calculated from the enthalpy of atomisation, the selected enthalpy of formation of Se(g), the enthalpy of formation of Ce(g), (423 ± 10(estimated)) kJ·mol^{-1}, in [82WAG/EVA], to be $\Delta_f H_m^\circ$ (CeSe, g, 298.15 K) =

[82WAG/EVA], to be $\Delta_f H_m^\circ$ (CeSe, g, 298.15 K) = (165 ± 23) kJ·mol^{-1}. Since the accuracy of the employed thermal data cannot be ascertained the review adopts a rounded value for the enthalpy of formation of CeSe(g) with increased error bars:

$$\Delta_f H_m^\circ \text{(CeSe, g, 298.15 K)} = (165 \pm 25)\ \text{kJ·mol}^{-1}.$$

The datum is included in Appendix E since it was calculated with non-TDB auxiliary data.

V.12.14.2 Cerium selenites

Two determinations of the solubility product of cerium(III) selenite are available. Little attention has been paid to the composition and quality of the solid phase, which will be denoted from lack of information as $Ce_2(SeO_3)_3$(s) only.

Chukhlantsev and Tomashevsky [57CHU/TOM] prepared cerium(III) selenite by mixing 0.1 M solutions of selenious acid and a cerium(III) salt. The mixture was neutralised with ammonia at 333 K in order to precipitate cerium selenite. The separated selenite was dried at 313 K. No information on its quality is available but it is most likely amorphous. Chemical analysis showed a 2:3 ratio between Ce(III) and Se(IV). The solubility of the specimen in dilute solution of nitric or sulphuric acid was measured at 293 K. The experiments were performed and the data recalculated as outlined in Appendix A, [56CHU]. The result for:

$$Ce_2(SeO_3)_3(s) \rightleftharpoons 2Ce^{3+} + 3SeO_3^{2-} \qquad (V.163)$$

was $\log_{10} K_{s,0}^\circ$ ((V.163), 293.15 K) = – (27.07 ± 0.37). The reported conditional constant is – (24.46 ± 0.10). Aleksandrovich and Serebrennikov [62ALE/SER] measured the solubility of $Ce_2(SeO_3)_3$(s) at (298.15 ± 0.05) K in HCl, HNO_3, and H_2SO_4 of unspecified concentration. The experiments and their evaluation are described in Appendix A. The result is $\log_{10} K_{s,0}^\circ$ ((V.163), 298.15 K) = – (31.17 ± 0.30). The reported conditional value is – 28.4.

A comparison between the primary data in [57CHU/TOM] and [62ALE/SER] reveals that the measured solubility is much the same, while the reported pH of the equilibrium solution is about one unit less in the latter work. No selection of the solubility product of cerium(III) selenite can be made due to the widely differing experimental results and vague information on the composition and quality of the solid phase.

Maier, Suponitskii, and Karapet'yants [67MAI/SUP] made a calorimetric determination of the enthalpy change of the reaction between an aqueous solution of cerium(III) chloride and Na_2SeO_3(cr). The cerium selenite formed was amorphous. As no other datum exists for the enthalpy of formation of this compound the result of the investigation is included in the review but not selected. The calorimetric measurement is re-evaluated in Appendix A on the assumption that the selenite contains 10 molecules of water of crystallisation. The result is $\Delta_f H_m^\circ$ ($Ce_2(SeO_3)_3 \cdot 10H_2O$, am, 298.15 K) = – (5732.8 ± 5.3) kJ·mol^{-1}.

V.12.15 Lanthanum compounds

V.12.15.1 Lanthanum selenides

V.12.15.1.1 LaSe(cr)

The heat capacity of LaSe(cr) was measured in the temperature range 90 to 390 K by Tikhonov, Golubkov, and Smirnov [67TIK/GOL]. The value obtained at 298.15 K is selected:

$$C^{\circ}_{p,\mathrm{m}}\,(\mathrm{LaSe, cr, 298.15\ K}) = (49.2 \pm 1.0)\ \mathrm{J{\cdot}K^{-1}{\cdot}mol^{-1}}.$$

No entropy value for LaSe(cr) at 298.15 K was selected from the measurements in [67TIK/GOL] because of the missing experimental information in the temperature range 0 to 90 K.

The enthalpy of atomisation of LaSe(cr) was measured to be $\Delta_{\mathrm{at}}H^{\circ}_{\mathrm{m}}$ (LaSe, g, 0 K) = (1062 ± 19) kJ·mol^{-1} by Nagai, Shinmei, and Yokokawa [74NAG/SHI] by measuring the partial pressures of the species in the gas phases using mass spectrometry and Knudsen effusion cells. The temperature range of the measurements was not mentioned and the extrapolation of the results from high temperatures to 0 K was made using undocumented estimated Gibbs energy functions. In the absence of complete low temperature heat capacity data of LaSe(cr), the atomisation enthalpy is estimated by the review to be 3 kJ·mol^{-1} larger at 298.15 K than at 0 K yielding $\Delta_{\mathrm{at}}H^{\circ}_{\mathrm{m}}$ (LaSe, cr, 298.15 K) = (1065 ± 20) kJ·mol^{-1}. The enthalpy of formation of LaSe(cr) was calculated from the enthalpy of atomisation, the selected enthalpy of formation of Se(g), the enthalpy of formation of La(g), (431.0 ± 3.0(estimated)) kJ·mol^{-1}, in [82WAG/EVA], to be:

$$\Delta_{\mathrm{f}}H^{\circ}_{\mathrm{m}}\,(\mathrm{LaSe, cr, 298.15\ K}) = -\,(398.5 \pm 20.6)\ \mathrm{kJ{\cdot}mol^{-1}}.$$

The datum is included in Appendix E since it was calculated with non-TDB auxiliary data.

V.12.15.1.2 LaSe(g)

The entropy of LaSe(g) at 298.15 K was calculated to be S°_{m} (LaSe, g, 298.15 K) = 263.8 J·K^{-1}·mol^{-1} by Kovalevskaya, Sidorenko, Lysenko, and Fesenko [75KOV/SID]. The entropy value is not selected because all parameters required for the calculation using statistical mechanics were estimated.

The dissociation enthalpy of LaSe(g) was determined by Ni and Wahlbeck [72NI/WAH] by measuring the partial pressures of the gas species in the temperature range 1965 to 2214 K using mass spectrometry and effusion cells. The results were evaluated to be $\Delta_{\mathrm{at}}H^{\circ}_{\mathrm{m}}$ (LaSe, g, 0 K) = (456.5 ± 42.3) kJ·mol^{-1} from the second law and $\Delta_{\mathrm{at}}H^{\circ}_{\mathrm{m}}$ (LaSe, g, 0 K) = (474.1 ± 3.4) kJ·mol^{-1} from the third law, respectively, using Gibbs energy functions of LaSe(g) calculated from estimated molecular parameters. The weighted average, $\Delta_{\mathrm{at}}H^{\circ}_{\mathrm{m}}$ (LaSe, g, 0 K) = (474.0 ± 3.4) kJ·mol^{-1}, was recalculated to

298.15 K by the review employing molecular parameters for LaSe(g) identical to those in [72NI/WAH], the selected properties of Se(g), and the properties of La(g) in [82WAG/EVA] yielding $\Delta_{at}H_m^\circ$ (LaSe, g, 298.15 K) = (476.9 ± 3.4) kJ·mol^{-1}. The enthalpy of formation of LaSe(g) was calculated from this value, the selected enthalpy of Se(g), and the enthalpy of La(g) in [82WAG/EVA] to be $\Delta_f H_m^\circ$ (LaSe, g, 298.15 K) = (190.2 ± 5.5) kJ·mol^{-1}.

Nagai, Shinmei, and Yokokawa [74NAG/SHI] also determined the dissociation enthalpy of LaSe(g) by measuring the partial pressures of the gas species using mass spectrometry and Knudsen effusion cells and reported $\Delta_{at}H_m^\circ$ (LaSe, g, 0 K) = (478.2 ± 14.6) kJ·mol^{-1}. The experimental conditions and the assumptions made for the recalculation of the high temperature results to 0 K are poorly documented. A complete recalculation by the review is not possible since, among other details, temperature values are missing. However, the dissociation enthalpy at 298.15 K is calculated by the review to be 2.9 kJ·mol^{-1} larger than the value at 0 K using the same procedure as for the results in [72NI/WAH] above, yielding $\Delta_{at}H_m^\circ$ (LaSe, g, 298.15 K) = (481.1 ± 14.6) kJ·mol^{-1}. The value was combined with the enthalpy of La(g) in [82WAG/EVA] and the selected enthalpy of Se(g) to yield $\Delta_f H_m^\circ$ (LaSe, g, 298.15 K) = (186.0 ± 15.2) kJ·mol^{-1}.

The enthalpy change of the reaction

$$\text{LaSe(g)} + \text{V(g)} \rightleftharpoons \text{La(g)} + \text{VSe(g)} \qquad \text{(V.164)}$$

was determined by Bergman, Coppens, Drowart, and Smoes [70BER/COP] in mass spectrometric measurements of ion the intensities of the gaseous species in the temperature range 1945 to 2400 K. The results were evaluated to be $\Delta_r H_m^\circ$ ((V.164), 0 K) = (130.1 ± 1.9) kJ·mol^{-1} from the second law and $\Delta_r H_m^\circ$ ((V.164), 0 K) = (130.2 ± 0.8) kJ·mol^{-1} from the third law, respectively, using Gibbs energy functions of LaSe(g) and VSe(g) calculated from estimated molecular parameters. Unfortunately, the molecular data used for VSe(g) were not mentioned and the weighted average of the reaction enthalpy, $\Delta_r H_m^\circ$ ((V.164), 0 K) = (130.2 ± 0.7) kJ·mol^{-1}, was recalculated to 298.15 K by the review employing a value of 9.5 kJ·mol^{-1} for the enthalpy increment of VSe(g) from 0 to 298.15 K, molecular parameters of LaSe(g) identical to those in [70BER/COP], and the properties of La(g) and V(g) in [82WAG/EVA] yielding $\Delta_r H_m^\circ$ ((V.164), 298.15 K) = (128.6 ± 0.9) kJ·mol^{-1}. The enthalpy of formation of LaSe(g) is calculated to be $\Delta_f H_m^\circ$ (LaSe, g, 298.15 K) = (193.2 ± 17.0) kJ·mol^{-1} by combining the reaction enthalpy with the enthalpy values of La(g) and V(g) in [82WAG/EVA] and the Appendix E value for the enthalpy of formation of VSe(g). The corresponding atomisation enthalpy of LaSe(g) at 298.15 K is (473.9 ± 17.6) kJ·mol^{-1}.

The weighted average of the determinations of the enthalpy of formation of LaSe(g) in [72NI/WAH], [74NAG/SHI], and [70BER/COP] is

$$\Delta_f H_m^\circ \text{(LaSe, g, 298.15 K)} = (190.0 \pm 5.0)\ \text{kJ·mol}^{-1}.$$

The datum is included in Appendix E since it was calculated with non-TDB auxiliary data.

V.12.15.1.3 La_2Se_3(cr)

Paukov, Nogteva, and Yarembash [66PAU/NOG] measured the heat capacity of La_2Se_3(cr) in the temperature range 13 to 295 K. The heat capacity and entropy values at 298.15 K derived from these measurements are selected:

$$C^{\circ}_{p,\mathrm{m}}(\mathrm{La_2Se_3, cr, 298.15\ K}) = (125.6 \pm 1.0)\ \mathrm{J{\cdot}K^{-1}{\cdot}mol^{-1}}$$

and

$$S^{\circ}_{\mathrm{m}}(\mathrm{La_2Se_3, cr, 298.15\ K}) = (202.2 \pm 1.8)\ \mathrm{J{\cdot}K^{-1}{\cdot}mol^{-1}}.$$

The heat content of La_2Se_3(cr) was measured in the temperature range 467 to 2209 K using drop calorimetry by Bolgar, Kopan, and Kriklya [2001BOL/KOP]. The result was described analytically by the expression

$$C^{\circ}_{p,\mathrm{m}}(\mathrm{La_2Se_3, cr, (298.15 - 2000)\ K}) = (116.45 + 27.64 \times 10^{-3}\ T + 8.0631 \times 10^{4}\ T^{-2})\ \mathrm{J{\cdot}K^{-1}{\cdot}mol^{-1}}.$$

V.12.15.2 Lanthanum selenites

Aleksandrovich and Serebrennikov [62ALE/SER] measured the solubility of $La_2(SeO_3)_3$(s) at (298.15 ± 0.05) K in HCl, HNO_3, and H_2SO_4 of unspecified concentration. The experiments and the evaluation of the solubility product of the reaction

$$\mathrm{La_2(SeO_3)_3(s)} \rightleftharpoons 2\mathrm{La^{3+}} + 3\mathrm{SeO_3^{2-}} \qquad (V.165)$$

are described in Appendix A. The result is $\log_{10} K^{\circ}_{s,0}$ ((V.165), 298.15 K) = – (31.49 ± 0.58). The reported conditional value is – 28.5. The measurements most likely refer to a hydrated amorphous phase. Reference [82WAG/EVA] reports $\Delta_f G^{\circ}_m$ ($La_2(SeO_3)_3$, cr, 298.15 K) = – 2633.7 kJ mol^{-1} which corresponds to $\log_{10} K^{\circ}_{s,0}$ = – 31.4. This is close to the value evaluated from the measurements in [62ALE/SER] by the review. Since the other results obtained in [62ALE/SER] are not included in [82WAG/EVA], the datum may originate from an unidentified source. The review accepts the solubility product $\log_{10} K^{\circ}_{s,0}$ ((V.165), 298.15 K) = – (31.0 ± 2.0) for freshly precipitated lanthanum selenite although its quality cannot be assessed. This datum is not selected.

Maier, Suponitskii, and Karapet'yants [71MAI/SUP] prepared anhydrous crystalline lanthanum selenite and measured its enthalpy of dissolution in HCl(aq, 1:18.5) in a calorimeter. The data are evaluated in Appendix A and yield

$$\Delta_f H^{\circ}_{\mathrm{m}}(\mathrm{La_2(SeO_3)_3, cr, 298.15\ K}) = -(2852.2 \pm 6.0)\ \mathrm{kJ{\cdot}mol^{-1}}$$

which is included in Appendix E, since it was calculated with non-TDB auxiliary data. The non-referenced value reported in [82WAG/EVA] is – 2879.43 kJ mol^{-1}.

V.12.15.3 Lanthanum selenates

The measurements of Serebrennikov and Tsybukova [79SER/TSY] indicate that $La_2(SeO_4)_3{\cdot}5H_2O$ is the stable lanthanum selenate in contact with water at 298.15 K. The reported solubility at this temperature is accepted:

$$m(La_2(SeO_4)_3{\cdot}5H_2O(cr), \text{aq sat}, 298.15\ K) = 0.639\ \text{mol}(La_2(SeO_4)_3){\cdot}\text{kg}^{-1}.$$

No uncertainty can be assigned to this value.

Smolyakova, Efremov, and Serebrennikov [73SMO/EFR] determined the enthalpy change of the reaction between $La_2(SeO_4)_3{\cdot}5H_2O(cr)$ and a solution of barium chloride with formation of $BaSeO_4(cr)$ by a calorimetric measurement. The review calculated:

$$\Delta_f H_m^\circ (La_2(SeO_4)_3{\cdot}5H_2O, cr, 298.15\ K) = -(4555.0 \pm 17.5)\ kJ{\cdot}mol^{-1}$$

from this measurement, see Appendix A. The result is adopted although the source is of uncertain quality and included in Appendix E, since it was calculated with non-TDB auxiliary data.

No other thermodynamic data have been found for lanthanum selenates.

V.13 Actinide compounds and complexes

V.13.1 Curium compounds

V.13.1.1 Curium selenides

The compounds CmSe(cr) and Cm_2Se_3(cr) have been reported, but no thermodynamic information is available.

V.13.2 Americium compounds

V.13.2.1 Americium selenides

The compounds AmSe(cr), $AmSe_2$(cr) and Am_3Se_4(cr) have been reported, but no thermodynamic information is available.

V.13.3 Plutonium compounds

V.13.3.1 Plutonium selenides

The compounds PuSe(cr), Pu_2Se_3(cr), and Pu_3Se_4(cr) have been reported. Experimental thermodynamic information is only available for PuSe(cr).

V.13.3.1.1 PuSe(cr)

The heat capacity of PuSe(cr) was measured in the temperature range 7 to 300 K by Hall, Mortimer, Harding and Spirlet in [91HAL/MOR] using a transient method up to 20 K and an adiabatic technique in the range 13 to 300 K. The heat capacity and entropy values determined at 298.15 K are selected as in the previous NEA-TDB reviews [2001LEM/FUG] and [2003GUI/FAN]:

$$C^{\circ}_{p,\mathrm{m}}(\mathrm{PuSe, cr, 298.15\ K}) = (59.7 \pm 1.2)\ \mathrm{J{\cdot}K^{-1}{\cdot}mol^{-1}}$$

and

$$S^{\circ}_{\mathrm{m}}(\mathrm{PuSe, cr, 298.15\ K}) = (92.1 \pm 1.8)\ \mathrm{J{\cdot}K^{-1}{\cdot}mol^{-1}}.$$

V.13.4 Neptunium compounds

V.13.4.1 Neptunium selenides

The compounds NpSe(cr), $NpSe_3$(cr), Np_2Se_3(cr), Np_2Se_5(cr), Np_3Se_4(cr), and Np_3Se_5(cr) have been reported, but no experimental thermodynamic information is available.

V.13.5 Uranium compounds

V.13.5.1 Uranium selenides

The phase equilibria in the system uranium-selenium was studied by differential thermal (DTA), chemical, and X-ray diffraction analysis by Ellert, Slovyanskikh, and Sevastyanov in [71ELL/SLO] and [75ELL/SEV]. The phases USe(cr), U_3Se_4(cr), U_2Se_3(cr), U_3Se_5(cr), $USe_{1.88}$(cr), β-USe_2, γ-USe_2, and USe_3(cr) were reported. The phase $USe_{1.88}$(cr) has been denoted α-USe_2, but it is not a polymorphic modification of uranium diselenide. The stable form of the composition USe_2 is β-USe_2, which crystallises in an orthorhombic structure. γ-USe_2 is metastable.

Thermodynamic data for uranium selenides were previously assessed in the book on uranium compounds and complexes [92GRE/FUG]. Comparisons are frequently made with this source in the following sections but a number of selections made there are not made by the review since they were based entirely on estimates.

V.13.5.1.1 USe(cr)

The heat capacity of USe(cr) has been measured in the temperature range 5 to 350 K by Takahashi and Westrum [65TAK/WES]. The reported values of the heat capacity and entropy at 298.15 K are selected:

$$C^{\circ}_{p,\mathrm{m}}(\mathrm{USe, cr, 298.15\ K}) = (54.8 \pm 0.3)\ \mathrm{J{\cdot}K^{-1}{\cdot}mol^{-1}}$$

and

$$S^\circ_m(\text{USe, cr, 298.15 K}) = (96.5 \pm 1.5)\ \text{J·K}^{-1}\text{·mol}^{-1}.$$

The expression for the heat capacity of USe(cr) in the temperature range 298.15 to 800 K selected in [92GRE/FUG] and [2003GUI/FAN] is based on an estimate made in [84GRO/DRO] from a comparison with data for US(cr). No $C^\circ_{p,m}(T)$ function for USe(cr) is selected by the review.

The enthalpies of formation of USe(cr) and U_3Se_4(cr) were measured by Baskin and Smith [70BAS/SMI] using direct synthesis calorimetry. However, the reactions were incomplete and the sample aimed at the compound USe(cr) consisted of approximately 25% of U_3Se_4(cr) in addition to USe(cr) after the reaction. Similarly, the sample aimed at the compound U_3Se_4(cr) consisted of 45% of U_3Se_4(cr), the rest being made up of 30% of USe(cr) and 25% of β-USe_2 after reaction. The β-USe_2 phase is a minority product and present only in one of the samples, and by estimating its enthalpy of formation to be – 426.8 kJ·mol^{-1}, Baskin and Smith evaluated the enthalpy of formation of USe(cr) and U_3Se_4(cr) to be – (275.7 ± 14.6) kJ·mol^{-1} and – (982.8 ± 85.4) kJ·mol^{-1}, respectively. The evaluation of the data in [70BAS/SMI] was repeated by the review by considering the constraints set by the condition that the three phases are all stable at room temperature instead of using an estimated value of the enthalpy of formation of β-USe_2. The procedure results in extreme limits for the enthalpy of formation of the phases, $-477\ \text{kJ·mol}^{-1} < \Delta_f H^\circ_m(\text{USe}_2, \beta, 298.15\ \text{K}) < -359\ \text{kJ·mol}^{-1}$, $-1115\ \text{kJ·mol}^{-1} < \Delta_f H^\circ_m(\text{U}_3\text{Se}_4, \text{cr}, 298.15\ \text{K}) < -931\ \text{kJ·mol}^{-1}$, and $-254\ \text{kJ·mol}^{-1} < \Delta_f H^\circ_m(\text{USe, cr, 298.15 K}) < -289\ \text{kJ·mol}^{-1}$. The enthalpy ranges are consistent with the values evaluated by Baskin and Smith and the review selects the mean of the extreme values of the enthalpy of formation of USe(cr):

$$\Delta_f H^\circ_m(\text{USe, cr, 298.15 K}) = -(272 \pm 18)\ \text{kJ·mol}^{-1}.$$

The error limits include the uncertainties due to the incomplete reaction.

The Gibbs energy of formation is calculated from the selected enthalpy of formation and the standard entropies to be:

$$\Delta_f G^\circ_m(\text{USe, cr, 298.15 K}) = -(273.3 \pm 18.0)\ \text{kJ·mol}^{-1}.$$

The selected values of the thermodynamic quantities are in close agreement with those selected in [92GRE/FUG], retained in [2003GUI/FAN].

V.13.5.1.2 $USe_{1.88}$(cr)

The compound $USe_{1.88}$(cr) crystallises with tetragonal symmetry and has also been denoted α-USe_2. However, according to the X-ray and phase diagram studies by Ellert, Slovyanskikh, and Sevastyanov in [71ELL/SLO] and [75ELL/SEV], $USe_{1.88}$(cr) is a separate phase and not a polymorphic modification of uranium diselenide. This condition has not been known to all investigators of the system and has lead to confusing results for the phases $USe_{1.88}$(cr) (α-USe_2) and β-USe_2. The latter is orthorhombic and of exact stoichiometry.

Westrum and Grønvold [70WES/GRO] measured the low temperature heat capacity of a phase anticipated to be α-USe_2. The phase was not identified using X-ray diffraction, but the composition was determined to be exactly stoichiometric USe_2 by chemical analysis. Considering the synthesis method and the results in [71ELL/SLO] and [75ELL/SEV], the phase investigated was most probably β-USe_2. The results of the measurements in [70WES/GRO] are therefore attributed to β-USe_2 rather than $USe_{1.88}$(cr) by the review. However, the heat capacity and entropy at 298.15 K derived from the measurements in [70WES/GRO] were selected for the α-USe_2 phase in [92GRE/FUG]. Baskin and Smith [70BAS/SMI] estimated the enthalpy of formation of USe_2(cr) and the value was accepted for the α-USe_2 phase in [92GRE/FUG].

Attributing the measurements in [70WES/GRO] to β-USe_2, no experimental thermodynamic information is available for the $USe_{1.88}$(cr) phase. The data for the compound denoted by α-USe_2 in [92GRE/FUG] should therefore most likely be allotted to β-USe_2 instead.

V.13.5.1.3 β-USe_2

The heat capacity of USe_2(cr) was measured in the temperature range 5 to 350 K by Westrum and Grønvold [70WES/GRO]. The authors anticipated the phase to be tetragonal α-USe_2 without identifying the phase using X-ray diffraction. However, as discussed in the previous section, the phase investigated was most likely β-USe_2. The results of the investigation are therefore attributed to β-USe_2 by the review. No anomalies were found in the heat capacity except for a λ-type transformation with $\Delta_{trs}S_m$ = 0.79 $J\cdot K^{-1}\cdot mol^{-1}$ at 13.1 K. The heat capacity and entropy at 298.15 K evaluated from the measurements are selected:

$$C^{o}_{p,m}(USe_2, \beta, 298.15\ K) = (79.2 \pm 0.3)\ J\cdot K^{-1}\cdot mol^{-1}$$

and

$$S^{o}_{m}(USe_2, \beta, 298.15\ K) = (133.8 \pm 1.5)\ J\cdot K^{-1}\cdot mol^{-1}.$$

The error limits of the entropy were increased in order to incorporate a possible zero-point entropy of 1.17 $J\cdot K^{-1}\cdot mol^{-1}$. These selections agree with the selections in [92GRE/FUG] but there considered to apply to α-USe_2.

Baskin and Smith [70BAS/SMI] measured the enthalpy of formation USe(cr) and U_3Se_4(cr) using direct reaction calorimetry and found minor amounts of β-USe_2 as a by-product in one of their samples. In order to evaluate their results, the enthalpy of formation of β-USe_2 was estimated to be – 426.8 $kJ\cdot mol^{-1}$. The measurements were re-evaluated by the review as discussed in Section V.13.5.1.1 without using an estimated value, yielding extreme limits for the enthalpy of formation of β-USe_2, – 477 $kJ\cdot mol^{-1}$ <

$\Delta_f H^\circ_m$ (USe_2, β, 298.15 K) < – 359 kJ·mol^{-1}. No value is selected because the β-USe_2 phase was only found in small quantities as a by-product. The enthalpy estimated in [70BAS/SMI] was accepted in [92GRE/FUG] and there considered to apply to α-USe_2.

V.13.5.1.4 USe_3(cr)

The selenium vapour pressure over a mixture of USe_2(cr) and USe_3(cr) was measured in the temperature range 1023 to 1178 K by Sevastyanov, Slovyanskikh, and Ellert in [71SEV/SLO]. The results were combined with an estimated value of the enthalpy of formation of USe_2(cr) in [92GRE/FUG] to yield an accepted value of the enthalpy of formation of USe_3(cr). Similarly, the reviews in [92GRE/FUG] and [2003GUI/FAN] accepted the value of the entropy of USe_3(cr) estimated by Westrum and Grønvold [62WES/GRO].

The review does not derive a value for the enthalpy of formation of USe_3(cr) since no experimental value for the enthalpy of formation of USe_2(cr) is available.

V.13.5.1.5 U_2Se_3(cr)

The heat capacity of U_2Se_3(cr) was measured in the temperature range 2 to 293 K by Lagnier, Suski, and Wojakowski [75LAG/SUS]. Several anomalies and peculiarities were found and no numerical values of the entropy and heat capacity were reported. The values $C^\circ_{p,m}$ (U_2Se_3, cr, 298.15 K) = (84.4 ± 3.0) J·K^{-1}·mol^{-1} and S°_m (U_2Se_3, cr, 298.15 K) = (131.6 ± 1.5) J·K^{-1}·mol^{-1} were evaluated by the review from the graph in Figure 1 of the paper. The values are similar to those of β-USe_2 and consequently too small and in error. The IAEA review [84GRO/DRO] reports an entropy which is twice as large, S°_m (U_2Se_3, cr, 298.15 K) = (261.4 ± 1.7) J·K^{-1}·mol^{-1}, referring to the same experimental data. The resulting entropy of formation is $\Delta_f S^\circ_m$ (U_2Se_3, cr, 298.15 K) = 34.73 J·K^{-1}·mol^{-1} which is out of line with the corresponding values of USe(cr) and β-USe_2, 4.21 J·K^{-1}·mol^{-1} and – 0.58 J·K^{-1}·mol^{-1}, respectively. No values of the heat capacity and entropy at 298.15 K were selected. The review concludes that the values selected in [92GRE/FUG] are likely to be in error.

The review in [92GRE/FUG] accepted the value of the enthalpy of formation of U_2Se_3(cr) estimated by Mills [74MIL].

V.13.5.1.6 U_3Se_4(cr)

The enthalpy of formation of USe(cr) and U_3Se_4(cr) was measured by Baskin and Smith [70BAS/SMI] using direct reaction calorimetry. However, the reactions were incomplete and a minor amount of β-USe_2 was found as a by-product in one of the samples. In order to correct for this, the enthalpy of formation of β-USe_2 was estimated to be – 426.8 kJ·mol^{-1} resulting in $\Delta_f H^\circ_m$ (U_3Se_4, cr, 298.15 K) = – (983 ± 56) kJ·mol^{-1}. The measurements were re-evaluated by the review as discussed in Section V.13.5.1.1 without using an estimated value, yielding extreme limits for the enthalpy of formation of

U_3Se_4(cr), – 1115 kJ·mol^{-1} < $\Delta_f H_m^\circ$ (U_3Se_4, cr, 298.15 K) < – 931 kJ·mol^{-1}. The range is consistent with the value obtained by Baskin and Smith. Their result was selected in [92GRE/FUG]. The review prefers a value that does not include an estimate and selects for the enthalpy of formation of U_3Se_4(cr) the mean value of the extremes of the range:

$$\Delta_f H_m^\circ (U_3Se_4, \text{cr}, 298.15\text{ K}) = -(1023 \pm 92)\text{ kJ·mol}^{-1}.$$

The error limits reflect the uncertainties due to the incomplete reaction.

The reviews in [92GRE/FUG] and [2003GUI/FAN] accepted the value of the entropy of U_3Se_4(cr) estimated by Westrum and Grønvold [62WES/GRO].

V.13.5.1.7 U_3Se_5(cr)

No experimental thermodynamic information is available for U_3Se_5(cr). The reviews in [92GRE/FUG] and [2003GUI/FAN] accepted the value of the entropy of U_3Se_5(cr) estimated by Westrum and Grønvold [62WES/GRO] and the value of the enthalpy of formation estimated by Mills [74MIL].

V.13.5.2 Uranium selenites

No phase diagram has been found for the system UO_2SeO_3-H_2O-SeO_2.

Krylov and Chukhlantsev [57KRY/CHU] prepared dioxouranium(VI) selenite at 323 K by mixing 1 M solutions of dioxouranium(VI) acetate and selenious acid. The precipitate was aged for several days in the mother liquor at this temperature, then separated and dried at 373 K. Chemical analysis confirmed the 1:1 ratio between U and Se(IV). The specimen was anhydrous. No X-ray diffraction measurements were performed. The information presented by Claude [60CLA] indicates that no water of crystallisation is present in dioxouranium(VI) selenite precipitated from aqueous solution although the amorphous product tenaciously withholds water. The solubility of the specimen in dilute solution of nitric or sulphuric acid was measured at 293 K. The data have been recalculated by the review, *cf.* Appendix A, [56CHU]. The result for:

$$UO_2SeO_3(s) \rightleftharpoons UO_2^{2+} + SeO_3^{2-} \qquad (V.166)$$

was $\log_{10} K_{s,0}^\circ$ ((V.166), 293.15 K) = – (11.10 ± 0.18). The reported conditional value is – (10.41 ± 0.12). This is the only determination of the solubility of uranium(VI) selenite found. The estimate $\log_{10} K_{s,0}^\circ$ ((V.166), 298.15 K) = – (11.0 ± 2.0) has been included in the review and might be applicable to an aged, amorphous uranium(VI) selenite phase. This datum is not selected.

Cordfunke and Ouweltjes [77COR/OUW2] determined the enthalpy of formation of dioxouranium(VI) selenite using a specimen obtained by a solid-state synthesis at 630 K. The selected value:

$$\Delta_f H_m^\circ (UO_2SeO_3, \text{cr}, 298.15\text{ K}) = -(1522.2 \pm 1.1)\text{ kJ·mol}^{-1}$$

differs only from the value given in [77COR/OUW2] and that selected in [92GRE/FUG] by a small difference due to the selected enthalpy of formation of SeO_2(cr).

V.13.5.3 Uranium selenates

The phase diagram studies by Serezhkina *et al.* show that the stable solid phase in contact with water at 298.15 K is $UO_2SeO_4{\cdot}4H_2O$. The solubility is:

$$m(UO_2SeO_4{\cdot}4H_2O(cr), \text{aq sat}, 298.15\text{ K}) = 5.31\text{ mol}(UO_2SeO_4){\cdot}kg^{-1}.$$

A phase diagram and related information are available for the system: UO_2SeO_4-H_2SeO_4-H_2O at 298.15 K in [88SER/BLA]; UO_2SeO_4-$BeSeO_4$-H_2O at 298.15 K in [83SER/KUD]; UO_2SeO_4-$MgSeO_4$-H_2O at 298.15 K in [84SER/SER]; UO_2SeO_4-$(NH_4)_2SeO_4$-H_2O at 298.15 K in [94SER]; UO_2SeO_4-Li_2SeO_4-H_2O at 298.15 K in [89TAT/SER]; UO_2SeO_4-Na_2SeO_4-H_2O at 298.15 K in [90TAT/SER]; UO_2SeO_4-K_2SeO_4-H_2O at 298.15 K in [93SER/KUC]; UO_2SeO_4-Rb_2SeO_4-H_2O at 298.15 K in [89KUC/SHT]; UO_2SeO_4-Cs_2SeO_4-H_2O at 298.15 K in [87SER/SER].

The value of the solubility product of $UO_2SeO_4{\cdot}4H_2O$(cr),

$$UO_2SeO_4{\cdot}4H_2O(cr) \rightleftharpoons UO_2^{2+} + SeO_4^{2-} + 4H_2O(l) \qquad (V.167)$$

at 298.15 K has been reported by Kumov and Batyreva [90KUM/BAT] to be $\log_{10} K^{\circ}_{s,0}$ ((V.167), 298.15 K) = – (2.25 ± 0.04), see Appendix A. As the reviewers failed to assess the accuracy of this datum, it will not be included amongst the selected data.

Complex formation in the dioxouranium(VI)-selenate system has been studied by Lubal and Havel [97LUB/HAV] by spectrophotometric and potentiometric methods at (298.2 ± 0.5) K. The evidence for the formation of UO_2SeO_4(aq) and $UO_2(SeO_4)_2^{2-}$ is convincing, see Appendix A. The following equilibrium constants of Reactions (V.168) and (V.169):

$$UO_2^{2+} + SeO_4^{2-} \rightleftharpoons UO_2SeO_4(aq) \qquad (V.168)$$

$$UO_2^{2+} + 2SeO_4^{2-} \rightleftharpoons UO_2(SeO_4)_2^{2-} \qquad (V.169)$$

were determined by spectrophotometry in 3 M $Na^+(ClO_4^-, SeO_4^{2-})$ medium, $\log_{10} \beta_1$ ((V.168), I = 3 M, 298.15 K) = (1.50 ± 0.02) and $\log_{10} \beta_2$ ((V.169), I = 3 M, 298.15 K) = (2.29 ± 0.02). The authors obtained $\log_{10} \beta_1^{\circ}$ ((V.168), I = 0, 298.15 K) = (2.64 ± 0.01) from the potentiometric measurements at low ionic strength. Djogic, Pizeta, and Zelic [99DJO/PIZ] made a careful voltammetric study of the same system at (296 ± 2) K in sodium perchlorate medium. From the shift of the peak potential with selenate concentration they obtained $\log_{10} \beta_1$ ((V.168), I = 3 M, 296 K) = (1.50 ± 0.02) in agreement with the result in [97LUB/HAV]. However, no evidence was found for the formation of $UO_2(SeO_4)_2^{2-}$. At I = 0.1 M, $\log_{10} \beta_1$ ((V.168), I = 0.1 M, 296 K) = (2.0 ± 0.1) is reported together with evidence for the formation of a second complex but its formation constant could not be determined due to the short accessible concentration range of the ligand.

Extrapolation of $\log_{10} \beta_1$ ((V.168), I = 3 M) by the SIT approach with $\Delta\varepsilon$ = – (0.34 ± 0.07) kg·mol^{-1} as for the corresponding sulphate complex, [92GRE/FUG], yields $\log_{10} \beta_1^{\circ}$ = (2.31 ± 0.25) and similarly $\log_{10} \beta_1^{\circ}$ = (2.84 ± 0.15) from $\log_{10} \beta_1$ ((V.168), I = 0.1 M). The mean of the latter value and the potentiometric result will be selected:

$$\log_{10} \beta_1^{\circ} \text{ ((V.168), 298.15 K)} = (2.74 \pm 0.25).$$

The correspondingly extrapolated value of β_2 is $\log_{10} \beta_2^{\circ}$ = (3.10 ± 0.50). This value cannot be selected due to lack of data supporting the long extrapolation. The potentiometric data in [97LUB/HAV] indicated that $\log_{10} \beta_2^{\circ}$ does not exceed 3.4. This selection yields:

$$\Delta_f G_m^{\circ} (\mathrm{UO_2SeO_4}, \text{aq}, 298.15 \text{ K}) = -(1407.7 \pm 2.7) \text{ kJ·mol}^{-1}.$$

The enthalpy of formation of dioxouranium(VI) selenate selected in [92GRE/FUG] and retained in [2003GUI/FAN] is accepted:

$$\Delta_f H_m^{\circ} (\mathrm{UO_2SeO_4}, \text{cr}, 298.15 \text{ K}) = -(1539.3 \pm 3.3) \text{ kJ·mol}^{-1}.$$

No value of the standard enthalpy of formation of $UO_2SeO_4 \cdot 4H_2O$(cr) has been found.

V.13.6 Thorium compounds

V.13.6.1 Thorium selenides

The compounds ThSe(cr), Th_2Se_3(cr), Th_2Se_5(cr), and Th_7Se_{12}(cr) have been reported, but no experimental thermodynamic information is available.

V.13.6.2 Thorium selenites

Krylov and Chukhlantsev [57KRY/CHU] prepared thorium selenite at 323 K by adding a concentrated solution of selenious acid to a 1 M solution of thorium chloride. The precipitate was aged for several days in the mother liquor at this temperature, then separated and dried at 373 K. Chemical analysis confirmed the 1:2 ratio between Th and Se(IV) and the specimen was anhydrous. No X-ray diffraction measurements were performed.

The solubility of the specimen in dilute solution of nitric or sulphuric acid was measured at 293 K. The data at pH 3 or below have been recalculated by the review, *cf.* Appendix A, [56CHU]. A correction for the hydrolysis of the Th^{4+} ion was applied. The result for:

$$\mathrm{Th(SeO_3)_2(s)} \rightleftharpoons \mathrm{Th^{4+}} + 2\mathrm{SeO_3^{2-}} \quad \text{(V.170)}$$

was $\log_{10} K^{\circ}_{s,0}$ ((V.170), 293.15 K) = – (21.92 ± 0.67). The reported conditional value is – (19.87 ± 0.62). The estimate $\log_{10} K^{\circ}_{s,0}$ ((V.170), 298.15 K) = – (22.0 ± 2.0) has been included in the review as the only datum available for the solubility product of thorium selenite but not selected. Its applicability to equilibrium calculations is somewhat doubtful, as the composition of the solid phase is not well defined. Selenious acid is known to precipitate amorphous hydrates from Th^{4+} solution, which turn into basic salts when pH increases. According to the results in [99NES/GLY] discussed in the next paragraph, the anhydrous $Th(SeO_3)_2$ used in the solubility measurements should change to $Th(SeO_3)_2{\cdot}H_2O$ in contact with water.

Nesterenko and Glybin [99NES/GLY] prepared crystalline $Th(SeO_3)_2{\cdot}H_2O$ from aqueous solution. The vapour pressure of the salt:

$$Th(SeO_3)_2{\cdot}H_2O(cr) \rightleftharpoons Th(SeO_3)_2(cr) + H_2O(g) \qquad (V.171)$$

was measured with a membrane manometer in the temperature range 428 to 515 K and found to follow:

$$\log_{10} p_{H_2O} = (9.8 \pm 0.3) - (4862 \pm 121)\, T^{-1}$$

The extrapolated data estimated in the paper are:

$$\Delta_r H^{\circ}_m ((V.171), 298.15\ K) = (94.1 \pm 2.9)\ kJ{\cdot}mol^{-1},$$

$$\Delta_r S^{\circ}_m ((V.171), 298.15\ K) = (191.2 \pm 4.5)\ J{\cdot}K^{-1}{\cdot}mol^{-1}.$$

The auxiliary data used are not mentioned. The result is selected, as the extrapolation is comparatively short.

The paper also reports $\Delta_f H^{\circ}_m$ ($Th(SeO_3)_2$, cr, 298.15 K) = – (1702.5 ± 5.0) $kJ{\cdot}mol^{-1}$. This datum will not be selected as it was obtained by extra-thermodynamic methods.

V.14 Group 2 compounds and complexes

V.14.1 Beryllium compounds

V.14.1.1 Beryllium selenides

No experimental thermodynamic information is available for beryllium selenides.

V.14.1.1.1 BeSe(cr)

Mills [74MIL] estimated the enthalpy of formation and entropy of BeSe(cr) to be $\Delta_f H^{\circ}_m$ (BeSe, cr, 298.15 K) = – (167.4 ± 42.0) $kJ{\cdot}mol^{-1}$ and S°_m (BeSe, cr, 298.15 K) = (46.0 ± 8.4) $J{\cdot}K^{-1}{\cdot}mol^{-1}$, respectively. No values are selected by the review.

V.14.1.2 Beryllium selenites

Yanitskii and Stashene [85YAN/STA] studied the phase diagram of the system Be-SeO_3-H_2SeO_3-H_2O by solubility measurements at (298.15 ± 0.1) K over the interval 0 to 80 mass-% H_2SeO_3 in the liquid phase. The tie lines and chemical analysis of the solids showed the presence of $BeSeO_3\cdot 2H_2O$, $BeSeO_3\cdot H_2O$, $Be(HSeO_3)_2\cdot H_2O$, and $Be(HSeO_3)_2\cdot H_2SeO_3$ as solid phases. Infrared spectra indicated formation of hydrogen selenites and absence of the diselenite ion.

The authors also measured the solubility of $BeSeO_3\cdot 2H_2O$, $Be(HSeO_3)_2\cdot H_2O$, and $Be(HSeO_3)_2\cdot H_2SeO_3$ in water between about 293 and 353 K. No pH measurements were performed. The results are only presented in a small graph and will not be accepted by the review as quantitative information. The graph indicates that $BeSeO_3\cdot 2H_2O$ is slightly soluble whereas the acid salts are readily soluble.

Ripan and Vericeanu [68RIP/VER] studied the solubility of beryllium selenite ($BeSeO_3\cdot 2H_2O$?) by conductivity measurements as described in Appendix A. The Be^{2+} ion forms strong hydroxo complexes. A calculation based on the assumptions made in the Appendix together with the total concentration, mean value 1.0×10^{-4} M, in the paper leads to the conclusion that the major part of Be(II) would be present as hydroxo complexes at equilibrium. Hence no solubility product can be calculated from the data in [68RIP/VER].

No other experimental studies associated with thermodynamic data of beryllium selenites have been found.

V.14.1.3 Beryllium selenates

The stable form of beryllium selenate in contact with water at 298.15 K is $BeSeO_4\cdot 4H_2O$. The selected solubility is based on the measurements in [26SID/LEW] and [83SER/KUD]:

$$m(BeSeO_4\cdot 4H_2O(cr), \text{aq sat}, 298.15\text{ K}) = (3.73 \pm 0.02)\text{ mol}(BeSeO_4)\cdot\text{kg}^{-1}.$$

The value of the solubility product of $BeSeO_4\cdot 4H_2O(cr)$,

$$BeSeO_4\cdot 4H_2O(cr) \rightleftharpoons Be^{2+} + SeO_4^{2-} + 4H_2O(l) \qquad (V.172)$$

at 298.15 K has been reported by Kumov and Batyreva [90KUM/BAT] to be $\log_{10} K^{\circ}_{s,0}$ ((V.172), 298.15 K) = – (2.94 ± 0.06), see Appendix A. As the reviewers failed to assess the accuracy of this datum, it will not be included amongst the selected data.

Selivanova and Shneider [61SEL/SHN3] measured the enthalpy change of the reaction $BeSeO_4\cdot 4H_2O(cr) + BaCl_2(aq, 1:900) \rightarrow BaSeO_4(cr) + BeCl_2(aq, 1:900) + 4H_2O(l)$ in a calorimeter. Similarly they determined the integral enthalpy of dissolution of $BeSeO_4\cdot 4H_2O(cr)$ and $BeSeO_4\cdot 2H_2O(cr)$ to $BeSeO_4$(aq, 1:900), and the enthalpy of dissolution (1:3600) in 1 M KOH of $BeSeO_4{\cdot}2H_2O(cr)$ and $BeSeO_4(cr)$. The review obtained from the experimental data, see Appendix A:

$$\Delta_f H_m^\circ (BeSeO_4{\cdot}4H_2O, cr, 298.15 K) = -(2107.4 \pm 6.8) kJ{\cdot}mol^{-1},$$

$$\Delta_f H_m^\circ (BeSeO_4{\cdot}2H_2O, cr, 298.15 K) = -(1501.9 \pm 6.8) kJ{\cdot}mol^{-1},$$

$$\Delta_f H_m^\circ (BeSeO_4, cr, 298.15 K) = -(884.6 \pm 6.9) kJ{\cdot}mol^{-1}.$$

These results, which agree with [82WAG/EVA] within the uncertainties, are included in Appendix E since they were calculated with non-TDB auxiliary data.

V.14.2 Magnesium compounds

V.14.2.1 Magnesium selenides

No experimental thermodynamic information is available for magnesium selenides.

V.14.2.1.1 MgSe(cr)

Mills [74MIL] estimated the enthalpy of formation and entropy of MgSe(cr) to be $\Delta_f H_m^\circ$(MgSe, cr, 298.15 K) = $-(293 \pm 42)$ kJ·mol^{-1} and S_m°(MgSe, cr, 298.15 K) = (62.8 ± 8.4) J·K^{-1}·mol^{-1}, respectively. No values are selected by the review.

V.14.2.2 Magnesium selenites

Ebert and Havlicek [80EBE/HAV] established the phase diagram of the system $MgSeO_3$-H_2SeO_3-H_2O at 298.15 K from solubility measurements. Crystallisation fields of $MgSeO_3{\cdot}6H_2O$ and $Mg(HSeO_3)_2{\cdot}4H_2O$ are present in the diagram. The latter compound, which is incongruently soluble in water, was shown by its infrared spectrum to be a hydrogen selenite and not a diselenite. No primary solubility data are published except the compositions of the two invariant points. Stashene and Yanitskii [86STA/YAN] made a similar study also at 298.15 K, which confirmed the formation of $MgSeO_3{\cdot}6H_2O$ and $Mg(HSeO_3)_2{\cdot}4H_2O$. In addition $Mg(HSeO_3)_2{\cdot}3H_2O$ was observed to form at concentrations of selenious acid greater than 45.5 mass-%. As no primary data are presented in [80EBE/HAV] a closer comparison between the two studies cannot be made. The phase diagram at 348.15 K was also established.

V.14.2.2.1 MgSeO₃·6H₂O(cr)

The solubility product of magnesium selenite has been determined in a number of studies in which the solid phase has been denoted as $MgSeO_3$. The review assumes that the determinations pertain to the hexahydrate and writes:

$$MgSeO_3{\cdot}6H_2O(cr) \rightleftharpoons Mg^{2+} + SeO_3^{2-} + 6H_2O(l) \qquad (V.173)$$

Chukhlantsev [56CHU] measured the solubility in dilute solution of nitric or hydrochloric acid at 293 K. The data have been recalculated as described in Appendix A, [56CHU]. The result was $\log_{10} K^{\circ}_{s,0}$ ((V.173), 293.15 K) = – (5.52 ± 0.24). The reported conditional constant is – (4.88 ± 0.10). Ripan and Vericeanu [68RIP/VER] studied the solubility of the salt by conductivity measurements as described in Appendix A. Hydrolysis can be neglected for Mg^{2+}, [76BAE/MES], and the original data, mean solubility 1.38×10^{-3} M, were assumed to provide a measure of the solubility in pure water. The solubility product calculated from the data with a correction for the protolysis of the selenite ion and extrapolated to $I = 0$ by the SIT expression ($\Delta\varepsilon = 0$ kg·mol^{-1}) is $\log_{10} K^{\circ}_{s,0}$ ((V.173), 291 K) = – (6.01 ± 0.25).

Savenko [95SAV] measured the solubility of crystalline magnesium selenite in water. The pH was adjusted between 8.2 and 7.5 by small additions of strong acid or base. Both the selenium(IV) and magnesium concentrations were determined at equilibrium. The experimental temperature was 293.15 K. The data have been recalculated with the protonation constant of the selenite ion selected here and activity coefficients were found from the SIT expression with $\varepsilon = 0$ kg·mol^{-1}. The result was $\log_{10} K^{\circ}_{s,0}$ ((V.173), 293.15 K) = – (5.88 ± 0.05). No trend in the value of the solubility product with pH was observed. Sharmasakar, Reddy, and Vance [96SHA/RED] obtained $\log_{10} K^{\circ}_{s,0}$ ((V.173), 298.2 K) = – 7.56 from measurements of the solubility of crystalline $MgSeO_3$ in aqueous media. For reasons presented in Appendix A, this result has been given zero weight.

The review selects the result obtained by Savenko [95SAV] as the most reliable and calculates at standard conditions:

$$\log_{10} K^{\circ}_{s,0}\,((\text{V.173}), 298.15\ \text{K}) = -(5.82 \pm 0.25).$$

The uncertainty is estimated by the review.

Leshchinskaya and Selivanova [66LES/SEL] measured the enthalpy change of the reaction between Na_2SeO_3(cr) and a solution of magnesium sulphate with formation of $MgSeO_3\cdot 6H_2O$(cr). Their data have been re-evaluated in Appendix A with new auxiliary data and a minor correction for incomplete conversion in the studied reaction. The result is:

$$\Delta_f H^{\circ}_m (\text{MgSeO}_3\cdot 6\text{H}_2\text{O, cr, 298.15 K}) = -(2707.21 \pm 1.30)\ \text{kJ}\cdot\text{mol}^{-1}.$$

The selected data yield:

$$\Delta_f G^{\circ}_m (\text{MgSeO}_3\cdot 6\text{H}_2\text{O, cr, 298.15 K}) = -(2273.83 \pm 2.60)\ \text{kJ}\cdot\text{mol}^{-1},$$

$$S^{\circ}_m (\text{MgSeO}_3\cdot 6\text{H}_2\text{O, cr, 298.15 K}) = (328.5 \pm 9.8)\ \text{J}\cdot\text{K}^{-1}\cdot\text{mol}^{-1}.$$

V.14.2.2.2 $MgSeO_3$(cr)

Leshchinskaya and Selivanova [66LES/SEL] prepared $MgSeO_3$(cr) and measured the enthalpy of dissolution of $MgSeO_3$(cr), – (70.00 ± 0.17), and $MgSeO_3\cdot 6H_2O$(cr),

(21.88 ± 0.33) kJ·mol^{-1} in 8% HNO_3. The data combine with $\Delta_f H_m^\circ$ ($MgSeO_3 \cdot 6H_2O$, cr, 298.15 K) to give $\Delta_f H_m^\circ$ ($MgSeO_3$, cr, 298.15 K) = – (900.34 ± 1.35) kJ·mol^{-1}.

Bakeeva, Buketov, and Pashinkin [69BAK/BUK] investigated the equilibrium:

$$MgSeO_3(cr, l) \rightleftharpoons MgO(cr) + SeO_2(g) \qquad (V.174)$$

by the dew point technique in the temperature range 921 to 1175 K. Magnesium selenite melts at about 1060 K. The vapour pressure relationship used to calculate p_{SeO_2} is not presented in the paper. From data in [71PAS/ARO] it could be deduced that the vapour pressure expression in [62MAR/GET] had been used for this purpose. The published pressures have been recalculated accordingly with the selected expression for the saturated vapour pressure of SeO_2(cr), see [71PAS/ARO] in Appendix A. The result for $MgSeO_3$(cr) and the temperature interval 921 to 1060 K is $\log_{10} p(SeO_2, g\ T) = (9.786 \pm 0.172) - (11600 \pm 167)\, T^{-1}$ and for $MgSeO_3$(l) $\log_{10} p(SeO_2, g\ T) = (8.710 \pm 0.199) - (10455 \pm 225)\, T^{-1}$ in the temperature interval 1067 to 1175 K. The uncertainties are entered here as twice the standard deviations from the least-squares calculation. The data yield T_{fus} = 1063 K and:

$$\Delta_{fus} H_m^\circ (MgSeO_3, cr, 1063\ K) = (21.9 \pm 5.4)\ kJ \cdot mol^{-1}.$$

The high temperature data have been extrapolated to standard conditions with $C_{p,m}^\circ$ values in [73BAR/KNA] (no phase transitions assumed) leading to $\Delta_r C_{p,m}^\circ$ = $(25.1 - 41.0 \times 10^{-3}\, T - 21.3 \times 10^5\, T^{-2})$ J·K^{-1}·mol^{-1} and the following estimates, $\Delta_f H_m^\circ$ ($MgSeO_3$, cr, 298.15 K) = – (938.0 ± 4.6) kJ·mol^{-1} and S_m° ($MgSeO_3$, cr, 298.15 K) = (93.4 ± 4.5) J·K^{-1}·mol^{-1}. The corresponding values calculated in [71PAS/ARO] with unknown auxiliary data are – (941.4 ± 5.9) kJ·mol^{-1} and (86.2 ± 5.9) J·K^{-1}·mol^{-1}, respectively. Pashinkin and Gospodinov [94PAS/GOS] obtained S_m° ($MgSeO_3$, cr, 298.15 K) = (84.2 ± 9.2) J·K^{-1}·mol^{-1} from their reconsideration of the data. By subtracting the entropy difference between $MgSO_3 \cdot 6H_2O$(cr) and $MgSO_3$(cr), 234.3 J·K^{-1}·mol^{-1}, in [82WAG/EVA] from S_m° ($MgSeO_3 \cdot 6H_2O$, cr, 298.15 K) the review obtains S_m° ($MgSeO_3$, cr, 298.15 K) = 95.4 J·K^{-1}·mol^{-1}. The original results in [69BAK/BUK] were $\Delta_f H_m^\circ$ ($MgSeO_3$, cr, 298.15 K) = – 913.8 kJ·mol^{-1} and S_m° ($MgSeO_3$, cr, 298.15 K) = 94.6 J·K^{-1}·mol^{-1}.

The review concludes that the revised data lead to an unresolved discrepancy between the enthalpy of formation of $MgSeO_3$(cr) obtained from the calorimetric (– 900 kJ·mol^{-1}) and vapour pressure measurements (– 940 kJ·mol^{-1}). The calorimetric value is selected with the uncertainty assigned by the review:

$$\Delta_f H_m^\circ (MgSeO_3, cr, 298.15\ K) = -(900.3 \pm 2.5)\ kJ \cdot mol^{-1}.$$

The unselected mean value of the entropy of $MgSeO_3$(cr) from the various estimates is S_m° ($MgSeO_3$, cr, 298.15 K) = (90.0 ± 10.0) J·K^{-1}·mol^{-1}.

V.14.2.3 Magnesium selenates

Klein [40KLE] studied the solubility of magnesium selenate in water as a function of temperature. A heptahydrate, $MgSeO_4{\cdot}7H_2O$, is the stable phase below 276.8 K. In the temperature range 276.8 to 362.7 K, $MgSeO_4{\cdot}6H_2O$ represents the stable phase and above 362.7 K this compound transforms to $MgSeO_4{\cdot}4.5H_2O$. The solubilities reported in a similar study by Meyer and Aulich [28MEY/AUL] appear to be in error, but the reference agrees on the formation of the hepta- and hexahydrates.

The solubility of $MgSeO_4{\cdot}6H_2O$(cr) in water at 298.15 K has been measured in many papers. The selected solubility:

$$m(MgSeO_4{\cdot}6H_2O(cr), \text{aq sat}, 298.15\ K) = (3.29 \pm 0.05)\ \text{mol}(MgSeO_4){\cdot}kg^{-1}$$

is based on [40HIL/SOT], [84SER/SER], [89VOJ/EBE], [95STO/OJK] reporting a solubility of 3.325, 3.305, 3.245, 3.280 mol($MgSeO_4$)·kg^{-1}, respectively. The references are chosen so that only one contribution is included from each research group. The selected value agrees with the interpolated value in [40KLE].

A phase diagram and related information are available for the system: $MgSeO_4$-H_2SeO_4-H_2O at 303.15 K in [44FUR/KIN]; $MgSeO_4$-$(NH_4)_2SeO_4$-H_2O at 303.15 K in [38LAW/KIN] and at 298.15 K in [89VOJ/EBE]; $MgSeO_4$-Li_2SeO_4-H_2O at 298.15 K in [89VOJ/EBE], [93OJK/MIC]; $MgSeO_4$-Na_2SeO_4-H_2O at 298.15 K in [28MEY/AUL]; $MgSeO_4$-K_2SeO_4-H_2O at 298.15 K in [28MEY/AUL], [40HIL/SOT]; $MgSeO_4$-Rb_2SeO_4-H_2O at 298.15 K in [87VOJ/EBE2]; $MgSeO_4$-Cs_2SeO_4-H_2O at 298.15 K in [87VOJ/EBE2].

The value of solubiliy product of $MgSeO_4{\cdot}6H_2O$(cr),

$$MgSeO_4{\cdot}6H_2O(cr) \rightleftharpoons Mg^{2+} + SeO_4^{2-} + 6H_2O(l) \qquad (V.175)$$

at 298.15 K has been reported by Christov [97CHR] to be $\log_{10} K^{\circ}_{s,0} = -1.133$ and by Kumov and Batyreva [90KUM/BAT] to be – 1.20, see Appendix A. The results are based on the same isopiestic data by Ojkova and Staneva [89OJK/STA]. The former value:

$$\log_{10} K^{\circ}_{s,0}\ ((V.175), 298.15\ K) = -(1.133 \pm 0.044)$$

will be selected as it has been evaluated with the Pitzer ion interaction approach. The uncertainty added by the review corresponds to an uncertainty in the mean activity coefficient of 5%.

Parker, Tice, and Thomson [97PAR/TIC] cite the value of the equilibrium constant of the reaction:

$$Mg^{2+} + SeO_4^{2-} \rightleftharpoons MgSeO_4(aq) \qquad (V.176)$$

$\log_{10} \beta^{\circ}_1$ (V.176) = 2.2 from the thesis by Quinn. This work does not seem to have been published elsewhere and has not been available to the review. The authors attempted to verify this result by a measurement similar to that described in Section V.14.3.3 for Ca^{2+}. However, they were not able to obtain data of sufficiently high quality to either to refute or verify the reported value. As no other value is available and it is consistent with the value of the stability constant of the sulphate complex, the review selects:

$$\log_{10} \beta_1^\circ \text{ ((V.176), 298.15 K)} = (2.2 \pm 0.2).$$

The uncertainty has been added by the review.

Selivanova, Shneider, and Ryabova [61SEL/SHN] measured the enthalpy change of the reaction $MgSeO_4{\cdot}6H_2O(cr) + 2AgNO_3(aq, 1{:}800) \rightarrow Ag_2SeO_4(cr) + Mg(NO_3)_2(aq, 1{:}1600) + 6H_2O(l)$ in a calorimeter. They also determined the integral enthalpies of dissolution to (aq, 1:1500) of $MgSeO_4{\cdot}6H_2O(cr)$, $MgSeO_4{\cdot}4H_2O(cr)$, $MgSeO_4{\cdot}H_2O(cr)$, and $MgSeO_4(cr)$. The data were recalculated with new auxiliary data in Appendix A and the selected result:

$$\Delta_f H_m^\circ (MgSeO_4{\cdot}6H_2O, \text{cr, 298.15 K}) = -(2781.4 \pm 2.4) \text{ kJ}{\cdot}\text{mol}^{-1},$$

$$\Delta_f H_m^\circ (MgSeO_4{\cdot}4H_2O, \text{cr, 298.15 K}) = -(2192.67 \pm 2.65) \text{ kJ}{\cdot}\text{mol}^{-1},$$

$$\Delta_f H_m^\circ (MgSeO_4{\cdot}H_2O, \text{cr, 298.15 K}) = -(1298.24 \pm 2.58) \text{ kJ}{\cdot}\text{mol}^{-1},$$

$$\Delta_f H_m^\circ (MgSeO_4, \text{cr, 298.15 K}) = -(971.32 \pm 2.62) \text{ kJ}{\cdot}\text{mol}^{-1}$$

was obtained. The present selection agrees with [82WAG/EVA].

The standard Gibbs energy of formation of $MgSeO_4{\cdot}6H_2O(cr)$ is calculated from the Gibbs energy of Reaction (V.175) and the selected values of $\Delta_f G_m^\circ$ for the ions and water to be:

$$\Delta_f G_m^\circ (MgSeO_4{\cdot}6H_2O, \text{cr, 298.15 K}) = -(2324.20 \pm 2.00) \text{ kJ}{\cdot}\text{mol}^{-1}.$$

The standard entropy of $MgSeO_4{\cdot}6H_2O$ then becomes

$$S_m^\circ (MgSeO_4{\cdot}6H_2O, \text{cr, 298.15 K}) = (351.1 \pm 10.5) \text{ J}{\cdot}\text{K}^{-1}{\cdot}\text{mol}^{-1}.$$

V.14.3 Calcium compounds

V.14.3.1 Calcium selenides

The phase CaSe(cr) is the only compound reported in the system calcium-selenium.

V.14.3.1.1 CaSe(cr)

Fabre [1887FAB] determined the enthalpy of formation of CaSe(cr) to be $\Delta_f H_m^\circ$ (CaSe, cr, 298.15 K) = – 284 $\text{kJ}{\cdot}\text{mol}^{-1}$ from solution calorimetry measurements. The review has re-assessed the work by Fabre in Appendix A and obtained $\Delta_f H_m^\circ$ (CaSe, cr, 298.15/K) = – (383.1 ± 3.4) $\text{kJ}{\cdot}\text{mol}^{-1}$. This value is more uncertain than indicated by the experimental accuracy since the specimen of CaSe(cr) was of somewhat doubtful quality.

Marx and Petzel [92MAR/PET] measured the total pressure of the vapour in equilibrium with stoichiometric CaSe(cr) in the temperature range 1832 to 2138 K using the Knudsen effusion weight-loss method. The vapour was assumed to predominantly contain Ca(g) and Se(g) based on results from calculations using estimated thermodynamic properties of CaSe(g). The enthalpy change for the reaction

$$CaSe(cr) \rightleftharpoons Ca(g) + Se(g)$$

was determined to be $\Delta_{at}H^{\circ}_{m}$ (CaSe, cr, 298.15 K) = (830.0 ± 28.0) kJ·mol^{-1} from the second law and to be (858.0 ± 43.7) kJ·mol^{-1} from the third law using an estimated heat capacity and Gibbs energy function of CaSe(cr). The second law entropy change was determined to be $\Delta_{at}S^{\circ}_{m}$ (CaSe, cr, 298.15 K) = (248.3 ± 22.7) J·K^{-1}·mol^{-1}. The corresponding values of the enthalpy of formation and the entropy of CaSe(cr) are calculated to be $\Delta_{f}H^{\circ}_{m}$ (CaSe,cr, 298.15 K) = – (416.1 ± 28.1) kJ·mol^{-1} from the second law and – (444.1 ± 43.7) kJ·mol^{-1} from the third law, and S°_{m} (CaSe, cr, 298.15 K) = (83.3 ± 22.7) J·K^{-1}·mol^{-1} employing the selected values of Se(g) and the data of Ca(g) in [89COX/WAG]. The selected value of the enthalpy of formation is the mean of the second and third law values and the calorimetric result with the uncertainty limits covering the range of data:

$$\Delta_{f}H^{\circ}_{m}\text{ (CaSe, cr 298.15 K)} = -(414.4 \pm 35.0)\text{ kJ·mol}^{-1}.$$

Mills [74MIL] estimated the entropy of CaSe(cr) to be S°_{m} (CaSe, cr, 298.15 K) = (69.0 ± 6.0) J·K^{-1}·mol^{-1}. No value was selected for the entropy of CaSe(cr) because of the large difference between the experimental value in [92MAR/PET] and the estimated value in [74MIL].

V.14.3.2 Calcium selenites

Dumm and Brown [97DUM/BRO] studied phase formation in the system CaO-SeO_2-H_2O at 296.2 and 353.2 K by solubility measurements. Only results at the lower temperature will be cited, since the study at 353.2 K is incomplete. The calcium containing phases formed are $CaSeO_3 \cdot H_2O$ and $Ca(HSeO_3)_2 \cdot H_2O$. The same compounds were found by Ebert and Havlicek [81EBE/HAV] to be the calcium containing phases formed in the system $CaSeO_3$-SeO_2-H_2O system at 298 K.

The solid phases studied in the solubility experiments are assumed by the review to have the composition $CaSeO_3 \cdot H_2O$ and the solubility equilibrium is written:

$$CaSeO_3 \cdot H_2O(cr) \rightleftharpoons Ca^{2+} + SeO_3^{2-} + H_2O(l) \qquad (V.177)$$

Chukhlantsev [56CHU] prepared calcium selenite by mixing 0.1 M solutions of calcium chloride and sodium selenite solution in stoichiometric amounts. Chemical analysis confirmed the 1:1 ratio between Ca(II) and Se(IV). No X-ray diffraction measurements were performed. The solubility of the specimen in dilute solution of nitric or hydrochloric acid was measured at 293 K. The data have been recalculated as described in Appendix A, [56CHU], to yield $\log_{10} K^{\circ}_{s,0}$ ((V.177), 293.15 K) = – (6.15 ± 0.22). The reported conditional constant is $-(5.50 \pm 0.14)$. Ripan and Vericeanu [68RIP/VER] studied the solubility of $CaSeO_3(s)$ by conductivity measurements as described in Appendix A. Hydrolysis can be neglected for Ca^{2+}, [76BAE/MES], and the original data, mean value 1.33×10^{-3} M, were assumed to provide a measure of the solubility of $CaSeO_3{\cdot}H_2O(cr)$ in pure water. The solubility product calculated from the data with a correction for the protolysis of the selenite ion and extrapolated to $I = 0$ by the SIT expression ($\Delta\varepsilon = 0$ kg·mol^{-1}) is $\log_{10} K^{\circ}_{s,0}$ ((V.177), 291 K) = $-(6.04 \pm 0.35)$. Savenko [95SAV] measured the solubility of crystalline calcium selenite in water. The pH was adjusted between 8.1 and 7.3 by small additions of strong acid or base. Both the total selenium(IV) and calcium concentrations were determined at equilibrium. The experimental temperature was 293.15 K. The data have been recalculated with the protonation constant of the selenite ion selected here and activity coefficients were found from the SIT expression with $\varepsilon = 0$ kg·mol^{-1}. The result was $\log_{10} K^{\circ}_{s,0}$ ((V.177), 293.15 K) = $-(6.55 \pm 0.03)$. No trend in the value of the solubility product with pH was observed.

Sharmasakar, Reddy, and Vance [96SHA/RED] obtained $\log_{10} K^{\circ}_{s,0}$ ((V.177), 298.2 K) = -7.76 from measurements of the solubility of crystalline $CaSeO_3{\cdot}H_2O$ in aqueous media. For reasons presented in Appendix A, this result has been given zero weight.

The review considers the measurements by Savenko to be the most satisfactory and selects:

$$\log_{10} K^{\circ}_{s,0}\ ((\text{V.177}),\ 298.15\ \text{K}) = -(6.40 \pm 0.25).$$

Leshchinskaya and Selivanova [63LES/SEL3] made a calorimetric measurement of the reaction between $Ca(NO_3)_2{\cdot}4H_2O(cr)$ and a solution of sodium selenite. The data were evaluated with $CaSeO_3{\cdot}2H_2O$ as reaction product, while equilibrium studies show that the stable phase is $CaSeO_3{\cdot}H_2O$ under the conditions of the calorimetric measurement. Their data have been re-evaluated in Appendix A with new auxiliary data and the assumption that $CaSeO_3{\cdot}H_2O$ is indeed formed. The result is $\Delta_f H^{\circ}_m$ ($CaSeO_3{\cdot}H_2O$, cr, 298.15 K) = $-(1324.83 \pm 2.81)$ kJ·mol^{-1}. With $CaSeO_3{\cdot}2H_2O$ as reaction product $\Delta_f H^{\circ}_m$ ($CaSeO_3{\cdot}2H_2O$, cr, 298.15 K) = $-(1610.6 \pm 2.8)$ kJ·mol^{-1} is obtained, which is close to the value found for this compound in [82WAG/EVA].

The review adopts:

$$\Delta_f H^{\circ}_m\ (\text{CaSeO}_3{\cdot}\text{H}_2\text{O, cr, 298.15 K}) = -(1324.83 \pm 2.81)\ \text{kJ·mol}^{-1}$$

and from the solubility product:

$$\Delta_f G^{\circ}_m\ (\text{CaSeO}_3{\cdot}\text{H}_2\text{O, cr, 298.15 K}) = -(1188.90 \pm 2.50)\ \text{kJ·mol}^{-1}.$$

These results yield:

$$S^{\circ}_m\ (\text{CaSeO}_3{\cdot}\text{H}_2\text{O, cr, 298.15 K}) = (168.8 \pm 12.6)\ \text{J·K}^{-1}\text{·mol}^{-1}.$$

This value appears to be somewhat too large. The review therefore concludes that additional measurements would be desirable, preferably with better control of the solid phase. Note that the enthalpy value was calculated with non-TDB auxiliary data and therefore included in Appendix E.

V.14.3.3 Calcium selenates

The solubility of $CaSeO_4{\cdot}2H_2O(cr)$ in water between 270 and 375 K was studied by Selivanova and Shneider [59SEL/SHN4], see Appendix A. The result at 298.15 K was 0.3557 m. This value differs considerably from the value of 0.404 m obtained by Meyer and Aulich [28MEY/AUL] in a similar study. The data from the latter source are consistently 10 to 15% larger. The measurement at 303 K by Welton and King [39WEL/KIN] agrees best with [28MEY/AUL]. The mean value is selected:

$$m(CaSeO_4{\cdot}2H_2O(cr), \text{aq sat}, 298.15\text{ K}) = (0.380 \pm 0.025)\ \text{mol}(CaSeO_4){\cdot}\text{kg}^{-1}.$$

A phase diagram or related information is available for the system: $CaSeO_4$-$(NH_4)_2SeO_4$-H_2O at 303.15 K in [39WEL/KIN]; $CaSeO_4$-Na_2SeO_4-H_2O at 298.15 K in [28MEY/AUL]; $CaSeO_4$-K_2SeO_4-H_2O at 298.15 K in [28MEY/AUL].

The selected solubility will be used for an estimate of the solubility product of calcium selenate dihydrate:

$$CaSeO_4{\cdot}2H_2O(cr) \rightleftharpoons Ca^{2+} + SeO_4^{2-} + 2H_2O(l) \qquad \text{(V.178)}$$

The mean activity coefficients of the sulphates of Cd, Cu, Mg, Mn, Ni, and Zn are quite similar [50HAR/OWE]. The variation at a concentration of 0.5 m is less than 10%. Ojkova and Staneva [89OJK/STA] made isopiestic measurement on aqueous solutions of the selenates of Co, Cu, Mg, Ni, and Zn. The mean activity coefficients obtained from these measurements also show a moderate variation between the salts. The value of $\gamma_\pm$ for $MgSeO_4$, $\gamma_\pm = 0.12$, will be used to extrapolate the conditional solubility product to $I = 0$. The selected result is:

$$\log_{10} K^{\circ}_{s,0}\,((\text{V.178}), 298.15\text{ K}) = -\,(2.68 \pm 0.25).$$

The estimated uncertainty corresponds to an uncertainty in $\gamma_\pm$ of 25%.

Parker, Tice, and Thomason [97PAR/TIC] measured the Ca^{2+} activity in solutions containing total Ca^{2+} concentrations in the range 20×10^{-6} to 200×10^{-6} M and the selenate concentration 0.004, 0.01, or 0.03 M with a carefully calibrated ion-selective Ca^{2+} electrode. Activity coefficient corrections were made by Davies' equation. The equilibrium constant of the reaction:

$$Ca^{2+} + SeO_4^{2-} \rightleftharpoons CaSeO_4(aq) \qquad \text{(V.179)}$$

was found by trial and error. The stability constant was varied systematically and the agreement between calculated and measured Ca^{2+} activities tested. The best agreement was found for:

$$\log_{10} \beta_1^\circ \text{ ((V.179), 298.15 K)} = (2.00 \pm 0.10).$$

This value is selected and the uncertainty has been estimated by the review. The quality of the work is supported by the parallel determination of the value of the constant for the sulphate complex, which agreed with literature data.

Selivanova and Shneider [59SEL/SHN3] determined the enthalpy change of the reaction $CaSeO_4{\cdot}2H_2O(cr) + 2AgNO_3(aq, 1{:}800) \rightarrow Ag_2SeO_4(cr) + Ca(NO_3)_2(aq, 1{:}1600) + 2H_2O(l)$ by a calorimetric measurement. They also determined the integral enthalpies of dissolution of $CaSeO_4{\cdot}2H_2O$(cr) and $CaSeO_4$(cr) to $CaSeO_4$(aq, 1:1600). The data were recalculated in Appendix A to provide the selected values:

$$\Delta_f H_m^\circ (CaSeO_4{\cdot}2H_2O, \text{cr, 298.15 K}) = -(1709.0 \pm 2.6)\ \text{kJ}{\cdot}\text{mol}^{-1},$$

$$\Delta_f H_m^\circ (CaSeO_4, \text{cr, 298.15 K}) = -(1112.1 \pm 2.6)\ \text{kJ}{\cdot}\text{mol}^{-1}$$

and from the stability constant of Reaction (V.179) :

$$\Delta_f G_m^\circ (CaSeO_4, \text{aq, 298.15 K}) = -(1003.7 \pm 1.9)\ \text{kJ}{\cdot}\text{mol}^{-1}.$$

The temperature variation of the solubility product of $CaSeO_4{\cdot}2H_2O$(cr) measured in [59SEL/SHN4] results in $\Delta_r H_m^\circ$ ((V.178), 298.15 K) = $-(5.31 \pm 0.59)$ kJ·mol^{-1}, see Appendix A. With accepted auxiliary data this value leads to $\Delta_f H_m^\circ$ ($CaSeO_4{\cdot}2H_2O$, cr, 298.15 K) = $-(1712.85 \pm 3.69)$ kJ·mol^{-1}. The calorimetric determination of this quantity has been preferred and it agrees with [82WAG/EVA].

The standard Gibbs energy of formation of calcium selenate dihydrate is calculated from the Gibbs energy of Reaction (V.178) and the selected values of $\Delta_f G_m^\circ$ for the ions and water to be:

$$\Delta_f G_m^\circ (CaSeO_4{\cdot}2H_2O, \text{cr, 298.15 K}) = -(1481.9 \pm 2.3)\ \text{kJ}{\cdot}\text{mol}^{-1}.$$

The value in [82WAG/EVA] is $-$ 1486.8 kJ·mol^{-1}. This result apparently originates from $\log_{10} K_{s,0}^\circ$ ((V.178), 298.15 K) = $-$ 3.08 reported in [59SEL/SHN4] and calculated with an activity coefficient for $MgSO_4$(aq), see Appendix A. The standard entropy of $CaSeO_4{\cdot}2H_2O$ becomes from the selections:

$$S_m^\circ (CaSeO_4{\cdot}2H_2O, \text{cr, 298.15 K}) = (198.8 \pm 11.7)\ \text{J}{\cdot}\text{K}^{-1}{\cdot}\text{mol}^{-1}.$$

V.14.4 Strontium compounds

V.14.4.1 Strontium selenides

The phases $SrSe_3$(cr), $SrSe_2$(cr), Sr_2Se_3(cr), and SrSe(s) are stable compounds according to the phase diagram reported by Lyskova and Vakhobov [75LYS/VAK]. Thermodynamic information is available for SrSe(cr) and SrSe(g).

V.14.4.1.1 SrSe(cr)

The enthalpy of formation of SrSe(cr) was determined to be $\Delta_f H^\circ_m$(SrSe, cr, 298.15 K) = – 285.1 kJ·mol^{-1} in the early work of Fabre [1887FAB] using solution calorimetry. The review has re-assessed the work by Fabre in Appendix A and obtained $\Delta_f H^\circ_m$(SrSe, cr, 298.15 K) = – (395.8 ± 3.3) kJ·mol^{-1}. Berkowitz and Chupka [66BER/CHU] determined the partial pressures of both Sr(g) and Se(g) to be (1.2 ± 0.3) × 10^{-5} bar in equilibrium with SrSe(cr) at 1853 K using mass spectrometry. A third law evaluation made by the review as discussed in Appendix A yielded $\Delta_f H^\circ_m$(SrSe, cr, 298.15 K) = – (415.8 ± 65.0) kJ·mol^{-1} from these measurements. Marx and Petzel [92MAR/PET] measured the saturated vapour pressure in equilibrium with SrSe(cr) in the temperature range 1862 to 2122 K using the Knudsen effusion weight-loss method. It was assumed (see Section V.14.3.1.1) that SrSe(s) vaporises according to reaction

$$\text{SrSe(cr)} \rightleftharpoons \text{Sr(g)} + \text{Se(g)}.$$

The enthalpy of reaction was evaluated to be $\Delta_r H^\circ_m$(298.15 K) = (893.7 ± 28.0) kJ·mol^{-1} from the second law and (851.0 ± 42.0) kJ·mol^{-1} from the third law using an estimated heat capacity and Gibbs energy function of SrSe(cr). The entropy of reaction was determined to be $\Delta_r S^\circ_m$(298.15 K) = (281.8 ± 22.9) J·K^{-1}·mol^{-1}. On combining the reaction quantities with the data of Sr(g) in [98CHA], $\Delta_f H^\circ_m$(Sr, g, 298.15 K) = (164.0 ± 1.7) kJ·mol^{-1}, and the selected values of Se(g), the corresponding enthalpy of formation values $\Delta_f H^\circ_m$(SrSe, cr, 298.15 K) = – (493.6 ± 28.0) kJ·mol^{-1} and $\Delta_f H^\circ_m$(SrSe, cr, 298.15 K) = – (450.9 ± 42.0) kJ·mol^{-1} and entropy S°_m(SrSe, cr, 298.15 K) = (59.5 ± 22.9) J·K^{-1}·mol^{-1} are obtained. However, according to the phase diagram in [75LYS/VAK], SrSe(cr) melts at 1873 K. The vaporisation reaction in the experiments in [92MAR/PET] was therefore presumably

$$\text{SrSe(l)} \rightleftharpoons \text{Sr(g)} + \text{Se(g)}$$

and the enthalpies and entropies at 298.15 K should be corrected accordingly. The enthalpy of fusion of SrSe(cr) is estimated by the review to be of the order of 50 kJ·mol^{-1}, rendering the values of the enthalpy of formation of SrSe(cr) still more negative by the same amount. The second law entropy of SrSe(s) after correction for the entropy of fusion is approximately 32 J·K^{-1}·mol^{-1} which can be compared with the value S°_m(SrSe, cr, 298.15 K) = (80.8 ± 6.3) J·K^{-1}·mol^{-1} estimated by Mills [74MIL]. The review assigns zero weight to the results in [92MAR/PET] and selects the mean of the other two determinations with a large uncertainty:

$$\Delta_f H^\circ_m(\text{SrSe, cr, 298.15 K}) = -(406.0 \pm 35.0)\ \text{J·mol}^{-1}.$$

V.14.4.1.2 SrSe(g)

The dissociation energy of SrSe(g) at 1853 K was determined by Berkowitz and Chupka [66BER/CHU] in vaporisation studies at this temperature. A recalculation of the value by the review using the selected properties of Se(g) and Se_2(g), and the Gibbs energy function of SrSe(g) calculated from molecular parameters estimated in [66BER/CHU] to the standard temperature yielded $\Delta_{at}H^{o}_{m}$ (SrSe, g, 298.15 K) = (282 ± 25) kJ·mol^{-1}. No value of the enthalpy of formation of SrSe(g) has been selected from this result since it is based on only one experimental point.

V.14.4.2 Strontium selenites

Ebert and Havlicek [82EBE/HAV] investigated the system $SrSeO_3$-H_2SeO_3-H_2O by solubility measurements at 298.15 K. The solids identified in the phase diagram were $SrSeO_3 \cdot nH_2O$ n = 0.6 to 1.0, $Sr(HSeO_3)_2$ and H_2SeO_3. The formation of a hydrogen selenite was confirmed by its infrared spectrum. It is incongruently soluble. No solubility data are included in the paper except the compositions of the invariant points.

The solid phase denoted $SrSeO_3 \cdot nH_2O$ n = 0.6 to 1.0 in [82EBE/HAV] is not accompanied by any analytical data. Strontium selenite was prepared by reacting equivalent amounts of strontium chloride and sodium selenite solutions in [65LES/AVE], and $SrCl_2 \cdot 6H_2O$(cr) with a 0.055 M solution of Na_2SeO_3 in [63SEL/LES]. The strontium selenite formed was crystalline in both instances. Chemical analyses of Sr and Se are lower than expected for the composition $SrSeO_3$. However, the values of n calculated from the analytical data are n = 0.13 from the determination of Sr and n = 0.06 from the determination of Se. These values are thus much lower than those reported in [82EBE/HAV]. Based on the Russian work, the review will assume that the solid phase present in the solubility and calorimetric measurements discussed below was $SrSeO_3$(cr).

Chukhlantsev [56CHU] prepared strontium selenite by mixing 0.1 M solution of strontium chloride and sodium selenite solution in stoichiometric amounts. Chemical analysis confirmed the 1:1 ratio between Sr and Se(IV). No X-ray diffraction measurements were performed. The solubility of the specimen in dilute solution of nitric or hydrochloric acid was measured at 293 K. The data have been recalculated as described in Appendix A, [56CHU]. The result for:

$$SrSeO_3(cr) \rightleftharpoons Sr^{2+} + SeO_3^{2-} \qquad (V.180)$$

was $\log_{10} K^{o}_{s,0}$ ((V.180), 293.15 K) = – (6.33 ± 0.19). The reported conditional constant is – (5.68 ± 0.08). The solubility in water was also measured. The data have been corrected for the hydrolysis of the selenite ion and extrapolated to I = 0. The result was $\log_{10} K^{o}_{s,0}$ ((V.180), 293.15 K) = – (6.32 ± 0.05). The reported conditional constant is – (6.07 ± 0.05). Ripan and Vericeanu [68RIP/VER] studied the solubility of $SrSeO_3$(s) by conductivity measurements as described in Appendix A. Hydrolysis can be neglected for Sr^{2+}, [76BAE/MES], and the original solubility data, mean value 2.82 × 10^{-3} M, were assumed to provide a measure of the solubility of $SrSeO_3$(s) in pure water. The solubility product calculated from the data with a correction for the protolysis of the selenite ion and extrapolated to I = 0 by the SIT expression ($\Delta\varepsilon$ = 0 kg·mol^{-1}) is $\log_{10} K^{\circ}_{s,0}$ ((V.180), 291 K) = – (5.48 ± 0.05).

The review selects the result by Chukhlantsev and proposes:

$$\log_{10} K^{\circ}_{s,0}\ ((\text{V.180}),\ 298.15\ \text{K}) = -(6.30 \pm 0.50).$$

Selivanova and Leshchinskaya [63SEL/LES] measured the enthalpy change of the reaction between equimolar quantities of $SrCl_2·6H_2O$ and Na_2SeO_3(aq, 1:1000) with formation of crystalline $SrSeO_3$. Their data have been re-evaluated with new auxiliary data in Appendix A. The selected result is:

$$\Delta_f H^{\circ}_m (\text{SrSeO}_3,\ \text{cr},\ 298.15\ \text{K}) = -(1051.9 \pm 2.7)\ \text{kJ·mol}^{-1}.$$

Reference [82WAG/EVA] reports – 1047.7 kJ·mol^{-1}. The difference between the two values can be accounted for by the change in the data for strontium made in [92GRE/FUG]. The selected data further yield:

$$\Delta_f G^{\circ}_m (\text{SrSeO}_3,\ \text{cr},\ 298.15\ \text{K}) = -(962.2 \pm 3.4)\ \text{kJ·mol}^{-1},$$

$$S^{\circ}_m (\text{SrSeO}_3,\ \text{cr},\ 298.15\ \text{K}) = (104.7 \pm 14.6)\ \text{J·K}^{-1}\text{·mol}^{-1}.$$

V.14.4.3 Strontium selenates

Selivanova and Zubova [58SEL/ZUB], [59SEL/ZUB3] measured the solubility of $SrSeO_4$(cr) at temperatures between 273 and 373 K, see [59SEL/ZUB3] in Appendix A. The solubility at 298.15 K was (1.937 ± 0.070) × 10^{-2} M. The solubility product obtained from this fairly high equilibrium concentration is substantially affected by the selection of the activity coefficient. The value of $\gamma_{\pm}$ = 0.40 was calculated by SIT with $\varepsilon(Sr^{2+}, SeO_4^{2-}) \approx 0$ kg·mol^{-1} as suggested by the data for $MgSeO_4$ in [89OJK/STA]. The same datum for sulphates of divalent metal ions is 0.30 from [50HAR/OWE]. The value of the solubility product of $SrSeO_4$(cr):

$$\text{SrSeO}_4(\text{cr}) \rightleftharpoons \text{Sr}^{2+} + \text{SeO}_4^{2-} \quad (\text{V.181})$$

has been calculated with $\gamma_{\pm}$ = 0.30 and 0.40, respectively, and the mean is $\log_{10} K^{\circ}_{s,0}$ ((V.181), 298.15 K) = – (4.35 ± 0.15).

The enthalpy change of Reaction (V.181) was studied in [59SEL/ZUB3]. Two values were evaluated in Appendix A from the data, namely $\Delta_r H^{\circ}_m$ ((V.181), 298.15 K) = – (0.98 ± 0.17) kJ·mol^{-1} from the temperature variation of the solubility product and –(2.49 ± 0.86) kJ·mol^{-1} from the calorimetric measurements. The review therefore adopts $\Delta_r H^{\circ}_m$ ((V.181), 298.15 K) = – (1.70 ± 1.50) kJ·mol^{-1}. The value combines with the ionic enthalpies to yield $\Delta_f H^{\circ}_m$ ($SrSeO_4$, cr, 298.15 K) = – (1152.70 ± 3.84) kJ·mol^{-1}.

The value in [82WAG/EVA] is – 1142.7 kJ·mol^{-1}. Most of the difference, 9.5 kJ·mol^{-1}, is probably caused by the selection of new auxiliary data.

The standard Gibbs energy of formation of strontium selenate is obtained from the Gibbs energy of Reaction ((V.181)) and the selected values of $\Delta_f G_m^\circ$ for the ions to be $\Delta_f G_m^\circ$ ($SrSeO_4$, cr, 298.15 K) = – (1028.20 ± 1.90) $kJ \cdot mol^{-1}$.

The formation data yield S_m° ($SrSeO_4$, cr, 298.15 K) = (90.5 ± 14.4) $J \cdot K^{-1} \cdot mol^{-1}$.

The value of the standard entropy is smaller than expected from a comparison with data for alkaline earth sulphates in [82WAG/EVA] and indicates a systematic error in the formation data, which could not be traced from the experimental data available. No formation values for strontium selenate have therefore been selected.

V.14.5 Barium compounds

V.14.5.1 Barium selenides

The phases $BaSe_3$(cr), $BaSe_2$(cr), Ba_2Se_3(cr), and BaSe(cr) are stable compounds according to the phase diagram reported by Lyskova and Vakhobov [75LYS/VAK2]. Thermodynamic information is available for BaSe(cr).

V.14.5.1.1 BaSe(cr)

The early solution calorimetry work of Fabre [1887FAB] has been re-assessed in Appendix A yielding $\Delta_f H_m^\circ$ (BaSe, cr, 298.15 K) = – (378.7 ± 4.1) $kJ \cdot mol^{-1}$. Marx and Petzel [92MAR/PET] supposedly measured the saturated vapour pressure in equilibrium with BaSe(cr) in the temperature range 1860 to 2158 K using the Knudsen effusion weight-loss method. It was assumed (see Section V.14.3.1.1) that BaSe(cr) vaporises according to reaction

$$\text{BaSe(cr)} \rightleftharpoons \text{Ba(g)} + \text{Se(g)}$$

under these conditions. The enthalpy of reaction was evaluated to be $\Delta_r H_m^\circ$ (298.15 K) = (887.3 ± 25.4) $kJ \cdot mol^{-1}$ from the second law and (881.5 ± 43.5) $kJ \cdot mol^{-1}$ from the third law using an estimated heat capacity and Gibbs energy function of BaSe(cr). The entropy of reaction was determined to be $\Delta_r S_m^\circ$ (298.15 K) = (258.8 ± 21.7) $J \cdot K^{-1} \cdot mol^{-1}$. However, according to the phase diagram in [75LYS/VAK2], BaSe(cr) melts at 2053 K. The vaporisation reaction was therefore presumably

$$\text{BaSe(l)} \rightleftharpoons \text{Ba(g)} + \text{Se(g)}$$

in the upper part of the temperature range. Hence the evaluation was made across the fusion transition making the results doubtful. A correct treatment of the fusion would lead to a still larger value of the atomisation enthalpy of BaSe(cr). Since the difference between the high and low temperature results is almost 100 $kJ \cdot mol^{-1}$, the review does not select a value of the standard enthalpy of formation of BaSe(cr).

V.14.5.2 Barium selenites

Neal and McCrosky [38NEA/CRO] established the phase diagram of the system $BaSeO_3$-SeO_2-H_2O at 273.15, 298.15, and 323.15 K by solubility measurement in the range 0 to about 70 mass-% SeO_2 in the liquid phase. The solids $BaSeO_3$ and $BaSe_2O_5$ were formed at all temperatures.

Ripan and Vericeanu [68RIP/VER] studied the solubility of $BaSeO_3$(s) by conductivity measurements as described in Appendix A. Hydrolysis can be neglected for Ba^{2+} [76BAE/MES] and the original data, mean solubility value 7.92×10^{-4} M, were assumed to provide a measure of the solubility of $BaSeO_3$(s) in pure water. The equilibrium constant (solubility product) of the reaction:

$$BaSeO_3(cr) \rightleftharpoons Ba^{2+} + SeO_3^{2-} \qquad (V.182)$$

calculated from the data with a correction for the protolysis of the selenite ion and extrapolated to $I = 0$ by the SIT expression ($\Delta\varepsilon = 0$ kg·mol^{-1}) is $\log_{10} K^{\circ}_{s,0}$ ((V.182), 291 K) = – (6.43 ± 0.10). Leshchinskaya, Averbukh, and Selivanova [65LES/AVE] prepared crystalline barium selenite and measured its solubility in water at 298.15 K. The time of equilibration was about two weeks with constant stirring. The solubility product has been recalculated with allowance for the hydrolysis of the selenite ion and extrapolated to $I = 0$ with the SIT expression. The result was $\log_{10} K^{\circ}_{s,0}$ ((V.182), 298.15 K) = – 6.58. No uncertainty can be assigned to this value as there was only one experiment performed. The conditional constant reported in the paper is – 6.37.

The review selects:

$$\log_{10} K^{\circ}_{s,0}\ ((V.182),\ 298.15\ K) = -(6.50 \pm 0.25).$$

Leshchinskaya, Selivanova, and Strel'tsov [63LES/SEL] measured the enthalpy change of the reaction between equimolar quantities of $BaCl_2 \cdot 2H_2O$(cr) and Na_2SeO_3(aq, 1:1000) with formation of crystalline $BaSeO_3$. Their data have been re-evaluated with new auxiliary data in Appendix A. The selected result is:

$$\Delta_f H^{\circ}_m\ (BaSeO_3,\ cr,\ 298.15\ K) = -(1036.7 \pm 2.9)\ kJ \cdot mol^{-1}.$$

Reference [82WAG/EVA] reports – 1040.6 kJ·mol^{-1}. The major part of the difference between the two values is due to the change in the data for barium made in [92GRE/FUG]. The selected data further yield:

$$\Delta_f G^{\circ}_m\ (BaSeO_3,\ cr,\ 298.15\ K) = -(957.2 \pm 3.4)\ kJ \cdot mol^{-1},$$

$$S^{\circ}_m\ (BaSeO_3,\ cr,\ 298.15\ K) = (145.6 \pm 15.0)\ J \cdot K^{-1} \cdot mol^{-1}.$$

V.14.5.3 Barium selenates

The solubility of $BaSeO_4$(cr) in water in the temperature interval 288 to 368 K was determined by Selivanova and Shneider [58SEL/SHN3], see Appendix A. The solubility is

$(1.87 \pm 0.01) \times 10^{-4}$ M at 298.15 K. From this selected value the solubility product of $BaSeO_4(cr)$,

$$BaSeO_4(cr) \rightleftharpoons Ba^{2+} + SeO_4^{2-} \qquad (V.183)$$

is calculated to be:

$$\log_{10} K^{\circ}_{s,0} ((V.183), 298.15\ K) = -(7.56 \pm 0.10),$$

with $\gamma_{\pm} = 0.884$ from SIT.

Selivanova, Kapustinskii, and Zubova [59SEL/KAP] determined the enthalpy of reaction of $BaCl_2$(aq, 1:500) + H_2SeO_4(7.07 M) → $BaSeO_4$(cr) + 2HCl(aq, 1: 250). Their measurements are evaluated in Appendix A resulting in $\Delta_r H^{\circ}_m$(V.183) = (5.70 ± 2.54) kJ·mol^{-1}. With the selected enthalpies of formation of the ions, the standard enthalpy of formation of $BaSeO_4$(cr) becomes:

$$\Delta_f H^{\circ}_m (BaSeO_4, cr, 298.15\ K) = -(1144.0 \pm 5.0)\ kJ\cdot mol^{-1}.$$

Reference [82WAG/EVA] reports – 1146.4 kJ·mol^{-1} and the value reported in the paper is – 1168.2 kJ·mol^{-1}.

The standard Gibbs energy of formation of barium selenate is calculated from the Gibbs energy of Reaction (V.183) and the selected values of $\Delta_f G^{\circ}_m$ for the ions to be:

$$\Delta_f G^{\circ}_m (BaSeO_4, cr, 298.15\ K) = -(1040.3 \pm 3.0)\ kJ\cdot mol^{-1}.$$

The value in [82WAG/EVA] is – 1044.7 kJ·mol^{-1}. It has apparently been calculated from the data in [58SEL/SHN3]. The difference from the result of the review is caused by the new ionic Gibbs energies introduced.

The selected data yield:

$$S^{\circ}_m (BaSeO_4, cr, 298.15\ K) = (167.0 \pm 19.6)\ J\cdot K^{-1}\cdot mol^{-1}.$$

V.14.6 Radium compounds

V.14.6.1 Radium selenides

No experimental thermodynamic information is available for radium selenides. Estimated thermodynamic properties of RaSe(cr) are available in [85LOW].

V.15 Ammonium and group 1 compounds and complexes

V.15.1 Ammonium compounds

V.15.1.1 $NH_4HSe(cr)$

Ammonium hydrogen selenide, $NH_4HSe(cr)$, is at room temperature a white crystalline solid, which can be prepared by reacting gaseous ammonia with gaseous hydrogen selenide. It is highly susceptible to oxidation and decomposes in the presence of traces of air [49MIK/POS].

Mikus and Poss [49MIK/POS] studied the reaction:

$$NH_4HSe(cr) \rightleftharpoons NH_3(g) + H_2Se(g);\ K_p = p_{NH_3} \times p_{H_2Se} \qquad (V.184)$$

by means of vapour pressure measurements over the temperature range 288 to 303 K. The measurements were performed both at stoichiometric conditions, as well as in the presence of excess ammonia or hydrogen selenide gas, and yielded consistent results.

The plot of $\log_{10} K_p$ versus $1/T$ resulted in a straight line

$$\log_{10} K_p\,(V.184) = (16.14 \pm 0.61) - (6011 \pm 182)\ T^{-1}$$

from which the selected data:

$$\Delta_r H_m^\circ\,((V.184), 298.15\ K) = (115.1 \pm 3.5)\ kJ\cdot mol^{-1}$$

$$\Delta_r G_m^\circ\,((V.184), 298.15\ K) = (23.0 \pm 4.9)\ kJ\cdot mol^{-1}$$

$$\Delta_r S_m^\circ\,((V.184), 298.15\ K) = (309.0 \pm 11.7)\ J\cdot K^{-1}\cdot mol^{-1}$$

are derived.

These data combine with CODATA [89COX/WAG] for $NH_3(g)$ and the selected data for $H_2Se(g)$ to the selected data:

$$\Delta_f H_m^\circ\,(NH_4HSe, cr, 298.15\ K) = -(132.0 \pm 4.0)\ kJ\cdot mol^{-1}$$

$$S_m^\circ\,(NH_4HSe, cr, 298.15\ K) = (102.8 \pm 11.7)\ J\cdot K^{-1}\cdot mol^{-1}$$

$$\Delta_f G_m^\circ\,(NH_4HSe, cr, 298.15\ K) = -(24.2 \pm 5.3)\ kJ\cdot mol^{-1}.$$

V.15.1.2 Ammonium selenites

Micka, Pikal, and Ebert [85MIC/PIK] studied the system $(NH_4)_2SeO_3$-H_2SeO_3-H_2O at 298.15 and 303.15 K and established the phase diagrams from solubility measurements. The crystalline phases $(NH_4)_2SeO_3\cdot H_2O$, NH_4HSeO_3, and $NH_4H_3(SeO_3)_2$ were identified at 298.15 K, while at 303.15 K $(NH_4)_2SeO_3$, $(NH_4)_2Se_2O_5$, and $NH_4H_3(SeO_3)_2$ were reported. The intermediate phases are congruently soluble, but no solubility data are presented. The diagrams indicate that two phase transitions, *viz.*,

$$2NH_4HSeO_3(cr) \rightarrow (NH_4)_2Se_2O_5(cr) + H_2O(l)$$

and

$$(NH_4)_2SeO_3{\cdot}H_2O(cr) \rightarrow (NH_4)_2SeO_3(cr) + H_2O(l)$$

take place between 298 and 303 K.

V.15.1.2.1 $(NH_4)_2SeO_3{\cdot}H_2O(cr)$

Janickis [34JAN] determined the solubility of ammonium selenite in water over the temperature interval 253.2 to 343.2 K and found that $(NH_4)_2SeO_3{\cdot}H_2O(cr)$ was the only solid phase formed. Above about 343 K oxidation of NH_4^+ to N_2 begins. The solubility of $(NH_4)_2SeO_3{\cdot}H_2O(cr)$ in water at 298.15 K was reported to be 7.40 $mol((NH_4)_2SeO_3){\cdot}kg^{-1}$.

Measured distances in the small phase diagrams in [85MIC/PIK] yielded the same solubility at 298.15 and 303.15 K, namely 7.1 $mol((NH_4)_2SeO_3){\cdot}kg^{-1}$. The approximate agreement with the measurements by Janickis might therefore indicate the presence of metastabile phases as he used $(NH_4)_2SeO_3{\cdot}H_2O$ as starting material, while $(NH_4)_2SeO_3$ was used in [85MIC/PIK]. The review selects:

$$m((NH_4)_2SeO_3{\cdot}H_2O(cr), \text{aq sat}, 298.15\text{ K}) = (7.4 \pm 0.2)\ mol((NH_4)_2SeO_3){\cdot}kg^{-1}.$$

V.15.1.2.2 $NH_4HSeO_3(cr)$ and $(NH_4)_2Se_2O_5(cr)$

Janickis [34JAN] could not prepare NH_4HSeO_3 and the various efforts at its preparation always yielded the diselenite. The solubility of ammonium diselenite in water was measured over the temperature range 253.2 to 343.3 K. The compound crystallises with three moles of water below about 305 K and is anhydrous above this temperature. The solubility of $(NH_4)_2Se_2O_5{\cdot}3H_2O(cr)$ was 9.97 $mol((NH_4)_2Se_2O_5){\cdot}kg^{-1}$ at 298.15 K. The transition temperature agrees with the phase diagrams in [85MIC/PIK] in so far as these indicate that $(NH_4)_2Se_2O_5$ is the stable phase above 303 K. Micka, Pikal, and Ebert obtained $NH_4HSeO_3(cr)$ at 298 K and the composition found by chemical analysis was confirmed by the infrared spectrum, which did not contain frequencies from the –Se–O–Se– group. The review concludes that the phase relationship between the hydrogen selenite and diselenite cannot be settled at present and no figure for the solubility will be selected.

V.15.1.2.3 $NH_4H_3(SeO_3)_2(cr)$

Janickis [34JAN] prepared ammonium trihydrogen selenite by crystallisation from an aqueous mixture of selenious acid and ammonia in the molar ratio 2:1. He studied the solubility of the compound over the temperature interval from 258.4 to 303.2 K. Only anhydrous $NH_4H_3(SeO_3)_2(cr)$ is present as solid and melting occurs at 313.5 K.

The solubility in water estimated by interpolation in [34JAN] is selected:

$$m(NH_4H_3(SeO_3)_2(cr), \text{aq sat}, 298.15\text{ K}) = (32 \pm 1)\ mol(NH_4H_3(SeO_3)_2){\cdot}kg^{-1}.$$

No enthalpies or Gibbs energies of formation of ammonium selenites have been found.

V.15.1.3 Ammonium selenates

The solubility of ammonium selenate, $(NH_4)_2SeO_4(cr)$, at 298.15 K has been recorded in [45KIN/BEC], [80KAD/TAN], [89VOJ/EBE], [89VOJ/EBE2], [94SER] to be 6.565, 6.378, 6.712, 6.686, and 6.613 mol·kg^{-1}, respectively, after recalculation from mass-% to molality. The mean of the three most recent measurements will be selected yielding:

$$m((NH_4)_2SeO_4(cr), \text{aq sat}, 298.15\ K) = (6.67 \pm 0.02)\ \text{mol}((NH_4)_2SeO_4)\cdot kg^{-1}.$$

A phase diagram or similar information is available for the system: $(NH_4)_2SeO_4$-H_2SeO_4-H_2O at 303.15 K in [37KIN]; $(NH_4)_2SeO_4$-Li_2SeO_4-H_2O at 303.15 K in [95HAV/MIC]; $(NH_4)_2SeO_4$-Na_2SeO_4-H_2O at 303.15 K in [77ORL].

The solubility product of $(NH_4)_2SeO_4(cr)$:

$$(NH_4)_2SeO_4(cr) \rightleftharpoons 2NH_4^+ + SeO_4^{2-} \qquad (V.185)$$

was determined at 298.15 K by Ojkova, Christov, and Mihov [99OJK/CHR] to be $\log_{10} K^\circ_{s,0} = 0.920$ and by Kumok and Batyreva [90KUM/BAT] to be 0.45, see Appendix A. The former value is adopted by the review as it is based on isopiestic data treated by the Pitzer ion interaction approach. It was calculated using a solubility of $(NH_4)_2SeO_4(cr)$ = 6.715 mol·kg^{-1} in [99OJK/CHR]. A recalculation with the selected solubility results in the selected value:

$$\log_{10} K^\circ_{s,0}((V.185), 298.15\ K) = (0.911 \pm 0.065).$$

The uncertainty added by the review corresponds to an uncertainty in the mean activity coefficient of 5%.

Selivanova, Prymova, and Kravchenko [71SEL/PRY] measured the enthalpy change of the reaction $(NH_4)_2SeO_4(cr) + Pb(NO_3)_2(aq, 1{:}1600) \rightarrow PbSeO_4(cr) + 2NH_4NO_3(aq, 1{:}800)$ and the enthalpy of dissolution of $(NH_4)_2SeO_4(cr)$ to $(NH_4)_2SeO_4(aq, 1{:}1600)$ in a calorimeter. The review has recalculated the data to standard conditions and obtained for:

$$(NH_4)_2SeO_4(cr) + Pb^{2+} \rightarrow PbSeO_4(cr) + 2NH_4^+ \qquad (V.186)$$

$$\Delta_r H^\circ_m((V.186), 298.15\ K) = (3.67 \pm 1.51)\ kJ\cdot mol^{-1}$$

and for Reaction (V.185):

$$\Delta_r H^\circ_m((V.185), 298.15\ K) = (8.48 \pm 0.52)\ kJ\cdot mol^{-1}.$$

The standard enthalpy of formation of $(NH_4)_2SeO_4(cr)$ found from the enthalpies of reaction and selected enthalpies of formation is $-(878.41 \pm 4.62)$ and $-(878.50 \pm 3.57)$ kJ·mol^{-1}, respectively. The average will be selected:

$$\Delta_f H_m^\circ ((NH_4)_2SeO_4, cr, 298.15 K) = -(878.45 \pm 3.50) kJ \cdot mol^{-1}.$$

The value in [82WAG/EVA] is – 874.5 and in the paper – 881.6 kJ·mol^{-1}.

The standard Gibbs energy of formation of ammonium selenate is calculated from the Gibbs energy of Reaction (V.185) and the selected values of $\Delta_f G_m^\circ$ for the ions to be:

$$\Delta_f G_m^\circ ((NH_4)_2SeO_4, cr, 298.15 K) = -(593.10 \pm 1.60) kJ \cdot mol^{-1}.$$

The selected data yield:

$$S_m^\circ ((NH_4)_2SeO_4, cr, 298.15 K) = (209.6 \pm 12.9) J \cdot K^{-1} \cdot mol^{-1}.$$

V.15.2 Lithium compounds

V.15.2.1 Lithium selenides

Only anhydrous lithium selenides have been considered.

The phase diagram of the lithium-selenium system was assessed by Sangster and Pelton [97SAN/PEL3]. Li_2Se(cr) is the only intermediary compound and it melts at 1575 K.

V.15.2.1.1 Li_2Se(cr)

There is no experimental information available on the heat capacity and entropy of Li_2Se(cr). The heat capacity was estimated to be $C_{p,m}^\circ$ (Li_2Se, cr, (298.15 – 700) K) = (74.81 + 10.293 × 10^{-3} T) J·K^{-1}·mol^{-1} by Smith and Moser [76SMI/MOS] and $C_{p,m}^\circ$ (Li_2Se, cr, (298.15 – 1575) K) = (66.1 + 18.0 × 10^{-3} T) J·K^{-1}·mol^{-1} by Barin [95BAR]. The entropy at 298.15 K has been estimated by Ader [74ADE] to be S_m° (Li_2Se, cr, 298.15 K) = (96.2 ± 21.0) J·K^{-1}·mol^{-1} using Latimer's rule, by Voronin [70VOR] to be S_m° (Li_2Se, cr, 298.15 K) = 69.9 J·K^{-1}·mol^{-1} using a semi-empirical relationship involving vibrational contributions to the entropy and by Mills [74MIL] to be S_m° (Li_2Se, cr, 298.15 K) = (71.1 ± 12.6) J·K^{-1}·mol^{-1} by comparison with Li_2S(cr). No values of the entropy and heat capacity are selected by the review.

The Gibbs energy of formation of Li_2Se(cr) was determined at 633 K by Cairns, Kucera and Cunningham [73CAI/KUC] and at 693 K by [78CHE/MOR] using electrochemical cells. The values obtained were $\Delta_f G_m^\circ$ (Li_2Se, cr, 633 K) = – 393.3 kJ·mol^{-1} and $\Delta_f G_m^\circ$ (Li_2Se, scr, 693 K) = – 402.4 kJ·mol^{-1}, respectively. The Gibbs energy values cannot be recalculated to 298.15 K with sufficient accuracy for a selection by the review because reliable thermal functions are unavailable.

The enthalpy of formation of Li_2Se(cr) was determined to be $\Delta_f H_m^\circ$ (Li_2Se, cr, 298.15 K) = – (425.9 ± 9.0) kJ·mol^{-1} in [74ADE] using bomb calorimetry. The value is less negative than the value $\Delta_f H_m^\circ$ (Li_2Se, cr, 298.15 K) = – (442.3 ± 4.0) kJ·mol^{-1} recalculated from the solution calorimetry measurements of Fabre [1887FAB] as discussed in Appendix A. The average of the two measurements is selected with error bars estimated by the review:

$$\Delta_f H_m^\circ(\text{Li}_2\text{Se, cr, 298.15 K}) = -(434.1 \pm 15.0) \text{ kJ}\cdot\text{mol}^{-1}.$$

V.15.2.2 Lithium selenites

A phase diagram and related information of the system Li_2SeO_3-H_2SeO_3-H_2O at 298.15 K are available in [86MIC/EBE], [86MIC/EBE2]. The crystalline compounds $Li_2SeO_3\cdot H_2O$, $LiHSeO_3$, and $LiH_3(SeO_3)_2$ were found. No solubility data are provided of the intermediate phases.

V.15.2.2.1 Li_2SeO_3(cr) and $Li_2SeO_3\cdot H_2O$(cr)

Rosenheim and Krause [21ROS/KRA] determined the solubility of lithium selenite in water in the temperature range 273.2 to 373.2 K. The composition of the solid phase was $Li_2SeO_3\cdot\frac{3}{4}H_2O$ over the whole temperature interval according to a chemical analysis of Li and Se. Klushina, Selivanova, and Poltavtseva [67KLU/SEL] prepared the salt from a concentrated aqueous solution of LiOH and H_2SeO_3 in stoichiometric proportions. The lithium and selenium contents of the product agreed well with the composition $Li_2SeO_3\cdot H_2O$. The same composition was found by Micka and co-workers [86MIC/EBE]. The review therefore adopts the formula $Li_2SeO_3\cdot H_2O$ for the lithium selenite phase in contact with water.

The solubility of $Li_2SeO_3\cdot H_2O$ in water at 298.15 K is reported to be 1.43 mol(Li_2SeO_3)·kg^{-1} in [21ROS/KRA] and 1.92 mol(Li_2SeO_3)·kg^{-1} in [86MIC/EBE2]. The review selects:

$$m(\text{Li}_2\text{SeO}_3\cdot\text{H}_2\text{O(cr), aq sat, 298.15 K}) = (1.92 \pm 0.10) \text{ mol}(\text{Li}_2\text{SeO}_3)\cdot\text{kg}^{-1}.$$

Klushina, Selivanova, and Poltavtseva [67KLU/SEL] measured the enthalpy change of the reaction between $Li_2SeO_3\cdot H_2O$(cr) and a solution of lead nitrate. Their data have been corrected and evaluated to furnish the enthalpy change of the reaction $Li_2SeO_3\cdot H_2O(cr) + Pb^{2+} \rightarrow PbSeO_3(cr) + 2Li^+ + H_2O(l)$, *cf.* Appendix A. The result is $-(47.17 \pm 1.61)$ kJ·mol^{-1}, which combines with selected standard enthalpies of formation to the selected value:

$$\Delta_f H_m^\circ(\text{Li}_2\text{SeO}_3\cdot\text{H}_2\text{O, cr, 298.15 K}) = -(1328.60 \pm 2.43) \text{ kJ}\cdot\text{mol}^{-1}.$$

The value in [82WAG/EVA] is $-$ 1334.3 kJ·mol^{-1}. This value is probably based on the data in [67KLU/SEL], but the mistake made there (see Appendix A) was apparently not detected.

Klushina, Selivanova, and Artem'eva [68KLU/SEL3] determined the integral heats of dissolution to aq(1:800) of $Li_2SeO_3\cdot H_2O$, $-(20.25 \pm 0.21)$ kJ·mol^{-1}, and Li_2SeO_3, $-(31.21 \pm 0.33)$ kJ·mol^{-1}. These data yield the selected value:

$$\Delta_f H_m^\circ(\text{Li}_2\text{SeO}_3\text{, cr, 298.15 K}) = -(1031.81 \pm 2.46) \text{ kJ}\cdot\text{mol}^{-1}.$$

V.15.2.2.2 $Li_2Se_2O_5$(cr)

Roshchina and Selivanova [68ROS/SEL2] prepared crystalline $Li_2Se_2O_5$(cr) from $LiHSeO_3$(cr) by heating. The specimen was reacted with a solution of $Pb(NO_3)_2$ with formation of crystalline $PbSeO_3$ in a calorimeter. The calorimetric data were used to estimate the enthalpy change of the reaction $Li_2Se_2O_5(cr) + 2Pb^{2+} + H_2O(l) \rightarrow 2PbSeO_3(cr) + 2Li^+ + 2H^+$ in Appendix A to be – (26.24 ± 3.00) kJ·mol^{-1}. The standard enthalpy of formation of $Li_2Se_2O_5$(cr) is calculated from this datum to the selected value:

$$\Delta_f H_m^\circ (Li_2Se_2O_5, cr, 298.15 K) = -(1310.87 \pm 4.70)\ kJ\cdot mol^{-1}$$

with selected auxiliary data.

V.15.2.2.3 $LiHSeO_3$(cr) and $LiH_3(SeO_3)_2$(cr)

Roshchina and Selivanova [68ROS/SEL] reacted a solution of $LiHSeO_3$ with crystalline $Pb(NO_3)_2$ and measured the enthalpy change of the reaction in which $PbSeO_3$(cr) is formed. The result together with the measured enthalpy of dissolution of $LiHSeO_3$(cr) in water was used in Appendix A to estimate the enthalpy change of the reaction $LiHSeO_3(cr) + Pb^{2+} \rightarrow PbSeO_3(cr) + Li^+ + H^+$ to be (6.38 ± 0.25) kJ·mol^{-1}. This datum combines with selected standard enthalpies of formation to the selected value:

$$\Delta_f H_m^\circ (LiHSeO_3, cr, 298.15 K) = -(817.85 \pm 1.84)\ kJ\cdot mol^{-1}.$$

The same authors [70ROS/SEL2] made an analogous study of $LiH_3(SeO_3)_2$(cr). The experimental results obtained are used in Appendix A to estimate the enthalpy change of the reaction $LiH_3(SeO_3)_2(cr) + 2Pb^{2+} \rightarrow 2PbSeO_3(cr) + Li^+ + 3H^+$ to be (27.23 ± 0.53) kJ·mol^{-1}. This value then yields the selected value:

$$\Delta_f H_m^\circ (LiH_3(SeO_3)_2, cr, 298.15 K) = -(1371.7 \pm 3.7)\ kJ\cdot mol^{-1}$$

with selected auxiliary standard enthalpies of formation.

Water vapour measurements of saturated solutions in the system Li_2SeO_3-H_2SeO_3-H_2O were used in [86MIC/EBE2] to obtain the selected Gibbs energies of the reactions:

$$LiHSeO_3(cr) + H_2SeO_3(cr) \rightleftharpoons LiH_3(SeO_3)_2(cr) \quad (V.187)$$

$$\Delta_r G_m^\circ ((V.187), 298.15 K) = -(15.35 \pm 0.90)\ kJ\cdot mol^{-1}.$$

$$Li_2SeO_3{\cdot}H_2O(cr) + LiH_3(SeO_3)_2(cr) \rightleftharpoons 3LiHSeO_3(cr) + H_2O(l) \quad (V.188)$$

$$\Delta_r G_m^\circ ((V.188), 298.15 K) = -(0.93 \pm 0.10)\ kJ\cdot mol^{-1}.$$

V.15.2.3 Lithium selenates

The review selects the consistent result of the solubility of the stable lithium selenate, $Li_2SeO_4{\cdot}H_2O$, at standard conditions obtained in [89TAT/SER], [89VOJ/EBE], [98OJK/CHR]:

$$m(Li_2SeO_4{\cdot}H_2O(cr), \text{aq sat}, 298.15\text{ K}) = (4.275 \pm 0.020)\text{ mol}(Li_2SeO_4){\cdot}kg^{-1}.$$

A phase diagram or similar information is available for the systems: Li_2SeO_4-H_2SeO_4-H_2O at 303.15 K in [95HAV/MIC]; Li_2SeO_4-Cs_2SeO_4-H_2O at 303.15 K in [2000HAV/MIC].

The solubility product of $Li_2SeO_4{\cdot}H_2O(cr)$,

$$Li_2SeO_4{\cdot}H_2O(cr) \rightleftharpoons 2Li^+ + SeO_4^{2-} + H_2O(l) \qquad (V.189)$$

was determined at 298.15 K by Ojkova, Christov, and Mihov [98OJK/CHR] to be $\log_{10} K^{\circ}_{s,0} = 1.762$ and by Kumok and Batyreva [90KUM/BAT] to be 2.05, see Appendix A. The former value:

$$\log_{10} K^{\circ}_{s,0}\,((V.189), 298.15\text{ K}) = (1.762 \pm 0.087)$$

is selected by the review as it based on isopiestic data treated by the Pitzer ion interaction approach. The uncertainty added by the review corresponds to an uncertainty in the mean activity coefficient of 5%.

Selivanova, Shneider and Sazykina [62SEL/SHN] measured the enthalpy change of the reaction $Li_2SeO_4{\cdot}H_2O(cr) + BaCl_2(aq, 1{:}800) \rightarrow BaSeO_4(cr) + 2LiCl(aq, 1{:}400) + H_2O(l)$ in a calorimeter. With the same calorimeter the enthalpies of dissolution of $Li_2SeO_4{\cdot}H_2O(cr)$ and $Li_2SeO_4(cr)$ to $Li_2SeO_4(aq, 1{:}800)$ were also determined. Their results were recalculated with new auxiliary data in Appendix A to the selected values:

$$\Delta_f H^{\circ}_m(Li_2SeO_4{\cdot}H_2O, cr, 298.15\text{ K}) = -(1420.0 \pm 5.6)\text{ kJ}{\cdot}mol^{-1},$$

$$\Delta_f H^{\circ}_m(Li_2SeO_4, cr, 298.15\text{ K}) = -(1127.7 \pm 5.6)\text{ kJ}{\cdot}mol^{-1},$$

which overlap the same quantities in [82WAG/EVA].

The standard Gibbs energy of formation of lithium selenate is calculated from the Gibbs energy of Reaction (V.189) and the selected values of $\Delta_f G^{\circ}_m$ for the ions and water to be:

$$\Delta_f G^{\circ}_m(Li_2SeO_4{\cdot}H_2O, cr, 298.15\text{ K}) = -(1252.4 \pm 1.5)\text{ kJ}{\cdot}mol^{-1}.$$

The selected data yield:

$$S^{\circ}_m(Li_2SeO_4{\cdot}H_2O, cr, 298.15\text{ K}) = (181.8 \pm 19.4)\text{ J}{\cdot}K^{-1}{\cdot}mol^{-1}.$$

V.15.3 Sodium compounds

V.15.3.1 Sodium selenides

Only anhydrous sodium selenides have been considered.

The phase diagram of the sodium-selenium system was assessed by Sangster and Pelton [97SAN/PEL4]. The intermediary compounds Na_2Se(cr), Na_2Se_2(cr), Na_2Se_3(cr), Na_2Se_4(cr), Na_2Se_5(cr) and Na_2Se_6(cr) have been reported, but the existence of the last two compounds is questionable. Na_2Se(cr) and Na_2Se_2(cr) are the only sodium selenides which have been characterised by X-ray diffraction and for which thermodynamic information is available.

V.15.3.1.1 Na_2Se(cr)

The entropy at 298.15 K was estimated by Voronin [70VOR] to be S°_{m} (Na_2Se, cr, 298.15 K) = 104.2 $J{\cdot}K^{-1}{\cdot}mol^{-1}$ using a semi-empirical relationship involving vibrational contributions to the entropy and by Mills [74MIL] to be S°_{m} (Na_2Se, cr, 298.15 K) = (92.0 ± 12.6) $J{\cdot}K^{-1}{\cdot}mol^{-1}$ by comparison with Na_2S(cr). No value for the entropy is selected.

The enthalpy of formation of Na_2Se(cr) was determined to be $\Delta_f H^{\circ}_{m}$ (Na_2Se, cr, 265 K) = – 343.1 $kJ{\cdot}mol^{-1}$ by Mulder and Schmidt [51MUL/SCH] using solution calorimetry in liquid ammonia. The value is reasonably close to $\Delta_f H^{\circ}_{m}$ (Na_2Se, cr, 298.15 K) = – (332.2 ± 5.9) $kJ{\cdot}mol^{-1}$ recalculated from the solution calorimetry measurements of Fabre [1887FAB] as discussed in Appendix A. The average of the two measurements is selected with error bars estimated by the review:

$$\Delta_f H^{\circ}_{m} (Na_2Se, cr, 298.15 K) = -(337.7 \pm 10.0)\ kJ{\cdot}mol^{-1}.$$

V.15.3.1.2 Na_2Se_2(cr)

The enthalpy of the reaction 2Na(in NH_3(l)) + 2Se(trigonal) → Na_2Se_2(in NH_3(l)) was determined to be $\Delta_r H^{\circ}_{m}$ (265 K) = – 418.4 $kJ{\cdot}mol^{-1}$ in [51MUL/SCH]. Mills [74MIL] estimated the enthalpy of dissolution of Na_2Se_2(cr) in liquid ammonia to be $\Delta_{sol} H^{\circ}_{m}$ (Na_2Se_2(cr), NH_3(l), 298.15 K) = 18.4 $kJ{\cdot}mol^{-1}$ by comparison with the analogous reaction for Na_2Te_2(cr) and employed the experimental value of the enthalpy of dissolution of sodium in liquid ammonia in [51MUL/SCH] to calculate $\Delta_f H^{\circ}_{m}$ (Na_2Se_2, cr, 298.15 K) = – (388.3 ± 42.0) $kJ{\cdot}mol^{-1}$. Sangster and Pelton [97SAN/PEL4] erroneously assigned the enthalpy of the reaction above to the enthalpy of formation of Na_2Se_2(cr) in their review. No value is selected for the enthalpy of formation of Na_2Se_2(cr) since it involves an estimated value for the enthalpy of dissolution in ammonia.

The entropy at 298.15 K was estimated by Mills [74MIL] to be S°_{m} (Na_2Se_2, cr, 298.15 K) = (125.5 ± 20.9) $J \cdot K^{-1} \cdot mol^{-1}$ by comparison with the values for Na_2S_2(cr) and Na_2Se(cr). No value for the entropy is selected.

V.15.3.2 Sodium selenites

Sabbah and Périnet [65SAB/PER], and Micka and Ebert [86MIC/EBE2] determined the phase diagram and related data of the system Na_2SeO_3-H_2SeO_3-H_2O at 298.15 K. The two data sets agree well. The salts $Na_2SeO_3 \cdot 5H_2O$, Na_2SeO_3, $NaHSeO_3$, $NaH_3(SeO_3)_2$, and $NaH_7(SeO_3)_4$ were isolated as solid phases.

Solubility data for the system Na_2SeO_3-Na_2SeO_4-H_2O at 273.2, 293.2, 298.2, 303.2, and 313.2 K are presented in [70YAN/PAT]. No intermediate phases are formed.

V.15.3.2.1 Na_2SeO_3(cr), $Na_2SeO_3 \cdot 5H_2O$(cr), and $Na_2SeO_3 \cdot 8H_2O$(cr)

Masson *et al.* have reviewed the system Na_2SeO_3-H_2O in [86MAS/LUT] and found published data to be in reasonable agreement. Three compounds, $Na_2SeO_3 \cdot 8H_2O$(cr), $Na_2SeO_3 \cdot 5H_2O$(cr), and Na_2SeO_3(cr), have been identified in the temperature range 253 to 393 K as stable phases in contact with water. According to [32JAN] the transitions $Na_2SeO_3 \cdot 8H_2O$(cr) $\rightarrow$ $Na_2SeO_3 \cdot 5H_2O$(cr) and $Na_2SeO_3 \cdot 5H_2O$(cr) $\rightarrow$ Na_2SeO_3(cr) occur at (264.5 ± 0.3) K and (309.1 ± 0.2) K, respectively. Masson derived the following relationships for the temperature dependence of the solubility, S, in mass-% of Na_2SeO_3. The symbol s(S) denotes the standard deviation of S about the regression line.

$Na_2SeO_3 \cdot 8H_2O$ (253 – 264) K: $S = 48.2 + 0.821\,(T - 273.2) + 0.0176\,(T - 273.2)^2$; s($S$) = 0.19,

$Na_2SeO_3 \cdot 5H_2O$ (264 – 310) K: $S = 43.2 + 0.108\,(T - 273.2) + 0.00205\,(T - 273.2)^2$; s($S$) = 0.15,

Na_2SeO_3 (310 – 373) K: $S = 54.0 - 0.133\,(T - 273.2) + 0.000483\,(T - 273.2)^2$; s($S$) = 0.21.

The calculated solubility of $Na_2SeO_3 \cdot 5H_2O$(cr) at 298.15 K is:

$$m(Na_2SeO_3 \cdot 5H_2O\text{(cr), aq sat, 298.15 K}) = (5.15 \pm 0.30)\ \text{mol}(Na_2SeO_3) \cdot \text{kg}^{-1}.$$

Amitin, Klushina, Lebedeva, and Paukuv [80AMI/KLU] determined the heat capacity of Na_2SeO_3(cr) in the temperature range 6 to 300 K. The selected results are:

$$C^{\circ}_{p,m}(Na_2SeO_3\text{, cr, 298.15 K}) = (126.1 \pm 0.3)\ \text{J} \cdot \text{K}^{-1} \cdot \text{mol}^{-1},$$

$$S^{\circ}_{m}(Na_2SeO_3\text{, cr, 298.15 K}) = (147.4 \pm 0.3)\ \text{J} \cdot \text{K}^{-1} \cdot \text{mol}^{-1},$$

$$H^{\circ}_{m}(Na_2SeO_3\text{, cr, 298.15 K}) - H^{\circ}_{m}(Na_2SeO_3\text{, cr, 0 K}) = (23.00 \pm 0.05)\ \text{kJ} \cdot \text{mol}^{-1}.$$

The standard enthalpy of formation of Na_2SeO_3(cr) is evaluated from the data by Thomsen [1882THO] in Appendix A to be – (959.62 ± 0.95) $kJ \cdot mol^{-1}$. This value agrees reasonably well with the same datum in [82WAG/EVA], -958.6 kJ·mol^{-1}, probably also derived from [1882THO]. The selected values is:

$$\Delta_f H_m^\circ (Na_2SeO_3, cr, 298.15 K) = -(959.62 \pm 0.95) kJ\cdot mol^{-1}.$$

Carre *et al.* [79CAR/GER] determined the enthalpy of formation of Na_2SeO_3(aq, 1 M, 298.15 K) to be $-(984.7 \pm 3.4)$ kJ·mol^{-1} from the enthalpy of reaction of SeO_2(cr) with a sodium hydroxide solution. This implies a heat of dissolution to 1 M solution of $-(25.1 \pm 3.5)$ kJ·mol^{-1}, which is compatible with known heats of solution of sodium selenite.

Klushina, Selivanova, Lapin, and Artem'eva [69KLU/SEL] measured the heat of dissolution of Na_2SeO_3(cr) and $Na_2SeO_3\cdot 5H_2O$(cr) in water to the same concentration (aq, 1:800). The standard enthalpy of formation of the pentahydrate is calculated from the known standard enthalpy of formation of the anhydrous salt in Appendix A. The selected result is:

$$\Delta_f H_m^\circ (Na_2SeO_3\cdot 5H_2O, cr, 298.15 K) = -(2404.67 \pm 1.00) kJ\cdot mol^{-1},$$

which is almost identical with the value in [82WAG/EVA].

Shkodin, Malyshev, and Makhmetov [71SHK/MAL] measured the vapour pressure of SeO_2(g) above solid and molten sodium selenite by the dew point method. The paper is discussed in Appendix A. No thermodynamic quantities will be derived from it.

V.15.3.2.2 $Na_2Se_2O_5$(cr)

The solubility of $Na_2Se_2O_5$(cr) has been studied by Janitzki [32JAN]. The compound is stable only above (366.2 ± 0.5) K in the aqueous environment. Below this temperature it transforms to hydrogen selenites. The solubility at the transition temperature is about 11.0 mol·kg^{-1}.

Roshchina and Selivanova [69ROS/SEL4] prepared $Na_2Se_2O_5$(cr) and reacted it with a solution of $Pb(NO_3)_2$ in a calorimeter. The enthalpy change of the reaction $Na_2Se_2O_5(cr) + H_2O(l) + 2Pb^{2+} \rightarrow 2PbSeO_3(cr) + 2Na^+ + 2H^+$ is estimated from the calorimetric data to be $-(41.21 \pm 3.00)$ kJ·mol^{-1} in Appendix A. This datum combines with selected standard enthalpies of formation to give the selected value:

$$\Delta_f H_m^\circ (Na_2Se_2O_5, cr, 298.15 K) = -(1219.64 \pm 4.71) kJ\cdot mol^{-1}.$$

Shkrodin, Makhmetov, and Malyshev [75SHK/MAK] studied the thermal behaviour of sodium diselenite. The compound melts at (653 ± 10) K and dissociates above the melting point according to the reaction:

$$Na_2Se_2O_5(l) \rightleftharpoons Na_2SeO_3(cr) + SeO_2(g) \qquad (V.190)$$

The vapour pressure of SeO_2(g) was measured by the dew point method in the temperature range 703 to 963 K yielding $\log_{10} p_{SeO_2} = (8.239 \pm 0.500) - (7848 \pm 500)$

T/T^{-1}. The uncertainties have been added by the review as no primary data are furnished. No experimental data are available for an extrapolation to the standard state.

V.15.3.2.3 $NaHSeO_3$(cr) and $NaH_3(SeO_3)_2$(cr)

Janitzki [32JAN] studied the system $NaHSeO_3$-H_2O in the temperature interval 263.9 to 382.7 K. Three solid phases were identified, $NaHSeO_3{\cdot}3H_2O$(cr), $NaHSeO_3$(cr), and $Na_2Se_2O_5$(cr). The transition $NaHSeO_3{\cdot}3H_2O$(cr) $\rightarrow$ $NaHSeO_3$(cr) occurs at (300.3 ± 0.3) K and the transition $NaHSeO_3$(cr) $\rightarrow$ $Na_2Se_2O_5$(cr) at (366.2 ± 0.5) K. The solubility of $NaHSeO_3{\cdot}3H_2O$(cr) in water at 298.15 K is calculated by interpolation from data in [32JAN] to be:

$$m(\text{NaHSeO}_3{\cdot}3\text{H}_2\text{O(cr), aq sat, 298.15 K}) = (10.6 \pm 0.2)\ \text{mol(NaHSeO}_3)\cdot\text{kg}^{-1}.$$

Janitzki [32JAN] investigated the system $NaH_3(SeO_3)_2$-H_2O in the temperature range 266.3 to 361.2 K. Only one solid phase, $NaH_3(SeO_3)_2$(cr), is present. Its solubility in water at 298.15 K is calculated by interpolation from the data in [32JAN] to be:

$$m(\text{NaH}_3(\text{SeO}_3)_2\text{(cr), aq sat, 298.15 K}) = (4.31 \pm 0.15)\ \text{mol(NaH}_3(\text{SeO}_3)_2)\cdot\text{kg}^{-1}.$$

Thomsen [1882THO] measured the heat of neutralisation of selenious acid to give sodium hydrogen selenite. The result is used to estimate the standard enthalpy of formation of $NaHSeO_3$(cr) to be – (758.20 ± 1.10) $kJ\cdot mol^{-1}$ in Appendix A. Selivanova and Roshchina [68SEL/ROS] measured the enthalpy of reaction between $NaHSeO_3$(cr) and a lead nitrate solution with formation of lead selenite. Their data were used to estimate the enthalpy change of the reaction $NaHSeO_3$(cr) + Pb^{2+} $\rightarrow$ $PbSeO_3$(cr) + H^+ + Na^+ to be – (11.48 ± 1.50) $kJ\cdot mol^{-1}$ in Appendix A. This datum yields with selected standard enthalpies of formation $\Delta_f H_m^\circ$ ($NaHSeO_3$, cr, 298.15 K) = – (761.86 ± 2.35) $kJ\cdot mol^{-1}$. The value in [82WAG/EVA] is – 759.23 and believed to be based on the data in [1882THO]. The review selects:

$$\Delta_f H_m^\circ(\text{NaHSeO}_3\text{, cr, 298.15 K}) = -(760.10 \pm 2.50)\ \text{kJ}\cdot\text{mol}^{-1}.$$

The enthalpies of dissolution of $NaHSeO_3$(cr) and $NaHSeO_3{\cdot}3H_2O$(cr) in water to aq(1:800) have been measured to be (5.67 ± 0.01) $kJ\cdot mol^{-1}$ in [68SEL/ROS] and (30.63 ± 0.04) $kJ\cdot mol^{-1}$ in [69ROS/SEL], respectively. These data yield in combination with the standard enthalpy of formation of the anhydrous salt the selected value:

$$\Delta_f H_m^\circ(\text{NaHSeO}_3{\cdot}3\text{H}_2\text{O, cr, 298.15 K}) = -(1642.55 \pm 2.50)\ \text{kJ}\cdot\text{mol}^{-1}.$$

Roshchina, Selivanova, and Anan'ina [69ROS/SEL2] prepared $NaH_3(SeO_3)_2$ and reacted it with an equivalent amount of a lead nitrate solution in a calorimeter. The enthalpy data were used to calculate the enthalpy change of the reaction $NaH_3(SeO_3)_2$(cr) + $2Pb^{2+}$ $\rightarrow$ $2PbSeO_3$(cr) + Na^+ + $3H^+$ to be (10.30 ± 3.00) $kJ\cdot mol^{-1}$ in Appendix A. Combined with selected enthalpies of formation this datum results in the selected value:

$$\Delta_f H_m^\circ(\text{NaH}_3(\text{SeO}_3)_2, \text{cr}, 298.15 \text{ K}) = -(1316.64 \pm 4.71) \text{ kJ·mol}^{-1}.$$

Micka and Ebert [86MIC/EBE2] measured the water vapour pressure of the saturated solutions in the system Na_2SeO_3-H_2SeO_3-H_2O. From these measurements they calculated the selected Gibbs energies of the reactions:

$$\text{NaH}_3(\text{SeO}_3)_2(\text{cr}) + 2\text{H}_2\text{SeO}_3(\text{cr}) \rightleftharpoons \text{NaH}_7(\text{SeO}_3)_4(\text{cr}) \quad (\text{V.191})$$

$$\Delta_r G_m^\circ((\text{V.191}), 298.15 \text{ K}) = -(6.20 \pm 0.40) \text{ kJ·mol}^{-1},$$

$$\text{NaHSeO}_3(\text{cr}) + \text{H}_2\text{SeO}_3(\text{cr}) \rightleftharpoons \text{NaH}_3(\text{SeO}_3)_2(\text{cr}) \quad (\text{V.192})$$

$$\Delta_r G_m^\circ((\text{V.192}), 298.15 \text{ K}) = -(18.05 \pm 1.10) \text{ kJ·mol}^{-1},$$

$$\text{Na}_2\text{SeO}_3(\text{cr}) + \text{NaH}_3(\text{SeO}_3)_2(\text{cr}) \rightleftharpoons 3\text{NaHSeO}_3(\text{cr}) \quad (\text{V.193})$$

$$\Delta_r G_m^\circ((\text{V.193}), 298.15 \text{ K}) = -(6.33 \pm 0.40) \text{ kJ·mol}^{-1}.$$

V.15.3.3 Sodium selenates

Funk [00FUN], Meyer and Aulich [28MEY/AUL], and Yanitskii and Patkauskas [70YAN/PAT] determined the solubility of sodium selenate in water as a function of temperature with concordant results. $Na_2SeO_4 \cdot 10H_2O$(cr) is the stable phase below 305.0 K and Na_2SeO_4(cr) above this temperature.

A phase diagram or similar information is available for the system: Na_2SeO_4-H_2SeO_4-H_2O at 298.15 K in [71SEL/SOB]; Na_2SeO_4-NaOH-H_2O at 291.15 K in [51WIN/STO]; Na_2SeO_4-Na_2SO_4-H_2O at 298.15 K in [51WIN/STO]; Na_2SeO_4-Na_2SeO_3-H_2O at 273 to 333 K in [70YAN/PAT]; Na_2SeO_4-K_2SeO_4-H_2O at 298.15 K in [28MEY/AUL].

V.15.3.3.1 Na_2SeO_4(cr)

Selivanova and Sazykina [64SEL/SAZ] determined the enthalpy change of the reaction Na_2SeO_4(cr) + $BaCl_2$(aq, 1:800) → $BaSeO_4$(cr) + 2NaCl(aq, 1:400) by a calorimetric measurement. Their enthalpy of reaction has been recalculated with new auxiliary data to be – (15.53 ± 0.24) kJ·mol^{-1} at standard conditions in Appendix A. This value was combined with selected standard enthalpies of formation to obtain $\Delta_f H_m^\circ$ (Na_2SeO_4, cr, 298.15 K) = – (1074.35 ± 5.60) kJ·mol^{-1}.

The value of $\Delta_f H_m^\circ$ (Na_2SeO_4, aq, 298.15 K) calculated from the selected ionic enthalpies is – (1084.18 ± 3.50) kJ·mol^{-1}. This implies a total enthalpy of dissolution of Na_2SeO_4(cr) to the aqueous standard state of – 9.8 kJ·mol^{-1}. The heat of dissolution of the salt to Na_2SeO_4(aq, 1:800) determined in [62SEL/SAZ3] is – (4.06 ± 0.17) kJ·mol^{-1}. The enthalpy of dilution of Na_2SeO_4(aq, 1:800) to the standard state would then be about – 5.7 kJ·mol^{-1}, which appears too large in comparison with the corresponding value for Na_2SO_4, – 1.02 kJ·mol^{-1}. This comparison thus indicates that the value of $\Delta_f H_m^\circ$ (Na_2SeO_4, cr, 298.15 K) should be somewhat more exotermic than calculated from the data in [64SEL/SAZ]. On the other hand, the enthalpies of formation in [82WAG/EVA] yield – 8.4, respectively, – 10.1 kJ·mol^{-1} for the enthalpies of dissolution of Na_2SeO_4(cr) to (aq, 1:500) and (aq, 1:5000). These figures thus support the result in [64SEL/SAZ]. The review calculates $\Delta_f H_m^\circ$ (Na_2SeO_4, cr, 298.15 K) = – (1079.10 ± 3.50) kJ·mol^{-1} from the ionic enthalpies, the heat of dissolution in [62SEL/SAZ3] and the estimated enthalpy of dilution to the aqueous standard state and selects:

$$\Delta_f H_m^\circ (\text{Na}_2\text{SeO}_4, \text{cr}, 298.15\ \text{K}) = -(1076.50 \pm 3.50)\ \text{kJ·mol}^{-1}.$$

The value in [82WAG/EVA] is – 1069.0 kJ·mol^{-1}.

Vasilev and Kruchina [91VAS/KRU] determined the heat capacity of solutions of sodium selenate (m < 1.50 mol·kg^{-1}) at 298.15 K:

$$C_{p,m}^\circ (\text{Na}_2\text{SeO}_4, \text{aq}, 298.15\ \text{K}) = (4.1795 - 0.8386 \times m + 0.3143 \times m^2 - 0.0679 \times m^3)\ \text{J·K}^{-1}\text{·mol}^{-1}.$$

V.15.3.3.2 $Na_2SeO_4·10H_2O$(cr)

The selected value of the solubility of $Na_2SeO_4·10H_2O$(cr) in water at 298.15 K:

$$m(\text{Na}_2\text{SeO}_4\text{·10H}_2\text{O(cr), aq sat}, 298.15\ \text{K}) = (3.05 \pm 0.05)\ \text{mol(Na}_2\text{SeO}_4\text{)·kg}^{-1}$$

is based on work in [28MEY/AUL], [71SEL/SOB], [90TAT/SER], [98CHR/OJK] reporting 3.030, 3.074, 3.070, and 3.029 mol·kg^{-1}, respectively.

The solubility product of $Na_2SeO_4·10H_2O$(cr),

$$\text{Na}_2\text{SeO}_4\text{·10H}_2\text{O(cr)} \rightleftharpoons 2\text{Na}^+ + \text{SeO}_4^{2-} + 10\text{H}_2\text{O(l)} \qquad \text{(V.194)}$$

was determined at 298.15 K by Christov, Ojkova, and Mihov [98CHR/OJK] to be $\log_{10} K_{s,0}^\circ$ = – 0.690 and by Kumok and Batyreva [90KUM/BAT] to be – 0.41. The former value is selected by the review as it based on isopiestic data treated by the Pitzer ion interaction approach. With the selected value of the solubility the value is changed to be:

$$\log_{10} K_{s,0}^\circ ((\text{V.194}), 298.15\ \text{K}) = -(0.681 \pm 0.087).$$

The uncertainty added by the review corresponds to an uncertainty in the mean activity coefficient of 5%.

The standard enthalpy of formation of $Na_2SeO_4·10H_2O$(cr) is calculated from the standard enthalpy of formation of Na_2SeO_4(cr) and the enthalpies of dissolution of the two salts to Na_2SeO_4(aq, 1:800). The two latter quantities were determined by Selivanova and Sazykina [62SEL/SAZ3] to be – (4.06 ± 0.17) kJ·mol^{-1} for the anhydrous salt and (74.73 ± 0.63) kJ·mol^{-1} for the hydrate. The selected result is:

$$\Delta_f H_m^\circ (\text{Na}_2\text{SeO}_4\text{·10H}_2\text{O}, \text{cr}, 298.15\ \text{K}) = -(4013.6 \pm 3.6)\ \text{kJ·mol}^{-1}.$$

The standard Gibbs energy of formation of $Na_2SeO_4\cdot10H_2O$ is calculated from the Gibbs energy of Reaction (V.194) and the selected values of $\Delta_f G^\circ_m$ for the ions and water to be:

$$\Delta_f G^\circ_m (Na_2SeO_4\cdot10H_2O, cr, 298.15 K) = -(3338.70 \pm 1.60) kJ\cdot mol^{-1}.$$

The selected data yield:

$$S^\circ_m (Na_2SeO_4\cdot10H_2O, cr, 298.15 K) = (624.0 \pm 13.2) J\cdot K^{-1}\cdot mol^{-1}.$$

V.15.3.3.3 $Na_2Se_2O_7$(cr)

Sobol and Selivanova [70SOB/SEL] prepared crystalline sodium diselenate, $Na_2Se_2O_7$, by gradual heating of $NaHSeO_4$ to 475 K. Its standard enthalpy of formation was obtained from calorimetric measurements and three thermochemical cycles, see Appendix A. The mean:

$$\Delta_f H^\circ_m (Na_2Se_2O_7, cr, 298.15 K) = -(1322.0 \pm 5.0) kJ\cdot mol^{-1}$$

is selected.

V.15.3.3.4 $NaHSeO_4$(cr)

Sobol and Selivanova [73SOB/SEL] measured the enthalpy change of the reaction $NaHSeO_4$(cr) + $BaCl_2$(aq, 1:800) → $BaSeO_4$(cr) + NaCl(aq, 1:800) + HCl(aq, 1:800) in a calorimeter. The result has been recalculated with new auxiliary data in Appendix A to be:

$$\Delta_f H^\circ_m (NaHSeO_4, cr, 298.15 K) = -(822.23 \pm 5.59) kJ\cdot mol^{-1}.$$

The selected value agrees with the value in [82WAG/EVA] within the error limits.

V.15.4 Potassium compounds

V.15.4.1 Potassium selenides

Only anhydrous potassium selenides have been considered.

The phase diagram of the potassium-selenium system was assessed by Sangster and Pelton [97SAN/PEL2]. The intermediary compounds K_2Se(cr), K_2Se_2(cr), K_2Se_3(cr), K_2Se_4(cr), and K_2Se_5(cr) have been reported, but thermodynamic information is only available for K_2Se(cr).

V.15.4.1.1 K_2Se(cr)

The entropy of K_2Se(cr) was estimated by Voronin [70VOR] to be S°_m(K_2Se, cr, 298.15 K) = 117.15 $J\cdot K^{-1}\cdot mol^{-1}$ using a semi-empirical relationship involving vibrational contributions to the entropy and by Mills [74MIL] to be S°_m(K_2Se, cr,

298.15 K) = (125.5 ± 12.6) J·K^{-1}·mol^{-1} by comparison with K_2S(cr). No value for the entropy is selected.

The enthalpy of formation of K_2Se(cr) at 298.15 K was determined by Fabre [1887FAB]. The measurements were recalculated by the review using recent data for some reactions as discussed in Appendix A. The value obtained is selected with an uncertainty estimated by the review:

$$\Delta_f H_m^\circ (K_2Se, cr, 298.15 K) = -(398.7 \pm 10.0) kJ·mol^{-1}.$$

V.15.4.2 Potassium selenites

Sabbah and Périnet [66SAB/PER] and Micka and Ebert [86MIC/EBE2] determined the phase diagram and related data at 298.15 K of the system K_2SeO_3-H_2SeO_3-H_2O. The two studies agree reasonably well except around 50 mass-% H_2SeO_3, where large changes in the concentration of K_2SeO_3 occur. The agreement is only moderate between the two sets of data in this region. The crystalline phases K_2SeO_3 $K_5H_3(SeO_3)_4$, $KHSeO_3$, and $KH_3(SeO_3)_2$ were isolated in both studies. The phase diagrams at 293.15 and 303.15 K are available in [86MIC/EBE3], but the only solubility data reported are the compositions at the invariant points. The diselenite, $K_2Se_2O_5$, forms at 303.15 K instead of the hydrogen selenite.

V.15.4.2.1 K_2SeO_3(cr) and $K_2SeO_3·4H_2O$(cr)

Masson *et al.* have reviewed the system K_2SeO_3-H_2O in [86MAS/LUT] and found published data to be in reasonable agreement. Two solid phases, $K_2SeO_3·4H_2O$(cr) and K_2SeO_3(cr), have been identified as stable phases in contact with water in the temperature range 253 to 393 K. The transition $K_2SeO_3·4H_2O$(cr) → K_2SeO_3(cr) occurs at (297.5 ± 0.3) K according to Janitzki [32JAN]. Masson derived the following relationships for the temperature dependence of the solubility, *S*, in mass-% of K_2SeO_3. The symbol s(*S*) denotes the standard deviation of *S* about the regression line:

$K_2SeO_3·4H_2O$ (253 – 297) K: $S = 62.8 + 0.182 (T - 273.2) + 0.00198 (T - 273.2)^2$; $s(S) = 0.11$,

K_2SeO_3 (297 – 353) K: $S = 68.42 + 0.0000749 (T - 273.2)^2$; $s(S) = 0.04$.

The calculated solubility of K_2SeO_3(cr) at 298.15 K is:

$$m(K_2SeO_3(cr), aq sat, 298.15 K) = (10.58 \pm 0.10) mol(K_2SeO_3)·kg^{-1}$$

which agrees with the value of 10.66 mol(K_2SeO_3)·kg^{-1} in [86MIC/EBE2]. (This datum was not included in the regression by Masson *et al.*).

Klushina, Selivanova, Lapin, and Novikova [68KLU/SEL] prepared crystalline anhydrous potassium selenite. The compound was reacted with a lead nitrate solution in a calorimeter. The measured enthalpy change of the reaction was used in Appendix A to calculate the enthalpy change of the reaction $K_2SeO_3(cr) + Pb^{2+} \rightarrow PbSeO_3(cr) + 2K^+$ to be – (56.19 ± 1.50) $kJ \cdot mol^{-1}$. From this value:

$$\Delta_f H^\circ_m (K_2SeO_3, cr, 298.15 K) = -(981.09 \pm 2.36) kJ \cdot mol^{-1}$$

is obtained with selected standard enthalpies of formation. The selected result agrees with [82WAG/EVA] within the estimated uncertainty.

V.15.4.2.2 $K_2Se_2O_5(cr)$ and $KHSeO_3(cr)$

Janitzki [32JAN] studied the system $KHSeO_3$-H_2O in the temperature interval 252.6 to 376.0 K. Two solid phases were identified, $KHSeO_3(cr)$ and $K_2Se_2O_5(cr)$. The transition $KHSeO_3(cr) \rightarrow K_2Se_2O_5(cr)$ occurs at (298 ± 1) K in agreement with the phase diagrams established in [86MIC/EBE2], [86MIC/EBE3] and is very sluggish. The solubility of $K_2Se_2O_5(cr)$ in water at 298.15 K is calculated by interpolation from data in [32JAN] to be:

$$m(K_2Se_2O_5(cr), aq sat, 298.15 K) = (10.6 \pm 0.2) mol(K_2Se_2O_5) \cdot kg^{-1}.$$

Roshchina and Selivanova [70ROS/SEL] reacted a solution of $KHSeO_3$ with crystalline $Pb(NO_3)_2$ and measured the enthalpy change of the reaction in which $PbSeO_3(cr)$ is formed. The result together with the measured enthalpy of dissolution of $KHSeO_3(cr)$ in water was used in Appendix A to estimate the enthalpy change of the reaction $KHSeO_3(cr) + Pb^{2+} \rightarrow PbSeO_3(cr) + K^+ + H^+$ to be – (11.65 ± 0.25) $kJ \cdot mol^{-1}$. This datum combines with selected standard enthalpies of formation to give the selected value:

$$\Delta_f H^\circ_m (KHSeO_3, cr, 298.15 K) = -(773.49 \pm 1.84) kJ \cdot mol^{-1}.$$

In the same reference the authors prepared crystalline $K_2Se_2O_5(cr)$ from $KHSeO_3(cr)$ by heating. The specimen was reacted with a solution of $Pb(NO_3)_2$ in a calorimeter with formation of crystalline $PbSeO_3$. The calorimetric data were used to calculate the enthalpy change of the reaction $K_2Se_2O_5(cr) + 2Pb^{2+} + H_2O(l) \rightarrow 2PbSeO_3(cr) + 2K^+ + 2H^+$ in Appendix A to be – (34.40 ± 3.00) $kJ \cdot mol^{-1}$. The standard enthalpy of formation of $K_2Se_2O_5(cr)$ is calculated from this datum to the selected value:

$$\Delta_f H^\circ_m (K_2Se_2O_5, cr, 298.15 K) = -(1250.05 \pm 4.72) kJ \cdot mol^{-1}$$

with selected auxiliary data.

V.15.4.2.3 $KH_3(SeO_3)_2(cr)$

Janitzki [32JAN] investigated the system $KH_3(SeO_3)_2$-H_2O in the temperature range 266.3 to 332.6 K. Only one solid phase, $KH_3(SeO_3)_2(cr)$, is present. The compound melts at about 350 K and its solubility in water at 298.15 K is estimated by interpolation from the data in [32JAN] to be:

$$m(KH_3(SeO_3)_2(cr), \text{aq sat}, 298.15 \text{ K}) = (8.75 \pm 0.25) \text{ mol}(KH_3(SeO_3)_2)\cdot\text{kg}^{-1}.$$

Roshchina and Selivanova [70ROS/SEL3] prepared and reacted $KH_3(SeO_3)_2$ with an equivalent amount of a lead nitrate solution in a calorimeter. The enthalpy data were used to calculate the enthalpy change of the reaction $KH_3(SeO_3)_2(cr) + 2Pb^{2+} \rightarrow 2PbSeO_3(cr) + K^+ + 3H^+$ to be (10.94 ± 3.00) kJ·mol^{-1} in Appendix A. Combined with selected enthalpies of formation this datum results in the selected value:

$$\Delta_f H_m^\circ (KH_3(SeO_3)_2, \text{cr}, 298.15 \text{ K}) = -(1329.08 \pm 4.71) \text{ kJ}\cdot\text{mol}^{-1}.$$

Micka and Ebert [86MIC/EBE2] measured the water vapour pressure of saturated solutions in the system K_2SeO_3-H_2SeO_3-H_2O. From these measurements they calculated the selected Gibbs energy of the reaction:

$$K_2SeO_3(cr) + KH_3(SeO_3)_2(cr) \rightleftharpoons 3KHSeO_3(cr) \qquad \text{(V.195)}$$

$$\Delta_r G_m^\circ ((\text{V.195}), 298.15 \text{ K}) = -(6.24 \pm 0.40) \text{ kJ}\cdot\text{mol}^{-1}.$$

V.15.4.3 Potassium selenates

Meyer and Aulich [28MEY/AUL] studied the solubility of potassium selenate in water as a function of temperature up to about 335 K and found the anhydrous salt, $K_2SeO_4(cr)$, to be the stable phase. A similar study by Friend [29FRI] between 273.2 and 370.0 K yielded a 1 to 2% higher solubility in the overlapping range.

The selected value of the solubility of $K_2SeO_4(cr)$ in water at 298.15 K:

$$m(K_2SeO_4(cr), \text{aq sat}, 298.15 \text{ K}) = (5.14 \pm 0.02) \text{ mol}(K_2SeO_4)\cdot\text{kg}^{-1}$$

is based on the measurements in [28MEY/AUL], [40HIL/SOT], [89VOJ/EBE2], [93SER/KUC] reporting 5.142, 5.152, 5.130, and 5.140 mol·kg^{-1}, respectively.

The solubility product of $K_2SeO_4(cr)$,

$$K_2SeO_4(cr) \rightleftharpoons 2K^+ + SeO_4^{2-} \qquad \text{(V.196)}$$

was determined by and Ojkova, Christov, and Mihov [99OJK/CHR], see Appendix A, to be $\log_{10} K_{s,0}^\circ = 0.902$. Kumov and Batyreva [90KUM/BAT] obtained $-(0.73 \pm 0.04)$ for this datum from their isopiestic measurements, which differ greatly from those in [99OJK/CHR]. As the measurements by Christov and co-workers of osmotic coefficients of other alkali metal selenates are close to those obtained for potassium, the result of Kumov and Batyreva is not considered. The solubility selected by the review changes the value of the solubility product to be:

$$\log_{10} K_{s,0}^\circ ((\text{V.196}), 298.15 \text{ K}) = (0.904 \pm 0.065).$$

The uncertainty added by the review to the selected value corresponds to an uncertainty in the mean activity coefficient of 5%.

The standard Gibbs energy of formation of potassium selenate is calculated from the Gibbs energy of Reaction (V.196) and the selected values of $\Delta_f G^\circ_m$ for the ions to be:

$$\Delta_f G^\circ_m (K_2SeO_4, cr, 298.15 K) = -(999.3 \pm 1.5) kJ\cdot mol^{-1}$$

The unreferenced value in [82WAG/EVA] is – 1002.8 kJ·mol^{-1}.

Atake *et al.* [83ATA/NOM2] measured the heat capacity of K_2SeO_4(cr) between 3 and 300 K. The thermodynamic functions are given in a table for this temperature range and contains the selected values:

$$C^\circ_{p,m} (K_2SeO_4, cr, 298.15 K) = (142.64 \pm 1.00) J\cdot K^{-1}\cdot mol^{-1}$$

$$S^\circ_m (K_2SeO_4, cr, 298.15 K) = (202.1 \pm 1.0) J\cdot K^{-1}\cdot mol^{-1}.$$

The uncertainties have been estimated by the review. The selected data yield $\Delta_f H^\circ_m$ (K_2SeO_4, cr, 298.15 K) = – (1112.5 ± 1.5) kJ·mol^{-1}.

An experimental value of standard enthalpy of formation of K_2SeO_4(cr) was calculated in Appendix A from the calorimetric measurement by Sazikina and Selivanova [63SAZ/SEL] of the reaction K_2SeO_4(cr) + 2$AgNO_3$(aq, 1:400) → Ag_2SeO_4(cr) + 2KNO_3(aq, 1:400) to be – (1121.29 ± 2.15) kJ·mol^{-1}. The experimental value is thus considerably more negative than the calculated value. The enthalpy of dissolution of K_2SeO_4(cr) to K_2SeO_4(aq, 1:440) is reported to be (6.82 ± 0.04) kJ·mol^{-1} in [62SEL/ZUB]. The enthalpies of dilution of Na_2SO_4, K_2SO_4, and Rb_2SO_4 from (aq, 1:440) to the aqueous standard state are all about – 1 kJ·mol^{-1}. If the enthalpy of dilution of K_2SeO_4(aq, 1:440) to the aqueous standard state is assumed to be the same as for the sulphate, the ionic enthalpies yield $\Delta_f H^\circ_m$ (K_2SeO_4, cr, 298.15 K) = – (1113.62 ± 3.50) kJ·mol^{-1}. The calorimetric result thus appears to be in error and it is not accepted.

For consistency,

$$\Delta_f H^\circ_m (K_2SeO_4, cr, 298.15 K) = -(1112.5 \pm 1.5) kJ\cdot mol^{-1}$$

is selected. The value reported in [82WAG/EVA] is – 1110.02 kJ·mol^{-1} and of unknown origin.

V.15.5 Rubidium compounds

V.15.5.1 Rubidium selenides

The phase diagram of the rubidium-selenium system was assessed by Sangster and Pelton [97SAN/PEL5]. The intermediary compounds Rb_2Se(cr), Rb_2Se_2(cr), Rb_2Se_3(cr), Rb_2Se_4(cr), and Rb_2Se_5(cr) have been reported, but Rb_2Se(cr) is the only rubidium selenide for which thermodynamic information is available.

V.15.5.1.1 $Rb_2Se(cr)$

The heat capacity of $Rb_2Se(cr)$ has been measured in the temperature range 310 to 800 K using differential scanning calorimetry by Kohli [97KOH]. The heat capacity was described by the expression $C^{\circ}_{p,m}$ (Rb_2Se, cr, (298.15 – 800) K) = (66.801 + 25.85 × 10^{-3} T + 0.129 × 10^{5} T^{-2}) $J \cdot K^{-1} \cdot mol^{-1}$ and the value at 298.15 K calculated from this expression is selected:

$$C^{\circ}_{p,m}(\text{Rb}_2\text{Se, cr, 298.15 K}) = (74.7 \pm 2.0)\ \text{J}\cdot\text{K}^{-1}\cdot\text{mol}^{-1}.$$

The enthalpy of formation of $Rb_2Se(cr)$ was given as $\Delta_f H^{\circ}_m$ (Rb_2Se, cr, 298.15 K) = – (326.4 ± 12) $kJ \cdot mol^{-1}$ by Yatsenko, Kuznetsov and Chuntov [77YAT/KUZ], referring to unpublished results. The experimental technique was not mentioned. No value for the enthalpy of formation of $Rb_2Se(cr)$ has been selected.

V.15.5.2 Rubidium selenites

Micka and Ebert have investigated the system Rb_2SeO_3-H_2SeO_3-H_2O at 298.15 K. The solubility measurement is presented in [86MIC/EBE2] and the phase diagram in [87MIC/EBE]. The crystalline phases $RbH_3(SeO_3)_2$, $RbHSeO_3$, and Rb_2SeO_3 were identified. The solubility of $Rb_2SeO_3(cr)$ is reported to be:

$$m(\text{Rb}_2\text{SeO}_3\text{(cr), aq sat, 298.15 K}) = 7.86\ \text{mol}(\text{Rb}_2\text{SeO}_3)\cdot\text{kg}^{-1}.$$

Klushina, Selivanova, Lapin, and Fedyainov [69KLU/SEL3] reacted $Rb_2SeO_3(cr)$ with a lead nitrate solution in a calorimeter with formation of crystalline lead selenite. The measured enthalpy change was used in Appendix A to estimate the enthalpy change of the reaction $Rb_2SeO_3(cr) + Pb^{2+} \rightarrow PbSeO_3(cr) + 2Rb^{+}$ to be – (62.07 ± 1.56) $kJ \cdot mol^{-1}$. This value combines with selected standard enthalpies of formation to give the selected value:

$$\Delta_f H^{\circ}_m(\text{Rb}_2\text{SeO}_3\text{, cr, 298.15 K}) = -(973.17 \pm 2.40)\ \text{kJ}\cdot\text{mol}^{-1}.$$

Within the estimated uncertainty this value agrees with [82WAG/EVA].

Micka and Ebert [86MIC/EBE2] measured the water vapour pressure of saturated solutions in the system Rb_2SeO_3-H_2SeO_3-H_2O. From these measurements they calculated the selected Gibbs energy of the reaction:

$$\text{Rb}_2\text{SeO}_3\text{(cr)} + \text{RbH}_3(\text{SeO}_3)_2\text{(cr)} \rightleftharpoons 3\text{RbHSeO}_3\text{(cr)} \qquad \text{(V.197)}$$

$$\Delta_r G^{\circ}_m((\text{V.197}),\ 298.15\ \text{K}) = -(8.55 \pm 0.50)\ \text{kJ}\cdot\text{mol}^{-1}.$$

V.15.5.3 Rubidium selenates

Kadoshnikova, Deichman, and Tananaev [77KAD/DEI] determined the solubility of $Rb_2SeO_4(cr)$ in water at 298.15 K to be 5.448 $mol \cdot kg^{-1}$. Vojtisek and Ebert [87VOJ/EBE2] report 5.20 $mol \cdot kg^{-1}$ for the same datum. This value is confirmed in [2003BAR/CHR] and the review selects:

$$m(\mathrm{Rb_2SeO_4}, \text{aq sat}, 298.15\ \mathrm{K}) = (5.20 \pm 0.05)\ \mathrm{mol(Rb_2SeO_4)\cdot kg^{-1}}.$$

Barkov, Christov, and Ojkova [2003BAR/CHR] determined the solubility product of:

$$\mathrm{Rb_2SeO_4(cr)} \rightleftharpoons 2\mathrm{Rb^+} + \mathrm{SeO_4^{2-}} \quad (V.198)$$

to be $\log_{10} K^\circ_{s,0}$ ((V.198), 298.15 K) = (0.430 ± 0.065) and Kumok and Batyreva [90KUM/BAT] report $\log_{10} K^\circ_{s,0}$ = – 0.97, see Appendix A. The result obtained in [2003BAR/CHR] is selected and corresponds to $\Delta_f G^\circ_m$ (Rb_2SeO_4, cr, 298.15 K) = – 1005.10 kJ·mol^{-1} calculated with the selected values of $\Delta_f G^\circ_m$ for the ions.

The standard enthalpy of formation of Rb_2SeO_4(cr) was evaluated from the measurement by Selivanova and Sazikina [63SEL/SAZ] of the reaction Rb_2SeO_4(cr) + 2$AgNO_3$(aq, 1:400) → Ag_2SeO_4(cr) + 2$RbNO_3$(aq, 1:400) in Appendix A. The value obtained was $\Delta_f H^\circ_m$ (Rb_2SeO_4, cr, 298.15 K) = – 1116.50 kJ·mol^{-1} and agrees with [82WAG/EVA].

The data yield S°_m (Rb_2SeO_4, cr, 298.15 K) = (232.3 ± 9.0) J·K^{-1}·mol^{-1}. This entropy value appears to be too large compared with the accepted value for K_2SeO_4(cr) (202.1 ± 1.0) J·K·mol^{-1}, and Cs_2SeO_4(cr), (220.7 ± 13.2) J·K·mol^{-1}, for reasons that cannot be settled with the experimental information at hand. The values selected are thus:

$$\Delta_f H^\circ_m (\mathrm{Rb_2SeO_4}, \text{cr}, 298.15\ \mathrm{K}) = -(1116.5 \pm 2.2)\ \mathrm{kJ\cdot mol^{-1}},$$

$$\Delta_f G^\circ_m (\mathrm{Rb_2SeO_4}, \text{cr}, 298.15\ \mathrm{K}) = -(1005.1 \pm 1.5)\ \mathrm{kJ\cdot mol^{-1}},$$

$$S^\circ_m (\mathrm{Rb_2SeO_4}, \text{cr}, 298.15\ \mathrm{K}) = (232.3 \pm 9.0)\ \mathrm{J\cdot K^{-1}\cdot mol^{-1}}.$$

V.15.6 Caesium compounds

V.15.6.1 Caesium selenides

The phase diagram of the caesium-selenium system was assessed by Sangster and Pelton [97SAN/PEL]. A number of intermediary selenides exist but thermodynamic information is only available for Cs_2Se(cr).

V.15.6.1.1 Cs_2Se(cr)

The heat capacity of Cs_2Se(cr) has been measured in the temperature range 310 to 800 K using differential scanning calorimetry by Kohli [97KOH]. The heat capacity was described by the expression $C^\circ_{p,m}$ (Cs_2Se, cr, (298.15 – 800) K) = (67.975 + 26.49 × 10^{-3} T + 0.143 × 10^5 T^{-2}) J·K^{-1}·mol^{-1} and the value at 298.15 K calculated from this expression is selected:

$$C^\circ_{p,m} (\mathrm{Cs_2Se}, \text{cr}, 298.15\ \mathrm{K}) = (76.0 \pm 2.0)\ \mathrm{J\cdot K^{-1}\cdot mol^{-1}}.$$

V.15.6.2 Caesium selenites

Micka and Ebert have investigated the system Cs_2SeO_3-H_2SeO_3-H_2O at 298.15 K. The solubility measurement is presented in [86MIC/EBE2] and the phase diagram in [87MIC/EBE]. The crystalline phases $CsH_5(SeO_3)_3$, $CsH_3(SeO_3)_2$, $CsHSeO_3$, and $Cs_2SeO_3{\cdot}H_2O$ were identified. The solubility of $Cs_2SeO_3{\cdot}H_2O$(cr) is reported to be:

$$m(Cs_2SeO_3{\cdot}H_2O\text{, aq sat, 298.15 K}) = 9.88\ \text{mol}(Cs_2SeO_3){\cdot}\text{kg}^{-1}.$$

Klushina *et al.* [70KLU/SEL] prepared $Cs_2SeO_3{\cdot}H_2O$(cr) and reacted it with a lead nitrate solution in a calorimeter. The enthalpy data have been used in Appendix A to estimate the enthalpy change of the reaction $Cs_2SeO_3{\cdot}H_2O(cr) + Pb^{2+} \rightarrow PbSeO_3(cr) + 2Cs^+ + H_2O(l)$. The value obtained was – (23.00 ± 1.50) kJ·mol^{-1}, which furnishes the standard enthalpy of formation of caesium selenite:

$$\Delta_f H^\circ_m (Cs_2SeO_3{\cdot}H_2O\text{, cr, 298.15 K}) = -(1311.83 \pm 2.55)\ \text{kJ}{\cdot}\text{mol}^{-1}$$

with selected standard enthalpies of formation. The selected result differs somewhat from the value in [82WAG/EVA], 1314.61 kJ·mol^{-1}.

Micka and Ebert [86MIC/EBE2] measured the water vapour pressure of saturated solutions in the system Cs_2SeO_3-H_2SeO_3-H_2O. From these measurements they calculated the selected Gibbs energy changes of the reactions:

$$CsH_3(SeO_3)_2(cr) + H_2SeO_3(cr) \rightleftharpoons CsH_5(SeO_3)_3(cr) \qquad (V.199)$$

$$\Delta_r G^\circ_m ((V.199)\text{, 298.15 K}) = -(7.70 \pm 0.50)\ \text{kJ}{\cdot}\text{mol}^{-1},$$

$$CsHSeO_3(cr) + H_2SeO_3(cr) \rightleftharpoons CsH_3(SeO_3)_2(cr) \qquad (V.200)$$

$$\Delta_r G^\circ_m ((V.200)\text{, 298.15 K}) = -(19.25 \pm 1.20)\ \text{kJ}{\cdot}\text{mol}^{-1},$$

$$Cs_2SeO_3{\cdot}H_2O(cr) + CsH_3(SeO_3)_2(cr) \rightleftharpoons 3CsHSeO_3(cr) + H_2O(l) \qquad (V.201)$$

$$\Delta_r G^\circ_m ((V.201)\text{, 298.15 K}) = -(14.46 \pm 0.90)\ \text{kJ}{\cdot}\text{mol}^{-1}.$$

V.15.6.3 Caesium selenates

Information on the solubility of Cs_2SeO_4(cr) in water at 298.15 K can be found in [77KAD/DEI3], [87VOJ/EBE2], [2001BAR/CHR], which report 6.449, 6.340, and 6.353 mol·kg^{-1}, respectively. The review selects the mean of the three results:

$$m(Cs_2SeO_4\text{(cr),aq sat, 298.15 K}) = (6.38 \pm 0.07)\ \text{mol}(Cs_2SeO_4){\cdot}\text{kg}^{-1}.$$

A phase diagram and related information are available for the system Cs_2SeO_4-H_2SeO_4-H_2O at 303.15 K in [2000HAV/MIC].

The solubility product of Cs_2SeO_4(cr):

$$Cs_2SeO_4(cr) \rightleftharpoons 2Cs^+ + SeO_4^{2-} \qquad (V.202)$$

was determined by Barkov, Christov, and Ojkova [2001BAR/CHR], see Appendix A, to be $\log_{10} K^{\circ}_{s,0} = 0.630$. With the selected value of the solubility the value of $K^{\circ}_{s,0}$ is changed to the selected value:

$$\log_{10} K^{\circ}_{s,0}\ ((\text{V.202}), 298.15\ \text{K}) = (0.636 \pm 0.065).$$

The uncertainty added by the review corresponds to an uncertainty in the mean activity coefficient of 5%.

The standard Gibbs energy of formation of caesium selenate is calculated from the Gibbs energy of Reaction (V.202) and the selected values of $\Delta_f G^{\circ}_m$ for the ions to be:

$$\Delta_f G^{\circ}_m\ (\text{Cs}_2\text{SeO}_4, \text{cr}, 298.15\ \text{K}) = -(1018.8 \pm 1.8)\ \text{kJ}\cdot\text{mol}^{-1}.$$

Selivanova and Sazykina [62SEL/SAZ5] measured the enthalpy changes of the reactions Cs_2SeO_4(aq, 1:1000) + $BaCl_2\cdot 2H_2O$(cr) $\rightarrow$ $BaSeO_4$(cr) + $2CsCl_2$(aq, 1:500) + $2H_2O$(l) and Cs_2SeO_4(cr) + aq $\rightarrow$ Cs_2SeO_4(aq, 1:1000) in a calorimetric study. The results were recalculated with new auxiliary data and discussed in Appendix A. The review selects:

$$\Delta_f H^{\circ}_m\ (\text{Cs}_2\text{SeO}_4, \text{cr}, 298.15\ \text{K}) = -(1138.7 \pm 3.5)\ \text{kJ}\cdot\text{mol}^{-1}.$$

The selected data yield:

$$S^{\circ}_m\ (\text{Cs}_2\text{SeO}_4, \text{cr}, 298.15\ \text{K}) = (220.7 \pm 13.2)\ \text{J}\cdot\text{K}^{-1}\cdot\text{mol}^{-1}.$$

Chapter VI

Discussion of auxiliary data selection

VI.1 Group 17 auxiliary species

VI.1.1 Chlorine auxiliary species

VI.1.1.1 Hypochlorous acid and the hypochlorite anion

Neumann and Müller [29NEU/MUL] extensively studied the enthalpy of the reaction

$$Cl_2(g) + 2OH^- \rightleftharpoons ClO^- + Cl^- + H_2O(l) \qquad (VI.1)$$

by calorimetry in lithium, sodium, potassium, calcium, strontium, or barium hydroxide medium. The mean value of the enthalpy change was $-(101.04 \pm 0.34)$ kJ·mol^{-1}. The temperature of the experiments is not mentioned but is most likely to be in the range 291 to 293 K. The mole ratio ClO^- to $H_2O(l)$ varied between 1:750 and 1:3200 in the experiments. Concurrently the enthalpy of the reaction

$$HClO(aq) + OH^- \rightleftharpoons ClO^- + H_2O(l) \qquad (VI.2)$$

was measured using the above alkali and alkaline earth hydroxides. The mean value of $\Delta_r H^\circ_m$((VI.2), 291 K) was $-(37.40 \pm 0.12)$ kJ·mol^{-1}. The review calculates from these data $\Delta_f H^\circ_m$(ClO^-, 291 K) $= -(108.17 \pm 0.37)$ kJ·mol^{-1} and $\Delta_f H^\circ_m$(HClO, aq, 291 K) $= -(126.58 \pm 0.38)$ kJ·mol^{-1}. The same reactions had earlier been studied by Thomsen [1882THO] employing sodium hydroxide solutions. The review obtains from his data $\Delta_f H^\circ_m$(ClO^-, 291 K) $= -(110.2 \pm 0.8)$ kJ·mol^{-1} and $\Delta_f H^\circ_m$(HClO, aq, 291 K) $= -(124.3 \pm 1.0)$ kJ·mol^{-1}.

The enthalpy of the reaction $ClO^- + H_2O_2(aq) \rightleftharpoons Cl^- + O_2(g) + H_2O(l)$ was measured by Flis, Mishchenko, and Pakhomova [58FLI/MIS] in calorimetric experiments. The review finds from their data $\Delta_f H^\circ_m$(ClO^-, 298.15 K) $= -(107.22 \pm 0.50)$ kJ·mol^{-1}.

The enthalpy of formation of HClO(aq) has also been evaluated from the enthalpy change of the reaction HClO(aq) + H^+ + $2I^-$ $\rightleftharpoons$ I_2(cr) + Cl^- + H_2O(l), $\Delta_r H^\circ_m$(291 K) = – (215.2 ± 0.8) kJ·mol^{-1} measured by Thomsen [1882THO] and of the reaction Cl_2(aq) + H_2O(l) $\rightleftharpoons$ HClO(aq) +H^+ + Cl^-, $\Delta_r H^\circ_m$(298.15 K) = (21.1 ± 0.5) kJ·mol^{-1}, measured by Connick and Chia [59CON/CHI]. The former quantity was determined by calorimetry and the latter from the temperature variation of the equilibrium constant of the reaction determined by conductometric measurements. The enthalpies of formation calculated from the data were – (124.1 ± 0.8) and – (123.4 ± 0.7) kJ·mol^{-1}, respectively.

The review selects the mean of all data

$$\Delta_f H^\circ_m(\text{HClO, aq, 298.15 K}) = -(125.7 \pm 1.5)\ \text{kJ·mol}^{-1}$$

$$\Delta_f H^\circ_m(\text{ClO}^-, \text{298.15 K}) = -(108.3 \pm 1.5)\ \text{kJ·mol}^{-1}$$

The error bars have been increased to include the uncertainty from the influence of temperature. The selections lead to an enthalpy change of the reaction

$$H^+ + ClO^- \rightleftharpoons HClO(aq) \quad \text{(VI.3)}$$

amounting to $\Delta_r H^\circ_m$((VI.3), 298.15 K) = – (17.4 ± 2.1) kJ·mol^{-1}, which is compatible with the previously selected value – (19 ± 9) kJ·mol^{-1} in [92GRE/FUG].

The determinations of the enthalpy of dilution of HClO(aq) in [29NEU/MUL] indicate that it is negligible for solutions more dilute than HClO(aq, 1:200).

The selection of $\Delta_f H^\circ_m$(ClO^-, 298.15 K) = – (108.3 ± 1.5) kJ·mol^{-1} agrees with the value in [82WAG/EVA], which is – 107.1 kJ·mol^{-1}, whereas the selection $\Delta_f H^\circ_m$(HClO, aq, 298.15 K) = – (125.7 ± 1.5) kJ·mol^{-1} disagrees with the NBS value of – 120.9 kJ·mol^{-1}. Since the NBS data are not referenced the origin of the discrepancy cannot be identified. The selection in [82WAG/EVA] corresponds to an enthalpy change of Reaction (VI.3) of – 13.8 kJ·mol^{-1}. The evaluation of this reaction enthalpy in [92GRE/FUG] found that the data fell into two groups at (15 ± 2) kJ·mol^{-1} and (22.2 ± 6) kJ·mol^{-1}, respectively, and the value cited above (19 ± 9) kJ·mol^{-1} was selected. This evaluation did not include the extensive measurements in [29NEU/MUL], which result in an enthalpy change of Reaction (VI.3) of – 18.41·kJ·mol^{-1}. The determinations of the enthalpy of neutralisation of HClO(aq) thus indicate that the enthalpy of formation of HClO(aq) should be more negative than shown by the selections in [82WAG/EVA]. This conclusion is supported by the two determinations of $\Delta_f H^\circ_m$(HClO, aq, 298.15 K) in [1882THO] and [59CON/CHI] that do not involve the enthalpy change of Reaction (VI.3).

VI.2 Group 14 auxiliary species

VI.2.1 Carbon auxiliary species

VI.2.1.1 Hydrocyanic acid and the cyanide ion

The evaluation of the thermodynamic quantities of HCN(aq, 298.15 K) and CN^-(aq, 298.15 K) will start from the corresponding quantities for HCN(g, 298.15 K).

Thomsen [1882THO] measured the enthalpy change of the reaction $HCN(g) + 2.5O_2(g) \rightarrow CO_2(g) + 0.5N_2(g) + H_2O(l)$ by burning a stream of HCN(g) in oxygen at 293 K. The measured enthalpy of reaction, corrected to 298.15 K by the review, was $-(663.77 \pm 1.67)$ kJ·mol^{-1}. Gordon [37GOR] suggested that a correction amounting to -3.15 kJ·mol^{-1} should be applied to Thomsen's result from the formation of some nitric oxide during the combustion. A second correction stems from the slight association of the hydrogen cyanide molecules in the gaseous phase. Data for the polymerisation reactions have been provided by Giauque and Ruehrwein [39GIA/RUE]. The partial pressure of HCN(g) in the combustion mixture is estimated to be 0.3 bar by the review and the calculated correction for polymerisation is -0.27 kJ·mol^{-1}. The two corrections have been applied and the calorimetric measurement results in $\Delta_f H_m^\circ$(HCN, g, 298.15 K) = (130.76 ± 1.67) kJ·mol^{-1} with auxiliary data from [89COX/WAG].

The value of $\Delta_f H_m^\circ$(HCN, g, 298.15 K) listed in the JANAF Tables [98CHA] and NBS [82WAG/EVA] is 135.14 kJ·mol^{-1}. It is based on Thomsen's experiment. The reason for the large difference from the value obtained here is unknown, but might indicate that the combustion reaction was considered to have been carried out as a bomb calorimetric measurement at constant volume.

Badger [24BAD] studied the reaction $NH_3(g) + C(cr) \rightleftharpoons HCN(g) + H_2(g)$ and Horiuchi, Yano, and Kanai [51HOR/YAN] investigated the reaction $2CO(g) + NH_3(g) \rightleftharpoons HCN(g) + CO_2(g) + H_2(g)$. The data were re-evaluated in [98CHA] and the results obtained there are accepted. Badger's data yielded an enthalpy of reaction at 298.15 K that differed by about 50 kJ·mol^{-1} between the evaluations by the second and by the third law, respectively. Hence his results will not be considered further. The data for the second equilibrium reaction lead to an enthalpy of reaction at 298.15 K of (12.1 ± 10.9) kJ·mol^{-1} from the second law and (4.60 ± 2.10) kJ·mol^{-1} from the third law. From the last result $\Delta_f H_m^\circ$(HCN, g, 298.15 K) = (131.11 ± 2.16) kJ·mol^{-1} is obtained.

Several evaluations of $\Delta_f H_m^\circ$(HCN, g, 298.15 K) from spectroscopic determinations of the enthalpy of atomisation of HCN(g) and from quantum mechanical *ab initio* calculations of enthalpies of reactions are available. Martin [96MAR] calculated $\Delta_f H_m^\circ$(HCN, g, 298.15 K) = (128.24 ± 2.54) kJ·mol^{-1} from experimental spectroscopic data. From *ab initio* calculations of the enthalpy of atomisation of HCN(g) at 0 K and the enthalpy of the reaction $2HCN(g) \rightarrow (CN)_2(g) + N_2(g)$ he obtained with auxiliary data $\Delta_f H^\circ_m$(HCN, g, 298.15 K) = (130.0 ± 0.2) kJ·mol^{-1} and (129.50 ± 0.50) kJ·mol^{-1}, respectively. Finch *et al.* [93FIN/GAR] report evaluations of spectroscopic data and quantum mechanical calculations yielding $\Delta_f H^\circ_m$(HCN, g, 298.15 K) = (129.8 ± 2.2) kJ·mol^{-1} and (129 ± 7) kJ·mol^{-1}, respectively. Gardner [92GAR] found $\Delta_f H^\circ_m$(HCN, g, 298.15 K) = (137 ± 10) kJ·mol^{-1} from an *ab initio* calculation of the enthalpy of the reaction H_2(g) + $(CN)_2$(g) → 2HCN(g) and auxiliary data.

The review selects with equal weight the estimates of the standard enthalpy of formation of HCN(g) from [1882THO], [51HOR/YAN], the spectroscopic data in [96MAR] and [93FIN/GAR], and the quantum mechanical calculations in [96MAR] and obtains

$$\Delta_f H^\circ_m(\text{HCN, g, 298.15 K}) = (129.90 \pm 2.50)\ \text{kJ·mol}^{-1}$$

The review selects from the concordant calculations of the entropy of HCN(g) at 298.15 K by statistical mechanics the value in [61MCB/GOR]

$$S^\circ_m(\text{HCN, g, 298.15 K}) = (201.71 \pm 0.10)\ \text{J·K}^{-1}\text{·mol}^{-1}$$

The Gibbs energy of formation is hence

$$\Delta_f G^\circ_m(\text{HCN, g, 298.15 K}) = (119.52 \pm 2.50)\ \text{kJ·mol}^{-1}$$

The equilibrium constant of the reaction

$$\text{HCN(aq)} \rightleftharpoons \text{HCN(g)} \qquad \text{(VI.4)}$$

has been determined by static and dynamic methods. No direct measurement of the enthalpy of the reaction has been made and it has been obtained from the temperature dependence of the equilibrium constant. The available information is contained in Table VI-1 after recalculation of the data in [51HOR/TAN] and [74KOT/LEB2] to the unit bar·kg·mol^{-1}.

The data in [79ROD/ROB] have not been considered since they differ substantially from those in the other references. The enthalpy change given by the authors in [51HOR/TAN] is (21.3 ± 1.7) kJ·mol^{-1}. This result has also been ignored.

The association of HCN(g) was not considered by the authors in the derivation of the equilibrium constant. However, the measurements in [83AVE/SPI] were carried out at low pressures where the association is negligible and the extrapolation procedure in [92RUM/MAU] would at least remove most of the influence of the association. No experimental information is available in the other references for an application of a correction, which is estimated to be 10% at most. The review selects the mean of the values in the Table and obtains

$$\log_{10} K_p((\text{VI.4}), 298.15\ \text{K}) = -(0.902 \pm 0.050).$$

Table VI-1: Thermodynamic data for Reaction (VI.4) at 298.15 K. The quantities were obtained by inter- or extrapolation of experimental results or functions in the cited papers.

Reference	K_p((VI.4), 298.15 K) (bar·kg·mol^{-1})	$\Delta_r H^\circ_m$ ((VI.4), 298.15) (kJ·mol^{-1})	Remark
[51HOR/TAN]	0.122	26.70	$\log_{10} K_p((\text{VI.4}), T) = f(1/T)$ (284 – 313) K
[74KOT/LEB2]	0.132	24.36	$\log_{10} K_p((\text{VI.4}), T) = f(1/T)$ (293 – 368) K
[83AVE/SPI]	0.116	-	298.15 K
[92RUM/MAU]	0.124	27.29	$\ln K_p((\text{VI.4}), T) = f(1/T, T)$ (313 – 413) K
[92RUM/MAU]	0.132	23.48	$\log_{10} K_p((\text{VI.4}), T) = f(1/T)$ (313 – 353) K

The vapour pressure of HCN(l) is close to 1 bar at the standard temperature. Thus the Gibbs energy of formation of HCN(l) and HCN(g) are almost equal at 298.15 K. The enthalpy of vaporisation to HCN(g) can be calculated from the entropy of HCN(l), 112.9 J·K^{-1}·mol^{-1}, in [39GIA/RUE] and of HCN(g) selected here to be $\Delta_{vap} H^\circ_m$ (HCN, l, 298.15 K) = 26.47 kJ·mol^{-1}. This result is somewhat larger than the experimental value, 25.22 kJ·mol^{-1}, obtained for the evaporation to a total hydrogen cyanide pressure of 1 atm in [39GIA/RUE]. Combined with the enthalpy of dissolution of HCN(l) in water in [1875BER], – 1.6 kJ·mol^{-1}, $\Delta_r H^\circ_m$((VI.4), 298.15 K) = 28.1 kJ·mol^{-1} is obtained. This datum is combined with the mean from Table VI-1, (25.5 ± 2.50) kJ·mol^{-1} in the weight ratio 1:3 to provide the selected result

$$\Delta_r H^\circ_m((\text{VI.4}), 298.15\ \text{K}) = (26.15 \pm 2.50)\ \text{kJ·mol}^{-1}.$$

The selected reaction data yield the selected thermodynamic quantities for undissociated HCN in aqueous solution

$$\Delta_f H^\circ_m(\text{HCN, aq, 298.15 K}) = (103.75 \pm 3.54)\ \text{kJ·mol}^{-1}$$

$$\Delta_f G^\circ_m(\text{HCN, aq, 298.15 K}) = (114.37 \pm 2.52)\ \text{kJ·mol}^{-1}$$

$$S^\circ_m(\text{HCN, aq, 298.15 K}) = (131.3 \pm 8.4)\ \text{J·K}^{-1}\text{·mol}^{-1}.$$

For the protonation of the cyanide ion

$$\text{H}^+ + \text{CN}^- \rightleftharpoons \text{HCN(aq)} \qquad \text{(VI.5)}$$

the consistent result for the enthalpy of reaction from calorimetric determinations in [62IZA/CHR], [70CHR/JOH], and [96SOL/MAY] is selected

$$\Delta_r H_m^\circ((\text{VI.5}), 298.15\text{ K}) = -(43.60 \pm 0.20)\text{ kJ}\cdot\text{mol}^{-1}.$$

The IUPAC recommended value of the equilibrium constant at infinite dilution evaluated by Beck in [87BEC] is selected

$$\log_{10} K((\text{VI.5}), 298.15\text{ K}) = (9.21 \pm 0.02)$$

$$\Delta_r G_m^\circ((\text{VI.5}), 298.15\text{ K}) = -(52.57 \pm 0.11)\text{ kJ}\cdot\text{mol}^{-1}.$$

The selections result in:

$$\Delta_f H_m^\circ(\text{CN}^-, 298.15\text{ K}) = (147.35 \pm 3.55)\text{ kJ}\cdot\text{mol}^{-1}$$

$$\Delta_f G_m^\circ(\text{CN}^-, 298.15\text{ K}) = (166.94 \pm 2.52)\text{ kJ}\cdot\text{mol}^{-1}$$

$$S_m^\circ(\text{CN}^-, 298.15\text{ K}) = (101.2 \pm 8.4)\text{ J}\cdot\text{K}^{-1}\cdot\text{mol}^{-1}.$$

Finch *et al.* [93FIN/GAR] determined the enthalpy of formation of the cyanide ion from the enthalpies of combustion of tetramethyl- and tetraethylammonium cyanide. The molar enthalpies of formation so obtained were combined with experimental determinations of the enthalpies of dissolution of the salts and the enthalpies of formation of $(CH_3)_4N^+$ and $(C_2H_5)_4N^+$ as auxiliary data. The consistent results so obtained yielded $\Delta_f H_m^\circ(\text{CN}^-, 298.15\text{ K}) = (140.3 \pm 2.2)\text{ kJ}\cdot\text{mol}^{-1}$ as the mean. As this result is significantly smaller than the result obtained from the hydrogen cyanide route it was not included in the selection.

Appendix A

Discussion of selected references

This appendix comprises discussions relating to a number of key publications, which contain experimental information cited in this review. These discussions are fundamental in explaining the accuracy of the data concerned and the interpretation of the experiments, but they are too lengthy or are related to too many different Sections to be included in the main text. The notation used in this appendix is consistent with that used in the present book, and not necessarily consistent with that used in the publication under discussion.

[1882THO]

The only experiment, which leads to a determination of the standard enthalpy of formation of sodium selenite, appears to be that of Thomsen [1882THO]. He measured the enthalpy of the reaction between one mole of selenious acid and two moles of sodium hydroxide in solution. The data are used to calculate the standard enthalpy of formation of $NaSeO_3(cr)$ in Table A-1.

Table A-1: Evaluation of the standard enthalpy of formation of $NaSeO_3(cr)$. The sources of the data are indicated.

#	Reaction	$\Delta_r H_m$ (kJ·mol^{-1})	Reference
1	$2NaOH(aq, 1:200) + H_2SeO_3(aq, 1:400) \rightarrow 2H_2O(l) + Na_2SeO_3(aq, 1:800)$	– (113.07 ± 0.30)	[1882THO]
2	$NaOH(aq, 1:200) + aq \rightarrow Na^+ + OH^-$	– (0.50 ± 0.05)	[82WAG/EVA]
3	$H_2SeO_3(cr) + aq \rightarrow H_2SeO_3(aq, 1:400)$	(17.90 ± 0.60)	[61SEL/PAK]
4	$Se(cr) + H_2(g) + 1.5O_2(g) \rightarrow H_2SeO_3(cr)$	– (524.72 ± 0.65)	(review)
5	$Na(cr) + 0.5O_2(g) + 0.5H_2(g) + aq \rightarrow Na^+ + OH^-$	– (470.36 ± 0.07)	[89COX/WAG]
6	$Na_2SeO_3(cr) + aq \rightarrow Na_2SeO_3(aq, 1:800)$	– (28.33 ± 0.08)	[69KLU/SEL]
7	$H_2(g) + 0.5O_2(g) \rightarrow H_2O(l)$	– (285.83 ± 0.04)	[89COX/WAG]
8	$2Na(cr) + Se(cr) + 1.5O_2(g) \rightarrow Na_2SeO_3(cr)$	– (959.62 ± 0.95)	(result)

$\Delta_r H_8 = \Delta_f H^\circ_m(Na_2SeO_3, cr, 298.15 K) = \Delta_r H_1 - 2\Delta_r H_2 + \Delta_r H_3 + \Delta_r H_4 + 2\Delta_r H_5 - \Delta_r H_6 - 2\Delta_r H_7$

Thomsen also measured the enthalpy of reaction between one mole of selenious acid and one mole of sodium hydroxide in solution. His data are used in Table A-2 to estimate the standard enthalpy of formation of $NaHSeO_3(cr)$.

Table A-2: Estimate of the standard enthalpy of formation of $NaHSeO_3(cr)$. The sources of the data are indicated.

#	Reaction	$\Delta_r H_m$ (kJ·mol^{-1})	Reference
1	$NaOH(aq, 1:200) + H_2SeO_3(aq, 1:400) \rightarrow H_2O(l) + NaHSeO_3(aq, 1:600)$	– (61.81 ± 0.30)	[1882THO]
2	$NaOH(aq, 1:200) + aq \rightarrow Na^+ + OH^-$	– (0.50 ± 0.05)	[82WAG/EVA]
3	$H_2SeO_3(cr) + aq \rightarrow H_2SeO_3(aq, 1:400)$	(17.90 ± 0.60)	[61SEL/PAK]
4	$Se(cr) + H_2(g) + 1.5O_2(g) \rightarrow H_2SeO_3(cr)$	– (524.72 ± 0.65)	(review)
5	$Na(cr) + 0.5O_2(g) + 0.5H_2(g) + aq \rightarrow Na^+ + OH^-$	– (470.36 ± 0.07)	[89COX/WAG]
6	$NaHSeO_3(cr) + aq \rightarrow NaHSeO_3(aq, 1:800) \cong NaHSeO_3(aq, 1:600)$	(5.67 ± 0.50)	[68SEL/ROS]
7	$H_2(g) + 0.5O_2(g) \rightarrow H_2O(l)$	– (285.83 ± 0.04)	[89COX/WAG]
8	$Na(cr) + Se(cr) + 0.5H_2(g) + 1.5O_2(g) \rightarrow NaHSeO_3(cr)$	– (758.20 ± 1.10)	(result)

$\Delta_r H_8 = \Delta_f H^\circ_m(\text{NaHSeO}_3\text{, cr, 298.15 K}) = \Delta_r H_1 - \Delta_r H_2 + \Delta_r H_3 + \Delta_r H_4 + \Delta_r H_5 - \Delta_r H_6 - \Delta_r H_7$

[1887FAB]

Fabre determined the enthalpy of formation of the metal selenides $Li_2Se(s)$, $Na_2Se(s)$, $K_2Se(s)$, $CaSe(s)$, $SrSe(s)$, $BaSe(s)$, $FeSe(s)$, $MnSe(s)$, $NiSe(s)$, $CoSe(s)$, $ZnSe(s)$, $CdSe(s)$, $CuSe(s)$, $Tl_2Se(s)$, $PbSe(s)$, $HgSe(s)$, and $Ag_2Se(s)$ using aqueous solution calorimetry. The work was excellent at that time but the characterisation of the materials does not meet the standards of today. The investigation by Fabre is still the only source of experimental thermochemical information for some systems. Recalculations of the results are necessary since more accurate auxiliary data are now available. Reassessments of Fabre's work have been made by Mills in [74MIL] and by Rossini, Wagman, Evans, and Levine in [52ROS/WAG]. In the present evaluation, the alkaline and alkaline earth selenides are considered to be crystalline.

The enthalpies of formation of the alkali metal selenides were re-evaluated by the review from the enthalpy changes of Reactions #1 and #2 exemplified for lithium in Table A-3. The temperature of the experiments was about 288 K. The reactions combine to Reaction #3 from which the enthalpy of formation of the alkali metal selenide was calculated with selected auxiliary data at 298.15 K. A correction for the formation of some Se^{2-} in Reaction #2 cannot be made due to lack of data. This also pertains to the correction of the data from the experimental temperature to 298.15 K. The results are assumed to be valid at standard conditions within the error limits and the following values were obtained: $\Delta_f H_m^\circ$ (Li_2Se, cr, 298.15 K) = – (442.3 ± 4.0) kJ·mol^{-1}, $\Delta_f H_m^\circ$ (Na_2Se, cr, 298.15 K) = – (332.2 ± 5.9) kJ·mol^{-1}, and $\Delta_f H_m^\circ$ (K_2Se, cr, 298.15 K) = – (398.7 ± 5.0) kJ·mol^{-1}.

Table A-3: Evaluation of the standard enthalpy of formation of alkali metal (Reactions #1 to #3) and alkaline earth metal (Reaction #4) selenides. The sources of the data are indicated.

#	Reaction	$\Delta_f H_m^\circ$ (kJ·mol^{-1})	Reference
1	$Li_2Se(cr) + H_2O(l) + aq \rightarrow 2Li^+ + HSe^- + OH^-$	– (44.6 ± 2.5)	(paper)
2	$H_2Se(g) + OH^- \rightarrow HSe^- + H_2O(l)$	– (70.6 ± 2.5)	(paper)
3	$H_2Se(g) + 2OH^- + 2Li^+ \rightarrow Li_2Se(cr) + 2H_2O(l)$	– (26.0 ± 3.5)	(calculated)
4	$CaSe(cr) + 2H^+ \rightarrow Ca^{2+} + H_2Se(aq)$	– (145.6 ± 2.5)	(paper)

The enthalpies of formation of the alkaline earth metal selenides can be calculated from Reaction #4 of Table A-3 exemplified for CaSe. The dissolution was carried out with 0.25 M HCl at about 288 K. A simplified evaluation with selected auxiliary data yields: $\Delta_f H_m^\circ$ (CaSe, cr, 298.15 K) = – (383.1 ± 3.4) kJ·mol^{-1}, $\Delta_f H_m^\circ$ (SrSe, cr, 298.15 K) = – (395.8 ± 3.3) kJ·mol^{-1}, and $\Delta_f H_m^\circ$ (BaSe, cr, 298.15 K) = – (378.7 ± 4.1) kJ·mol^{-1}.

The enthalpies of dissolution in water saturated with bromine and in the presence of Br_2(l) were measured for the other metal selenides mentioned. Since the information on experimental details is scarce, no re-evaluation of these measurements was attempted.

Concordant results were obtained for Reaction #2 with the lithium, sodium, and potassium hydroxides. The data yield with selected auxiliary data the following estimate of the standard enthalpy of formation of the hydrogen selenide ion, $\Delta_f H_m^\circ$ (HSe^-, 298.15 K) = (14.3 ± 3.2) kJ·mol^{-1}.

[13BRU]

The experimental technique used in [13BRU] and [23HLA] to evaluate the first acidity constant of H_2Se(aq) is identical. The ionic conductivity of HSe^- is first determined from a series of NaHSe(aq) solutions, prepared by neutralising NaOH(aq) with H_2Se(g). This value (Λ_∞ = 70, no unit provided) is then used, in combination with the ionic conductivity of H^+ and the measured conductivity of H_2Se(aq) solutions, to deduce the degree of dissociation in these solutions. The acidity constant was calculated from these data with an assumed concentration of H_2Se(aq) of 0.1 M at p_{H_2Se} = 1 atm.

The procedure has been accepted by the review, but the data have been recalculated with the presently evaluated value of 0.0838 M for the sum of the concentration of $H_2Se(aq)$ and HSe^- at p_{H_2Se} = 1 bar. The results are presented in Chapter V.

[13PRE/BRO]

The vapour pressure of liquid selenium was measured with a Bourdon gage manometer in the temperature range 663 to 983 K. A third law evaluation of the enthalpy of formation of $Se_2(g)$ at 298.15 K was made by the review using the thermal functions of Se(l), Se(trig) and $Se_2(g)$ and mole fractions of $Se_2(g)$ estimated from the selected selenium data. The enthalpy of formation derived was $\Delta_f H_m^\circ$ (Se_2, g, 298.15 K) = (136.7 ± 2.5) $kJ \cdot mol^{-1}$.

[28SCH/SWI]

The equilibrium constant of the reaction:

$$Se(cr) + 2I_2(cr) + 3H_2O(l) \rightleftharpoons H_2SeO_3(aq) + 4H^+ + 4I^- \quad (A.1)$$

has been determined from a chemical analysis of the equilibrium mixture. Five mixtures of the components were employed with Se(cr) and $I_2(cr)$ present from the start. The initial compositions of the test solutions are contained in Table A-4.

Table A-4: Initial concentrations (mM) of the test solutions.

#	KI	$HClO_4$	H_2SeO_3
1	4.15	50.15	0.00
2	87.88	39.67	10.00
3	0.00	0.00	0.00
4	0.00	9.87	14.00
5	58.88	26.50	25.00

The equilibrium was always approached from the left to the right.

Chemical analysis:

Although the analytical methods used are described at some length, details are lacking.

Total iodine, *i.e.*, $I_t(0) = [I_2(aq)] + [I_3^-]$ was determined by titration with $N_2H_5^+$ (called hydrazine here and in the paper for convenience) after neutralisation. Then the solution was acidified, an excess of hydrazine added, and the mixture heated. This treatment reduces Se(IV) to Se(0), which was filtered off. The evaluator believes that the selenium was collected and weighed and thus that the determination of selenium was made in this way, although it is not explicitly mentioned in the paper.

After removal of Se(0) the solution was treated with hypochlorous acid and liberated iodine titrated with hydrazine. This was intended to yield total iodine plus iodide in the equilibrium mixture. This analytical step is not understood. It may be mentioned, however, that this determination was not used in the calculation of the equilibrium constant or elsewhere in the paper.

The analytical concentration of strong acid was determined by a back-titration of an added excess of standard hydrogen carbonate with standard hydrochloric acid. Iodine was first reduced by thiosulphate in order to allow visual indication of the end point in the titration. The end point corresponds to the titration of strong acid plus almost all of the selenious acid. Hence, a blank titration was carried out on a solution of selenious acid of the same concentration as the test solution.

Calculations of the equilibrium concentrations:

The results of the calculation of the concentrations at equilibrium from the analytical data are found in Table II of the paper. The same notations as used there will be used here.

- $\Sigma I_3(deter)$ = $[I_2(aq)] + [I_3^-]$ is the total iodine concentration determined by titration with hydrazine.
- $(I_3^-)(calcd.)$ is $\Sigma I_3(deter) - [I_2(aq)]$. $[I_2(aq)]$ is the solubility of iodine in water and a value of 1.33 mM is mentioned in the text. This value has been used in experiments 2, 4, and 5. Why slightly different values were used in experiments 1 and 3 is not known.
- $(I^-)(calcd.)$ has been found from $(I_3^-)(calcd.)$ using the known extent of the reaction $I_2(cr) + I^- \rightleftharpoons I_3^-$. The equilibrium constants used in the calculation of the iodine and iodide concentrations were close to more recent values in [55KAT/GEB] and a recalculation makes little or no difference.
- *Strong acid(deter.)* is the analytical concentration of strong acid determined by titration:

  $$(H+)(calcd.) \text{ is } Strong\ acid(deter.) + \left[HSeO_3^-\right]$$

  and

  $$(\Sigma H_2SeO_3)(deter.) - (H_2SeO_3)(calc.) = \left[HSeO_3^-\right].$$

The concentrations of the Se(IV) species were calculated without corrections for activity coefficients with $\log_{10} K_{1,2} = 2.57$. A recalculation with $\log_{10} K_{1,2} = 2.64$ and introduction of activity coefficients resulted in only minor changes (< 3%) in the concentrations of H^+ and H_2SeO_3. Since the solutions are strongly acid almost all Se(IV) is present as H_2SeO_3.

Stoichiometry:

The observed changes in concentration from the initial solution to the equilibrium solution are entered in Table A-5.

The expected changes in the concentrations of H^+ and I(– I) are 4Δ[Se(IV)]. Agreement with the expectation is found only for experiment 3. The other experiments deviate (Dev.) substantially and oxidation of iodide by oxygen was proposed as the cause of the discrepancy.

Table A-5: Concentration changes during equilibration, mM.

#	Δ[Se(IV)]	4Δ[Se(IV)]	Δ[H^+]	Dev. [H^+]	Δ[I(– I)]	Dev.[I(–I)]
1	9.33	37.32	24.19	13.13	36.4	1.0
2	1.65	6.60	– 5.65	12.25	– 2.3	8.9
3	13.03	52.12	51.62	0.50	52.2	0.1
4	12.62	50.48	40.90	9.57	42.7	7.8
5	5.82	23.28	5.44	17.84	8.5	14.8

[32SHE/LYO]

The emf of the cell:

$$\text{Pt} \mid \text{quinhydrone}, H_2SO_4\ (\text{m}) \mid H_2SeO_4\ (\text{m}), \text{quinhydrone} \mid \text{Pt}$$

$$E = g \ln \frac{a_{\mathrm{H}^+}^{(\mathrm{Se})}}{a_{\mathrm{H}^+}^{(\mathrm{S})}} + E_j$$

was measured at (25 ± 0.05)°C and at equal concentrations of the acids. $a_{\mathrm{H}^+}^{(\mathrm{Se})}$ and $a_{\mathrm{H}^+}^{(\mathrm{S})}$ represent the activity of the hydrogen ion in the selenic acid and sulphuric acid solutions, respectively, and g is the Nernstian coefficient. The concentration of acid was varied from 0.01 m to 0.4 m at seven different levels. The electrode potential of the selenium half-cell became unstable above the latter concentration. The emf of the cell seldom differed from zero by more than one or two tenths of a millivolt and the authors concluded that the activities of hydrogen ion in solutions of selenic and of sulphuric acid of the same concentration are very nearly equal.

This result is at variance with the accepted values of the second dissociation constants of the acids. Reference [92GRE/FUG] selects $\log_{10} K_2^\circ = -1.80$ for selenic and $\log_{10} K_2^\circ = -1.98$ for sulphuric acid. The expected potential difference would therefore be about 1 to 2 mV from these data if the liquid junction potential (E_j) is ignored, and not a few tenths of a millivolt. The contradiction can probably be traced to the neglect of E_j. An estimate of E_j by the Henderson equation indicates that the first term in the above expression for E is largely cancelled by the second term. The same conclusion is reached if the cell is regarded as a pseudo-concentration cell. The much larger mobility of H^+ in comparison with the other ions present reduces the expected potential difference to the order of a few tenths of a millivolt as observed.

The review thus concludes that, despite the simplicity of the experiments, the results of the investigation do not show that the second dissociation constants of selenic and sulphuric acid have the same value.

[42GEL/KIN]

The emf of the cell (Q, QH_2 = quinhydrone):

$$\text{Pt} \mid \text{Q}, \text{QH}_2, C \text{ m } \text{H}_2\text{SeO}_4(\text{sat'd Q}, \text{QH}_2) \mid C \text{ m } \text{H}_2\text{SeO}_4(\text{sat'd Ag}_2\text{SeO}_4) \mid \text{Ag}_2\text{SeO}_4,\text{Ag}$$

$$E = E^{\circ}_{\text{Ag/Ag}_2\text{SeO}_4} - E^{\circ}_{\text{Q,QH}_2} - 0.5\, g\, \log_{10}\left\{\text{H}^+\right\}^2_- \left\{\text{SeO}_4^{2-}\right\}_+ = E_\text{o} - 0.5\, g\, \log_{10}\left\{\text{H}^+\right\}^2_- \left\{\text{SeO}_4^{2-}\right\}_+$$

was measured at 298.15 K. Curled brackets represent activities and $g = RTF^{-1}\ln 10$. The + and – signs are introduced here to stress that the activities pertain to different parts of the galvanic cell. The authors derived equation (9) in the paper from the stoichiometry of the test solution and the expression for the emf of the cell and obtained:

$$E = E_\text{o} - 0.5\, g\, [2 \log_{10}(1 + \alpha) + \log_{10}(c_\text{S}/C + \alpha) + \log_{10}\beta + 3 \log_{10} C]$$

α is the degree of protolysis of the hydrogen selenate ion, β the activity coefficient correction calculated from the Debye-Hückel equation, and c_S the solubility of silver selenate at the appropriate selenic acid concentration. The last quantity was found in a separate measurement in which the solubility of Ag_2SeO_4 in selenic acid solution with concentrations between 0.00 and 0.1218 M was determined. This was the range in C covered by the potentiometric measurements.

The value of the protonation constant of the selenate ion, K_1, was varied systematically until a (reasonably) constant value of E_o, independent of C, was obtained from the above equation. The result was $\log_{10} K_1 = 2.0$ and $E_\text{o} = -0.1295$ V. However, the equation is not quite correct as the degree of dissociation of the hydrogen selenate ion is substantially different in the two half-cells, *cf.* [35HAR/HAM]. The value obtained for K_1 will therefore not be accepted. A recalculation of the data by the review has not been attempted.

[52BRO]

The vapour pressure of liquid selenium was measured with Bourdon gauges in the temperature range 536 to 961 K. A third law evaluation of the enthalpy of formation of $Se_2(g)$ at 298.15 K was made by the review using the vapour pressure at 950 K and the thermal functions of Se(l), Se(trig), $Se_2(g)$. A mole fraction of $Se_2(g)$, $x_{\text{Se}_2(\text{g})} = 0.64$ was estimated from the selected selenium data. The enthalpy of formation derived was $\Delta_\text{f} H^{\circ}_\text{m}(\text{Se}_2, \text{g}, 298.15\text{ K}) = (140.1 \pm 1.6)\text{ kJ·mol}^{-1}$.

[56CHU]

The solubility products of a number of metal selenites were determined from solubility measurements in dilute mineral acids at 293.2 K.

The selenites were generally prepared by mixing a 0.1 to 0.2 M solution of the metal ion and a 0.1 M solution of sodium selenite in stoichiometric proportions. In some cases the precipitate was aged in the mother liquor before separation and drying. The specimens were analysed but only the ratio metal:selenium is reported. X-ray diffraction patterns were not registered. It is therefore not clearly established whether the preparations were crystalline or (aged) amorphous phases.

The solubility in a 1×10^{-3} to 1×10^{-2} M solution of sulphuric, nitric, or hydrochloric acid was determined after equilibration for 8 hours with continuous stirring. Two acids were used for each selenite with two or three experiments performed with each acid. The solid and solution were separated by centrifugation and the metal ion concentration in the saturated solution was determined. The total selenite concentration was calculated from the total metal ion concentration. The pH values of the start and equilibrium solutions were measured with a glass electrode calibrated against the quinhydrone electrode. The pH scale is somewhat undefined but can probably be regarded as an activity scale.

The SeO_3^{2-} concentration was calculated from the total selenite concentration using the protonation constants from Rumpf [33RUM]. No activity coefficient or side reaction corrections were included in the calculation of the solubility product. The results are presented in the appropriate Sections in Chapter V.

The procedure outlined above is fairly typical for a number of determinations of solubility products of metal selenites. Data from such investigations have been re-evaluated by the review with the accepted protonation constants of the selenite ion, corrected for the hydrolysis of the metal ion when necessary, and the value of the solubility product extrapolated to standard state conditions. It has been observed that the initial and final pH values in cases are in conflict. This has been ignored and the calculations have been based solely on the data for the equilibrium solution. Complexation of the metal ion by the anions and temperature effects were neglected, which probably introduces a negligible error compared with other sources of error. Activity coefficients were calculated by the SIT expression with $\varepsilon = 0$ kg·mol^{-1}, which is a reasonable simplification due to the low ionic concentrations. The ionic strength was obtained by an iterative procedure from knowledge of the total metal concentration and the pH of the equilibrium solution. The results of the recalculations are entered in Chapter V.

[57ILL/LAP]

The vapour pressure of liquid selenium was measured by a static method in the temperature range 823 to 1173 K. The total vapour pressure at equilibrium was given as $\log_{10}(p/\text{bar}) = 5.203 - 4987\ T^{-1}$. A third law evaluation by this review of the enthalpy of formation of $Se_2(g)$ at 298.15 K was made using the vapour pressure at 1150 K and the selected data for the heat capacities and entropies of Se(l) and $Se_2(g)$. A mole fraction of $Se_2(g)$, $x_{Se_2(g)}$ = 0.80 was estimated from the selected selenium data. The enthalpy of formation derived was $\Delta_f H_m^\circ$ (Se_2, g, 298.15 K) = (141.1 ± 2.1) kJ·mol^{-1}.

[57KIU/WAG]

The Gibbs energy of formation of $Ag_2Se(cr)$ was obtained from emf measurements in the temperature range 433 to 643 K using the cell:

(–) Ag(cr) | AgI(cr) | β-Ag_2Se, Se(cr,l) (+).

The enthalpy and entropy change for the reaction:

$$2Ag(cr) + Se(l) \rightleftharpoons Ag_2Se(cr) \quad (A.2)$$

at 500 K were evaluated to be $\Delta_r H_m^\circ$((A.2), 500 K) = – 40.836 kJ·mol^{-1} and $\Delta_r S_m^\circ$((A.2), 500 K) = 30.962 J·K^{-1}·mol^{-1}, respectively.

The above values were recalculated to 298.15 K by the review using the second and third laws and necessary selected data. The value of $\Delta_f H_m^\circ$ (Ag_2Se, α, 298.15 K) obtained from the second law was – 43.6 and – 40.0 kJ·mol^{-1} from the third law. The result for the entropy S_m° (Ag_2Se, α, 298.15 K) was 148.5 J·K^{-1}·mol^{-1}.

[57TOR]

Toropova measured the solubility of a mercury(II) selenite specimen, $HgSeO_3$(cr?), in acidified 1 M $NaNO_3$ at (298.15 ± 0.05) K. The experimental results are entered in Table A-6.

At the low pH values used in the investigation the dissolution equilibrium is best described by the reaction:

$$HgSeO_3(cr) + 2H^+ \rightleftharpoons Hg^{2+} + H_2SeO_3(aq). \quad (A.3)$$

Table A-6: Experimental results and the equilibrium constant of Reaction (A.3).

pH	$\log_{10} [Hg^{2+}]$	$\log_{10} K$
1.00	– 3.04	– 4.08
1.26	– 3.30	– 4.08
1.30	– 3.33	– 4.07
1.80	– 3.80	– 4.07

The mixed conditional equilibrium constant of this reaction was calculated with neglect of complexation of Hg^{2+} by NO_3^-. A correction for the presence of small amounts of hydrogen selenite was made with $\log_{10} K_{1,2} = 2.36$ [77FOW/STR], [88OZE/YAG] valid in 1 M Na^+ medium. The result is:

$$\log_{10} K\ ((\text{A.3}),\ 1\ \text{M NaNO}_3,\ 298.15\ \text{K}) = \log_{10}\left(\frac{m_{\text{Hg}^{2+}}\, m_{\text{H}_2\text{SeO}_4}}{a^2_{\text{H}^+}}\right) = -(4.07 \pm 0.02).$$

Extrapolation to $I = 0$ was made by the SIT expression with $\varepsilon(\text{Hg}^{2+}, \text{NO}_3^-) = -(0.1 \pm 0.1)\ \text{kg·mol}^{-1}$ [92GRE/FUG]. The result is $\log_{10} K^\circ((\text{A.3}),\ 298.15\ \text{K}) = -(4.98 \pm 0.10)$. In combination with the accepted protonation constants of the selenite ion this datum yields for the solubility product of mercury(II) selenite:

$$\text{HgSeO}_3(\text{cr}) \rightleftharpoons \text{Hg}^{2+} + \text{SeO}_3^{2-} \qquad (\text{A.4})$$

$$\log_{10} K^\circ_{s,0}((\text{A.4}),\ 298.15\ \text{K}) = -(15.98 \pm 0.30).$$

[58KUL/SHA]

The vapour pressure of solid selenium presented in this paper is about ten times higher than in other investigations and the equation for the vapour pressure does not fit the line in Figure 2 of the paper. The results of this investigation are therefore rejected.

[58SEL]

The original data used by Selivanova for the calculation of the standard Gibbs energy and enthalpy of formation of $Tl_2SeO_4(cr)$ are contained in Table A-7.

Table A-7: Experimental values of the solubility of $Tl_2SeO_4(cr)$.

t (°C)	Solubility g Tl_2SeO_4/100 g H_2O	Reference
9.3	2.13	[07TUT]
12	2.4	[07TUT]
20	2.8	[10GLA]
80	8.5	[10GLA]
100	10.86	[07TUT]

The data in Table A-7 were used to construct Table A-8.

Table A-8: Interpolated values of the solubility of $Tl_2SeO_4(cr)$.

t (°C)	$m_{Tl_2SeO_4}$
10	0.0390
20	0.0500
25	0.0570
30	0.0645
40	0.0800

[58SEL/SHN3]

The concentration of barium in water solutions saturated with $BaSeO_4(cr)$ was determined by a polarographic method ($E_{1/2} = -1.8$ V with 0.1 M LiCl as supporting electrolyte). The measured solubilities are reported in Table A-9. There is no description of the method used to test for equilibrium and no thermodynamic data were calculated in the paper.

Table A-9: Solubility of barium selenate in water, results from two parallel experiments.

Temperature K (±0.2 K)	Run number	Time of stirring, days	Results g/L
288	1	36	0.0482
	2	36	0.0482
298	1	20	0.0523
	2	20	0.0526
303	1	15	0.0543
	2	15	0.0533
313	1	12	0.0426
	2	12	0.0421
323	1	10	0.0367
	2	10	0.0375
348	1	8	0.0324
	2	8	0.0330
368	1	4	0.0280
	2	4	0.0280

[58TRZ]

This paper contains no new experimental data but cites the otherwise unavailable experimental results of Tuma [57TUM].

Trzil calculates the enthalpy of formation of selenic acid. The first step comprises the thermochemical cycle in Table A-10. The table contains, however, the auxiliary data used by the review and not the data used by Trzil. Changes in some of the $\Delta_r H_m$ values between 293 and 298 K have been ignored as the necessary quantities for temperature corrections are not available.

Table A-10: Determination of $\Delta_f H^\circ_m$ (H_2SeO_4, aq 1:1200, 298.15 K). The sources of the data are indicated.

#	Reaction	$\Delta_r H_m$ (kJ·mol^{-1})	Reference
1	SeO_2(aq, 1:500) + HClO(aq, 1:700) → H_2SeO_4(aq, 1:1200) + HCl(aq, 1:1200)	– (125.03 ± 0.40)	[1882THO]
2	$0.5H_2$(g) + $0.5Cl_2$(g) + aq → HCl(aq, 1:1200)	– (166.68 ± 0.05)	[89COX/WAG] [82WAG/EVA]
3	$0.5H_2$(g) + $0.5Cl_2$(g) + $0.5O_2$(g) + aq → HClO(aq, 1:700)	– (125.70 ± 1.50)	(review)
4	H_2(g) + $0.5O_2$(g) → H_2O(l)	– (285.83 ± 0.04)	[89COX/WAG]
5	SeO_2(cr) + aq → SeO_2(aq, 1:500)	(4.47 ± 0.17)	[61SEL/PAK]
6	Se(cr) + O_2(g) → SeO_2(cr)	– (225.39 ± 0.60)	(review)

$\Delta_f H^\circ_m$ (H_2SeO_4, aq 1:1200, 298.15 K) = $\Delta_r H_1 - \Delta_r H_2 + \Delta_r H_3 + \Delta_r H_4 + \Delta_r H_5 + \Delta_r H_6$ = – (590.80 ± 1.67) kJ·mol^{-1}

The value found by Trzil was – 601.6 kJ·mol^{-1}. The earlier accepted enthalpy of formation of SeO_2(cr), which was about 10 kJ·mol^{-1} too exothermic, mainly causes the difference.

The integral enthalpy of dissolution of H_2SeO_4(cr) to H_2SeO_4(aq, 1:2000) was determined as well as the enthalpies of dilution of H_2SeO_4(aq, 1:750). The latter measurements are only presented as a small graph. As this reference may be difficult to find some enthalpies of dissolution of H_2SeO_4(cr) estimated from the data are presented here: (dilution, $\Delta_{dis} H^\circ_m$ (H_2SeO_4(cr), aq dilution)) kJ·mol^{-1} (750, – 58.6); (1000, – 59.6); (1200, – 60.3); (2000, – 62.3); (5000, – 66.0); (7000, – 67.0). The dilution 1:7000 was the largest one measured. No error limits are provided in the paper but are estimated here to be ± 0.8 kJ·mol^{-1}.

[58WOO]

Three different methods are used in the paper to determine the second dissociation constant, K_{a2}, of H_2Se: (1) ordinary potentiometric titration employing KOH(aq) and a glass electrode, (2) addition of small amounts of H_2Se(aq) to KOH(aq) and subsequent pH determination with a KOH(aq)-calibrated glass electrode, and (3) the determination of the solubility of Na_2Se(cr) in solutions of known initial concentrations of sodium hydroxide/acetate. For each technique only a very limited number of data (4, 2, and 5, respectively) are presented.

The potentiometric titration data (295 K?, pH-buffer scale) were all collected at an analytical composition of the solution very close to HSe^-, which make them highly vulnerable to experimental errors. The evaluated value of pK_{a2}, 11.7, is also claimed to be "much too low because of the formation of polyselenide ions".

The second series of data (273.6 K, I = 0.2 M) uses a two-point KOH(aq)-calibrated glass electrode to determine the concentration of OH^- and, from that information, the fractions of HSe^- and Se^{2-} are calculated. The equilibrium constant, K_H, of the reaction $Se^{2-} + H_2O \rightleftharpoons HSe^- + OH^-$ is determined to be (1.1 ± 0.4) and "usual thermodynamic equations", not defined in the paper, are used to reach a pK_{a2} value of (14.5 ± 0.4) at 295 K and I = 0.

The third set of data (295 K, I = 4.2 M) is used to simultaneously evaluate the solubility product of Na_2Se(cr) and K_H and the values obtained were (0.040 ± 0.002) M^3 and (0.25 ± 0.09) M, respectively. Again, the application of undefined "usual thermodynamic equations" is referred to and the K_H value is claimed to correspond to a pK_{a2} value of (15.9 ± 0.8) at I = 0.

In view of the much more precise investigation presented by Levy and Myers [90LEV/MYE], these data are given zero weight.

[59GOL/KOS3]

Golub and Kosmatyi [59GOL/KOS3] studied the complex formation between Pb^{2+} and $SeCN^-$ by potentiometric measurements, and Golub [59GOL] performed solubility measurements on $Pb(SeCN)_2$(cr). The concentration of selenocyanate was varied between 0.7 and 3.6 M and the experiments were carried out at 293 to 303 K.

In one series of the potentiometric measurements an undefined amount of $NaNO_3$ was added "to maintain the ionic strength constant". The value of the ionic strength is not mentioned, but obviously the ionic environment varied substantially since a sodium and a potassium salt were mixed. Other experiments were carried out without the addition of a supporting ionic medium.

The lead amalgam electrode was standardised in a 0.005 M $Pb(NO_3)_2$ solution and supposed to provide a measure of the Pb^{2+} concentration in the test solutions. Total aqueous lead and selenium were determined by chemical analysis in the solubility study.

The potentiometric measurements were used to calculate the equilibrium constant of the reaction:

$$Pb^{2+} + 6SeCN^- \rightleftharpoons Pb(SeCN)_6^{4-} \quad (A.5)$$

to be $\log_{10} K$ ((A.5), 293 K) = 3.85 and $\log_{10} K$ ((A.5), 303 K) = 2.97.

The solubility measurements resulted in the equilibrium constant $\log_{10} K$ ((A.6), 293 K) = – 2.63 for the reaction:

$$Pb(SeCN)_2(cr) + 4SeCN^- \rightleftharpoons Pb(SeCN)_6^{4-} \quad (A.6)$$

and corresponds to $\log_{10} K_{s,0} = -6.48$ for $Pb(SeCN)_2$(cr). This value is not in harmony with an observation in [59GOL] that the solubility of $Pb(SeCN)_2$(cr) in water at 293 K is 0.00239 M.

In both papers the stoichiometry of the predominating aqueous species, $Pb(SeCN)_6^{4-}$, was made probable from the approximate constancy of the equilibrium constants evaluated. However, the lack of ionic medium control in the experiments, which significantly influences the activities of appearing species, seriously affects the validity of this conclusion. Furthermore, the calibration routine of the lead amalgam electrode is inadequate for measuring free Pb^{2+} concentrations in concentrated solutions of $SeCN^-$. The formation constant of $Pb(SeCN)_6^{4-}$ is comparatively small. From a statistical point of view it therefore appears very unlikely that this species would predominate over the whole concentration range of the ligand from 0.7 to 3.6 M. Hence, these data cannot be accepted by this review.

[59GUL/PET]

The heat capacities of α-HgSe and α-Ag_2Se were measured in the temperature range 80 to 300 K. The values at 298.15 K were not stated. The values $C^{\circ}_{p,m}$ (HgSe, α, 298.15 K) = (49.79 ± 2.50) $J \cdot K^{-1} \cdot mol^{-1}$ and $C^{\circ}_{p,m}$ (Ag_2Se, α, 298.15 K) = (79.2 ± 5.0) $J \cdot K^{-1} \cdot mol^{-1}$ were interpolated by the review from data in Tables 6 and 4 of the paper.

[59MIT]

Mityureva determined the solubilities of thallium selenite, $Tl_2(SeO_3)_3$(cr), and indium selenite, $In_2(SeO_3)_3 \cdot 6H_2O$(cr), in hydrochloric, nitric, and sulphuric acids at 293.2 K.

The indium concentration was determined by polarographic and colorimetric methods after previous removal of the selenium, and the thallium concentration by a volumetric titration. The attainment of a constant solubility was checked.

Only the protolysis of the selenite ion was considered in the evaluation of the measurements, which are presented in Table A-11 and Table A-12. The protonation constants published by Rumpf [33RUM] were used without introduction of activity coefficients.

Table A-11: Solubility product of indium selenite.

Solvent	pH of saturated solution	Concentration of In^{3+} (M)	$-\log_{10}[In^{3+}]^2[SeO_3^{2-}]^3$
0.1 M H_2SO_4	1.36	9.49×10^{-3}	32.54
0.05 M HCl	1.81	2.74×10^{-3}	32.69

Table A-12: Solubility product of thallium selenite.

Solvent	pH of saturated solution	Concentration of Tl^{3+} (M)	$-\log_{10}[Tl^{3+}]^2[SeO_3^{2-}]^3$
0.1 M H_2SO_4	1.29	6.47×10^{-4}	38.78
0.1 M HNO_3	1.28	6.84×10^{-4}	38.66

Side reactions of the metal ions were not considered in the paper. In the case of In(III) a major part of the dissolved In(III) would be expected to be present as complexes with the ions of the medium, while Tl(III) would be present mainly as hydroxo complexes. The review has not found it possible to derive solubility products from the limited data at hand due to uncertainties about the extent of the side reactions and the large corrections from the activity coefficients involved.

[59SEL/KAP]

Selivanova, Kapustinskii, and Zubova measured the solubility of $PbSeO_4(cr)$ at 5 temperatures between 273.15 and 323.15 K. Their data have been transformed to the molality scale and $K^\circ_{s,0}$ calculated for reaction:

$$PbSeO_4(cr) \rightleftharpoons Pb^{2+} + SeO_4^{2-}, \qquad (A.7)$$

using the SIT equation with $\Delta\varepsilon = 0$ kg·mol^{-1}. From the temperature dependence of $K^\circ_{s,0}$, $\Delta_r H^\circ_m$((A.7), 298.15 K) = (15.29 ± 0.52) kJ·mol^{-1} was calculated by the review. The enthalpy change of the reaction was also determined from a calorimetric measurement at 298.15 K of the heat of precipitation of Pb^{2+} by an equivalent amount of selenic acid according to the reaction:

$$Pb(NO_3)_2(aq, 1{:}850) + H_2SeO_4(7.07\ M) \rightarrow PbSeO_4(cr) + 2HNO_3(aq, 1{:}425).$$

The lead selenate was shown to be crystalline by X-ray diffraction.

The experimental results and the evaluation of $\Delta_r H^\circ_m$((A.7), 298.15 K) from the experimental data are presented in Table A-13. The evaluation requires a value of the enthalpy of dilution of $Pb(NO_3)_2$(aq, 1:850) to the aqueous standard state. The value of the integral enthalpy of dissolution of $Pb(NO_3)_2(cr)$ to the aqueous standard state, $\Delta_{sol}H^{\circ}_{m}$ ($Pb(NO_3)_2$, cr, 298.15 K), calculated from data in [82WAG/EVA] is 35.60 $kJ \cdot mol^{-1}$. This figure appears to be too large. The standard enthalpy of formation of $Pb(NO_3)_2(cr)$ determined by Steele *et al.* [73STE/CAR] together with the ionic enthalpies of Pb^{2+} and NO_3^- result in $\Delta_{sol}H^{\circ}_{m}$ ($Pb(NO_3)_2$, cr, 298.15 K) = (27.88 ± 1.04) $kJ \cdot mol^{-1}$. The two values are thus significantly different. Muldrow and Hepler [58MUL/HEP] determined the integral enthalpy of dissolution of $Pb(NO_3)_2(cr)$ to $Pb(NO_3)_2$(aq, 1:3200 – 4500, 298.15 K) to be (32.9 ± 0.5) $kJ \cdot mol^{-1}$. By a similar measurement Payne [95PAY] obtained $\Delta_{sol}H^{\circ}_{m}$ ($Pb(NO_3)_2$, cr, aq 1:7500 – 18000, 298.15 K) = (33.76 ± 0.13) $kJ \cdot mol^{-1}$. In analogy with the behaviour of strontium and barium nitrate on dissolution, $\Delta_{sol}H^{\circ}_{m}$ ($Pb(NO_3)_2$, cr, 298.15 K) is assumed to be very close to this result. From the data in [82WAG/EVA] the enthalpy of dissolution of $Pb(NO_3)_2(cr)$ to (aq, 1:850) is estimated to be (31.90 ± 1.50) $kJ \cdot mol^{-1}$. This results in an enthalpy of dilution of $Pb(NO_3)_2$(aq, 1:850) to the aqueous standard state amounting to (1.85 ± 1.50) $kJ \cdot mol^{-1}$.

Table A-13: Evaluation of $\Delta_r H^{\circ}_{m}$((A.7), 298.15 K). The sources of the data are indicated.

#	Reaction	$\Delta_r H_m$ ($kJ \cdot mol^{-1}$)	Reference
1	H_2SeO_4(7.07 M) + aq → H_2SeO_4(aq, 1:850)	– (24.77 ± 0.04)	(paper)
2	$Pb(NO_3)_2$(aq, 1:850) + H_2SeO_4(7.07 M) → $PbSeO_4$(cr) + $2HNO_3$(aq, 1: 425)	– (40.84 ± 0.29)	(paper)
3	H_2SeO_4(aq, 1:850) + aq → $2H^+$ + SeO_4^{2-}	– (14.00 ± 2.00)	(review)
4	$Pb(NO_3)_2$(aq, 1:850) + aq → Pb^{2+} + 2 NO_3^-	(1.85 ± 1.50)	(review, see text)
5	HNO_3(aq, 1:425) + aq → H^+ + NO_3^-	– (0.40 ± 0.10)	[82WAG/EVA]
6	$PbSeO_4$(cr) → Pb^{2+} + SeO_4^{2-}	(4.72 ± 2.53)	(result)

$\Delta_r H_6 = \Delta_r H_1 - \Delta_r H_2 + \Delta_r H_3 + \Delta_r H_4 - 2\,\Delta_r H_5$

The value of $\Delta_r H_6$ = $\Delta_r H^{\circ}_{m}$((A.7), 298.15 K) thus disagrees with the value obtained from the temperature dependence of the solubility product. The calorimetric value will be preferred.

The authors also determined the enthalpy of reaction:

$$BaSeO_4(cr) \rightleftharpoons Ba^{2+} + SeO_4^{2-} \qquad (A.8)$$

from a calorimetric measurement of the reaction: $BaCl_2$(aq, 1:500) + H_2SeO_4(7.07 M) → $BaSeO_4$(cr) + 2HCl(aq, 1: 250).

The experimental data have been re-evaluated in Table A-14.

Table A-14: Evaluation of $\Delta_r H_m^\circ$((A.8), 298.15 K). The sources of the data are indicated.

#	Reaction	$\Delta_r H_m$ (kJ·mol^{-1})	Reference
1	H_2SeO_4(7.07 M) + aq → H_2SeO_4(aq, 1:500)	– (24.06 ± 0.04)	(paper)
2	$BaCl_2$(aq, 1:500) + H_2SeO_4(7.07 M) → $BaSeO_4$(cr) + 2HCl(aq, 1:250)	– (45.98 ± 0.21)	(paper)
3	H_2SeO_4(aq, 1:500) + aq → $2H^+$ + SeO_4^{2-}	– (16.00 ± 2.50)	(review)
4	HCl(aq, 1:250) + aq → H^+ + Cl^-	– (0.81 ± 0.19)	[82WAG/EVA]
5	$BaCl_2$(aq, 1:500) + aq → Ba^{2+} + $2Cl^-$	– (1.84 ± 0.10)	[82WAG/EVA]
6	Ba^{2+} + SeO_4^{2-} → $BaSeO_4$(cr)	– (5.70 ± 2.54)	(result)

$\Delta_r H_6 = -\Delta_r H_1 + \Delta_r H_2 - \Delta_r H_3 + 2\,\Delta_r H_4 - \Delta_r H_5$

[59SEL/SHN3]

Selivanova and Shneider determined the enthalpy change of the reaction $CaSeO_4 \cdot 2H_2O$(cr) + $2AgNO_3$(aq, 1:800) → Ag_2SeO_4(cr) + $Ca(NO_3)_2$(aq, 1:1600) + $2H_2O$(l) by a calorimetric measurement. The result has been recalculated to standard conditions in Table A-15.

Table A-15: Enthalpy change of the reaction $CaSeO_4 \cdot 2H_2O$(cr) + $2Ag^+$ → Ag_2SeO_4(cr) + Ca^{2+} + $2H_2O$(l). The sources of the data are indicated.

#	Reaction	$\Delta_r H_m$ (kJ·mol^{-1})	Reference
1	$CaSeO_4 \cdot 2H_2O$(cr) + $2AgNO_3$(aq, 1:800) → Ag_2SeO_4(cr) + $Ca(NO_3)_2$(aq, 1:1600) + $2H_2O$(l)	– (38.74 ± 1.00)	(paper)
2	$AgNO_3$(aq, 1:800) + aq → Ag^+ + NO_3^-	(0.05 ± 0.10)	[82WAG/EVA]
3	$Ca(NO_3)_2$(aq, 1:1600) + aq → Ca^{2+} + 2 NO_3^-	– (0.92 ± 0.10)	[82WAG/EVA]
4	$CaSeO_4 \cdot 2H_2O$(cr) + $2Ag^+$ → Ag_2SeO_4(cr) + Ca^{2+} + $2H_2O$(l)	– (39.76 ± 1.01)	(result)

$\Delta_r H_4 = \Delta_r H_1 - 2\,\Delta_r H_2 + \Delta_r H_3$

The integral enthalpies of dissolution to ($CaSeO_4$, aq 1:1600) of $CaSeO_4 \cdot 2H_2O$(cr) and $CaSeO_4$(cr) were also measured and found to be – (8.49 ± 0.21) and – (33.68 ± 0.21) kJ·mol^{-1}, respectively. These results combine with accepted standard enthalpies of formation to $\Delta_f H_m^\circ$($CaSeO_4 \cdot 2H_2O$, cr, 298.15 K) = – (1708.99 ± 2.56) kJ·mol^{-1} and $\Delta_f H_m^\circ$($CaSeO_4$, cr, 298.15 K) = – (1112.14 ± 2.58) kJ·mol^{-1}.

[59SEL/SHN4]

The solubility of $CaSeO_4 \cdot 2H_2O(cr)$ in water was determined between 270.6 and 375.4 K. The calcium concentration was found by the oxalate-permanganate method and by a complexometric titration. As the solubility data are not readily available they are reproduced in Table A-16.

Table A-16: Solubility of $CaSeO_4 \cdot 2H_2O(cr)$ in water.

T (K)	Time of stirring (days)	Density of the saturated solution ($g \cdot cm^{-3}$)	Solubility ($mol \cdot kg^{-1}$)	Solubility ($mol \cdot L^{-1}$)
270.6	–	–	–	0.4405
283.15	8	1.093	0.4159	0.4224
288.15	6	1.087	0.3890	0.3947
293.15	6	1.082	0.3759	0.3805
298.15	6	1.076	0.3557	0.3593
303.15	6	1.073	0.3358	0.3395
313.15	4	1.062	0.3275	0.3281
333.15	4	1.053	0.2847	0.2849
243.15	3	1.048	0.2482	0.2488
353.15	2	1.042	0.2229	0.2232
363.15	1.5	1.038	0.1955	0.1959
375.35	1.5	1.035	0.1757	0.1762

The original molalities have been recalculated as an erroneous molecular weight was used.

The treatment of the solubility data in the paper for the determination of the standard enthalpy of dissolution could not be fully understood and the following treatment was resorted to by the evaluator. The data in the temperature interval 283 to 313 K were selected. Approximate activity coefficients were taken from the data for $MgSO_4$ in [50HAR/OWE]. A second order polynomial was fitted to these data and mean activity coefficients for $CaSeO_4$ in the saturated solutions obtained by interpolation. No attempt was made to correct for the temperature variation of the activity coefficient.

For the equilibrium:

$$CaSeO_4 \cdot 2H_2O(cr) \rightleftharpoons Ca^{2+} + SeO_4^{2-} + 2H_2O(l) \qquad (A.9)$$

the solubility product $\log_{10} K^{\circ}_{s,0}$ ((A.9), 298.15 K) = – 3.09 and the enthalpy of reaction $\Delta_r H^{\circ}_m$ ((A.9), 298.15 K) = – (5.31 ± 0.59) $kJ \cdot mol^{-1}$ were obtained.

[59SEL/ZUB]

Solubility measurements

Ag_2SeO_4 was prepared by mixing stoichiometric quantities of silver nitrate and selenic acid. The salt contained 60.04% Ag which is close to the theoretical value of 60.14%. The preparation was shown to be crystalline by X-ray diffraction. The solubility was measured after 15 to 20 days in a thermostat. No increase in solubility was observed between day 15 and day 20. The silver concentration was determined by a turbidimetric method.

The solubilities are presented in a table under the heading Ag_2SeO_4 without stating the concentration scale. The mean value is $(1.24 \pm 0.01) \times 10^{-3}$. If this value, c, refers to Ag_2SeO_4, then the concentration solubility product, L_p, would be $4c^3$, which is four times the value presented in the table, 1.91×10^{-9}. If, on the other hand, the concentration figure refers to that of the silver ion, L_p becomes $c^3/2$, which is half of the value in the table. It is hence concluded that some mistake has been made.

The solubility product is calculated as $L_a = \gamma_{\pm}^3 L_p$, which is correct only if $L_p = 4c^3$. The mean activity coefficent $\gamma_{\pm}$ was obtained from the limiting Debye-Hückel law as $\gamma_{\pm} = 0.866$. This value is consistent with the solubility data only if the concentration refers to Ag_2SeO_4 and is expressed in $mol{\cdot}L^{-1}$ (M). The solubility product is therefore most likely 4 times larger than reported.

Calorimetric measurement

The heat of precipitation silver selenate was measured by mixing 0.04241 moles of silver nitrate dissolved in 325 g of water with 0.02122 moles of selenic acid in an isothermal calorimeter equipped with a sensitive thermometer. The acid (7.07 M) was contained initially in a bulb that was broken in order to start the reaction. The Ag_2SeO_4 formed was shown to be crystalline by X-ray diffraction. The water equivalent was determined after each run. The experimental results and the evaluation of the standard enthalpy of formation of Ag_2SeO_4(cr) are shown in Table A-17.

Table A-17: Evaluation of the standard enthalpy of formation of $Ag_2SeO_4(cr)$. The sources of the data are indicated.

#	Reaction	$\Delta_r H_m$ (kJ·mol^{-1})	Reference
1	$H_2SeO_4(7.07\ M) + aq \rightarrow H_2SeO_4(aq, 1:850)$	– (24.85 ± 0.08)	(paper)
2	$2AgNO_3(aq, 1:425) + H_2SeO_4(7.07\ M) \rightarrow Ag_2SeO_4(cr) + 2HNO_3(aq, 1:425)$	– (68.24 ± 0.10)	(paper)
3	$H_2(g) + 2O_2(g) + Se(cr) + aq \rightarrow H_2SeO_4(aq, 1:850)$	– (589.50 ± 2.10)	(review)
4	$AgNO_3(aq, 1:425) + aq \rightarrow Ag^+ + NO_3^-$	(0.20 ± 0.05)	[82WAG/EVA]
5	$HNO_3(aq, 1:425) + aq \rightarrow H^+ + NO_3^-$	– (0.40 ± 0.10)	[82WAG/EVA]
6	$Ag(cr) + H^+ \rightarrow Ag^+ + 0.5H_2(g)$	(105.79 ± 0.08)	[89COX/WAG]
7	$2Ag(cr) + Se(cr) + 2O_2(g) \rightarrow Ag_2SeO_4(cr)$	– (422.51 ± 2.12)	(result)

$\Delta_r H_7 = -\Delta_r H_1 + \Delta_r H_2 + \Delta_r H_3 - 2\,\Delta_r H_4 + \Delta_r H_5 + 2\,\Delta_r H_6$

[59SEL/ZUB3]

The solubility of $SrSeO_4(cr)$ in water has been determined over the temperature interval 273 to 373 K. Crystalline strontium selenate was prepared by mixing stoichiometric amounts of strontium nitrate and selenic acid. Under the light microscope the crystals appear as very thin needles. The strontium concentration in saturated solution was obtained by polarographic (half-wave potential at – 2.1 V) and gravimetric measurements. Both methods gave concordant results, Table A-18.

Table A-18: Solubility of strontium selenate in water at 298.15 K.

Time of stirring (hours)	Solubility (g($SrSeO_4$)·L^{-1})	
	Polarographic method	Gravimetric method
13	4.2826	–
90	4.4210	4.540
13	4.668	–
104	4.5130	4.514

The solubility data reported in Table A-19 are taken from [58SEL/ZUB].

The enthalpy change of the reaction:

$$SrSeO_4(cr) \rightleftharpoons Sr^{2+} + SeO_4^{2-} \qquad (A.10),$$

at temperatures around 300 K was estimated from the variation of the solubility product with temperature to be – (0.98 ± 0.17) kJ·mol^{-1} by the review.

Table A-19: Solubility of strontium selenate in water in the temperature interval 273 to 373 K. Equilibration times 3 to 33 days.

Temperature (K)	Solubility (g($SrSeO_4$)·L^{-1})
273.15	2.031×10^{-2}
288.15	2.031×10^{-2}
303.15	1.932×10^{-2}
313.15	1.798×10^{-2}
325.15	1.617×10^{-2}
338.15	1.258×10^{-2}
358.15	9.984×10^{-3}
373.15	7.767×10^{-3}

The authors determined the standard enthalpy of formation of $SrSeO_4$(cr) from a measurement of the enthalpy change of the reaction $SrCl_2\cdot 6H_2O$(cr) + K_2SeO_4(aq, 1:240) → $SrSeO_4$(cr) + 2KCl(aq, 1:120) + $6H_2O$(l). The data have been recalculated with new auxiliary data in Table A-20.

Table A-20: Calculation of the enthalpy change of the reaction Sr^{2+} + $SeO_4^{2-} \rightleftharpoons SrSeO_4$(cr). The sources of the data are indicated.

#	Reaction	$\Delta_r H_m$ (kJ·mol^{-1})	Reference
1	$SrCl_2\cdot 6H_2O$(cr) + K_2SeO_4(aq, 1:240) → $SrSeO_4$(cr) + 2KCl(aq, 1:120) + $6H_2O$(l)	(30.92 ± 0.59)	(paper)
2	$SrCl_2\cdot 6H_2O$(cr) + aq → $SrCl_2$(aq, 1:240) + $6H_2O$(l)	(31.51 ± 0.29)	(paper)
3	$SrCl_2$(aq, 1:240) + aq → Sr^{2+} + $2Cl^-$	– (2.30 ± 0.10)	[82WAG/EVA]
4	KCl(aq, 1:120) + aq → K^+ + Cl^-	– (0.26 ± 0.10)	[82WAG/EVA]
5	K_2SeO_4(aq, 1:240) + aq → $2K^+$ + SeO_4^{2-}	– (1.30 ± 0.50)	(review)
6	Sr^{2+} + SeO_4^{2-} → $SrSeO_4$(cr).	(2.49 ± 0.86)	(result)

$\Delta_r H_6 = \Delta_r H_1 - \Delta_r H_2 - \Delta_r H_3 + 2\,\Delta_r H_4 - \Delta_r H_5$

The review observed that the reaction in the calorimeter is not complete. Equivalent amounts of the reactants were mixed and at the ionic strength of the final solution, $I \simeq 0.45$ M, a rough calculation indicates that only about 80% of the selenate had reacted. Since $\Delta_r H_m^\circ$ (A.10) is close to zero, the correction for the incompleteness of the reaction is small and was neglected in Table A-20.

[59ZLO/POP]

The evaporation of PbSe(cr) was studied in the temperature range 774 to 941 K using the Langmuir method and in the range 914 to 991 K using the Knudsen effusion weight loss technique. The original evaluation employed auxiliary data different from those used by the review and the vapour pressure expression was therefore re-evaluated using the selected values of the entropies and heat capacities of PbSe(cr) and PbSe(g). The results from the two series of measurements agree well and the new evalution was based on a combination of both series of measurements. The second law entropy of PbSe(g) and the enthalpy of sublimation of PbSe(cr) were calculated to be S°_m(PbSe, g, 298.15 K) = (259.8 ± 9.0) $J \cdot K^{-1} \cdot mol^{-1}$ and $\Delta_{sub}H^\circ_m$(PbSe, cr, 298.15 K) = (222.6 ± 9.0) $kJ \cdot mol^{-1}$, respectively. The third law enthalpy of sublimation was calculated to be $\Delta_{sub}H^\circ_m$(PbSe, cr, 298.15 K) = (226.0 ± 3.0) $kJ \cdot mol^{-1}$.

[60FIN/WAG]

The Gibbs energy of formation of PbSe(cr) was measured in the temperature range 473 to 573 K using an electrochemical cell. The reaction studied was

$$\text{Pb(cr)} + \text{Se(l)} \rightleftharpoons \text{PbSe(cr)}. \qquad \text{(A.11)}$$

The enthalpy and entropy of reaction were calculated to be $\Delta_r H^\circ_m$((A.11), 523 K) = – (106.4 ± 3.0) $kJ \cdot mol^{-1}$ and $\Delta_r S^\circ_m$((A.11), 523 K) = – (15.9 ± 6.0) $J \cdot K^{-1} \cdot mol^{-1}$ by the review from Eq. (II) of the paper. A recalculation to 298.15 K employing the selected heat capacity of PbSe(cr), the selected data of selenium and the data for Pb(cr) in [89COX/WAG] yielded $\Delta_f H^\circ_m$(PbSe, cr, 298.15 K) = – (99.5 ± 3.0) $kJ \cdot mol^{-1}$ and $\Delta_f S^\circ_m$(PbSe, cr, 298.15 K) = (1.8 ± 6.0) $J \cdot K^{-1} \cdot mol^{-1}$. The corresponding entropy of PbSe(cr) was calculated to be S°_m(PbSe, cr, 298.15 K) = (108.7 ± 6.0) $J \cdot K^{-1} \cdot mol^{-1}$ using the selected entropy of selenium and the entropy of Pb(cr) in [89COX/WAG].

[60KOR/SOK]

The saturated vapour pressure of α-ZnSe and α-CdSe were measured using Knudsen effusion cells in the temperature range 913 to 1093 K and 813 to 1013 K, respectively. Apparent sublimation enthalpies for the average temperatures were derived but no evaluations of thermodynamic properties at 298.15 K were made. The experimental data were therefore evaluated by the review using the second and the third law, and assuming the reactions α-ZnSe $\rightleftharpoons$ Zn(g) + ½Se_2(g) and α-CdSe $\rightleftharpoons$ Cd(g) + ½Se_2(g). The results are given in Table A-21. The evaluations used the selected thermodynamic functions of selenium, the data for Zn and Cd in [89COX/WAG], the heat capacity expression of α-ZnSe in Section V.9.1.1.1, the selected heat capacity of α-CdSe and the selected entropy of α-ZnSe and α-CdSe at 298.15 K in the case of the third law.

Table A-21: Enthalpy of formation and entropy of α-ZnSe and α-CdSe at 298.15 K evaluated by the review from the data in [60KOR/SOK].

Compound	$\Delta_f H^\circ_m$ (298.15 K) (kJ·mol^{-1})		S°_m (298.15 K) (J·K^{-1}·mol^{-1})	Temperature (K)
	2nd law	3rd law		
α-ZnSe	– (215.9 ± 35.0)	– (175.2 ± 12.0)	(31.6 ± 20.0)	913 – 1093
α-CdSe	– (139.0 ± 15.0)	– (141.9 ± 8.0)	(90.9 ± 10.0)	813 – 1013

[60POR/SPE]

The equilibrium constant of the reaction

$$Bi_2(g) + Se_2(g) \rightleftharpoons 2BiSe(g) \qquad (A.12)$$

was determined at 1050 K to be $\log_{10} K$ ((A.12), 1050 K) = (40 ± 10) using mass spectroscopy and Knudsen effusion from a Bi-Se mixture of varying composition. The result corresponds to $\Delta_f G^\circ_m$ (1050 K) = – (32.2 ± 8.1) kJ·mol^{-1} and using the selected properties of BiSe(g) and Se_2(g) and thermodynamic properties of Bi_2(g) calculated from the molecular constants given in [96BAR/TAH], the enthalpy of reaction at 298.15 K is calculated to be $\Delta_r H^\circ_m$ ((A.12), 298.15 K) = – (24.4 ± 8.0) kJ·mol^{-1} from the third law. This yields $\Delta_f H^\circ_m$ (BiSe, g, 298.15 K) = (167.5 ± 9.0) kJ·mol^{-1} using $\Delta_f H^\circ_m$ (Bi_2, g, 298.15 K) = 219.7 kJ·mol^{-1} in [82WAG/EVA] and the selected enthalpy of formation of Se_2(g).

[61HAM]

Hamada studied the reaction

$$SeCN^- + H^+ \rightleftharpoons Se(s, red) + HCN(aq). \qquad (A.13)$$

Dilute solutions, 0.001 to 0.05 M, of hydrochloric acid were added to aliquots of a solution of KSeCN in flasks. The reaction flasks were purged with nitrogen, capped and left in a thermostated bath for one day to reach equilibrium. The precipitate of red selenium was weighed, which together with a pH determination yields sufficient information for the calculation of the equilibrium constant.

Checks for the reversibility of the reaction and equilibration times are not mentioned.

Measurements were carried out at four temperatures in the narrow temperature range of 273.5 to 285.7 K. Low concentrations were used in the experiments and the results are considered to refer to standard conditions. The enthalpy of reaction was evaluated from the temperature variation of the equilibrium constant to be:

$$\Delta_r H^\circ_m ((A.13), 298.15 K) = 29.7 kJ·mol^{-1}.$$

The equilibrium constant determined at 280.5 K was then extrapolated to 298.15 K to yield $\log_{10} K^{\circ}$ ((A.13), 298.15 K) = 3.57 equivalent to:

$$\Delta_r G_m^{\circ}((\text{A.13}), 298.15\text{ K}) = -20.4\text{ kJ·mol}^{-1}.$$

These values were finally combined to yield:

$$\Delta_r S_m^{\circ}((\text{A.13}), 298.15\text{ K}) = 168.0\text{ J·K}^{-1}\text{·mol}^{-1}.$$

By combining these data with the accepted data for Se(monoclinic, 298.15 K) ($\Delta_f H_m^{\circ}$ = 2.14 kJ·mol^{-1}, $\Delta_f G_m^{\circ}$ = 1.28 kJ·mol^{-1}, and S_m° = 44.97 J·K^{-1}·mol^{-1}) and the data evaluated for HCN(aq) ($\Delta_f H_m^{\circ}$ = (103.75 ± 3.54) kJ·mol^{-1}, $\Delta_f G_m^{\circ}$ = (114.37 ± 2.52) kJ·mol^{-1}, and S_m° = (131.3 ± 8.4) J·K^{-1}·mol^{-1}),

$$\Delta_f H_m^{\circ}(\text{SeCN}^-, 298.15\text{ K}) = 76.2\text{ kJ·mol}^{-1},$$

$$\Delta_f G_m^{\circ}(\text{SeCN}^-, 298.15\text{ K}) = 136.1\text{ kJ·mol}^{-1},$$

$$\text{and } S_m^{\circ}(\text{SeCN}^-, 298.15\text{ K}) = 8.3\text{ J·K}^{-1}\text{·mol}^{-1},$$

can be calculated.

However, this value of the entropy for $SeCN^-$ appears to be inconsistent with the corresponding values in [82WAG/EVA] for OCN^- and SCN^-, 106.7 and 144.3 J·K^{-1}·mol^{-1}, respectively, and cannot, therefore, be accepted. It is highly probable that this inconsistency predominantly originates from an erroneously determined value of the reaction enthalpy and hence the enthalpy of formation for $SeCN^-$ will not be accepted. The determination of the equilibrium constant for the reaction can be judged as being less influenced by experimental uncertainties and is, as such, accepted. To account for the uncertainties of the determination, and of the temperature extrapolation made, it is, however, ascribed an uncertainty of 0.5 logarithmic units. This results in the selected value:

$$\Delta_f G_m^{\circ}(\text{SeCN}^-, 298.15\text{ K}) = (136.1 \pm 3.8)\text{ kJ·mol}^{-1}.$$

[61LIN/PAN]

Measurements were made at 298.15 K of the cells:

Hg,Hg_2Cl_2(s) | KCl (sat.) | KNO_3 (m) | $NaHSeO_3$ (m), Ag_2SeO_3 (sat) | Ag_2SeO_3, Ag (I)

Hg,Hg_2Cl_2(s) | KCl (sat.) | KNO_3 (m) | $NaHSeO_3$ (m), Q, QH_2 | Pt (II)

and combined to yield the emf of the cell:

Pt | QH_2, Q, $NaHSeO_3$ (m), Ag_2SeO_3 (sat.) | Ag_2SeO_3, Ag (III)

that could not be measured directly, probably due to a reaction between quinhydrone and silver selenite. The emf of cell (I) was stable for 12 to 24 hours and the reproducibility was about 1 mV. The corresponding figures for cell (II) were 4 to 6 hours and 0.5 mV, respectively. It can be noted that cell (II) is unbuffered. The combination of cells (I) and (II) to represent cell (III) assumes that the solubility of silver selenite is small.

Silver selenite was prepared by addition of a selenious acid solution to a solution of silver nitrate. The product was recrystallised three times from nitric acid (?). Sodium hydrogen selenite solution was prepared by neutralisation of recrystallised selenious acid by carbonate-free sodium hydroxide. The silver-silver selenite electrodes were obtained by first electroplating silver onto a platinum foil and then converting some of the silver to silver selenite by further anodic electrolysis. The procedure was apparently the same as used for the preparation of silver-silver chloride electrodes. The experiments were carried out in the dark or in subdued daylight.

The authors report $\log_{10} K_{s,0}$ (298.15 K) = – 14.74, for the reaction:

$$Ag_2SeO_3(cr) \rightleftharpoons 2Ag^+ + SeO_3^{2-}. \qquad (A.14)$$

This value cannot be accepted since a number of mistakes were made in the evaluation of the constant and a recalculation has therefore been attempted.

The emf of cell (I) as written can be expressed by the relationship:

$$E_I = E^o_{Ag/Ag_2SeO_3} - E_{ref} - \frac{RT}{2F}\ln a_{SeO_3^{2-}}.$$

However, the measured emf of the cell increases with m, which would indicate a reversed sign of the cell potential. E_I was therefore combined with the emf of cell (II):

$$E_{II} = E^o_{Q,QH_2} - E_{ref} + \frac{RT}{F}\ln a_{H^+}$$

in two ways:

$$E_{III} = E_I - E_{II} = E^o_{Ag/Ag_2SeO_3} - E^o_{Q,QH_2} - \frac{RT}{2F}\ln a^2_{H^+}\, a_{SeO_3^{2-}},$$

$$E_{IV} = E_I + E_{II} = E^o_{Q,QH_2} - E^o_{Ag/Ag_2SeO_3} + \frac{RT}{2F}\ln a^2_{H^+}\, a_{SeO_3^{2-}}.$$

The activity product was calculated with the accepted protonation constants of the selenite ion using the SIT approach for the activity coefficient correction. The value of E^o_{Ag/Ag_2SeO_3} obtained from E_{III} was (309.5 ± 18.9) mV with a systematic trend in the value with m. The value from E_{IV} was – (137.4 ± 3.5) mV with no significant trend. Hence a reversal of the sign of the potentials from cell (I) is indicated. The values were calculated with E^o_{Q,QH_2} = 699.4 mV. With $E^o_{Ag^+/Ag}$ = 799.1 mV the corresponding solubility products are $\log_{10} K_{s,0}$ ((A.14), 298.15 K) = – (16.6 ± 0.6) and – (31.7 ± 0.1). These findings were confirmed by a similar data treatment considering the silver/silver selenite electrode as a probe of the silver ion activity. This evaluation used data from cell (I) in conjunction with a value of 241.5 mV for the potential of the reference electrode.

The review finds that it is not possible to derive a reliable value of the solubility product from the data. The value $\log_{10} K_{s,0}$ ((A.14), 298.15 K)= – 31.7 is numerically too negative and the other value, which is closer to literature values, shows an unacceptable trend. Apparently Masson *et al.* [86MAS/LUT] experienced the same problems as they write "attempted recalculations gave nonsensical figures, so the work is rejected".

[61NES/PAS]

The total pressure of the vapour in equilibrium with β-SnSe was measured in the temperature range 862 to 920 K using the Knudsen effusion technique. The enthalpy and entropy of sublimation according to the reaction α-SnSe $\rightleftharpoons$ SnSe(g) were calculated by the review from the reported vapour pressure expression and the selected heat capacities of SnSe(g) and α-SnSe, the enthalpy of transformation α-SnSe → β-SnSe being 1.28 kJ·mol^{-1} (cf V.7.4.1.2), and a heat capacity expression of β-SnSe being identical to that of α-SnSe, yielding $\Delta_{\text{sub}}H_{\text{m}}^{\circ}$ (SnSe, α, 298.15 K) = (188.2 ± 20.0) kJ·mol^{-1} and $\Delta_{\text{sub}}S_{\text{m}}^{\circ}$ (SnSe, α, 298.15 K) = (133.0 ± 25.0) J·K^{-1}·mol^{-1}, respectively.

The error limits are rather large in this investigation due to the small temperature range.

[61SEL/PAK]

Selivanova and Pakhorukov measured the integral heat of dissolution of H_2SeO_3(cr) in a conventional calorimeter. The heat equivalent of the equipment was obtained by electrical calibration. The accuracy of the instrumentation was checked by a measurement of the enthalpy of dissolution of KCl to KCl(aq, 1:200).

The results are entered in Table A-22, as they are not readily available. The uncertainty of the data is estimated to be ± 150 J·mol^{-1}.

Table A-22: Integral enthalpies of solution of H_2SeO_3(cr) in water.

Dilution	$\Delta_{\text{sol}}H_{\text{m}}$ (kJ·mol^{-1})	Dilution	$\Delta_{\text{sol}}H_{\text{m}}$ (kJ·mol^{-1})	Dilution	$\Delta_{\text{sol}}H_{\text{m}}$ (kJ·mol^{-1})
1:100	16.88	1:1000	18.20	1:2000	18.47
1:200	17.68	1:1200	18.21	1:2500	18.51
1:400	17.90	1:1400	18.23	1:3000	18.56
1:600	18.03	1:1600	18.31	1:3500	18.59
1:800	18.10	1:1800	18.38	1:4000	18.64

The integral enthalpies of dissolution of SeO_2(cr) in water can be calculated by adding the enthalpy change of the reaction SeO_2(cr) + H_2O(l) $\rightleftharpoons$ H_2SeO_3(cr), – (13.50 ± 0.18) kJ·mol^{-1}, adopted in Section V.3.2.1.

[61SEL/SHN]

Selivanova, Shneider, and Ryabova measured the enthalpy change of the reaction: $MgSeO_4{\cdot}6H_2O(cr) + 2AgNO_3(aq, 1{:}800) \rightarrow Ag_2SeO_4(cr) + Mg(NO_3)_2(aq, 1{:}1600) + 6H_2O(l)$ in a calorimeter. The result has been recalculated to standard conditions in Table A-23.

Table A-23: Enthalpy change of the reaction $MgSeO_4{\cdot}6H_2O(cr) + 2Ag^+ \rightarrow Ag_2SeO_4(cr) + Mg^{2+} + 6H_2O(l)$. The sources of the data are indicated.

#	Reaction	$\Delta_r H_m$ (kJ·mol^{-1})	Reference
1	$MgSeO_4{\cdot}6H_2O(cr) + 2AgNO_3(aq, 1{:}800) \rightarrow Ag_2SeO_4(cr) + Mg(NO_3)_2(aq, 1{:}1600) + 6H_2O(l)$	– (33.30 ± 1.00)	(paper)
2	$AgNO_3(aq, 1{:}800) + aq \rightarrow Ag^+ + NO_3^-$	(0.05 ± 0.10)	[82WAG/EVA]
3	$Mg(NO_3)_2(aq, 1{:}1600) + aq \rightarrow Mg^{2+} + 2\,NO_3^-$	– (1.29 ± 0.10)	[82WAG/EVA]
4	$MgSeO_4{\cdot}6H_2O(cr) + 2Ag^+ \rightarrow Ag_2SeO_4(cr) + Mg^{2+} + 6H_2O(l)$	– (34.69 ± 1.01)	(result)

$\Delta_r H_4 = \Delta_r H_1 - 2\,\Delta_r H_2 + \Delta_r H_3$

The enthalpies of dissolution to (aq, 1:1500) were measured for $MgSeO_4{\cdot}6H_2O(cr)$, – (0.40 ± 0.20) kJ·mol^{-1}, $MgSeO_4{\cdot}4H_2O(cr)$, – (17.45 ± 1.05) kJ·mol^{-1}, $MgSeO_4{\cdot}H_2O(cr)$, – (54.39 ± 0.84) kJ·mol^{-1}, and $MgSeO_4(cr)$, – (95.48 ± 0.96) kJ·mol^{-1}. The experimental data in combination with accepted auxiliary data yield:

$$\Delta_f H_m^\circ(MgSeO_4{\cdot}6H_2O, cr, 298.15\ K) = -(2781.38 \pm 2.43)\ kJ\cdot mol^{-1},$$

$$\Delta_f H_m^\circ(MgSeO_4{\cdot}4H_2O, cr, 298.15\ K) = -(2192.67 \pm 2.65)\ kJ\cdot mol^{-1},$$

$$\Delta_f H_m^\circ(MgSeO_4{\cdot}H_2O, cr, 298.15\ K) = -(1298.24 \pm 2.58)\ kJ\cdot mol^{-1},$$

and $\Delta_f H_m^\circ(MgSeO_4, cr, 298.15\ K) = -(971.32 \pm 2.62)\ kJ\cdot mol^{-1}$.

[61SEL/SHN3]

Selivanova and Shneider measured the enthalpy of the reaction $BeSeO_4{\cdot}4H_2O(cr) + BaCl_2(aq, 1{:}900) \rightarrow BaSeO_4(cr) + BeCl_2(aq, 1{:}900) + 4H_2O(l)$ in a calorimeter. They also measured the enthalpy of dissolution of $BeSeO_4{\cdot}4H_2O(cr)$ and $BeSeO_4{\cdot}2H_2O(cr)$ in water (1:900) to be – (9.96 ± 0.33) and – (43.76 ± 0.41) kJ·mol^{-1}, respectively, as well as the enthalpies of dissolution of $BeSeO_4{\cdot}2H_2O(cr)$ and $BeSeO_4(cr)$ in 1 M KOH (1:3600) to be – (109.75 ± 0.63) and – (155.44 ± 0.84) kJ·mol^{-1}, respectively. The experimental results were used to calculate the standard enthalpy of formation of $BeSeO_4(cr)$ and its hydrates. The data have been recalculated by the review with new auxiliary data in Table A-24.

The value of $\Delta_f H_m^\circ(BeCl_2$, aq 1:900, 298.15 K) was obtained from the enthalpy of formation of $BeCl_2(cr)$ and its enthalpy of dissolution in water. The enthalpy of formation was reported to be – (493.85 ± 2.35) kJ·mol^{-1} from direct synthesis calorimetry in [61JOH/GIL] and to be – (494.76 ± 2.10) kJ·mol^{-1} from the enthalpy of dissolution of beryllium metal in hydrochloric acid in [62THO/SIN]. The mean value, $\Delta_f H_m^\circ$ (BeCl$_2$, cr, 298.15 K) = – (494.30 ± 3.30) kJ·mol^{-1}, is adopted.

Table A-24: Evaluation of the standard enthalpy of formation of $BeSeO_4 \cdot 4H_2O$(cr). The sources of the data are indicated.

#	Reaction	$\Delta_r H_m$ (kJ·mol^{-1})	Reference
1	$BeSeO_4 \cdot 4H_2O$(cr) + $BaCl_2$(aq, 1:900) → $BaSeO_4$(cr) + $BeCl_2$(aq, 1:900) + $4H_2O$(l)	– (21.25 ± 0.33)	(paper)
2	Ba(cr) + Cl_2(g) + aq → $BaCl_2$(aq, 1:900)	– (866.97 ± 2.50)	(see text)
3	Be(cr) + Cl_2(g) + aq → $BeCl_2$(aq, 1:900)	– (708.3 ± 3.9)	(see text)
4	Ba(cr) + Se(cr) + $2O_2$(g) → $BaSeO_4$(cr)	– (1144.00 ± 5.00)	(review)
5	H_2(g) + $0.5O_2$(g) → H_2O(l)	– (285.83 ± 0.04)	[89COX/WAG]
6	Be(cr) + Se(cr) + $4H_2$(g) + $4O_2$(g) → $BeSeO_4 \cdot 4H_2O$(cr)	– (2107.0 ± 6.8)	(result)

$\Delta_r H_6 = \Delta_f H_m^\circ$ ($BeSeO_4 \cdot 4H_2O$, cr, 298.15 K) $= -\Delta_r H_1 - \Delta_r H_2 + \Delta_r H_3 + \Delta_r H_4 + 4\,\Delta_r H_5$

The enthalpies of solution of the two polymorphs of $BeCl_2$(cr) in 0.01 M HCl to a dilution of 1:6000, – (216.7 ± 0.1) kJ·mol^{-1} for α-BeCl$_2$ and – (213.8 ± 0.1) kJ·mol^{-1} for β-BeCl$_2$, were determined in [95SPU/FIN]. Since the formation data apparently refer to mixtures of the two forms [65DON/OET] the mean of the two dissolution data, – (215.3 ± 1.5) kJ·mol^{-1}, is used. Matignon and Marchal [26MAT/MAR] obtained – 213.8 kJ·mol^{-1} for dilutions to 1:900 and 1:2700. Dilution effects in the range 1:900 to 1:6000 thus appear to be small and the review adopts – (214.0 ± 2.0) kJ·mol^{-1} for the enthalpy of dissolution to aq 1:900 which gives $\Delta_f H_m^\circ$ (BeCl$_2$, aq 1:900, 298.15 K) = – (708.3 ± 3.9) kJ·mol^{-1}.

The value of $\Delta_f H_m^\circ$ (BaCl$_2$, aq 1:900, 298.15 K) is found from the standard enthalpy of formation of $BaCl_2$(cr) in [92GRE/FUG] and its enthalpy of dissolution in [82WAG/EVA] to be – (866.97 ± 2.50) kJ·mol^{-1}.

The standard enthalpies of formation of the other two selenates are found from the enthalpies of dissolution to be $\Delta_f H_m^\circ$ ($BeSeO_4 \cdot 2H_2O$, cr, 298.15 K) = – (1501.9 ± 6.8) kJ·mol^{-1} and $\Delta_f H_m^\circ$ (BeSeO$_4$, cr, 298.15 K) = – (884.6 ± 6.9) kJ·mol^{-1}.

[61SOM]

The saturated vapour pressure of α-CdSe was measured in the temperature range 975 to 1210 K using Bourdon gauge and dew point techniques. The two methods gave identical results. The total vapour pressure was described by the expression $\log_{10}(p/\text{bar}) = 6.92 - 10020\ T^{-1}$. No evaluations of thermodynamic properties at 298.15 K were made.

The pressure relation was therefore used by the review in deriving second and third law quantities assuming the reaction α-CdSe(s) $\rightleftharpoons$ Cd(g) + ½Se_2(g). The evaluations used the selected thermodynamic functions of selenium, the data for Cd in [89COX/WAG], the selected heat capacity of α-CdSe, and the selected entropy of α-CdSe at 298.15 K in the case of the third law. The results were $\Delta_f H_m^\circ$ (CdSe, α, 298.15 K) = – (121.1 ± 8.0) kJ·mol^{-1}, $\Delta_f H_m^\circ$ (CdSe, α, 298.15 K) = – (119.4 ± 6.0) kJ·mol^{-1} and S_m° (CdSe, α, 298.15 K) = (84.5 ± 8.0) J·K^{-1}·mol^{-1} for the second and third law enthalpy of formation and the entropy, respectively.

[61WOS]

The saturated vapour pressure of α-CdSe was measured in the temperature range 1016 to 1170 K using a gas flow method. Values for the entropy and the enthalpy of formation of α-CdSe at 298.15 were derived using the second law and an estimated heat capacity of α-CdSe. The experimental results were therefore re-evaluated by the review using both the second and the third law, the selected thermodynamic functions of selenium, the data for Cd in [89COX/WAG], the selected heat capacity of α-CdSe, and the selected entropy of α-CdSe in the case of the third law. The vaporisation was assumed to occur according to the reaction α-CdSe $\rightleftharpoons$ Cd(g) + ½Se_2(g). The results were $\Delta_f H_m^\circ$ (CdSe, α, 298.15 K) = – (158.0 ± 15.0) kJ·mol^{-1}, $\Delta_f H_m^\circ$ (CdSe, α, 298.15 K) = – (146.2 ± 3.0) kJ·mol^{-1}, and S_m° (CdSe, α, 298.15 K) = (75.4 ± 15.0) J·K^{-1}·mol^{-1} for the second and third law enthalpy of formation and the entropy, respectively.

[62ALE/SER]

Aleksandrovich and Serebrennikov measured the solubility of the selenites of Ce^{3+}, La^{3+}, Nd^{3+}, Pr^{3+} and Sm^{3+} in HCl, HNO_3, and H_2SO_4 of unspecified concentration. The quality of the solid phases is not described. The experiments were carried out in closed vessels at (298.15 ± 0.05) K in a nitrogen atmosphere. The phases were separated after 8 hours by centrifugation and the total metal ion concentration and pH were measured in the aqueous phase. The primary data were evaluated according to [56CHU] in this Appendix and the results are entered in Chapter V.

[62KUD/UST]

The total vapour pressure of liquid selenium at equilibrium was measured by a static method in the temperature range 715 to 958 K and given as $\log_{10}(p/\text{bar}) = 5.230 - 5011\,T^{-1}$. A third law evaluation of the enthalpy of formation of Se_2(g) at 298.15 K was made by this review using the vapour pressure at 950 K and the selected data for the heat capacities and entropies of Se(l) and Se_2(g). A mole fraction of Se_2(g), $x_{Se_2(g)}$ = 0.64 was estimated from the selected selenium data. The enthalpy of formation derived was $\Delta_f H_m^\circ$ (Se_2, g, 298.15 K) = (140.1 ± 1.6) kJ·mol^{-1}.

[62SEL/LES]

Crystalline Ag_2SeO_3 was prepared by mixing 0.1 M solutions of $AgNO_3$ and Na_2SeO_3 in stoichiometric proportions. The solubility of the specimen in water at (298.15 ± 0.10) K was determined. Periods of mixing between 6 to 10 days were employed and the solubility changed little during this period and equilibrium is thus assumed to prevail. The solubility was determined by a turbidimetric method to be $(8.31 \pm 0.03) \times 10^{-6}$ M. No measurement of the pH of the equilibrium solution was made. From this datum the review calculated $\log_{10} K^{\circ}_{s,0}$ ((A.14), 298.15 K) = – (15.87 ± 0.02) after allowance for side reactions, see [68RIP/VER] in this Appendix.

The authors also measured the enthalpy change of the reaction between $AgNO_3$(cr) and a solution of sodium selenite with formation of crystalline silver selenite as verified by chemical analysis and X-ray diffraction. The calorimeter was calibrated by electrical resistance heating. The calorimetric data are evaluated in Table A-25.

Table A-25: Calculation of the standard enthalpy of formation of Ag_2SeO_3(cr). The sources of the data are indicated.

#	Reaction	$\Delta_r H_m$ (kJ·mol^{-1})	Reference
1	$2AgNO_3(cr) + Na_2SeO_3(aq, 1{:}800) \rightarrow Ag_2SeO_3(cr) + 2NaNO_3(aq, 1{:}400)$	– (22.13 ± 0.21)	(paper)
2	$AgNO_3(cr) + aq \rightarrow Ag^+ + NO_3^-$	(22.73 ± 0.04)	[89COX/WAG]
3	$NaNO_3(aq, 1{:}400) \rightarrow Na^+ + NO_3^-$	– (0.08 ± 0.10)	[82WAG/EVA]
4	$Na_2SeO_3(cr) + aq \rightarrow Na_2SeO_3(aq, 1{:}800)$	– (28.33 ± 0.08)	[69KLU/SEL]
5	$2Na(cr) + Se(cr) + 1.5O_2(g) \rightarrow Na_2SeO_3(cr)$	– (959.62 ± 0.95)	(review)
6	$Ag(cr) + Na^+ \rightarrow Ag^+ + Na(cr)$	– (346.13 ± 0.10)	[89COX/WAG]
7	$2Ag(cr) + Se(cr) + 1.5O_2(g) \rightarrow Ag_2SeO_3(cr)$	– (363.44 ± 1.02)	(result)

$\Delta_r H_7 = \Delta_f H^{\circ}_m(Ag_2SeO_3, cr, 298.15\ K) = \Delta_r H_1 + 2\Delta_r H_2 + 2\Delta_r H_3 + \Delta_r H_4 + \Delta_r H_5 + 2\Delta_r H_6$

[62SEL/LES3]

The enthalpy change of the reaction between Na_2SeO_3(cr) and a lead nitrate solution with formation of crystalline lead selenite was measured in a calorimeter. The data have been used to calculate the standard enthalpy change of the reaction $Na_2SeO_3(cr) + Pb^{2+} \rightarrow PbSeO_3(cr) + 2Na^+$ in Table A-26.

The value of the standard enthalpy of formation of lead selenite obtained from $\Delta_r H_4$ and accepted standard enthalpies of formation is:

$$\Delta_f H^{\circ}_m(PbSeO_3, cr, 298.15\ K) = -(532.08 \pm 1.80)\ kJ\cdot mol^{-1}.$$

[82WAG/EVA] reports – 537.6 kJ·mol^{-1}. Changes in the accepted standard enthalpies of formation of the ions account for 2.5 kJ·mol^{-1} of the difference. The remainder of the difference is most likely due mainly to the change in the values of the heat of dilution of $Pb(NO_3)_2$(aq) made by the review and discussed in Appendix A, [59SEL/KAP].

Table A-26: Calculation of the standard enthalpy change of the reaction Na_2SeO_3(cr) + Pb^{2+} → $PbSeO_3$(cr) + 2Na^+. The sources of the data are indicated.

#	Reaction	$\Delta_r H_m$ (kJ·mol^{-1})	Reference
1	Na_2SeO_3(cr) + $Pb(NO_3)_2$(aq, 1:1200) → $PbSeO_3$(cr) + 2$NaNO_3$(aq, 1:600)	– (52.26 ± 0.17)	(paper)
2	$Pb(NO_3)_2$(aq, 1:1200) + aq → Pb^{2+} + 2 NO_3^-	(1.50 ± 1.50)	(review)
3	$NaNO_3$(aq, 1:600) + aq → Na^+ + NO_3^-	– (0.15 ± 0.05)	[82WAG/EVA]
4	Na_2SeO_3(cr) + Pb^{2+} → $PbSeO_3$(cr) + 2Na^+	– (54.06 ± 1.50)	(result)

$\Delta_r H_4 = \Delta_r H_1 - \Delta_r H_2 + 2\,\Delta_r H_3$

[62SEL/SAM]

Selivanova, Samplavskaya, and Maier [62SEL/SAM] determined the enthalpy change of the reaction of a copper selenate solution with solid silver nitrate by a calorimetric measurement. They similarly measured the heat of dissolution of $CuSeO_4\cdot5H_2O$(cr) and $CuSeO_4$(cr) to $CuSeO_4$(aq, 1:700). The data have been recalculated with new auxiliary data to obtain the enthalpies of formation of the copper selenates as shown in Table A-27. The enthalpy of formation of $AgNO_3$(cr) was calculated from the enthalpy of dissolution of $AgNO_3$(cr) to the aqueous standard state in [89COX/WAG] and the standard ionic enthalpies of formation in the same reference.

Table A-27: Evaluation of the standard enthalpy of formation of $CuSeO_4$(cr). The sources of the data are indicated.

#	Reaction	$\Delta_r H_m$ (kJ·mol^{-1})	Reference
1	2$AgNO_3$(cr) + $CuSeO_4$(aq, 1:700) → Ag_2SeO_4(cr) + $Cu(NO_3)_2$(aq, 1:700)	(8.58 ± 0.13)	(paper)
2	$CuSeO_4$(cr) + aq → $CuSeO_4$(aq, 1:700)	– (54.98 ± 0.33)	(paper)
3	2Ag(cr) + Se(cr) + 2O_2(g) → Ag_2SeO_4(cr)	– (422.51 ± 2.12)	(review)
4	Ag(cr) + 0.5N_2(g) + 1.5O_2(g) → $AgNO_3$(cr)	– (123.79 ± 0.41)	(see text)
5	Cu(cr) + N_2(g) + 3O_2(g) +aq → $Cu(NO_3)_2$(aq,1:700)	– (351.80 ± 1.62)	[89COX/WAG] [82WAG/EVA]
6	Cu(cr) + Se(cr) + 2O_2(g) → $CuSeO_4$(cr)	– (480.33 ± 2.81)	(result)

$\Delta_r H_6 = -\Delta_r H_1 - \Delta_r H_2 + \Delta_r H_3 - 2\,\Delta_r H_4 + \Delta_r H_5$

The result from the table is $\Delta_f H_m^\circ$ ($CuSeO_4$, cr, 298.15 K) = – (480.33 ± 2.81) $kJ\cdot mol^{-1}$.

The enthalpy of dissolution of $CuSeO_4\cdot 5H_2O$(cr) was determined to be (8.87 ± 0.13) $kJ\cdot mol^{-1}$. The enthalpy of hydration of $CuSeO_4$(cr) is thus – (63.85 ± 0.35) $kJ\cdot mol^{-1}$, which yields $\Delta_f H_m^\circ$ ($CuSeO_4\cdot 5H_2O$, cr, 298.15 K) = – (1973.33 ± 2.83) $kJ\cdot mol^{-1}$.

The present results differ from those in the paper by about 20 $kJ\cdot mol^{-1}$, which is mainly caused by the new value of the standard enthalpy of formation of Ag_2SeO_4(cr).

[62SEL/SAZ5]

The enthalpy changes of the reactions Cs_2SeO_4(aq, 1:1000) + $BaCl_2\cdot 2H_2O$(cr) → $BaSeO_4$(cr) + $2CsCl_2$(aq, 1:500) + $2H_2O$(l) and Cs_2SeO_4(cr) + aq → Cs_2SeO_4(aq, 1:1000) were determined by Selivanova and Sazykina by calorimetric measurements. Their data have been recalculated with new auxiliary data in Table A-28. The standard enthalpy of formation of $BaCl_2\cdot 2H_2O$(cr) needed was obtained from the standard enthalpy of $BaCl_2$ in [92GRE/FUG] and the enthalpy of hydration of $BaCl_2$ from [82WAG/EVA].

Table A-28: Calculation of the standard enthalpy of formation of Cs_2SeO_4(cr). The sources of the data are indicated.

#	Reaction	$\Delta_r H_m$ ($kJ\cdot mol^{-1}$)	Reference
1	Cs_2SeO_4(aq, 1:1000) + $BaCl_2\cdot 2H_2O$(cr) → $BaSeO_4$(cr) + $2CsCl_2$(aq, 1:500) + $2H_2O$(l)	(14.27 ± 0.63)	(paper)
2	Cs_2SeO_4(cr) + aq → Cs_2SeO_4(aq,1:1000)	(17.49 ± 0.42)	(paper)
3	Ba(cr) + Se(cr) +$2O_2$(g) → $BaSeO_4$(cr)	– (1144.00 ± 5.00)	(review)
4	Cs(cr) + $0.5Cl_2$(g) + aq → CsCl(aq,1:500)	– (424.85 ± 0.50)	[89COX/WAG] [82WAG/EVA]
5	Ba(cr) + Cl_2(g) + $2H_2$(g) + O_2(g) → $BaCl_2\cdot 2H_2O$(cr)	– (1456.70 ± 2.50)	(see text)
6	H_2(g) + $0.5O_2$(g) → H_2O(l)	– (285.83 ± 0.04)	[89COX/WAG]
7	2Cs(cr) + Se(cr) + $2O_2$(g) → Cs_2SeO_4(cr)	– (1140.42 ± 5.82)	(result)

$\Delta_r H_7 = -\Delta_r H_1 - \Delta_r H_2 + \Delta_r H_3 + 2\Delta_r H_4 - \Delta_r H_5 + 2\Delta_r H_6$

The enthalpy of formation has also been estimated from Reaction #2 on the assumption that the enthalpy of dilution of Cs_2SeO_4(aq, 1:1000) to the aqueous standard state is close to zero as for the sulphate. With the accepted ionic enthalpies the result is

– (1137.00 ± 2.50) kJ·mol^{-1}. The review selects the mean of the two estimates – (1138.70 ± 3.50) kJ·mol^{-1}.

[62SEL/SHN]

Selivanova, Shneider, and Sazykina measured the enthalpy change of the reaction $Li_2SeO_4 \cdot H_2O$(cr) + $BaCl_2$(aq, 1:800) → $BaSeO_4$(cr) + 2LiCl(aq, 1:400) + H_2O(l) in a calorimeter. With the same calorimeter the enthalpies of dissolution of $Li_2SeO_4 \cdot H_2O$(cr) and Li_2SeO_4(cr) to Li_2SeO_4(aq, 1:800) were determined to be – (25.02 ± 0.25) and – (31.51 ± 0.29) kJ·mol^{-1}, respectively. The enthalpy of reaction has been recalculated with new auxiliary data to standard conditions in Table A-29.

Table A-29: Enthalpy change of the reaction $Li_2SeO_4 \cdot H_2O$(cr) + Ba^{2+} → $BaSeO_4$(cr) + $2Li^+$ + H_2O(l). The sources of the data are indicated.

#	Reaction	$\Delta_r H_m$ (kJ·mol^{-1})	Reference
1	$Li_2SeO_4 \cdot H_2O$(cr) + $BaCl_2$(aq, 1:800) → $BaSeO_4$(cr) + 2LiCl(aq, 1:400) + H_2O(l)	– (32.59 ± 0.17)	(paper)
2	$BaCl_2$(aq, 1:800) + aq → Ba^{2+} + $2Cl^-$	– (1.63 ± 0.10)	[82WAG/EVA]
3	LiCl(aq, 1:400) + aq → Li^+ + Cl^-	– (0.50 ± 0.10)	[82WAG/EVA]
4	$Li_2SeO_4 \cdot H_2O$(cr) + Ba^{2+} → $BaSeO_4$(cr) + $2Li^+$ + H_2O(l)	– (31.96 ± 0.28)	(result)

$\Delta_r H_4 = \Delta_r H_1 - \Delta_r H_2 + 2\,\Delta_r H_3$

The experimental data in combination with accepted standard enthalpies of formation result in $\Delta_f H_m^\circ$ ($Li_2SeO_4 \cdot H_2O$, cr, 298.15 K) = – (1420.01 ± 5.60) kJ·mol^{-1} and $\Delta_f H_m^\circ$ (Li_2SeO_4, cr, 298.15 K) = – (1127.69 ± 5.60) kJ·mol^{-1}.

The calculated enthalpy of formation of Li_2SeO_4 in the aqueous standard state is – (1160.44 ± 3.50) kJ·mol^{-1}. On the assumption that the enthalpy of dilution of Li_2SeO_4(aq, 1:800 to Li_2SeO_4(aq, 1:∞) is the same as for the corresponding sulphate (– 1.89 kJ·mol^{-1} [82WAG/EVA]) the review calculates $\Delta_f H_m^\circ$ (Li_2SeO_4, cr, 298.15 K) = – (1127.04 ± 3.50) kJ·mol^{-1} from the data. The calorimetric result is selected.

[62WOS/GEE]

The saturated vapour pressure of α-ZnSe was measured in the temperature range 1060 to 1393 K using a gas flow method. Values for the entropy and the enthalpy of formation of α-ZnSe at 298.15 were derived using the second law and an estimated heat capacity of α-ZnSe. The experimental results were therefore re-evaluated by the review using both the second and the third law, the selected thermodynamic functions of selenium, the data for Zn in [89COX/WAG], the heat capacity expression of α-ZnSe in Section V.9.1.1.1, and the selected entropy of α-ZnSe in the case of the third law. The vaporisation was assumed to occur according to the reaction α-ZnSe ⇌ Zn(g) + ½Se_2(g).

The results were $\Delta_f H_m^\circ$(ZnSe, α, 298.15 K) = – (198.6 ± 19.0) kJ·mol^{-1}, $\Delta_f H_m^\circ$(ZnSe, α, 298.15 K) = – (182.8 ± 6.0) kJ·mol^{-1}, and S_m°(ZnSe, α, 298.15 K) = (58.9 ± 15.0) J·K^{-1}·mol^{-1} for the second and third law enthalpy of formation and the entropy, respectively.

[63HIR/ICH]

The total pressure of the vapour in equilibrium with β-SnSe was measured in the temperature range 773 to 898 K using the Knudsen effusion technique. The enthalpy of and entropy of sublimation according to the reaction α-SnSe ⇌ SnSe(g) were calculated by the review from the reported vapour pressure expression and the selected heat capacities of SnSe(g) and α-SnSe, the enthalpy of transformation α-SnSe → β-SnSe being 1.28 kJ·mol^{-1} (cf V.7.4.1.2), and a heat capacity expression of β-SnSe being identical to that of α-SnSe yielding $\Delta_{sub} H_m^\circ$(SnSe, α, 298.15 K) = (210.8 ± 12.0) kJ·mol^{-1} and $\Delta_{sub} S_m^\circ$(SnSe, α, 298.15 K) = (158.9 ± 13.0) J·K^{-1}·mol^{-1}, respectively.

[63LES/SEL]

The authors reacted $BaCl_2·2H_2O$(cr) with a solution of sodium selenite in an electrically calibrated calorimeter. Crystalline barium selenite was formed. The measured enthalpy change is combined with new auxiliary data in Table A-30 to estimate the standard enthalpy of formation of $BaSeO_3$(cr). The standard enthalpy of formation of $BaCl_2·2H_2O$(cr) was calculated as described in this Appendix, [62SEL/SAZ5].

Table A-30: Estimation of the standard enthalpy of formation of $BaSeO_3$(cr). The sources of the data are indicated.

#	Reaction	$\Delta_r H_m$ (kJ·mol^{-1})	Reference
1	$BaCl_2·2H_2O$(cr) + Na_2SeO_3(aq, 1:1000) → $BaSeO_3$(cr) + 2NaCl(aq, 1:500) + $2H_2O$(l)	(23.51 ± 0.13)	(paper)
2	Ba(cr) + Cl_2(g) + $2H_2$(g) + O_2(g) → $BaCl_2·2H_2O$(cr)	– (1456.70 ± 2.70)	(see text)
3	Na_2SeO_3(cr) + aq → Na_2SeO_3(aq, 1:1000)	– (29.66 ± 0.13)	[63SEL/LES]
4	2Na(cr) + Se(cr) + $1.5O_2$(g) → Na_2SeO_3(cr)	– (959.62 ± 0.95)	(review)
5	Na(cr) + $0.5Cl_2$(g) + aq → NaCl(aq, 1:500)	– (407.07 ± 0.15)	[89COX/WAG] [82WAG/EVA]
6	H_2(g) + $0.5O_2$(g) → H_2O(l)	– (285.83 ± 0.04)	[89COX/WAG]
7	Ba(cr) + Se(cr) + $1.5O_2$(g) → $BaSeO_3$(cr)	– (1036.67 ± 2.89)	(result)

$\Delta_r H_7 = \Delta_f H_m^\circ(BaSeO_3, cr, 298.15\ K) = \Delta_r H_1 + \Delta_r H_2 + \Delta_r H_3 + \Delta_r H_4 - 2\,\Delta_r H_5 - 2\,\Delta_r H_6$

[63LES/SEL3]

The authors determined the enthalpy change of the reaction between crystalline $Ca(NO_3)_2{\cdot}4H_2O$ and a solution of sodium selenite in an electrically calibrated calorimeter. The reaction product was crystalline and assigned the composition $CaSeO_3{\cdot}2H_2O$ from chemical analysis. According to [81EBE/HAV], [97DUM/BRO] and other sources, however, the stable phase has the composition $CaSeO_3{\cdot}H_2O$. A comparison of the diffraction pattern of the reaction product in the calorimeter with that of $CaSeO_3{\cdot}H_2O$ in [81EBE/HAV] seems to indicate that the monohydrate was formed in the calorimetric experiment. At least this possibility cannot be ruled out. The calorimetric data have therefore been evaluated on the assumption that $CaSeO_3{\cdot}H_2O$ was formed.

The rather large solubility product of calcium selenite suggests that the reaction in the calorimeter was not complete. It is estimated that a few per cent of the components remained unreacted. No correction has been applied for this fact. The calculation of the standard enthalpy of formation of $CaSeO_3{\cdot}H_2O$ with new auxiliary data is shown in Table A-31.

Table A-31: Calculation of the standard enthalpy of formation of $CaSeO_3{\cdot}H_2O$. The sources of the data are indicated.

#	Reaction	$\Delta_r H_m$ ($kJ{\cdot}mol^{-1}$)	Reference
1	$Ca(NO_3)_2{\cdot}4H_2O(cr)$ + Na_2SeO_3(aq, 1:1000) → $CaSeO_3{\cdot}H_2O(cr)$ + $2NaNO_3$(aq, 1:500) + $3H_2O(l)$	(45.15 ± 0.13)	(paper)
2	$Na_2SeO_3(cr)$ + aq → Na_2SeO_3(aq, 1:1000)	– (29.66 ± 0.13)	[63SEL/LES]
3	$2Na(cr) + Se(cr) + 1.5O_2(g) \rightarrow Na_2SeO_3(cr)$	– (959.62 ± 0.95)	(review)
4	$Ca(cr) + 4H_2(g) + 5O_2(g) + N_2(g) \rightarrow Ca(NO_3)_2{\cdot}4H_2O(cr)$	– (2132.33 ± 2.50)	[82WAG/EVA]
5	$Na(cr) + 1.5O_2(g) + 0.5N_2(g)$ + aq → $NaNO_3$(aq, 1:500)	– (447.07 ± 0.42)	[89COX/WAG] [82WAG/EVA]
6	$H_2(g) + 0.5O_2(g) \rightarrow H_2O(l)$	– (285.83 ± 0.04)	[89COX/WAG]
7	$Ca(cr) + Se(cr) + H_2(g) + 2O_2(g) \rightarrow CaSeO_3{\cdot}H_2O(cr)$	– (1324.83 ± 2.81)	(result)

$\Delta_r H_7 = \Delta_f H_m^\circ(CaSeO_3{\cdot}H_2O, cr, 298.15\ K) = \Delta_r H_1 + \Delta_r H_2 + \Delta_r H_3 + \Delta_r H_4 - 2\,\Delta_r H_5 - 3\,\Delta_r H_6$

[63LES/SEL4]

The authors made a calorimetric determination of the enthalpy change of the reaction between $Co(NO_3)_2{\cdot}6H_2O(cr)$ and a solution of sodium selenite with formation of $CoSeO_3{\cdot}2H_2O(cr)$ as established by chemical analysis and X-ray diffraction. The standard enthalpy of formation of cobalt selenite is calculated in Table A-32 with new auxiliary data.

Table A-32: Calculation of the standard enthalpy of formation of $CoSeO_3{\cdot}2H_2O(cr)$. The sources of the data are indicated.

#	Reaction	$\Delta_r H_m$ ($kJ{\cdot}mol^{-1}$)	Reference
1	$Co(NO_3)_2{\cdot}6H_2O(cr) + Na_2SeO_3(aq, 1{:}1200) \rightarrow CoSeO_3{\cdot}2H_2O(cr) + 2NaNO_3(aq, 1{:}600) + 4H_2O(l)$	(47.57 ± 0.13)	(paper)
2	$Co(cr) + 6H_2(g) + 6O_2(g) + N_2(g) \rightarrow Co(NO_3)_2{\cdot}6H_2O(cr)$	– (2211.20 ± 2.50)	[82WAG/EVA]
3	$Na(cr) + 1.5O_2(g) + 0.5N_2(g) + aq \rightarrow NaNO_3(aq, 1{:}600)$	– (447.04 ± 0.42)	[89COX/WAG] [82WAG/EVA]
4	$2Na(cr) + Se(cr) + 1.5O_2(g) \rightarrow Na_2SeO_3(cr)$	– (959.62 ± 0.95)	(review)
5	$Na_2SeO_3(cr) + aq \rightarrow Na_2SeO_3(aq, 1{:}1200)$	– (29.71 ± 0.15)	[63SEL/LES5]
6	$H_2(g) + 0.5O_2(g) \rightarrow H_2O(l)$	– (285.83 ± 0.04)	[89COX/WAG]
7	$Co(cr) + Se(cr) + 2H_2(g) + 2.5O_2(g) \rightarrow CoSeO_3{\cdot}2H_2O(cr)$	– (1115.56 ± 2.81)	(result)

$\Delta_r H_7 = \Delta_f H^\circ_m(CoSeO_3{\cdot}2H_2O, cr, 298.15 K) = \Delta_r H_1 + \Delta_r H_2 - 2\,\Delta_r H_3 + \Delta_r H_4 + \Delta_r H_5 - 4\,\Delta_r H_6$

[63MIJ]

Mijlhoff calculated the enthalpy of formation of $SeO_3(cr)$ by a procedure that starts from the enthalpy change of reaction:

$$H_2SeO_3(aq, 1{:}500) + HClO(aq, 1{:}700) \rightarrow H_2SeO_4(aq, 1{:}1200) + HCl(aq, 1{:}1200)$$

as determined by Thomsen [1882THO]. This quantity was combined with auxiliary enthalpy data all taken from [1882THO] except the heat of dissolution of $SeO_3(cr)$, which was obtained from [62JER/MEN]. The outcome of the calculation was:

$$\Delta_f H^\circ_m(SeO_3, cr, 298.15 K) = -184.5 \text{ kJ}{\cdot}\text{mol}^{-1}.$$

This result will not be considered since the auxiliary data used have been superseded by new data.

[63SAZ/SEL]

The enthalpy change of the reaction $K_2SeO_4(cr) + 2AgNO_3(aq, 1{:}400) \rightarrow Ag_2SeO_4(cr) + 2KNO_3(aq, 1{:}400)$ was measured by Sazykina and Selvanova. Their results have been recalculated to standard conditions with new auxiliary data in Table A-33.

In combination with accepted standard enthalpies of formation $\Delta_r H_4$ yields $\Delta_f H^\circ_m(K_2SeO_4, cr, 298.15 K) = -(1121.29 \pm 2.15)$ $kJ{\cdot}mol^{-1}$.

Table A-33: Calculation of the enthalpy change of the reaction $K_2SeO_4(cr) + 2Ag^+ \rightarrow Ag_2SeO_4(cr) + 2K^+$. The sources of the data are indicated.

#	Reaction	$\Delta_r H_m$ (kJ·mol^{-1})	Reference
1	$K_2SeO_4(cr) + 2AgNO_3(aq, 1{:}400) \rightarrow Ag_2SeO_4(cr) + 2KNO_3(aq, 1{:}400)$	– (17.28 ± 0.13)	(paper)
2	$AgNO_3(aq, 1{:}400) + aq \rightarrow Ag^+ + NO_3^-$	(0.20 ± 0.10)	[82WAG/EVA]
3	$KNO_3(aq, 1{:}400) + aq \rightarrow K^+ + NO_3^-$	(0.30 ± 0.10)	[82WAG/EVA]
4	$K_2SeO_4(cr) + 2Ag^+ \rightarrow Ag_2SeO_4(cr) + 2K^+$	– (17.08 ± 0.31)	(result)

$\Delta_r H_4 = \Delta_r H_1 - 2\,\Delta_r H_2 + 2\,\Delta_r H_3$

[63SEL/LES]

The authors reacted $SrCl_2{\cdot}6H_2O(cr)$ with a solution of sodium selenite in an electrically calibrated calorimeter. Crystalline strontium selenite was formed. The integral enthalpy of dilution of Na_2SeO_3 to (aq, 1:1000) was also measured. The measured enthalpy changes are combined with new auxiliary data in Table A-34 to estimate the standard enthalpy of formation of $SrSeO_3(cr)$. The standard enthalpy of formation of $SrCl_2{\cdot}6H_2O(cr)$ used was calculated from the enthalpy of hydration of $SrCl_2(cr)$ in [82WAG/EVA] and its enthalpy of formation in [92GRE/FUG].

Table A-34: Estimation of the standard enthalpy of formation of $SrSeO_3(cr)$. The sources of the data are indicated.

#	Reaction	$\Delta_r H_m$ (kJ·mol^{-1})	Reference
1	$SrCl_2{\cdot}6H_2O(cr) + Na_2SeO_3(aq, 1{:}1000) \rightarrow SrSeO_3(cr) + 2NaCl(aq, 1{:}500) + 6H_2O(l)$	(37.07 ± 0.13)	(paper)
2	$Sr(cr) + Cl_2(g) + 6H_2(g) + 3O_2(g) \rightarrow SrCl_2{\cdot}6H_2O(cr)$	– (2628.74 ± 0.50)	(see text)
3	$Na_2SeO_3(cr) + aq \rightarrow Na_2SeO_3(aq, 1{:}1000)$	– (29.66 ± 0.13)	(paper)
4	$2Na(cr) + Se(cr) + 1.5O_2(g) \rightarrow Na_2SeO_3(cr)$	– (959.62 ± 0.95)	(review)
5	$Na(cr) + 0.5Cl_2(g) + aq \rightarrow NaCl(aq, 1{:}500)$	– (407.07 ± 0.15)	[89COX/WAG] [82WAG/EVA]
6	$H_2(g) + 0.5O_2(g) \rightarrow H_2O(l)$	– (285.83 ± 0.04)	[89COX/WAG]
7	$Sr(cr) + Se(cr) + 1.5O_2(g) \rightarrow SrSeO_3(cr)$	– (1051.83 ± 2.70)	(result)

$\Delta_r H_7 = \Delta_f H^\circ_m(SrSeO_3, cr, 298.15\ K) = \Delta_r H_1 + \Delta_r H_2 + \Delta_r H_3 + \Delta_r H_4 - 2\,\Delta_r H_5 - 6\,\Delta_r H_6$

[63SEL/LES5]

The authors reacted a solution of cadmium chloride with $Na_2SeO_3(cr)$ in an electrically calibrated calorimeter. The precipitate of $CdSeO_3$ was amorphous. Additional measurements were made in [63LES/SEL6] of the enthalpies of dissolution of $CdSeO_3$(am) and $CdSeO_3$(cr) to the same concentration in nitric acid (aq, 1:50 and denoted sln). The data are combined in Table A-35 and used to calculate the enthalpy of formation of cadmium selenite.

Table A-35: Calculation of the standard enthalpy of formation of $CdSeO_3$. The sources of the data are indicated.

#	Reaction	$\Delta_r H_m$ ($kJ\cdot mol^{-1}$)	Reference
1	Na_2SeO_3(cr) + $CdCl_2$(aq, 1:1000) → $CdSeO_3$(am) + 2NaCl(aq, 1:500)	– (22.59 ± 0.13)	(paper)
2	Na(cr) + $0.5Cl_2$(g) + aq → NaCl(aq, 1:500)	– (407.07 ± 0.15)	[89COX/WAG] [82WAG/EVA]
3	Cd(cr) + Cl_2(g) + aq → $CdCl_2$(aq, 1:1000)	– (406.53 ± 0.66)	[89COX/WAG] [82WAG/EVA]
4	2Na(cr) + Se(cr) + $1.5O_2$(g) → Na_2SeO_3(cr)	– (959.62 ± 0.95)	(review)
5	Cd(cr) + Se(cr) + $1.5O_2$(g) → $CdSeO_3$(am)	– (574.60 ± 1.17)	(result)
6	$CdSeO_3$(am) + sln → $CdSeO_3$(sln)	– (9.10 ± 0.10)	[63LES/SEL6]
7	$CdSeO_3$(cr) + sln → $CdSeO_3$(sln)	– (7.70 ± 0.12)	[63LES/SEL6]
8	Cd(cr) + Se(cr) + $1.5O_2$(g) → $CdSeO_3$(cr)	– (576.00 ± 1.18)	(result)

$\Delta_r H_5 = \Delta_f H^\circ_m$ ($CdSeO_3$, am, 298.15 K) $= \Delta_r H_1 - 2\,\Delta_r H_2 + \Delta_r H_3 + \Delta_r H_4$

$\Delta_r H_8 = \Delta_f H^\circ_m$ ($CdSeO_3$, cr, 298.15 K) $= \Delta_r H_5 + \Delta_r H_6 - \Delta_r H_7$

[63SEL/MAI]

Selivanova, Maier, and Luk'yanova [63SEL/MAI] prepared $CdSeO_4\cdot H_2O$(cr) and $CdSeO_4$(cr) and measured their heats of dissolution in water (1:1200) in a calorimeter. They also measured the enthalpy change of the reaction between the monohydrate and a barium chloride solution with formation of $BaSeO_4$(cr). The result of the last reaction has been re-evaluated in Table A-36.

The standard enthalpy of formation of $CdSeO_4\cdot H_2O$(cr) is calculated from $\Delta_r H_4$ and accepted enthalpies of formation to be – (943.05 ± 5.62) $kJ\cdot mol^{-1}$. The enthalpies of dissolution of the salts were – (23.89 ± 0.21) and – (46.99 ± 0.25) $kJ\cdot mol^{-1}$, respectively. With these enthalpies the standard enthalpy of formation of $CdSeO_4$(cr) becomes – (634.12 ± 5.63) $kJ\cdot mol^{-1}$.

The standard enthalpies of formation obtained by the review differ from those in the paper by about 20 $kJ\cdot mol^{-1}$ mainly due to a change in the value of $\Delta_f H^\circ_m$ ($BaSeO_4$, cr, 298.15 K).

Table A-36: Enthalpy change of the reaction $CdSeO_4{\cdot}H_2O(cr) + Ba^{2+} \rightarrow BaSeO_4(cr) + Cd^{2+} + H_2O(l)$. The sources of the data are indicated.

#	Reaction	$\Delta_r H_m$ (kJ·mol^{-1})	Reference
1	$CdSeO_4{\cdot}H_2O(cr) + BaCl_2(aq, 1:800) \rightarrow BaSeO_4(cr) + CdCl_2(aq, 1:800) + H_2O(l)$	– (25.82 ± 0.08)	(paper)
2	$BaCl_2(aq, 1:800) + aq \rightarrow Ba^{2+} + 2Cl^-$	– (1.63 ± 0.10)	[82WAG/EVA]
3	$CdCl_2(aq, 1:800) + aq \rightarrow Cd^{2+} + 2Cl^-$	– (3.71 ± 0.10)	[82WAG/EVA]
4	$CdSeO_4{\cdot}H_2O(cr) + Ba^{2+} \rightarrow BaSeO_4(cr) + Cd^{2+} + H_2O(l)$	– (27.90 ± 0.16)	(result)

$\Delta_r H_4 = \Delta_r H_1 - \Delta_r H_2 + \Delta_r H_3$

[63SEL/MAI3]

Selivanova, Maier, and Luk'yanova prepared $ZnSeO_4{\cdot}6H_2O(cr)$, $ZnSeO_4{\cdot}H_2O(cr)$, and $ZnSeO_4(cr)$ and measured their heats of dissolution in water (1:1200) in a calorimeter. They also measured the enthalpy change of the reaction between the monohydrate and a barium chloride solution with formation of $BaSeO_4(cr)$. The result of the last reaction has been re-evaluated in Table A-37.

Table A-37: Enthalpy change of the reaction $ZnSeO_4{\cdot}H_2O(cr) + Ba^{2+} \rightarrow BaSeO_4(cr) + Zn^{2+} + H_2O(l)$. The sources of the data are indicated.

#	Reaction	$\Delta_r H_m$ (kJ·mol^{-1})	Reference
1	$ZnSeO_4{\cdot}H_2O(cr) + BaCl_2(aq, 1:1200) \rightarrow BaSeO_4(cr) + ZnCl_2(aq, 1:1200) + H_2O(l)$	– (64.43 ± 0.38)	(paper)
2	$BaCl_2(aq, 1:1200) + aq \rightarrow Ba^{2+} + 2Cl^-$	– (1.43 ± 0.10)	[82WAG/EVA]
3	$ZnCl_2(aq, 1:1200) + aq \rightarrow Zn^{2+} + 2Cl^-$	– (2.35 ± 0.10)	[82WAG/EVA]
4	$ZnSeO_4{\cdot}H_2O(cr) + Ba^{2+} \rightarrow BaSeO_4(cr) + Zn^{2+} + H_2O(l)$	– (65.35 ± 0.40)	(result)

$\Delta_r H_4 = \Delta_r H_1 - \Delta_r H_2 + \Delta_r H_3$

The standard enthalpy of formation of $ZnSeO_4{\cdot}H_2O(cr)$ is calculated from $\Delta_r H_4$ and accepted enthalpies of formation to be – (983.07 ± 5.61) kJ·mol^{-1}. The enthalpies of dissolution of the salts were (5.94 ± 0.04), – (40.38 ± 0.17), and – (72.89 ± 0.42) kJ·mol^{-1}, respectively. With these enthalpies the standard enthalpies of formation of $ZnSeO_4{\cdot}6H_2O(cr)$ and $ZnSeO_4(cr)$ become – (2458.54 ± 5.61) and – (664.73 ± 5.63) kJ·mol^{-1}.

The standard enthalpies of formation obtained by the review differ from those in the paper by about 20 kJ·mol^{-1} mainly due to the change in the accepted value of $\Delta_f H_m^\circ$ ($BaSeO_4$, cr, 298.15 K).

[63SEL/SAZ]

The enthalpy change f the reaction $Rb_2SeO_4(cr) + 2AgNO_3(aq, 1:400) \rightarrow Ag_2SeO_4(cr) + 2RbNO_3(aq, 1:400)$ was measured by Selvanova and Sazykina. Their results have been recalculated to standard conditions with new auxiliary data in Table A-38.

Table A-38: Calculation of the enthalpy change of the reaction $Rb_2SeO_4(cr) + 2Ag^+ \rightarrow Ag_2SeO_4(cr) + 2Rb^+$. The sources of the data are indicated.

#	Reaction	$\Delta_r H_m$ (kJ·mol^{-1})	Reference
1	$Rb_2SeO_4(cr) + 2AgNO_3(aq, 1:400) \rightarrow Ag_2SeO_4(cr) + 2RbNO_3(aq, 1:400)$	– (20.21 ± 0.08)	(paper)
2	$AgNO_3(aq, 1:400) + aq \rightarrow Ag^+ + NO_3^-$	(0.20 ± 0.10)	[82WAG/EVA]
3	$RbNO_3(aq, 1:400) + aq \rightarrow Rb^+ + NO_3^-$	(0.39 ± 0.10)	[82WAG/EVA]
4	$Rb_2SeO_4(cr) + 2Ag^+ \rightarrow Ag_2SeO_4(cr) + 2Rb^+$	– (19.83 ± 0.29)	(result)

$\Delta_r H_4 = \Delta_r H_1 - 2\,\Delta_r H_2 + 2\,\Delta_r H_3$

In combination with accepted standard enthalpies of formation $\Delta_r H_4$ yields $\Delta_f H_m^\circ$ (Rb_2SeO_4, cr, 298.15 K) = – (1116.50 ± 2.20) kJ·mol^{-1}.

[64BON/SHU]

The total vapour pressure of liquid selenium at equilibrium was measured by a static method in the temperature range 573 to 1167 K and given as $\log_{10}(p/\text{bar}) = 5.722 - 5468\,T^{-1}$. A third law evaluation of the enthalpy of formation of $Se_2(g)$ at 298.15 K was made by this review using the vapour pressure at 1150 K and the selected data for the heat capacities and entropies of Se(l) and $Se_2(g)$. A mole fraction of $Se_2(g)$, $x_{Se_2(g)} = 0.80$ was estimated from the selected selenium data. The enthalpy of formation derived was $\Delta_f H_m^\circ$ (Se_2, g, 298.15 K) = (139.9 ± 3.4) kJ·mol^{-1}.

[64COL/DRO]

The partial pressures of the gaseous species in equilibrium with β-SnSe were measured in the temperature range 792 to 979 K using mass spectroscopy and the Knudsen effusion technique. The observed species were Sn(g), Se(g), $Se_2(g)$, SnSe(g), and $Sn_2Se_2(g)$ and thermodynamic properties were derived for SnSe(g) and $Sn_2Se_2(g)$. The enthalpy of the reaction β-SnSe $\rightleftharpoons$ SnSe(g) was calculated in the paper by combining the measurements there with those reported in [61NES/PAS] and [63HIR/ICH]. The measurements in these references are treated separately by the review. In order to obtain results representing the measurements in [64COL/DRO] only, the enthalpy and entropy of the reaction were re-evaluated by the review from the experimental values reported in Table 3 of the paper. The selected heat capacities of SnSe(g) and α-SnSe were employed, the enthalpy of transformation α-SnSe → β-SnSe was set to 1.28 kJ·mol^{-1} (cf V.7.4.1.2), and the heat capacity expression of β-SnSe was assumed to be identical to that of α-SnSe yielding $\Delta_r H_m^\circ(298.15 \text{ K}) = (150.8 \pm 45.0) \text{ kJ·mol}^{-1}$ and $\Delta_r S_m^\circ(298.15 \text{ K}) = (92.7 \pm 40.0) \text{ J·K}^{-1}\text{·mol}^{-1}$ from the second law. The authors calculated a third law value to $\Delta_r H_m^\circ(298.15 \text{ K}) = (220.1 \pm 12.0) \text{ kJ·mol}^{-1}$ using estimated entropies and heat capacities of SnSe(s) and SnSe(g). The difference between the second and third law results is thus very large. The third law result is likely to be more correct and agrees much better with the results obtained in the other investigations.

The authors also evaluated the enthalpy of the dimerisation reaction $2\text{SnSe(g)} \rightleftharpoons \text{Sn}_2\text{Se}_2\text{(g)}$ to be $\Delta_r H_m^\circ(298.15 \text{ K}) = -(194.6 \pm 22.0) \text{ kJ·mol}^{-1}$ in a third law analysis of measurements made in the temperature range 936 to 991 K. No evaluation was made according to the second law because of the short temperature range. The third law evaluation employed Gibbs energy functions calculated from parameters estimated by comparison with the isoelectronic molecules As_4(g) and Sb_4(g). In the absence of better values of the molecular parameters, no recalculation was made by the review.

[64LES/SEL]

The authors measured the enthalpy change of the reaction between $ZnSO_4·7H_2O$(cr) and a sodium selenite solution in an electrically calibrated calorimeter. The reaction product is assigned the composition $ZnSeO_3·H_2O$ from the X-ray diffraction pattern and a chemical analysis, which shows moderate agreement with the expected composition of the compound. The standard enthalpy of formation of the monohydrate is evaluated in Table A-39.

Table A-39: Calculation of the standard enthalpy of formation of $ZnSeO_3·H_2O$. The sources of the data are indicated.

#	Reaction	$\Delta_r H_m$ (kJ·mol^{-1})	Reference
1	$ZnSO_4·7H_2O$(cr) + Na_2SeO_3(aq, 1:1000) → $ZnSeO_3·H_2O$(cr) + Na_2SO_4(aq, 1:1000) + $6H_2O$(l)	(32.05 ± 0.21)	(paper)
2	2Na(cr) + S(cr) + $2O_2$(g) + aq → Na_2SO_4(aq, 1:1000)	– (1388.99 ± 0.43)	[89COX/WAG] [82WAG/EVA]
3	Zn(cr) + S(cr) + $7H_2$(g) + $5.5O_2$(g) → $ZnSO_4·7H_2O$(cr)	– (3077.75 ± 2.50)	[82WAG/EVA]
4	Na_2SeO_3(cr) + aq → Na_2SeO_3(aq, 1:1000)	– (29.66 ± 0.13)	[63SEL/LES]
5	2Na(cr) + Se(cr) + $1.5O_2$(g) → Na_2SeO_3(cr)	– (959.62 ± 0.95)	(review)
6	H_2(g) + $0.5O_2$(g) → H_2O(l)	– (285.83 ± 0.04)	[89COX/WAG]
7	Zn(cr) + Se(cr) + H_2(g) + $2O_2$(g) → $ZnSeO_3·H_2O$(cr)	– (931.01 ± 2.73)	(result)

$\Delta_r H_7 = \Delta_f H_m^\circ(\text{ZnSeO}_3\text{·H}_2\text{O, cr, 298.15 K}) = \Delta_r H_1 - \Delta_r H_2 + \Delta_r H_3 + \Delta_r H_4 + \Delta_r H_5 - 6\,\Delta_r H_6$

[64MAS/WEI]

The temperature dependence of the magnetic susceptibilities of liquid selenium and its saturated vapour was measured in the temperature range 800 to 1100 K. The partial pressure of $Se_2(g)$ was calculated assuming $Se_2(g)$ to be the only paramagnetic species in the system. The results were only presented graphically and the expression $\log_{10}(p_{Se_2}/\text{bar}) = 10.824 - 11064\ T^{-1}$ was evaluated by the review from the graph in Fig 2 of the paper. The enthalpy of formation at 298.15 K of $Se_2(g)$ was calculated from this expression using the second and the third laws and the selected thermal functions of Se(l), Se(trig), and $Se_2(g)$ yielding the values $\Delta_f H^\circ_m$ (Se_2, g, 298.15 K) = (121.0 ± 4.6) kJ·mol^{-1} and $\Delta_f H^\circ_m$ (Se_2, g, 298.15 K) = (143.5 ± 3.4) kJ·mol^{-1}, respectively.

[64MIJ/BLO]

The vapour pressure of $SeO_3(g)$ over $SeO_3(cr)$ was measured in the temperature range 350 to 390 K by a membrane null manometer. A few measurements were also made with $SeO_3(l)$. Several series of experiments were performed but the reproducibility was poor, which inevitably casts some doubts on the thermodynamic data obtained. The authors selected two series of experiments (C_1 and C_2) that showed concordant vapour pressure curves for the calculations.

A linear regression of $\log_{10} p_{SeO_3}$ on $1/T$ was carried out by the review on the data below 390 K, a temperature close to the melting point. The calculation yielded $\log_{10} p_{SeO_3} = (4.84 \pm 0.53) - (2783 \pm 197)\ T^{-1}$ corresponding, for the reaction:

$$SeO_3(cr) \rightleftharpoons SeO_3(g), \qquad (A.15)$$

to $\Delta_{sub} H^\circ_m$ ((A.15), 370 K) = (53.3 ± 3.8) kJ·mol^{-1} and $\Delta_{sub} S^\circ_m$ ((A.15), 370 K) = (92.7 ± 1.0) J·K^{-1}·mol^{-1}.

The $\log_{10} p_{SeO_3}$ *versus* $1/T$ curve is, however, clearly non-linear. Inclusion of $\Delta_r C^\circ_{p,m}$ in the regression results in an excellent fit but an unreasonable value of $\Delta_r C^\circ_{p,m}$. If the experimental data below 380 K are used, then $\log_{10} p_{SeO_3} = (2.67 \pm 0.38) - (1977 \pm 139)\ T^{-1}$ is obtained, corresponding to $\Delta_{sub} H^\circ_m$ ((A.15), 365 K) = (37.8 ± 2.7) kJ mol^{-1} and $\Delta_{sub} S^\circ_m$ ((A.15), 365 K) = (51.1 ± 7.3) J·K^{-1}·mol^{-1}. The experimental points fall within twice the estimated standard deviation in $\log_{10} p_{SeO_3}$.

In the paper, the curvature of the $\log_{10} p_{SeO_3} = f(1/T)$ graph was interpreted as an indication of a tetramer-monomer equilibrium in the gaseous phase. The data were analysed under this assumption and produced the following enthalpy values (350 K < T < 391 K):

$$(SeO_3)_4(s) \rightleftharpoons (SeO_3)_4(g) \qquad \Delta_r H^\circ_m = 37.7 \text{ kJ·mol}^{-1};$$

$$(SeO_3)_4(s) \rightleftharpoons 4SeO_3(g) \qquad \Delta_r H^\circ_m = 536 \text{ kJ·mol}^{-1}.$$

This interpretation needs confirmation and for the time being the data obtained from the linear approximation will be adopted. This position was also taken by Mills [74MIL].

The few data above the melting point yield $\log_{10} p_{SeO_3}$ = (5.19 ± 0.28) – (2886 ± 111) T^{-1} corresponding, for the reaction:

$$SeO_3(l) \rightleftharpoons SeO_3(g), \qquad (A.16)$$

to $\Delta_{vap}H^{\circ}_{m}$ ((A.16), 398 K) = (55.3 ± 2.1) kJ·mol^{-1} and $\Delta_{vap}S^{\circ}_{m}$ ((A.16), 398 K) = (99.3 ± 5.3) J·K^{-1}·mol^{-1}.

[64SEL/LES]

Nickel selenite was prepared by mixing 0.1 M solutions of sodium selenite and nickel nitrate in stoichiometric proportions at room temperature. The amorphous precipitate crystallised on boiling the reaction mixture for 2 hours. X-ray diffraction showed that the preparation was crystalline and chemical analysis agreed with the formula $NiSeO_3 \cdot 2H_2O$.

The calorimetric measurements comprised determinations of the enthalpies of dissolution in 70 mass-% H_2SO_4, H_2SO_4(aq, 1:2.33), as shown in Table A-40. This solution will be denoted by sln. The partial standard molar enthalpies of formation of H_2O and H_2SO_4 were estimated from data in [82WAG/EVA].

Table A-40: Calculation of the standard enthalpy of formation of $NiSeO_3 \cdot 2H_2O$. The sources of the data are indicated.

#	Reaction	$\Delta_r H_m$ (kJ mol^{-1})	Reference
1	$NiSeO_3 \cdot 2H_2O$(cr) + H_2SO_4(sln) + sln → $NiSO_4$(sln, 1:1100) + H_2SeO_3(sln, 1:1100) + $2H_2O$(l, sln)	– (38.61 ± 0.10)	(paper)
2	$NiSO_4 \cdot 7H_2O$(cr) + sln → $NiSO_4$(sln, 1:1100) + $7H_2O$(l, sln)	(31.61 ± 0.22)	(paper)
3	H_2SeO_3(cr) + sln → H_2SeO_3(sln, 1:1100)	(11.21 ± 0.10)	(paper)
4	S(cr) + H_2(g) + $2O_2$(g) + sln → H_2SO_4(sln)	– (842.30 ± 0.50)	[82WAG/EVA]
5	Se(cr) + H_2(g) + $1.5O_2$(g) → H_2SeO_3(cr)	– (524.72 ± 0.65)	(review)
6	Ni(cr) + S(cr) + $7H_2$(g) + $5.5O_2$(g) → $NiSO_4 \cdot 7H_2O$(cr)	– (2977.33 ± 0.98)	[2005GAM/BUG]
7	H_2(g) + $0.5O_2$(g) + sln → H_2O(l, sln)	– (293.80 ± 0.50)	[82WAG/EVA]
8	Ni(cr) + Se(cr) + $2H_2$(g) + $2.5O_2$(g) → $NiSeO_3 \cdot 2H_2O$(cr)	– (1109.33 ± 2.83)	(result)

$\Delta_r H_8 = -\Delta_r H_1 + \Delta_r H_2 + \Delta_r H_3 - \Delta_r H_4 + \Delta_r H_5 + \Delta_r H_6 - 5\,\Delta_r H_7$

The result in [82WAG/EVA] is – 1134.32 kJ mol^{-1} and has probably been calculated from the same basic data with apparent molar enthalpies of formation.

[64SEL/SAZ]

Selivanova and Sazykina determined the enthalpy change of the reaction $Na_2SeO_4(cr)$ + $BaCl_2(aq, 1:800) \rightarrow BaSeO_4(cr) + 2NaCl(aq, 1:400)$ by a calorimetric measurement. The enthalpy change at standard conditions is evaluated in Table A-41.

Table A-41: Enthalpy change of the reaction $Na_2SeO_4(cr) + Ba^{2+} \rightarrow BaSeO_4(cr) + 2Na^+$. The sources of the data are indicated.

#	Reaction	$\Delta_r H_m$ (kJ·mol^{-1})	Reference
1	$Na_2SeO_4(cr) + BaCl_2(aq, 1:800) \rightarrow BaSeO_4(cr) + 2NaCl(aq, 1:400)$	$-(16.44 \pm 0.08)$	(paper)
2	$BaCl_2(aq, 1:800) + aq \rightarrow Ba^{2+} + 2Cl^-$	$-(1.63 \pm 0.10)$	[82WAG/EVA]
3	$NaCl(aq, 1:400) + aq \rightarrow Na^+ + Cl^-$	$-(0.36 \pm 0.10)$	[82WAG/EVA]
4	$Na_2SeO_4(cr) + Ba^{2+} \rightarrow BaSeO_4(cr) + 2Na^+$	$-(15.53 \pm .24)$	(result)

$\Delta_r H_4 = \Delta_r H_1 - \Delta_r H_2 + 2\,\Delta_r H_3$

[64TER/RAT]

The Gibbs energy of formation of α-CdSe was measured in the temperature range 600 to 700 K using the electrochemical cell

$$(-)\ Cd(l)\ |\ 5\ \text{weight-\%}\ CdBr_2(\text{in LiBr + KBr melt})\ |\ \alpha\text{-CdSe, Se(l)}\ (+).$$

The cell reaction is $Cd(l) + Se(l) \rightleftharpoons \alpha\text{-CdSe}$. The Gibbs energy of formation was evaluated by linear regression of the emf on temperature. The Gibbs energy of formation in the temperature range was described by $\Delta_f G^\circ_m(\text{CdSe}, \alpha, T) = (-157318 + 34.73\,T)$ J·mol^{-1}. Thermodynamic quantities at 298.15 K were calculated but a set of auxiliary data different from that used by the review was applied. The Gibbs energy expression was therefore recalculated to 298.15 K by the review according to the second and the third law using the selected thermodynamic functions of selenium, the data of Cd in [89COX/WAG], the selected heat capacity of α-CdSe, and the selected entropy of α-CdSe in the case of the third law. The results were $\Delta_f H^\circ_m(\text{CdSe}, \alpha, 298.15\text{ K}) = -(136.6 \pm 4.0)$ kJ·mol^{-1}, $\Delta_f H^\circ_m(\text{CdSe}, \alpha, 298.15\text{ K}) = -(136.4 \pm 2.0)$ kJ·mol^{-1}, and $S^\circ_m(\text{CdSe}, \alpha, 298.15\text{ K}) = (86.3 \pm 2.0)$ J·K·mol^{-1} for the second and third law enthalpy of formation and the entropy, respectively.

[65AND/KAR]

Andreeva and Karapet'yants mixed a solution of sodium selenite and $Bi(NO_3)_3 \cdot 5H_2O(cr)$ in an electrically calibrated calorimeter and measured the enthalpy change of the reaction. The reaction product was identified as crystalline $(BiO)_2SeO_3$. Its standard enthalpy of formation is estimated in Table A-42.

Table A-42: Calculation of the standard enthalpy of formation of $(BiO)_2SeO_3$. The sources of the data are indicated.

#	Reaction	$\Delta_r H_m$ (kJ·mol^{-1})	Reference
1	$2Bi(NO_3)_3{\cdot}5H_2O(cr) + Na_2SeO_3(aq, 1{:}3000) \rightarrow (BiO)_2SeO_3(cr) + 2NaNO_3(aq, 1{:}1500) + 4HNO_3(aq, 1{:}750) + 8H_2O(l)$	(31.84 ± 0.08)	(paper)
2	$Bi(cr) + 5H_2(g) + 7O_2(g) + 1.5N_2(g) \rightarrow Bi(NO_3)_3{\cdot}5H_2O(cr)$	– (2005.05 ± 2.80)	(paper, review)
3	$2Na(cr) + Se(cr) + 1.5O_2(g) + aq \rightarrow Na_2SeO_3(aq, 1{:}3000) \cong 2Na^+ + SeO_3^{2-}$	– (987.85 ± 1.14)	(review) [89COX/WAG]
4	$Na(cr) + 0.5N_2(g) + 1.5O_2(g) + aq \rightarrow NaNO_3(aq, 1{:}1500)$	– (447.01 ± 0.40)	[89COX/WAG] [82WAG/EVA]
5	$0.5H_2(g) + 1.5O_2(g) + 0.5N_2(g) + aq \rightarrow HNO_3(aq, 1{:}750)$	– (206.49 ± 0.40)	[89COX/WAG] [82WAG/EVA]
6	$H_2(g) + 0.5O_2(g) \rightarrow H_2O(l)$	– (285.83 ± 0.04)	[89COX/WAG]
7	$2Bi(cr) + Se(cr) + 2.5O_2(g) \rightarrow (BiO)_2SeO_3(cr)$	– (959.49 ± 5.99)	(result)

$\Delta_r H_7 = \Delta_f H_m^\circ((BiO)_2SeO_3, cr, 298.15\ K) = \Delta_r H_1 + 2\,\Delta_r H_2 + \Delta_r H_3 - 2\,\Delta_r H_4 - 4\,\Delta_r H_5 - 8\,\Delta_r H_6$

The standard enthalpy of formation of $Bi(NO_3)_3{\cdot}5H_2O(cr)$ determined in the paper has been re-evaluated in Table A-43.

Table A-43: Calculation of the standard enthalpy of formation of $Bi(NO_3)_3{\cdot}5H_2O(cr)$. The sources of the data are indicated.

#	Reaction	$\Delta_r H_m$ (kJ mol^{-1})	Reference
1	$Bi(NO_3)_3{\cdot}5H_2O(cr) + 3NaOH(aq, 1{:}1000) \rightarrow Bi(OH)_3(cr) + 3NaNO_3(aq, 1{:}1000) + 5H_2O(l)$	– (66.40 ± 0.42)	(paper)
2	$Bi(cr) + 1.5H_2(g) + 1.5O_2(g) \rightarrow Bi(OH)_3(cr)$	– (711.3 ± 2.50)	[82WAG/EVA]
3	$Na(cr) + 0.5O_2(g) + 0.5H_2(g) + aq \rightarrow NaOH(aq, 1{:}1000)$	– (470.01 ± 0.09)	[89COX/WAG] [82WAG/EVA]
4	$Na(cr) + 0.5N_2(g) + 1.5O_2(g) + aq \rightarrow NaNO_3(aq, 1{:}1000)$	– (447.01 ± 0.40)	[89COX/WAG] [82WAG/EVA]
5	$H_2(g) + 0.5O_2(g) \rightarrow H_2O(l)$	– (285.83 ± 0.04)	[89COX/WAG]
6	$Bi(cr) + 5H_2(g) + 7O_2(g) + 1.5N_2(g) \rightarrow Bi(NO_3)_3{\cdot}5H_2O(cr)$	– (2005.05 ± 2.80)	(result)

$\Delta_r H_6 = \Delta_f H_m^\circ(Bi(NO_3)_3{\cdot}5H_2O, cr, 298.15\ K) = -\,\Delta_r H_1 + \Delta_r H_2 - 3\,\Delta_r H_3 + 3\,\Delta_r H_4 + 5\,\Delta_r H_5$

The calculated value of the standard enthalpy of formation of $Bi(NO_3)_3{\cdot}5H_2O$(cr) is $\Delta_f H^\circ_m$ ($Bi(NO_3)_3{\cdot}5H_2O$, cr, 298.15 K) = – (2005.05 ± 2.80) kJ·mol^{-1} and will be accepted. The value in the paper is – 2002.9 kJ·mol^{-1}, calculated with another set of auxiliary data.

[65BRE]

The vapour pressure of Hg(g) in equilibrium with congruently vaporising α-HgSe was measured in the temperature range 623 to 873 K using optical density techniques. Thermodynamic quantities for α-HgSe were derived using auxiliary data for the gaseous selenium species and assuming stoichiometric composition of the gas phase. The auxiliary data applied to the selenium species differ from those selected by the review and the results were therefore re-evaluated using the selected set of thermodynamic data of the selenium species.

The equilibrium Hg(g) partial pressure was reported to be $\log_{10}(p_{Hg}/\text{bar})$ = 6.226 – 5900 T^{-1}. The evaluation made by the review involved the matching of the concentrations of the elements mercury and selenium in the gas phase assuming a stoichiometric composition. All species in the series Se(g)-Se_8(g) and Hg(g) were considered. The selected thermodynamic data for the selenium species and the CODATA [89COX/WAG] values for mercury were applied. The mole fractions of the most important selenium species are given in Table A-44.

Table A-44: Mole fractions of the most important gaseous selenium species in equilibrium with α-HgSe.

T (K)	Se_2(g)	Se_5(g)	Se_6(g)
723	0.63	0.25	0.09
873	0.76	0.17	0.04

The expression $\Delta_r G^\circ_m(T)$ = (175.3 – 0.1801 T) kJ·mol^{-1} for the reaction α-HgSe $\rightleftharpoons$ Hg(g) + ½Se_2(g) was derived from the calculated partial pressures. No error limits were assigned because no information about the accuracy of the measurements was given. Recalculation to 298.15 K using the selected heat capacity of α-HgSe yields $\Delta_f H^\circ_m$ (HgSe, α, 298.15 K) = – 51.2 kJ·mol^{-1} and S°_m (HgSe, α, 298.15 K) = 103.3 J·K^{-1}·mol^{-1}. The corresponding third law value of the enthalpy is $\Delta_f H^\circ_m$ (HgSe, α, 298.15 K) = – 55.0 kJ·mol^{-1} and was calculated using the selected values for the heat capacity and entropy of α-HgSe.

[65KEL/CRU]

The heat capacity of α-HgSe was measured in the temperature range 293 to 513 K using a heat pulse method. The result is only presented in a graph and the heat capacity at

298.15 K, $C^{\circ}_{p,m}$(HgSe, α, 298.15 K) = (55.15 ± 1.00) J·K^{-1}·mol^{-1}, and the heat capacity expression $C^{\circ}_{p,m}$(HgSe, α, (293 – 513) K) = (52.31 + 9.52 × 10^{-3} T) J·K^{-1}·mol^{-1} were evaluated by the review from Figure 3 of the paper.

[65LES/SEL]

The authors reacted Na_2SeO_3(cr) with a copper sulphate solution in an electrically calibrated calorimeter and measured the enthalpy change of the reaction. The product was $CuSeO_3 \cdot 2H_2O$(cr) as shown by chemical analysis and X-ray diffraction. Crystalline anhydrous copper selenite was also prepared and the integral enthalpies of dissolution of the two selenites in 8% HNO_3 (HNO_3, aq 1:40, and denoted sln below) were determined. The data have been used in Table A-45 to calculate standard enthalpies of formation of the copper selenites.

Table A-45: Calculation of the standard enthalpies of formation of $CuSeO_3 \cdot 2H_2O$(cr) and $CuSeO_3$(cr). The sources of the data are indicated.

#	Reaction	$\Delta_r H_m$ (kJ·mol^{-1})	Reference
1	Na_2SeO_3(cr) + $CuSO_4$(aq, 1:800) + $2H_2O$ → $CuSeO_3 \cdot 2H_2O$(cr) + Na_2SO_4(aq, 1:800)	– (16.40 ± 0.13)	(paper)
2	Cu(cr) + S(cr) + $2O_2$(g) + aq →$CuSO_4$(aq, 1:800)	– (839.92 ± 1.08)	[89COX/WAG] [82WAG/EVA]
3	2Na(cr) + S(cr) + $2O_2$(g) + aq → Na_2SO_4(aq, 1:800)	– (1389.00 ± 0.43)	[89COX/WAG] [82WAG/EVA]
4	2Na(cr) + Se(cr) + $1.5O_2$(g) → Na_2SeO_3(cr)	– (959.62 ± 0.95)	(review)
5	H_2(g) + $0.5O_2$(g) → H_2O(l)	– (285.83 ± 0.04)	[89COX/WAG]
6	Cu(cr) + Se(cr) + $2H_2$(g) + $2.5O_2$(g) → $CuSeO_3 \cdot 2H_2O$(cr)	– (998.60 ± 1.51)	(result)
7	$CuSeO_3 \cdot 2H_2O$(cr) + sln → $CuSeO_3$(sln, 0.05 M) + $2H_2O$(l, sln)	(15.86 ± 0.08)	(paper)
8	$CuSeO_3$(cr) + sln → $CuSeO_3$(sln, 0.05 M)	– (24.18 ± 0.08)	(paper)
9	H_2O(l) + sln → H_2O(l, sln)	(0.00 ± 0.05)	[82WAG/EVA]
10	Cu(cr) + Se(cr) + $1.5O_2$(g) → $CuSeO_3$(cr)	– (386.90 ± 1.52)	(result)

$\Delta_r H_6 = \Delta_f H^{\circ}_m(CuSeO_3 \cdot 2H_2O, cr, 298.15 K) = \Delta_r H_1 + \Delta_r H_2 - \Delta_r H_3 + \Delta_r H_4 + 2\,\Delta_r H_5$

$\Delta_r H_{10} = \Delta_f H^{\circ}_m(CuSeO_3, cr, 298.15 K) = -2\,\Delta_r H_5 + \Delta_r H_6 + \Delta_r H_7 - \Delta_r H_8 - 2\,\Delta_r H_9$

[65SIL/KAR]

The apparent vapour pressure of α-HgSe was measured in the temperature range 613 to 753 K using a gas flow method. The thermodynamic quantities were evaluated by the review because no values were reported in the original paper.

The result of the measurements was given in the form of an apparent total vapour pressure corresponding to the total pressure which would be obtained if no dissociation occurred in the gas phase. The expression $\log_{10}(p_{app}/\text{bar}) = 6.898 - 6505\ T^{-1}$ was derived from the data given in Table 2. The evaluation involved the matching of observed and calculated apparent pressures by varying the total pressure of the mercury and selenium species and assuming a stoichiometric composition of the gas phase. All species in the series Se(g)-Se_8(g) and Hg(g) were considered. The selected thermodynamic data of the selenium species and the CODATA [89COX/WAG] values for mercury were employed. The mole fractions of the most important selenium species are given in Table A-46.

Table A-46: Mole fractions of the most important gaseous selenium species in equilibrium with α-HgSe.

T (K)	Se_2(g)	Se_5(g)	Se_6(g)
613	0.67	0.22	0.09
753	0.77	0.16	0.05

The expression $\Delta_r G^\circ_m(T) = (189.0 - 0.1919\ T)$ kJ·mol^{-1} for the reaction α-HgSe $\rightleftharpoons$ Hg(g) + ½Se_2(g) was derived from the calculated partial pressures. No error limits were assigned because of the limited information about the experiments. A recalculation to 298.15 K using the selected heat capacity of HgSe yields $\Delta_f H^\circ_m$ (HgSe, α, 298.15 K) = – 62.9 kJ·mol^{-1} and S°_m (HgSe, α, 298.15 K) = 94.4 J·K^{-1}·mol^{-1}. The corresponding third law value of the enthalpy is $\Delta_f H^\circ_m$ (HgSe, α, 298.15 K) = – 60.0 kJ·mol^{-1} and was calculated using the selected heat capacity and entropy of α-HgSe.

[66AND/KAR]

Andreeva and Karapet'yants prepared crystalline bismuth selenite and measured in a calorimeter its enthalpy of dissolution in HNO_3(aq, 1:9) denoted sln. The enthalpies of dissolution of bismuth nitrate and selenious acid in the same solvent were also determined. The numerical values used in the thermochemical cycle in the paper indicate that the compound denoted there by $Bi(NO_3)_3$(cr) is in fact $Bi(NO_3)_3·5H_2O$ (cr). The data were evaluated on this assumption in Table A-47. The partial standard molar enthalpies of formation of H_2O and HNO_3 were estimated from [82WAG/EVA].

The result in the paper is – 1403.7 kJ·mol^{-1}, calculated with a different set of auxiliary data.

Table A-47: Calculation of the standard enthalpy of formation of $Bi_2(SeO_3)_3$(cr). The sources of the data are indicated.

#	Reaction	$\Delta_r H_m$ (kJ·mol^{-1})	Reference
1	$Bi_2(SeO_3)_3$(cr) + 6HNO_3(sln) + sln → 2$Bi(NO_3)_3$(sln, aq 1:1000) + 3H_2SeO_3(sln, aq 1:700)	(19.41 ± 0.08)	(paper)
2	$Bi(NO_3)_3 \cdot 5H_2O$(cr) + sln → $Bi(NO_3)_3$(sln, aq 1:1000) + 5H_2O(sln)	(37.57 ± 0.38)	(paper)
3	H_2SeO_3(cr) + sln → H_2SeO_3(sln, aq 1:700)	(15.94 ± 0.13)	(paper)
4	Se(cr) + H_2(g) + 1.5O_2(g) → H_2SeO_3(cr)	– (524.72 ± 0.65)	(review)
5	Bi(cr) + 5H_2(g) + 1.5N_2(g) + 7O_2(g) → $Bi(NO_3)_3 \cdot 5H_2O$(cr)	– (2005.05 ± 2.80)	[65AND/KAR]
6	0.5H_2(g) + 1.5O_2(g) + 0.5N_2(g) + sln → HNO_3(sln)	– (203.10 ± 0.50)	[89COX/WAG] [82WAG/EVA]
7	H_2(g) + 0.5O_2(g) + sln → H_2O(sln)	– (286.03 ± 0.10)	[89COX/WAG] [82WAG/EVA]
8	2Bi(cr) + 3Se(cr) + 4.5O_2(g) → $Bi_2(SeO_3)_3$(cr)	– (1401.81 ± 6.77)	(result)

$\Delta_r H_8 = \Delta_f H^\circ_m$ ($Bi_2(SeO_3)_3$, cr, 298.15 K) $= -\Delta_r H_1 + 2\,\Delta_r H_2 + 3\,\Delta_r H_3 + 3\,\Delta_r H_4 + 2\,\Delta_r H_5 - 6\,\Delta_r H_6 - 10\,\Delta_r H_7$

[66BER/CHU]

The compositions of the gaseous phases in equilibrium with SrSe(cr), CdSe(cr), HgSe(cr), and trigonal and liquid selenium were studied by mass-spectrometry using Knudsen cells. In the equilibrium with solid or liquid selenium all the ions Se^+(g)- Se_8^+ (g) were observed, but the temperature dependence could not be obtained for the Se^+(g) and Se_4^+ (g) ions because of their weak intensities. The primary results for the equilibria are given in the Table A-48.

Table A-48: Reaction enthalpies and temperatures used for the calculation of thermodynamic quantities of gaseous selenium molecules at standard conditions.

Reaction no	Reaction	Temperature (K)	$\Delta_r H^\circ_m$ (T) (kJ·mol^{-1})
1	2Se_3(g) ⇌ 3Se_2(g)	850	107.5
2	Se_6(g) ⇌ 3Se_2(g)	475	272.0
3	5Se_6(g) ⇌ 6Se_5(g)	500	231.0
4	6Se_7(g) ⇌ 7Se_6(g)	500	58.9
5	3Se_8(g) ⇌ 4Se_6(g)	500	69.0
6	6Se(cr) ⇌ Se_6(g)	460	141.0
7	6Se(l) ⇌ Se_6(g)	525	83.3

From Reactions #6 and #7 the authors calculated the enthalpy of fusion to be $\Delta_{fus}H^{\circ}_{m}$ (Se, trigonal) = 9.623 kJ·mol^{-1} and noted that it is considerably larger than the generally accepted value $\Delta_{fus}H^{\circ}_{m}$ (Se, trigonal) = 5.439 kJ·mol^{-1} at that time. The selected value of $\Delta_{fus}H^{\circ}_{m}$ (Se, trigonal) is 6.159 kJ·mol^{-1}. No explanation for the discrepancy was given, but it was concluded that the difference was much larger than the experimental error. The data for the reactions above were recalculated to 298.15 K by the present review using the second and third laws with the selected values for heat capacities and entropies of the Se_n(g) molecules and Se(s, l). Because Reactions #6 and #7 lead to different values for the enthalpy of formation of Se_6(g), the evaluation has been made separately for the vaporisation of the solid and the liquid, using the appropriate enthalpies of formation of Se_6(g). The results are given in the Table A-49.

Table A-49: Standard entropy and enthalpy of formation of gaseous selenium species evaluated from data in [66BER/CHU].

Species	Evaluation state	$\Delta_f H^{\circ}_{m}$ (298.15 K) (kJ·mol^{-1})		S°_{m} (298.15 K) (J·K^{-1}·mol^{-1})
		2nd law	3rd law	
Se_2(g)	Trigonal	(139.4 ± 0.7)	(141.2 ± 1.4)	(243.9 ± 1.4)
Se_2(g)	Liquid	(133.8 ± 1.7)	(141.2 ± 1.4)	(232.6 ± 3.2)
Se_3(g)	Trigonal	(157.3 ± 2.1)	(170.0 ± 7.0)	(292.2 ± 4.3)
Se_3(g)	Liquid	(149.0 ± 5.1)	(170.0 ± 7.0)	(275.4 ± 9.6)
Se_5(g)	Trigonal	(160.4 ± 1.4)	(145.3 ± 3.1)	(430.6 ± 3.1)
Se_5(g)	Liquid	(146.7 ± 4.1)	(145.3 ± 3.1)	(402.5 ± 7.8)
Se_6(g)	Trigonal	(145.5 ± 0.8)	(138.3 ± 0.4)	(444.3 ± 1.7)
Se_6(g)	Liquid	(128.9 ± 2.4)	(138.3 ± 0.5)	(410.6 ± 4.6)
Se_7(g)	Trigonal	(159.4 ± 1.9)	(151.2 ± 0.9)	(505.1 ± 4.2)
Se_7(g)	Liquid	(140.1 ± 5.7)	(151.2 ± 1.3)	(465.8 ± 10.9)
Se_8(g)	Trigonal	(169.8 ± 2.2)	(157.0 ± 1.2)	(553.7 ± 4.8)
Se_8(g)	Liquid	(147.8 ± 6.6)	(157.0 ± 1.6)	(508.7 ± 12.4)

The difference between the results obtained by the second law for a particular gaseous selenium species from the equilibria with solid and liquid selenium, respectively, is of the order of 10 to 20 kJ·mol^{-1}, while the third law evaluation made by review brings the difference down to less than 0.1 kJ·mol^{-1}. The temperature ranges of the experiments therefore seem to be too small to provide accurate values using the second law.

An inconsistency was noted by Brebrick [68BRE] in the equilibrium constants given in Table X of the paper and the partial pressures of Se_2(g) given in column 4 of

Table VIII of the paper. This inconsistency was due to a calculation error and was corrected in [68BER/CHU]. These changes do not affect the results obtained by this review.

The dissociation energy of $Se_2(g)$ at 298.15 K obtained from the measurements on the Sr-Se system was re-evaluated by this review using the third law and the selected heat capacities and entropies for the Se(g) and $Se_2(g)$ molecules. The value obtained, 319.1 kJ·mol^{-1}, differs from the original value in the paper, 299.6 kJ·mol^{-1}, due to the different heat capacities and entropies used.

The enthalpy of formation of SrSe(cr) was calculated by the review from the observation that the partial pressures of Sr(g) and Se(g) were both determined to be $(1.2 \pm 0.3) \times 10^{-5}$ bar in equilibrium with SrSe(cr) at 1853 K. The third law evaluation was made according to the reaction SrSe(cr) $\rightleftharpoons$ Sr(g) + Se(g) employing the selected thermodynamic properties of Se(g), the data of Sr(g) in [98CHA], the entropy value S°_{m}(SrSe, cr, 298.15 K) = (80.8 ± 6.0) J·K^{-1}·mol^{-1} estimated by Mills [74MIL] and the heat capacity expression

$$C^{\circ}_{p,m}(\text{SrSe, cr, (298.15 – 2000) K}) = (49.9709 + 8.534 \times 10^{-3}\, T - 0.9429 \times 10^{-6}\, T^2 - 2.9351 \times 10^{5}\, T^{-2})\ \text{J·K}^{-1}\text{·mol}^{-1}.$$

which was estimated by assuming the heat capacity to be equal to that of SrS(cr) as reported in the JANAF Tables [98CHA] plus the difference in heat capacity between EuSe(cr) and EuS(cr) as obtained by McMasters and Gschneidner [74MCM/GSC]. The calculated value of the enthalpy of formation of SrSe(cr) was $\Delta_f H^{\circ}_{m}$(SrSe, cr, 298.15 K) = – (416 ± 65) kJ·mol^{-1}.

[66FUJ/WES]

The mass spectrum of the species $Se_2(g)$-$Se_8(g)$ in equilibrium with condensed selenium was measured in the temperature range 375 to 460 K using a Knudsen cell and a mass spectrometer. It was concluded that at the experimental conditions only the ions $Se_5^+(g)$ - $Se_8^+(g)$ were formed as the direct result of ionisation of the corresponding parent molecules. The apparatus used did not allow for the measurement of absolute vapour pressures and the evaluation of the experiments involved a normalisation against the vapour pressure of $Se_6(g)$ as given by Stull and Sinke [56STU/SIN]. The enthalpies of formation at 448 K in Table III of the paper were recalculated by the present review to 298.15 K using the selected data of the heat capacities to provide for a comparison with other investigations. Similarly, third law enthalpies of formation and second law entropies were evaluated by combining the data of Table III of the paper, the total pressure measurements of Niwa and Shibata [40NIW/SHI], and the selected values for heat capacities and entropies of $Se_5(g)$-$Se_8(g)$ and Se(trig). The results are summarised in the Table A-50.

Table A-50: Entropies and enthalpies of formation of the species $Se_5(g)$-$Se_8(g)$.

Species	$\Delta_f H_m$ (448 K) (kJ·mol^{-1})	$\Delta_f H^\circ_m$ (298.15 K) (kJ·mol^{-1})		S°_m (298.15 K) (J·K^{-1}·mol^{-1})
	Original values	2nd law	3rd law	
$Se_5(g)$	(151.5 ± 4.1)	(155.5 ± 4.1)	(145.3 ± 4.1)	(421.0 ± 9.2)
$Se_6(g)$	(141.4 ± 4.1)	(145.6 ± 4.1)	(137.4 ± 4.1)	(446.9 ± 9.2)
$Se_7(g)$	(160.7 ± 4.1)	(165.2 ± 4.1)	(150.7 ± 4.1)	(519.3 ± 9.2)
$Se_8(g)$	(166.9 ± 5.7)	(171.8 ± 5.7)	(155.8 ± 5.7)	(561.7 ± 12.8)

[66LES/SEL]

The authors reacted 9×10^{-3} moles of $MgSO_4$(aq, 1:2000) with an equivalent amount of Na_2SeO_3(cr) in 325 grams of water in an electrically calibrated calorimeter. Crystalline $MgSeO_3\cdot 6H_2O$ was formed. From the solubility product of magnesium selenite the review estimates that about 6% of the Mg^{2+} remained in solution after the reaction. A correction has been applied, which changes the enthalpy of Reaction #1 in Table A-51 from – (47.49 ± 0.13) in the paper to – (48.25 ± 0.25) kJ·mol^{-1}.

The authors also determined the enthalpies of dissolution of $MgSeO_3\cdot 6H_2O$(cr) and $MgSeO_3$(cr) in nitric acid.

Table A-51: Calculation of the standard enthalpy of formation of $MgSeO_3\cdot 6H_2O$. The sources of the data are indicated.

#	Reaction	$\Delta_r H_m$ (kJ·mol^{-1})	Reference
1	$MgSO_4$(aq, 1:2000) + Na_2SeO_3(cr) + $6H_2O$(l) → $MgSeO_3\cdot 6H_2O$(cr) + Na_2SO_4(aq, 1:2000)	– (48.25 ± 0.25)	(see text)
2	Mg(cr) + S(cr) + $2O_2$(g) + aq → $MgSO_4$(aq, 1:2000)	– (1373.42 ± 0.73)	[89COX/WAG] [82WAG/EVA]
3	2Na(cr) + Se(cr) + $1.5O_2$(g) → Na_2SeO_3(cr)	– (959.62 ± 0.95)	(review)
4	2Na(cr) + S(cr) + $2O_2$(g) + aq → Na_2SO_4(aq, 1:2000)	– (1389.06 ± 0.43)	[89COX/WAG] [82WAG/EVA]
5	H_2(g) + $0.5O_2$(g) → H_2O(l)	– (285.83 ± 0.04)	[89COX/WAG]
6	Mg(cr) + Se(cr) + $6H_2$(g) + $4.5O_2$(g) → $MgSeO_3\cdot 6H_2O$(cr)	– (2707.21 ± 1.30)	(result)

$\Delta_r H_6 = \Delta_f H^\circ_m$ ($MgSeO_3\cdot 6H_2O$, cr, 298.15 K) $= \Delta_r H_1 + \Delta_r H_2 + \Delta_r H_3 - \Delta_r H_4 + 6\,\Delta_r H_5$

[66LES/SEL3]

Leshchinskaya and Selivanova [66LES/SEL] measured the enthalpy of dissolution of $MnSeO_3\cdot 2H_2O$(cr), $MnSO_4$(cr), and H_2SeO_3 in H_2SO_4(aq, 1:30) denoted sln. Their data have been recalculated with new auxiliary data in Table A-52. The partial standard enthalpies of formation of H_2O and H_2SO_4 were estimated from [82WAG/EVA].

Table A-52: Calculation of the standard enthalpy of formation of $MnSeO_3{\cdot}2H_2O(cr)$. The sources of the data are indicated.

#	Reaction	$\Delta_r H_m$ (kJ·mol^{-1})	Reference
1	$MnSeO_3{\cdot}2H_2O(cr) + H_2SO_4(sln) + sln \rightarrow MnSO_4(sln, 1{:}1000) + H_2SeO_3(sln, 1{:}1000) + 2H_2O(l, sln)$	– (22.71 ± 0.13)	(paper)
2	$MnSO_4(cr) + sln \rightarrow MnSO_4(sln, 1{:}1000)$	– (44.10 ± 0.17)	(paper)
3	$H_2SeO_3(cr) + sln \rightarrow H_2SeO_3(sln, 1{:}1000)$	(17.82 ± 0.17)	(paper)
4	$Se(cr) + H_2(g) + 1.5O_2(g) \rightarrow H_2SeO_3(cr)$	– (524.72 ± 0.65)	(review)
5	$S(cr) + H_2(g) + 2O_2(g) + sln \rightarrow H_2SO_4(sln)$	– (884.60 ± 0.30)	[89COX/WAG] [82WAG/EVA]
6	$Mn(cr) + S(cr) + 2O_2(g) \rightarrow MnSO_4(cr)$	– (1065.25 ± 2.50)	[82WAG/EVA]
7	$H_2(g) + 0.5O_2(g) + sln \rightarrow H_2O(l, sln)$	– (285.88 ± 0.04)	[89COX/WAG]
8	$Mn(cr) + Se(cr) + 2H_2(g) + 2.5O_2(g) \rightarrow MnSeO_3{\cdot}2H_2O(cr)$	– (1280.70 ± 2.62)	(result)

$\Delta_r H_8 = \Delta_f H^\circ_m(MnSeO_3{\cdot}2H_2O, cr, 298.15\ K) = -\Delta_r H_1 + \Delta_r H_2 + \Delta_r H_3 + \Delta_r H_4 - \Delta_r H_5 + \Delta_r H_6 + 2\,\Delta_r H_7$

[66PYA/DUR]

Pyatniskii and Durdyev measured the solubility of $CoSeO_3{\cdot}2H_2O$ in water and as a function of the total selenite concentration at a pH of about 8.5. The temperature of the experiments is not mentioned. The primary data of the water solubility, which included the pH of the equilibrium solutions, have been used in a recalculation that yielded $\log_{10} K^\circ_{s,0} = -(8.17 \pm 0.05)$, for the reaction $CoSeO_3{\cdot}2H_2O(cr) \rightleftharpoons Co^{2+} + SeO_3^{2-} + 2H_2O$. The result in the paper is – 7.64 calculated with an unspecified value of the protonation constant of the selenite ion.

The study of the variation of the solubility with the selenite concentration is stated to have been carried at the constant ionic strengths 0.01 and 0.3 M. How this was accomplished was not clear from the information in the paper. The data in the tables rather seem to indicate that the ionic strength varied and reached 0.03 and 0.5 M, respectively, in the solution with the highest selenite concentrations. The analysis of the data was made with an equilibrium model that comprised the solubility equilibrium and the formation of the complex $Co(SeO_3)_2^{2-}$. Thus the formation of $CoSeO_3(aq)$ was not included in the model. The analysis led to values of the solubility product at the two ionic strengths that appear to be inconsistent with the value obtained from the solubility in water. This result together with the improbable model made the review reject the outcome of the equilibrium analysis.

The variation of the solubility of $CoSeO_3 \cdot 2H_2O$ with selenite concentration is only moderate. A more likely model of the system would therefore include $CoSeO_3(aq)$ instead of (or in addition to) $Co(SeO_3)_2^{2-}$. The vague description of the experiments and compositions of the test solutions detracted the review from further analysis of the data in the paper.

[66RAW/TOG]

Rawling and Toguri studied the formation of $H_2Se(g)$ via the transpiration method, where $H_2(g)$ was slowly bubbled through Se(l) in the temperature range of 800 to 900 K. By varying the gas flow, it was ascertained that equilibrium was attained and by positioning two selenium crucibles in series with the first one at a higher temperature, the reversibility of the reaction was verified.

The gases leaving the furnace were first passed through a cold trap where elemental selenium was condensed. The partial pressure of $H_2Se(g)$ was then determined via the absorption in an alkaline potassium bromate solution and analysis, while the partial pressure of $H_2(g)$ was calculated from the volume of hydrogen passing the system and the weight of the selenium analysed in the bromate solution.

A total number of 8 determinations of the equilibrium constant K is reported and the uncertainty in the individual determination seem to be in the range of $\pm$ 0.02.

[66SAD/SEM]

The Gibbs energy of formation of PbSe(cr) was measured in the temperature range 673 to 873 K using an electrochemical cell. Both selenium and lead are liquids at these temperatures and the reaction studied was

$$\text{Pb(l)} + \text{Se(l)} \rightleftharpoons \text{PbSe(cr)}. \qquad \text{(A.17)}$$

The enthalpy and entropy of reaction, $\Delta_r H_m^\circ$((A.17), 723 K) = – (108.8 ± 3.0) kJ·mol^{-1} and $\Delta_r S_m^\circ$((A.17), 723 K) = – (25.1 ± 4.0) J·K^{-1}·mol^{-1}, were determined by the review from Eq. (II) of the paper. A recalculation to 298.15 K employing the selected heat capacity of PbSe(cr), the selected data of selenium and the data for Pb(cr, l) in [89COX/WAG] yielded $\Delta_f H_m^\circ$(PbSe, cr, 298.15 K) = – (95.1 ± 3.0) kJ·mol^{-1} and $\Delta_f S_m^\circ$(PbSe, cr, 298.15 K) = – (0.1 ± 4.0) J·K^{-1}·mol^{-1}. The corresponding entropy of PbSe(cr) was calculated to be S_m°(PbSe, cr, 298.15 K) = (106.7 ± 4.0) J·K^{-1}·mol^{-1} using the selected entropy of selenium and the entropy of Pb(cr) in [89COX/WAG].

[66SAP/MAR]

Sapozhnikov and Markovskii measured at 298.15 K the enthalpy of dissolution of a number of zinc, cadmium, and mercury selenites in mineral acid in a calorimeter. Sulphuric acid (aq, 1:50) was used for the zinc and cadmium compounds, while hydrochloric acid (aq, 1:35) was used for the mercury salts. The experimental details are few and it is estimated by the review that the concentrations of the dissolved salts were in a dilution range of about (aq, 1:500).

The enthalpy of dissolution of SeO_2(cr) to (aq, 1:500 – 1:600) in the acids (sln) was determined to be (2.68 ± 0.04) kJ·mol^{-1} from which the review calculates an enthalpy of formation of H_2SeO_3(sln, 1:500) of – (508.5 ± 1.0) kJ·mol^{-1}. The evaluation in the paper used – 514.6 kJ·mol^{-1}. The enthalpies of formation of the metal sulphates and mercury chloride in the media were not determined and the values at infinite dilution were used. The review employs enthalpies of formation at a dilution of (aq, 1:500) and assumes that $HgCl_4^{2-}$ is formed rather than $HgCl_2$(aq). This introduces a correction of – 3.4 kJ·mol^{-1}. The results are entered in Table A-53.

Table A-53: Measured enthalpies of dissolution of zinc and cadmium selenites in H_2SO_4(aq, 1:50) and mercury selenites in HCl(aq, 1:35). The enthalpies of formation have been estimated from these data. The uncertainties stem from the original publication. The data for the compounds provided with an asterisk are given for information only as they do not seem to form under normal aqueous conditions (*cf.* phase diagram studies reported in Chapter V).

Substance	$\Delta_{dis}H_m^o$ (kJ·mol^{-1})	$-\Delta_f H_m^o$ (kJ·mol^{-1})
$ZnSeO_3 \cdot 2H_2O$*	– (8.91 ± 0.42)	(1243.5 ± 10.5)
$ZnSeO_3 \cdot H_2O$	– (22.67 ± 0.42)	(943.9 ± 10.5)
α-$ZnSeO_3$	– (43.89 ± 1.26)	(636.8 ± 11.3)
β-$ZnSeO_3$	– (28.07 ± 0.63)	(652.7 ± 10.9)
$Zn(HSeO_3)_2 \cdot 2H_2O$	(14.43 ± 0.63)	(1775.3 ± 12.6)
$ZnSe_2O_5$	– (15.10 ± 0.63)	(888.3 ± 12.6)
$CdSeO_3 \cdot 2/3H_2O$*	– (6.69 ± 0.42)	(785.0 ± 6.3)
$CdSeO_3 \cdot ½H_2O$*	– (7.61 ± 0.63)	(736.4 ± 6.7)
α-$CdSeO_3$	– (17.61 ± 0.63)	(583.5 ± 6.7)
β-, γ-$CdSeO_3$	– (14.48 ± 0.84)	(586.7 ± 7.1)
$Cd_3H_2(SeO_3)_4$	– (6.27 ± 1.88)	(2305.6 ± 22.6)
$CdSe_2O_5$	(3.89 ± 0.63)	(827.7 ± 8.4)
α-$HgSeO_3$	– (27.86 ± 1.05)	(375.0 ± 5.4)
β-$HgSeO_3$	– (30.46 ± 0.63)	(372.4 ± 5.0)
γ-$HgSeO_3$	– (29.62 ± 0.63)	(373.3 ± 5.0)
$Hg_3H_2(SeO_3)_4$	– (64.38 ± 1.89)	(1652.8 ± 16.3)

The uncertainties are transferred from the paper. The enthalpies of formation are approximate, but comparisons with other data suggest that the correct values fall within twice the stated uncertainties.

[66SED/AGU]

The total pressure was measured in systems containing α-ZnSe and I_2(cr) as starting materials in the temperature range 900 to 1200 K using a Bourdon gauge. The results were analysed assuming the equilibria α-ZnSe + I_2(g) $\rightleftharpoons$ ZnI_2(g) + ½Se_2(g) and I_2(g) $\rightleftharpoons$ 2I(g) to be dominant. The calculation of thermodynamic quantities at 298.15 K from the high temperature data involved an estimated heat capacity and entropy of ZnI_2(g). The obtained results were adjusted by the review to the selected entropy and enthalpy of formation of Se_2(g) yielding $\Delta_f H^\circ_m$ (ZnSe, α, 298.15 K) = – (185.8 ± 16.0) $kJ\cdot mol^{-1}$ and S°_m (ZnSe, α, 298.15 K) = (62.2 ± 20.0) $J\cdot K^{-1}\cdot mol^{-1}$. The error limits of the entropy and the enthalpy of formation were obtained by considering the standard deviation of the experiments and the uncertainties in the thermodynamic properties of solid and gaseous ZnI_2 to be the main error contributions.

[67CHA/BEV]

The enthalpy of formation of the supposed compound Bi_2Se(s) was determined using metal solution calorimetry. However, Bi_2Se(s) is not a stable compound in the bismuth-selenium system and the results are not considered by the review.

[67KLU/SEL]

Klushina, Selivanova, and Poltavtseva prepared crystalline $Li_2SeO_3\cdot H_2O$ and measured the enthalpy of the reaction of the salt with a lead nitrate solution in an electrically calibrated calorimeter. Crystalline lead selenite is formed. In the evaluation of the calorimetric measurements the authors divided the heat evolved by the number of moles of $Li_2SeO_3\cdot H_2O$ taken. This is incorrect as the amount of $Pb(NO_3)_2$ used limits the amount of lead selenite that can be formed. The data have been recalculated to find the enthalpy change of the reaction $Li_2SeO_3\cdot H_2O$(cr) + Pb^{2+} → $PbSeO_3$(cr) + 2Li^+ + H_2O(l) in Table A-54.

Table A-54: Calculation of the enthalpy change of the reaction $Li_2SeO_3{\cdot}H_2O(cr)$ + $Pb^{2+} \rightarrow PbSeO_3(cr) + 2Li^+ + H_2O(l)$. The sources of the data are indicated.

#	Reaction	$\Delta_r H_m$ (kJ·mol^{-1})	Reference
1	$Li_2SeO_3{\cdot}H_2O(cr) + Pb(NO_3)_2(aq, 1{:}800) \rightarrow PbSeO_3(cr) + 2LiNO_3(aq, 1{:}400) + H_2O(l)$	– (44.44 ± 0.59)	(paper)
2	$Pb(NO_3)_2(aq, 1{:}800) + aq \rightarrow Pb^{2+} + 2\,NO_3^-$	(1.85 ± 1.50)	(review)
3	$LiNO_3(aq, 1{:}400) + aq \rightarrow Li^+ + NO_3^-$	– (0.44 ± 0.05)	[82WAG/EVA]
4	$Li_2SeO_3{\cdot}H_2O(cr) + Pb^{2+} \rightarrow PbSeO_3(cr) + 2Li^+ + H_2O(l)$	– (47.17 ± 1.61)	(review)

$\Delta_r H_4 = \Delta_r H_1 - \Delta_r H_2 + 2\,\Delta_r H_3$

[67KOL/IVA]

The complex formation between Sc^{3+} and SeO_4^{2-} was studied by measurement of the distribution of Sc^{3+} between the aqueous phase and a cation exchanger at selenate concentrations in the range 0 to 0.167 M. Sodium perchlorate was added to keep the ionic strength at I = 0.5 M. The radioactive nuclide ^{46}Sc was used to follow the distribution ratio at a total scandium concentration of 2×10^{-5} M. Measurements were carried out at 298.15, 308.15, and 318.15 K and the pH was adjusted to 4.0 to 4.5. No disturbance from the hydrolysis of Sc^{3+} occurs at this pH according to the paper although this would be expected from the data in [76BAE/MES].

In each experiment 2.40 ml (V) of Sc^{3+} solution, activity n_o, were equilibrated with 10 to 30 mg (m) of the resin. The loading of the resin is low. The activity (n_L) of the aqueous phase was measured and the distribution coefficient calculated as:

$$K_D = \frac{[R_3Sc]}{[Sc^{3+}]_{aq,tot}} = \frac{n_o - n_L}{n_L}\frac{V}{m}.$$

Preliminary calculations indicated that the resin only sorbs Sc^{3+} and the stability constants were calculated from:

$$(K_D^o / K_D - 1)/[SeO_4^{2-}] = \beta_1 + \beta_2\,[SeO_4^{2-}] + \cdots$$

K_D^o is the distribution coefficient in the absence of the ligand, *i e.* in 0.5 M $NaClO_4$.The results presented for 298.15 K are $\beta_1 = (60.8 \pm 2.4)\ M^{-1}$ and $\beta_2 = (435 \pm 17)\ M^{-2}$.

The review suspects that the equation used to calculate β_1 and β_2 requires that a_{Na^+} is kept constant. This is not the case as $[Na^+]$ varies from 0.5 to 0.33 M with increasing concentration of the ligand. The value of β_2 is therefore not accepted and the value of β_1 evaluated from the data at selenate concentrations below 0.025 M. The result is $\log_{10} \beta_1$ (0.5 M $NaClO_4$, 298.15 K) = (1.65 ± 0.20).

The extent of side-reactions from the hydrolysis of Sc^{3+} increases with temperature and may be the reason for the anomalous quotient between the stepwise stability constants obtained at 308.15 and 318.15 K. The quotient is 0.95 at 318.15 K. The results for these temperatures are not accepted.

The related paper by the same research group [71KOL/TET] does not contain quantitative information on the complexation.

[67MAI/SUP]

The authors reacted a solution of cerium(III) or neodymium chloride with Na_2SeO_3(cr) in an electrically calibrated calorimeter. The selenites precipitated were amorphous and contained 10 and 8 molecules of water, respectively, when dried at 298 K. These numbers have been used in the re-evaluation of the measurements. The recalculation used the values of the standard enthalpies of formation of Ce^{3+}(aq) and Nd^{3+}(aq) proposed in the critical evaluation by Cordfunke and Konings [2001COR/KON], see Appendix E, and of Cl^-(aq) in [89COX/WAG]. A correction for the enthalpy of dilution from (aq 1:1500) to the aqueous standard state of – (3.5 ± 1.0) kJ·mol^{-1} was applied.

Table A-55: Calculation of the enthalpies of formation of $Ce_2(SeO_3)_3·10H_2O$(am) and $Nd_2(SeO_3)_3·8H_2O$(am). The sources of the data are indicated.

#	Reaction	$\Delta_r H_m$ (kJ·mol^{-1})	Reference
1a	$2CeCl_3$(aq, 1:1500) + $3Na_2SeO_3$(cr) + $10H_2O$(l) → $Ce_2(SeO_3)_3·10H_2O$(am) + 6NaCl(aq, 1:500)	– (37.95 ± 0.33)	(paper)
1b	$2NdCl_3$(aq, 1:1500) + $3Na_2SeO_3$(cr) + $8H_2O$(l) → $Nd_2(SeO_3)_3·8H_2O$(am) + 6NaCl(aq, 1:500)	– (33.93 ± 0.38)	(paper)
2	2Na(cr) + Se(cr) + $1.5O_2$(g) → Na_2SeO_3(cr)	– (959.62 ± 0.95)	(review)
3a	Ce + $1.5Cl_2$ + aq → $CeCl_3$(aq, 1:1500)	– (1200.1 ± 2.2)	(see text)
3b	Nd + $1.5Cl_2$ + aq → $NdCl_3$(aq, 1:1500)	– (1192.5 ± 2.2)	(see text)
4	Na(cr) + $0.5Cl_2$(g) + aq → NaCl(aq, 1:500)	– (407.09 ± 0.15)	[89COX/WAG] [82WAG/EVA]
5	H_2(g) + $0.5O_2$(g) → H_2O(l)	– (285.83 ± 0.04)	[89COX/WAG]
6a	2Ce(cr) + 3Se(cr) + $10H_2$(g) + $9.5O_2$(g) → $Ce_2(SeO_3)_3·10H_2O$(am)	– (5732.8 ± 5.3)	(result)
6b	2Nd(cr) + 3Se(cr) + $8H_2$(g) + $8.5O_2$(g) → $Nd_2(SeO_3)_3·8H_2O$(am)	– (5141.9 ± 5.3)	(result)

$\Delta_r H_6 = \Delta_r H_1 + 3\,\Delta_r H_2 + 2\,\Delta_r H_3 - 6\,\Delta_r H_4 + 10(8)\,\Delta_r H_5$

[67RAU2]

Measurements of the selenium vapour density in the temperature range 712 to 1147 K are presented. The results were re-evaluated by the author in [74RAU] using new information about the species in the gas phase.

[67TER/RAT]

The Gibbs energy of formation of α-ZnSe was measured in the temperature range 360 to 420 K using the cell

$$(-)\ \mathrm{Zn(cr)}\ |\ \mathrm{ZnCl_2(in\ glycerol)}\ |\ \alpha\text{-}\mathrm{ZnSe,\ Se(trig)}\ (+)$$

and the reaction Zn(cr) + Se(cr) $\rightleftharpoons$ α-ZnSe. The enthalpy and entropy of formation at 298.15 K were evaluated by linear regression on emf *versus* temperature. The entropy of α-ZnSe at 298.15 K was calculated by the review from the entropy of formation and the selected entropy of selenium and the entropy of Zn(cr) at 298.15 K in [89COX/WAG] yielding S°_{m} (ZnSe, α, 298.15 K) = (69.9 ± 6.0) J·K^{-1}·mol^{-1}. The value of the enthalpy of formation $\Delta_f H^{\circ}_{m}$ (ZnSe, α, 298.15 K) = – (166.1 ± 9.0) kJ·mol^{-1} was retained. The error limits of the entropy and the enthalpy of formation were considered to be those of the original paper multiplied by three.

[68BAK]

The vapour pressure of liquid selenium was measured by a static method in the temperature range 663 to 1693 K and the total vapour pressure was given as $\log_{10}(p/\mathrm{bar}) = 5.271 - 5043\ T^{-1}$. A third law evaluation of the enthalpy of formation of Se_2(g) at 298.15 K was made by this review using the vapour pressure at 1500 K and the selected data for the heat capacities and entropies of Se(cr, l) and Se_2(g). A mole fraction of Se_2(g), $x_{\mathrm{Se_2(g)}}$, = 0.90 was estimated from the selected selenium data. The enthalpy of formation derived was $\Delta_f H^{\circ}_{m}$ (Se_2, g, 298.15 K) = (140.5 ± 2.2) kJ·mol^{-1}.

[68BER/CHU]

The calculated partial pressures presented in Table X of [66BER/CHU] are corrected. No changes were made to the thermodynamic properties derived in the previous paper.

[68BON/PAS]

The saturated vapour pressure of Bi_2Se_3(cr) was measured in the temperature range 735 to 879 K using the Knudsen effusion technique. The total vapour pressure was described by the expression $\log_{10}(p/\mathrm{bar}) = (9.347 \pm 1.036) - (11890 \pm 818)\ T^{-1}$. Thermodynamic data were evaluated according to the reaction Bi_2Se_3(cr) $\rightleftharpoons$ BiSe(g) + Se_2(g). However, the mass spectrometric investigation by Porter and Spencer [60POR/SPE] shows that selenium-rich bismuth-selenium alloys evaporate primarily by giving off Se_2(g) leading to an accumulation of bismuth in the condensed phase. The correct reaction should therefore involve Bi_2Se_3(cr), BiSe(cr) of the composition at the selenium-rich limit of the homogeneity range (approximately $Bi_{0.45}Se_{0.55}$), and Se_2(g). The thermodynamic values derived for Bi_2Se_3(cr) in this paper are inappropriate and rejected.

[68COL/DRO]

Mass spectrometry and effusion cells were used to measure the partial pressures of In(g), Se(g), Se_2(g), InSe(g), In_2Se(g), and In_2Se_2(g) in In-Se gas mixtures. The temperature ranges of the reactions studied are summarised in Table A-56. The evaluation used estimated free energy functions of the indium selenides and because better values are available, the experiments were re-evaluated by the review. The procedure utilized the partial pressures listed in Table 5 of the paper, the selected properties of the selenium species, entropies and heat capacities of the indium selenides calculated from the molecular parameters obtained by Erkoc, Katircioglu, and Yilmaz [2001ERK/KAT] in quantum mechanical studies of In-Se clusters, the enthalpy of In(g) in [82WAG/EVA], and the entropy and heat capacity of In(g) calculated from the atomic data listed by Moore [71MOO2]. The results are presented in Table A-57 and compared with the original results in Table A-58. The relatively large difference between the original enthalpy of reaction of the third reaction and the new value originates in significantly lower values of the Gibbs energy function of In_2Se_2(g) in [68COL/DRO] than used by the review.

Table A-56: Reactions and temperature ranges studied in [68COL/DRO].

Reaction	Temperature range (K)
InSe(g) $\rightleftharpoons$ In(g) + Se(g)	1495 – 1815
In_2Se(g) $\rightleftharpoons$ 2In(g) + Se(g)	1495 – 1815
In_2Se_2(g) $\rightleftharpoons$ In_2Se(g) + ½Se_2(g)	1030 – 1134

Table A-57: Thermodynamic properties of the gaseous indium selenides at standard conditions as re-evaluated by the review from the investigation in [68COL/DRO]. The enthalpy values of In_2Se_2(g) were calculated using the third law value of In_2Se(g).

Species	S^{o}_{m} (298.15 K), 2nd law ($J\cdot K^{-1}\cdot mol^{-1}$)	$\Delta_f H^{o}_{m}$ (298.15 K), 2nd law ($kJ\cdot mol^{-1}$)	$\Delta_f H^{o}_{m}$ (298.15 K), 3rd law ($kJ\cdot mol^{-1}$)
InSe(g)	251.7	216.7	226.6
In_2Se(g)	334.8	167.6	147.3
In_2Se_2(g)	371.6	139.8	159.4

Table A-58: Comparison between the re-evaluation made by the review and the original results for the reaction enthalpies.

Reaction	$\Delta_r H^\circ_m$ (298.15 K), 2nd law (kJ·mol^{-1})	$\Delta_r H^\circ_m$ (298.15 K), 3rd law (kJ·mol^{-1})	$\Delta_r H^\circ_m$ (298.15 K), original value (kJ·mol^{-1})
$InSe(g) \rightleftharpoons In(g) + Se(g)$	262.7	252.8	(242.7 ± 14.6)
$In_2Se(g) \rightleftharpoons 2In(g) + Se(g)$	555.1	575.4	(556.5 ± 16.7)
$In_2Se_2(g) \rightleftharpoons In_2Se(g) + \frac{1}{2} Se_2(g)$	78.0	58.4	(129.7 ± 16.7)

[68FIC/HAS]

The partial pressures of the species in gaseous aluminium-selenium mixtures were determined in the temperature range 1232 to 1352 K using mass spectrometry and Knudsen effusion cells. The measured ion intensities were converted to partial pressures by normalising the ion intensity of Al(g) in equilibrium with Al(l) to the known total vapour pressure. The derived thermodynamic quantities were recalculated by the review using the selected thermodynamic properties of $Se_2(g)$, the CODATA [89COX/WAG] properties of Al(g), and the entropies and heat capacity expressions of the aluminium selenides given in Sections V.8.2.1.1 to V.8.2.1.3. The results are summarised in Table A-59.

Table A-59: Second and third law enthalpies of reactions involving the species AlSe(g), $Al_2Se(g)$, and $Al_2Se_2(g)$.

Reaction	$\Delta_r H^\circ_m$ (298.15 K) 2nd law, (kJ·mol^{-1})	$\Delta_r H^\circ_m$ (298.15 K) 3rd law, (kJ·mol^{-1})	$\Delta_r S^\circ_m$ (298.15 K) 2nd law, (J·K^{-1}·mol^{-1})
$Al(g) + \frac{1}{2}Se_2(g) \rightleftharpoons AlSe(g)$	– (220 ± 13)	– (138 ± 9)	– (115 ± 7)
$2\,Al(g) + \frac{1}{2}Se_2(g) \rightleftharpoons Al_2Se(g)$	– (507 ± 30)	– (433 ± 25)	– (220 ± 13)
$2\,Al(g) + Se_2(g) \rightleftharpoons Al_2Se_2(g)$	– (768 ± 45)	– (978 ± 60)	(58 ± 19)

[68KLU/SEL]

Crystalline K_2SeO_3 was prepared and chemical analysis of the very hygroscopic substance agreed reasonably well with the expected composition for K_2SeO_3. A specimen of the preparation was reacted with a lead nitrate solution in an electrically calibrated calorimeter. Crystalline $PbSeO_3$ was formed. The calorimetric data have been used to estimate the enthalpy change of the reaction $K_2SeO_3(cr) + Pb^{2+} \rightarrow PbSeO_3(cr) + 2K^+$ in Table A-60.

Table A-60: Calculation of the enthalpy change of the reaction $K_2SeO_3(cr) + Pb^{2+} \rightarrow PbSeO_3(cr) + 2K^+$. The sources of the data are indicated.

#	Reaction	$\Delta_r H_m$ (kJ·mol^{-1})	Reference
1	$K_2SeO_3(cr) + Pb(NO_3)_2(aq, 1{:}800) \rightarrow PbSeO_3(cr) + 2KNO_3(aq, 1{:}400)$	– (54.94 ± 0.04)	(paper)
2	$Pb(NO_3)_2(aq, 1{:}800) + aq \rightarrow Pb^{2+} + 2\,NO_3^-$	(1.85 ± 1.50)	(review)
3	$KNO_3(aq, 1{:}400) + aq \rightarrow K^+ + NO_3^-$	(0.30 ± 0.05)	[82WAG/EVA]
4	$K_2SeO_3(cr) + Pb^{2+} \rightarrow PbSeO_3(cr) + 2K^+$	– (56.19 ± 1.50)	(result)

$\Delta_r H_4 = \Delta_r H_1 - \Delta_r H_2 + 2\,\Delta_r H_3$

[68MAI/SUP]

The enthalpies of formation of $Sc_2(SeO_3)_3$(am) and $Y_2(SeO_3)_3$(cr) have been determined from calorimetric experiments and auxiliary data. The primary data are available in four tables containing the amount of substance used, the voltage, current and time used in the electrical calibration of the calorimeter, and the observed temperature changes during reaction and calibration. For Sc^{3+} the review could not obtain an agreement between the reported enthalpy changes of the reactions and the primary data in the tables. For the reactions involving yttrium agreement was found.

The scandium selenite formed in the reaction between an aqueous solution of $ScCl_3$ and Na_2SeO_3(cr) was amorphous and contained 10 molecules of water. The enthalpy of formation of $Sc_2(SeO_3)_3\cdot10H_2O$(am) is evaluated in Table A-61 both with the reported enthalpy changes and those calculated by the review from the primary data.

The largest discrepancy between the primary data and the reported enthalpy change occurs for Reaction #4. The review has exchanged the result for this reaction with auxiliary data by Leonidov and Furkaluk [99LEO/FUR] in Table A-62.

Crystalline yttrium selenite was prepared and its enthalpy of dissolution in the solvent HCl(aq, 1:18.5), denoted sln, as well as the enthalpy of dissolution of selenious acid in the same solvent were determined. The enthalpy of dissolution in the medium of yttrium chloride, which is needed as an auxiliary datum, was not measured. The review has therefore estimated the enthalpy change of Reaction #3 in Table A-63 from dissolution data of YCl_3(cr) in HCl(aq) in [88XIA/ZHU] and [97OPP/EHR]. The evaluation also uses the mean value of the enthalpy of formation of YCl_3(cr) in these references.

The result presented in the paper is $\Delta_f H_m^\circ$ ($Y_2(SeO_3)_3$, cr, 298.15 K) = – 3235 kJ·mol^{-1} but the evaluation contains a number of errors. The result was later changed to – 2808 kJ·mol^{-1} in a footnote in the VINITI document No. 2968-71 without comments. Reference [82WAG/EVA] reports – 3020.0 kJ·mol^{-1} of unknown origin. This value appears to be too negative in view of the results for some of the lanthanide selenites.

Table A-61: Calculation of the enthalpy of formation of $Sc_2(SeO_3)_3\cdot 10H_2O(am)$. The sources of the data are indicated.

#	Reaction	$\Delta_r H_m$ (kJ·mol^{-1})	Reference
1	$2ScCl_3(aq, 1:2500) + 3Na_2SeO_3(cr) + 10H_2O(l) \rightarrow$	– (65.19 ± 0.46)	(paper)
	$Sc_2(SeO_3)_3\cdot 10H_2O(am) + 6NaCl(aq, 1:850)$	– (71.55 ± 0.46)	(review)
2	$2Na(cr) + Se(cr) + 1.5O_2(g) \rightarrow Na_2SeO_3(cr)$	– (959.62 ± 0.95)	(review)
3	$Sc + 6H_2(g) + 3O_2(g) + 1.5Cl_2(g) \rightarrow ScCl_3\cdot 6H_2O(cr)$	– (2789.10 ± 2.50)	[82WAG/EVA]
4	$ScCl_3\cdot 6H_2O(cr) + aq \rightarrow ScCl_3(aq, 1:2500) + 6H_2O(l)$	– (29.54 ± 0.10)	(paper)
		– (50.63 ± 0.15)	(review)
5	$Na(cr) + 0.5Cl_2(g) + aq \rightarrow NaCl(aq, 1:850)$	– (407.10 ± 0.15)	[89COX/WAG]
			[82WAG/EVA]
6	$H_2(g) + 0.5O_2(g) \rightarrow H_2O(l)$	– (285.83 ± 0.04)	[89COX/WAG]
7	$2Sc(cr) + 3Se(cr) + 10H_2(g) + 9.5O_2(g) \rightarrow$	– (5567.10 ± 5.90)	(paper)
	$Sc_2(SeO_3)_3\cdot 10H_2O(am)$	– (5615.60 ± 5.90)	(result)

$$\Delta_r H_7 = \Delta_r H_1 + 3\,\Delta_r H_2 + 2\,\Delta_r H_3 + 2\,\Delta_r H_4 - 6\,\Delta_r H_5 - 2\,\Delta_r H_6$$

Table A-62: Calculation of the enthalpy of formation of $Sc_2(SeO_3)_3\cdot 10H_2O(am)$ with additional auxiliary data. The sources of the data are indicated.

#	Reaction	$\Delta_r H_m$ (kJ·mol^{-1})	Reference
1	$2ScCl_3(aq, 1:2500) + 3Na_2SeO_3(cr) + 10H_2O(l) \rightarrow$	– (68.37 ± 4.00)	(paper, review)
	$Sc_2(SeO_3)_3\cdot 10H_2O(am) + 6NaCl(aq, 1:850)$		
2	$2Na(cr) + Se(cr) + 1.5O_2(g) \rightarrow Na_2SeO_3(cr)$	– (959.62 ± 0.95)	(review)
3	$Sc + 1.5Cl_2(g) \rightarrow ScCl_3(cr)$	– (944.8 ± 0.6)	[99LEO/FUR]
4	$ScCl_3(cr) + aq \rightarrow ScCl_3(aq, 1:2500)$	– (197.0 ± 3.0)	[99LEO/FUR]
5	$Na(cr) + 0.5Cl_2(g) + aq \rightarrow NaCl(aq, 1:850)$	– (407.10 ± 0.15	[89COX/WAG]
			[82WAG/EVA]
6	$H_2(g) + 0.5O_2(g) \rightarrow H_2O(l)$	– (285.83 ± 0.04)	[89COX/WAG]
7	$2Sc(cr) + 3Se(cr) + 10H_2(g) + 9.5O_2(g) \rightarrow$	– (5646.5 ± 7.9)	(result)
	$Sc_2(SeO_3)_3\cdot 10H_2O(am)$		

$$\Delta_r H_7 = \Delta_r H_1 + 3\,\Delta_r H_2 + 2\,\Delta_r H_3 + 2\,\Delta_r H_4 - 6\,\Delta_r H_5 + 10\,\Delta_r H_6$$

Table A-63: Calculation of the standard enthalpy of formation of $Y_2(SeO_3)_3$(cr). The sources of the data are indicated. The concentrations employed are small and denoted as dilute (dil) only.

#	Reaction	$\Delta_r H_m$ (kJ·mol^{-1})	Reference
1	$Y_2(SeO_3)_3$(cr) + 6HCl(sln) + sln → 2YCl_3(sln, dil) + 3H_2SeO_3(sln, dil)	– (71.13 ± 0.84)	(paper)
2	H_2SeO_3(cr) + sln → H_2SeO_3(sln, dil)	(18.62 ± 0.13)	(paper)
3	YCl_3(cr) + sln → YCl_3(sln, dil)	– (197.0 ± 3.0)	(see text)
4	Se(cr) + H_2(g) + 1.5O_2(g) → H_2SeO_3(cr)	– (524.72 ± 0.65)	(review)
5	Y(cr) + 1.5Cl_2(g) → YCl_3(cr)	– (1012.1 ± 9.0)	(see text)
6	0.5H_2(g) + 0.5Cl_2(g) + sln → HCl(sln)	– (160.90 ± 0.05)	[2000RAN/FUG]
7	2Y(cr) + 3Se(cr) + 4.5O_2(g) → $Y_2(SeO_3)_3$(cr)	– (2900.0 ± 19.1)	(result)

$\Delta_r H_7 = -\Delta_r H_1 + 3\,\Delta_r H_2 + 2\,\Delta_r H_3 + 3\,\Delta_r H_4 + 2\,\Delta_r H_5 - 6\,\Delta_r H_6$

[68RAT/TER]

The Gibbs energy of formation of α-HgSe was measured in the temperature range 320 to 420 K using a galvanic cell employing the reaction Hg(l) + Se(cr) ⇌ α-HgSe. The enthalpy and entropy of formation at 298.15 K were evaluated by linear regression of the emf on temperature. The entropy of α-HgSe at 298.15 K was calculated by the review from the entropy of formation of α-HgSe, the selected entropy of selenium and the entropy of Hg(l) at 298.15 K in [89COX/WAG] to be S°_m(HgSe, α, 298.15 K) = (93.7 ± 9.0) J·K^{-1}·mol^{-1}. The value of the enthalpy of formation $\Delta_f H^\circ_m$(HgSe, α, 298.15 K) = – (45.2 ± 9.0) kJ·mol^{-1} was retained. The error limits of the entropy and the enthalpy of formation are those of the original paper multiplied by two.

[68REZ/ISA]

The vapour pressure of α-HgSe was measured in the temperature range 573 to 823 K using a gas flow method. No thermodynamic quantities were evaluated in the paper and this was therefore performed by the review.

The result of the measurements was given in the form of an apparent total vapour pressure corresponding to the total vapour pressure which would have been obtained if no dissociation occurred in the gas phase. The temperature dependence of the apparent pressure was $\log_{10}(p_{app}/\text{bar})$ = (4.444 ± 0.150) – 4681 T^{-1}. The evaluation involved the matching of observed and calculated apparent pressures by varying the total pressure of the mercury and selenium species and assuming stoichiometric composition of the gas phase. All species in the series Se(g)-Se_8(g) and Hg(g) were considered. The selected thermodynamic data of the selenium species and the CODATA [89COX/WAG] values of mercury were employed. The mole fractions of the most important selenium species are given in Table A-64.

Table A-64: Mole fractions of the most important gaseous selenium species in equilibrium with α-HgSe.

T (K)	Se_2(g)	Se_5(g)	Se_6(g)
573	0.32	0.36	0.26
823	0.88	0.09	0.02

The expression $\Delta_r G^\circ_m$ (T)= (143.8 – 0.1305 T) kJ·mol^{-1} for the reaction α-HgSe ⇌ Hg(g) + ½Se_2(g) was derived from the calculated partial pressures. No error limits were assigned because of the limited information about the experiments. A recalculation to 298.15 K using the selected heat capacity of α-HgSe yields $\Delta_f H^\circ_m$ (HgSe, α, 298.15 K) = – 17.9 kJ·mol^{-1} and S°_m (HgSe, α, 298.15 K) = 155.5 J·K^{-1}·mol^{-1}. The corresponding third law value of the enthalpy is $\Delta_f H^\circ_m$ (HgSe, α, 298.15 K) = – 57.6 kJ·mol^{-1} and was calculated using the selected heat capacity and entropy of α-HgSe. The large discrepancy between the second and third law values indicate serious systematic errors.

[68RIP/VER]

Ripan and Vericeanu [68RIP/VER] attempted to determine the solubility product of the selenites of Zn, Cd, Cu, Ni, Co, Be, Mg, Ca, Sr, and Ba from conductivity measurements of saturated aqueous solutions. The experimental temperature was (291.2 ± 0.5) K. The evaluation of the data is based on the thorough presentation of the paper by Masson *et al.* in [86MAS/LUT].

The ionic conductivity of the selenite ion in water was determined by measuring the molar conductivities of a series of aqueous solutions of lithium selenite with concentrations below 2 × 10^{-3} M. The molar conductivity at infinite dilution was found by extrapolation to be (172.0 ± 2.9) ohm^{-1}·cm^2·mol^{-1}. With the ionic conductivity of Li^+ equal to 33.4 ohm^{-1}·cm^2·mol^{-1}, the ionic conductivity of the selenite ion becomes 105.2 ohm^{-1}·cm^2·mol^{-1}. This value was then combined with the ionic conductivity of the metal ion to find the molar conductivity of the dissolved selenite.

The selenite salts were prepared by precipitation from aqueous solution and well washed. Whether the specimen so obtained was crystalline or amorphous is not mentioned. The selenite was mixed with water and agitated intermittently for 1 h and the conductivity measured. The sampling and measurement were repeated until a constant value of the conductivity was obtained. No pH measurements were apparently made.

Masson points out that the protolysis of the selenite ion was not taken into account in the determination of the ionic conductivity of the selenite ion or in the evaluation of the solubility product. An entirely satisfactory correction cannot be found for this omission. However, as pointed out by Masson, because the calibration and sample solutions had concentrations of about the same magnitude, the errors would cancel to some extent. The review agrees with Masson and adds another source of uncertainty caused by the hydrolysis of the metal ion. In order to obtain a measure of the extent of the side reactions calculations were made on the assumption that the selenite had been dissolved in pure water free from carbon dioxide to the concentration calculated in [68RIP/VER]. Allowance was made for the protolysis of the selenite ion and the hydrolysis of the metal ion. Equilibrium constants from the review and [76BAE/MES] were used without a temperature correction. Activity coefficients were estimated by the SIT expression with $\Delta\varepsilon = 0\ \text{kg·mol}^{-1}$. The results are entered at the appropriate Sections in Chapter V.

[68ROS/SEL]

Crystalline lithium hydrogen selenite was prepared and checked by X-ray and chemical analysis. The enthalpy of the reaction between a solution of the salt and $Pb(NO_3)_2(cr)$ was determined as well as the heat of dissolution of $LiHSeO_3$. The experimental findings are furnished in Table A-65, in which the enthalpy change of the reaction $LiHSeO_3(cr) + Pb^{2+} \rightarrow PbSeO_3(cr) + Li^+ + H^+$ is calculated.

Table A-65: Calculation of the enthalpy change of the reaction $LiHSeO_3(cr) + Pb^{2+} \rightarrow PbSeO_3(cr) + Li^+ + H^+$. The sources of the data are indicated.

#	Reaction	$\Delta_r H_m$ (kJ·mol^{-1})	Reference
1	$LiHSeO_3(aq, 1{:}800) + Pb(NO_3)_2(cr) \rightarrow PbSeO_3(cr) + LiNO_3(aq, 1{:}800) + HNO_3(aq, 1{:}800)$	(22.58 ± 0.01)	(paper)
2	$Pb(NO_3)_2(cr) + aq \rightarrow Pb^{2+} + 2\ NO_3^-$	(33.80 ± 0.25)	[95PAY]
3	$LiNO_3(aq, 1{:}\ 800) + aq \rightarrow Li^+ + NO_3^-$	– (0.35 ± 0.05)	[82WAG/EVA]
4	$HNO_3(aq, 1{:}800) + aq \rightarrow H^+ + NO_3^-$	– (0.36 ± 0.05)	[82WAG/EVA]
5	$LiHSeO_3(cr) + aq \rightarrow LiHSeO_3(aq, 1{:}800)$	(18.31 ± 0.01)	(paper)
6	$LiHSeO_3(cr) + Pb^{2+} \rightarrow PbSeO_3(cr) + Li^+ + H^+$	(6.38 ± 0.25)	(result)

$\Delta_r H_6 = \Delta_r H_1 - \Delta_r H_2 + \Delta_r H_3 + \Delta_r H_4 + \Delta_r H_5$

[68ROS/SEL2]

Crystalline lithium diselenite was obtained by heating $LiHSeO_3(cr)$ to constant weight at 373 K. Chemical analysis and X-ray diffraction confirmed the crystalline state and composition of the salt. A specimen of the compound was reacted in an electrically calibrated calorimeter with a solution of $Pb(NO_3)_2$. Crystalline $PbSeO_3$ was formed. The measured enthalpy change was used to calculate the enthalpy change of the reaction $Li_2Se_2O_5(cr) + 2Pb^{2+} + H_2O(l) \rightarrow 2PbSeO_3(cr) + 2Li^+ + 2H^+$ in Table A-66.

Table A-66: Calculation of the enthalpy change of the reaction $Li_2Se_2O_5(cr) + 2Pb^{2+} + H_2O(l) \rightarrow 2PbSeO_3(cr) + 2Li^+ + 2H^+$. The sources of the data are indicated.

#	Reaction	$\Delta_r H_m$ (kJ·mol^{-1})	Reference
1	$Li_2Se_2O_5(cr) + 2Pb(NO_3)_2(aq, 1{:}400) + H_2O(l) \rightarrow 2PbSeO_3(cr) + 2LiNO_3(aq, 1{:}400) + 2HNO_3(aq, 1{:}400)$	– (19.54 ± 0.04)	(paper)
2	$Pb(NO_3)_2(aq, 1{:}400) + aq \rightarrow Pb^{2+} + 2\ NO_3^-$	(2.50 ± 1.50)	(review)
3	$LiNO_3(aq, 1{:}400) + aq \rightarrow Li^+ + NO_3^-$	– (0.44 ± 0.05)	[82WAG/EVA]
4	$HNO_3(aq, 1{:}400) + aq \rightarrow H^+ + NO_3^-$	– (0.41 ± 0.05)	[82WAG/EVA]
5	$Li_2Se_2O_5(cr) + 2Pb^{2+} + H_2O(l) \rightarrow 2PbSeO_3(cr) + 2Li^+ + 2H^+$	– (26.24 ± 1.50)	(result)

$\Delta_r H_5 = \Delta_r H_1 - 2\,\Delta_r H_2 + 2\,\Delta_r H_3 + 2\,\Delta_r H_4$

[68SEL/ROS]

Sodium hydrogen selenite was prepared and the specimen checked by X-ray and chemical analysis. The enthalpy of reaction between solid $NaHSeO_3$ and a solution of lead nitrate with formation of crystalline lead selenite was measured in an electrically calibrated calorimeter. The result is used in Table A-67 to calculate the enthalpy of the reaction $NaHSeO_3(cr) + Pb^{2+} \rightarrow PbSeO_3(cr) + H^+ + Na^+$.

Table A-67: Calculation of the enthalpy change of the reaction $NaHSeO_3(cr) + Pb^{2+} \rightarrow PbSeO_3(cr) + H^+ + Na^+$. The sources of the data are indicated.

#	Reaction	$\Delta_r H_m$ (kJ·mol^{-1})	Reference
1	$NaHSeO_3(cr) + Pb(NO_3)_2(aq, 1{:}800) \rightarrow PbSeO_3(cr) + NaNO_3(aq, 1{:}800) + HNO_3(aq, 1{:}800)$	– (9.10 ± 0.01)	(paper)
2	$Pb(NO_3)_2(aq, 1{:}800) + aq \rightarrow Pb^{2+} + 2\ NO_3^-$	(1.85 ± 1.50)	(review)
3	$NaNO_3(aq, 1{:}800) + aq \rightarrow Na^+ + NO_3^-$	– (0.17 ± 0.05)	[82WAG/EVA]
4	$HNO_3(aq, 1{:}800) + aq \rightarrow H^+ + NO_3^-$	– (0.36 ± 0.05)	[82WAG/EVA]
5	$NaHSeO_3(cr) + Pb^{2+} \rightarrow PbSeO_3(cr) + H^+ + Na^+$	– (11.48 ± 1.50)	(result)

$\Delta_r H_5 = \Delta_r H_1 - \Delta_r H_2 + \Delta_r H_3 + \Delta_r H_4$

[68TAN/GOR]

Tananaev *et al.* prepared a crystalline hydrate of gallium selenate by evaporating a solution of gallium selenate until a solid began to form. The crystals formed upon cooling were washed with 70% ethanol and left to dry in air. The salt was analysed for selenium and gallium. The water of crystallisation was obtained by difference on the basis of the result of the chemical analyses. The findings were, in mass-% with the expected result for 18 molecules of water within parentheses, Ga 15.40 (15.62); Se 26.62 (26.53); H_2O 36.42 (36.33). The agreement with the formulae $Ga_2(SeO_4)_3{\cdot}18H_2O$ is thus good. No discussion is made of the stability of the hydrate. The review assumes that the specimen used in the calorimetric measurements had the assigned composition and hence that no loss of water had occurred.

The authors mixed $Ga_2(SeO_4)_3{\cdot}18H_2O(cr)$ with a solution of barium chloride in a calorimeter and measured the enthalpy of the reaction: $Ga_2(SeO_4)_3{\cdot}18H_2O(cr)$ + $3BaCl_2$(aq, 1:1600) → $3BaSeO_4(cr)$ + $2GaCl_3$(aq, 1:2400) + $18H_2O(l)$

In their evaluation of the result, the authors did not consider the hydrolysis of Ga^{3+}, which is not negligible. Although the details of the hydrolysis of the ion are not well understood, the equilibrium constants in [76BAE/MES] can be used to estimate with fair reliability that about 20% of the Ga(III) will be converted to hydroxo complexes under the conditions of the calorimetric experiment. The reaction mixture should then be about 4.3×10^{-3} M in H^+, but the effect of the low pH on the solubility of $BaSeO_4(cr)$ can be ignored. From the enthalpy data in the same reference, the enthalpy change per mole of OH^- bound is estimated to be (38.0 ± 8.0) kJ. The evaluation of the data by the review is shown in Table A-68. The enthalpy of dilution of $GaCl_3$(aq, 1:2400) is assumed here to be the same as for $AlCl_3$(aq, 1:2400). It was neglected in the paper.

Table A-68: Calculation of the enthalpy change of the reaction $Ga_2(SeO_4)_3{\cdot}18H_2O(cr)$ + $3Ba^{2+}$ → $3BaSeO_4(cr)$ + $2Ga^{3+}$ + $18H_2O(l)$. The sources of the data are indicated.

#	Reaction	$\Delta_r H_m$ (kJ·mol^{-1})	Reference
1	$Ga_2(SeO_4)_3{\cdot}18H_2O(cr)$ + $3BaCl_2$(aq, 1:1600) → $3BaSeO_4(cr)$ + $2GaCl_3$(aq, 1:2400, hydrolysed) + $18H_2O(l)$	– (60.04 ± 0.33)	(paper)
2	$GaCl_3$(aq, 1:2400, hydrolysed) → $GaCl_3$(aq, 1:2400)	– (7.00 ± 1.50)	(review)
3	$GaCl_3$(aq, 1:2400) + aq → Ga^{3+} + $3Cl^-$	(2.50 ± 0.50)	(review)
4	$BaCl_2$(aq, 1:1600) + aq → Ba^{2+} + $2Cl^-$	– (1.38 ± 0.10)	[82WAG/EVA]
5	$Ga_2(SeO_4)_3{\cdot}18H_2O(cr)$ + $3Ba^{2+}$ → $3BaSeO_4(cr)$ + $2Ga^{3+}$ + $18H_2O(l)$	– (64.90 ± 3.19)	(result)

$$\Delta_r H_5 = \Delta_r H_1 + 2\,\Delta_r H_2 + 2\,\Delta_r H_3 - 3\,\Delta_r H_4$$

The standard enthalpy of formation of $Ga_2(SeO_4)_3{\cdot}18H_2O(cr)$ is found from $\Delta_r H_5$ and accepted enthalpies of formation to be – (7331.1 ± 17.8) kJ·mol^{-1}. The result in the paper is – 7401.0 kJ·mol^{-1}. An analysis reveals that the main cause of the discrepancy between the two results originates from the selection of the standard heat of formation of $BaSeO_4(cr)$. The paper uses the original value in [59SEL/KAP], – 1168.2 kJ·mol^{-1}, which differs from the accepted value by 24.2 kJ·mol^{-1}, see Section V.14.5.3.

The enthalpies of dissolution of $Ga_2(SeO_4)_3{\cdot}18H_2O(cr)$ and $Ga_2(SeO_4)_3(cr)$ in 0.2 M KOH to aq 1:7200 were also measured and resulted in an enthalpy of hydration of – (277.80 ± 1.48) $kJ \cdot mol^{-1}$. From this datum the standard enthalpy of formation of the anhydrous salt becomes – (1908.4 ± 17.9) $kJ \cdot mol^{-1}$.

[68UY/MUE]

The vapour pressure of Ga_2Se_3(cr) was measured in the temperature range 931 to 1094 K using mass spectrometry and Knudsen effusion cells. It was found that the vaporisation reaction is:

$$Ga_2Se_3(cr) \rightleftharpoons Ga_2Se(g) + Se_2(g) \qquad (A.18)$$

with an average second law enthalpy of reaction of $\Delta_r H^\circ_m$((A.18), 1010 K) = (627.6 ± 17.0) $kJ \cdot mol^{-1}$. No primary data or information on the temperature dependence of the total pressure was reported. The enthalpy of reaction was recalculated to 298.15 K by the review using the selected data of selenium and the heat capacities of Ga_2Se(g) and Ga_2Se_3(cr) in Sections V.8.3.1.2 and V.8.3.1.3, yielding $\Delta_r H^\circ_m$((A.18), 298.15 K) = (649.6 ± 20.0) $kJ \cdot mol^{-1}$.

[68WEL/SAL]

Wells and Salam have made extensive studies of the kinetics of the reaction between Fe^{2+} and hydrogen peroxide in the presence of various anions. It was observed that the oxidation rate was unaffected by perchlorate and nitrate ions, whereas a large number of other anions including selenate changed the rate. These anions were all potential ligands and the following reaction scheme, somewhat simplified here, was proposed to account for the findings. It is written for SeO_4^{2-} but the same scheme applies to other anions as well:

$$Fe^{2+} + SeO_4^{2-} \xrightleftharpoons{\beta_1} FeSeO_4(aq),$$

$$Fe^{2+} + H_2O_2 \xrightarrow{k_1} Fe^{3+} + OH^- + OH\cdot,$$

$$FeSeO_4(aq) + H_2O_2 \xrightarrow{k_2} FeSeO_4^+ + OH^- + OH\cdot,$$

$$Fe(II) + OH\cdot \xrightarrow{rapid} Fe(III) + OH^-.$$

The rate of formation of Fe(III) measured by spectrophotometry followed:

$$\frac{d[Fe(III)]}{dt} = 2k_o[Fe(II)][H_2O_2]$$

which yields with the proposed mechanism:

$$k_o = \frac{k_1 + k_2\beta_1\,[SeO_4^{2-}]}{1 + \beta_1\,[SeO_4^{2-}]}.$$

The value of β_1 determined in a medium containing 4×10^{-3} M $HClO_4$ and I = 1 M adjusted with $NaClO_4$ was 12 M^{-1} at 298.15 K.

No primary data are provided, which does not permit an assessment of the result. Such data are, however, provided in [64WEL/SAL] for the sulphate ion, which apparently behaves very similar to the selenate ion with $\beta_1 = 10\ M^{-1}$. The data are reproduced in Table A-69 and Figure A-1. The value of k_1 was determined to be 49.5 $mol^{-1}\cdot L\cdot s^{-1}$ from the value of k_o at $[SO_4^{2-}] = 0$ and k_2 was assigned to the limiting value of k_o at high sulphate concentrations, 74 $mol^{-1}\cdot L\cdot s^{-1}$.

Table A-69: Primary data for the reaction between Fe^{2+} and SO_4^{2-}. k_o in $mol^{-1}\cdot L\cdot s^{-1}$.

$[SO_4^{2-}]$ / M	k_o	$[SO_4^{2-}]$ / M	k_o
0	49.5	0.122	68.2
0.024	53.0	0.147	73.0
0.0482	57.9	0.197	74.4
0.0733	60.0	0.299	72.5
0.0975	65.7	0.347	73.0

Figure A-1: Comparison between the experimental value of the rate constant taken from Table A-69 (+) and the rate constant calculated from the parameters given in [64WEL/SAL] (line) at different sulphate concentrations.

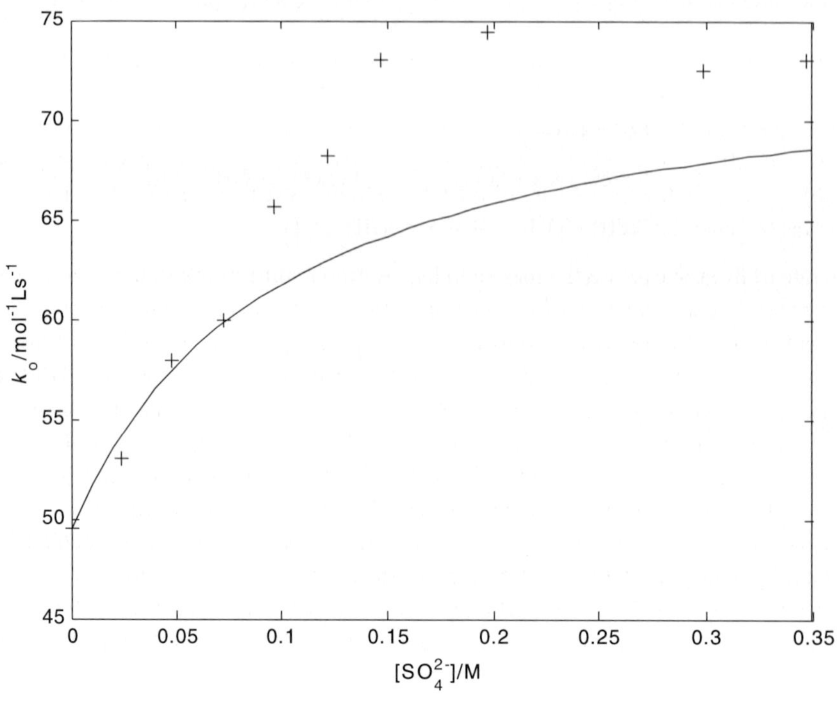

The value of the stability constant was calculated by iteration (i is the number of the iteration) by plotting $y_i = k_o(1 + \beta_{1(i-1)}\,[SO_4^{2-}])$ against $x = [SO_4^{2-}]$. Starting from $\beta_{1(0)} = 0$, $\beta_{1(i)}$ was found from the slope of $y_i = f(x)$ and k_2. In this way the authors found $\beta_1 = 10.2\ M^{-1}$. However, with this value of the equilibrium constant $\beta_1\,[SO_4^{2-}]$ is too small for k_o to reach its limiting value. This is demonstrated in Figure A-1, which compares the k_o values calculated with the parameters presented and the experimental results. Hence k_2 must be in error.

The authors state that the iteration method used converges to an accurate value of β_1. In fact, the iteration method does not converge and cannot be used to find β_1. Other methods must be used such as non-linear least squares optimisation, but a good fit can hardly be expected between the model and the experimental data over the whole range of sulphate concentrations.

The data for selenate have been evaluated in the same way as described for sulphate with similar numerical results. With no primary data available and in view of the objections to the method used in the sulphate calculations, the reported value $\beta_1 = 12\ M^{-1}$ at 298.15 K and $I = 1$ will not be accepted.

[68WIE/GOY]

The total pressure of the saturated vapour in equilibrium with solid MnSe was measured in the temperature range 1414 to 1610 K using the Knudsen effusion method. The vaporisation process can be described by the reaction

$$\alpha\text{-MnSe} \rightarrow \text{Mn(g)} + x\text{Se(g)} + 0.5(1-x)\text{Se}_2\text{(g)} \qquad \text{(A.19)}$$

leading to congruent effusion, which was verified by X-ray powder diffractograms of the residues after vaporisation. Se(g) and Se_2(g) were both considered in the evaluation of the thermodynamic properties of α-MnSe by employing auxiliary data for the Mn(g), Se(g), and Se_2(g) molecules. The thermodynamic properties of the species Se(g) and Se_2(g) selected by the review are different from those used in the paper and the evaluation was therefore remade.

The vapour pressure expression $\log_{10}(p/\text{bar}) = 9.5579 - 26420\ T^{-1}$ was calculated from a least-squares fit of the data in Table I of the paper. The total vapour pressure calculated from the reactions α-MnSe ⇌ Mn(g) + Se(g) and α-MnSe ⇌ Mn(g) + ½Se_2(g) was subsequently matched to the pressure expression at 1414 K and 1610 K by adjusting the Gibbs energy of formation of α-MnSe while maintaining a correct Se(g)/Se_2(g) equilibrium. The new values of x in Reaction (A.19), $x = 0.43$ at 1414 K and $x = 0.44$ at 1610 K, agree well with the values of about $x = 0.45$ obtained in the paper for the whole temperature range. Second law entropy and enthalpy values were evaluated from the Gibbs energy values at 1414 K and 1610 K and recalculated to 298.15 K yielding S°_m (MnSe, α, 298.15 K) = (94.3 ± 9.0) $J\cdot K^{-1}\cdot mol^{-1}$ and $\Delta_f H^\circ_m$ (MnSe, α, 298.15 K) = – (169.5 ± 10.0) $kJ\cdot mol^{-1}$. The selected heat capacity of α-MnSe, the selected values of the selenium species and the data of Mn(g) in [82WAG/EVA] were employed in the calculations. The corresponding third law enthalpy of formation was evaluated to be $\Delta_f H^\circ_m$ (MnSe, α, 298.15 K) = – (174.7 ± 8.0) kJ·mol^{-1} from the Gibbs energy at the mean temperature of 1512 K.

[68YAM/POR]

The composition of the selenium vapour evaporating from liquid selenium was studied in the temperature range 578 to 667 K using a mass spectrometer and Knudsen effusion cells. The ions Se_3^+ (g) and Se_4^+ (g) were found to be formed as a result of the fragmentation of Se_5(g) and Se_6(g) molecules and were excluded from the thermodynamic evaluation. It was also found that at lower temperatures, the Se_2^+ (g) ion intensities contained contributions from the fragmentation of Se_5(g) (< 14%) and Se_6(g) (< 29%) molecules. The evaluation of the data in the paper did not correct for this and the result for the Se_2(g) molecules is more uncertain than for the other molecules as shown by the large discrepancy between the second and third law values in the Table A-70.

In the present review, the enthalpies of formation at 625 K in Table III of the paper were recalculated to 298.15 K using the selected heat capacities of Se(cr, l), Se_2(g), and Se_5(g)-Se_8(g) to provide for a comparison with other investigations. Similarly, third law enthalpies of formation and second law entropies were evaluated by combining the data of Table III in the paper and the total pressure measurements by the Knudsen torsion-effusion method in the same paper with the selected values for the heat capacities and entropies. The results are summarised in the Table A-70.

Table A-70: Entropies and enthalpies of formation of gaseous selenium species.

Species	$\Delta_f H_m$ (625 K) (kJ·mol^{-1})	$\Delta_f H^\circ_m$ (298.15 K) (kJ·mol^{-1})		S°_m (298.15 K) (J·K^{-1}·mol^{-1})
	original values	2nd law	3rd law	
Se_2(g)	(95.4 ± 8.9)	(113.1 ± 6.3)	(131.5 ± 6.3)	(216.5 ± 10.5)
Se_5(g)	(96.7 ± 11.8)	(140.2 ± 8.4)	(142.7 ± 8.4)	(394.2 ± 13.9)
Se_6(g)	(83.7 ± 11.8)	(134.8 ± 8.4)	(134.4 ± 8.4)	(429.1 ± 13.9)
Se_7(g)	(84.5 ± 11.8)	(143.3 ± 8.4)	(149.1 ± 8.4)	(477.4 ± 13.9)
Se_8(g)	(90.4 ± 11.8)	(156.8 ± 8.4)	(155.6 ± 8.4)	(528.2 ± 13.9)

[69BER/STR]

The vapour pressure of Ga_2Se_3(cr) was measured in the temperature range 970 to 1180 K using mass spectrometry and Knudsen cells. It was found that the vaporisation reaction is

$$Ga_2Se_3(cr) \rightleftharpoons Ga_2Se(g) + Se_2(g). \quad (A.20)$$

Both the $Ga_2Se(g)$ and $Se_2(g)$ partial pressures were recorded resulting in the expression $\log_{10} K_p = 19.945 - 34055\ T^{-1}$. The enthalpy and entropy of reaction were evaluated by the review using the second and the third law and the selected data of selenium and the heat capacities of $Ga_2Se(g)$ and $Ga_2Se_3(cr)$ in Sections V.8.3.1.2 and V.8.3.1.3. The changes in the thermodynamic quantities of Reaction (A.20) at 298.15 K were $\Delta_r H^\circ_m$((A.20), 298.15 K) = (678.6 ± 42.0) kJ·mol^{-1} and $\Delta_r S^\circ_m$((A.20), 298.15 K) = (421.2 ± 40.0) J·K^{-1}·mol^{-1} from the second law and $\Delta_r H^\circ_m$((A.20), 298.15 K) = (636.9 ± 30.0) kJ·mol^{-1} from the third law.

The vapour pressure of $In_2Se_3(cr)$ was measured in the temperature range 900 to 1120 K using mass spectrometry and Knudsen cells. It was found that the vaporisation reaction is

$$In_2Se_3(cr) \rightleftharpoons In_2Se(g) + Se_2(g). \qquad (A.21)$$

Both the $In_2Se(g)$ and $Se_2(g)$ partial pressures were recorded resulting in the expression $\log_{10} K_p = 5.816 - 19069\ T^{-1}$. The enthalpy and entropy of reaction were evaluated by the review using the second and the third law , the selected properties of $Se_2(g)$, the heat capacity and entropy of $In_2Se(g)$ calculated by the review from the molecular parameters obtained by Erkoc, Katircioglu, and Yilmaz [2001ERK/KAT] in quantum mechanical studies of In-Se clusters, and the heat capacity, transition enthalpies, and entropy of solid In_2Se_3 given in Section V.8.4.1.5. The changes in the thermodynamic quantities of Reaction (A.21) at 298.15 K were $\Delta_r H^\circ_m$((A.21), 298.15 K) = 400.2 kJ·mol^{-1} and $\Delta_r S^\circ_m$((A.21), 298.15 K) = 168.4 J·K^{-1}·mol^{-1} from the second law and $\Delta_r H^\circ_m$((A.21), 298.15 K) = 598.4 kJ·mol^{-1} from the third law. The second and third law values differ wildly indicating serious systematic errors and the anticipated congruent vaporisation reaction is probably not correct. The results are therefore rejected.

[69BOE/BEN]

The total pressure of evaporating species during the congruent vaporisation of α-ZnSe was measured at the temperatures 1423 K, 1473 K, and 1513 K using a transpiration method with Ar(g) as the carrier gas. The vaporisation reaction 2α-ZnSe $\rightleftharpoons$ 2Zn(g) + $Se_2(g)$ was assumed and the relation $\log_{10}(K_p/\text{atm}^3) = 13.510 - 31078\ T^{-1}$ was derived from the measurements using a vapour pressure of $Se_2(g)$ equal to one half that of Zn(g). The Gibbs energy of formation $\Delta_f G^\circ_m$(ZnSe, α, 298.15 K) = – 149.3 kJ·mol^{-1} was evaluated from the second law and the enthalpy of formation $\Delta_f H^\circ_m$(ZnSe, α, 298.15 K) = – 156.7 kJ·mol^{-1} was then calculated by estimating the entropy of formation of α-ZnSe. The review was not able to reproduce those values from the data and the expressions given in the paper. The vapour pressure reported in Table 2 of the paper was therefore evaluated by the review yielding the expression $\log_{10}(K_p/\text{bar}^3) = 16.784 - 29263\ T^{-1}$ for the vaporisation reaction. A recalculation to 298.15 K using the selected thermodynamic functions of selenium, the data for Zn in [89COX/WAG], the heat capacity expression of α-ZnSe in Section V.9.1.1.1, and the selected entropy of α-ZnSe in the case of the third law, yielded the values $\Delta_f H_m^\circ$(ZnSe, α, 298.15 K) = – (383 ± 51) kJ·mol^{-1}, $\Delta_f H_m^\circ$(ZnSe, α, 298.15 K) = – (193.0 ± 18.0) kJ·mol^{-1}, and S_m°(ZnSe, α, 298.15 K) = – (58 ± 43) J·K^{-1}·mol^{-1} for the second and third law enthalpy of formation and the entropy, respectively. The large discrepancy between the second and third law enthalpy of formation and the negative entropy indicate serious errors in the measurements. The results were therefore rejected.

[69BOE/BEN2]

The total pressure of evaporating species during the congruent vaporisation of α-CdSe was measured in the temperature range 1094 to 1240 K using a transpiration method with Ar(g) as the carrier gas. The vaporisation reaction 2α-CdSe $\rightleftharpoons$ 2Cd(g) + Se_2(g) was assumed and the relation $\log_{10}(K_p/\text{bar}^3)$ = 13.53 – 27928 T^{-1} was derived from the measurements using a vapour pressure of Se_2(g) equal to one half that of Cd(g). The enthalpy of formation $\Delta_f H_m^\circ$(CdSe, α, 298.15 K) = – 132.9 kJ·mol^{-1} was evaluated from the second law.

The vapour pressure measurements are of very low quality as shown by the condition that the pressures measured at 1473 K vary by a factor of 18. Furthermore, the pressure expression above does not reproduce the measured vapour pressures listed in Table 2 of the paper. The pressures in this table were used by the review in a re-evaluation which yielded completely unreasonable results. The work is therefore rejected.

[69FLO]

The equilibrium vapour pressure of Zn(g) was measured during the congruent vaporisation of α-ZnSe at the temperatures 1173 K, 1273 K, and 1413 K by a transpiration method using N_2(g) as a carrier gas. The vaporisation reaction 2α-ZnSe $\rightleftharpoons$ 2Zn(g) + Se_2(g) was assumed and the relation $\log_{10}(K_p/\text{atm}^3)$ = 22.373 – 39978 T^{-1} was derived from the measurements using a vapour pressure of Se_2(g) equal to one half that of Zn(g). No thermodynamic quantities for α-ZnSe were calculated. The vapour pressure reported in Table 1a of the paper was therefore evaluated by the review according to the second and the third law using the selected thermodynamic functions of selenium, the data for Zn in [89COX/WAG], the heat capacity expression of α-ZnSe in Section V.9.1.1.1, and the selected entropy of α-ZnSe in the case of the third law. The values obtained were $\Delta_f H_m^\circ$(ZnSe, α, 298.15 K) = – (192.5 ± 23.0) kJ·mol^{-1}, $\Delta_f H_m^\circ$(ZnSe, α, 298.15 K) = – (170.5 ± 4.9) kJ·mol^{-1}, and S_m°(ZnSe, α, 298.15 K) = (54.9 ± 12.0) J·K^{-1}·mol^{-1} for the second and third law enthalpy of formation and the entropy, respectively.

The measurement method was also applied to the α-CdSe system. The results were re-evaluated by the review in the same manner as for the α-ZnSe system yielding $\Delta_f H_m^\circ$(CdSe, α, 298.15 K) = – (143.8 ± 15.0) kJ·mol^{-1}, $\Delta_f H_m^\circ$(CdSe, α, 298.15 K) =

– (142.3 ± 3.0) kJ·mol^{-1}, and S°_{m} (CdSe, α, 298.15 K) = (85.0 ± 12.0) J·K^{-1}·mol^{-1} for the second and third law enthalpy of formation and the entropy, respectively.

[69HAN/MUN]

The evaporation of PbSe(cr) was studied in the temperature range 887 to 948 K using the torsion-Langmuir method and in the range 936 to 1106 K using the torsion-effusion technique. The original evaluation employed auxiliary data different from those used by the review. The vapour pressure expression was therefore re-evaluated using the selected values of the entropies and heat capacities of PbSe(cr) and PbSe(g). The free surface results agree less well with other studies ([59ZLO/POP], [69SOK/PAS], [93BRU/PIA]) and the new evaluation was based on the torsion-effusion measurements only. The second law entropy of PbSe(g) and the second law enthalpy of sublimation of PbSe(cr) were calculated to be S°_{m} (PbSe, g, 298.15 K) = (274.7 ± 7.0) J·K^{-1}·mol^{-1} and $\Delta_{sub}H^{\circ}_{m}$ (PbSe, cr, 298.15 K) = (242.6 ± 7.0) kJ·mol^{-1}, respectively. The third law enthalpy of sublimation was calculated to be $\Delta_{sub}H^{\circ}_{m}$ (PbSe, cr, 298.15 K) = (231.0 ± 4.0) kJ·mol^{-1}.

[69KLU/SEL]

Crystalline (checked by X-ray diffraction) Na_2SeO_3 and $Na_2SeO_3 \cdot 5H_2O$ were prepared and the compositions confirmed by chemical analysis. The integral enthalpies of dissolution of the salts to the same concentration were measured in a calorimeter calibrated electrically. The results are used in Table A-71 to calculate the enthalpy of formation of $Na_2SeO_3 \cdot 5H_2O$ from the corresponding quantity for Na_2SeO_3.

Table A-71: Evaluation of the standard enthalpy of formation of $Na_2SeO_3 \cdot 5H_2O$. The sources of the data are indicated.

#	Reaction	Δ_rH_m (kJ·mol^{-1})	Reference
1	Na_2SeO_3(cr) + aq → Na_2SeO_3(aq, 1:800)	– (28.33 ± 0.08)	(paper)
2	$Na_2SeO_3 \cdot 5H_2O$(cr) + aq → Na_2SeO_3(aq, 1:800) + $5H_2O$(l)	– (12.43 ± 0.21)	(paper)
3	H_2(g) + $0.5O_2$(g) → H_2O(l)	– (285.83 ± 0.04)	[89COX/WAG]
4	Na(cr) + Se(cr) + $1.5O_2$(g) → Na_2SeO_3(cr)	– (959.62 ± 0.95)	(review)
5	2Na(cr) + Se(cr) + $5H_2$(g) + $4O_2$(g) → $Na_2SeO_3 \cdot 5H_2O$(cr)	– (2404.67 ± 1.00)	(result)

$\Delta_rH_5 = \Delta_fH^{\circ}_{m}$ ($Na_2SeO_3 \cdot 5H_2O$, cr, 298.15 K) = $\Delta_rH_1 - \Delta_rH_2 + 5\,\Delta_rH_3 + \Delta_rH_4$

[69KLU/SEL3]

Rubidium selenite was prepared and the very hygroscopic substance was dried in a vacuum desiccator over phosphorus pentoxide. The crystalline state was controlled by X-ray diffraction and chemical analyses agreed with the expected composition for Rb_2SeO_3. A specimen of the preparation was reacted with a lead nitrate solution in an electrically calibrated calorimeter. Crystalline $PbSeO_3$ was formed. The calorimetric data have been used to estimate the enthalpy change of the reaction $Rb_2SeO_3(cr) + Pb^{2+} \rightarrow PbSeO_3(cr) + 2Rb^+$ in Table A-72.

Table A-72: Calculation of the enthalpy change of the reaction $Rb_2SeO_3(cr) + Pb^{2+} \rightarrow PbSeO_3(cr) + 2Rb^+$. The sources of the data are indicated.

#	Reaction	$\Delta_r H_m$ (kJ·mol^{-1})	Reference
1	$Rb_2SeO_3(cr) + Pb(NO_3)_2(aq, 1:800) \rightarrow PbSeO_3(cr) + 2RbNO_3(aq, 1:400)$	– (61.00 ± 0.42)	(paper)
2	$Pb(NO_3)_2(aq, 1:800) + aq \rightarrow Pb^{2+} + 2\,NO_3^-$	(1.85 ± 1.50)	(review)
3	$RbNO_3(aq, 1:400) + aq \rightarrow Rb^+ + NO_3^-$	(0.39 ± 0.05)	[82WAG/EVA]
4	$Rb_2SeO_3(cr) + Pb^{2+} \rightarrow PbSeO_3(cr) + 2Rb^+$	– (62.07 ± 1.56)	(result)

$\Delta_r H_4 = \Delta_r H_1 - \Delta_r H_2 + 2\,\Delta_r H_3$

[69LAF/CER]

The system cobalt-selenium was studied in the temperature range 673 to 873 K using the solid state electrochemical cells

$$(-)\ Co(cr), CoCl_2(cr) \mid BaCl_2(cr) \mid CoCl_2(cr), Co_{1-x}Se(cr)\ (+)$$

$$\text{and}\ (-)\ Co(cr), CoCl_2(cr) \mid BaCl_2(cr) \mid CoCl_2(cr), CoSe_2(cr)\ (+).$$

Thermodynamic quantities were only derived at the temperatures of the measurements by the authors and not recalculated to standard conditions. The values reported for $CoSe_2(cr)$ were $\Delta_f H_m^\circ(CoSe_2, cr, 678\ K) = -179.9$ kJ·mol^{-1} and $\Delta_f S_m^\circ(CoSe_2, cr, 678\ K) = -154.4$ J·K^{-1}·mol^{-1}. The value of the entropy change is much too negative and the enthalpy of formation is about twice as negative as in other investigations. Mills [74MIL] reassessed the measurements and calculated the enthalpy of formation to $\Delta_f H_m^\circ(CoSe_2, cr, 298.15\ K) = -(93.7 \pm 8.4)$ kJ·mol^{-1} using the third law and an estimated value of $(G_m^\circ(CoSe_2, cr, 678\ K) - H_m^\circ(CoSe_2, cr, 298.15\ K))/678\ K = (118.4 \pm 12.6)$ J·K^{-1}·mol^{-1}. No attempt was made by the review to recalculate the value by Mills because the change would be very small, a proper heat capacity of $CoSe_2(cr)$ is unavailable above 298.15 K, and the results would still be dubious.

The values reported for the enthalpy and entropy of formation of $Co_{1-x}Se(s)$ in table 3 of the paper are also unrealistic and similar to those of $CoSe_2(cr)$. The measurements were re-evaluated by Mills [74MIL] yielding $\Delta_f H_m^\circ(Co_{0.90}Se, cr, 298.15\ K) = -58.8$ kJ·mol^{-1}.

[69MAI/SOL]

The compound $Pr_2(SeO_3)_3{\cdot}H_2SeO_3{\cdot}5H_2O$ was prepared by mixing 10% solutions of $PrCl_3$ and H_2SeO_3 (the amount of H_2SeO_3 taken exceeded the stoichiometric ratio by a factor of three) at room temperature. Crystalline $Pr_2(SeO_3)_3{\cdot}H_2SeO_3{\cdot}5H_2O$ formed on standing. The chemical analysis of Pr and Se agreed with the formula, but water was not determined. $Pr_2(SeO_3)_3{\cdot}H_2SeO_3$ was obtained by heating the hydrate to 525 K. The product was apparently crystalline and the composition determined by chemical analysis was in fair agreement with the calculated composition.

The calorimetric measurements comprised the determination of the enthalpies of dissolution of $Pr_2(SeO_3)_3{\cdot}H_2SeO_3{\cdot}5H_2O$, $Pr_2(SeO_3)_3{\cdot}H_2SeO_3$, and H_2SeO_3 in HCl(aq, 1:18.5) and denoted sln. The standard enthalpies of formation of the two compounds are evaluated in Table A-73. The origin of the value of $\Delta_f H_m^\circ$ ($PrCl_3{\cdot}7H_2O$, cr, 298.15 K) used in Table A-73 is found in this Appendix, [71MAI/SUP].

Table A-73: Calculation of the standard enthalpies of formation of $Pr_2(SeO_3)_3{\cdot}H_2SeO_3{\cdot}5H_2O$ and $Pr_2(SeO_3)_3{\cdot}H_2SeO_3$. The sources of the data are indicated.

#	Reaction	$\Delta_r H_m$ (kJ·mol^{-1})	Reference
1	$Pr_2(SeO_3)_3{\cdot}H_2SeO_3{\cdot}5H_2O$(cr) + 6HCl(sln) + sln → $2PrCl_3$(sln, aq 1:2000) + $4H_2SeO_3$(sln, aq 1:1000) + $5H_2O$(l, sln)	(9.46 ± 0.17)	(paper)
2	$PrCl_3{\cdot}7H_2O$(cr) + sln → $PrCl_3$(sln, aq 1:2000) + $7H_2O$(l, sln)	– (8.79 ± 0.88)	[71MAI/SUP]
3	Pr(cr) + $7H_2$(g) + $3.5O_2$(g) + $1.5Cl_2$(g) → $PrCl_3{\cdot}7H_2O$(cr)	– (3181.0 ± 2.0)	(see text)
4	H_2SeO_3(cr) + sln → H_2SeO_3(sln, aq 1:1000))	(21.05 ± 0.17)	(paper)
5	Se(cr) + H_2(g) + $1.5O_2$(g) → H_2SeO_3(cr)	– (524.72 ± 0.65)	(review)
6	H_2(g) + $0.5O_2$(g) + sln → H_2O(l, sln)	– (285.98 ± 0.04)	[2000RAN/FUG]
7	$0.5H_2$(g) + $0.5Cl_2$(g) + sln → HCl(sln)	– (160.90 ± 0.10)	[2000RAN/FUG]
8	2Pr(cr) + 4Se(cr) + $6H_2$(g) + $8.5O_2$(g) → $Pr_2(SeO_3)_3{\cdot}H_2SeO_3{\cdot}5H_2O$(cr)	– (4864.5 ± 5.2)	(result)
9	$Pr_2(SeO_3)_3{\cdot}H_2SeO_3$(cr) + 6HCl(sln) + sln → $2PrCl_3$(sln, aq 1:2000) + $4H_2SeO_3$(sln, aq 1:1000)	– (58.95 ± 0.54)	(paper)
10	2Pr(cr) + 4Se(cr) + H_2(g) + $6O_2$(g) → $Pr_2(SeO_3)_3{\cdot}H_2SeO_3$(cr)	– (3366.2 ± 5.2)	(result)

$\Delta_r H_8 = \Delta_f H_m^\circ$ ($Pr_2(SeO_3)_3{\cdot}H_2SeO_3{\cdot}5H_2O$, cr, 298.15 K) $= -\Delta_r H_1 + 2\,\Delta_r H_2 + 2\,\Delta_r H_3 + 4\,\Delta_r H_4 + 4\,\Delta_r H_5 - 9\,\Delta_r H_6 - 6\,\Delta_r H_7$

$\Delta_r H_{10} = \Delta_f H_m^\circ$ ($Pr_2(SeO_3)_3{\cdot}H_2SeO_3$, cr, 298.15 K) $= -\Delta_r H_9 + 2\,\Delta_r H_2 + 2\,\Delta_r H_3 + 4\,\Delta_r H_4 + 4\,\Delta_r H_5 - 14\,\Delta_r H_6 - 6\,\Delta_r H_7$

The results reported in the paper and in [82WAG/EVA] are – 4833.9 and 4880.1 kJ·mol^{-1}, respectively, for $Pr_2(SeO_3)_3{\cdot}H_2SeO_3{\cdot}5H_2O$, and – 3336.2 and – 3382.6

kJ·mol^{-1} for $Pr_2(SeO_3)_3·H_2SeO_3$. Most of the discrepancies between the three sets of results are caused by different selections of auxiliary data for praseodymium and the solvent. The present evaluation uses the experimental heat of dissolution of $PrCl_3·7H_2O$(cr) in the solvent as determined in [71MAI/SUP] and the standard enthalpy of formation of $PrCl_3·7H_2O$(cr). The other evaluations used the enthalpy of formation of $PrCl_3$(aq, 1:2000). Further partial instead of apparent molar enthalpies have been used here. The two changes cause the difference of about 25 kJ·mol^{-1} from the result in [82WAG/EVA].

[69MAM]

This thesis was not available to the review. Mills [74MIL] used the galvanic cell data reported for Ga_2Se(cr), GaSe(cr), and Ga_2Se_3(cr) in second and third law evaluations and corrected several serious errors. It can be noted that the alleged Ga_2Se(cr) phase does not exist. The values reported by the review rely on the work of Mills.

[69MEH/GUB]

This section also includes a discussion of Reference [68MEH].

The solubility of Ag_2SeO_3(cr) at 298.15 K in aqueous solution was studied as a function of pH and selenite concentration. The ionic strength of the test solutions is stated to have been kept at 1.0 M by the addition of $NaClO_4$. From an examination of the original data in [68MEH] the reviewers rather believe that the solutions contained 1 M $NaClO_4$ in addition to the sodium selenite concentration. Three series of experiments were conducted.

In the first series solutions of silver nitrate and sodium selenite were mixed to the total concentrations 5.0×10^{-3} and 48.0×10^{-3} M, respectively. The pH was adjusted by addition of $HClO_4$ or NaOH to values between 1.2 and 13.1 (53 experimental points). After equilibration for 6 days, the phases were separated. The pH and pAg were obtained from potentiometric measurements with a glass and a silver electrode. The latter electrode was calibrated on the concentration scale. The following equilibrium constants were evaluated from the data on the molar scale:

$$Ag_2SeO_3(cr) \rightleftharpoons 2Ag^+ + SeO_3^{2-} \quad (A.22)$$

$$\log_{10} K_{s,0}((A.22), 298.15\text{ K}) = -(15.58 \pm 0.12),$$

$$SeO_3^{2-} + H^+ \rightleftharpoons HSeO_3^- \quad (A.23)$$

$$\log_{10} K_1((A.23), 298.15\text{ K}) = (8.12 \pm 0.28),$$

$$HSeO_3^- + H^+ \rightleftharpoons H_2SeO_3 \quad (A.24)$$

$$\log_{10} K_{1,2}((A.24), 298.15\text{ K}) = (2.26 \pm 0.09).$$

Closely agreeing values were obtained by the review from the same data. The protonation constants obtained are consistent with those reported in Table V-5. These constants are mixed conditional constants.

In series 2 a silver nitrate solution spiked with the radioactive nuclide ^{110}Ag was mixed with a sodium selenite solution. The pH was again adjusted from about 1.2 to 12.5 in the mixtures. After equilibration the phases were separated and the pH determined. The total concentration of dissolved silver was found by scintillation spectrometry. The solubility determinations were performed at total concentrations of selenite at 6.45×10^{-3}, 0.10, and 0.20 M.

The design of the experiment is not well suited for an equilibrium analysis, but definitely showed that Ag_2SeO_3(cr) is dissolved by an excess of SeO_3^{2-}. No equilibrium constants were derived in the paper from these data. The review calculated $K_{s,0}$ ((A.22), 298.15 K) from the data at the lowest total selenite concentration and pH < 6. The protonation constants from the first series were used in the calculation. The result was $\log_{10} K_{s,0}$ ((A.22), 298.15 K) = – (15.20 ± 0.15). Hence there is a significant difference between the values of the solubility product determined by the two techniques, which cannot be accounted for and apparently not observed by the authors.

In series 3 mixtures containing a total silver concentration of 4.0×10^{-4} M and a total concentration of selenite between 0.001 and 0.90 M were prepared. The pH was adjusted to about 9.5. After equilibration and phase separation the total dissolved silver concentration was measured by scintillation spectrometry. A plot of $\log_{10} [Ag^+]$ against $\log_{10} [SeO_3^{2-}]$ has two linear branches with slopes 0.5 and 1.5 suggesting the formation of $AgSeO_3^-$ (β_1) and $Ag(SeO_3)_2^{3-}$ (β_2). The values of the formation constants calculated from the data were $\log_{10} \beta_1 = (2.42 \pm 0.12)$ and $\log_{10} \beta_2 = (3.76 \pm 0.05)$.

The β_2 was calculated from data obtained at selenite concentrations in the range 0.40 to 0.90 M. Thus the ionic strength could not have been kept constant at 1 M as stated in the paper. The value of the constant is not accepted.

The value of β_1 has been recalculated by the review using the data obtained below 0.03 M selenite where the correction for the presence of $Ag(SeO_3)_2^{3-}$ is moderate. The concentration of Ag^+ was calculated with $\log_{10} K_{s,0} = -15.58$ from the first series of experiment and – 15.20 obtained from the measurement of the total solubility in the second series. The result was $\log_{10} \beta_1 = (2.46 \pm 0.10)$ and (2.16 ± 0.10), respectively. The first value agrees with the paper in which $\log_{10} K_{s,0} = -15.58$ was used to find the silver ion concentration.

As there is no obvious reason to prefer one of the values of the solubility product the mean is chosen, which results in $\log_{10} K_{s,0}$ ((A.22), 1 M $NaClO_4$, 298.15 K) = – (15.40 ± 0.35) and $\log_{10} \beta_1$ (1 M $NaClO_4$, 298.15 K) = (2.30 ± 0.25). Note that it is not possible to derive values of the protonation constants from the solubility measurements.

[69PAS/UST]

The saturated vapour pressure of α-HgSe was measured in the temperature range 756 to 951 K using a quartz manometer. The results were combined with previous investigations made by the same authors and evaluated considering the species $Se_2(g)$, $Se_4(g)$, $Se_6(g)$, $Se_8(g)$ and Hg(g). However, the thermodynamic data used for the selenium species were different from those selected by the review and the results were therefore re-evaluated.

The total vapour pressure expression given in the paper is $\log_{10}(p/\text{bar}) = (6.735 \pm 0.050) - (6445 \pm 24)\,T^{-1}$ for the temperature range 613 to 1043 K and includes the result of previous investigations. The results of the manometer measurements are only given in a graph and it is not possible to separate them from previous results. However, it can be concluded that the expression satisfactorily describes the new measurements.

The re-evaluation involved the matching of observed and calculated pressures considering all species in the series Se(g)-$Se_8(g)$ and Hg(g). The selected thermodynamic data of the selenium species and the CODATA [89COX/WAG] values of mercury were used in the procedure. Stoichiometric composition of the gas phase was assumed. The calculated mole fractions of $Se_2(g)$ were 0.80 and 0.87 at the temperatures 756 K and 951 K, respectively. The expression $\Delta_r G^\circ_m(T) = ((186.4 \pm 8.0) - (0.1862 \pm 0.008)\,T)$ kJ·mol^{-1} for the reaction α-HgSe $\rightleftharpoons$ Hg(g) + ½$Se_2(g)$ was derived from the calculated partial pressures. The error limits correspond to those given in the vapour expression multiplied by a factor of two and increased to account for the reduced number of measurements actually stemming from the manometer measurements. Recalculation to 298.15 K using the selected heat capacity of α-HgSe yields $\Delta_f H^\circ_m$ (HgSe, α, 298.15 K) = – (63.3 ± 10.0) kJ·mol^{-1} and S°_m (HgSe, α, 298.15 K) = (96.0 ± 10.0) J·K^{-1}·mol^{-1}. The corresponding third law value of the enthalpy is $\Delta_f H^\circ_m$ (HgSe, α, 298.15 K) = – (61.1 ± 10.0) kJ·mol^{-1} and was calculated using the selected heat capacity and entropy of α-HgSe.

[69ROS/SEL2]

Crystalline sodium trihydrogen selenite was prepared and reacted in stoichiometric proportions with a solution of lead nitrate in an electrically calibrated calorimeter. The lead selenite formed was crystalline and the reaction was quantitative. The calorimetric data have been used to calculate the enthalpy change of the reaction $NaH_3(SeO_3)_2(cr) + 2Pb^{2+} \rightarrow 2PbSeO_3(cr) + Na^+ + 3H^+$ in Table A-74.

Table A-74: Calculation of the enthalpy change of the reaction $NaH_3(SeO_3)_2(cr)$ + $2Pb^{2+} \rightarrow 2PbSeO_3(cr) + Na^+ + 3H^+$. The sources of the data are indicated.

#	Reaction	$\Delta_r H_m^o$ (kJ·mol^{-1})	Reference
1	$NaH_3(SeO_3)_2(cr) + 2Pb(NO_3)_2(aq, 1{:}400) \rightarrow 2PbSeO_3(cr)$ + $NaNO_3(aq, 1{:}800) + 3HNO_3(aq, 1{:}267)$	(16.82 ± 0.02)	(paper)
2	$Pb(NO_3)_2(aq, 1{:}400) + aq \rightarrow Pb^{2+} + 2\,NO_3^-$	(2.50 ± 1.50)	(review)
3	$NaNO_3(aq, 1{:}800) + aq \rightarrow Na^+ + NO_3^-$	– (0.17 ± 0.05)	[82WAG/EVA]
4	$HNO_3(aq, 1{:}267) + aq \rightarrow H^+ + NO_3^-$	– (0.45 ± 0.05)	[82WAG/EVA]
5	$NaH_3(SeO_3)_2(cr) + 2Pb^{2+} \rightarrow 2PbSeO_3(cr) + Na^+ + 3H^+$	(10.30 ± 3.00)	(result)

$\Delta_r H_5 = \Delta_r H_1 - 2\,\Delta_r H_2 + \Delta_r H_3 + 3\,\Delta_r H_4$

[69ROS/SEL4]

Sodium diselenite was prepared from $NaHSeO_3$ by heating at 383 K. The quality of the specimen was controlled by X-ray and chemical analysis. The salt was reacted with a solution of lead nitrate in an electrically calibrated calorimeter with formation of crystalline lead selenite. The enthalpy change of the reaction $Na_2Se_2O_5(cr) + H_2O(l)$ + $2Pb^{2+} \rightarrow 2PbSeO_3(cr) + 2Na^+ + 2H^+$ is estimated in Table A-75.

Table A-75: Calculation of the enthalpy change of the reaction $Na_2Se_2O_5(cr) + H_2O(l)$ + $2Pb^{2+} \rightarrow 2PbSeO_3(cr) + 2Na^+ + 2H^+$. The sources of the data are indicated.

#	Reaction	$\Delta_r H_m^o$ (kJ·mol^{-1})	Reference
1	$Na_2Se_2O_5(cr) + 2Pb(NO_3)_2(aq, 1{:}400) + H_2O(l) \rightarrow 2PbSeO_3(cr)$ + $2NaNO_3(aq, 1{:}400) + 2HNO_3(aq, 1{:}400)$	– (35.23 ± 0.01)	(paper)
2	$Pb(NO_3)_2(aq, 1{:}400) + aq \rightarrow Pb^{2+} + 2\,NO_3^-$	(2.50 ± 1.50)	(review)
3	$NaNO_3(aq, 1{:}400) + aq \rightarrow Na^+ + NO_3^-$	– (0.08 ± 0.05)	[82WAG/EVA]
4	$HNO_3(aq, 1{:}400) + aq \rightarrow H^+ + NO_3^-$	– (0.41 ± 0.05)	[82WAG/EVA]
5	$Na_2Se_2O_5(cr) + H_2O(l) + 2Pb^{2+} \rightarrow 2PbSeO_3(cr) + 2Na^+ + 2H^+$	– (41.21 ± 3.00)	(result)

$\Delta_r H_5 = \Delta_r H_1 - 2\,\Delta_r H_2 + 2\,\Delta_r H_3 + 2\,\Delta_r H_4$

[69SEM/MEL]

The experimental results reported for $Bi_2Se_3(cr)$ are identical to those reported in [68MAL/SEM].

[69SOK/PAS]

The evaporation of PbSe(cr) was studied in the temperature range 936 to 1016 K using the Knudsen weight loss technique. The original evaluation employed auxiliary data different from those used by the review and the vapour pressure expression was therefore re-evaluated using the selected values of the entropies and heat capacities of PbSe(cr) and PbSe(g). The second law entropy of PbSe(g) and the second law enthalpy of sublimation of PbSe(cr) were calculated to be S°_m (PbSe, g, 298.15 K) = (270.5 ± 7.0) $J \cdot K^{-1} \cdot mol^{-1}$ and $\Delta_{sub}H^\circ_m$ (PbSe, cr, 298.15 K) = (234.9 ± 7.0) $kJ \cdot mol^{-1}$, respectively. The third law enthalpy of sublimation was calculated to be $\Delta_{sub}H^\circ_m$ (PbSe, cr, 298.15 K) = (227.1 ± 3.0) $kJ \cdot mol^{-1}$.

[69SUP/MAI]

Suponitskii *et al.* [69SUP/MAI] calculated enthalpies of formation of the selenates of Pr, Nd, and Sm (Ln) from a calorimetric determination of the enthalpy change of the reaction $Ln_2(SeO_4)_3 \cdot nH_2O(cr) + 3BaCl_2(aq) \rightarrow 3BaSeO_4(cr) + 2LnCl_3(aq) + nH_2O(l)$.

An inspection of the primary data suggests that the enthalpy of this reaction has been expressed per mole of selenate and not per mole of the salt. The published $\Delta_r H$ values should therefore be multiplied by a factor of three before they are entered into thermochemical cycles. This was not done in the paper. The correction is supported by the results for $Nd_2(SeO_4)_3 \cdot 5H_2O(cr)$ in [73SMO/EFR].

The details of the calculation of the standard enthalpies are not shown in the paper. The evaluation made by the review is exemplified for Pr in Table A-76. The recalculation used the values of the standard enthalpy of formation of Pr^{3+}(aq), Nd^{3+}(aq), and Sm^{3+}(aq) from the critical assessment of Cordfunke and Konings [2001COR/KON], see Appendix E, and of Cl^-(aq) in [89COX/WAG]. A correction for the enthalpy of dilution from (aq, 1:5000) to the aqueous standard state of – (2.4 ± 0.5) $kJ \cdot mol^{-1}$ was applied.

Table A-76: Evaluation of the standard enthalpy of formation of $Pr_2(SeO_4)_3 \cdot 5H_2O(cr)$. The sources of the data are indicated.

#	Reaction	$\Delta_r H^\circ_m$ ($kJ \cdot mol^{-1}$)	Reference
1	$Pr_2(SeO_4)_3 \cdot 5H_2O(cr) + 3BaCl_2(aq, 1:3300) \rightarrow 3BaSeO_4(cr) + 2PrCl_3(aq, 1:5000) + 5H_2O(l)$	– (121.60 ± 2.80)	(paper)
2	$Ba(cr) + Cl_2(g) + aq \rightarrow BaCl_2(aq, 1:3300)$	– (867.86 ± 2.50)	[92GRE/FUG] [82WAG/EVA]
3	$Pr(cr) + 1.5Cl_2(g) + aq \rightarrow PrCl_3(aq, 1:5000)$	– (1204.5 ± 2.1)	(see text)
4	$Ba(cr) + Se(cr) + 2O_2(g) \rightarrow BaSeO_4(cr)$	– (1144.00 ± 5.00)	(review)
5	$H_2(g) + 0.5O_2(g) \rightarrow H_2O(l)$	– (285.83 ± 0.04)	[89COX/WAG]
6	$2Pr(cr) + 3Se(cr) + 5H_2(g) + 8.5O_2(g) \rightarrow Pr_2(SeO_4)_3 \cdot 5H_2O(cr)$	– (4545.0 ± 17.6)	(result)

$\Delta_r H_6 = \Delta_f H^\circ_m (Pr_2(SeO_4)_3 \cdot 5H_2O, cr, 298.15 K) = - \Delta_r H_1 - 3\Delta_r H_2 + 2\Delta_r H_3 + 3\Delta_r H_4 + 5\Delta_r H_5$

Analogous calculations were made for the other two salts resulting in:

$$\Delta_f H_m^\circ (Nd_2(SeO_4)_3{\cdot}5H_2O, cr, 298.15 K) = -(4511.1 \pm 17.5) kJ{\cdot}mol^{-1}$$

and

$$\Delta_f H_m^\circ (Sm_2(SeO_4)_3{\cdot}8H_2O, cr, 298.15 K) = -(5395.0 \pm 17.8) kJ{\cdot}mol^{-1},$$

with $\Delta_r H_1 = -(132.70 \pm 2.00)$ and $-(96.60 \pm 3.90)$ kJ·mol^{-1}, respectively.

[69UY/DRO]

The exchange reaction

$$Pb(g) + BiSe(g) \rightleftharpoons PbSe(g) + Bi(g) \quad (A.25)$$

was studied in the temperature range 864 to 1020 K using Knudsen effusion and mass spectroscopy. The result was represented by the expression $\log_{10} K$ ((A.25), T) = $(0.715 \pm 0.243) + (5397 \pm 227)\ T^{-1}$. This corresponds to the mean values $\Delta_r H_m^\circ$ ((A.25), 942 K) = $-(39.5 \pm 1.9)$ kJ·mol^{-1} and $\Delta_r S_m^\circ$ ((A.25), 942 K) = (5.9 ± 2.0) J·K^{-1}·mol^{-1}. The values were recalculated to 298.15 K by the review using the selected properties of BiSe(g) and PbSe(g), the properties of Pb(g) in [89COX/WAG], and Bi(g) in [56STU/SIN] yielding $\Delta_f H_m^\circ$ (BiSe, g, 298.15 K) = (172.5 ± 7.2) kJ·mol^{-1} and $\Delta_f H_m^\circ$ (BiSe, g, 298.15 K) = (174.8 ± 7.2) kJ·mol^{-1} from the second and the third law, respectively.

[70BLA/MUN]

The total pressure of the vapour in equilibrium with β-SnSe was measured in the temperature range 789 to 975 K using the torsion effusion technique. The enthalpy of and entropy of sublimation according to the reaction α-SnSe $\rightleftharpoons$ SnSe(g) were calculated by the review from the reported vapour pressure expression and the selected heat capacities of SnSe(g) and α-SnSe, the enthalpy of transformation α-SnSe → β-SnSe being 1.28 kJ·mol^{-1} (cf V.7.4.1.2), and a heat capacity expression of β-SnSe being identical to that of α-SnSe yielding $\Delta_{sub} H_m^\circ$ (SnSe, α, 298.15 K) = (217.0 ± 7.0) kJ·mol^{-1} and $\Delta_{sub} S_m^\circ$ (SnSe, α, 298.15 K) = (164.7 ± 8.0) J·K^{-1}·mol^{-1}, respectively.

[70BOG/BEZ]

The heat capacity of α-HgSe was measured in the temperature range 15 to 290 K. The result is only presented in a graph and the heat capacity value $C_{p,m}^\circ$ (HgSe, α, 298.15 K) = (54.28 ± 0.50) J·K^{-1}·mol^{-1}, and the standard entropy S_m° (HgSe, α, 298.15 K) = (100.65 ± 2.00) J·K^{-1}·mol^{-1} were evaluated by the review from Figure 1 of the paper.

[70CHA/TIK]

The enthalpy of formation of α-ZnSe was determined by heating mixtures of the elements using a calibrated DTA apparatus. The reaction was complete at 800 K. The value $\Delta_f H^\circ_m$ (ZnSe, α, 298.15 K) = – (156 ± 4) kJ·mol^{-1} was determined as the result of nine measurements. The error limits were increased by the review to be ± 12 kJ·mol^{-1} to allow for systematic errors and a confidence interval of 95%.

[70GHO/NAI]

Ghosh and Nair made careful measurements of the cell:

$$H_2, Pt \mid S \mid AgCl, Ag.$$

Three main types of the test solution were studied: S_1; HCl(m_1), Na_2SeO_4(m_2), S_2; HCl(m_1), H_2SeO_4(m_2), S_3; HCl(m_1), H_2SeO_4(m_2), MCl_2(m_3) with M^{2+} = Co^{2+}, Mn^{2+}, or Ni^{2+}. Solutions S_1 and S_2 were used to determine the protonation constant of the selenate ion. The evaluation of these measurements is presented in Section V.3.3.2.

The equilibrium constant of the reaction:

$$M^{2+} + SeO_4^{2-} \rightleftharpoons MSeO_4(aq) \qquad (A.26)$$

was found from the measurements on S_3 at six different compositions. Typical concentrations contained in S_3 were: m_1 = 0.5 × 10^{-3} mol·kg^{-1}, m_2 = (2 – 6) × 10^{-3} mol·kg^{-1}, and m_3 = 5 × 10^{-3} mol·kg^{-1}. Measurements were made between 273.15 and 318.15 K including 298.15 K.

In the evaluation of the data, the complexation between M^{2+} and Cl^- was neglected, which is acceptable, and only the 1:1 complex assumed to form. The hydrogen ion concentration, m_{H^+}, was calculated from the measured potential as:

$$-\log_{10} m_{H^+} = (E - E^\circ) / (\log_{10} RT/F) + \log_{10} \gamma^2_{\pm(HCl)} + \log_{10} (m_1 + 2\, m_3),$$

by an iterative procedure. The activity coefficients were calculated by Davies' equation with the coefficient of the linear term in I equal to 0.3. The value of the protonation constant, K_H, employed in the calculations is not stated, but most likely the value $\log_{10} K_H$ = 1.70 was used as obtained from the measurements with S_1 and S_2. Very consistent values of the stability constant were obtained from the six compositions measured for each metal and temperature.

The stability constants at 298.15 K have been recalculated from the original data by the review with the accepted value of $\log_{10} K_H$ = 1.75 and various expressions for the estimate of the activity coefficients. The calculations resulted in mean values close to those reported in the paper but suggested that the uncertainty should be increased.

The enthalpy of Reaction (A.26) was found from the temperature variation of the stability constant. The plots of $\log_{10} \beta_1$ as a function of $1/T$ are linear and comprise 6 or 7 points. A least squares treatment of the data by the review resulted in slightly different values from those reported in the paper, but no revision of the data there has been made.

Nair and Nancollas determined by the same technique the equilibrium constants and enthalpy changes for ion-pair formation between the sulphate ion and Mg^{2+} and Zn^{2+} [58NAI/NAN2], and Co^{2+}, Mn^{2+}, and Ni^{2+} [59NAI/NAN]. Powell [73POW] has summarised and critically evaluated the enthalpy changes on formation of these ion pairs obtained from calorimetric measurements. Agreement between the results from the temperature variation of the equilibrium constant and calorimetry is observed only for Co^{2+}, while for the other ions the results are larger by about 6 to 15 $kJ \cdot mol^{-1}$ (average 9.7 $kJ \cdot mol^{-1}$). Hence it is concluded that the reported values for the enthalpy changes of the selenates are most likely too large and they have not been selected.

The numerical results from this reference are reported in Chapter V at the appropriate places.

[70KAR/MAI]

Praseodymium hydrogen selenite was prepared by mixing a 10% solution of praseodymium carbonate with a 40% solution of selenious acid. The amount of acid taken was eight times the stoichiometric amount. The crystalline product had the composition $Pr(HSeO_3)_3 \cdot 2H_2O$ as established by determinations of Pr and Se by chemical analysis, and water by drying the sample. The compounds $Pr(HSeO_3)_3$, $Pr_2(SeO_3)_3 \cdot 3SeO_2$, $Pr_2(SeO_3)_3 \cdot SeO_2$, $Pr_2(SeO_3)_3$, and $Pr_2O_3 \cdot SeO_2$ were prepared by heat treatment of $Pr(HSeO_3)_3 \cdot 2H_2O$ at increasingly higher temperatures. The compounds were analysed for Pr and Se and the ratios Pr:Se obtained were consistent with the above formulas. X-ray diffraction patterns are presented of $Pr(HSeO_3)_3 \cdot 2H_2O$, $Pr(HSeO_3)_3$, and $Pr_2(SeO_3)_3 \cdot 3SeO_2$. The number of lines listed is only three except for the parent compound. The quality of the preparations is hence doubtful.

The enthalpies of dissolution of the compounds in HCl(aq, 1:18.5), denoted sln, at 298.15 K were measured. The dilution was 1:2000 or 1:4000 (dil). The calorimeter was calibrated by electric resistance heating. The evaluator has been unable to reach the same figures of the heat equivalent of the calorimeter as the authors from the published *V*, *I*, *t* data. This problem has been encountered several times in Russian papers. It is believed that the heat equivalent is some constant fraction of the true heat equivalent and used only as an intermediate figure in the calculation of the enthalpy change. Hence the data will be accepted. The result for $Pr(HSeO_3)_3 \cdot 2H_2O$ is evaluated in Table A-77 with new auxiliary data. The results for the other compounds were evaluated similarly. As the compound $Pr_2(SeO_3)_3$ has been re-measured in [71MAI/SUP] it is not considered here. The origin of the value of $\Delta_f H_m^\circ$ ($PrCl_3 \cdot 7H_2O$, cr, 298.15 K) used in Table A-77 is found in this Appendix, [71MAI/SUP].

Table A-77: Calculation of the standard enthalpy of formation of $Pr(HSeO_3)_3 \cdot 2H_2O$. The sources of the data are indicated.

#	Reaction	$\Delta_r H_m^\circ$ (kJ·mol^{-1})	Reference
1	$Pr(HSeO_3)_3 \cdot 2H_2O(cr) + 3HCl(sln) + sln \rightarrow PrCl_3(sln, dil) + 3H_2SeO_3(sln, dil) + 2H_2O(l, sln)$	(28.12 ± 0.50)	(paper)
2	$PrCl_3 \cdot 7H_2O(cr) + sln \rightarrow PrCl_3(sln, dil) + 7H_2O(l, sln)$	– (8.79 ± 0.88)	[71MAI/SUP]
3	$Pr(cr) + 1.5Cl_2(g) + 7H_2(g) + 3.5O_2(g) \rightarrow PrCl_3 \cdot 7H_2O(cr)$	– (3181.0 ± 2.0)	(see text)
4	$H_2SeO_3(cr) + sln \rightarrow H_2SeO_3(sln, dil)$	(18.62 ± 0.13)	[68MAI/SUP]
5	$Se(cr) + H_2(g) + 1.5O_2(g) \rightarrow H_2SeO_3(cr)$	– (524.72 ± 0.65)	(review)
6	$H_2(g) + 0.5O_2(g) + sln \rightarrow H_2O(l, sln)$	– (285.98 ± 0.04)	[2000RAN/FUG]
7	$0.5H_2(g) + 0.5Cl_2(g) + sln \rightarrow HCl(sln)$	– (160.90 ± 0.10)	[2000RAN/FUG]
8	$Pr(cr) + 3Se(cr) + 3.5H_2(g) + 5.5O_2(g) \rightarrow Pr(HSeO_3)_3 \cdot 2H_2O(cr)$	– (2823.6 ± 3.0)	(result)

$\Delta_r H_8 = \Delta_f H_m^\circ(Pr(HSeO_3)_3 \cdot 2H_2O, cr, 298.15 K) = -\Delta_r H_1 + \Delta_r H_2 + \Delta_r H_3 + 3\,\Delta_r H_4 + 3\,\Delta_r H_5 - 5\,\Delta_r H_6 - 3\,\Delta_r H_7$

The results for the other compounds are:

$$\Delta_f H_m^\circ(Pr(HSeO_3)_3, cr, 298.15 K) = -(2212.5 \pm 2.9)\ kJ \cdot mol^{-1},$$

$$\Delta_f H_m^\circ(Pr_2(SeO_3)_3 \cdot 3SeO_2, cr, 298.15 K) = -(3545.2 \pm 4.5)\ kJ \cdot mol^{-1},$$

$$\Delta_f H_m^\circ(Pr_2(SeO_3)_3 \cdot SeO_2, cr, 298.15 K) = -(3108.3 \pm 3.5)\ kJ \cdot mol^{-1},$$

$$\Delta_f H_m^\circ(Pr_2O_3 \cdot SeO_2, cr, 298.15 K) = -(2302.4 \pm 3.1)\ kJ \cdot mol^{-1}.$$

[70KEL]

The vapour pressures of the gaseous species Se_2-Se_8 were measured in the temperature range 473 to 723 K using an electrochemical Knudsen cell and a mass spectrometer. The electrochemical cell has the advantage that the selenium activity can be varied and allows for an extraordinary large temperature range to be investigated. This study is unique because the activity of selenium can be set sufficiently low to provide conditions where $Se_3(g)$ and $Se_4(g)$ are the primary species and not the result of fragmentation.

Enthalpies of formation at 0 K for the species $Se_2(g)$-$Se_8(g)$ were derived using Gibbs energy functions calculated from estimated molecular parameters. In the present review, the experimental results presented in Tables 11 and 15 to 20 of the paper were re-evaluated using the second and third laws and the selected data for the heat capacities and entropies. The results are summarised in the Table A-78. No evaluation using the second law can be made for $Se_4(g)$ because the measurements were made at a single temperature only.

Table A-78: Entropies and enthalpies of formation of gaseous selenium species. The error limits at 298.15 K are based only on the experimental values and do not reflect contributions from systematic errors.

Species	$\Delta_f H_m$ (0 K) (kJ·mol^{-1}) original values	$\Delta_f H^\circ_m$ (298.15 K) (kJ·mol^{-1}) 2nd law	3rd law	S°_m (298.15 K) (J·K^{-1}·mol^{-1})
Se_2	(146.1 ± 1.7)	(145.8 ± 2.6)	(144.0 ± 1.0)	(250.0 ± 4.1)
Se_3	(184.5 ± 8.4)	(181.0 ± 2.1)	(178.0 ± 0.8)	(313.3 ± 3.0)
Se_4	(190.8 ± 8.4)		(163.0 ± 3.4)	
Se_5	(143.9 ± 10.5)	(145.2 ± 4.5)	(148.6 ± 1.6)	(392.6 ± 7.2)
Se_6	(140.2 ± 10.5)	(142.5 ± 8.3)	(134.3 ± 2.4)	(441.7 ± 14.2)
Se_7	(151.5 ± 10.5)	(158.8 ± 8.5)	(148.0 ± 2.4)	(505.2 ± 14.8)
Se_8	(160.7 ± 10.5)	(161.7 ± 7.7)	(155.4 ± 2.3)	(537.7 ± 14.4)

The selenium activity was adjusted by the EMF applied to the electrochemical cell Pt, Ag | AgI | Ag_2Se, Pt.

According to the Ag-Se phase diagram of [90KAR/THO] the silver selenide has a homogeneity range above 405 K spanning the compositions between Ag_2Se and $Ag_2Se_{0.98}$. In the evaluations by Keller, the ideal stoichiometry was assumed and the results can be slightly distorted. The total pressures calculated from the data by Keller are somewhat low at higher temperatures compared to most other investigations.

[70KLU/SEL]

The authors prepared crystalline $Cs_2SeO_3{\cdot}H_2O$ and measured the enthalpy of the reaction of the salt with a lead nitrate solution in an electrically calibrated calorimeter. Crystalline lead selenite is formed. The data have been recalculated to find the enthalpy change of the reaction $Cs_2SeO_3{\cdot}H_2O(cr) + Pb^{2+} \rightarrow PbSeO_3(cr) + 2Cs^+ + H_2O(l)$ in Table A-79.

Table A-79: Calculation of the enthalpy change of the reaction $Cs_2SeO_3{\cdot}H_2O(cr) + Pb^{2+} \rightarrow PbSeO_3(cr) + 2Cs^+ + H_2O(l)$. The sources of the data are indicated.

#	Reaction	$\Delta_r H^\circ_m$ (kJ·mol^{-1})	Reference
1	$Cs_2SeO_3{\cdot}H_2O(cr) + Pb(NO_3)_2(aq, 1{:}800) \rightarrow PbSeO_3(cr) + 2CsNO_3(aq, 1{:}400) + H_2O(l)$	– (20.13 ± 0.16)	(paper)
2	$Pb(NO_3)_2(aq, 1{:}800) + aq \rightarrow Pb^{2+} + 2\,NO_3^-$	(1.85 ± 1.50)	(review)
3	$CsNO_3(aq, 1{:}400) + aq \rightarrow Cs^+ + NO_3^-$	– (0.51 ± 0.05)	[82WAG/EVA]
4	$Cs_2SeO_3{\cdot}H_2O(cr) + Pb^{2+} \rightarrow PbSeO_3(cr) + 2Cs^+ + H_2O(l)$	– (23.00 ± 1.50)	(review)

$\Delta_r H_4 = \Delta_r H_1 - \Delta_r H_2 + 2\,\Delta_r H_3$

[70ROS/SEL]

Crystalline potassium hydrogen selenite was prepared and the product was checked by X-ray and chemical analysis. The enthalpy of the reaction between a solution of the salt and $Pb(NO_3)_2(cr)$ was determined as well as the heat of dissolution of $KHSeO_3$. The experimental findings are furnished in Table A-80, in which the enthalpy change of the reaction $KHSeO_3(cr) + Pb^{2+} \rightarrow PbSeO_3(cr) + K^+ + H^+$ is estimated.

Table A-80: Calculation of the enthalpy change of the reaction $KHSeO_3(cr) + Pb^{2+} \rightarrow PbSeO_3(cr) + K^+ + H^+$. The sources of the data are indicated.

#	Reaction	$\Delta_r H_m$ (kJ·mol^{-1})	Reference
1	$KHSeO_3$(aq, 1:800) + $Pb(NO_3)_2$(cr) → $PbSeO_3$(cr) + KNO_3(aq, 1:800) + HNO_3(aq, 1:800)	(12.43 ± 0.01)	(paper)
2	$Pb(NO_3)_2$(cr) + aq → Pb^{2+} + 2 NO_3^-	(33.80 ± 0.25)	[95PAY]
3	KNO_3(aq, 1: 800) + aq → K^+ + NO_3^-	(0.04 ± 0.05)	[82WAG/EVA]
4	HNO_3(aq, 1:800) + aq → H^+ + NO_3^-	– (0.36 ± 0.05)	[82WAG/EVA]
5	$KHSeO_3$(cr) + aq → $KHSeO_3$(aq, 1:800)	(10.04 ± 0.04)	(paper)
6	$KHSeO_3$(cr) + Pb^{2+} → $PbSeO_3$(cr) + K^+ + H^+	– (11.65 ± 0.25)	(result)

$\Delta_r H_6 = \Delta_r H_1 - \Delta_r H_2 + \Delta_r H_3 + \Delta_r H_4 + \Delta_r H_5$

Crystalline $K_2Se_2O_5$ was prepared by heating $KHSeO_3$ at 343 K. Chemical and X-ray analysis confirmed the composition and crystalline state of the preparation. The salt was reacted with a lead nitrate solution in an electrically calibrated calorimeter. The calorimetric result is used in Table A-81 to calculate the enthalpy change of the reaction $K_2Se_2O_5(cr) + 2Pb^{2+} + H_2O(l) \rightarrow 2PbSeO_3(cr) + 2K^+ + 2H^+$.

Table A-81: Calculation of the enthalpy change of the reaction $K_2Se_2O_5(cr) + 2Pb^{2+} + H_2O(l) \rightarrow 2PbSeO_3(cr) + 2K^+ + 2H^+$. The sources of the data are indicated.

#	Reaction	$\Delta_r H_m$ (kJ·mol^{-1})	Reference
1	$K_2Se_2O_5$(cr) + 2$Pb(NO_3)_2$(aq, 1:400) + H_2O(l) → 2$PbSeO_3$(cr) + 2KNO_3(aq, 1:400) + 2HNO_3(aq, 1:400)	– (29.18 ± 0.04)	(paper)
2	$Pb(NO_3)_2$(aq, 1:400) + aq → Pb^{2+} + 2 NO_3^-	(2.50 ± 1.50)	(review)
3	KNO_3(aq, 1:400) + aq → K^+ + NO_3^-	(0.30 ± 0.05)	[82WAG/EVA]
4	HNO_3(aq, 1:400) + aq → H^+ + NO_3^-	– (0.41 ± 0.05)	[82WAG/EVA]
5	$K_2Se_2O_5$(cr) + 2Pb^{2+} + H_2O(l) → 2$PbSeO_3$(cr) + 2K^+ + 2H^+	– (34.40 ± 3.00)	(result)

$\Delta_r H_5 = \Delta_r H_1 - 2\,\Delta_r H_2 + 2\,\Delta_r H_3 + 2\,\Delta_r H_4$

[70ROS/SEL2]

Crystalline lithium trihydrogen selenite was prepared and checked by X-ray and chemical analysis. The enthalpy of the reaction between a solution of the salt and $Pb(NO_3)_2(cr)$ was determined as well as the heat of dissolution of $LiH_3(SeO_3)_2(cr)$. The experimental findings are furnished in Table A-82, in which the enthalpy change of the reaction $LiH_3(SeO_3)_2(cr) + 2Pb^{2+} \rightarrow 2PbSeO_3(cr) + Li^+ + 3H^+$ is calculated.

Table A-82: Calculation of the enthalpy change of the reaction $LiH_3(SeO_3)_2(cr) + 2Pb^{2+} \rightarrow 2PbSeO_3(cr) + Li^+ + 3H^+$. The sources of the data are indicated.

#	Reaction	$\Delta_r H_m$ (kJ·mol^{-1})	Reference
1	$LiH_3(SeO_3)_2(aq, 1:800) + 2Pb(NO_3)_2(cr) \rightarrow 2PbSeO_3(cr) + LiNO_3(aq, 1:800) + 3HNO_3(aq, 1:267)$	(56.12 ± 0.04)	(paper)
2	$Pb(NO_3)_2(cr) + aq \rightarrow Pb^{2+} + 2\,NO_3^-$	(33.80 ± 0.25)	[95PAY]
3	$LiNO_3(aq, 1: 800) + aq \rightarrow Li^+ + NO_3^-$	– (0.35 ± 0.05)	[82WAG/EVA]
4	$HNO_3(aq, 1:267) + aq \rightarrow H^+ + NO_3^-$	– (0.46 ± 0.05)	[82WAG/EVA]
5	$LiH_3(SeO_3)_2(cr) + aq \rightarrow LiH_3(SeO_3)_2(aq, 1:800)$	(40.44 ± 0.08)	(paper)
6	$LiH_3(SeO_3)_2(cr) + 2Pb^{2+} \rightarrow 2PbSeO_3(cr) + Li^+ + 3H^+$	(27.23 ± 0.53)	(result)

$\Delta_r H_6 = \Delta_r H_1 - 2\,\Delta_r H_2 + \Delta_r H_3 + 3\,\Delta_r H_4 + \Delta_r H_5$

[70ROS/SEL3]

Crystalline potassium trihydrogen selenite was prepared and subjected to chemical and X-ray analysis for confirmation of the composition and crystalline state. The compound was reacted in stoichiometric proportions with a solution of lead nitrate in an electrically calibrated calorimeter. The lead selenite formed was crystalline and the reaction was quantitative. The calorimetric data have been used to calculate the enthalpy change of the reaction $KH_3(SeO_3)_2(cr) + 2Pb^{2+} \rightarrow 2PbSeO_3(cr) + K^+ + 3H^+$ in Table A-83.

Table A-83: Calculation of the enthalpy change of the reaction $KH_3(SeO_3)_2(cr) + 2Pb^{2+} \rightarrow 2PbSeO_3(cr) + K^+ + 3H^+$. The sources of the data are indicated.

#	Reaction	$\Delta_r H_m$ (kJ·mol^{-1})	Reference
1	$KH_3(SeO_3)_2(cr) + 2Pb(NO_3)_2(aq, 1:400) \rightarrow 2PbSeO_3(cr) + KNO_3(aq, 1:800) + 3HNO_3(aq, 1:267)$	(17.28 ± 0.02)	(paper)
2	$Pb(NO_3)_2(aq, 1:400) + aq \rightarrow Pb^{2+} + 2\,NO_3^-$	(2.50 ± 1.50)	(review)
3	$KNO_3(aq, 1:800) + aq \rightarrow K^+ + NO_3^-$	(0.04 ± 0.05)	[82WAG/EVA]
4	$HNO_3(aq, 1:267) + aq \rightarrow H^+ + NO_3^-$	– (0.46 ± 0.05)	[82WAG/EVA]
5	$KH_3(SeO_3)_2(cr) + 2Pb^{2+} \rightarrow 2PbSeO_3(cr) + K^+ + 3H^+$	(10.94 ± 1.50)	(result)

$\Delta_r H_5 = \Delta_r H_1 - 2\,\Delta_r H_2 + \Delta_r H_3 + 3\,\Delta_r H_4$

[70SOB/SEL]

Sobol and Selivanova prepared crystalline sodium diselenate by gradually heating crystalline $NaHSeO_4$ up to 475 K. The crystalline state and composition of the specimen were confirmed by X-ray diffraction and chemical analysis. The authors made a number of calorimetric measurements from which the standard enthalpy of formation of $Na_2Se_2O_7$(cr) can be derived. The measurements and the calculations by the review are summarised in the following three tables.

Table A-84: Determination of enthalpy change of the reaction $Na_2Se_2O_7$(cr) + $2Pb^{2+}$ + H_2O(l) → $2PbSeO_4$(cr) + $2Na^+$ + $2H^+$. The sources of the data are indicated.

#	Reaction	$\Delta_r H_m$ (kJ·mol^{-1})	Reference
1	$Na_2Se_2O_7$(cr) + $2Pb(NO_3)_2$(aq, 1:800) +H_2O(l) → $2PbSeO_4$(cr) + $2NaNO_3$(aq, 1:800) + $2HNO_3$(aq, 1:800)	– (79.39 ± 0.11)	(paper)
2	$NaNO_3$(aq, 1:800) + aq → Na^+ + NO_3^-	– (0.18 ± 0.10)	[82WAG/EVA]
3	HNO_3(aq, 1:800) + aq → H^+ + NO_3^-	– (0.36 ± 0.05)	[82WAG/EVA]
4	$Pb(NO_3)_2$ (aq, 1:800) + aq → Pb^{2+} + 2 NO_3^-	(1.85 ± 1.50)	(review)
5	$Na_2Se_2O_7$(cr) + $2Pb^{2+}$ +H_2O(l) → $2PbSeO_4$(cr) + $2Na^+$ + $2H^+$	– (84.17 ± 3.01)	(result)

$\Delta_r H_5 = \Delta_r H_1 + 2\,\Delta_r H_2 + 2\,\Delta_r H_3 - 2\,\Delta_r H_4$

The enthalpy of reaction, $\Delta_r H_5$ in Table A-84, combined with accepted standard enthalpies of formation and neglect of the enthalpy of mixing leads to $\Delta_f H_m^\circ$ ($Na_2Se_2O_7$, cr, 298.15 K) = – (1327.12 ± 9.18) kJ·mol^{-1}.

Table A-85: Determination of $\Delta_f H_m^\circ$ ($Na_2Se_2O_7$, cr, 298.15 K) from the enthalpy of dissolution of $Na_2Se_2O_7$(cr) in water. The sources of the data are indicated.

#	Reaction	$\Delta_r H_m$ (kJ·mol^{-1})	Reference
1	$Na_2Se_2O_7$(cr) + H_2O(l) + aq → $2NaHSeO_4$(aq, 1:800)	– (55.65 ± 0.12)	(paper)
2	$NaHSeO_4$(cr) + aq → $NaHSeO_4$(aq, 1:800)	– (8.10 ± 0.03)	[73SOB/SEL]
3	Na(cr) + Se(cr) + $0.5H_2$(g) + $2O_2$(g) → $NaHSeO_4$(cr)	– (822.23 ± 5.59)	(review)
4	H_2(g) + $0.5O_2$(g) → H_2O(l)	– (285.83 ± 0.04)	[89COX/WAG]
5	2Na(cr) + 2Se(cr) + $3.5O_2$(g) → $Na_2Se_2O_7$(cr)	– (1319.18 ± 11.18)	(result)

$\Delta_r H_5 = -\Delta_r H_1 + 2\,\Delta_r H_2 + 2\,\Delta_r H_3 - \Delta_r H_4$

Table A-86: Determination of $\Delta_f H_m^\circ$ ($Na_2Se_2O_7$, cr, 298.15 K) from the heat of neutralisation of $Na_2Se_2O_7$(cr) with NaOH(aq). The sources of the data are indicated.

#	Reaction	$\Delta_r H_m$ (kJ·mol^{-1})	Reference
1	$Na_2Se_2O_7$(cr) + 2NaOH(aq, 1:800) → $2Na_2SeO_4$(aq, 1:800) + H_2O(l)	– (186.82 ± 0.13)	(paper)
2	Na_2SeO_4(cr) + aq → Na_2SeO_4(aq, 1:800)	– (4.06 ± 0.17)	[62SEL/SAZ3]
3	2Na(cr) + Se(cr) + $2O_2$(g) → Na_2SeO_4(cr)	– (1076.50 ± 3.50)	(review)
4	Na + $0.5H_2$(g) + $0.5O_2$(g) + aq → NaOH(aq, 1:800)	– (469.99 ± 0.12)	[89COX/WAG] [82WAG/EVA]
5	H_2(g) + $0.5O_2$(g) → H_2O(l)	– (285.83 ± 0.04)	[89COX/WAG]
6	2Na(cr) + 2Se(cr) + $3.5O_2$(g) → $Na_2Se_2O_7$(cr)	– (1320.15 ± 7.01)	(result)

$\Delta_r H_6 = -\Delta_r H_1 + 2\,\Delta_r H_2 + 2\,\Delta_r H_3 - 2\,\Delta_r H_4 + \Delta_r H_5$

[71GLA/KOR]

The total pressure of the vapour in equilibrium with Cu_2Se(cr) was measured in the temperature range 1133 to 1293 K using the Knudsen effusion technique. The enthalpy and entropy of the assumed reaction $Cu_2Se(cr) \rightleftharpoons 2Cu(cr) + \frac{1}{2}Se_2(g)$ were evaluated at the mean temperature only and no recalculations to 298.15 K were made because of missing enthalpy and entropy increments of Cu_2Se(cr). The results were therefore recalculated by the review using the entropy and enthalpy increments of Cu_2Se(cr) in [78BLA/GUN], the selected thermodynamic properties of Se_2(g), and the data of Cu(cr) in [89COX/WAG] yielding the values $\Delta_f H_m^\circ$ (Cu_2Se, α, 298.15 K) = – (26.5 ± 10.0) kJ·mol^{-1} and S_m° (Cu_2Se, α, 298.15 K) = (90.1 ± 20.0) J·K^{-1}·mol^{-1}.

[71GRI/BOR]

The saturated vapour pressure of liquid InSe was measured at ten temperatures in the temperature range 1113 to 1253 K. On the assumption that the gas phase consisted of the molecules In(g), Se(g), Se_2(g), InSe(g), In_2Se(g), and In_2Se_2(g) the enthalpies and entropies of formation of InSe(g), In_2Se(g), and In_2Se_2(g), in all six parameters, were determined by a least-squares procedure. Considering the uncertainties in the pressure measurements and the neglect of the other Se_n(g) species, it is not possible to determine six unknown parameters from ten measured points with any degree of accuracy. The results of this investigation are therefore rejected.

[71KEL/RIC]

The paper is a shortened version of the dissertation by Keller [70KEL] and does not provide new information.

[71MAI/SUP]

Maier, Suponitskii, and Karapet'yants prepared anhydrous crystalline lanthanum, praseodymium, and neodymium selenite by keeping the amorphous salts at an elevated temperature in a sealed tube. The salts are denoted by $M_2(SeO_3)_3$. The standard enthalpies of formation of the compounds have been calculated from the experimental results and auxiliary data in Table A-87. The solvent (sln) employed in the calorimetric dissolution experiments was HCl(aq, 1:18.5). Dissolution was made to the appropriate concentrations and the composition of the solutions will be denoted simply as dilute (dil).

Table A-87: Calculation of the standard enthalpies of formation of the anhydrous selenites of La, Pr, and Nd. The sources of the data are indicated and entered in the order La, Pr, Nd. *n* is 7 for La and Pr, and 6 for Nd.

#	Reaction	$\Delta_r H_m$ (kJ·mol^{-1})	Reference
1	$M_2(SeO_3)_3(cr) + 6HCl(sln) + sln \rightarrow 2MCl_3(sln, dil) + 3H_2SeO_3(sln, dil)$	– (71.50 ± 2.13)	(paper)
		– (71.96 ± 1.13)	
		– (62.13 ± 0.63)	
2	$H_2SeO_3(cr) + sln \rightarrow H_2SeO_3(sln, dil)$	(18.62 ± 0.13)	[68MAI/SUP]
3	$MCl_3{\cdot}nH_2O(cr) + sln \rightarrow MCl_3(sln, dil) + nH_2O(l, sln)$	– (8.37 ± 0.67)	(paper)
		– (8.79 ± 0.88)	
		– (20.00 ± 0.21)	
4	$Se(cr) + H_2(g) + 1.5O_2(g) \rightarrow H_2SeO_3(cr)$	– (524.72 ± 0.65)	(review)
5	$M(cr) + 1.5Cl_2(g) + nH_2(g) + 0.5nO_2(g) \rightarrow MCl_3{\cdot}nH_2O(cr)$	– (3178.9 ± 2.5)	(see text)
		– (3181.0 ± 2.0)	
		– (2874.2 ± 2.0)	
6	$H_2(g) + 0.5O_2(g) + sln \rightarrow H_2O(l, sln)$	– (285.98 ± 0.04)	[2000RAN/FUG]
7	$0.5H_2(g) + 0.5Cl_2(g) + sln \rightarrow HCl(sln)$	– (160.90 ± 0.10)	[2000RAN/FUG]
8	$2M(cr) + 3Se(cr) + 4.5O_2(g) \rightarrow M_2(SeO_3)_3(cr)$	– (2852.2 ± 6.0)	(results)
		– (2856.8 ± 5.0)	
		– (2848.2 ± 4.6)	

$\Delta_r H_8 = \Delta_f H^\circ_m(M_2(SeO_3)_3, cr, 298.15 K) = -\Delta_r H_1 + 3\Delta_r H_2 + 2\Delta_r H_3 + 3\Delta_r H_4 + 2\Delta_r H_5 - 2n\Delta_r H_6 - 6\Delta_r H_7$

The value of $\Delta_f H^\circ_m$ ($NdCl_3{\cdot}6H_2O$, cr, 298.15 K) was calculated from the weighted mean of the enthalpy of dissolution in water of $NdCl_3{\cdot}6H_2O$(cr) to infinite dilution as determined by Spedding, Naumann, and Eberts [58SPE/NAU], – (38.9 ± 0.4) kJ·mol^{-1}, and by Seifert and Funke [98SEI/FUN], – (36.5 ± 0.2) kJ·mol^{-1}. The value of $\Delta_{sol} H^\circ_m$ ($NdCl_3{\cdot}6H_2O$, cr, 298.15 K) so obtained, – (36.8 ± 0.2) kJ·mol^{-1}, was combined with the standard enthalpy of formation of Nd^{3+}(aq) in [2001COR/KON] and of $Cl^-(aq)$ and $H_2O(l)$ in [89COX/WAG] to give $\Delta_f H_m^\circ$ ($NdCl_3 \cdot 6H_2O$, cr, 298.15 K). The standard enthalpies of formation of the other two salts were calculated in the same way. The values of $\Delta_{sol} H_m^\circ$ ($LaCl_3 \cdot 7H_2O$, cr, 298.15 K), $-(30.8 \pm 0.4)$ kJ·mol^{-1}, obtained from [70HIN/COB2] and $\Delta_{sol} H_m^\circ$ ($PrCl_3 \cdot 7H_2O$, cr, 298.15 K), $-(26.8 \pm 0.2)$ kJ·mol^{-1}, obtained from [70HIN/COB2] and [98SEI/FUN] were used.

The results from the table:

$$\Delta_f H_m^\circ (La_2(SeO_3)_3, \text{cr}, 298.15 \text{ K}) = -(2852.0 \pm 6.0) \text{ kJ mol}^{-1},$$

$$\Delta_f H_m^\circ (Pr_2(SeO_3)_3, \text{cr}, 298.15 \text{ K}) = -(2856.8 \pm 5.0) \text{ kJ mol}^{-1},$$

$$\Delta_f H_m^\circ (Nd_2(SeO_3)_3, \text{cr}, 298.15 \text{ K}) = -(2848.2 \pm 4.6) \text{ kJ mol}^{-1},$$

agree with the values in the paper, which were calculated in the same way, after allowance for the changes in the auxiliary data and the fact that partial molar enthalpies of formation were not used for H_2O and HCl. The auxiliary data used in the paper can be found in the deposited document VINITI No. 2968-71.

[71MEH/GUB]

In the early 1970's, Mehra and Gubeli published a series of papers [70MEH/GUB], [70MEH/GUB2], [71MEH/GUB] on selenide complexation and precipitation with Ag^+, Mn^{2+} and Hg^{2+}. The three works all originate from a thesis by M. C. Mehra, "Studies on the stabilities of some metal selenite, sulphide, and selenide complexes in solution" [68MEH], and constitute the only studies in the literature where soluble selenide complexes have been indicated.

The experimental set-up was common for all three metal ions. It included the mixing, under a nitrogen gas atmosphere, of a $(0.5 - 1) \times 10^{-3}$ M radioactively labelled metal ion solution with a 0.015 to 0.08 M solution of Se(– II) at different pH. After equilibration for five days the solid formed was separated from the aqueous phase by means of a sintered glass filter of undefined pore size. The aqueous phase was analysed with respect to pH (glass electrode), pMe (Ag^+ and Hg^{2+}, ion selective electrode), $[Me(aq)]_{tot}$ (γ-ray scintillation) and [Se(– II)] (iodometric titration).

The solubility products of the three metal selenides were calculated from the experimental data, but included the use of erroneous acidity constants for $H_2Se(aq)$. The review has accepted the measurements and recalculated the experimental data with selected values for the acidity constants of hydrogen selenide.

For all three systems, and in certain parts of the pH interval, the γ-ray scintillation data indicated significantly higher total metal solubility than expected from the solubility product. This finding was interpreted as being due to the formation of soluble complexes. An alternative explanation, however, is that the glass filter used was inefficient in separating out colloidally sized metal selenide particles. These measurements might therefore not prove the existence of true aqueous complexes. This is a concern which is especially valid for the regions where low 3×10^{-7} to 1×10^{-8} M and pH-independent solubilities were recorded and interpreted as due to the existence of uncharged metal selenide complexes. These complexes and their corresponding equilibrium constants have therefore not been accepted by the review.

In the Hg^{2+}-Se^{2-} system, and at pH ≥ 7, a significant solubilisation of HgSe(s) was recorded. These data cannot be questioned by the experimental uncertainty discussed above, and clearly indicate the formation of a soluble complex, ($HgSe_2^{2-}$). This species has therefore been accepted with an equilibrium constant recalculated by the review.

Also in the Ag^+-Se^{2-} system a significant solubilisation of Ag_2Se(s) was recorded, although only at a pH well above 11. The species suggested to explain the increased solubility, $Ag(Se)_2(OH)^{4-}$, is, however, highly questionable in view of the well-known tendency of Ag^+ for linear coordination. This complex and its corresponding equilibrium constant have therefore not been accepted by the review.

[71PAS/ARO]

The vapour pressures obtained from dew point measurements will depend on the relationship used for the temperature dependence of the saturated vapour pressure. A number of relationships, $p_{SeO_2} = f(T)$, have been proposed for the equilibrium $SeO_2(cr) \rightleftharpoons SeO_2(g)$, see Section V.2.2.3. in Chapter V. In order to obtain consistency it is necessary to revise previous vapour pressures calculated from the dew point temperature, T_{dew}, to conform with the accepted $p_{SeO_2} = f(T)$ function. The problem met in this revision has been the fact that in some instances only p_{SeO_2} values have been published, while the dew point temperatures and the relationship used to obtain p_{SeO_2} are absent.

Pashinkin *et al.* reported revised values of $\Delta_f H_m^\circ$ and S_m° for a number of selenites. The values were derived from previously published vapour pressures in [64BAK/BUK], [68BAK/BUK], [68BAK/BUK3], [69BAK/BUK], [69BAK/BUK2]. The new values together with data deposited in VINITI (No. 2710-71) made it possible to deduce that the vapour pressure expression in [62MAR/GET] had been used in the original publications. The re-evaluation of the original data starts with a calculation of T_{dew} from published vapour pressures and the relationship in [62MAR/GET]. The revised pressures are obtained from T_{dew} and the selected relationship; $\log_{10}(p_{SeO_2}/\text{bar}) = 15.893 - 6250.2 \times T^{-1} - 1.960 \times \log_{10} T$. The correction is substantial. For instance, the enthalpy change of the reaction $MgSeO_3(cr) \rightarrow MgO(cr) + SeO_2(g)$ is altered from 189.1 to 222.0 $kJ\cdot mol^{-1}$.

[71SEA/MUN]

The saturated vapour pressure of α-CdSe was measured in the temperature range 942 to 1041 K using the Knudsen torsion effusion technique. A value for the enthalpy of formation of α-CdSe at 298.15 K was derived using an undocumneted set of auxiliary data.

The experimental results given in Table 3 of the paper were therefore re-evaluated by the review using both the second and the third law, the selected thermodynamic functions of selenium, the data for Cd in [89COX/WAG], the selected heat capacity of α-CdSe, and the selected entropy of α-CdSe in the case of the third law. The vaporisation was assumed to occur according to the reaction α-CdSe $\rightleftharpoons$ Cd(g) + ½Se_2(g). The results were $\Delta_f H_m^\circ$(CdSe, α, 298.15 K) = – (144.1 ± 12.0) kJ·mol^{-1}, $\Delta_f H_m^\circ$(CdSe, α, 298.15 K) = – (142.0 ± 2.0) kJ·mol^{-1}, and S_m°(CdSe, α, 298.15 K) = (84.1 ± 8.0) J·K^{-1}·mol^{-1} for the second and third law enthalpy of formation and the entropy, respectively.

[71SEL/PRY]

Selivanova, Prymova, and Kravchenko [71SEL/PRY] measured the enthalpy change of the reaction:

$$(NH_4)_2SeO_4(cr) + Pb(NO_3)_2(aq, 1{:}1600) \rightarrow PbSeO_4(cr) + 2NH_4NO_3(aq,1{:}800),$$

in a calorimeter. They also measured the integral enthalpy of dissolution of $(NH_4)_2SeO_4$(cr) to $(NH_4)_2SeO_4$(aq, 1:1600). Their data have been re-evaluated in Table A-88 and Table A-89.

Table A-88: Enthalpy change of the reaction $(NH_4)_2SeO_4$(cr) + Pb^{2+} → $PbSeO_4$(cr) + 2 NH_4^+. The sources of the data are indicated.

#	Reaction	$\Delta_r H_m$ (kJ·mol^{-1})	Reference
1	$(NH_4)_2SeO_4$(cr) + $Pb(NO_3)_2$(aq, 1:1600) → $PbSeO_4$(cr) + $2NH_4NO_3$(aq,1:800)	(5.61 ± 0.08)	(paper)
2	$Pb(NO_3)_2$(aq, 1:1600) + aq → Pb^{2+} + 2 NO_3^-	(1.50 ± 1.50)	(review)
3	NH_4NO_3(aq,1:800) + aq → NH_4^+ + NO_3^-	– (0.22 ± 0.10)	[82WAG/EVA]
4	$(NH_4)_2SeO_4$(cr) + Pb^{2+} → $PbSeO_4$(cr) + 2 NH_4^+	(3.67 ± 1.51)	(result)

$\Delta_r H_4 = \Delta_r H_1 - \Delta_r H_2 + 2\,\Delta_r H_3$

Table A-89: Integral enthalpy of dissolution of $(NH_4)_2SeO_4$(cr) to the standard state. The sources of the data are indicated.

#	Reaction	$\Delta_r H_m$ (kJ·mol^{-1})	Reference
1	$(NH_4)_2SeO_4$(cr) + aq → $(NH_4)_2SeO_4$(aq, 1:1600)	(10.33 ± 0.13)	(paper)
2	$(NH_4)_2SeO_4$(aq, 1:1600) + aq → 2 NH_4^+ + SeO_4^{2-}	– (1.85 ± 0.50)	(estimated)
3	$(NH_4)_2SeO_4$(cr) + aq → 2 NH_4^+ + SeO_4^{2-}	(8.48 ± 0.52)	(result)

$\Delta_r H_3 = \Delta_r H_1 + \Delta_r H_2$

The standard enthalpy of formation of $(NH_4)_2SeO_4(cr)$ is found from $\Delta_r H_4$ and accepted enthalpies of formation to be – (879.33 ± 4.64) $kJ\cdot mol^{-1}$.

The dissolution data were evaluated on the assumption that the enthalpy of dilution of the ammonium selenate solution was the same as for the corresponding sulphate solution with data from [82WAG/EVA].

The heat of dissolution to the aqueous standard state and the accepted enthalpies of the ions combine to give – (878.50 ± 3.57) $kJ\cdot mol^{-1}$ as an estimate of the enthalpy of formation of $(NH_4)_2SeO_4(cr)$.

[71SHK/MAL]

Shkodin, Malyshev, and Makhmetov measured the vapour pressure of $SeO_2(g)$ above solid and molten sodium selenite by the dew point method. The vapour pressure for $Na_2SeO_3(cr)$ and the temperature interval 967 to 1009 K was $\log_{10} p_{SeO_2} = (15.87 \pm 1.10) - (17700 \pm 1120)\,T^{-1}$ and for $Na_2SeO_3(l)$ and the temperature interval 1015 to 1171 K $\log_{10} p_{SeO_2} = (13.01 \pm 1.02) - (14810 \pm 1110)\,T^{-1}$. The uncertainties have been calculated by the review from an evaluation of the experimental data presented in the deposited VINITI document (No. 3040-71). The review failed to obtain a reasonable agreement between the published primary data and the vapour pressure equation for molten sodium selenite.

The authors write "the dissociation of sodium selenite occurs stepwise" and describe the process by the expression $x\mathrm{Na_2SeO_3(cr, l)} \rightarrow x\mathrm{Na_2O}(x-y)\mathrm{SeO_2(cr)} + y\mathrm{SeO_2(g)}$. The decomposition of Na_2SeO_3 is also discussed in [67KLU/SEL3].

The review does not find it possible to extract thermodynamic quantities from this paper.

[71UY/DRO]

The relative abundances of the species in gaseous aluminium-selenium mixtures were studied in the temperature range 1520 to 1758 K by measuring the ion intensities of the gaseous species in a mass spectrometer. The derived thermodynamic quantities were recalculated by the review using the selected thermodynamic properties of Se(g) and $Se_2(g)$, the CODATA [89COX/WAG] properties of Al(g), and the entropies and heat capacity expressions of the aluminium selenides given in Sections V.8.2.1.1 to V.8.2.1.3. The results are summarised in Table A-90.

Table A-90: Second and third law enthalpies of reactions involving the species AlSe(g), Al_2Se(g), and Al_2Se_2(g).

Reaction	$\Delta_r H_m^\circ$(298.15 K) 2nd law, (kJ·mol^{-1})	$\Delta_r H_m^\circ$(298.15 K) 3rd law, (kJ·mol^{-1})
AlSe(g) + Se(g) $\rightleftharpoons$ Al(g) + Se_2(g)	– (6 ± 25)	– (39 ± 20)
Al_2Se(g) + Se(g) $\rightleftharpoons$ 2AlSe(g)	(3 ± 13)	(14 ± 36)
Al_2Se_2(g) + 2Se(g) $\rightleftharpoons$ 2AlSe(g) + Se_2(g)	(58 ± 38)	(61 ± 50)

[72CHE/KUZ]

The total vapour pressure of liquid selenium at equilibrium was measured by a static method in the temperature range 673 to 958 K and given as $\log_{10}(p/\text{bar}) = 5.217 - 4991\, T^{-1}$. A third law evaluation of the enthalpy of formation of Se_2(g) at 298.15 was made by this review using the vapour pressure at 940 K and the selected data for the heat capacities and entropies of Se(l) and Se_2(g). A mole fraction of Se_2(g), $x_{Se_2(g)}$ = 0.63 was estimated from the selected selenium data. The enthalpy of formation derived was $\Delta_f H_m^\circ$ (Se_2, g, 298.15 K) = (140.8 ± 1.5) kJ·mol^{-1}.

[72FLO]

In two papers, [69FLO] and [72FLO], Flögel exposed powders of HgSe(cr), CdSe(cr), and ZnSe(cr) to slowly passing nitrogen and hydrogen gas at 1 atm pressure. By measuring the flow of gas, and the weight loss of the powders per time unit, the partial pressures of the metal and selenium vapours were calculated. These partial pressures were found to be independent of the flow rate of the gas over the range of 1 to 6 dm^3/h, and therefore a state of equilibrium could be assumed.

The nitrogen gas experiments were used to evaluate equilibrium constants for the reaction 2MeSe(cr) $\rightleftharpoons$ 2Me(g) + Se_2(g).

In the hydrogen gas experiments, a partial formation of H_2Se(g) was assumed to take place according to the reaction MeSe(cr) + $(1-\alpha)H_2$(g) $\rightleftharpoons$ Me(g) + $(1-\alpha)H_2Se$(g) + $0.5\alpha Se_2$(g). The data were used, in combination with the nitrogen gas data, to deduce an equilibrium constant for the reaction H_2(g) + ½Se_2(g) $\rightleftharpoons$ H_2Se(g). In these calculations, the approximations $p_{Me} >> p_{Se_2}$ and $p_{H_2} >> p_{Me} + p_{Se_2}$ were made.

Due to the different properties of the selenide phases studied, different temperature spans were employed, *viz.* 620 to 810 K for HgSe(cr), 970 to 1370 K for CdSe(cr), and 1170 to 1410 K for ZnSe(cr). The expression for the temperature-dependent equilibrium constant presented in [72FLO] is based on the combined information from all three systems studied.

[72GRI/BOR]

The saturated vapour pressure in equilibrium with In_2Se_3(cr) was studied in the temperature range 600 to 1200 K using a Bourdon gauge manometer. The enthalpy of formation of γ-In_2Se_3 was determined to be $\Delta_f H_m^\circ$ (In_2Se_3, γ, 298.15 K) = – 216.3 kJ·mol^{-1} from the the vapour pressure in the temperature range 923 to 1013 K assuming the vaporisation reaction:

$$\gamma\text{-}In_2Se_3 \rightleftharpoons In_2Se(g) + Se_2(g). \qquad (A.27)$$

The vapour pressure data given in Table 3 of the paper were re-evaluated by the review using the selected properties of Se_2(g), the heat capacity and entropy of In_2Se(g) calculated by the review from the molecular parameters obtained by Erkoc, Katircioglu, and Yilmaz [2001ERK/KAT], and the heat capacity, transition enthalpies, and entropy of solid In_2Se_3 given in Section V.8.4.1.5, yielding $\Delta_r H_m^\circ$((A.27), 298.15 K) = 451.8 kJ·mol^{-1} and $\Delta_r S_m^\circ$((A.27), 298.15 K) = 373.6 J·K^{-1}·mol^{-1} from the second law and $\Delta_r H_m^\circ$((A.27), 298.15 K) = 444.8 kJ·mol^{-1} from the third law.

[72MAE/YOK]

The enthalpy of mixing in the liquid state was measured at the temperatures 736 K, 828 K, and 998 K for various compositions in system bismuth-selenium. No enthalpies of formation at 298.15 K were determined in the paper. The enthalpy of formation of Bi_2Se_3(cr) at 298.15 K was therefore calculated by the review from the value reported for the enthalpy of mixing of $Bi_{0.4}Se_{0.6}$ at 998 K, $\Delta_f H_m^\circ$ ($Bi_{0.4}Se_{0.6}$, l, 998 K) = – 22.43 kJ·mol^{-1}, the selected value of the enthalpy of fusion and the heat capacity of Bi_2Se_3(cr), the selected properties of selenium, and the properties of bismuth in [56STU/SIN] yielding $\Delta_f H_m^\circ$ (Bi_2Se_3, cr, 298.15 K) = – 188.8 kJ·mol^{-1}. The value is about 40 kJ·mol^{-1} more negative than in other investigations for unknown reasons.

Analogous measurements were also made in the system antimony-selenium at 935 K yielding $\Delta_{mix} H_m^\circ$ ($Sb_{0.40}Se_{0.60}$, l, 935 K) = – (29.3 ± 0.5) kJ·mol^{-1}. No enthalpy of formation of Sb_2Se_3(cr) was derived but by combining the value of the enthalpy of mixing with the heat capacities of solid and liquid Sb_2Se_3, the enthalpy of fusion of Sb_2Se_3(cr), the data for antimony in [73HUL/DES], and the selected data of selenium, the enthalpy of formation of Sb_2Se_3(cr) was calculated to be $\Delta_f H_m^\circ$ (Sb_2Se_3, cr, 298.15 K) = – (133.0 ± 8.0) kJ·mol^{-1} by the review.

[72MAT/GEI]

The thermodynamic properties of the γ phase $Co_{1-x}Se$(cr) was studied as a function of composition in the temperature range 633 to 818 K using the electrochemical cell

$$(-)\ Co(cr)\ |\ 0.5\ \text{weight-\%}\ CoCl_2\ (\text{in KCl + LiCl melt})\ |\ Co_{1-x}Se(cr)\ (+).$$

The reported values of the enthalpy and entropy of formation were only tabulated at a temperature of 700 K and a recalculation to 298.15 K has been made by the review.

Two compositions, approximately corresponding to the limits of the homogeneity range of $Co_{1-x}Se$(cr), were considered. The recalculation employed the selected properties of selenium, the data of cobalt in [98CHA], and the estimated mean heat capacity of $Co_{1-x}Se$(cr) in the temperature range 298.15 to 700 K. The calculated values are tabulated in Table A-91.

Table A-91: Enthalpy and entropy of formation of $Co_{1-x}Se$(cr). The last column gives the estimated value of the mean heat capacity in the temperature interval 298.15 to 700 K.

Composition	$\Delta_f H_m^\circ$ ($Co_{1-x}Se$, cr, 298.15 K) (kJ·mol^{-1})	$\Delta_f S_m^\circ$ ($Co_{1-x}Se$, cr, 298.15 K) (J·K^{-1}·mol^{-1})	$C_{p,m}^\circ$ ($Co_{1-x}Se$, cr, (298.15 – 700) K) (J·K^{-1}·mol^{-1})
$Co_{0.75}Se$	– (54.5 ± 4.0)	(2.3 ± 6.0)	48
$Co_{0.93}Se$	– (52.7 ± 4.0)	(8.3 ± 6.0)	53

[72MAT/GEI2]

The thermodynamic properties of $Co_{0.72}Se$(cr) were studied in the temperature range 633 to 821 K using the electrochemical cell

(–) Co(cr) | 0.5 weight-% $CoCl_2$ (in KCl + LiCl melt) | $Co_{0.72}Se$(cr) + $CoSe_2$(cr) (+).

The values reported for the enthalpy and entropy of formation at 700 K were $\Delta_f H_m^\circ$ ($Co_{0.72}Se$, cr, 700 K) = – (61.1 ± 3.0) kJ·mol^{-1} and $\Delta_f S_m^\circ$ ($Co_{0.72}Se$, cr, 700 K) = – (13.4 ± 4.0) J·K^{-1}·mol^{-1}. These values were recalculated to 298.15 K by the review using the selected properties of selenium, the data of cobalt in [98CHA], and an estimated effective heat capacity in the temperature range 298.15 to 700 K of $C_{p,m}^\circ$ ($Co_{0.72}Se$, cr, (298.15 – 700) K) = 48 J·K^{-1}·mol^{-1}. The calculated values are $\Delta_f H_m^\circ$ ($Co_{0.72}Se$, cr, 298.15 K) = – (53.8 ± 4.0) kJ·mol^{-1} and $\Delta_f S_m^\circ$ ($Co_{0.72}Se$, cr, 298.15 K) = (0.6 ± 4.0) J·K^{-1}·mol^{-1}.

[72MUN/MES]

The vapour pressure of α-HgSe was measured in the temperature range 530 to 570 K using the torsion-effusion technique. The results were analysed considering all the species in the series Se(g)-Se_8(g) and Hg(g). However, the thermodynamic data applied for the selenium species were different from those selected by the review and the results were therefore re-evaluated.

The total vapour pressure measured was given as $\log_{10}(p/\text{bar})$ = (8.242 ± 0.402) – (6963 ± 220) T^{-1}. The expression was used in the re-evaluation by matching the observed and calculated total pressures considering all species in the series Se(g)-Se_8(g) and Hg(g). The selected data of the selenium species and the CODATA

[89COX/WAG] values for mercury were used. The composition of the gas phase was assumed to correspond to congruent evaporation of stoichiometric α-HgSe. The mole fractions of the most important selenium species are given in Table A-92.

Table A-92: Mole fractions of the most important gaseous selenium species in equilibrium with α-HgSe.

T (K)	Se_2(g)	Se_5(g)	Se_6(g)
530	0.37	0.33	0.26
570	0.39	0.34	0.23

The expression $\Delta_r G^\circ_m(T) = ((201.8 \pm 27.0) - (0.2274 \pm 0.021)\,T)$ kJ·mol^{-1} for the reaction α-HgSe $\rightleftharpoons$ Hg(g) + ½Se_2(g) was derived for the temperature range 530 to 570 K from the calculated partial pressures. A recalculation to 298.15 K using the selected heat capacity of HgSe yields $\Delta_f H^\circ_m$ (HgSe, α, 298.15 K) = – (73.5 ± 20.0) kJ·mol^{-1} and S°_m (HgSe, α, 298.15 K) = (62.6 ± 15.0) J·K^{-1}·mol^{-1}. The corresponding third law value of the enthalpy is $\Delta_f H^\circ_m$ (HgSe, α, 298.15 K) = – (57.0 ± 20.0) kJ·mol^{-1} and was calculated using the selected values for the heat capacity and entropy of α-HgSe. The error limits correspond to those given in the vapour expression multiplied by a factor of two.

[72SIG/WIE]

The saturated vapour pressure of α-CdSe was measured in the temperature range 866 to 1035 K using the Knudsen effusion technique. The congruency of the sublimation was studied in detail and it was found that Se_2(g) evaporates preferentially in the initial stages of the evaporation until the stoichiometry becomes $CdSe_{0.998}$. The selenide then evaporates congruently without further change in stoichiometry. Thermodynamic properties at 298.15 K were evaluated but a set of auxiliary data different from that used by the review was applied. The experimental results given in Table 1 and 3 of the paper were therefore re-evaluated by the review using both the second and the third law, the selected thermodynamic functions of selenium, the data for Cd in [89COX/WAG], the selected heat capacity of α-CdSe, and the selected entropy of α-CdSe in the case of the third law. The vaporisation was assumed to occur according to the reaction α-CdSe $\rightleftharpoons$ Cd(g) + ½Se_2(g). The results were $\Delta_f H^\circ_m$ (CdSe, α, 298.15 K) = – (147.7 ± 5.0) kJ·mol^{-1}, $\Delta_f H^\circ_m$ (CdSe, α, 298.15 K) = – (147.1 ± 2.0) kJ·mol^{-1}, and S°_m (CdSe, α, 298.15 K) = (85.6 ± 5.0) J·K^{-1}·mol^{-1} for the second and third law enthalpy of formation and the entropy, respectively.

[73HAS/MUN]

The saturated vapour pressure of α-ZnSe was measured in the temperature range 952 to 1209 K using the torsion-effusion Knudsen cell technique. The results were evaluated according to the second law assuming the reaction 2α-ZnSe $\rightleftharpoons$ 2Zn(g) + Se_2(g) giving the dissociation enthalpy $\Delta_{sub}H^{\circ}_{m}$(ZnSe, α, 298.15 K) = (710 ± 12) $kJ\cdot mol^{-1}$. No thermodynamic quantities were derived for α-ZnSe. The vapour pressure expression given in Eq. (1) of the paper was therefore evaluated by the review according to the second and the third law using the selected thermodynamic functions of selenium, the data for Zn in [89COX/WAG], the heat capacity expression of α-ZnSe in Section V.9.1.1.1, and the selected entropy of α-ZnSe in the case of the third law. The values obtained were $\Delta_f H^{\circ}_{m}$(ZnSe, α, 298.15 K) = – (157.8 ± 13.0) $kJ\cdot mol^{-1}$, $\Delta_f H^{\circ}_{m}$(ZnSe, α, 298.15 K) = – (175.9 ± 6.6) $kJ\cdot mol^{-1}$, and S°_{m}(ZnSe, α, 298.15 K) = (88.7 ± 12.0) $J\cdot K^{-1}\cdot mol^{-1}$ for the second and third law enthalpy of formation and the entropy, respectively.

[73PIN/NES]

The authors prepared zirconium(IV) selenite by reacting a 0.1 M solution of $ZrOCl_2$ with a large excess of 2.5 M selenious acid. The composition was checked by chemical analysis, but as no X-ray diffraction measurement is reported the specimen will be denoted as $Zr(SeO_3)_2$(s). The enthalpies of dissolution of ZrO_2(cr), SeO_2(cr), and $Zr(SeO_3)_2$(s) in 11.28 M HF (sln) were measured. The resulting solutions will be denoted as dilute (dil) as concentrations used were somewhat variable. The results are used in Table A-93 to calculate the standard enthalpy of formation of $Zr(SeO_3)_2$(s).

Table A-93: Calculation of the standard enthalpy of formation of $Zr(SeO_3)_2$(s). The sources of the data are indicated.

#	Reaction	$\Delta_r H_m$ ($kJ\cdot mol^{-1}$)	Reference
1	ZrO_2(cr) + 4HF(sln) + sln → ZrF_4(sln, dil) + $2H_2O$(l, sln)	– (131.80 ± 2.00)	(paper)
2	SeO_2(cr) + H_2O(l, sln) + sln → H_2SeO_3(sln, dil)	(4.39 ± 0.38)	(paper)
3	$Zr(SeO_3)_2$(s) + 4HF(sln) + sln → ZrF_4(sln, dil) + $2H_2SeO_3$(sln, dil)	– (64.85 ± 1.30)	(paper)
4	Zr(cr) + O_2(g) → ZrO_2(cr)	– (1100.60 ± 0.70)	[2004BRO]
5	Se(cr) + O_2(g) → SeO_2(cr)	– (225.39 ± 0.60)	(review)
6	Zr(cr) + 2Se(cr) + $3O_2$(g) → $Zr(SeO_3)_2$(s)	– (1609.55 ± 2.69)	(result)

$\Delta_r H_6 = \Delta_f H^{\circ}_{m}(Zr(SeO_3)_2, s, 298.15\ K) = \Delta_r H_1 + 2\,\Delta_r H_2 - \Delta_r H_3 + \Delta_r H_4 + 2\,\Delta_r H_5$

The result in the paper is – 1603.3 $kJ\cdot mol^{-1}$. The difference between the two estimates is caused by a change in the enthalpy of formation of ZrO_2(cr).

[73SEL/ZAL]

Selivanova, Zalogina, and Gorokhov mixed $Al_2(SeO_4)_3\cdot18H_2O(cr)$ with a solution of barium chloride in a calorimeter and measured the enthalpy change of the reaction $Al_2(SeO_4)_3\cdot18H_2O(cr) + 3BaCl_2(aq, 1:1600) \rightarrow 3BaSeO_4(cr) + 2AlCl_3(aq, 1:2400) + 18H_2O(l)$.

The hydrolysis of Al^{3+} was not considered in the evaluation of the result in [73SEL/ZAL]. The equilibrium constants in [76BAE/MES] were used to estimate the concentrations of the hydroxo complexes formed under the conditions of the calorimetric experiment. Thus only formation of $AlOH^{2+}$, $Al_2(OH)_2^{4+}$, and $Al_3(OH)_4^{5+}$ was considered. Of the total Al(III), Al^{3+} comprised 98.4%. The enthalpy change per mole of OH^- bound is about (40.0 ± 5.0) kJ according to the same Reference. The evaluation of the experimental data by the review is shown in Table A-94.

Table A-94: Enthalpy change of the reaction $Al_2(SeO_4)_3\cdot18H_2O(cr) + 3Ba^{2+} \rightarrow 3BaSeO_4(cr) + 2Al^{3+} + 18H_2O(l)$. The sources of the data are indicated.

#	Reaction	$\Delta_r H_m$ ($kJ\cdot mol^{-1}$)	Reference
1	$Al_2(SeO_4)_3\cdot18H_2O(cr) + 3BaCl_2(aq, 1:1600) \rightarrow 3BaSeO_4(cr) + 2AlCl_3(aq, 1:2400, hydrolysed) + 18H_2O(l)$	– (87.26 ± 0.50)	(paper)
2	$AlCl_3(aq, 1:2400, hydrolysed) \rightarrow AlCl_3(aq, 1:2400)$	– (0.60 ± 0.20)	(review)
3	$AlCl_3(aq, 1:2400) + aq \rightarrow Al^{3+} + 3Cl^-$	(2.50 ± 0.25)	[82WAG/EVA]
4	$BaCl_2(aq, 1:1600) + aq \rightarrow Ba^{2+} + 2Cl^-$	– (1.38 ± 0.10)	[82WAG/EVA]
5	$Al_2(SeO_4)_3\cdot18H_2O(cr) + 3Ba^{2+} \rightarrow 3BaSeO_4(cr) + 2Al^{3+} + 18H_2O(l)$	– (79.32 ± 0.87)	(result)

$\Delta_r H_5 = \Delta_r H_1 + 2\,\Delta_r H_2 + 2\,\Delta_r H_3 - 3\,\Delta_r H_4$

The standard enthalpy of formation of $Al_2(SeO_4)_3\cdot18H_2O(cr)$ is found from $\Delta_r H_5$ and accepted enthalpies of formation to be – (7970.02 ± 17.06) $kJ\cdot mol^{-1}$.

The authors also prepared anhydrous aluminium selenate by heating $AlH(SeO_4)_2\cdot nH_2O$ to 600 K. The product will be denoted as solid only, as a check of its crystalline state is not mentioned. The enthalpy of dissolution of the two aluminium selenates in 0.5 M KOH to the same unspecified concentration was measured to be – (178.90 ± 0.92) $kJ\cdot mol^{-1}$ and – (442.30 ± 1.02) $kJ\cdot mol^{-1}$ with the result for the anhydrous compound mentioned last. From these data the standard enthalpy of formation of $Al_2(SeO_4)_3(s)$ is calculated to be – (2561.68 ± 17.11) $kJ\cdot mol^{-1}$.

[73SHA/MIS]

The Gibbs energy of formation of PbSe(cr) was measured in the temperature ranges 490 to 600 K and 650 to 858 K using electrochemical cells. The determined enthalpies and entropies of reaction at the mean temperatures of the temperature intervals are given in Table A-95. The quantities were recalculated to 298.15 K by the review and are tabulated in Table A-96. In addition, the corresponding values of the absolute entropy of PbSe(cr) at 298.15 K were calculated. All calculations employed the selected heat capacity of PbSe(cr), the selected thermodynamic properties of selenium, and the data for Pb(cr, l) in [89COX/WAG].

Table A-95: Enthalpies and entropies of formation of PbSe(cr) at the mean reaction temperatures.

Temperature (K)	Reaction	$\Delta_f H_m^\circ$ (PbSe, cr) (kJ·mol^{-1})	$\Delta_f S_m^\circ$ (PbSe, cr) (J·K^{-1}·mol^{-1})
545	Pb(cr) + Se(l) $\rightleftharpoons$ PbSe(cr)	– (107.5 ± 3.0)	– (17.2 ± 4.0)
754	Pb(l) + Se(l) $\rightleftharpoons$ PbSe(cr)	– (108.8 ± 3.0)	– (26.5 ± 4.0)

Table A-96: Standard enthalpies and entropies of formation and the standard entropy of PbSe(cr).

Temperature (K)	$\Delta_f H_m^\circ$ (PbSe, cr, 298.15 K) (kJ·mol^{-1})	$\Delta_f S_m^\circ$ (PbSe, cr, 298.15 K) (J·K^{-1}·mol^{-1})	S_m° (PbSe, cr, 298.15 K) (J·K^{-1}·mol^{-1})
545	– (100.3 ± 3.0)	(2.7 ± 4.0)	(109.6 ± 4.0)
754	– (94.9 ± 4.0)	(1.0 ± 4.0)	(107.9 ± 4.0)

[73SMO/EFR]

Smolyakova, Efremov, and Serebrennikov calculated the enthalpies of formation of the selenates of La, Nd, Eu, Gd, Tb, and Dy (Ln) from a determination of the enthalpy change of the reaction $Ln_2(SeO_4)_3{\cdot}nH_2O$(cr) + $3BaCl_2$(aq) $\rightarrow$ $3BaSeO_4$(cr) + $2LnCl_3$(aq) + nH_2O(l) at 298.15 K, $\Delta_r H$. No primary experimental results or other details are presented in the paper and the standard enthalpies of formation were assigned the wrong sign.

The deposited material in VINITI (No. 6103-73) contains a table of the enthalpies of the precipitation reaction but no primary experimental data. These enthalpies carry a positive sign, which contradicts the findings in [69SUP/MAI] for neodymium that is common to both investigations. Since the primary experimental data there show that the precipitation reaction is exothermic, the review has changed the sign of $\Delta_r H$

before entering it into the thermochemical calculations. An inspection of the VINITI document further revealed that the standard enthalpies of formation have been calculated with auxiliary enthalpies of formation of $BaCl_2$(cr) and $LnCl_3$(cr) instead of the appropriate values for aqueous solution. The published results are therefore unacceptable.

A recalculation of $\Delta_f H^\circ_m$ ($Ln_2(SeO_4)_3\cdot nH_2O$, cr, 298.15 K) was made from the relationship $\Delta_f H^\circ_m$ ($Ln_2(SeO_4)_3\cdot nH_2O$(cr)) = $3\Delta_f H^\circ_m$ ($BaSeO_4$(cr)) + $2\Delta_f H^\circ_m$ (Ln^{3+}) + $n\Delta_f H^\circ_m$ (H_2O, l) − $3\Delta_f H^\circ_m$ (Ba^{2+}) − $\Delta_r H^\circ_m$. This approximation is estimated to cause only a minor systematic error. Auxiliary data were taken from [2001COR/KON] and the review. The results are summarised in Table A-97.

Table A-97: Results of and data used for the calculation of the standard enthalpy of formation of $Ln_2(SeO_4)_3\cdot nH_2O$(cr).

Salt	$-\Delta_f H^\circ_m$ (Ln^{3+}) (kJ·mol^{-1})	$-\Delta_r H^\circ_m$ (kJ·mol^{-1})	$-\Delta_f H^\circ_m$ (salt) (kJ·mol^{-1})
$La_2(SeO_4)_3\cdot 5H_2O$(cr)	(707.6 ± 2.5)	(117.0 ± 0.80)	(4555.0 ± 17.5)
$Nd_2(SeO_4)_3\cdot 5H_2O$(cr)	(694.8 ± 2.0)	(122.2 ± 0.54)	(4524.2 ± 17.5)
$Eu_2(SeO_4)_3\cdot 8H_2O$(cr)	(605.4 ± 4.0)	(89.83 ± 1.59)	(5235.2 ± 17.5)
$Gd_2(SeO_4)_3\cdot 8H_2O$(cr)	(698.4 ± 2.0)	(66.52 ± 1.30)	(5444.5 ± 17.5)
$Tb_2(SeO_4)_3\cdot 8H_2O$(cr)	(699.8 ± 4.0)	(87.70 ± 1.13)	(5426.1 ± 17.5)
$Dy_2(SeO_4)_3\cdot 8H_2O$(cr)	(700.2 ± 3.0)	(143.2 ± 0.59)	(5371.4 ± 17.5)

[73SOB/SEL]

Sobol and Selivanova measured the enthalpy change of the reaction $NaHSeO_4$(cr) + $BaCl_2$(aq, 1:800) → $BaSeO_4$(cr) + NaCl(aq, 1:800) + HCl(aq, 1:800) in a calorimeter. The result has been recalculated to standard conditions with new auxiliary data in Table A-98. The enthalpy of mixing was evaluated separately.

With accepted standard enthalpies of formation $\Delta_r H_6$ yields:

$$\Delta_f H^\circ_m (\text{NaHSeO}_4, \text{cr}, 298.15\text{ K}) = -(822.23 \pm 5.59)\text{ kJ·mol}^{-1}.$$

Table A-98: Determination of the enthalpy change of the reaction: $NaHSeO_4$(cr) + Ba^{2+} → $BaSeO_4$(cr) + Na^+ + H^+ under standard conditions. The sources of the data are indicated.

#	Reaction	$\Delta_r H_m$ (kJ·mol^{-1})	Reference
1	$NaHSeO_4$(cr) + $BaCl_2$(aq, 1:800) → $BaSeO_4$(cr) + {NaCl(aq, 1:800) + HCl(aq, 1:800)}	– (27.41 ± 0.13)	(paper)
2	$BaCl_2$(aq, 1:800) + aq → Ba^{2+} + 2Cl^-	– (1.63 ± 0.10)	[82WAG/EVA]
3	NaCl(aq, 1:800) + aq → Na^+ + Cl^-	– (0.32 ± 0.10)	[82WAG/EVA]
4	HCl(aq, 1:800) + aq → H^+ + Cl^-	– (0.47 ± 0.05)	[82WAG/EVA]
5	NaCl(aq, 1:800) + HCl(aq, 1:800) → {NaCl(aq, 1:800) + HCl(aq, 1:800)}	(0.74 ± 0.10)	(paper)
6	$NaHSeO_4$(cr) + Ba^{2+} → $BaSeO_4$(cr) + Na^+ + H^+	– (27.31 ± 0.22)	(result)

$\Delta_r H_6 = \Delta_r H_1 - \Delta_r H_2 + \Delta_r H_3 + \Delta_r H_4 - \Delta_r H_5$

[74JEL/KOM]

The thermodynamic properties of the γ phase $Co_{1-x}Se$(cr) was studied as a function of composition in the temperature range 873 to 1273 K by measuring the vapour pressure of selenium using an isopiestic method. The reported values of the enthalpy and entropy of formation were tabulated at a temperature of 873 K and a recalculation to 298.15 K was made by review. Two compositions, approximately corresponding to the limits of the homogeneity range of $Co_{1-x}Se$(cr), were considered. The recalculation employed the selected properties of selenium, the data for cobalt in [98CHA], and the estimated mean heat capacity of $Co_{1-x}Se$(cr) in the temperature range 298.15 to 873 K. The calculated values are tabulated in Table A-99.

Table A-99: Enthalpy and entropy of formation of $Co_{1-x}Se$. The last column gives the estimated value of the mean heat capacity in the temperature interval 298.15 to 873 K.

Composition	$\Delta_f H^\circ_m$ ($Co_{1-x}Se$, cr, 298.15 K) (kJ·mol^{-1})	$\Delta_f S^\circ_m$ ($Co_{1-x}Se$, cr, 298.15 K) (J·K^{-1}·mol^{-1})	$C^\circ_{p,m}$ ($Co_{1-x}Se$, cr, (298.15 – 873) K) (J·K^{-1}·mol^{-1})
$Co_{0.75}Se$	– 71.4	– 2.1	49
$Co_{0.92}Se$	– 57.4	20.9	55

[74KUL2]

Kullberg measured the heat evolved when small amounts of red HgO(cr) were dissolved in solutions of 0.3 and 1 M KSeCN in a reaction solution calorimeter. The HgO(cr) was contained in a thin-walled glass ampoule, which could be broken inside the calorimeter and the initial temperature of the solution in the calorimeter was adjusted to give an almost constant temperature after the reaction. Due to the slowness of the reaction, this was a necessary precaution to obtain good data. To ascertain that the effects registered were not due to the presence of contaminating CN^- ions, a suitable amount of Hg^{2+} ions was added to the solution immediately prior to the experiments.

The heats evolved in a series of 8 experimental runs were found to be directly proportional to the amount of HgO(cr) dissolved and independent of the $SeCN^-$ concentration. Based on the work of Toropova [56TOR], Kullberg assumed that $Hg(SeCN)_4^{2-}$ was formed and calculated the reaction enthalpy for

$$HgO(cr) + 4SeCN^- + H_2O \rightarrow Hg(SeCN)_4^{2-} + 2OH^-$$

to be $-(102.5 \pm 1.4)$ kJ·mol^{-1}.

By using the same experimental set-up, Kullberg also determined the reaction enthalpy for the reaction $HgO(cr) + 2H^+ \rightarrow Hg^{2+} + H_2O$ in 1 M $HClO_4$, $-(22.8 \pm 0.4)$ kJ·mol^{-1}. He then combined the two values with the reaction enthalpy for the formation of water at $I = 1$ M, -56.11 kJ·mol^{-1}, to yield the reaction enthalpy for

$$Hg^{2+} + 4SeCN^- \rightleftharpoons Hg(SeCN)_4^{2-}\ ; \Delta_r H = -(191.9 \pm 1.5)\ \text{kJ·mol}^{-1}.$$

[74RAU]

New density and total pressure measurements of selenium vapour are presented. A previous investigation by the same author, [67RAU], is re-evaluated and combined with the new data in an evaluation of the thermodynamic properties of the molecules $Se_2(g)$-$Se_8(g)$. The enthalpies of formation and entropies were varied in a procedure to obtain a good fit to the observed densities. The procedure contains many parameters that can be varied and different sets may fit the experimental data almost equally well. The thermodynamic properties derived for the species $Se_2(g)$-$Se_8(g)$ are too dependent on the choices made by the author in order to be considered by the review. However, the new measurements of the total vapour pressure in the temperature range 1073 to 1373 K were evaluated by the review for the thermodynamic properties of $Se_2(g)$ using the third law and mole fractions of $Se_2(g)$ estimated from the selected selenium data. The enthalpy of formation derived was $\Delta_f H_m^\circ$ (Se_2, g, 298.15 K) = (138.9 ± 1.9) kJ·mol^{-1}.

[75HOA/REY]

The vapour pressures of the species $Se_2(g)$, and $Se_5(g)$-$Se_8(g)$ in equilibrium with selenium were determined for the temperature range 429 to 575 K using a mass spectrometer and a Knudsen cell. The data were divided into two sets, one for the equilibrium with solid selenium and one for the equilibrium with liquid selenium. In the evaluation of the experimental data in Figure 3 of the paper, the straight lines should intersect at the melting point of selenium. However, the intersection is at 479 K and not at the true melting point (494 K) indicating either an offset in the temperature values or a mistake made in the selection of the experimental points for the linear regressions. In the latter case, points from the first set have been incorporated in the second set and used in the least squares calculations on the equilibrium with liquid selenium. The misplaced data would then be missing in the corresponding calculation on the equilibrium with solid selenium. It can also be noted that the heat of fusion (10.878 kJ·mol^{-1}) obtained from the slopes of the lines in Figure 3 is almost twice as large as the correct value, 6.159 kJ·mol^{-1}. These observations taken together indicate systematic errors in the data. Indeed, the enthalpies obtained from the second and third law evaluations made by the review show an unacceptable large difference for the equilibria with the liquid phase, while the difference is reasonable for the equilibria with solid selenium (see Table A-100 and Table A-101).

Table A-100: Second and third law evaluations of equilibria with solid selenium.

Species	$\Delta_f H_m$ (445 K) (kJ·mol^{-1}) original values	$\Delta_f H^\circ_m$ (298.15 K) (kJ·mol^{-1}) 2nd law	3rd law	S°_m (298.15 K) (J·K^{-1}·mol^{-1})
$Se_5(g)$	(147.3 ± 8.4)	(151.2 ± 8.4)	(145.5 ± 5.9)	(411.2 ± 18.8)
$Se_6(g)$	(136.4 ± 8.4)	(140.5 ± 8.4)	(137.8 ± 5.9)	(434.6 ± 18.8)
$Se_7(g)$	(147.3 ± 8.4)	(151.7 ± 8.4)	(148.6 ± 5.9)	(494.1 ± 18.8)

Table A-101: Second and third law evaluations of equilibria with liquid selenium.

Species	$\Delta_f H_m$ (525 K) (kJ·mol^{-1}) original values	$\Delta_f H^\circ_m$ (298.15 K) (kJ·mol^{-1}) 2nd law	3rd law	S°_m (298.15 K) (J·K^{-1}·mol^{-1})
$Se_2(g)$	(123.4 ± 8.4)	(138.9 ± 8.4)		
$Se_5(g)$	(90.8 ± 8.4)	(129.3 ± 8.4)	(142.5 ± 8.4)	(373.2 ± 15.9)
$Se_6(g)$	(73.6 ± 8.4)	(118.9 ± 8.4)	(136.2 ± 8.4)	(395.6 ± 15.9)
$Se_7(g)$	(69.5 ± 8.4)	(121.7 ± 8.4)	(147.0 ± 8.4)	(438.7 ± 15.9)
$Se_8(g)$	(60.7 ± 8.4)	(119.8 ± 8.4)		

The results obtained at 445 K and 525 K were recalculated to 298.15 K using the second and third laws and the heat capacities and entropies selected by this review. For the third law evaluation, relative ion intensities were taken from Figure 3 of the paper and combined with the total pressures reported by Niwa and Shibata [40NIW/SHI] for the lower and by Yamdagni and Porter [68YAM/POR] for the higher temperature range. A third law evaluation for $Se_2(g)$ and $Se_8(g)$ could not be made because of the lack of partial pressure data.

[76CHA/CHE]

Chao and Cheng [76CHA/CHE] studied the determination of a number of anions, single or in mixtures, by a stepwise potentiometric titration with silver nitrate at pH = 11. The silver ion activity was measured with a silver ion selective electrode based on silver sulphide. The data were also used to estimate the solubility products of the silver salts formed during a titration. The method is only sketched in the paper but appears to have proceeded along the following course. The potential of incipient precipitation (E_{prec}) was estimated graphically from the shape of the titration curve. E_{prec} would thus be a measure of the silver ion activity at the nominal and known concentration of the anion in the presence of its silver salt.

The calibration of the electrode was accomplished by mixing the sample with a known concentration of sulphide ion. The first (Ag_2S, S^{2-}) and last (Ag_2O, pH = 11) points of the titration were then used to calibrate the electrode by calculating the corresponding pAg values from the relevant solubility products (no details presented). Intermediate pAg values were calculated from the measured E_{prec} by assuming a linear relationship between the electrode potential and pAg between the two calibration points. The numerical value of the slope of the calibration curve is not mentioned. The procedure is described in the Ph. D. thesis by one of the authors, but this source was not consulted by the review.

Although the method of the paper is of doubtful accuracy most of the presented solubility products agree within an order of magnitude or better with published values. The solubility product of Ag_2SeO_3 was reported to be $\log_{10} K_{s,0}$ (Ag_2SeO_3(s), 293.15 K) = – (15.45 ± 0.15) at I = 0.1 (medium not stated). The value at I = 0 is estimated to be – (15.85 ± 0.30) by the SIT approach.

[76GLA/MIK]

The vapour pressure of Se_2(g) in equilibrium with $MoSe_2$(cr) and $TaSe_2$(cr) were measured in this study. Only the measurements on $MoSe_2$(cr) are discussed here. The results for $TaSe_2$(cr) are discussed in Section V.11.18. The vapour pressure of Se_2(g) in equilibrium with $MoSe_2$(cr) and Mo(cr) was determined in the temperature range 1233 to 1353 K using mass spectrometry and Knudsen effusion cells. The results were evaluated according to the reaction

$$MoSe_2(cr) \rightleftharpoons Mo(cr) + Se_2(g).$$

In [69SPI/ROU] it was shown that the homogeneity range of $MoSe_2$ extends to $MoSe_{1.91}$ and, therefore, it is more appropriate to make the evaluation using the reaction $MoSe_{1.91}(cr) \rightleftharpoons Mo(cr) + 0.955Se_2(g)$ as basis. This was done in a similar study by Viksman, Gordienko, Yanaki, and Klochkov [86VIK/GOR].

The results presented in Table 1 of the paper were re-evaluated by the present review using both the second and third laws. The necessary data were taken from the selected data in this review except for molybdenum for which data in [98CHA] were used. The selected heat capacity of $MoSe_2$(cr) was rescaled to the composition $MoSe_{1.91}$ according to the stoichiometry. The second law results are $\Delta_f H^\circ_m$ ($MoSe_{1.91}$, cr, 298.15 K) = – (264.1 ± 13.0) kJ·mol^{-1} and S°_m ($MoSe_{1.91}$, cr, 298.15 K) = (42.0 ± 10.0) J·K^{-1}·mol^{-1}. The third law enthalpy of formation is $\Delta_f H^\circ_m$ ($MoSe_{1.91}$, cr, 298.15 K) = – (209.4 ± 13.0) kJ·mol^{-1}. The error limits correspond to one standard deviation in the linear regression on the experimental data.

[76KIR/RAV]

The saturated vapour pressure of α-ZnSe was measured in the temperature range 923 to 1123 K using a weight loss technique and mass spectrometry. The results were presented as $\log_{10} K_p$ for the reaction 2α-ZnSe $\rightleftharpoons$ 2Zn(g) + Se_2(g). No thermodynamic quantities were derived for α-ZnSe. The values of $\log_{10} K_p$ reported in Table 1 of the paper were therefore evaluated by the review according to the second and the third law using the selected thermodynamic functions of selenium, the data for Zn in [89COX/WAG], the heat capacity expression of α-ZnSe in Section V.9.1.1.1, and the selected entropy of α-ZnSe in the case of the third law. The values obtained were $\Delta_f H^\circ_m$ (ZnSe, α, 298.15 K) = – (164.5 ± 6.0) kJ·mol^{-1}, $\Delta_f H^\circ_m$ (ZnSe, α, 298.15 K) = – (179.1 ± 4.2) kJ·mol^{-1}, and S°_m (ZnSe, α, 298.15 K) = (86.2 ± 9.0) J·K^{-1}·mol^{-1} for the second and third law enthalpy of formation and the entropy, respectively.

[76PET/KOF]

The heat capacity of α-CdSe was measured in the temperature range 4.2 to 300 K. The heat capacity showed no anomalies except in the range 45 to 50 K where a small deviation from smooth behaviour was observed. The absolute entropy was presented only in a graph of poor resolution and no value was given at 298.15 K. The entropy was therefore evaluated by the review from the heat capacity values given in Table 1 of the paper. The resulting absolute entropy at 298.15 K was S°_m (CdSe, α, 298.15 K) = (86.76 ± 1.00) J·K^{-1}·mol^{-1}. The error limit was estimated from the general accuracy for this type of measurements and includes the effect of estimated systematic errors.

[76PIA/BAR]

The vapour pressure of GaSe(cr) was measured in the temperature range 1053 to 1143 K using mass spectrometry, thermogravimetry, and torsion-effusion Knudsen cells. It was found that the vaporization reaction is

$$4GaSe(cr) \rightleftharpoons Ga_2Se(g) + Ga_2Se_3(cr) \qquad (A.28)$$

and the total pressure was reported to be $\log_{10}(p/\text{bar}) = (12.61 \pm 0.5) - (19800 \pm 500)\, T^{-1}$. The expression was re-evaluated by the review using the second and the third law and the data for the gallium selenides given in Sections V.8.3.1.1, V.8.3.1.2, and V.8.3.1.3. The changes in the thermodynamic quantities for Reaction (A.28) at 298.15 K were $\Delta_r H^\circ_m$ ((A.28), 298.15 K) = 400.8 kJ·mol^{-1} and $\Delta_r S^\circ_m$ ((A.28),

298.15 K) = 241.4 J·K^{-1}·mol^{-1} from the second law and $\Delta_r H^\circ_m$((A.28), 298.15 K) = 334.3 kJ·mol^{-1} from the third law. The large difference between the second and third law values of the enthalpy of formation indicates systematic errors.

[76TAN/VOL]

Tananaev *et al.* [76TAN/VOL] prepared crystalline $GaH(SeO_3)_2 \cdot 2H_2O$ and $GaH(SeO_3)_2$. The enthalpies of dissolution of the salts in HCl(aq, 1:8), denoted sln, were measured at 298.15 K and with auxiliary data used to calculate the standard enthalpies of formation of the gallium acid selenites in Table A-102.

Table A-102: Calculation of the standard enthalpies of formation of $GaH(SeO_3)_2 \cdot 2H_2O$ and $GaH(SeO_3)_2$. The sources of the data are indicated.

#	Reaction	$\Delta_r H_m$ (kJ·mol^{-1})	Reference
1	$GaH(SeO_3)_2 \cdot 2H_2O$(cr) + 3HCl(sln) + sln → $GaCl_3$(sln, aq 1:3100) + $2H_2SeO_3$(sln, aq 1:1550) + $2H_2O$(l, sln)	(47.24 ± 0.71)	(paper)
2	$GaH(SeO_3)_2$(cr) + 3HCl(sln) + sln → $GaCl_3$(sln, aq 1:3100) + $2H_2SeO_3$(sln, aq 1:1550)	(8.95 ± 1.25)	(paper)
3	$GaCl_3$(cr) + sln → $GaCl_3$(sln, aq 1:3100)	– (103.22 ± 1.67)	[76TAN/VOL2]
4	H_2SeO_3(cr) + sln → H_2SeO_3(sln, aq 1:2070)	(24.98 ± 0.42)	[76TAN/VOL2]
5	$0.5H_2 + 0.5Cl_2$ + sln → HCl(sln)	– (153.21 ± 0.11)	[2000RAN/FUG]
6	Se(cr) + H_2(g) + $1.5O_2$(g) → H_2SeO_3(cr)	– (524.72 ± 0.65)	(review)
7	Ga(cr) + $1.5Cl_2$(g) → $GaCl_3$(cr)	– (524.70 ± 2.50)	[82WAG/EVA]
8	H_2(g) + $0.5O_2$(g) + sln → H_2O(l, sln)	– (286.67 ± 0.04)	[2000RAN/FUG]
9	Ga(cr) + 2Se(cr) + (n+0.5)H_2(g) + (0.5n+3)O_2(g) → $GaH(SeO_3)_2 \cdot nH_2O$(cr)	– (1788.35 ± 3.47) – (1176.72 ± 3.62)	(result)

$\Delta_r H_9 = -\Delta_r H_{1\text{-}2} + \Delta_r H_3 + 2\,\Delta_r H_4 - 3\,\Delta_r H_5 + 2\,\Delta_r H_6 + \Delta_r H_7 + n\,\Delta_r H_8$

The results:

$$\Delta_f H^\circ_m(GaH(SeO_3)_2 \cdot 2H_2O, \text{cr}, 298.15\text{ K}) = -(1788.4 \pm 3.5)\text{ kJ·mol}^{-1},$$

$$\Delta_f H^\circ_m(GaH(SeO_3)_2, \text{cr}, 298.15\text{ K}) = -(1176.7 \pm 3.5)\text{ kJ·mol}^{-1},$$

obtained here differ substantially from those in the paper, which did not use partial molar enthalpies of formation for H_2O and HCl in the calculations. Furthermore, the published value for the anhydrous salt disagrees with the primary data.

[76TAN/VOL2]

Tananaev *et al.* [76TAN/VOL2] prepared crystalline $Ga_2(SeO_3)_3 \cdot 6H_2O$, $Ga_2(SeO_3)_3 \cdot H_2O$, and $Ga_2(SeO_3)_3$. The enthalpies of dissolution of the salts in HCl(aq, 1:8), denoted sln, were measured at 298.15 K. The enthalpies of dissolution of $GaCl_3$(cr) and H_2SeO_3(cr) in the same medium were also determined. The standard enthalpies of formation of the gallium selenites are calculated from the measurements with new auxiliary data in Table A-103.

The results are $\Delta_f H_m^\circ$ ($Ga_2(SeO_3)_3 \cdot 6H_2O$, cr, 298.15 K) = – (3682.3 ± 6.6) $kJ \cdot mol^{-1}$, $\Delta_f H_m^\circ$ ($Ga_2(SeO_3)_3 \cdot H_2O$, cr, 298.15 K) = – (2138.4 ± 6.6) $kJ \cdot mol^{-1}$, $\Delta_f H_m^\circ$ ($Ga_2(SeO_3)_3$, cr, 298.15 K) = – (1823.7 ± 6.5) $kJ \cdot mol^{-1}$. They differ substantially from those in the paper, which did not use partial molar enthalpies of formation for H_2O and HCl in the calculations.

Table A-103: Calculation of the standard enthalpies of formation of $Ga_2(SeO_3)_3 \cdot 6H_2O$, $Ga_2(SeO_3)_3 \cdot H_2O$, and $Ga_2(SeO_3)_3$. The sources of the data are indicated.

#	Reaction	$\Delta_r H_m$ ($kJ \cdot mol^{-1}$)	Reference
1	$Ga_2(SeO_3)_3 \cdot 6H_2O$(cr) + 6HCl(sln) + sln → $2GaCl_3$(sln, aq 1:3100) + $3H_2SeO_3$(sln, aq 1:2070) + $6H_2O$(l, sln)	(126.44 ± 1.34)	(paper)
2	$Ga_2(SeO_3)_3 \cdot H_2O$(cr) + 6HCl(sln) + sln → $2GaCl_3$(sln, aq 1:3100) + $3H_2SeO_3$(sln, aq 1:2070) + H_2O(l, sln)	(15.94 ± 1.25)	(paper)
3	$Ga_2(SeO_3)_3$(cr) + 6HCl(sln) + sln → $2GaCl_3$(sln, aq 1:3100) + $3H_2SeO_3$(sln, aq 1:2070)	– (12.09 ± 0.25)	(paper)
4	$GaCl_3$(cr) + sln → $GaCl_3$(sln, aq 1:3100)	– (103.22 ± 1.67)	(paper)
5	H_2SeO_3(cr) + sln → H_2SeO_3(sln, aq 1:2070)	(24.98 ± 0.42)	(paper)
6	$0.5H_2 + 0.5Cl_2$ + sln → HCl(sln)	– (153.21 ± 0.11)	[2000RAN/FUG]
7	Se(cr) + H_2(g) + $1.5O_2$(g) → H_2SeO_3(cr)	– (524.72 ± 0.65)	(review)
8	Ga(cr) + $1.5Cl_2$(g) → $GaCl_3$(cr)	– (524.70 ± 2.50)	[82WAG/EVA]
9	H_2(g) + $0.5O_2$(g) + sln → H_2O(l, sln)	– (286.67 ± 0.04)	[2000RAN/FUG]
10	2Ga(cr) + 3Se(cr) + nH_2(g) + $0.5(n+9)O_2$(g) → $Ga_2(SeO_3)_3 \cdot nH_2O$(cr)	– (3682.26 ± 6.62) – (2138.37 ± 6.59) – (1823.71 ± 6.48)	(result)

$\Delta_r H_{10} = - \Delta_r H_{1\text{-}3} + 2\,\Delta_r H_4 + 3\,\Delta_r H_5 - 6\,\Delta_r H_6 + 3\,\Delta_r H_7 + 2\,\Delta_r H_8 + n\,\Delta_r H_9$

[77DUT/TIS]

Mass spectrometry and Knudsen effusion cells were used to measure the partial pressure of Se_2(g) in equilibrium with solid and liquid Ag_2Se. For this work, the results were re-evaluated using the second and third laws, the selected values for the heat capacity and entropy of Ag_2Se, and the selected data for selenium and silver. The result is given in Table A-104.

Table A-104: Re-evaluated enthalpy of formation and entropy of α-Ag_2Se at 298.15 K. The evaluations were made at 1100 K and 1200 K in the lower and higher temperature range, respectively. The uncertainties are based on the standard deviations of the coefficients in the linear regressions.

$\Delta_f H_m^\circ$ (298.15 K) (kJ·mol^{-1})		S_m° (298.15 K) (J·K^{-1}·mol^{-1})	Temperature (K)
2nd law	3rd law		
– (36.7 ± 8.0)	– 40.9	(153.1 ± 6.0)	1070 – 1170
– (14.5 ± 8.0)	– 40.8	(175.8 ± 6.0)	1170 – 1270

[77MIK/GLA]

The pressure of the vapour in equilibrium with WSe_2(cr) was studied in the temperature range 1138 to 1503 K using mass spectroscopy and a Knudsen cell. The ion intensities were recorded in the temperature range 1138 to 1428 K and the enthalpy change of the reaction

$$WSe_2(cr) \rightleftharpoons W(cr) + Se_2(g) \quad (A.29)$$

was derived from the change in the ion intensity with temperature to be $\Delta_r H_m^\circ$((A.29), 1300 K) = (259.4 ± 9.0) kJ·mol^{-1}. In contrast, the absolute pressures listed for the range 1268 to 1503 K in Table 1 of the paper yield a value of $\Delta_r H_m^\circ$((A.29), 1400 K) = (109.3 ± 25.0) kJ·mol^{-1} for the same reaction. The former enthalpy of reaction results in an enthalpy of formation at 298.15 K of $\Delta_f H_m^\circ$(WSe_2, cr, 298.15 K) = – (134.1 ± 10.0) kJ·mol^{-1} and the latter yields a value of $\Delta_f H_m^\circ$(WSe_2, cr, 298.15 K) = (14.0 ± 25.0) kJ·mol^{-1}. The selected thermodynamic values for Se_2(g) and W(cr), and the heat capacity of WSe_2(cr) were employed in the calculations. Similarly, the standard entropy at 298.15 K becomes S_m°(WSe_2, cr, 298.15 K) = (245.9 ± 20.0) J·K^{-1}·mol^{-1} from the vapour pressure measurements. A third law evaluation results in $\Delta_f H_m^\circ$(WSe_2, cr, 298.15 K) = – (209.7 ± 10.0) kJ·mol^{-1}. The large inconsistencies between the values make this investigation highly questionable.

[77OGA/NIS]

The heat capacity of $CoSe_2$(cr) was measured in the temperature range 0 to 282 K. No values of the heat capacity or entropy at 298.15 K were given in the paper. These properties were therefore evaluated by the review from the graph in Figure 3 of the paper. The values derived from the slightly extrapolated heat capacity data were $C_{p,m}^\circ$($CoSe_2$, cr, 298.15 K) = (75.6 ± 1.0) J·K^{-1}·mol^{-1} and S_m°($CoSe_2$, cr, 298.15 K) = (97.2 ± 2.0) J·K^{-1}·mol^{-1}.

[77PIS/MIK]

The vapour pressure of $WSe_2(cr)$ was determined thermogravimetrically using a Knudsen cell. The vapour pressure was summarised by the formula $\log_{10}(p/\text{mm Hg}) = 13.550 - 19850\ T^{-1}$ where the decimal point in the first constant is missing in the original paper. The pressure calculated from this expression is smaller by a factor of 10 compared to that presented in Figures 1 and 2 in the paper. The equation was assumed to be correct by this review and entropy and enthalpy values at 298.15 K were derived using the selected thermodynamic properties for $Se_2(g)$ and W(cr), and the heat capacity of $WSe_2(cr)$. The second law results were $\Delta_f H^\circ_m(WSe_2, cr, 298.15\ K) = -252.4\ kJ\cdot mol^{-1}$ and $S^\circ_m(WSe_2, cr, 298.15\ K) = 40.8\ J\cdot K^{-1}\cdot mol^{-1}$. The third law enthalpy of formation was $\Delta_f H^\circ_m(WSe_2, cr, 298.15\ K) = -198.3\ kJ\cdot mol^{-1}$. Considering the general uncertainties in this investigation no error limits were assigned.

[78ARU]

Aruga [78ARU] determined the enthalpy of the reactions:

$$Zn^{2+} + SeO_4^{2-} \rightleftharpoons ZnSeO_4(aq) \quad (A.30)$$

$$Cd^{2+} + SeO_4^{2-} \rightleftharpoons CdSeO_4(aq) \quad (A.31)$$

from calorimetric titrations at 298.15 K. Successive amounts of a 0.17 M tetraethylammonium selenate solution were added to a 0.17 M metal nitrate solution. The measurements were made with an LKB Precision Calorimeter System, but only three points were measured for each reaction.

The data were evaluated on the assumption that only the 1:1 complex was formed, which is reasonable. The stability constants determined by Banks [34BAN] were used and recalculated to $I = 0.5$ M by Davies' equation with the coefficient of the linear term equal to 0.2. The result was $\beta_1(ZnSeO_4) = 8.13\ M^{-1}$ and $\beta_1(CdSeO_4) = 9.77\ M^{-1}$. Since the complex formation is rather weak, the calculated concentration changes during the titration will be sensitive to the β_1-values employed. The uncertainty in these values is estimated to be $\pm 2\ M^{-1}$.

The enthalpy changes obtained were:

$$\Delta_r H_m((A.30), I = 0.5\ M, 298.15\ K) = (0.20 \pm 0.04)\ kJ\cdot mol^{-1},$$

$$\Delta_r H_m((A.31), I = 0.5\ M, 298.15\ K) = (3.90 \pm 0.40)\ kJ\cdot mol^{-1}.$$

The uncertainty for the cadmium complex has been increased considerably by the review due to the uncertainty in the value of the equilibrium constant. Aruga concurrently determined the corresponding quantities for the sulphate complexes and calculated theoretically the difference between the enthalpy change at $I = 0$ and 0.5 M to be $4.05\ kJ\cdot mol^{-1}$. The correction was estimated by the review to be $(4.40 \pm 0.20)\ kJ\cdot mol^{-1}$. A comparison of the sulphate data extrapolated to standard conditions with other calorimetric data in [73POW] indicates that the results reported by Aruga are low by about

4 kJ·mol^{-1}. With a correction to standard conditions amounting to 4.40 kJ·mol^{-1} as for sulphate, the following selected results are found for the selenates:

$$\Delta_r H^\circ_m((A.30), 298.15\ K) = (4.6 \pm 4.0)\ kJ\cdot mol^{-1},$$

$$\Delta_r H^\circ_m((A.31), 298.15\ K) = (8.3 \pm 4.0)\ kJ\cdot mol^{-1}.$$

The assigned error limits reflect a possible systematic error in the evaluation of the data.

[78BLA/GUN]

The heat content of $Ag_2Se(cr)$ and $Cu_2Se(cr)$ were measured in the temperature range 350 to 1500 K using a drop calorimeter. Tables with thermodynamic properties at different temperatures were presented but no analytical expressions for the temperature dependence of the heat capacity were given. The following expressions were derived by this review from least-squares refinements using the data in Tables 1 and 3 of the paper,

$$C^\circ_{p,m}(Cu_2Se, \alpha, (298.15 - 395)\ K) = (-\ 10.6639 + 286.3173 \times 10^{-3}\ T)\ J\cdot K^{-1}\cdot mol^{-1},$$

$$C^\circ_{p,m}(Cu_2Se, \beta, (395 - 1390)\ K) = (75.9761 + 5.5020 \times 10^{-3}\ T + 10.1346 \times 10^{5}\ T^{-2})\ J\cdot K^{-1}\cdot mol^{-1},$$

$$C^\circ_{p,m}(Ag_2Se, \alpha, (298.15 - 407)\ K) = (37.709 + 107.819 \times 10^{-3}\ T + 9.85077 \times 10^{5}\ T^{-2})\ J\cdot K^{-1}\cdot mol^{-1},$$

and

$$C^\circ_{p,m}(Ag_2Se, \beta, (407 - 1170)\ K) = (82.4368 + 7.003 \times 10^{-3}\ T)\ J\cdot K^{-1}\cdot mol^{-1}.$$

[78GER/GLA]

The paper is a compilation of vapour pressure expressions for a large number of elements. No new experimental information on selenium is given.

[78ROL/HOC]

The saturated vapour pressure of α-ZnSe was measured in the temperature range 875 to 1100 K using Knudsen cells and mass spectrometry. The results were evaluated by applying the second and third law and assuming the reaction α-ZnSe $\rightleftharpoons$ Zn(g) + ½Se_2(g). The sublimation enthalpies $\Delta_{sub}H^\circ_m$(298.15 K) = 408 kJ·mol^{-1} and $\Delta_{sub}H^\circ_m$(298.15 K) = 416 kJ·mol^{-1} were derived for the second and third law enthalpy of reaction, respectively. No quantities were derived for α-ZnSe at 298.15 K and no vapour pressure expression was given. The experimental vapour pressure data reported in Table 1 of the paper were therefore evaluated by the review according to the second and the third law using the selected thermodynamic functions of selenium, the data for Zn in [89COX/WAG], the heat capacity expression of α-ZnSe in Section V.9.1.1.1, and the selected entropy of α-ZnSe in the case of the third law. The values obtained were $\Delta_f H^\circ_m$(ZnSe, α, 298.15 K) = – (215.5 ± 19.0) kJ·mol^{-1}, $\Delta_f H^\circ_m$(ZnSe, α, 298.15 K) =

– (169.7 ± 15.0) kJ·mol^{-1}, and S°_{m} (ZnSe, α, 298.15 K) = (25.4 ± 16.0) J·K^{-1}·mol^{-1} for the second and third law enthalpy of formation and the entropy, respectively. The large difference between the second and third law values indicate serious systematic errors and the entropy value is much too low. The result of this investigation was disregarded while selecting thermodynamic values for α-ZnSe.

[78SMO/PAT]

This is a thorough study of the evaporation of α-MnSe in the temperature range 1225 to 1780 K using the mass-spectrometric Knudsen-cell method. The predominant vaporisation process is

$$\alpha\text{-MnSe} \rightarrow \text{Mn(g)} + x\text{Se(g)} + 0.5(1-x)\text{Se}_2\text{(g)}$$

leading to congruent effusion which was verified by running the experiments to complete vaporisation, observing no residues or reaction with the crucible, and noting that all ion intensities decreased abruptly at the end of the experiments. It was found that the difference δ in the composition of $MnSe_{1+\delta}$ was less than 1×10^{-4} for a temperature change from 1330 K to 1550 K. Ion intensities originating from the gaseous molecules Mn(g), Se(g), Se_2(g), MnSe(g), Mn_2Se(g), $MnSe_2$(g) and Mn_2Se_2(g) were measured and transformed into partial pressure values. The second and third laws were used to evaluate enthalpies of formation and dissociation energies. The heat capacity of α-MnSe was calculated from the expression

$$C^{\circ}_{p,m}\text{(MnSe, α, (298.15 – 1600) K)} = (49.036 + 6.736 \times 10^{-3}\ T)\ \text{J·K}^{-1}\text{·mol}^{-1}$$

estimated by Wiedemeier and Goyette [68WIE/GOY]. The Gibbs energy functions and the enthalpy functions were calculated from estimated molecular parameters for the gaseous manganese selenides. The values of these functions in Table 3 of the paper were used by the review to calculate the standard entropy values at 298.15 K in Table A-105. The estimated molecular constants in Table 3 of the paper were used by the review to calculate heat capacities using statistical mechanics and to derive heat capacity expressions from least-squares regressions. The expressions are listed in Table A-106.

Table A-105: Standard entropies calculated from the enthalpy and Gibbs energy functions in Table 3 of [78SMO/PAT]. The error limits have been set by the review.

Species	S°_{m} (298.15 K) (J·K^{-1}·mol^{-1})
MnSe(g)	(262 ± 5)
Mn_2Se(g)	(307 ± 8)
$MnSe_2$(g)	(326 ± 10)
Mn_2Se_2(g)	(365 ± 12)

Table A-106: Heat capacity expressions derived by least-squares regressions on heat capacity values calculated using statistical mechanics and estimated molecular parameters. The expressions describe the results from statistical mechanics to within ± 0.04 $J \cdot K^{-1} \cdot mol^{-1}$ in the temperature range 298.15 to 2000 K.

Species	Heat capacity expression ($J \cdot K^{-1} \cdot mol^{-1}$)
MnSe(g)	$C^{o}_{p,m} = 37.215 + 0.7302 \times 10^{-3}\ T - 6.9 \times 10^{-8}\ T^{2} - 1.48257 \times 10^{5}\ T^{-2}$
$Mn_2Se(g)$	$C^{o}_{p,m} = 57.792 + 0.4718 \times 10^{-3}\ T - 1.42 \times 10^{-7}\ T^{2} - 3.29577 \times 10^{5}\ T^{-2}$
$MnSe_2(g)$	$C^{o}_{p,m} = 57.795 + 0.4579 \times 10^{-3}\ T - 1.41 \times 10^{-7}\ T^{2} - 3.25757 \times 10^{5}\ T^{-2}$
$Mn_2Se_2(g)$	$C^{o}_{p,m} = 82.846 + 0.3437 \times 10^{-3}\ T - 1.033 \times 10^{-7}\ T^{2} - 3.90602 \times 10^{5}\ T^{-2}$

The atomisation energies of the gaseous species were recalculated from 0 to 298.15 K using the values of $H^{\circ}(T) - H^{\circ}(0\ K)$ given in Table 3 of the paper. Enthalpies of formation were calculated from the atomisation energies at 298.15 K and the selected data of Se(g) and the enthalpy of formation of Mn(g) in [82WAG/EVA]. The results are given in Table A-107.

Table A-107: Atomisation energies and enthalpies of formation derived from measurements in [78SMO/PAT].

Species	$\Delta_{at}H^{\circ}_{m}$ (0 K) ($kJ \cdot mol^{-1}$)	$\Delta_{at}H^{\circ}_{m}$ (298.15 K) ($kJ \cdot mol^{-1}$)	$\Delta_{f}H^{\circ}_{m}$ (298.15 K) ($kJ \cdot mol^{-1}$)
α-MnSe	(697.9 ± 10.0)	(697.5 ± 10.0)	– (180.7 ± 10.0)
MnSe(g)	(235.6 ± 10.0)	(238.4 ± 10.0)	(278.4 ± 10.0)
$Mn_2Se(g)$	(528.0 ± 20.0)	(533.2 ± 20.0)	(264.3 ± 20.0)
$MnSe_2(g)$	(516.3 ± 20.0)	(521.4 ± 20.0)	(231.5 ± 20.0)
$Mn_2Se_2(g)$	(892.4 ± 25.0)	(898.1 ± 25.0)	(135.5 ± 25.0)

[78SMO/SER]

The authors measured the enthalpy change of the reaction between aqueous solutions of the chlorides of Lu, Yb, Er, Ho, Dy, Tb, Gd, and Nd (Ln) and Na_2SeO_3(cr). No experimental details of the calorimetric measurements are presented. From other sources it is concluded that the selenites formed were amorphous. The chemical formulas of the selenites are written with 8 molecules of water except for neodymium selenite, which is written with 4 molecules of water. No analytical results are, however, provided in the paper.

Maier, Suponitski, and Karapet'yants [67MAI/SUP] studied the same reaction for neodymium and state that the amorphous selenite formed contained 8 molecules of water. The enthalpy change of the reaction found by the two groups is the same. Neodymium selenite will therefore also be written with 8 molecules of water. As the experimental conditions are unknown in [78SMO/SER] the data have been recalculated assuming that they were similar to those in [67MAI/SUP], see Appendix A. The recalculation used the values of the enthalpies of formation of $LnCl_3(aq)$ obtained from the values of the enthalpies of formation of the aqueous lanthanide ions proposed in the critical assessment by Cordfunke and Konings [2001COR/KON], see Appendix E, and the enthalpy of formation of $Cl^-(aq)$ in [89COX/WAG]. The value of $\Delta_f H^\circ_m$ ($LnCl_3$, aq 1:1500) was obtained by applying a correction for the enthalpy of dilution to the standard state of – (3.5 ± 1.0) kJ·mol^{-1}.

Table A-108: Calculation of the enthalpy of formation of $Ln_2(SeO_3)_3\cdot 8H_2O(am)$. The uncertainties are those found in the paper.

Lanthanide	$\Delta_r H^\circ_m$ (kJ·mol^{-1})	$\Delta_f H^\circ_m$ ($Ln_2(SeO_3)_3$, am, 298.15 K) (kJ·mol^{-1})
Lu	– (51.04 ± 2.60)	– (5176.0 ± 11.5)
Yb	– (43.60 ± 0.80)	– (5114.6 ± 6.7)
Er	– (41.13 ± 1.00)	– (5175.9 ± 10.9)
Ho	– (40.38 ± 0.94)	– (5174.2 ± 8.7)
Dy	– (39.58 ± 0.82)	– (5158.4 ± 9.4)
Tb	– (38.12 ± 0.92)	– (5156.1 ± 8.5)
Gd	– (52.84 ± 0.10)	– (5168.0 ± 6.7)
Nd	– (33.43 ± 0.36)	– (5141.4 ± 8.0)

[79BEN/PAS]

The vapour pressure of $Ga_2Se_3(cr)$ was measured in the temperature range 1121 to 1269 K using the weight-loss Knudsen method. The measured total vapour pressure was given by the expression $\log_{10}(p/\text{bar}) = (9.435 \pm 1.000) - (16352 \pm 1200)\, T^{-1}$. This result was evaluated by the review using the second and third laws, the selected data of selenium and the heat capacities of $Ga_2Se(g)$ and $Ga_2Se_3(cr)$ in Sections V.8.3.1.2 and V.8.3.1.3, and assuming the congruent vaporisation reaction:

$$Ga_2Se_3(cr) \rightleftharpoons Ga_2Se(g) + Se_2(g). \quad (A.32)$$

The thermodynamic quantities for the reaction at 298.15 K derived from the second law were $\Delta_r H^\circ_m$ ((A.32), 298.15 K) = (657.6 ± 46.0) kJ·mol^{-1} and $\Delta_r S^\circ_m$ ((A.32), 298.15 K) = (393.3 ± 38.0) J·K^{-1}·mol^{-1}. The corresponding third law value of the enthalpy of reaction was $\Delta_r H^\circ_m$ ((A.32), 298.15 K) = (645.6 ± 30.0) kJ·mol^{-1}.

[79KUL/ZLO]

The partial pressures of the gaseous species in equilibrium with $SnSe_2$(cr) were measured in the temperature range 550 to 690 K using mass spectroscopy and effusion cells. The observed species were Sn(g), Se(g), Se_2(g), Se_3(g), SnSe(g), $SnSe_2$(g), Sn_2Se(g), Sn_2Se_2(g), and Sn_2Se_3(g). Thermodynamic quantities were derived for reactions involving $SnSe_2$(cr), SnSe(g), $SnSe_2$(g), and Sn_2Se_2(g) from the second and third laws and employing estimated thermal functions. It can be noted that the reported second law enthalpy of the dimerisation reaction 2SnSe(g) $\rightleftharpoons$ Sn_2Se_2(g), $\Delta_r H^\circ_m(620\ \text{K}) = -(193.8 \pm 16.7)\ \text{kJ·mol}^{-1}$, was only calculated at the mean temperature of the measurements and not recalculated to 298.15 K. The third law value, $\Delta_r H^\circ_m(298.15\ \text{K}) = -(196.4 \pm 12.8)\ \text{kJ·mol}^{-1}$, was obtained using the free energy functions calculated from molecular parameters estimated by Colin and Drowart [64COL/DRO]. No attempt was made by the review to recalculate the results for the dimerisaton reaction because improved thermal functions only exist for SnSe(g).

[79SUL/PRU]

This is a thorough study of the thermodynamics of the gaseous species in the antimony-selenium system using mass spectroscopy and effusion cells in the temperature range 765 to 885 K. The results are tested and cross-examined for reasonableness in systematic comparisons and appear to be reliable. Previous investigations of the gas phase are discussed and compared with the current results. However, no quantities were calculated at 298.15 K and they were therefore evaluated by the review in the following manner. The enthalpies of the reactions at 825 K in Table 4 of the paper were rearranged such that the enthalpies of formation of the various species (at 298.15 K) could be calculated from the enthalpies of reaction and the enthalpies of formation of Sb_2Se_3(cr) and Se_2(g) only. The enthalpies of formation at 298.15 K of SbSe(g), Sb_2Se_2(g), Sb_3Se_3(g), and Sb_4Se_4(g) were then calculated using the selected data of selenium and Sb_2Se_3(cr) and enthalpy functions calculated from the molecular data given in Tables 5, 8, and 10 of the paper using the common methods of statistical mechanics. The enthalpy of formation at 298.15 K of the other antimony selenides were calculated in the same manner, but with enthalpy increments interpolated from those of the SbSe(g) polymers, since molecular parameters were not available. The procedure adds some minor errors to the enthalpies of formation at 298.15 K, compared to those at 825 K and the error limits were adjusted accordingly. The results are summarised in Table A-109.

A similar procedure could be applied to obtain absolute entropies of the species at 298.15 K, but such calculations were omitted in this review because of the large uncertainties of the heat capacities.

Table A-109: Enthalpies of formation derived by the review from the measurements in [79SUL/PRU].

Species	$\Delta_f H_m^\circ$ (g, 298.15 K) (kJ·mol^{-1})
SbSe(g)	(177.0 ± 20.0)
Sb_2Se_2(g)	(127.2 ± 30.0)
Sb_2Se_3(g)	(83.5 ± 30.0)
Sb_2Se_4(g)	(58.3 ± 30.0)
Sb_3Se(g),	(171.3 ± 40.0)
Sb_3Se_2(g)	(141.4 ± 30.0)
Sb_3Se_3(g)	(88.6 ± 30.0)
Sb_4Se_3(g)	(81.2 ± 40.0)
Sb_4Se_4(g)	(35.6 ± 40.0)

[79VEN/PAS]

The saturated vapour pressure in equilibrium with In_2Se_3(cr) was studied in the temperature range 1055 to 1167 K using Knudsen cells and mass spectroscopy. It was found that the vaporisation reaction is

$$In_2Se_3(cr) \rightleftharpoons In_2Se(g) + Se_2(g). \quad (A.33)$$

Both the In_2Se(g) and Se_2(g) partial pressures were recorded resulting in the expression $\log_{10} K = (13.388 \pm 2.880) - (25844 \pm 3240)\,T^{-1}$. The enthalpy and entropy of reaction were evaluated by the review using the second and the third law , the selected properties of Se_2(g), the heat capacity and entropy of In_2Se(g) calculated by the review from the molecular parameters obtained by Erkoc, Katircioglu, and Yilmaz [2001ERK/KAT] in quantum mechanical studies of In-Se clusters, and the heat capacity, transition enthalpies, and entropy of solid In_2Se_3 given in Section V.8.4.1.5. The calculated changes in the thermodynamic quantities of Reaction (A.33) at 298.15 K were $\Delta_r H_m^\circ$(298.15 K) = (536.2 ± 62.0) kJ·mol^{-1} and $\Delta_r S_m^\circ$(298.15 K) = (319.4 ± 55.0) J·K^{-1}·mol^{-1} from the second law and $\Delta_r H_m^\circ$(298.15 K) = (588.1 ± 40.0) kJ·mol^{-1} from the third law.

[80SIR/PET]

The heat capacity of α-ZnSe was measured in the temperature range 4.2 to 300 K. The heat capacity showed no anomalies in the studied temperature range. The absolute entropy was presented only in a graph of poor resolution and no value was given at 298.15 K. The entropy was therefore evaluated by the review from the heat capacity values given in Table 1 of the paper. The resulting absolute entropy at 298.15 K was S_m°(ZnSe, α, 298.15 K) = (71.94 ± 1.00) J·K^{-1}·mol^{-1}. The error limit was estimated from the general accuracy for this type of measurements and includes systematic errors.

[80UML/WAL]

The polarographic behaviour of the selenite ion in slightly acidic solution was studied in the presence of a number of metal ions that form slightly soluble selenides. In the absence of such cations the selenite ion is reduced in an irreversible step according to the reaction $H_2SeO_3(aq) + 4H^+ + 4e^- \rightarrow HgSe(s) + 3H_2O(l)$. At more negative potentials the mercury selenide formed is reduced in a reversible step, $HgSe(s) + 2H^+ + 2e^- \rightleftharpoons Hg(l) + H_2Se(aq)$. In presence of the studied metal ions the position of the second reduction step is shifted towards more positive potentials and the reduction is assumed to follow (for a divalent metal ion) $HgSe(s) + Me^{2+} + 2e^- \rightleftharpoons MeSe(s) + Hg(l)$.

The authors derived the following equation for the half-wave potential, $E_{1/2}$, on the assumption that the process was reversible

$$E_{1/2} = E^{o}_{Hg^{2+}/Hg} + (g/2)\log_{10} K_{s,0}(HgSe(s)) - (g/2q)\log_{10} K_{s,0}(Me_pSe_q(s)) + (g\,p/2q)\log_{10}([Me^{z+}]/2)$$

$E^{o}_{Hg^{2+}/Hg}$ is the standard potential of the Hg^{2+}/Hg electrode and g equals $(RT \ln 10)/F$ and p and q represent the stoichiometric composition of the $Me_pSe_q(s)$ solid phase.

The paper does not indicate that the derived relationship was thoroughly tested, but it is accepted by the review. Only one composition of the test solution appears to have been measured for each metal ion. The solubility products obtained from the data and calculated with the accepted solubility product of α-HgSe, $\log_{10} K_{s,0}(\alpha\text{-HgSe}) = -(60 \pm 2)$, are entered in Chapter V. No recalculations of the polarographic measurements to standard conditions from the ionic strength 0.1 or 0.3 M have been made due to the large uncertainties involved in the primary results.

[81KOS/GRI]

The heat capacities of $\alpha\text{-In}_2Se_3$ and $\beta\text{-In}_2Se_3$ were measured in the temperature range 5 to 303 K and the absolute entropies were evaluated. There is an error in the integration of the heat capacity of $\alpha\text{-In}_2Se_3$ in the temperature range 90 to 100 K leading to a minor error in the entropy above these temperatures (see Table 3 of the paper). The entropy value at 298.15 K should read $S^{o}_{m}(In_2Se_3, \alpha, 298.15\ K) = (203.2 \pm 0.6)\ J\cdot K^{-1}\cdot mol^{-1}$. The supercooling of the $\beta\text{-In}_2Se_3$ phase below 330 K has not been observed in any other investigation and Boehnke, Kühn, Berezovskii, and Spassov [87BOE/KUH] suggested that the $\beta\text{-In}_2Se_3$ phase used in the measurements actually was the rhombohedral form of $\alpha\text{-In}_2Se_3$.

[81STE]

It was observed that the values of both the heat capacities and the entropies of the sulphur ring compounds $S_6(g)$, $S_7(g)$, $S_8(g)$, and $S_{12}(g)$ vary linearly with the number of atoms in the molecule. Even the $S_2(g)$ molecule fits the linear relationship if its entropy is lowered by the contribution of the two unpaired electrons. Assuming an analogous relationship to be valid for gaseous selenium ring compounds, the heat capacities and entropies were calculated for the species Se_5(g), Se_6(g), Se_7(g), Se_9(g), Se_{10}(g), Se_{11}(g), and Se_{12}(g) from a linear relationship based on the properties of Se_2(g) and Se_8(g). The structures of the species (except Se_2(g)) have not been experimentally verified and the properties of the Se_8(g) molecule were derived from data of the Se_8 molecule in the solid state. Considering these circumstances, the proposed values of the heat capacities and entropies can be expected to suffer from systematic errors that are difficult to estimate.

[82GRI/GRI]

Information on the fragmentation and ionisation, which result from the bombardment of selenium vapour in the ion source of a mass spectrometer, was obtained for the temperature range 463 to 483 K. All ions from Se^+(g) to Se_8^+(g) were observed. However, it was concluded that the intensities of the ions Se^+(g) - Se_4^+(g) contained substantial contributions from fragmentation. The evaluation of thermodynamic quantities was therefore limited in the original paper to what was considered as "apparent" enthalpies of formation. The results are summarised in the Table A-110 and have been converted to 298.15 K using the values of the heat capacities selected in this review. Third law enthalpies of formation were evaluated by the review by combining the "apparent" enthalpies of formation with (*i*) the ions intensities in Table I of the paper and (*ii*) the supposedly corrected mole fractions in Table V of the paper together with the total pressure measurements of Niwa and Shibata [40NIW/SHI] and the adopted values of the heat capacities and entropies of the molecules Se(g)-Se_8(g). Second law entropies were calculated similarly. The enthalpy obtained from the ion intensities for the molecules Se(g)-Se_4(g) can be considered as extreme lower limit values.

Table A-110: Standard entropies and enthalpies of formation of gaseous selenium species. The second value in a column is calculated from the mole fractions in Table V of the paper. All other values are derived from the "apparent" enthalpy of sublimation and the relative intensities in Table I of the paper.

Species	$\Delta_f H_m$ (483 K) (kJ·mol^{-1})	$\Delta_f H_m^\circ$ (298.15 K) (kJ·mol^{-1})		S_m° (298.15 K) (J·K^{-1}·mol^{-1})
	original values	2nd law	3rd law	
Se(g)	(130.5 ± 4.1)	(133.1 ± 4.1)	(140.9 ± 4.1)	(160.5± 8.5)
Se_2(g)	(133.5 ± 4.1)	(135.5 ± 4.1)	(136.5 ± 4.1), (143.3 ± 4.1)	(245.2 ± 8.5), (231.1 ± 8.5)
Se_3(g)	(139.3 ± 4.1)	(143.6 ± 4.1)	(146.7 ± 4.1)	(302.6 ± 8.5)
Se_4(g)	(128.9 ± 4.1)	(133.7 ± 4.1)	(141.9 ± 4.1), (148.6 ± 4.1)	(323.7 ± 8.5), (309.8 ± 8.5)
Se_5(g)	(143.9 ± 4.1)	(149.1 ± 4.1)	(147.0 ± 4.1), (143.5 ± 4.1)	(402.5 ± 8.5), 409.8 ± 8.5)
Se_6(g)	(130.1 ± 4.1)	(135.5 ± 4.1)	(139.3 ± 4.1), (138.3 ± 4.1)	(420.6 ± 8.5), (422.8 ± 8.5)
Se_7(g)	(143.5 ± 4.1)	(149.3 ± 4.1)	(152.8 ± 4.1), (153.7 ± 4.1)	(479.8 ± 8.5), (477.8 ± 8.5)
Se_8(g)	(144.8 ± 4.1)	(150.9 ± 4.1)	(158.3 ± 4.1), (159.4 ± 4.1)	(511.0 ± 8.5), (508.6 ± 8.5)

[82RAO]

This paper presents the FORTRAN code of a small computer program used for the calculation of partial pressures of selenium species in saturated selenium vapour. The data used for the calculations are mainly those in [74RAU] and no new information is given.

[82SCH/FRA]

The incongruent vaporisation of WSe_2(cr) was investigated using a weight-loss Knudsen technique and mass spectrometry. The temperature range 1290 to 1720 K was studied. The simultaneous formation of Se_2(g) and Se(g) was considered and for the reaction:

$$\tfrac{1}{2}WSe_2(cr) \rightleftharpoons \tfrac{1}{2}W(cr) + Se(g) \qquad (A.34)$$

the results $\Delta_r H_m^\circ$((A.34), 1500 K) = (317 ± 7) kJ·mol^{-1} and $\Delta_r S_m^\circ$((A.34), 1500 K) = (159 ± 5) J·K^{-1}·mol^{-1} were obtained. The thermodynamic properties of WSe_2(cr) at 298.15 K were calculated by this review from these quantities and the thermodynamic values for Se(g) and W(cr), and the heat capacity of WSe_2(cr). The second law results were $\Delta_f H_m^\circ$(WSe_2, cr, 298.15 K) = – (175.0 ± 12.0) kJ·mol^{-1} and S_m°(WSe_2, cr,

298.15 K) = (53.2 ± 5.0) $J\cdot K^{-1}\cdot mol^{-1}$ and the third law enthalpy of formation was $\Delta_f H^\circ_m$ (WSe_2, cr, 298.15 K) = – (125.6 ± 12.0) $kJ\cdot mol^{-1}$.

[82VEN/GOR]

This is a very accurate calculation of thermodynamic functions of Se_2(g) from spectroscopic data using the wavenumber $\Delta\nu_e$ = 509.95 cm^{-1} for the important splitting of ground electronic state $X^3\Sigma^+$ into the components XO^+_g and $X1_g$. Calculated values were tabulated at 100 K intervals in the temperature range 100 to 3000 K. The presented analytical relationship of the temperature dependence of the heat capacity does not extend all the way down to 300 K due to the curvature of the heat capacity at low temperatures. Corresponding expressions were evaluated by this review from the tabulated values divided into three different temperature intervals, whereby the whole temperature range 298.15 to 3000 K could be described. The three expressions are:

$$C^\circ_{p,m}(\mathrm{Se_2, g, (298.15-700)\ K}) = (54.7611 - 38.147\times10^{-3}\,T + 26.0538\times10^{-6}\,T^2 - 0.35225\times10^{5}\,T^{-2})\ \mathrm{J\cdot K^{-1}\cdot mol^{-1}},$$

$$C^\circ_{p,m}(\mathrm{Se_2, g, (700-1700)\ K}) = (34.2410 + 7.2440\times10^{-3}\,T - 1.9539\times10^{-6}\,T^2 + 0.85196\times10^{5}\,T^{-2})\ \mathrm{J\cdot K^{-1}\cdot mol^{-1}},$$

$$C^\circ_{p,m}(\mathrm{Se_2, g, (1700-3000)\ K}) = (45.4581 - 2.4932\times10^{-3}\,T + 0.4510\times10^{-6}\,T^2 - 3.78247\times10^{5}\,T^{-2})\ \mathrm{J\cdot K^{-1}\cdot mol^{-1}}.$$

[83WIE/PUL]

The total pressure of the vapour in equilibrium with SnSe(cr) was measured in the temperature range 736 to 967 K using the Knudsen effusion technique. The enthalpy of and entropy of sublimation according to the reaction SnSe(cr) $\rightleftharpoons$ SnSe(g) were calculated by the review from the reported vapour pressure expression and the selected heat capacity expressions for SnSe(g) and SnSe(cr) yielding $\Delta_{sub}H^\circ_m$(SnSe, α, 298.15 K) = (219.7 ± 10.0) $kJ\cdot mol^{-1}$ and $\Delta_{sub}S^\circ_m$(SnSe, α, 298.15 K) = (166.7 ± 7.0) $J\cdot K^{-1}\cdot mol^{-1}$, respectively.

[83ZEL/KOL]

The enthalpy of formation of WSe_2(cr) was measured by direct synthesis calorimetry, and by the Knudsen effusion method in the temperature range 1173 to 1423 K. The calorimetric result was $\Delta_f H^\circ_m$ (WSe_2, cr, 298.15 K) = – (154.2 ± 2.0) $kJ\cdot mol^{-1}$. The estimated error is based on the standard deviation of six experiments. The reaction was incomplete and 10 to 15% of the selenium remained unreacted.

The vapour pressure measurements gave an enthalpy of reaction $\Delta_r H^\circ_m$(1300 K) = (259.6 ± 4.0) $kJ\cdot mol^{-1}$ for the reaction WSe_2(cr) $\rightleftharpoons$ W(cr) + Se_2(g). This value does not agree with the partial pressures given in Table 2 of the paper. The review re-evaluated the data and derived the thermodynamic properties of WSe_2(cr) at 298.15 K using the thermodynamic values for Se_2(g) and W(cr), and the heat capacity of

WSe_2(cr). The second law results were $\Delta_f H_m^\circ$ (WSe_2, cr, 298.15 K) = – (98.0 ± 50.0) kJ·mol^{-1} and S_m° (WSe_2, cr, 298.15 K) = (146.3 ± 30.0) J·K^{-1}·mol^{-1} and the third law enthalpy of formation was $\Delta_f H_m^\circ$ (WSe_2, cr, 298.15 K) = – (190.2 ± 10.0) kJ·mol^{-1}.

[84DRO/SMO]

The partial pressures of the species Se_5(g), Se_6(g), and Se_7(g) in equilibrium with trigonal selenium were determined for the temperature range 430 to 490 K using a mass spectrometer and a Knudsen cell. The results obtained at 473 K were recalculated to 298.15 K by this review using the second and third laws and the selected heat capacities and entropies of Se(trig), Se_5(g), Se_6(g), and Se_7(g). The results are summarised in Table A-111.

Table A-111: Standard entropies and enthalpies of formation of gaseous selenium species.

Species	$\Delta_f H_m$ (473 K) (kJ·mol^{-1})	$\Delta_f H_m^\circ$ (298.15 K) (kJ·mol^{-1})		S_m° (298.15 K) (J·K^{-1}·mol^{-1})
	original values	2nd law	3rd law	
Se_5(g)	(150.6 ± 4.2)	(155.4 ± 4.2)	(145.3 ± 4.2)	(419.7 ± 8.8)
Se_6(g)	(138.9 ± 3.8)	(144.0 ± 3.8)	(137.0 ± 3.8)	(443.3 ± 8.0)
Se_7(g)	(149.4 ± 5.4)	(154.8 ± 5.4)	(151.0 ± 5.4)	(495.0 ± 11.5)

[84HUA/GIL]

The partial pressures of the gaseous species in equilibrium with liquid selenium were determined in the temperature range 485 to 503 K using a mass spectrometer and a Knudsen cell. The results were evaluated at 500 K and are summarised in Table A-112. The values were recalculated by this review to 298.15 K using the second and third laws and the selected heat capacities and entropies for the species Se_2(g)-Se_8(g) and Se(cr, l). No fragmentation was considered. Since this has been a problem in other investigations, particularly for Se_3(g) and Se_4(g), the results for those species could be in error. The values given here may be considered as lower limits for the enthalpies of formation because the recorded pressures are probably too high.

Table A-112: Standard entropies and enthalpies of formation of gaseous selenium species.

Species	$\Delta_f H_m$ (500 K) (kJ·mol^{-1})	$\Delta_f H^\circ_m$ (298.15 K) (kJ·mol^{-1})		S°_m (298.15 K) (J·K^{-1}·mol^{-1})
	original values	2nd law	3rd law	
$Se_2(g)$	(133.9 ± 7.1)	(148.6 ± 7.1)	(140.6 ± 10.0)	(263.6 ± 14.2)
$Se_3(g)$	(115.5 ± 4.2)	(138.8 ± 4.2)	(149.3 ± 7.7)	(288.2 ± 8.4)
$Se_4(g)$	(104.2 ± 3.8)	(134.4 ± 3.8)	(141.4 ± 11.4)	(326.9 ± 8.0)
$Se_5(g)$	(115.5 ± 4.2)	(152.2 ± 4.2)	(140.1 ± 8.0)	(422.8 ± 8.8)
$Se_6(g)$	(100.0 ± 3.3)	(143.2 ± 3.3)	(133.0 ± 7.1)	(449.3 ± 7.1)
$Se_7(g)$	(109.6 ± 3.8)	(159.5 ± 3.8)	(146.2 ± 6.2)	(513.5 ± 8.0)
$Se_8(g)$	(105.9 ± 3.3)	(162.4 ± 3.3)	(154.9 ± 5.8)	(541.2 ± 7.1)

[84IMP/PIA]

The vapour pressure of α-HgSe was measured in the temperature range 541 to 602 K using the torsion-effusion technique and in the temperature range 538 to 568 K using the Knudsen-effusion technique. The results were analysed assuming that $Se_2(g)$ was the only selenium species in the vapour phase. Therefore, the results have been re-evaluated by the review.

The total vapour pressure measured by the torsion-effusion method was given as $\log_{10}(p/\text{bar}) = (6.84 \pm 0.12) - (6350 \pm 70)\ T^{-1}$. The expression was re-evaluated by matching observed and calculated total pressures considering all species in the series $Se(g)$-$Se_8(g)$ and $Hg(g)$. The selected thermodynamic data of the selenium species and the CODATA [89COX/WAG] values of mercury were used. A stoichiometric composition of the gas phase was assumed. The mole fractions of the most important selenium species are given in Table A-113.

Table A-113: Mole fractions of the most important gaseous selenium species in equilibrium with α-HgSe.

T (K)	$Se_2(g)$	$Se_5(g)$	$Se_6(g)$
541	0.49	0.29	0.19
602	0.57	0.27	0.13

The expression $\Delta_r G^\circ_m(T) = ((186.5 \pm 6.0) - (0.1927 \pm 0.010)\ T)$ kJ·mol^{-1} for the reaction α-HgSe $\rightleftharpoons$ Hg(g) + ½$Se_2(g)$ was derived for the temperature range 541 to 602 K from the calculated partial pressures. A recalculation to 298.15 K using the selected heat capacity of α-HgSe yields $\Delta_f H^\circ_m$(HgSe, α, 298.15 K) = – (58.5 ± 5.5)

$kJ\cdot mol^{-1}$ and S°_{m}(HgSe, α, 298.15 K) = (96.6 ± 8.0) $J\cdot K^{-1}\cdot mol^{-1}$. The corresponding third law value of the enthalpy is $\Delta_f H^{\circ}_{m}$(HgSe, α, 298.15 K) = – (57.0 ± 5.5) $kJ\cdot mol^{-1}$ using the selected heat capacity and entropy of α-HgSe. The error limits correspond to those given in the vapour expression multiplied by a factor of two.

[84PUP/RUS]

The authors performed a theoretical analysis of the available structural data of molecules of the group 16 elements using the MOLCAO model and motivated a simple scaling procedure for the calculation of the vibrational frequencies and moments of inertia of the species $Se_3(g)$-$Se_8(g)$ from those of the corresponding sulphur species. Subsequently, the entropy and heat content of the selenium species were calculated using the common statistical mechanics expressions except for $Se_5(g)$ for which the vibrational contribution was interpolated from that of $Se_4(g)$ and $Se_6(g)$. The thermodynamic functions were tabulated at 100 K intervals in the temperature range 300 to 1500 K. No heat capacities were calculated. Expressions describing the temperature dependence of the heat capacity were evaluated by the review by fitting the standard heat capacity expression (*cf.* Section II.6) to heat capacities calculated from the molecular parameters given in Table 1 of the paper. The expressions are given in Section V.1.

[84SLA/POP]

The authors prepared $Ge(SeO_3)_2(s)$, $Sn(SeO_3)_2(s)$, and $PbSeO_3(s)$ and measured their solubilities in dilute mineral acids at 293 K. The initial pH of the acids ranged from 1.2 to 2.1. This is the only primary datum furnished in the paper.

The solubility products for germanium(IV) ($\log_{10} K_s = -25.5$) and tin(IV) ($\log_{10} K_s = -30.6$) will not be accepted as the equilibrium reactions that define the solubility products are absent. These reactions are not self-evident.

The result for lead selenite is accepted, but a recalculation with the protonation constants of the selenite ion and the activity coefficient expression selected by the review is not possible.

The published solubilities in water are clearly wrong. They have been obtained as the square root of the solubility product and, in the case of tin and germanium, by a further division by a factor of two. This observation further supports the rejection of the proposed solubility products of $Ge(SeO_3)_2$ and $Sn(SeO_3)_2$.

[84SMO/DRO]

Smoes and Drowart studied the equilibrium reaction:

$$SeO(g) + SeS(g) \rightleftharpoons SO(g) + Se_2(g). \quad (A.35)$$

The equilibrium was established over the temperature range 1300 to 1800 K in a Knudsen cell containing a mixture of MnSe(cr), EuS(cr), and Fe_2O_3(cr). The partial pressures of the species partaking in the reaction were taken as proportional to the ion beam intensities in the mass spectrometer. Linear regression of $\log_{10} K^\circ$ (A.35) on $1/T$ yielded $\Delta_r H^\circ_m$((A.35), T) = – (52.7 ± 1.5) kJ·mol^{-1}. The enthalpy and Gibbs functions for SeO(g), SeS(g), SO(g), and Se_2(g) were calculated by statistical-mechanical methods and tabulated in the temperature interval 298.15 to 2000 K. These data yield $\Delta_r H^\circ_m$((A.35), 0 K) = – (53.14 ± 5.40) kJ·mol^{-1} by the second-law approach. The estimated inaccuracy in this value is introduced by the authors and includes errors in ionisation cross-sections and electron multiplier yields. The same data yielded $\Delta_r H^\circ_m$((A.35), 0 K) = – (52.64 ± 5.20) kJ·mol^{-1} by the third-law approach. The statistical uncertainty in the value was ± 0.71 kJ·mol^{-1}. The authors selected the third-law value of $\Delta_r H^\circ_m$((A.35), 0 K).

The primary aim of the investigation was a determination of the dissociation energy of SeO(g). The selected enthalpy of reaction was combined with selected values of the dissociation energies of SO(g), SeS(g), and Se_2(g) in order to arrive at this datum, see Table A-114.

Table A-114: Thermochemical cycle for the calculation of the dissociation energy of SeO(g), T = 0 K.

#	Reaction	$\Delta_r H_m$ (kJ·mol^{-1})	Reference
1	SO(g) → S(g) + O(g)	(517.1 ± 0.4)	Sources selected in [84SMO/DRO]
2	SeS(g) → Se(g) + S(g)	(366.9 ± 6.7)	Sources selected in [84SMO/DRO]
3	Se_2(g) → 2Se(g)	(329.07 ± 0.25)	Sources selected in [84SMO/DRO]
4	SeO(g) + SeS(g) → Se_2(g) + SO(g)	– (52.64 ± 0.71)	[84SMO/DRO]
5	SeO(g) → Se(g) + O(g)	(426.63 ± 6.30)	(result)

$\Delta_{at} H^\circ_m$ (SeO, g, 0 K) = $\Delta_r H_1 - \Delta_r H_2 + \Delta_r H_3 + \Delta_r H_4$ = 426.63 kJ·mol^{-1}

The thermochemical result was compared with results from spectroscopic measurements. After a long discussion, the authors concluded that the best estimate of the dissociation energy of SeO(g) could be made from the spectroscopic information available and adopted $\Delta_{at} H^\circ_m$ (SeO, g, 0 K) = ($424.7^{+0.6}_{-6.0}$) kJ·mol^{-1}. This result is accepted by the review with symmetrical error bars, thus: $\Delta_{at} H^\circ_m$ (SeO, g, 0 K) = (424.7 ± 6.0) kJ·mol^{-1}.

The enthalpy of formation of SeO(g) under standard conditions is found from the data in the following table.

Table A-115: Calculation of $\Delta_f H_m^\circ$ (SeO, g, 298.15 K). The sources of the data are indicated.

#	Reaction	$\Delta_r H_m$ (kJ·mol^{-1})	Reference
1	Se(g, 0 K) + O(g, 0 K) → SeO(g, 0 K)	– (424.70 ± 6.00)	(review)
2	Se(g, 0 K) → Se(g, 298.15 K)	(6.199 ± 0.010)	[82WAG/EVA]
3	Se(cr, 298.15 K) → Se(g, 298.15 K)	(236.1 ± 1.6)	(review)
4	O(g, 0 K) → O(g, 298.15 K)	(6.725 ± 0.001)	[89COX/WAG]
5	0.5O_2(g, 298.15 K) → O(g, 298.15 K)	(249.18 ± 0.10)	[89COX/WAG]
6	SeO(g, 0 K) → SeO(g, 298.15 K)	9.746	[84SMO/DRO]
7	Se(cr) + 0.5O_2(g) → SeO(g)	(57.40 ± 6.21)	(result)

$\Delta_f H_m^\circ$ (SeO, g, 298.15 K) = $\Delta_r H_1 - \Delta_r H_2 + \Delta_r H_3 - \Delta_r H_4 + \Delta_r H_5 + \Delta_r H_6$

[86GLO/LOM]

Selenium electrodes were obtained by electroplating selenium onto various substrates from a 10.45 M solution of potassium selenite. They were measured against a calomel electrode and the substrate did not affect the potentials. The emf of the selenium-calomel cell was measured with test solutions of potassium selenite of concentrations 0.1, 1, 2, 4, 6, 8, and 10.45 M. The solutions also contained 1 M KOH. The dependence of the emf on pH was studied in another series of experiments.

The main analysis of the data centers on the half-cell reaction:

$$SeO_3^{2-} + 3H_2O(l) + 4e^- \rightleftharpoons Se(s) + 6OH^- \quad (A.36)$$

$$E = E_o + \frac{RT}{4F}\ln\frac{a_{SeO_3^{2-}}\,a_{H_2O}^{\ 3}}{a_{OH^-}^{\ 6}}.$$

Plots of $E = f(\log_{10}[SeO_3^{2-}])_{pH}$ and $E = f(pH)_{Se(IV)}$ at 293 K were linear and had the slopes 14.5 and 87.5 mV, respectively. These values are close to the Nernstian slopes required by Reaction (A.36) and the authors concluded that the half-cell was reversible. The potential, recalculated to the standard hydrogen electrode scale, measured at $[SeO_3^{2-}]$ = 1 M and $[OH^-]$ = 1 M was – 0.364 V and assigned as the standard potential of the redox–couple in alkaline solution. The authors noted this value is close to the value in [56LAT], but they did obviously not know that this datum is in error by about 0.03V.

The review finds it difficult to assess this paper as experimental conditions and primary data are lacking. There can be no doubt that the authors have used concentrations instead of activities in their calculations. The activity of water was set equal to 1 despite the high concentration of solutes and any liquid junction potentials have been neglected. It appears fortuitous that Nernstian slopes were observed. The large concentrations used in the investigation and the way the experimental data have been treated preclude the incorporation of the results of this paper in the review.

[86LYO/YOU]

Lyons and Young employed elemental high-purity selenium to electrolytically produce a 0.01 M solution of HSe^-/Se^{2-} in 1 M KOH. Part of this solution was then saturated with elemental selenium, which was found to result in a solution of the average composition $Se_{3.8}^{2-}$. The two stock solutions were finally mixed in various proportions ($n = 26$). The UV/VIS-spectra at 220 to 540 nm and redox-potentials *vs.* the Saturated Calomel Electrode of the mixtures were measured at 300 and 304 K, respectively.

The spectrophotometric data were first used to evaluate the number of species formed in the system and it was found that four significant components were needed to reproduce the data.

The redox-potential data were then modelled under the assumption that Se^{2-}, Se_2^{2-}, Se_3^{2-}, and Se_4^{2-} ions were gradually formed with increasing proportions of Se(0). A good fit to the measured potentials was obtained with the data presented in Chapter V and the root-mean square error was 0.0031V.

Finally, the spectrophotometric data were analysed under the same assumption in order to obtain the absorption spectra of the different species and their distribution as a function of the mean composition of the solution. The result of this analysis was found to be in good agreement with the evaluation from the redox-potential data.

It was also shown in the paper by ESR measurements that none of the species formed contain unpaired electrons, and ^{77}Se-NMR data indicate that the exchange kinetics between the species is fast on the NMR time scale.

[86POP/SLA]

The authors prepared $Al_2(SeO_3)_3$(s), $Ga_2(SeO_3)_3$(s), $In_2(SeO_3)_3$(s), and $Tl_2(SeO_3)_3$(s). The solubilities of the salts were measured in nitric, perchloric, or sulphuric acid at 298 K. No primary experimental data are furnished and the review assumes that the pH values, 1.7 to 2.0, reported refer to the saturated solutions. The solubility products were calculated from the analytical concentration of dissolved metal ion with corrections for the protonation of the selenite ion and the activity coefficients. No recalculation with the values of these quantities accepted by the review is possible.

Hydrolysis of the metal ions was neglected. This is acceptable except for Tl(III). For this ion, the hydrolysis is so extensive at pH 2 [76BAE/MES] that the value of the solubility product of $Tl_2(SeO_3)_3$(cr) cannot be accepted. The results for the other salts are entered in the appropriate Sections in Chapter V.

The solubilities in water of the salts published in the paper are not correct.

[86VIK/GOR]

The vapour pressures of the species in equilibrium with $MoSe_{1.91}$(cr) were measured in the temperature range 1459 to 1570 K using mass spectrometry and Knudsen cells. The results were evaluated assuming that the decomposition followed the reaction scheme $MoSe_{1.91}(cr) \rightleftharpoons Mo(cr) + 0.955Se_2(g)$. The authors obtained an enthalpy of reaction of $\Delta_r H_m^\circ$ ($MoSe_{1.91}$, cr, 298.15 K) = – (404.4 ± 30.0) kJ·mol^{-1} using the third law. The second law was not used. The result does not agree with the pressures presented in Table 1 of the paper. The experimental data were re-evaluated by the present review using both the second and third laws. Necessary data were taken from the selected data in this review except for molybdenum for which data in [98CHA] were used. The selected heat capacity of $MoSe_2$ was rescaled to the composition $MoSe_{1.91}$ according to the stoichiometry. The second law results are $\Delta_f H_m^\circ$ ($MoSe_{1.91}$, cr, 298.15 K) = – 70.2 kJ·mol^{-1} and S_m° ($MoSe_{1.91}$, cr, 298.15 K) = 210.2 J·K^{-1}·mol^{-1}. The third law enthalpy of formation is $\Delta_f H_m^\circ$ ($MoSe_{1.91}$, cr, 298.15 K) = – (259.2 ± 35.0) kJ·mol^{-1}. The second law values are very much in error and no error limits were assigned. The error limits of the third law value are retained from the paper. The total pressure measured for $MoSe_{1.91}$ is of the right order of magnitude, but its temperature dependence is apparently in error. Similarly, the vapour pressure of the species in equilibrium with $WSe_{1.87}$(cr) were measured in the temperature range 1512 to 1556 K. The temperature range is too narrow and the measured values too few and scattered to make a second law evaluation meaningful. The third law evaluation made by this review yields an enthalpy of formation $\Delta_f H_m^\circ$ ($WSe_{1.87}$, cr, 298.15 K) = – (265.2 ± 35.0) kJ·mol^{-1} using the thermodynamic functions for Se_2(g) and W(cr), and the heat capacity of WSe_2(cr). The enthalpy of WSe_2(cr) was obtained to $\Delta_f H_m^\circ$ (WSe_2, cr, 298.15 K) = – (283.6 ± 35.0) kJ·mol^{-1} by rescaling the value of $WSe_{1.87}$(cr) according to stoichiometry.

[87BOE/KUH]

The heat capacity of hexagonal α_H-In_2Se_3 was measured in the temperature range 7 to 285 K and of both hexagonal α_H-In_2Se_3 and rhombohedral α_R-In_2Se_3 in the temperature range 333 to 463 K. No expressions for the temperature dependence of the heat capacity were given and no thermodynamic properties at 298.15 K were evaluated. The following quantities were therefore calculated by the review. The heat capacity at 298.15 K was calculated to be $C_{p,m}^\circ$ (In_2Se_3, α_H, 298.15 K) = (121.1 ± 0.4) J·K^{-1}·mol^{-1} by extrapolating the values in Table 1 of the paper. Similarly, the entropy was calculated to be S_m° (In_2Se_3, α, 298.15 K) = (203.0 ± 0.6) J·K^{-1}·mol^{-1} by a third law integration of the data in the same table. The heat capacity at 298.15 K is about 4 J·K^{-1}·mol^{-1} smaller in this investigation than in the investigations in [75DEM/KOS], [76MIL], and [81KOS/GRI], and it was therefore not considered meaningful to derive heat capacity expressions from the values listed in Table 2 of the paper.

[87BOL/TRO]

The enthalpy of $NbSe_2$(cr), $NbSe_{1.75}$(cr), and $MoSe_2$(cr) were measured in the temperature range 1180 to 1656 K, 1162 to 1738 K, and 1219 to 1806 K, respectively. The composition $NbSe_{1.75}$ is not a single phase and is not considered here. Analytical expressions for the enthalpy and heat capacity as a function of temperature were derived by a least-squares refinement on the enthalpy data. Unfortunately, the expressions presented in the paper do not reproduce the experimental results.

The enthalpies given in Table 2 of the paper were used in a new least-squares procedure by this review to evaluate heat capacities using the analytical expression $C^{\circ}_{p,\mathrm{m}}(T) = (a + b\,T + c\,T^{2} + d\,T^{-2})$ J·K^{-1}·mol^{-1}. The heat capacities at 298.15 K were fixed to the selected values. The set of parameters in the heat capacity expression, which best fitted the experimental enthalpy values for $NbSe_2$(cr), led to a physically impossible behaviour of the temperature dependence of the heat capacity because of the scatter in the enthalpy values. A rescaling of the heat capacity expression for WSe_2(cr) in [90BOL/TRO] to match the heat capacity of $NbSe_2$(cr) at 298.15 K did not give an acceptable result. The selected expression for $NbSe_2$(cr) was derived by slightly modifying the least–squares result obtained using only the parameters a and b by the addition of a small contribution from a d term. In this way, the shape of the $C^{\circ}_{p,\mathrm{m}} = \mathrm{f}(T)$ curve is adequate and a reasonable fit to the enthalpy at high temperatures is maintained. The resulting expression is

$$C^{\circ}_{p,\mathrm{m}}(\mathrm{NbSe_2, cr, (298.15 - 1700)\ K}) = (68.763 + 0.0235\,T - 2.0 \times 10^{5}\,T^{-2})\ \mathrm{J{\cdot}K^{-1}{\cdot}mol^{-1}}.$$

The enthalpy of $MoSe_2$(cr) was analysed using the same procedure as for $NbSe_2$(cr) and gave the same type of problems. In this case, the heat capacity expression for WSe_2(cr) in [90BOL/TRO] gave a reasonable fit to the enthalpy at high temperatures after rescaling to match the selected heat capacity of $MoSe_2$(cr) at 298.15 K. The resulting heat capacity expression for $MoSe_2$(cr) is

$$C^{\circ}_{p,\mathrm{m}}(\mathrm{MoSe_2, cr, (298.15 - 1800)\ K}) = (69.112 + 0.014221\,T - 2.4621 \times 10^{5}\,T^{-2})\ \mathrm{J{\cdot}K^{-1}{\cdot}mol^{-1}}.$$

[87HON/IID]

The specific heat of Ag_2Se(cr) was measured in the temperature range 120 to 520 K. The experimental heat capacity is shown only in figures and a value at 298.15 K was derived from Figure 1 of the paper, $C^{\circ}_{p,\mathrm{m}}(\mathrm{Ag_2Se}, \alpha, 298.15\ \mathrm{K}) = (83.68 \pm 3.00)$ J·K^{-1}·mol^{-1}. The uncertainty has been estimated by this review because no error analysis was made by the authors nor were any estimated experimental errors given. The entropy change for the α to β transition was found to be 16.48 J·K^{-1}·mol^{-1} and corresponds to an enthalpy of transformation of (6.69 ± 1.00) kJ·mol^{-1}. The uncertainty has been assessed by this review.

[87SRI/EDW]

This is a careful study of the vaporisation chemistry of the indium-selenium system using simultaneous Knudsen effusion and torsion effusion methods, by X-ray powder diffraction, and by sealed tube and closed cell methods. It was found that $In_2Se_3(cr)$ vaporises incongruently according to the simultaneous reactions $4In_2Se_3(cr) \rightleftharpoons 2In_4Se_5(l) + Se_2(g)$ and $2In_4Se_5(l) \rightleftharpoons 3In_2Se_3(cr) + In_2Se(g)$ at the temperatures of the measurements 1042 to 1156 K. However, the actual composition of the liquid phase varies with temperature. The latter condition poses a special problem in evaluating thermochemical data from the reactions and the agreement between second and third law results was poor. To circumvent the problem, the results for two sets of reactions, (*i*) $4In_2Se_3(cr) \rightleftharpoons 2In_4Se_5(l) + Se_2(g)$, $10In_4Se_5(l) \rightleftharpoons 8In_5Se_6(l) + Se_2(g)$, and $4/17In_5Se_6(l) \rightleftharpoons 10/17In_2Se(g) + 7/17Se_2(g)$, and (*ii*) $10/3In_2Se_3(cr) \rightleftharpoons 4/3In_5Se_6(l) + Se_2(g)$ and $4/17In_5Se_6(l) \rightleftharpoons 10/17In_2Se(g) + 7/17Se_2(g)$, were individually combined to form the hypothetical reaction $In_2Se_3(cr) \rightleftharpoons In_2Se(g) + Se_2(g)$ which does not involve liquids of variable stoichiometry. In this way, errors for the reactions involving the liquids cancel to a large extent and the results from second and third law evaluations agree well. The results for the two reactions were summarised to $\Delta_r H_m^\circ(298.15\ K) = (604.2 \pm 2.4)\ kJ\cdot mol^{-1}$ and $\Delta_r H_m^\circ(298.15\ K) = (600.4 \pm 2.4)\ kJ\cdot mol^{-1}$), respectively. Unfortunately, any numerical values were only given for the thermodynamic quantities as recalculated to 298.15 K without specifying the employed auxiliary data. The experimental data actually measured at the high temperatures can therefore not be re-evaluated using the auxiliary data set employed by the review.

[88BAR/IER]

The saturated vapour pressure of α-CdSe was measured in the temperature range 873 to 1115 K using the Knudsen effusion and torsion-effusion techniques. The enthalpy and entropy of sublimation at 298.15 K were merely derived from the measurements using a set of auxiliary data different from that used by the review. The results given in Table 2 of the paper were therefore re-evaluated by the review using both the second and the third law, the selected thermodynamic functions of selenium, the data for Cd in [89COX/WAG], the selected heat capacity of α-CdSe, and the selected entropy of α-CdSe in the case of the third law. The vaporisation was assumed to occur according to the reaction $\alpha\text{-CdSe(s)} \rightleftharpoons Cd(g) + \frac{1}{2}Se_2(g)$. The results were $\Delta_f H_m^\circ(\text{CdSe}, \alpha, 298.15\ K) = -(142.6 \pm 3.0)\ kJ\cdot mol^{-1}$, $\Delta_f H_m^\circ(\text{CdSe}, \alpha, 298.15\ K) = -(146.8 \pm 2.0)\ kJ\cdot mol^{-1}$, and $S_m^\circ(\text{CdSe}, \alpha, 298.15\ K) = (90.5 \pm 4.0)\ J\cdot K^{-1}\cdot mol^{-1}$ for the second and third law enthalpy of formation and the entropy, respectively.

[88BAR/PIA]

The vaporisation of $In_2Se_3(cr)$ was studied by measuring the total vapour pressure in the temperature range 984 to 1159 K using the Knudsen effusion and torsion effusion techniques. The results were evaluated assuming the reaction $In_2Se_3(cr) \rightleftharpoons In_2Se(g) +$

$Se_2(g)$ and using auxiliary data different from those used by the review. The results were therefore re-evaluated using the selected properties of $Se_2(g)$, the heat capacity and entropy of $In_2Se(g)$ calculated by the review from the molecular parameters obtained by Erkoc, Katircioglu, and Yilmaz [2001ERK/KAT], and the heat capacity, transition enthalpies, and entropy of solid In_2Se_3 given in Section V.8.4.1.5, yielding $\Delta_r H^\circ_m(298.15\ K) = 549\ kJ\cdot mol^{-1}$ and $\Delta_r S^\circ_m(298.15\ K) = 332\ J\cdot K^{-1}\cdot mol^{-1}$ from the second law and $\Delta_r H^\circ_m(298.15\ K) = 596\ kJ\cdot mol^{-1}$ from the third law.

[88OZE/YAG]

Ozeki, Yagasaki, Ichida and Sasaki studied the $H^+ - MoO_4^{2-} - SeO_3^{2-}$ system at 298.15 K by potentiometric and ^{77}Se NMR measurements. The data covered the ranges $5.8 > -\log_{10} h\ (\mathrm{p}h) > 2.0$, $58.9 > B/\mathrm{mM} > 33.2$, $48.8 > C/\mathrm{mM} > 4.3$, and $0.8 < B/C < 10.0$, where h is the concentration of free hydrogen ions. B stands for the total concentration of molybdate and C that of selenite. The equilibria will be written as:

$$p\mathrm{H}^+ + q\,\mathrm{MoO_4^{2-}} + r\,\mathrm{SeO_3^{2-}} \rightleftharpoons (\mathrm{H}^+)_p(\mathrm{MoO_4^{2-}})_q(\mathrm{SeO_3^{2-}})_r.$$

At $B/C < 2.5$ the data were well explained by the species $Se_2Mo_5O_{21}^{4-}$; $(p,q,r) = (10,5,2)$. This anion has been found in the solid state and its structure has been determined. In the following complexes will be identified by their (p, q, r) values. Minor amounts of (11,5,2) formed at low ph.

At $B/C > 2.5$ it was necessary to introduce additional complexes. The equilibrium analysis suggested several possible sets of complexes. The simplest model, which satisfactorily explained the experimental data, added the species (14,8,2) and (15,8,2). This model was selected and yielded the formation constants in the Table A-116.

Table A-116: Formation constants of heteropoly anions in the system $H^+ - MoO_4^{2-} - SeO_3^{2-}$.

(p,q,r)	$\log_{10}\beta_{pqr}$
(10,5,2)	(71.38 ± 0.02)
(11,5,2)	(72.73 ± 0.04)
(14,8,2)	(97.97 ± 0.02)
(15,8,2)	(101.04 ± 0.03)

This work appears to be the only equilibrium study of heteropoly anions containing selenium.

[88PAS/ZHA]

The saturated vapour pressure in equilibrium with $In_2Se_3(cr)$ was studied in the temperature range 980 to 1100 K using the Knudsen effusion method. The vapour pressure was given by the expression $\log_{10}(p/\text{bar}) = (8.05 \pm 0.75) - (13189 \pm 790)\ T^{-1}$ and thermodynamic quantities were evaluated assuming the reaction

$$In_2Se_3(cr) \rightleftharpoons In_2Se(g) + Se_2(g) \qquad (A.37)$$

The vapour pressure expression was re-evaluated by the review using the selected properties of $Se_2(g)$, the heat capacity and entropy of $In_2Se(g)$ calculated by the review from the molecular parameters obtained by Erkoc, Katircioglu, and Yilmaz [2001ERK/KAT], and the heat capacity, transition enthalpies, and entropy of solid In_2Se_3 given in Section V.8.4.1.5. in order to obtain a consistent set of evaluations. The calculated changes in the thermodynamic quantities of Reaction (A.37) at 298.15 K were $\Delta_r H^\circ_m$((A.37), 298.15 K) = (546.4 ± 16.0) kJ·mol^{-1} and $\Delta_r S^\circ_m$((A.37), 298.15 K) = 360 J·K^{-1}·mol^{-1} from the second law and $\Delta_r H^\circ_m$((A.37), 298.15 K) = (563.2 ± 14.0) kJ·mol^{-1} from the third law.

[90BAR/TRI]

The saturated vapour pressure of α-ZnSe was measured using the thermogravimetric Knudsen effusion method in the temperature range 1025 to 1288 K. The results were evaluated by applying the second and third law and assuming the reaction α-ZnSe ⇌ Zn(g) + ½Se_2(g). The sublimation enthalpies $\Delta_{sub} H^\circ_m$(ZnSe, α, 298.15 K) = (377 ± 4) kJ·mol^{-1} and $\Delta_{sub} H^\circ_m$(ZnSe, α, 298.15 K) = (332 ± 4) kJ·mol^{-1} were derived for the second and third law enthalpy of reaction, respectively. No quantities were derived for α-ZnSe at 298.15 K. The experimental data given by the vapour pressure expression $\log_{10}(p/\text{kPA}) = (9.15 \pm 0.06) - (12798 \pm 73)\ T^{-1}$ were therefore re-evaluated by the review according to the second and the third law using the selected thermodynamic functions of selenium, the data for Zn in [89COX/WAG], the heat capacity expression of α-ZnSe in Section V.9.1.1.1, and the selected entropy of α-ZnSe in the case of the third law. The values obtained were $\Delta_f H^\circ_m$(ZnSe, α, 298.15 K) = – (183.4 ± 14.0) kJ·mol^{-1}, $\Delta_f H^\circ_m$(ZnSe, α, 298.15 K) = – (181.8 ± 7.0) kJ·mol^{-1}, and S°_m(ZnSe, α, 298.15 K) = (66.1 ± 9.0) J·K^{-1}·mol^{-1} for the second and third law enthalpy of formation and the entropy, respectively.

[90BOL/TRO]

There are some errors in the equations. Equations (1) and (3) of the paper should read:

$$H^\circ_m(T) - H^\circ_m(298.15\ \text{K}) = (71.086\ T + 7.313 \times 10^{-3}\ T^2 + 253246\ T^{-1} - 22693)\ \text{kJ·mol}^{-1}$$

and

$C^\circ_{p,m}(T) = (71.086 + 14.627 \times 10^{-3}\ T - 253246\ T^{-2})$ J·K^{-1}·mol^{-1}, respectively.

[90KUM/BAT]

Kumok and Batyreva have published calculations of the solubility product of a considerable number of selenates at 298.15 K. The variation of the mean activity coefficient of a salt was approximated by an equation proposed by Kusik and Meissner [78KUS/MEI]. This equation contains only one adjustable parameter, q. q-values are presented for a number of selenates, but details of their calculation and references to the experimental data are generally lacking.

The way in which the solubility products were evaluated from the primary data could not be followed in such detail that a critical evaluation could be made. This comment applies in particular to the calculations performed on data from ternary solubility diagrams. The results obtained by Kumok and Batyreva are therefore mentioned in Chapter V but not accepted.

[90LEV/MYE]

To study the second dissociation constant of $H_2Se(aq)$, Levy and Myers utilised the difference in the UV-absorbance of HSe^- ($\lambda_{max} \approx 250$ nm; $\varepsilon_{max} = 8030$ $M^{-1}\cdot cm^{-1}$) and Se^{2-} ($\lambda_{max} \approx 270$ nm; $\varepsilon_{max} = 12290$ $M^{-1}\cdot cm^{-1}$). Also considered in the evaluation was the observation that the values of λ_{max} were somewhat blue-shifted when the ionic strength of the solution increased.

The samples were prepared by dissolving measured volumes of $H_2Se(g)$ in degassed solutions of NaOH, which were held in 0.1 mm quartz cells. The cells were then filled with a slight overpressure of $N_2(g)$ and a colourless sample was taken to indicate that no selenide oxidation had occurred.

The experimental data cover the range from 0.12 to 2.5 molal NaOH at [Se(–II)] = 0.01 M (n = 12) and in each point, the ratio $m_{Se^{2-}} / m_{HSe^-}$ and the position of λ_{max} for Se^{2-} were optimised in an iterative manner. Together with the known initial concentration of OH^-, these ratios were used to evaluate the equilibrium constant, K_m, of the reaction $HSe^- + OH^- \rightleftharpoons Se^{2-} + H_2O$.

Linear regression of $\log_{10} K_m$ on $m^{½}$, *i.e.* the square root of the molality of NaOH, yielded $\log_{10} K_m = -0.997 + 0.793\cdot m^{½}$, which fitted the data well. It was also found that the evaluated frequency shift of the Se^{2-} spectrum varied linearly with the square root of the molality of NaOH, which lends credibility to the evaluation made.

The value of K_m at $m = 0$ was further extrapolated to $I = 0$ by applying activity coefficient functions for γ_{NaHSe} and γ_{Na_2Se} according to Guggenheim. In this way, a "best" value of $K_m(I = 0) = (0.0877 \pm 0.004)$ was obtained which, combined with $\log_{10} K^\circ_w = -13.997$ results in $\log_{10} K^\circ(HSe^- \rightleftharpoons Se^{2-} + H^+, 298.15 \text{ K}) = -(15.05 \pm 0.04)$.

[90NAS/SHA]

The Gibbs energy of formation of α-ZnSe was measured in the temperature range 635 to 825 K using the cell

$$(-)\ \text{Zn(cr, l)} \mid 5\ \text{weight-\%}\ \text{ZnCl}_2\ (\text{in LiCl + KCl melt}) \mid \alpha\text{-ZnSe, Se(l)}\ (+).$$

The cell reaction was Zn(cr, l) + Se(l) $\rightleftharpoons$ α-ZnSe. Zinc melts at 693 K within the experimental temperature range and the data were separated into two different temperature ranges. The Gibbs energy of formation was evaluated by linear regression of the emf on temperature. Equation 4 of the paper for the high temperature range (equilibrium with molten zinc) is in error and does not fit the data given in Figure 1 of the paper, neither does the Gibbs energy of formation given in Eqs.(3, 4) of the paper match at the melting point of zinc. A new expression for the Gibbs energy of formation yielding $\Delta_f G^\circ_m$ (ZnSe, α, T) = (– 191479 + 36.62 T) J·mol^{-1}, was therefore derived by the review from the graph in Figure 1 of the paper. This equation and Equation 3 of the paper were recalculated to 298.15 K by the review according to the second and the third law using the selected thermodynamic functions of selenium, the entropy of Zn(cr) at 298.15 K in [89COX/WAG], the heat capacity and enthalpy of fusion of Zn in [98CHA], the the heat capacity expression of α-ZnSe in Section V.9.1.1.1, and the selected entropy of α-ZnSe in the case of the third law. The results are given in Table A-117.

The enthalpy of formation of α-ZnSe was also measured direcly in a quantitative DTA calorimeter by the reaction of the powdered elements. The value obtained at 707 K was recalculated by the review to 298.15 K using the selected thermodynamic functions of selenium, the heat capacity of Zn in [89COX/WAG], and the heat capacity expression of α-ZnSe in Section V.9.1.1.1 because a different set of auxiliary data was employed for the calculations in the paper. The value obtained was $\Delta_f H^\circ_m$ (ZnSe, α, 298.15 K) = – (177.1 ± 8.0) kJ·mol^{-1}.

Table A-117. Enthalpy of formation and entropy of α-ZnSe at 298.15 evaluated by the review from the data in [90NAS/SHA].

$\Delta_f H^\circ_m$ (298.15 K) (kJ·mol^{-1})		S°_m (298.15 K) (J·K·mol^{-1})	Temperature (K)
2nd law	3rd law		
– (177.5 ± 6.0)	– (177.3 ± 4.2)	(59.4 ± 12.0)	635 – 693
– (175.0 ± 6.0)	– (177.5 ± 4.2)	(63.9 ± 9.0)	693 – 825

[90SCA/PIA]

The temperature dependence of the total vapour pressure in equilibrium with Ag_2Se was measured by the torsion Knudsen technique. The data were evaluated assuming the reaction to be $Ag_2Se(cr, l) \rightleftharpoons 2Ag(g) + \frac{1}{2}Se_2(g)$. The partial pressure of silver was assumed to be the same as for pure silver because the vaporisation is incongruent and metallic silver is formed. The measurements were re-evaluated using the second and third laws, the selected heat capacity and entropy of Ag_2Se, and the selected data of selenium and silver. The result is given in Table A-118.

Table A-118: Re-evaluated enthalpy of formation and entropy of α-Ag_2Se at 298.15 K. The evaluations were made at 1100 K and 1200 K in the lower and higher temperature range, respectively. The error limits are based on the standard deviations of the coefficients in the linear regressions.

$\Delta_f H^\circ_m$ (298.15 K) (kJ·mol^{-1})		S°_m (298.15 K) (J·K^{-1}·mol^{-1})	Temperature (K)
2nd law	3rd law		
– (58.9 ± 8.0)	– 38.7	(131.9 ± 5.0)	1045 – 1166
– (67.9 ± 6.0)	– 40.9	(127.6 ± 4.0)	1187 – 1305

[92SEM/KUL]

The vapour pressure of $MoSe_2(cr)$ was studied in the temperature range 1600 to 1750 K but the decomposition products are unknown. Therefore, the measured vapour pressures cannot be used for an evaluation of thermodynamic properties.

The evaporation of $WSe_2(cr)$ was investigated using a gas flow reactor in the temperature range 1000 to 1600 K. The vapour pressure in the temperature range 1460 to 1600 K was described by the equation $\log_{10}(p/\text{Pa}) = (13.189 \pm 1.729) - (1.858 \pm 0.264) \times 10^4\, T^{-1}$. The thermodynamic properties of the reaction $WSe_2(cr) \rightleftharpoons W(cr) + Se_2(g)$ at 298.15 K were evaluated from this expression by this review using the thermodynamic properties of Se(g), $Se_2(g)$, and W(cr) and heat capacity of $WSe_2(cr)$. At equilibrium the partial pressure of Se(g) is 22% of the $Se_2(g)$ partial pressure at the relevant temperatures and total pressures and this was taken into account in the evaluation. The second law results were $\Delta_f H^\circ_m$ (WSe_2, cr, 298.15 K) = – (234.7 ± 50.0) kJ·mol^{-1} and S°_m (WSe_2, cr, 298.15 K) = (103.2 ± 33.0) J·K^{-1}·mol^{-1} and the third law enthalpy of formation was $\Delta_f H^\circ_m$ (WSe_2, cr, 298.15 K) = – (256.1 ± 50.0) kJ·mol^{-1}.

[93BRU/PIA]

The evaporation of PbSe(cr) was studied in the temperature range 835 to 1047 K using the torsion effusion technique. The original evaluation employed auxiliary data different from those used by the review and the vapour pressure expression was therefore re-evaluated using the selected values of the entropies and heat capacities of PbSe(cr) and PbSe(g). The second law entropy of PbSe(g) and the second law enthalpy of sublimation of PbSe(cr) were calculated to be S°_m (PbSe, g, 298.15 K) = (260.6 ± 7.0)

$J\cdot K^{-1}\cdot mol^{-1}$ and $\Delta_{sub}H^{\circ}_{m}$ (PbSe, cr, 298.15 K) = (224.4 ± 7.0) $kJ\cdot mol^{-1}$, respectively. The third law enthalpy of sublimation was calculated to be $\Delta_{sub}H^{\circ}_{m}$ (PbSe, cr, 298.15 K) = (227.1 ± 3.0) $kJ\cdot mol^{-1}$.

[95RAI/FEL3]

This paper presents a serious attempt to determine the solubility product of iron(III) selenite.

The experiments show that $Fe_2(SeO_3)_3\cdot 6H_2O$ is the stable form of iron(III) selenite at pH < 4. Above this pH other, but not identified, solid phases are formed. The specimen, prepared from solutions of $FeCl_3$ and Na_2SeO_3 and aged at 360 K, was well crystallised as demonstrated by the sharp X-ray diffraction pattern.

Several series of solubility measurements were conducted at (296 ± 2) K. In one series the solubility of the specimen in solutions of HCl with a pH ranging from 0.77 to 5.1 was followed. Another series was conducted at a constant pH of about 1.4 and with the total selenite concentration varied from 0.0002 to 0.01 M. In two experiments equilibrium was approached from supersaturation achieved by a pH change. The test solutions were continuously agitated and analysed after intervals ranging from 1 to about 1700 days. The phases were separated by centrifugation and the supernatants passed through a 0.0018 μm pore size filter before instrumental analysis. Solubility equilibrium is slowly established and the results indicate that more than 30 days are needed to reach a steady state. The measured concentrations show that $Fe_2(SeO_3)_3\cdot 6H_2O$ dissolves congruently below pH 4. No change in the oxidation state of selenium was observed.

The equilibrium constant (solubility product) of the reaction

$$Fe_2(SeO_3)_3\cdot 6H_2O(cr) \rightleftharpoons 2Fe^{3+} + 3SeO_3^{2-} + 6H_2O(l) \qquad (A.38)$$

was calculated by a non-linear least squares program based on a minimisation of the Gibbs energy change of the reaction. Known side-reactions of Fe(III) and Se(IV) were included in the model and the activity coefficients were calculated from the Pitzer model. The standard Gibbs energy of formation of the known species and the ion-interaction parameters are included in the paper as well as the primary data. In order to obtain a reasonably good fit to the experimental data at increasing selenite concentration it was necessary to introduce a complex between Fe(III) and Se(IV) into the equilibrium model. The simplest model that best represented the experimental findings included $FeSeO_3^+$. The standard Gibbs energies of formation found for this model yielded:

$$\log_{10} K^{\circ}_{s,0}\ ((A.38), 296\ K) = -(41.58 \pm 0.11)$$

and

$$\log_{10} \beta^{\circ}_{1}\ ((A.38), 296\ K) = (11.15 \pm 0.11).$$

A calculation without $FeSeO_3^+$ included in the model on 4 data points at low pH and low total selenite concentrations yielded $\log_{10} K^\circ_{s,0}$ (A.38) = – 41.04, which indicates that the value of the solubility product is not critically dependent on the equilibrium model.

The review noted that an erroneous value of $\Delta_f G^\circ_m(SeO_3^{2-})$ was used. No recalculation has been attempted.

[95ZOC/PIA]

The total pressure of the vapour in equilibrium with SnSe(cr) was measured in the temperature range 739 to 1040 K using the Knudsen torsion effusion technique in this accurate study. The enthalpy and entropy of sublimation according to the reaction SnSe(cr) $\rightleftharpoons$ SnSe(g) were calculated by the review from the reported vapour pressure expression and the selected heat capacity expressions for SnSe(g) and SnSe(cr) yielding $\Delta_{sub}H^\circ_m$ (SnSe, α, 298.15 K) = (222.5 ± 3.4) kJ·mol^{-1} and $\Delta_{sub}S^\circ_m$ (SnSe, α, 298.15 K) = (168.9 ± 4.2) J·K^{-1}·mol^{-1}, respectively.

[96BRE/LIU2]

This is a thorough study of the vapour pressures of Zn(g) and Se_2(g) in equilibrium with α-ZnSe using optical density techniques. The Gibbs energy for the reaction

$$\alpha\text{-ZnSe} \rightleftharpoons \text{Zn(g)} + \tfrac{1}{2}\text{Se}_2\text{(g)} \qquad \text{(A.39)}$$

was evaluated to be $\Delta_r G^\circ_m$ ((A.39), T) = (– 361807 + 191.485 T) J·mol^{-1}, in the temperature range 1260 to 1410 K. The enthalpy of formation of α-ZnSe at 298.15 K was evaluated to be $\Delta_f H^\circ_m$ (ZnSe, α, 298.15 K) = – (177.8 ± 2.1) kJ·mol^{-1} using the third law, but applying a set of auxiliary data differing from the values selected by the review. The expression for the Gibbs energy of reaction was therefore recalculated according to the second and the third law using the selected thermodynamic functions of selenium, the data for Zn in [89COX/WAG], the heat capacity expression of α-ZnSe in Section V.9.1.1.1, and the selected entropy of α-ZnSe in the case of the third law. The values obtained were $\Delta_f H^\circ_m$ (ZnSe, α, 298.15 K) = – (175.4 ± 8.0) kJ·mol^{-1}, $\Delta_f H^\circ_m$ (ZnSe, α, 298.15 K) = – (178.2 ± 3.6) kJ·mol^{-1}, and S°_m (ZnSe, α, 298.15 K) = (74.0 ± 6.0) J·K^{-1}·mol^{-1} for the second and third law enthalpy of formation and the entropy, respectively.

[96SCH/FRE]

The saturated vapour pressure of α-ZnSe was measured using Knudsen effusion cells in the temperature range 1190 to 1310 K. The results were evaluated by applying the second and third law and assuming the reaction

$$\alpha\text{-ZnSe} \rightleftharpoons \text{Zn(g)} + \tfrac{1}{2}\text{Se}_2\text{(g)}. \qquad \text{(A.40)}$$

The reaction quantities $\Delta_r S^\circ_m$ ((A.40), 298.15 K) = (215.0 ± 2.0) J·K^{-1}·mol^{-1}, $\Delta_r H^\circ_m$ ((A.40), 298.15 K) = (382.0 ± 3.0) kJ·mol^{-1}, and $\Delta_r H^\circ_m$ ((A.40), 298.15 K) = (369.4 ± 0.5) kJ·mol^{-1} were obtained for the entropy of reaction and the second and third law enthalpy of reaction, respectively. No quantities were derived for α-ZnSe at 298.15 K. The experimental data given by the vapour pressure expression $\log_{10}(p/\text{kPa}) = (9.19 \pm 0.08) - (12896 \pm 96)\,T^{-1}$ were therefore re-evaluated by the review according to the second and the third law using the selected thermodynamic functions of selenium, the data for Zn in [89COX/WAG], the heat capacity expression of α-ZnSe in Section V.9.1.1.1, and the selected entropy of α-ZnSe in the case of the third law. The values obtained were $\Delta_f H^\circ_m$ (ZnSe, α, 298.15 K) = – (182.4 ± 14.0) kJ·mol^{-1}, $\Delta_f H^\circ_m$ (ZnSe, α, 298.15 K) = – (177.6 ± 3.8) kJ·mol^{-1}, and S°_m (ZnSe, α, 298.15 K) = (68.1 ± 14.0) J·K^{-1}·mol^{-1} for the second and third law enthalpy of formation and the entropy, respectively.

[96SHA/RED]

The solubility of the crystalline phases $CaSeO_3{\cdot}H_2O$, $MgSeO_3$, $MnSeO_3$, and $ZnSeO_3{\cdot}H_2O$ were determined at (298.2 ± 1.0) K in water, 0.01 M NaCl, 0.03 M NaCl, and 0.02 M Na_2SeO_3. The mixture of solid and solution was agitated continuously for 40 weeks. Samples were withdrawn at regular intervals and analysed for selenite. The determinations showed that equilibrium had been established after 40 weeks. The final separation of the aqueous phase was made by centrifugation followed by filtration through a 0.45 μm membrane filter. The filtrate was analysed for total soluble cations and anions but the analytical techniques used are not revealed. The pH of the equilibrium solution was not measured.

The primary solubility data and the calculation of the solubility products, defined in the usual way, are presented. This presentation contains some unexpected results. The total solubility of the metal ion and selenite are approximately equal in water and in the inert salt solutions for the magnesium and manganese selenites. This is the expected result for a simple dissolution reaction. For calcium selenite, the metal ion concentration was about 100 times greater than the total selenite concentration whereas for zinc selenite the opposite was found. There is no comment in the paper on these results, which contradict the equilibrium reactions used in the paper to define the reported solubility products for calcium and zinc selenite. The review also noted that the calculation of the magnesium and selenite activities from the total concentrations introduces activity coefficients between 0.1 to 0.01 at moderate ionic strengths. Thus the values of these coefficients appear unreasonably small. On the whole, the activity coefficient corrections introduced appear to vary in an erratic way between the various systems studied.

Another unexpected result is the decrease in the total selenite concentration with time in the experiments with dissolution in 0.02 M Na_2SeO_3 solution. The explanation given by the authors is not tenable. The statement in the paper that control of pH is unnecessary because H^+ does not enter into the solubility expression, together with the remarks made above, forces the review to attach zero weight to the results obtained in the paper.

[97BAH/PAR]

Ion interaction coefficients, $\varepsilon_{j,k}$, for the selenocyanate ion and its metal ion complexes are not available. Data for the thiocyanate ion have therefore been used as an approximation for extrapolation by SIT of selenocyanate equilibrium constants to standard conditions.

Bahta, Parker, and Tuck [97BAH/PAR] have made a critical evaluation of the stability constants of complexes of the thiocyanate ion. Their scrutiny of the literature leaves rather few data as "Recommended" or "Tentatively recommended". Sufficient data for an evaluation of $\varepsilon_{j,k}$ values are found only for Cd^{2+} in $NaClO_4$ medium. These data are entered in Table A-119 as mean values at the appropriate $NaClO_4$ concentration. No data below I = 1 M are in general available. The small values of the stability constants require rather high concentrations of the ligand to be added to the test solutions, which in turn requires a high concentration of the inert salt background.

Table A-119: Stability constants of cadmium thiocyanate complexes (mean values) in $NaClO_4$ medium with an extrapolation to I = 0 by the SIT method. The uncertainties correspond to twice the standard deviation from the least-squares calculation.

I (M)	$\log_{10}\beta_1$	$\log_{10}\beta_2$	$\log_{10}\beta_3$	$\log_{10}\beta_4$
(0)	(1.89)	(2.78)	(2.85)	(2.3)
1	1.35	1.92	2.14	2.13
2	1.26	2.14	1.9	2.25
3	1.39	2.02	2.32	2.33
4	1.26	2.38	2.73	2.91
$\log_{10}\beta_n^o$	(2.13 ± 0.19)	(3.01 ± 0.31)	(2.94 ± 0.44)	(2.63 ± 0.32)
$\Delta\varepsilon$	– (0.026 ± 0.058)	– (0.145 ± 0.094)	– (0.202 ± 0.134)	– (0.173 ± 0.099)

The evaluation of the $\Delta\varepsilon$ followed the method outlined in [92GRE/FUG]. The data presented for I = 0 in [97BAH/PAR], which were generated via a different extrapolation algorithm, were not included in the linear regression. Due to the lack of data, the results obtained here will be assumed to be applicable to other divalent metal ions and to other ionic media as well.

[97CHR]

Christov calculated the parameters in the Pitzer ion interaction model from isopiestic measurements at 298.15 K by Ojkova and Staneva [89OJK/STA]. This reference contains (interpolated) osmotic coefficients of zinc, magnesium, cobalt, and nickel selenate solutions from 0.1 mol·kg^{-1} to saturated solution. Sodium chloride standards were used and the agreement between duplicate determinations was 0.2% or better.

These parameters (see Table A-120) were used to find the mean activity coefficient of the saturated solution and the solubility product. The results were:

$\log_{10} K^{\circ}_{s,0}$ ($ZnSeO_4$) = – 1.572, solubility 3.05 mol·kg^{-1};

$\log_{10} K^{\circ}_{s,0}$ ($CoSeO_4$) = – 1.759, solubility 2.82 mol·kg^{-1};

$\log_{10} K^{\circ}_{s,0}$ ($NiSeO_4$) = – 1.486, solubility 1.73 mol·kg^{-1};

$\log_{10} K^{\circ}_{s,0}$ ($MgSeO_4$) = – 1.133, solubility 3.28 mol·kg^{-1}.

The result for nickel selenate has been revised in later publications [98OJK/CHR], [2001BAR/CHR].

Table A-120: Pitzer coefficients of Zn, Co, Ni, and Mg selenate solutions at 298.15 K.

Salt	α_1 fixed	α_2 fixed	$\beta^{(0)}_{MX}$	$\beta^{(1)}_{MX}$	C^{ϕ}_{MX}	$\sigma(\phi)$
$ZnSeO_4$	1.4	12	0.2324	3.8156	0.0224	0.0056
$CoSeO_4$	1.4	12	0.4409	1.4440	– 0.303	0.0045
$NiSeO_4$	1.4	12	0.5517	3.2103	– 0.0271	0.0050
$MgSeO_4$	1.4	12	0.3010	4.2720	0.0163	0.0050

[97FER/FIN]

The interaction between the selenite or selenate ion with Cu^{2+} and Pb^{2+} was studied by classical polarography. The measurements were carried out in unbuffered solution with 0.15 M $NaNO_3$ added as supporting electrolyte. The temperature is not mentioned. The concentration of Cu^{2+} was 1×10^{-4} M and that of Pb^{2+} 1×10^{-5} M. The ligand concentration was varied over the interval 2×10^{-5} to 1×10^{-2} M.

With selenate as ligand no decrease in the limiting current or shift in the half-wave potential, $E_{1/2}$, was observed. It was concluded that no precipitate was formed and that the complex formation was weak. The review assigns the accuracy of the potential measurement to be ± 2 mV, which leads to an upper value of the stability constant of the 1:1 complex of about 20 M^{-1} in 0.15 M $NaNO_3$.

With selenite as ligand a decrease in the limiting current started already at an added ligand concentration of 2×10^{-5} M. For Cu^{2+} no shift in $E_{1/2}$ was observed, while a shift was present for Pb^{2+}. The primary data are presented only as small graphs. The authors calculated from their observations a solubility product for

$$CuSeO_3{\cdot}2H_2O(cr) \rightleftharpoons Cu^{2+} + SeO_3^{2-} + 2H_2O(l) \qquad (A.41)$$

$$\log_{10} K_{s,0}\,(A.41) = -\,8.42$$

and a stability constant for:

$$Pb^{2+} + SeO_3^{2-} \rightleftharpoons PbSeO_3(aq) \qquad (A.42)$$

$$\log_{10} \beta_1\,(A.42) = 4.69$$

The absence of primary data and any details of the calculation of the constants make the assessment of the correctness of the constants almost impossible. As no pH values were apparently measured, the concentration used for the selenite ion must have been its total concentration. As discussed in [68RIP/VER] in this Appendix, a major part of the copper(II) in solution would be present as hydroxo complexes. Hence the solubility product reported in the paper would be expected to be greater than the true value. The greater activity of a microscopic phase as compared to a bulk phase also leads to this conclusion.

The presence of a precipitate in the Pb^{2+} - SeO_3^{2-} system makes the evaluation of the data difficult and somewhat arbitrary with the data at hand. As no details of the evaluation procedure is disclosed, the data for this system cannot be accepted. A discrepancy between the observed and calculated potential shifts noted by the review contributed to this decision.

[97LUB/HAV]

Lubal and Havel studied the complex formation between UO_2^{2+} and SeO_4^{2-} by spectrophotometric and potentiometric measurements at (298.2 ± 0.5) K. The spectrophotometric measurements were carried out in 3 M Na(ClO_4, SeO_4) medium. The total selenate concentration was varied between 0 and 0.6 M and the UO_2^{2+} concentration was kept constant at 0.0331 M. The potentiometric work used metal ion concentrations in the range (1 – 5) × 10^{-4} M and selenate ion concentrations in the range 0 to 0.3 M. No salt background was added, since the ion–selective dioxouranium(VI) electrode did not function in the presence of perchlorate ions.

Spectrophotometric work in the $-\log_{10}[H^+]$ range 1.4 to 3.0 gave no evidence for the formation of protonated species. Data collected in the $-\log_{10}[H^+]$ range 2.26 to 2.86 (spectrophotometry) and 3.5 (potentiometry) were evaluated by least squares methods. Factor analysis of the spectrophotometric data strongly indicated the presence of three absorbing species. Hence the data were interpreted by an equilibrium model including UO_2^{2+}, $UO_2SeO_4(aq)$, and $UO_2(SeO_4)_2^{2-}$. The equilibrium analysis of the 0 - 4 derivative spectra yielded consistent results. The following equilibrium constants, recalculated here to the molality scale, were reported:

$$UO_2^{2+} + SeO_4^{2-} \rightleftharpoons UO_2SeO_4(aq) \quad (A.43)$$

$$\log_{10} \beta_1 \,((A.43), I = 3\text{ M}, 298.15\text{ K}) = (1.50 \pm 0.02),$$

$$\log_{10} \beta_1^{\circ} \,((A.43), I = 0, 298.15\text{ K}) = (2.64 \pm 0.01),$$

$$UO_2^{2+} + 2\,SeO_4^{2-} \rightleftharpoons UO_2(SeO_4)_2^{2-} \quad (A.44)$$

$$\log\beta_2((A.44), I = 3\text{ M}, 298.15\text{ K}) = (2.29 \pm 0.02).$$

These results will be selected although it is not clear if a correction for the formation of hydrogen selenate was applied to the spectrophotometric data and the variation of the medium is considerable. The behaviour of the ion-selective electrode also poses questions on the reliability of the potentiometric data and hence the value of $\log_{10} \beta_1^{\circ}$ at $I = 0$.

The cited value of $\varepsilon(UO_2^{2+}, ClO_4^-)$ refers to $\varepsilon(UO_2^+, ClO_4^-)$, which affects the conclusions drawn in the paper from the application of the SIT approach to the data.

[98OJK/CHR]

The osmotic coefficient of Li_2SeO_4 solutions in the concentration range 0.927 to 3.708 $mol{\cdot}kg^{-1}$ was determined by isopiestic measurements at 298.15 K. Sodium chloride solutions were employed as reference standards. An equilibration period of 20 days yielded an agreement within 0.2% between duplicates.

The osmotic coefficients were used to find the parameters in the Pitzer ion interaction model. These parameters (see Table A-121) were then employed to find the activity coefficients in saturated solution.

Table A-121: Pitzer coefficients of Li_2SeO_4 solutions.

α_1	α_2	$\beta^{(0)}_{MX}$	$\beta^{(1)}_{MX}$	C^{ϕ}_{MX}	$\sigma(\phi)$
2 (fixed)	1 (fixed)	0.18252	1.62841	– 0.00283	0.00476

With the solubility of $Li_2SeO_4{\cdot}H_2O(cr)$, 4.275 $mol{\cdot}kg^{-1}$, determined in the paper, the corresponding solubility product was calculated to be $\log_{10} K^{\circ}_{s,0} = 1.762$.

The paper also contains a calculation of the Pitzer coefficients of nickel selenate solutions from isopiestic measurements by Ojkova and Staneva [89OJK/STA]. The results contained in Table A-122 also appear in [98CHR/OJK], [2001BAR/CHR].

With the solubility of $NiSeO_4{\cdot}6H_2O(cr)$, 1.93 $mol{\cdot}kg^{-1}$, determined in [98CHR/OJK] $\log_{10} K^{\circ}_{s,0}$ is calculated to be – 1.381.

The Pitzer model could successfully reproduce the solubility in the ternary system Li_2SeO_4-$NiSeO_4$-H_2O.

Table A-122: Pitzer coefficients of $NiSeO_4$ solutions.

t	α_2	$\beta^{(0)}_{MX}$	$\beta^{(1)}_{MX}$	C^{ϕ}_{MX}	$\sigma(\phi)$
1.4(fixed)	12 (fixed)	0.58712	2.98530	– 0.004540	0.00243

[99OJK/CHR]

The osmotic coefficient of $(NH_4)_2SeO_4$ solutions in the concentration range 2.74 to 5.50 mol·kg^{-1} and of K_2SeO_4 solutions between 1.09 and 4.74 mol·kg^{-1} were determined by isopiestic measurements at 298.15 K. Sodium chloride solutions were employed as reference standards. An equilibration period of 20 days yielded an agreement within 0.2% between duplicates.

The osmotic coefficients were used to find the parameters in the Pitzer ion interaction model (see Table A-123). These parameters were then employed to find the activity coefficients in saturated solution.

Table A-123: Pitzer coefficients of $(NH_4)_2SeO_4$ and K_2SeO_4 solutions.

System	α	$\beta^{(0)}_{MX}$	$\beta^{(1)}_{MX}$	C^{ϕ}_{MX}	$\sigma(\phi)$
$(NH_4)_2SeO_4$–H_2O	2 (fixed)	0.04155	0.77095	0.00312	0.00351
K_2SeO_4–H_2O	2 (fixed)	0.10911	0.71354	– 0.00325	0.00698

With the solubility of $(NH_4)_2SeO_4$(cr), 6.715 mol·kg^{-1}, and K_2SeO_4(cr), 5.13 mol·kg^{-1}, both extracted from the literature the corresponding solubility products were calculated to be $\log_{10} K^{\circ}_{s,0}$ = 0.920 and 0.902, respectively.

[99OPP/GOB]

Oppermann *et al.* studied the ternary system Bi-O-Se. Six crystalline phases were prepared by high temperature synthesis from Bi_2O_3(cr) and SeO_2(cr). The phases were characterised by X-ray powder diffraction and IR spectroscopy. The standard enthalpies of formation and entropies of the compounds were obtained from the temperature dependence of the equilibrium constants of the decomposition reactions. The standard enthalpies of formation were also found from solution calorimetric work in which the enthalpies of dissolution of the compounds, Bi_2O_3(cr), and SeO_2(cr) in 4 M HCl were measured. It should benoted that the results in the paper were obtained with a correct value of the enthalpy of formation of SeO_2(cr) and not the erroneous value in Table II of tha paper.

The results from the calorimetric work for the compounds $Bi_2Se_3O_9$ of the same stoichiometry as $Bi_2(SeO_3)_3$, and Bi_2SeO_5 of the same stoichiometry as $(BiO)_2SeO_3$ are $\Delta_f H^{\circ}_m$ ($Bi_2Se_3O_9$, cr, 298.15 K) = – (1404.6 ± 7.1) kJ·mol^{-1} and

$\Delta_f H_m^\circ$ (Bi_2SeO_5, cr, 298.15 K) = – (902.5 ± 5.4) kJ·mol^{-1}. The results from the decomposition reactions $Bi_2Se_3O_9$(cr) ⇌ Bi_2SeO_5(cr) + $2SeO_2$(g) and $8Bi_2SeO_5$(cr) ⇌ $Bi_{16}Se_5O_{34}$(cr) + $3SeO_2$(g) were $\Delta_f H_m^\circ$ ($Bi_2Se_3O_9$, cr, 298.15 K) = – (1389.9 ± 33.5) kJ·mol^{-1}, S_m° ($Bi_2Se_3O_9$, cr, 298.15 K) = (392 ± 38) J·K^{-1}·mol^{-1}, and $\Delta_f H_m^\circ$ (Bi_2SeO_5, cr, 298.15 K) = – (884.5 ± 12.6) kJ·mol^{-1}, S_m° (Bi_2SeO_5, cr, 298.15 K) = (240 ± 13) J·K^{-1}·mol^{-1}.

[2001BAR/CHR]

The osmotic coefficient of Cs_2SeO_4 solutions in the concentration range 1.09 to 4.59 mol·kg^{-1} was determined by isopiestic measurements at 298.15 K with NaCl solutions as reference standard. An equilibration period of 20 days yielded an agreement within 0.2% between duplicates.

The osmotic coefficients were used to find the parameters in the Pitzer ion interaction model (see Table A-124). These parameters were then employed to find the activity coefficient in saturated solution.

Table A-124: Pitzer coefficients of Cs_2SeO_4 solutions

α	$\beta^{(0)}_{MX}$	$\beta^{(1)}_{MX}$	C^{ϕ}_{MX}	$\sigma(\phi)$
2 (fixed)	0.18234	– 1.64961	– 0.01407	0.0041

With the solubility of Cs_2SeO_4(cr) determined in the paper, 6.353 mol·kg^{-1}, the solubility product was calculated to be $\log_{10} K^\circ_{s,0} = 0.630$.

[2001OPP/SCH]

Oppermann *et al.* studied some of the equilibria in the system Se-O-Br by total pressure measurements. Thermodynamic quantities were derived from the temperature dependence of the equilibrium constant. The extrapolations of the results to standard conditions do not use experimental thermal functions but employ various estimates.

The vapour pressure of $SeOBr_2$(l) was determined at temperatures below 475 K. The enthalpy of vaporisation $\Delta_{vap} H_m^\circ$ ($SeOBr_2$, l, 298.15 K) = (50.6 ± 1.3) kJ·mol^{-1} was obtained. No formation data can be derived from this measurement since selected data for the enthalpy of fusion of $SeOBr_2$(cr) and the enthalpy of formation of $SeOBr_2$(g) are lacking.

The gas phase reaction $2SeOBr_2$(g) ⇌ SeO_2(g) + $SeBr_2$(g) +Br_2(g) was measured in the temperature range 570 to 770 K. No primary data are presented and it is not understood why an iterative process was necessary for the calculation of the equilibrium constant. The data were evaluated by the third law to yield $\Delta_f H_m^\circ$ ($SeOBr_2$, g,

298.15 K) = – (92.9 ± 13.8) kJ·mol^{-1}. A considerable number of estimates of thermochemical data are involved in the calculation of this value.

The equilibrium $Se_2Br_2(l) \rightleftharpoons Se(cr) + SeBr_2(g)$ was measured in the temperature range 325 to 475 K. The data derived in the paper correspond to an enthalpy of reaction at standard conditions of $\Delta_r H_m^\circ$(298.15 K) = (49 ± 16) kJ·mol^{-1}. This value is in reasonable agreement with $\Delta_r H_m^\circ$(298.15 K) = (63 ± 20) kJ·mol^{-1} selected by the review from the measurements of the same reaction in [70HOG/LUN]. The latter value is preferred since it was obtained in the temperature range 293 to 348 K and hence the result calculated for standard conditions is less affected by the uncertainties in estimated thermal quantities. However, the values of the standard enthalpy of formation of $SeBr_2$(g) derived in the paper and in the review are widely different, – (23.8 ± 12.6) and (32 ± 20) kJ·mol^{-1}, respectively. The difference originates mainly in the value of $\Delta_f H_m^\circ$ (Se_2Br_2, l, 298.15 K) employed in the evaluation. The review selected an experimental value from [78TSV], – (31 ± 3) kJ·mol^{-1}, while Oppermann *et al.* used a value adopted in Mills [74MIL], – 72.8 kJ·mol^{-1} together with estimated temperature corrections. This value was calculated by Mills from the measurements in [70HOG/LUN] and a not referenced NBS value for $\Delta_f H_m^\circ$ ($SeBr_2$, g, 298.15 K) = – 21 kJ·mol^{-1}.

[2001OPP/ZHA]

Oppermann and Zhang-Preβe studied the ternary system Sm-O-Se. A number of crystalline phases were prepared by high temperature synthesis from mixtures of Sm_2O_3(cr) and SeO_2(cr) as starting materials. The new phases were characterised by X-ray powder diffraction and IR spectrometry. Enthalpies of formation, entropies and heat capacities as well as a phase diagram are presented.

The standard enthalpy of formation of the phase $Sm_2Se_3O_9$(cr), which has the same stoichiometric composition as $Sm_2(SeO_3)_3$, was determined to be – (2921 ± 41) kJ·mol^{-1} from the temperature dependence of the equilibrium constant of the reaction $2Sm_2Se_3O_9(cr) \rightleftharpoons 2Sm_2Se_{1.5}O_6(cr) + 3SeO_2(g)$ in the temperature range 875 to 1125 K and experimental C_p values.

The authors used a value for $\Delta_f H_m^\circ$ (SeO_2, g, 298.15 K) that is 16.2 kJ·mol^{-1} more negative than the selected value. They also used values of $H_m^\circ(T) - H_m^\circ$(298.15 K) for SeO_2(g) that, independent of T, are 16.2 kJ·mol^{-1} more positive than those calculated by the selected expression for the heat capacity of SeO_2(g). The two differences cancel and no recalculation is necessary.

[2001SEB/POT]

Séby *et al.* attempted to extrapolate K_1 and $K_{1,2}$ data for the protonation of the selenite ion from Barcza and Sillén [71BAR/SIL] to I = 0 by the "modified Bromley methodology", which is shortly presented in the paper and contains interaction parameters reminiscent of the SIT approach. It is not clear to the review if only data at high ionic strengths were used in the fitting procedure or whether they were supplemented by data at I = 0. Anyhow, the extrapolated values, $\log_{10} K^{\circ}_{1,2}$ = (2.70 ± 0.06) and $\log_{10} K^{\circ}_{1}$ = (8.54 ± 0.04), are quite close to the accepted values.

The results obtained have not been selected for two reasons. If the extrapolation was based mainly on the data in [71BAR/SIL] the constants would be affected by experimental artifacts as discussed in [76BAE/MES], [92GRE/FUG]. If, on the other hand, the data in [71BAR/SIL] are supplemented by data at I = 0, the values of the extrapolated constants would not represent independent information.

[2002OPP/ZHA]

Oppermann *et al.* studied the ternary system Nd-O-Se. A number of crystalline phases were prepared by high temperature synthesis from mixtures of Nd_2O_3(cr) and SeO_2(cr) as starting materials. The new phases were characterised by X-ray powder diffraction and IR spectrometry. Enthalpies of formation, entropies and heat capacities as well as a phase diagram are presented.

The standard enthalpy of formation of the phase $Nd_2Se_3O_9$(cr), which has the same stoichiometric composition as $Nd_2(SeO_3)_3$, was determined to be – (2846 ± 41) kJ·mol^{-1} from the temperature dependence of the equilibrium constant of the reaction $2Nd_2Se_3O_9(cr) \rightleftharpoons 2Nd_2Se_{1.5}O_6(cr) + 3SeO_2(g)$ in the range 875 to 1175 K and experimental C_p values.

The authors used a value for $\Delta_f H^{\circ}_m$ (SeO_2, g, 298.15 K) that is 16.2 kJ·mol^{-1} more negative than the selected value. They also used values of $H^{\circ}_m(T) - H^{\circ}_m$(298.15 K) for SeO_2(g) that, independent of T, are 16.2 kJ·mol^{-1} more positive than those calculated by the selected expression for the heat capacity of SeO_2(g). The two differences cancel and no recalculation is necessary.

[2002OPP/ZHA2]

This paper contains a thermochemical study of the system Yb_2O_3-SeO_2. The experimental approach and the thermochemical quantities derived are the same as those presented in [2001OPP/ZHA] and [2002OPP/ZHA]. The standard enthalpy of formation of the phase $Yb_2Se_3O_9$(cr), which is of the same stoichiometric composition as $Yb_2(SeO_3)_3$, was determined to be – (2779 ± 42) kJ·mol^{-1} from the temperature dependence of the equilibrium constant of the reaction $Yb_2Se_3O_9(cr) \rightleftharpoons Yb_2SeO_5(cr) + 2SeO_2(g)$ in the temperature range 825 to 1050 K.

[2002ZHA/OPP]

Zhang-Preβe and Oppermann prepared crystalline phases of the general composition $Me_2Se_xO_{3+2x}$ for Me = Nd, Sm, and Y by high temperature synthesis using Me_2O_3(cr) and SeO_2(cr) as starting materials. These materials were mixed in stoichiometric amounts corresponding to the x values 1, 1.5, 3, 3.5, and 4. The details of the preparation and characterisation of the compounds are described, for instance, in References [2001OPP/ZHA] and [2002OPP/ZHA].

The standard enthalpies of formation of the compounds were found from solution calorimetric measurements of the enthalpy of dissolution in 4 M HCl of $Me_2Se_xO_{3+2x}$, Me_2O_3(cr) and SeO_2(cr), respectively. The results obtained for $Me_2Se_3O_9$(cr), which has the same stoichiometric composition as $Me_2(SeO_3)_3$, are; $\Delta_f H_m^\circ$ ($Y_2Se_3O_9$, cr, 298.15 K) = – (2828.0 ± 8.4) kJ·mol^{-1}, $\Delta_f H_m^\circ$ ($Sm_2Se_3O_9$, cr, 298.15 K) = – (2827.1 ± 5.9) kJ·mol^{-1}, $\Delta_f H_m^\circ$ ($Nd_2Se_3O_9$, cr, 298.15 K) = – (2833.8 ± 5.4) kJ·mol^{-1}.

[2002ZHA/OPP2]

This paper contains a thermochemical study of the system Y_2O_3-SeO_2. The experimental approach and the thermochemical quantities derived are the same as those presented in [2001OPP/ZHA] and [2002OPP/ZHA]. The standard enthalpy of formation of the phase $Y_2Se_3O_9$(cr), which is of the same stoichiometric composition as $Y_2(SeO_3)_3$, was determined to be – (2893 ± 50) kJ·mol^{-1} from the temperature dependence of the equilibrium constant of the reaction $Y_2Se_3O_9$(cr) $\rightleftharpoons$ Y_2SeO_5(cr) + $2SeO_2$(g) in the temperature range 875 to 1100 K.

The authors used a value for $\Delta_f H_m^\circ$ (SeO_2, g, 298.15 K) that is 16.2 kJ·mol^{-1} more negative than the selected value. They also used values of $H_m^\circ(T) - H_m^\circ(298.15\ \text{K})$ for SeO_2(g) that, independent of T, are 16.2 kJ·mol^{-1} more positive than those calculated by the selected expression for the heat capacity of SeO_2(g). The two differences cancel and no recalculation is necessary.

[2003BAR/CHR]

The osmotic coefficient of Rb_2SeO_4 solutions in the concentration range 1.07 to 5.20 mol·kg^{-1} was determined by isopiestic measurements at 298.15 K with NaCl solutions as reference standard. An equilibration period of 20 days yielded an agreement within 0.2% between duplicates.

The osmotic coefficients were used to find the parameters in the Pitzer ion interaction model (see Table A-125). These parameters were then employed to find the activity coefficient in saturated solution.

Table A-125: Pitzer coefficients of Rb_2SeO_4 solutions

α	$\beta^{(0)}_{MX}$	$\beta^{(1)}_{MX}$	C^{ϕ}_{MX}	$\sigma(\phi)$
2 (fixed)	0.12672	– 0.25052	– 0.00860	0.0034

With the solubility of Rb_2SeO_4(cr) determined in the paper, 5.20 mol·kg^{-1}, the solubility product was calculated to be $\log_{10} K^{\circ}_{s,0} = 0.430$.

Appendix B

Ionic strength corrections[1]

Thermodynamic data always refer to a selected standard state. The definition given by IUPAC [82LAF] is adopted in this review as outlined in Section II.3.1. According to this definition, the standard state for a solute B in a solution is a hypothetical solution, at the standard state pressure, in which $m_B = m^\circ = 1$ mol·kg^{-1}, and in which the activity coefficient γ_B is unity. However, for many reactions, measurements cannot be made accurately (or at all) in dilute solutions from which the necessary extrapolation to the standard state would be simple. This is invariably the case for reactions involving ions of high charge. Precise thermodynamic information for these systems can only be obtained in the presence of an inert electrolyte of sufficiently high concentration, ensuring that activity factors are reasonably constant throughout the measurements. This appendix describes and illustrates the method used in this review for the extrapolation of experimental equilibrium data to zero ionic strength.

The activity factors of all the species participating in reactions in high ionic strength media must be estimated in order to reduce the thermodynamic data obtained from the experiment to the state $I = 0$. Two alternative methods can be used to describe the ionic medium dependence of equilibrium constants:

- One method takes into account the individual characteristics of the ionic media by using a medium dependent expression for the activity coefficients of the species involved in the equilibrium reactions. The medium dependence is described by virial or ion interaction coefficients as used in the Pitzer equations [73PIT] and in the specific ion interaction theory.

---

[1] This Appendix contains essentially the text written by Grenthe and Wanner [2000GRE/WAN], earlier versions of which have been printed in the previous NEA-TDB reviews [92GRE/FUG], [95SIL/BID], [99RAR/RAN], [2001LEM/FUG], [2003GUI/FAN], and [2005GAM/BUG]. The equations presented here are an essential part to the review procedure and are required to use the selected thermodynamic values. The contents of Tables B.4 and B.5 have been revised.

- The other method uses an extended Debye-Hückel expression in which the activity coefficients of reactants and products depend only on the ionic charge and the ionic strength, but it accounts for the medium specific properties by introducing ionic pairing between the medium ions and the species involved in the equilibrium reactions. Earlier, this approach has been used extensively in marine chemistry, *cf.* Refs. [79JOH/PYT], [79MIL], [79PYT], [79WHI2].

The activity factor estimates are thus based on the use of Debye-Hückel type equations. The "extended" Debye-Hückel equations are either in the form of specific ion interaction methods or the Davies equation [62DAV]. However, the Davies equation should in general not be used at ionic strengths larger than 0.1 $mol\cdot kg^{-1}$. The method preferred in the NEA Thermochemical Data Base review is a medium-dependent expression for the activity coefficients, which is the specific ion interaction theory in the form of the Brønsted-Guggenheim-Scatchard approach. Other forms of specific ion interaction methods (the Pitzer and Brewer "B-method" [61LEW/RAN] and the Pitzer virial coefficient method [79PIT]) are described in the NEA Guidelines for the extrapolation to zero ionic strength [2000GRE/WAN].

The specific ion interaction methods are reliable for intercomparison of experimental data in a given concentration range. In many cases this includes data at rather low ionic strengths, I = 0.01 to 0.1 M, *cf.* Figure B-1, while in other cases, notably for cations of high charge ($\geq +4$ and ≤ -4), the lowest available ionic strength is often 0.2 M or higher, see for example Figures V.12 and V.13 in [92GRE/FUG]. It is reasonable to assume that the extrapolated equilibrium constants at $I = 0$ are more precise in the former than in the latter cases. The extrapolation error is composed of two parts, one due to experimental errors, and the other due to model errors. The model errors seem to be rather small for many systems, less than 0.1 units in $\log_{10} K^\circ$. For reactions involving ions of high charge, which may be extensively hydrolysed, one cannot perform experiments at low ionic strengths. Hence, it is impossible to estimate the extrapolation error. This is true for all methods used to estimate activity corrections. Systematic model errors of this type are not included in the uncertainties assigned to the selected data in this review.

It should be emphasised that the specific ion interaction model is *approximate*. Modifying it, for example by introducing the equations suggested by Ciavatta ([90CIA], Eqs. (8–10), *cf.* Section B.1.4), would result in slightly different ion interaction coefficients and equilibrium constants. Both methods provide an internally consistent set of values. However, their absolute values may differ somewhat. Grenthe *et al.* [92GRE/FUG] estimate that these differences in general are less than 0.2 units in $\log_{10} K^\circ$, *i.e.*, approximately 1 $kJ\cdot mol^{-1}$ in derived $\Delta_f G^\circ_m$ values.

B.1 The specific ion interaction equations

B.1.1 Background

The Debye-Hückel term, which is the dominant term in the expression for the activity coefficients in dilute solution, accounts for electrostatic, non-specific long-range interactions. At higher concentrations short range, non-electrostatic interactions have to be taken into account. This is usually done by adding ionic strength dependent terms to the Debye-Hückel expression. This method was first outlined by Brønsted [22BRO], [22BRO2], and elaborated by Scatchard [36SCA] and Guggenheim [66GUG]. The two basic assumptions in the specific ion interaction theory are described below.

- **Assumption 1:** The activity coefficient γ_j of an ion j of charge z_j in the solution of ionic strength I_m may be described by Eq. (B.1):

$$\log_{10}\gamma_j = -z_j^2 D + \sum_k \varepsilon(j,k,I_m)\, m_k \quad \text{(B.1)}$$

D is the Debye-Hückel term:

$$D = \frac{A\sqrt{I_m}}{1 + Ba_j\sqrt{I_m}} \quad \text{(B.2)}$$

where I_m is the molal ionic strength:

$$I_m = \frac{1}{2}\sum_i m_i z_i^2$$

A and B are constants which are temperature and pressure dependent, and a_j is an ion size parameter ("distance of closest approach") for the hydrated ion j. The Debye-Hückel limiting slope, A, has a value of (0.509 ± 0.001) $\text{kg}^{½}\cdot\text{mol}^{-½}$ at 25°C and 1 bar, (*cf.* Section B.1.2). The term Ba_j in the denominator of the Debye-Hückel term has been assigned a value of Ba_j = 1.5 $\text{kg}^{½}\cdot\text{mol}^{-½}$ at 25°C and 1 bar, as proposed by Scatchard [76SCA] and accepted by Ciavatta [80CIA]. This value has been found to minimise, for several species, the ionic strength dependence of $\varepsilon(j, k, I_m)$ between I_m = 0.5 m and I_m = 3.5 m. It should be mentioned that some authors have proposed different values for Ba_j ranging from Ba_j = 1.0 [35GUG] to Ba_j = 1.6 [62VAS]. However, the parameter Ba_j is empirical and as such correlated to the value of $\varepsilon(j, k, I_m)$. Hence, this variety of values for Ba_j does not represent an uncertainty range, but rather indicates that several different sets of Ba_j and $\varepsilon(j, k, I_m)$ may describe equally well the experimental mean activity coefficients of a given electrolyte. The ion interaction coefficients at 25°C listed in Table B-4, Table B-5 and Table B-6 have thus to be used with Ba_j = 1.5 $\text{kg}^{½}\cdot\text{mol}^{-½}$.

The summation in Eq. (B.1) extends over all ions k present in solution. Their molality is denoted m_k, and the specific ion interaction parameters, $\varepsilon(j, k, I_m)$, in general depend only slightly on the ionic strength. The concentrations of the ions of the ionic medium are often very much larger than those of the reacting species. Hence, the ionic medium ions will make the main contribution to the value of $\log_{10}\gamma_j$ for the reacting ions. This fact often makes it possible to simplify the summation $\sum_k \varepsilon(j,k,I_m)m_k$ so that only ion interaction coefficients between the participating ionic species and the ionic medium ions are included, as shown in Eqs. (B.4) to (B.8).

- **Assumption 2:** The ion interaction coefficients $\varepsilon(j, k, I_m)$ are zero for ions of the same charge sign and for uncharged species. The rationale behind this is that ε, which describes specific short-range interactions, must be small for ions of the same charge since they are usually far from one another due to electrostatic repulsion. This holds to a lesser extent also for uncharged species.

Eq. (B.1) will allow fairly accurate estimates of the activity coefficients in mixtures of electrolytes if the ion interaction coefficients are known. Ion interaction coefficients for simple ions can be obtained from tabulated data of mean activity coefficients of strong electrolytes or from the corresponding osmotic coefficients. Ion interaction coefficients for complexes can either be estimated from the charge and size of the ion or determined experimentally from the variation of the equilibrium constant with the ionic strength.

Ion interaction coefficients are not strictly constant but may vary slightly with the ionic strength. The extent of this variation depends on the charge type and is small for 1:1, 1:2 and 2:1 electrolytes for molalities less than 3.5 m. The concentration dependence of the ion interaction coefficients can thus often be neglected. This point was emphasised by Guggenheim [66GUG], who has presented a considerable amount of experimental material supporting this approach. The concentration dependence is larger for electrolytes of higher charge. In order to accurately reproduce their activity coefficient data, concentration dependent ion interaction coefficients have to be used, *cf.* Lewis *et al.* [61LEW/RAN], Baes and Mesmer [76BAE/MES], or Ciavatta [80CIA]. By using a more elaborate virial expansion, Pitzer and co-workers [73PIT], [73PIT/MAY], [74PIT/KIM], [74PIT/MAY], [75PIT], [76PIT/SIL], [78PIT/PET], [79PIT] have managed to describe measured activity coefficients of a large number of electrolytes with high precision over a large concentration range. Pitzer's model generally contains three parameters as compared to one in the specific ion interaction theory. The use of the theory requires the knowledge of all these parameters. The derivation of Pitzer coefficients for many complexes such as those of the actinides would require a very large amount of additional experimental work, since few data of this type are currently available.

The way in which the activity coefficient corrections are performed in this review according to the specific ion interaction theory is illustrated below for a general case of a complex formation reaction. Charges are omitted for brevity.

$$m\,\mathrm{M} + q\,\mathrm{L} + n\,\mathrm{H_2O(l)} \rightleftharpoons \mathrm{M}_m\mathrm{L}_q(\mathrm{OH})_n + n\,\mathrm{H}^+$$

The formation constant of $M_mL_q(OH)_n$, ${}^*\beta_{q,n,m}$, determined in an ionic medium (1:1 salt NX) of the ionic strength I_m, is related to the corresponding value at zero ionic strength, ${}^*\beta^{\circ}_{q,n,m}$ by Eq.(B.3):

$$\begin{aligned}\log_{10} {}^*\beta_{q,n,m} = \log_{10} {}^*\beta^{\circ}_{q,n,m} + m\log_{10}\gamma_M + q\log_{10}\gamma_L + n\log_{10} a_{H_2O} \\ -\log_{10}\gamma_{q,n,m} - n\log_{10}\gamma_{H^+}\end{aligned} \tag{B.3}$$

The subscript (q, n, m) denotes the complex ion $M_mL_q(OH)_n$. If the concentrations of N and X are much greater than the concentrations of M, L, $M_mL_q(OH)_n$ and H^+, only the molalities m_N and m_X have to be taken into account for the calculation of the term $\sum_k \varepsilon(j,k,I_m)m_k$ in Eq. (B.1). For example, for the activity coefficient of the metal cation M, γ_M, Eq. (B.4) is obtained at 25°C and 1 bar.

$$\log_{10}\gamma_M = \frac{-z_M^2\, 0.509\sqrt{I_m}}{1+1.5\sqrt{I_m}} + \varepsilon(M,X,I_m)m_X \tag{B.4}$$

Under these conditions, $I_m \approx m_X = m_N$ Substituting the $\log_{10}\gamma_j$ values in Eq. (B.3) with the corresponding forms of Eq. (B.4) and rearranging leads to:

$$\log_{10} {}^*\beta_{q,n,m} - \Delta z^2 D - n\log_{10} a_{H_2O} = \log_{10} {}^*\beta^{o}_{q,n,m} - \Delta\varepsilon\, I_m \tag{B.5}$$

where, at 25°C and 1 bar:

$$\Delta z^2 = (m z_M - q z_L - n)^2 + n - m z_M^2 - q z_L^2 \tag{B.6}$$

$$D = \frac{0.509\sqrt{I_m}}{1+1.5\sqrt{I_m}} \tag{B.7}$$

$$\Delta\varepsilon = \varepsilon(q, n, m, \text{N or X}) + n\varepsilon(\text{H, X}) - q\varepsilon(\text{N, L}) - m\varepsilon(\text{M, X}) \tag{B.8}$$

Here $(m z_M - q z_L - n)$, z_M and z_L are the charges of the complex $M_mL_q(OH)_n$, the metal ion M and the ligand L, respectively.

Equilibria involving $H_2O(l)$ as a reactant or product require a correction for the activity of water, a_{H_2O}. The activity of water in an electrolyte mixture can be calculated as:

$$\log_{10} a_{H_2O} = \frac{-\phi\sum_k m_k}{\ln(10)\times 55.51} \tag{B.9}$$

where ϕ is the osmotic coefficient of the mixture and the summation extends over all solute species k with molality m_k present in the solution. In the presence of an ionic medium NX in dominant concentration, Eq. (B.9) can be simplified by neglecting the contributions of all minor species, *i.e.*, the reacting ions. Hence, for a 1:1 electrolyte of ionic strength $I_m \approx m_{NX}$, Eq. (B.9) becomes:

$$\log_{10} a_{H_2O} = \frac{-2\, m_{NX}\, \phi}{\ln(10) \times 55.51} \tag{B.10}$$

Alternatively, water activities can be taken from Table B-1. These have been calculated for the most common ionic media at various concentrations applying Pitzer's ion interaction approach and the interaction parameters given in [91PIT]. Data in *italics* have been calculated for concentrations beyond the validity of the parameter set applied. These data are therefore extrapolations and should be used with care.

Table B-1: Water activities a_{H_2O} for the most common ionic media at various concentrations applying Pitzer's ion interaction approach and the interaction parameters given in [91PIT]. Data in *italics* have been calculated for concentrations beyond the validity of the parameter set applied. These data are therefore extrapolations and should be used with care.

Water activities a_{H_2O}								
c (*m*)	$HClO_4$	$NaClO_4$	$LiClO_4$	NH_4ClO_4	$Ba(ClO_4)_2$	HCl	NaCl	LiCl
0.10	0.9966	0.9966	0.9966	0.9967	0.9953	0.9966	0.9966	0.9966
0.25	0.9914	0.9917	0.9912	0.9920	0.9879	0.9914	0.9917	0.9915
0.50	0.9821	0.9833	0.9817	0.9844	0.9740	0.9823	0.9833	0.9826
0.75	0.9720	0.9747	0.9713	0.9769	0.9576	0.9726	0.9748	0.9731
1.00	0.9609	0.9660	0.9602	0.9694	0.9387	0.9620	0.9661	0.9631
1.50	0.9357	0.9476	0.9341	0.9542	0.8929	0.9386	0.9479	0.9412
2.00	0.9056	0.9279	0.9037		*0.8383*	0.9115	0.9284	0.9167
3.00	0.8285	0.8840	0.8280		*0.7226*	0.8459	0.8850	0.8589
4.00	0.7260	0.8331	*0.7309*			0.7643	0.8352	0.7991
5.00	*0.5982*	*0.7744*				0.6677	0.7782	0.7079
6.00	*0.4513*	*0.7075*				*0.5592*		*0.6169*
c (*m*)	KCl	NH_4Cl	$MgCl_2$	$CaCl_2$	NaBr	HNO_3	$NaNO_3$	$LiNO_3$
0.10	0.9966	0.9966	0.9953	0.9954	0.9966	0.9966	0.9967	0.9966
0.25	0.9918	0.9918	0.9880	0.9882	0.9916	0.9915	0.9919	0.9915
0.50	0.9836	0.9836	0.9744	0.9753	0.9830	0.9827	0.9841	0.9827
0.75	0.9754	0.9753	0.9585	0.9605	0.9742	0.9736	0.9764	0.9733
1.00	0.9671	0.9669	0.9399	0.9436	0.9650	0.9641	0.9688	0.9635
1.50	0.9500	0.9494	0.8939	0.9024	0.9455	0.9439	0.9536	0.9422
2.00	0.9320	0.9311	0.8358	0.8507	0.9241	0.9221	0.9385	0.9188

(Continued on next page)

Table B-1: (continued)

Water activities a_{H_2O}								
c (*m*)	$HClO_4$	$NaClO_4$	$LiClO_4$	NH_4ClO_4	$Ba(ClO_4)_2$	HCl	NaCl	LiCl
3.00	0.8933	0.8918	0.6866	0.7168	0.8753	0.8737	0.9079	0.8657
4.00	0.8503	0.8491	0.5083	0.5511	*0.8174*	0.8196	0.8766	0.8052
5.00		*0.8037*		*0.3738*	*0.7499*	0.7612	0.8446	0.7390
6.00					*0.6728*	*0.7006*	*0.8120*	*0.6696*
c (*m*)	NH_4NO_3	Na_2SO_4	$(NH_4)_2SO_4$	Na_2CO_3	K_2CO_3	NaSCN		
0.10	0.9967	0.9957	0.9958	0.9956	0.9955	0.9966		
0.25	0.9920	0.9900	0.9902	0.9896	0.9892	0.9915		
0.50	0.9843	0.9813	0.9814	0.9805	0.9789	0.9828		
0.75	0.9768	0.9732	0.9728	0.9720	0.9683	0.9736		
1.00	0.9694	0.9653	0.9640	0.9637	0.9570	0.9641		
1.50	0.9548	0.9491	0.9455	0.9467	0.9316	0.9438		
2.00	0.9403		0.9247	0.9283	0.9014	0.9215		
3.00	0.9115		0.8735		0.8235	0.8708		
4.00	0.8829		0.8050		0.7195	*0.8115*		
5.00	*0.8545*				*0.5887*	*0.7436*		
6.00	*0.8266*					*0.6685*		

Values of osmotic coefficients for single electrolytes have been compiled by various authors, *e.g.*, Robinson and Stokes [59ROB/STO]. The activity of water can also be calculated from the known activity coefficients of the dissolved species. In the presence of an ionic medium $N_{\nu_+}X_{\nu_-}$ of a concentration much larger than those of the reacting ions, the osmotic coefficient can be calculated according to Eq. (B.11) (*cf.* Eqs. (23-39), (23-40) and (A4-2) in [61LEW/RAN]).

$$1-\phi = \frac{A\ln(10)|z_+z_-|}{I_m(B\,a_j)^3}\left[1+B\,a_j\sqrt{I_m}-2\ln(1+B\,a_j\sqrt{I_m})-\frac{1}{1+B\,a_j\sqrt{I_m}}\right] - \ln(10)\;\varepsilon(\mathrm{N,X})\,m_{\mathrm{NX}}\left(\frac{\nu_+\nu_-}{\nu_++\nu_-}\right) \qquad \text{(B.11)}$$

where ν_+ and ν_- are the number of cations and anions in the salt formula ($\nu_+z_+ = \nu_-z_-$) and in this case:

$$I_m = \frac{1}{2}|z_+z_-|\,m_{\mathrm{NX}}(\nu_+ + \nu_-)$$

The activity of water is obtained by inserting Eq. (B.11) into Eq. (B.10). It should be mentioned that in mixed electrolytes with several components at high concentrations, it might be necessary to use Pitzer's equation to calculate the activity of water. On the other hand, a_{H_2O} is nearly constant in most experimental studies of equilibria in dilute aqueous solutions, where an ionic medium is used in large excess with respect to the reactants. The medium electrolyte thus determines the osmotic coefficient of the solvent.

In natural waters the situation is similar; the ionic strength of most surface waters is so low that the activity of $H_2O(l)$ can be set equal to unity. A correction may be necessary in the case of seawater, where a sufficiently good approximation for the osmotic coefficient may be obtained by considering NaCl as the dominant electrolyte.

In more complex solutions of high ionic strengths with more than one electrolyte at significant concentrations, *e.g.*, (Na^+, Mg^{2+}, Ca^{2+}) (Cl^-, SO_4^{2-}), Pitzer's equation (*cf.* [2000GRE/WAN]) may be used to estimate the osmotic coefficient; the necessary interaction coefficients are known for most systems of geochemical interest.

Note that in all ion interaction approaches, the equation for mean activity coeffi-cients can be split up to give equations for conventional single ion activity coefficients in mixtures, *e.g.*, Eq. (B.1). The latter are strictly valid only when used in combinations which yield electroneutrality. Thus, while estimating medium effects on standard potentials, a combination of redox equilibria with $H^+ + e^- \rightleftharpoons \frac{1}{2}H_2(g)$ is necessary (*cf.* Example B.3).

B.1.2 Ionic strength corrections at temperatures other than 298.15 K

Values of the Debye-Hückel parameters *A* and *B* in Eqs. (B.2) and (B.11) are listed in Table B-2 for a few temperatures at a pressure of 1 bar below 100°C and at the steam saturated pressure for $t \geq 100$°C. The values in Table B-2 may be calculated from the static dielectric constant and the density of water as a function of temperature and pressure, and are also found for example in Refs. [74HEL/KIR], [79BRA/PIT], [81HEL/KIR], [84ANA/ATK], [90ARC/WAN].

The term Ba_j in the denominator of the Debye-Hückel term *D*, *cf.* Eq. (B.2), has been assigned in this review a value of 1.5 $kg^{1/2}\cdot mol^{-1/2}$ at 25°C and 1 bar, *cf.* Section B.1.1 At temperatures and pressures other than the reference and standard state, the following possibilities exist:

- The value of Ba_j is calculated at each temperature assuming that ion sizes are independent of temperature and using the values of *B* listed in Table B-2.
- The value Ba_j is kept constant at 1.5 $kg^{1/2}\cdot mol^{-1/2}$. Due to the variation of *B* with temperature, *cf.* Table B-2, this implies a temperature dependence for ion size parameters. Assuming that ion sizes are in reality constant, then it is seen that

this simplification introduces an error in D, which increases with temperature and ionic strength (this error is less than ± 0.01 at $t \leq 100°C$ and $I < 6$ m, and less than ±0.006 at $t \leq 50°C$ and $I \leq 4$ m).

- The value of Ba_j is calculated at each temperature assuming a given temperature variation for a_j and using the values of B listed in Table B-2. For example, in the aqueous ionic model of Helgeson and co-workers ([88TAN/HEL], [88SHO/HEL], [89SHO/HEL], [89SHO/HEL2]) ionic sizes follow the relation: $a_j(T) = a_j(298.15\text{ K, 1 bar}) + |z_j| g(T, p)$ [90OEL/HEL], where $g(T, p)$ is a temperature and pressure function which is tabulated in [88TAN/HEL], [92SHO/OEL], and is approximately zero at temperatures below 175°C.

Table B-2: Debye-Hückel constants as a function of temperature at a pressure of 1 bar below 100°C and at the steam saturated pressure for $t \geq 100°C$. The uncertainty in the A parameter is estimated in [2001LEM/FUG] to be ±0.001 at 25°C, and ± 0.006 at 300°C, while for the B parameter the estimated uncertainty ranges from ±0.0003 at 25°C to ± 0.001 at 300°C.

t(°C)	P (bar)	A ((kg·mol^{-1})^{½})	$B \times 10^{-10}$ (kg^{½}·mol$^{-½}$·m^{-1})
0	1.00	0.491	0.3246
5	1.00	0.494	0.3254
10	1.00	0.498	0.3261
15	1.00	0.501	0.3268
20	1.00	0.505	0.3277
25	1.00	0.509	0.3284
30	1.00	0.513	0.3292
35	1.00	0.518	0.3300
40	1.00	0.525	0.3312
50	1.00	0.534	0.3326
75	1.00	0.564	0.3371
100	1.013	0.600	0.3422
125	2.32	0.642	0.3476
150	4.76	0.690	0.3533
175	8.92	0.746	0.3593
200	15.5	0.810	0.365
250	29.7	0.980	0.379
300	85.8	1.252	0.396

The values of $\varepsilon(j, k, I_m)$, obtained with the methods described in Section B.1.3 at temperatures other than 25°C will depend on the value adopted for Ba_j. As long as a consistent approach is followed, values of ε(*j, k, I_m*) absorb the choice of Ba_j , and for moderate temperature intervals (between 0 and 200°C) the choice $Ba_j = 1.5$ kg^{½}·mol$^{-½}$ is the simplest one.

The variation of ε(*j, k, I_m*) with temperature is discussed by Lewis *et al.* [61LEW/RAN], Millero [79MIL], Helgeson *et al.* [81HEL/KIR], [90OEL/HEL], Giffaut *et al.* [93GIF/VIT2] and Grenthe and Plyasunov [97GRE/PLY]. The absolute values for the reported ion interaction parameters differ in these studies due to the fact that the Debye-Hückel term used by these authors is not exactly the same. Nevertheless, common to all these studies is the fact that values of $(\partial\varepsilon/\partial T)_p$ are usually ≤ 0.005 kg·mol^{-1}·K^{-1} for temperatures below 200°C. Therefore, if values of ε(*j, k, I_m*) obtained at 25°C are used in the temperature range 0 to 50°C to perform ionic strength corrections, the error in $\log_{10}\gamma_j / I_m$ will be ≤ 0.13. It is clear that in order to reduce the uncertainties on solubility calculations at $t \neq 25$°C, studies on the variation of ε(*j, k, I_m*) values with temperature should be undertaken.

B.1.3 Estimation of ion interaction coefficients

B.1.3.1 Estimation from mean activity coefficient data

Example B.1:

The ion interaction coefficient $\varepsilon(H^+, Cl^-)$ can be obtained from published values of $\gamma_{\pm,\mathrm{HCl}}$ *versus* m_{HCl}:

$$\begin{aligned} 2\log_{10}\gamma_{\pm,\mathrm{HCl}} &= \log_{10}\gamma_{\mathrm{H}^+} + \log_{10}\gamma_{\mathrm{Cl}^-} \\ &= -D + \varepsilon(\mathrm{H}^+,\mathrm{Cl}^-)\,m_{\mathrm{Cl}^-} - D + \varepsilon(\mathrm{Cl}^-,\mathrm{H}^+)\,m_{\mathrm{H}^+} \\ \log_{10}\gamma_{\pm,\mathrm{HCl}} &= -D + \varepsilon(\mathrm{H}^+,\mathrm{Cl}^-)\,m_{\mathrm{HCl}} \end{aligned}$$

By plotting $(\log_{10}\gamma_{\pm,\mathrm{HCl}} + D)$ *versus* m_{HCl} a straight line with the slope $\varepsilon(H^+, Cl^-)$ is obtained. The degree of linearity should in itself indicate the range of validity of the specific ion interaction approach. Osmotic coefficient data can be treated in an analogous way.

B.1.3.2 Estimations based on experimental values of equilibrium constants at different ionic strengths

Example B.2:

Equilibrium constants are given in Table B-3 for the reaction:

$$UO_2^{2+} + Cl^- \rightleftharpoons UO_2Cl^+ \qquad (B.12)$$

The following formula is deduced from Eq. (B.5) for the extrapolation to $I = 0$:

$$\log_{10}\beta_1 + 4D = \log_{10}\beta_1^\circ - \Delta\varepsilon\, I_m \qquad (B.13)$$

The linear regression is done as described in Appendix C. The following results are obtained:

$$\log_{10} \beta_1^\circ = (0.170 \pm 0.021)$$

$$\Delta\varepsilon(\text{B.12}) = -(0.248 \pm 0.022)\,\text{kg}\cdot\text{mol}^{-1}.$$

The experimental data are depicted in Figure B-1, where the dashed area represents the uncertainty range that is obtained by using the results in $\log_{10} \beta_1^\circ$ and $\Delta\varepsilon$ and correcting back to $I \neq 0$.

Table B-3: The preparation of the experimental equilibrium constants for the extrapolation to I = 0 with the specific ion interaction method at 25°C and 1 bar, according to Reaction (B.12). The linear regression of this set of data is shown in Figure B-1

I_c	I_m	$\log_{10} \beta_1$ (exp) [(a)]	$\log_{10} \beta_{1,m}$ [(b)]	$\log_{10} \beta_{1,m} + 4D$
0.100	0.101	– (0.17 ± 0.10)	– 0.173	(0.265 ± 0.100)
0.200	0.202	– (0.25 ± 0.10)	– 0.254	(0.292 ± 0.100)
0.254	0.258	– (0.35 ± 0.04)	– 0.356	(0.230 ± 0.040)
0.303	0.308	– (0.39 ± 0.04)	– 0.398	(0.219 ± 0.040)
0.403	0.412	– (0.41 ± 0.04)	– 0.419	(0.246 ± 0.040)
0.500	0.513	– (0.32 ± 0.10)	– 0.331	(0.372 ± 0.100)
0.552	0.569	– (0.42 ± 0.04)	– 0.433	(0.288 ± 0.040)
0.651	0.674	– (0.34 ± 0.04)	– 0.355	(0.394 ± 0.040)
0.852	0.886	– (0.42 ± 0.04)	– 0.437	(0.358 ± 0.040)
1.000	1.052	– (0.31 ± 0.10)	– 0.332	(0.491 ± 0.100)
1.000	1.052	– (0.277 ± 0.260)	– 0.298	(0.524 ± 0.260)
1.500	1.617	– (0.24 ± 0.10)	– 0.273	(0.618 ± 0.100)
2.000	2.212	– (0.15 ± 0.10)	– 0.194	(0.743 ± 0.100)
2.000	2.212	– (0.12 ± 0.10)	– 0.164	(0.773 ± 0.100)
2.500	2.815	– (0.06 ± 0.10)	– 0.112	(0.860 ± 0.100)
3.000	3.503	(0.04 ± 0.10)	– 0.027	(0.973 ± 0.100)

a Equilibrium constants for Reaction (B.12) with assigned uncertainties, corrected to 25°C where necessary.

b Equilibrium constants corrected from molarity to molality units, as described in Section II.2

Figure B-1: Plot of $\log_{10} \beta_1 + 4D$ *versus* I_m for Reaction (B.12), at 25°C and 1 bar. The straight line shows the result of the weighted linear regression, and the dotted lines represent the uncertainty range obtained by propagating the resulting uncertainties at $I = 0$ back to $I = 4$ m.

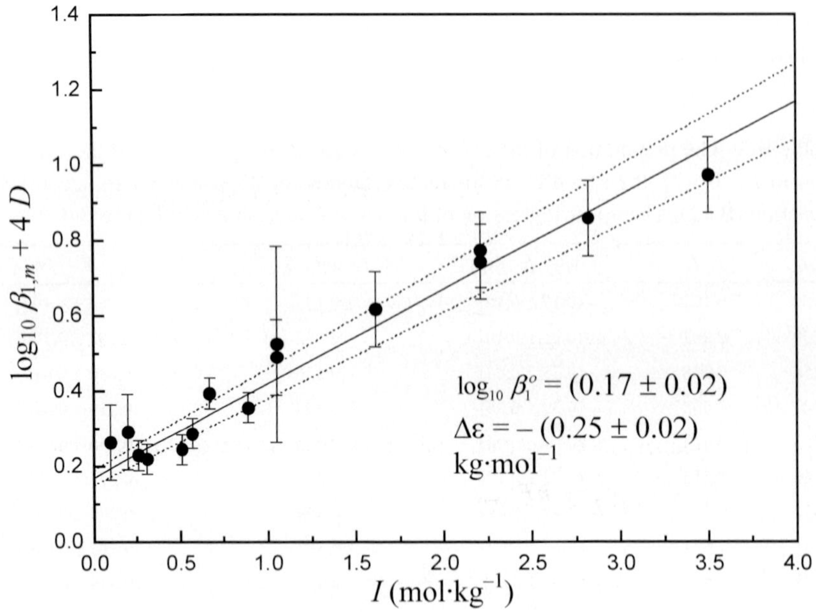

Example B.3:

When using the specific ion interaction theory, the relationship between the redox potential of the couple PuO_2^{2+}/Pu^{4+} in a medium of ionic strength I_m and the corresponding quantity at $I = 0$ should be calculated in the following way. The reaction in the galvanic cell:

$$Pt \mid H_2(g, r) \mid H^+(r) \mid PuO_2^{2+}, Pu^{4+}, H^+, H_2O(l) \mid Pt$$

is:

$$PuO_2^{2+} + H_2(g, r) + 4H^+ - 2H^+(r) \rightleftharpoons Pu^{4+} + 2H_2O(l) \qquad (B.14)$$

where "r" is used to indicate that $H_2(g)$ and H^+ are at the chemical conditions in the reference electrode compartment, *i.e.*, standard conditions when the reference electrode is the SHE. However, H^+, $H_2O(l)$ and the PuO_2^{2+}/Pu^{4+} are the conditions of the experimental measurements (*i.e.*, non-standard conditions, usually high ionic strength to improve the accuracy of the measurement).

For Reaction (B.14):

$$\log_{10} K^\circ = \log_{10}\left(\frac{a_{Pu^{4+}} \times a^2_{H_2O} \times a_{H^+(r)}}{a_{PuO_2^{2+}} \times a^4_{H^+} \times f_{H_2(r)}}\right).$$

Since by definition of the SHE, $f_{H_2(r)} = 1$ and $a_{H^+(r)} = 1$,

$$\log_{10} K^o = \log_{10} K + \log_{10}\gamma_{Pu^{4+}} - \log_{10}\gamma_{PuO_2^{2+}} - 4\log_{10}\gamma_{H^+} + 2\log_{10} a_{H_2O},$$

and

$$\log_{10}\gamma_{Pu^{4+}} = -16D + \varepsilon(Pu^{4+}, ClO_4^-)\ m_{ClO_4^-}$$

$$\log_{10}\gamma_{PuO_2^{2+}} = -4D + \varepsilon(PuO_2^{2+}, ClO_4^-)\ m_{ClO_4^-}$$

$$\log_{10}\gamma_{H^+} = -D + \varepsilon(H^+, ClO_4^-)\ m_{ClO_4^-}$$

Hence, $E^{o'}$

$$\log_{10} K^\circ = \log_{10} K - 8D + (\varepsilon(Pu^{4+}, ClO_4^-) - \varepsilon(PuO_2^{2+}, ClO_4^-) - 4\varepsilon(H^+, ClO_4^-))\, m_{ClO_4^-} + 2\log_{10} a_{H_2O} \qquad (B.15)$$

The relationship between the equilibrium constant and the redox potential is:

$$\ln K = \frac{nF}{RT} E^{o'} \qquad (B.16)$$

$$\ln K^\circ = \frac{nF}{RT} E^\circ. \qquad (B.17)$$

$E^{o'}$ is the redox potential in a medium of ionic strength I, E° is the corresponding standard potential at $I = 0$, and n is the number of transferred electrons in the reaction considered. Combining Eqs. (B.15), (B.16) and (B.17) and rearranging them leads to Eq.(B.18):

$$E^{o'} - (8D - 2\log_{10} a_{H_2O})\left(\frac{RT\ln(10)}{nF}\right) = E^o - \Delta\varepsilon\, m_{ClO_4^-}\left(\frac{RT\ln(10)}{nF}\right) \qquad (B.18)$$

For $n = 2$ in the present example and $T = 298.15$ K, Eq.(B.18) becomes:

$$E^{o'}[\text{mV}] - 236.6\ D + 59.16\log_{10} a_{H_2O} = E^o[\text{mV}] - 29.48\,\Delta\varepsilon\ m_{ClO_4^-}$$

where

$$\Delta\varepsilon = (\varepsilon(Pu^{4+}, ClO_4^-) - \varepsilon(PuO_2^{2+}, ClO_4^-) - 4\varepsilon(H^+, ClO_4^-))$$

The value of a_{H_2O} can be taken from experimental data or calculated from Eqs. (B.10) and (B.11).

In general, formal potentials are reported with reference to the standard hydrogen electrode, *cf.* Section II.1.6.5, as exemplified in Tables V.2 and V.3 of the uranium NEA review [92GRE/FUG]. In that case, the H^+ appearing in the reduction reaction is already at standard conditions. For example, experimental data are available on the formal potentials for reactions:

$$PuO_2^{2+} + 4H^+ + 2e^- \rightleftharpoons Pu^{4+} + 2H_2O(l) \qquad (B.19)$$

and

$$PuO_2^{2+} + e^- \rightleftharpoons PuO_2^+ \qquad (B.20).$$

While Reaction (B.19) corresponds to (B.14), reaction (B.20) is equivalent to:

$$PuO_2^{2+} + \tfrac{1}{2}H_2(g) \rightleftharpoons PuO_2^+ + H^+ \qquad (B.21)$$

where the designator "(r)" has been omitted, since in these equations only the H^+ in the reference compartment is relevant.

The cations in Reaction (B.14) represent aqueous species in the ionic media used during the experiments. In Reaction (B.21) H^+ represents the cation in the standard hydrogen electrode, and therefore it is already in standard conditions, and its activity coefficient must not be included in any extrapolation to $I = 0$ of experimental values for Reaction (B.20).

B.1.4 On the magnitude of ion interaction coefficients

Ciavatta [80CIA] made a compilation of ion interaction coefficients for a large number of electrolytes. Similar data for complexations of various kinds were reported by Spahiu [83SPA] and Ferri *et al.* [83FER/GRE]. These and some other data for 25°C and 1 bar have been collected and are listed in Section B.3.

It is obvious from the data in these tables that the charge of an ion is of great importance for the magnitude of the ion interaction coefficient. Ions of the same charge type have similar ion interaction coefficients with a given counter-ion. Based on the tabulated data, Grenthe *et al.* [92GRE/FUG] proposed that it is possible to estimate, with an error of at most ± 0.1 in ε, ion interaction coefficients for cases where there are insufficient experimental data for an extrapolation to $I = 0$. The error that is made by this approximation is estimated to ± 0.1 in Δε in most cases, based on comparison with $\Delta\varepsilon$ values of various reactions of the same charge type.

Ciavatta [90CIA] has proposed an alternative method to estimate values of ε for a first or second complex, ML or ML_2, in an ionic media NX, according to the following relationships:

$$\varepsilon(\text{ML, N or X}) \approx \left(\varepsilon(\text{M,X}) + \varepsilon(\text{L,N})\right)/2 \qquad (B.22)$$

$$\varepsilon(\text{ML}_2\text{, N or X}) \approx \left(\varepsilon(\text{M,X}) + 2\varepsilon(\text{L,N})\right)/3 \qquad (B.23)$$

Ciavatta obtained [90CIA] an average deviation of ± 0.05 kg·mol^{-1} between ε estimates according to Eqs. (B.22) and (B.23) and the ε values at 25°C obtained from ionic strength dependency of equilibrium constants.

B.2 Ion interaction coefficients *versus* equilibrium constants for ion pairs

It can be shown that the virial type of activity coefficient equations and the ionic pairing model are equivalent provided that the ionic pairing is weak. In these cases the distinction between complex formation and activity coefficient variations is difficult or even arbitrary unless independent experimental evidence for complex formation is available, *e.g.*, from spectroscopic data, as is the case for the weak uranium(VI) chloride complexes. It should be noted that the ion interaction coefficients evaluated and tabulated by Ciavatta [80CIA] were obtained from experimental mean activity coefficient data without taking into account complex formation. However, it is known that many of the metal ions listed by Ciavatta form weak complexes with chloride and nitrate ion. This fact is reflected by ion interaction coefficients that are smaller than those for the non-complexing perchlorate ion, *cf* Table B-4. In [2001LEM/FUG] chloride and nitrate complex formation was taken into account when these ions are part of the ionic medium and the value of the ion interaction coefficient $\varepsilon(\mathrm{M}^{n+},\mathrm{ClO}_4^-)$ was used as a substitute for $\varepsilon(\mathrm{M}^{n+},\mathrm{Cl}^-)$ and $\varepsilon(\mathrm{M}^{n+},\mathrm{NO}_3^-)$. In this way, the medium dependence of the activity coefficients is described with a combination of a specific ion interaction model and an ion pairing model. It is evident that the use of NEA recommended data with ionic strength correction models that differ from those used in the evaluation procedure can lead to inconsistencies in the results of the speciation calculations.

It should be mentioned that complex formation may also occur between negatively charged complexes and the cation of the ionic medium. An example is the stabilization of the complex ion $\mathrm{UO_2(CO_3)_3^{5-}}$ at high ionic strength, see for example Section V.7.1.2.1.d (p. 322) in the uranium review [92GRE/FUG].

B.3 Tables of ion interaction coefficients

Table B-4, Table B-5 and Table B-6 contain the selected specific ion interaction coefficients used in this review, according to the specific ion interaction theory described. Table B-4 contains cation interaction coefficients with $\mathrm{Cl^-}$, $\mathrm{ClO_4^-}$ and $\mathrm{NO_3^-}$. Table B-5 anion interaction coefficients with $\mathrm{Li^+}$, with $\mathrm{Na^+}$ or $\mathrm{NH_4^+}$ and with $\mathrm{K^+}$. The coefficients have the units of kg·mol^{-1} and are valid for 298.15 K and 1 bar. The species are ordered by charge and appear, within each charge class, in standard order of arrangement, *cf.* Section II.1.8.

In some cases, the ionic interaction can be better described by assuming ion interaction coefficients as functions of the ionic strength rather than as constants. Ciavatta [80CIA] proposed the use of Eq. (B.24) for cases where the uncertainties in Table B-4 and Table B-5 are ± 0.03 kg·mol^{-1} or greater.

$$\varepsilon = \varepsilon_1 + \varepsilon_2 \log_{10} I_m \quad \text{(B.24)}$$

For these cases, and when the uncertainty can be improved with respect to the use of a constant value of ε, the values ε_1 and ε_2 given in Table B-6 should be used.

It should be noted that ion interaction coefficients tabulated in Table B-4, Table B-5 and Table B-6 may also involve ion pairing effects, as described in Section B.2. In direct comparisons of ion interaction coefficients, or when estimates are made by analogy, this aspect must be taken into account.

Table B-4: Ion interaction coefficients ε(*j, k*) (kg·mol^{-1}) for cations *j* with $k = Cl^-$, ClO_4^- and NO_3^-, taken from Ciavatta [80CIA], [88CIA] unless indicated otherwise. The uncertainties represent the 95% confidence level. The ion interaction coefficients marked with † can be described more accurately with an ionic strength dependent function, listed in Table B-6. The coefficients $\varepsilon(M^{n+}, Cl^-)$ and $\varepsilon(M^{n+}, NO_3^-)$ reported by Ciavatta [80CIA] were evaluated without taking chloride and nitrate complexation into account, as discussed in Section B. 2.

j *k*→ ↓	Cl^-	ClO_4^-	NO_3^-
H^+	(0.12 ± 0.01)	(0.14 ± 0.02)	(0.07 ± 0.01)
NH_4^+	−(0.01 ± 0.01)	−(0.08 ± 0.04)†	−(0.06 ± 0.03)†
H_2gly^+	−(0.06 ± 0.02)		
Tl^+		−(0.21 ± 0.06)†	
$ZnHCO_3^+$	0.2(a)		
$CdCl^+$		(0.25 ± 0.02)	
CdI^+		(0.27 ± 0.02)	
$CdSCN^+$		(0.31 ± 0.02)	
$HgCl^+$		(0.19 ± 0.02)	
Cu^+		(0.11 ± 0.01)	
Ag^+		(0.00 ± 0.01)	−(0.12 ± 0.05)†
$NiOH^+$	−(0.01 ± 0.07)(K)	(0.14 ± 0.07)(L)	
NiF^+		(0.34 ± 0.08)(M)	
$NiCl^+$		(0.47 ± 0.06)(N)	
$NiNO_3^+$		(0.44 ± 0.14)(O)	
$NiBr^+$		(0.59 ± 0.10)(P)	
$NiHS^+$		−(0.85 ± 0.39)(Q)	

(Continued on next page)

Table B-4 (continued)

$j \downarrow$ $k \rightarrow$	Cl^-	ClO_4^-	NO_3^-
$NiSCN^+$		$(0.31 \pm 0.04)^{(R)}$	
YCO_3^+		$(0.17 \pm 0.04)^{(b)}$	
$Am(OH)_2^+$	$-(0.27 \pm 0.20)^{(q)}$	$(0.17 \pm 0.04)^{(c)}$	
AmF_2^+		$(0.17 \pm 0.04)^{(c)}$	
$AmSO_4^+$		$(0.22 \pm 0.08)^{(d)}$	
$AmCO_3^+$	$(0.01 \pm 0.05)^{(r)}$	$(0.17 \pm 0.04)^{(c)}$	
PuO_2^+		$(0.24 \pm 0.05)^{(e)}$	
PuO_2F^+		$(0.29 \pm 0.11)^{(f)}$	
PuO_2Cl^+		$(0.50 \pm 0.09)^{(g)}$	
NpO_2^+	(0.09 ± 0.05)	$(0.25 \pm 0.05)^{(h)}$	
NpO_2OH^+		$-(0.06 \pm 0.40)^{(i)}$	
$(NpO_2)_3(OH)_5^+$		(0.45 ± 0.20)	
NpO_2F^+		$(0.29 \pm 0.12)^{(j)}$	
NpO_2Cl^+		$(0.50 \pm 0.14)^{(k)}$	
$NpO_2IO_3^+$		$(0.33 \pm 0.04)^{(l)}$	
$Np(SCN)_3^+$		$(0.17 \pm 0.04)^{(m)}$	
UO_2^+		$(0.26 \pm 0.03)^{(n)}$	
UO_2OH^+		$-(0.06 \pm 0.40)^{(n)}$	$(0.51 \pm 1.4)^{(n)}$
$(UO_2)_3(OH)_5^+$	$(0.81 \pm 0.17)^{(n)}$	$(0.45 \pm 0.15)^{(n)}$	$(0.41 \pm 0.22)^{(n)}$
UF_3^+	$(0.1 \pm 0.1)^{(o)}$	$(0.1 \pm 0.1)^{(o)}$	
UO_2F^+	$(0.04 \pm 0.07)^{(p)}$	(0.28 ± 0.04)	
UO_2Cl^+		$(0.33 \pm 0.04)^{(n)}$	
$UO_2ClO_3^+$		$(0.33 \pm 0.04)^{(o)}$	
UO_2Br^+		$(0.24 \pm 0.04)^{(o)}$	
$UO_2BrO_3^+$		$(0.33 \pm 0.04)^{(o)}$	
$UO_2IO_3^+$		$(0.33 \pm 0.04)^{(o)}$	
$UO_2N_3^+$		$(0.3 \pm 0.1)^{(o)}$	
$UO_2NO_3^+$		$(0.33 \pm 0.04)^{(o)}$	
UO_2SCN^+		$(0.22 \pm 0.04)^{(o)}$	
Pb^{2+}		(0.15 ± 0.02)	$-(0.20 \pm 0.12)^{\dagger}$
$AlOH^{2+}$	$0.09^{(s)}$	$0.31^{(s)}$	
$Al_2CO_3(OH)_2^{2+}$	$0.26^{(s)}$		
Zn^{2+}		(0.33 ± 0.03)	(0.16 ± 0.02)
$ZnCO_3^{2+}$	$(0.35 \pm 0.05)^{(a)}$		

(Continued on next page)

Table B-4 (continued)

$j \downarrow$ $k \rightarrow$	Cl^-	ClO_4^-	NO_3^-
Cd^{2+}			(0.09 ± 0.02)
Hg^{2+}		(0.34 ± 0.03)	$-(0.1 \pm 0.1)^{\dagger}$
Hg_2^{2+}		(0.09 ± 0.02)	$-(0.2 \pm 0.1)^{\dagger}$
Cu^{2+}	(0.08 ± 0.01)	(0.32 ± 0.02)	(0.11 ± 0.01)
Ni^{2+}	(0.17 ± 0.02)	$(0.370 \pm 0.032)^{(S)}$	$(0.182 \pm 0.010)^{(T)}$
Co^{2+}	(0.16 ± 0.02)	(0.34 ± 0.03)	(0.14 ± 0.01)
$FeOH^{2+}$		$0.38^{(b)}$	
$FeSCN^{2+}$		$0.45^{(b)}$	
Mn^{2+}	(0.13 ± 0.01)		
$YHCO_3^{2+}$		$(0.39 \pm 0.04)^{(b)}$	
$AmOH^{2+}$	$-(0.04 \pm 0.07)^{(q)}$	$(0.39 \pm 0.04)^{(c)}$	
AmF^{2+}		$(0.39 \pm 0.04)^{(c)}$	
$AmCl^{2+}$		$(0.39 \pm 0.04)^{(c)}$	
AmN_3^{2+}		$(0.39 \pm 0.04)^{(c)}$	
$AmNO_2^{2+}$		$(0.39 \pm 0.04)^{(c)}$	
$AmNO_3^{2+}$		$(0.39 \pm 0.04)^{(c)}$	
$AmH_2PO_4^{2+}$		$(0.39 \pm 0.04)^{(c)}$	
$AmSCN^{2+}$		$(0.39 \pm 0.04)^{(c)}$	
PuO_2^{2+}		$(0.46 \pm 0.05)^{(t)}$	
PuF_2^{2+}		$(0.36 \pm 0.17)^{(j)}$	
$PuCl^{2+}$		$(0.39 \pm 0.16)^{(u)}$	
PuI^{2+}		$(0.39 \pm 0.04)^{(v)}$	
NpO_2^{2+}		$(0.46 \pm 0.05)^{(w)}$	
$(NpO_2)_2(OH)_2^{2+}$		(0.57 ± 0.10)	
NpF_2^{2+}		$(0.38 \pm 0.17)^{(j)}$	
$NpSO_4^{2+}$		(0.48 ± 0.11)	
$Np(SCN)_2^{2+}$		$(0.38 \pm 0.20)^{(j)}$	
UO_2^{2+}	$(0.21 \pm 0.02)^{(x)}$	(0.46 ± 0.03)	$(0.24 \pm 0.03)^{(v)}$
$(UO_2)_2(OH)_2^{2+}$	$(0.69 \pm 0.07)^{(n)}$	$(0.57 \pm 0.07)^{(n)}$	$(0.49 \pm 0.09)^{(n)}$
$(UO_2)_3(OH)_4^{2+}$	$(0.50 \pm 0.18)^{(n)}$	$(0.89 \pm 0.23)^{(n)}$	$(0.72 \pm 1.0)^{(n)}$
UF_2^{2+}		$(0.3 \pm 0.1)^{(o)}$	
USO_4^{2+}		$(0.3 \pm 0.1)^{(o)}$	
$U(NO_3)_2^{2+}$		$(0.49 \pm 0.14)^{(y)}$	
Mg^{2+}	(0.19 ± 0.02)	(0.33 ± 0.03)	(0.17 ± 0.01)

(Continued on next page)

Table B-4 (continued)

$j \downarrow$ $k \rightarrow$	Cl^-	ClO_4^-	NO_3^-
Ca^{2+}	(0.14 ± 0.01)	(0.27 ± 0.03)	(0.02 ± 0.01)
Ba^{2+}	(0.07 ± 0.01)	(0.15 ± 0.02)	−(0.28 ± 0.03)
Al^{3+}	(0.33 ± 0.02)		
Ni_2OH^{3+}		(0.59 ± 0.15)[(U)]	
Fe^{3+}		(0.56 ± 0.03)	(0.42 ± 0.08)
Cr^{3+}	(0.30 ± 0.03)		(0.27 ± 0.02)
La^{3+}	(0.22 ± 0.02)	(0.47 ± 0.03)	
$La^{3+} \rightarrow Lu^{3+}$		0.47 → 0.52[(b)]	
Am^{3+}	(0.23 ± 0.02)[(G)]	(0.49 ± 0.03)[(c)]	
Pu^{3+}		(0.49 ± 0.05)[(z)]	
$PuOH^{3+}$		(0.50 ± 0.05)[(i)]	
PuF^{3+}		(0.56 ± 0.11)[(i)]	
$PuCl^{3+}$		(0.85 ± 0.09)[(#)]	
$PuBr^{3+}$		(0.58 ± 0.16)[(A)]	
$PuSCN^{3+}$		(0.39 ± 0.04)[(B)]	
Np^{3+}		(0.49 ± 0.05)[(z)]	
$NpOH^{3+}$		(0.50 ± 0.05)[(i)]	
NpF^{3+}		(0.58 ± 0.07)[(C)]	
$NpCl^{3+}$		(0.81 ± 0.09)[(D)]	
NpI^{3+}		(0.77 ± 0.26)[(E)]	
$NpSCN^{3+}$		(0.76 ± 0.12)[(j)]	
U^{3+}		(0.49 ± 0.05)[(y)]	
UOH^{3+}		(0.48 ± 0.08)[(y)]	
UF^{3+}		(0.48 ± 0.08)[(o)]	
UCl^{3+}		(0.50 ± 0.10)[(k)]	
UBr^{3+}		(0.52 ± 0.10)[(o)]	
UI^{3+}		(0.55 ± 0.10)[(o)]	
UNO_3^{3+}		(0.62 ± 0.08)[(y)]	
Be_2OH^{3+}		(0.50 ± 0.05)[(F)]	
$Be_3(OH)_3^{3+}$	(0.30 ± 0.05)[(F)]	(0.51 ± 0.05)[(y)]	(0.29 ± 0.05)[(F)]
$Al_3CO_3(OH)_4^{4+}$	0.41[(S)]		
$Ni_4(OH)_4^{4+}$		(1.08 ± 0.08)[(V)]	
$Fe_2(OH)_2^{4+}$		0.82[(b)]	
$Y_2CO_3^{4+}$		(0.80 ± 0.04)[(b)]	

(Continued on next page)

Table B-4: (continued)

$j \downarrow$ $k \rightarrow$	Cl^-	ClO_4^-	NO_3^-
Pu^{4+}		$(0.82 \pm 0.07)^{(H)}$	
Np^{4+}		$(0.84 \pm 0.06)^{(I)}$	
U^{4+}		$(0.76 \pm 0.06)^{(J)(o)}$	
Th^{4+}	(0.25 ± 0.03)		(0.11 ± 0.02)
$Al_3(OH)_4^{5+}$	$0.66^{(s)}$	$1.30^{(s)}$	

a Taken from Ferri *et al.* [85FER/GRE].

b Taken from Spahiu [83SPA].

c Estimated in [95SIL/BID].

d Evaluated in [95SIL/BID].

e Derived from $\Delta\varepsilon = \varepsilon(PuO_2^{2+},ClO_4^-) - \varepsilon(PuO_2^+,ClO_4^-) = (0.22 \pm 0.03)$ kg·mol^{-1} [95CAP/VIT]. In [92GRE/FUG], $\varepsilon(PuO_2^+,ClO_4^-) = (0.17 \pm 0.05)$ kg·mol^{-1} was tabulated based on [89ROB], [89RIG/ROB] and [90RIG]. Capdevila and Vitorge's data [92CAP], [94CAP/VIT] and [95CAP/VIT] were unavailable at that time.

f Estimated in [2001LEM/FUG] by analogy with Δε of the corresponding Np(IV) reaction.

g From Δε evaluated by Giffaut [94GIF].

h As in [92GRE/FUG], derived from $\Delta\varepsilon = \varepsilon(NpO_2^{2+},ClO_4^-) - \varepsilon(NpO_2^+,ClO_4^-) = (0.21 \pm 0.03)$ kg·mol^{-1} [87RIG/VIT], [89RIG/ROB] and [90RIG].

i Estimated in [2001LEM/FUG].

j Estimated in [2001LEM/FUG] by analogy with Δε of the corresponding U(IV) reaction.

k Estimated in [2001LEM/FUG] by analogy with Δε of the corresponding P(VI) reaction.

l Estimated in [2001LEM/FUG] by assuming $\varepsilon(NpO_2IO_3^+,ClO_4^-) \approx \varepsilon(UO_2IO_3^+,ClO_4^-)$.

m Estimated in [2001LEM/FUG] by assuming $\varepsilon(Np(SCN)_3^+,ClO_4^-) \approx \varepsilon(AmF_2^+,ClO_4^-)$.

n Evaluated in the uranium review [92GRE/FUG], using $\varepsilon(UO_2^{2+},X) = (0.46 \pm 0.03)$ kg·mol^{-1}, where X = Cl^-, ClO_4^- and NO_3^-.

o Estimated in the uranium review [92GRE/FUG].

p Taken from Riglet *et al.* [89RIG/ROB], where the following assumptions were made: $\varepsilon(Np^{3+},ClO_4^-) \approx \varepsilon(Pu^{3+},ClO_4^-) = 0.49$ kg·mol^{-1} as for other (M^{3+}, ClO_4^-) interactions, and $\varepsilon(NpO_2^{2+},ClO_4^-) \approx \varepsilon(PuO_2^{2+},ClO_4^-) \approx \varepsilon(UO_2^{2+},ClO_4^-) = 0.46$ kg·mol^{-1}.

q Evaluated in [2003GUI/FAN] (*cf.* Section 12.3.1.1) from Δε (in NaCl solution) for the reactions $An^{3+} + nH_2O(l) \rightleftharpoons An(OH)_n^{(3-n)} + nH^+$.

r Evaluated in [2003GUI/FAN] (*cf.* Section 12.6.1.1.1) from Δε (in NaCl solution) for the reactions $An^{3+} + nCO_3^{2-} \rightleftharpoons An(CO_3)_n^{(3-n)}$ (based on $\varepsilon(Am^{3+},Cl^-) = (0.23 \pm 0.02)$ kg·mol^{-1} and $\varepsilon(Na^+,CO_3^{2-}) = -(0.08 \pm 0.03)$ kg·mol^{-1}.

s Taken from Hedlund [88HED].

t By analogy with $\varepsilon(UO_2^{2+},ClO_4^-)$ as derived from isopiestic measurements in [92GRE/FUG]. The uncertainty is increased because the value is estimated by analogy.

u Estimated in [2001LEM/FUG] by analogy with Δε of the corresponding Am(III) reaction.

v Estimated in [2001LEM/FUG] by assuming $\varepsilon(\mathrm{PuI}^{2+},\mathrm{ClO}_4^-) \approx \varepsilon(\mathrm{AmSCN}^{2+},\mathrm{ClO}_4^-)$ and $\varepsilon(\mathrm{I}^-,\mathrm{NH}_4^+) \approx \varepsilon(\mathrm{SCN}^-,\mathrm{Na}^+)$.

w By analogy with $\varepsilon(\mathrm{UO}_2^{2+},\mathrm{ClO}_4^-)$ as derived from isopiestic measurements noted in [92GRE/FUG]. The uncertainty is increased because the value is estimated by analogy.

x These coefficients were not used in the NEA-TDB uranium review [92GRE/FUG] because they were evaluated by Ciavatta [80CIA] without taking chloride and nitrate complexation into account. Instead, Grenthe *et al.* used $\varepsilon(\mathrm{UO}_2^{2+},\mathrm{X}) = (0.46 \pm 0.03)$ kg·mol^{-1}, for X = Cl^-, ClO_4^- and NO_3^-.

y Evaluated in the uranium review [92GRE/FUG] using $\varepsilon(\mathrm{U}^{4+},\mathrm{X}) = (0.76 \pm 0.06)$ kg·mol^{-1}.

z Estimated by analogy with $\varepsilon(\mathrm{Ho}^{3+},\mathrm{ClO}_4^-)$ [83SPA] as in previous books in this series [92GRE/FUG], [95SIL/BID]. The uncertainty is increased because the value is estimated by analogy.

\# Derived from the $\Delta\varepsilon$ evaluated in [2001LEM/FUG].

A Estimated in [2001LEM/FUG] by analogy with $\Delta\varepsilon$ of the corresponding U(IV) reaction, and by assuming $\varepsilon(\mathrm{Br}^-,\mathrm{H}^+) \approx \varepsilon(\mathrm{Br}^-,\mathrm{Na}^+)$.

B Estimated in [2001LEM/FUG] by assuming $\varepsilon(\mathrm{PuSCN}^{2+},\mathrm{ClO}_4^-) \approx \varepsilon(\mathrm{AmSCN}^{2+},\mathrm{ClO}_4^-)$.

C Evaluated in [2001LEM/FUG].

D Derived from the $\Delta\varepsilon$ selected in [2001LEM/FUG].

E Estimated in [2001LEM/FUG] by analogy with $\Delta\varepsilon$ of the corresponding Np(IV) chloride reaction, and by assuming $\varepsilon(\mathrm{I}^-,\mathrm{H}^+) \approx \varepsilon(\mathrm{I}^-,\mathrm{Na}^+)$.

F Taken from Bruno [86BRU], where the following assumptions were made: $\varepsilon(\mathrm{Be}^{2+},\mathrm{ClO}_4^-) = 0.30$ kg·mol^{-1} as for other $\varepsilon(\mathrm{M}^{2+},\mathrm{ClO}_4^-)$; $\varepsilon(\mathrm{Be}^{2+},\mathrm{Cl}^-) = 0.17$ kg·mol^{-1} as for other $\varepsilon(\mathrm{M}^{2+},\mathrm{Cl}^-)$ and $\varepsilon(\mathrm{Be}^{2+},\mathrm{NO}_3^-) = 0.17$ kg·mol^{-1} as for other $\varepsilon(\mathrm{M}^{2+},\mathrm{NO}_3^-)$.

G The ion interaction coefficient $\varepsilon(\mathrm{An}^{3+},\mathrm{Cl}^-)$ for An = Am and Cm is assumed to equal to $\varepsilon(\mathrm{Nd}^{3+},\mathrm{Cl}^-)$ which is calculated from trace activity coefficients of Nd^{3+} ion in 0 to 4 m NaCl. These trace activity coefficients are based on the ion interaction Pitzer parameters evaluated in [97KON/FAN] from osmotic coefficients in aqueous $\mathrm{NdCl_3}$-NaCl and $\mathrm{NdCl_3}$-$\mathrm{CaCl_2}$.

H Derived from $\Delta\varepsilon = \varepsilon(\mathrm{Pu}^{4+},\mathrm{ClO}_4^-) - \varepsilon(\mathrm{Pu}^{3+},\mathrm{ClO}_4^-) = (0.33 \pm 0.035)$ kg·mol^{-1} [95CAP/VIT]. Uncertainty estimated in [2001LEM/FUG] (see Appendix A). In the first book of this series [92GRE/FUG], $\varepsilon(\mathrm{Pu}^{3+},\mathrm{ClO}_4^-) = (1.03 \pm 0.05)$ kg·mol^{-1} was tabulated based on references [89ROB], [89RIG/ROB], [90RIG]. Capdevila and Vitorge's data [92CAP], [94CAP/VIT] and [95CAP/VIT] were unavailable at that time.

I Derived from $\Delta\varepsilon = \varepsilon(\mathrm{Np}^{4+},\mathrm{ClO}_4^-) - \varepsilon(\mathrm{Np}^{3+},\mathrm{ClO}_4^-) = (0.35 \pm 0.03)$ kg·mol^{-1} [89ROB], [89RIG/ROB], [90RIG].

J Using the measured value of $\Delta\varepsilon = \varepsilon(\mathrm{U}^{4+},\mathrm{ClO}_4^-) - \varepsilon(\mathrm{U}^{3+},\mathrm{ClO}_4^-) = (0.35 \pm 0.06)$ kg·mol^{-1} p.89 [90RIG], where the uncertainty is recalculated in [2001LEM/FUG] from the data given in this thesis, and $\varepsilon(\mathrm{U}^{3+},\mathrm{ClO}_4^-) = (0.49 \pm 0.05)$ kg·mol^{-1} (see footnote (y)), a value for $\varepsilon(\mathrm{U}^{4+},\mathrm{ClO}_4^-)$ can be calculated in the same way as is done for $\varepsilon(\mathrm{Np}^{4+},\mathrm{ClO}_4^-)$ and $\varepsilon(\mathrm{Pu}^{4+},\mathrm{ClO}_4^-)$. This value, $\varepsilon(\mathrm{U}^{4+},\mathrm{ClO}_4^-) = (0.84 \pm 0.06)$ kg·mol^{-1} is consistent with that tabulated $\varepsilon(\mathrm{U}^{4+},\mathrm{ClO}_4^-) = (0.76 \pm 0.06)$ kg·mol^{-1}, since the uncertainties overlap. The authors of [2001LEM/FUG] do not believe that a change in the previously selected value for $\varepsilon(\mathrm{U}^{4+},\mathrm{ClO}_4^-)$ is justified at present.

K Evaluated in [2005GAM/BUG] (*cf.* [2005GAM/BUG], Section V.3.1.1) from $\Delta\varepsilon$ in chloride media for the reaction $\mathrm{Ni}^{2+} + \mathrm{H_2O} \rightleftharpoons \mathrm{NiOH}^+ + \mathrm{H}^+$.

L Evaluated in [2005GAM/BUG] (*cf.* [2005GAM/BUG], Section V.3.1.1) from $\Delta\varepsilon$ in perchlorate media for the reaction $\mathrm{Ni}^{2+} + \mathrm{H_2O} \rightleftharpoons \mathrm{NiOH}^+ + \mathrm{H}^+$.

M Derived from $\Delta\varepsilon = \varepsilon(\mathrm{NiF}^+,\mathrm{ClO}_4^-) - \varepsilon(\mathrm{Ni}^{2+},\mathrm{ClO}_4^-) - \varepsilon(\mathrm{Na}^+,\mathrm{F}^-) = -(0.049 \pm 0.060)$ kg·mol^{-1} in [2005GAM/BUG] (*cf.* Section V.4.2.3).

N Evaluated in [2005GAM/BUG].

O Evaluated in [2005GAM/BUG]. See details in [2005GAM/BUG], Section V.6.1.2, specially sub-section V.6.1.2.1 for an alternative treatment of this system.

P Evaluated in [2005GAM/BUG]. See details in [2005GAM/BUG], Section V.4.2.5, specially sub-section V.4.2.5.1 for an alternative treatment of this system.

Q Evaluated in [2005GAM/BUG].

R Derived from $\Delta\varepsilon = \varepsilon(\text{NiSCN}^-, \text{Na}^+) - \varepsilon(\text{SCN}^-, \text{Na}^+) - \varepsilon(\text{Ni}^{2+}, \text{ClO}_4^-) = -(0.109 \pm 0.025)\ \text{kg·mol}^{-1}$ in [2005GAM/BUG] (*cf.* [2005GAM/BUG], Section V.7.1.3.1).

S Derived from the ionic strength dependence of the osmotic and mean activity coefficient of $Ni(ClO_4)_2$ solution in [2005GAM/BUG] (*cf.* [2005GAM/BUG], Section V.4.3).

T Derived from the ionic strength dependence of the osmotic and mean activity coefficient of $Ni(NO_3)_2$ solution in [2005GAM/BUG] (*cf.* [2005GAM/BUG], Section V.6.1.2.1).

U Derived by assuming $\varepsilon(\text{Ni}_2\text{OH}^{3+}, \text{ClO}_4^-) \approx \varepsilon(\text{Be}_2\text{OH}^{3+}, \text{ClO}_4^-)$ in [2005GAM/BUG] (*cf.* [2005GAM/BUG], Section V.3.1.1.)

V Derived from $\Delta\varepsilon = 4\cdot\varepsilon(\text{H}^+, \text{ClO}_4^-) - \varepsilon(\text{Ni}_4\text{OH}_4^{4+}, \text{ClO}_4^-) - 4\cdot\varepsilon(\text{Ni}^{2+}, \text{ClO}_4^-) = (0.16 \pm 0.05)\ \text{kg·mol}^{-1}$ in [2005GAM/BUG] (*cf.* [2005GAM/BUG], Section V.3.1.1.1).

Table B-5: Ion interaction coefficients $\varepsilon(j, k)$ ($kg \cdot mol^{-1}$) for anions j with k =Li, Na and K, taken from Ciavatta [80CIA], [88CIA] unless indicated otherwise. The uncertainties represent the 95% confidence level. The ion interaction coefficients marked with † can be described more accurately with an ionic strength dependent function, listed in Table B-6.

j $k \rightarrow$ ↓	Li^+	Na^+	K^+
OH^-	$-(0.02 \pm 0.03)$†	(0.04 ± 0.01)	(0.09 ± 0.01)
F^-		(0.02 ± 0.02)[a]	(0.03 ± 0.02)
HF_2^-		$-(0.11 \pm 0.06)$[a]	
Cl^-	(0.10 ± 0.01)	(0.03 ± 0.01)	(0.00 ± 0.01)
ClO_3^-		$-(0.01 \pm 0.02)$	
ClO_4^-	(0.15 ± 0.01)	(0.01 ± 0.01)	
Br^-	(0.13 ± 0.02)	(0.05 ± 0.01)	(0.01 ± 0.02)
BrO_3^-		$-(0.06 \pm 0.02)$	
I^- [p]	(0.16 ± 0.01)	(0.08 ± 0.02)	(0.02 ± 0.01)
IO_3^-		$-(0.06 \pm 0.02)$[b]	
HSO_4^-		$-(0.01 \pm 0.02)$	
$HSeO_4^-$		$-(0.01 \pm 0.02)$[y]	
N_3^-		(0.0 ± 0.1)[b]	
NO_2^-	(0.06 ± 0.04)†	(0.00 ± 0.02)	$-(0.04 \pm 0.02)$
NO_3^-	(0.08 ± 0.01)	$-(0.04 \pm 0.03)$†	$-(0.11 \pm 0.04)$†
$H_2PO_4^-$		$-(0.08 \pm 0.04)$†	$-(0.14 \pm 0.04)$†
HCO_3^-		(0.00 ± 0.02)[d]	$-(0.06 \pm 0.05)$[i]
CN^-		(0.07 ± 0.03)[x]	
SCN^-		(0.05 ± 0.01)	$-(0.01 \pm 0.01)$
$SeCN^-$		(0.05 ± 0.01)[z]	
$HCOO^-$		(0.03 ± 0.01)	
CH_3COO^-	(0.05 ± 0.01)	(0.08 ± 0.01)	(0.09 ± 0.01)
$SiO(OH)_3^-$		$-(0.08 \pm 0.03)$[a]	
$Si_2O_2(OH)_5^-$		$-(0.08 \pm 0.04)$[b]	
$B(OH)_4^-$		$-(0.07 \pm 0.05)$†	
$Ni(SCN)_3^-$		(0.66 ± 0.13)[t]	
$Am(SO_4)_2^-$		$-(0.05 \pm 0.05)$[c]	

(Continued on next page)

Table B-5 (continued)

j ↓ k →	Li^+	Na^+	K^+
$Am(CO_3)_2^-$		$-(0.14 \pm 0.06)^{(r)}$	
$PuO_2CO_3^-$		$-(0.18 \pm 0.18)^{(o)}$	
$NpO_2(OH)_2^-$		$-(0.01 \pm 0.07)^{(q)}$	
$NpO_2CO_3^-$		$-(0.18 \pm 0.15)^{(f)}$	
$(NpO_2)_2CO_3(OH)_3^-$		$(0.00 \pm 0.05)^{(k)}$	
$UO_2(OH)_3^-$		$-(0.09 \pm 0.05)^{(b)}$	
$UO_2F_3^-$		$-(0.14 \pm 0.05)^{(w)}$	
$UO_2(N_3)_3^-$		$(0.0 \pm 0.1)^{(b)}$	
$(UO_2)_2CO_3(OH)_3^-$		$(0.00 \pm 0.05)^{(b)}$	
SO_3^{2-}		$-(0.08 \pm 0.05)$†	
SO_4^{2-}	$-(0.03 \pm 0.04)$†	$-(0.12 \pm 0.06)$†	$-(0.06 \pm 0.02)$
SeO_4^{2-}		$-(0.12 \pm 0.06)^{(\#)}$	
$S_2O_3^{2-}$		$-(0.08 \pm 0.05)$†	
HPO_4^{2-}		$-(0.15 \pm 0.06)$†	$-(0.10 \pm 0.06)$†
CO_3^{2-}		$-(0.08 \pm 0.03)^{(d)}$	(0.02 ± 0.01)
$SiO_2(OH)_2^{2-}$		$-(0.10 \pm 0.07)^{(a)}$	
$Si_2O_3(OH)_4^{2-}$		$-(0.15 \pm 0.06)^{(b)}$	
$Ni(CN)_4^{2-}$		$(0.185 \pm 0.081)^{(u)}$	
CrO_4^{2-}		$-(0.06 \pm 0.04)$†	$-(0.08 \pm 0.04)$†
$NpO_2(HPO_4)_2^{2-}$		$-(0.1 \pm 0.1)$	
$NpO_2(CO_3)_2^{2-}$		$-(0.02 \pm 0.14)^{(k)}$	
$UO_2F_4^{2-}$		$-(0.30 \pm 0.06)^{(w)}$	
$UO_2(SO_4)_2^{2-}$		$-(0.12 \pm 0.06)^{(b)}$	
$UO_2(N_3)_4^{2-}$		$-(0.1 \pm 0.1)^{(b)}$	
$UO_2(CO_3)_2^{2-}$		$-(0.02 \pm 0.09)^{(d)}$	
$(UO_2)_2(OH)_2(SO_4)_2^{2-}$		$-(0.14 \pm 0.22)$	
PO_4^{3-}		$-(0.25 \pm 0.03)$†	$-(0.09 \pm 0.02)$
$Si_3O_6(OH)_3^{3-}$		$-(0.25 \pm 0.03)^{(b)}$	
$Si_3O_5(OH)_5^{3-}$		$-(0.25 \pm 0.03)^{(b)}$	
$Si_4O_7(OH)_5^{3-}$		$-(0.25 \pm 0.03)^{(b)}$	
$Ni(CN)_5^{3-}$		$(0.25 \pm 0.14)^{(v)}$	
$Am(CO_3)_3^{3-}$		$-(0.23 \pm 0.07)^{(r)}$	

(Continued on next page)

Table B-5 (continued)

j $k\rightarrow$ ↓	Li^+	Na^+	K^+
$Np(CO_3)_3^{3-}$			$-(0.15 \pm 0.07)^{(n)}$
$NpO_2(CO_3)_2^{3-}$		$-(0.33 \pm 0.17)^{(f)}$	
$P_2O_7^{4-}$		$-(0.26 \pm 0.05)$	$-(0.15 \pm 0.05)$
$Fe(CN)_6^{4-}$			$-(0.17 \pm 0.03)$
$NpO_2(CO_3)_3^{4-}$		$-(0.40 \pm 0.19)^{(e)}$	$-(0.62 \pm 0.42)^{(g)(h)}$
$NpO_2(CO_3)_2OH^{4-}$		$-(0.40 \pm 0.19)^{(m)}$	
$U(CO_3)_4^{4-}$		$-(0.09 \pm 0.10)^{(b)(d)}$	
$UO_2(CO_3)_3^{4-}$		$-(0.01 \pm 0.11)^{(d)}$	
$(UO_2)_3(OH)_4(SO_4)_3^{4-}$		(0.6 ± 0.6)	
$NpO_2(CO_3)_3^{5-}$		$-(0.53 \pm 0.19)^{(f)}$	$-(0.22 \pm 0.03)^{(s)}$
$UO_2(CO_3)_3^{5-}$		$-(0.62 \pm 0.15)^{(d)}$	
$Np(CO_3)_5^{6-}$			$-(0.73 \pm 0.68)^{(j)}$
$(NpO_2)_3(CO_3)_6^{6-}$		$-(0.46 \pm 0.73)^{(e)}$	
$U(CO_3)_5^{6-}$		$-(0.30 \pm 0.15)^{(d)}$	$-(0.70 \pm 0.31)^{(i)}$
$(UO_2)_3(CO_3)_6^{6-}$		$(0.37 \pm 0.11)^{(d)}$	
$(UO_2)_2NpO_2(CO_3)_6^{6-}$		$(0.09 \pm 0.71)^{(l)}$	
$(UO_2)_5(OH)_8(SO_4)_4^{6-}$		(1.10 ± 0.5)	
$(UO_2)_4(OH)_7(SO_4)_4^{7-}$		(2.80 ± 0.7)	

a Evaluated in the NEA-TDB uranium review [92GRE/FUG].

b Estimated in the NEA-TDB uranium review [92GRE/FUG].

c Estimated in the NEA-TDB americium review [95SIL/BID].

d These values differ from those reported in the NEA-TDB uranium review. See the discussion in [95GRE/PUI]. Values for CO_3^{2-} and HCO_3^- are based on [80CIA].

e Calculated in [2001LEM/FUG] (Section 12.1.2.1.2).

f Calculated in [2001LEM/FUG] (Section 12.1.2.1.3).

g Calculated in [2001LEM/FUG] (Section 12.1.2.2.1).

h $\varepsilon(NpO_2(CO_3)_3^{4-}, NH_4^+) = -(0.78 \pm 0.25)$ kg·mol^{-1} is calculated in [2001LEM/FUG] (Section 12.1.2.2.1).

i Calculated in [2001LEM/FUG] from Pitzer coefficients [98RAI/FEL].

j Calculated in [2001LEM/FUG] (Section 12.1.2.1.4).

k Estimated by analogy in [2001LEM/FUG] (Section 12.1.2.1.2).

l Estimated by analogy in [2001LEM/FUG] (Section 12.1.2.2.1).

m Estimated in [2001LEM/FUG] by analogy with $NpO_2(CO_3)_3^{4-}$.

n Estimated by analogy in [2001LEM/FUG] (Section 12.1.2.1.5).

o Estimated in [2001LEM/FUG] by analogy with $\varepsilon(NpO_2CO_3^-, Na^+)$.

p $\varepsilon(I^-, NH_4^+) \approx \varepsilon(SCN^-, Na^+) = (0.05 \pm 0.01)$ kg·mol^{-1}.

q Estimated in [2001LEM/FUG] (Section 8.1.3).

r Evaluated in [2003GUI/FAN], Section 12.6.1.1.1 from $\Delta\varepsilon_n$ (in NaCl solution) for the reactions $An^{3+} + n\,CO_3^{2-} \rightleftharpoons An(CO_3)_n^{(3-n)}$ (based on $\varepsilon(Am^{3+}, Cl^-)$ = (0.23 ± 0.02) kg·mol^{-1} and $\varepsilon(Na^+, CO_3^{2-}) = -(0.08 \pm 0.03)$ kg·mol^{-1}.

s Evaluated in [2005GAM/BUG] (*cf.* Appendix A), discussion of [98ALM/NOV] from $\Delta\varepsilon$ for the reactions $KNpO_2CO_3(s) + 2\,CO_3^{2-} \rightleftharpoons NpO_2(CO_3)_3^{5-} + K^+$ (in K_2CO_3-KCl solution) and $K_3NpO_2(CO_3)_2(s)$ + $CO_3^{2-} \rightleftharpoons NpO_2(CO_3)_3^{5-} + 3\ K^+$ (in K_2CO_3 solution) (based on $\varepsilon(K^+, CO_3^{2-}) = (0.02 \pm 0.01)$ kg·mol^{-1}).

t Evaluated in [2005GAM/BUG] (*cf.* Section V.7.1.3.1) from $\Delta\varepsilon_n$ in perchlorate media for the reaction $Ni^{2+} + 3\,SCN^- \rightleftharpoons Ni(SCN)_3^-$.

u Evaluated in [2005GAM/BUG] (*cf.* Section V.7.1.2.1.1) from $\Delta\varepsilon_n$ in perchlorate media for the reaction $Ni^{2+} + 4\,CN^- \rightleftharpoons Ni(CN)_4^{2-}$.

v Evaluated in [2005GAM/BUG] (*cf.* Section V.7.1.2.1.1) from $\Delta\varepsilon_n$ in perchlorate media for the reaction $Ni^{2+} + 5\,CN^- \rightleftharpoons Ni(CN)_5^{3-}$.

w Evaluated in [2003GUI/FAN], (*cf.* Section 9.4.2.2.1.1).

x As reported in [92BAN/BLI].

y Estimated in this review by analogy with $\varepsilon(Na^+, HSO_4^-)$.

z Estimated in this review by analogy with $\varepsilon(Na^+, SCN^-)$.

\# Estimated in this review by analogy with $\varepsilon(Na^+, SO_4^{2-})$.

Table B-6: Ion interaction coefficients $\varepsilon_1(j, k)$ ($kg \cdot mol^{-1}$) and $\varepsilon_2(j, k)$ ($kg \cdot mol^{-1}$) for cations j with $k = Cl^-$, ClO_4^- and NO_3^- (first part), and for anions j with $k = Li^+$, Na^+ and K^+ (second part), according to the relationship $\varepsilon = \varepsilon_1 + \varepsilon_2 \log_{10} I_m$. The data are taken from Ciavatta [80CIA], [88CIA]. The uncertainties represent the 95% confidence level.

j $k \rightarrow$	Cl^-		ClO_4^-		NO_3^-	
$\downarrow$	ε_1	ε_2	ε_1	ε_2	ε_1	ε_2
NH_4^+			–(0.088 ± 0.002)	(0.095 ± 0.012	–(0.075 ± 0.001)	(0.057 ± 0.004)
Tl^+			–(0.18 ± 0.02)	(0.09 ± 0.02)		
Ag^+					–(0.1432 ± 0.0002)	(0.0971 ±0.000
Pb^{2+}					–(0.329 ± 0.007)	(0.288 ± 0.018)
Hg^{2+}					–(0.145 ± 0.001)	(0.194 ± 0.002)
Hg_2^{2+}					–(0.2300 ± 0.0004)	(0.194 ± 0.002)
j $k \rightarrow$	Li^+		Na^+		K^+	
$\downarrow$	ε_1	ε_2	ε_1	ε_2	ε_1	ε_2
OH^-	–(0.039 ± 0.002)	(0.072 ± 0.006)				
NO_2^-	(0.02 ± 0.01)	(0.11 ± 0.01)				
NO_3^-			–(0.049 ± 0.001)	(0.044 ± 0.002	–(0.131 ± 0.002)	(0.082 ± 0.006)
$H_2PO_4^-$			–(0.109 ± 0.001)	(0.095 ± 0.003	–(0.1473 ± 0.0008)	(0.121 ± 0.004)
$B(OH)_4^-$			–(0.092 ± 0.002)	(0.103 ± 0.005		
SO_3^{2-}			–(0.125 ± 0.008)	(0.106 ± 0.009		
SO_4^{2-}	–(0.068 ± 0.003)	(0.093 ± 0.007)	–(0.184 ± 0.002)	(0.139 ± 0.006		
$S_2O_3^{2-}$			–(0.125 ± 0.008)	(0.106 ± 0.009		
HPO_4^{2-}			–(0.19 ± 0.01)	(0.11 ± 0.03)	–(0.152 ± 0.007)	(0.123 ± 0.016)
CrO_4^{2-}			–(0.090 ± 0.005)	(0.07 ± 0.01)	–(0.123 ± 0.003)	(0.106 ± 0.007)
PO_4^{3-}			–(0.29 ± 0.02)	(0.10 ± 0.01)		

Appendix C

Assigned uncertainties [1]

This Appendix describes the origin of the uncertainty estimates that are given in the TDB Tables of selected data. The original text in [92GRE/FUG] has been retained in [95SIL/BID], [99RAR/RAN], [2001LEM/FUG], [2003GUI/FAN] and [2005GAM/BUG] except for some minor changes. Because of the importance of the uncertainty estimates an extension to the Appendix was added in [2003GUI/FAN]. The extension (essentially Section C.1) is included here with minor changes.

C.1 The general problem

The focus of this section is on the uncertainty estimates of equilibria in solution, where the key problem is analytical, *i.e.* the determination of the stoichiometric composition and equilibrium constants of complexes that are in rapid equilibrium with one another. We can formulate analysis of the experimental data in the following way: From N measurements, y_i, of the variable y we would like to determine a set of n equilibrium constants k_r, $r = 1, 2,..., n$, assuming that we know the functional relationship:

$$y = f(k_1, k_2, ...k_r...k_n; a_1, a_2,....) \quad \text{(C.1)}$$

where a_1, a_2 *etc.* are quantities that can be varied but whose values (a_{1i}, a_{2i}, *etc.*) are assumed to be known accurately in each experiment from the sets of data (y_i, a_{1i}, a_{2i},...), $i = 1, 2, ...N$. The functional relationship (C.1) is obtained from the chemical model proposed and in general several different models have to be tested before the "best" one is selected. Details of the procedures are given in Rossotti and Rossotti [61ROS/ROS].

[1] This Appendix contains essentially the text of the TDB-3 guideline, also printed in the uranium NEA-TDB review as Appendix C [92GRE/FUG]. Because of its importance in the selection of data and to guide the users of the values in Chapters III and IV, the text is reproduced here with minor revisions.

When selecting the functional relationship (C.1) and determining the set of equilibrium constants that best describes the experiments one often uses a least-squares method. Within this method, the "best" description is the one that will minimise the residual sum of squares, U:

$$U = \sum_i w_i \left[y_i - f(k_1...k_n;\ a_{1i},\ a_{2i}...) \right]^2 \quad \text{(C.2)}$$

where w_i is the weight of each experimental measurement y_i.

The minimum of the function (C.2) is obtained by solving a set of normal equations:

$$\frac{\partial U}{\partial k_r} = 0,\ r = 1,.....n \quad \text{(C.3)}$$

A "true" minimum is only obtained if:

- the functional relationship (C.1) is correct, *i.e.* if the chemical model is correct.
- all errors are random errors in the variable y, in particular there are no systematic errors.
- the random errors in y follow a Gaussian (normal) distribution.
- the weight $w_i(y_i,\ a_{1i},\ a_{2i},....)$ of an experimental determination is an exact measure of its inherent accuracy.

To ascertain that the first condition is fulfilled requires chemical insight, such as information of coordination geometry, relative affinity between metal ions and various donor atoms, etc. It is in particular important to test if the chemical equilibrium constants of complexes that occur in small amounts are chemically reasonable. Too many experimentalists seem to look upon the least-squares refinement of experimental data more as an exercise in applied mathematics than as a chemical venture. One of the tasks in the review of the literature is to check this point. An erroneous chemical model is the most serious type of systematic error.

The experimentalist usually selects the variable that he/she finds most appropriate to fulfill the second condition. If the estimated errors in a_{1i}, a_{2i} ... are smaller than the error in y_i the second condition is reasonably well fulfilled. The choice of error-carrying variable is a matter of choice based on experience, but one must be aware that it has implications, especially in the estimated uncertainty.

The presence of systematic errors is, potentially, the most important source of uncertainty. There is no possibility to handle systematic errors using statistics; statistical methods may indicate their presence, no more. Systematic errors in the chemical model have been mentioned. In addition there may be systematic errors in the methods used. By comparing experimental data obtained with different experimental methods one can obtain an indication of the presence and magnitude of such errors. The systematic errors of this type are accounted for both in the review of the literature and when taking the average of data obtained with different experimental methods. The electrode calibration, especially the conversion between measured pH and $-\log_{10}[H^+]$ is an important source of systematic error.

The assumption of a normal distribution of the random errors is a choice made in the absence of better alternatives.

Finally, a comment on the weights used in least-squares refinements; this is important because it influences the uncertainty estimate of the equilibrium constants. The weights of individual experimental points can be obtained by repeating the experiment several times and then calculating the average and standard deviation of these data. This procedure is rarely used, instead most experimentalists seem to use unit weight when making least-squares analysis of their data. However, also in this case there is a weighting of the data by the number of experimental determinations in the parameter range where the different complexes are formed. In order to have comparable uncertainty estimates for the different complexes one should therefore try to have the same number of experimental data points in the concentration ranges where these complexes are predominant, a procedure very rarely used.

As indicated above the assignment of uncertainties to equilibrium constants is not a straightforward procedure and it is complicated further when there is lack of primary experimental data. The uncertainty estimates given for the individual equilibrium constants reported by the authors and for some cases re-estimated by this review are given in the tables of this and previous reviews. The procedure used to obtain these estimates is given in the original publications and in the Appendix A discussions. However, this uncertainty is still an estimate and to a large extent based on "expert judgment".

C.2 Uncertainty estimates in the selected thermodynamic data

The uncertainty estimate in the selected thermodynamic data is based on the uncertainty of the individual equilibrium constants or other thermodynamic data, calculated as described in the following sections. A weighted average of the individual $\log_{10} K$ values is calculated using the estimated uncertainty of the individual experimental values to assign its weight. The uncertainty in this average is then calculated using the formulae given in the following text. This uncertainty depends on the number of experimental data – for N data with the same estimated uncertainty σ the uncertainty in the average is $\sigma/\sqrt{N}$. The average and the associated uncertainty reported in the tables of selected data is reported with many more digits than justified only in order to allow the users to back-track the calculations. The reported uncertainty is much smaller than the estimated experimental uncertainty and the users of the tables should look at the discussion of the selected constants in order to get a better estimate of the uncertainty in an experimental determination using a specific method.

One of the objectives of the NEA Thermochemical Data Base (TDB) project is to provide an idea of the uncertainties associated with the data selected in this review. As a rule, the uncertainties define the range within which the corresponding data can be reproduced with a probability of 95% at any place and by any appropriate method. In many cases, statistical treatment is limited or impossible due to the availability of only one or few data points. A particular problem has to be solved when significant discrepancies occur between different source data. This appendix outlines the statistical procedures which were used for fundamentally different problems and explains the philosophy used in this review when statistics were inapplicable. These rules are followed consistently throughout the series of reviews within the TDB Project. Four fundamentally different cases are considered:

1. One source datum available
2. Two or more independent source data available
3. Several data available at different ionic strengths
4. Data at non-standard conditions: Procedures for data correction and recalculation.

C.3 One source datum

The assignment of an uncertainty to a selected value that is based on only one experimental source is a highly subjective procedure. In some cases, the number of data points the selected value is based on allows the use of the "root mean square" [82TAY] deviation of the data points X_i to describe the standard deviation s_X associated with the average $\overline{X}$:

$$s_X = \sqrt{\frac{1}{N-1}\sum_{i=1}^{N}(X_i-\overline{X})^2} \qquad \text{(C.4)}$$

The standard deviation s_X is thus calculated from the dispersion of the equally weighted data points X_i around the average $\overline{X}$, and the probability is 95% that an X_i is within $\overline{X}$ $\pm$ 1.96 s_X, see Taylor [82TAY] (pp.244 – 245). The standard deviation s_X is a measure of the precision of the experiment and does not include any systematic errors.

Many authors report standard deviations s_X calculated with Eq. (C.4) (but often not multiplied by 1.96), but these do not represent the quality of the reported values in absolute terms. It is thus important not to confuse the standard deviation s with the uncertainty σ. The latter reflects the reliability and reproducibility of an experimental value and also includes all kinds of systematic errors s_j that may be involved. The uncertainty σ can be calculated with Eq. (C.5), assuming that the systematic errors are independent.

$$\sigma_X = \sqrt{s_X^2 + \sum_j (s_j^2)} \qquad \text{(C.5)}$$

The estimation of the systematic errors s_j (which, of course, have to relate to $\overline{X}$ and be expressed in the same unit) can only be made by a person who is familiar with the experimental method. The uncertainty σ has to correspond to the 95% confidence level preferred in this review. It should be noted that for all the corrections and recalculations made *(e.g.*, temperature or ionic strength corrections) the rules of the propagation of errors have to be followed, as outlined in Section C.6.2.

More often, the determination of s_X is not possible because either only one or two data points are available, or the authors did not report the individual values. The uncertainty σ in the resulting value can still be estimated using Eq. (C.5) assuming that s_X^2 is much smaller than $\sum_j (s_j^2)$, which is usually the case anyway.

C.4 Two or more independent source data

Frequently, two or more experimental data sources are available, reporting experimental determinations of the desired thermodynamic data. In general, the quality of these determinations varies widely, and the data have to be weighted accordingly for the calculation of the mean. Instead of assigning weight factors, the individual source data X_i are provided with an uncertainty σ_i that also includes all systematic errors and represents the 95% confidence level, as described in Section C.3. The weighted mean $\overline{X}$ and its uncertainty $\sigma_{\overline{X}}$ are then calculated according to Eqs. (C.6) and (C.7).

$$\overline{X} \equiv \frac{\sum_{i=1}^{N}\left(\frac{X_i}{\sigma_i^2}\right)}{\sum_{i=1}^{N}\left(\frac{1}{\sigma_i^2}\right)} \tag{C.6}$$

$$\sigma_{\overline{X}} = \sqrt{\frac{1}{\sum_{i=1}^{N}\left(\frac{1}{\sigma_i^2}\right)}} \tag{C.7}$$

Eqs. (C.6) and (C.7) may only be used if all the X_i belong to the same parent distribution. If there are serious discrepancies among the X_i, one proceeds as described below under Section C.4.1. It can be seen from Eq. (C.7) that σ_X is directly dependent on the absolute magnitude of the σ_i values, and not on the dispersion of the data points around the mean. This is reasonable because there are no discrepancies among the X_i, and because the σ_i values already represent the 95% confidence level. The selected uncertainty $\sigma_{\overline{X}}$ will therefore also represent the 95% confidence level.

In cases where all the uncertainties are equal $\sigma_i = \sigma$, Eqs. (C.6) and (C.7) reduce to Eqs. (C.8) and (C.9).

$$\overline{X} = \frac{1}{N}\sum_{i=1}^{N} X_i \tag{C.8}$$

$$\sigma_{\overline{X}} = \frac{\sigma}{\sqrt{N}} \tag{C.9}$$

Example C.1:

Five data sources report values for the thermodynamic quantity *X*. The reviewer has assigned uncertainties that represent the 95% confidence level as described in Section C.3.

i	X_i	σ_i
1	25.3	0.5
2	26.1	0.4
3	26.0	0.5
4	24.85	0.25
5	25.0	0.6

According to Eqs.(C.6) and (C.7), the following result is obtained:

$$\overline{X} = (25.3 \pm 0.2).$$

The calculated uncertainty $\sigma_{\overline{X}} = 0.2$ appears relatively small but is statistically correct, for the values are assumed to follow a Gaussian distribution. As a consequence of Eq. (C.7), $\sigma_{\overline{X}}$ will always come out smaller than the smallest σ_i. Assuming $\sigma_4 = 0.10$ instead of 0.25 would yield $\overline{X} = (25.0 \pm 0.1)$ and $\sigma_4 = 0.60$ would result in $\overline{X} = (25.6 \pm 0.2)$. In fact, the values $(X_i \pm \sigma_i)$ in this example are at the limit of consistency, that is, the range $(X_4 \pm \sigma_4)$ does not overlap with the ranges $(X_2 \pm \sigma_2)$ and $(X_3 \pm \sigma_3)$. There might be a better way to solve this problem. Three possible alternatives seem more reasonable:

i. The uncertainties σ_i are reassigned because they appear too optimistic after further consideration. Some assessments may have to be reconsidered and the uncertainties reassigned. For example, multiplying all the σ_i by 2 would yield $\overline{X} = (25.3 \pm 0.3)$.

ii. If reconsideration of the previous assessments gives no evidence for reassigning the X_i and σ_i (95% confidence level) values listed above, the statistical conclusion will be that all the X_i do not belong to the same parent distribution and cannot therefore be treated in the same group (*c.f.* item iii below for a non–statistical explanation). The values for $i = 1$, 4 and 5 might be considered as belonging to Group A and the values for $i = 2$ and 3 to Group B. The weighted average of the values in Group A is X_A ($i = 1, 4, 5$) = (24.95 ± 0.21) and of those in Group B, X_B ($i = 2, 3$) = (26.06 ± 0.31), the second digit after the decimal point being carried over to avoid loss of information. The selected value is now determined as described below under "Discrepancies" (Section C.4.1), Case I. X_A and X_B are averaged (straight average, there is no reason for giving X_A a larger weight than X_B), and $\sigma_{\bar{X}}$ is chosen in such a way that it covers the complete ranges of expectancy of X_A and X_B. The selected value is then $\bar{X} = (25.5 \pm 0.9)$.

iii. Another explanation could be that unidentified systematic errors are associated with some values. If this seems likely to be the case, there is no reason for splitting the values up into two groups. The correct way of proceeding would be to calculate the unweighted average of all the five points and assign an uncertainty that covers the whole range of expectancy of the five values. The resulting value is then $\bar{X} = (25.45 \pm 1.05)$, which is rounded according to the rules in Section C.6.3 to $\bar{X} = (25.4 \pm 1.1)$.

C.4.1 Discrepancies

Two data are called discrepant if they differ significantly, *i.e.*, their uncertainty ranges do not overlap. In this context, two cases of discrepancies are considered. Case I: Two significantly different source data are available. Case II: Several, mostly consistent source data are available, one of them being significantly different, *i.e.*, an "outlier".

Case I. Two discrepant data: This is a particularly difficult case because the number of data points is obviously insufficient to allow the preference of one of the two values. If there is absolutely no way of discarding one of the two values and selecting the other, the only solution is to average the two source data in order to obtain the selected value, because the underlying reason for the discrepancy must be unrecognised systematic errors. There is no point in calculating a weighted average, even if the two source data have been given different uncertainties, because there is obviously too little information to give even only limited preference to one of the values. The uncertainty $\sigma_{\bar{X}}$ assigned to the selected mean $\bar{X}$ has to cover the range of expectation of both source data X_1, X_2, as shown in Eq.(C.10),

$$\sigma_{\bar{X}} = |X_i - \bar{X}| + \sigma_{\max} \tag{C.10}$$

where $i = 1, 2$, and $\sigma_{\max}$ is the larger of the two uncertainties σ_i, see Example C.1.ii and Example C.2.

Example C.2:

The following credible source data are given:

$$X_1 = (4.5 \pm 0.3)$$

$$X_2 = (5.9 \pm 0.5).$$

The uncertainties have been assigned by the reviewer. Both experimental methods are satisfactory, and there is no justification to discard one of the data. The selected value is then:

$$\overline{X} = (5.2 \pm 1.2).$$

Figure C-1: Illustration for Example C.2

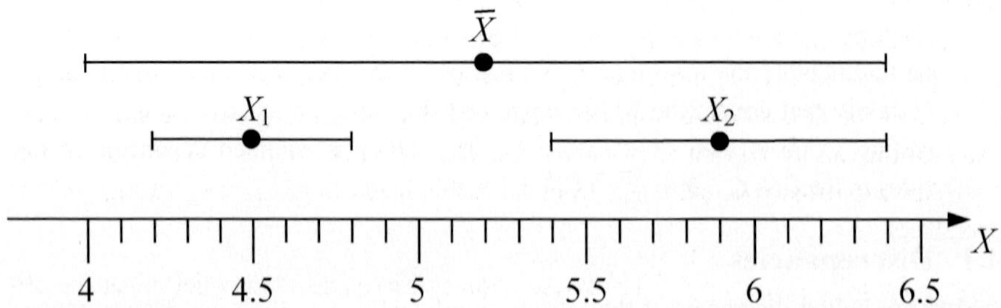

Case II. Outliers: This problem can often be solved by either discarding the outlying data point, or by providing it with a large uncertainty to lower its weight. If, however, the outlying value is considered to be of high quality and there is no reason to discard all the other data, this case is treated in a way similar to Case I. Example C.3 illustrates the procedure.

Example C.3:

The following data points are available. The reviewer has assigned the uncertainties and sees no justification for any change.

i	X_i	σ_i
1	4.45	0.35
2	5.9	0.5
3	5.7	0.4
4	6.0	0.6
5	5.2	0.4

There are two sets of data that, statistically, belong to different parent distributions, A and B. According to Eqs. (C.6) and (C.7), the following average values are found for the two groups: X_A (i =1) = (4.45 ± 0.35) and X_B (i = 2, 3, 4, 5) =

(5.62 ± 0.23). The selected value will be the straight average of X_A and X_B, analogous to Example C.1:

$$\overline{X} = (5.0 \pm 0.9).$$

C.5 Several data at different ionic strengths

The extrapolation procedure used in this review is the specific ion interaction model outlined in Appendix B. The objective of this review is to provide selected data sets at standard conditions, *i.e.,* among others, at infinite dilution for aqueous species. Equilibrium constants determined at different ionic strengths can, according to the specific ion interaction equations, be extrapolated to $I = 0$ with a linear regression model, yielding as the intercept the desired equilibrium constant at $I = 0$, and as the slope the stoichiometric sum of the ion interaction coefficients, $\Delta\varepsilon$. The ion interaction coefficient of the target species can usually be extracted from $\Delta\varepsilon$ and is listed in the corresponding table of Appendix B.

The available source data may sometimes be sparse or may not cover a sufficient range of ionic strengths to allow a proper linear regression. In this case, the correction to $I = 0$ should be carried out according to the procedure described in Section C.6.1.

If sufficient data are available at different ionic strengths and in the same inert salt medium, a weighted linear regression will be the appropriate way to obtain both the constant at $I = 0$, $\overline{X}^\circ$, and $\Delta\varepsilon$. The first step is the conversion of the ionic strength from the frequently used molar ($\text{mol}\cdot\text{dm}^{-3}$, M) to the molal ($\text{mol}\cdot\text{kg}^{-1}$, m) scale, as described in Section II.2. The second step is the assignment of an uncertainty σ_i, to each data point X_i at the molality $m_{k,i}$, according to the rules described in Section C.3. A large number of commercial and public domain computer programs and routines exist for weighted linear regressions. The subroutine published by Bevington [69BEV] (*pp.*104–105) has been used for the calculations in the examples of this appendix. Eqs. (C.11) through (C.15) present the equations that are used for the calculation of the intercept $\overline{X}^\circ$ and the slope $-\Delta\varepsilon$:

$$\overline{X}^\circ = \frac{1}{\Delta}\left(\sum_{i=1}^{N}\frac{m_{k,i}^2}{\sigma_i^2}\sum_{i=1}^{N}\frac{X_i}{\sigma_i^2} - \sum_{i=1}^{N}\frac{m_{k,i}}{\sigma_i^2}\sum_{i=1}^{N}\frac{m_{k,i}X_i}{\sigma_i^2}\right) \tag{C.11}$$

$$-\Delta\varepsilon = \frac{1}{\Delta}\left(\sum_{i=1}^{N}\frac{1}{\sigma_i^2}\sum_{i=1}^{N}\frac{m_{k,i}X_i}{\sigma_i^2} - \sum_{i=1}^{N}\frac{m_{k,i}}{\sigma_i^2}\sum_{i=1}^{N}\frac{X_i}{\sigma_i^2}\right) \tag{C.12}$$

$$\sigma_{\overline{X}^\circ} = \sqrt{\frac{1}{\Delta}\sum_{i=1}^{N}\frac{m_{k,i}^2}{\sigma_i^2}} \tag{C.13}$$

$$\sigma_{\Delta\varepsilon} = \sqrt{\frac{1}{\Delta}\sum_{i=1}^{N}\frac{1}{\sigma_i^2}} \tag{C.14}$$

where $$\Delta = \sum_{i=1}^{N}\frac{1}{\sigma_i^2}\sum_{i=1}^{N}\frac{m_{k,i}^2}{\sigma_i^2} - \left(\sum_{i=1}^{N}\frac{m_{k,i}}{\sigma_i^2}\right)^2 . \qquad \text{(C.15)}$$

In this way, the uncertainties σ_i are not only used for the weighting of the data in Eqs. (C.11) and (C.12), but also for the calculation of the uncertainties $\sigma_{\bar{X}^\circ}$ and $\sigma_{\Delta\varepsilon}$ in Eqs. (C.13) and (C.14). If the σ_i represents the 95% confidence level, $\sigma_{\bar{X}^\circ}$ and $\sigma_{\Delta\varepsilon}$ will also do so. In other words, the uncertainties of the intercept and the slope do not depend on the dispersion of the data points around the straight line but rather directly on their absolute uncertainties σ_i.

Example C.4:

Ten independent determinations of the equilibrium constant $\log_{10}{}^*\beta$ for the reaction:

$$UO_2^{2+} + HF(aq) \rightleftharpoons UO_2F^+ + H^+ \qquad \text{(C.16)}$$

are available in $HClO_4$/$NaClO_4$ media at different ionic strengths. Uncertainties that represent the 95% confidence level have been assigned by the reviewer. A weighted linear regression, ($\log_{10}{}^*\beta + 2D$) *vs.* m_k, according to the formula $\log_{10}{}^*\beta$ (C.16) + $2D$ = $\log_{10}{}^*\beta^\circ$(C.16) – $\Delta\varepsilon\ m_k$, will yield the correct values for the intercept $\log_{10}{}^*\beta^\circ$(C.16) and the slope $\Delta\varepsilon$. In this case, m_k corresponds to the molality of ClO_4^-. D is the Debye-Hückel term, *cf.* Appendix B.

i	$m_{ClO_4^-,i}$	$\log_{10}{}^*\beta + 2D$	σ_i.
1	0.05	1.88	0.10
2	0.25	1.86	0.10
3	0.51	1.73	0.10
4	1.05	1.84	0.10
5	2.21	1.88	0.10
6	0.52	1.89	0.11
7	1.09	1.93	0.11
8	2.32	1.78	0.11
9	2.21	2.03	0.10
10	4.95	2.00	0.32

The results of the linear regression are:

intercept = $(1.837 \pm 0.054) = \log_{10}{}^*\beta^\circ$(C.16)

slope = $(0.029 \pm 0.036) = -\Delta\varepsilon$.

Calculation of the ion interaction coefficient $\varepsilon(UO_2F^+,ClO_4^-) = \Delta\varepsilon + \varepsilon(UO_2^{2+},ClO_4^-) - \varepsilon(H^+,ClO_4^-)$: from $\varepsilon(UO_2^{2+},ClO_4^-) = (0.46 \pm 0.03)$ kg·mol^{-1}, $\varepsilon(H^+,ClO_4^-) = (0.14 \pm 0.02)$ kg·mol^{-1} (see Appendix B) and the slope of the linear regression, $\Delta\varepsilon = -(0.03 \pm 0.04)$ kg·mol^{-1}, it follows that $\varepsilon(UO_2F^+,ClO_4^-) = (0.29 \pm 0.05)$ kg·mol^{-1}. Note that the uncertainty (± 0.05) kg·mol^{-1} is obtained based on the rules of error propagation as described in Section C.6.2:

$$\sigma = \sqrt{(0.04)^2 + (0.03)^2 + (0.02)^2}$$

The resulting selected values are thus:

$$\log_{10} {}^*\beta^\circ(\text{C.16}) = (1.84 \pm 0.05)$$

$$\varepsilon(UO_2F^+,\ ClO_4^-) = (0.29 \pm 0.05)\ \text{kg·mol}^{-1}.$$

C.5.1 Discrepancies or insufficient number of data points

Discrepancies are principally treated as described in Section C.4. Again, two cases can be defined. Case I: Only two data are available. Case II: An "outlier" cannot be discarded. If only one data point is available, the procedure for correction to zero ionic strength outlined in Section C.6 should be followed.

Case I. Too few molalities: If only two source data are available, there will be no straightforward way to decide whether or not these two data points belong to the same parent distribution unless either the slope of the straight line is known or the two data refer to the same ionic strength. Drawing a straight line right through the two data points is an inappropriate procedure because all the errors associated with the two source data would accumulate and may lead to highly erroneous values of $\log_{10} K^\circ$ and $\Delta\varepsilon$. In this case, an ion interaction coefficient for the key species in the reaction in question may be selected by analogy (charge is the most important parameter), and a straight line with the slope $\Delta\varepsilon$ as calculated may then be drawn through each data point. If there is no reason to discard one of the two data points based on the quality of the underlying experiment, the selected value will be the unweighted average of the two standard state data obtained by this procedure, and its uncertainty must cover the entire range of expectancy of the two values, analogous to Case I in Section C.4. It should be mentioned that the ranges of expectancy of the corrected values at $I = 0$ are given by their uncertainties which are based on the uncertainties of the source data at $I \neq 0$ and the uncertainty in the slope of the straight line. The latter uncertainty is not an estimate but is calculated from the uncertainties in the ion interaction coefficients involved, according to the rules of error propagation outlined in Section C.6.2. The ion interaction coefficients estimated by analogy are listed in the table of selected ion interaction coefficients (Appendix B), but they are flagged as estimates.

Case II. Outliers and inconsistent data sets: This case includes situations where it is difficult to decide whether or not a large number of points belong to the same parent distribution. There is no general rule on how to solve this problem, and decisions are left to the judgment of the reviewer. For example, if eight data points follow a straight line reasonably well and two lie way out, it may be justified to discard the "outliers". If, however, the eight points are scattered considerably and two points are just a bit further out, one can probably not consider them as "outliers". It depends on the particular case and on the judgment of the reviewer whether it is reasonable to increase the uncertainties of the data to reach consistency, or whether the slope Δε of the straight line should be estimated by analogy.

Example C.5:

Six reliable determinations of the equilibrium constant $\log_{10}\beta$ of the reaction:

$$UO_2^{2+} + SCN^- \rightleftharpoons UO_2SCN^+ \quad (C.17)$$

are available in different electrolyte media:

I_c = 0.1 M (KNO_3): $\log_{10}\beta$ (C.17) = (1.19 ± 0.03)

I_c = 0.33 M (KNO_3): $\log_{10}\beta$ (C.17) = (0.90 ± 0.10)

I_c = 1.0 M ($NaClO_4$): $\log_{10}\beta$ (C.17) = (0.75 ± 0.03)

I_c = 1.0 M ($NaClO_4$): $\log_{10}\beta$ (C.17) = (0.76 ± 0.03)

I_c = 1.0 M ($NaClO_4$): $\log_{10}\beta$ (C.17) = (0.93 ± 0.03)

I_c = 2.5 M ($NaNO_3$): $\log_{10}\beta$ (C.17) = (0.72 ± 0.03).

The uncertainties are assumed to represent the 95% confidence level. From the values at I_c = 1 M, it can be seen that there is a lack of consistency in the data, and that a linear regression like in Example C.4 would not be appropriate. Instead, the use of Δε values from reactions of the same charge type is encouraged. Analogies with Δε are more reliable than analogies with single ε values due to canceling effects. For the same reason, the dependency of Δε on the type of electrolyte is often smaller than for single ε values.

A reaction of the same charge type as Reaction (C.17), and for which Δε is well known, is:

$$UO_2^{2+} + Cl^- \rightleftharpoons UO_2Cl^+ . \quad (C.18).$$

The value of Δε (C.18) = – (0.25 ± 0.02) was obtained from a linear regression using 16 experimental data between I_c = 0.1 M and I_c = 3 M $Na(Cl,ClO_4)$ [92GRE/FUG]. It is thus assumed that:

$$\Delta\varepsilon\,(C.17) = \Delta\varepsilon\,(C.18) = -(0.25 \pm 0.02).$$

The correction of $\log_{10}\beta$ (C.17) to I_c = 0 is done using the specific ion interaction equation, *cf.* TDB-2, which uses molal units:

$$\log_{10} \beta + 4D = \log_{10} \beta^{\circ} - \Delta\varepsilon\, I_m. \qquad (C.19)$$

D is the Debye-Hückel term in molal units and I_m the ionic strength converted to molal units by using the conversion factors listed in [76BAE/MES] (p.439). The following list gives the details of this calculation. The resulting uncertainties in $\log_{10} \beta$ are obtained based on the rules of error propagation as described in Section C.6.2.

Table C-1: Details of the calculation of equilibrium constant corrected to $I = 0$, using (C.19).

I_m	electrolyte	$\log_{10} \beta$	$4D$	$\Delta\varepsilon\, I_m$	$\log_{10} \beta^{\circ}$
0.101	KNO_3	(1.19 ± 0.03)	0.438	– 0.025	(1.68 ± 0.03)[a]
0.335	KNO_3	(0.90 ± 0.10)	0.617	– 0.084	(1.65 ± 0.10)[a]
1.050	$NaClO_4$	(0.75 ± 0.03)	0.822	– 0.263	(1.31 ± 0.04)
1.050	$NaClO_4$	(0.76 ± 0.03)	0.822	– 0.263	(1.32 ± 0.04)
1.050	$NaClO_4$	(0.93 ± 0.03)	0.822	– 0.263	(1.49 ± 0.04)
2.714	$NaNO_3$	(0.72 ± 0.03)	0.968	– 0.679	(1.82 ± 0.13)[a]

a These values were corrected for the formation of the nitrate complex $UO_2NO_3^+$ by using $\log_{10} K(UO_2NO_3^+) = (0.30 \pm 0.15)$ [92GRE/FUG].

As was expected, the resulting values $\log_{10} \beta^{\circ}$ are inconsistent and have therefore to be treated as described in Case I of Section C.4. That is, the selected value will be the unweighted average of $\log_{10} \beta^{\circ}$, and its uncertainty will cover the entire range of expectancy of the six values. A weighted average would only be justified if the six values of $\log_{10} \beta^{\circ}$ were consistent. The result is:

$$\log_{10} \beta^{\circ} = (1.56 \pm 0.39).$$

C.6 Procedures for data handling

C.6.1 Correction to zero ionic strength

The correction of experimental data to zero ionic strength is necessary in all cases where a linear regression is impossible or appears inappropriate. The method used throughout the review is the specific ion interaction equations described in detail in Appendix B. Two variables are needed for this correction, and both have to be provided with an uncertainty at the 95% confidence level: the experimental source value, $\log_{10} K$ or $\log_{10} \beta$, and the stoichiometric sum of the ion interaction coefficients, $\Delta\varepsilon$. The ion interaction coefficients (see Tables B-4, B-5 and B-6 of Appendix B) required to calculate $\Delta\varepsilon$ may not all be known. Missing values therefore need to be estimated. It is recalled that the electric charge has the most significant influence on the magnitude of the ion interaction coefficients, and that it is in general more reliable to estimate $\Delta\varepsilon$ from known reactions of the same charge type, rather than to estimate single ε values. The uncertainty of the corrected value at I = 0 is calculated by taking into account the propagation of errors, as described below. It should be noted that the ionic strength is frequently given in moles per dm^3 of solution (molar, M) and has to be converted to moles per kg H_2O (molal, m), as the model requires. Conversion factors for the most common inert salts are given in Table II-5.

Example C.6:

For the equilibrium constant of the reaction:

$$M^{3+} + 2H_2O(l) \rightleftharpoons M(OH)_2^+ + 2H^+, \qquad (C.20)$$

only one credible determination in 3 M $NaClO_4$ solution is known, $\log_{10}{}^*\beta$ (C.20) = – 6.31, to which an uncertainty of ± 0.12 has been assigned. The ion interaction coefficients are as follows:

$$\varepsilon(M^{3+}, ClO_4^-) = (0.56 \pm 0.03) \text{ kg·mol}^{-1},$$

$$\varepsilon(M(OH)_2^+, ClO_4^-) = (0.26 \pm 0.11) \text{ kg·mol}^{-1},$$

$$\varepsilon(H^+, ClO_4^-) = (0.14 \pm 0.02) \text{ kg·mol}^{-1}.$$

The values of ε and $\sigma_{\Delta\varepsilon}$ can be obtained readily (*cf.* Eq. (C.22)):

$$\Delta\varepsilon = \varepsilon(M(OH)_2^+, ClO_4^-) + 2\varepsilon(H^+, ClO_4^-) - \varepsilon(M^{3+}, ClO_4^-) = -0.02 \text{ kg·mol}^{-1},$$

$$\sigma_{\Delta\varepsilon} = \sqrt{(0.11)^2 + (2\times 0.02)^2 + (0.03)^2} = 0.12 \text{ kg·mol}^{-1}.$$

The two variables are thus:

$$\log_{10}{}^*\beta\text{ (C.20)} = -(6.31 \pm 0.12),$$

$$\Delta\varepsilon = -(0.02 \pm 0.12) \text{ kg·mol}^{-1}.$$

According to the specific ion interaction model the following equation is used to correct for ionic strength for the reaction considered here:

$$\log_{10}{}^*\beta\text{ (C.20)} + 6D = \log_{10}{}^*\beta^\circ\text{ (C.20)} - \Delta\varepsilon\, m_{ClO_4^-}.$$

D is the Debye-Hückel term:

$$D = \frac{0.509\sqrt{I_m}}{(1+1.5\sqrt{I_m})}.$$

The ionic strength I_m and the molality $m_{ClO_4^-}$ ($I_m \approx m_{ClO_4^-}$) have to be expressed in molal units, 3 M $NaClO_4$ corresponding to 3.5 m $NaClO_4$ (see Section II.2), giving D = 0.25. This results in:

$$\log_{10}{}^*\beta^\circ\text{ (C.20)} = -4.88.$$

The uncertainty in $\log_{10} {}^*\beta^\circ$ is calculated from the uncertainties in $\log_{10} {}^*\beta$ and $\Delta\varepsilon$ (*cf.* Eq. (C.22)):

$$\sigma_{\log_{10} {}^*\beta^\circ} = \sqrt{\sigma^2_{\log_{10} {}^*\beta} + (m_{\mathrm{ClO_4^-}}\,\sigma_{\Delta\varepsilon})^2} = \sqrt{(0.12)^2 + (3.5\times 0.12)^2} = 0.44.$$

The selected, rounded value is:

$$\log_{10} {}^*\beta^\circ\,(\mathrm{C.20}) = -\,(4.9 \pm 0.4).$$

C.6.2 Propagation of errors

Whenever data are converted or recalculated, or other algebraic manipulations are performed that involve uncertainties, the propagation of these uncertainties has to be taken into account in a correct way. A clear outline of the propagation of errors is given by Bevington [69BEV]. A simplified form of the general formula for error propagation is given by Eq.(C.21), supposing that X is a function of $Y_1, Y_2, \ldots, Y_N$.

$$\sigma_X^2 = \sum_{i=1}^{N}\left(\frac{\partial X}{\partial Y_i}\sigma_{Y_i}\right)^2 \quad \text{(C.21)}$$

Eq. (C.21) can be used only if the variables $Y_1, Y_2, \ldots, Y_N$ are independent or if their uncertainties are small, that is the covariances can be disregarded. One of these two assumptions can almost always be made in chemical thermodynamics, and Eq. (C.21) can thus almost universally be used in this review. Eqs. (C.22) through (C.26) present explicit formulas for a number of frequently encountered algebraic expressions, where c, c_1, c_2 are constants.

$X = c_1Y_1 \pm c_2Y_2$: $$\sigma_X^2 = (c_1\sigma_{Y_1})^2 + (c_2\sigma_{Y_2})^2 \quad \text{(C.22)}$$

$X = \pm\, cY_1Y_2$ and $X = \pm\dfrac{cY_1}{Y_2}$: $$\left(\frac{\sigma_X}{X}\right)^2 = \left(\frac{\sigma_{Y_1}}{Y_1}\right)^2 + \left(\frac{\sigma_{Y_2}}{Y_2}\right)^2 \quad \text{(C.23)}$$

$X = c_1Y^{\pm c_2}$: $$\frac{\sigma_X}{X} = c_2\frac{\sigma_Y}{Y} \quad \text{(C.24)}$$

$X = c_1e^{\pm c_2Y}$: $$\frac{\sigma_X}{X} = c_2\sigma_Y \quad \text{(C.25)}$$

$X = c_1\ln(\pm c_2Y)$: $$\sigma_X = c_1\frac{\sigma_Y}{Y} \quad \text{(C.26)}$$

Example C.7:

A few simple calculations illustrate how these formulas are used. The values have not been rounded.

$$\text{Eq. (C.22): } \Delta_r G_m = 2\cdot[-(277.4 \pm 4.9)]\ \text{kJ}\cdot\text{mol}^{-1} - [-(467.3 \pm 6.2)]\ \text{kJ}\cdot\text{mol}^{-1}$$

$$= -(87.5 \pm 11.6)\ \text{kJ}\cdot\text{mol}^{-1}.$$

$$\text{Eq. (C.23): } K = \frac{(0.038 \pm 0.002)}{(0.0047 \pm 0.0005)} = (8.09 \pm 0.96)$$

$$\text{Eq. (C.24): } K = 4\cdot(3.75 \pm 0.12)^3 = (210.9 \pm 20.3)$$

$$\text{Eq. (C.25): } K^\circ = e^{\frac{-\Delta_r G_m^\circ}{RT}}; \qquad \Delta_r G_m^\circ = -(2.7 \pm 0.3)\ \text{kJ}\cdot\text{mol}^{-1}$$

$$R = 8.3145\ \text{J}\cdot\text{K}^{-1}\cdot\text{mol}^{-1}$$

$$T = 298.15\ \text{K}$$

$$K^\circ = (2.97 \pm 0.36).$$

Note that powers of 10 have to be reduced to powers of *e*, *i.e.*, the variable has to be multiplied by ln(10), *e.g.*,

$$\log_{10} K = (2.45 \pm 0.10);\ K = 10^{\log_{10} K} = e^{(\ln(10)\cdot\log_{10} K)} = (282 \pm 65).$$

$$\text{Eq. (C.26): } \Delta_r G_m^\circ = -RT \ln K^\circ; \quad K^\circ = (8.2 \pm 1.2) \times 10^6$$

$$R = 8.3145\ \text{J}\cdot\text{K}^{-1}\cdot\text{mol}^{-1}$$

$$T = 298.15\ \text{K}$$

$$\Delta_r G_m^\circ = -(39.46 \pm 0.36)\ \text{kJ}\cdot\text{mol}^{-1}$$

$$\ln K^\circ = (15.92 \pm 0.15)$$

$$\log_{10} K^\circ = \ln K^\circ / \ln(10) = (6.91 \pm 0.06).$$

Again, it can be seen that the uncertainty in $\log_{10} K^\circ$ cannot be the same as in $\ln K^\circ$. The constant conversion factor of ln(10) = 2.303 is also to be applied to the uncertainty.

C.6.3 Rounding

The standard rules to be used for rounding are:

1. When the digit following the last digit to be retained is less than 5, the last digit retained is kept unchanged.
2. When the digit following the last digit to be retained is greater than 5, the last digit retained is increased by 1.
3. When the digit following the last digit to be retained is 5 and

a) there are no digits (or only zeroes) beyond the 5, an odd digit in the last place to be retained is increased by 1 while an even digit is kept unchanged.

b) other non–zero digits follow, the last digit to be retained is increased by 1, whether odd or even.

This procedure avoids introducing a systematic error from always dropping or not dropping a 5 after the last digit retained.

When adding or subtracting, the result is rounded to the number of decimal places (not significant digits) in the term with the least number of places. In multiplication and division, the results are rounded to the number of significant digits in the term with the least number of significant digits.

In general, all operations are carried out in full, and only the final results are rounded, in order to avoid the loss of information from repeated rounding. For this reason, several additional digits are carried in all calculations until the final selected set of data is developed, and only then are data rounded.

C.6.4 Significant digits

The uncertainty of a value basically defines the number of significant digits a value should be given.

Example: (3.478 ± 0.008)
(3.48 ± 0.01)
(2.8 ± 0.4)
(10 ± 1)
(105 ± 20).

In the case of auxiliary data or values that are used for later calculations, it is often not convenient to round to the last significant digit. In the value (4.85 ± 0.26), for example, the "5" is close to being significant and should be carried along a recalculation path in order to avoid loss of information. In particular cases, where the rounding to significant digits could lead to slight internal inconsistencies, digits with no significant meaning in absolute terms are nevertheless retained. The uncertainty of a selected value always contains the same number of digits after the decimal point as the value itself.

Appendix D

Selenium minerals

This Appendix presents two lists of naturally occurring compounds containing selenium. The first list contains minerals ratified by the International Mineralogical Association (IMA). The minerals in this table are subdivided into the groups; Native element, Oxides, Selenides, Selenites, and Selenates. Within each subgroup the minerals are presented in alphabetical order according to their assigned names. The second list contains unnamed minerals and naturally occurring compounds, which have been incompletely investigated. For instance, the space group is usually missing. At present some of the minerals in this list have not been ratified by IMA, or ratified as unnamed minerals, or denoted as naturally occurring compounds incompletely investigated. They are arranged in alphabetical order according to the chemical formulae assigned to a particular phase. Minerals and mineral varieties containing selenium in which selenium is present as an isomorphic admixture (1 to 2 mass %) were not considered.

In order to keep the number of references within bounds, only one reference is provided for each entry in the tables. In general this reference is to the most recent work available. That is why the earliest reference is dated as late as 1937. In cases of conflicting data, preference has been given to the results published in regular mineralogical journals and reviews such as "The American Mineralogist", "The Canadian Mineralogist", and "Zapiski Vserossiiskogo Mineralogicheskogo Obshchestva".

The lists do not pretend to be critical compilations of the available information and they should rather be understood to inform that published material exists on the listed minerals and naturally occurring compounds.

No thermodynamic data were found during the search of the mineralogical literature.

Table D-1: Minerals of selenium ratified by the International Mineralogical Association.

Mineral name	Chemical formula	Space Group	Reference
	NATIVE ELEMENT		
Selenium	Se	$P3_121$	[83KRU/RYA]
	OXIDES		
Downeyite	SeO_2	$P4_2/mbc$	[77FIN/MRO]
	SELENIDES		
Achavalite	FeSe	$P6_3/mmc$	[59SIN]
Aguilarite	Ag_4SeS	$P2_12_12_1$	[74PET/OWE]
Antimonselite	Sb_2Se_3	Pnma	[95MIN/LI]
Athabascaite	Cu_5Se_4	$P2_12_12$	[70HAR/CAB]
Babkinite	$Pb_2Bi_2(S,Se)_3$	$P\bar{3}$ or $P\bar{3}m1$	[96BRY/SPI]
Bambollaite	$Cu(Te,Se)_2$	$I4_1/amd$	[72HAR/NUF]
Bellidoite	Cu_2Se	$P4_2/n$	[75MON]
Berzelianite	Cu_2Se	$F\bar{4}3m$	[50EAR]
Bohdanowiczite	$AgBiSe_2$	$P\bar{3}m1$	[79BAN/ATK]
Bornhardtite	Co_3Se_4	Fd3m	[55RAM/SCH]
Brodtkorbite	Cu_2HgSe_2	$P2_1/c$	[2002PAA/TOP]
Bukovite	$Cu_3Tl_2FeSe_4$	Tetragonal	[71JOH/KVA2]
Cadmoselite	CdSe	$P6_3/mc$	[57BUR/KOV]
Chameanite	$(Cu,Fe)_4As(Se,S)_4$	$I\bar{4}3m$	[82JOH/PIC]
Chrisstanleyite	$Ag_2Pd_3Se_4$	$P2_1/m$ or $P2_1$	[98PAA/ROB]
Clausthalite	PbSe	Fm3m	[89NIK/AND]
Crookesite	$(Cu,Tl,Ag)_2Se$ or	I4/mmm	[50EAR]
	Cu_7TlSe_4	I4, I4/m or $I\bar{4}$	[87JOH]
Drysdallite	$Mo(Se,S)_2$	$P6_3/mmc$	[73CEC/RIE]
Dzharkenite	$FeSe_2$	Pa3	[95YAS/RYA]
Eskebornite	$CuFeSe_2$	$P4_2/mmc$	[88JOH]
Eucairite	AgCuSe	Pmmn	[50EAR]
Ferroselite	$FeSe_2$	Pnnm or P2nn	[56BUR]
Fischesserite	Ag_3AuSe_2	$I4_132$ or I432	[71JOH/PIC2]
Freboldite	CoSe	$P6_3/mmc$	[55RAM/SCH]
Geffroyite	$(Cu,Fe,Ag)_9(S,Se)_8$	Fm3m	[82JOH/PIC]
Giraudite	$(Cu,Zn,Ag)_{12}(As,Sb)_4(Se,S)_{13}$	$I\bar{4}3m$	[82JOH/PIC]
Guanajuatite	$Bi_2(Se,S)_3$	Pnma	[50EAR]
Hakite	$(Cu,Hg)_{12}Sb_4(Se,S)_{13}$	$I\bar{4}3m$	[71JOH/KVA]
Hastite	$CoSe_2$	Pnnm	[55RAM/SCH]
Jolliffeite	NiAsSe	Pa3	[91CAB/LAF]

(Continued on next page)

Table D-1: (continued)

Mineral name	Chemical formula	Space Group	Reference
	SELENIDES		
Junoite	$Cu_2Bi_8Pb_3(S,Se)_{16}$	C2/m	[75MUM]
Kawazulite	Bi_2Te_2Se	$R\bar{3}m$, R3m or R32	[81MIL]
Kitkaite	NiTeSe	$P\bar{3}m1$	[65HAK/VUO]
Klockmannite	CuSe	$P6_3/mmc$	[49EAR]
Krutaite	$CuSe_2$	Pa3	[72JOH/PIC]
Kullerudite	$NiSe_2$	Pnnm	[64VUO/HUH]
Laitakarite	Bi_4Se_2S	$R\bar{3}m$, R32 or R3m	[60VOR]
Laphamite	$As_2(Se,S)_3$	$P2_1/c$	[86DUN/PEA]
Luberoite	Pt_5Se_4	$P2_1/c$	[92JED/CER]
Makinenite	γ-NiSe	R3m	[64VUO/HUH]
Mgriite	$(Cu,Fe)_3AsSe_3$	Fd3m	[82DYM/LOS]
Naumannite	Ag_2Se	$P2_12_12_1$	[74PET/OWE]]
Nevskite	Bi(Se,S)	$P\bar{3}m1$, P3m1 or P321	[84NEC/CHI]
Nordstromite	$Pb_3CuBi_7(S,Se)_{14}$	$P2_1/m$	[80MUM]
Oosterboschite	$(Pd,Cu)_7Se_5$	Orth	[70JOH/PIC]
Padmaite	PdBiSe	$P4_132$ or $P4_332$	[91POL/VOL]
Palladseite	$Pd_{17}Se_{15}$	Pm3m	[77DAV/CLA]
Paraguanajuatite	Bi_2Se_3	$R\bar{3}m$	[89NIK/AND]
Penroseite	$(Ni,Co,Cu)Se_2$	Pa3	[37BAN/HEY]
Permingeatite	Cu_3SbSe_4	$I\bar{4}2m$	[71JOH/PIC]
Petrovicite	$PbHgCu_3BiSe_5$	$Pna2_1$ or Pnma	[76JOH/KVA]
Poubaite	$PbBi_2(Se,Te,S)_4$	$R\bar{3}m$	[78CEC/VAV]
Proudite	$Cu_{0-1}Pb_{7.5}Bi_{9.67}(S,Se)_{22}$	C2/m	[76MUM]
Sabatierite	Cu_6TlSe_4	Orth	[78JOH/KVA]
Sederholmite	β-NiSe	$P6_3/mmc$	[64VUO/HUH]
Selenostephanite	$Ag_5Sb(Se,S)_4$	$P2_12_12_1$	[85BOT/SAN]
Skippenite	Bi_2TeSe_2	R3m or $R\bar{3}m$ or R32	[87JOH/PIC]
Soucekite	$PbCuBi(S,Se)_3$	$Pn2_1m$	[79CEC/VAV]
Stilleite	ZnSe	$F\bar{4}3m$	[56RAM]
Sudovikovite	$PtSe_2$	$P\bar{3}m1$	[97POL/TAR]
Telluronevskite	Bi_3TeSe_2	$P\bar{3}m1$	[2001RID/SKA]
Tiemannite	HgSe	$F\bar{4}3m$	[50EAR]
Trogtalite	$CoSe_2$	Pa3	[55RAM/SCH]
Trustedtite	Ni_3Se_4	Fd3m	[64VUO/HUH]
Tyrrellite	$(Cu,Co,Ni)_3Se_4$	Fd3m	[70HAR]
Umangite	Cu_3Se_2	$P42_1m$	[66MOR/KOT]

(Continued on next page)

Table D-1: (continued)

Mineral name	Chemical formula	Space Group	Reference
	SELENITES		
Verbeekite	$PdSe_2$	C2/m	[2002ROB/PAA]
Watkinsonite	$Cu_2PbBi_4(Se,S)_8$	P2, Pm or P2/m	[87JOH/PIC]
Weibullite	$Pb_6Bi_8(S,Se)_{18}$	Pnma	[80MUM]
Wilkmanite	Ni_3Se_4	C2/m	[64VUO/HUH]
Wittite	$Pb_9Bi_{12}(S,Se)_{27}$	P2/m and C2/m	[80MUM]
Ahlfeldite	$NiSeO_3{\cdot}2H_2O$	$P2_1/c$	[94MAN]]
Chalcomenite	$CuSeO_3{\cdot}2H_2O$	$P2_12_12_1$	[94MAN]
Chloromenite	$Cu_9O_2(SeO_3)_4Cl_6$	C2/m	[99VER/KRI]
Clinochalcomenite	$CuSeO_3{\cdot}2H_2O$	$P2_1/c$	[94MAN]
Cobaltomenite	$CoSeO_3{\cdot}2H_2O$	$P2_1/c$	[94MAN]
Demesmaekerite	$Pb_2Cu_5(UO_2)_2(SeO_3)_6(OH)_6{\cdot}2H_2O$	$P\bar{1}$ or P1	[65CES/BAC]
Derriksite	$Cu_4(UO_2)(SeO_3)_2(OH)_6{\cdot}H_2O$	Pnmm or $Pmn2_1$	[71CES/PIE]
	$Cu_4(UO_2)(SeO_3)_2(OH)_6$	$Pn2_1m$	[83GIN/CES]
Francisite	$Cu_3Bi(SeO_3)_2O_2Cl$	Pmmn	[90PRI/GAT]
Georgbokiite	$Cu_5O_2(SeO_3)_2Cl_2$	$P2_1/c$	[99VER/SEM]
Guilleminite	$Ba(UO_2)_3(SeO_3)_2(OH)_4{\cdot}3H_2O$	Pnma	[65PIE/TOU]
	$Ba(UO_2)_3(SeO_3)_2\ O_2{\cdot}3H_2O$	$Pmn2_1$	[95COO/HAW]
Haynesite	$(UO_3)_2(OH)_2(SeO_3)_2{\cdot}5H_2O$	Pnc2 or Pmna	[91DEL/PIR]
Ilinskite	$NaCu_5O_2(SeO_3)_2Cl_3$	Pnma or $Pna2_1$	[97VER/SEM]
Mandarinoite	$Fe_2Se_3O_9{\cdot}4H_2O$	$P2_1/c$	[78DUN/PEA]
Marthozite	$Cu(UO_2)_4(SeO_3)_4(OH)_2{\cdot}10H_2O$	Pnma or $Pna2_1$	[69CES/OOS]
Molybdomenite	$PbSeO_3$	$P2_1/m$ or $P2_1$	[65MAN]
Orlandiite	$Pb_3Cl_4(SeO_3)\ {\cdot}H_2O$	$P\bar{1}$	[99CAM/GRA]
Piretite	$Ca(UO_2)_3(SeO_3)_2(OH)_4{\cdot}4H_2O$	$Pmn2_1$ or Pmmn	[96VOC/BLA]
Sophiite	$Zn_2(SeO_3)Cl_2$	Pccn	[89VER/FIL]
	SELENATES		
Carlosruizite	$K_6(Na,K)_4Na_6Mg_{10}(SeO_4)_{12}(IO_3)_{12}$ ${\cdot}12H_2O$	$P\bar{3}c1$	[94KON/EVA]
Olsacherite	$Pb_2(SO_4)(SeO_4)$	$P222_1$	[69HUR/ARI]
Schmiederite	$Pb_2Cu_2(SeO_3)(SeO_4)(OH)_4$	$P2_1/m$	[87EFF]

Table D-2: Minerals of selenium not specified by name and naturally occurring selenium compounds, which have been incompletely investigated.

Formula	Reference
$(Ag,Bi)_2Se$	[88PAR]
$(Ag,Cu)Bi_3(Se,S)_5$	[89KOV/NEK]
$Ag_{11}FeAs_4(Se,S)_{12.5}$	[89KOV/NEK]
Ag_3BiSe_3	[98PRI]
$(Bi_{2.47}Pb_{0.90}Cu_{0.57}Ag_{0.03}Fe_{0.01})_{3.98}(S_{3.11}Se_{1.89})_{5.00}$	[76THO/PRI]
$Bi_{3.76}Pb_{0.74}Cu_{0.02}Fe_{0.02}Ag_{0.01}Se_{2.78}S_{1.00}$	[76THO/PRI]
$(Bi,Pb)_2(Se,Te,S)_3$	[87JOH/PIC]
Bi_2SeS	[83SPI/BAD]
$Bi_4Se_3Te_3$	[81MIL]
$Bi_2(Se,Te,S)_2$	[87JOH/PIC]
$Bi_3(Te,Se,S)_4$	[95SHI/SCH]
Bi_4Se_3	[92PIE]
$Bi_3(Se_2TeS)_4$	[88EFI/NEN]
$Bi_9Cu_4Pb_2Se_{18}$	[81MIL]
Bi_4Te_2Se	[2001GU/WAT]
$Bi_6(Te,Se)_3$	[2001GU/WAT]
$(Co,Fe)Se_2$	[74MON]
CoAsSe	[91CAB/LAF]
$CuSbSe_2$	[95CHE/LI]
CuPbSe	[83DYM/LOS]
$(Cu,Hg)_{1-x}Sb_x(Se,S)$	[95CHE/LI]
$HgPd_2Se_3$	[98PAA/ROB]
$Ir(As,Se,S)_2$	[97TOL/KRI]
Mo_3Se_4	[93PAS/SUL]
$Ni_3As_3(S,Se)_4$	[95LIU/ZHE]
$NiSe_2$	[85FRA/WET]
$(Pb,Bi)_3(S,Se)_4$	[93SAK/BRY]
$Pb_2Bi_2(S,Se)_3$	[93SAK/BRY]
$PbBi_2(Se,S)_3$	[99HOL/SOD]
Pd_3Se_2	[95RUD/KNA]
PtBiSe	[97POL/TAR]
SeS_x $(x = 1 - 7)$	[97SER/BUS]
Te,Se	[95SPI/SPI]
Te_3Se_4	[89KOV/NEK]
$KCdCu_7O_2(SeO_3)_2Cl_9$	[2001GRI/FER]
$Pb_4CuCl_3(SeO_3)_3(OH)$	[99CAM/GRA]

Appendix E

Additional values of thermodynamic properties for Se compounds

This Appendix contains two tables.

Table E-1 lists the values of the properties for a large number of selenium species which have been evaluated using auxiliary data (themselves listed in Table E-2) that have **not** been critically reviewed following the NEA-TDB Guidelines. The original experimental sources reporting the thermodynamic cycles leading to the values listed in Table E-1 have been submitted to a critical review following the NEA-TDB Guidelines, as discussed in Chapter V, and have been found to be acceptable according to such criteria. While the data contained in Tables E-1 and E-2 **are not selected values**, it is considered that they are the best values currently available, and further critical reviews would benefit from the information they provide.

Table E-1. Values of the properties for selenium species at 298.15 K and 0.1 MPa pressure, which have been evaluated using auxiliary data (listed in Table E-2) that have **not** been critically reviewed following the NEA-TDB Guidelines. Thus, the data contained are not selected data, and should be used with this in mind.

Compound	$\Delta_f G_m^\circ$ (kJ·mol^{-1})	$\Delta_f H_m^\circ$ (kJ·mol^{-1})	S_m° (J·K^{-1}·mol^{-1})
TeSe(g)	*104.681* *± 3.396*	*154.350* [(a)] *± 3.391*	
As_2Se_3(cr)	*− 83.941* *± 4.222*	*− 81.100* *± 4.100*	
BiSe(g)		*172.200* *± 4.400*	

(Continued on next page)

Table E-1: (continued)

Compound	$\Delta_f G_m^\circ$ (kJ·mol⁻¹)	$\Delta_f H_m^\circ$ (kJ·mol⁻¹)	S_m° (J·K⁻¹·mol⁻¹)
$Bi_2(SeO_3)_3$(cr)		*− 1403.200 ± 5.000*	
BSe(g)		*342.500 ± 20.000*	
$Al_2(SeO_4)_3$(cr)		*− 2561.700 ± 17.100*	
$Al_2(SeO_4)_3 \cdot 18H_2O$(cr)		*− 7970.000 ± 17.100*	
GaSe(cr)	*− 141.200 ± 15.000*		
$Ga_2(SeO_3)_3$(cr)		*− 1823.700 ± 6.500*	
$Ga_2(SeO_3)_3 \cdot H_2O$(cr)		*− 2138.400 ± 6.600*	
$Ga_2(SeO_3)_3 \cdot 6H_2O$(cr)	*− 3039.200* (a) *± 14.700*	*− 3682.300 ± 6.600*	*373.682 ± 54.056*
$Ga_2(SeO_4)_3$(cr)		*− 1908.400 ± 17.900*	
$Ga_2(SeO_4)_3 \cdot 18H_2O$(cr)		*− 7331.100 ± 17.800*	
$GaH(SeO_3)_2$(cr)		*− 1176.700 ± 3.500*	
$GaH(SeO_3)_2 \cdot 2H_2O$(cr)		*− 1788.400 ± 3.500*	
InSe(g)		*221.600 ± 20.000*	
α-In_2Se_3	*− 306.000 ± 20.000*		
α-TlSe	*− 55.700 ± 7.100*	*− 56.700 ± 7.100*	
Tl_2Se(cr)	*− 88.500 ± 9.800*	*− 87.200 ± 9.800*	
Tl_2SeO_4(cr)		*− 635.800* (a) *± 6.400*	*214.400* (a) *± 22.300*
$ZnSeO_3 \cdot H_2O$(cr)		*− 931.000 ± 2.700*	

(Continued on next page)

Table E-1: (continued)

Compound	$\Delta_f G_m^o$ (kJ·mol^{-1})	$\Delta_f H_m^o$ (kJ·mol^{-1})	S_m^o (J·K^{-1}·mol^{-1})
AuSe(g)		*344.800* *±14.600*	
$Co_{0.84}Se(cr)$			*70.300* *±8.000*
$CoSe_2(cr)$	*– 100.400* *±15.000*		
$CoSeO_3·2H_2O(cr)$	*– 936.280* *±4.800*	*– 1115.56* *±2.810*	*245.100* *±18.600*
$CoSeO_4(aq)$	*–509.407* [(a)] *±4.173*		
$CoSeO_4·6H_2O(cr)$	*–1926.877* [(a)] *±4.178*		
$CoSeCN^+$	*72.979* [(a)] *±5.715*		
$FeSe_2(cr)$	*– 101.300* *±15.000*		
β-$Fe_{1.04}Se$	*– 70.100* *±4.000*		
γ-Fe_3Se_4	*– 244.000* *±30.000*		
α-Fe_7Se_8	*– 489.000* *±20.000*		
$Fe_2(SeO_3)_3·6H_2O(cr)$	*–2756.750* [(a)] *±7.300*		
$RuSe_2(cr)$		*– 175.300* *±8.600*	
$OsSe_2(cr)$		*– 128.600* *±11.000*	
α-MnSe	*– 182.900* *±12.000*	*– 177.900* *±12.000*	
MnSe(g)		*278.400* *±10.000*	
$MnSe_2(cr)$	*– 187.700* *±10.600*	*– 180.500* *±10.000*	*140.300* *±12.000*
$MnSeO_3·2H_2O(cr)$	*– 1108.200* [(a)] *±6.700*	*– 1280.700* *±2.600*	*269.300* *±24.100*
$MnSeO_4(aq)$	*–681.460* [(a)] *±3.340*		

(Continued on next page)

Table E-1: (continued)

Compound	$\Delta_f G_m^o$ (kJ·mol^{-1})	$\Delta_f H_m^o$ (kJ·mol^{-1})	S_m^o (J·K^{-1}·mol^{-1})
γ-Cr_2Se_3	*–247.900* *±20.000*	*–246.700* *±20.000*	
$MoSe_2$(cr)		*–234.400* *±3.500*	
Mo_6Se_6(cr)		*–478.400* *±8.600*	
WSe_2(cr)		*–185.500* *±5.500*	
VSe(g)		*405.000* *±16.000*	
$Zr(SeO_3)_2$(cr)		*1609.600* *±2.700*	
ScSe(g)		*228.100* *±13.700*	
YSe(g)		*226.300* *±15.600*	
LuSe(g)		*240.200* *±14.900*	
YbSe(cr)		*–443.900* *±31.300*	
$Yb_2(SeO_3)_3$(cr)		*–2780.000* *±50.000*	
HoSe(g)		*206.000* *±20.000*	
$Dy_2(SeO_4)_3\cdot 8H_2O$(cr)		*–5371.400* *±17.500*	
$Tb_2(SeO_4)_3\cdot 8H_2O$(cr)		*–5426.100* *±17.500*	
GdSe(g)		*198.200* *±14.900*	
$Gd_2(SeO_4)_3\cdot 8H_2O$(cr)		*–5444.500* *±17.500*	
EuSe(cr)		*–450.400* *±10.000*	*99.600* *±7.300*
EuSe(g)		*108.900* *±15.600*	
$Eu_2(SeO_4)_3\cdot 8H_2O$(cr)		*–5235.200* *±17.500*	

(Continued on next page)

Table E-1: (continued)

Compound	$\Delta_f G_m^\circ$ (kJ·mol^{-1})	$\Delta_f H_m^\circ$ (kJ·mol^{-1})	S_m° (J·K^{-1}·mol^{-1})
SmSe(cr)		*– 447.000 ± 20.000*	
SmSe(g)		*112.000 ± 20.000*	
$Sm_2(SeO_3)_3$(cr)		*– 2827.000 ± 6.000*	
$Sm_2(SeO_4)_3 \cdot 8H_2O$(cr)		*– 5395.000 ± 17.800*	
NdSe(cr)		*– 409.900 ± 19.800*	
NdSe(g)		*168.400 ± 15.000*	
$Nd_2(SeO_3)_3$(cr)		*– 2841.000 ± 15.000*	
$Nd_2(SeO_4)_3 \cdot 5H_2O$(cr)		*– 4517.700 ± 15.000*	
PrSe(cr)		*– 428.600 ± 22.700*	
PrSe(g)		*145.000 ± 20.000*	
$Pr_2(SeO_3)_3$(cr)		*– 2856.800 ± 5.000*	
$Pr(HSeO_3)_3$(cr)		*– 2212.500 ± 2.900*	
$Pr(HSeO_3)_3 \cdot 2H_2O$(cr)		*– 2823.500 ± 3.000*	
$Pr_2(SeO_4)_3 \cdot 5H_2O$(cr)		*– 4545.000 ± 17.600*	
$Pr_2H_2(SeO_3)_4$(cr)		*– 3366.200 ± 5.200*	
$Pr_2H_2(SeO_3)_4 \cdot 5H_2O$(cr)		*– 4864.500 ± 5.200*	
CeSe(cr)		*– 414.000 ± 22.000*	
CeSe(g)		*165.000 ± 25.000*	
LaSe(cr)		*– 398.500 ± 20.600*	

(Continued on next page)

Table E-1: (continued)

Compound	$\Delta_f G_m^o$ (kJ·mol^{-1})	$\Delta_f H_m^o$ (kJ·mol^{-1})	S_m^o (J·K^{-1}·mol^{-1})
LaSe(g)		*190.000* *±5.000*	
$La_2(SeO_3)_3$(cr)		*– 2852.200* *±6.000*	
$La_2(SeO_4)_3 \cdot H_2O$(cr)		*– 4555.000* *±17.500*	
$BeSeO_4$(cr)		*– 884.600* *±6.900*	
$BeSeO_4 \cdot 2H_2O$(cr)		*– 1501.900* *±6.800*	
$BeSeO_4 \cdot 4H_2O$(cr)		*– 2107.400* *±6.800*	
$CaSeO_3 \cdot H_2O$(cr)		*– 1324.800* *±2.800*	*168.754* *±12.589*

a Value calculated from an NEA-TDB selected reaction data.

Table E-2. This table lists the values of auxiliary data that have been used to calculate the values reported in Table E-1. They are given for information purposes only, since these data have not been critically reviewed following the NEA-TDB Guidelines. All data are deemed to correspond to the reference temperature of 298.15 K and to the standard state, *i.e.*, a pressure of 0.1 MPa and, for the aqueous species, infinite dilution (I = 0). The error bars are the uncertainties used in the calculation of the propagated error in the data presented in Table E-1.

Property and species	Value	Reference	Compound(s) to which it applies
$\Delta_f H_m^\circ$ (Te_2, g)	(163.2 ± 1.0) (kJ·mol^{-1})	[95DAV/RAN]	TeSe(g)
$\Delta_f H_m^\circ$ (AsF_5, g)	– (1234.2 ± 0.7) (kJ·mol^{-1})	[93OHA]	As_2Se_3(cr)
$\Delta_f H_m^\circ$ (Bi, g)	(198.7 ± 4.0) (kJ·mol^{-1})	[56STU/SIN]	BiSe(g)
$\Delta_f H_m^\circ$ ($Bi(OH)_3$, cr)	– (711.3 ± 2.5) (kJ·mol^{-1})	[82WAG/EVA]	$Bi_2(SeO_3)_3$(cr)
$\Delta_r H_m^\circ$ (Al^{3+}, hydrolysis)	(40.0 ± 5.0) (kJ·mol^{-1})	[76BAE/MES]	$Al_2(SeO_4)_3$(cr) $Al_2(SeO_4)_3 \cdot 18H_2O$(cr)
S_m° (Ga, cr)	(40.88 ± 0.30) (J·K^{-1}·mol^{-1})	[82WAG/EVA]	GaSe(cr)
$\Delta_f G_m^\circ$ (Ga^{3+})	– (159.0 ± 4.0) (kJ·mol^{-1})	[82WAG/EVA]	$Ga_2(SeO_3)_3 \cdot 6H_2O$(cr)
$\Delta_f H_m^\circ$ (Ga^{3+})	– (211.7 ± 2.5) (kJ·mol^{-1})	[82WAG/EVA]	$Ga_2(SeO_4)_3$(cr) $Ga_2(SeO_4)_3 \cdot 18H_2O$(cr)
$\Delta_f H_m^\circ$ ($GaCl_3$, cr)	– (524.70 ± 2.50) (kJ·mol^{-1})	[82WAG/EVA]	$Ga_2(SeO_3)_3$(cr) $Ga_2(SeO_3)_3 \cdot H_2O$(cr) $Ga_2(SeO_3)_3 \cdot 6H_2O$(cr) $GaH(SeO_3)_2$(cr) $GaH(SeO_3)_2 \cdot 2H_2O$(cr)
$\Delta_f H_m^\circ$ (In, g)	(243.30 ± 1.30) (kJ·mol^{-1})	[82WAG/EVA]	InSe(g)
S_m° (In, cr)	(57.82 ± 0.30) (J·K^{-1}·mol^{-1})	[82WAG/EVA]	α-In_2Se_3
$\Delta_f H_m^\circ$ (Tl, l)	(3.72 ± 0.10) (kJ·mol^{-1})	[91DIN]	α-TlSe, Tl_2Se(cr)
S_m° (Tl, cr)	(64.18 ± 0.30) (J·K^{-1}·mol^{-1})	[82WAG/EVA]	α-TlSe, Tl_2Se(cr)
$\Delta_f H_m^\circ$ (Tl^+)	(5.36 ± 1.00) (kJ·mol^{-1})	[82WAG/EVA]	Tl_2SeO_4(cr)

(Continued on next page)

Table E-2: (continued).

Property and species	Value	Reference	Compound(s) to which it applies
S°_m (Tl^+)	(125.5 ± 3.5) ($J\cdot K^{-1}\cdot mol^{-1}$)	[82WAG/EVA]	Tl_2SeO_4(cr)
$\Delta_f H^\circ_m$ ($ZnSO_4\cdot 7H_2O$, cr)	– (3077.75 ± 2.50) ($kJ\cdot mol^{-1}$)	[82WAG/EVA]	$ZnSeO_3\cdot H_2O$
$\Delta_f H^\circ_m$ (Au, g)	(366.1 ± 2.5) ($kJ\cdot mol^{-1}$)	[82WAG/EVA]	AuSe(g)
S°_m (Co, cr)	(30.07 ± 0.10) ($J\cdot K^{-1}\cdot mol^{-1}$)	[98CHA]	$Co_{0.84}Se$(cr) $CoSe_2$(cr) $CoSeO_3\cdot 2H_2O$(cr)
$\Delta_f H^\circ_m$ ($Co(NO_3)_3\cdot 6H_2O$, cr)	– (2211.20 ± 2.50) ($kJ\cdot mol^{-1}$)	[82WAG/EVA]	$CoSeO_3\cdot 2H_2O$(cr)
$\Delta_f G^\circ_m$ (Co^{2+})	– (54.4 ± 3.9) ($kJ\cdot mol^{-1}$)	[82WAG/EVA]	$CoSeO_3\cdot 2H_2O$(cr) $CoSeO_4\cdot 6H_2O$(cr) $CoSeO_4$(aq) $CoSeCN^+$
S°_m (Fe, cr)	(27.32 ± 0.13) ($J\cdot K^{-1}\cdot mol^{-1}$)	[98CHA]	All iron selenides
S°_m (Ru, cr)	(28.61 ± 0.25) ($J\cdot K^{-1}\cdot mol^{-1}$)	[73HUL/DES]	$RuSe_2$(cr) $OsSe_2$(cr)
$\Delta_f G^\circ_m$ (Fe^{3+})	– (4.70 ± 2.50) ($kJ\cdot mol^{-1}$)	[82WAG/EVA]	$Fe_2(SeO_3)_3\cdot 6H_2O$(cr)
S°_m (Mn, cr)	(32.01 ± 0.40) ($J\cdot K^{-1}\cdot mol^{-1}$)	[82WAG/EVA]	α-MnSe $MnSe_2$(cr) $MnSeO_3\cdot 2H_2O$(cr)
$\Delta_f G^\circ_m$ (Mn^{2+})	– (228.1 ± 3.0) ($kJ\cdot mol^{-1}$)	[82WAG/EVA]	α-MnSe $MnSeO_4$(aq) $MnSeO_3\cdot 2H_2O$(cr)
$\Delta_f H^\circ_m$ (Mn, g)	(280.7 ± 1.7) ($kJ\cdot mol^{-1}$)	[82WAG/EVA]	α-MnSe MnSe(g)
$\Delta_f H^\circ_m$ ($MnSO_4$, cr)	– (1065.25 ± 2.50) ($kJ\cdot mol^{-1}$)	[82WAG/EVA]	$MnSeO_3\cdot 2H_2O$(cr)
S°_m (Cr, cr)	(23.77 ± 0.21) ($J\cdot K^{-1}\cdot mol^{-1}$)	[82WAG/EVA]	γ-Cr_2Se_3
$\Delta_f H^\circ_m$ (MoF_6, g)	– (1557.9 ± 0.9) ($kJ\cdot mol^{-1}$)	[87OHA/TAS]	$MoSe_2$(cr) Mo_6Se_6(cr)
$\Delta_f H^\circ_m$ (WF_6, g)	– (1721.7 ± 1.7) ($kJ\cdot mol^{-1}$)	[88OHA/LEW2]	WSe_2(cr)
$\Delta_f H^\circ_m$ (V, g)	(514.21 ± 1.30) ($kJ\cdot mol^{-1}$)	[82WAG/EVA]	VSe(g) HoSe(g) LaSe(g)

(Continued on next page)

Table E-2: (continued).

Property and species	Value	Reference	Compound(s) to which it applies
$\Delta_f H_m^\circ$ (ZrO_2, cr)	– (1100.5 ± 0.7) ($kJ \cdot mol^{-1}$)	[2004BRO]	$Zr(SeO_3)_2$(cr)
$\Delta_f H_m^\circ$ (Sc, g)	(377.8 ± 2.1) ($kJ \cdot mol^{-1}$)	[82WAG/EVA]	ScSe(g) LuSe(g) GdSe(g) NdSe(g)
$\Delta_f H_m^\circ$ (Y, g)	(421.3 ± 2.1) ($kJ \cdot mol^{-1}$)	[82WAG/EVA]	YSe(g) BSe(g)
$\Delta_f H_m^\circ$ (Lu, g)	(427.6 ± 9.0) ($kJ \cdot mol^{-1}$)	[82WAG/EVA]	LuSe(g)
$\Delta_f H_m^\circ$ (Yb, g)	(152.3 ± 3.0) ($kJ \cdot mol^{-1}$)	[82WAG/EVA]	YbSe(cr)
$\Delta_f H_m^\circ$ (Yb_2O_3, cr)	(1814.6 ± 2.5) ($kJ \cdot mol^{-1}$)	[82WAG/EVA]	$Yb_2(SeO_3)_3$(cr)
$\Delta_f H_m^\circ$ (Ho, g)	(300.8 ± 9.0) ($kJ \cdot mol^{-1}$)	[82WAG/EVA]	HoSe(g)
$\Delta_f H_m^\circ$ (Dy^{3+})	– (700.2 ± 3.0) ($kJ \cdot mol^{-1}$)	[2001COR/KON]	$Dy_2(SeO_4)_3 \cdot 8H_2O$(cr)
$\Delta_f H_m^\circ$ (Tb^{3+})	– (699.8 ± 4.0) ($kJ \cdot mol^{-1}$)	[2001COR/KON]	$Tb_2(SeO_4)_3 \cdot 8H_2O$(cr)
$\Delta_f H_m^\circ$ (Gd, g)	(397.5 ± 9.0) ($kJ \cdot mol^{-1}$)	[82WAG/EVA]	GdSe(g)
$\Delta_f H_m^\circ$ (Gd^{3+})	(698.4 ± 2.0) ($kJ \cdot mol^{-1}$)	[2001COR/KON]	$Gd_2(SeO_4)_3 \cdot 8H_2O$(cr)
$\Delta_f H_m^\circ$ (Eu, g)	(175.3 ± 4.0) ($kJ \cdot mol^{-1}$)	[82WAG/EVA]	EuSe(cr) EuSe(g)
$\Delta_f H_m^\circ$ (Eu^{3+})	– (605.4 ± 4.0) ($kJ \cdot mol^{-1}$)	[2001COR/KON]	$Eu_2(SeO_4)_3 \cdot 8H_2O$(cr)
$\Delta_f H_m^\circ$ (Sm, g)	(206.7 ± 4.0) ($kJ \cdot mol^{-1}$)	[82WAG/EVA]	SmSe(cr) SmSe(g)
$\Delta_f H_m^\circ$ (Sm_2O_3, cr)	– (1823.0 ± 2.5) ($kJ \cdot mol^{-1}$)	[82WAG/EVA]	$Sm_2(SeO_3)_3$(cr)
$\Delta_f H_m^\circ$ (Sm^{3+})	– (690.0 ± 2.0) ($kJ \cdot mol^{-1}$)	[2001COR/KON]	$Sm_2(SeO_4)_3 \cdot 8H_2O$(cr)
$\Delta_f H_m^\circ$ (Nd, g)	(327.6 ± 4.0) ($kJ \cdot mol^{-1}$)	[82WAG/EVA]	NdSe(cr) NdSe(g)
$\Delta_f H_m^\circ$ ($NdCl_3 \cdot 6H_2O$, cr)	– (2874.2 ± 2.0) ($kJ \cdot mol^{-1}$)	See Appendix A [71MAI/SUP]	$Nd_2(SeO_3)_3$(cr)

(Continued on next page)

Table E-2: (continued).

Property and species	Value	Reference	Compound(s) to which it applies
$\Delta_f H_m^\circ$ (Nd_2O_3, cr)	– (1807.9 ± 2.5) ($kJ\cdot mol^{-1}$)	[82WAG/EVA]	$Nd_2(SeO_3)_3$(cr)
$\Delta_f H_m^\circ$ (Nd^{3+})	– (694.8 ± 2.0) ($kJ\cdot mol^{-1}$)	[2001COR/KON]	$Nd_2(SeO_4)_3\cdot 5H_2O$(cr)
$\Delta_f H_m^\circ$ (Pr, g)	(355.6 ± 9.0) ($kJ\cdot mol^{-1}$)	[82WAG/EVA]	PrSe(cr) PrSe(g)
$\Delta_f H_m^\circ$ ($PrCl_3\cdot 7H_2O$, cr)	– (3180.0 ± 2.0) ($kJ\cdot mol^{-1}$)	See Appendix A [71MAI/SUP]	Praseodymium selenites
$\Delta_f H_m^\circ$ (Pr^{3+})	– (705.7 ± 2.0) ($kJ\cdot mol^{-1}$)	[2001COR/KON]	$Pr_2(SeO_4)_3\cdot 5H_2O$(cr)
$\Delta_f H_m^\circ$ (Ce, g)	(423 ± 10) ($kJ\cdot mol^{-1}$)	[82WAG/EVA]	CeSe(cr) CeSe(g)
$\Delta_f H_m^\circ$ (La, g)	(431.0 ± 3.0) ($kJ\cdot mol^{-1}$)	[82WAG/EVA]	LaSe(cr) LaSe(g)
$\Delta_f H_m^\circ$ ($LaCl_3\cdot 7H_2O$, cr)	– (3178.9 ± 2.5) ($kJ\cdot mol^{-1}$)	See Appendix A [71MAI/SUP]	$La_2(SeO_3)_3$(cr)
$\Delta_f H_m^\circ$ (La^{3+})	– (707.6 ± 2.5) ($kJ\cdot mol^{-1}$)	[2001COR/KON]	$La_2(SeO_4)_3\cdot 5H_2O$(cr)
$\Delta_f H_m^\circ$ ($BeCl_2$, aq 1:900)	– (708.3 ± 3.9) ($kJ\cdot mol^{-1}$)	See Appendix A [61SEL/SHN3]	$BeSeO_4\cdot 4H_2O$(cr) $BeSeO_4\cdot 2H_2O$(cr) $BeSeO_4$(cr)
$\Delta_f H_m^\circ$ ($Ca(NO_3)_2\cdot 4H_2O$, cr)	– (2132.33 ± 2.50) ($kJ\cdot mol^{-1}$)	[82WAG/EVA]	$CaSeO_3\cdot H_2O$(cr)

Appendix F

Solubility of metal selenates and alkali metal selenites in water

This Appendix contains the solubility in water of metal selenates and alkali metal selenites. The solubility of most metal selenites in water is a less well-defined quantity due to the protolysis of the selenite ion and as a consequence metal-hydroxide ion complexation and, in cases, transformation of the normal solid phase to a basic selenite phase occur. Most solubilities of metal selenites have therefore been determined in solutions with an adjusted pH value, mainly in the range pH 2 to 4. Such data have been used to calculate the solubility product of the usually slightly or moderately soluble metal selenites as described in Chapter V and Appendix A.

Table F-1. Solubility of metal selenates and alkali metal selenites in water. The comment " accepted" indicates that only one experimental source was available. The footnote (a) indicates that the data were not obtained at 298.15 K.

Compound	Solubility in water	Comment
$PbSeO_4(cr)$	$(4.7 \pm 1.0) \times 10^{-4}$ mol($PbSeO_4$)·kg^{-1}	Selected
$Al_2(SeO_4)_3 \cdot 16H_2O(cr)$	0.603 mol($Al_2(SeO_4)_3$)·kg^{-1}	Accepted
$In_2(SeO_4)_3 \cdot 9H_2O(cr)$ [(a)]	1.87 mol($In_2(SeO_4)_3$)·kg^{-1}	Accepted
$Tl_2SeO_4(cr)$	(0.058 ± 0.010) mol(Tl_2SeO_4)·kg^{-1}	Selected
$ZnSeO_4 \cdot 6H_2O(cr)$	(3.17 ± 0.20) mol($ZnSeO_4$)·kg^{-1}	Selected
$CdSeO_4 \cdot 2H_2O(cr)$	(2.56 ± 0.02) mol($CdSeO_4$)·kg^{-1}	Selected
$CdSeO_4 \cdot H_2O(cr)$	(2.42 ± 0.05) mol($CdSeO_4$)·kg^{-1}	Selected
$CuSeO_4 \cdot 5H_2O(cr)$	(0.922 ± 0.033) mol($CuSeO_4$)·kg^{-1}	Selected

(Continued on next page)

Table F-1: (continued)

Compound	Solubility in water	Comment
$Ag_2SeO_4(cr)$	$(1.8 \pm 0.6) \times 10^{-3}$ mol(Ag_2SeO_4)·kg^{-1}	Selected
$NiSeO_4 \cdot 6H_2O(cr)$	(1.93 ± 0.05) mol($NiSeO_4$)·kg^{-1}	Selected
$CoSeO_4 \cdot 6H_2O(cr)$	(2.82 ± 0.02) mol($CoSeO_4$)·kg^{-1}	Selected
$MnSeO_4 \cdot 5H_2O(cr)$	(2.86 ± 0.08) mol($MnSeO_4$)·kg^{-1}	Selected
$Yb_2(SeO_4)_3 \cdot 8H_2O(cr)$	1.289 mol($Yb_2(SeO_4)_3$)·kg^{-1}	Accepted
$Eu_2(SeO_4)_3 \cdot 8H_2O$	0.608 mol($Eu_2(SeO_4)_3$)·kg^{-1}	Accepted
$Sm_2(SeO_4)_3 \cdot 8H_2O(cr)$	0.582 mol($Sm_2(SeO_4)_3$)·kg^{-1}	Accepted
$Pr_2(SeO_4)_3 \cdot 5H_2O(cr)$	0.459 mol($Pr_2(SeO_4)_3$)·kg^{-1}	Accepted
$La_2(SeO_4)_3 \cdot 5H_2O(cr)$	0.639 mol($La_2(SeO_4)_3$)·kg^{-1}	Accepted
$UO_2SeO_4 \cdot 4H_2O(cr)$	5.31 mol(UO_2SeO_4)·kg^{-1}	Accepted
$BeSeO_4 \cdot 4H_2O(cr)$	(3.73 ± 0.02) mol($BeSeO_4$)·kg^{-1}	Selected
$MgSeO_4 \cdot 6H_2O(cr)$	(3.29 ± 0.05) mol($MgSeO_4$)·kg^{-1}	Selected
$CaSeO_4 \cdot 2H_2O(cr)$	(0.380 ± 0.025) mol($CaSeO_4$)·kg^{-1}	Selected
$SrSeO_4(cr)$	$(1.94 \pm 0.07) \times 10^{-2}$ mol($SrSeO_4$)·kg^{-1}	Selected
$BaSeO_4(cr)$	$(1.87 \pm 0.01) \times 10^{-4}$ mol($BaSeO_4$)·kg^{-1}	Selected
$(NH_4)_2SeO_3 \cdot H_2O(cr)$	(7.4 ± 0.2) mol($(NH_4)_2SeO_3$)·kg^{-1}	Selected
$NH_4H_3(SeO_3)_2(cr)$	(32 ± 1) mol($NH_4H_3(SeO_3)_2$)·kg^{-1}	Selected
$(NH_4)_2SeO_4(cr)$	(6.67 ± 0.02) mol($(NH_4)_2SeO_4$)·kg^{-1}	Selected
$Li_2SeO_3 \cdot H_2O(cr)$	(1.92 ± 0.10) mol(Li_2SeO_3)·kg^{-1}	Selected
$Li_2SeO_4 \cdot H_2O(cr)$	(4.275 ± 0.020) mol(Li_2SeO_4)·kg^{-1}	Selected
$Na_2SeO_3 \cdot 5H_2O(cr)$	(5.15 ± 0.30) mol(Na_2SeO_3)·kg^{-1}	Selected
$NaHSeO_3 \cdot 3H_2O(cr)$	(10.6 ± 0.2) mol($NaHSeO_3$)·kg^{-1}	Selected
$NaH_3(SeO_3)_2(cr)$	(4.31 ± 0.15) mol($NaH_3(SeO_3)_2$)·kg^{-1}	Selected
$Na_2SeO_4 \cdot 10H_2O(cr)$	(3.05 ± 0.05) mol(Na_2SeO_4)·kg^{-1}	Selected
$K_2SeO_3(cr)$	(10.58 ± 0.10) mol(K_2SeO_3)·kg^{-1}	Selected
$K_2Se_2O_5(cr)$	(10.6 ± 0.2) mol($K_2Se_2O_5$)·kg^{-1}	Selected
$KH_3(SeO_3)_2(cr)$	(8.75 ± 0.25) mol($KH_3(SeO_3)_2$)·kg^{-1}	Selected
$K_2SeO_4(cr)$	(5.14 ± 0.02) mol(K_2SeO_4)·kg^{-1}	Selected
$Rb_2SeO_3(cr)$	7.86 mol(Rb_2SeO_3)·kg^{-1}	Accepted
Rb_2SeO_4	(5.20 ± 0.05) mol(Rb_2SeO_4)·kg^{-1}	Selected

(Continued on next page)

Table F-1: (continued)

Compound	Solubility in water	Comment
$Cs_2SeO_3 \cdot H_2O$	9.88 $mol(Cs_2SeO_3) \cdot kg^{-1}$	Accepted
$Cs_2SeO_4(cr)$	(6.38 ± 0.07) $mol(Cs_2SeO_4) \cdot kg^{-1}$	Selected

a Data obtained at 293.15 K.

Bibliography

[1875BER] Berthelot, M., Recherches thermochimiques sur la série du cyanogène, *Ann. Chim. Phys.*, **5**, (1875), 433-493, in French. Cited on page: 435.

[1882THO] Thomsen, J., *Thermochemishe Untersuchungen*, J. A. Barth, Leipzig, (1882), in German. Cited on pages: 127, 142, 144, 149, 167, 170, 417, 418, 419, 431, 432, 433, 434, 437, 438, 448, 472.

[1884BER/VIE] Berthelot, M., Vieille, P., Sur le séléniure d'azote, *Ann. Chim. Phys.*, (1884), 91, in French. Cited on page: 186.

[1887FAB] Fabre, M. C., Recherches thermiques sur les séléniures, *Ann. Chim. Phys.*, **10**, (1887), 472-550, in French. Cited on pages: 135, 398, 403, 406, 412, 416, 423, 438.

[1891PET] Petersen, E., Über die allotropen Zustände einiger Elemente, *Z. Phys. Chem. (Leipzig)*, (1891), 601-623, in German. Cited on page: 167.

[1898PEL] Pelabon, M. H., Ueber die Dissociation des Selenwasserstoffs, *Z. Phys. Chem. (Munich)*, **26**, (1898), 659-689, in German. Cited on page: 132.

[1899BOD] Bodenstein, M., Gasreaktionen in der chemischen Kinetik. IV. Bildung und Zersetzung von Selenwasserstoff, *Z. Phys. Chem. (Munich)*, **29**, (1899), 429-448, in German. Cited on page: 132.

[00FUN] Funk, R., Über die Natriumsalze einiger, der Schwefelsäure analoger zweibasischer Säuren. Studien über die Löslichkeit der Salze.VI, *Ber. Dtsch. Chem. Ges.*, **33**, (1900), 3696-3698, in German. Cited on page: 420.

[02RIN] Ringer, W. E., Mischkrystalle von Schwefel und Selen, *Z. Anorg. Chem.*, **32**, (1902), 183-218, in German. Cited on page: 178.

[07TUT] Tutton, A. E. H., The relation of thallium to the alkali metals: a study of thallium sulphate and selenate, *Proc. R. Soc. London, Ser. A*, **79**, (1907), 351-382. Cited on page: 446.

[08KRE/HOF] Kremann, R., Hofmeier, F., Über die Hydrate der Selensäure, *Monatsh. Chem.*, **29**, (1908), 1111-1118, in German. Cited on page: 144.

[09ROS/PRI] Rosenheim, A., Pritze, M., Über einige komplexe selenhaltige Anionen, *Z. Anorg. Chem.*, **63**, (1909), 275-281, in German. Cited on page: 284.

[10GLA] Glauser, R. T., Über Thalloselenat, (Tl_2SeO_4), *Z. Anorg. Allg. Chem.*, **66**, (1910), 437-438, in German. Cited on page: 446.

[13BRU] Bruner-Krakau, L., Über Selen- und Tellurwasserstoff als Säuren, *Z. Elektrochem.*, **21**, (1913), 861, in German. Cited on pages: 134, 439.

[13JAN/MEY] Jannek, J., Meyer, J., Eine neue Bestimmung des Atomgewichtes des Selens, *Z. Anorg. Allg. Chem.*, **83**, (1913), 51-96, in German. Cited on pages: 137, 138.

[13PRE/BRO] Preuner, G., Brockmöller, I., Gasdruckmessungen mit Spiralmanometer aus Quarzglas. Isothermen von Selen, Schwefel, Arsen, Phosfor; Dissociation des Kupfersulfids und des Selenwasserstoffs., *Z. Phys. Chem. (Leipzig)*, **81**, (1913), 129-170, in German. Cited on pages: 98, 99, 440.

[21ROS/KRA] Rosenheim, A., Krause, L., Über selenige Säure und Heteropolyselenite, *Z. Anorg. Allg. Chem.*, **118**, (1921), 177-192, in German. Cited on page: 413.

[22BRO] Brønsted, J. N., Studies of solubility: IV. The principle of specific interaction of ions, *J. Am. Chem. Soc.*, **44**, (1922), 877-898. Cited on page: 589.

[22BRO2] Brønsted, J. N., Calculation of the osmotic and activity functions in solutions of uni-univalent salts, *J. Am. Chem. Soc.*, **44**, (1922), 938-948. Cited on page: 589.

[22MAN/ORT] Manchot, W., Ortner, K., Über die Hydratbildung des Selendioxydes, *Z. Anorg. Allg. Chem.*, **120**, (1922), 300-308, in German. Cited on pages: 137, 138, 139.

[23HLA] de Hlasko, M., Sur la dissociation électrolytique de l'hydrogène sélénié et de l'hydrogène telluré , *J. Chim. Phys. Phys.-Chim. Biol.*, **20**, (1923), 167-172, in French. Cited on pages: 134, 439.

[23PEA/FOR] Pearce, J. N., Fortsch, A. R., The free energy of dilution and the activity of the ions of hydrogen iodide in aqueous solution, *J. Am. Chem. Soc.*, **45**, (1923), 2852-2857. Cited on pages: 150, 151.

[24BAD] Badger, R. M., The ammonia, carbon, hydrogen cyanide, hydrogen equilibrium and the free energy of hydrogen cyanide, *J. Am. Chem. Soc.*, **46**, (1924), 2166-2172. Cited on page: 433.

[25AMI/FEL] McAmis, A. J., Felsing, W. A., The solubility of hydrogen selenide, *J. Am. Chem. Soc.*, **47**, (1925), 2633-2637. Cited on pages: 133, 134.

[25LEN/KAO] Lenher, V., Kao, C. H., The preparation of selenic acid and of certain selenates, *J. Am. Chem. Soc.*, **47**, (1925), 1521-1522. Cited on page: 305.

[25WIL/PRI] Willcox, J. S., Prideaux, E. B. R., The dissociation constants of selenious acid, *J. Chem. Soc.*, (1925), 1543-1546. Cited on page: 141.

[26MAT/MAR] Matignon, C., Marchal, G., Thermochimie du glucinium, *Bull. Soc. Chim. Fr.*, **39**, (1926), 167-190, in French. Cited on page: 464.

[26SID/LEW] Sidgwick, N. V., Lewis, N. B., The solubility of beryllium oxide in solutions of its salts, *J. Chem. Soc.*, (1926), 1287-1302. Cited on page: 393.

[27KRA/FRI] Krauss, F., Fricke, A., Über das Aluminiumsulfat und seine Hydrate. Zur Kenntnis der Doppelsulfate und ihrer Komponenten II., *Z. Anorg. Chem.*, **166**, (1927), 170-176, in German. Cited on page: 232.

[28MEY/AUL] Meyer, J., Aulich, W., Zur Kenntnis der Doppelsalze der Selensäure, *Z. Anorg. Allg. Chem.*, **172**, (1928), 321-343, in German. Cited on pages: 144, 397, 401, 420, 421, 425.

[28SCH/SWI] Schott, H. F., Swift, E. H., Yost, D. M., The reduction potential of selenium and the free energy of aqueous selenous acid, *J. Am. Chem. Soc.*, **50**, (1928), 721-727. Cited on pages: 150, 440.

[28SHE/IZA] Sherrill, M. S., Izard, E. F., The reduction potential of selenous acid and the free energy of aqueous selenic acid, *J. Am. Chem. Soc.*, **50**, (1928), 1665-1675. Cited on pages: 152, 153.

[29FRI] Friend, J. A. N., The solubility of potassium selenate in water between 0° and 100°, *J. Chem. Soc.*, (1929), 2782-2783. Cited on page: 425.

[29NEU/MUL] Neumann, B., Müller, G., Zur Thermochemie der unterchlorigen Säure und einiger ihrer Salze in Wässriger Lösung, *Z. Anorg. Chem.*, **182**, (1929), 235-253, in German. Cited on pages: 431, 432.

[30BIR/HUT] Birckenbach, L., Huttner, K., Über die Cyan-Abkömmlinge der flüchtigen Hydride, *Z. Anorg. Allg. Chem.*, **190**, (1930), 1-26, in German. Cited on page: 307.

[30YOS/KIR] Yost, D. M., Kircher, C. E., The vapor pressures of selenium tetrachloride. The existence of selenium dichloride, *J. Am. Chem. Soc.*, **52**, (1930), 4680-4685. Cited on page: 169.

[31BRI/ROB] Britton, H. T. S., Robinson, R. A., The use of the antimony-antimonious oxide electrode in the determination of the concentration of hydrogen ions and in potentiometric titrations. The Prideaux-Ward universal buffer mixture, *J. Chem. Soc.*, (1931), 458-473, Contains report of measurements by W. L. German. Cited on page: 141.

[31FRI] Friend, J. A. N., The selenates of neodynium and their solubilities in water and in selenic acid solutions, *J. Chem. Soc.*, (1931), 1802-1809. Cited on page: 375.

[32ALP/SAY] Macalpine, G. B., Sayce, L. A., The system cupric selenate-water-selenic acid at 25°, *J. Chem. Soc.*, (1932), 1560-1565. Cited on pages: 146, 295.

[32BRI/ROB] Britton, H. T. S., Robinson, R. A., The use of the glass electrode in titrimetric work and precipitation reactions. The application of the principle of the solubility product to basic precipitates, *Trans. Faraday Soc.*, **28**, (1932), 531-545. Cited on page: 141.

[32JAN] Janitzki, J., Über die selenigsauren Salze des Natriums und des Kaliums, *Z. Anorg. Allg. Chem.*, **205**, (1932), 49-76, in German. Cited on pages: 417, 418, 419, 423, 424.

[32MEY/HIN] Meyer, J., Hinke, W., Zur Kenntnis der Selensäureester, *Z. Anorg. Allg. Chem.*, **204**, (1932), 29-32. Cited on page: 305.

[32SHE/LYO] Sherrill, M. S., Lyons, Jr., E. H., The comparative ionization of selenic and sulfuric acids, *J. Am. Chem. Soc.*, **54**, (1932), 979-984. Cited on pages: 146, 442.

[33RUM] Rumpf, P., Sur le titrage électrométrique des acides sulfureux, sélénieux et alfa-oxyalcoylsulfoniques, *Compt. Rend.*, **197**, (1933), 686-689, in French. Cited on pages: 141, 444, 451.

[34BAN] Banks, W. H., The dissociation of the selenates of zinc and cadmium in water, *J. Chem. Soc.*, (1934), 1010-1012. Cited on pages: 264, 265, 275, 549.

[34HAR/EMB] Harned, H. S., Embree, N. D., The temperature variation of ionization constants in aqueous solutions, *J. Am. Chem. Soc.*, **56**, (1934), 1050-1053. Cited on page: 143.

[34JAN] Janickis, J., Über die selenigsauren Salze des Ammoniums, *Z. Anorg. Allg. Chem.*, **218**, (1934), 89-103, in German. Cited on page: 410.

[35GUG] Guggenheim, E. A., The specific thermodynamic properties of aqueous solutions of strong electrolytes, *Philos. Mag.*, **19**, (1935), 588-643. Cited on page: 589.

[35HAR/HAM] Harned, H. S., Hamer, W. J., The thermodynamics of aqueous sulfuric acid solutions from electromotive force measurements, *J. Am. Chem. Soc.*, **57**, (1935), 27-33. Cited on page: 443.

[36HAR/KES] Harned, H. S., Keston, A. S., Donelson, J. G., The thermodynamics of hybrobromic acid in aqueous solution from electromotive force measurements, *J. Am. Chem. Soc.*, **58**, (1936), 989-994. Cited on page: 153.

[36MER/SCH] Merten, H., Schlüter, H., Über die Verbrennungswärmen einiger organischer Selen-Verbindungen, *Ber. Dtsch. Chem. Ges.*, **69**, (1936), 1364-1366, in German. Cited on page: 206.

[36SCA] Scatchard, G., Concentrated solutions of strong electrolytes, *Chem. Rev.*, **19**, (1936), 309-327. Cited on page: 589.

[37AKE/TEA] Åkerlöf, G., Teare, J. W., Thermodynamics of concentrated aqueous solutions of hydrochloric acid, *J. Am. Chem. Soc.*, **59**, (1937), 1855-1868. Cited on page: 153.

[37BAN/HEY] Bannister, F. A., Hey, M. H., The identity of penroseite and blockite, *Am. Mineral.*, **22**, (1937), 319-324. Cited on page: 635.

[37GOR] Gordon, A. R., The free energy of hydrogen cyanide from spectroscopic data, *J. Chem. Phys.*, **5**, (1937), 30-32. Cited on page: 433.

[37KIN] King, G. B., The system ammonia-selenic acid-water at 30°C, *J. Phys. Chem.*, **41**, (1937), 797-801. Cited on page: 411.

[37KRU/CLU] Kruis, A., Clusius, K., Über die Molwärmen und Umwandlungen der kondensierten Hydride und Deuteride des Schwefels und Selens, *Z. Phys. Chem., Abt. B*, **38**, (1937), 156-176, in German. Cited on page: 131.

[38ISH/ABE] Ishikawa, F., Abe, H., Studies of selenious acid, *Sci. Pap. Inst. Phys. Chem. Res. (Jpn.)*, **34**, (1938), 775-786. Cited on pages: 137, 138, 139.

[38LAW/KIN] Lawrence, R. W., King, G. B., The system ammonium selenate-magnesium selenate-water at 30°, *J. Am. Chem. Soc.*, **60**, (1938), 1987-1988. Cited on page: 397.

[38NEA/CRO] Neal, Jr., J. L., McCrosky, C. R., The ternary system selenium dioxide-barium selenite-water at 0, 25, and 50°, *J. Am. Chem. Soc.*, **60**, (1938), 911-914. Cited on pages: 139, 407.

[39GIA/RUE] Giauque, W. F., Ruehrwein, R. A., The entropy of hydrogen cyanide. Heat capacity, heat of vaporization and vapor pressure. Hydrogen bond polymerization of the gas in chains of infinite length, *J. Am. Chem. Soc.*, **61**, (1939), 2626-2633. Cited on pages: 433, 435.

[39HAG] Hagisawa, H., Dissociation constants of selenious acid, *Bull. Inst. Phys. Chem. Res. (Tokyo)*, **18**, (1939), 648-656. Cited on page: 141.

[39KEL] Kelley, K. K., The specific heats at low temperatures of manganese, manganous selenide, and manganous telluride, *J. Am. Chem. Soc.*, **61**, (1939), 203-207. Cited on pages: 337, 338.

[39KLE/SOD] Klemm, W., Sodomann, H., Langmesser, P., Beiträge zur Kenntnis der Alkalimetallchalkogenide, *Z. Anorg. Allg. Chem.*, **241**, (1939), 281-304, in German. Cited on page: 154.

[39WEL/KIN] Welton, R. C., King, G. B., The system calcium selenate-ammonium selenate-water at 30°, *J. Am. Chem. Soc.*, **61**, (1939), 1251-1252. Cited on page: 401.

[40HIL/SOT] Hill, A. E., Soth, G. C., Ricci, J. E., Some further solid solutions of picromerites at 25°, *J. Am. Chem. Soc.*, **62**, (1940), 2717-2723. Cited on pages: 397, 425.

[40KLE] Klein, A., Etude sur les sélénates des métaux de la série magnésienne, *Ann. Chim. (Paris)*, **14**, (1940), 263-317, in French. Cited on pages: 263, 274, 275, 295, 314, 321, 397.

[40NIW/SHI] Niwa, K., Shibata, Z., Molecular weight determination of vapours of low vapour pressure substances. I. Determination of vapour pressure and molecular weight of solid and fused selenium, *J. Fac. Sci., Hokkaido Univ., Ser.3*, **3**, (1940), 53-73. Cited on pages: 487, 543, 557.

[41HAG] Hagisawa, H., Dissociation constants of hydrogen selenide, *Bull. Inst. Phys. Chem. Res. (Tokyo)*, **20**, (1941), 384-389, in Japanese. Cited on page: 134.

[41HAR/ROB] Harned, H. S., Robinson, R. A., The activity coefficient of hydriodic acid at 25°. From isopiestic vapour pressure measurements, *Trans. Faraday Soc.*, **37**, (1941), 302-307. Cited on page: 151.

[42GEL/KIN] Gelbach, R. W., King, G. B., The secondary ionization and activity coefficients of selenic acid, *J. Am. Chem. Soc.*, **64**, (1942), 1054-1057. Cited on pages: 146, 305, 443.

[44AME/BEL] Amelin, A. G., Belyakov, M. I., Saturated vapour pressure of SeO_2, *Zh. Fiz. Khim.*, **18**, (1944), 466-468, in Russian. Cited on pages: 123, 124, 261.

[44FUR/KIN] Furukawa, H., King, G. B., The system magnesium selenate-selenic acid-water at 30°C, *J. Phys. Chem.*, **48**, (1944), 174-178. Cited on page: 397.

[45KIN/BEC] King, G. B., Beckman, W. A., The system ammonium selenate-cupric selenate-water at 25°, *J. Am. Chem. Soc.*, **67**, (1945), 857-859. Cited on pages: 295, 411.

[46WAI/SHU] Waitkins, G. R., Shutt, R., Aluminium selenide and hydrogen selenide, *Inorganic Synthesis, vol. II*, Fernelius, W. C., Ed., pp.183-186, McGraw-Hill, New York, USA, (1946). Cited on page: 131.

[47IVE/PIT] Ives, D. J. G., Pittman, R. W., Wardlaw, W., The preparation, properties, and chlorination products of carbon diselenide, *J. Chem. Soc.*, (1947), 1080-1083. Cited on pages: 205, 206.

[48LIN/NIE] Lingane, J. J., Niedrach, L. W., Polarography of selenium and tellurium. I. The -2 states, *J. Am. Chem. Soc.*, **70**, (1948), 4115-4120. Cited on pages: 134, 281.

[49EAR] Earley, J. W., Studies of natural and artificial selenides: I-Klockmannite, CuSe, *Am. Mineral.*, **34**, (1949), 435-440. Cited on page: 635.

[49MIK/POS] Mikus, F. F., Poss, F. J., Equilibrium between ammonia and hydrogen selenide, *J. Am. Chem. Soc.*, **71**, (1949), 429-431. Cited on page: 409.

[49PAP/HAI] Pappas, A. , Haissinsky, M., Sur le système H_2Se-Se-Na_2CO_3, *Bull. Soc. Chim. Fr.*, (1949), 645-647, in French. Cited on page: 154.

[50BAS] Bassett, H., The determination of selenate, and the preparation of some aluminium selenates, *J. Chem. Soc.*, (1950), 1191-1194. Cited on page: 231.

[50EAR] Earley, J. W., Description and synthesis of the selenide minerals, *Am. Mineral.*, **35**, (1950), 337-364. Cited on pages: 634, 635.

[50HAR/OWE] Harned, H. S., Owen, B. B., *The Physical Chemistry of Electrolytic Solutions, 2nd ed.*, Reinhold, New York, (1950). Cited on pages: 401, 405, 454.

[50HER] Herzberg, G., *Molecular spectra and molecular structure. I. Spectra of diatomic molecules*, 2nd. Edition, Van Nostrand Reinhold, New York, (1950). Cited on pages: 210, 215, 221, 228.

[50PAM/AGA] Pamfilov, A. V., Agafonova, A. L., Dissociation constant of selenic acid, *Zh. Fiz. Khim.*, **24**, (1950), 1147-1151, in Russian. Cited on pages: 146, 148.

[50ROH/FRO] Rohrer, C. S., Froning, H. R., A study on heterogeneous equilibria in aqueous solutions of the nickel selenates at 30°, *J. Am. Chem. Soc.*, **72**, (1950), 4656-4659. Cited on page: 314.

[51HOR/TAN] Horiuchi, J., Tanabe, K., Solubility of prussic acid in water, *J. Res. Inst. Catal., Hokkaido Univ.*, **1**, (1951), 117-131. Cited on pages: 434, 435.

[51HOR/YAN] Horiuchi, J., Yano, T., Kanai, K., Synthesis of hydrocyanic acid. IV. The kinetics of the synthetic reaction and the equilibrium, *Shokubai*, 7, (1951), 8-16, in Japanese. Cited on pages: 433, 434.

[51KAP/GOL] Kapustinskii, A. F., Golutvin, Y. M., The method of limiting heats. II. The heat of formation of iron selenide, *Zh. Fiz. Khim.*, **25**, (1951), 729-731, in Russian. Cited on page: 231.

[51KAP/ZHD] Kapustinskii, A. F., Zhdanova, A. N., Investigation of aqueous solutions of selenic acid by the specific-gravity method, *Zh. Prikl. Khim. (S.-Peterburg)*, **24**, (1951), 583-589, in Russian, English translation in [51KAP/ZHD2]. Cited on page: 146.

[51KAP/ZHD2] Kapustinskii, A. F., Zhdanova, A. N., Investigation of aqueous solutions of selenic acid by the specific-gravity method, *Russ. J. Appl. Chem.*, **24**, (1951), 649-656. From a citation in this Bibliography

[51KAP/ZHD3] Kapustinskii, A. F., Zhdanova, A. N., The thermal and vicosimetric analysis of the selenic acid-water system, *Zh. Prikl. Khim. (S.-Peterburg)*, **24**, (1951), 658-661, in Russian, English translation in [51KAP/ZHD4]. Cited on page: 144.

[51KAP/ZHD4] Kapustinskii, A. F., Zhdanova, A. N., The thermal and viscosimetric analysis of the selenic acid-water system, *Russ. J. Appl. Chem.*, **24**, (1951), 739-743. From a citation in this Bibliography

[51MUL/SCH] Mulder, H. D., Schmidt, F. C., Heats of solution and reaction in liquid ammonia. VIII., *J. Am. Chem. Soc.*, **73**, (1951), 5575-5577. Cited on page: 416.

[51TOU/DOS] Toul, F., Dostál, K., Darstellung und Eigenschaften des Selentrioxyds, *Collect. Czech. Chem. Commun.*, **16**, (1951), 531-541, in German. Cited on page: 126.

[51WIN/STO] Windmaisser, F., Stöckl, F., Basiche Alkaliselenate, *Monatsh. Chem.*, **82**, (1951), 287-294, in German. Cited on page: 420.

[52AYN/PEA] Aynsley, E. E., Peacock, R. D., Robinson, P. L., The preparation and properties of selenium tetrafluoride and oxyfluoride, *J. Chem. Soc.*, (1952), 1231-1234. Cited on page: 163.

[52BRO] Brooks, L. S., The vapor pressures of tellurium and selenium, *J. Am. Chem. Soc.*, **74**, (1952), 227-229. Cited on pages: 99, 443.

[52GAB/ALV] Gabriel, H., Alvarez-Tostado, C., Silicon disulfide and silicon diselenide, *J. Am. Chem. Soc.*, **74**, (1952), 262-264. Cited on page: 208.

[52HAR] Hart, D., The periodicity of chemical thermodynamic functions, *J. Phys. Chem.*, **56**, (1952), 202-214. Cited on page: 126.

[52LEH/KRU] Lehmann, H.-A., Krüger, G., Darstellung von reinem Selentrioxyd, *Z. Anorg. Allg. Chem.*, **267**, (1952), 324-333, in German. Cited on page: 126.

[52ROS/WAG] Rossini, F. D., Wagman, D. D., Evans, W. H., Levine, S., Jaffe, I., Selected values of chemical thermodynamic properties, NBS Circular, Report 500, (1952). Cited on page: 438.

[53PEA] Peacock, R. D., Some properties of selenium tetrafluoride, *J. Chem. Soc.*, (1953), 3617-3619. Cited on pages : 161, 162.

[54PAR/QUA] Parkinson, D. H., Quarrington, J. E., The molar heats of lead sulphide, selenide and telluride in the temperature range 20°K to 260°K, *Proc. Phys. Soc., London, Sect. A*, **67**, (1954), 569-579. Cited on pages: 218, 219, 220.

[54SCH/GAT2] Schneider, A., Gattow, G., Beitrag zum System Aluminium-Selen, *Z. Anorg. Allg. Chem.*, **277**, (1954), 49-59, in German. Cited on page: 231.

[54SHA] Sharma, C. B., The absorption spectrum of bismuth selenide and bismuth telluride in the region 2900-2200 Å, *Proc. Phys. Soc., London,*, **A67**, (1954), 935-938. Cited on page: 197.

[55DOS] Dostál, K., Thermische Analyse des Systems SeO_3-H_2O, *Collect. Czech. Chem. Commun.*, **20**, (1955), 1033-1040, in German. Cited on page: 145.

[55GAT/SCH] Gattow, G., Schneider, A., Die Bildungswärmen der Wismutchalkogenide, *Angew. Chem.*, **67**, (1955), 306-307, in German. Cited on pages: 201, 202.

[55KAT/GEB] Katzin, L. I., Gebert, E., The iodide-iodine-triiodide equilibrium and ion activity coefficient ratios, *J. Am. Chem. Soc.*, **77**, (1955), 5814-5819. Cited on page: 441.

[55RAM/SCH] Ramdohr, P., Schmidt, M., Vier neue naturliche Kobaltselenide von Steinbruch Trogtal bei Lautenthal im Harz, *Neues Jahrb. Mineral., Monatsh.*, **6**, (1955), 133-142, in German. Cited on pages: 634, 635.

[55SEL/BOG] Selivanova, N. M., Boguslavskii, R. Y., Solubility of lead selenate in aqueous solutions of alkali metal selenates, *Zh. Fiz. Khim.*, **29**, (1955), 128-134, in Russian. Cited on page: 224.

[56BUR] Bur'yanova, E. Z., On ferroselite, *Zap. Vses. Mineral. O-va.*, **85**, (1956), 286-291, in Russian. Cited on page: 634.

[56CHU] Chukhlantsev, V. G., The solubility products of some metal selenites, *Zh. Neorg. Khim.*, **1**, (1956), 2300-2305, in Russian, English translation in [56CHU2]. Cited on pages: 223, 262, 263, 272, 274, 283, 293, 294, 303, 312, 319, 330, 341, 380, 389, 391, 394, 395, 399, 404, 444, 465.

[56CHU2] Chukhlantsev, V. G., The solubility products of some metal selenites, *Russ. J. Inorg. Chem.*, **1**, (1956), 132-138. From a citation in this Bibliography

[56ERD/PAU] Erdey, L., Paulik, F., Paulik, J., Ein neues thermisches Verfahren: Die Derivationsthermogravimetrie, *Acta Chim. Acad. Sci. Hung.*, **10**, (1956), 61-97, in German. Cited on page: 232.

[56GAT/SCH] Gattow, G., Schneider, A., Die Bildungsenthalpien im System Kupfer-Selen, *Z. Anorg. Allg. Chem.*, **286**, (1956), 296-306, in German. Cited on pages: 289, 290, 291, 292.

[56HAH/BUR] Hahn, H., Burow, F., Über die Bildungsenthalpien der Sulfide, Selenide und Telluride des Galliums und Indiums, *Angew. Chem.*, **68**, (1956), 382, in German. Cited on pages: 234, 236, 240, 243, 244.

[56LAT] Latimer, W. M., *The Oxidation States of the Elements and their Potentials in Aqueous Solutions, 2nd ed., 3rd printing*, Prentice-Hall Inc., Englewood Cliffs, N. J., (1956). Cited on pages: 151, 564.

[56RAM] Ramdohr, P., Stilleit, ein neues Mineral, naturliches Zinkselenid, von Shinkolobwe, Geotektonisches Symposium zu Ehren von Hans Stille (1956), cited from Am. Mineral. 42 (1957) 584. Cited on page: 635.

[56STU/SIN] Stull, D. R., Sinke, G. C., *Thermodynamic properties of the elements*, American Chemical Society, Washington, D.C., (1956). Cited on pages: 93, 95, 487, 519, 534, 645.

[56TOR] Toropova, V. F., A study of selenocyanate complexes of certain metals, *Zh. Neorg. Khim.*, **1**, (1956), 243-250, in Russian, English translation in [56TOR2]. Cited on pages: 276, 277, 286, 287, 297, 306, 307, 322, 542.

[56TOR2] Toropova, V. F., A study of selenocyanate complexes of certain metals, *Russ. J. Inorg. Chem.*, **1**, (1956), 56-63. From a citation in this Bibliography

[56VUI] Vuillard, G., Le système binaire eau-acide sélénique, *Compt. Rend.*, **242**, (1956), 1326-1329, in French. Cited on page: 144.

[57BUR/KOV] Bur'yanova, E. Z., Kovalev, G. A., Kolikov, A. I., Cadmoselite, a new mineral, *Zap. Vses. Mineral. O-va.*, **86**, (1957), 626-628, in Russian. Cited on page: 634.

[57CHU/TOM] Chukhlantsev, V. G., Tomashevsky, G. P., The solubility of the selenites of certain metals, *Zh. Anal. Khim.*, **12**, (1957), 296-302, in Russian, English translation in [57CHU/TOM2]. Cited on pages: 223, 283, 312, 319, 320, 330, 341, 380.

[57CHU/TOM2] Chukhlantsev, V. G., Tomashevsky, G. P., The solubility of the selenites of certain metals, *J. Anal. Chem. USSR*, **12**, (1957), 303-309. From a citation in this Bibliography

[57ILL/LAP] Illarionov, V. V., Lapina, L. M., Association states of selenium in the gas phase, *Dokl. Akad. Nauk SSSR*, **114**, (1957), 1021-1024, in Russian. Cited on pages: 98, 99, 444.

[57KIU/WAG] Kiukkola, K., Wagner, C., Measurements on galvanic cells involving solid electrolytes, *J. Electrochem. Soc.*, **104**, (1957), 379-387. Cited on pages: 299, 300, 445.

[57KRY/CHU] Krylov, E. I., Chukhlantsev, V. G., Uranyl and thorium selenites, *Zh. Anal. Khim.*, **12**, (1957), 451-456, in Russian, English translation in [57KRY/CHU2]. Cited on pages: 389, 391.

[57KRY/CHU2] Krylov, E. I., Chukhlantsev, V. G., Uranyl and thorium selenites, *J. Anal. Chem. USSR*, **12**, (1957), 469-473. From a citation in this Bibliography

[57TOR] Toropova, V. F., Investigation of the stability of complex compounds with addenda containing selenium, *Zh. Neorg. Khim.*, **2**, (1957), 515-522, in Russian, English translation in [57TOR2]. Cited on pages: 272, 284, 285, 286, 445.

[57TOR2] Toropova, V. F., Investigation of the stability of complex compounds with addenda containing selenium, *Russ. J. Inorg. Chem.*, **2**, (1957), 63-76. From a citation in this Bibliography

[57TUM] Tuma, J., Enthalpy of dissolution of selenic acids, Ph. D. Thesis, Faculty of science, Masaryk University, (1957), Brno, in Czech. Cited on pages : 126, 448.

[57VUI] Vuillard, G., Contribution à l'étude de l'état vitreux et de la cristallisation des solutions aqueuses, *Ann. Chim. (Paris)*, **2**, (1957), 233-297, in French. Cited on page : 144.

[58FLI/MIS] Flis, I. E., Mishchenko, K. P., Pakhomova, N. V., Calculation of fundamental thermodynamic properties of ClO_{aq}^- and $HClO_{aq}$ at 25, *Zh. Neorg. Khim.*, **3**, (1958), 1781-1784. Cited on page: 431.

[58JAN/DOE] Jander, J., Doetsch, V., Untersuchungen am Selenstickstoff, *Angew. Chem.*, **70**, (1958), 704, in German. Cited on page: 186.

[58KUL/SHA] Kuliev, A. A., Shakhtakhtinskii, M. G., Saturated vapor pressure of selenium below its boiling point, *Dokl. Akad. Nauk SSSR*, **120**, (1958), 1284-1286, in Russian. Cited on page: 446.

[58MUL/HEP] Muldrow, Jr., C. N., Hepler, L. G., Heats of precipitation and formation of lead and calcium molybdates, *J. Phys. Chem.*, **62**, (1958), 982-984. Cited on page: 452.

[58NAI/NAN] Nair, V. S. K., Nancollas, G. H., Thermodynamic of ion association. Part V. Dissociation of the bisulphate ion, *J. Chem. Soc.*, (1958), 4144-4147. Cited on page : 149.

[58NAI/NAN2] Nair, V. S. K., Nancollas, G. H., Thermodynamics of ion association. Part VI. Magnesium and zinc sulphates, *J. Chem. Soc.*, (1958), 3706-3710. Cited on page: 521.

[58SEL/SHN3] Selivanova, N. M., Shneider, V. A., On some physico-chemical properties of barium selenate, *Nauchn. Dokl. Vyssh. Shk., Khim. Khim. Tekhnol.*, **2**, (1958), 216-220, in Russian. Cited on pages: 407, 408, 447.

[58SEL/ZUB] Selivanova, N. M., Zubova, G. A., Physico-chemical properties of strontium selenate, *Izv. Vyssh. Uchebn. Zaved., Khim. Khim. Tekhnol.*, **1**, (1958), 27-33, in Russian. Cited on pages : 405, 456.

[58SEL] Selivanova, N. M., On some thermodynamic properties of thallium(I) selenate, *Zh. Fiz. Khim.*, **32**, (1958), 1277-1279, in Russian. Cited on pages: 253, 446.

[58SPE/NAU] Spedding, F. H., Naumann, A. W., Eberts, R. E., Heats of dilution and related thermodynamic properties of aqueous rare earth salt solutions at 25°; integral heats of solution of $NdCl_3.6H_2O$, *J. Am. Chem. Soc.*, **81**, (1958), 23-28. Cited on page: 528.

[58TRZ] Trzil, J., Thermodynamic constants of selenium oxide, *Sb. Ved. Pr. Vys. Sk. Banske Ostrave*, **4**, (1958), 277-281, in Czech. Cited on pages: 126, 127, 131, 144, 149, 448.

[58WOO] Wood, R. H., The second ionization constant of hydrogen selenide, *J. Am. Chem. Soc.*, **80**, (1958), 1559-1562. Cited on pages: 135, 449.

[59CON/CHI] Connick, R. E., Chia, Y. T., The hydrolysis of chlorine and its variation with temperature, *J. Am. Chem. Soc.*, **81**, (1959), 1280-1284. Cited on page: 432.

[59GOL/KOS3] Golub, A. M., Kosmatyi, Y. V., A potentiometric study of lead selenocyanate, *Zh. Neorg. Khim.*, **4**, (1959), 1347-1352, in Russian, English translation in [59GOL/KOS4]. Cited on pages: 225, 226, 449.

[59GOL/KOS4] Golub, A. M., Kosmatyi, Y. V., A potentiometric study of lead selenocyanate, *Russ. J. Inorg. Chem.*, **4**, (1959), 606-608. From a citation in this Bibliography

[59GOL/POM] Golub, A. M., Pomerants, G. B., Complex silver selenocyanates, *Zh. Neorg. Khim.*, **4**, (1959), 769-775, in Russian, English translation in [59GOL/POM2]. Cited on page: 307.

[59GOL/POM2] Golub, A. M., Pomerants, G. B., Complex silver selenocyates, *Russ. J. Inorg. Chem.*, **4**, (1959), 349-352. From a citation in this Bibliography

[59GOL] Golub, A. M., A study of selenocyanato complexes of lead by the solubility method, *Russ. J. Inorg. Chem.*, **4**, (1959), 711-713. Cited on pages: 225, 449, 450.

[59GRO/WES] Grønvold, F., Westrum, Jr., E. F., Chou, C., Heat capacities and thermodynamic properties of the pyrrhotites FeS and Fe0.877S from 5 to 350 K, *J. Chem. Phys.*, **30**, (1959), 528-531. Cited on pages: 325, 326, 327.

[59GUL/PET] Gul'tyaev, P. V., Petrov, A. V., The specific heats of a number of semiconductors, *Sov. Phys. Solid State*, **1**, (1959), 330-334. Cited on pages: 279, 299, 450.

[59MIT] Mityureva, T. T., Determination of solubility products of the selenites of indium and thallium-III, *Dopov. Akad. Nauk Ukr. RSR*, **2**, (1959), 166-168, in Russian. Cited on pages: 246, 252, 450.

[59NAI/NAN] Nair, V. S. K., Nancollas, G. H., Thermodynamics of ion association. Part VI. Some transition metal sulphates, *J. Chem. Soc.*, (1959), 3934-3939. Cited on page: 521.

[59ROB/STO] Robinson, R. A., Stokes, R. H., *Electrolyte solutions*, 2nd. Edition, Butterworths, London, (1959, revised 1970), 571 pp.. Cited on pages: 151, 593.

[59SEL/KAP] Selivanova, N. M., Kapustinskii, A. F., Zubova, G. A., Thermochemical properties of slightly soluble selenates and the entropy of the selenate ion in aqueous solution, *Izv. Akad. Nauk SSSR, Ser. Khim.*, (1959), 187-194, in Russian. Cited on pages: 225, 408, 451, 467, 504.

[59SEL/SHN3] Selivanova, N. M., Shneider, V. A., Physico-chemical properties of selenates: V. Heat of formation of calcium selenate from the elements, *Izv. Vyssh. Uchebn. Zaved., Khim. Khim. Tekhnol.*, **2**, (1959), 475-479, in Russian. Cited on pages: 402, 453.

[59SEL/SHN4] Selivanova, N. M., Shneider, V. A., Physico-chemical properties of selenates: VI. Solubility of calcium selenate in water at various temperatures, *Izv. Vyssh. Uchebn. Zaved., Khim. Khim. Tekhnol.*, **2**, (1959), 651-656, in Russian. Cited on pages: 401, 402, 454.

[59SEL/ZUB] Selivanova, N. M., Zubova, G. A., Finkel'shtein, E. I., Thermodynamic properties of silver selenate, *Zh. Fiz. Khim.*, **33**, (1959), 2365-2369, in Russian, English translation in [59SEL/ZUB2]. Cited on pages: 305, 306, 455.

[59SEL/ZUB2] Selivanova, N. M., Zubova, G. A., Finkel'shtein, E. I., Thermodynamic properties of silver selenate, *Russ. J. Phys. Chem.*, **33**, (1959), 430-432. From a citation in this Bibliography

[59SEL/ZUB3] Selivanova, N. M., Zubova, G. A., The thermodynamic properties of strontium selenate, *Zh. Fiz. Khim.*, **33**, (1959), 141-146, in Russian. Cited on pages: 405, 456.

[59SIN] Sindeeva, N. D., *Mineralogy, types of deposits, and basic pattern of selenium and tellurium geochemistry*, p. 46, Izdatel. Akad. Nauk SSSR, Moscow, (1959), in Russian. Cited on page: 634.

[59ZLO/POP] Zlomanov, V. P., Popovkin, B. A., Novoselova, A. V., Determination of the saturated vapour pressure of solid lead selenide, *Russ. J. Inorg. Chem.*, **4**, (1959), 1231-1233. Cited on pages : 222, 458, 511.

[60CLA] Claude, R., Préparation et propriétés des sélénito-uranates, *Ann. Chim. (Paris)*, **5**, (1960), 165-206, in French. Cited on page: 389.

[60FIN/WAG] Finch, C. B., Wagner, Jr., J. B., Determination of the standard free energy of formation of lead selenide, *J. Electrochem. Soc.*, **107**, (1960), 932-933. Cited on pages: 220, 458.

[60GAD/SHA] Gadzhiev, S. N., Sharifov, K. A., Determination of heat of formation of tin selenide formed in the calorimeter bomb, *Dokl.-Akad. Nauk Az. SSR*, **16**, (1960), 659-662, in Russian. Cited on page: 214.

[60GOL/SKO] Golub, A. M., Skopenko, V. V., Copper selenocyanates, *Zh. Neorg. Khim.*, **5**, (1960), 1973-1976, in Russian, English translation in [60GOL/SKO2]. Cited on page: 296.

[60GOL/SKO2] Golub, A. M., Skopenko, V. V., Copper selenocyanates, *Russ. J. Phys. Chem.*, **5**, (1960), 961-963. From a citation in this Bibliography

[60GRO/THU] Grønvold, F., Thurmann-Moe, T., Westrum, Jr., E. F., Levitin, N. E., Low temperature heat capacities and thermodynamic properties of the $Ni_{1-x}Se$-Phase, *Acta Chem. Scand.*, **14**, (1960), 634-640. Cited on page: 310.

[60JAN/DOE] Jander, J., Doetsch, V., Ein Beitrag zur Kenntnis des Stickstoffselenides (Selenstickstoffs), *Chem. Ber.*, **93**, (1960), 561-569, in German. Cited on page: 186.

[60KEL] Kelley, K. K., Contributions to the data on theoretical metallurgy: XIII. High-temperature heat-content, heat-capacity and entropy data for the elements and inorganic compounds, *Bull. - U. S., Bur. Mines*, **584**, (1960), 1-232. Cited on page: 221.

[60KLE/OST] Klebanov, G. S., Ostapkevich, N. A., Reactions of selenium with aqueous solutions of alkali-metal sulfites, *Zh. Prikl. Khim. (S.-Peterburg)*, **33**, (1960), 1957-1961, in Russian, English translation in [60KLE/OST2]. Cited on page: 182.

[60KLE/OST2] Klebanov, G. S., Ostapkevich, N. A., Reactions of selenium with aqueous solutions of alkali-metal sulfites, *Russ. J. Appl. Chem.*, **33**, (1960), 1938-1941. From a citation in this Bibliography

[60KOR/SOK] Korneeva, I. V., Sokolov, V. V., Novoselova, A. V., Saturated-vapour pressure of zinc and cadmium selenides in the solid state, *Russ. J. Inorg. Chem.*, **5**, (1960), 117-119. Cited on pages: 257, 258, 269, 270, 458, 459.

[60POR/SPE] Porter, R. F., Spencer, C. W., Stabilities of the gaseous molecules, BiSe, BiTe, and SbTe, *J. Chem. Phys.*, **32**, (1960), 943-944. Cited on pages: 198, 200, 459, 495.

[60SIM/PAE2] Simon, A., Paetzold, R., Untersuchungen an Selen-Sauerstoff-Verbindungen: IV. Über den Zustand wässriger Alkalihydrogen- und Alkalipyroselenitlösungen, *Z. Anorg. Allg. Chem.*, **303**, (1960), 46-52, in German. Cited on page: 141.

[60TOB] Tobisawa, S., Magnetic susceptibilities of sulfur, selenium and tellurium from room temperature to 1100°C, *Bull. Chem. Soc. Jpn.*, **33**, (1960), 889-893. Cited on page: 97.

[60VOR] Vorma, A., Laitakarite, a new Bi-Se-mineral, *Bull. Comm. Geol. Finl.*, **188**, (1960), 1-10. Cited on page: 635.

[61GAD/SHA] Gadzhiev, S. N., Sharifov, K. A., Thermochemical study of the chalcogenides of gallium, *Vopr. Metall. Fiz. Poluprovodn., Akad. Nauk SSSR, Tr. 4-go[Chetvertogo] Soveshch.*, **4**, (1961), 43-45, in Russian. Cited on page: 236.

[61HAM] Hamada, S., Decomposition equilibrium of selenocyanate ion, *Nippon Kagaku Kaishi (J. Chem. Soc. Japan)*, **82**, (1961), 1327-1330, in Japanese, English translation in [62HAM]. Cited on pages: 207, 459.

[61JOH/GIL] Johnson, W. H., Gilliland, A. A., Heat of formation of beryllium chloride, *J. Res. Natl. Bur. Stand.*, **65A**, (1961), 59-61. Cited on page: 464.

[61KEL/KIN] Kelley, K. K., King, E. G., Contributions to the data on theoretical metallurgy: XIV. Entropies of the elements and inorganic compounds, *Bull. - U. S., Bur. Mines*, **592**, (1961). Cited on pages: 92, 221.

[61LEW/RAN] Lewis, G. N., Randall, M., *Thermodynamics, as revised by Pitzer, K. S. and Brewer, L.*, 2nd. Edition, McGraw-Hill, New York, (1961), 723 pp.. Cited on pages: 22, 23, 588, 590, 593, 596.

[61LIN/PAN] Lin, J.-L., Pan, K., A study of the silver-silver selenite electrode, *J. Chin. Chem. Soc. (Taipei, Taiwan)*, **8**, (1961), 14-22. Cited on pages: 303, 460.

[61MCB/GOR] McBride, B. J., Gordon, S., Thermodynamic functions of several triatomic molecules in the ideal gas state, *J. Chem. Phys.*, **35**, (1961), 2198-2206. Cited on page : 434.

[61NES/PAS] Nesterova, Y. M., Pashinkin, A. S., Novoselova, A. V., Determination of the saturated vapour pressure of solid tin selenide and telluride, *Russ. J. Inorg. Chem.*, **6**, (1961), 1031-1033. Cited on pages: 216, 462, 476.

[61ROS/ROS] Rossotti, F. J. C., Rossotti, H., *The determination of stability constants and other equilibrium constants in solution*, McGraw-Hill, New York, (1961). Cited on pages : 22, 615.

[61SEL/PAK] Selivanova, N. M., Pakhorukov, N. I., On the heats of solution of selenious acid, *Izv. Vyssh. Uchebn. Zaved., Khim. Khim. Tekhnol.*, **4**, (1961), 355-358, in Russian. Cited on pages: 130, 136, 137, 143, 437, 438, 448, 462.

[61SEL/SHN] Selivanova, N. M., Shneider, V. A., Ryabova, R. I., Heat of formation of $MgSeO_4$ from its elements, *Zh. Neorg. Khim.*, **6**, (1961), 27-33, in Russian, English translation in [61SEL/SHN2]. Cited on pages: 398, 463.

[61SEL/SHN2] Selivanova, N. M., Shneider, V. A., Ryabova, R. I., Heat of formation of magnesium selenate from its elements, *Russ. J. Inorg. Chem.*, **6**, (1961), 13-16. From a citation in this Bibliography

[61SEL/SHN3] Selivanova, N. M., Shneider, V. A., Physicochemical properties of selenates: X. Heat of formation of beryllium selenate, *Zh. Fiz. Khim.*, **35**, (1961), 574-579, in Russian, English translation in [61SEL/SHN4]. Cited on pages: 393, 463, 648.

[61SEL/SHN4] Selivanova, N. M., Shneider, V. A., Physicochemical properties of selenates: X. Heat of formation of beryllium selenate, *Russ. J. Phys. Chem.*, **35**, (1961), 280-282. From a citation in this Bibliography

[61SOM] Somorjai, G. A., Vapor pressure and solid-vapor equilibrium of CdSe (Cadmium Selenide), *J. Phys. Chem.*, **65**, (1961), 1059-1061. Cited on pages: 269, 270, 464.

[61WOS] Wösten, W. J., The vapor pressure of cadmium selenide, *J. Phys. Chem.*, **65**, (1961), 1949-1951. Cited on pages: 269, 270, 465.

[62ALE/SER] Aleksandrovich, A. M., Serebrennikov, V. V., Solubility and solubility product of selenites and sulphites of light rare earth elements, *Tr. Tomsk. Gos. Univ.*, **154**, (1962), 105-109, in Russian. Cited on pages: 370, 374, 377, 380, 383, 465.

[62BAE/BUS] Baer, Y., Busch, G., Fröhlich, C., Steigmeier, E., Wärmeleitfähigkeit, elektrische Leitfähigkeit, Hall-effekt, Thermospannung und spezifische Wärme von Ag_2Se, *Z. Naturforsch.*, **17a**, (1962), 886-889, in German. Cited on page: 299.

[62CHU/PAS] Ch'ün-hua, L., Pashinkin, A. S., Novoselova, A. V., Determination of the vapour pressures of solid germanium selenide and telluride, *Russ. J. Inorg. Chem.*, **7**, (1962), 496-498. Cited on page: 210.

[62DAG] Dagron, C., Sur la préparation et quelques propriétés des tétrafluorures de sélénium et de tellure et de l'oxyfluorure $SeOF_2$, *Compt. Rend.*, **255**, (1962), 122-124, in French. Cited on pages: 161, 165.

[62DAV] Davies, C. W., *Ion association*, Butterworths, London, (1962), 190 pp.. Cited on page: 588.

[62GAT] Gattow, G., Die Bildungsenthalpie von SeO_2 und die Umwandlungswärmen der Selen-Modifikationen, *Z. Anorg. Allg. Chem.*, **317**, (1962), 245-260, in German. Cited on pages: 120, 121.

[62GEN/NES] Genov, L. K., Nesmeyanov, A. N., Priselkov, Y. A., The rate of evaporation of thallium selenide, *Vestn. Mosk. Univ. Ser. 2: Khim.*, **17**, (1962), 34-35, in Russian. Cited on page: 247.

[62GOL/AND] Golub, A. M., Andreichenko, O. E., Selenocyanato-complexes of cadmium, *Zh. Neorg. Khim.*, **7**, (1962), 549-555, in Russian, English translation in [62GOL/AND2]. Cited on pages: 276, 277, 278.

[62GOL/AND2] Golub, A. M., Andreichenko, O. E., Selenocyanato-complexes of cadmium, *Russ. J. Inorg. Chem.*, 7, (1962), 279-283. From a citation in this Bibliography

[62GOL/SKO] Golub, A. M., Skopenko, V. V., Selenocyanatocobalt complexes, *Zh. Neorg. Khim.*, **7**, (1962), 1012-1021, in Russian, English translation in [62GOL/SKO2]. Cited on page: 322.

[62GOL/SKO2] Golub, A. M., Skopenko, V. V., Selenocyanatocobalt complexes, *Russ. J. Inorg. Chem.*, **7**, (1962), 520-525. From a citation in this Bibliography

[62GRO/WES] Grønvold, F., Westrum, Jr., E. F., Heat capacities and thermodynamic functions of iron disulphide (pyrite), iron diselenide, and nickel diselenide from 5 to 350°K. The estimation of standard entropies of transition metal chalcogenides, *Inorg. Chem.*, **1**, (1962), 36-48. Cited on pages: 311, 328.

[62HAM] Hamada, S., Decomposition equilibrium of selenocyanate ion, *Chem. Abstr.*, **57**, (1962). From a citation in this Bibliography

[62IZA/CHR] Izatt, R. M., Christensen, J. J., Pack, R. T., Bench, R., Thermodynamics of metal-cyanide coordination. I. pK, ΔH^{o}, and ΔS^{o} values as a function of temperature for hydrocyanic acid dissociation in aqueous solution, *Inorg. Chem.*, **1**, (1962), 828-831. Cited on page: 435.

[62JER/MEN] Jerschkewitz, H. G., Menning, K., Über die thermische Zersetzung des Selen(VI)-oxids und die Verbindung Se_2O_5, *Z. Anorg. Allg. Chem.*, **319**, (1962), 82-93, in German. Cited on pages: 126, 127, 130, 131, 472.

[62KUD/UST] Kudryavtsev, A. A., Ustyugov, G. P., Determination of the pressure of the saturated vapors of selenium and tellurium, *Tr. Inst.-Mosk. Khim.-Tekhnol. Inst. im. D. I. Mendeleeva*, **38**, (1962), 42-46, in Russian. Cited on pages: 99, 465.

[62MAR/GET] Margulis, E. V., Getskin, L. S., Mil'skaya, N. S., Vapour pressure of SeO_2, *Zh. Neorg. Khim.*, **7**, (1962), 729-731, in Russian, English translation in [62MAR/GET2]. Cited on pages: 123, 124, 313, 320, 396, 530.

[62MAR/GET2] Margulis, E. V., Getskin, L. S., Mil'skaya, N. S., Vapour pressure of SeO_2, *Russ. J. Inorg. Chem.*, 7, (1962), 375-377. From a citation in this Bibliography

[62PAP] Papousek, D., Thermodynamische Funktionen des anharmonischen Oszillators und schwingenden Rotators II. Rotationsverteilungs-funktion zweiatomiger Moleküle in Multiplett-Sigma-Elektro-nenzuständen, *Collect. Czech. Chem. Commun.*, **27**, (1962), 1-8, in German. Cited on page: 119.

[62RAT/RIC] Ratchford, R. J., Rickert, H., Eine electrochemische Knudsenzelle zur Untersuchung des Thermodynamik von Selendampf und über die Zersetzung von Ag_2Se, *Z. Elektrochem.*, **66**, (1962), 497-502, in German. Cited on pages: 95, 97.

[62RUM/ZHI] Rumyantsev, Y. V., Zhiteneva, G. M., Bolondz, F. M., Thermal stabilities of selenides and tellurides of iron and nickel, *Tr. Vost.-Sib. Fil., Akad. Nauk SSSR*, **41**, (1962), 114-120, in Russian. Cited on page : 326.

[62SEL/LES] Selivanova, N. M., Leshchinskaya, Z. L., Klushina, T. V., Physicochemical properties of selenites. I. Thermodynamic properties of silver selenite, *Zh. Fiz. Khim.*, **36**, (1962), 1349-1352, in Russian, English translation in [62SEL/LES2]. Cited on pages : 303, 304, 305, 466.

[62SEL/LES2] Selivanova, N. M., Leshchinskaya, Z. L., Klushina, T. V., Physicochemical properties of selenites. I. Thermodynamic properties of silver selenite, *Russ. J. Phys. Chem.*, **36**, (1962), 719-722. From a citation in this Bibliography

[62SEL/LES3] Selivanova, N. M., Leshchinskaya, Z. L., Physical-chemical properties of selenites. Thermodynamic properties of lead selenite, *Tr. Inst.-Mosk. Khim.-Tekhnol. Inst. Im. D. I. Mendeleeva*, **38**, (1962), 37-42, in Russian. Cited on pages: 223, 466.

[62SEL/SAM] Selivanova, N. M., Samplavskaya, K. K., Maier, A. I., Thermochemical properties of selenates. Heat of formation of copper selenates, *Tr. Inst.-Mosk. Khim.-Tekhnol. Inst. im. D. I. Mendeleeva*, **38**, (1962), 30-36, in Russian. Cited on pages: 296, 467.

[62SEL/SAZ3] Selivanova, N. M., Sazykina, T. A., Heat of formation of sodium selenate decahydrate, *Zh. Neorg. Khim.*, 7, (1962), 537-541, in Russian, English translation in [62SEL/SAZ4]. Cited on pages: 420, 421, 527.

[62SEL/SAZ4] Selivanova, N. M., Sazykina, T. A., Heat of formation of sodium selenate decahydrate, *Russ. J. Inorg. Chem.*, 7, (1962), 273-275. From a citation in this Bibliography

[62SEL/SAZ5] Selivanova, N. M., Sazykina, T. A., Physicochemical properties of selenates. Heat of formation of caesium selenate, *Tr. Inst.-Mosk. Khim.-Tekhnol. Inst. im. D. I. Mendeleeva*, **38**, (1962), 26-29, in Russian. Cited on pages: 430, 468, 470.

[62SEL/SHN] Selivanova, N. M., Shneider, V. A., Sazykina, T. A., Physicochemical properties of selenates. XII. Heat of formation from the elements of lithium selenate, *Izv. Vyssh. Uchebn. Zaved., Khim. Khim. Tekhnol.*, **5**, (1962), 183-187, in Russian. Cited on pages: 415, 469.

[62SEL/ZUB] Selivanova, N. M., Zubova, G. A., Abramov, I. I., Kalinina, A. V., Sazykina, T. A., Physico-chemical properties of selenates. Properties of potassium selenate, *Tr. Inst.-Mosk. Khim.-Tekhnol. Inst. im. D. I. Mendeleeva*, **38**, (1962), 21-25, in Russian. Cited on page: 426.

[62THO/SIN] Thompson, C. J., Sinke, G. C., Stull, D. R., Heat of formation of beryllium chloride, *J. Chem. Eng. Data*, 7, (1962), 380-381. Cited on page: 464.

[62VAS] Vasil'ev, V. P., Influence of ionic strength on the instability constants of complexes, *Russ. J. Inorg. Chem.*, 7, (1962), 924-927. Cited on page: 589.

[62WAL/ART] Walsh, P. N., Art, E. W., White, D., The heat capacity of the silver chalcogenides, $Ag_{1.99}S$, $Ag_{1.99}Se$, and $Ag_{1.88}Te$ from 16 to 300° K, *J. Phys. Chem.*, **66**, (1962), 1546-1549. Cited on page: 299.

[62WAL] Walrafen, G. E., Raman spectral studies of molten selenious acid, *J. Chem. Phys.*, **37**, (1962), 1468-1472. Cited on page: 140.

[62WAL2] Walrafen, G. E., Raman spectral studies of aqueous solutions of selenious acid, *J. Chem. Phys.*, **36**, (1962), 90-93. Cited on page: 140.

[62WES/GRO] Westrum, Jr., E. F., Grønvold, F., Chemical thermodynamics of the actinide element chalcogenides, Thermodynamics of nuclear materials, Proc. Int. Symp., held in Vienna, Austria,21-25 May 1962, pp.3-37, International Atomic Energy Agency, Vienna, Austria, (1962). Cited on pages: 388, 389.

[62WOS/GEE] Wösten, W. J., Geers, M. G., The vapour pressure of zinc selenide, *J. Phys. Chem.*, **66**, (1962), 1252-1253. Cited on pages: 257, 258, 469.

[63DEV/UMI] Devyatykh, G. G., Umilin, V. A., Runovskaya, L. V., Liquid-vapour equilibrium in the sulphur-selenium system at elevated pressure, *Zh. Neorg. Khim.*, **8**, (1963), 149-152, in Russian, English translation in [63DEV/UMI2]. Cited on page: 180.

[63DEV/UMI2] Devyatykh, G. G., Umilin, V. A., Runovskaya, L. V., Liquid-vapour equilibrium in the sulphur-selenium system at elevated pressure, *Russ. J. Inorg. Chem.*, **8**, (1963), 76-78. From a citation in this Bibliography

[63GOL/JEU] Goldfinger, P., Jeunehomme, M., Mass spectrometric and Knudsen-Cell vaporization studies of group 2B-6B compounds, *Trans. Faraday Soc.*, **59**, (1963), 2851-2867. Cited on pages: 117, 118, 279.

[63HIR/ICH] Hirayama, C., Ichikawa, Y., DeRoo, A. M., Vapor pressures of tin selenide and tin telluride, *J. Phys. Chem.*, **67**, (1963), 1039-1042. Cited on pages: 216, 470, 476.

[63LES/SEL] Leshchinskaya, Z. L., Selivanova, N. M., Strel'tsov, I. S., Heat of formation of barium selenite, *Zh. Neorg. Khim.*, **8**, (1963), 763-764, in Russian, English translation in [63LES/SEL2]. Cited on pages: 407, 470.

[63LES/SEL2] Leshchinskaya, Z. L., Selivanova, N. M., Strel'tsov, I. S., Heat of formation of barium selenite, *Russ. J. Inorg. Chem.*, **8**, (1963), 387-388. From a citation in this Bibliography

[63LES/SEL3] Leshchinskaya, Z. L., Selivanova, N. M., Thermodynamic properties of calcium selenite, *Tr. Inst.-Mosk. Khim.-Tekhnol. Inst. im. D. I. Mendeleeva*, **44**, (1963), 37-40, in Russian. Cited on pages: 400, 471.

[63LES/SEL4] Leshchinskaya, Z. L., Selivanova, N. M., Maier, A. I., Strel'tsov, I. S., Muzalev, E. Y., Heat of formation of nickel and cobalt selenites, *Zh. Vses. Khim. O-va. im. D. I. Mendeleeva*, **8**, (1963), 577-578, in Russian. Cited on pages: 312, 320, 471.

[63LES/SEL5] Leshchinskaya, Z. L., Selivanova, N. M., Thermodynamic properties of cadmium selenite, *Tr. Inst.-Mosk. Khim.-Tekhnol. Inst. im. D. I. Mendeleeva*, **41**, (1963), 18-21, in Russian. Cited on page: 273.

[63LES/SEL6] Leshchinskaya, Z. L., Selivanova, N. M., On the heat of transformation of non equilibrium (amorphous) cadmium selenite to the equilibrium form, *Tr. Inst.-Mosk. Khim.-Tekhnol. Inst. im. D. I. Mendeleeva*, **41**, (1963), 22-25, in Russian. Cited on pages: 273, 474.

[63MIJ/GER] Mijlhoff, F. C., Gerding, H., The binary system sulphur trioxide-selenium trioxide, *Recl. Trav. Chim. Pays-Bas*, **82**, (1963), 807-813. Cited on page: 127.

[63MIJ] Mijlhoff, F. C., On the heat of formation of selenium trioxide, *Recl. Trav. Chim. Pays-Bas*, **82**, (1963), 822. Cited on pages: 127, 472.

[63NAG2] Nagarajan, G., Mean amplitudes of vibration and thermodynamic functions of the dioxides of chlorine, sulphur and selenium, *Bull. Soc. Chim. Belg.*, **72**, (1963), 524-536. Cited on page: 122.

[63NES] Nesmeyanov, A. N., *Vapor pressure of the chemical elements*, Elsevier Publishing Company, Amsterdam/London/New York, (1963). Cited on page: 98.

[63RUS/MAR] Rustamov, P. G., Mardakhev, B. N., Investigation of the solubility in the system SeO_2-Na_2O-H_2O at 25°C, *Azerb. Khim. Zh.*, **4**, (1963), 131-139, in Russian. Cited on page: 139.

[63SAZ/SEL] Sazykina, T. A., Selivanova, N. M., On the heat of formation of potassium selenate and some properties of selenates and sulphates of alkali metals, *Tr. Inst.-Mosk. Khim.-Tekhnol. Inst. Im. D. I. Mendeleeva*, **41**, (1963), 14-17, in Russian. Cited on pages : 426, 472.

[63SEL/LES] Selivanova, N. M., Leshchinskaya, Z. L., Thermodynamic properties of strontium selenite, *Zh. Neorg. Khim.*, **8**, (1963), 563-566, in Russian, English translation in [63SEL/LES2]. Cited on pages : 404, 405, 470, 471, 473, 477.

[63SEL/LES2] Selivanova, N. M., Leshchinskaya, Z. L., Thermodynamic properties of strontium selenite, *Russ. J. Inorg. Chem.*, **8**, (1963), 286-288. From a citation in this Bibliography

[63SEL/LES3] Selivanova, N. M., Leshchinskaya, Z. L., Maier, A. I., Strel'tsov, I. S., Muzalev, E. Y., Thermodynamic properties of nickel selenite dihydrate, *Zh. Fiz. Khim.*, **37**, (1963), 1563-1567, in Russian, English translation in [63SEL/LES4]. Cited on page : 312.

[63SEL/LES4] Selivanova, N. M., Leshchinskaya, Z. L., Maier, A. I., Strel'tsov, I. S., Muzalev, E. Y., Thermodynamic properties of nickel selenite dihydrate, *Russ. J. Phys. Chem.*, **37**, (1963), 837-839. From a citation in this Bibliography

[63SEL/LES5] Selivanova, N. M., Leshchinskaya, Z. L., Strel'tsov, I. S., Heat of formation of cadmium selenite, *Zh. Fiz. Khim.*, **37**, (1963), 668-670, in Russian, English translation in [63SEL/LES6]. Cited on pages : 273, 472, 473.

[63SEL/LES6] Selivanova, N. M., Leshchinskaya, Z. L., Strel'tsov, I. S., Heat of formation of cadmium selenite, *Russ. J. Phys. Chem.*, **37**, (1963), 346-347. From a citation in this Bibliography

[63SEL/MAI] Selivanova, N. M., Maier, A. I., Luk'yanova, T. A., Physicochemical properties of selenates: XIX. Heat of formation of cadmium selenate, *Zh. Fiz. Khim.*, **37**, (1963), 1588-1592, in Russian, English translation in [63SEL/MAI2]. Cited on pages: 276, 474.

[63SEL/MAI2] Selivanova, N. M., Maier, A. I., Luk'yanova, T. A., Physicochemical properties of selenates: XIX. Heat of formation of cadmium selenate, *Russ. J. Phys. Chem.*, **37**, (1963), 851-853. From a citation in this Bibliography

[63SEL/MAI3] Selivanova, N. M., Maier, A. I., Luk'yanova, T. A., Heat of formation of zinc selenate, *Zh. Neorg. Khim.*, **8**, (1963), 2428-2433, in Russian, English translation in [63SEL/MAI4]. Cited on pages: 265, 475.

[63SEL/MAI4] Selivanova, N. M., Maier, A. I., Luk'yanova, T. A., Heat of formation of zinc selenate, *Russ. J. Inorg. Chem.*, **8**, (1963), 1273-1276. From a citation in this Bibliography

[63SEL/SAM] Selivanova, N. M., Samplavskaya, K. K., Strel'tsov, I. S., Mazepova, V. I., Thermal decomposition of aluminium selenate, *Zh. Neorg. Khim.*, **8**, (1963), 1645-1653, in Russian, English translation in [63SEL/SAM2]. Cited on page: 231.

[63SEL/SAM2] Selivanova, N. M., Samplavskaya, K. K., Strel'tsov, I. S., Mazepova, V. I., Thermal decomposition of aluminium selenate, *Russ. J. Inorg. Chem.*, **8**, (1963), 857-861. From a citation in this Bibliography

[63SEL/SAZ] Selivanova, N. M., Sazykina, T. A., Physicochemical properties of selenates. XXI. Heat of formation of rubidium selenate, *Izv. Vyssh. Uchebn. Zaved., Khim. Khim. Tekhnol.*, **6**, (1963), 531-533, in Russian. Cited on pages: 428, 476.

[63SHA/KUL] Shakhtakhtinskii, M. G., Kuliev, A. A., Adbullaev, G. B., Investigation of the saturated vapor pressure of some selenides with radioisotopes, *Vopr. Metall. Fiz. Poluprovodn., Akad. Nauk SSSR, Tr. 4-go[Chetvertogo] Soveshch.*, **1961**, (1963), 38-42, in Russian. Cited on page: 247.

[63WAL] Walrafen, G. E., Raman spectral studies of aqueous solutions of selenic acid, *J. Chem. Phys.*, **39**, (1963), 1479-1492. Cited on page: 146.

[64BAK/BUK] Bakeeva, S. S., Buketov, E. A., Bakeev, M. I., On the dissociation pressure of zinc selenite, *Tr. Inst. Metall. Obogashch. Akad. Nauk Kaz. SSR*, **11**, (1964), 163-167, in Russian. Cited on pages : 261, 262, 530.

[64BON/SHU] Bonilla, C. F., Shulman, G., Selenium: possible rankine-cycle fluid, *Nucleonics*, **22**, (1964), 58-60. Cited on pages: 99, 476.

[64BUK/PAS] Buketov, E. A., Pashinkin, A. S., Ugorets, M. Z., Muldagalieva, R. A., Sapozhnikov, R. A., The thermal stability of silver selenite, *Russ. J. Inorg. Chem.*, **9**, (1964), 1456-1458. Cited on page: 304.

[64COL/DRO] Colin, R., Drowart, J., Thermodynamic study of tin selenide and tin telluride using a mass spectrometer, *Trans. Faraday Soc.*, **60**, (1964), 673-683. Cited on pages: 214, 215, 216, 217, 218, 476, 554.

[64DRO] Drowart, J., Études thermochimiques par spectrométrie de masse, *Bull. Soc. Chim. Belg.*, **73**, (1964), 451-458, in French. Cited on pages: 204, 208.

[64GLE/MUL] Glemser, O., Müller, A., Stöcker, U., Dampfdrucke von SeO_2, CrO_3 und Re_2O_7, *Z. Anorg. Allg. Chem.*, **333**, (1964), 25-28, in German. Cited on pages: 123, 124.

[64GUN] Gunn, S. R., The heats of formation of H_2Se and H_2Te. Correlations between simple covalent hydrides, *J. Phys. Chem.*, **68**, (1964), 949-952. Cited on page: 132.

[64HOW/MIS] Howlett, B. W., Misra, S., Bever, M. B., On the thermodynamic properties of the compounds Sb_2Se_3, Bi_2Se_3, Sb_2Te_3, and Bi_2Te_3, *Trans. Metall. Soc. AIME*, **230**, (1964), 1367-1372. Cited on pages : 194, 202.

[64LES/SEL] Leshchinskaya, Z. L., Selivanova, N. M., Heat of formation of zinc selenite ($ZnSeO_3.H_2O$), *Zh. Fiz. Khim.*, **38**, (1964), 972-974, in Russian, English translation in [64LES/SEL2]. Cited on pages: 262, 263, 477.

[64LES/SEL2] Leshchinskaya, Z. L., Selivanova, N. M., Heat of formation of zinc selenite ($ZnSeO_3.H_2O$), *Russ. J. Phys. Chem.*, **38**, (1964), 527-528. From a citation in this Bibliography

[64MAR/SAP] Markovskii, L. Y., Sapozhnikov, Y. P., Boev, E. I., Bismuth selenites, *Zh. Neorg. Khim.*, **9**, (1964), 856-866, in Russian, English translation in [64MAR/SAP2]. Cited on page: 203.

[64MAR/SAP2] Markovskii, L. Y., Sapozhnikov, Y. P., Boev, E. I., Bismuth selenites, *Russ. J. Inorg. Chem.*, **9**, (1964), 472-477. From a citation in this Bibliography

[64MAS/WEI] Massen, C. H., Weijts, A. G. L. M., Poulis, J. A., Magnetic susceptibility of liquid and gaseous selenium, *Trans. Faraday Soc.*, **60**, (1964), 317-322. Cited on pages: 96, 97, 98, 478.

[64MIJ/BLO] Mijlhoff, F. C., Block, R., Vapour pressure of selenium trioxide, *Recl. Trav. Chim. Pays-Bas*, **83**, (1964), 799-809. Cited on pages: 128, 129, 478.

[64NAI] Nair, V. S. K., Dissociation of the biselenate ion, *J. Inorg. Nucl. Chem.*, **26**, (1964), 1911-1917. Cited on pages: 146, 147, 148, 149.

[64SEL/LES] Selivanova, N. M., Leshchinskaya, Z. L., Heat of transformation of the non-equilibrium (amorphous) form of nickel selenite into the equilibrium form, *Zh. Neorg. Khim.*, **9**, (1964), 259-263, in Russian, English translation in [64SEL/LES2]. Cited on pages: 313, 479.

[64SEL/LES2] Selivanova, N. M., Leshchinskaya, Z. L., Heat of transformation of the non-equilibrium (amorphous) form of nickel selenite into the equilibrium form, *Russ. J. Inorg. Chem.*, **9**, (1964), 143-145. From a citation in this Bibliography

[64SEL/SAZ] Selivanova, N. M., Sazykina, T. A., Heat of formation of anhydrous crystalline sodium selenate, *Zh. Prikl. Khim. (S.-Peterburg)*, **37**, (1964), 514-517, in Russian. Cited on pages: 420, 421, 480.

[64SIL/MAR] Sillén, L. G., Martell, A. E., *Stability constants of metal ion complexes*, Special Publication, No. 17, Chemical Society, London, (1964), 754 pp.. Cited on page: 14.

[64TER/RAT] Terpilowski, J., Ratajczak, E., Thermodynamic properties of CdSe and CdTe, *Bull. Acad. Pol. Sci., Ser. Sci. Chim.*, **12**, (1964), 355-358. Cited on pages: 269, 270, 480.

[64VUO/HUH] Vuorelainen, Y., Huhma, A., Häkli, A., Sederholmite, wilkmanite, kullerudite, mäkinenite and trüstedtite, five new nickel selenide minerals, *Bull. Comm. Geol. Finl.*, **215**, (1964), 113-125. Cited on pages: 635, 636.

[64WEL/SAL] Wells, C. F., Salam, M. A., A kinetic approach to the nature of ferrous ions in aqueous solution, *Nature (London)*, **203**, (1964), 751-752. Cited on page: 506.

[65AND/KAR] Andreeva, L. L., Karapet'yants, M. K., Heat of formation of bismuth oxide selenite, *Zh. Fiz. Khim.*, **39**, (1965), 2410-2412, in Russian, English translation in [65AND/KAR2]. Cited on pages: 202, 480, 485.

[65AND/KAR2] Andreeva, L. L., Karapet'yants, M. K., Heat of formation of bismuth oxide selenite, *Russ. J. Phys. Chem.*, **39**, (1965), 1287-1288. From a citation in this Bibliography

[65AND/KUD] Andreeva, L. L., Kudryavtsev, A. A., Heat of formation of bismuth triselenide (Bi_2Se_3), *Tr. Inst.-Mosk. Khim.-Tekhnol. Inst. im. D. I. Mendeleeva*, **49**, (1965), 25-27, in Russian. Cited on page: 202.

[65BAR/COR] Barker, C. K., Cordes, A. W., Margrave, J. L., The heats of formation of solid tetrasulfur tetranitride and tetraselenium tetranitride, *J. Phys. Chem.*, **69**, (1965), 334-335. Cited on page: 186.

[65BRE] Brebrick, R. F., Partial pressure of Se_2 and optical density of selenium vapor in the visible and ultraviolet, *J. Chem. Phys.*, **43**, (1965), 3031-3036. Cited on pages: 280, 281, 482.

[65CES/BAC] Cesbron, F., Bachet, B., Oosterbosch, R., La demesmaekerite, sélénite hydraté d'uranium, cuivre et plomb, *Bull. Soc. Fr. Minéral. Cristallogr.*, **88**, (1965), 422-425, in French. Cited on page: 636.

[65CLE/WES] Clever, H. L., Westrum, Jr., E. F., Cordes, A. W., Heat capacities and thermodynamic properties of globular molecules. XIV. Tetraphosphorus trisulfide, tetraphosphorus triselenide, and tetraphosphorus decasulfide between 5 and 350°K, *J. Phys. Chem.*, **69**, (1965), 1214-1219. Cited on page: 187.

[65COV/DOB] Covington, A. K., Dobson, J. V., The dissociation constant of the biselenate ion at 25°C, *J. Inorg. Nucl. Chem.*, **27**, (1965), 1435-1436. Cited on page: 148.

[65DEV/UMI] Devyatykh, G. G., Umilin, V. A., Agliulov, N. K., Kutsepin, V. F., Liquid-vapour equilibrium in the sulfur-selenium system, *Russ. J. Inorg. Chem.*, **10**, (1965), 898-901. Cited on page: 180.

[65DON/OET] McDonald, R. A., Oetting, F. L., The thermodynamic properties and allotropy of beryllium chloride between 13 and 715 K, *J. Phys. Chem.*, **69**, (1965), 3839-3845. Cited on page: 464.

[65EFE/KAR] Efendiev, G. K., Karaev, Z. S., Synthesis and study of rare earth sulfo selenogallates, *Khal'kogenidy*, **1**, (1965), 133-140, in Russian. Cited on page: 236.

[65ELL] Elliott, R. P., *Constitution of binary alloys, 1st supplement*, McGraw-Hill, New York, (1965), 877 pp.. Cited on pages: 353, 354.

[65GAD/DOB] Gadzhiev, S. M., Dobrotin, R. B., Suvorov, A. V., Thermodynamic properties of sulphur, selenium and tellurium oxides and hydrates, *Uch. Zap. Azerb. Gos. Univ. Ser. Khim. Nauk*, **2**, (1965), 81-89, in Russian. Cited on pages: 123, 129.

[65GOL/SKO] Golub, A. M., Skopenko, V. V., Metal selenocyanates and their properties, *Russ. Chem. Rev.*, **34**, (1965), 901-908. Cited on page: 207.

[65HAK/VUO] Häkli, T. A., Vuorelainen, Y., Sahama, T. G., Kitkaite (NiTeSe), a new mineral from Kuusamo, Northeast Finland, *Am. Mineral.*, **50**, (1965), 581-586. Cited on page: 635.

[65KEL/CRU] Kelemen, F., Cruceanu, E., Niculescu, D., Untersuchung einiger thermischer Eigenschaften der Verbindungen HgSe, HgTe und ZnTe, *Phys. Status Solidi*, **11**, (1965), 865-872, in German. Cited on pages: 279, 482.

[65LES/AVE] Leshchinskaya, Z. L., Averbukh, M. A., Selivanova, N. M., Solubility and thermodynamic properties of barium and strontium selenites, *Zh. Fiz. Khim.*, **39**, (1965), 2036-2038, in Russian, English translation in [65LES/AVE2]. Cited on pages: 404, 407.

[65LES/AVE2] Leshchinskaya, Z. L., Averbukh, M. A., Selivanova, N. M., Solubility and thermodynamic properties of barium and strontium selenites, *Russ. J. Phys. Chem.*, **39**, (1965), 1082-1083. From a citation in this Bibliography

[65LES/SEL] Leshchinskaya, Z. L., Selivanova, N. M., Thermodynamic properties of copper selenites, *Zh. Fiz. Khim.*, **39**, (1965), 2430-2434, in Russian, English translation in [65LES/SEL2]. Cited on pages: 294, 483.

[65LES/SEL2] Leshchinskaya, Z. L., Selivanova, N. M., Thermodynamic properties of copper selenites, *Russ. J. Phys. Chem.*, **39**, (1965), 1297-1299. From a citation in this Bibliography

[65MAN] Mandarino, J. A., Molybdomenite from the Ranwick uranium mine, Montreal River Area, Ontario, *Can. Mineral.*, **8**, (1965), 149-158. Cited on page: 636.

[65MAR/GOL] De Maria, G., Goldfinger, P., Malaspina, L., Piacente, V., Mass-spectrometric study of gaseous molecules ZnS, ZnSe and ZnTe, *Trans. Faraday Soc.*, **61**, (1965), 2146-2152. Cited on page: 260.

[65MIJ] Mijlhoff, F. C., The crystal structure of tetragonal selenium trioxide, *Acta Crystallogr., Sect. B: Struct. Sci.*, **18**, (1965), 795-798. Cited on page: 127.

[65MIJ2] Mijlhoff, F. C., An electron diffraction study of selenium trioxide, *Recl. Trav. Chim. Pays-Bas*, **84**, (1965), 74-80. Cited on page: 127.

[65PAE/AMO] Paetzold, R., Amoulong, H., Untersuchungen an Selen-Sauerstoff-Verbindungen. XXVIII. Die Schwingungsspektren von kristallisiertem, flüssigem und gasförmigem Selentrioxid, *Z. Anorg. Allg. Chem.*, **337**, (1965), 225-237, in German. Cited on page: 128.

[65PIE/TOU] Pierrot, R., Toussaint, J., Verbeek, T., La guilleminite, une nouvelle espèce minérale, *Bull. Soc. Fr. Minéral. Cristallogr.*, **88**, (1965), 132-135, in French. Cited on page: 636.

[65SAB/PER] Sabbah, R., Perinet, G., Préparation et stabilité thermique des sélénites de sodium obtenus par voie aqueuse, *J. Chim. Phys. Phys.-Chim. Biol.*, **62**, (1965), 929-936, in French. Cited on pages: 139, 417.

[65SAP/MAR] Sapozhnikov, Y. P., Markovskii, L. Y., Composition and some properties of mercury selenite, *Zh. Neorg. Khim.*, **10**, (1965), 1399-1402, in Russian, English translation in [65SAP/MAR2]. Cited on page: 285.

[65SAP/MAR2] Sapozhnikov, Y. P., Markovskii, L. Y., Composition and some properties of mercury selenite, *Russ. J. Inorg. Chem.*, **10**, (1965), 760-762. From a citation in this Bibliography

[65SIL/KAR] Silina, E. Y., Karapet'yants, M. K., Dissociation of mercury selenide vapour, *Russ. J. Phys. Chem.*, **39**, (1965), 1611-1614. Cited on pages: 280, 281, 483.

[65TAK/WES] Takahashi, Y., Westrum, Jr., E. F., Uranium monoselenide. Heat capacity and thermodynamic properties from 5 to 350°K, *J. Phys. Chem.*, **69**, (1965), 3618-3621. Cited on page: 385.

[66AND/KAR] Andreeva, L. L., Karapet'yants, M. K., Heats of formation of bismuth selenite and tellurite, *Zh. Fiz. Khim.*, **40**, (1966), 470-473, in Russian, English translation in [66AND/KAR2]. Cited on pages: 202, 484.

[66AND/KAR2] Andreeva, L. L., Karapet'yants, M. K., Heats of formation of bismuth selenite and tellurite, *Russ. J. Phys. Chem.*, **40**, (1966), 253-254. From a citation in this Bibliography

[66BAR/CHA] Barrow, R. F., Chandler, G. G., Meyer, C. B., The $B(^3\Sigma_u^-)$-$X(^3\Sigma_g^-)$ band system of the Se_2 molecule, *Philos. Trans. R. Soc. London*, **260**, (1966), 395-456. Cited on pages: 93, 94.

[66BER/CHU] Berkowitz, J., Chupka, W. A., Equilibrium composition of selenium vapor; the thermodynamics of the vaporization of HgSe, CdSe, and SrSe, *J. Chem. Phys.*, **45**, (1966), 4289-4302. Cited on pages: 95, 96, 97, 99, 100, 102, 105, 107, 108, 110, 111, 112, 114, 115, 116, 117, 279, 403, 404, 485, 486, 495.

[66BOU/KEL] Boughton, J. H., Keller, R. N., Dissociation constants of hydropseudohalic acids, *J. Inorg. Nucl. Chem.*, **28**, (1966), 2851-2859. Cited on page: 207.

[66FUJ/WES] Fujisaki, H., Westmore, J. B., Tickner, A. W., Mass spectrometric study of subliming selenium, *Can. J. Chem.*, **44**, (1966), 3063-3071. Cited on pages: 105, 107, 108, 110, 112, 114, 115, 117, 118, 487.

[66GAD] Gadzhiev, S. N., Enthalpy of formation of selenium dioxide, *Zh. Fiz. Khim.*, **40**, (1966), 241-243, in Russian, English translation in [66GAD2]. Cited on pages: 120, 121.

[66GAD2] Gadzhiev, S. N., Enthalpy of formation of selenium dioxide, *Russ. J. Phys. Chem.*, **40**, (1966), 128-129. From a citation in this Bibliography

[66GAD/DOB] Gadzhiev, S. M., Dobrotin, R. B., Suvorov, A. V., Thermodynamics of the vaporisation of selenium trioxide, *Zh. Neorg. Khim.*, **11**, (1966), 2155-2156, in Russian, English translation in [66GAD/DOB2]. Cited on page: 129.

[66GAD/DOB2] Gadzhiev, S. M., Dobrotin, R. B., Suvorov, A. V., Thermodynamics of the vaporisation of selenium trioxide, *Russ. J. Inorg. Chem.*, **11**, (1966), 1154-1155. From a citation in this Bibliography

[66GAT/DRA] Gattow, G., Dräger, M., Thermochemie des Selens. IV. Über die Bildungsenthalpie von SeO_2 und die Umwandlungswärmen der Selen-Modifikationen, *Z. Anorg. Allg. Chem.*, **343**, (1966), 55-57, in German. Cited on pages: 120, 121, 205.

[66GAT/DRA2] Gattow, G., Dräger, M., Über Chalkogenocarbonate. XVI. Das Kohlenstoffdiselenid. 1. Darstellung und physikalische Eigenschaften des Cse_2, *Z. Anorg. Allg. Chem.*, **343**, (1966), 11-24, in German. Cited on pages: 204, 205.

[66GAT/DRA3] Gattow, G., Dräger, M., Über Chalkogenocarbonate. XVII. Das Kohlenstoffdiselenid. 2. Thermochemie des Cse_2, *Z. Anorg. Allg. Chem.*, **343**, (1966), 232-243, in German. Cited on pages: 205, 206.

[66GLA/COR] Glatz, A. C., Cordo, K. E., High-temperature enthalpy studies of bismuth trisulfide and antimony triselenide, *J. Phys. Chem.*, **70**, (1966), 3757-3760. Cited on pages: 194, 201.

[66GOR/KRE] Gorbov, S. I., Krestovnikov, A. N., Thermodynamic properties of the gaseous chalcogenides of group V elements, *Russ. J. Phys. Chem.*, **40**, (1966), 505-507. Cited on pages: 197, 198.

[66GUG] Guggenheim, E. A., *Applications of Statistical Mechanics*, Clarendon Press, Oxford, (1966). Cited on pages: 589, 590.

[66HAJ] Hajiev, S. N., Über die Bildungsenthalpie von SeO_2, *Z. Anorg. Allg. Chem.*, **343**, (1966), 50-54, in German. Cited on pages: 120, 121.

[66KAL/YAR] Kalitin, V. I., Yarembash, E. I., Luzhnaya, N. P., Phase diagram of the system praseodymium-selenium, *Inorg. Mater.*, **2**, (1966), 1672-1679. Cited on page : 376.

[66LES/SEL] Leshchinskaya, Z. L., Selivanova, N. M., Thermodynamic properties of magnesium selenites, *Zh. Neorg. Khim.*, **11**, (1966), 260-265, in Russian, English translation in [66LES/SEL2]. Cited on pages: 395, 488.

[66LES/SEL2] Leshchinskaya, Z. L., Selivanova, N. M., Thermodynamic properties of magnesium selenites, *Russ. J. Inorg. Chem.*, **11**, (1966), 143-146. From a citation in this Bibliography

[66LES/SEL3] Leshchinskaya, Z. L., Selivanova, N. M., Thermodynamic properties of manganese(II) selenite, *Izv. Vyssh. Uchebn. Zaved., Khim. Khim. Tekhnol.*, **9**, (1966), 523-527, in Russian. Cited on pages: 341, 342, 488.

[66MOR/KOT] Morimoto, N., Koto, K., Crystal structure of umangite, Cu_3Se_2, *Science*, **152**, (1966), 345. Cited on page: 635.

[66NAG/MUL] Nagarajan, G., Müller, A., Thermodynamische Funktionen und molekulare Polarisierbarkeiten von Thionyl- und Seleninylhalogeniden, *Z. Naturforsch., B: Chem. Sci.*, **21**, (1966), 612-617, in German. Cited on pages: 165, 171.

[66OHA/KOZ] Ohashi, T., Kozuka, Z., Moriyama, J., Vapor pressure measurements on bismuth-selenium alloys, *Nippon Kinzoku Gakkaishi*, **30**, (1966), 785-788, in Japanese. Cited on page: 201.

[66OHA/SET] O'Hare, P. A. G., Settle, J. L., Hubbard, W. N., Fluorine bomb calorimetry. Part 14. Enthalpies of formation of the hexafluorides of sulphur, selenium, and tellurium and their thermodynamic properties to 1500°K, *Trans. Faraday Soc.*, **62**, (1966), 558-565. Cited on page: 164.

[66OSM/VAG] Osman-Zade, S. D., Vagramyan, A. T., Equilibrium potential of the selenium electrode in selenious acid solution, *Elektrokhimiya*, **2**, (1966), 85-87, in Russian. Cited on page: 151.

[66PAU/NOG] Paukov, I. E., Nogteva, V. V., Yarembash, E. I., The true heat capacity of rare-earth chalcogenides at low temperatures. I. The true low-temperature heat capacity, entropy, and enthalpy of lanthanum selenide, *Russ. J. Phys. Chem.*, **40**, (1966), 1661-1662. Cited on page: 383.

[66PYA/DUR] Pyatnitskii, I. V., Durdyev, M., Investigation of the solubility of cobalt selenite, *Ukr. Khim. Zh. (Russ. Ed.)*, **32**, (1966), 77-81, in Russian. Cited on pages: 319, 320, 489.

[66RAW/TOG] Rawling, J. R., Toguri, J. M., The thermodynamic properties of hydrogen selenide, *Can. J. Chem.*, **44**, (1966), 451-456. Cited on pages : 132, 490.

[66SAB/CAR] Sabbah, R., Carpeni, G., Etude électrométrique des systèmes H_2SeO_3-Na_2SeO_3 (ou K_2SeO_3) en milieux Cl^- (Na^+ ou K^+) = 3M à 25°C, *J. Chim. Phys. Phys.-Chim. Biol.*, **63**, (1966), 1549-1554, in French. Cited on pages: 140, 141.

[66SAB/PER] Sabbah, R., Perinet, G., Préparation, comportement thermique et identification aux rayons X des sélénites de potassium obtenus par voie aqueuse, *J. Chim. Phys. Phys.-Chim. Biol.*, **63**, (1966), 332-336, in French. Cited on page: 423.

[66SAD/SEM] Sadykov, K. B., Semenkovich, S. A., Study of the thermodynamic properties of lead selenide by the electromotive force method, *Izv. Akad. Nauk Turkm. SSR, Ser. Fiz.-Tekh. Khim. Geol. Nauk*, **3**, (1966), 25-28, in Russian. Cited on pages: 220, 490.

[66SAP/MAR] Sapozhnikov, Y. P., Markovskii, L. Y., Determination of the enthalpy of formation of different phases of zinc, cadmium and mercury selenites, *Tr. –Gos. Inst. Prikl. Khim.*, **53**, (1966), 168-174, in Russian. Cited on pages: 260, 261, 263, 272, 273, 274, 285, 286, 490.

[66SED/AGU] Sedgwick, T. O., Agule, B. J., Bourdon gauge determination of equilibrium in the ZnSe(s)-I_2(g) system, *J. Electrochem. Soc.*, **113**, (1966), 54-57. Cited on pages: 257, 258, 492.

[67BAR/STO] Barrow, R. F., Stobart, O. V., Vaughan, H., Absorption bands of gaseous BiS and BiSe, *Proc. Phys. Soc., London,*, **90**, (1967), 555-561. Cited on page: 198.

[67CHA/BEV] Chaudhari, P., Bever, M. B., A calorimetric investigation of the bismuth-rich region of the system bismuth-selenium, *Trans. Metall. Soc. AIME*, **239**, (1967), 501-504. Cited on page: 492.

[67COO/CUL] Cooper, R., Culka, J. V., Interchalcogen compounds-I. The sulfur-selenium system, *J. Inorg. Nucl. Chem.*, **29**, (1967), 1217-1224. Cited on page: 179.

[67GRE/ZAI] Greiver, T. N., Zaitseva, I. G., Interaction of selenium with sulfide solutions and of sulfur with selenide solutions, *Russ. J. Appl. Chem.*, **40**, (1967), 1623-1626. Cited on page: 181.

[67HUM/BON] Humffray, A. A., Bond, A. M., Forrest, J. S., Polarographic study of the selenocyanate complexes of zinc and cadmium, *J. Electroanal. Chem.*, **15**, (1967), 67-71. Cited on pages: 266, 276, 277.

[67KAR/PAS] Karakhanova, M. I., Pashinkin, A. S., Novoselova, A. V., Determining the dissociation pressure of solid tin diselenide, *Proc. Acad. Sci. USSR*, **3**, (1967), 1352-1355. Cited on page: 217.

[67KAW/MAN] Kawassiades, C. T., Manoussakis, G. E., Tossidis, J. A., Study of selenous acid dissociation in mixed aqueous-alcoholic solvents, *J. Inorg. Nucl. Chem.*, **29**, (1967), 401-405. Cited on pages: 141, 142.

[67KLU/SEL] Klushina, T. V., Selivanova, N. M., Poltavtseva, L. S., Heat of formation of lithium selenite monohydrate, *Zh. Neorg. Khim.*, **12**, (1967), 2546-2549, in Russian, English translation in [67KLU/SEL2]. Cited on pages: 413, 492.

[67KLU/SEL2] Klushina, T. V., Selivanova, N. M., Poltavtseva, L. S., Heat of formation of lithium selenite monohydrate, *Russ. J. Inorg. Chem.*, **12**, (1967), 1344-1346. From a citation in this Bibliography

[67KLU/SEL3] Klushina, T. V., Selivanova, N. M., Lapin, V. V., Behaviour of sodium selenite on heating, *Zh. Neorg. Khim.*, **12**, (1967), 2297-2304, in Russian, English translation in [67KLU/SEL4]. Cited on page: 532.

[67KLU/SEL4] Klushina, T. V., Selivanova, N. M., Lapin, V. V., Behaviour of sodium selenite on heating, *Russ. J. Inorg. Chem.*, **12**, (1967), 1213-1217. From a citation in this Bibliography

[67KOL/IVA] Kolosov, I. V., Ivanov-Emin, B. N., Korotaeva, L. G., Tetsu, K., Study of the complex formation of scandium with selenate and sulfate ions, *Radiokhimiya*, **9**, (1967), 473-478, in Russian, English translation in [67KOL/IVA2]. Cited on pages: 355, 493.

[67KOL/IVA2] Kolosov, I. V., Ivanov-Emin, B. N., Korotaeva, L. G., Tetsu, K., Study of the complex formation of scandium with selenate and sulfate ions, *Sov. Radiochem.*, **9**, (1967), 457-461. From a citation in this Bibliography

[67KRE/GOR] Krestovnikov, A. N., Gorbov, S. I., Vapour pressure of bismuth selenide, *Russ. J. Phys. Chem.*, **41**, (1967), 376-378. Cited on page: 201.

[67MAI/SUP] Maier, A. I., Suponitskii, Y. L., Karapet'yants, M. K., Heats of formation of cerium and neodynium selenites, *Izv. Akad. Nauk SSSR, Neorg. Mater.*, **3**, (1967), 1511-1513, in Russian, English translation in [67MAI/SUP2]. Cited on pages: 374, 380, 494, 552, 553.

[67MAI/SUP2] Maier, A. I., Suponitskii, Y. L., Karapet'yants, M. K., Heats of formation of cerium and neodynium selenites, *Inorg. Mater.*, **3**, (1967), 1320-1322. From a citation in this Bibliography

[67MAM/KER] Mamedov, K. K., Kerimov, I. G., Kostryukov, V. N., Guseinov, G. D., Heat capacity and entropy of indium monoselenide, *Russ. J. Phys. Chem.*, **41**, (1967), 691-692. Cited on page: 239.

[67MAM/KER2] Mamedov, K. K., Kerimov, I. G., Kostryukov, V. N., Mekhtiev, M. I., Heat capacity of gallium selenide and thallium selenide, *Fiz. Tekh. Poluprovodn. (S.-Peterburg)*, **1**, (1967), 441-442, in Russian. Cited on page: 233.

[67MEL/SEM] Melekh, B. T., Semenkovich, S. A., The specific heat of bismuth telluride and selenide over the temperature range 773-398 and at 298°K, *Inorg. Mater.*, **3**, (1967), 1727-1729. Cited on pages: 198, 199.

[67MUR/HEY] Murray, J. J., Heyding, R. D., Decomposition pressures and enthalpies of formation of some transition metal diarsenides and diselenides, *Can. J. Chem.*, **45**, (1967), 2675-2687. Cited on pages: 333, 334, 335, 336.

[67RAU] Rau, H., Das Bildungsgleichgewicht des Selenwasserstoffs, *Ber. Bunsen-Ges.*, **71**, (1967), 716-719, in German. Cited on pages: 132, 336, 542.

[67RAU2] Rau, H., Thermodynamik des Selendampfes, *Ber. Bunsen-Ges.*, **71**, (1967), 711-715, in German. Cited on page: 495.

[67SHA/AZI] Sharifov, K. A., Azizov, T. K., Standard enthalpies of formation of gallium and indium selenides (Ga_2Se_3 and In_2Se_3), *Russ. J. Phys. Chem.*, **41**, (1967), 645-646. Cited on pages: 236, 243, 244.

[67TER/RAT] Terpilowski, J., Ratajczak, E., Wlasnosci termodynamiczne selenku cynku, *Rocz. Chem.*, **41**, (1967), 429-433, in Polish. Cited on pages: 256, 257, 258, 495.

[67TIK/GOL] Tikhonov, V. V., Golubkov, A. V., Smirnov, I. A., Specific heat of NdS, LaSe, and LaTe, *Sov. Phys. Solid State*, **8**, (1967), 2860-2863. Cited on page: 381.

[67VAS/NIK] Vasil'ev, V. P., Nikol'skaya, A. V., Bachinskaya, A. G., Gerasimov, Y. I., Thermodynamic properties of thallium monoselenide, *Dokl. Akad. Nauk SSSR*, **176**, (1967), 1335-1338, in Russian. Cited on pages : 249, 250, 251.

[67ZAI/GRE] Zaitseva, I. G., Greiver, T. N., Sodium polyselenides, *Russ. J. Appl. Chem.*, **40**, (1967), 1862-1865. Cited on page: 154.

[68BAK] Baker, E. H., The vapour pressure and resistivity of selenium at high temperatures, *J. Chem. Soc. A*, (1968), 1089-1092. Cited on pages: 99, 495.

[68BAK/BUK] Bakeeva, S. S., Buketov, E. A., Pashinkin, A. S., Dissociation pressure of silver selenite, *Zh. Neorg. Khim.*, **13**, (1968), 32-35, in Russian, English translation in [68BAK/BUK2]. Cited on pages: 304, 305, 530.

[68BAK/BUK2] Bakeeva, S. S., Buketov, E. A., Pashinkin, A. S., Dissociation pressure of silver selenite, *Russ. J. Inorg. Chem.*, **13**, (1968), 16-17. From a citation in this Bibliography

[68BAK/BUK3] Bakeeva, S. S., Buketov, E. A., Pashinkin, A. S., Dissociation pressure of cadmium selenite, *Zh. Neorg. Khim.*, **13**, (1968), 1779-1781, in Russian, English translation in [68BAK/BUK4]. Cited on pages: 273, 530.

[68BAK/BUK4] Bakeeva, S. S., Buketov, E. A., Pashinkin, A. S., Dissociation pressure of cadmium selenite, *Russ. J. Inorg. Chem.*, **13**, (1968), 927-928. From a citation in this Bibliography

[68BER/CHU] Berkowitz, J., Chupka, W. A., Comment on the composition of selenium vapor, *J. Chem. Phys.*, **48**, (1968), 5743-5744. Cited on pages: 487, 495.

[68BON/PAS] Boncheva-Mladenova, Z., Pashinkin, A. S., Novoselova, A. V., Determination of the saturated vapor pressure of solid bismuth selenide, *Inorg. Mater.*, **4**, (1968), 904-907. Cited on pages: 200, 201, 202, 495.

[68BOW/BRO] Bowater, I. C., Brown, R. D., Burden, F. R., The microwave spectrum, dipole moment, and structure analysis of selenium tetrafluoride, *J. Mol. Spectrosc.*, **28**, (1968), 454-460. Cited on page: 162.

[68BOW/BRO2] Bowater, I. C., Brown, R. D., Burden, F. R., The microwave spectrum, dipole moment, and structure analysis of seleninyl fluoride, *J. Mol. Spectrosc.*, **28**, (1968), 461-470. Cited on page: 166.

[68BRE] Brebrick, R. F., Partial pressure of Se_2(g) in selenium vapor, *J. Chem. Phys.*, **48**, (1968), 5741-5743. Cited on page: 486.

[68BRO/WHA] Brown, A. J., Whalley, E., Poly(carbon diselenide), *Inorg. Chem.*, **7**, (1968), 1254-1255. Cited on page: 205.

[68COL/DRO] Colin, R., Drowart, J., Mass spectrometric determination of dissociation energies of gaseous indium sulfides, selenides, and tellurides, *Trans. Faraday Soc.*, **64**, (1968), 2611-2621. Cited on pages: 240, 241, 496.

[68DOB/SUV] Dobrotin, R. B., Suvorov, A. V., Gadzhiev, S. M., Thermodynamics of gaseous forms of selenium oxides and hydrates, *Probl. Sovrem. Khim. Koord. Soedin.*, **2**, (1968), 23-52, in Russian. Cited on pages: 123, 124, 129, 137, 138.

[68DYA/KHO] D'yachkova, I. B., Khodakovskiy, I. L., Vernadskiy, V. I., Thermodynamic equilibria in the systems S-H_2O, Se-H_2O, and Te-H_2O in the 25-300°C temperature range and their geochemical interpretations, *Geochem. Int.*, **11**, (1968), 1108-1125. Cited on page: 136.

[68FIC/HAS] Ficalora, P. J., Hastie, J. W., Margrave, J. L., Mass spectrometric studies at high temperatures. XXVII. The reactions of aluminum vapor with S_2(g), Se_2(g), and Te_2(g), *J. Phys. Chem.*, **72**, (1968), 1660-1663. Cited on pages: 228, 229, 230, 497.

[68GOR/KRE] Gorbov, S. I., Krestovnikov, A. N., Vapour pressure of diarsenic triselenide, *Russ. J. Inorg. Chem.*, **13**, (1968), 779-782. Cited on page: 189.

[68GRO] Grønvold, F., Heat capacities and thermodynamic properties of the iron selenides $Fe_{1.04}Se$, Fe_7Se_8, and Fe_3Se_4 from 298 to 1050°K, *Acta Chem. Scand.*, **22**, (1968), 1219-1240. Cited on pages: 325, 326, 327, 329.

[68JOS/KUB] Jost, W., Kubaschewski, P., Spezifische Wärmen von Silber- und Kupfer(I)-Chalkogeniden von -70°C bis zu 550°C, *Z. Phys. Chem. (Munich)*, **60**, (1968), 69-78, in German. Cited on page: 299.

[68KAR/ZLO] Karbanov, S. G., Zlomanov, V. P., Novoselova, A. V., Saturated vapor pressures of solid and liquid germanium monosulfide and monoselenide, *Inorg. Mater.*, **4**, (1968), 1634-1637. Cited on pages: 210, 212.

[68KLU/SEL] Klushina, T. V., Selivanova, N. M., Lapin, V. V., Novikova, A. A., Heat of formation of potassium selenite, *Zh. Neorg. Khim.*, **13**, (1968), 2917-2921, in Russian, English translation in [68KLU/SEL2]. Cited on pages: 423, 497.

[68KLU/SEL2] Klushina, T. V., Selivanova, N. M., Lapin, V. V., Novikova, A. A., Heat of formation of potassium selenite, *Russ. J. Inorg. Chem.*, **13**, (1968), 1502-1505. From a citation in this Bibliography

[68KLU/SEL3] Klushina, T. V., Selivanova, N. M., Artem'eva, L. N., Heat of formation of anhydrous lithium selenite, *Izv. Vyssh. Uchebn. Zaved., Khim. Khim. Tekhnol.*, **11**, (1968), 861-866, in Russian. Cited on page: 413.

[68KNO/BAN] Knox, B. E., Ban, V. S., Mass spectrometric studies of laser-induced vaporization. III. The arsenic-selenium system, *Mater. Res. Bull.*, **3**, (1968), 885-894. Cited on page: 189.

[68LUN/LEL] Lundkvist, M., Lellep, M., Conductance and equilibrium studies on the system Se-Cl. II. Equilibrium studies on the system Se-Cl, *Acta Chem. Scand.*, **22**, (1968), 291-298. Cited on pages : 167, 175.

[68MAI/SUP] Maier, A. I., Suponitskii, Y. L., Karapet'yants, M. K., Enthalpy of formation of scandium and yttrium selenites, *Izv. Akad. Nauk SSSR, Neorg. Mater.*, **4**, (1968), 921-924, in Russian, English translation in [68MAI/SUP2]. Cited on pages : 354, 356, 498, 522, 528.

[68MAI/SUP2] Maier, A. I., Suponitskii, Y. L., Karapet'yants, M. K., Enthalpy of formation of scandium and yttrium selenites, *Inorg. Mater.*, **4**, (1968), 808-811. From a citation in this Bibliography

[68MAL/PAS] Mal'tsev, A. K., Pashinkin, A. S., Bakeeva, S. S., Zhdanov, V. M., Low-temperature heat capacity and thermodynamic functions of selenium dioxide, *Zh. Fiz. Khim.*, **42**, (1968), 2615-2617, in Russian, English translation in [68MAL/PAS2]. Cited on page : 120.

[68MAL/PAS2] Mal'tsev, A. K., Pashinkin, A. S., Bakeeva, S. S., Zhdanov, V. M., Low-temperature heat capacity and thermodynamic functions of selenium dioxide, *Russ. J. Phys. Chem.*, **42**, (1968), 1381-1382. From a citation in this Bibliography

[68MAL/SEM] Malekh, B. T., Semenkovich, S. A., Thermodynamic properties of Bi(III) telluride and selenide, *Inorg. Mater.*, **4**, (1968), 1180-1182. Cited on pages: 199, 202, 517.

[68MEH] Mehra, M. C., Studies on the stabilities of some metal selenite, sulphide and selenide complexes in solution, Ph. D. Thesis, Laval University, Quebec, Canada, (1968). Cited on pages: 135, 282, 301, 302, 303, 304, 338, 341, 514, 529.

[68NOG/PAU] Nogteva, V. V., Paukov, I. E., Yarembash, E. I., The true heat capacity of rare-earth chalcogenides at low temperatures. II. The true heat capacity of neodymium selenide in the range 12.1-229°K and its standard entropy and enthalpy, *Russ. J. Phys. Chem.*, **42**, (1968), 121-122. Cited on page: 373.

[68OHA] O'Hare, P. A. G., The thermodynamic properties of some chalcogen fluorides, *Argonne Natl. Lab., [Rep.] ANL*, **7315**, (1968), 1-95. Cited on pages: 161, 162, 163.

[68RAT/TER] Ratajczak, E., Terpilowski, J., Thermodynamic properties of HgSe, *Rocz. Chem.*, **42**, (1968), 433-436, in Polish. Cited on pages: 280, 281, 500.

[68REZ/ISA] Reznyakov, A. A., Isakova, R. A., Thermal dissociation of mercury selenide vapours, *Russ. J. Inorg. Chem.*, **13**, (1968), 326-329. Cited on pages: 280, 281, 500.

[68RIP/VER] Ripan, R., Vericeanu, G., Conductimetric determination of the solubility product of some selenites, *Stud. Univ. Babes-Bolyai, Chem.*, **13**, (1968), 31-37, in Romanian. Cited on pages: 262, 272, 293, 312, 319, 393, 395, 400, 404, 407, 466, 501, 502, 579.

[68ROS/SEL] Roshchina, Z. V., Selivanova, N. M., Heat of formation of lithium hydrogen selenite $LiHSeO_3$, *Tr. Inst.-Mosk. Khim.-Tekhnol. Inst. im. D. I. Mendeleeva*, **58**, (1968), 7-10, in Russian. Cited on pages: 414, 502.

[68ROS/SEL2] Roshchina, Z. V., Selivanova, N. M., Synthesis and thermochemical study of lithium pyroselenite, *Tr. Inst.-Mosk. Khim.-Tekhnol. Inst. im. D. I. Mendeleeva*, **58**, (1968), 13-15, in Russian. Cited on pages: 414, 502.

[68SAV/TAN] Savchenko, G. S., Tananaev, I. V., Volodina, A. N., The formation, composition and some properties of aluminium, gallium and indium selenites, *Izv. Akad. Nauk SSSR, Neorg. Mater.*, **4**, (1968), 369-380, in Russian, English translation in [68SAV/TAN2]. Cited on pages: 237, 246.

[68SAV/TAN2] Savchenko, G. S., Tananaev, I. V., Volodina, A. N., The formation, composition and some properties of aluminium, gallium and indium selenites, *Inorg. Mater.*, **4**, (1968), 310-318. From a citation in this Bibliography

[68SEK/IWA] Sekine, T., Iwaki, H., Sakairi, M., Shimada, F., Inarida, M., Studies on the liquid-liquid partition systems: VI. Solvent extraction study of the dissociation of sulfurous, selenious, and tellurous acids in acid perchlorate media, *Bull. Chem. Soc. Jpn.*, **41**, (1968), 1-7. Cited on pages: 141, 142.

[68SEL/ROS] Selivanova, N. M., Roshchina, Z. V., Heat of formation of sodium hydrogen selenite $NaHSeO_3$, *Zh. Neorg. Khim.*, **13**, (1968), 630-634, in Russian, English translation in [68SEL/ROS2]. Cited on pages: 419, 438, 503.

[68SEL/ROS2] Selivanova, N. M., Roshchina, Z. V., Heat of formation of sodium hydrogen selenite $NaHSeO_3$, *Russ. J. Inorg. Chem.*, **13**, (1968), 329-331. From a citation in this Bibliography

[68TAN/GOR] Tananaev, I. V., Gorokhov, V. K., Selivanova, N. M., Bol'shakova, N. K., Heats of formation and thermal stability of neutral gallium selenates, *Izv. Akad. Nauk SSSR, Neorg. Mater.*, **4**, (1968), 2124-2128, in Russian, English translation in [68TAN/GOR2]. Cited on pages: 238, 503.

[68TAN/GOR2] Tananaev, I. V., Gorokhov, V. K., Selivanova, N. M., Bol'shakova, N. K., Heats of formation and thermal stability of neutral gallium selenides, *Inorg. Mater.*, **4**, (1968), 1847-1851. From a citation in this Bibliography

[68TAR/ZHD] Tarasov, V. V., Zhdanov, V. M., Mal'tsev, A. K., Low-temperature heat capacity, enthalpy, and entropy of germanium disulfide and germanium diselenide, *Russ. J. Phys. Chem.*, **42**, (1968), 685-687. Cited on page: 210.

[68TER/ZAL] Terpilowski, J., Zaleska, E., Gawel, W., Electrochemical investigations on thallium-selenium solid system, *Rocz. Chem.*, **42**, (1968), 1845-1848, in Polish. Cited on pages: 247, 250, 252.

[68UST/VIG] Ustyugov, G. P., Vigdorovich, E. N., Kudryavtsev, A. A., Molecular composition of vapor in tellurium-selenium system, *Inorg. Mater.*, **4**, (1968), 1566-1567. Cited on page: 184.

[68UY/MUE] Uy, O. M., Muenow, D. W., Ficalora, P. J., Margrave, J. L., Mass spectrometric studies at high temperatures. Part 30.-Vaporizaton of Ga_2S_3, Ga_2Se_3 and Ga_2Te_3, and stabilities of the gaseous gallium chalcogenides, *Trans. Faraday Soc.*, **64**, (1968), 2998-3005. Cited on pages: 233, 234, 235, 505.

[68VAS/SOM] Vasil'ev, V. P., Somov, A. P., Nikol'skaya, A. V., Gerasimov, Y. I., Thermodynamic properties of bismuth selenide investigated by the e.m.f. method, *Russ. J. Phys. Chem.*, **42**, (1968), 355-356. Cited on pages: 199, 202.

[68WEL/SAL] Wells, C. F., Salam, M. A., Complex formation between iron(II) and inorganic anions. Part II. The effect of oxyanions on the reaction of iron(II) with hydrogen peroxide, *J. Chem. Soc. A*, (1968), 308-315. Cited on pages: 332, 505.

[68WIE/GOY] Wiedemeier, H., Goyette, W. J., Evaluation of Knudsen measurements of the sublimation of manganese(II) selenide, *J. Chem. Phys.*, **48**, (1968), 2936-2941. Cited on pages: 337, 338, 507, 551.

[68YAM/POR] Yamdagni, R., Porter, R. F., Mass spectrometric and torsion effusion studies of the evaporation of liquid selenium, *J. Electrochem. Soc.*, **115**, (1968), 601-604. Cited on pages : 94, 95, 96, 97, 98, 105, 107, 108, 110, 112, 114, 115, 117, 118, 508, 543.

[68ZHD/MAL] Zhdanov, V. M., Mal'tsev, A. K., Low-temperature heat capacity, enthalpy, and entropy of crystalline and vitreous arsenic sesquiselenide and of crystalline arsenic sesquitelluride, *Russ. J. Phys. Chem.*, **42**, (1968), 1078-1080. Cited on pages: 190, 191.

[69BAK/BUK] Bakeeva, S. S., Buketov, E. A., Pashinkin, A. S., Dissociation pressure of nickel and magnesium selenites, *Tr. Khim.-Metall. Inst. Akad. Nauk Kaz. SSR*, **9**, (1969), 16-23, in Russian. Cited on pages: 313, 396, 530.

[69BAK/BUK2] Bakeeva, S. S., Buketov, E. A., Pashinkin, A. S., Dissociation pressure of cobalt selenite, *Tr. Khim.-Metall. Inst. Akad. Nauk Kaz. SSR*, **7**, (1969), 49-52, in Russian. Cited on pages: 320, 530.

[69BER/STR] Berger, L. I., Strel'chenko, S. S., Bondar, S. A., Molodyk, A. D., Balanevskaya, A. E., Lebedev, V. V., Dissociation pressures of binary semiconductor compounds of the $B_2^{III}C_3^{VI}$ type, *Inorg. Mater.*, **5**, (1969), 743-745. Cited on pages: 233, 234, 235, 243, 244, 508.

[69BEV] Bevington, P. R., *Data reduction and error analysis for the physical sciences*, McGraw-Hill, New York, (1969), 336 pp.. Cited on pages: 623, 629.

[69BOE/BEN] Boev, E. I., Benderskii, L. A., Mil'kov, G. A., Thermal dissociation of zinc group chalcogenides. I. Thermal dissociation of zinc sulfide and selenide, *Russ. J. Phys. Chem.*, **43**, (1969), 776-778. Cited on pages: 257, 258, 509.

[69BOE/BEN2] Boev, E. I., Benderskii, L. A., Minaeva, N. V., Bunin, A. M., Thermal dissociation of zinc subgroup chalcogenides. II. Thermal dissociation of cadmium sulphide and selenide, *Russ. J. Phys. Chem.*, **43**, (1969), 1252-1254. Cited on pages: 270, 510.

[69CES/OOS] Cesbron, F., Oosterbosch, R., Pierrot, R., Une nouvelle espèce minérale: la marthozite. Uranyl-sélénite de cuivre hydraté , *Bull. Soc. Fr. Minéral. Cristallogr.*, **92**, (1969), 278-283, in French. Cited on page: 636.

[69COM] Comité International des Poids et des Mesures, The International Practical Temperature Scale of 1968, *Metrologia*, **5**, (1969), 35-47. Cited on page: 33.

[69DEM/MAL] Demidenko, A. F., Mal'tsev, A. K., Heat capacity of zinc telluride in the range 56-300°K. Entropy and enthalpy of ZnTe, CdS, CdSe, CdTe, *Inorg. Mater.*, **5**, (1969), 130-132. Cited on pages: 267, 268.

[69DEM] Demidenko, A. F., Heat capacities of cadmium sulfide, selenide, and telluride between 55 and 300°K, *Inorg. Mater.*, **5**, (1969), 210-212. Cited on pages: 267, 268.

[69DYA/VIG] D'yachkova, N. N., Vigdorovich, E. N., Ustyugov, G. P., Kudryavtsev, A. A., Saturated vapor pressure and molecular composition of selenium dioxide in the gaseous phase, *Izv. Akad. Nauk SSSR, Neorg. Mater.*, **5**, (1969), 2219-2220, in Russian, English translation in [69DYA/VIG2]. Cited on pages : 123, 124.

[69DYA/VIG2] D'yachkova, N. N., Vigdorovich, E. N., Ustyugov, G. P., Kudryavtsev, A. A., Saturated vapor pressure and molecular composition of selenium dioxide in the gaseous phase, *Inorg. Mater.*, **5**, (1969), 1900-1901. From a citation in this Bibliography

[69FIC/THO] Ficalora, P. J., Thompson, J. C., Margrave, J. L., Mass spectroscopic studies at high temperatures: XXVI. The sublimation of SeO_2 and SeO_3, *J. Inorg. Nucl. Chem.*, **31**, (1969), 3771-3774. Cited on pages: 123, 124, 128, 129.

[69FLO] Flögel, P., Zum Gleichgewicht zwishen Selen und Wasserstoff bei 1000°C, *Z. Anorg. Allg. Chem.*, **370**, (1969), 16-30, in German. Cited on pages: 132, 257, 258, 269, 270, 510, 533.

[69FOF/KIT] Fofanov, G. M., Kitaev, G. A., Analysis of the conditions for the precipitation of metal selenides from aqueous solutions with sodium selenosulphate, *Russ. J. Inorg. Chem.*, **14**, (1969), 322-324. Cited on page: 183.

[69GAD/SAT] Gadzhiev, S. M., Sattar-Zade, I. S., Thermodynamic investigation of selenium trioxide, *Uch. Zap. Azerb. Gos. Univ. Ser. Khim. Nauk*, **1**, (1969), 35-37, in Russian. Cited on page: 129.

[69GME] *Gmelins Handbuch der anorganischen Chemie: Blei, Teil C – Lieferung 2*, Hantke, G., Ed., Verlag Chemie, Weinheim/Bergstr., (1969). Cited on page: 224.

[69GOR] Gordon, J. S., Thermodynamic functions of hydrogen selenide, *J. Chem. Eng. Data*, **14**, (1969), 460. Cited on page: 132.

[69GRO/WES] Grønvold, F., Westrum, Jr., E. F., Radebaugh, R., Tetrapalladium sulfide and tetrapalladium selenide. Heat capacities and thermodynamic properties from 5° to 350°K, *J. Chem. Eng. Data*, **14**, (1969), 205-207. Cited on page: 332.

[69HAN/MUN] Hansen, E. E., Munir, Z. A., Mitchell, M. J., Sublimation pressure and sublimation coefficient of single-crystal lead selenide, *J. Am. Ceram. Soc.*, **52**, (1969), 610-612. Cited on pages: 222, 511.

[69HUR/ARI] Hurlbut, Jr., C. S., Aristarain, L. F., Olsacherite, $Pb_2(SO_4)(SeO_4)$, a new mineral from Bolivia, *Am. Mineral.*, **54**, (1969), 1519-1527. Cited on page: 636.

[69KLU/SEL] Klushina, T. V., Selivanova, N. M., Lapin, V. V., Artem'eva, L. N., Heat of formation of crystalline sodium selenite pentahydrate from the elements, *Zh. Prikl. Khim. (S.-Peterburg)*, **42**, (1969), 1510-1516, in Russian, English translation in [69KLU/SEL2]. Cited on pages: 418, 437, 466, 511.

[69KLU/SEL2] Klushina, T. V., Selivanova, N. M., Lapin, V. V., Artem'eva, L. N., Heat of formation of crystalline sodium selenite pentahydrate from the elements, *Russ. J. Appl. Chem.*, **42**, (1969), 1426-1431. From a citation in this Bibliography

[69KLU/SEL3] Klushina, T. V., Selivanova, N. M., Lapin, V. V., Fedyainov, N. V., Preparation of rubidium selenite and its heat of formation, *Zh. Fiz. Khim.*, **43**, (1969), 1398-1402, in Russian, English translation in [69KLU/SEL4]. Cited on pages: 427, 511.

[69KLU/SEL4] Klushina, T. V., Selivanova, N. M., Lapin, V. V., Fedyainov, N. V., Preparation of rubidium selenite and its heat of formation, *Russ. J. Phys. Chem.*, **43**, (1969), 779-781. From a citation in this Bibliography

[69KOZ/VAN] Kozhitov, L. V., Vanyukov, A. V., Liquid-vapor equilibria in tellurium-selenium and tellurium-sulfur systems, *Russ. J. Appl. Chem.*, **42**, (1969), 1661-1667. Cited on pages: 183, 184.

[69LAF/CER] Laffitte, M., Cerclier, O., Etude thermodynamique du système cobalt-selénium, *High Temp. High Pressures*, **1**, (1969), 449-455, in French. Cited on pages: 317, 512.

[69LEA/WAH] Leary, Jr., H. J., Wahlbeck, P. G., The vaporization process and rate of effusion of crystalline Sc_2Se_3, *High Temp. Sci.*, **1**, (1969), 277-286. Cited on page : 354.

[69MAI/SOL] Maier, A. I., Soldatova, T. A., Karapet'yants, M. K., Thermal properties of $Pr_2(SeO_3)_3.H_2SeO_3.5H_2O$, *Izv. Akad. Nauk SSSR, Neorg. Mater.*, **5**, (1969), 1712-1716, in Russian, English translation in [69MAI/SOL2]. Cited on pages : 378, 513.

[69MAI/SOL2] Maier, A. I., Soldatova, T. A., Karapet'yants, M. K., Thermal properties of $Pr_2(SeO_3)_3.H_2SeO_3.5H_2O$, *Inorg. Mater.*, **5**, (1969), 1451-1454. From a citation in this Bibliography

[69MAM] Mamedov, K. N., Dissert for Examn. Of Academic Degree for a Candidate in Tech. Sciences, Report , (1969). Cited from [74MIL] on pages: 236, 514.

[69MEH/GUB] Mehra, M. C., Gubeli, A. O., Complexing characteristics of insoluble selenites I. Silver selenite, *Radiochem. Radioanal. Lett.*, **2**, (1969), 61-68. Cited on pages: 303, 304, 514.

[69PAS/UKH] Pashinkin, A. S., Ukhlinov, G. A., Novoselova, A. V., Saturated vapor pressure of solid germanium monoselenide, *Moscow Univ. Chem. Bull.*, **24**, (1969), 80-81. Cited on page: 210.

[69PAS/UST] Pashinkin, A. S., Ustyugov, G. P., Vigdorovich, E. N., Silina, E. Y., Klinchikova, S. A., A study of the evaporation of mercury selenide, *Inorg. Mater.*, **5**, (1969), 400-405. Cited on pages : 280, 281, 516.

[69PIA/MAL] Piacente, V., Malaspina, L., Bardi, G., Étude thermodynamique de l'évaporation de l'oxyde de sélénium, *Rev. Int. Hautes Temp. Refract.*, **6**, (1969), 25-28, in French. Cited on pages : 123, 124.

[69ROS] Rossotti, H., *Chemical applications of potentiometry*, D. Van Nostrand, Princeton N. J., (1969), 229 pp.. Cited on pages: 21, 22.

[69ROS/PER] Rosso, B., Perret, R., Identification cristallographique des phases cristallisées hydratées et anhydres des séléniates d'aluminium et de gallium, *C. R. Séances Acad. Sci., Ser. C*, **268**, (1969), 487-490, in French. Cited on page: 231.

[69ROS/SEL] Roshchina, Z. V., Selivanova, N. M., Anan'ina, M. S., Physicochemical study of sodium hydrogen selenite trihydrate, *Izv. Vyssh. Uchebn. Zaved., Khim. Khim. Tekhnol.*, **12**, (1969), 535-538, in Russian. Cited on page: 419.

[69ROS/SEL2] Roshchina, Z. V., Selivanova, N. M., Anan'ina, M. S., Synthesis and heat of formation of sodium trihydroselenite, *Izv. Akad. Nauk SSSR, Neorg. Mater.*, **5**, (1969), 530-535, in Russian, English translation in [69ROS/SEL3]. Cited on pages: 419, 516.

[69ROS/SEL3] Roshchina, Z. V., Selivanova, N. M., Anan'ina, M. S., Synthesis and heat of formation of sodium trihydroselenite, *Inorg. Mater.*, **5**, (1969), 443-448. From a citation in this Bibliography

[69ROS/SEL4] Roshchina, Z. V., Selivanova, N. M., Preparation and thermochemical investigation of sodium pyroselenite, ($Na_2Se_2O_5$), *Zh. Neorg. Khim.*, **14**, (1969), 877-879, in Russian, English translation in [69ROS/SEL5]. Cited on pages: 418, 517.

[69ROS/SEL5] Roshchina, Z. V., Selivanova, N. M., Preparation and thermochemical investigation of sodium pyroselenite, ($Na_2Se_2O_5$), *Russ. J. Inorg. Chem.*, **14**, (1969), 456-458. From a citation in this Bibliography

[69SAL/HAK] Salomaa, P., Hakala, R., Vesala, S., Aalto, T., Solvent deuterium isotope effects on acid-base reactions: Part III. Relative acidity constants of inorganic oxyacids in light and heavy water. Kinetic applications, *Acta Chem. Scand.*, **23**, (1969), 2116-2126. Cited on pages: 141, 142.

[69SCH/KAR] Schöneshöfer, M., Karmann, W., Henglein, A., Die Pulsradiolyse des Selenwasserstoffes in wässriger Lösung, *Int. J. Radiat. Phys. Chem.*, **1**, (1969), 407-423, in German. Cited on page: 154.

[69SEM/MEL] Semenkovich, S. A., Melekh, B. T., Thermodynamic properties of Bi_2Te_3, Bi_2Se_3 and Sb_2Te_3, Sb_2Se_3, *Khim. Svyaz Krist.*, (1969), 277-281, in Russian. Cited on page: 517.

[69SHA/KOS] Shaulov, I. K., Kostina, E. N., Heat capacity and entropy of cadmium selenide at liquid nitrogen temperatures, *Russ. J. Phys. Chem.*, **43**, (1969), 549-550. Cited on pages: 267, 268.

[69SOK/PAS] Sokolov, V. V., Pashinkin, A. S., Novoselova, A. V., Ryazantsev, A. A., Dolgikh, V. A., Klinchikova, S. A., Saturated vapor pressures of tin and lead tellurides and lead selenide, *Inorg. Mater.*, **5**, (1969), 12-15. Cited on pages: 222, 511, 517.

[69SON/NOV] Sonin, V. I., Novikov, G. I., Polyachenok, O. G., Thermodynamic investigation of the sublimation of selenium dioxide, *Zh. Fiz. Khim.*, **43**, (1969), 2980, in Russian, English translation in [69SON/NOV2]. Cited on pages: 123, 124, 203, 309, 331.

[69SON/NOV2] Sonin, V. I., Novikov, G. I., Polyachenok, O. G., Thermodynamic investigation of the sublimation of selenium dioxide, *Russ. J. Phys. Chem.*, **43**, (1969), 1680. From a citation in this Bibliography

[69SPI/ROU] Spiesser, M., Rouxel, J., Kerriou, M., Goureaux, M. G., Caractérisation et étude physico-chimique de séléniures et tellurures non stoechiométriques de molybdène, *Bull. Soc. Chim. Fr.*, **5**, (1969), 1427-1431, in French. Cited on pages: 345, 544.

[69SUP/MAI] Suponitskii, Y. L., Maier, A. I., Fedorov, V. V., Kravchenko, L. K., Doroshenko, N. A., Karapet'yants, M. K., Enthalpy of formation of selenates of praseodymium, neodymium, and samarium, *Zh. Fiz. Khim.*, **43**, (1969), 1284, in Russian, English translation in [69SUP/MAI2]. Cited on pages: 371, 375, 378, 518, 539.

[69SUP/MAI2] Suponitskii, Y. L., Maier, A. I., Fedorov, V. V., Kravchenko, L. K., Doroshenko, N. A., Karapet'yants, M. K., Enthalpy of formation of selenates of praseodymium, neodymium, and samarium, *Russ. J. Phys. Chem.*, **43**, (1969), 711. From a citation in this Bibliography

[69UST/VIG] Ustyugov, G. P., Vigdorovich, E. N., Timoshin, I. A., The saturated vapor pressure of bismuth selenide and telluride, *Inorg. Mater.*, **5**, (1969), 138-139. Cited on page: 201.

[69UST/VIG3] Ustyugov, G. P., Vigdorovich, E. N., Thermal stability of hydrogen selenide and hydrogen telluride, *Inorg. Mater.*, **5**, (1969), 135-137. Cited on page: 132.

[69UST/VIG4] Ustyugov, G. P., Vigdorovich, E. N., Bezobrazov, E. G., Liquid-vapor phase equilibrium in binary systems of tellurium with impurities, *Inorg. Mater.*, **5**, (1969), 300-301. Cited on page: 183.

[69UY/DRO] Uy, O. M., Drowart, J., Mass spectrometric determination of the dissociation energies of the molecules BiO, BiS, BiSe, and BiTe, *Trans. Faraday Soc.*, **65**, (1969), 3221-3230. Cited on pages: 197, 198, 221, 519.

[69YUS/DEV] Yushin, A. S., Devyatykh, G. G., Equilibrium constant for heat of formation of hydrogen selenide from the elements, *Russ. J. Phys. Chem.*, **43**, (1969), 550-552. Cited on page: 132.

[69ZHD] Zhdanov, V. M., Low-temperature specific heat, enthalpy, and entropy of germanium monoselenide, *Russ. J. Phys. Chem.*, **43**, (1969), 1468-1470. Cited on page: 209.

[70BAS/SMI] Baskin, Y., Smith, S. D., Enthalpy-of-formation data on compounds of uranium with groups VA and VIA elements, *J. Nucl. Mater.*, **37**, (1970), 209-222. Cited on pages: 386, 387, 388.

[70BER/COP] Bergman, C., Coppens, P., Drowart, J., Smoes, S., Mass spectrometric determination of the dissociation energies of the gaseous rare earth monoselenides and monotellurides, *Trans. Faraday Soc.*, **66**, (1970), 800-808. Cited on pages: 349, 354, 357, 360, 361, 362, 363, 365, 367, 369, 372, 373, 379, 382.

[70BLA/MUN] Blair, R. C., Munir, Z. A., Sublimation coefficient of tin selenide, *J. Am. Ceram. Soc.*, **53**, (1970), 301-303. Cited on pages: 216, 519.

[70BLA/SCH] Blachnik, R., Schneider, A., Schmelzwärmen von III/V-, IV/V- und V/VI-Verbindungen, *Z. Anorg. Allg. Chem.*, **372**, (1970), 314-324, in German. Cited on pages: 191, 194, 201.

[70BOD/PLA] Bodewig, F. G., Plambeck, J. A., Electrochemical behavior of selenium and tellurium in fused LiCl-KCl eutectic, *J. Electrochem. Soc.*, **117**, (1970), 618-621. Cited on page: 168.

[70BOG/BEZ] Bogdanov, V. I., Bezborodova, V. M., Temperature dependence of the specific heat of HgSe, *Fiz. Tverd. Tela (Leningrad)*, **12**, (1970), 1849-1850, in Russian. Cited on pages : 279, 280, 519.

[70CHA/TIK] Charlot, C., Tikhomiroff, N., Laffitte, M., Le séléniure de zinc : préparation et enthalpie de formation, *Bull. Soc. Chim. Fr.*, **2**, (1970), 459-463, in French. Cited on pages: 257, 258, 519.

[70CHR/JOH] Christensen, J. J., Johnston, H. D., Izatt, R. M., Thermodynamics of proton ionization in aqueous solution. Part XII. ΔH°, ΔS°, and ΔC_p° values for hydrocyanic acid dissociation at 10, 25, and 40°, *J. Chem. Soc. A*, (1970), 454-455. Cited on page: 435.

[70GHO/NAI] Ghosh, R., Nair, V. S. K., Studies on metal complexes in aqueous solution: III. The biselenate ion and transition metal selenates, *J. Inorg. Nucl. Chem.*, **32**, (1970), 3041-3051. Cited on pages: 146, 148, 149, 315, 322, 342, 520.

[70GOS/PAS] Gospodinov, G. G., Pashinkin, A. S., Boncheva-Mladenova, Z., Novoselova, A. V., Determination of the saturated vapor pressure of solid antimony selenide, *Inorg. Mater.*, **6**, (1970), 639-643. Cited on pages: 193, 201.

[70GRO] Grønvold, F., Heat capacities and thermodynamic properties of the $Ni_{1-x}Se$-Phase from 298 to 1050°K, *Acta Chem. Scand.*, **24**, (1970), 1036-1050. Cited on page: 311.

[70HAJ] Hajiev, S. N., Advances in experimental thermochemistry. 1. The determination of enthalpies of binary semiconducting compounds by direct synthesis in a bomb calorimeter, *J. Chem. Thermodyn.*, **2**, (1970), 765-773. Cited on page: 220.

[70HAR] Harris, D. C., New data on tyrrellite, *Can. Mineral.*, **10**, (1970), 731-736. Cited on page: 635.

[70HAR/CAB] Harris, D. C., Cabri, L. J., Kaiman, S., Athabascaite: a new copper selenide mineral from Martin Lake, Saskatchewan, *Can. Mineral.*, **10**, (1970), 207-215. Cited on page: 634.

[70HIN/COB2] Hinchey, R. J., Cobble, J. W., Standard-state entropies for the aqueous trivalent lanthanide and yttrium ions, *Inorg. Chem.*, **9**, (1970), 917-921. Cited on page: 529.

[70HOG/LUN] Högberg, K., Lundqvist, M., Conductance and equilibrium studies on the system Se-Br, *Acta Chem. Scand.*, **24**, (1970), 255-260. Cited on pages: 175, 583.

[70JOH/PIC] Johan, Z., Picot, P., Pierrot, R., Verbeek, T., L'oosterboschite (Pd, Cu)$_7$Se$_5$ une nouvelle espèce minérale et la trogtalite cupro-palladifère de Musonoï (Katange), *Bull. Soc. Fr. Minéral. Cristallogr.*, **93**, (1970), 476-481, in French. Cited on page: 635.

[70KAR/MAI] Karapet'yants, M. K., Maier, A. I., Soldatova, T. A., Enthalpies of formation of certain praseodymium selenites, *Izv. Akad. Nauk SSSR, Neorg. Mater.*, **6**, (1970), 1118-1122, in Russian, English translation in [70KAR/MAI2]. Cited on pages: 378, 521.

[70KAR/MAI2] Karapet'yants, M. K., Maier, A. I., Soldatova, T. A., Enthalpies of formation of certain praseodymium selenites, *Inorg. Mater.*, **6**, (1970), 976-979. From a citation in this Bibliography

[70KEL] Keller, H., Zur Thermodynamic von Selendampf. Massenspektrometrische Untersuchungen mit der elektrochemischen Knudsenzelle, Ph. D. Thesis, University of Dortmund, (1970), Germany, in German. Cited on pages: 95, 96, 97, 99, 100, 101, 102, 103, 104, 105, 107, 108, 110, 111, 112, 114, 115, 117, 522, 527.

[70KLU/SEL] Klushina, T. V., Selivanova, N. M., Karapet'yants, M. K., Lapin, V. V., Fedyainov, N. V., Heats of formation of caesium and francium selenites, *Termodin. Termokhim. Konstanty,* Astakhov, K. V., Ed., pp. 152-158, Akademiya Nauk SSSR, Izd. Nauka, Moscow, (1970). Cited on pages: 429, 523.

[70MAK/PEC] Makatun, V. N., Pechkovskii, V. V., Vapour pressure over solid and liquid selenium dioxide and parameters of its triple point, *Zh. Fiz. Khim.*, **44**, (1970), 2667, in Russian, English translation in [70MAK/PEC2]. Cited on pages: 121, 123, 124.

[70MAK/PEC2] Makatun, V. N., Pechkovskii, V. V., Vapour pressure over solid and liquid selenium dioxide and parameters of its triple point, *Russ. J. Phys. Chem.*, **44**, (1970), 1522. From a citation in this Bibliography

[70MAM/KER] Mamedov, K. K., Kerimov, I. G., Kostryukov, V. N., Mekhtiev, M. I., Specific heat, entropy, and enthalpy of gallium, indium, and thallium selenides in the 12-300°K range, *Termodin. Termokhim. Konstanty,* Astakhov, K. V., Ed., pp. 217-223, Akademiya Nauk SSSR, Izd. Nauka, Moscow, (1970). Cited on pages: 233, 239, 240.

[70MEH/GUB] Mehra, M. C., Gubeli, A. O., The complexing characteristics of insoluble selenides. 1. Silver selenide, *Can. J. Chem.*, **48**, (1970), 3491-3497. Cited on pages: 134, 135, 301, 302, 529.

[70MEH/GUB2] Mehra, M. C., Gubeli, A. O., The complexing characteristics of insoluble selenides. 2. Manganese selenide, *J. Less-Common Met.*, **22**, (1970), 281-285. Cited on pages: 338, 529.

[70MEL/WAH] Melucci, R. C., Wahlbeck, P. G., Mass spectroscopic observations of gaseous boron selenides, *Inorg. Chem.*, **9**, (1970), 1065-1068. Cited on page: 226.

[70OZI/VOE] Ozin, G. A., Vander Voet, A., A gas-phase Raman study of the dissociation of selenium tetrachloride and the vibrational spectrum and molecular structure of selenium dichloride, *Chem. Commun.*, (1970), 896-897. Cited on page: 169.

[70ROS/SEL] Roshchina, Z. V., Selivanova, N. M., Heat of formation of potassium hydrogen selenite and pyroselenite, *Zh. Neorg. Khim.*, **15**, (1970), 3181-3184, in Russian, English translation in [70ROS/SEL4]. Cited on pages: 424, 524.

[70ROS/SEL4] Roshchina, Z. V., Selivanova, N. M., Heat of formation of potassium hydrogen selenite and pyroselenite, *Russ. J. Inorg. Chem.*, **15**, (1970), 1657-1659. From a citation in this Bibliography

[70ROS/SEL2] Roshchina, Z. V., Selivanova, N. M., Heat of formation of lithium trihydrogen selenite, *Izv. Vyssh. Uchebn. Zaved., Khim. Khim. Tekhnol.*, **13**, (1970), 3-5, in Russian. Cited on pages: 414, 525.

[70ROS/SEL3] Roshchina, Z. V., Selivanova, N. M., Heat of formation of potassium trihydrogen selenite, *Izv. Vyssh. Uchebn. Zaved., Khim. Khim. Tekhnol.*, **13**, (1970), 926-928, in Russian. Cited on pages: 425, 525.

[70SOB/SEL] Sobol, L. G., Selivanova, N. M., Heat of formation of sodium pyroselenate ($Na_2Se_2O_7$)., *Izv. Vyssh. Uchebn. Zaved., Khim. Khim. Tekhnol.*, **13**, (1970), 1728-1732, in Russian. Cited on pages: 422, 526.

[70TRE/WIL] Treinin, A., Wilf, J., Electronic spectra of the oxyanions of selenium in solution, *J. Phys. Chem.*, **74**, (1970), 4131-4137. Cited on page: 148.

[70UST/BEZ] Ustyugov, G. P., Bezobrazov, E. G., Mareev, Y. M., The saturated vapor pressure of silicon selenide, *Inorg. Mater.*, **6**, (1970), 1918. Cited on page: 208.

[70UY/DRO] Uy, O. M., Drowart, J., Mass spectrometric determination of the dissociation energies of the boron monochalcogenides, *High Temp. Sci.*, **2**, (1970), 293-298. Cited on page: 226.

[70VOR] Voronin, G. F., Estimation of standard entropies of chemical compounds, *Russ. J. Phys. Chem.*, **44**, (1970), 1717-1719. Cited on pages: 412, 416, 422.

[70WES/GRO] Westrum, Jr., E. F., Grønvold, F., Uranium chalcogenides: III. Heat capacities and thermodynamic properties of $US_{1.9}$ and Use_2 from 5 to 350°K, *J. Inorg. Nucl. Chem.*, **32**, (1970), 2169-2177. Cited on page: 387.

[70YAN/PAT] Yanitskii, I. V., Patkauskas, R. M., Solubility and crystallization in the system sodium selenate-sodium selenite-water, *Zh. Prikl. Khim. (S.-Peterburg)*, **43**, (1970), 522-527, in Russian, English translation in [70YAN/PAT2]. Cited on pages: 417, 420.

[70YAN/PAT2] Yanitskii, I. V., Patkauskas, R. M., Solubility and crystallization in the system sodium selenate-sodium selenite-water, *Russ. J. Appl. Chem.*, **43**, (1970), 530-535. From a citation in this Bibliography

[71BAK/BUK] Bakeeva, S. S., Buketov, E. A., Pashinkin, A. S., Dissociation pressure of lead selenite, *Izv. Akad. Nauk SSSR, Neorg. Mater.*, **7**, (1971), 1261-1262, in Russian, English translation in [71BAK/BUK2]. Cited on page: 224.

[71BAK/BUK2] Bakeeva, S. S., Buketov, E. A., Pashinkin, A. S., Dissociation pressure of lead selenite, *Inorg. Mater.*, 7, (1971), 1120-1121. From a citation in this Bibliography

[71BAR/SIL] Barcza, L., Sillén, L. G., Equilibrium studies of polyanions: 19. Polyselenite equilibria in various ionic media, *Acta Chem. Scand.*, **25**, (1971), 1250-1260. Cited on pages: 140, 141, 583, 584.

[71BLA/SCH] Blachnik, R., Schneider, A., High-temperature enthalpies of As_2Te_3, Sb_2Se_3, Bi_2Se_3, GeAs, $GeAs_2$, and SnAs, *J. Chem. Thermodyn.*, **3**, (1971), 227-233. Cited on pages: 194, 198, 199.

[71CES/PIE] Cesbron, F., Pierrot, R., Verbeek, T., La derriksite, $Cu_4(UO_2)(SeO_3)_2(OH)_6.H_2O$, une nouvelle espèce minérale, *Bull. Soc. Fr. Minéral. Cristallogr.*, **94**, (1971), 534-537, in French. Cited on page: 636.

[71DUB/SIS] Dubeau, C., Sisi, J.-C., Ozanne, N., The solubility of hydrogen selenide gas in water, *J. Chem. Eng. Data*, **16**, (1971), 78-79. Cited on pages: 133, 134.

[71ELL/SLO] Ellert, G. V., Slovyanskikh, V. K., Sevast'yanov, V. G., The U-S and U-Se systems, *Russ. J. Inorg. Chem.*, **16**, (1971), 910-914. Cited on pages: 385, 386, 387.

[71GLA/KOR] Glazov, V. M., Korenchuk, N. M., Vapour pressure and thermodynamic properties of copper and silver chalcogenides, *Russ. J. Phys. Chem.*, **45**, (1971), 1520-1521. Cited on pages: 291, 527.

[71GLA/KUL] Glazov, V. M., Kuliev, R. A., Krestovnikov, A. N., Entropies and heats of fusion of antimony chalcogenides, *Russ. J. Phys. Chem.*, **45**, (1971), 1516-1517. Cited on page: 194.

[71GOR/FES] Gordienko, S. P., Fesenko, V. V., Fenochka, B. V., Lysenko, A. A., Mass-spectrometric investigation of evaporation of rare-earth metal monochalcogenides, *Russ. J. Phys. Chem.*, **45**, (1971), 1100-1101. Cited on pages: 367, 368, 369, 370, 371, 372, 373.

[71GRE/ZAI] Greiver, T. N., Zaitseva, I. G., Sal'dau, E. P., Sodium sulfoselenides, *Russ. J. Appl. Chem.*, **44**, (1971), 1713-1719. Cited on page: 181.

[71GRI/BOR] Grinberg, Y. K., Boryakova, V. A., Shevel'kov, V. F., Evaporation of InSe, *Inorg. Mater.*, **7**, (1971), 669-672. Cited on page: 527.

[71JEN] Jensen, K. A., *Nomenclature of inorganic chemistry*, IUPAC Commission on Nomenclature of Inorganic Compounds, Pergamon Press, Oxford, (1971), 110 pp.. Cited on page: 11.

[71JOH/KVA] Johan, Z., Kvacek, M., La hakite, un nouveau minéral du groupe de la tétraédrite, *Bull. Soc. Fr. Minéral. Cristallogr.*, **94**, (1971), 45-48, in French. Cited on page: 634.

[71JOH/KVA2] Johan, Z., Kvacek, M., La bukovite, $Cu_{3+x}Tl_2FeSe_{4-x}$, une nouvelle espèce minérale, *Bull. Soc. Fr. Minéral. Cristallogr.*, **94**, (1971), 529-533, in French. Cited on page: 634.

[71JOH/PIC] Johan, Z., Picot, P., Pierrot, R., Kvacek, M., La permingeatite Cu_3SbSe_4, un nouveau minéral du groupe de la luzonite, *Bull. Soc. Fr. Minéral. Cristallogr.*, **94**, (1971), 162-165, in French. Cited on page: 635.

[71JOH/PIC2] Johan, Z., Picot, P., Pierrot, R., Kvacek, M., La fischesserite, Ag_3AuSe_2, premier séléniure d'or, isotype de la petzite, *Bull. Soc. Fr. Minéral. Cristallogr.*, **94**, (1971), 381-384, in French. Cited on page: 634.

[71KEL/RIC] Keller, H., Rickert, H., Detry, D., Drowart, J., Goldfinger, P., Zur Thermodynamik von Selendampf. Massenspektrometrische Untersuchungen mit der elektrochemischen Knudsen-Zelle, *Z. Phys. Chem. (Munich)*, **75**, (1971), 273-286, in German. Cited on page: 527.

[71KOL/TET] Kolosov, I. V., Tetsu, K., Ivanov-Emin, B. N., Korotaeva, L. G., An anion-exchange study of the interaction between scandium and selenate ions, *Zh. Neorg. Khim.*, **16**, (1971), 3107-3109, in Russian, English translation in [71KOL/TET2]. Cited on page: 494.

[71KOL/TET2] Kolosov, I. V., Tetsu, K., Ivanov-Emin, B. N., Korotaeva, L. G., An anion-exchange study of the interaction between scandium and selenate ions, *Russ. J. Inorg. Chem.*, **16**, (1971), 1649-1650. From a citation in this Bibliography

[71MAE/YOK] Maekawa, T., Yokokawa, T., Niwa, K., Enthalpies of mixing in the liquid state II. Tl + Se and Tl + S, *J. Chem. Thermodyn.*, **3**, (1971), 707-710. Cited on pages: 249, 250, 251.

[71MAI/SUP] Maier, A. I., Suponitskii, Y. L., Karapet'yants, M. K., Enthalpies of formation of crystalline lanthanum, praseodymium, and neodymium selenites, *Zh. Fiz. Khim.*, **45**, (1971), 2377-2378, in Russian, English translation in [71MAI/SUP2]. Cited on pages: 375, 378, 383, 513, 514, 521, 522, 528, 647, 648.

[71MAI/SUP2] Maier, A. I., Suponitskii, Y. L., Karapet'yants, M. K., Enthalpies of formation of crystalline lanthanum, praseodymium, and neodymium selenites, *Russ. J. Phys. Chem.*, **45**, (1971), 1349. From a citation in this Bibliography

[71MAK/RAT] Makatun, V. N., Rat'kovskii, I. A., Pechkovskii, V. V., The dissociation pressures of neutral and basic copper selenites, *Zh. Fiz. Khim.*, **45**, (1971), 1591, in Russian, English translation in [71MAK/RAT2]. Cited on page: 294.

[71MAK/RAT2] Makatun, V. N., Rat'kovskii, I. A., Pechkovskii, V. V., The dissociation pressures of neutral and basic copper selenites, *Russ. J. Phys. Chem.*, **45**, (1971), 904. From a citation in this Bibliography

[71MEH/GUB] Mehra, M. C., Gubeli, A. O., The complexing characteristics of insoluble selenides. 3. Mercuric selenide, *J. Less-Common Met.*, **25**, (1971), 221-224. Cited on pages: 134, 135, 282, 301, 338, 529.

[71MEL/STE] Lelekh, B. T., Spepanova, N. B., Fomina, T. A., Semenkovich, S. A., Thermodynamic properties of compounds in the tin-selenium system, *Zh. Fiz. Khim.*, **45**, (1971), 2018-2020, in Russian. Cited on pages: 214, 216, 217.

[71MOO2] Moore, C. E., *Atomic energy levels*, III, *Natl. Bur. Stand. Ref. Data Ser.*, No. 35/V.III, U. S. National Bureau of Standards, Washington, D.C., USA, (1971), 282 pp.. Cited on page: 496.

[71NAG/BRI] Nagarajan, G., Brinkley, D. C., Statistical thermodynamics: Enthalpy, free energy, entropy, and heat capacity of some hexafluorides of octahedral symmetry, *Z. Naturforsch.*, **26A**, (1971), 1658-1666. Cited on page: 164.

[71PAS/ARO] Pashinkin, A. S., Aron, B. M., Bakeeva, S. S., Buketov, E. A., Karyakin, N. V., Krylova, G. P., Rustembekov, K. G., Ugorets, M. Z., Measurement and calculation of certain thermodynamic and physicochemical constants of selenium dioxide, *Zh. Fiz. Khim.*, **45**, (1971), 1599-1600, in Russian, English translation in [71PAS/ARO2]. Cited on pages: 120, 121, 122, 273, 274, 304, 305, 313, 320, 321, 396, 530.

[71PAS/ARO2] Pashinkin, A. S., Aron, B. M., Bakeeva, S. S., Buketov, E. A., Karyakin, N. V., Krylova, G. P., Rustembekov, K. G., Ugorets, M. Z., Measurement and calculation of certain thermodynamic and physicochemical constants of selenium dioxide, *Russ. J. Phys. Chem.*, **45**, (1971), 909-910. From a citation in this Bibliography

[71RAB/RAU] Rabenau, A., Rau, H., Rosenstein, G., Phase relationships in the gold-selenium system, *J. Less-Common Met.*, **24**, (1971), 291-299. Cited on page: 308.

[71RAM/JAY] Ramaswamy, K., Jayaraman, S., Molecular force field for some X_2Y_2 molecules of C_2 symmetry, *Acta Phys. Pol.*, **A40**, (1971), 883-889. Cited on pages: 168, 174.

[71SAD/YAR] Sadovskaya, O. A., Yarembash, E. I., Naginaev, E. N., Europium selenides, *Inorg. Mater.*, **7**, (1971), 284-285. Cited on page: 366.

[71SEA/MUN] Seacrist, L., Munir, Z. A., Studies on the sublimation of IIB-VIA compounds. IV. The anisotropic behavior of the sublimation of basal-plane oriented crystals of cadmium selenide, *High Temp. Sci.*, **3**, (1971), 340-348. Cited on pages: 269, 270, 530.

[71SEL/PRY] Selivanova, N. M., Prymova, L. A., Kravchenko, L. K., Physicochemical study of ammonium selenate, *Izv. Vyssh. Uchebn. Zaved., Khim. Khim. Tekhnol.*, **14**, (1971), 647-650, in Russian. Cited on pages: 411, 531.

[71SEL/SOB] Selivanova, N. M., Sobol, L. G., Gorokhov, V. K., The sodium selenate-selenic acid-water system at 25°C, *Zh. Neorg. Khim.*, **16**, (1971), 2800-2803, in Russian, English translation in [71SEL/SOB2]. Cited on pages: 420, 421.

[71SEL/SOB2] Selivanova, N. M., Sobol, L. G., Gorokhov, V. K., The sodium selenate-selenic acid-water system at 25°C, *Russ. J. Inorg. Chem.*, **16**, (1971), 1491-1493. From a citation in this Bibliography

[71SEV/SLO] Sevast'yanov, V. G., Slovyanskikh, V. K., Ellert, G. V., Equilibria in the US_x-Br_2 and USe_x-Br_2 system, *Russ. J. Inorg. Chem.*, **16**, (1971), 1776-1779. Cited on page: 388.

[71SHK/MAL] Shkodin, V. G., Malyshev, V. P., Makhmetov, M. Z., Polukarov, A. N., Dissociation pressure of sodium selenite, *Zh. Fiz. Khim.*, **45**, (1971), 2689, in Russian, English translation in [71SHK/MAL2]. Cited on pages: 418, 532.

[71SHK/MAL2] Shkodin, V. G., Malyshev, V. P., Makhmetov, M. Z., Polukarov, A. N., Dissociation pressure of sodium selenite, *Russ. J. Phys. Chem.*, **45**, (1971), 1528. From a citation in this Bibliography

[71SIL/MAR] Sillén, L. G., Martell, A. E., *Stability constants of metal ion complexes, Suppl. No. 1*, Special Publication, No. 25, Chemical Society, London, (1971), 865 pp.. Cited on page: 14.

[71SON/POL] Sonin, V. I., Polyachenok, O. G., Thermal stability of gaseous selenium dioxide, *Vestsi Akad. Navuk BSSR, Ser. Khim. Navuk*, (1971), 121-123, in Russian. Cited on page: 123.

[71UY/DRO] Uy, O. M., Drowart, J., Determination by the mass spectrometric Knudsen cell method of the atomization energies of the gaseous aluminium chalcogenides, Al_2, AlCu, AlCuS and $AlCuS_2$, *Trans. Faraday Soc.*, **67**, (1971), 1293-1301. Cited on pages: 228, 229, 230, 532.

[71VAS/NIK] Vasil'ev, V. P., Nikol'skaya, A. V., Gerasimov, Y. I., Thermodynamic study of thallium-selenium alloys by the electromotive force method, *Russ. J. Phys. Chem.*, **45**, (1971), 1169-1170. Cited on page: 251.

[71ZAI/GRE] Zaitseva, I. G., Greiver, T. N., Sal'dau, E. P., Sodium chalcogenotellurites, *Russ. J. Appl. Chem.*, **44**, (1971), 1502-1506. Cited on page: 185.

[71ZHD] Zhdanov, V. M., Low-temperature heat capacity, enthalpy, and entropy of antimony selenide, *Russ. J. Phys. Chem.*, **45**, (1971), 1356-1357. Cited on page: 194.

[72ADA/DOW] Adams, C. J., Downs, A. J., The vibrational spectra and structures of selenium tetrafluoride and tellurium tetrafluoride: including a matrix-isolation study, *Spectrochim. Acta*, **28A**, (1972), 1841-1854. Cited on page: 162.

[72AGA/SHA] Agaev, A. B., Sharifov, K. A., Standard enthalpy of formation of Gd_2Se_3 and Gd_2Te_3, *Uch. Zap. Azerb. Gos. Univ. Ser. Khim. Nauk*, **3**, (1972), 72-74, in Russian. Cited on page: 364.

[72ALE/BEA] Alexander, L. E., Beattie, I. R., A sapphire cell for high temperature Raman studies of reactive gases. The vibrational spectra of SeF_4, $SeOF_2$, TiF_4, and SbF_3, *J. Chem. Soc. Dalton Trans.*, (1972), 174-1750. Cited on page: 166.

[72ARN/BAR] Arnek, R., Barcza, L., Thermochemical studies of hydrolytic reactions: 11. Polyselenite equilibria in various ionic media, *Acta Chem. Scand.*, **26**, (1972), 213-217. Cited on pages: 140, 142.

[72BAK/BUK] Bakeeva, S. S., Buketov, E. A., Pashinkin, A. S., Bakeev, M. I., Muldagalieva, R. A., Dissociation pressure of copper selenite, *Tr. Khim.-Metall. Inst. Akad. Nauk Kaz. SSR*, **19**, (1972), 136-142, in Russian. Cited on page: 294.

[72BAK/PAS] Bakeeva, S. S., Pashinkin, A. S., Bakeev, M. I., Dissociation pressure of bismuth selenite, *Tr. Khim.-Metall. Inst. Akad. Nauk Kaz. SSR*, **19**, (1972), 130-135, in Russian. Cited on page: 203.

[72CHE/KUZ] Chernozubov, Y. S., Kuznetsov, B. P., Klimenko, A. A., Podmogil'nyi, E. V., Vapour pressures of chalcogens, *Russ. J. Phys. Chem.*, **46**, (1972), 168. Cited on pages: 99, 533.

[72CHR/SCH] Christe, K. O., Schack, C. J., Curtis, E. C., Selenium pentafluoride chloride, SeF_5Cl. Vibrational spectrum, force constants, and thermodynamic properties, *Inorg. Chem.*, **11**, (1972), 583-586. Cited on pages : 164, 165.

[72DEI/TAN] Deichman, E. N., Tananaev, I. V., Kadoshnikova, N. V., The $In_2(SeO_4)_3$-H_2SeO_4-H_2O system at 20°C, *Zh. Neorg. Khim.*, **17**, (1972), 222-225, in Russian, English translation in [72DEI/TAN2]. Cited on page : 247.

[72DEI/TAN2] Deichman, E. N., Tananaev, I. V., Kadoshnikova, N. V., The $In_2(SeO_4)_3$-H_2SeO_4-H_2O system at 20°C, *Russ. J. Inorg. Chem.*, **17**, (1972), 117-118. From a citation in this Bibliography

[72EMO/THE] Emons, H.-H., Theisen, L., Dampfdruckmessungen an Siliciummonochalkogeniden, *Monatsh. Chem.*, **103**, (1972), 62-71, in German. Cited on page: 208.

[72FLO] Flögel, P., Zur Kristallzüchtung von Cadmiumsulfid und anderen II- VI-Verbindungen. IV. Zum Gleichgewicht zwischen Selen und Wasserstoff bei 400°C, *Z. Anorg. Allg. Chem.*, **388**, (1972), 218-228, in German. Cited on pages: 132, 533.

[72GLA/KIR] Gladyshev, V. P., Kireeva, E. P., Electrode reactions on a mercury electrode in which the anions Se^{2-} and Te^{2-} participate, *Sov. Electrochem.*, **8**, (1972), 1068. Cited on page: 281.

[72GRI/BOR] Grinberg, Y. K., Boryakova, V. A., Shevel'kov, V. F., Medvedeva, Z. S., p-T-x phase diagram and thermodynamic properties of In_2Se_3, *Inorg. Mater.*, **8**, (1972), 1846-1853. Cited on page: 244, 534.

[72GRO] Grønvold, F., High-temperature reaction calorimeter. Enthalpy of formation of iron selenides and nickel selenides at 1050 K, *Acta Chem. Scand.*, **26**, (1972), 2085-2099. Cited on pages: 311, 325, 326, 327.

[72HAR/NUF] Harris, D. C., Nuffield, E. W., Bambollaite, a new copper telluro-selenide, *Can. Mineral.*, **11**, (1972), 738-742. Cited on page: 634.

[72HIL/CUE] Hillel, R., Cueilleron, J., Etude du séléniure de bore : B_2Se_3, *Bull. Soc. Chim. Fr.*, **1**, (1972), 98-103, in French. Cited on page: 226.

[72HOA/CAB] Hoareau, A., Cabaud, B., Uzan, R., Etude d'un composé du type SbSe par spectrométrie de masse, *C. R. Séances Acad. Sci., Ser. B*, **275**, (1972), 705-707, in French. Cited on page: 193.

[72JOH/PIC] Johan, Z., Picot, P., Pierrot, R., Kvacek, M., La krutaïte, $CuSe_2$, un nouveau minéral du groupe de la pyrite, *Bull. Soc. Fr. Minéral. Cristallogr.*, **95**, (1972), 475-481, in French. Cited on page: 635.

[72KOM/WES] Komarek, K. L., Wessely, K., Übergangsmetall-Chalkogensysteme, 1. Mitt.: Das System Kobalt-Selen, *Monatsh. Chem.*, **103**, (1972), 896-906, in German. Cited on page: 317.

[72MAE/YOK] Maekawa, T., Yokokawa, T., Niwa, K., Enthalpies of mixing in the liquid state. IV. Bi + Se and Sb + Se, *J. Chem. Thermodyn.*, **4**, (1972), 873-878. Cited on pages: 194, 201, 202, 534.

[72MAT/GEI] Matlasevich, O. B., Geiderikh, V. A., Gerassimov, J. I., Thermodynamic investigation of nonstoichiometric phases in Co-Te and Co-Se systems, *Rev. Chim. Minér.*, **9**, (1972), 191-204. Cited on pages: 317, 534.

[72MAT/GEI2] Matlasevich, O. B., Geiderikh, V. A., Thermodynamic properties of cobalt selenides ($CoSe_2$ and $Co_{0.72}Se$), *Russ. J. Phys. Chem.*, **46**, (1972), 1048-1049. Cited on pages: 317, 318, 535.

[72MUN/MES] Munir, Z. A., Meschi, D. J., Pound, G. M., The partial pressures of Hg(g) and Se_n(g) in equilibrium with crystalline mercury selenide, *J. Cryst. Growth*, **15**, (1972), 263-267. Cited on pages: 280, 281, 535.

[72MUR/TAK] Murayama, T., Takayanagi, A., Determination of the stability constants of the bis(selenocyanato)mercury(II) and tris(selenocyanato)mercurate(II) complexes by a polarographic method, *Bull. Chem. Soc. Jpn.*, **45**, (1972), 3549-3551. Cited on pages: 286, 287.

[72NI/WAH] Ni, R. Y., Wahlbeck, P. G., Dissociation energies of gaseous ScSe, YSe, and LaSe, *High Temp. Sci.*, **4**, (1972), 326-346. Cited on pages: 353, 354, 355, 356, 381, 382.

[72NOV/ZLO] Novoselova, A. V., Zlomanov, V. P., Karbanov, S. G., Matveyev, O. V., Gas'kov, A. M., Physico-chemical study of the germanium, tin, lead chalcogenides, *Prog. Solid State Chem.*, **7**, (1972), 85-115. Cited on page: 218.

[72PAS/UST] Pashinkin, A. S., Ustyugov, G. P., Bakeeva, S. S., Taraskin, V. V., Bakeev, M. I., On the evaporation of mercury selenite, *Tr. Khim.-Metall. Inst. Akad. Nauk Kaz. SSR*, **19**, (1972), 126-129, in Russian. Cited on page: 285.

[72POL/MAK] Polukarov, A. N., Makhmetov, M. Z., Dissociation pressure of zinc selenite, *Zh. Fiz. Khim.*, **46**, (1972), 1059, in Russian, English translation in [72POL/MAK2]. Cited on pages: 261, 262.

[72POL/MAK2] Polukarov, A. N., Makhmetov, M. Z., Dissociation pressure of zinc selenite, *Russ. J. Phys. Chem.*, **46**, (1972), 622. From a citation in this Bibliography

[72SID/KON] Sidorova, E. E., Kondrat'ev, S. N., Mochalov, K. N., Blagoveshchenskaya, G. I., Pribylov, K. P., Thermographic study of the decomposition of the higher oxides of selenium, *Zh. Neorg. Khim.*, **17**, (1972), 299-301, in Russian, English translation in [72SID/KON2]. Cited on pages: 129, 130.

[72SID/KON2] Sidorova, E. E., Kondrat'ev, S. N., Mochalov, K. N., Blagoveshchenskaya, G. I., Pribylov, K. P., Thermographic study of the decomposition of the higher oxides of selenium, *Russ. J. Inorg. Chem.*, **17**, (1972), 157-158. From a citation in this Bibliography

[72SIG/WIE] Sigai, A. G., Wiedemeier, H., Solid-gas phase equilibria and thermodynamic properties of cadmium selenide, *J. Electrochem. Soc.*, **119**, (1972), 910-914. Cited on pages: 267, 268, 269, 270, 536.

[72SMO/MAN] Smoes, S., Mandy, F., Vander Auwera-Mahieu, A., Drowart, J., Determination by the mass spectrometric Knudsen cell method of the dissociation energies of the group IB chalcogenides, *Bull. Soc. Chim. Belg.*, **81**, (1972), 45-56. Cited on pages: 290, 292, 297, 298, 308, 309.

[72SVE] Svendsen, S. R., Decomposition pressures and standard enthalpy of formation for the iron selenides FeSe, Fe_7Se_8, Fe_3Se_4 and $FeSe_2$, *Acta Chem. Scand.*, **26**, (1972), 3757-3774. Cited on pages: 326, 327, 329.

[72SVE2] Svendsen, S. R., The enthalpy of iron diselenide from 300 to 853 K by drop calorimetry, *Acta Chem. Scand.*, **26**, (1972), 3834-3838. Cited on page: 329.

[72WHI/MCC] White, H. W., McCollum, D. C., Specific heat of EuSe between 0.5 and 16°K, *J. Appl. Phys.*, **43**, (1972), 1225-1226. Cited on page: 366.

[73BAK/PAS] Bakeeva, S. S., Pashinkin, A. S., Bakeev, M. I., Buketov, E. A., Dissociation pressure of gold selenite, *Zh. Fiz. Khim.*, **47**, (1973), 481, in Russian, English translation in [73BAK/PAS3]. Cited on page: 309.

[73BAK/PAS3] Bakeeva, S. S., Pashinkin, A. S., Bakeev, M. I., Buketov, E. A., Dissociation pressure of gold selenite, *Russ. J. Phys. Chem.*, **47**, (1973), 275. From a citation in this Bibliography

[73BAK/PAS2] Bakeeva, S. S., Pashinkin, A. S., Bakeev, M. I., Buketov, E. A., Dissociation pressure of iron(III) selenites, *Zh. Fiz. Khim.*, **47**, (1973), 481, in Russian, English translation in [73BAK/PAS4]. Cited on page: 331.

[73BAK/PAS4] Bakeeva, S. S., Pashinkin, A. S., Bakeev, M. I., Buketov, E. A., Dissociation pressure of iron(III) selenites, *Russ. J. Phys. Chem.*, **47**, (1973), 276. From a citation in this Bibliography

[73BAR/KNA] Barin, I., Knacke, O., *Thermochemical properties of inorganic substances*, Springer Verlag, Berlin, (1973), 921 pp.. Cited on pages: 224, 261, 274, 285, 295, 313, 321, 396.

[73BAR/MOR] Barnes, D. S., Mortimer, C. T., Enthalpies of combustion of selenium and diphenyl selenide, *J. Chem. Thermodyn.*, **5**, (1973), 371-377. Cited on page: 121.

[73BAT] Bates, R. G., *Determination of pH, theory and practice*, John Wiley & Sons, New York, (1973), 479 pp.. Cited on page: 21.

[73BRE/MEI] Brekow, G., Mei, M., Scheiba, M., Tausend, A., Wobig, D., Specific heat of thallium monoselenide between 3 and 600 K, *J. Phys. C: Solid State Phys.*, **6**, (1973), L462-L464. Cited on pages : 248, 249.

[73CAI/KUC] Cairns, E. J., Kucera, G. H., Cunningham, P. T., Thermodynamic studies of the lithium-selenium system by an emf method, *J. Electrochem. Soc.*, **120**, (1973), 595-597. Cited on page: 412.

[73CEC/RIE] Cech, F., Rieder, M., Vrana, S., Drysdallite, $MoSe_2$, a new mineral, *Neues Jahrb. Mineral., Monatsh.*, **10**, (1973), 433-442. Cited on page : 634.

[73DEI/TAN] Deichman, E. N., Tananaev, I. V., Kadoshnikova, N. V., Sodium selenatoindates, *Zh. Neorg. Khim.*, **18**, (1973), 2085-2089, in Russian, English translation in [73DEI/TAN2]. Cited on page : 247.

[73DEI/TAN2] Deichman, E. N., Tananaev, I. V., Kadoshnikova, N. V., Sodium selenatoindates, *Russ. J. Inorg. Chem.*, **18**, (1973), 1103-1105. From a citation in this Bibliography

[73GAN/KUZ] Ganelina, E. S., Kuz'micheva, V. P., Krasnopol'skaya, M. B., State of the selenite ion in aqueous solution, *Zh. Neorg. Khim.*, **18**, (1973), 1321-1325, in Russian, English translation in [73GAN/KUZ2]. Cited on page: 140.

[73GAN/KUZ2] Ganelina, E. S., Kuz'micheva, V. P., Krasnopol'skaya, M. B., State of the selenite ion in aqueous solution, *Russ. J. Inorg. Chem.*, **18**, (1973), 698-700. From a citation in this Bibliography

[73GRE/BOR] Greenberg, J. H., Borjakova, V. A., Shevelkov, V. F., Thermodynamic properties of In_2Se, *J. Chem. Thermodyn.*, **5**, (1973), 233-237. Cited on page: 241.

[73GRO2] Grønvold, F., Heat capacities and thermodynamic properties of hexagonal and liquid selenium in the range 298 to 1000 K. Enthalpy and temperature of fusion, *J. Chem. Thermodyn.*, **5**, (1973), 525-531. Cited on page: 91.

[73HAS/MUN] Hassan, M. S., Munir, Z. A., Studies on the sublimation of IIB-VIA compounds. VI. Thermodynamics of the dissociation of zinc selenide, *High Temp. Sci.*, **5**, (1973), 34-39. Cited on pages: 257, 258, 537.

[73HUL/DES] Hultgren, R., Desai, P. D., Hawkins, D. T., Gleiser, M., Kelley, K. K., Wagman, D. D., *Selected values of the thermodynamic properties of the elements*, American Society for Metals, Metals Park, Ohio, USA, (1973), 636 pp.. Cited on pages: 93, 95, 194, 308, 366, 367, 534, 646.

[73KLU/LES] Klushina, T. V., Leshchinskaya, Z. L., Selivanova, N. M., Energies of crystal lattices and specific heats of selenites, *Izv. Vyssh. Uchebn. Zaved., Khim. Khim. Tekhnol.*, **16**, (1973), 362-364, in Russian. Cited on page: 224.

[73KUB] Kubaschewski, P., Spezifische Wärmen und thermische Fehlordnung von Kupferchalkogeniden. Teil II: Die Systeme Cu_2S-CuS und Cu_2Se-CuSe, *Ber. Bunsen-Ges.*, **77**, (1973), 74-80, in German. Cited on page: 291.

[73LEB/BOS] Lebreton, J., Bosser, G., Marsigny, L., New emission spectrum of carbon monoselenide, *J. Phys. B: At. Mol. Phys.*, **6**, (1973), L226-L227. Cited on page: 204.

[73MAE/YOK] Maekawa, T., Yokokawa, T., Niwa, K., Enthalpies of binary mixtures of chalcogen elements, *Bull. Chem. Soc. Jpn.*, **46**, (1973), 761-765. Cited on pages: 180, 183.

[73NAZ/SHI] Nazarenko, V. A., Shitareva, G. G., Poluektova, E. N., The ionisation constants of tellurous and selenous acids, *Zh. Neorg. Khim.*, **18**, (1973), 1155-1159, in Russian, English translation in [73NAZ/SHI2]. Cited on page: 141.

[73NAZ/SHI2] Nazarenko, V. A., Shitareva, G. G., Poluektova, E. N., The ionisation constants of tellurous and selenous acids, *Russ. J. Inorg. Chem.*, **18**, (1973), 609-611. From a citation in this Bibliography

[73NES/PIN] Nesterenko, V. P., Pinaev, G. F., Pechkovskii, V. V., Dissociation presure of zirconium diselenite, *Zh. Fiz. Khim.*, **47**, (1973), 1057-1058, in Russian, English translation in [73NES/PIN2]. Cited on page: 352.

[73NES/PIN2] Nesterenko, V. P., Pinaev, G. F., Pechkovskii, V. V., Dissociation presure of zirconium diselenite, *Russ. J. Phys. Chem.*, **47**, (1973), 603. From a citation in this Bibliography

[73OKU/OKA] Okumura, A., Okazaki, N., Kinetics of oxygen exchange between selenite ions and water, *Bull. Chem. Soc. Jpn.*, **46**, (1973), 1084-1088. Cited on pages: 140, 141.

[73PIN/NES] Pinaev, G. F., Nesterenko, V. P., Glybin, V. P., Pechkovskii, V. V., Standard enthalpy of formation of zirconium diselenite, *Dokl. Akad. Nauk BSSR*, **17**, (1973), 446-449, in Russian. Cited on pages: 352, 537.

[73PIT/MAY] Pitzer, K. S., Mayorga, G., Thermodynamics of electrolytes. II. Activity and osmotic coefficients for strong electrolytes with one or both ions univalent, *J. Phys. Chem.*, **77**, (1973), 2300-2308. Cited on page: 590.

[73PIT] Pitzer, K. S., Thermodynamics of electrolytes: I. Theoretical basis and general equations, *J. Phys. Chem.*, **77**, (1973), 268-277. Cited on pages: 587, 590.

[73POL/ZAK] Poltavtsev, Y. G., Zakharov, V. P., Protas, I. M., Remizovich, T. V., Chugaev, V. N., Molecular composition of vapors and condensate structure during sputtering of some arsenic chalcogenides by laser emission, *Ukr. Fiz. Zh. (Russ. Ed.)*, **18**, (1973), 752-755, in Russian. Cited on page: 189.

[73POW] Powell, H. K. J., Entropy titrations: A reassessment of data for the reaction of the sulphate ion with protons and with bivalent metals, *J. Chem. Soc. Dalton Trans.*, (1973), 1947-1951. Cited on pages: 521, 549.

[73RAM/THU] Ramakrishna, R. S., Thuraisingham, R., Polarographic studies of the thiocyanates and selenocyanates of thallium, indium and gallium, *J. Inorg. Nucl. Chem.*, **35**, (1973), 2805-2812. Cited on page: 254.

[73SEL/ZAL] Selivanova, N. M., Zalogina, E. A. , Gorokhov, V. K., Heat of formation of aluminium selenate, *Zh. Fiz. Khim.*, **47**, (1973), 2713-2714, in Russian, English translation in [73SEL/ZAL2]. Cited on pages : 232, 538.

[73SEL/ZAL2] Selivanova, N. M., Zalogina, E. A. , Gorokhov, V. K., Heat of formation of aluminium selenate, *Russ. J. Phys. Chem.*, **47**, (1973), 1530. From a citation in this Bibliography

[73SHA/AZI] Sharifov, K., Azizov, T., Aliev, I., Abbasov, A., Sixth All-Union Conf. On Calorimetry. Expanded Summaries of Reports, (1973). Cited from [77BUR] on page: 234.

[73SHA/MIS] Shamsuddin, Misra, S., Thermodynamic properties of lead selenide, *Scr. Metall.*, **7**, (1973), 547-554. Cited on pages: 220, 539.

[73SMO/EFR] Smolyakova, K. E., Efremov, A. A., Serebrennikov, V. V., Enthalpies of formation of selenates of the rare-earth metals, *Zh. Fiz. Khim.*, **47**, (1973), 2452-2453, in Russian, English translation in [73SMO/EFR2]. Cited on pages: 363, 364, 366, 368, 375, 384, 518, 539.

[73SMO/EFR2] Smolyakova, K. E., Efremov, A. A., Serebrennikov, V. V., Enthalpies of formation of selenates of the rare-earth metals, *Russ. J. Phys. Chem.*, **47**, (1973), 1389. From a citation in this Bibliography

[73SOB/SEL] Sobol, L. G., Selivanova, N. M., Heat of formation of sodium hydrogen selenate $NaHSeO_4$., *Izv. Vyssh. Uchebn. Zaved., Khim. Khim. Tekhnol.*, **16**, (1973), 1493-1495, in Russian. Cited on pages: 422, 526, 540.

[73STE/CAR] Steele, W. V., Carson, A. S., Laye, P. G., Spencer, J. A., The enthalpy of formation of lead nitrate, *J. Chem. Thermodyn.*, **5**, (1973), 477-479. Cited on page: 452.

[73ZIG/ORL] Zigel, U. U., Orlova, G. M., Low-temperature heat capacity of glassy arsenic selenides, *Zh. Prikl. Khim. (S.-Peterburg)*, **46**, (1973), 721-724, in Russian. Cited on page: 190.

[74ADE] Ader, M., Enthalpy of formation of lithium monoselenide, *J. Chem. Thermodyn.*, **6**, (1974), 587-597. Cited on page: 412.

[74AHR/AVS] Ahrland, S., Avsar, E., Kullberg, L., Thermodynamics of metal complex formation in aqueous solution. VI. Equilibrium and enthalpy measurements on the zinc and cadmium selenocyanate systems, *Acta Chem. Scand.*, A 28, (1974), 855-865. Cited on pages: 206, 266, 276, 277.

[74BEH/LEM] Behrens, R. G., Lemons, R. S., Rosenblatt, G. M., Vapor pressure and thermodynamics of selenium dioxide. The enthalpy of atomization of $SeO_2(g)$, *J. Chem. Thermodyn.*, **6**, (1974), 457-466. Cited on pages: 120, 122, 123, 124.

[74BLA/IGE2] Blachnik, R., Igel, R., Thermodynamische Eigenschaften von IV-VI-Verbindungen: Bleichalkogenide, *Z. Naturforsch.*, **29b**, (1974), 625-629, in German. Cited on pages: 214, 219.

[74DAS/SAH] Das, R. C., Sahu, G., Satyanarayana, D., Misra, S. N., Silver-silver selenocyanate electrode: determination of standard electrode potential, solubility product of AgSeCN and various thermodynamic parameters at 35, 40, 45 and 50°C, *Electrochim. Acta*, **19**, (1974), 887-890. Cited on page: 307.

[74GOS/MOL] Gospodinov, G., Molodyk, A. D., Pashinkin, A. S., Strel'chenko, S. S., Determination of the saturated vapor pressure of solid AsSe, *Inorg. Mater.*, **10**, (1974), 962-963. Cited on page: 189.

[74HAR/EIC] Hariharan, A. V., Eick, H. A., Sublimation thermodynamics of EuSe, *J. Chem. Thermodyn.*, **6**, (1974), 373-378. Cited on pages: 366, 367.

[74HEL/KIR] Helgeson, H. C., Kirkham, D. H., Theoretical prediction of the thermodynamic behavior of aqueous electrolytes at high pressures and temperatures: I. Summary of the thermodynamic/electrostatic properties of the solvent, *Am. J. Sci.*, **274**, (1974), 1089-1198. Cited on page: 594.

[74IRW/LAC] Irwin, J. C., LaCombe, J., Specific heats of ZnTe, ZnSe, and GaP, *J. Appl. Phys.*, **45**, (1974), 567-573. Cited on page: 255.

[74JEL/KOM] Jelinek, H., Komarek, K. L., Thermodynamic properties of cobalt-selenium alloys, *Monatsh. Chem.*, **105**, (1974), 689-711. Cited on pages: 317, 318, 541.

[74KOT/LEB2] Kotlik, S. B., Lebedeva, G. N., Equilibrium pressures of HCN and NH_3 over aqueous solutions, *Russ. J. Appl. Chem.*, **47**, (1974), 440-441. Cited on pages: 434, 435.

[74KUL] Kullberg, L., Thermodynamics of metal complex formation in aqueous solution. VII. Equilibrium and enthalpy measurements on the nickel(II)-selenocyanate system, *Acta Chem. Scand.*, A 28, (1974), 897-902. Cited on pages: 297, 315, 316.

[74KUL2] Kullberg, L., Thermodynamics of metal complex formation in aqueous solution. VIII. A calorimetric study of the mercury(II) thiocyanate, selenocyanate, and thiosulfate systems, *Acta Chem. Scand.*, A 28, (1974), 979-985. Cited on pages: 287, 541.

[74MCM/GSC] McMasters, O. D., Gschneidner, K. A., Kaldis, E., Sampietro, G., High-temperature enthalpies and standard Gibbs free energies of formation of the europium chalcogenides: EuO, EuS, EuSe, and EuTe, *J. Chem. Thermodyn.*, **6**, (1974), 845-857. Cited on pages: 358, 366, 487.

[74MIL] Mills, K. C., *Thermodynamic data for inorganic sulphides, selenides and tellurides*, Butterworths, London, (1974), 845 pp.. Cited on pages: 88, 92, 93, 95, 99, 100, 101, 102, 103, 104, 105, 106, 108, 109, 111, 112, 113, 114, 115, 116, 118, 120, 121, 122, 127, 132, 133, 163, 169, 170, 175, 192, 230, 233, 234, 235, 236, 252, 317, 332, 336, 343, 349, 352, 353, 355, 388, 389, 392, 394, 399, 403, 412, 416, 417, 422, 438, 479, 487, 512, 514, 583.

[74MON] De Montreuil, L. A., An unknown cobalt-iron selenide, *El Ingeniero, Geol. Univ. Nacl. Mayor de San Marcos, Geol.*, (1974), 65-70. Cited on page: 637.

[74MOR/SEK] Moriya, H., Sekine, T., A solvent-extraction study of zinc(II) complexes with several divalent anions of carboxylic and inorganic acids, *Bull. Chem. Soc. Jpn.*, **47**, (1974), 747-748. Cited on page: 264.

[74MOR/VLA] Morozova, M. P., Vladimirova, V. A., Stolyarova, T. A., Nepomnyashchaya, V. I., Enthalpy of formation of nickel selenides, *Russ. J. Phys. Chem.*, **48**, (1974), 446. Cited on pages: 310, 311.

[74MUS/AZI] Mustafaev, F. M., Azizov, T. K., Aliev, I. Y., Abbasov, A. S., Thermodynamic properties of the indium-selenium system, *Azerb. Khim. Zh.*, **2**, (1974), 93-96, in Russian. Cited on pages: 239, 243.

[74NAG/SHI] Nagai, S.-I., Shinmei, M., Yokokawa, T., Heats of atomization, dissociation energies and heats of sublimation of several rare earth monochalcogenides, *J. Inorg. Nucl. Chem.*, **36**, (1974), 1904-1905. Cited on pages: 367, 369, 370, 372, 373, 376, 377, 379, 381, 382.

[74PAR] Parsons, R., Manual of Symbols and Terminology for Physicochemical Quantities and Units. Appendix III. Electrochemical nomenclature, *Pure Appl. Chem.*, **37**, (1974), 503-516. Cited on page: 20.

[74PET/OWE] Petruk, W., Owens, D. R., Stewart, J. M., Murray, E. J., Observations on acanthite, aquilarite and naumannite, *Can. Mineral.*, **12**, (1974), 365-369. Cited on pages: 634, 635.

[74PIN/GOR] Pinaev, G. F., Gornostaeva, L. V., Volkova, V. P., Glybin, V. P., A calorimetric study of iron(III) selenites, *Zh. Fiz. Khim.*, **48**, (1974), 2713-2714, in Russian, English translation in [74PIN/GOR2]. Cited on pages: 330, 331.

[74PIN/GOR2] Pinaev, G. F., Gornostaeva, L. V., Volkova, V. P., Glybin, V. P., A calorimetric study of iron(III) selenites, *Russ. J. Phys. Chem.*, **48**, (1974), 1705. From a citation in this Bibliography

[74PIT/KIM] Pitzer, K. S., Kim, J. J., Thermodynamics of electrolytes. IV. Activity and osmotic coefficients for mixed electrolytes, *J. Am. Chem. Soc.*, **96**, (1974), 5701-5707. Cited on page: 590.

[74PIT/MAY] Pitzer, K. S., Mayorga, G., Thermodynamics of electrolytes. III. Activity and osmotic coefficients for 2-2 electrolytes, *J. Solution Chem.*, **3**, (1974), 539-546. Cited on page: 590.

[74RAU] Rau, H., Vapour composition and critical constants of selenium, *J. Chem. Thermodyn.*, **6**, (1974), 525-535. Cited on pages: 88, 99, 100, 101, 103, 105, 106, 109, 113, 116, 495, 542, 558.

[74WAS/NOU] Wasif, S., Nour, M. M., Hussein, M. A., Solutions in selenic acid: Part 7. Cryoscopic studies of selenic acid. The SeO_3 + H_2O system., *J. Chem. Soc. A*, **70**, (1974), 929-935. Cited on pages: 145, 146.

[74WIE/IRE] Wiedemeier, H., Irene, E. A., Knudsen measurements of the sublimation and the heat of formation of GeSe, *Z. Anorg. Allg. Chem.*, **404**, (1974), 299-307. Cited on pages: 210, 212.

[75BIR] Birch, J. A., Heat capacities of ZnS, ZnSe and CdTe below 25 K, *J. Phys. C: Solid State Phys.*, **8**, (1975), 2043-2047. Cited on page: 255.

[75BLE/MIJ] Bleijerveld, R. H. T., Mijlhoff, F. C., Vapour pressure and structure of selenium trioxide, *Recl. Trav. Chim. Pays-Bas*, **94**, (1975), 190-191. Cited on pages: 128, 129.

[75BRE/MEI] Brekow, G., Mei, M., Scheiba, M., Tausend, A., Wobig, D., Specific heat of Tl_2Se between 3 and 640 K, *J. Phys. C: Solid State Phys.*, **8**, (1975), L456-L460. Cited on pages: 250, 251.

[75DEM/KOS] Demidenko, A. F., Koshchenko, V. I., Grinberg, Y. K., Boryakova, V. A., Gastev, S. V., Thermodynamic and optical properties of In_2Se_3, *Inorg. Mater.*, **11**, (1975), 1839-1841. Cited on pages: 242, 243, 566.

[75ELL/SEV] Ellert, G. V., Sevast'yanov, V. G., Slovyanskikh, V. K., The Se-U and Te-U systems, *Russ. J. Inorg. Chem.*, **20**, (1975), 120-124. Cited on pages: 385, 386, 387.

[75GRO/KVE] Grønvold, F., Kveseth, N. J., Sveen, A., Vacancy disorder in $Ni_{1-x}Se$. Heat capacities and thermodynamic properties of 10 samples in the range 300 to about 750 K, *J. Chem. Thermodyn.*, 7, (1975), 617-632. Cited on page: 311.

[75GRO] Grønvold, F., Heat capacities and thermodynamic properties of iron diselenide and nickel diselenide from 300 to 1000 K, *J. Chem. Thermodyn.*, 7, (1975), 645-654. Cited on pages: 311, 329.

[75HOA/REY] Hoareau, A., Reymond, J.-M., Cabaud, B., Uzan, R., Etude des agrégats de sélénium en phase vapeur en fonction de leur maille. I. Résultats expérimentaux par spectrométrie de masse, *J. Phys. (Paris)*, **36**, (1975), 737-743, in French. Cited on pages: 97, 105, 106, 107, 108, 110, 112, 113, 114, 116, 117, 118, 542.

[75KAD/DEI] Kadoshnikova, N. V., Deichman, E. N., Tananaev, I. V., Kharitonov, Y. Y., Potassium selenatoindates, *Zh. Neorg. Khim.*, **20**, (1975), 2066-2070, in Russian, English translation in [75KAD/DEI2]. Cited on page: 247.

[75KAD/DEI2] Kadoshnikova, N. V., Deichman, E. N., Tananaev, I. V., Kharitonov, Y. Y., Potassium selenatoindates, *Russ. J. Inorg. Chem.*, **20**, (1975), 1150-1152. From a citation in this Bibliography

[75KIW/WES3] Kiwia, H. L., Westrum, Jr., E. F., Low-temperature heat capacities of molybdenum diselenide and ditelluride, *J. Chem. Thermodyn.*, **7**, (1975), 683-691. Cited on page: 345.

[75KOV/SID] Kovalevskaya, E. I., Sidorenko, A. Y., Lysenko, A. A., Fesenko, V. V., Thermodynamic functions of gaseous monoselenides of the cerium subgroup, *Russ. J. Phys. Chem.*, **49**, (1975), 791. Cited on pages: 365, 367, 369, 372, 376, 379, 381.

[75LAG/SUS] Lagnier, R., Suski, W., Wojakowski, A., Heat capacity of the uranium sesquiselenide, *Phys. Status Solidi*, **A29**, (1975), K51-K54. Cited on page: 388.

[75LYS/VAK] Lyskova, Y. B., Vakhobov, A. V., Phase diagrams of the systems Sr-Se and Sr-Te, *Inorg. Mater.*, **11**, (1975), 304-305. Cited on pages: 402, 403.

[75LYS/VAK2] Lyskova, Y. B., Vakhobov, A. V., Phase diagrams of the systems Ba-Se and Ba-Te, *Inorg. Mater.*, **11**, (1975), 1784-1785. Cited on page: 406.

[75MON] De Montreuil, L. A., Bellidoite: a new copper selenide, *Econ. Geol.*, **70**, (1975), 384-387. Cited on page: 634.

[75MUM] Mumme, W. G., Junoite, $Cu_2Pb_3Bi_8(S,Se)_{16}$ a new sulfosalt from Tennant Creek, Australia: its crystal structure and relationship with other bismuth sulfosalts, *Am. Mineral.*, **60**, (1975), 548-558. Cited on page: 635.

[75MUR/HEY] Murray, R. M., Heyding, R. D., The copper-selenium system at temperatures to 850 K and pressures to 50 kbar, *Can. J. Chem.*, **53**, (1975), 878-887. Cited on page: 292.

[75NOR] Norbury, A. H., Coordination chemistry of the cyanate, thiocyanate, and selenocyanate ions, *Adv. Inorg. Chem. Radiochem.*, **17**, (1975), 231-386. Cited on page: 206.

[75PIT] Pitzer, K. S., Thermodynamics of electrolytes. V. Effects of higher-order electrostatic terms, *J. Solution Chem.*, **4**, (1975), 249-265. Cited on page: 590.

[75PRE/PIE] Predel, B., Piehl. J., Pool, M. J., Beitrag zur Kenntnis der thermodynamischen Eigenschafte flüssiger Thallium-Selen-, Wismut-Selen-, und Antimon-Selen-Legierungen, *Z. Metallkd.*, **66**, (1975), 388-395, in German. Cited on pages: 250, 251.

[75SAT/SAH] Satyanarayana, D., Sahu, G., Das, R. C., Studies on ion association: thermodynamics of formation of monoselenocyanato-complexes of cobalt(II), nickel(II), and cadmium(II), *J. Chem. Soc. Dalton Trans.*, (1975), 2236-2239. Cited on pages: 276, 277, 278, 315, 316, 323.

[75SHK/MAK] Shkodin, V. G., Makhmetov, M. Z., Malyshev, V. P., Dissociation pressure of sodium pyroselenite, VINITI, (1975), in Russian. Cited on page: 418.

[75VES/KOS] Vesala, A., Koskinen, V., The first dissociation constant of selenous acid in light and heavy water determined by potentiostatic titration, *Finn. Chem. Lett.*, (1975), 145-148. Cited on pages: 141, 142.

[76ASK/ALI] Askerova, K. A., Alieva, N. A., Azizov, T. K., Abbasov, A. S., Mustafayev, F. M., Investigation of the thermodynamic properties of copper selenides, *Izv. Akad. Nauk Az. SSR, Ser. Fiz.-Tekh. Mat. Nauk*, **1976**, (1976), 137-139, in Russian. Cited on pages: 289, 290, 291, 292.

[76BAB/KUL] Babanli, M. B., Kuliev, A. A., Evaluation of the melting heat and melting entropy of the chalcogenide of gallium and indium of the $A_2^{III}B_3^{VI}$ type, *Izv. Akad. Nauk Az. SSR, Ser. Fiz.-Tekh. Mat. Nauk*, **1976**, (1976), 145-147, in Russian. Cited on page: 236.

[76BAE/MES] Baes, Jr., C. F., Mesmer, R. E., *The hydrolysis of cations*, Wiley & Sons, New York, (1976), 489 pp.. Cited on pages : 26, 141, 148, 224, 262, 272, 355, 395, 400, 404, 407, 493, 502, 504, 538, 565, 584, 590, 627, 645.

[76BRO/CAS] Bros, H., Castanet, R., Kehiaian, H. V., Etude calorimétrique du système bismuth-sélénium, *High Temp. High Pressures*, **8**, (1976), 271-278, in French. Cited on pages : 197, 201, 202.

[76CHA/CHE] Chao, E. E., Cheng, K. L., Stepwise titration of some anion mixtures and determination of K_{sp} of silver precipitates with silver ion selective electrode, *Anal. Chem.*, **48**, (1976), 267-271. Cited on pages: 303, 544.

[76COO/WOO] Cooper, J. N., Woods, M., Sullivan, J. C., Deutsch, E., Kinetics of the oxidation of selenium(IV) by neptunium(VII) in aqueous acidic perchlorate media, *Inorg. Chem.*, **15**, (1976), 2862-2864. Cited on page: 140.

[76GLA/MIK] Glazunov, M. P., Mikhailov, E. S., Piskarev, N. V., Chupakhin, M. S., A mass-spectrometric investigation of the evaporation of molybdenum and tantalum diselenides, *Sov. Powder Metallurgy Met. Ceram.*, **15**, (1976), 866-868. Cited on pages: 345, 346, 351, 544.

[76GRE/BAR] Greenwood, D. J., Barrow, R. F., A medium-resolution study of fluorescence in $^{80}Se_2$ excited by lines of the argon-ion laser, *J. Phys. B: At. Mol. Phys.*, **9**, (1976), 2123-2126. Cited on page: 94.

[76JOH/KVA] Johan, Z., Kvacek, M., Picot, P., La petrovicite, $Cu_3HgPbBiSe_5$, un nouveau minéral, *Bull. Soc. Fr. Minéral. Cristallogr.*, **99**, (1976), 310-313, in French. Cited on page: 635.

[76KIR/RAV] Kirk, D. L., Raven, M. S., A theory dealing with the vacuum evaporation and deposition of binary semiconducting compounds, with special reference to the thermal evaporation of zinc selenide, *J. Phys. D: Appl. Phys.*, **9**, (1976), 2015-2026. Cited on pages: 257, 258, 545.

[76MIL] Mills, K. C., Molar heat capacities and enthalpies of transition for InSe(c), $InSe_{1.2}$(c) and In_2Se_3(c), *High Temp. High Pressures*, **8**, (1976), 225-230. Cited on pages: 239, 242, 243, 566.

[76MUM] Mumme, W. G., Proudite from Tennant Creek, Northern Territory, Australia: its crystal structure and relationship with weibullite and wittite, *Am. Mineral.*, **61**, (1976), 839-852. Cited on page: 635.

[76PET/KOF] Petrova, Z. K., Kofman, N. A., Heat capacity and some thermodynamic properties of cadmium selenide, *Dokl. Akad. Nauk BSSR*, **20**, (1976), 688-690, in Russian. Cited on pages: 267, 268, 545.

[76PIA/BAR] Piacente, V., Bardi, G., Di Paolo, V., Ferro, D., The vapour pressure over Ga_2S_2 and Ga_2Se_2, *J. Chem. Thermodyn.*, **8**, (1976), 391-401. Cited on pages: 233, 234, 235, 545.

[76PIN/VOL] Pinaev, G. F., Volkova, V. P., Crystallisation fields of iron selenites in the Fe_2O_3-SeO_2-H_2O system at 120-240°C, *Zh. Neorg. Khim.*, **21**, (1976), 1341-1345, in Russian, English translation in [76PIN/VOL2]. Cited on page: 330.

[76PIN/VOL2] Pinaev, G. F., Volkova, V. P., Crystallisation fields of iron selenites in the Fe_2O_3-SeO_2-H_2O system at 120-240°C, *Russ. J. Inorg. Chem.*, **21**, (1976), 734-736. From a citation in this Bibliography

[76PIT/SIL] Pitzer, K. S., Silvester, L. F., Thermodynamics of electrolytes. VI. Weak electrolytes including H_3PO_4, *J. Solution Chem.*, **5**, (1976), 269-278. Cited on page: 590.

[76RAS/MED] Rasulov, S. M., Medzhidov, R. A., Thermodynamic properties of GeSe in solid and liquid states, *Inzh.-Fiz. Zh.*, **31**, (1976), 739-740, in Russian. Cited on page: 209.

[76SCA] Scatchard, G., *Equilibrium in solutions: Surface and colloid chemistry*, Harvard University Press, Cambridge, Massachusetts, (1976), 306 pp.. Cited on page: 589.

[76SMI/MOS] Smith, J. F., Moser, Z., Thermodynamic properties of binary lithium systems – a review, *J. Nucl. Mater.*, **59**, (1976), 158-174. Cited on page: 412.

[76TAN/VOL] Tananaev, I. V., Volodina, A. N., Selivanova, N. M., Bol'shakova, N. K., Heats of formation of gallium hydrogen selenites, *Zh. Neorg. Khim.*, **21**, (1976), 1138-1140, in Russian, English translation in [76TAN/VOL3]. Cited on pages: 237, 546.

[76TAN/VOL3] Tananaev, I. V., Volodina, A. N., Selivanova, N. M., Bol'shakova, N. K., Heats of formation of gallium hydrogen selenites, *Russ. J. Inorg. Chem.*, **21**, (1976), 619-620. From a citation in this Bibliography

[76TAN/VOL2] Tananaev, I. V., Volodina, A. N., Selivanova, N. M., Bol'shakova, N. K., Heats of formation of normal gallium selenites, *Zh. Neorg. Khim.*, **21**, (1976), 1135-1138, in Russian, English translation in [76TAN/VOL4]. Cited on pages: 237, 546, 547.

[76TAN/VOL4] Tananaev, I. V., Volodina, A. N., Selivanova, N. M., Bol'shakova, N. K., Heats of formation of normal gallium selenites, *Russ. J. Inorg. Chem.*, **21**, (1976), 617-619. From a citation in this Bibliography

[76THO/PRI] Thorpe, R. I., Pringle, G. J., Plant, A. G., Occurrence of selenide and sulphide minerals in bornite ore of the Kidd Creek massive sulphide deposit, Timmins, Ontario, *Pap.-Geol. Surv. Can.*, **76-1A**, (1976), 311-317. Cited on page: 637.

[77AZI/ABB] Azizov, T. K., Abbasov, A. S., Mustafaev, F. M., Aliev, I. Y., Determination of the enthalpy of formation of Ce_2Te_3, Dy_2S_3, Dy_2Se_3, Dy_2Te_3, Yb_2S_3, Yb_2Se_3, *Vses. Konf. Kalorim., [Rasshir. Tezisy Dokl.], 7th*, **1**, (1977), 30-32, in Russian. Cited on pages: 358, 362.

[77BUR] Burylev, B. P., Determination of the standard enthalpy of formation of Ga_2Se_3, *Inorg. Mater.*, **13**, (1977), 754-755. Cited on page: 236.

[77COR/OUW2] Cordfunke, E. H. P., Ouweltjes, W., Standard enthalpies of formation of uranium compounds: IV. α-UO_2SO_4, β-UO_2SeO_4, and UO_2SeO_3, *J. Chem. Thermodyn.*, **9**, (1977), 1057-1062. Cited on pages: 149, 389, 390.

[77DAV/CLA] Davis, R. J., Clark, A. M., Criddle, A. J., Palladseite, a new mireral from Itabira, Minas Gerais, Brazil, *Mineral. Mag.*, **41**, (1977), M10-M13. Cited on page: 635.

[77DRO/SMO] Drowart, J., Smoes, S., Determination by the mass spectrometric Knudsen cell method and discussion of the dissociation energies of the molecules Se_2(g), SSe(g) and SeTe(g), *J. Chem. Soc. Faraday Trans. 2*, **73**, (1977), 1755-1767. Cited on pages: 93, 181, 184.

[77DUT/TIS] Dutchak, Y. I., Tishchenko, I. A., Korenchuk, N., Mass-spectrometric investigation of the vapor pressure above solid and liquid silver chalcogenides, *Inorg. Mater.*, **13**, (1977), 1586-1588. Cited on pages: 299, 300, 547.

[77FER] Fernelius, W. C., *How to name an inorganic substance*, IUPAC Commission on Nomenclature of Inorganic Compounds, Pergamon Press, Oxford, (1977), 36 pp.. Cited on page: 11.

[77FIN/MRO] Finkelman, R. B., Mrose, M. E., Downeyite, the first verified natural occurrence of SeO_2, *Am. Mineral.*, **62**, (1977), 316-320. Cited on page: 634.

[77FOW/STR] Fowless, A. D., Stranks, D. R., Selenitometal complexes: 1. Synthesis and characterization of selenito complexes of cobalt(III) and their equilibrium properties in solution, *Inorg. Chem.*, **16**, (1977), 1271-1276. Cited on pages: 140, 141, 142, 446.

[77FOW/STR2] Fowless, A. D., Stranks, D. R., Selenitometal complexes: 3. Kinetics and mechanism of the reaction of hydroxocobalt(III) and hydroxorhodium(III) cations with monomeric and dimeric selenite anions, *Inorg. Chem.*, **16**, (1977), 1282-1286. Cited on page: 140.

[77KAD/DEI] Kadoshnikova, N. V., Deichman, E. N., Tananaev, I. V., Rubidium selenatoindates, *Zh. Neorg. Khim.*, **22**, (1977), 924-927, in Russian, English translation in [77KAD/DEI2]. Cited on pages: 247, 427.

[77KAD/DEI2] Kadoshnikova, N. V., Deichman, E. N., Tananaev, I. V., Rubidium selenatoindates, *Russ. J. Inorg. Chem.*, **22**, (1977), 512-514. From a citation in this Bibliography

[77KAD/DEI3] Kadoshnikova, N. V., Deichman, E. N., Tananaev, I. V., Caesium selenatoindates, *Zh. Neorg. Khim.*, **22**, (1977), 675-678, in Russian, English translation in [77KAD/DEI4]. Cited on pages: 247, 429.

[77KAD/DEI4] Kadoshnikova, N. V., Deichman, E. N., Tananaev, I. V., Caesium selenatoindates, *Russ. J. Inorg. Chem.*, **22**, (1977), 372-374. From a citation in this Bibliography

[77KOV/SID] Kovalevskaya, E. I., Sidorenko, A. Y., Lysenko, A. A., Fesenko, V. V., Thermodynamic functions of gaseous yttrium-subgroup lanthanide monoselenides, *Russ. J. Phys. Chem.*, **51**, (1977), 607-608. Cited on pages: 357, 358, 360, 361, 362, 363.

[77KUL/ZLO] Kulyukhina, E. A., Zlomanov, V. P., Novoselova, A. V., p-T projection of the phase diagram of the system SnSe-Se, *Inorg. Mater.*, **13**, (1977), 200-203. Cited on page: 217.

[77LAK/PEL] Lakeenkov, V. M., Pelevin, O. V., The heat and entropy of fusion of zinc selenide, *Russ. J. Phys. Chem.*, **51**, (1977), 453-454. Cited on page: 259.

[77MIK/GLA] Mikhailov, E. S., Glazunov, M. P., Piskarev, N. V., Chupakhin, M. S., The evaporation of tungsten diselenide and tungsten niobium diselenide, *Russ. J. Phys. Chem.*, **51**, (1977), 422-423. Cited on pages: 348, 548.

[77MIK/PIS] Mikhailov, E. S., Piskarev, N. V., Chupakhin, M. S., Thermal dissociation of $NbSe_2$, *Inorg. Mater.*, **13**, (1977), 803-806. Cited on page: 350.

[77MOR/VLA] Morozova, M. P., Vladimirova, V. A., Grudetskii, A. V., Malysheva, E. A., Enthalpy of formation of cobalt selenides, *Russ. J. Phys. Chem.*, **51**, (1977), 927. Cited on pages: 317, 318.

[77OGA/NIS] Ogawa, S., Nishihara, Y., Specific heat of $CoSe_2$ and Mossbauer effect of ^{57}Fe doped in $CoSe_2$, *J. Phys. Soc. Jpn.*, **42**, (1977), 343-344. Cited on pages: 318, 548.

[77ORL] Orlyanskii, Y. N., The $(NH_4)_2SeO_4$-Na_2SeO_4-H_2O system at 30°C, *Zh. Neorg. Khim.*, **22**, (1977), 2274-2276, in Russian, English translation in [77ORL2]. Cited on page: 411.

[77ORL2] Orlyanskii, Y. N., The $(NH_4)_2SeO_4$-Na_2SeO_4-H_2O system at 30°C, *Russ. J. Inorg. Chem.*, **22**, (1977), 1230-1232. From a citation in this Bibliography

[77PIS/MIK] Piskarev, I. V., Mikhailov, E. S., Chupakhin, M. S., Determination of the vapour pressure of WSe_2, *Russ. J. Inorg. Chem.*, **22**, (1977), 1121-1123. Cited on pages: 348, 549.

[77RAS/MED] Rasulov, S. M., Medzhidov, R. A., Thermodynamic properties of glassy, crystalline, and liquid arsenic selenide, *Russ. J. Phys. Chem.*, **51**, (1977), 969-970. Cited on page: 190.

[77SHA/KOE] Sharp, K. W., Koehler, W. H., Synthesis and characterization of sodium polyselenides in liquid ammonia solution, *Inorg. Chem.*, **16**, (1977), 2258-2265. Cited on page: 154.

[77SMO/DRO] Smoes, S., Drowart, J., Determination of the atomization energies of the molecules CSe(g) and CSe_2(g) by the mass spectrometric Knudsen cell method, *J. Chem. Soc. Faraday Trans. 2*, **73**, (1977), 1746-1754. Cited on page: 204.

[77YAT/KUZ] Yatsenko, S. P., Kuznetsov, A. N., Chuntonov, K. A., Phase equilibria in the system Se-Rb, *Inorg. Mater.*, **13**, (1977), 602-603. Cited on page: 427.

[78ARU] Aruga, R., Thermodynamics of association of selenate and sulfate with bivalent metals. Calculation, by calorimetric data, of the water molecules displaced from a cationic sphere, *Inorg. Chem.*, **17**, (1978), 2503-2505. Cited on pages: 265, 275, 549.

[78BLA/GUN] Blachnik, R., Gunia, P. G., Enthalpien von Kupfer- und Silberchalkogeniden, *Z. Naturforsch.*, **33a**, (1978), 190-196, in German. Cited on pages: 291, 292, 298, 299, 527, 550.

[78CEC/VAV] Cech, F., Vavrin, I., Poubaite, $PbBi_2(Se,Te,S)_4$, a new mineral, *Neues Jahrb. Mineral., Monatsh.*, **1**, (1978), 9-19. Cited on page: 635.

[78CHE/MOR] Chekoev, N. G., Morachevskii, A. G., Thermodynamic properties of lithium-selenium system melts, *Zh. Prikl. Khim. (S.-Peterburg)*, **51**, (1978), 2348-2350, in Russian. Cited on page: 412.

[78CRA/SAL] Craven, R. A., Di Salvo, F. J., Hsu, F. S. L., Mechanisms for the 200 K transition in $TiSe_2$: A measurement of the specific heat, *Solid State Commun.*, **25**, (1978), 39-42. Cited on page: 352.

[78DUN/PEA] Dunn, P. J., Peacor, D. R., Sturman, B. D., Mandarinoite, a new ferric-iron selenite from Bolivia, *Can. Mineral.*, **16**, (1978), 605-609. Cited on page: 636.

[78GER/GLA] Gerasimov, Y. I., Glazov, V. M., Gasarev, V. G., Sharov, V. V., Evaporation and sublimation equations for simple subtances, *Dokl. Akad. Nauk SSSR*, **242**, (1978), 868-871, in Russian. Cited on page: 550.

[78HAY/NAK] Hayashi, K., Nakahira, M., Stability and the equilibrium selenium vapor pressure of the Vse_2 phase, *J. Solid State Chem.*, **24**, (1978), 153-161. Cited on page: 349.

[78JOH/KVA] Johan, Z., Kvacek, M., Picot, P., La sabatierite, un nouveau séléniure de cuivre et de thallium, *Bull. Minéral.*, **101**, (1978), 557-560, in French. Cited on page: 635.

[78KUS/MEI] Kusik, C. L., Meissner, H. P., Electrolyte activity coefficients in inorganic processing, *AlChE Symp. Ser.*, **74**, (1978), 14-20. Cited on page: 571.

[78LAI/NII] Laitinen, R., Niinistö , L., Thermal behaviour of the monoclinic allotropes γ-sulfur and α-selenium and some structurally related sulfur-selenium compounds, *J. Therm. Anal.*, **13**, (1978), 99-104. Cited on pages: 179, 180.

[78LUK/POL] Lukashenko, G. M., Polotskaya, R. I., Thermodynamic properties of manganese selenides, *Inorg. Mater.*, **14**, (1978), 1203-1205. Cited on pages: 337, 338, 339.

[78MAM/ALD] Mamedov, K. K., Aldzhanov, M. A., Kerimov, I. G., Mekhtiev, M. I., Specific heat and moments of the vibrational spectrum of gallium monochalcogenides, *Fiz. Tverd. Tela (Leningrad)*, **20**, (1978), 42-47, in Russian. Cited on page: 233.

[78PIT/PET] Pitzer, K. S., Peterson, J. R., Silvester, L. F., Thermodynamics of electrolytes: IX. Rare earth chlorides, nitrates, and perchlorates, *J. Solution Chem.*, **7**, (1978), 45-56. Cited on page: 590.

[78RAS/MED] Rasulov, S. M., Medzhidov, R. A., Enthalpy and heat capacity of lead selenide in the temperature interval 300-1480 K, *Teplofiz. Vys. Temp.*, **16**, (1978), 304-309, in Russian, English translation in [78RAS/MED2]. Cited on page: 219.

[78RAS/MED2] Rasulov, S. M., Medzhidov, R. A., Enthalpy and heat capacity of lead selenide in the temperature interval 300-1480°K, *High Temp.*, **16**, (1978), 251-256. From a citation in this Bibliography

[78ROL/HOC] Rolinski, E. J., Hoch, M., A mass spectrometric study of the vaporization of zinc selenide, *Adv. Mass Spectrom.*, **7A**, (1978), 588-592. Cited on pages: 257, 258, 550.

[78SER/BAT] Serebrennikov, V. V., Batyreva, V. A., Tsybukova, T. N., The $Nd(BrO_3)_3$-$Pr(BrO_3)_3$-H_2O and $Nd_2(SeO_4)_3$-$Pr_2(SeO_4)_3$-H_2O systems at 25°C, *Zh. Neorg. Khim.*, **23**, (1978), 1947-1951, in Russian, English translation in [78SER/BAT2]. Cited on pages: 375, 378.

[78SER/BAT2] Serebrennikov, V. V., Batyreva, V. A., Tsybukova, T. N., The $Nd(BrO_3)_3$-$Pr(BrO_3)_3$-H_2O and $Nd_2(SeO_4)_3$-$Pr_2(SeO_4)_3$-H_2O systems at 25°C, *Russ. J. Inorg. Chem.*, **23**, (1978), 1070-1072. From a citation in this Bibliography

[78SMO/PAT] Smoes, S., Pattje, W. R., Drowart, J., Thermodynamic study of the vaporization of manganese metal and manganese selenide by Knudsen-cell mass spectrometry, *High Temp. Sci.*, **10**, (1978), 109-130. Cited on pages: 338, 339, 340, 551, 552.

[78SMO/SER] Smolyakova, K. E., Serebrennikov, V. V., Enthalpy of the formation of some rare earth element selenites, *Reaktsionnie Sposobnosti Veshchestv*, Serebrennikov, V. V., Ed., pp.12-14, Izddatel'stvo Tomskogo Universiteta, (1978). Cited on pages: 357, 359, 360, 362, 363, 365, 374, 552, 553.

[78TSV] Tsvetkov, V. G., Enthalpies of the reaction of selenium and tellurium with bromine, *Russ. J. Inorg. Chem.*, **23**, (1978), 1086-1087. Cited on pages: 174, 176, 583.

[78WES/MAK] Westland, A. D., Makhija, R., Enthalpies of formation of crystalline selenium(IV) chloride and bromide, *Can. J. Chem.*, **56**, (1978), 1586-1588. Cited on pages: 170, 176.

[78WIE/SIE] Wiedemeier, H., Siemers, P., Gaur, U., Wunderlich, B., Heat capacity measurements of GeS, GeSe and GeTe, *Thermochim. Acta*, **27**, (1978), 223-231. Cited on page: 209.

[79BAN/ATK] Banas, M., Atkin, D., Bowles, J. F. W., Simpson, P. R., Definitive data on bohdanowiczite, a new silver bismuth selenide, *Mineral. Mag.*, **43**, (1979), 131-133. Cited on page: 634.

[79BEN/PAS] Bendrikh, N. F., Pashinkin, A. S., Evaporation of Ga_2Se_3, *Inorg. Mater.*, **15**, (1979), 326-329. Cited on pages: 234, 235, 553.

[79BLA/HOP] Blachnik, R., Hoppe, A., Präparation und thermochemische Untersuchung von Verbindungen der Systeme Phosphor-Schwefel und Phosphor-Selen, *Z. Anorg. Allg. Chem.*, **457**, (1979), 91-104, in German. Cited on pages: 186, 187.

[79BOR/KNI] Born, P., Kniep, R., Mootz, D., Phasenbeziehungen im System Se-Br und die Kristallstrukturen des dimorphen $SeBr_4$, *Z. Anorg. Allg. Chem.*, **451**, (1979), 12-24, in German. Cited on page: 175.

[79BRA/PIT] Bradley, D. J., Pitzer, K. S., Thermodynamics of electrolytes. 12. Dielectric properties of water and Debye-Hückel parameters to 350°C and 1 kbar, *J. Phys. Chem.*, **83**, (1979), 1599-1603. Cited on page: 594.

[79CAR/GER] Carre, J., Germain, P., Kollmannsberger, M., Perachon, G., Thourey, J., Détermination de l'enthalpie de formation standard de l'oxydifluorure de sélenium $SeOF_2$ et du tetrafluorure de sélenium SeF_4 par hydrolyse alcaline, *J. Fluorine Chem.*, **13**, (1979), 365-374, in French. Cited on pages: 162, 166, 418.

[79CEC/VAV] Cech, F., Vavrin, I., Soucekite $CuPbBi(S,Se)_3$, a new mineral of the bournonite group, *Neues Jahrb. Mineral., Monatsh.*, **7**, (1979), 289-295, in German. Cited on page: 635.

[79EYS/SUN] Eysel, H. H., Sunder, S., Molecular chalcogenide, tetraselenium tetrasulphide: Some comments, *Indian J. Chem.*, **18A**, (1979), 447. Cited on page: 179.

[79HAA/WIL] Haacke, D. F., Williams, P. A., The aqueous chemistry of uranium minerals: Part I. Divalent cation zippeite, *Mineral. Mag.*, **43**, (1979), 539-541. Cited on page: 160.

[79HUA/MES] Huang, M.-D., Meschi, D. J., Searcy, A. W., Thermodynamic properties of SeS gas, *J. Electrochem. Soc.*, **126**, (1979), 1825-1826. Cited on page: 181.

[79HUB/HER] Huber, K. P., Herzberg, G., *Molecular spectra and molecular structure IV, constants of diatomic molecules*, pp.562-563, Van Nostrand Reinhold Co., New York, (1979). Cited on pages: 208, 210, 215.

[79IGA/NIS] Igaki, K., Nishine, S., Thermodynamic properties and crystal structures of nonstoichiometric niobium selenides $NbSe_x$ for x=1.48-2.00, *Trans. Jpn. Inst. Met.*, **20**, (1979), 219-226. Cited on page: 350.

[79JOH/PYT] Johnson, K. S., Pytkowicz, R. M., Ion association and activity coefficients in multicomponent solutions, *Activity coefficients in electrolyte solutions*, Pytkowicz, R. M., Ed., II, pp.1-62, CRC Press, Boca Raton, Florida, (1979). Cited on page: 588.

[79KAR/CRE] Kargol, J. A., Crecely, R. W., Burmeister, J. L., Carbon-13 nuclear magnetic resonance study of coordinated thiocyanate, selenocyanate, and cyanate, *Inorg. Chem.*, **18**, (1979), 2532-2535. Cited on page: 206.

[79KUL/ZLO] Kulyukhina, E. A., Zlomanov, V. P., Novoselova, A. V., Sublimation of tin selenides, *Inorg. Mater.*, **15**, (1979), 921-923. Cited on pages: 215, 216, 217, 218, 554.

[79LAI/NII] Laitinen, R., Niinistö , L., Steudel, R., Structural studies on the sulfur-selenium binary system, *Acta Chem. Scand.*, **33A**, (1979), 737-747. Cited on page: 179.

[79LAV/PIN] Lavendy, H., Pinchemel, B., Rotational analysis of the $A^2\Sigma^+$ transitions of $^{27}Al^{80}Se$, *Can. J. Phys.*, **57**, (1979), 607-611. Cited on page: 227.

[79MIL] Millero, F. J., Effects of pressure and temperature on activity coefficients, *Activity coefficients in electrolyte solutions*, Pytkowicz, R. M., Ed., II, pp.63-151, CRC Press, Boca Raton, Florida, (1979). Cited on pages: 588, 596.

[79MIL/HAI] Milne, J., LaHaie, P., Chloroselenate(IV) equilibria in aqueous hydrochloric acid, *Inorg. Chem.*, **18**, (1979), 3180-3183. Cited on pages: 153, 172, 173.

[79PIT] Pitzer, K. S., Theory: Ion Interaction Approach, *Activity coefficients in electrolyte solutions*, Pytkowicz, R. M., Ed., vol. 1, pp.157-208, CRC Press, Boca Raton, Florida, (1979). Cited on pages: 588, 590.

[79POP/TON] Popovic, S., Tonejc, A., Grzeta-Plenkovic, B., Celustka, B., Trojko, R., Revised and new crystal data for indium selenides, *J. Appl. Crystallogr.*, **12**, (1979), 416-420. Cited on page: 242.

[79PRE/GER] Predel, B., Gerdes, F., Gerling, U., Berücksichtung der Assoziation in der Dampfphase bei Aktivitätsbestimmungen und Revision der Aktivitäten flüssiger Legierungen der Systeme Selen-Thallium, Selen-Wismut und Selen-Antimon, *Z. Metallkd.*, **70**, (1979), 109-112, in German. Cited on page: 250.

[79PYT] Pytkowicz, R. M., Activity coefficients, ionic media, and equilibria in solutions, *Activity coefficients in electrolyte solutions*, Pytkowicz, R. M., Ed., II, pp.301-305, CRC Press, Boca Raton, Florida, (1979). Cited on page: 588.

[79ROD/ROB] Rodkey, F. L., Robertson, R. F., The solubility of hydrogen cyanide in water, *J. Combust. Toxicol.*, **6**, (1979), 44-47. Cited on page: 434.

[79SER/TSY] Serebrennikov, V. V., Tsybukova, T. N., The $La_2(SeO_4)_3$-$Sm_2(SeO_4)_3$-H_2O and $Eu_2(SeO_4)_3$-$La_2(SeO_4)_3$-H_2O systems at 25°C, *Zh. Neorg. Khim.*, **24**, (1979), 2545-2548, in Russian, English translation in [79SER/TSY2]. Cited on pages: 368, 371, 375, 384.

[79SER/TSY2] Serebrennikov, V. V., Tsybukova, T. N., The $La_2(SeO_4)_3$-$Sm_2(SeO_4)_3$-H_2O and $Eu_2(SeO_4)_3$-$La_2(SeO_4)_3$-H_2O systems at 25°C, *Russ. J. Inorg. Chem.*, **24**, (1979), 1414-1417. From a citation in this Bibliography

[79SUL/PRU] Sullivan, C. L., Prusaczyk, J. E., Miller, R. A., Carlson, K. D., Thermodynamic properties of the molecular species vaporized from solid Sb_2Se_3, *High Temp. Sci.*, **11**, (1979), 95-122. Cited on pages: 193, 195, 554, 555.

[79VEN/PAS] Vendrikh, N. F., Pashinkin, A. S., Vapor pressure of solid In_2Se_3, *Inorg. Mater.*, **15**, (1979), 600-603. Cited on pages: 243, 244, 555.

[79WHI] Whitfield, M., Activity coefficients in natural waters, *Activity coefficients in electrolyte solutions*, Pytkowicz, R. M., Ed., II, pp.153-299, CRC Press, Boca Raton, Florida, (1979). Cited on pages: 9, 10.

[79WHI2] Whiffen, D. H., *Manual of symbols and terminology for physicochemical quantities and units*, IUPAC Commission on Physicochemical Symbols, Terminology and Units, Pergamon Press, Oxford, (1979), 41 pp.. Cited on page: 588.

[80AMI/KLU] Amitin, E. B., Klushina, T. V., Lebedeva, E. G., Paukov, I. E., Heat capacities of Na_2SeO_3 at low temperatures, entropy and enthalpy under standard conditions, *Izv. Sib. Otd. Akad. Nauk SSSR, Ser. Khim. Nauk*, **6**, (1980), 11-12, in Russian. Cited on page: 417.

[80AZI/AGA] Azizov, T. K., Agaev, A. B., Guseinov, A. G., Abbasov, A. S., Aliev, I. Y., Mustafaev, F. M., Investigation of Gd-Se system thermodynamical properties, *Dokl.-Akad. Nauk Az. SSR*, **36**, (1980), 36-38, in Russian. Cited on page: 364.

[80BLA/HOP] Blachnik, R., Hoppe, A., Wickel, U., Die Systeme Arsen-Schwefel und Arsen-Selen und die thermodynamischen Daten ihrer Verbindungen, *Z. Anorg. Allg. Chem.*, **463**, (1980), 78-90, in German. Cited on pages : 191, 192.

[80BRE/LAM] Brewer, L., Lamoreaux, R. H., Thermochemical properties, in: Molybdenum: Physico-chemical properties of its compounds and alloys, *At. Energy Rev., Spec. Issue*, (1980), 11-191. Cited on page: 345.

[80CIA] Ciavatta, L., The specific interaction theory in evaluating ionic equilibria, *Ann. Chim. (Rome)*, **70**, (1980), 551-567. Cited on pages: 589, 590, 600, 601, 602, 607, 609, 611, 613.

[80EBE/HAV] Ebert, M., Havlicek, D., Magnesium selenites, *Chem. Zvesti*, **34**, (1980), 441-451. Cited on page: 394.

[80GRA/HIR] Grade, M., Hirschwald, W., Equilibrium gas phase composition of IIB/VIA compounds and identification of gaseous MeX(g) molecules, *Z. Anorg. Allg. Chem.*, **460**, (1980), 106-114. Cited on page: 279.

[80JOH/MCC] Johanson, W. R., McCollum, D. C., Specific heats of EuTe and EuSe near their magnetic transitions, *Phys. Rev. B*, **22**, (1980), 2435-2438. Cited on page: 366.

[80KAD/TAN] Kadoshnikova, N. V., Tananaev, I. V., Ammonium selenatoindates(III), *Zh. Neorg. Khim.*, **25**, (1980), 2838-2842, in Russian, English translation in [80KAD/TAN2]. Cited on pages: 247, 411.

[80KAD/TAN2] Kadoshnikova, N. V., Tananaev, I. V., Ammonium selenatoindates(III), *Russ. J. Inorg. Chem.*, **25**, (1980), 1566-1569. From a citation in this Bibliography

[80LAI/STE] Laitinen, R., Steudel, R., Vibrational spectroscopic investigations of S_nSe_{8-n} molecules, *J. Mol. Struct.*, **68**, (1980), 19-32. Cited on page: 179.

[80MAM/ALD] Mamedov, K. K., Aldzhanov, M. A., Mekhtiev, M. I., Kerimov, I. G., Heat capacity of thallium sulfide, selenide, and telluride at low temperatures, *Inzh.-Fiz. Zh.*, **39**, (1980), 1005-1009, in Russian. Cited on pages: 248, 249.

[80MUM] Mumme, W. G., Seleniferrous lead-bismuth sulphosalts from Falun, Sweden: weibullite, wittite and nordströmite, *Am. Mineral.*, **65**, (1980), 789-796. Cited on pages: 635, 636.

[80OJK/GOS] Ojkova, T., Gospodinov, G., Untersuchung der Phasengleichgewichte im System Bi_2O_3-SeO_2-H_2O, *Z. Anorg. Allg. Chem.*, **470**, (1980), 227-232, in German. Cited on page: 202.

[80SHU/GAU] Shu, H.-C., Gaur, U., Wunderlich, B., Heat capacity and chemical equilibria of liquid selenium, *J. Polym. Sci, Polym. Phys. Ed.*, **18**, (1980), 449-456. Cited on page: 91.

[80SIR/PET] Sirota, N. N., Petrova, Z. K., Solovoskii, T. D., Heat capacity of zinc selenide in the 4,2-300 K region, *Dokl. Akad. Nauk BSSR*, **24**, (1980), 214-217, in Russian. Cited on pages: 255, 256, 555.

[80UML/WAL] Umland, F., Wallmeier, W., Polarographische Untersuchungen der Wechsel-Wirkungen von Selen und Tellur mit einigen Kationen in saurer Lösung, *J. Less-Common Met.*, **76**, (1980), 245-254, in German. Cited on pages: 195, 221, 252, 258, 271, 281, 556.

[80WES/ROS] Westphal, G. H., Rosenberg, F., Infrared spectrum of matrix-isolated $SeOCl_2$, *J. Mol. Spectrosc.*, **83**, (1980), 355-359. Cited on page: 171.

[80WOO] Woolf, A. A., The pattern of dissociation energies of selenium(IV) fluoride and its neighbouring main group fluorides, *J. Fluorine Chem.*, **15**, (1980), 533-537. Cited on pages: 163, 166.

[80WUN/SHU] Wunderlich, B., SHU, H.-C., The crystallization and melting of selenium, *J. Cryst. Growth*, **48**, (1980), 227-239. Cited on pages: 91, 92.

[80ZAK] Zak, Z., Crystal structure of diselenium pentoxide Se_2O_5, *Z. Anorg. Allg. Chem.*, **460**, (1980), 81-85. Cited on page: 130.

[80ZLO/NOV] Zlomanov, V. P., Novozhilov, A. F., Makarov, A. V., Sublimation of GeSe and $GeSe_2$, *Inorg. Mater.*, **16**, (1980), 419-423. Cited on page: 212.

[81BAL/LEG] Baldé, L., Legendre, B., Souleau, C., Khodadad, P., Capacité calorifique de l'alliage $Sn_{0,50}Sn_{0,50}$ à l'état solide entre 378 et 1135 K, *J. Less-Common Met.*, **80**, (1981), 45-50, in French. Cited on page: 214.

[81BOS/LEB] Bosser, G., Lebreton, J., Rotational analysis of the $a^3\Pi_{1,0+} \rightarrow X^1\Sigma^+$ system of carbon monoselenide, *J. Phys. B: At. Mol. Phys.*, **14**, (1981), 1445-1448. Cited on page: 204.

[81EBE/HAV] Ebert, M., Havlicek, D., Calcium selenites, *Collect. Czech. Chem. Commun.*, **46**, (1981), 1740-1747. Cited on pages: 399, 471.

[81GAU/SHU] Gaur, U., Shu, H.-C., Mehta, A., Wunderlich, B., Heat capacity and other thermodynamic properties of linear macromolecules. I. Selenium, *J. Phys. Chem. Ref. Data*, **10**, (1981), 89-117. Cited on pages: 88, 90, 91, 92.

[81HEL/KIR] Helgeson, H. C., Kirkham, D. H., Flowers, G. C., Theoretical prediction of the thermodynamic behavior of aqueous electrolytes at high pressures and temperatures: IV. Calculation of activity coefficients, osmotic coefficients, and apparent molal and standard and relative partial molal properties to 600°C and 5 kb, *Am. J. Sci.*, **281**, (1981), 1249-1516. Cited on pages: 594, 596.

[81KOS/GRI] Koshchenko, V. I., Grinberg, Y. K., Demidenko, A. F., Zhelagina, V. A., Temperature dependence of thermodynamic properties of In_2Se_3 at 5-300 K, *Inorg. Mater.*, **17**, (1981), 1471-1474. Cited on pages: 242, 243, 556, 566.

[81MIL] Miller, R., Kawazulite Bi_2Te_2Se, related bismuth minerals and selenian covellite from the Northwest Territories, *Can. Mineral.*, **19**, (1981), 341-348. Cited on pages: 635, 637.

[81MOF] Moffatt, W. G., *The Handbook of Binary Phase Diagrams*, General Electric Company, New York, (1981). Cited on page: 345.

[81OIK/GOS] Oikova, T., Gospodinov, G., Phase interactions in the PbO-SeO_2-H_2O and PbO-SeO_3-H_2O systems at 100°C, *Zh. Neorg. Khim.*, **26**, (1981), 491-494, in Russian, English translation in [81OIK/GOS2]. Cited on pages: 223, 224.

[81OIK/GOS2] Oikova, T., Gospodinov, G., Phase interactions in the PbO-SeO_2-H_2O and PbO-SeO_3-H_2O systems at 100°C, *Russ. J. Inorg. Chem.*, **26**, (1981), 266-268. From a citation in this Bibliography

[81SER/BAT] Serebrennikov, V. V., Batyreva, V. A., Tsybukova, T. N., The $Nd(BrO_3)_3$-$Yb(BrO_3)_3$-H_2O and $Nd_2(SeO_4)_3$-$Yb_2(SeO_4)_3$-H_2O systems at 25°C, *Zh. Neorg. Khim.*, **26**, (1981), 2837-2840, in Russian, English translation in [81SER/BAT2]. Cited on pages: 359, 375.

[81SER/BAT2] Serebrennikov, V. V., Batyreva, V. A., Tsybukova, T. N., The $Nd(BrO_3)_3$-$Yb(BrO_3)_3$-H_2O and $Nd_2(SeO_4)_3$-$Yb_2(SeO_4)_3$-H_2O systems at 25°C, *Russ. J. Inorg. Chem.*, **26**, (1981), 1518-1519. From a citation in this Bibliography

[81STE] Steudel, R., Entropies and heat capacities of gaseous selenium molecules Se_n (n=5...12), *Z. Naturforsch.*, **36a**, (1981), 408-409. Cited on pages: 104, 105, 106, 108, 109, 111, 112, 113, 114, 115, 116, 118, 556.

[81STU/MOR] Stumm, W., Morgan, J. J., *Aquatic Chemistry. An introduction emphasizing chemical equilibria in natural waters*, 2nd. Edition, John Wiley and Sons, New York, (1981), 780 pp.. Cited on page: 20.

[81TOM/BIL] Tomic, S., Biljakovic, K., Djurek, D., Cooper, J. R., Calorimetric study of the phase transitions in $NbSe_3$, *Chem. Scr.*, **17**, (1981), 189-190. Cited on page: 351.

[81TOM/BIL2] Tomic, S., Biljakovic, K., Djurek, D., Cooper, J. R., Monceau, P., Meerschaut, A., Calorimetric study of the phase transitions in niobium triselenide $NbSe_3$, *Solid State Commun.*, **38**, (1981), 109-112. Cited on page: 351.

[81WIE/PUL] Wiedermeier, H., Pultz, G., Gaur, U., Wunderlich, B., Heat capacity measurements of SnSe and $SnSe_2$, *Thermochim. Acta*, **43**, (1981), 297-303. Cited on pages: 214, 216.

[81WIL] Willner, H., Photochemische Erzeugung von SF_2 und SeF_2 in Argonmatrix, *Z. Anorg. Allg. Chem.*, **481**, (1981), 117-125, in German. Cited on page: 161.

[82CAS/JER] Castro, A., Jerez, A., Pico, C., Veiga, M. L., Preparation and characterization of mixed oxides of selenium and tellurium, *J. Chem. Soc. Dalton Trans.*, (1982), 733-736. Cited on pages: 185, 186.

[82DIE/SAN] Dieleman, J., Sanders, F. H. M., Van Dommelen, J. H. J., The phase diagram of the Ga-Se system, *Philips J. Res.*, **37**, (1982), 204-229. Cited on pages: 233, 234, 236.

[82DRE] Drever, J. I., *The geochemistry of natural waters*, Prentice-Hall, Englewood Cliffs, N. J., (1982), 388 pp.. Cited on page: 20.

[82DYM/LOS] Dymkov, Y. M., Loseva, T. I., Zavyalov, E. N., Ryzkov, B. I., Bochek, L. I., Mgriite (Cu, Fe)$_3$AsSe$_3$, a new mineral, *Zap. Vses. Mineral. O-va.*, **111**, (1982), 215-219, in Russian. Cited on page: 635.

[82EBE/HAV] Ebert, M., Havlicek, D., Strontium and barium selenites. Synthesis and some properties, *Collect. Czech. Chem. Commun.*, **47**, (1982), 1923-1930. Cited on page: 404.

[82EBE/MIC] Ebert, M., Micka, Z., Pekova, I., Cobalt(II) selenites, *Chem. Zvesti*, **36**, (1982), 169-177. Cited on page: 319.

[82EBE/MIC2] Ebert, M., Micka, Z., Pekova, I., Synthesis of nickel selenites, their solubility and bonding in them, *Collect. Czech. Chem. Commun.*, **47**, (1982), 2069-2076. Cited on page: 312.

[82FRA/ZUM] Frank, A. S., Zuman, P., Acid-base and photochemical reactions of selenious acid, *Anal. Lett.*, **15**, (1982), 1609-1619. Cited on page: 141.

[82FRI/KAL] Fritzler, B., Kaldis, E., Jilek, E., High temperature phase diagram and enthalpies of solution of TmSe, *Rare Earths Mod. Sci. Technol.*, **3**, (1982), 243-248. Cited on page: 359.

[82GLA/DUB] Gladyshev, V. P., Dubinina, L. K., The free energy of formation of polyselenide anions in aqueous solutions, *Russ. J. Phys. Chem.*, **56**, (1982), 259-260. Cited on page: 154.

[82GOR/FEN] Gordienko, S. P., Fenochina, B. V., Viskman, G. S., Klochkov, L. A., Mikhlina, T. M., Characteristics of the behavior of praseodymium monoselenide at high temperatures, *Proc. Acad. Sci. USSR*, **18**, (1982), 13-17. Cited on pages: 376, 377.

[82GRI/GRI] Grimley, R. T., Grindstaff, Q. G., DeMercurio, T. A., Forsman, J. A., A study of the selenium vapor system by angular distribution mass spectrometry, *J. Phys. Chem.*, **86**, (1982), 976-982. Cited on pages: 95, 96, 97, 99, 100, 102, 103, 104, 105, 107, 108, 110, 112, 114, 115, 117, 557.

[82HAM] Hamann, S. D., The influence of pressure on ionization equilibria in aqueous solutions, *J. Solution Chem.*, **11**, (1982), 63-68. Cited on page: 35.

[82JOH/PIC] Johan, Z., Picot, P., Ruhlmann, F., Evolution paragénétique de la minéralisation uranifère de Chaméane (Puy-de-Dome), France : chaméanite, geffroyite et giraudite, trois séléniures nouveaux de Cu, Fe, Ag et As, *TMPM, Tschermaks Mineral. Petrogr. Mitt.*, **29**, (1982), 151-167, in French. Cited on page: 634.

[82LAF] Laffitte, M., A report of IUPAC commission I.2 on thermodynamics: Notation for states and processes, significance of the word ''standard'' in chemical thermodynamics, and remarks on commonly tabulated forms of thermodynamic functions, *J. Chem. Thermodyn.*, **14**, (1982), 805-815. Cited on pages: 12, 13, 28, 30, 33, 587.

[82LYK/CHE] Lykov, S. N., Chernik, I. A., Heat capacity of lead chalcogenides in the 1-20 K, *Fiz. Tverd. Tela (Leningrad)*, **24**, (1982), 3102-3105, in Russian. Cited on pages : 219, 220.

[82MOR/LEG] Morgant, G., Legendre, B., Souleau, C., Capacités calorifiques molaires des composés TlSe et Tl_2Se dans les états solide et liquide, *Ann. Chim. (Paris)*, 7, (1982), 301-308, in French. Cited on pages: 249, 251.

[82NEM] Nemilov, S., Relation between excess thermodynamic functions of glassy arsenic(III) sulfide and arsenic selenide (As_2Se_3) at absolute zero and their structural characteristics, *Fiz. Khim. Stekla*, **8**, (1982), 465-471, in Russian. Cited on page: 191.

[82NIS/SOK] Nisel'son, L. A., Sokolova, T. D., Popov, V. A., Miklyaev, A. D., Phase equilibria in the Re_2O_7-SeO_2 system, *Zh. Neorg. Khim.*, **27**, (1982), 1335-1338, in Russian, English translation in [82NIS/SOK2]. Cited on page: 122.

[82NIS/SOK2] Nisel'son, L. A., Sokolova, T. D., Popov, V. A., Miklyaev, A. D., Phase equilibria in the Re_2O_7-SeO_2 system, *Russ. J. Inorg. Chem.*, **27**, (1982), 753-754. From a citation in this Bibliography

[82OJK/GOS] Ojkova, T., Gospodinov, G., Untersuchung der Phasengleichgewichte in den Systemen CuO-SeO_2-H_2O und Ag_2O-SeO_2-H_2O bei 100°C, *Z. Anorg. Allg. Chem.*, **484**, (1982), 235-240, in German. Cited on pages: 293, 294, 303.

[82PRI/VAS] Pribyl'skii, N. Y., Vasilieva, I. G., Gamidov, R. S., Phase equilibria of the Gd-Se system, *Mater. Res. Bull.*, **17**, (1982), 1147-1153. Cited on page: 364.

[82PRO/BAR] Prosser, S. J., Barrow, R. F., Effantin, C., D'Incan, J., Vergès, J., Laser-excited fluorescence of gaseous Se_2 studied by Fourier transform spectrometry, *J. Phys. B: At. Mol. Phys.*, **15**, (1982), 4151-4160. Cited on page: 94.

[82RAO] Rao, Y. K., Composition of liquid-saturated selenium vapor, *Metall. Trans.*, **14B**, (1982), 308-311. Cited on pages: 88, 558.

[82SCH/FRA] Schiffman, R. A., Franzen, H. F., Ziegler, R. J., Thermodynamics of incongruently vaporizing tungsten diselenide, *High Temp. Sci.*, **15**, (1982), 69-78. Cited on pages: 348, 558.

[82SHU/SEN] Shukla, A. K., Sen, P., Sarma, D. D., Transformations in $Ag_{2+\delta}S$ and $Ag_{2+\delta}Se$: A calorimetric study, *Ber. Bunsen-Ges.*, **86**, (1982), 198-202. Cited on pages: 299, 300.

[82TAN/KAD] Tananaev, I. V., Kadoshnikova, N. V., Double selenates of indium and univalent cations, *Zh. Neorg. Khim.*, **27**, (1982), 2113-2119, in Russian, English translation in [82TAN/KAD2]. Cited on page: 247.

[82TAN/KAD2] Tananaev, I. V., Kadoshnikova, N. V., Double selenates of indium and univalent cations, *Russ. J. Inorg. Chem.*, **27**, (1982), 1193-1197. From a citation in this Bibliography

[82TAY] Taylor, J. R., *An introduction to error analysis: The study of uncertainties in physical measurements*, University Science Books, Mill Valley, CA, USA, (1982), 270 pp.. Cited on page: 618.

[82VEN/GOR] Vendrikh, N. F., Gorbov, S. I., Pashinkin, A. S., Thermodynamic functions of diatomic selenium, *Inorg. Mater.*, **18**, (1982), 921-925. Cited on pages: 94, 95, 559.

[82WAG/EVA] Wagman, D. D., Evans, W. H., Parker, V. B., Schumm, R. H., Halow, I., Bailey, S. M., Churney, K. L., Nuttall, R. L., The NBS tables of chemical thermodynamic properties: Selected values for inorganic and C_1 and C_2 organic substances in (SI) units, *J. Phys. Chem. Ref. Data*, 11, Suppl. 2, (1982), 1-392. Cited on pages : 4, 23, 24, 29, 30, 36, 127, 131, 133, 136, 137, 142, 143, 144, 149, 152, 154, 164, 170, 172, 174, 185, 186, 191, 206, 223, 225, 234, 237, 243, 245, 250, 251, 252, 253, 254, 263, 265, 274, 276, 286, 294, 296, 298, 304, 305, 306, 308, 313, 320, 323, 338, 339, 344, 350, 354, 355, 356, 357, 358, 361, 365, 367, 368, 369, 370, 372, 373, 376, 377, 379, 380, 381, 382, 383, 394, 396, 398, 400, 402, 405, 406, 407, 408, 412, 413, 415, 418, 419, 421, 422, 424, 426, 427, 428, 429, 432, 433, 437, 438, 448, 452, 453, 456, 457, 459, 460, 463, 464, 466, 467, 468, 469, 470, 471, 472, 473, 474, 475, 476, 477, 479, 480, 481, 483, 484, 485, 488, 489, 493, 494, 496, 498, 499, 502, 503, 504, 507, 512, 513, 514, 517, 518, 523, 524, 525, 526, 527, 531, 532, 538, 541, 546, 547, 552, 564, 645, 646, 647, 648.

[83ATA/NOM2] Atake, T., Nomoto, K., Chaudhuri, B. K., Chihara, H., K_2SeO_4: Calorimetric studies of two successive phase transitions, *J. Chem. Thermodyn.*, **15**, (1983), 383-394. Cited on page: 426.

[83AVE/SPI] Avedesian, M. M., Spira, P., Kanduth, H., Stripping of HCN in a packed tower, *Can. J. Chem. Eng.*, **61**, (1983), 801-806. Cited on pages : 434, 435.

[83DIM/LAK] Dimitriev, Y., Lakov, L., Ivanova, Y., SeO_2-TeO_2 phase diagram, *J. Mater. Sci. Lett.*, **2**, (1983), 635-637. Cited on pages: 185, 186.

[83DYM/LOS] Dymkov, Y. M., Loseva, T. I., Boitsev, V. E., Alteration of Cu-Fe-sulphides in the process of formation of berzelianite on the deposit Rozhna, (Moraviya, ChSSR), *Nov. Dannye Miner.*, **31**, (1983), 41-51, in Russian. Cited on page: 637.

[83ELS/ABO] El-Sharkawy, A. A., Abou El-Azm, A. M., Kenawy, M. I., Hillal, A. S., Abu-Basha, H. M., Thermophysical properties of polycrystalline PbS, PbSe, and PbTe in the temperature range 300-700 K, *Int. J. Thermophys.*, **4**, (1983), 261-269. Cited on pages: 218, 219.

[83FER/GRE] Ferri, D., Grenthe, I., Salvatore, F., Studies on metal carbonate equilibria. 7. Reduction of the tris(carbonato)dioxouranate(VI) ion, $UO_2(CO_3)_3^{4-}$, in carbonate solutions, *Inorg. Chem.*, **22**, (1983), 3162-3165. Cited on page: 600.

[83FER/HAA] Fernholt, L., Haaland, A., Seip, R., Kniep, R., Korte, L., The molecular structure of selenium dichloride, $SeCl_2$, determined by gas electron diffraction, *Z. Naturforsch., B: Chem. Sci.*, **38b**, (1983), 1072-173. Cited on page: 169.

[83GIN/CES] Ginderow, D., Cesbron, F., Structure de la derriksite $Cu_4(UO_2)(SeO_3)_2(OH)_6$, *Acta Crystallogr., Sect. B: Struct. Sci.*, **C39**, (1983), 1605-1607, in French. Cited on page: 636.

[83GOS/BOG] Gospodinov, G. G., Bogdanov, B. G., Dimorphism of the anhydrous selenites of zinc, cadmium and mercury, *Thermochim. Acta*, **71**, (1983), 391-395. Cited on pages: 261, 262, 273, 274, 285.

[83KRU/RYA] Kruglova, V. G., Ryabeva, E. G., Crystals of native selenium from coal deposit, *Zap. Vses. Mineral. O-va.*, **112**, (1983), 51-57, in Russian. Cited on page: 634.

[83MAH/MIL] Mahadevan, M., Milne, J., The nature of selenium dissolved in hydrochloric acid containing selenium dioxide, *Inorg. Chem.*, **22**, (1983), 1648-1651. Cited on page: 173.

[83MIY/HAY] Miyauchi, K., Hayashi, K., Nakahira, M., Phase equilibrium of the VSe_2- V_3Se_4 system and the thermodynamic quantities of the V_3Se_4 phase, *Mater. Res. Bull.*, **18**, (1983), 757-764. Cited on page: 349.

[83NAU/RAK] Naumov, V. N., Rakhmenkulov, F. S., Fedorov, V. E., Mishchenko, A. V., Thermal properties of niobium diselenide at 4.7-314 K, *Russ. J. Phys. Chem.*, **57**, (1983), 18-20. Cited on page: 350.

[83SER/KUD] Serezhkina, L. B., Kudryashova, Y. G., Serezhkin, V. N., The $BeSeO_4$-UO_2SeO_4-H_2O system at 25°C, *Zh. Neorg. Khim.*, **28**, (1983), 1054-1056, in Russian, English translation in [83SER/KUD2]. Cited on pages: 390, 393.

[83SER/KUD2] Serezhkina, L. B., Kudryashova, Y. G., Serezhkin, V. N., The $BeSeO_4$-UO_2SeO_4-H_2O system at 25°C, *Russ. J. Inorg. Chem.*, **28**, (1983), 598-600. From a citation in this Bibliography

[83SER/SER] Serezhkina, L. B., Serezhkin, V. N., The $MnSeO_4$-UO_2SeO_4-H_2O system at 25°C, *Zh. Neorg. Khim.*, **28**, (1983), 2666-2669, in Russian, English translation in [83SER/SER2]. Cited on page: 342.

[83SER/SER2] Serezhkina, L. B., Serezhkin, V. N., The $MnSeO_4$-UO_2SeO_4-H_2O system at 25°C, *Russ. J. Inorg. Chem.*, **28**, (1983), 1513-1514. From a citation in this Bibliography

[83SPA] Spahiu, K., Carbonate complex formation in lanthanoid and actinoid systems, Ph. D. Thesis, The Royal Institute of Technology, (1983), Stockholm, Sweden. Cited on pages: 600, 606, 607.

[83SPI/BAD] Spiridonov, E. M., Badalov, A. S., New sulfurselentellurides and sulfoselenides of bismuth from volcanic deposit Kiragach (East Uzbekistan), *Uzb. Geol. Zh.*, **6**, (1983), 82-84, in Russian. Cited on page: 637.

[83STE/STR] Steudel, R., Strauss, E.-M., Chromatographische Trennung schwefelhaltiger Heterocyclen der Typen $(CH_2)_nS_m$ und Se_nS_{8-n} mittels HPLC, *Z. Naturforsch.*, **38b**, (1983), 719-725, in German. Cited on page: 179.

[83WIE/PUL] Wiedemeier, H., Pultz, G., Equilibrium sublimation and thermodynamic properties of SnSe and $SnSe_2$, *Z. Anorg. Allg. Chem.*, **499**, (1983), 130-144. Cited on pages: 216, 559.

[83ZEL/KOL] Zelikman, A. N., Kolchin, Y. O., Golutvin, Y. M., Enthalpy of formation of tungsten diselenide, *Russ. J. Phys. Chem.*, **57**, (1983), 519-521. Cited on pages: 348, 349, 559.

[84ANA/ATK] Ananthaswamy, J., Atkinson, G., Thermodynamics of concentrated electrolyte mixtures. 4. Pitzer-Debye-Hückel limiting slopes for water from 0 to 100°C and from 1 atm to 1 kbar, *J. Chem. Eng. Data*, **29**, (1984), 81-87. Cited on page: 594.

[84BAR/PET] Bart, J. C. J., Petrini, G., Compound formation in the systems TeO_2-SeO_2 and TeO_2-SeO_2-H_2O, *Z. Anorg. Allg. Chem.*, **509**, (1984), 183-191. Cited on pages: 185, 186.

[84CAS/PIC] Castro, A., Pico, C., Jerez, A., Veiga, M. L., The thermal behaviour of solid phases in the Te-Se-O_2 system, *Thermochim. Acta*, **82**, (1984), 381-386. Cited on pages: 185, 186.

[84CHI/PIA] Chirulli, G., Piacente, V., Standard sublimation enthalpy of condensed $GeSe_2$ by torsion effusion measurements, *J. Mater. Sci. Lett.*, **3**, (1984), 925-928. Cited on page: 212.

[84DRO/SMO] Drowart, J., Smoes, S., Vander Auwera-Mahieu, A. M., Vaporization of pure and alloyed production line selenium. A mass spectrometric study, Industrial uses of selenium and tellurium. 3rd International Symposium, pp.323-350, (1984). Cited on pages: 107, 108, 110, 112, 114, 189, 560.

[84EBE/MIC] Ebert, M., Micka, Z., Uchytilova, M., Zinc selenites: The solubility diagram and its use for the isolation of the compounds, their spectral features and thermal behaviour and the strength of bonds in them, *Collect. Czech. Chem. Commun.*, **49**, (1984), 1653-1659. Cited on page: 260.

[84FRE] Freeman, R. D., Conversion of standard (1 atm) thermodynamic data to the new standard state pressure, 1 bar (10^5 Pa), *J. Chem. Eng. Data*, **29**, (1984), 105-111. Cited on page: 30.

[84GOS] Gospodinov, G., Das Löslichkeitsdiagramm des Systems $ZnSeO_4$-H_2SeO_4-H_2O und einige Eigenschaften der erhaltenen Verbindungen, *Z. Anorg. Allg. Chem.*, **517**, (1984), 223-230, in German. Cited on pages: 263, 264.

[84GOS/BOG] Gospodinov, G. G., Bogdanov, B. G., Thermal and thermodynamic data concerning the selenites from group IB in the periodic table, *Thermochim. Acta*, **75**, (1984), 295-302. Cited on pages: 294, 295, 304, 305.

[84GOS2] Gospodinov, G., Untersuchung der Löslichkeitsisothermen im System $CuSeO_4$-H_2SeO_4-H_2O bei 25 und 100 °C, der Löslichkeitspolytherme im System $CuSeO_4$-H_2O und einiger Eigenschaften der erhaltenen Verbindungen, *Z. Anorg. Allg. Chem.*, **513**, (1984), 213-221, in German. Cited on page: 295.

[84GOS3] Gospodinov, G. G., Physico-chemical investigation of the system ZnO-SeO_2-H_2O and some properties of the compounds obtained, *Thermochim. Acta*, **77**, (1984), 439-444. Cited on page: 260.

[84GOS4] Gospodinov, G. G., Solubility isotherm and some properties of the phases of the Ga_2O_3-SeO_2-H_2O system, *Thermochim. Acta*, **82**, (1984), 375-380. Cited on page: 237.

[84GOS5] Gospodinov, G. G., Phases of the system Tl_2O_3-SeO_2-H_2O and their thermal stability, *Thermochim. Acta*, **77**, (1984), 445-450. Cited on page: 252.

[84GOS6] Gospodinov, G. G., Interactions in the three-component system In_2O_3-SeO_2-H_2O at 100°C, *Thermochim. Acta*, **82**, (1984), 367-374. Cited on page: 246.

[84GRO/DRO] Grønvold, F., Drowart, J., Westrum, Jr., E. F., *The chemical thermodynamics of actinide elements and compounds: Part 4, The actinide chalcogenides (excluding oxides)*, International Atomic Energy Agency, Vienna, (1984), 265 pp.. Cited on pages: 93, 94, 95, 96, 97, 98, 99, 386, 388.

[84HOS] Hostettler, J. D., Electrode electrons, aqueous electrons, and redox potentials in natural waters, *Am. J. Sci.*, **284**, (1984), 734-759. Cited on page: 20.

[84HUA/GIL] Huang, J. Y.-K., Gilles, P. W., Bennett, J. E., Kinetics of vaporization of molten selenium, *High Temp. Sci.*, **17**, (1984), 109-133. Cited on pages: 95, 97, 100, 102, 103, 104, 105, 107, 108, 110, 112, 114, 115, 117, 560.

[84IMP/PIA] Imperatori, P., Piacente, V., Vapour pressure of condensed HgSe from torsion-effusion measurements, *J. Chem. Thermodyn.*, **16**, (1984), 949-954. Cited on pages: 280, 281, 561.

[84KEN/WHE] Kennedy, S. J., Wheeler, J. C., Equilibrium copolymerization as a critical phenomenon, *J. Phys. Chem.*, **88**, (1984), 6595-6605. Cited on page: 180.

[84KON/STR] Kondrat'ev, S. N., Strizhov, B. V., Doronin, V. N., Heat capacity and thermodynamic functions of selenium trioxide, *Probl. Kalorim. Khim. Termodin., Dokl. Vses. Konf., 10th*, **2**, (1984), 433-435, in Russian. Cited on page: 126.

[84LAN/EDW] Lane, W. C., Edwards, T. H., Gillis, J. R., Bonomo, F. S., Murcray, F. J., Analysis of $?_2$ of H_2Se, *J. Mol. Spectrosc.*, **107**, (1984), 306-317. Cited on pages: 132, 133.

[84MAR/MES] Marshall, W. L., Mesmer, R. E., Pressure-density relationships and ionization equilibria in aqueous solutions, *J. Solution Chem.*, **13**, (1984), 383-391. Cited on page: 35.

[84MIC/UCH] Micka, Z., Uchytilova, M., Ebert, M., Cadmium(II) and mercury(II) selenites, *Chem. Zvesti*, **38**, (1984), 759-770. Cited on pages: 272, 284.

[84NEC/CHI] Nechelyustov, G. N., Chistyakova, N. I., Zavyalov, E. N., Nevskite Bi(Se, S) - a new bismuth selenide, *Zap. Vses. Mineral. O-va.*, **113**, (1984), 351-355, in Russian. Cited on page: 635.

[84OGA/KAP] Oganesyan, E. N., Kapantsyan, E. E., Sarkisyan, R. A., Nabivanets, B. I., Babayan, G. G., The state of selenium(IV) in perchloric acid solutions, *Zh. Neorg. Khim.*, **29**, (1984), 2546-2549, in Russian, English translation in [84OGA/KAP2]. Cited on page: 140.

[84OGA/KAP2] Oganesyan, E. N., Kapantsyan, E. E., Sarkisyan, R. A., Nabivanets, B. I., Babayan, G. G., The state of selenium(IV) in perchloric acid solutions, *Russ. J. Inorg. Chem.*, **29**, (1984), 1454-1456. From a citation in this Bibliography

[84PUP/RUS] Pupyshev, V. I., Rusin, A. D., Estimation of molecular constants of some molecules of Se_n row, *Vestn. Mosk. Univ. Ser. 2: Khim.*, **25**, (1984), 352-356, in Russian. Cited on pages: 88, 99, 100, 101, 102, 103, 104, 105, 106, 108, 109, 111, 112, 113, 114, 115, 116, 562.

[84SER/SER] Serezhkina, L. B., Serezhkin, V. N., The $MgSeO_4$-UO_2SeO_4-H_2O system at 25°C, *Zh. Neorg. Khim.*, **29**, (1984), 796-799, in Russian, English translation in [84SER/SER2]. Cited on pages : 390, 397.

[84SER/SER2] Serezhkina, L. B., Serezhkin, V. N., The $MgSeO_4$-UO_2SeO_4-H_2O system at 25°C, *Russ. J. Inorg. Chem.*, **29**, (1984), 459-460. From a citation in this Bibliography

[84SER/TSY] Serebrennikov, V. V., Tsybukova, T. N., Velikov, A. A., Solid solutions in the $Nd_2(SeO_4)_3$-$Sm_2(SeO_4)_3$-H_2O system, *Zh. Neorg. Khim.*, **29**, (1984), 2392-2395, in Russian, English translation in [84SER/TSY2]. Cited on pages: 371, 375.

[84SER/TSY2] Serebrennikov, V. V., Tsybukova, T. N., Velikov, A. A., Solid solutions in the $Nd_2(SeO_4)_3$-$Sm_2(SeO_4)_3$-H_2O system, *Russ. J. Inorg. Chem.*, **29**, (1984), 1368-1370. From a citation in this Bibliography

[84SLA/POP] Slavtscheva, J., Popova, E., Gospodinov, G., Untersuchung von Löslichkeit und Löslichkeitsprodukt bei Seleniten der Elemente der IV. Gruppe des Periodensystems, *Z. Chem.*, **24**, (1984), 105-106, in German. Cited on pages: 213, 218, 223, 562.

[84SMO/DRO] Smoes, S., Drowart, J., Determination of the dissociation energy of selenium monoxide by the mass-spectrometric Knudsen-cell method, *J. Chem. Soc. Faraday Trans.*, **80**, (1984), 1171-1180. Cited on pages: 118, 119, 562, 563, 564.

[85BAR/PAR] Bard, A. J., Parsons, R., Jordan, J., *Standard Potentials in Aqueous Solution*, Bard, A. J., Parsons, R., Jordan, J., Eds., International Union of Pure and Applied Chemistry. Marcel Dekker, Inc., New York, (1985), 834 pp.. Cited on page: 20.

[85BOT/SAN] Botova, M. M., Sandomirskaya, S. M., Chuvikina, N. G., Selenostephanite - $Ag_5Sb(Se, S)_4$ - a new mineral, *Zap. Vses. Mineral. O-va.*, **114**, (1985), 627-630, in Russian. Cited on page: 635.

[85FAK] Fakhruddin, A., Fluorescence spectra of matrix-isolated BiS, BiSe, and BiTe, *J. Chem. Phys.*, **82**, (1985), 4401-4404. Cited on page: 198.

[85FER/GRE] Ferri, D., Grenthe, I., Hietanen, S., Nàer-Neumann, E., Salvatore, F., Studies on metal carbonate equilibria: 12. Zinc(II) carbonate complexes in acid solution, *Acta Chem. Scand.*, **A39**, (1985), 347-353. Cited on page: 606.

[85FRA/WET] Franz, E. D., Wetzenschtein, W., Natürlisches kubisches $NiSe_2$ von Caracoles/Bolivien, *Naturwissenschaften*, **72**, (1985), 655-656, in German. Cited on page: 637.

[85GOS] Gospodinov, G., Löslichkeitsisothermen und thermisches Verhalten der Phasen im System $CdSeO_4$-H_2SeO_4-H_2O bei 100°C, *Z. Anorg. Allg. Chem.*, **525**, (1985), 237-240, in German. Cited on page: 275.

[85JOH/NII] Johansson, G., Niinistö, L., Wakita, H., Coordination of the trivalent lanthanoids in aqueous selenate and perchlorate solutions, *Acta Chem. Scand.*, **39A**, (1985), 359-361. Cited on page: 361.

[85JUD/WUN] Judovits, L., Wunderlich, B., On the entropies of glassy and monoclinic selenium at 0 K, *J. Therm. Anal.*, **30**, (1985), 895-899. Cited on page: 90.

[85LOW] Lowson, R. T., The thermochemistry of radium, *Thermochim. Acta*, **91**, (1985), 185-212. Cited on page: 408.

[85MES/MAR] Mestres, L., Martinez, M. L., Rodriguez, A., Solans, X., Font-Altaba, M., Study on the system $CoSeO_4$-$NiSeO_4$-H_2O at 30°C. Crystal structure of $CoSeO_4.5H_2O$, *Z. Anorg. Allg. Chem.*, **528**, (1985), 183-190, in German. Cited on page: 314.

[85MIC/PIK] Micka, Z., Pikal, P., Ebert, M., Ammonium selenites, *Chem. Pap.*, **39**, (1985), 617-628. Cited on pages: 409, 410.

[85MIL/HAI] Milne, J., LaHaie, P., Bromoselenate(IV) equilibria in aqueous hydrobromic acid, *Inorg. Chem.*, **24**, (1985), 840-844. Cited on pages: 153, 177.

[85MUL] Muller, A. B., (1985), Private communication, OECD Nuclear Energy Agency, Paris. Cited on page: 2.

[85OPP/HAN] Oppermann, H., Hanke, U., Kunze, G., Zur Bestimmung der Bildungsenthalpie von $SeCl_4$ und $SeOCl_2$, *Z. Anorg. Allg. Chem.*, **530**, (1985), 163-168, in German. Cited on pages: 170, 172.

[85SOH/NOV] Söhnel, O., Novotný, P., *Densities of aqueous solutions of inorganic substances*, Elsevier, Amsterdam, (1985), 335 pp.. Cited on pages: 26, 27.

[85SPY/KAL] Spychiger, H., Kaldis, E., Fritzler, B., Tm-Se: a system with valence fluctuations, *J. Less-Common Met.*, **110**, (1985), 61-73. Cited on page: 359.

[85VOJ/EBE] Vojtisek, P., Ebert, M., Löslichkeitsuntersuchung in den Systemen K_2SeO_4-Co(II)SeO_4-H_2O und K_2SeO_4-Ni(II)SeO_4-H_2O bei 25°C, *Z. Chem.*, **25**, (1985), 414-415, in German. Cited on pages: 314, 321.

[85YAN/STA] Yanitskii, I. V., Stashene, S. S., Solubility and crystallisation in the $BeSeO_3$-H_2SeO_3-H_2O system, *Zh. Neorg. Khim.*, **30**, (1985), 2149-2152, in Russian, English translation in [85YAN/STA2]. Cited on page: 393.

[85YAN/STA2] Yanitskii, I. V., Stashene, S. S., Solubility and crystallisation in the $BeSeO_3$-H_2SeO_3-H_2O system, *Russ. J. Inorg. Chem.*, **30**, (1985), 1220-1222. From a citation in this Bibliography

[86BRU] Bruno, J., Stoichiometric and structural studies on the Be^{2+}-H_2O-CO_2(g) system, Ph. D. Thesis, The Royal Institute of Technology, (1986), Stockholm, Sweden. Cited on page: 607.

[86COD] CODATA, The 1986 adjustment of the fundamental physical constants. A report of the CODATA Task Group on Fundamental Constants, prepared by Cohen, E. R. and Taylor, B. N., Pergamon Journals, Oxford, CODATA Bulletin, Report 63, (1986), 36 pp.. Cited on page: 33.

[86DUN/PEA] Dunn, P. J., Peacor, D. R., Criddle, A. J., Finkelman, R. B., Laphamite, an arsenic selenide analogue of orpiment, from burning anthracite deposits in Pennsylvania, *Mineral. Mag.*, **50**, (1986), 279-282. Cited on page: 635.

[86FER/PIA] Ferro, D., Piacente, V., Bardi, G., Vaporization study of solid germanium monoselenide and monotelluride, *High Temp. Sci.*, **21**, (1986), 69-83. Cited on page: 210.

[86GLO/LOM] Globa, A. V., Lomakina, T. P., Sklyarov, V. I., Electrode potential of selenium films in alkaline solutions, *Elektrokhimiya*, **22**, (1986), 1107-1108, in Russian, English translation in [86GLO/LOM2]. Cited on pages: 151, 564.

[86GLO/LOM2] Globa, A. V., Lomakina, T. P., Sklyarov, V. I., Electrode potential of selenium films in alkaline solutions, *Sov. Electrochem.*, **22**, (1986), 1039-1041. From a citation in this Bibliography

[86GON/LUK] Goncharuk, L. V., Lukashenko, G. M., Thermodynamic properties of chromium selenide Cr_2Se_3, *Russ. J. Phys. Chem.*, **60**, (1986), 1089-1090. Cited on page: 344.

[86LYO/YOU] Lyons, L. E., Young, T. L., Alkaline selenide, polyselenide electrolytes: Concentrations, absorption spectra and formal potentials, *Aust. J. Chem.*, **39**, (1986), 511-527. Cited on pages: 135, 154, 155, 156, 157, 565.

[86MAS] Massalski, T. B., *Binary Alloy Phase Diagrams*, vol. 1 and 2, American Society for Metals, 2224 pp.. Cited on pages: 288, 310, 337, 345.

[86MAS/LUT] Masson, M. R., Lutz, H. D., Engelen, B., *Sulfites, Selenites and Tellurites*, vol. 26, IUPAC Solubility Data Series, Pergamon Press, Oxford, (1986). Cited on pages: 262, 293, 303, 417, 423, 462, 501.

[86MIC/EBE] Micka, Z., Ebert, M., France, P., Ptacek, P., Lithium and thallium(I) selenites, *Chem. Pap.*, **40**, (1986), 309-322. Cited on pages: 252, 413.

[86MIC/EBE2] Micka, Z., Ebert, M., The change in the standard Gibbs energy during the formation of the acid selenites of the alkali metals, *Collect. Czech. Chem. Commun.*, **51**, (1986), 1933-1941. Cited on pages: 139, 413, 414, 417, 420, 423, 424, 425, 427, 429.

[86MIC/EBE3] Micka, Z., Ebert, M., Ptacek, P., Preparation of potassium selenites, their solubility and bonding conditions, *Collect. Czech. Chem. Commun.*, **51**, (1986), 2741-2750. Cited on pages: 423, 424.

[86MOR3] Morse, M. D., Clusters of transition metal atoms, *Chem. Rev.*, **86**, (1986), 1049-1109. Cited on page: 298.

[86MUR] Murray, J. L., The Se-Ti (selenium-titanium) system, *Bull. Alloy Phase Diagrams*, **7**, (1986), 163-165. Cited on page: 351.

[86NOR/MUN] Nordstrom, D. K., Munoz, J. L., *Geochemical Thermodynamics*, Blackwell Sci. Publ., Palo Alto, (1986). Cited on page: 20.

[86OHA] O'Hare, P. A. G., The standard molar enthalpy of formation at 298.15 K of crystalline germanium diselenide ($GeSe_2$). Thermochemical Ge-Se bond-dissociation enthalpies. Revised values for the standard molar enthalpies of formation of GeF_4(g) and SeF_6(g), *J. Chem. Thermodyn.*, **18**, (1986), 555-562. Cited on page: 211.

[86OPP/HAN] Oppermann, H., Hanke, U., Krabbes, G., Gleichgewichtsdruckmessungen im System Se/O/Cl, *Z. Anorg. Allg. Chem.*, **542**, (1986), 89-101, in German. Cited on pages: 167, 169, 171.

[86POP/SLA] Popova, E., Slavtscheva, J., Gospodinov, G., Untersuchung der Löslichkeit einiger Phasen des Systems M_2O_3-SeO_2-H_2O der Elemente der 3. Gruppe des Periodensystems der Elemente, *Z. Chem.*, **26**, (1986), 342-343, in German. Cited on pages: 231, 237, 246, 252, 565.

[86PRO/AYA] Provencher, R., Ayache, C., Jandl, S., Jay-Gerin, J.-P., Low-temperature specific heats of $ZrSe_3$ and ZrS_3, *Solid State Commun.*, **59**, (1986), 553-556. Cited on page: 352.

[86STA/YAN] Stashene, S. S., Yanitskii, I. V., The $MgSeO_3$-H_2SeO_3-H_2O system, *Zh. Neorg. Khim.*, **31**, (1986), 1056-1061, in Russian, English translation in [86STA/YAN2]. Cited on page: 394.

[86STA/YAN2] Stashene, S. S., Yanitskii, I. V., The $MgSeO_3$-H_2SeO_3-H_2O system, *Russ. J. Inorg. Chem.*, **31**, (1986), 602-606. From a citation in this Bibliography

[86STE/ALI] Steblevskii, A. V., Alikhanyan, A. S., Gorgoraki, V. I., Pashinkin, A. S., Vaporisation processes in the As-Se system, *Russ. J. Inorg. Chem.*, **31**, (1986), 474-476. Cited on pages: 189, 191.

[86VIK/GOR] Viksman, G. S., Gordienko, S. P., Yanaki, A. A., Klochkov, L. A., Dissociation of molybdenum and tungsten selenides, *Sov. Powder Metallurgy Met. Ceram.*, **1**, (1986), 64-66. Cited on pages: 345, 346, 348, 544, 566.

[87AMI/MIN] Amitin, E. B., Minenkov, Y. F., Nabutovskaya, O. A., Naumov, V. N., Paukov, I. E., Kondrat'ev, S. N., Thermodynamic properties of selenium dioxide at 7-304 K, *Zh. Fiz. Khim.*, **61**, (1987), 2611-2615, in Russian, English translation in [87AMI/MIN2]. Cited on page: 120.

[87AMI/MIN2] Amitin, E. B., Minenkov, Y. F., Nabutovskaya, O. A., Naumov, V. N., Paukov, I. E., Kondrat'ev, S. N., Thermodynamic properties of selenium dioxide at 7-304 K, *Russ. J. Phys. Chem.*, **61**, (1987), 1375-1378. From a citation in this Bibliography

[87BEC] Beck, M. T., Critical survey of stability constants of cyano complexes, *Pure Appl. Chem.*, **59**, (1987), 1703-1720. Cited on page: 436.

[87BOC/KUL] Boctor, N. Z., Kullerud, G., Phase relations and equilibrium copolymerization in the Se-S system, *J. Solid State Chem.*, **71**, (1987), 513-521. Cited on pages: 178, 179, 180.

[87BOE/KUH] Boehnke, U.-C., Kühn, G., Berezovskii, G. A., Spassov, T., Some aspects of the thermal behaviour of In_2Se_3, *J. Therm. Anal.*, **32**, (1987), 115-120. Cited on pages: 242, 243, 556, 566.

[87BOL/TRO] Bolgar, A. S., Trofimova, Z. A., Yanaki, A. A., Kopylova, L. I., Zaletilo, L. S., Enthalpy and specific heat of $NbSe_2$, $NbSe_{1.75}$, and $MoSe_2$ at high temperatures, *Inorg. Mater.*, **23**, (1987), 798-801. Cited on pages: 345, 350, 567.

[87BRA/EDW] Brassington, N. J., Edwards, H. G. M., Fawcett, V., The vibrational Raman spectra of selenium trioxide, *Spectrochim. Acta*, **43A**, (1987), 451-454. Cited on page: 128.

[87BRI/OGD] Brisdon, A. K., Ogden, J. S., Matrix isolation and mass spectrometric studies on selenium trioxide vapour: The characterization of monomeric SeO_3, *J. Mol. Struct.*, **157**, (1987), 141-153. Cited on page: 128.

[87CIA/IUL] Ciavatta, L., Iuliano, M., Porto, R., The hydrolysis of the La(III) ion in aqueous perchlorate solution at 60°C, *Polyhedron*, **6**, (1987), 1283-1290. Cited on page: 37.

[87EFF] Effenberger, H., Crystal structure and chemical formula of schmiederite, $Pb_2Cu_2(OH)_4(SeO_3)(SeO_4)$, with a comparison to linarite, $PbCu(OH)_2(SO_4)$, *Mineral. Petrol.*, **36**, (1987), 3-12. Cited on page: 636.

[87FIN/KRU] Fink, E. H., Kruse, H., Ramsay, D. A., Wang, D.-W., High resolution studies of the $b^1\Sigma^+$-$X^3\Sigma^-$ emission system of SeS, *Mol. Phys.*, **60**, (1987), 277-290. Cited on page: 180.

[87GAR/PAR] Garvin, D., Parker, V. B., White, Jr., H. J., *CODATA thermodynamic tables: Selection of some compounds of calcium and related mixtures: A prototype set of tables*, Springer-Verlag, Berlin, (1987), 356 pp.. Cited on page: 24.

[87HON/IID] Honma, K., Iida, K., Specific heat of superionic conductor Ag_2S, Ag_2Se and Ag_2Te in α-phase, *J. Phys. Soc. Jpn.*, **56**, (1987), 1828-1836. Cited on pages: 299, 567.

[87JOH/PIC] Johan, Z., Picot, P., Ruhlmann, F., The ore mineralogy of the Otish Mountains uranium deposit, Quebec: skippenite, Bi_2Se_2Te, and watkinsonite, $Cu_2PbBi_4(Se,S)_8$, two new mineral species, *Can. Mineral.*, **25**, (1987), 625-637. Cited on pages: 635, 636, 637.

[87JOH] Johan, Z., La crookesite, $TlCu_7Se_4$: nouvelles données et isotypie avec $NH_4Cu_7S_4$, *C. R. Séances Acad. Sci., Ser. B*, **304**, (1987), 1121-1124, in French. Cited on page: 634.

[87KEL] Kelly, R. L., Atomic and ionic spectrum lines below 2000 angstroms: hydrogen through krypton. Part II (Mn-Kr), *J. Phys. Chem. Ref. Data*, **16**, (1987), 651-1369. Cited on page: 92.

[87MIC/EBE] Micka, Z., Ebert, M., Rubidium and cesium acidic selenites, *Chem. Pap.*, **41**, (1987), 321-334. Cited on pages: 427, 429.

[87MIL] Milne, J., Haloselenate(IV) formation and selenous acid dissociation equilibria in hydrochloric and hydrofluoric acids, *Can. J. Chem.*, **65**, (1987), 316-321. Cited on pages: 141, 153, 166, 172, 173.

[87OHA/SUS] O'Hare, P. A. G., Susman, S., Volin, K. J., The energy difference between the crystalline and vitreous forms of germanium diselenide as determined by combustion calorimetry in fluorine. The Ge-Se bond energy, *J. Non-Cryst. Solids*, **89**, (1987), 24-30. Cited on page: 211.

[87OHA/TAS] O'Hare, P. A. G., Tasker, I. R., Tarascon, J. M., A fluorine-combustion calorimetric study of two molybdenum selenides: $MoSe_2$ and Mo_6Se_6, *J. Chem. Thermodyn.*, **19**, (1987), 61-68. Cited on pages: 345, 346, 646.

[87OJK/STA] Ojkova, T., Staneva, D., Untersuchung der Systeme $Al_2(SeO_4)_3$-$MnSeO_4$-H_2O und $Al_2(SeO_4)_3$-$MgSeO_4$-H_2O bei 25°C, *Z. Anorg. Allg. Chem.*, **554**, (1987), 227-230, in German. Cited on pages: 231, 232.

[87RIG/VIT] Riglet, C., Vitorge, P., Grenthe, I., Standard potentials of the (MO_2^{2+}/MO_2^{+}) systems for uranium and other actinides, *Inorg. Chim. Acta*, **133**, (1987), 323-329. Cited on page: 606.

[87RUN/LIN] Runnells, D. D., Lindberg, R. D., Kempton, J. H., Irrerversibility of Se(VI)/Se(IV) redox couple in synthetic basaltic ground water at 25°C and 75°C, *Mater. Res. Soc. Symp. Proc.*, **84**, (1987), 723-733. Cited on page: 150.

[87SER/SER] Serezhkina, L. B., Serezhkin, V. N., The Cs_2SeO_4-UO_2SeO_4-H_2O system at 25°C, *Zh. Neorg. Khim.*, **32**, (1987), 480-483, in Russian, English translation in [87SER/SER2]. Cited on page: 390.

[87SER/SER2] Serezhkina, L. B., Serezhkin, V. N., The Cs_2SeO_4-UO_2SeO_4-H_2O system at 25°C, *Russ. J. Inorg. Chem.*, **32**, (1987), 267-269. From a citation in this Bibliography

[87SRI/EDW] Srinivasa, R. S., Edwards, J. G., Phase equilibria, chemistry, and thermodynamics during vaporization of indium selenide, *J. Electrochem. Soc.*, **134**, (1987), 1811-1817. Cited on pages: 243, 244, 568.

[87SVE/GRO] Svendsen, S. R., Grønvold, F., Westrum, Jr., E. F., Thermodynamic properties of $RuSe_2$ from 5 to 1500 K, *J. Chem. Thermodyn.*, **19**, (1987), 1009-1022. Cited on pages: 333, 334, 335.

[87VOJ/EBE] Vojtisek, P., Ebert, M., Löslichkeitsuntersuchung in den Systemen $CuSeO_4$-K_2SeO_4-H_2O und $ZnSeO_4$-K_2SeO_4-H_2O bei 25°C, *Z. Chem.*, **27**, (1987), 229-230, in German. Cited on pages: 264, 295.

[87VOJ/EBE2] Vojtisek, P., Ebert, M., Löslichkeitsuntersuchung in den Systemen Rb_2SeO_4-$MgSeO_4$-H_2O und Cs_2SeO_4-$MgSeO_4$-H_2O bei 25°C, *Z. Chem.*, **27**, (1987), 266-267, in German. Cited on pages: 397, 427, 429.

[87VOJ/EBE3] Vojtisek, P., Ebert, M., Löslichkeitsuntersuchung in den Systemen $(NH_4)_2SeO_4$-M(II)SeO_4-H_2O (M(II) = Zn, Cd) und K_2SeO_4-$CdSeO_4$-H_2O bei 25°C, *Z. Chem.*, **27**, (1987), 345-346, in German. Cited on pages: 263, 264, 275.

[87VOJ/EBE4] Vojtisek, P., Ebert, M., Leblova, M., Löslichkeitsuntersuchung in den Systemen $(NH_4)_2SeO_4$-M(II)SeO_4-H_2O (M(II) = Co und Ni) bei 25°C, *Z. Chem.*, **27**, (1987), 375-376, in German. Cited on pages: 314, 321.

[88BAR/IER] Bardi, G., Ieronimakis, K., Trionfetti, G., Vaporization enthalpy of cadmium selenide and telluride obtained by vapour pressure measurements, *Thermochim. Acta*, **129**, (1988), 341-343. Cited on pages: 269, 270, 568.

[88BAR/PIA] Bardi, G., Piacente, V., Scardala, P., Vaporization enthalpy of indium sesquiselenide from its vapor pressure, *High Temp. Sci.*, **25**, (1988), 175-185. Cited on page: 244, 568.

[88BUR/WIL] Bürger, H., Willner, H., The high-resolution Fourier transform spectrum of monoisotopic $C^{80}Se_2$ in the ν_2, ν_3, and $\nu_1 + \nu_3$ regions: equilibrium, ground, and excited state parameters determined by rovibrational analyses, *J. Mol. Spectrosc.*, **128**, (1988), 221-235. Cited on page: 206.

[88CIA] Ciavatta, L., (1988), Università di Napoli, Naples, Italy, Private Communication, (from the citation in [92GRE/FUG]). Cited on pages: 602, 609, 613.

[88EFI/NEN] Efimov, A. V., Nenashev, S. N., Borodaev, Y. S., Mozgova, N. N., Sivtsov, A. V., On hypogene selenium mineralisation on tin ore deposit Nevskoe (North – East of USSR), *Nov. Dannye Miner.*, **35**, (1988), 128-151, in Russian. Cited on page: 637.

[88GHO/LUK] Ghosh, G., Lukas, H. L., Delaey, L., A thermodynamic assessment of the Se- Te system, *CALPHAD: Comput. Coupling Phase Diagrams Thermochem.*, **12**, (1988), 295-299. Cited on page: 183.

[88HED] Hedlund, T., Studies of complexation and precipitation equilibria in some aqueous aluminium(III) systems, Ph. D. Thesis, University of Umeå , (1988), Sweden. Cited on page: 606.

[88HIR/UED] Hirota, T., Ueda, Y., Kosuge, K., Phase diagram of the $TiSe_x$ system ($0.95 < x < 2.00$), *Mater. Res. Bull.*, **23**, (1988), 1641-1650. Cited on page: 351.

[88JOH] Johan, Z., Crystal symmetry of eskerbornite, $CuFeSe_2$, *Neues Jahrb. Mineral., Monatsh.*, **8**, (1988), 337-343. Cited on page: 634.

[88MAM/YAN] Mamedov, K. K., Yangirov, A. Y., Guseinov, A. G., Abdullaev, A. M., Heat capacity and phase transistions in highly anisotropic $A^{III}B^{IV}$-type semiconductors and their analogs at low temperatures, *Phys. Status Solidi*, **106**, (1988), 315-331. Cited on page: 250.

[88MIL/CVI] Mills, I., Cvitaš, T., Homann, K., Kallay, N., Kuchitsu, K., *Quantities, units and symbols in physical chemistry, IUPAC*, Blackwell Scientific Publications, Oxford, (1988), 134 pp.. Cited on pages: 9, 18, 22.

[88OHA/LEW2] O'Hare, P. A. G., Lewis, B. M., Parkinson, B. A., Standard molar enthalpy of formation by fluorine-combustion calorimetry of tungsten diselenide (WSe_2). Thermodynamics of the high-temperature vaporization of WSe_2. Revised value of the standard molar enthalpy of formation of molybdenite (MoS_2), *J. Chem. Thermodyn.*, **20**, (1988), 681-691. Cited on pages: 349, 646.

[88OZE/YAG] Ozeki, T., Yagasaki, A., Ichida, H., Sasaki, Y., Equilibria of the H^+-MoO_4^{2-}-SeO_3^{2-} system in aqueous 1 M Na(Cl) medium, *Polyhedron*, **7**, (1988), 1131-1134. Cited on pages: 140, 141, 142, 347, 446, 569.

[88PAR] Parnell, J., Mercury and silver-bismuth selenides at Alva, Scotland, *Mineral. Mag.*, **52**, (1988), 719-720. Cited on page: 637.

[88PAS/ZHA] Pashinkin, A. S., Zharov, V. V., Vapor pressure of solid In_2Se_3, *Inorg. Mater.*, **24**, (1988), 1247-1250. Cited on page: 244, 569.

[88RAS] Rasulov, S. M., Components of the heat capacity of the compounds $A^{IV}Se$ and $A_2{}^{V}Se_3$, *High Temp.*, **26**, (1988), 70-75. Cited on pages: 199, 209.

[88SER/BLA] Serezhkina, L. B., Blatov, V. A., Serezhkin, V. N., Hydroxonium selenatouranylates [Oxoselenatouranates], *Zh. Neorg. Khim.*, **33**, (1988), 1900-1902, in Russian, English translation in [88SER/BLA2]. Cited on page: 390.

[88SER/BLA2] Serezhkina, L. B., Blatov, V. A., Serezhkin, V. N., Hydroxonium selenatouranylates [Oxoselenatouranates], *Russ. J. Inorg. Chem.*, **33**, (1988), 1083-1084. From a citation in this Bibliography

[88SHO/HEL] Shock, E. L., Helgeson, H. C., Calculation of the thermodynamic and transport properties of aqueous species at high pressures and temperatures: Correlation algorithms for ionic species and equation of state predictions to 5 kb and 1000°C, *Geochim. Cosmochim. Acta*, **52**, (1988), 2009-2036. Cited on pages: 35, 595.

[88TAN/HEL] Tanger, IV, J. C., Helgeson, H. C., Calculation of the thermodynamic and transport properties of aqueous species at high pressures and temperatures: Revised equations of state for the standard partial molal properties of ions and electrolytes, *Am. J. Sci.*, **288**, (1988), 19-98. Cited on pages: 35, 595.

[88WAN] Wanner, H., The NEA Thermochemical Data Base Project, *Radiochim. Acta*, **44/45**, (1988), 325-329. Cited on page: 2.

[88XIA/ZHU] Xiang-Yun, W., Zhu, J. T., Goudiakas, J., Fuger, J., Thermodynamics of lanthanide elements IV. Molar enthalpies of formation of Y^{3+}(aq), YCl_3(cr), YBr_3(cr), and YI_3(cr), *J. Chem. Thermodyn.*, **20**, (1988), 1195-1202. Cited on page: 498.

[89COX/WAG] Cox, J. D., Wagman, D. D., Medvedev, V. A., *CODATA Key Values for Thermodynamics*, Hemisphere Publ. Corp., New York, (1989), 271 pp.. Cited on pages: 3, 29, 34, 36, 67, 68, 69, 137, 143, 149, 170, 172, 204, 208, 209, 210, 211, 216, 221, 223, 226, 227, 228, 229, 230, 231, 260, 289, 290, 298, 300, 399, 409, 433, 437, 438, 448, 456, 458, 464, 465, 466, 467, 468, 469, 470, 471, 472, 473, 474, 477, 480, 481, 482, 483, 484, 485, 488, 489, 490, 494, 495, 497, 499, 500, 509, 510, 511, 516, 518, 519, 526, 527, 529, 531, 532, 536, 537, 539, 545, 550, 553, 561, 564, 568, 570, 572, 575, 576.

[89FIN/SET] Fink, E. H., Setzer, K. D., Ramsay, D. A., Vervloet, M. V., Xu, G. Z., Rotational analysis of the 0-0 band of the $b0^+ \rightarrow X_10^+$ system of $^{130}Te^{80}Se$, *J. Mol. Spectrosc.*, **136**, (1989), 218-221. Cited on page: 184.

[89GHO/LUK] Ghosh, G., Lukas, H. L., Delaey, L., A thermodynamic assessment of the Sb-Se system, *Z. Metallkd.*, **80**, (1989), 663-668. Cited on page: 193.

[89GLA/KOV] Gladyshev, V. P., Kovaleva, S. V., Solubility products of mercury chalcogenides, *J. Gen. Chem. USSR*, **59**, (1989), 308-311. Cited on page: 281.

[89GLA/PAS] Glazov, V. M., Pashinkin, A. S., Malkova, A. S., Specific heat of cadmium selenide in the range 360-760 K, *Russ. J. Phys. Chem.*, **63**, (1989), 19-20. Cited on page: 268.

[89GOS/BOG] Gospodinov, G. G., Bogdanov, B. G., Thermal and thermodynamic properties of the selenites of the elements from group IV in the periodic table, *Thermochim. Acta*, **146**, (1989), 15-19. Cited on pages: 213, 218, 224.

[89GUS/ALD] Guseinov, D. A., Aldzhanov, M. A., Sadykhov, R. Z., Guseinov, N. G., Low-temperature heat capacity and the Gruneisen parameter of Fe_3Se_4, *Inorg. Mater.*, **25**, (1989), 1018-1020. Cited on page: 327.

[89KOV/NEK] Kovalenker, V. A., Nekrasov, I. Y., Sadomirskaya, S. M., Nekrasova, A. N., Malov, V. S., Danchenko, V. Y., Dmitrieva, M. T., Sulphide-Selenide-Telluride mineralization of epithermal manifestations in the Kuril-Kamchatsk volcanic belt, *Mineral. Zh.*, **11**, (1989), 3-18, in Russian. Cited on page: 637.

[89KUC/SHT] Kuchumova, N. V., Shtokova, I. P., Serezhkina, L. B., Serezhkin, V. N., Complex formation in the Rb_2SeO_4-UO_2SeO_4-H_2O system, *Zh. Neorg. Khim.*, **34**, (1989), 1029-1031, in Russian, English translation in [89KUC/SHT2]. Cited on page: 390.

[89KUC/SHT2] Kuchumova, N. V., Shtokova, I. P., Serezhkina, L. B., Serezhkin, V. N., Complex formation in the Rb_2SeO_4-UO_2SeO_4-H_2O system, *Russ. J. Inorg. Chem.*, **34**, (1989), 577-579. From a citation in this Bibliography

[89MIL] Milne, J., The nature of selenium dissolved in hydrobromic acid containing selenium dioxide, *Can. J. Chem.*, **67**, (1989), 1056-1060. Cited on page: 178.

[89NEV/RYB] Neverov, V. G., Rubakova, G. A., Panus, V. R., Thermodynamic study of alloys of the system As-Se, *Inorg. Mater.*, **25**, (1989), 191-193. Cited on page: 191.

[89NIK/AND] Nikitin, S. A., Andersen, E. B., Petrova, N. V., New data on platynite and isotopy of lead in associating selenides, *Zap. Vses. Mineral. O-va.*, **3**, (1989), 22-27, in Russian. Cited on pages: 634, 635.

[89OHA/SUS] O'Hare, P. A. G., Susman, S., Volin, K. J., Thermochemistry of germanium monoselenide, and the Ge-Se bond dissociation enthalpy, *J. Chem. Thermodyn.*, **21**, (1989), 827-836. Cited on page: 209.

[89OJK/STA] Ojkova, T., Staneva, D., Wasseraktivität und Aktivitätskoeffizienten in binären Lösungen von Zink-, Magnesium-, Kobalt-, Nickel- und Kupferselenaten, *Z. Phys. Chem. (Leipzig)*, **270**, (1989), 628-632, in German. Cited on pages: 264, 296, 314, 322, 397, 401, 405, 578, 580.

[89OPP/RIT] Oppermann, H., Rittner, P., Funke, K., Schadow, H., Zur calorimetrischen Bestimmung der Bildungsenthalpie von $SeBr_4$ und $SeOBr_2$, *Z. Anorg. Allg. Chem.*, **575**, (1989), 71-76, in German. Cited on page: 176.

[89PHI/BUC] Phillips, W. A., Buchenau, U., Nücker, N., Dianoux, A.-J., Petry, W., Dynamics of glassy and liquid selenium, *Phys. Rev. Lett.*, **63**, (1989), 2381-2384. Cited on page: 90.

[89RIG/ROB] Riglet, C., Robouch, P., Vitorge, P., Standard potentials of the (MO_2^{2+}/MO_2^{+}) and (M^{4+}/M^{3+}) redox systems for neptunium and plutonium, *Radiochim. Acta*, **46**, (1989), 85-94. Cited on pages : 606, 607.

[89ROB] Robouch, P., Contribution à la prévision du comportement de l'américium, du plutonium et du neptunium dans la géosphère ; données chimiques, Commissariat à l'Energie Atomique, Report CEA-R-5473, (1989). Cited on pages: 606, 607.

[89SHA/CHA] Sharma, R. C., Chang, Y. A., Thermodynamic analysis and phase equilibria calculations for the Cd-Te, Cd-Se, and Cd-S systems, *J. Electrochem. Soc.*, **136**, (1989), 1536-1542. Cited on page: 267.

[89SHO/HEL] Shock, E. L., Helgeson, H. C., Corrections to Shock and Helgeson Geochimica et Cosmochimica Acta, 52, (1988) 2009-2036, *Geochim. Cosmochim. Acta*, **53**, (1989), 215. Cited on pages: 35, 595.

[89SHO/HEL2] Shock, E. L., Helgeson, H. C., Sverjensky, D. A., Calculation of the thermodynamic and transport properties of aqueous species at high pressures and temperatures: Standard partial molal properties of inorganic neutral species, *Geochim. Cosmochim. Acta*, **53**, (1989), 2157-2183. Cited on pages: 35, 595.

[89TAT/SER] Tatarinova, E. E., Serezhkina, L. B., Serezhkin, V. N., Solubility in the Li_2XO_4-UO_2XO_4-H_2O (X = S, Se) systems at 25°C, *Zh. Neorg. Khim.*, **34**, (1989), 2157-2159, in Russian, English translation in [89TAT/SER2]. Cited on pages: 390, 415.

[89TAT/SER2] Tatarinova, E. E., Serezhkina, L. B., Serezhkin, V. N., Solubility in the Li_2XO_4-UO_2XO_4-H_2O (X = S, Se) systems at 25°C, *Russ. J. Inorg. Chem.*, **34**, (1989), 1227-1228. From a citation in this Bibliography

[89VER/FIL] Vergasova, L. P., Filatov, S. K., Semyonova, T. F., Filosophova, T. M., Sophiite, $Zn_2(SeO_3)Cl_2$, a new mineral from volcanic sublimates, *Zap. Vses. Mineral. O-va.*, **1**, (1989), 65-69, in Russian. Cited on page: 636.

[89VOJ/EBE] Vojtisek, P., Ebert, M., Löslichkeitsuntersuchung in den Systemen $M_2(I)SeO_4$-$MgSeO_4$-H_2O (M(I) = Li, NH_4 und Tl) bei 25°C, *Z. Chem.*, **29**, (1989), 220-221, in German. Cited on pages: 253, 397, 411, 415.

[89VOJ/EBE2] Vojtisek, P., Ebert, M., Sieglova, Z., Untersuchung in den Systemen $MnSeO_4$-$M_2(I)SeO_4$-H_2O (M(I) = K, NH_4, Rb) bei 25°C, *Z. Chem.*, **29**, (1989), 265-266, in German. Cited on pages: 342, 411, 425.

[90ARC/WAN] Archer, D. G., Wang, P., The dielectric constant of water and Debye-Hückel limiting law slopes, *J. Phys. Chem. Ref. Data*, **19**, (1990), 371-411. Cited on page: 594.

[90BAR/TRI] Bardi, G., Trionfetti, G., Vapour pressure and sublimation enthalpy of zinc selenide and zinc telluride by thermogravimetric Knudsen-effusion method, *Thermochim. Acta*, **157**, (1990), 287-294. Cited on pages: 257, 258, 570.

[90BEC/NAG] Beck, M. T., Nagypál, I., *Chemistry of complex equilibria, 2nd. Edition*, Horwood Limited Publishers, New York, (1990), 402 pp.. Cited on page: 22.

[90BIN] Binnewies, M., Bildung und Stabilität von gasförmigem BSe_2, *Z. Anorg. Allg. Chem.*, **589**, (1990), 115-221, in German. Cited on page: 227.

[90BOL/TRO] Bolgar, A. S., Trofimova, Z. A., Yanaki, A. A., Thermodynamic properties of tungsten diselenide in a broad temperature range, *Poroshk. Metall. (Kiev)*, **5**, (1990), 53-56, in Russian. Cited on pages: 348, 567, 570.

[90CIA] Ciavatta, L., The specific interaction theory in equilibrium analysis: Some empirical rules for estimating interaction coefficients of metal ion complexes, *Ann. Chim. (Rome)*, **80**, (1990), 255-263. Cited on pages: 588, 600, 601.

[90DAS/NAR] Dasgupta, P. K., Nara, O., Measurement of acid dissociation constants of weak acids by cation exchange and conductometry, *Anal. Chem.*, **62**, (1990), 1117-1122. Cited on pages: 141, 142.

[90KAR/THO] Karakaya, I., Thompson, W. T., The Ag-Se (silver-selenium) system, *Bull. Alloy Phase Diagrams*, **11**, (1990), 266-271. Cited on pages: 297, 523.

[90KOS/YAC] Koshchenko, V. I., Yachmenev, V. E., Zhegalina, V. A., Pashinkin, A. S., Danilenko, G. N., Low-temperature heat capacity of α-Cr_2Se_3, *Inorg. Mater.*, **26**, (1990), 2130-2133. Cited on pages: 343, 344.

[90KUM/BAT] Kumok, V. N., Batyreva, V. A., Interpretation of the solubility diagrams of selenate systems at 25°C, *Zh. Neorg. Khim.*, **35**, (1990), 2663-2667, in Russian, English translation in [90KUM/BAT2]. Cited on pages: 264, 275, 296, 314, 322, 342, 390, 393, 397, 411, 415, 421, 425, 428, 571.

[90KUM/BAT2] Kumok, V. N., Batyreva, V. A., Interpretation of the solubility diagrams of selenate systems at 25°C, *Russ. J. Inorg. Chem.*, **35**, (1990), 1514-1517. From a citation in this Bibliography

[90LEI] Leigh, G. J., *Nomenclature of inorganic chemistry, recommendations 1990, issued by the Commission on Nomenclature of Inorganic Chemistry*, IUPAC, Ed., Blackwell Sci. Publ., Oxford, (1990), 289 pp.. Cited on page: 11.

[90LEV/MYE] Levy, D. E., Myers, J. R., Spectroscopic determination of the second dissociation constant of H_2Se and the activity coefficients and spectral shifts of its ions, *J. Phys. Chem.*, **94**, (1990), 7842-7847. Cited on pages: 135, 155, 301, 449, 571.

[90MAR/PET] Marx, V., Petzel, T., The thermodynamics of sublimation and formation of ytterbium(II) selenide, *J. Less-Common Met.*, **163**, (1990), 199-207. Cited on page: 358.

[90MIC/CER] Micka, Z., Cermak, M., Niznansky, D., The solubility of cupric selenate and the bonding conditions in the compound $Cu(HseO_3)_2$, *Collect. Czech. Chem. Commun.*, **55**, (1990), 2441-2446. Cited on page: 293.

[90MIN/AMI] Minenkov, Y. F., Amitin, E. B., Naumov, V. N., Paukov, I. E., Krabbes, G., Lattice and electron components of specific heat of Mo_6Se_8 and Mo_6Te_8, *Mater. Sci. Forum*, **62/64**, (1990), 115-116. Cited on page : 346.

[90MON] Monnin, C., The influence of pressure on the activity coefficients of the solutes and on the solubility of minerals in the system Na-Ca-Cl-SO_4-H_2O to 200°C and 1 kbar, and to high NaCl concentration, *Geochim. Cosmochim. Acta*, **54**, (1990), 3265-3282. Cited on page: 35.

[90NAS/SHA] Nasar, A., Shamsuddin, M., Thermodynamic properties of zinc selenide, *Z. Metallkd.*, **81**, (1990), 244-246. Cited on pages: 256, 257, 258, 572.

[90NDI/CAR] Ndikumana, T., Carleer, M., Colin, R., Laser-induced fluorescence spectrum and radiative lifetimes of the gaseous PbSe molecule, *Mol. Phys.*, **69**, (1990), 229-239. Cited on page: 221.

[90OEL/HEL] Oelkers, E. H., Helgeson, H. C., Triple-ion anions and polynuclear complexing in supercritical electrolyte solutions, *Geochim. Cosmochim. Acta*, **54**, (1990), 727-738. Cited on pages: 595, 596.

[90OHA/LEW] O'Hare, P. A. G., Lewis, B. M., Susman, S., Volin, K. J., Standard molar enthalpies of formation and transition at the temperature 298.15 K and other thermodynamic properties of the crystalline and vitreous forms of arsenic sesquiselenide (As_2Se_3). Dissociation enthalpies of As-Se bonds, *J. Chem. Thermodyn.*, **22**, (1990), 1191-1206. Cited on pages: 189, 190, 191.

[90OJK/BAL] Ojkova, T., Balarew, C., Staneva, D., Thermodynamische Untersuchung des Systems $MgSeO_4$-$ZnSeO_4$-H_2O bei 25.0°C, *Z. Anorg. Allg. Chem.*, **584**, (1990), 217-224, in German. Cited on page: 264.

[90PRI/GAT] Pring, A., Gatehouse, B. M., Birch, W. D., Francisite, $Cu_3Bi(SeO_3)_2O_2Cl$, a new mineral from Iron Monarch, South Australia: description and crystal structure, *Am. Mineral.*, **75**, (1990), 1421-1425. Cited on page: 636.

[90RIG] Riglet, C., Chimie du neptunium et autres actinides en milieu carbonate, Commissariat à l'Energie Atomique, Report CEA-R-5535, (1990), 267 pp.. Cited on pages: 606, 607.

[90SCA/PIA] Scardala, P., Piacente, V., Ferro, D., Standard sublimation enthalpy of solid Ag_2Se from vapour pressure measurements, *J. Less-Common Met.*, **162**, (1990), 11-21. Cited on pages: 299, 300, 572.

[90STO/GAV] Stolyarova, T. A., Gavrilov, N. M., Nekrasov, I. Y., Thermodynamic properties of bismuth selenides, *Geokhimiya*, (1990), 1367-1374, in Russian. Cited on pages: 197, 198, 199, 202.

[90TAT/SER] Tatarinova, E. E., Serezhkina, L. B., Serezhkin, V. N., System Na_2SeO_4-UO_2SeO_4-H_2O at 25°C, *Radiokhimiya*, **32**, (1990), 1-3, in Russian, English translation in [90TAT/SER2]. Cited on pages: 390, 421.

[90TAT/SER2] Tatarinova, E. E., Serezhkina, L. B., Serezhkin, V. N., System Na_2SeO_4-UO_2SeO_4-H_2O at 25°C, *Sov. Radiochem.*, **32**, (1990), 303-305. From a citation in this Bibliography

[91AND/CAS] Anderson, G. M., Castet, S., Schott, J., Mesmer, R. E., The density model for estimation of thermodynamic parameters of reactions at high temperatures and pressures, *Geochim. Cosmochim. Acta*, **55**, (1991), 1769-1779. Cited on page: 35.

[91BAL/SAC] Balducci, G., Sacconi, L. V., Gigli, G., The dimerization energies of the Ge_2S_2, Ge_2Se_2, and Ge_2Te_2 molecules, *Ber. Bunsen-Ges.*, **95**, (1991), 1651-1658. Cited on page: 212.

[91BJO/SCH] Björgvinsson, M., Schrobilgen, G. J., Homo- and heteropolychalcogenide anions Ch^{2-}, HCh^-, Ch_2^{2-}, Ch_3^{2-}, and Ch_4^{2-} (Ch = Se and/or Te): Solution ^1H, ^{77}Se, ^{123}Te, and ^{125}Te NMR study, *Inorg. Chem.*, **30**, (1991), 2540-2547. Cited on page: 154.

[91CAB/LAF] Cabri, L. J., Laflamme, J. H. G., Roberts, A. C., Criddle, A. J., Hulbert, L. J., Jolliffeite and unnamed CoAsSe: two new arsenoselenides from the north shore of Lake Athabasca, Saskatchewan, *Can. Mineral.*, **29**, (1991), 411-418. Cited on pages: 634, 637.

[91DEL/PIR] Deliens, M., Piret, P., La haynesite, sélénite hydraté d'uranyle, nouvelle espèce minérale de la mine Repete, Comté de San Juan, Utah, *Can. Mineral.*, **29**, (1991), 561-564, in French. Cited on page: 636.

[91DIN] Dinsdale, A. T., SGTE data for pure elements, *CALPHAD: Comput. Coupling Phase Diagrams Thermochem.*, **15**, (1991), 317-425. Cited on pages: 29, 249, 336, 645.

[91GOS] Gospodinov, G. G., Regularities in the formation of selenites in the three-component system Al_2O_3-SeO_2-H_2O and their properties, *Thermochim. Acta*, **180**, (1991), 169-176. Cited on page: 231.

[91GOS/BAR] Gospodinov, G., Barkov, D., Phase equilibrium in the system $HgSeO_4$-H_2SeO_4-H_2O at 100°C, *Cryst. Res. Technol.*, **26**, (1991), 599-603. Cited on page: 286.

[91GOS/BAR2] Gospodinov, G., Barkov, D., Das System HgO-SeO_2-H_2O bei 25 und 100°C, *Z. Phys. Chem. (Munich)*, **171**, (1991), 241-247, in German. Cited on page: 284.

[91HAL/MOR] Hall, R. O. A., Mortimer, M. J., Harding, S. R., Spirlet, J. C., Low temperature heat capacity of plutonium selenide over the temperature range 7-300 K, AEA Technology report, Report AEA-FS-0048H, (1991). Cited on page: 385.

[91MOR/FEU] Morgant, G., Feutelais, Y., Schnitter, J., Legendre, B., Thermodynamic properties of binary liquid thallium-selenium solutions, *Thermochim. Acta*, **180**, (1991), 89-98. Cited on pages: 249, 250, 251.

[91OKA] Okamoto, H., The Fe-Se (iron-selenium) system, *J. Phase Equilib.*, **12**, (1991), 383-389. Cited on page: 324.

[91OYK/MIH] Oykova, T., Mihov, D., Pavlova, P., Phase interaction in the systems sodium selenate - copper selenate - water and sodium selenate - zinc selenate - water at 25°C, *Cryst. Res. Technol.*, **26**, (1991), 1071-1075. Cited on pages: 264, 295.

[91PIT] Pitzer, K. S., Ion interaction approach: theory and data correlation, *Activity coefficients in electrolyte solutions*, Pitzer, K. S., Ed., pp.75-153, CRC Press, Boca Raton, Florida, (1991). Cited on page: 592.

[91POL/VOL] Polekhovsky, Y. S., Voloshin, A. V., Tarasova, I. P., Nikitin, S. A., Pakhomovsky, Y. A., Menshikov, Y. P., Kretser, Y. L., Kolytcheva, T. I., Padmaite PdBiSe, a new palladium and bismuth selenide from metasomatites of South Karelia, *Zap. Vses. Mineral. O-va.*, **3**, (1991), 85-88, in Russian. Cited on page: 635.

[91RAS] Rasulov, S. M., Vacancy concentrations and formation energies for Sb_2Se_3, and SnSe, *Proc. Acad. Sci. USSR*, **27**, (1991), 914-916. Cited on page: 214.

[91VAS/KRU] Vasilev, V. A., Kruchina, T. I., Effect of anion hydration on the heat capacity and density of aqueous sodium selenate solutions at 298.15 K, *Izv. Vyssh. Uchebn. Zaved., Khim. Khim. Tekhnol.*, **34**, (1991), 43-46, in Russian. Cited on page: 421.

[91WAN] Wanner, H., On the problem of consistency of chemical thermodynamic data bases, Scientific basis for nuclear waste management XIV, vol. 212, pp.815-822, Materials Research Society, (1991). Cited on page: 2.

[92BAN/BLI] Bányai, I., Blixt, J., Glaser, J., Tóth, I., On the dissociation of hydrogen cyanide in aqueous solutions containing different ionic media. A combined potentiometric and carbon-13 NMR study, *Acta Chem. Scand.*, **46**, (1992), 138-141. Cited on page: 612.

[92BOO/KLE] Boone, S., Kleppa, O. J., Enthalpies of formation for group IV selenides ($GeSe_2$, $GeSe_2$(am), SnSe, $SnSe_2$, PbSe) by direct-combination drop calorimetry, *Thermochim. Acta*, **197**, (1992), 109-121. Cited on pages : 211, 214, 217, 220.

[92CAP] Capdevila, H., Données thermodynamiques sur l'oxydoréduction du plutonium en milieux acide et carbonate. Stabilité de Pu(V), Ph. D. Thesis, Université de Paris-Sud, Orsay, France, (1992), in French. Also published as CEA-R-5643 Commissariat à l'Energie Atomique, France, (1993). Cited on pages: 606, 607.

[92GAR] Gardner, P. J., The standard enthalpies of formation of hydrogen cyanide and tricyanophosphine by ab initio molecular orbital calculation, *Thermochim. Acta*, **205**, (1992), 65-73. Cited on page: 434.

[92GOP/MIL] Gopal, M., Milne, J., Spectroscopic evidence for selenium iodides in CS_2 solution: Se_3I_2, Se_2I_2, and SeI_2, *Inorg. Chem.*, **31**, (1992), 4530-4533. Cited on page: 178.

[92GOS/BAR] Gospodinov, G., Barkov, D., Phasenverhältnisse der Selenate im Dreistoffsystem Tl_2O_3-SeO_3-H_2O bei 100°C, *Monatsh. Chem.*, **123**, (1992), 701-704, in German. Cited on page: 254.

[92GRE/FUG] Grenthe, I., Fuger, J., Konings, R. J. M., Lemire, R. J., Muller, A. B., Nguyen-Trung, C., Wanner, H., *Chemical Thermodynamics of Uranium*, Wanner, H., Forest, I., Nuclear Energy Agency, Organisation for Economic Co-operation and Development, Eds., vol. 1, *Chemical Thermodynamics*, North Holland Elsevier Science Publishers B. V., Amsterdam, The Netherlands, (1992), 715 pp.. Cited on pages: 2, 3, 24, 26, 36, 67, 68, 69, 70, 71, 72, 73, 74, 75, 76, 77, 78, 79, 80, 81, 82, 83, 84, 85, 86, 121, 134, 141, 143, 147, 148, 151, 152, 153, 183, 191, 193, 195, 264, 385, 386, 387, 388, 389, 390, 391, 405, 407, 432, 442, 446, 464, 468, 473, 518, 577, 584, 587, 588, 600, 601, 606, 607, 611, 615, 626, 627.

[92JED/CER] Jedwab, J., Cervelle, B., Gouet, G., Hubaut, X., Piret, P., The new platinum selenide luberoite Pt_5Se_4 from the Lubero region (Kivu Province, Zaire), *Eur. J. Mineral.*, **4**, (1992), 683-692. Cited on page: 635.

[92KOS/IVA] Koshchenko, V. I., Ivanyas, A. F., Zhegalina, V. A., Danilenko, G. N., Thermodynamic properties of γ-chromium selenide (γ-Cr_2Se_3) in the temperature range 60-300 K, *Russ. J. Phys. Chem.*, **66**, (1992), 412-413. Cited on page: 344.

[92MAR/PET] Marx, V., Petzel, T., Über die Verdampfungsthermodynamik und die Bildungenthalpien von CaSe, SrSe und BaSe, *Z. Anorg. Allg. Chem.*, **614**, (1992), 7-11, in German. Cited on pages: 399, 403, 406.

[92OJK/STO] Ojkova, T., Stoilova, D., Staneva, D., Study on the $MgSeO_4$-$CoSeO_4$-H_2O system at 25°C, *Z. Phys. Chem. (Munich)*, **177**, (1992), 247-254. Cited on page: 321.

[92OYK/MIH] Oykova, T., Mihov, D., Study on the phase equilibrium in the systems sodium selenate - cadmium selenate - water and sodium selenate - manganese selenate - water at 25°C, *Cryst. Res. Technol.*, **27**, (1992), 697-701. Cited on pages: 275, 342.

[92PIA/SCA] Piacente, V., Scardala, P., Fontana, D., Decomposition pressure and standard enthalpies of sublimation and of formation of iron monoselenide, *J. Alloys Compd.*, **189**, (1992), 263-267. Cited on page: 326.

[92PIE] Piestrzynski, A., Bi_4Se_3 a new unnamed mineral from the Kupferschiefer, Polish Lowland: preliminary report, *Mineral. Pol.*, **23**, (1992), 35-42. Cited on page: 637.

[92RUM/MAU] Rumpf, B., Maurer, G., Solubilities of hydrogen cyanide and sulfur dioxide in water at temperatures from 293.15 to 413.15 K and pressures up to 2.5 MPa, *Fluid Phase Equilib.*, **81**, (1992), 241-260. Cited on pages: 434, 435.

[92SEM/KUL] Semenov-Kobzar, A. A., Kulikov, L. M., Bogatyr, A. S., Zaletilo, L. S., Thermal stability of polycrystalline molybdenum and tungsten diselenides in argon, *Inorg. Mater.*, **28**, (1992), 693-695. Cited on pages: 345, 348, 573.

[92SHA/CHA] Sharma, R. C., Chang, Y. A., Guminski, C., The Hg-Se (mercury-selenium) system, *J. Phase Equilib.*, **13**, (1992), 663-671. Cited on page: 279.

[92SHO/OEL] Shock, E. L., Oelkers, E. H., Johnson, J. W., Sverjensky, D. A., Helgeson, H. C., Calculation of the thermodynamic properties of aqueous species at high pressures and temperatures. Effective electrostatic radii, dissociation constants and standard partial molal properties to 1000°C and 5 kbar, *J. Chem. Soc. Faraday Trans.*, **88**, (1992), 803-826. Cited on page: 595.

[92SIR/GAV] Sirota, N. N., Gavaleshko, N. P., Novikova, V. V., Novikov, A. V., Heat capacities and thermodynamic functions of the $Cd_xHg_{1-x}Se$ (x > 0.5) alloys in the temperature range 5-300 K, *Russ. J. Phys. Chem.*, **66**, (1992), 433-435. Cited on pages: 267, 268.

[93BLA] Blachnik, R., The compounds P_4S_3, P_4Se_3, As_4S_3 and As_4Se_3 and the quaternary system P_4S_3-P_4Se_3-As_4S_3-As_4Se_3, *Thermochim. Acta*, **213**, (1993), 241-259. Cited on page: 186.

[93BRU/PIA] Brunetti, B., Piacente, V., Thermodynamic properties of lead monoselenide, *J. Mater. Sci. Lett.*, **12**, (1993), 1738-1740. Cited on pages: 222, 511, 573.

[93CHA] Chatillon, C., Critical analysis of the thermodynamic properties of the In-Se gaseous and solid phases, *J. Cryst. Growth*, **129**, (1993), 297-311. Cited on pages: 238, 239, 240, 241, 243, 246.

[93FIN/GAR] Finch, A., Gardner, P. J., Head, A. J., Xiaoping, W., Yefimov, M. E., Furkaluk, M. U., The standard enthalpy of formation of the aqueous cyanide ion, *J. Chem. Thermodyn.*, **25**, (1993), 1385-1390. Cited on pages: 434, 436.

[93GHO] Ghosh, G., The Sb-Se (antimony-selenium) system, *J. Phase Equilib.*, **14**, (1993), 753-763. Cited on page: 193.

[93GIF/VIT2] Giffaut, E., Vitorge, P., Capdevila, H., Corrections de température sur les coefficients d'activité calculés selon la TIS, Commissariat à l'Energie Atomique, Report CEA-N-2737, (1993), 29 pp.. Cited on page: 596.

[93GLA/SIN] Glass, R. S., Singh, W. P., Jung, W., Veres, Z., Scholz, T. D., Stadtman, T. C., Monoselenophosphate: synthesis, charaterization, and identity with the prokaryotic biological selenium donor, compound SePX, *Biochemistry*, **32**, (1993), 12555-12559. Cited on page: 187.

[93GOS] Gospodinov, G. G., Röntgenstruktur- und thermische Untersuchungen der Dreistoffverbindungen im System Sb-Se-O, *Monatsh. Chem.*, **124**, (1993), 343-347, in German. Cited on page: 195.

[93OHA] O'Hare, P. A. G., Calorimetric measurements of the specific energies of reaction of arsenic and of selenium with fluorine. Standard molar enthalpies of formation $\Delta_f H_m^o$ at the temperature 298.15 K of AsF_5, SeF_6, As_2Se_3, As_4S_4, and As_2S_3. Thermodynamic properties of AsF_5 and SeF_6 in the ideal-gas state. Critical assessment of $\Delta_f H_m^o(AsF_3$, l), and the dissociation enthalpies of As-F bonds, *J. Chem. Thermodyn.*, **25**, (1993), 391-402. Cited on pages: 164, 189, 190, 191, 645.

[93OJK/MIC] Ojkova, T., Michov, D., Jankova, R., Dreistoffsysteme Wasser-Salz mit Lithiumselenat, Natriumselenat, Cobaltselenat und Magnesiumselenat bei 25°C, *Monatsh. Chem.*, **124**, (1993), 349-354, in German. Cited on pages: 321, 397.

[93OJK/STO] Ojkova, T. G. , Stoilova, D. G., Staneva, D. I., Thermodynamical study of the zinc selenate-nickel selenate-water system at 25°C, *Bulg. Chem. Commun.*, **26**, (1993), 259-267. Cited on pages: 263, 264, 314.

[93PAS/SUL] Pasava, J., Sulovsky, P., Kovalova, M., Geochemistry and mineralogy of Proterozoic metal-rich black shales from the Bohemian massif, Czech Republic, with a description of possible new molybdenum selenide and telluride phases, *Can. Mineral.*, **31**, (1993), 745-754. Cited on page: 637.

[93SAK/BRY] Sakharova, M. S., Bryzgalov, I. A., Ryahovskaya, S. K., Mineralogy of selenides in deposits of volcanic belts, *Zap. Vseross. Mineral. O-va.*, **3**, (1993), 1-9, in Russian. Cited on page: 637.

[93SER/KUC] Serezhkina, L. B., Kuchumova, N. V., Serezhkin, V. N., Complexation in the system K_2SeO_4-UO_2SeO_4-H_2O, *Radiokhimiya*, **35**, (1993), 31-35, in Russian, English translation in [93SER/KUC2]. Cited on pages: 390, 425.

[93SER/KUC2] Serezhkina, L. B., Kuchumova, N. V., Serezhkin, V. N., Complexation in the system K_2SeO_4-UO_2SeO_4-H_2O, *Radiochemistry (Moscow)*, **35**, (1993), 640-643. From a citation in this Bibliography

[93SHA/CHA] Sharma, R. C., Chang, Y. A., Guminski, C., The Hg-Se (mercury-selenium) system. (Erratum to document cited in CA119(2):13115s), *J. Phase Equilib.*, **14**, (1993), 6. Cited on page: 279.

[93SLA/POP] Slavtscheva, J., Popova, E., Gospodinov, G., Untersuchung der Löslichkeit und Bestimmung des Löslichkeitsproduktes von Kupfer- und Silberselenit, *Monatsh. Chem.*, **124**, (1993), 1115-1118, in German. Cited on pages: 293, 294, 303.

[94CAP/VIT] Capdevila, H., Vitorge, P., Potentiels redox des couples PuO_2^{2+} /PuO_2^+ et Pu^{4+}/Pu^{3+} à force ionique et température variables. Entropie et capacité calorifique, Commissariat à l'Energie Atomique, Report CEA-N-2762, (1994), 73 pp.. Cited on pages: 606, 607.

[94CHI/SHE] Chizhevskaya, S. N., Shelimova, L. E., Zaitseva, I. A., Critical analysis and matching of data on the phase diagram of the Bi-Se system, *Inorg. Mater.*, **30**, (1994), 1285-1292. Cited on page: 196.

[94GIF] Giffaut, E., Influence des ions chlorure sur la chimie des actinides, Ph. D. Thesis, Université de Paris-Sud, Orsay, France, (1994), 259 pp., in French. Cited on page: 606.

[94GOS/STA] Gospodinov, G. G., Stancheva, M. G., A physicochemical study of the selenites of yttrium, *Cryst. Res. Technol.*, **29**, (1994), K83-K88. Cited on page: 356.

[94KON/EVA] Konnert, J. A., Evans, Jr., H. T., McGee, J. J., Ericksen, G. E., Mineralogical studies of the nitrate deposits of Chile: VII. Two new saline minerals with composition $K_6(Na,K)_4Na_6Mg_{10}(XO_4)_{12}(IO_3)_{12}.12H_2O$: fuenzalidaite (X = S) and carlosruizite (X = Se), *Am. Mineral.*, **79**, (1994), 1003-1008. Cited on page: 636.

[94MAN] Mandarino, J. A., Natural and synthetic selenites and selenates and their Gladstone-Dale compatibility, *Eur. J. Mineral.*, **6**, (1994), 337-349. Cited on page: 636.

[94OJK/STO] Ojkova, T., Stoilova, D., Staneva, D., Thermodynamical study on the zinc selenate-cobalt selenate-water system at 25°C, *Monatsh. Chem.*, **125**, (1994), 1067-1074. Cited on pages : 264, 321.

[94PAS/GOS] Pashinkin, A. S., Gospodinov, G. G., Standard entropies of some metal selenites, *Zh. Fiz. Khim.*, **68**, (1994), 1545-1549, in Russian, English translation in [94PAS/GOS2]. Cited on pages : 285, 305, 309, 313, 353, 396.

[94PAS/GOS2] Pashinkin, A. S., Gospodinov, G. G., Standard entropies of some metal selenites, *Russ. J. Phys. Chem.*, **68**, (1994), 1401-1404. From a citation in this Bibliography

[94PAS/ZLO] Pashinkin, A. S., Zlomanov, V. P., Malkova, A. S., On the heat capacity of lead selenide and lead telluride, *Inorg. Mater.*, **30**, (1994), 1036-1038. Cited on page: 219.

[94RIC/IPS] Richter, K. W., Ipser, H., Transition metal-chalcogen systems. XI: The platinum-selenium phase diagram, *J. Phase Equilib.*, **15**, (1994), 165-170. Cited on page: 335.

[94SER] Serezhkina, L. B., Complex formation in the system $(NH_4)_2SeO_4$-UO_2SeO_4-H_2O, *Zh. Neorg. Khim.*, **39**, (1994), 1020-1022, in Russian, English translation in [94SER2]. Cited on pages: 390, 411.

[94SER2] Serezhkina, L. B., Complex formation in the system $(NH_4)_2SeO_4$-UO_2SeO_4-H_2O, *Russ. J. Inorg. Chem.*, **39**, (1994), 976-978. From a citation in this Bibliography

[94TOM/SUS] Tomaszkiewicz, I., Susman, S., Volin, K. J., O'Hare, P. A. G., Fluorine-combustion calorimetric determinations of the standard molar enthalpy changes for the formation of $SiSe_2$(cr), $SiSe_{1.94}$(cr), and $SiSe_{1.94}$(vit), and for the transition: $SiSe_{1.94}$(vit) = $SiSe_{1.94}$(cr) at the temperature T = 298.15 K. Implications of the results for the enthalpies of dissociation D_m^o(Se-SiSe) and D_m^o(SiSe). Thermodynamic properties of SiSe(g), *J. Chem. Thermodyn.*, **26**, (1994), 1081-1093. Cited on pages: 207, 208.

[94YAM/KAM] Yamaguchi, K., Kameda, K., Takeda, Y., Itagaki, K., Measurements of high temperature heat content of the II-VI and IV-VI (II: Zn, Cd IV: Sn, Pb VI: Se, Te) compounds, *Mater. Trans., JIM*, **35**, (1994), 118-124. Cited on pages: 214, 219, 255, 259, 268.

[95BAL/MIL] Ball, S., Milne, J., Studies on the interaction of selenite and selenium with sulfur donors. Part 3. Sulfite, *Can. J. Chem.*, **73**, (1995), 716-724. Cited on page: 182.

[95BAR] Barin, I., *Thermochemical data of pure substances*, 3rd. Edition, VCH, Weinheim, (1995). Cited on pages: 169, 227, 412.

[95CAP/VIT] Capdevila, H., Vitorge, P., Redox potentials of PuO_2^{2+} / PuO_2^+ and Pu^{4+} / Pu^{3+} at different ionic strengths and temperatures. Entropy and heat capacity, *Radiochim. Acta*, **68**, (1995), 51-62. Cited on pages: 606, 607.

[95CHE/LI] Chen, L., Li, D., Wang, G., Zhang, Q., A study on two unnamed selenide minerals, *Kuangwu Xuebao*, **15**, (1995), 418-421, in Chinese. Cited on page: 637.

[95CLA/COM] Clavaguera-Mora, M. T., Comas, C., Clavaguera, N., Contributions to the modeling of the thermodynamic behaviour of Se-Te liquid solutions, *J. Alloys Compd.*, **220**, (1995), 39-47. Cited on page: 183.

[95COO/HAW] Cooper, M. A., Hawthorne, F. C., The crystal structure of guilleminite, a hydrated Ba-U-Se sheet structure, *Can. Mineral.*, **33**, (1995), 1103-1109. Cited on page: 636.

[95DAV/RAN] Davydov, A. V., Rand, M. H., Argent, B. B., Review of heat capacity data for tellurium, *CALPHAD: Comput. Coupling Phase Diagrams Thermochem.*, **19**, (1995), 375-387. Cited on pages: 185, 645.

[95GRE/PUI] Grenthe, I., Puigdomènech, I., Rand, M. H., Sandino, M. C. A., *Corrections to the Uranium NEA-TDB review, Appendix D, in: Chemical Thermodynamics of Americium (authored by Silva, R. J., Bidoglio, G., Rand, M. H. and Robouch, P. B., Wanner, H., Puigdomènech, I.)*, Nuclear Energy Agency, Organisation for Economic Co-operation and Development, Ed., pp.347-374, North Holland Elsevier Science Publishers B. V., Amsterdam, The Netherlands, (1995). Cited on page: 611.

[95HAV/MIC] Havlicek, D., Micka, Z., Boublikova, R., Lithium and ammonium selenates, *Collect. Czech. Chem. Commun.*, **60**, (1995), 969-976. Cited on pages: 411, 415.

[95LI/HAG] Li, D. X., Haga, Y., Shida, H., Suzuki, T., Fukama, H., Kasuya, T., Akimoto, R., Kobayashi, M., Susceptibility and specific heat of EuSe, *J. Alloys Compd.*, **21**, (1995), 211-214. Cited on page: 366.

[95LIC/FOR] Licht, S., Forouzan, F., Speciation analysis of aqueous polyselenide solutions, *J. Electrochem. Soc.*, **142**, (1995), 1546-1551. Cited on pages: 154, 155, 156.

[95LIU/ZHE] Liu, J., Zheng, M., Liu, J., A description of possible new Ni-As-S-Se mineral phase, *Kuangwu Xuebao*, **15**, (1995), 425-427, in Chinese. Cited on page: 637.

[95MIN/LI] Min, M., Li, D., Shi, N., Liu, Q., Cao, Y., Some new data on antimonselite, *Kuangwu Xuebao*, **15**, (1995), 303-304, in Chinese. Cited on page: 634.

[95PAY] Payne, J. R., Lead nitrate: a thermal solubility study, *Thermochim. Acta*, **259**, (1995), 207-213. Cited on pages: 452, 502, 524, 525.

[95RAI/FEL3] Rai, D., Felmy, A. R., Moore, D. A., The solubility product of crystalline ferric selenite hexahydrate and the complexation constant of $FeSeO_3^+$, *J. Solution Chem.*, **24**, (1995), 735-752. Cited on pages: 302, 329, 574.

[95RUD/KNA] Rudashevskii, N. C., Knauf, V. V., Chernyshov, H. M., Platinum group minerals from black shales KMA, *Dokl. Akad. Nauk*, **344**, (1995), 91-95, in Russian. Cited on page: 637.

[95SAV] Savenko, V. S., Solubility product of magnesium and calcium selenites, *Zh. Neorg. Khim.*, **40**, (1995), 1254-1256, in Russian, English translation in [95SAV2]. Cited on pages: 395, 400.

[95SAV2] Savenko, V. S., Solubility product of magnesium and calcium selenites, *Russ. J. Inorg. Chem.*, **40**, (1995), 1206-1208. From a citation in this Bibliography

[95SHI/SCH] Shimuzu, M., Schmidt, S. T., Stanley, C. J., Tsunoda, K., Kawazulite and unnamed $Bi_3(Te,Se,S)_4$ in Ag-Bi-Te-Se-S mineralization from the Suttsu mine, Hokkaido, Japan, *Neues Jahrb. Mineral., Abh.*, **169**, (1995), 305-308. Cited on page: 637.

[95SIL/BID] Silva, R. J., Bidoglio, G., Rand, M. H., Robouch, P., Wanner, H., Puigdomènech, I., *Chemical Thermodynamics of Americium*, Nuclear Energy Agency, Organisation for Economic Co-operation and Development, Ed., vol. 2, *Chemical Thermodynamics*, North Holland Elsevier Science Publishers B. V., Amsterdam, The Netherlands, (1995), 374 pp.. Cited on pages: 2, 587, 606, 607, 611, 615.

[95SPI/SPI] Spiridonov, E. M., Spiridonov, M. F., Yudin, A. V., Shscerbachov, D. K., New data of skipennite Bi_2TeSe_2, *Dokl. Akad. Nauk*, **343**, (1995), 229-232, in Russian. Cited on page : 637.

[95SPU/FIN] Spundflasche, E., Seifert, H. J., Zur Thermochemie und Struktur von Berylliumchlorid, *Z. Anorg. Allg. Chem.*, **621**, (1995), 1723-1726. Cited on page: 464.

[95STO/CHR] Stoilova, D., Christov, C., Ojkova, T., Staneva, D., Thermodynamical study on the magnesium selenate-nickel selenate-water system at 25°C, *Monatsh. Chem.*, **126**, (1995), 535-541. Cited on page: 314.

[95STO/OJK] Stoilova, D., Ojkova, T., Staneva, D., On the $(Mg_xNi_{1-x})SeO_4.6H_2O$ and $(Mg_xCu_{1-x})SeO_4.5H_2O$ mixed crystal formation, *Cryst. Res. Technol.*, **30**, (1995), 3-7. Cited on pages: 295, 314, 397.

[95YAS/RYA] Yashunsky, Y. V., Ryabeva, E. G., Abramov, M. V., Rasulova, S. D., Dzharkenite $FeSe_2$ – the new mineral, *Zap. Vseross. Mineral. O-va.*, **1**, (1995), 85-90, in Russian. Cited on page: 634.

[95ZOC/PIA] Zocchi, F., Piacente, V., Sublimation enthalpy of tin monoselenides, *J. Mater. Sci. Lett.*, **14**, (1995), 235-237. Cited on pages: 216, 575.

[96AMZ/GIL] Amzil, A., Gilbert, M., Bichara, C., Mathieu, J.-C., A thermodynamic and structural analysis of selenium-tellurium melts, *J. Phys.: Condens. Matter*, **8**, (1996), 5281-5293. Cited on page: 183.

[96BAD/FER] Badiello, R., Feroci, G., Fini, A., Interaction between trace elements: selenium and cadmium ions, *J. Trace Elem. Med. Biol.*, **10**, (1996), 156-162. Cited on page: 275.

[96BAR/TAH] Barrow, R. F., Taher, F., D'Incan, J., Effantin, C., Ross, A. J., Topouzkhanian, A., Wannous, G., Vergès, J., Electronic states of Bi_2, *Mol. Phys.*, **87**, (1996), 725-733. Cited on page: 459.

[96BRE/LIU] Brebrick, R. F., Liu, H., Analysis of the Zn-Se system, *J. Phase Equilib.*, **17**, (1996), 495-501. Cited on page: 255.

[96BRE/LIU2] Brebrick, R. F., Liu, H., Partial pressures of Zn and Se_2 over ZnSe(c) from optical density measurements, *High Temp. Mater. Sci.*, **35**, (1996), 215-237. Cited on pages: 257, 258, 575.

[96BRY/SPI] Bryzgalov, I. A., Spiridonov, E. M., Petrova, I. V., Sakharova, M. S., Babkinite $Pb_2Bi_2(S, Se)_3$ – a new mineral, *Dokl. Akad. Nauk*, **346**, (1996), 656-659, in Russian. Cited on page : 634.

[96FEU/MAJ] Feutelais, Y., Majid, M., Legendre, B., Fries, S. G., Phase diagram investigation and proposition of a thermodynamic evaluation of the tin-selenium system, *J. Phase Equilib.*, **17**, (1996), 40-49. Cited on pages: 213, 214, 215, 216, 217.

[96HEI/KOC] Heinemann, C., Koch, W., Lindner, G.-G., Reinen, D., Widmark, P.-O., Ground- and excited-state properties of neutral and anionic selenium dimers and trimers, *Phys. Rev.*, **54**, (1996), 1979-1993. Cited on page: 94.

[96KAZ/BOL] Kazenas, E. K., Bol'shih, M. A., Evaporation of SeO_2, TeO_2 oxides, *Metally*, **2**, (1996), 9-16, in Russian. Cited on pages: 123, 124.

[96MAR] Martin, J. M. L., The total atomization energy and heat of formation of HCN(g), *Chem. Phys. Lett.*, **259**, (1996), 679-682. Cited on pages: 433, 434.

[96OHA/ZYW] O'Hare, P. A. G., Zywocinski, A., Curtiss, L. A., Thermodynamics of (germanium + selenium): a review and critical assessment, *J. Chem. Thermodyn.*, **28**, (1996), 459-480. Cited on pages: 209, 210, 211, 212.

[96OHT/YAM] Ohta, T., Yamamuro, O., Matsuo, T., Suga, H., Calorimetric study of phase transitions in SeF_6 and WF_6 crystals, *J. Phys. Chem.*, **100**, (1996), 2353-2359. Cited on page: 164.

[96OKA/KAW] Okada, H., Kawanaka, T., Ohmoto, S., Study on the ZnSe phase diagram by differential thermal analysis, *J. Cryst. Growth*, **165**, (1996), 31-36. Cited on page: 255.

[96SCH/FRE] Schönherr, E., Freiberg, M., Siche, D., Hartmann, H., The vapor composition and vapor pressure of ZnSe from a modified Knudsen technique between 1190 and 1310 K, *Ber. Bunsen-Ges.*, **100**, (1996), 1766-1771. Cited on pages: 257, 258, 575.

[96SHA/CHA] Sharma, R. C., Chang, Y. A., The Se-Zn (selenium-zinc) system, *J. Phase Equilib.*, **17**, (1996), 155-160. Cited on page: 255.

[96SHA/CHA2] Sharma, R. C., Chang, Y. A., The S-Se (sulfur-selenium) system, *J. Phase Equilib.*, **17**, (1996), 148-150. Cited on page: 179.

[96SHA/CHA3] Sharma, R. C., Chang, Y. A., The Cd-Se (cadmium-selenium) system, *J. Phase Equilib.*, **17**, (1996), 140-145. Cited on page: 267.

[96SHA/RED] Sharmasarkar, S., Reddy, K. J., Vance, G. F., Preliminary quantification of metal selenite solubility in aqueous solutions, *Chem. Geol.*, **132**, (1996), 165-170. Cited on pages: 262, 341, 395, 400, 576.

[96SOL/MAY] Solis, J. S., May, P. M., Hefter, G., Cyanide thermodynamics. III. Enthalpies and entropies of ionization of water and hydrogen cyanide, *Aust. J. Chem.*, **49**, (1996), 651-657. Cited on page: 435.

[96STO/FJE] Stølen, S., Fjellvåg, H., Grønvold, F., Sipowska, J. T., Westrum, Jr., E. F., Heat capacity, structural and thermodynamic properties of synthetic klockmannite CuSe at temperatures from 5 K to 652.7 K. Enthalpy of decomposition, *J. Chem. Thermodyn.*, **28**, (1996), 753-766. Cited on page: 289.

[96VOC/BLA] Vochten, R., Blaton, N., Peeters, O., Deliens, M., Piretite, $Ca(UO_2)_3(SeO_3)_2(OH)_4.4H_2O$, a new calcium uranyl selenite from Shinkolobwe, Shaba, Zaire, *Can. Mineral.*, **34**, (1996), 1317-1322. Cited on page: 636.

[97ALL/BAN] Allard, B., Banwart, S. A., Bruno, J., Ephraim, J. H., Grauer, R., Grenthe, I., Hadermann, J., Hummel, W., Jakob, A., Karapiperis, T., Plyasunov, A. V., Puigdomènech, I., Rard, J. A., Saxena, S., Spahiu, K., *Modelling in Aquatic Chemistry*, Grenthe, I., Puigdomènech, I., Eds., Nuclear Energy Agency, Organisation for Economic Co-operation and Development, Paris, France, (1997), 724 pp.. Cited on page: 22.

[97BAH/PAR] Bahta, A., Parker, G. A., Tuck, D. G., Critical survey of stability constants of complexes of thiocyanate ion, *Pure Appl. Chem.*, **69**, (1997), 1489-1548. Cited on pages: 266, 277, 286, 315, 577.

[97CHI/SHE] Chizhevskaya, S. N., Shelimova, L. E., Se-Te phase diagram and structures of amorphous and crystalline $Se_{1-x}Te_x$ alloys: A critical review, *Russ. J. Inorg. Chem.*, **42**, (1997), 741-750. Cited on page: 183.

[97CHR] Christov, C., Thermodynamics of the formation of solid solutions of the type (Me,Me')$SeO_4.6H_2O$ where Me and Me' denote Mg, Co, Ni, or Zn from aqueous solutions, *J. Chem. Thermodyn.*, **29**, (1997), 481-489. Cited on pages: 264, 314, 322, 397, 578.

[97COM/MOR] Comas, C., Mora, M. T., A thermodynamic model for polymerized unary liquids, *J. Chem. Phys.*, **107**, (1997), 4668-4673. Cited on page: 183.

[97DEG/PEL] Degterov, S. A., Pelton, A. D., L'Ecuyer, J. D., Thermodynamic optimization of the selenium-arsenic (Se-As) system, *J. Phase Equilib.*, **18**, (1997), 357-368. Cited on pages: 188, 189, 190, 191, 192.

[97DUM/BRO] Dumm, J. Q., Brown, P. W., Phase formation in the system CaO-SeO_2-H_2O, *J. Am. Ceram. Soc.*, **80**, (1997), 2488-2494. Cited on pages: 399, 471.

[97FER/FIN] Feroci, G., Fini, A., Badiello, R., Breccia, A., Interaction between selenium derivatives and heavy metal ions: Cu^{2+} and Pb^{2+}, *Microchem. J.*, **57**, (1997), 379-388. Cited on pages: 225, 293, 296, 578.

[97GOS/STA] Gospodinov, G. G., Stancheva, M. G., A study of the compounds present in the three-component system Dy_2O_3-SeO_2-H_2O at 100°C, *J. Therm. Anal.*, **48**, (1997), 1351-1356. Cited on page: 362.

[97GRE/PLY] Grenthe, I., Plyasunov, A. V., On the use of semiempirical electrolyte theories for the modeling of solution chemical data, *Pure Appl. Chem.*, **69**, (1997), 951-958. Cited on page: 596.

[97KAM/GLA] Kaminski, R., Glass, R. S., Schroeder, T. B., Michalski, J., Skowronska, A., Monoselenophosphate: Its hydrolysis and its ability to phosphorylate alcohols and amines, *Bioorg. Chem.*, **25**, (1997), 247-259. Cited on page: 187.

[97KIL/WRI] Kilmartin, P. A., Wright, G. A., Copper selenocyanate formation on the copper anode, *Aust. J. Chem.*, **50**, (1997), 321-327. Cited on page: 297.

[97KOH] Kohli, R., The heat capacities of selected inorganic alkali metal compounds, *J. Therm. Anal.*, **49**, (1997), 1321-1326. Cited on pages: 427, 428.

[97KON/FAN] Könnecke, T., Fanghänel, T., Kim, J. I., Thermodynamics of trivalent actinides in concentrated electrolyte solutions. Modelling the chloride complexation of Cm(III), *Radiochim. Acta*, **76**, (1997), 131-135. Cited on page: 607.

[97LUB/HAV] Lubal, P., Havel, J., The study of complex equilibria of uranium(VI) with selenate, *Talanta*, **44**, (1997), 457-466. Cited on pages: 390, 391, 579.

[97MAN/GAL] Manceau, A., Gallup, D. L., Removal of selenocyanate in water by precipitation: characterization of copper-selenium precipitate by X-ray diffraction, infrared, and X-ray absorption spectroscopy, *Environ. Sci. Technol.*, **31**, (1997), 968-976. Cited on page: 297.

[97MAY/GAL] Mayer, P. M., Gal, J.-F., Radom, L., The heats of formation, gas-phase acidities, and related thermochemical properties of the third-row hydrides GeH_4, AsH_3, SeH_2 and HBr from G2 ab initio calculations, *Int. J. Mass Spectrom. Ion Processes*, **167/168**, (1997), 689-696. Cited on page: 133.

[97OPP/EHR] Oppermann, H., Ehrlich, S., Henning, C., Calorimetrie an Yttriumhalogeniden und Yttriumoxidhalogeniden, *Z. Naturforsch., B: Chem. Sci.*, **52**, (1997), 305-310, in German. Cited on page: 498.

[97PAN/ZVI] Pankratova, O. Y., Zvinchuk, R. A., Titanium monoselenide of variable composition, *Russ. J. Inorg. Chem.*, **42**, (1997), 1146-1149. Cited on pages: 351, 352.

[97PAR/TIC] Parker, D. R., Tice, K. R., Thomason, D. N., Effects of ion pairing with calcium and magnesium on selenate availability to higher plants, *Environ. Toxicol. Chem.*, **16**, (1997), 565-571. Cited on pages: 397, 401.

[97POL/TAR] Polekhovskii, Y. S., Tarasova, I. P., Nesterov, A. R., Pakhomovskiy, Y. A., Bakhchisaraitsev, A. Y., Sudovikovite $PtSe_2$ - new selenide of platinum from metasomatites of South Karelia, *Dokl. Akad. Nauk*, **354**, (1997), 82-85, in Russian. Cited on pages: 635, 637.

[97PUI/RAR] Puigdomènech, I., Rard, J. A., Plyasunov, A. V., Grenthe, I., Temperature corrections to thermodynamic data and enthalpy calculations, *Modelling in Aquatic Chemistry*, Grenthe, I., Puigdomènech, I., Eds., pp.427-493, Nuclear Energy Agency, Organisation for Economic Co-operation and Development, Paris, France, (1997). Cited on pages: 37, 38, 313, 321.

[97RAY/DIC] Raymond, C. C., Dick, D. L., Dorhout, P. K., Speciation of main-group metal ions by electrospray mass spectrometry. 1. Investigation of aqueous polyselenide species and effects of cations and pH, *Inorg. Chem.*, **36**, (1997), 2678-2681. Cited on page: 154.

[97SAN/PEL] Sangster, J., Pelton, A. D., The Cs-Se (cesium-selenium) system, *J. Phase Equilib.*, **18**, (1997), 173-176. Cited on page: 428.

[97SAN/PEL2] Sangster, J., Pelton, A. D., The K-Se (potassium-selenium) system, *J. Phase Equilib.*, **18**, (1997), 177-180. Cited on page: 422.

[97SAN/PEL3] Sangster, J., Pelton, A. D., The Li-Se (lithium-selenium) system, *J. Phase Equilib.*, **18**, (1997), 181-184. Cited on page: 412.

[97SAN/PEL4] Sangster, J., Pelton, A. D., The Na-Se (sodium-selenium) system, *J. Phase Equilib.*, **18**, (1997), 185-189. Cited on page: 416.

[97SAN/PEL5] Sangster, J., Pelton, A. D., The Rb-Se (rubidium-selenium) system, *J. Phase Equilib.*, **18**, (1997), 190-193. Cited on page: 426.

[97SER/BUS] Sergeev, N. B., Buslaeva, E. Y., Bulakhov, G. A., Kuz'mina, O. V., Selenium sulfides SeS_x (x = 1-7) - a new group of natural compounds from the supergene zone of the Gai deposit, the Southern Ural, *Dokl. Akad. Nauk*, **352**, (1997), 804-807, in Russian. Cited on page: 637.

[97TOL/KRI] Tolstykh, N. D., Krivenko, A. P., Pospelova, L. N., Unusual compounds of Ir, Os and Ru with Se, Te and As from the fields of the Zolotaya river in the Western Sayan, *Zap. Vseross. Mineral. O-va.*, **6**, (1997), 23-34, in Russian. Cited on page: 637.

[97VER/SEM] Vergasova, L. P., Semenova, T. F., Shuvalov, R. R., Filatov, S. K., Anan'ev, V. V., Ilinskite $NaCu_5O_2(SeO_3)_2Cl_3$ - a new mineral from volcanic sublimates, *Dokl. Akad. Nauk*, **353**, (1997), 641-644, in Russian. Cited on page: 636.

[98ALM/NOV] Al Mahamid, I., Novak, C. F., Becraft, K. A., Carpenter, S. A., Hakem, N., Solubility of Np(V) in K-Cl-CO_3 and Na-K-Cl-CO_3 solutions to high concentrations: measurements and thermodynamic model predictions, *Radiochim. Acta*, **81**, (1998), 93-101. Cited on page: 612.

[98CHA] Chase, Jr., M. W., Monograph No. 9. NIST-JANAF Thermochemical Tables, *J. Phys. Chem. Ref. Data*, (1998). Cited on pages: 290, 318, 319, 351, 403, 433, 487, 535, 541, 545, 566, 572, 646.

[98CHR/OJK] Christov, C., Ojkova, T., Mihov, D., Thermodynamic study of ($m_1Na_2SeO_4$ + m_2NiSeO_4)(aq) where m denotes molality at the temperature 298.15 K, *J. Chem. Thermodyn.*, **30**, (1998), 73-79. Cited on pages: 314, 421, 580.

[98OJK/CHR] Ojkova, T., Christov, C., Mihov, D., Investigation of the aqueous lithium and nickel selenate system, *Z. Phys. Chem. (Munich)*, **203**, (1998), 87-93. Cited on pages: 314, 415, 578, 580.

[98PAA/ROB] Paar, W. H., Roberts, A. C., Criddle, A. J., Topa, D., A new mineral, chrisstanleyite, $Ag_2Pd_3Se_4$ from Hope's Nose, Torquay, Devon, England, *Mineral. Mag.*, **62**, (1998), 257-264. Cited on pages: 634, 637.

[98PRI] Pring, A., Selenides and sulfides from Iron Monarch, South Australia, *Neues Jahrb. Mineral., Monatsh.*, **1**, (1998), 36-48. Cited on page: 637.

[98RAI/FEL] Rai, D., Felmy, A. R., Hess, N. J., Moore, D. A., Yui, M., A thermodynamic model for the solubility of UO_2(am) in the aqueous K^+-Na^+-HCO_3^--CO_3^{2-}-OH^--H_2O system, *Radiochim. Acta*, **82**, (1998), 17-25. Cited on page: 611.

[98SCH] Schlesinger, M. E., The Mn-Se (Manganese-Selenium) system, *J. Phase Equilib.*, **19**, (1998), 588-590. Cited on page: 337.

[98SEI/FUN] Seifert, H. J., Funke, S., Solution enthalpies of hydrates $LnCl_3.xH_2O$ (Ln=Ce-Lu), *Thermochim. Acta*, **320**, (1998), 1-7. Cited on pages: 528, 529.

[98VAS/DAO] Vassilev, G. P., Daouchi, B., Record, M.-C., Tedenac, J.-C., Thermodynamic studies of the In-Se system, *J. Alloys Compd.*, **269**, (1998), 107-115. Cited on pages: 238, 239, 242, 243, 246.

[99CAM/GRA] Campostrini, I., Gramaccioli, C. M., Demartin, F., Orlandiite, $Pb_3Cl_4(SeO_3).H_2O$, a new mineral species, and an associated lead-copper selenite chloride from the Baccu Locci mine, Sardinia, Italy, *Can. Mineral.*, **37**, (1999), 1493-1498. Cited on pages: 636, 637.

[99DJO/PIZ] Djogic, R., Pizeta, I., Zelic, M., Voltammetric study of uranyl-selenium interactions, *Anal. Chim. Acta*, **404**, (1999), 159-166. Cited on page: 390.

[99HOL/SOD] Holtstam, D., Söderhielm, J., The discreditation of platynite, *Can. Mineral.*, **37**, (1999), 1313-1315. Cited on page: 637.

[99KOL/STO] Koleva, V., Stoilova, D., DTA, DSC and X-ray studies on copper and manganese selenate hydrates, *Thermochim. Acta*, **342**, (1999), 89-95. Cited on page: 342.

[99LEO/FUR] Leonidov, V. Y., Furkaluk, M. Y., Thermochemistry of some aqueous ions, *Pure Appl. Chem.*, **71**, (1999), 1267-1274. Cited on pages: 498, 499.

[99NES/GLY] Nesterenko, V. P., Glybin, V. P., Crystalline thorium selenite monohydrate: Synthesis and thermodynamic research of dehydration process, *Czech. J. Phys.*, **49**, (1999), Suppl. S1, 903-908. Cited on page: 392.

[99OJK/CHR] Ojkova, T., Christov, C., Mihov, D., Thermodynamic study on aqueous solutions of $(NH_4)_2SeO_4$ and K_2SeO_4 at 298.15 K, *Monatsh. Chem.*, **130**, (1999), 1061-1065. Cited on pages: 411, 425, 581.

[99OPP/GOB] Oppermann, H., Göbel, H., Schadow, H., Schmidt, P., Hennig, C., Vassilev, V., Markova-Deneva, I., Thermochemische Untersuchungen zum ternären System Bi/Se/O. II. Das Teilsystem Bi_2O_3-SeO_2, *Z. Naturforsch., B: Chem. Sci.*, **54b**, (1999), 239-251, in German. Cited on pages: 203, 581.

[99RAR/RAN] Rard, J. A., Rand, M. H., Anderegg, G., Wanner, H., *Chemical Thermodynamics of Technetium*, Sandino, M. C. A., Östhols, E., Nuclear Energy Agency Data Bank, Organisation for Economic Co-operation and Development, Eds., vol. 3, *Chemical Thermodynamics*, North Holland Elsevier Science Publishers B. V., Amsterdam, The Netherlands, (1999). Cited on pages: 2, 67, 68, 77, 253, 255, 343, 587, 615.

[99STO/JOH] Stølen, S., Johnsen, H.-B., Abe, R., Atake, T., Grande, T., Heat capacity and thermodynamic properties of $GeSe_2$ at temperatures from T = (13 to 900)K. Structural phase transition at T = 423 K, *J. Chem. Thermodyn.*, **31**, (1999), 465-477. Cited on page: 210.

[99VER/KRI] Vergasova, L., Krivovichev, S., Semenova, T., Filatov, S., Ananiev, V., Chloromenite, $Cu_9O_2(SeO_3)_4Cl_6$, a new mineral from the Tolbachik volcano, Kamchatka, Russia, *Eur. J. Mineral.*, **11**, (1999), 119-123. Cited on page: 636.

[99VER/SEM] Vergasova, L. P., Semenova, T. F., Filatov, S. K., Krivovichev, S. V., Shuvalov, R. R., Anan'ev, V. V., Georgbokiite $Cu_5O_2(SeO_3)_2Cl_2$ – a new mineral from volcanic sublimates, *Dokl. Akad. Nauk*, **364**, (1999), 527-531, in Russian. Cited on page: 636.

[99WAN/OST] Wanner, H., Östhols, E., TDB-3: Guidelines for the assignment of uncertainties, Nuclear Energy Agency Data Bank, Organisation for Economic Co-operation and Development, Report , (1999), 17 pp.. Cited on page: 4.

[99WAN] Wanner, H., TDB-6: Guidelines for the independent peer review of TDB reports, Nuclear Energy Agency Data Bank, Organisation for Economic Co-operation and Development, Report , (1999), 8 pp.. Cited on pages: 4, 5.

[2000BRE] Brebrick, R. F., Vapor pressure of SeO_2(s) and optical density of SeO_2(g), *J. Phase Equilib.*, **21**, (2000), 235-240. Cited on pages: 122, 123, 124.

[2000GLA/PAS] Glazov, V. M., Pashinkin, A. S., Fedorov, V. A., Phase equilibria in the Cu-Se system, *Inorg. Mater.*, **36**, (2000), 641-652. Cited on page: 288.

[2000GRE/WAN] Grenthe, I., Wanner, H., Östhols, E., TDB-2: Guidelines for the extrapolation to zero ionic strength, OECD Nuclear Energy Agency, Data Bank, Report , (2000), 34 pp.. Cited on pages: 4, 587, 588, 594.

[2000HAV/MIC] Havlicek, D., Micka, Z., Barborak, V., Smejkal, P., Cesium and cesium-lithium selenates, *Collect. Czech. Chem. Commun.*, **65**, (2000), 167-178. Cited on pages: 415, 429.

[2000OJK/BAR] Ojkova, T., Barkov, D., Popov, A., Untersuchung der Phasengleichgewichte in den Systemen Cs_2SeO_4-$CoSeO_4$-H_2O und Cs_2SeO_4-$NiSeO_4$-H_2O bei 25.0°C, *Monatsh. Chem.*, **131**, (2000), 727-732, in German. Cited on pages: 314, 321.

[2000OST/WAN] Östhols, E., Wanner, H., TDB-0: The NEA Thermochemical Data Base Project, Nuclear Energy Agency Data Bank, Organisation for Economic Co-operation and Development, Report , (2000), 21 pp.. Cited on page: 4.

[2000RAN/FUG] Rand, M. H., Fuger, J., The thermodynamic properties of the transuranium halides: Part 1, Neptunium and plutonium halides, ITU, European Commission Joint Research Centre report, Report 17332 EN, (2000). Cited on pages: 500, 513, 522, 528, 546, 547.

[2000ROM/FEU] Römermann, F., Feutelais, Y., Fries S. G., Blachnik, R., Phase diagram experimental investigation and thermodynamic assessment of the thallium-selenium system, *Intermetallics*, **8**, (2000), 53-65. Cited on pages: 247, 248, 249, 250, 251, 252.

[2000WAN/OST] Wanner, H., Östhols, E., TDB-1: Guidelines for the review procedure and data selection, Nuclear Energy Agency Data Bank, Organisation for Economic Co-operation and Development, Report , (2000), 5 pp.. Cited on page: 4.

[2001ANT/RUD] Antipov, A. V., Rudnyi, E. B., Dobrokhotova, Zh. V., Thermodynamic evaluation of the Bi-Se system, *Inorg. Mater.*, **37**, (2001), 126-132. Cited on pages: 196, 197, 199, 201, 202.

[2001BAR/CHR] Barkov, D., Christov, C., Ojkova, T., Thermodynamic study of $(m_1Cs_2SeO_4 + m_2NiSeO_4)(aq)$ where m denotes molality at T = 298.15 K, *J. Chem. Thermodyn.*, **33**, (2001), 1073-1080. Cited on pages: 314, 429, 430, 578, 580, 582.

[2001BOL/KOP] Bolgar, O. S., Kopan, A. R., Kriklya, O. I., Thermodynamic properties of La_2Se_3 and Sm_2Se_3 in a wide temperature range, *Inzh.-Khim. Fak.*, **5**, (2001), 26-32, in Ukrainian. Cited on pages: 370, 383.

[2001COR/KON] Cordfunke, E. H. P., Konings, R. J. M., The enthalpies of formation of lanthanide compounds II. Ln^{3+}(aq), *Thermochim. Acta*, **375**, (2001), 51-64. Cited on pages: 494, 518, 528, 540, 553, 647, 648.

[2001ERK/KAT] Erkoc, S., Katircioglu, S., Yilmaz, T., Structural and electronic properties of In_mSe_n microclusters: density functional theory calculations, *J. Mol. Struct.*, **542**, (2001), 101-105. Cited on pages: 240, 241, 496, 509, 534, 555, 569, 570.

[2001GRI/FER] Grice, J. D., Ferraris, G., New minerals approved in 2000 by the Commission on new minerals and mineral names, International Mineralogical Association, *Can. Mineral.*, **39**, (2001), 917-923. Cited on page: 637.

[2001GU/WAT] Gu, X.-P., Watanabe, M., Hoshino, K., Shibata, Y., Mineral chemistry and associations of Bi-Te(S,Se) minerals from China, *Neues Jahrb. Mineral., Monatsh.*, 7, (2001), 289-309. Cited on page: 637.

[2001LEM/FUG] Lemire, R. J., Fuger, J., Nitsche, H., Potter, P. E., Rand, M. H., Rydberg, J., Spahiu, K., Sullivan, J. C., Ullman, W. J., Vitorge, P., Wanner, H., *Chemical Thermodynamics of Neptunium and Plutonium*, Nuclear Energy Agency Data Bank, Organisation for Economic Co-operation and Development, Ed., vol. 4, *Chemical Thermodynamics*, North Holland Elsevier Science Publishers B. V., Amsterdam, The Netherlands, (2001). Cited on pages: 2, 72, 74, 80, 81, 201, 385, 587, 595, 601, 606, 607, 611, 612, 615.

[2001OPP/SCH] Oppermann, H., Schadow, H., Funke, K., Schmidt, P., Glechgewichtdruckmessungen im System Se/O/Br, *Z. Anorg. Allg. Chem.*, **627**, (2001), 326-332, in German. Cited on pages: 175, 177, 582.

[2001OPP/ZHA] Oppermann, H., Zhang-Presse, M., Thermochemische Untersuchungen zu den Systemen SE_2O_3-SeO_2 II. Samarium-selenoxide auf dem Schnitt Sm_2O_3-SeO_2, *Z. Naturforsch., B: Chem. Sci.*, **56b**, (2001), 917-926, in German. Cited on pages: 370, 583, 584, 585.

[2001RID/SKA] Ridkosil, T., Skala, R., Johan, Z., Srein, V., Telluronevskite, Bi_3TeSe_2, a new mineral, *Eur. J. Mineral.*, **13**, (2001), 177-185. Cited on page: 635.

[2001SEB/POT] Seby, F., Potin-Gautier, M., Giffaut, E., Borge, G., Donard, O. F. X., A critical review of thermodynamic data for selenium species at 25°C, *Chem. Geol.*, **171**, (2001), 173-194. Cited on pages: 141, 158, 583.

[2002BOL/KOP] Bolgar, O. S., Kopan, A. R., Gorbachuk, N. P., Thermodynamic peoperties of Pr_3Se_4 and Nd_3Se_4 at low temperatures, *Ukr. Khim. Zh. (Russ. Ed.)*, **68**, (2002), 87-91, in Russian. Cited on pages: 374, 377.

[2002OPP/ZHA] Oppermann, H., Zhang-Presse, M., Weck, St., Liebig, S., Thermochemische Untersuchungen zu den Systemen SE_2O_3-SeO_3. I. Neodymselenoxide auf dem Schnitt Nd_2O_3-SeO_2, *Z. Anorg. Allg. Chem.*, **628**, (2002), 81-90, in German. Cited on pages: 375, 584, 585.

[2002OPP/ZHA2] Oppermann, H., Zhang-Presse, M., Schmidt, P., Thermochemische Untersuchungen zu den Systemen SE_2O_3-SeO_2. V. Ytterbiumselenoxide auf dem Schnitt Yb_2O_3-SeO_2, *Z. Naturforsch., B: Chem. Sci.*, **57b**, (2002), 868-876, in German. Cited on pages: 359, 584.

[2002PAA/TOP] Paar, W. H., Topa, D., Roberts, A. C., Criddle, A. J., Amann, G., Sureda, R. J., The new mineral species brodtkorbite, Cu_2HgSe_2, and the associated selenide assemblage from Tuminico, Sierra de Cacho, La Rioja, Argentina, *Can. Mineral.*, **40**, (2002), 225-237. Cited on page: 634.

[2002PAS/MAL] Pashinkin, A. S., Malkova, A. S., Mikhailova, M. S., The heat capacity of zinc and cadmium chalcogenides (ZnTe, CdSe, and CdTe), *Russ. J. Phys. Chem.*, **76**, (2002), 552-555. Cited on page: 272.

[2002ROB/PAA] Roberts, A. C., Paar, W. H., Cooper, M. A., Topa, D., Criddle, A. J., Jedwab, J., Verbeekite, monoclinic $PdSe_2$, a new mineral from the Musonoi Cu-Co-Mn-U mine, near Kolwezi, Shaba Province, Democratic Republic of Congo, *Mineral. Mag.*, **66**, (2002), 173-179. Cited on page: 636.

[2002ZHA/OPP] Zhang-Presse, M., Oppermann, H., Thermochemische Untersuchungen zu den Systemen SE_2O_3-SeO_2. IV. Lösungs-kalorimetrie der $SE_2Se_xO_{3+2x}$-Phasen (SE = Nd, Sm, Yb), *Z. Naturforsch., B: Chem. Sci.*, **57b**, (2002), 661-667, in German. Cited on pages: 356, 371, 375, 584.

[2002ZHA/OPP2] Zhang-Presse, M., Oppermann, H., Thermochemical investigation of RE_2O_3-SeO_2 Systems. III. Yttrium selenium oxides in the pseudo-binary system, *J. Therm. Anal. Calorim.*, **69**, (2002), 301-316. Cited on pages: 356, 585.

[2003BAR/CHR] Barkov, D., Christov, C., Ojkova, T., Thermodynamic study of aqueous rubidium and cobalt selenate system at the temperature 298.15 K, *J. Chem. Thermodyn.*, **35**, (2003), 689-697. Cited on pages: 321, 427, 428, 585.

[2003BOL/KOP] Bolgar, O. S., Kopan, A. R., The enthalpy of selenides of praseodymium and neodymium of M_3Se_4 composition, *Ukr. Khim. Zh. (Russ. Ed.)*, **69**, (2003), 34-37, in Russian. Cited on pages: 374, 377.

[2003GRO/STO] Grønvold, F., Stølen, S., Semenov, Y., Heat capacity and thermodynamic properties of silver(I) selenide, oP-Ag_2Se from 300 to 406 K and cI-Ag_2Se from 406 to 900 K: transitional behavior and formation properties, *Thermochim. Acta*, **399**, (2003), 213-224. Cited on pages: 298, 299.

[2003GUI/FAN] Guillaumont, R., Fanghänel, T., Fuger, J., Grenthe, I., Neck, V., Palmer, D. A., Rand, M. H., *Update on the chemical thermodynamics of Uranium, Neptunium, Plutonium, Americium and Technetium*, Nuclear Energy Agency Data Bank, Organisation for Economic Co-operation and Development, Ed., vol. 5, *Chemical Thermodynamics*, North Holland Elsevier Science Publishers B. V., Amsterdam, The Netherlands, (2003). Cited on pages: 2, 67, 68, 72, 78, 79, 80, 385, 386, 388, 389, 391, 587, 606, 612, 615.

[2003LI/REC] Li, J.-B., Record, M.-C., Tedenac, J.-C., A thermodynamic assessment of the In-Se system, *Z. Metallkd.*, **94**, (2003), 381-389. Cited on pages: 239, 240, 245, 246.

[2003PAS/MAL] Pashinkin, A. S., Malkova, A. S., Heat capacity of zinc selenide, *Russ. J. Phys. Chem.*, 77, (2003), 2068-2079. Cited on page: 259.

[2004BRO] Brown, P., Chairman NEA-TDB review on Zirconium, personal communication, (2004). Cited on pages: 537, 647.

[2005GAM/BUG] Gamsjäger, H., Bugajski, J., Gajda, T., Lemire, R. J., Preis, W., *Chemical thermodynamics of Nickel*, Nuclear Energy Agency Data Bank, Organisation for Economic Co-operation and Development, Ed., vol. 6, *Chemical Thermodynamics*, North Holland Elsevier Science Publishers B. V., Amsterdam, The Netherlands, (2005). Cited on pages: 2, 67, 68, 74, 78, 81, 311, 312, 316, 479, 587, 607, 608, 612, 615.

List of cited authors

This chapter contains an alphabetical list of the authors of the references cited in this book. The reference codes given with each name correspond to the publications of which the person is the author or a co-author. Note that inconsistencies may occur due to a variation in spelling between different publications. No attempt was made to correct for such inconsistencies in this volume.

Athors list

Author	Reference
Aalto, T.	[69SAL/HAK]
Abbasov, A.	[73SHA/AZI]
Abbasov, A. S.	[74MUS/AZI], [76ASK/ALI], [77AZI/ABB], [80AZI/AGA]
Abdullaev, A. M.	[88MAM/YAN]
Abe, H.	[38ISH/ABE]
Abe, R.	[99STO/JOH]
Abou El-Azm, A. M.	[83ELS/ABO]
Abramov, I. I.	[62SEL/ZUB]
Abramov, M. V.	[95YAS/RYA]
Abu-Basha, H. M.	[83ELS/ABO]
Adams, C. J.	[72ADA/DOW]
Adbullaev, G. B.	[63SHA/KUL]
Ader, M.	[74ADE]
Agaev, A. B.	[72AGA/SHA], [80AZI/AGA]
Agafonova, A. L.	[50PAM/AGA]
Agliulov, N. K.	[65DEV/UMI]
Agule, B. J.	[66SED/AGU]
Ahrland, S.	[74AHR/AVS]
Åkerlöf, G.	[37AKE/TEA]
Akimoto, R.	[95LI/HAG]
Al Mahamid, I.	[98ALM/NOV]
Aldzhanov, M. A.	[78MAM/ALD], [80MAM/ALD], [89GUS/ALD]
Aleksandrovich, A. M.	[62ALE/SER]

(Continued on next page)

Authors list (continued)

Author	Reference
Alexander, L. E.	[72ALE/BEA]
Aliev, I.	[73SHA/AZI]
Aliev, I. Y.	[74MUS/AZI], [77AZI/ABB], [80AZI/AGA]
Alieva, N. A.	[76ASK/ALI]
Alikhanyan, A. S.	[86STE/ALI]
Allard, B.	[97ALL/BAN]
Alvarez-Tostado, C.	[52GAB/ALV]
Amann, G.	[2002PAA/TOP]
Amelin, A. G.	[44AME/BEL]
Amitin, E. B.	[80AMI/KLU], [87AMI/MIN], [90MIN/AMI]
Amoulong, H.	[65PAE/AMO]
Amzil, A.	[96AMZ/GIL]
Anan'ev, V. V.	[97VER/SEM], [99VER/SEM]
Anan'ina, M. S.	[69ROS/SEL], [69ROS/SEL2]
Ananiev, V.	[99VER/KRI]
Ananthaswamy, J.	[84ANA/ATK]
Anderegg, G.	[99RAR/RAN]
Andersen, E. B.	[89NIK/AND]
Anderson, G. M.	[91AND/CAS]
Andreeva, L. L.	[65AND/KAR], [65AND/KUD], [66AND/KAR]
Andreichenko, O. E.	[62GOL/AND]
Antipov, A. V.	[2001ANT/RUD]
Archer, D. G.	[90ARC/WAN]
Argent, B. B.	[95DAV/RAN]
Aristarain, L. F.	[69HUR/ARI]
Arnek, R.	[72ARN/BAR]
Aron, B. M.	[71PAS/ARO]
Art, E. W.	[62WAL/ART]
Artem'eva, L. N.	[68KLU/SEL3], [69KLU/SEL]
Aruga, R.	[78ARU]
Askerova, K. A.	[76ASK/ALI]
Atake, T.	[83ATA/NOM2], [99STO/JOH]
Atkin, D.	[79BAN/ATK]
Atkinson, G.	[84ANA/ATK]
Aulich, W.	[28MEY/AUL]
Avedesian, M. M.	[83AVE/SPI]
Averbukh, M. A.	[65LES/AVE]
Avsar, E.	[74AHR/AVS]

(Continued on next page)

Authors list (continued)

Author	Reference
Ayache, C.	[86PRO/AYA]
Aynsley, E. E.	[52AYN/PEA]
Azizov, T.	[73SHA/AZI]
Azizov, T. K.	[67SHA/AZI], [74MUS/AZI], [76ASK/ALI], [77AZI/ABB], [80AZI/AGA]
Babanli, M. B.	[76BAB/KUL]
Babayan, G. G.	[84OGA/KAP]
Bachet, B.	[65CES/BAC]
Bachinskaya, A. G.	[67VAS/NIK]
Badalov, A. S.	[83SPI/BAD]
Badger, R. M.	[24BAD]
Badiello, R.	[96BAD/FER], [97FER/FIN]
Baer, Y.	[62BAE/BUS]
Baes, Jr., C. F.	[76BAE/MES]
Bahta, A.	[97BAH/PAR]
Bailey, S. M.	[82WAG/EVA]
Bakeev, M. I.	[64BAK/BUK], [72BAK/BUK], [72BAK/PAS], [72PAS/UST], [73BAK/PAS], [73BAK/PAS2]
Bakeeva, S. S.	[64BAK/BUK], [68BAK/BUK], [68BAK/BUK3], [68MAL/PAS], [69BAK/BUK], [69BAK/BUK2], [71BAK/BUK], [71PAS/ARO], [72BAK/BUK], [72BAK/PAS], [72PAS/UST], [73BAK/PAS], [73BAK/PAS2]
Baker, E. H.	[68BAK]
Bakhchisaraitsev, A. Y.	[97POL/TAR]
Balanevskaya, A. E.	[69BER/STR]
Balarew, C.	[90OJK/BAL]
Baldé, L.	[81BAL/LEG]
Balducci, G.	[91BAL/SAC]
Ball, S.	[95BAL/MIL]
Ban, V. S.	[68KNO/BAN]
Banas, M.	[79BAN/ATK]
Banks, W. H.	[34BAN]
Bannister, F. A.	[37BAN/HEY]
Banwart, S. A.	[97ALL/BAN]
Bányai, I.	[92BAN/BLI]
Barborak, V.	[2000HAV/MIC]
Barcza, L.	[71BAR/SIL], [72ARN/BAR]

(Continued on next page)

Authors list (continued)

Author	Reference
Bard, A. J.	[85BAR/PAR]
Bardi, G.	[69PIA/MAL], [76PIA/BAR], [86FER/PIA], [88BAR/IER], [88BAR/PIA], [90BAR/TRI]
Barin, I.	[73BAR/KNA], [95BAR]
Barker, C. K.	[65BAR/COR]
Barkov, D.	[91GOS/BAR], [91GOS/BAR2], [92GOS/BAR], [2000OJK/BAR], [2001BAR/CHR], [2003BAR/CHR]
Barnes, D. S.	[73BAR/MOR]
Barrow, R. F.	[66BAR/CHA], [67BAR/STO], [76GRE/BAR], [82PRO/BAR], [96BAR/TAH]
Bart, J. C. J.	[84BAR/PET]
Baskin, Y.	[70BAS/SMI]
Bassett, H.	[50BAS]
Bates, R. G.	[73BAT]
Batyreva, V. A.	[78SER/BAT], [81SER/BAT], [90KUM/BAT]
Beattie, I. R.	[72ALE/BEA]
Beck, M. T.	[87BEC], [90BEC/NAG]
Beckman, W. A.	[45KIN/BEC]
Becraft, K. A.	[98ALM/NOV]
Behrens, R. G.	[74BEH/LEM]
Belyakov, M. I.	[44AME/BEL]
Bench, R.	[62IZA/CHR]
Benderskii, L. A.	[69BOE/BEN], [69BOE/BEN2]
Bendrikh, N. F.	[79BEN/PAS]
Bennett, J. E.	[84HUA/GIL]
Berezovskii, G. A.	[87BOE/KUH]
Berger, L. I.	[69BER/STR]
Bergman, C.	[70BER/COP]
Berkowitz, J.	[66BER/CHU], [68BER/CHU]
Berthelot, M.	[1875BER], [1884BER/VIE]
Bever, M. B.	[64HOW/MIS], [67CHA/BEV]
Bevington, P. R.	[69BEV]
Bezborodova, V. M.	[70BOG/BEZ]
Bezobrazov, E. G.	[69UST/VIG4], [70UST/BEZ]
Bichara, C.	[96AMZ/GIL]
Bidoglio, G.	[95SIL/BID]
Biljakovic, K.	[81TOM/BIL], [81TOM/BIL2]
Binnewies, M.	[90BIN]

(Continued on next page)

Authors list (continued)

Author	Reference
Birch, J. A.	[75BIR]
Birch, W. D.	[90PRI/GAT]
Birckenbach, L.	[30BIR/HUT]
Björgvinsson, M.	[91BJO/SCH]
Blachnik, R.	[70BLA/SCH], [71BLA/SCH], [74BLA/IGE2], [78BLA/GUN], [79BLA/HOP], [80BLA/HOP], [93BLA], [2000ROM/FEU]
Blagoveshchenskaya, G. I.	[72SID/KON]
Blair, R. C.	[70BLA/MUN]
Blaton, N.	[96VOC/BLA]
Blatov, V. A.	[88SER/BLA]
Bleijerveld, R. H. T.	[75BLE/MIJ]
Blixt, J.	[92BAN/BLI]
Block, R.	[64MIJ/BLO]
Bochek, L. I.	[82DYM/LOS]
Boctor, N. Z.	[87BOC/KUL]
Bodenstein, M.	[1899BOD]
Bodewig, F. G.	[70BOD/PLA]
Boehnke, U.-C.	[87BOE/KUH]
Boev, E. I.	[64MAR/SAP], [69BOE/BEN], [69BOE/BEN2]
Bogatyr, A. S.	[92SEM/KUL]
Bogdanov, B. G.	[83GOS/BOG], [84GOS/BOG], [89GOS/BOG]
Bogdanov, V. I.	[70BOG/BEZ]
Boguslavskii, R. Y.	[55SEL/BOG]
Boitsev, V. E.	[83DYM/LOS]
Bol'shakova, N. K.	[68TAN/GOR], [76TAN/VOL], [76TAN/VOL2]
Bol'shih, M. A.	[96KAZ/BOL]
Bolgar, A. S.	[87BOL/TRO], [90BOL/TRO]
Bolgar, O. S.	[2001BOL/KOP], [2002BOL/KOP], [2003BOL/KOP]
Bolondz, F. M.	[62RUM/ZHI]
Boncheva-Mladenova, Z.	[68BON/PAS], [70GOS/PAS]
Bond, A. M.	[67HUM/BON]
Bondar, S. A.	[69BER/STR]
Bonilla, C. F.	[64BON/SHU]
Bonomo, F. S.	[84LAN/EDW]
Boone, S.	[92BOO/KLE]
Borge, G.	[2001SEB/POT]
Borjakova, V. A.	[73GRE/BOR]

(Continued on next page)

Authors list (continued)

Author	Reference
Born, P.	[79BOR/KNI]
Borodaev, Y. S.	[88EFI/NEN]
Boryakova, V. A.	[71GRI/BOR], [72GRI/BOR], [75DEM/KOS]
Bosser, G.	[73LEB/BOS], [81BOS/LEB]
Botova, M. M.	[85BOT/SAN]
Boublikova, R.	[95HAV/MIC]
Boughton, J. H.	[66BOU/KEL]
Bowater, I. C.	[68BOW/BRO], [68BOW/BRO2]
Bowles, J. F. W.	[79BAN/ATK]
Bradley, D. J.	[79BRA/PIT]
Brassington, N. J.	[87BRA/EDW]
Brebrick, R. F.	[65BRE], [68BRE], [96BRE/LIU], [96BRE/LIU2], [2000BRE]
Breccia, A.	[97FER/FIN]
Brekow, G.	[73BRE/MEI], [75BRE/MEI]
Brewer, L.	[80BRE/LAM]
Brinkley, D. C.	[71NAG/BRI]
Brisdon, A. K.	[87BRI/OGD]
Britton, H. T. S.	[31BRI/ROB], [32BRI/ROB]
Brockmöller, I.	[13PRE/BRO]
Brooks, L. S.	[52BRO]
Bros, H.	[76BRO/CAS]
Brown, A. J.	[68BRO/WHA]
Brown, P.	[2004BRO]
Brown, P. W.	[97DUM/BRO]
Brown, R. D.	[68BOW/BRO], [68BOW/BRO2]
Bruner-Krakau, L.	[13BRU]
Brunetti, B.	[93BRU/PIA]
Bruno, J.	[86BRU], [97ALL/BAN]
Bryzgalov, I. A.	[93SAK/BRY], [96BRY/SPI]
Brønsted, J. N.	[22BRO], [22BRO2]
Buchenau, U.	[89PHI/BUC]
Bugajski, J.	[2005GAM/BUG]
Buketov, E. A.	[64BAK/BUK], [64BUK/PAS], [68BAK/BUK], [68BAK/BUK3], [69BAK/BUK], [69BAK/BUK2], [71BAK/BUK], [71PAS/ARO], [72BAK/BUK], [73BAK/PAS], [73BAK/PAS2]
Bulakhov, G. A.	[97SER/BUS]

(Continued on next page)

Authors list (continued)

Author	Reference
Bunin, A. M.	[69BOE/BEN2]
Bur'yanova, E. Z.	[56BUR], [57BUR/KOV]
Burden, F. R.	[68BOW/BRO], [68BOW/BRO2]
Bürger, H.	[88BUR/WIL]
Burmeister, J. L.	[79KAR/CRE]
Burow, F.	[56HAH/BUR]
Burylev, B. P.	[77BUR]
Busch, G.	[62BAE/BUS]
Buslaeva, E. Y.	[97SER/BUS]
Cabaud, B.	[72HOA/CAB], [75HOA/REY]
Cabri, L. J.	[70HAR/CAB], [91CAB/LAF]
Cairns, E. J.	[73CAI/KUC]
Campostrini, I.	[99CAM/GRA]
Cao, Y.	[95MIN/LI]
Capdevila, H.	[92CAP], [93GIF/VIT2], [94CAP/VIT], [95CAP/VIT]
Carleer, M.	[90NDI/CAR]
Carlson, K. D.	[79SUL/PRU]
Carpeni, G.	[66SAB/CAR]
Carpenter, S. A.	[98ALM/NOV]
Carre, J.	[79CAR/GER]
Carson, A. S.	[73STE/CAR]
Castanet, R.	[76BRO/CAS]
Castet, S.	[91AND/CAS]
Castro, A.	[82CAS/JER], [84CAS/PIC]
Cech, F.	[73CEC/RIE], [78CEC/VAV], [79CEC/VAV]
Celustka, B.	[79POP/TON]
Cerclier, O.	[69LAF/CER]
Cermak, M.	[90MIC/CER]
Cervelle, B.	[92JED/CER]
Cesbron, F.	[65CES/BAC], [69CES/OOS], [71CES/PIE], [83GIN/CES]
Ch'ün-hua, L.	[62CHU/PAS]
Chandler, G. G.	[66BAR/CHA]
Chang, Y. A.	[89SHA/CHA], [92SHA/CHA], [93SHA/CHA], [96SHA/CHA], [96SHA/CHA2], [96SHA/CHA3]
Chao, E. E.	[76CHA/CHE]
Charlot, C.	[70CHA/TIK]
Chase, Jr., M. W.	[98CHA]

(Continued on next page)

Authors list (continued)

Author	Reference
Chatillon, C.	[93CHA]
Chaudhari, P.	[67CHA/BEV]
Chaudhuri, B. K.	[83ATA/NOM2]
Chekoev, N. G.	[78CHE/MOR]
Chen, L.	[95CHE/LI]
Cheng, K. L.	[76CHA/CHE]
Chernik, I. A.	[82LYK/CHE]
Chernozubov, Y. S.	[72CHE/KUZ]
Chernyshov, H. M.	[95RUD/KNA]
Chia, Y. T.	[59CON/CHI]
Chihara, H.	[83ATA/NOM2]
Chirulli, G.	[84CHI/PIA]
Chistyakova, N. I.	[84NEC/CHI]
Chizhevskaya, S. N.	[94CHI/SHE], [97CHI/SHE]
Chou, C.	[59GRO/WES]
Christe, K. O.	[72CHR/SCH]
Christensen, J. J.	[62IZA/CHR], [70CHR/JOH]
Christov, C.	[95STO/CHR], [97CHR], [98CHR/OJK], [98OJK/CHR], [99OJK/CHR], [2001BAR/CHR], [2003BAR/CHR]
Chugaev, V. N.	[73POL/ZAK]
Chukhlantsev, V. G.	[56CHU], [57CHU/TOM], [57KRY/CHU]
Chuntonov, K. A.	[77YAT/KUZ]
Chupakhin, M. S.	[76GLA/MIK], [77MIK/GLA], [77MIK/PIS], [77PIS/MIK]
Chupka, W. A.	[66BER/CHU], [68BER/CHU]
Churney, K. L.	[82WAG/EVA]
Chuvikina, N. G.	[85BOT/SAN]
Ciavatta, L.	[80CIA], [87CIA/IUL], [88CIA], [90CIA]
Clark, A. M.	[77DAV/CLA]
Claude, R.	[60CLA]
Clavaguera, N.	[95CLA/COM]
Clavaguera-Mora, M. T.	[95CLA/COM]
Clever, H. L.	[65CLE/WES]
Clusius, K.	[37KRU/CLU]
Cobble, J. W.	[70HIN/COB2]
Colin, R.	[64COL/DRO], [68COL/DRO], [90NDI/CAR]
Comas, C.	[95CLA/COM], [97COM/MOR]

(Continued on next page)

Authors list (continued)

Author	Reference
Connick, R. E.	[59CON/CHI]
Cooper, J. N.	[76COO/WOO]
Cooper, J. R.	[81TOM/BIL], [81TOM/BIL2]
Cooper, M. A.	[95COO/HAW], [2002ROB/PAA]
Cooper, R.	[67COO/CUL]
Coppens, P.	[70BER/COP]
Cordes, A. W.	[65BAR/COR], [65CLE/WES]
Cordfunke, E. H. P.	[77COR/OUW2], [2001COR/KON]
Cordo, K. E.	[66GLA/COR]
Covington, A. K.	[65COV/DOB]
Cox, J. D.	[89COX/WAG]
Craven, R. A.	[78CRA/SAL]
Crecely, R. W.	[79KAR/CRE]
Criddle, A. J.	[77DAV/CLA], [86DUN/PEA], [91CAB/LAF], [98PAA/ROB], [2002PAA/TOP], [2002ROB/PAA]
Cruceanu, E.	[65KEL/CRU]
Cueilleron, J.	[72HIL/CUE]
Culka, J. V.	[67COO/CUL]
Cunningham, P. T.	[73CAI/KUC]
Curtis, E. C.	[72CHR/SCH]
Curtiss, L. A.	[96OHA/ZYW]
Cvitaš, T.	[88MIL/CVI]
D'Incan, J.	[82PRO/BAR], [96BAR/TAH]
D'yachkova, I. B.	[68DYA/KHO]
D'yachkova, N. N.	[69DYA/VIG]
Dagron, C.	[62DAG]
Danchenko, V. Y.	[89KOV/NEK]
Danilenko, G. N.	[90KOS/YAC], [92KOS/IVA]
Daouchi, B.	[98VAS/DAO]
Das, R. C.	[74DAS/SAH], [75SAT/SAH]
Dasgupta, P. K.	[90DAS/NAR]
Davies, C. W.	[62DAV]
Davis, R. J.	[77DAV/CLA]
Davydov, A. V.	[95DAV/RAN]
De Hlasko, M.	[23HLA]
De Maria, G.	[65MAR/GOL]
De Montreuil, L. A.	[74MON], [75MON]
DeMercurio, T. A.	[82GRI/GRI]

(Continued on next page)

Authors list (continued)

Author	Reference
DeRoo, A. M.	[63HIR/ICH]
Degterov, S. A.	[97DEG/PEL]
Deichman, E. N.	[72DEI/TAN], [73DEI/TAN], [75KAD/DEI], [77KAD/DEI], [77KAD/DEI3]
Delaey, L.	[88GHO/LUK], [89GHO/LUK]
Deliens, M.	[91DEL/PIR], [96VOC/BLA]
Demartin, F.	[99CAM/GRA]
Demidenko, A. F.	[69DEM/MAL], [69DEM], [75DEM/KOS], [81KOS/GRI]
Desai, P. D.	[73HUL/DES]
Detry, D.	[71KEL/RIC]
Deutsch, E.	[76COO/WOO]
Devyatykh, G. G.	[63DEV/UMI], [65DEV/UMI], [69YUS/DEV]
Di Paolo, V.	[76PIA/BAR]
Di Salvo, F. J.	[78CRA/SAL]
Dianoux, A.-J.	[89PHI/BUC]
Dick, D. L.	[97RAY/DIC]
Dieleman, J.	[82DIE/SAN]
Dimitriev, Y.	[83DIM/LAK]
Dinsdale, A. T.	[91DIN]
Djogic, R.	[99DJO/PIZ]
Djurek, D.	[81TOM/BIL], [81TOM/BIL2]
Dmitrieva, M. T.	[89KOV/NEK]
Dobrokhotova, Zh. V.	[2001ANT/RUD]
Dobrotin, R. B.	[65GAD/DOB], [66GAD/DOB], [68DOB/SUV]
Dobson, J. V.	[65COV/DOB]
Doetsch, V.	[58JAN/DOE], [60JAN/DOE]
Dolgikh, V. A.	[69SOK/PAS]
Donard, O. F. X.	[2001SEB/POT]
Donelson, J. G.	[36HAR/KES]
Dorhout, P. K.	[97RAY/DIC]
Doronin, V. N.	[84KON/STR]
Doroshenko, N. A.	[69SUP/MAI]
Dostál, K.	[51TOU/DOS], [55DOS]
Downs, A. J.	[72ADA/DOW]
Dräger, M.	[66GAT/DRA], [66GAT/DRA2], [66GAT/DRA3]
Drever, J. I.	[82DRE]

(Continued on next page)

Authors list (continued)

Author	Reference
Drowart, J.	[64COL/DRO], [64DRO], [68COL/DRO], [69UY/DRO], [70BER/COP], [70UY/DRO], [71KEL/RIC], [71UY/DRO], [72SMO/MAN], [77DRO/SMO], [77SMO/DRO], [78SMO/PAT], [84DRO/SMO], [84GRO/DRO], [84SMO/DRO]
Dubeau, C.	[71DUB/SIS]
Dubinina, L. K.	[82GLA/DUB]
Dumm, J. Q.	[97DUM/BRO]
Dunn, P. J.	[78DUN/PEA], [86DUN/PEA]
Durdyev, M.	[66PYA/DUR]
Dutchak, Y. I.	[77DUT/TIS]
Dymkov, Y. M.	[82DYM/LOS], [83DYM/LOS]
Earley, J. W.	[49EAR], [50EAR]
Ebert, M.	[80EBE/HAV], [81EBE/HAV], [82EBE/HAV], [82EBE/MIC], [82EBE/MIC2], [84EBE/MIC], [84MIC/UCH], [85MIC/PIK], [85VOJ/EBE], [86MIC/EBE], [86MIC/EBE2], [86MIC/EBE3], [87MIC/EBE], [87VOJ/EBE], [87VOJ/EBE2], [87VOJ/EBE3], [87VOJ/EBE4], [89VOJ/EBE], [89VOJ/EBE2]
Eberts, R. E.	[58SPE/NAU]
Edwards, H. G. M.	[87BRA/EDW]
Edwards, J. G.	[87SRI/EDW]
Edwards, T. H.	[84LAN/EDW]
Efendiev, G. K.	[65EFE/KAR]
Effantin, C.	[82PRO/BAR], [96BAR/TAH]
Effenberger, H.	[87EFF]
Efimov, A. V.	[88EFI/NEN]
Efremov, A. A.	[73SMO/EFR]
Ehrlich, S.	[97OPP/EHR]
Eick, H. A.	[74HAR/EIC]
El-Sharkawy, A. A.	[83ELS/ABO]
Ellert, G. V.	[71ELL/SLO], [71SEV/SLO], [75ELL/SEV]
Elliott, R. P.	[65ELL]
Embree, N. D.	[34HAR/EMB]
Emons, H.-H.	[72EMO/THE]
Engelen, B.	[86MAS/LUT]
Ephraim, J. H.	[97ALL/BAN]

(Continued on next page)

Authors list (continued)

Author	Reference
Erdey, L.	[56ERD/PAU]
Ericksen, G. E.	[94KON/EVA]
Erkoc, S.	[2001ERK/KAT]
Evans, Jr., H. T.	[94KON/EVA]
Evans, W. H.	[52ROS/WAG], [82WAG/EVA]
Eysel, H. H.	[79EYS/SUN]
Fabre, M. C.	[1887FAB]
Fakhruddin, A.	[85FAK]
Fanghänel, T.	[97KON/FAN], [2003GUI/FAN]
Fawcett, V.	[87BRA/EDW]
Fedorov, V. A.	[2000GLA/PAS]
Fedorov, V. E.	[83NAU/RAK]
Fedorov, V. V.	[69SUP/MAI]
Fedyainov, N. V.	[69KLU/SEL3], [70KLU/SEL]
Felmy, A. R.	[95RAI/FEL3], [98RAI/FEL]
Felsing, W. A.	[25AMI/FEL]
Fenochina, B. V.	[82GOR/FEN]
Fenochka, B. V.	[71GOR/FES]
Fernelius, W. C.	[77FER]
Fernholt, L.	[83FER/HAA]
Feroci, G.	[96BAD/FER], [97FER/FIN]
Ferraris, G.	[2001GRI/FER]
Ferri, D.	[83FER/GRE], [85FER/GRE]
Ferro, D.	[76PIA/BAR], [86FER/PIA], [90SCA/PIA]
Fesenko, V. V.	[71GOR/FES], [75KOV/SID], [77KOV/SID]
Feutelais, Y.	[91MOR/FEU], [96FEU/MAJ], [2000ROM/FEU]
Ficalora, P. J.	[68FIC/HAS], [68UY/MUE], [69FIC/THO]
Filatov, S.	[99VER/KRI]
Filatov, S. K.	[89VER/FIL], [97VER/SEM], [99VER/SEM]
Filosophova, T. M.	[89VER/FIL]
Finch, A.	[93FIN/GAR]
Finch, C. B.	[60FIN/WAG]
Fini, A.	[96BAD/FER], [97FER/FIN]
Fink, E. H.	[87FIN/KRU], [89FIN/SET]
Finkel'shtein, E. I.	[59SEL/ZUB]
Finkelman, R. B.	[77FIN/MRO], [86DUN/PEA]
Fjellvåg, H.	[96STO/FJE]
Flögel, P.	[69FLO], [72FLO]

(Continued on next page)

Authors list (continued)

Author	Reference
Flis, I. E.	[58FLI/MIS]
Flowers, G. C.	[81HEL/KIR]
Fofanov, G. M.	[69FOF/KIT]
Fomina, T. A.	[71MEL/STE]
Font-Altaba, M.	[85MES/MAR]
Fontana, D.	[92PIA/SCA]
Forouzan, F.	[95LIC/FOR]
Forrest, J. S.	[67HUM/BON]
Forsman, J. A.	[82GRI/GRI]
Fortsch, A. R.	[23PEA/FOR]
Fowless, A. D.	[77FOW/STR], [77FOW/STR2]
Fröhlich, C.	[62BAE/BUS]
France, P.	[86MIC/EBE]
Frank, A. S.	[82FRA/ZUM]
Franz, E. D.	[85FRA/WET]
Franzen, H. F.	[82SCH/FRA]
Freeman, R. D.	[84FRE]
Freiberg, M.	[96SCH/FRE]
Fricke, A.	[27KRA/FRI]
Friend, J. A. N.	[29FRI], [31FRI]
Fries S. G.	[2000ROM/FEU]
Fries, S. G.	[96FEU/MAJ]
Fritzler, B.	[82FRI/KAL], [85SPY/KAL]
Froning, H. R.	[50ROH/FRO]
Fuger, J.	[88XIA/ZHU], [92GRE/FUG], [2000RAN/FUG], [2001LEM/FUG], [2003GUI/FAN]
Fujisaki, H.	[66FUJ/WES]
Fukama, H.	[95LI/HAG]
Funk, R.	[00FUN]
Funke, K.	[89OPP/RIT], [2001OPP/SCH]
Funke, S.	[98SEI/FUN]
Furkaluk, M. U.	[93FIN/GAR]
Furkaluk, M. Y.	[99LEO/FUR]
Furukawa, H.	[44FUR/KIN]
Gabriel, H.	[52GAB/ALV]
Gadzhiev, S. M.	[65GAD/DOB], [66GAD/DOB], [68DOB/SUV], [69GAD/SAT]
Gadzhiev, S. N.	[60GAD/SHA], [61GAD/SHA], [66GAD]

(Continued on next page)

Authors list (continued)

Author	Reference
Gajda, T.	[2005GAM/BUG]
Gal, J.-F.	[97MAY/GAL]
Gallup, D. L.	[97MAN/GAL]
Gamidov, R. S.	[82PRI/VAS]
Gamsjäger, H.	[2005GAM/BUG]
Ganelina, E. S.	[73GAN/KUZ]
Gardner, P. J.	[92GAR], [93FIN/GAR]
Garvin, D.	[87GAR/PAR]
Gas'kov, A. M.	[72NOV/ZLO]
Gasarev, V. G.	[78GER/GLA]
Gastev, S. V.	[75DEM/KOS]
Gatehouse, B. M.	[90PRI/GAT]
Gattow, G.	[54SCH/GAT2], [55GAT/SCH], [56GAT/SCH], [62GAT], [66GAT/DRA], [66GAT/DRA2], [66GAT/DRA3]
Gaur, U.	[78WIE/SIE], [80SHU/GAU], [81GAU/SHU], [81WIE/PUL]
Gavaleshko, N. P.	[92SIR/GAV]
Gavrilov, N. M.	[90STO/GAV]
Gawel, W.	[68TER/ZAL]
Gebert, E.	[55KAT/GEB]
Geers, M. G.	[62WOS/GEE]
Geiderikh, V. A.	[72MAT/GEI], [72MAT/GEI2]
Gelbach, R. W.	[42GEL/KIN]
Genov, L. K.	[62GEN/NES]
Gerasimov, Y. I.	[67VAS/NIK], [68VAS/SOM], [71VAS/NIK], [78GER/GLA]
Gerassimov, J. I.	[72MAT/GEI]
Gerdes, F.	[79PRE/GER]
Gerding, H.	[63MIJ/GER]
Gerling, U.	[79PRE/GER]
Germain, P.	[79CAR/GER]
Getskin, L. S.	[62MAR/GET]
Ghosh, G.	[88GHO/LUK], [89GHO/LUK], [93GHO]
Ghosh, R.	[70GHO/NAI]
Giauque, W. F.	[39GIA/RUE]
Giffaut, E.	[93GIF/VIT2], [94GIF], [2001SEB/POT]
Gigli, G.	[91BAL/SAC]

(Continued on next page)

Authors list (continued)

Author	Reference
Gilbert, M.	[96AMZ/GIL]
Gilles, P. W.	[84HUA/GIL]
Gilliland, A. A.	[61JOH/GIL]
Gillis, J. R.	[84LAN/EDW]
Ginderow, D.	[83GIN/CES]
Gladyshev, V. P.	[72GLA/KIR], [82GLA/DUB], [89GLA/KOV]
Glaser, J.	[92BAN/BLI]
Glass, R. S.	[93GLA/SIN], [97KAM/GLA]
Glatz, A. C.	[66GLA/COR]
Glauser, R. T.	[10GLA]
Glazov, V. M.	[71GLA/KOR], [71GLA/KUL], [78GER/GLA], [89GLA/PAS], [2000GLA/PAS]
Glazunov, M. P.	[76GLA/MIK], [77MIK/GLA]
Gleiser, M.	[73HUL/DES]
Glemser, O.	[64GLE/MUL]
Globa, A. V.	[86GLO/LOM]
Glybin, V. P.	[73PIN/NES], [74PIN/GOR], [99NES/GLY]
Göbel, H.	[99OPP/GOB]
Goldfinger, P.	[63GOL/JEU], [65MAR/GOL], [71KEL/RIC]
Golub, A. M.	[59GOL/KOS3], [59GOL/POM], [59GOL], [60GOL/SKO], [62GOL/AND], [62GOL/SKO], [65GOL/SKO]
Golubkov, A. V.	[67TIK/GOL]
Golutvin, Y. M.	[51KAP/GOL], [83ZEL/KOL]
Goncharuk, L. V.	[86GON/LUK]
Gopal, M.	[92GOP/MIL]
Gorbachuk, N. P.	[2002BOL/KOP]
Gorbov, S. I.	[66GOR/KRE], [67KRE/GOR], [68GOR/KRE], [82VEN/GOR]
Gordienko, S. P.	[71GOR/FES], [82GOR/FEN], [86VIK/GOR]
Gordon, A. R.	[37GOR]
Gordon, J. S.	[69GOR]
Gordon, S.	[61MCB/GOR]
Gorgoraki, V. I.	[86STE/ALI]
Gornostaeva, L. V.	[74PIN/GOR]
Gorokhov, V. K.	[68TAN/GOR], [71SEL/SOB], [73SEL/ZAL]

(Continued on next page)

Authors list (continued)

Author	Reference
Gospodinov, G.	[74GOS/MOL], [80OJK/GOS], [81OIK/GOS], [82OJK/GOS], [84GOS], [84GOS2], [84SLA/POP], [85GOS], [86POP/SLA], [91GOS/BAR], [91GOS/BAR2], [92GOS/BAR], [93SLA/POP]
Gospodinov, G. G.	[70GOS/PAS], [83GOS/BOG], [84GOS/BOG], [84GOS3], [84GOS4], [84GOS5], [84GOS6], [89GOS/BOG], [91GOS], [93GOS], [94GOS/STA], [94PAS/GOS], [97GOS/STA]
Goudiakas, J.	[88XIA/ZHU]
Gouet, G.	[92JED/CER]
Goureaux, M. G.	[69SPI/ROU]
Goyette, W. J.	[68WIE/GOY]
Grade, M.	[80GRA/HIR]
Gramaccioli, C. M.	[99CAM/GRA]
Grande, T.	[99STO/JOH]
Grauer, R.	[97ALL/BAN]
Greenberg, J. H.	[73GRE/BOR]
Greenwood, D. J.	[76GRE/BAR]
Greiver, T. N.	[67GRE/ZAI], [67ZAI/GRE], [71GRE/ZAI], [71ZAI/GRE]
Grenthe, I.	[83FER/GRE], [85FER/GRE], [87RIG/VIT], [92GRE/FUG], [95GRE/PUI], [97ALL/BAN], [97GRE/PLY], [97PUI/RAR], [2000GRE/WAN], [2003GUI/FAN]
Grice, J. D.	[2001GRI/FER]
Grimley, R. T.	[82GRI/GRI]
Grinberg, Y. K.	[71GRI/BOR], [72GRI/BOR], [75DEM/KOS], [81KOS/GRI]
Grindstaff, Q. G.	[82GRI/GRI]
Grønvold, F.	[59GRO/WES], [60GRO/THU], [62GRO/WES], [62WES/GRO], [68GRO], [69GRO/WES], [70GRO], [70WES/GRO], [72GRO], [73GRO2], [75GRO/KVE], [75GRO], [84GRO/DRO], [87SVE/GRO], [96STO/FJE], [2003GRO/STO]
Grudetskii, A. V.	[77MOR/VLA]
Grzeta-Plenkovic, B.	[79POP/TON]
Gschneidner, K. A.	[74MCM/GSC]
Gu, X.-P.	[2001GU/WAT]

(Continued on next page)

Authors list (continued)

Author	Reference
Gubeli, A. O.	[69MEH/GUB], [70MEH/GUB], [70MEH/GUB2], [71MEH/GUB]
Guggenheim, E. A.	[35GUG], [66GUG]
Guillaumont, R.	[2003GUI/FAN]
Gul'tyaev, P. V.	[59GUL/PET]
Guminski, C.	[92SHA/CHA], [93SHA/CHA]
Gunia, P. G.	[78BLA/GUN]
Gunn, S. R.	[64GUN]
Guseinov, A. G.	[80AZI/AGA], [88MAM/YAN]
Guseinov, D. A.	[89GUS/ALD]
Guseinov, G. D.	[67MAM/KER]
Guseinov, N. G.	[89GUS/ALD]
Haacke, D. F.	[79HAA/WIL]
Haaland, A.	[83FER/HAA]
Hadermann, J.	[97ALL/BAN]
Haga, Y.	[95LI/HAG]
Hagisawa, H.	[39HAG], [41HAG]
Hahn, H.	[56HAH/BUR]
Haissinsky, M.	[49PAP/HAI]
Hajiev, S. N.	[66HAJ], [70HAJ]
Hakala, R.	[69SAL/HAK]
Hakem, N.	[98ALM/NOV]
Häkli, A.	[64VUO/HUH]
Häkli, T. A.	[65HAK/VUO]
Hall, R. O. A.	[91HAL/MOR]
Halow, I.	[82WAG/EVA]
Hamada, S.	[61HAM]
Hamann, S. D.	[82HAM]
Hamer, W. J.	[35HAR/HAM]
Hanke, U.	[85OPP/HAN], [86OPP/HAN]
Hansen, E. E.	[69HAN/MUN]
Harding, S. R.	[91HAL/MOR]
Hariharan, A. V.	[74HAR/EIC]
Harned, H. S.	[34HAR/EMB], [35HAR/HAM], [36HAR/KES], [41HAR/ROB], [50HAR/OWE]
Harris, D. C.	[70HAR/CAB], [70HAR], [72HAR/NUF]
Hart, D.	[52HAR]
Hartmann, H.	[96SCH/FRE]

(Continued on next page)

Authors list (continued)

Author	Reference
Hassan, M. S.	[73HAS/MUN]
Hastie, J. W.	[68FIC/HAS]
Havel, J.	[97LUB/HAV]
Havlicek, D.	[80EBE/HAV], [81EBE/HAV], [82EBE/HAV], [95HAV/MIC], [2000HAV/MIC]
Hawkins, D. T.	[73HUL/DES]
Hawthorne, F. C.	[95COO/HAW]
Hayashi, K.	[78HAY/NAK], [83MIY/HAY]
Head, A. J.	[93FIN/GAR]
Hedlund, T.	[88HED]
Hefter, G.	[96SOL/MAY]
Heinemann, C.	[96HEI/KOC]
Helgeson, H. C.	[74HEL/KIR], [81HEL/KIR], [88SHO/HEL], [88TAN/HEL], [89SHO/HEL], [89SHO/HEL2], [90OEL/HEL], [92SHO/OEL]
Henglein, A.	[69SCH/KAR]
Hennig, C.	[99OPP/GOB]
Henning, C.	[97OPP/EHR]
Hepler, L. G.	[58MUL/HEP]
Herzberg, G.	[50HER], [79HUB/HER]
Hess, N. J.	[98RAI/FEL]
Hey, M. H.	[37BAN/HEY]
Heyding, R. D.	[67MUR/HEY], [75MUR/HEY]
Hietanen, S.	[85FER/GRE]
Hill, A. E.	[40HIL/SOT]
Hillal, A. S.	[83ELS/ABO]
Hillel, R.	[72HIL/CUE]
Hinchey, R. J.	[70HIN/COB2]
Hinke, W.	[32MEY/HIN]
Hirayama, C.	[63HIR/ICH]
Hirota, T.	[88HIR/UED]
Hirschwald, W.	[80GRA/HIR]
Hoareau, A.	[72HOA/CAB], [75HOA/REY]
Hoch, M.	[78ROL/HOC]
Hofmeier, F.	[08KRE/HOF]
Högberg, K.	[70HOG/LUN]
Holtstam, D.	[99HOL/SOD]
Homann, K.	[88MIL/CVI]

(Continued on next page)

Authors list (continued)

Author	Reference
Honma, K.	[87HON/IID]
Hoppe, A.	[79BLA/HOP], [80BLA/HOP]
Horiuchi, J.	[51HOR/TAN], [51HOR/YAN]
Hoshino, K.	[2001GU/WAT]
Hostettler, J. D.	[84HOS]
Howlett, B. W.	[64HOW/MIS]
Hsu, F. S. L.	[78CRA/SAL]
Huang, J. Y.-K.	[84HUA/GIL]
Huang, M.-D.	[79HUA/MES]
Hubaut, X.	[92JED/CER]
Hubbard, W. N.	[66OHA/SET]
Huber, K. P.	[79HUB/HER]
Huhma, A.	[64VUO/HUH]
Hulbert, L. J.	[91CAB/LAF]
Hultgren, R.	[73HUL/DES]
Humffray, A. A.	[67HUM/BON]
Hummel, W.	[97ALL/BAN]
Hurlbut, Jr., C. S.	[69HUR/ARI]
Hussein, M. A.	[74WAS/NOU]
Huttner, K.	[30BIR/HUT]
Ichida, H.	[88OZE/YAG]
Ichikawa, Y.	[63HIR/ICH]
Ieronimakis, K.	[88BAR/IER]
Igaki, K.	[79IGA/NIS]
Igel, R.	[74BLA/IGE2]
Iida, K.	[87HON/IID]
Illarionov, V. V.	[57ILL/LAP]
Imperatori, P.	[84IMP/PIA]
Inarida, M.	[68SEK/IWA]
Ipser, H.	[94RIC/IPS]
Irene, E. A.	[74WIE/IRE]
Irwin, J. C.	[74IRW/LAC]
Isakova, R. A.	[68REZ/ISA]
Ishikawa, F.	[38ISH/ABE]
Itagaki, K.	[94YAM/KAM]
Iuliano, M.	[87CIA/IUL]
Ivanov-Emin, B. N.	[67KOL/IVA], [71KOL/TET]
Ivanova, Y.	[83DIM/LAK]

(Continued on next page)

Authors list (continued)

Author	Reference
Ivanyas, A. F.	[92KOS/IVA]
Ives, D. J. G.	[47IVE/PIT]
Iwaki, H.	[68SEK/IWA]
Izard, E. F.	[28SHE/IZA]
Izatt, R. M.	[62IZA/CHR], [70CHR/JOH]
Jaffe, I.	[52ROS/WAG]
Jakob, A.	[97ALL/BAN]
Jander, J.	[58JAN/DOE], [60JAN/DOE]
Jandl, S.	[86PRO/AYA]
Janickis, J.	[34JAN]
Janitzki, J.	[32JAN]
Jankova, R.	[93OJK/MIC]
Jannek, J.	[13JAN/MEY]
Jay-Gerin, J.-P.	[86PRO/AYA]
Jayaraman, S.	[71RAM/JAY]
Jedwab, J.	[92JED/CER], [2002ROB/PAA]
Jelinek, H.	[74JEL/KOM]
Jensen, K. A.	[71JEN]
Jerez, A.	[82CAS/JER], [84CAS/PIC]
Jerschkewitz, H. G.	[62JER/MEN]
Jeunehomme, M.	[63GOL/JEU]
Jilek, E.	[82FRI/KAL]
Johan, Z.	[70JOH/PIC], [71JOH/KVA], [71JOH/KVA2], [71JOH/PIC], [71JOH/PIC2], [72JOH/PIC], [76JOH/KVA], [78JOH/KVA], [82JOH/PIC], [87JOH/PIC], [87JOH], [88JOH], [2001RID/SKA]
Johanson, W. R.	[80JOH/MCC]
Johansson, G.	[85JOH/NII]
Johnsen, H.-B.	[99STO/JOH]
Johnson, J. W.	[92SHO/OEL]
Johnson, K. S.	[79JOH/PYT]
Johnson, W. H.	[61JOH/GIL]
Johnston, H. D.	[70CHR/JOH]
Jordan, J.	[85BAR/PAR]
Jost, W.	[68JOS/KUB]
Judovits, L.	[85JUD/WUN]
Jung, W.	[93GLA/SIN]

(Continued on next page)

Authors list (continued)

Author	Reference
Kadoshnikova, N. V.	[72DEI/TAN], [73DEI/TAN], [75KAD/DEI], [77KAD/DEI], [77KAD/DEI3], [80KAD/TAN], [82TAN/KAD]
Kaiman, S.	[70HAR/CAB]
Kaldis, E.	[74MCM/GSC], [82FRI/KAL], [85SPY/KAL]
Kalinina, A. V.	[62SEL/ZUB]
Kalitin, V. I.	[66KAL/YAR]
Kallay, N.	[88MIL/CVI]
Kameda, K.	[94YAM/KAM]
Kaminski, R.	[97KAM/GLA]
Kanai, K.	[51HOR/YAN]
Kanduth, H.	[83AVE/SPI]
Kao, C. H.	[25LEN/KAO]
Kapantsyan, E. E.	[84OGA/KAP]
Kapustinskii, A. F.	[51KAP/GOL], [51KAP/ZHD], [51KAP/ZHD3], [59SEL/KAP]
Karaev, Z. S.	[65EFE/KAR]
Karakaya, I.	[90KAR/THO]
Karakhanova, M. I.	[67KAR/PAS]
Karapet'yants, M. K.	[65AND/KAR], [65SIL/KAR], [66AND/KAR], [67MAI/SUP], [68MAI/SUP], [69MAI/SOL], [69SUP/MAI], [70KAR/MAI], [70KLU/SEL], [71MAI/SUP]
Karapiperis, T.	[97ALL/BAN]
Karbanov, S. G.	[68KAR/ZLO], [72NOV/ZLO]
Kargol, J. A.	[79KAR/CRE]
Karmann, W.	[69SCH/KAR]
Karyakin, N. V.	[71PAS/ARO]
Kasuya, T.	[95LI/HAG]
Katircioglu, S.	[2001ERK/KAT]
Katzin, L. I.	[55KAT/GEB]
Kawanaka, T.	[96OKA/KAW]
Kawassiades, C. T.	[67KAW/MAN]
Kazenas, E. K.	[96KAZ/BOL]
Kehiaian, H. V.	[76BRO/CAS]
Kelemen, F.	[65KEL/CRU]
Keller, H.	[70KEL], [71KEL/RIC]
Keller, R. N.	[66BOU/KEL]

(Continued on next page)

Authors list (continued)

Author	Reference
Kelley, K. K.	[39KEL], [60KEL], [61KEL/KIN], [73HUL/DES]
Kelly, R. L.	[87KEL]
Kempton, J. H.	[87RUN/LIN]
Kenawy, M. I.	[83ELS/ABO]
Kennedy, S. J.	[84KEN/WHE]
Kerimov, I. G.	[67MAM/KER], [67MAM/KER2], [70MAM/KER], [78MAM/ALD], [80MAM/ALD]
Kerriou, M.	[69SPI/ROU]
Keston, A. S.	[36HAR/KES]
Kharitonov, Y. Y.	[75KAD/DEI]
Khodadad, P.	[81BAL/LEG]
Khodakovskiy, I. L.	[68DYA/KHO]
Kilmartin, P. A.	[97KIL/WRI]
Kim, J. I.	[97KON/FAN]
Kim, J. J.	[74PIT/KIM]
King, E. G.	[61KEL/KIN]
King, G. B.	[37KIN], [38LAW/KIN], [39WEL/KIN], [42GEL/KIN], [44FUR/KIN], [45KIN/BEC]
Kircher, C. E.	[30YOS/KIR]
Kireeva, E. P.	[72GLA/KIR]
Kirk, D. L.	[76KIR/RAV]
Kirkham, D. H.	[74HEL/KIR], [81HEL/KIR]
Kitaev, G. A.	[69FOF/KIT]
Kiukkola, K.	[57KIU/WAG]
Kiwia, H. L.	[75KIW/WES3]
Klebanov, G. S.	[60KLE/OST]
Klein, A.	[40KLE]
Klemm, W.	[39KLE/SOD]
Kleppa, O. J.	[92BOO/KLE]
Klimenko, A. A.	[72CHE/KUZ]
Klinchikova, S. A.	[69PAS/UST], [69SOK/PAS]
Klochkov, L. A.	[82GOR/FEN], [86VIK/GOR]
Klushina, T. V.	[62SEL/LES], [67KLU/SEL], [67KLU/SEL3], [68KLU/SEL], [68KLU/SEL3], [69KLU/SEL], [69KLU/SEL3], [70KLU/SEL], [73KLU/LES], [80AMI/KLU]
Knacke, O.	[73BAR/KNA]
Knauf, V. V.	[95RUD/KNA]

(Continued on next page)

Authors list (continued)

Author	Reference
Kniep, R.	[79BOR/KNI], [83FER/HAA]
Knox, B. E.	[68KNO/BAN]
Kobayashi, M.	[95LI/HAG]
Koch, W.	[96HEI/KOC]
Koehler, W. H.	[77SHA/KOE]
Kofman, N. A.	[76PET/KOF]
Kohli, R.	[97KOH]
Kolchin, Y. O.	[83ZEL/KOL]
Koleva, V.	[99KOL/STO]
Kolikov, A. I.	[57BUR/KOV]
Kollmannsberger, M.	[79CAR/GER]
Kolosov, I. V.	[67KOL/IVA], [71KOL/TET]
Kolytcheva, T. I.	[91POL/VOL]
Komarek, K. L.	[72KOM/WES], [74JEL/KOM]
Kondrat'ev, S. N.	[72SID/KON], [84KON/STR], [87AMI/MIN]
Konings, R. J. M.	[92GRE/FUG]
Könnecke, T.	[97KON/FAN]
Konnert, J. A.	[94KON/EVA]
Konnings, R. J. M.	[2001COR/KON]
Kopan, A. R.	[2001BOL/KOP], [2002BOL/KOP], [2003BOL/KOP]
Kopylova, L. I.	[87BOL/TRO]
Korenchuk, N.	[77DUT/TIS]
Korenchuk, N. M.	[71GLA/KOR]
Korneeva, I. V.	[60KOR/SOK]
Korotaeva, L. G.	[67KOL/IVA], [71KOL/TET]
Korte, L.	[83FER/HAA]
Koshchenko, V. I.	[75DEM/KOS], [81KOS/GRI], [90KOS/YAC], [92KOS/IVA]
Koskinen, V.	[75VES/KOS]
Kosmatyi, Y. V.	[59GOL/KOS3]
Kostina, E. N.	[69SHA/KOS]
Kostryukov, V. N.	[67MAM/KER], [67MAM/KER2], [70MAM/KER]
Kosuge, K.	[88HIR/UED]
Kotlik, S. B.	[74KOT/LEB2]
Koto, K.	[66MOR/KOT]
Kovalenker, V. A.	[89KOV/NEK]
Kovalev, G. A.	[57BUR/KOV]
Kovaleva, S. V.	[89GLA/KOV]

(Continued on next page)

Authors list (continued)

Author	Reference
Kovalevskaya, E. I.	[75KOV/SID], [77KOV/SID]
Kovalova, M.	[93PAS/SUL]
Kozhitov, L. V.	[69KOZ/VAN]
Kozuka, Z.	[66OHA/KOZ]
Krabbes, G.	[86OPP/HAN], [90MIN/AMI]
Krasnopol'skaya, M. B.	[73GAN/KUZ]
Krause, L.	[21ROS/KRA]
Krauss, F.	[27KRA/FRI]
Kravchenko, L. K.	[69SUP/MAI], [71SEL/PRY]
Kremann, R.	[08KRE/HOF]
Krestovnikov, A. N.	[66GOR/KRE], [67KRE/GOR], [68GOR/KRE], [71GLA/KUL]
Kretser, Y. L.	[91POL/VOL]
Kriklya, O. I.	[2001BOL/KOP]
Krivenko, A. P.	[97TOL/KRI]
Krivovichev, S.	[99VER/KRI]
Krivovichev, S. V.	[99VER/SEM]
Kruchina, T. I.	[91VAS/KRU]
Krüger, G.	[52LEH/KRU]
Kruglova, V. G.	[83KRU/RYA]
Kruis, A.	[37KRU/CLU]
Kruse, H.	[87FIN/KRU]
Krylov, E. I.	[57KRY/CHU]
Krylova, G. P.	[71PAS/ARO]
Kubaschewski, P.	[68JOS/KUB], [73KUB]
Kucera, G. H.	[73CAI/KUC]
Kuchitsu, K.	[88MIL/CVI]
Kuchumova, N. V.	[89KUC/SHT], [93SER/KUC]
Kudryashova, Y. G.	[83SER/KUD]
Kudryavtsev, A. A.	[62KUD/UST], [65AND/KUD], [68UST/VIG], [69DYA/VIG]
Kühn, G.	[87BOE/KUH]
Kuliev, A. A.	[58KUL/SHA], [63SHA/KUL], [76BAB/KUL]
Kuliev, R. A.	[71GLA/KUL]
Kulikov, L. M.	[92SEM/KUL]
Kullberg, L.	[74AHR/AVS], [74KUL], [74KUL2]
Kullerud, G.	[87BOC/KUL]
Kulyukhina, E. A.	[77KUL/ZLO], [79KUL/ZLO]

(Continued on next page)

Authors list (continued)

Author	Reference
Kumok, V. N.	[90KUM/BAT]
Kunze, G.	[85OPP/HAN]
Kusik, C. L.	[78KUS/MEI]
Kutsepin, V. F.	[65DEV/UMI]
Kuz'micheva, V. P.	[73GAN/KUZ]
Kuz'mina, O. V.	[97SER/BUS]
Kuznetsov, A. N.	[77YAT/KUZ]
Kuznetsov, B. P.	[72CHE/KUZ]
Kvacek, M.	[71JOH/KVA], [71JOH/KVA2], [71JOH/PIC], [71JOH/PIC2], [72JOH/PIC], [76JOH/KVA], [78JOH/KVA]
Kveseth, N. J.	[75GRO/KVE]
L'Ecuyer, J. D.	[97DEG/PEL]
LaCombe, J.	[74IRW/LAC]
LaHaie, P.	[79MIL/HAI], [85MIL/HAI]
Laffitte, M.	[69LAF/CER], [70CHA/TIK], [82LAF]
Laflamme, J. H. G.	[91CAB/LAF]
Lagnier, R.	[75LAG/SUS]
Laitinen, R.	[78LAI/NII], [79LAI/NII], [80LAI/STE]
Lakeenkov, V. M.	[77LAK/PEL]
Lakov, L.	[83DIM/LAK]
Lamoreaux, R. H.	[80BRE/LAM]
Lane, W. C.	[84LAN/EDW]
Langmesser, P.	[39KLE/SOD]
Lapin, V. V.	[67KLU/SEL3], [68KLU/SEL], [69KLU/SEL], [69KLU/SEL3], [70KLU/SEL]
Lapina, L. M.	[57ILL/LAP]
Latimer, W. M.	[56LAT]
Lavendy, H.	[79LAV/PIN]
Lawrence, R. W.	[38LAW/KIN]
Laye, P. G.	[73STE/CAR]
Leary, Jr., H. J.	[69LEA/WAH]
Lebedev, V. V.	[69BER/STR]
Lebedeva, E. G.	[80AMI/KLU]
Lebedeva, G. N.	[74KOT/LEB2]
Leblova, M.	[87VOJ/EBE4]
Lebreton, J.	[73LEB/BOS], [81BOS/LEB]

(Continued on next page)

Authors list (continued)

Author	Reference
Legendre, B.	[81BAL/LEG], [82MOR/LEG], [91MOR/FEU], [96FEU/MAJ]
Lehmann, H.-A.	[52LEH/KRU]
Leigh, G. J.	[90LEI]
Lelekh, B. T.	[71MEL/STE]
Lellep, M.	[68LUN/LEL]
Lemire, R. J.	[92GRE/FUG], [2001LEM/FUG], [2005GAM/BUG]
Lemons, R. S.	[74BEH/LEM]
Lenher, V.	[25LEN/KAO]
Leonidov, V. Y.	[99LEO/FUR]
Leshchinskaya, Z. L.	[62SEL/LES], [62SEL/LES3], [63LES/SEL], [63LES/SEL3], [63LES/SEL4], [63LES/SEL5], [63LES/SEL6], [63SEL/LES], [63SEL/LES3], [63SEL/LES5], [64LES/SEL], [64SEL/LES], [65LES/AVE], [65LES/SEL], [66LES/SEL], [66LES/SEL3], [73KLU/LES]
Levine, S.	[52ROS/WAG]
Levitin, N. E.	[60GRO/THU]
Levy, D. E.	[90LEV/MYE]
Lewis, B. M.	[88OHA/LEW2], [90OHA/LEW]
Lewis, G. N.	[61LEW/RAN]
Lewis, N. B.	[26SID/LEW]
Li, D.	[95CHE/LI], [95MIN/LI]
Li, D. X.	[95LI/HAG]
Li, J.-B.	[2003LI/REC]
Licht, S.	[95LIC/FOR]
Liebig, S.	[2002OPP/ZHA]
Lin, J.-L.	[61LIN/PAN]
Lindberg, R. D.	[87RUN/LIN]
Lindner, G.-G.	[96HEI/KOC]
Lingane, J. J.	[48LIN/NIE]
Liu, H.	[96BRE/LIU], [96BRE/LIU2]
Liu, J.	[95LIU/ZHE], [95LIU/ZHE]
Liu, Q.	[95MIN/LI]
Lomakina, T. P.	[86GLO/LOM]
Loseva, T. I.	[82DYM/LOS], [83DYM/LOS]
Lowson, R. T.	[85LOW]
Lubal, P.	[97LUB/HAV]

(Continued on next page)

Authors list (continued)

Author	Reference
Luk'yanova, T. A.	[63SEL/MAI], [63SEL/MAI3]
Lukas, H. L.	[88GHO/LUK], [89GHO/LUK]
Lukashenko, G. M.	[78LUK/POL], [86GON/LUK]
Lundkvist, M.	[68LUN/LEL]
Lundqvist, M.	[70HOG/LUN]
Lutz, H. D.	[86MAS/LUT]
Luzhnaya, N. P.	[66KAL/YAR]
Lykov, S. N.	[82LYK/CHE]
Lyons, Jr., E. H.	[32SHE/LYO]
Lyons, L. E.	[86LYO/YOU]
Lysenko, A. A.	[71GOR/FES], [75KOV/SID], [77KOV/SID]
Lyskova, Y. B.	[75LYS/VAK], [75LYS/VAK2]
Macalpine, G. B.	[32ALP/SAY]
Maekawa, T.	[71MAE/YOK], [72MAE/YOK], [73MAE/YOK]
Mahadevan, M.	[83MAH/MIL]
Maier, A. I.	[62SEL/SAM], [63LES/SEL4], [63SEL/LES3], [63SEL/MAI], [63SEL/MAI3], [67MAI/SUP], [68MAI/SUP], [69MAI/SOL], [69SUP/MAI], [70KAR/MAI], [71MAI/SUP]
Majid, M.	[96FEU/MAJ]
Makarov, A. V.	[80ZLO/NOV]
Makatun, V. N.	[70MAK/PEC], [71MAK/RAT]
Makhija, R.	[78WES/MAK]
Makhmetov, M. Z.	[71SHK/MAL], [72POL/MAK], [75SHK/MAK]
Mal'tsev, A. K.	[68MAL/PAS], [68TAR/ZHD], [68ZHD/MAL], [69DEM/MAL]
Malaspina, L.	[65MAR/GOL], [69PIA/MAL]
Malekh, B. T.	[68MAL/SEM]
Malkova, A. S.	[89GLA/PAS], [94PAS/ZLO], [2002PAS/MAL], [2003PAS/MAL]
Malov, V. S.	[89KOV/NEK]
Malyshev, V. P.	[71SHK/MAL], [75SHK/MAK]
Malysheva, E. A.	[77MOR/VLA]
Mamedov, K. K.	[67MAM/KER], [67MAM/KER2], [70MAM/KER], [78MAM/ALD], [80MAM/ALD], [88MAM/YAN]
Mamedov, K. N.	[69MAM]
Manceau, A.	[97MAN/GAL]
Manchot, W.	[22MAN/ORT]

(Continued on next page)

Authors list (continued)

Author	Reference
Mandarino, J. A.	[65MAN], [94MAN]
Mandy, F.	[72SMO/MAN]
Manoussakis, G. E.	[67KAW/MAN]
Marchal, G.	[26MAT/MAR]
Mardakhev, B. N.	[63RUS/MAR]
Mareev, Y. M.	[70UST/BEZ]
Margrave, J. L.	[65BAR/COR], [68FIC/HAS], [68UY/MUE], [69FIC/THO]
Margulis, E. V.	[62MAR/GET]
Markova-Deneva, I.	[99OPP/GOB]
Markovskii, L. Y.	[64MAR/SAP], [65SAP/MAR], [66SAP/MAR]
Marshall, W. L.	[84MAR/MES]
Marsigny, L.	[73LEB/BOS]
Martell, A. E.	[64SIL/MAR], [71SIL/MAR]
Martin, J. M. L.	[96MAR]
Martinez, M. L.	[85MES/MAR]
Marx, V.	[90MAR/PET], [92MAR/PET]
Massalski, T. B.	[86MAS]
Massen, C. H.	[64MAS/WEI]
Masson, M. R.	[86MAS/LUT]
Mathieu, J.-C.	[96AMZ/GIL]
Matignon, C.	[26MAT/MAR]
Matlasevich, O. B.	[72MAT/GEI], [72MAT/GEI2]
Matsuo, T.	[96OHT/YAM]
Matveyev, O. V.	[72NOV/ZLO]
Maurer, G.	[92RUM/MAU]
May, P. M.	[96SOL/MAY]
Mayer, P. M.	[97MAY/GAL]
Mayorga, G.	[73PIT/MAY], [74PIT/MAY]
Mazepova, V. I.	[63SEL/SAM]
McAmis, A. J.	[25AMI/FEL]
McBride, B. J.	[61MCB/GOR]
McCollum, D. C.	[72WHI/MCC], [80JOH/MCC]
McCrosky, C. R.	[38NEA/CRO]
McDonald, R. A.	[65DON/OET]
McGee, J. J.	[94KON/EVA]
McMasters, O. D.	[74MCM/GSC]
Medvedev, V. A.	[89COX/WAG]

(Continued on next page)

Authors list (continued)

Author	Reference
Medvedeva, Z. S.	[72GRI/BOR]
Medzhidov, R. A.	[76RAS/MED], [77RAS/MED], [78RAS/MED]
Meerschaut, A.	[81TOM/BIL2]
Mehra, M. C.	[68MEH], [69MEH/GUB], [70MEH/GUB], [70MEH/GUB2], [71MEH/GUB]
Mehta, A.	[81GAU/SHU]
Mei, M.	[73BRE/MEI], [75BRE/MEI]
Meissner, H. P.	[78KUS/MEI]
Mekhtiev, M. I.	[67MAM/KER2], [70MAM/KER], [78MAM/ALD], [80MAM/ALD]
Melekh, B. T.	[67MEL/SEM], [69SEM/MEL]
Melucci, R. C.	[70MEL/WAH]
Menning, K.	[62JER/MEN]
Menshikov, Y. P.	[91POL/VOL]
Merten, H.	[36MER/SCH]
Meschi, D. J.	[72MUN/MES], [79HUA/MES]
Mesmer, R. E.	[76BAE/MES], [84MAR/MES], [91AND/CAS]
Mestres, L.	[85MES/MAR]
Meyer, C. B.	[66BAR/CHA]
Meyer, J.	[13JAN/MEY], [28MEY/AUL], [32MEY/HIN]
Michalski, J.	[97KAM/GLA]
Michov, D.	[93OJK/MIC]
Micka, Z.	[82EBE/MIC], [82EBE/MIC2], [84EBE/MIC], [84MIC/UCH], [85MIC/PIK], [86MIC/EBE], [86MIC/EBE2], [86MIC/EBE3], [87MIC/EBE], [90MIC/CER], [95HAV/MIC], [2000HAV/MIC]
Mihov, D.	[91OYK/MIH], [92OYK/MIH], [98CHR/OJK], [98OJK/CHR], [99OJK/CHR]
Mijlhoff, F. C.	[63MIJ/GER], [63MIJ], [64MIJ/BLO], [65MIJ], [65MIJ2], [75BLE/MIJ]
Mikhailov, E. S.	[76GLA/MIK], [77MIK/GLA], [77MIK/PIS], [77PIS/MIK]
Mikhailova, M. S.	[2002PAS/MAL]
Mikhlina, T. M.	[82GOR/FEN]
Miklyaev, A. D.	[82NIS/SOK]
Mikus, F. F.	[49MIK/POS]
Mil'kov, G. A.	[69BOE/BEN]
Mil'skaya, N. S.	[62MAR/GET]

(Continued on next page)

Authors list (continued)

Author	Reference
Miller, R.	[81MIL]
Miller, R. A.	[79SUL/PRU]
Millero, F. J.	[79MIL]
Mills, I.	[88MIL/CVI]
Mills, K. C.	[74MIL], [76MIL]
Milne, J.	[79MIL/HAI], [83MAH/MIL], [85MIL/HAI], [87MIL], [89MIL], [92GOP/MIL], [95BAL/MIL]
Min, M.	[95MIN/LI]
Minaeva, N. V.	[69BOE/BEN2]
Minenkov, Y. F.	[87AMI/MIN], [90MIN/AMI]
Mishchenko, A. V.	[83NAU/RAK]
Mishchenko, K. P.	[58FLI/MIS]
Misra, S.	[64HOW/MIS], [73SHA/MIS]
Misra, S. N.	[74DAS/SAH]
Mitchell, M. J.	[69HAN/MUN]
Mityureva, T. T.	[59MIT]
Miyauchi, K.	[83MIY/HAY]
Mochalov, K. N.	[72SID/KON]
Moffatt, W. G.	[81MOF]
Molodyk, A. D.	[69BER/STR], [74GOS/MOL]
Monceau, P.	[81TOM/BIL2]
Monnin, C.	[90MON]
Moore, C. E.	[71MOO2]
Moore, D. A.	[95RAI/FEL3], [98RAI/FEL]
Mootz, D.	[79BOR/KNI]
Mora, M. T.	[97COM/MOR]
Morachevskii, A. G.	[78CHE/MOR]
Morgan, J. J.	[81STU/MOR]
Morgant, G.	[82MOR/LEG], [91MOR/FEU]
Morimoto, N.	[66MOR/KOT]
Moriya, H.	[74MOR/SEK]
Moriyama, J.	[66OHA/KOZ]
Morozova, M. P.	[74MOR/VLA], [77MOR/VLA]
Morse, M. D.	[86MOR3]
Mortimer, C. T.	[73BAR/MOR]
Mortimer, M. J.	[91HAL/MOR]
Moser, Z.	[76SMI/MOS]
Mozgova, N. N.	[88EFI/NEN]

(Continued on next page)

Authors list (continued)

Author	Reference
Mrose, M. E.	[77FIN/MRO]
Muenow, D. W.	[68UY/MUE]
Muldagalieva, R. A.	[64BUK/PAS], [72BAK/BUK]
Mulder, H. D.	[51MUL/SCH]
Muldrow, Jr., C. N.	[58MUL/HEP]
Muller, A. B.	[85MUL], [92GRE/FUG]
Müller, A.	[64GLE/MUL], [66NAG/MUL]
Müller, G.	[29NEU/MUL]
Mumme, W. G.	[75MUM], [76MUM], [80MUM]
Munir, Z. A.	[69HAN/MUN], [70BLA/MUN], [71SEA/MUN], [72MUN/MES], [73HAS/MUN]
Munoz, J. L.	[86NOR/MUN]
Murayama, T.	[72MUR/TAK]
Murcray, F. J.	[84LAN/EDW]
Murray, E. J.	[74PET/OWE]
Murray, J. J.	[67MUR/HEY]
Murray, J. L.	[86MUR]
Murray, R. M.	[75MUR/HEY]
Mustafaev, F. M.	[74MUS/AZI], [77AZI/ABB], [80AZI/AGA]
Mustafayev, F. M.	[76ASK/ALI]
Muzalev, E. Y.	[63LES/SEL4], [63SEL/LES3]
Myers, J. R.	[90LEV/MYE]
Nabivanets, B. I.	[84OGA/KAP]
Nabutovskaya, O. A.	[87AMI/MIN]
Nàer-Neumann, E.	[85FER/GRE]
Nagai, S.-I.	[74NAG/SHI]
Nagarajan, G.	[63NAG2], [66NAG/MUL], [71NAG/BRI]
Naginaev, E. N.	[71SAD/YAR]
Nagypál, I.	[90BEC/NAG]
Nair, V. S. K.	[58NAI/NAN], [58NAI/NAN2], [59NAI/NAN], [64NAI], [70GHO/NAI]
Nakahira, M.	[78HAY/NAK], [83MIY/HAY]
Nancollas, G. H.	[58NAI/NAN], [58NAI/NAN2], [59NAI/NAN]
Nara, O.	[90DAS/NAR]
Nasar, A.	[90NAS/SHA]
Naumann, A. W.	[58SPE/NAU]
Naumov, V. N.	[83NAU/RAK], [87AMI/MIN], [90MIN/AMI]
Nazarenko, V. A.	[73NAZ/SHI]

(Continued on next page)

Authors list (continued)

Author	Reference
Ndikumana, T.	[90NDI/CAR]
Neal, Jr., J. L.	[38NEA/CRO]
Nechelyustov, G. N.	[84NEC/CHI]
Neck, V.	[2003GUI/FAN]
Nekrasov, I. Y.	[89KOV/NEK], [90STO/GAV]
Nekrasova, A. N.	[89KOV/NEK]
Nemilov, S.	[82NEM]
Nenashev, S. N.	[88EFI/NEN]
Nepomnyashchaya, V. I.	[74MOR/VLA]
Nesmeyanov, A. N.	[62GEN/NES], [63NES]
Nesterenko, V. P.	[73NES/PIN], [73PIN/NES], [99NES/GLY]
Nesterov, A. R.	[97POL/TAR]
Nesterova, Y. M.	[61NES/PAS]
Neumann, B.	[29NEU/MUL]
Neverov, V. G.	[89NEV/RYB]
Nguyen-Trung, C.	[92GRE/FUG]
Ni, R. Y.	[72NI/WAH]
Niculescu, D.	[65KEL/CRU]
Niedrach, L. W.	[48LIN/NIE]
Niinistö , L.	[78LAI/NII], [79LAI/NII], [85JOH/NII]
Nikitin, S. A.	[89NIK/AND], [91POL/VOL]
Nikol'skaya, A. V.	[67VAS/NIK], [68VAS/SOM], [71VAS/NIK]
Nisel'son, L. A.	[82NIS/SOK]
Nishihara, Y.	[77OGA/NIS]
Nishine, S.	[79IGA/NIS]
Nitsche, H.	[2001LEM/FUG]
Niwa, K.	[40NIW/SHI], [71MAE/YOK], [72MAE/YOK], [73MAE/YOK]
Niznansky, D.	[90MIC/CER]
Nogteva, V. V.	[66PAU/NOG], [68NOG/PAU]
Nomoto, K.	[83ATA/NOM2]
Norbury, A. H.	[75NOR]
Nordstrom, D. K.	[86NOR/MUN]
Nour, M. M.	[74WAS/NOU]
Novak, C. F.	[98ALM/NOV]
Novikov, A. V.	[92SIR/GAV]
Novikov, G. I.	[69SON/NOV]
Novikova, A. A.	[68KLU/SEL]

(Continued on next page)

Authors list (continued)

Author	Reference
Novikova, V. V.	[92SIR/GAV]
Novoselova, A. V.	[59ZLO/POP], [60KOR/SOK], [61NES/PAS], [62CHU/PAS], [67KAR/PAS], [68BON/PAS], [68KAR/ZLO], [69PAS/UKH], [69SOK/PAS], [70GOS/PAS], [72NOV/ZLO], [77KUL/ZLO], [79KUL/ZLO]
Novotný, P.	[85SOH/NOV]
Novozhilov, A. F.	[80ZLO/NOV]
Nücker, N.	[89PHI/BUC]
Nuffield, E. W.	[72HAR/NUF]
Nuttall, R. L.	[82WAG/EVA]
Oelkers, E. H.	[90OEL/HEL], [92SHO/OEL]
Oetting, F. L.	[65DON/OET]
Oganesyan, E. N.	[84OGA/KAP]
Ogawa, S.	[77OGA/NIS]
Ogden, J. S.	[87BRI/OGD]
O'Hare, P. A. G.	[66OHA/SET], [68OHA], [86OHA], [87OHA/SUS], [87OHA/TAS], [88OHA/LEW2], [89OHA/SUS], [90OHA/LEW], [93OHA], [94TOM/SUS], [96OHA/ZYW]
Ohashi, T.	[66OHA/KOZ]
Ohmoto, S.	[96OKA/KAW]
Ohta, T.	[96OHT/YAM]
Oikova, T.	[81OIK/GOS]
Ojkova, T.	[80OJK/GOS], [82OJK/GOS], [87OJK/STA], [89OJK/STA], [90OJK/BAL], [92OJK/STO], [93OJK/MIC], [94OJK/STO], [95STO/CHR], [95STO/OJK], [98CHR/OJK], [98OJK/CHR], [99OJK/CHR], [2000OJK/BAR], [2001BAR/CHR], [2003BAR/CHR]
Ojkova, T. G.	[93OJK/STO]
Okada, H.	[96OKA/KAW]
Okamoto, H.	[91OKA]
Okazaki, N.	[73OKU/OKA]
Okumura, A.	[73OKU/OKA]
Oosterbosch, R.	[65CES/BAC], [69CES/OOS]

(Continued on next page)

Authors list (continued)

Author	Reference
Oppermann, H.	[85OPP/HAN], [86OPP/HAN], [89OPP/RIT], [97OPP/EHR], [99OPP/GOB], [2001OPP/SCH], [2001OPP/ZHA], [2002OPP/ZHA], [2002OPP/ZHA2], [2002ZHA/OPP], [2002ZHA/OPP2]
Orlova, G. M.	[73ZIG/ORL]
Orlyanskii, Y. N.	[77ORL]
Ortner, K.	[22MAN/ORT]
Osman-Zade, S. D.	[66OSM/VAG]
Ostapkevich, N. A.	[60KLE/OST]
Östhols, E.	[99WAN/OST], [2000GRE/WAN], [2000OST/WAN], [2000WAN/OST]
Ouweltjes, W.	[77COR/OUW2]
Owen, B. B.	[50HAR/OWE]
Owens, D. R.	[74PET/OWE]
Oykova, T.	[91OYK/MIH], [92OYK/MIH]
Ozanne, N.	[71DUB/SIS]
Ozeki, T.	[88OZE/YAG]
Ozin, G. A.	[70OZI/VOE]
Paar, W. H.	[98PAA/ROB], [2002PAA/TOP], [2002ROB/PAA]
Pack, R. T.	[62IZA/CHR]
Paetzold, R.	[60SIM/PAE2], [65PAE/AMO]
Pakhomova, N. V.	[58FLI/MIS]
Pakhomovskiy, Y. A.	[97POL/TAR]
Pakhomovsky, Y. A.	[91POL/VOL]
Pakhorukov, N. I.	[61SEL/PAK]
Palmer, D. A.	[2003GUI/FAN]
Pamfilov, A. V.	[50PAM/AGA]
Pan, K.	[61LIN/PAN]
Pankratova, O. Y.	[97PAN/ZVI]
Panus, V. R.	[89NEV/RYB]
Papousek, D.	[62PAP]
Pappas, A.	[49PAP/HAI]
Parker, D. R.	[97PAR/TIC]
Parker, G. A.	[97BAH/PAR]
Parker, V. B.	[82WAG/EVA], [87GAR/PAR]
Parkinson, B. A.	[88OHA/LEW2]
Parkinson, D. H.	[54PAR/QUA]
Parnell, J.	[88PAR]

(Continued on next page)

Authors list (continued)

Author	Reference
Parsons, R.	[74PAR], [85BAR/PAR]
Pasava, J.	[93PAS/SUL]
Pashinkin, A. S.	[61NES/PAS], [62CHU/PAS], [64BUK/PAS], [67KAR/PAS], [68BAK/BUK], [68BAK/BUK3], [68BON/PAS], [68MAL/PAS], [69BAK/BUK], [69BAK/BUK2], [69PAS/UKH], [69PAS/UST], [69SOK/PAS], [70GOS/PAS], [71BAK/BUK], [71PAS/ARO], [72BAK/BUK], [72BAK/PAS], [72PAS/UST], [73BAK/PAS], [73BAK/PAS2], [74GOS/MOL], [79BEN/PAS], [79VEN/PAS], [82VEN/GOR], [86STE/ALI], [88PAS/ZHA], [89GLA/PAS], [90KOS/YAC], [94PAS/GOS], [94PAS/ZLO], [2000GLA/PAS], [2002PAS/MAL], [2003PAS/MAL]
Patkauskas, R. M.	[70YAN/PAT]
Pattje, W. R.	[78SMO/PAT]
Paukov, I. E.	[66PAU/NOG], [68NOG/PAU], [80AMI/KLU], [87AMI/MIN], [90MIN/AMI]
Paulik, F.	[56ERD/PAU]
Paulik, J.	[56ERD/PAU]
Pavlova, P.	[91OYK/MIH]
Payne, J. R.	[95PAY]
Peacock, R. D.	[52AYN/PEA], [53PEA]
Peacor, D. R.	[78DUN/PEA], [86DUN/PEA]
Pearce, J. N.	[23PEA/FOR]
Pechkovskii, V. V.	[70MAK/PEC], [71MAK/RAT], [73NES/PIN], [73PIN/NES]
Peeters, O.	[96VOC/BLA]
Pekova, I.	[82EBE/MIC], [82EBE/MIC2]
Pelabon, M. H.	[1898PEL]
Pelevin, O. V.	[77LAK/PEL]
Pelton, A. D.	[97DEG/PEL], [97SAN/PEL], [97SAN/PEL2], [97SAN/PEL3], [97SAN/PEL4], [97SAN/PEL5]
Perachon, G.	[79CAR/GER]
Perinet, G.	[65SAB/PER], [66SAB/PER]
Perret, R.	[69ROS/PER]
Petersen, E.	[1891PET]
Peterson, J. R.	[78PIT/PET]

(Continued on next page)

Authors list (continued)

Author	Reference
Petrini, G.	[84BAR/PET]
Petrov, A. V.	[59GUL/PET]
Petrova, I. V.	[96BRY/SPI]
Petrova, N. V.	[89NIK/AND]
Petrova, Z. K.	[76PET/KOF], [80SIR/PET]
Petruk, W.	[74PET/OWE]
Petry, W.	[89PHI/BUC]
Petzel, T.	[90MAR/PET], [92MAR/PET]
Phillips, W. A.	[89PHI/BUC]
Piacente, V.	[65MAR/GOL], [69PIA/MAL], [76PIA/BAR], [84CHI/PIA], [84IMP/PIA], [86FER/PIA], [88BAR/PIA], [90SCA/PIA], [92PIA/SCA], [93BRU/PIA], [95ZOC/PIA]
Pico, C.	[82CAS/JER], [84CAS/PIC]
Picot, P.	[70JOH/PIC], [71JOH/PIC], [71JOH/PIC2], [72JOH/PIC], [76JOH/KVA], [78JOH/KVA], [82JOH/PIC], [87JOH/PIC]
Piehl. J.	[75PRE/PIE]
Pierrot, R.	[65PIE/TOU], [69CES/OOS], [70JOH/PIC], [71CES/PIE], [71JOH/PIC], [71JOH/PIC2], [72JOH/PIC]
Piestrzynski, A.	[92PIE]
Pikal, P.	[85MIC/PIK]
Pinaev, G. F.	[73NES/PIN], [73PIN/NES], [74PIN/GOR], [76PIN/VOL]
Pinchemel, B.	[79LAV/PIN]
Piret, P.	[91DEL/PIR], [92JED/CER]
Piskarev, I. V.	[77PIS/MIK]
Piskarev, N. V.	[76GLA/MIK], [77MIK/GLA], [77MIK/PIS]
Pittman, R. W.	[47IVE/PIT]
Pitzer, K. S.	[73PIT/MAY], [73PIT], [74PIT/KIM], [74PIT/MAY], [75PIT], [76PIT/SIL], [78PIT/PET], [79BRA/PIT], [79PIT], [91PIT]
Pizeta, I.	[99DJO/PIZ]
Plambeck, J. A.	[70BOD/PLA]
Plant, A. G.	[76THO/PRI]
Plyasunov, A. V.	[97ALL/BAN], [97GRE/PLY], [97PUI/RAR]
Podmogil'nyi, E. V.	[72CHE/KUZ]

(Continued on next page)

Authors list (continued)

Author	Reference
Polekhovskii, Y. S.	[97POL/TAR]
Polekhovsky, Y. S.	[91POL/VOL]
Polotskaya, R. I.	[78LUK/POL]
Poltavtsev, Y. G.	[73POL/ZAK]
Poltavtseva, L. S.	[67KLU/SEL]
Poluektova, E. N.	[73NAZ/SHI]
Polukarov, A. N.	[71SHK/MAL], [72POL/MAK]
Polyachenok, O. G.	[69SON/NOV], [71SON/POL]
Pomerants, G. B.	[59GOL/POM]
Pool, M. J.	[75PRE/PIE]
Popov, A.	[2000OJK/BAR]
Popov, V. A.	[82NIS/SOK]
Popova, E.	[84SLA/POP], [86POP/SLA], [93SLA/POP]
Popovic, S.	[79POP/TON]
Popovkin, B. A.	[59ZLO/POP]
Porter, R. F.	[60POR/SPE], [68YAM/POR]
Porto, R.	[87CIA/IUL]
Pospelova, L. N.	[97TOL/KRI]
Poss, F. J.	[49MIK/POS]
Potin-Gautier, M.	[2001SEB/POT]
Potter, P. E.	[2001LEM/FUG]
Poulis, J. A.	[64MAS/WEI]
Pound, G. M.	[72MUN/MES]
Powell, H. K. J.	[73POW]
Predel, B.	[75PRE/PIE], [79PRE/GER]
Preis, W.	[2005GAM/BUG]
Preuner, G.	[13PRE/BRO]
Pribyl'skii, N. Y.	[82PRI/VAS]
Pribylov, K. P.	[72SID/KON]
Prideaux, E. B. R.	[25WIL/PRI]
Pring, A.	[90PRI/GAT], [98PRI]
Pringle, G. J.	[76THO/PRI]
Priselkov, Y. A.	[62GEN/NES]
Pritze, M.	[09ROS/PRI]
Prosser, S. J.	[82PRO/BAR]
Protas, I. M.	[73POL/ZAK]
Provencher, R.	[86PRO/AYA]
Prusaczyk, J. E.	[79SUL/PRU]

(Continued on next page)

Authors list (continued)

Author	Reference
Prymova, L. A.	[71SEL/PRY]
Ptacek, P.	[86MIC/EBE], [86MIC/EBE3]
Puigdomènech, I.	[95GRE/PUI], [95SIL/BID], [97ALL/BAN], [97PUI/RAR]
Pultz, G.	[81WIE/PUL], [83WIE/PUL]
Pupyshev, V. I.	[84PUP/RUS]
Pyatnitskii, I. V.	[66PYA/DUR]
Pytkowicz, R. M.	[79JOH/PYT], [79PYT]
Quarrington, J. E.	[54PAR/QUA]
Rabenau, A.	[71RAB/RAU]
Radebaugh, R.	[69GRO/WES]
Radom, L.	[97MAY/GAL]
Rai, D.	[95RAI/FEL3], [98RAI/FEL]
Rakhmenkulov, F. S.	[83NAU/RAK]
Ramakrishna, R. S.	[73RAM/THU]
Ramaswamy, K.	[71RAM/JAY]
Ramdohr, P.	[55RAM/SCH], [56RAM]
Ramsay, D. A.	[87FIN/KRU], [89FIN/SET]
Rand, M. H.	[95DAV/RAN], [95GRE/PUI], [95SIL/BID], [99RAR/RAN], [2000RAN/FUG], [2001LEM/FUG], [2003GUI/FAN]
Randall, M.	[61LEW/RAN]
Rao, Y. K.	[82RAO]
Rard, J. A.	[97ALL/BAN], [97PUI/RAR], [99RAR/RAN]
Rasulov, S. M.	[76RAS/MED], [77RAS/MED], [78RAS/MED], [88RAS], [91RAS]
Rasulova, S. D.	[95YAS/RYA]
Rat'kovskii, I. A.	[71MAK/RAT]
Ratajczak, E.	[64TER/RAT], [67TER/RAT], [68RAT/TER]
Ratchford, R. J.	[62RAT/RIC]
Rau, H.	[67RAU], [67RAU2], [71RAB/RAU], [74RAU]
Raven, M. S.	[76KIR/RAV]
Rawling, J. R.	[66RAW/TOG]
Raymond, C. C.	[97RAY/DIC]
Record, M.-C.	[98VAS/DAO], [2003LI/REC]
Reddy, K. J.	[96SHA/RED]
Reinen, D.	[96HEI/KOC]
Remizovich, T. V.	[73POL/ZAK]

(Continued on next page)

Authors list (continued)

Author	Reference
Reymond, J.-M.	[75HOA/REY]
Reznyakov, A. A.	[68REZ/ISA]
Ricci, J. E.	[40HIL/SOT]
Richter, K. W.	[94RIC/IPS]
Rickert, H.	[62RAT/RIC], [71KEL/RIC]
Ridkosil, T.	[2001RID/SKA]
Rieder, M.	[73CEC/RIE]
Riglet, C.	[87RIG/VIT], [89RIG/ROB], [90RIG]
Ringer, W. E.	[02RIN]
Ripan, R.	[68RIP/VER]
Rittner, P.	[89OPP/RIT]
Roberts, A. C.	[91CAB/LAF], [98PAA/ROB], [2002PAA/TOP], [2002ROB/PAA]
Robertson, R. F.	[79ROD/ROB]
Robinson, P. L.	[52AYN/PEA]
Robinson, R. A.	[31BRI/ROB], [32BRI/ROB], [41HAR/ROB], [59ROB/STO]
Robouch, P.	[89RIG/ROB], [89ROB], [95SIL/BID]
Rodkey, F. L.	[79ROD/ROB]
Rodriguez, A.	[85MES/MAR]
Rohrer, C. S.	[50ROH/FRO]
Rolinski, E. J.	[78ROL/HOC]
Römermann, F.	[2000ROM/FEU]
Rosenberg, F.	[80WES/ROS]
Rosenblatt, G. M.	[74BEH/LEM]
Rosenheim, A.	[09ROS/PRI], [21ROS/KRA]
Rosenstein, G.	[71RAB/RAU]
Roshchina, Z. V.	[68ROS/SEL], [68ROS/SEL2], [68SEL/ROS], [69ROS/SEL], [69ROS/SEL2], [69ROS/SEL4], [70ROS/SEL], [70ROS/SEL2], [70ROS/SEL3]
Ross, A. J.	[96BAR/TAH]
Rossini, F. D.	[52ROS/WAG]
Rosso, B.	[69ROS/PER]
Rossotti, F. J. C.	[61ROS/ROS]
Rossotti, H.	[61ROS/ROS], [69ROS]
Rouxel, J.	[69SPI/ROU]
Rubakova, G. A.	[89NEV/RYB]
Rudashevskii, N. C.	[95RUD/KNA]

(Continued on next page)

Authors list (continued)

Author	Reference
Rudnyi, E. B.	[2001ANT/RUD]
Ruehrwein, R. A.	[39GIA/RUE]
Ruhlmann, F.	[82JOH/PIC], [87JOH/PIC]
Rumpf, B.	[92RUM/MAU]
Rumpf, P.	[33RUM]
Rumyantsev, Y. V.	[62RUM/ZHI]
Runnells, D. D.	[87RUN/LIN]
Runovskaya, L. V.	[63DEV/UMI]
Rusin, A. D.	[84PUP/RUS]
Rustamov, P. G.	[63RUS/MAR]
Rustembekov, K. G.	[71PAS/ARO]
Ryabeva, E. G.	[83KRU/RYA], [95YAS/RYA]
Ryabova, R. I.	[61SEL/SHN]
Ryahovskaya, S. K.	[93SAK/BRY]
Ryazantsev, A. A.	[69SOK/PAS]
Rydberg, J.	[2001LEM/FUG]
Ryzkov, B. I.	[82DYM/LOS]
Sabbah, R.	[65SAB/PER], [66SAB/CAR], [66SAB/PER]
Sacconi, L. V.	[91BAL/SAC]
Sadomirskaya, S. M.	[89KOV/NEK]
Sadovskaya, O. A.	[71SAD/YAR]
Sadykhov, R. Z.	[89GUS/ALD]
Sadykov, K. B.	[66SAD/SEM]
Sahama, T. G.	[65HAK/VUO]
Sahu, G.	[74DAS/SAH], [75SAT/SAH]
Sakairi, M.	[68SEK/IWA]
Sakharova, M. S.	[93SAK/BRY], [96BRY/SPI]
Sal'dau, E. P.	[71GRE/ZAI], [71ZAI/GRE]
Salam, M. A.	[64WEL/SAL], [68WEL/SAL]
Salomaa, P.	[69SAL/HAK]
Salvatore, F.	[83FER/GRE], [85FER/GRE]
Sampietro, G.	[74MCM/GSC]
Samplavskaya, K. K.	[62SEL/SAM], [63SEL/SAM]
Sanders, F. H. M.	[82DIE/SAN]
Sandino, M. C. A.	[95GRE/PUI]
Sandomirskaya, S. M.	[85BOT/SAN]
Sangster, J.	[97SAN/PEL], [97SAN/PEL2], [97SAN/PEL3], [97SAN/PEL4], [97SAN/PEL5]

(Continued on next page)

Authors list (continued)

Author	Reference
Sapozhnikov, R. A.	[64BUK/PAS]
Sapozhnikov, Y. P.	[64MAR/SAP], [65SAP/MAR], [66SAP/MAR]
Sarkisyan, R. A.	[84OGA/KAP]
Sarma, D. D.	[82SHU/SEN]
Sasaki, Y.	[88OZE/YAG]
Sattar-Zade, I. S.	[69GAD/SAT]
Satyanarayana, D.	[74DAS/SAH], [75SAT/SAH]
Savchenko, G. S.	[68SAV/TAN]
Savenko, V. S.	[95SAV]
Saxena, S.	[97ALL/BAN]
Sayce, L. A.	[32ALP/SAY]
Sazykina, T. A.	[62SEL/SAZ3], [62SEL/SAZ5], [62SEL/SHN], [62SEL/ZUB], [63SAZ/SEL], [63SEL/SAZ], [64SEL/SAZ]
Scardala, P.	[88BAR/PIA], [90SCA/PIA], [92PIA/SCA]
Scatchard, G.	[36SCA], [76SCA]
Schack, C. J.	[72CHR/SCH]
Schadow, H.	[89OPP/RIT], [99OPP/GOB], [2001OPP/SCH]
Scheiba, M.	[73BRE/MEI], [75BRE/MEI]
Schiffman, R. A.	[82SCH/FRA]
Schlüter, H.	[36MER/SCH]
Schlesinger, M. E.	[98SCH]
Schmidt, F. C.	[51MUL/SCH]
Schmidt, M.	[55RAM/SCH]
Schmidt, P.	[99OPP/GOB], [2001OPP/SCH], [2002OPP/ZHA2]
Schmidt, S. T.	[95SHI/SCH]
Schneider, A.	[54SCH/GAT2], [55GAT/SCH], [56GAT/SCH], [70BLA/SCH], [71BLA/SCH]
Schnitter, J.	[91MOR/FEU]
Scholz, T. D.	[93GLA/SIN]
Schöneshöfer, M.	[69SCH/KAR]
Schönherr, E.	[96SCH/FRE]
Schott, H. F.	[28SCH/SWI]
Schott, J.	[91AND/CAS]
Schrobilgen, G. J.	[91BJO/SCH]
Schroeder, T. B.	[97KAM/GLA]
Schumm, R. H.	[82WAG/EVA]
Seacrist, L.	[71SEA/MUN]

(Continued on next page)

Authors list (continued)

Author	Reference
Searcy, A. W.	[79HUA/MES]
Seby, F.	[2001SEB/POT]
Sedgwick, T. O.	[66SED/AGU]
Seifert, H. J.	[95SPU/FIN], [98SEI/FUN]
Seip, R.	[83FER/HAA]
Sekine, T.	[68SEK/IWA], [74MOR/SEK]
Selivanova, N. M.	[55SEL/BOG], [58SEL/SHN3], [58SEL/ZUB], [58SEL], [59SEL/KAP], [59SEL/SHN3], [59SEL/SHN4], [59SEL/ZUB], [59SEL/ZUB3], [61SEL/PAK], [61SEL/SHN], [61SEL/SHN3], [62SEL/LES], [62SEL/LES3], [62SEL/SAM], [62SEL/SAZ3], [62SEL/SAZ5], [62SEL/SHN], [62SEL/ZUB], [63LES/SEL], [63LES/SEL3], [63LES/SEL4], [63LES/SEL5], [63LES/SEL6], [63SAZ/SEL], [63SEL/LES], [63SEL/LES3], [63SEL/LES5], [63SEL/MAI], [63SEL/MAI3], [63SEL/SAM], [63SEL/SAZ], [64LES/SEL], [64SEL/LES], [64SEL/SAZ], [65LES/AVE], [65LES/SEL], [66LES/SEL], [66LES/SEL3], [67KLU/SEL], [67KLU/SEL3], [68KLU/SEL], [68KLU/SEL3], [68ROS/SEL], [68ROS/SEL2], [68SEL/ROS], [68TAN/GOR], [69KLU/SEL], [69KLU/SEL3], [69ROS/SEL], [69ROS/SEL2], [69ROS/SEL4], [70KLU/SEL], [70ROS/SEL], [70ROS/SEL2], [70ROS/SEL3], [70SOB/SEL], [71SEL/PRY], [71SEL/SOB], [73KLU/LES], [73SEL/ZAL], [73SOB/SEL], [76TAN/VOL], [76TAN/VOL2]
Semenkovich, S. A.	[66SAD/SEM], [67MEL/SEM], [68MAL/SEM], [69SEM/MEL], [71MEL/STE]
Semenov, Y.	[2003GRO/STO]
Semenov-Kobzar, A. A.	[92SEM/KUL]
Semenova, T.	[99VER/KRI]
Semenova, T. F.	[97VER/SEM], [99VER/SEM]
Semyonova, T. F.	[89VER/FIL]
Sen, P.	[82SHU/SEN]

(Continued on next page)

Authors list (continued)

Author	Reference
Serebrennikov, V. V.	[62ALE/SER], [73SMO/EFR], [78SER/BAT], [78SMO/SER], [79SER/TSY], [81SER/BAT], [84SER/TSY]
Serezhkin, V. N.	[83SER/KUD], [83SER/SER], [84SER/SER], [87SER/SER], [88SER/BLA], [89KUC/SHT], [89TAT/SER], [90TAT/SER], [93SER/KUC]
Serezhkina, L. B.	[83SER/KUD], [83SER/SER], [84SER/SER], [87SER/SER], [88SER/BLA], [89KUC/SHT], [89TAT/SER], [90TAT/SER], [93SER/KUC], [94SER]
Sergeev, N. B.	[97SER/BUS]
Settle, J. L.	[66OHA/SET]
Setzer, K. D.	[89FIN/SET]
Sevast'yanov, V. G.	[71ELL/SLO], [71SEV/SLO], [75ELL/SEV]
Shakhtakhtinskii, M. G.	[58KUL/SHA], [63SHA/KUL]
Shamsuddin	[73SHA/MIS]
Shamsuddin, M.	[90NAS/SHA]
Sharifov, K.	[73SHA/AZI]
Sharifov, K. A.	[60GAD/SHA], [61GAD/SHA], [67SHA/AZI], [72AGA/SHA]
Sharma, C. B.	[54SHA]
Sharma, R. C.	[89SHA/CHA], [92SHA/CHA], [93SHA/CHA], [96SHA/CHA], [96SHA/CHA2], [96SHA/CHA3]
Sharmasarkar, S.	[96SHA/RED]
Sharov, V. V.	[78GER/GLA]
Sharp, K. W.	[77SHA/KOE]
Shaulov, I. K.	[69SHA/KOS]
Shelimova, L. E.	[94CHI/SHE], [97CHI/SHE]
Sherrill, M. S.	[28SHE/IZA], [32SHE/LYO]
Shevel'kov, V. F.	[71GRI/BOR], [72GRI/BOR]
Shevelkov, V. F.	[73GRE/BOR]
Shi, N.	[95MIN/LI]
Shibata, Y.	[2001GU/WAT]
Shibata, Z.	[40NIW/SHI]
Shida, H.	[95LI/HAG]
Shimada, F.	[68SEK/IWA]
Shimuzu, M.	[95SHI/SCH]
Shinmei, M.	[74NAG/SHI]
Shitareva, G. G.	[73NAZ/SHI]

(Continued on next page)

Authors list (continued)

Author	Reference
Shkodin, V. G.	[71SHK/MAL], [75SHK/MAK]
Shneider, V. A.	[58SEL/SHN3], [59SEL/SHN3], [59SEL/SHN4], [61SEL/SHN], [61SEL/SHN3], [62SEL/SHN]
Shock, E. L.	[88SHO/HEL], [89SHO/HEL], [89SHO/HEL2], [92SHO/OEL]
Shscerbachov, D. K.	[95SPI/SPI]
Shtokova, I. P.	[89KUC/SHT]
Shu, H.-C.	[80SHU/GAU], [80WUN/SHU], [81GAU/SHU]
Shukla, A. K.	[82SHU/SEN]
Shulman, G.	[64BON/SHU]
Shutt, R.	[46WAI/SHU]
Shuvalov, R. R.	[97VER/SEM], [99VER/SEM]
Siche, D.	[96SCH/FRE]
Sidgwick, N. V.	[26SID/LEW]
Sidorenko, A. Y.	[75KOV/SID], [77KOV/SID]
Sidorova, E. E.	[72SID/KON]
Sieglova, Z.	[89VOJ/EBE2]
Siemers, P.	[78WIE/SIE]
Sigai, A. G.	[72SIG/WIE]
Silina, E. Y.	[65SIL/KAR], [69PAS/UST]
Sillén, L. G.	[64SIL/MAR], [71BAR/SIL], [71SIL/MAR]
Silva, R. J.	[95SIL/BID]
Silvester, L. F.	[76PIT/SIL], [78PIT/PET]
Simon, A.	[60SIM/PAE2]
Simpson, P. R.	[79BAN/ATK]
Sindeeva, N. D.	[59SIN]
Singh, W. P.	[93GLA/SIN]
Sinke, G. C.	[56STU/SIN], [62THO/SIN]
Sipowska, J. T.	[96STO/FJE]
Sirota, N. N.	[80SIR/PET], [92SIR/GAV]
Sisi, J.-C.	[71DUB/SIS]
Sivtsov, A. V.	[88EFI/NEN]
Skala, R.	[2001RID/SKA]
Sklyarov, V. I.	[86GLO/LOM]
Skopenko, V. V.	[60GOL/SKO], [62GOL/SKO], [65GOL/SKO]
Skowronska, A.	[97KAM/GLA]
Slavtscheva, J.	[84SLA/POP], [86POP/SLA], [93SLA/POP]
Slovyanskikh, V. K.	[71ELL/SLO], [71SEV/SLO], [75ELL/SEV]

(Continued on next page)

Authors list (continued)

Author	Reference
Smejkal, P.	[2000HAV/MIC]
Smirnov, I. A.	[67TIK/GOL]
Smith, J. F.	[76SMI/MOS]
Smith, S. D.	[70BAS/SMI]
Smoes, S.	[70BER/COP], [72SMO/MAN], [77DRO/SMO], [77SMO/DRO], [78SMO/PAT], [84DRO/SMO], [84SMO/DRO]
Smolyakova, K. E.	[73SMO/EFR], [78SMO/SER]
Sobol, L. G.	[70SOB/SEL], [71SEL/SOB], [73SOB/SEL]
Söderhielm, J.	[99HOL/SOD]
Sodomann, H.	[39KLE/SOD]
Söhnel, O.	[85SOH/NOV]
Sokolov, V. V.	[60KOR/SOK], [69SOK/PAS]
Sokolova, T. D.	[82NIS/SOK]
Solans, X.	[85MES/MAR]
Soldatova, T. A.	[69MAI/SOL], [70KAR/MAI]
Solis, J. S.	[96SOL/MAY]
Solovoskii, T. D.	[80SIR/PET]
Somorjai, G. A.	[61SOM]
Somov, A. P.	[68VAS/SOM]
Sonin, V. I.	[69SON/NOV], [71SON/POL]
Soth, G. C.	[40HIL/SOT]
Souleau, C.	[81BAL/LEG], [82MOR/LEG]
Spahiu, K.	[83SPA], [97ALL/BAN], [2001LEM/FUG]
Spassov, T.	[87BOE/KUH]
Spedding, F. H.	[58SPE/NAU]
Spencer, C. W.	[60POR/SPE]
Spencer, J. A.	[73STE/CAR]
Spepanova, N. B.	[71MEL/STE]
Spiesser, M.	[69SPI/ROU]
Spira, P.	[83AVE/SPI]
Spiridonov, E. M.	[83SPI/BAD], [95SPI/SPI], [96BRY/SPI]
Spiridonov, M. F.	[95SPI/SPI]
Spirlet, J. C.	[91HAL/MOR]
Spundflasche, E.	[95SPU/FIN]
Spychiger, H.	[85SPY/KAL]
Srein, V.	[2001RID/SKA]
Srinivasa, R. S.	[87SRI/EDW]

(Continued on next page)

Authors list (continued)

Author	Reference
Stöcker, U.	[64GLE/MUL]
Stadtman, T. C.	[93GLA/SIN]
Stancheva, M. G.	[94GOS/STA], [97GOS/STA]
Staneva, D.	[87OJK/STA], [89OJK/STA], [90OJK/BAL], [92OJK/STO], [94OJK/STO], [95STO/CHR], [95STO/OJK]
Staneva, D. I.	[93OJK/STO]
Stanley, C. J.	[95SHI/SCH]
Stashene, S. S.	[85YAN/STA], [86STA/YAN]
Steblevskii, A. V.	[86STE/ALI]
Steele, W. V.	[73STE/CAR]
Steigmeier, E.	[62BAE/BUS]
Steudel, R.	[79LAI/NII], [80LAI/STE], [81STE], [83STE/STR]
Stewart, J. M.	[74PET/OWE]
Stobart, O. V.	[67BAR/STO]
Stoilova, D.	[92OJK/STO], [94OJK/STO], [95STO/CHR], [95STO/OJK], [99KOL/STO]
Stoilova, D. G.	[93OJK/STO]
Stokes, R. H.	[59ROB/STO]
Stolyarova, T. A.	[74MOR/VLA], [90STO/GAV]
Stranks, D. R.	[77FOW/STR], [77FOW/STR2]
Strauss, E.-M.	[83STE/STR]
Strel'chenko, S. S.	[69BER/STR], [74GOS/MOL]
Strel'tsov, I. S.	[63LES/SEL], [63LES/SEL4], [63SEL/LES3], [63SEL/LES5], [63SEL/SAM]
Strizhov, B. V.	[84KON/STR]
Stull, D. R.	[56STU/SIN], [62THO/SIN]
Stumm, W.	[81STU/MOR]
Sturman, B. D.	[78DUN/PEA]
St/olen, S.	[99STO/JOH]
Stöckl, F.	[51WIN/STO]
Stølen, S.	[96STO/FJE], [2003GRO/STO]
Suga, H.	[96OHT/YAM]
Sullivan, C. L.	[79SUL/PRU]
Sullivan, J. C.	[76COO/WOO], [2001LEM/FUG]
Sulovsky, P.	[93PAS/SUL]
Sunder, S.	[79EYS/SUN]

(Continued on next page)

Authors list (continued)

Author	Reference
Suponitskii, Y. L.	[67MAI/SUP], [68MAI/SUP], [69SUP/MAI], [71MAI/SUP]
Sureda, R. J.	[2002PAA/TOP]
Suski, W.	[75LAG/SUS]
Susman, S.	[87OHA/SUS], [89OHA/SUS], [90OHA/LEW], [94TOM/SUS]
Suvorov, A. V.	[65GAD/DOB], [66GAD/DOB], [68DOB/SUV]
Suzuki, T.	[95LI/HAG]
Sveen, A.	[75GRO/KVE]
Svendsen, S. R.	[72SVE], [72SVE2], [87SVE/GRO]
Sverjensky, D. A.	[89SHO/HEL2], [92SHO/OEL]
Swift, E. H.	[28SCH/SWI]
Taher, F.	[96BAR/TAH]
Takahashi, Y.	[65TAK/WES]
Takayanagi, A.	[72MUR/TAK]
Takeda, Y.	[94YAM/KAM]
Tanabe, K.	[51HOR/TAN]
Tananaev, I. V.	[68SAV/TAN], [68TAN/GOR], [72DEI/TAN], [73DEI/TAN], [75KAD/DEI], [76TAN/VOL], [76TAN/VOL2], [77KAD/DEI], [77KAD/DEI3], [80KAD/TAN], [82TAN/KAD]
Tanger, IV, J. C.	[88TAN/HEL]
Tarascon, J. M.	[87OHA/TAS]
Taraskin, V. V.	[72PAS/UST]
Tarasov, V. V.	[68TAR/ZHD]
Tarasova, I. P.	[91POL/VOL], [97POL/TAR]
Tasker, I. R.	[87OHA/TAS]
Tatarinova, E. E.	[89TAT/SER], [90TAT/SER]
Tausend, A.	[73BRE/MEI], [75BRE/MEI]
Taylor, J. R.	[82TAY]
Teare, J. W.	[37AKE/TEA]
Tedenac, J.-C.	[98VAS/DAO], [2003LI/REC]
Terpilowski, J.	[64TER/RAT], [67TER/RAT], [68RAT/TER], [68TER/ZAL]
Tetsu, K.	[67KOL/IVA], [71KOL/TET]
Theisen, L.	[72EMO/THE]
Thomason, D. N.	[97PAR/TIC]
Thompson, C. J.	[62THO/SIN]

(Continued on next page)

Authors list (continued)

Author	Reference
Thompson, J. C.	[69FIC/THO]
Thompson, W. T.	[90KAR/THO]
Thomsen, J.	[1882THO]
Thorpe, R. I.	[76THO/PRI]
Thourey, J.	[79CAR/GER]
Thuraisingham, R.	[73RAM/THU]
Thurmann-Moe, T.	[60GRO/THU]
Tice, K. R.	[97PAR/TIC]
Tickner, A. W.	[66FUJ/WES]
Tikhomiroff, N.	[70CHA/TIK]
Tikhonov, V. V.	[67TIK/GOL]
Timoshin, I. A.	[69UST/VIG]
Tishchenko, I. A.	[77DUT/TIS]
Tobisawa, S.	[60TOB]
Toguri, J. M.	[66RAW/TOG]
Tolstykh, N. D.	[97TOL/KRI]
Tomashevsky, G. P.	[57CHU/TOM]
Tomaszkiewicz, I.	[94TOM/SUS]
Tomic, S.	[81TOM/BIL], [81TOM/BIL2]
Tonejc, A.	[79POP/TON]
Topa, D.	[98PAA/ROB], [2002PAA/TOP], [2002ROB/PAA]
Topouzkhanian, A.	[96BAR/TAH]
Toropova, V. F.	[56TOR], [57TOR]
Tossidis, J. A.	[67KAW/MAN]
Tóth, I.	[92BAN/BLI]
Toul, F.	[51TOU/DOS]
Toussaint, J.	[65PIE/TOU]
Treinin, A.	[70TRE/WIL]
Trionfetti, G.	[88BAR/IER], [90BAR/TRI]
Trofimova, Z. A.	[87BOL/TRO], [90BOL/TRO]
Trojko, R.	[79POP/TON]
Trzil, J.	[58TRZ]
Tsunoda, K.	[95SHI/SCH]
Tsvetkov, V. G.	[78TSV]
Tsybukova, T. N.	[78SER/BAT], [79SER/TSY], [81SER/BAT], [84SER/TSY]
Tuck, D. G.	[97BAH/PAR]
Tuma, J.	[57TUM]

(Continued on next page)

Authors list (continued)

Author	Reference
Tutton, A. E. H.	[07TUT]
Uchytilova, M.	[84EBE/MIC], [84MIC/UCH]
Ueda, Y.	[88HIR/UED]
Ugorets, M. Z.	[64BUK/PAS], [71PAS/ARO]
Ukhlinov, G. A.	[69PAS/UKH]
Ullman, W. J.	[2001LEM/FUG]
Umilin, V. A.	[63DEV/UMI], [65DEV/UMI]
Umland, F.	[80UML/WAL]
Ustyugov, G. P.	[62KUD/UST], [68UST/VIG], [69DYA/VIG], [69PAS/UST], [69UST/VIG], [69UST/VIG3], [69UST/VIG4], [70UST/BEZ], [72PAS/UST]
Uy, O. M.	[68UY/MUE], [69UY/DRO], [70UY/DRO], [71UY/DRO]
Uzan, R.	[72HOA/CAB], [75HOA/REY]
Vagramyan, A. T.	[66OSM/VAG]
Vakhobov, A. V.	[75LYS/VAK], [75LYS/VAK2]
Van Dommelen, J. H. J.	[82DIE/SAN]
Vance, G. F.	[96SHA/RED]
Vander Auwera-Mahieu, A.	[72SMO/MAN]
Vander Auwera-Mahieu, A. M.	[84DRO/SMO]
Vander Voet, A.	[70OZI/VOE]
Vanyukov, A. V.	[69KOZ/VAN]
Vasil'ev, V. P.	[62VAS], [67VAS/NIK], [68VAS/SOM], [71VAS/NIK]
Vasilev, V. A.	[91VAS/KRU]
Vasilieva, I. G.	[82PRI/VAS]
Vassilev, G. P.	[98VAS/DAO]
Vassilev, V.	[99OPP/GOB]
Vaughan, H.	[67BAR/STO]
Vavrin, I.	[78CEC/VAV], [79CEC/VAV]
Veiga, M. L.	[82CAS/JER], [84CAS/PIC]
Velikov, A. A.	[84SER/TSY]
Vendrikh, N. F.	[79VEN/PAS], [82VEN/GOR]
Verbeek, T.	[65PIE/TOU], [70JOH/PIC], [71CES/PIE]
Veres, Z.	[93GLA/SIN]
Vergès, J.	[82PRO/BAR], [96BAR/TAH]
Vergasova, L.	[99VER/KRI]
Vergasova, L. P.	[89VER/FIL], [97VER/SEM], [99VER/SEM]

(Continued on next page)

Authors list (continued)

Author	Reference
Vericeanu, G.	[68RIP/VER]
Vernadskiy, V. I.	[68DYA/KHO]
Vervloet, M. V.	[89FIN/SET]
Vesala, A.	[75VES/KOS]
Vesala, S.	[69SAL/HAK]
Vieille, P.	[1884BER/VIE]
Vigdorovich, E. N.	[68UST/VIG], [69DYA/VIG], [69PAS/UST], [69UST/VIG], [69UST/VIG3], [69UST/VIG4]
Viksman, G. S.	[86VIK/GOR]
Viskman, G. S.	[82GOR/FEN]
Vitorge, P.	[87RIG/VIT], [89RIG/ROB], [93GIF/VIT2], [94CAP/VIT], [95CAP/VIT], [2001LEM/FUG]
Vladimirova, V. A.	[74MOR/VLA], [77MOR/VLA]
Vochten, R.	[96VOC/BLA]
Vojtisek, P.	[85VOJ/EBE], [87VOJ/EBE], [87VOJ/EBE2], [87VOJ/EBE3], [87VOJ/EBE4], [89VOJ/EBE], [89VOJ/EBE2]
Volin, K. J.	[87OHA/SUS], [89OHA/SUS], [90OHA/LEW], [94TOM/SUS]
Volkova, V. P.	[74PIN/GOR], [76PIN/VOL]
Volodina, A. N.	[68SAV/TAN], [76TAN/VOL], [76TAN/VOL2]
Voloshin, A. V.	[91POL/VOL]
Vorma, A.	[60VOR]
Voronin, G. F.	[70VOR]
Vrana, S.	[73CEC/RIE]
Vuillard, G.	[56VUI], [57VUI]
Vuorelainen, Y.	[64VUO/HUH], [65HAK/VUO]
Wagman, D. D.	[52ROS/WAG], [73HUL/DES], [82WAG/EVA], [89COX/WAG]
Wagner, C.	[57KIU/WAG]
Wagner, Jr., J. B.	[60FIN/WAG]
Wahlbeck, P. G.	[69LEA/WAH], [70MEL/WAH], [72NI/WAH]
Waitkins, G. R.	[46WAI/SHU]
Wakita, H.	[85JOH/NII]
Wallmeier, W.	[80UML/WAL]
Walrafen, G. E.	[62WAL], [62WAL2], [63WAL]
Walsh, P. N.	[62WAL/ART]
Wang, D.-W.	[87FIN/KRU]

(Continued on next page)

Authors list (continued)

Author	Reference
Wang, G.	[95CHE/LI]
Wang, P.	[90ARC/WAN]
Wanner, H.	[88WAN], [91WAN], [92GRE/FUG], [95SIL/BID], [99RAR/RAN], [99WAN/OST], [99WAN], [2000GRE/WAN], [2000OST/WAN], [2000WAN/OST], [2001LEM/FUG]
Wannous, G.	[96BAR/TAH]
Wardlaw, W.	[47IVE/PIT]
Wasif, S.	[74WAS/NOU]
Watanabe, M.	[2001GU/WAT]
Weck, S.	[2002OPP/ZHA]
Weijts, A. G. L. M.	[64MAS/WEI]
Wells, C. F.	[64WEL/SAL], [68WEL/SAL]
Welton, R. C.	[39WEL/KIN]
Wessely, K.	[72KOM/WES]
Westland, A. D.	[78WES/MAK]
Westmore, J. B.	[66FUJ/WES]
Westphal, G. H.	[80WES/ROS]
Westrum, Jr., E. F.	[59GRO/WES], [60GRO/THU], [62GRO/WES], [62WES/GRO], [65CLE/WES], [65TAK/WES], [69GRO/WES], [70WES/GRO], [75KIW/WES3], [84GRO/DRO], [87SVE/GRO], [96STO/FJE]
Wetzenschtein, W.	[85FRA/WET]
Whalley, E.	[68BRO/WHA]
Wheeler, J. C.	[84KEN/WHE]
Whiffen, D. H.	[79WHI2]
White, D.	[62WAL/ART]
White, H. W.	[72WHI/MCC]
White, Jr., H. J.	[87GAR/PAR]
Whitfield, M.	[79WHI]
Wickel, U.	[80BLA/HOP]
Widmark, P.-O.	[96HEI/KOC]
Wiedemeier, H.	[68WIE/GOY], [72SIG/WIE], [74WIE/IRE], [78WIE/SIE], [81WIE/PUL], [83WIE/PUL]
Wilf, J.	[70TRE/WIL]
Willcox, J. S.	[25WIL/PRI]
Williams, P. A.	[79HAA/WIL]
Willner, H.	[81WIL], [88BUR/WIL]

(Continued on next page)

Authors list (continued)

Author	Reference
Windmaisser, F.	[51WIN/STO]
Wobig, D.	[73BRE/MEI], [75BRE/MEI]
Wojakowski, A.	[75LAG/SUS]
Wood, R. H.	[58WOO]
Woods, M.	[76COO/WOO]
Woolf, A. A.	[80WOO]
Wösten, W. J.	[61WOS], [62WOS/GEE]
Wright, G. A.	[97KIL/WRI]
Wunderlich, B.	[78WIE/SIE], [80SHU/GAU], [80WUN/SHU], [81GAU/SHU], [81WIE/PUL], [85JUD/WUN]
Xiang-Yun, W.	[88XIA/ZHU]
Xiaoping, W.	[93FIN/GAR]
Xu, G. Z.	[89FIN/SET]
Yachmenev, V. E.	[90KOS/YAC]
Yagasaki, A.	[88OZE/YAG]
Yamaguchi, K.	[94YAM/KAM]
Yamamuro, O.	[96OHT/YAM]
Yamdagni, R.	[68YAM/POR]
Yanaki, A. A.	[86VIK/GOR], [87BOL/TRO], [90BOL/TRO]
Yangirov, A. Y.	[88MAM/YAN]
Yanitskii, I. V.	[70YAN/PAT], [85YAN/STA], [86STA/YAN]
Yano, T.	[51HOR/YAN]
Yarembash, E. I.	[66KAL/YAR], [66PAU/NOG], [68NOG/PAU], [71SAD/YAR]
Yashunsky, Y. V.	[95YAS/RYA]
Yatsenko, S. P.	[77YAT/KUZ]
Yefimov, M. E.	[93FIN/GAR]
Yilmaz, T.	[2001ERK/KAT]
Yokokawa, T.	[71MAE/YOK], [72MAE/YOK], [73MAE/YOK], [74NAG/SHI]
Yost, D. M.	[28SCH/SWI], [30YOS/KIR]
Young, T. L.	[86LYO/YOU]
Yudin, A. V.	[95SPI/SPI]
Yui, M.	[98RAI/FEL]
Yushin, A. S.	[69YUS/DEV]
Zaitseva, I. A.	[94CHI/SHE]
Zaitseva, I. G.	[67GRE/ZAI], [67ZAI/GRE], [71GRE/ZAI], [71ZAI/GRE]

(Continued on next page)

Authors list (continued)

Author	Reference
Zak, Z.	[80ZAK]
Zakharov, V. P.	[73POL/ZAK]
Zaleska, E.	[68TER/ZAL]
Zaletilo, L. S.	[87BOL/TRO], [92SEM/KUL]
Zalogina, E. A.	[73SEL/ZAL]
Zavyalov, E. N.	[82DYM/LOS], [84NEC/CHI]
Zelic, M.	[99DJO/PIZ]
Zelikman, A. N.	[83ZEL/KOL]
Zhang, Q.	[95CHE/LI]
Zhang-Presse, M.	[2001OPP/ZHA], [2002OPP/ZHA], [2002OPP/ZHA2], [2002ZHA/OPP], [2002ZHA/OPP2]
Zharov, V. V.	[88PAS/ZHA]
Zhdanov, V. M.	[68MAL/PAS], [68TAR/ZHD], [68ZHD/MAL], [69ZHD], [71ZHD]
Zhdanova, A. N.	[51KAP/ZHD], [51KAP/ZHD3]
Zhegalina, V. A.	[90KOS/YAC], [92KOS/IVA]
Zhelagina, V. A.	[81KOS/GRI]
Zheng, M.	[95LIU/ZHE]
Zhiteneva, G. M.	[62RUM/ZHI]
Zhu, J. T.	[88XIA/ZHU]
Ziegler, R. J.	[82SCH/FRA]
Zigel, U. U.	[73ZIG/ORL]
Zlomanov, V. P.	[59ZLO/POP], [68KAR/ZLO], [72NOV/ZLO], [77KUL/ZLO], [79KUL/ZLO], [80ZLO/NOV], [94PAS/ZLO]
Zocchi, F.	[95ZOC/PIA]
Zubova, G. A.	[58SEL/ZUB], [59SEL/KAP], [59SEL/ZUB], [59SEL/ZUB3], [62SEL/ZUB]
Zuman, P.	[82FRA/ZUM]
Zvinchuk, R. A.	[97PAN/ZVI]
Zywocinski, A.	[96OHA/ZYW]